Principles of Chemistry

A Molecular Approach

SECOND EDITION

NIVALDO J. TRO

Westmont College

PEARSON

Boston   Columbus   Indianapolis   New York   San Francisco   Upper Saddle River
Amsterdam   Cape Town   Dubai   London   Madrid   Milan   Munich   Paris   Montréal   Toronto
Delhi   Mexico City   São Paulo   Sydney   Hong Kong   Seoul   Singapore   Taipei   Tokyo

Library of Congress Cataloging-in-Publication Data

Tro, Nivaldo J.
  Principles of chemistry : a molecular approach / Nivaldo J. Tro. -- 2nd ed.
    p. cm.
  Includes index.
  ISBN 978-0-321-75090-7 (hardcover)
  1. Chemistry, Physical and theoretical. I. Title.
  QD453.3.T76 2013
  540--dc23
                                    2011044143

Editor in Chief: Adam Jaworski
Acquisitions Editor: Terry Haugen
Senior Marketing Manager: Jonathan Cottrell
Senior Project Editor: Jennifer Hart
VP/Executive Director, Development: Carol Trueheart
Development Editor: Erin Mulligan
Media Producer: Kristin Mayo
Assistant Editor: Lisa Pierce
Editorial Assistant: Erin Kneuer
Marketing Assistant: Nicola Houston
Managing Editor, Chemistry and Geosciences: Gina M. Cheselka
Senior Project Manager, Production: Shari Toron
Full Service/Composition: GEX Publishing Services
Senior Technical Art Specialist: Connie Long
Art Studio: Precision Graphics
Photo Manager: Maya Melenchuk
Photo Researcher: Stacey Stambaugh
Text Permissions Manager: Beth Wollar
Text Permissions Researcher: Natalie Giboney Turner
Design Manager: Mark Ong
Interior and Cover Design: Tamara Newnam
Operations Specialist: Jeff Sargent
Cover and Chapter Opener Illustrations: Quade Paul

Credits and acknowledgments borrowed from other sources and reproduced, with permission, in this textbook appear on the appropriate page within the text or on p. PC-1.

1 2 3 4 5 6 7 8 9 10—**CRK**—15 14 13 12 11

ISBN-10: 0-321-75090-X; ISBN-13: 978-0-321-75090-7 (Student Edition)
ISBN-10: 0-321-79997-6; ISBN-13: 978-0-321-79997-5 (Instructor's Review Copy)

www.pearsonhighered.com

List of Elements with Their Symbols and Atomic Masses

Element	Symbol	Atomic Number	Atomic Mass	Element	Symbol	Atomic Number	Atomic Mass
Actinium	Ac	89	227.03[a]	Mercury	Hg	80	200.59
Aluminum	Al	13	26.98	Molybdenum	Mo	42	95.94
Americium	Am	95	243.06[a]	Neodymium	Nd	60	144.24
Antimony	Sb	51	121.76	Neon	Ne	10	20.18
Argon	Ar	18	39.95	Neptunium	Np	93	237.05[a]
Arsenic	As	33	74.92	Nickel	Ni	28	58.69
Astatine	At	85	209.99[a]	Niobium	Nb	41	92.91
Barium	Ba	56	137.33	Nitrogen	N	7	14.01
Berkelium	Bk	97	247.07[a]	Nobelium	No	102	259.10[a]
Beryllium	Be	4	9.012	Osmium	Os	76	190.23
Bismuth	Bi	83	208.98	Oxygen	O	8	16.00
Bohrium	Bh	107	264.12[a]	Palladium	Pd	46	106.42
Boron	B	5	10.81	Phosphorus	P	15	30.97
Bromine	Br	35	79.90	Platinum	Pt	78	195.08
Cadmium	Cd	48	112.41	Plutonium	Pu	94	244.06[a]
Calcium	Ca	20	40.08	Polonium	Po	84	208.98[a]
Californium	Cf	98	251.08[a]	Potassium	K	19	39.10
Carbon	C	6	12.01	Praseodymium	Pr	59	140.91
Cerium	Ce	58	140.12	Promethium	Pm	61	145[a]
Cesium	Cs	55	132.91	Protactinium	Pa	91	231.04
Chlorine	Cl	17	35.45	Radium	Ra	88	226.03[a]
Chromium	Cr	24	52.00	Radon	Rn	86	222.02[a]
Cobalt	Co	27	58.93	Rhenium	Re	75	186.21
Copernicium	Cn	112	285	Rhodium	Rh	45	102.91
Copper	Cu	29	63.55	Roentgenium	Rg	111	272[a]
Curium	Cm	96	247.07[a]	Rubidium	Rb	37	85.47
Darmstadtium	Ds	110	271[a]	Ruthenium	Ru	44	101.07
Dubnium	Db	105	262.11[a]	Rutherfordium	Rf	104	261.11[a]
Dysprosium	Dy	66	162.50	Samarium	Sm	62	150.36
Einsteinium	Es	99	252.08[a]	Scandium	Sc	21	44.96
Erbium	Er	68	167.26	Seaborgium	Sg	106	266.12[a]
Europium	Eu	63	151.96	Selenium	Se	34	78.96
Fermium	Fm	100	257.10[a]	Silicon	Si	14	28.09
Fluorine	F	9	19.00	Silver	Ag	47	107.87
Francium	Fr	87	223.02[a]	Sodium	Na	11	22.99
Gadolinium	Gd	64	157.25	Strontium	Sr	38	87.62
Gallium	Ga	31	69.72	Sulfur	S	16	32.07
Germanium	Ge	32	72.64	Tantalum	Ta	73	180.95
Gold	Au	79	196.97	Technetium	Tc	43	98[a]
Hafnium	Hf	72	178.49	Tellurium	Te	52	127.60
Hassium	Hs	108	269.13[a]	Terbium	Tb	65	158.93
Helium	He	2	4.003	Thallium	Tl	81	204.38
Holmium	Ho	67	164.93	Thorium	Th	90	232.04
Hydrogen	H	1	1.008	Thulium	Tm	69	168.93
Indium	In	49	114.82	Tin	Sn	50	118.71
Iodine	I	53	126.90	Titanium	Ti	22	47.87
Iridium	Ir	77	192.22	Tungsten	W	74	183.84
Iron	Fe	26	55.85	Uranium	U	92	238.03
Krypton	Kr	36	83.80	Vanadium	V	23	50.94
Lanthanum	La	57	138.91	Xenon	Xe	54	131.293
Lawrencium	Lr	103	262.11[a]	Ytterbium	Yb	70	173.04
Lead	Pb	82	207.2	Yttrium	Y	39	88.91
Lithium	Li	3	6.941	Zinc	Zn	30	65.41
Lutetium	Lu	71	174.97	Zirconium	Zr	40	91.22
Magnesium	Mg	12	24.31	*[b]		113	284[a]
Manganese	Mn	25	54.94	*[b]		114	289[a]
Meitnerium	Mt	109	268.14[a]	*[b]		115	288[a]
Mendelevium	Md	101	258.10[a]	*[b]		116	292[a]

[a]Mass of longest-lived or most important isotope.

[b]The names of these elements have not yet been decided.

– To Michael, Ali, Kyle, and Kaden –

About the Author

Nivaldo Tro is a Professor of Chemistry at Westmont College in Santa Barbara, California, where he has been a faculty member since 1990. He received his Ph.D. in chemistry from Stanford University for work on developing and using optical techniques to study the adsorption and desorption of molecules to and from surfaces in ultrahigh vacuum. He then went on to the University of California at Berkeley, where he did postdoctoral research on ultrafast reaction dynamics in solution. Since coming to Westmont, Professor Tro has been awarded grants from the American Chemical Society Petroleum Research Fund, from Research Corporation, and from the National Science Foundation to study the dynamics of various processes occurring in thin adlayer films adsorbed on dielectric surfaces. He has been honored as Westmont's outstanding teacher of the year three times and has also received the college's outstanding researcher of the year award. Professor Tro lives in Santa Barbara with his wife, Ann, and their four children, Michael, Ali, Kyle, and Kaden. In his leisure time, Professor Tro enjoys surfing, biking, being outdoors with his family, and reading good books to his children.

Brief Contents

Contents

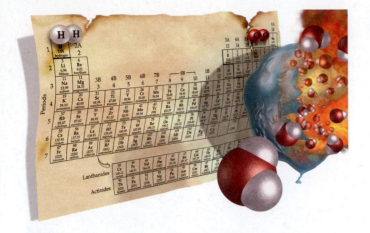

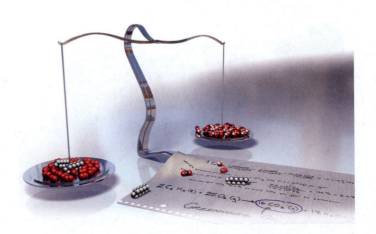

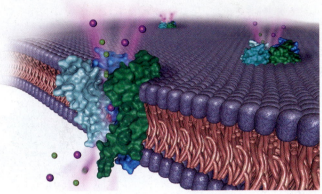

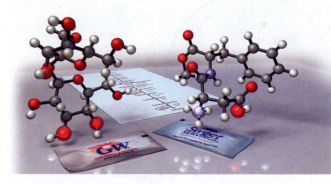

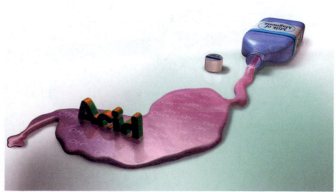

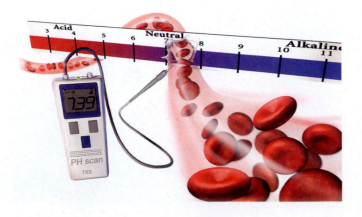

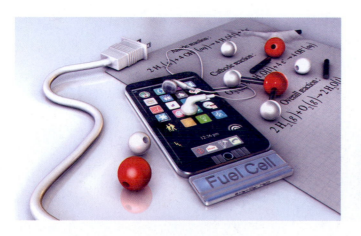

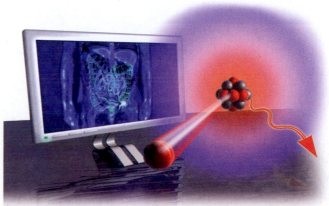

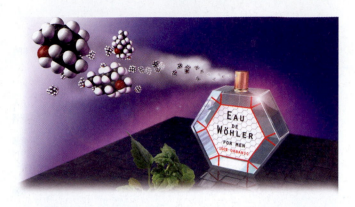

Preface

To the Student

As you begin this course, I invite you to think about your reasons for enrolling in it. Why are you taking general chemistry? More generally, why are you pursuing a college education? If you are like most college students taking general chemistry, part of your answer is probably that this course is required for your major and that you are pursuing a college education so you can get a good job some day. While these are good reasons, I would like to suggest a better one. I think the primary reason for your education is to prepare you to *live a good life*. You should understand chemistry—not for what it can *get* you—but for what it can *do* for you. Understanding chemistry, I believe, is an important source of happiness and fulfillment. Let me explain.

Understanding chemistry helps you to live life to its fullest for two basic reasons. The first is *intrinsic*: through an understanding of chemistry, you gain a powerful appreciation for just how rich and extraordinary the world really is. The second reason is *extrinsic*: understanding chemistry makes you a more informed citizen—it allows you to engage with many of the issues of our day. In other words, understanding chemistry makes *you* a deeper and richer person and makes your country and the world a better place to live. These reasons have been the foundation of education from the very beginnings of civilization.

How does chemistry help prepare you for a rich life and conscientious citizenship? Let me explain with two examples. My first one comes from the very first page of Chapter 1 of this book. There, I ask the following question: What is the most important idea in all of scientific knowledge? My answer to that question is this: **the behavior of matter is determined by the properties of molecules and atoms**. That simple statement is the reason I love chemistry. We humans have been able to study the substances that compose the world around us and explain their behavior by reference to particles so small that they can hardly be imagined. If you have never realized the remarkable sensitivity of the world we *can* see to the world we *cannot*, you have missed out on a fundamental truth about our universe. To have never encountered this truth is like never having read a play by Shakespeare or seen a sculpture by Michelangelo—or, for that matter, like never having discovered that the world is round. It robs you of an amazing and unforgettable experience of the world and the human ability to understand it.

My second example demonstrates how science literacy helps you to be a better citizen. Although I am largely sympathetic to the environmental movement, a lack of science literacy within some sectors of that movement, and the resulting anti-environmental backlash, creates confusion that impedes real progress and opens the door to what could be misinformed policies. For example, I have heard conservative pundits say that volcanoes emit more carbon dioxide—the most significant greenhouse gas—than does petroleum combustion. I have also heard a liberal environmentalist say that we have to stop using hairspray because it is causing holes in the ozone layer that will lead to global warming. Well, the claim about volcanoes emitting more carbon dioxide than petroleum combustion can be refuted by the basic tools you will learn to use in Chapter 4 of this book. We can easily show that volcanoes emit only 1/50th as much carbon dioxide as petroleum combustion. As for hairspray depleting the ozone layer and thereby leading to global warming: the chlorofluorocarbons that deplete ozone have been banned from hairspray since 1978, and ozone depletion has nothing to do with global warming anyway. People with special interests or axes to grind can conveniently distort the truth before an ill-informed public, which is why we all need to be knowledgeable.

So this is why I think you should take this course. Not just to satisfy the requirement for your major, and not just to get a good job some day, but to help you to lead a fuller life and to make the world a little better for everyone. I wish you the best as you embark on the journey to understand the world around you at the molecular level. The rewards are well worth the effort.

To the Professor

First and foremost, thanks to all of you who adopted this book in its first edition. You helped to make this book successful right out of the gate. I am grateful beyond words. Second, I have listened carefully to your feedback on the first edition. The changes you see in this edition are a direct result of your input, as well as my own experience in using the book in my general chemistry courses. If you have acted as a reviewer, or have contacted me directly, you are likely to see your suggestions reflected in the changes I have made. My hope is that the second edition is clearer, cleaner, and crisper than the first edition. In spite of the changes I just mentioned, the goal of the book remains the same: *to present a rigorous and accessible treatment of general chemistry in the context of relevance.*

Teaching general chemistry would be much easier if all of our students had exactly the same level of preparation and ability. But alas, that is not the case. Even though I teach at a relatively selective institution, my courses are populated with students with a range of backgrounds and abilities in chemistry. The challenge of successful teaching, in my opinion, is therefore figuring out how to instruct and challenge the best students while not losing those with lesser backgrounds and abilities. My strategy has always been to set the bar relatively high, while at the same time providing the motivation and support necessary to reach the high bar. That is exactly the philosophy of this book. We do not have to compromise away rigor in order to make chemistry accessible to our students. In this book, I have worked hard to combine rigor with accessibility—to create a book that does not dilute the content, yet can be used and understood by any student willing to put in the necessary effort.

Principles of Chemistry: A Molecular Approach is first and foremost a *student-oriented book*. My main goal is to motivate students and get them to achieve at the highest possible level. As we all know, many students take general chemistry because it is a requirement; they do not see the connection between chemistry and their lives or their intended careers. *Principles of Chemistry: A Molecular Approach* strives to make those connections consistently and effectively. Unlike other books, which often teach chemistry as something that happens only in the laboratory or in industry, this book teaches chemistry in the context of relevance. It shows students *why* chemistry is important to them, to their future careers, and to their world.

Principles of Chemistry: A Molecular Approach is secondly a *pedagogically driven book*. In seeking to develop problem-solving skills, a consistent approach (Sort, Strategize, Solve, and Check) is applied, usually in a two- or three-column format. In the two-column format, the left column shows the student how to analyze the problem and devise a solution strategy. It also lists the steps of the solution, explaining the rationale for each one, while the right column shows the implementation of each step. In the three-column format, the left column outlines a general procedure for solving an important category of problems that is then applied to two side-by-side examples. This strategy allows students to see both the general pattern and the slightly different ways in which the procedure may be applied in differing contexts. The aim is to help students understand both the *concept of the problem* (through the formulation of an explicit conceptual plan for each problem) and the *solution to the problem.*

Principles of Chemistry: A Molecular Approach is thirdly a *visual book*. Wherever possible, images are used to deepen the student's insight into chemistry. In developing chemical principles, multipart images help to show the connection between everyday processes visible to the unaided eye and what atoms and molecules are actually doing. Many of these images have three parts: macroscopic, molecular, and symbolic. This combination helps students to see the relationships between the formulas they write down on paper (symbolic), the world they see around them (macroscopic), and the atoms and molecules that compose that world (molecular). In addition, most figures are designed to teach rather than just to illustrate. They are rich with annotations and labels intended to help the student grasp the most important processes and the principles that underlie them. The resulting images contain significant amounts of information, but are also uncommonly clear and quickly understood.

Principles of Chemistry: A Molecular Approach is fourthly a *"big picture" book*. At the beginning of each chapter, a short paragraph helps students to see the key relationships between the different topics they are learning. Through focused and concise narrative, I strive to make the basic ideas of every chapter clear to the student. Interim summaries are provided at selected spots in the narrative, making it easier to grasp (and review) the main points of important discussions. And to make sure that students never lose sight of the forest for the trees, each chapter includes several *Conceptual Connections*, which ask them to think about concepts and

solve problems without doing any math. I want students to learn the concepts, not just plug numbers into equations to churn out the right answer.

Principles of Chemistry: A Molecular Approach is lastly a book that delivers the core of the standard chemistry curriculum, without sacrificing depth of coverage. Through our research, we have determined the topics that most faculty do not teach and eliminated them; but we have not made significant cuts on the topics that they do teach. When writing a brief book, the temptation is great to cut out the sections that show the excitement and relevance of chemistry; *we have not done that here.* Instead, we have cut out pet topics that are often included in books simply to satisfy a small minority of the market. We have also eliminated extraneous material that does not seem central to the discussion. The result is a lean book that covers core topics in depth, while still demonstrating the relevance and excitement of these topics.

I hope that this book supports you in your vocation of teaching students chemistry. I am increasingly convinced of the importance of our task. Please feel free to email me with any questions or comments about the book.

Nivaldo J. Tro
tro@westmont.edu

What's New in This Edition?

The second edition has been extensively revised and contains many more small changes than I can detail here. Below is a list of the most significant changes from the first edition to the second edition.

- Approximately 150 new end-of-chapter problems, most of them in the cumulative and challenge categories, have been added. As a result, the book now has about 15% more cumulative problems and about 40% more challenge problems.
- New Chapter opening sections for Chapter 6 (Thermochemistry) and Chapter 7 (The Quantum-Mechanical Model of the Atom) have been created that are more student accessible and relevant.
- Approximately 50 new conceptual connections have been added, including some that ask the students to draw matter or processes at the particulate level. As a result, the book now has over 50% more conceptual connections than the previous edition.
- New worked examples have been added to Chapter 3 (balancing chemical equations that contain polyatomic ions), Chapter 6 (thermal energy transfer) and Chapter 12 (calculating the vapor pressure of a solution containing an ionic solute).
- Data has been updated throughout and content on the termination of the Yucca Mountain Nuclear Waste Repository and on the accident at the Fukushima Daiichi nuclear power plant has been added.
- A new design was created that increases readability, clarity, and visual appeal. You will find the two- and three-column in-text examples easier to navigate and work through. I think the result is a cleaner, crisper presentation of the material.

- A section on thermal energy transfer was added to Chapter 6 (Thermochemistry). The addition of this material (in Section 6.4), as well as the corresponding new example and conceptual connection, gives students a better foundation for understanding calorimetry.
- Coverage of Coulomb's law from Chapter 9 (Chemical Bonding I: Lewis Theory) was moved to Chapter 8 (Periodic Properties of the Elements). The earlier introduction of Coulomb's law in Section 8.3 allows me to better explain key concepts in Chapter 8, including penetration, shielding, sub-level energy splitting in multielectron atoms, and effective nuclear charge.
- New content was added to Chapter 12 (Solutions) on colligative properties and vapor pressure, including a new example on how to calculate the vapor pressure of a solution containing an ionic solute (Example 12.12).
- A new section was added to Chapter 17 (Free Energy and Thermodynamics) that better describes the concepts of microstates and macrostates. This addition to Section 17.3 along with two new figures helps lay the foundation for a better understanding of the concept of entropy.

Acknowledgments

The book you hold in your hands bears my name on the cover, but I am really only one member of a large team that carefully crafted this book. Most importantly, I thank my editor, Terry Haugen, who has become a friend and colleague. Terry is new to this edition of the book, but he has already proven to be a skilled and competent editor. He has given me direction, inspiration, and most importantly, loads of support. I am just as grateful for my project editor, Jennifer Hart, who has now worked with me on multiple editions of several books. Jennifer, your guidance, organizational skills, and wisdom are central to the success of my projects, and I am eternally grateful. I am also grateful to Catherine Diamond who helped with organizing reviews, as well as numerous other tasks associated with keeping the team running smoothly. I also thank Erin Mulligan, who has now worked with me on several projects. Erin is an outstanding developmental editor who not only worked with me on crafting and thinking through every word, but also became a friend and fellow foodie in the process. I am also grateful to Adam Jaworski. I have only worked with Adam for a short time and on a limited basis, but I am already impressed by his skills and competence. And of course, I am continually grateful for Paul Corey, with whom I have now worked for over 10 years and 8 projects. Paul is a man of incredible energy and vision, and it is my great privilege to work with him. Paul told me many years ago (when he first signed me on to the Pearson team) to dream big, and then he provided the resources I needed to make those dreams come true. *Thanks, Paul.* I would also like to thank my first editor at Pearson, Kent Porter-Hamann. Kent and I spent many good years together writing books, and I continue to miss her presence in my work.

I am also grateful to my marketing manager Erin Gardner who has helped to develop a great marketing campaign and become a good friend. I am deeply grateful to Tamara Newnam for crafting the design of this text. I would like to thank Shari Toron, Gina Cheselka, and the rest of the Pearson production team. I also thank Kelly Morrison and her co-workers at GEX. This is the first time that I have worked with this team, and I am grateful that they put up with my continual nitpicking on endless details. I am grateful to Connie Long and to Andrew Troutt and his colleagues at Precision Graphics. I am also greatly indebted to my copy editor, Donna Young, for her dedication and professionalism, and to Stacey Stambaugh, for her exemplary photo research. I owe a special debt of gratitude to Quade and Emiko Paul, who continue to make my ideas come alive in their art.

I would like to acknowledge the help of my colleagues Allan Nishimura, Kristi Lazar, Steve Contakes, David Marten, and Carrie Hill, who have supported me in my department while I worked on this book. I am also grateful to those who have supported me personally. First on that list is my wife, Ann. Her love rescued a broken man twelve years ago and without her, none of this would have been possible. I am also indebted to my children, Michael, Ali, Kyle, and Kaden, whose smiling faces and love of life always inspire me. I come from a large Cuban family whose closeness and support most people would envy. Thanks to my parents, Nivaldo and Sara; my siblings, Sarita, Mary, and Jorge; my siblings-in-law, Jeff, Nachy, Karen, and John; my nephews and nieces, Germain, Danny, Lisette, Sara, and Kenny. These are the people with whom I celebrate life.

I would like to thank all of the general chemistry students who have been in my classes throughout my years as a professor at Westmont College. You have taught me much about teaching that is now in this book. I am especially grateful to Catherine Olson and Rose Corcoran, who worked with me in preparing the manuscript for this book. I would also like to express my appreciation to Michael Tro, who also helped in manuscript development, proofreading, and working new problems.

Lastly, I am indebted to the many reviewers whose ideas are imbedded throughout this book. They have corrected me, inspired me, and sharpened my thinking on how best to teach this subject we call chemistry. I deeply appreciate their commitment to this project. Thanks also to Frank Lambert for helping us all to think more clearly about entropy and for his review of the entropy sections of the book. Last but by no means least, I would like to record my gratitude to Becky Barlag, Angela Hoffman, Richard Langley, Sarah Siegel, and Steven Socol whose alertness, keen eyes, and scientific astuteness help make this a much better book.

Reviewers

T.J. Anderson, *Francis Marion University*
Yiyan Bai, *Houston Community College*
Maria Ballester, *Nova Southeastern University*
Shuhsien Batamo, *Houston Community College (Central Campus)*
Charles Benesh, *Wesleyan College*
Justin Briggle, *East Texas Baptist University*
Ron Briggs, *Arizona State University*

Katherine Burton, *Northern Virginia Community College*
Michael L. Denniston, *Georgia Perimeter College*
Ajit S. Dixit, *Wake Technical Community College*
David K. Erwin, *Rose-Hulman Institute of Technology*
Giga Geme, *University of Central Missouri*
Vincent P. Giannamore, *Nicholls State University*
Susan Hendrickson, *University of Colorado (Boulder)*
Angela Hoffman, *University of Portland*
Jason C. Jones, *Francis Marion University*
Dinty J. Musk, Jr., *Ohio Dominican University*
Changyong Qin, *Benedict College*
William Quintana, *New Mexico State University*
Melinda S. Ripper, *Butler County Community College*
Jamie Schneider, *University of Wisconsin (River Falls)*
John P. Scovill, *Temple University*
Dennis Swauger, *Ulster County Community College*
Chris Syvinski, *University of New England*
Tommaso A. Vannelli, *Western Washington University*
Kristofoland Varazo, *Francis Marion University*
Joshua Wallach, *Old Dominion University*
Edward P. Zovinka, *Saint Francis University*

Accuracy Reviewers

Rebecca Barlag, *Ohio University*
Angela Hoffman, *University of Portland*
Richard Langley, *Stephen F. Austin State University*
Sarah Siegel, *Gonzaga University*
Steven Socol, *McHenry County College*

Focus Group Participants

Andrew Price, *Temple University*
Cathrine Reck, *Indiana University*
Jason Kautz, *University of Nebraska (Lincoln)*
Lin Zhu, *Indiana University–Purdue University Indianapolis*
Michael Mueller, *Rose-Hulman Institute of Technology*
Sarah Siegel, *Gonzaga University*
Shusien Wang-Batamo, *Houston Community College*
Silas Blackstock, *University of Alabama*
Tom Pentecost, *Grand Valley State University*
Yiyan Bai, *Houston Community College*

Reviewers of the 1st Edition

Patricia G. Amateis, *Virginia Tech*
Paul Badger, *Robert Morris University*
Rebecca Barlag, *Ohio University*

Craig A. Bayse, *Old Dominion University*
Maria Benavides, *University of Houston, Downtown*
Silas C. Blackstock, *University of Alabama*
David A. Carter, *Angelo State University*
Linda P. Cornell, *Bowling Green State University, Firelands*
Charles T. Cox, Jr., *Georgia Institute of Technology*
David Cunningham, *University of Massachusetts, Lowell*
Pete Golden, *Sandhills Community College*
Robert A. Gossage, *Acadia University*
Angela Hoffman, *University of Portland*
Andrew W. Holland, *Idaho State University*
Narayan S. Hosmane, *Northern Illinois University*
Jason A. Kautz, *University of Nebraska, Lincoln*
Chulsung Kim, *Georgia Gwinnett College*
Scott Kirkby, *East Tennessee State University*
Richard H. Langley, *Stephen F. Austin State University*
Christopher Lovallo, *Mount Royal College*
Eric Malina, *University of Nebraska, Lincoln*
David H. Metcalf, *University of Virginia*
Edward J. Neth, *University of Connecticut*
MaryKay Orgill, *University of Nevada, Las Vegas*
Gerard Parkin, *Columbia University*
BarJean Phillips, *Idaho State University*
Nicholas P. Power, *University of Missouri*
Valerie Reeves, *University of New Brunswick*
Dawn J. Richardson, *Collin College*
Thomas G. Richmond, *University of Utah*
Jason Ritchie, *The University of Mississippi*
Christopher P. Roy, *Duke University*
Thomas E. Sorensen, *University of Wisconsin, Milwaukee*
Vinodhkumar Subramaniam, *East Carolina University*
Ryan Sweeder, *Michigan State University*
Dennis Taylor, *Clemson University*
David Livingstone Toppen, *California State University, Northridge*
Harold Trimm, *Broome Community College*
Susan Varkey, *Mount Royal College*
Clyde L. Webster, *University of California, Riverside*
Wayne Wesolowski, *University of Arizona*
Kurt Winkelmann, *Florida Institute of Technology*

Accuracy Reviewers of the 1st Edition

Margaret Asirvatham, *University of Colorado, Boulder*
Louis Kirschenbaum, *University of Rhode Island*
Richard H. Langley, *Stephen F. Austin State University*
Kathleen Thrush Shaginaw, *Particular Solutions, Inc.*

Supplements

For the Instructor

MasteringChemistry®

(http://www.masteringchemistry.com)

MasteringChemistry is the most effective, widely used online tutorial, homework, and assessment system for chemistry. It helps instructors maximize class time with customizable, easy to assign, and automatically graded assessments that motivate students to arrive prepared for class. These assessments can easily be customized and personalized by instructors to suit their individual teaching style. The powerful gradebook provides unique insight into student and class performance even before the first test. As a result, instructors can spend class time where students need it most.

Instructor Resource Center DVD (0-321-75180-9)

This DVD provides an integrated collection of resources designed to help you make efficient and effective use of your time. This DVD features all the art from the text, including figures and tables in PDF format for high-resolution printing, as well as four pre-built PowerPoint™ presentations. The first presentation contains the images/figures/tables embedded within the PowerPoint slides, while the second includes a complete modifiable lecture outline. The final two presentations contain worked 'in chapter' sample exercises and questions to be used with Classroom Response Systems. This DVD also contains movies and animations, as well as the TestGen version of the Printed Test Bank, which allows you to create and tailor exams to your needs.

Instructor Resource Manual (0-321-75182-5)

Organized by chapter, this useful guide includes objectives, lecture outlines, references to figures and solved problems, as well as teaching tips.

Printed Test Bank (0-321-75181-7)

The Printed Test Bank contains more than 1500 multiple choice, true/false, and short-answer questions.

Solutions Manual (0-321-76545-1)

This manual contains step-by-step solutions to all end-of-chapter exercises. With instructor permission, this manual may be made available to students.

Blackboard® and WebCT®

Practice and assessment materials are available upon request in these course management platforms.

For the Student

MasteringChemistry®

(http://www.masteringchemistry.com)

MasteringChemistry provides you with two learning systems: an extensive self-study area with an interactive eText and the most widely used chemistry homework and tutorial system (if your instructor chooses to make online homework part of your course).

Pearson eText

The integration of the Pearson eText within MasteringChemistry gives students with new books, easy access to the electronic text when they are logged into MasteringChemistry. Pearson eText pages look exactly like the printed text, offering powerful new functionality for students and instructors. Users can create notes, highlight text in different colors, create bookmarks, zoom, view in single-page or two-page view, etc. Pearson eText also links students to associated media files, including worked example animations, enabling them to view an animation as they read the text.

Selected Solutions Manual (0-321-75183-3)

This manual contains complete, step-by-step solutions to selected odd numbered end-of-chapter problems.

How This Text Will Help You Learn Chemistry

Niva Tro's true passion is teaching chemistry. Every day he finds new and innovative ways of presenting chemical concepts to his students.

Principles of Chemistry, **Second Edition** was written from Niva's classroom experience to address the challenges his students face. Niva's writing is known for a rigorous, yet accessible treatment of general chemistry—always in the context of its *relevance to everyday life*.

Chemistry is presented visually through macroscopic, molecular and symbolic representations, allowing you to see the connections among the formulas (symbolic), the world around you (macroscopic), and the atoms and molecules that make up the world (molecular).

Among other revisions, the **Second Edition** offers a crisp new design, adds more challenging problems, and significantly revises the coverage of electrochemistry. Like the first edition, the **Second Edition** is available with MasteringChemistry®—the premier online homework and assessment tool that will truly deepen your understanding of chemistry and will enhance your success in this course.

The following pages present some of the main features of this text:

Relevance and Writing Style
Chemistry becomes both meaningful and understandable. *See page xxi of this walkthrough.*

Pioneering Artwork
Superlative images help students grasp concepts. *See pages xxii–xxiii of this walkthrough.*

Problem-Solving
Proven strategies foster students' problem-solving skills. *See pages xxiv–xxv of this walkthrough.*

End-of-Chapter Material
Helpful learning aids reinforce concepts. *See pages xxvi–xxvii of this walkthrough.*

MasteringChemistry®
Improve students' understanding of chemistry and class performance with this online homework and assessment tool. *See pages xviii–xxix of this walkthrough.*

Relevance and Writing Style

Chemistry is relevant to every process occurring around us, at every second. Niva Tro helps you understand this connection by weaving specific, vivid examples throughout the text that tell the story of chemistry. Every chapter begins with a brief story showing chemistry is relevant to all people, at every moment.

How does a gecko walk across a ceiling? See Chapter 11 to find out how to use chemistry to understand how it can support its entire weight by a single toe in contact with a surface.

Have you ever wondered how a baby in the womb gets oxygen? See Chapter 14 to learn more about fetal hemoglobin and equilibrium.

These examples make the material more accessible by *contextualizing* the chemistry and grounding it in the world we live in.

Pioneering Artwork Makes Concepts Clear

Annotated Molecular Art

Many illustrations have three parts: a macroscopic image (what you can see with your eyes); a molecular image (what the molecules are doing); and a symbolic representation (how chemists represent the process with symbols and equations).

The goal is for you to connect what you see and experience (the macroscopic world) with the molecules responsible for that world, and with the way chemists represent those molecules. This is, after all, what chemistry is all about.

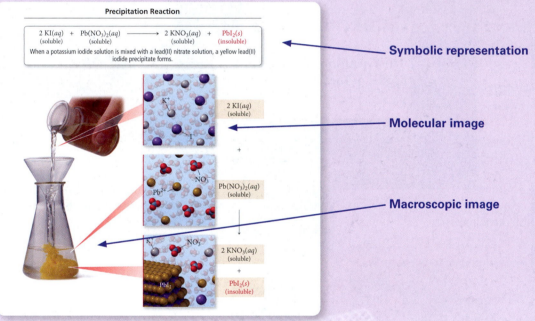

Precipitation Reaction

$$2 \, KI(aq) + Pb(NO_3)_2(aq) \longrightarrow 2 \, KNO_3(aq) + PbI_2(s)$$
(soluble)   (soluble)   (soluble)   (insoluble)

When a potassium iodide solution is mixed with a lead(II) nitrate solution, a yellow lead(II) iodide precipitate forms.

Symbolic representation

$2 \, KI(aq)$ (soluble)

Molecular image

$Pb(NO_3)_2(aq)$ (soluble)

Macroscopic image

$2 \, KNO_3(aq)$ (soluble)

$PbI_2(s)$ (insoluble)

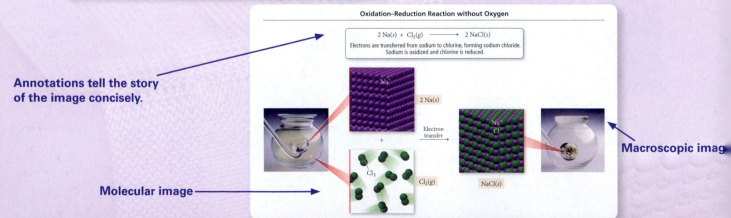

Oxidation–Reduction Reaction without Oxygen

$$2 \, Na(s) + Cl_2(g) \longrightarrow 2 \, NaCl(s)$$

Electrons are transferred from sodium to chlorine, forming sodium chloride. Sodium is oxidized and chlorine is reduced.

Annotations tell the story of the image concisely.

$2 \, Na(s)$

Electron transfer

$Cl_2(g)$

$NaCl(s)$

Molecular image

Macroscopic imag

Multipart images

Multipart images make connections among graphical representations, molecular processes, and the macroscopic world.

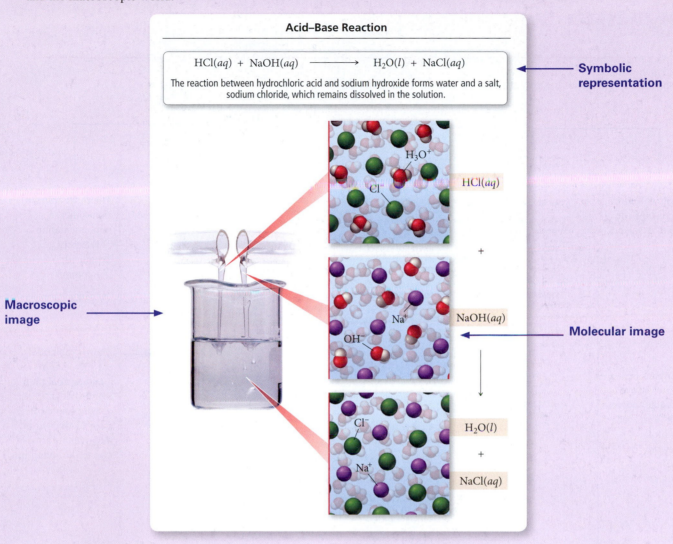

Acid–Base Reaction

$$HCl(aq) + NaOH(aq) \longrightarrow H_2O(l) + NaCl(aq)$$

The reaction between hydrochloric acid and sodium hydroxide forms water and a salt, sodium chloride, which remains dissolved in the solution.

Symbolic representation

H_3O^+

Cl^-

$HCl(aq)$

$+$

Na^+

OH^-

$NaOH(aq)$

Molecular image

Cl^-

Na^+

$H_2O(l)$

$+$

$NaCl(aq)$

Macroscopic image

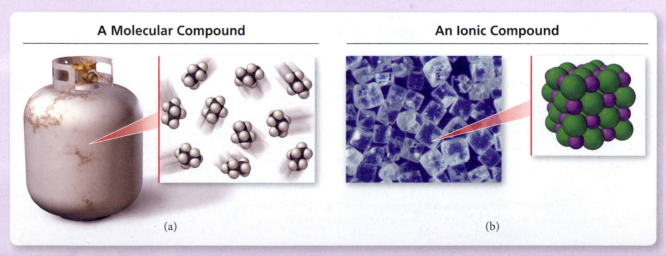

A Molecular Compound

(a)

An Ionic Compound

(b)

Consistent Strategies Help You Solve Problems

A consistent approach to problem solving is used throughout the book.

Two-Column Example

The **left column** explains how the problem is solved.

A four-part structure (**"Sort, Strategize, Solve, Check"**) provides you with a framework for analyzing and solving problems.

The **right column** shows the implementation of the steps explained in the left column.

Many problems are solved with a **conceptual plan** that provides a visual outline of the steps leading from the given information to the solution.

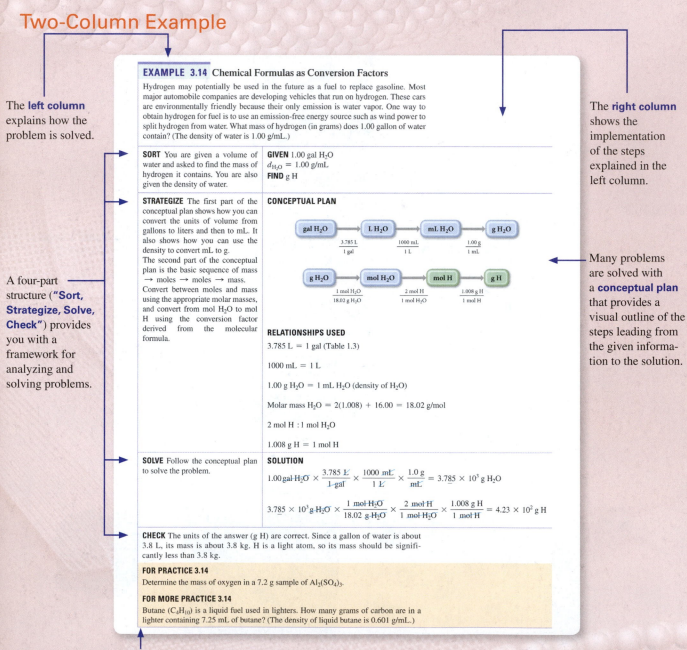

EXAMPLE 3.14 Chemical Formulas as Conversion Factors

Hydrogen may potentially be used in the future as a fuel to replace gasoline. Most major automobile companies are developing vehicles that run on hydrogen. These cars are environmentally friendly because their only emission is water vapor. One way to obtain hydrogen for fuel is to use an emission-free energy source such as wind power to split hydrogen from water. What mass of hydrogen (in grams) does 1.00 gallon of water contain? (The density of water is 1.00 g/mL.)

SORT You are given a volume of water and asked to find the mass of hydrogen it contains. You are also given the density of water.

GIVEN 1.00 gal H_2O
$d_{H_2O} = 1.00$ g/mL
FIND g H

STRATEGIZE The first part of the conceptual plan shows how you can convert the units of volume from gallons to liters and then to mL. It also shows how you can use the density to convert mL to g.
The second part of the conceptual plan is the basic sequence of mass → moles → moles → mass. Convert between moles and mass using the appropriate molar masses, and convert from mol H_2O to mol H using the conversion factor derived from the molecular formula.

CONCEPTUAL PLAN

gal H_2O → L H_2O → mL H_2O → g H_2O

$\dfrac{3.785\ L}{1\ gal}$ $\dfrac{1000\ mL}{1\ L}$ $\dfrac{1.00\ g}{1\ mL}$

g H_2O → mol H_2O → mol H → g H

$\dfrac{1\ mol\ H_2O}{18.02\ g\ H_2O}$ $\dfrac{2\ mol\ H}{1\ mol\ H_2O}$ $\dfrac{1.008\ g\ H}{1\ mol\ H}$

RELATIONSHIPS USED

3.785 L = 1 gal (Table 1.3)

1000 mL = 1 L

1.00 g H_2O = 1 mL H_2O (density of H_2O)

Molar mass H_2O = 2(1.008) + 16.00 = 18.02 g/mol

2 mol H : 1 mol H_2O

1.008 g H = 1 mol H

SOLVE Follow the conceptual plan to solve the problem.

SOLUTION

$1.00\ \text{gal}\ H_2O \times \dfrac{3.785\ L}{1\ gal} \times \dfrac{1000\ mL}{1\ L} \times \dfrac{1.0\ g}{mL} = 3.78\underline{5} \times 10^3\ \text{g}\ H_2O$

$3.78\underline{5} \times 10^3\ \text{g}\ H_2O \times \dfrac{1\ mol\ H_2O}{18.02\ g\ H_2O} \times \dfrac{2\ mol\ H}{1\ mol\ H_2O} \times \dfrac{1.008\ g\ H}{1\ mol\ H} = 4.23 \times 10^2\ \text{g H}$

CHECK The units of the answer (g H) are correct. Since a gallon of water is about 3.8 L, its mass is about 3.8 kg. H is a light atom, so its mass should be significantly less than 3.8 kg.

FOR PRACTICE 3.14
Determine the mass of oxygen in a 7.2 g sample of $Al_2(SO_4)_3$.

FOR MORE PRACTICE 3.14
Butane (C_4H_{10}) is a liquid fuel used in lighters. How many grams of carbon are in a lighter containing 7.25 mL of butane? (The density of liquid butane is 0.601 g/mL.)

Every worked Example is followed by "For Practice" Problems that you can try to solve on your own. Answers to "For Practice" Problems are in Appendix IV.

Three-Column Example

Problem-Solving Procedure Boxes for important categories of problems enable you to see how the same reasoning applies to different problems.

The **general procedure** is shown in the left column.

Two worked Examples, side by side, make it easy to see how differences are handled.

PROCEDURE FOR... Determining an Empirical Formula from Combustion Analysis	**EXAMPLE 3.18** Determining an Empirical Formula from Combustion Analysis	**EXAMPLE 3.19** Determining an Empirical Formula from Combustion Analysis
	Upon combustion, a compound containing only carbon and hydrogen produced 1.83 g CO_2 and 0.901 g H_2O. Find the empirical formula of the compound.	Upon combustion, a 0.8233 g sample of a compound containing only carbon, hydrogen, and oxygen produced 2.445 g CO_2 and 0.6003 g H_2O. Find the empirical formula of the compound.
1. Write down as *given* the masses of each combustion product and the mass of the sample (if given).	**GIVEN** 1.83 g CO_2, 0.901 g H_2O **FIND** empirical formula	**GIVEN** 0.8233 g sample, 2.445 g CO_2, 0.6003 g H_2O **FIND** empirical formula
2. Convert the masses of CO_2 and H_2O from step 1 to moles by using the appropriate molar mass for each compound as a conversion factor.	$1.83 \text{ g } CO_2 \times \dfrac{1 \text{ mol } CO_2}{44.01 \text{ g } CO_2}$ $= 0.0416 \text{ mol } CO_2$ $0.901 \text{ g } H_2O \times \dfrac{1 \text{ mol } H_2O}{18.02 \text{ g } H_2O}$ $= 0.0500 \text{ mol } H_2O$	$2.445 \text{ g } CO_2 \times \dfrac{1 \text{ mol } CO_2}{44.01 \text{ g } CO_2}$ $= 0.05556 \text{ mol } CO_2$ $0.6003 \text{ g } H_2O \times \dfrac{1 \text{ mol } H_2O}{18.01 \text{ g } H_2O}$ $= 0.03331 \text{ mol } H_2O$
3. Convert the moles of CO_2 and moles of H_2O from step 2 to moles of C and moles of H using the conversion factors inherent in the chemical formulas of CO_2 and H_2O.	$0.0416 \text{ mol } CO_2 \times \dfrac{1 \text{ mol C}}{1 \text{ mol } CO_2}$ $= 0.0416 \text{ mol C}$ $0.0500 \text{ mol } H_2O \times \dfrac{2 \text{ mol H}}{1 \text{ mol } H_2O}$ $= 0.100 \text{ mol H}$	$0.05556 \text{ mol } CO_2 \times \dfrac{1 \text{ mol C}}{1 \text{ mol } CO_2}$ $= 0.05556 \text{ mol C}$ $0.03331 \text{ mol } H_2O \times \dfrac{2 \text{ mol H}}{1 \text{ mol } H_2O}$ $= 0.06662 \text{ mol H}$
4. If the compound contains an element other than C and H, find the mass of the other element by subtracting the sum of the masses of C and H (obtained in step 3) from the mass of the sample. Finally, convert the mass of the other element to moles.	This sample contains no elements besides C and H, so proceed to next step.	$\text{Mass C} = 0.05556 \text{ mol C} \times \dfrac{12.01 \text{ g C}}{\text{mol C}}$ $= 0.6673 \text{ g C}$ $\text{Mass H} = 0.06662 \text{ mol H} \times \dfrac{1.008 \text{ g H}}{\text{mol H}}$ $= 0.06715 \text{ g H}$ $\text{Mass O} = 0.8233 \text{ g } -$ $(0.6673 \text{ g} + 0.06715 \text{ g}) = 0.0889 \text{ g}$ $\text{Mol O} = 0.0889 \text{ g O} \times \dfrac{\text{mol O}}{16.00 \text{ g O}}$ $= 0.00556 \text{ mol O}$
5. Write down a pseudoformula for the compound using the number of moles of each element (from steps 3 and 4) as subscripts.	$C_{0.0416}H_{0.100}$	$C_{0.05556}H_{0.06662}O_{0.00556}$
6. Divide all the subscripts in the formula by the smallest subscript. (Round all subscripts that are within 0.1 of a whole number.)	$C_{\frac{0.0416}{0.0416}} H_{\frac{0.100}{0.0416}} \rightarrow C_1 H_{2.4}$	$C_{\frac{0.05556}{0.00556}} H_{\frac{0.06662}{0.00556}} O_{\frac{0.00556}{0.00556}} \rightarrow C_{10}H_{12}O_1$

You can find answers to Conceptual Connections at the end of the chapter.

 **Conceptual Connection 3.1** **Structural Formulas**

In water, the oxygen atom is connected to each hydrogen atom by a single bond. Write a structural formula for water.

◀ **Conceptual Connections** are strategically placed to reinforce conceptual understanding of the most complex concepts. The book has over 50% more Conceptual Connections than the previous edition and answers can be found at the end of each chapter.

End-of-Chapter Material

The end-of-chapter review section helps you study the chapter's concepts and skills in a systematic way that is ideal for test preparation. Approximately 150 new end-of-chapter problems have been added, most of these in the cumulative and challenge categories.

End-of-Chapter Review Section

Key Terms

Section 3.2
ionic bond (75)
covalent bond (75)

Section 3.3
chemical formula (75)
empirical formula (75)
molecular formula (75)
structural formula (76)
ball-and-stick model (77)
space-filling molecular model (77)

Section 3.4
atomic element (77)

molecular element (77)
molecular compound (79)
ionic compound (79)
formula unit (79)
polyatomic ion (80)

Section 3.5
common name (82)
systematic name (82)
binary compound (83)
oxyanion (86)
hydrate (87)

Section 3.6
acid (89)

binary acid (90)
oxyacid (90)

Section 3.7
formula mass (91)

Section 3.8
mass percent composition (mass percent) (94)

Section 3.9
empirical formula molar mass (100)
combustion analysis (101)

Section 3.10
chemical reaction (103)
combustion reaction (103)
chemical equation (103)
reactants (103)
products (103)
balanced chemical equation (103)

Section 3.11
organic compound (107)
hydrocarbon (107)

▲ **Key Terms** list all of the chapter's boldfaced terms, organized by section in order of appearance, with page references. Definitions are found in the Glossary.

Key Concepts

Chemical Bonds (3.2)

► Chemical bonds, the forces that hold atoms together in compounds, arise from the interactions between nuclei and electrons in atoms.
► In an ionic bond, one or more electrons have been *transferred* from one atom to another, forming a cation (positively charged) and an anion (negatively charged). The two ions are drawn together by the attraction between the opposite charges.
► In a covalent bond, one or more electrons are *shared* between two atoms. The atoms are held together by the attraction between their nuclei and the shared electrons.

Representing Molecules and Compounds (3.3, 3.4)

► A compound is represented with a chemical formula, which indicates the elements present and the number of atoms of each.
► An empirical formula gives only the *relative* number of atoms, while a molecular formula gives the *actual* number present in the molecule.
► Structural formulas show how the atoms are bonded together, while molecular models show the geometry of the molecule.
► Compounds can be divided into two types: molecular compounds, formed between two or more covalently bonded nonmetals; and ionic compounds, usually formed between a metal ionically bonded to one or more nonmetals. The smallest identifiable unit of a molecular compound is a molecule, and the smallest identifiable unit of an ionic compound is a formula unit: the smallest electrically neutral collection of ions.
► Elements can also be divided into two types: molecular elements—which occur as (mostly diatomic) molecules—and atomic elements—which occur as individual atoms.

Naming Inorganic Ionic and Molecular Compounds and Acids (3.5, 3.6)

► A flowchart for naming simple inorganic compounds is provided at the end of this section. Use this chart to name inorganic compounds.

Formula Mass and the Mole Concept for Compounds (3.7)

► The formula mass of a compound is the sum of the atomic masses of all the atoms in the chemical formula. Like the atomic masses of elements, the formula mass characterizes the average mass of a molecule (or a formula unit).
► The mass of one mole of a compound (in grams) is the molar mass and is numerically equal to its formula mass (in amu).

Chemical Composition (3.8, 3.9)

► The mass percent composition of a compound is each element's percentage of the total compound's mass. The mass percent composition can be determined from the compound's chemical formula and the molar masses of its elements.
► The chemical formula of a compound provides the relative number of atoms (or moles) of each element in a compound, and can therefore be used to determine numerical relationships between moles of the compound and moles of its constituent elements.
► If the mass percent composition and molar mass of a compound are known, its empirical and molecular formulas can be determined.

Writing and Balancing Chemical Equations (3.10)

► In chemistry, we represent chemical reactions with chemical equations. The substances on the left side of a chemical equation are the reactants and the substances on the right side are the products.
► Chemical equations are balanced when the number of each type of atom on the left side of the equation is equal to the number on the right side.

Organic Compounds (3.11)

► Organic compounds—originally derived only from living organisms but now readily synthesized in the laboratory—are composed of carbon, hydrogen, and a few other elements such as nitrogen, oxygen, and sulfur.
► The simplest organic compounds are hydrocarbons, composed of only carbon and hydrogen.

▲ **The Key Concepts** section summarizes the chapter's most important ideas.

Key Equations and Relationships

Formula Mass (3.7)

$$\left(\begin{array}{c}\text{\# atoms of 1st element}\\ \text{in chemical formula}\end{array} \times \begin{array}{c}\text{atomic mass}\\ \text{of 1st element}\end{array}\right) + \left(\begin{array}{c}\text{\# atoms of 2nd element}\\ \text{in chemical formula}\end{array} \times \begin{array}{c}\text{atomic mass}\\ \text{of 2nd element}\end{array}\right) + \ldots$$

Mass Percent Composition (3.8)

$$\text{Mass \% of element X} = \frac{\text{mass of X in 1 mol compound}}{\text{mass of 1 mol compound}} \times 100\%$$

Empirical Formula Molar Mass (3.9)

$$\text{Molecular formula} = n \times \text{(empirical formula)}$$

$$n = \frac{\text{molar mass}}{\text{empirical formula molar mass}}$$

Inorganic Nomenclature Summary Chart

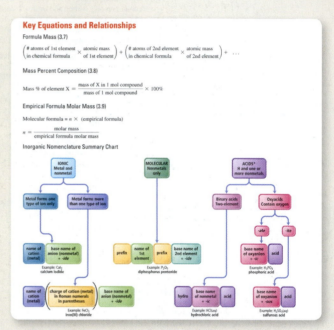

▲ **The Key Equations and Relationships** section lists each of the key equations and important quantitative relationships from the chapter.

Key Learning Objectives

Chapter Objectives	Assessment		
Writing Molecular and Empirical Formulas (3.3)	Example 3.1	For Practice 3.1	Exercises 1–4
Classifying Substances as Atomic Elements, Molecular Elements, Molecular Compounds, or Ionic Compounds (3.4)	Example 3.2	For Practice 3.2	Exercises 5–10
Writing Formulas for Ionic Compounds (3.5)	Examples 3.3, 3.4	For Practice 3.3, 3.4	Exercises 11–14, 23, 24
Naming Ionic Compounds (3.5)	Examples 3.5, 3.6	For Practice 3.5, 3.6	For More Practice 3.5, 3.6 Exercises 15–20
Naming Ionic Compounds Containing Polyatomic Ions (3.5)	Example 3.7	For Practice 3.7	For More Practice 3.7 Exercises 21–24
Naming Molecular Compounds (3.6)	Example 3.8	For Practice 3.8	For More Practice 3.8 Exercises 27–30
Naming Acids (3.6)	Examples 3.9, 3.10	For Practice 3.9, 3.10	For More Practice 3.10 Exercises 31–34
Calculating Formula Mass (3.7)	Example 3.11	For Practice 3.11	Exercises 35, 36
Using Formula Mass to Count Molecules by Weighing (3.7)	Example 3.12	For Practice 3.12	Exercises 39–44
Calculating Mass Percent Composition (3.8)	Example 3.13	For Practice 3.13	Exercises 45–50
Using Chemical Formulas as Conversion Factors (3.8)	Example 3.14	For Practice 3.14	For More Practice 3.14 Exercises 57, 58

▲ **NEW! Key Learning Objectives** list the concepts that you should know after reading the chapter and are linked to in-chapter and end-of-chapter examples that show mastery of those skills.

End-of-Chapter Exercises

Problems by Topic

Note: Answers to all odd-numbered Problems, numbered in blue, can be found in Appendix III. Exercises in the Problems by Topic section are paired, with each odd-numbered problem followed by a similar even-numbered problem. Exercises in the Cumulative Problems section are also paired, but somewhat more loosely. (Challenge Problems and Conceptual Problems, because of their nature, are unpaired.)

Chemical Formulas and Molecular View of Elements and Compounds

1. Determine the number of each type of atom in each formula:
   a. $Ca_3(PO_4)_2$ b. $SrCl_2$ c. KNO_3 d. $Mg(NO_2)_2$
2. Determine the number of each type of atom in each formula:
   a. $Ba(OH)_2$ b. NH_4Cl c. $NaCN$ d. $Ba(HCO_3)_2$
3. Write a chemical formula for each molecular model. (See Appendix IIA for color codes.)

(a)   (b)   (c)

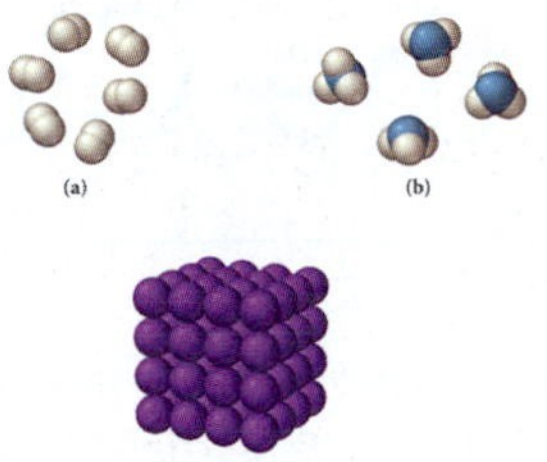

(c)

10. Based on the molecular views, classify each substance as an atomic element, a molecular element, an ionic compound, or a molecular compound.

▲ **Problems By Topic** are paired, with answers to the odd-numbered questions appearing in the appendix.

Conceptual Problems

119. When molecules are represented by molecular models, what does each sphere represent? How big is the nucleus of an atom in comparison to the sphere used to represent an atom in a molecular model?
120. Without doing any calculations, determine which element in each compound will have the highest mass percent composition.
    a. CO   b. N_2O
    c. $C_6H_{12}O_6$ d. NH_3
121. Explain the problem with the following statement and correct it. "The chemical formula for ammonia (NH_3) indicates that ammonia contains three grams of hydrogen to each gram of nitrogen."
122. Explain the problem with the following statement and correct it. "When a chemical equation is balanced, the number of molecules of each type on both sides of the equation will be equal."
123. Without doing any calculations, arrange the elements in H_2SO_4 in order of decreasing mass percent composition.
124. Element A is an atomic element and element B is a diatomic molecular element. Using circles to represent atoms of A and squares to represent atoms of B, draw molecular-level views of each element.

▲ **Conceptual Problems** let you test your grasp of key chapter concepts, often through reasoning that involves little or no math.

Cumulative Problems

85. How many molecules of ethanol (C_2H_5OH) (the alcohol in alcoholic beverages) are present in 165 mL of ethanol? The density of ethanol is 0.789 g/cm³.
86. A drop of water has a volume of approximately 0.05 mL. How many water molecules does it contain? The density of water is 1.0 g/cm³.
87. Determine the chemical formula of each compound and then use it to calculate the mass percent composition of each constituent element.
    a. potassium chromate   b. lead(II) phosphate
    c. sulfurous acid   d. cobalt(II) bromide

88. Determine the chemical formula of each compound and then use it to calculate the mass percent composition of each constituent element.
    a. perchloric acid   b. phosphorus pentachloride
    c. nitrogen triiodide   d. carbon dioxide
89. A Freon leak in the air conditioning system of an older car releases 32 g of CF_2Cl_2 per month. What mass of chlorine is emitted into the atmosphere each year by this car?
90. A Freon leak in the air-conditioning system of a large building releases 17 kg of CHF_2Cl per month. If the leak continues, how many kilograms of Cl will be emitted into the atmosphere each year?

▲ **Cumulative Problems** combine material from different parts of the chapter, and often from previous chapters as well, allowing you to see how well you can integrate the course material.

Challenge Problems

108. A mixture of NaCl and NaBr has a mass of 2.00 g and contains 0.75 g of Na. What is the mass of NaBr in the mixture?
109. Three pure compounds form when 1.00 g samples of element X combine with, respectively, 0.472 g, 0.630 g, and 0.789 g of element Z. The first compound has the formula X_2Z_3. Find the empirical formulas of the other two compounds.
110. A mixture of $CaCO_3$ and $(NH_4)_2CO_3$ is 61.9% CO_2 by mass. Find the mass percent of $CaCO_3$ in the mixture.
111. A mixture of 50.0 g of S and 1.00×10^2 g of Cl_2 reacts completely to form S_2Cl_2 and SCl_2. Find the mass of S_2Cl_2 formed.
112. Because of increasing evidence of damage to the ozone layer, chlorofluorocarbon (CFC) production was banned in 1996. However, there are about 100 million auto air conditioners that still use CFC-12 (CF_2Cl_2). These air conditioners are recharged from stockpiled supplies of CFC-12. If each of the 100 million automobiles contains 1.1 kg of CFC-12 and leaks 25% of its CFC-12 into the atmosphere per year, how much chlorine, in kg, enters the atmosphere each year due to auto air conditioners? (Assume two significant figures in your calculations.)

113. A particular coal contains 2.55% sulfur by mass. When the coal is burned, it produces SO_2 emissions which combine with rainwater to produce sulfuric acid. Use the formula of sulfuric acid to calculate the mass percent of S in sulfuric acid. Then determine how much sulfuric acid (in metric tons) is produced by the combustion of 1.0 metric ton of this coal. (A metric ton is 1000 kg.)
114. Lead is found in Earth's crust in several different lead ores. Suppose a certain rock is composed of 38.0% PbS (galena), 25.0% $PbCO_3$ (cerussite), and 17.4% $PbSO_4$ (anglesite). The remainder of the rock is composed of substances containing no lead. How much of this rock (in kg) must be processed to obtain 5.0 metric tons of lead? (A metric ton is 1000 kg.)
115. A 2.52 g sample of a compound containing only carbon, hydrogen, nitrogen, oxygen, and sulfur was burned in excess O_2 to yield 4.23 g of CO_2 and 1.01 g of H_2O. Another sample of the same compound, of mass 4.14 g, yielded 2.11 g of SO_3. A reaction to determine the nitrogen content was carried out on a third sample with a mass of 5.66 g and yielded

▲ **Challenge Problems** are designed to challenge stronger students.

Answers to Conceptual Connections

Structural Formulas

3.1 H—O—H

Representing Molecules

3.2 The spheres represent the electron cloud of the atom. It would be nearly impossible to draw a nucleus to scale on any of the space-filling molecular models—on this scale, the nucleus would be too small to see.

A Molecular View of Elements and Compounds

3.3

[diagram: A₂B and CD]

Ionic and Molecular Compounds

3.4 Choice (a) best describes the difference between ionic and molecular compounds. The (b) answer is incorrect because there are no "new" forces in bonding (just rearrangements that result in lower potential energy), and because ions do not group together in pairs in the solid phase. The (c) answer is incorrect because the main difference between ionic and molecular compounds is the way that the atoms bond. The (d) answer is incorrect because ionic compounds do not contain molecules.

Nomenclature

3.5 This conceptual connection addresses one of the main errors you can make in nomenclature: the failure to correctly categorize the compound. Remember that you must first determine whether the compound is an ionic compound, a molecular compound, or an acid, and then name it accordingly. NCl_3 is a molecular compound (two or more nonmetals), and therefore prefixes indicate the number of each type of atom—so NCl_3 is nitrogen trichloride. The compound $AlCl_3$, however, is an ionic compound (metal and nonmetal) and therefore does not require prefixes—so $AlCl_3$ is aluminum chloride.

Molecular Models and the Size of Molecules

3.6 Answer: (c) Atomic radii range in the hundreds of picometers, while the spheres in these models have radii of less than a centimeter. The scaling factor is therefore about 10^8 (100 million).

▲ **Conceptual Connections** are strategically placed to reinforce conceptual understanding of the most complex concepts. The book has over 50% more Conceptual Connections than the previous edition and answers can be found at the end of each chapter.

MasteringChemistry® Extend Learning Beyond the Classroom

MasteringChemistry® is designed with a single purpose: to help students reach the moment of understanding. The Mastering online homework and tutoring system delivers self-paced tutorials that provide students with individualized coaching set to your course objectives. MasteringChemistry helps students arrive better prepared for lecture and lab.

Engaging Experiences

MasteringChemistry® promotes interactivity and active learning. Research shows that Mastering's immediate feedback and tutorial assistance helps students understand and master concepts and skills in Chemistry—allowing them to retain more knowledge and perform better in this course and beyond.

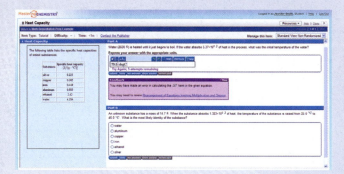

Student Tutorials

MasteringChemistry is the only system to provide instantaneous feedback specific to the most common wrong answers. Students can submit an answer and receive immediate, error-specific feedback. Simpler sub-problems—hints—are provided upon request.

Reading quizzes

Reading Quizzes give instructors the opportunity to assign reading, and test students on their comprehension of chapter content

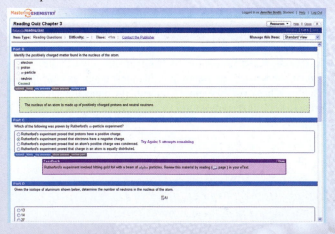

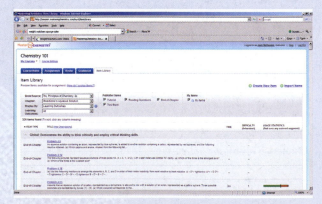

Learning Objectives

NEW! Learning Objectives appear at the end of each chapter and are linked to end-of-chapter problems in the text and in MasteringChemistry®. Problems in MasteringChemistry can be sorted by each learning objective increasing efficiencies in teaching and learning.

Make Learning Part of the Grade®

www.masteringchemistry.com

A Trusted Partner

The mastering platform was developed by scientists for science students and instructors, and has a proven history with more than 10 years of student use. Mastering currently has more than 1.5 million active registrations with active users in 50 states and 41 countries. The Mastering platform has 99.8% server reliability.

Pearson eText

Pearson eText provides access to the text when and wherever students have access to the Internet. eText pages look exactly like the printed text, offering powerful new functionality. Users can create notes, highlight the text in different colors, create bookmarks, zoom, click hyperlinked words and phrases to view definitions, view as single or two-pages. etc, etc.

Gradebook

Every assignment is automatically graded. Shades of red highlight vulnerable students and challenging assignments.

Gradebook Diagnostics

This screen provides you with your favorite diagnostics. With a single click, charts summarize the most difficult problems, vulnerable students, grade distribution, and even score improvement over the course.

Proven Results

The Mastering platform is the only homework system with research showing that it improves student learning. A wide variety of published papers, based on NSF-sponsored research and tests, illustrate the benefits of the Mastering program. Results documented in scientifically valid efficacy papers are available at **www.masteringchemistry.com/site/results**.

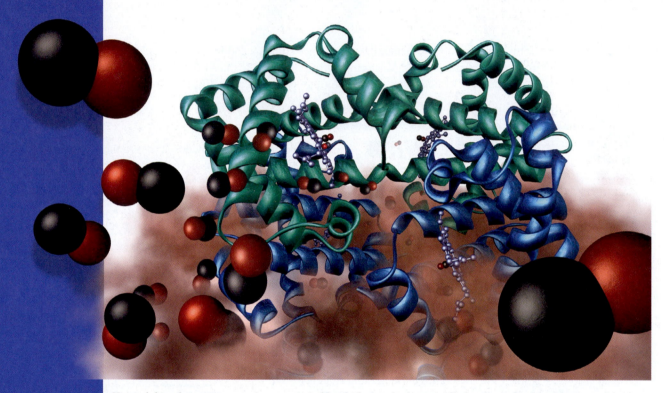

Matter, Measurement, and Problem Solving

The most incomprehensible thing about the universe is that it is comprehensible.
—Albert Einstein (1879–1955)

Hemoglobin, the oxygen-carrying protein in blood (depicted schematically here), can bind carbon monoxide molecules (the linked red and black spheres) as well as oxygen.

WHAT DO YOU THINK is the most important idea in all of human knowledge? There are, of course, many possible answers to this question—some practical, some philosophical, and some scientific. If we limit ourselves only to scientific answers, mine would be this: **the properties of matter are determined by the properties of molecules and atoms**. Atoms and molecules determine how matter behaves—if they were different, matter would be different. The properties of water molecules, for example, determine how water behaves; the properties of sugar molecules determine how sugar behaves;

and the molecules that compose our bodies determine how our bodies behave. The understanding of matter at the molecular level gives us unprecedented control over that matter. Our understanding of the details of the molecules that compose living organisms has made possible the revolution that has occurred in biology over the last 50 years.

1.1   Atoms and Molecules

The air over most U.S. cities, including my own, contains at least some pollution. A significant component of that pollution is carbon monoxide, a colorless gas emitted in the exhaust of cars and trucks. Carbon monoxide *gas* is composed of carbon monoxide *molecules*, each of which contains a carbon atom and an oxygen atom held together by a chemical bond. **Atoms** are the submicroscopic particles that constitute the fundamental building blocks of ordinary matter. However, free atoms are not common in nature; instead they bind together in specific geometric arrangements to form **molecules**.

The properties of the substances around us depend on the atoms and molecules that compose them, so the properties of carbon monoxide *gas* depend on the properties of carbon monoxide *molecules*. Carbon monoxide molecules happen to be just the right size and shape, and happen to have just the right chemical properties, to fit neatly into cavities within hemoglobin—the oxygen-carrying molecule in blood—that normally carry oxygen molecules (**Figure 1.1▼**). Consequently, carbon monoxide diminishes the oxygen-carrying capacity of blood. Breathing air containing too much carbon monoxide (greater than 0.04% by volume) can lead to unconsciousness and even death because not enough oxygen reaches the brain. Carbon monoxide deaths have occurred, for example, as a result of running an automobile in a closed garage or using a propane burner in an enclosed space for too long. In smaller amounts, carbon monoxide causes the heart and lungs to work harder and can result in headache, dizziness, weakness, and confusion.

Cars and trucks emit another closely related molecule, called carbon dioxide, in far greater quantities than carbon monoxide. The only difference between carbon dioxide and carbon monoxide is that carbon dioxide molecules contain two oxygen atoms instead of just one. This extra oxygen atom dramatically affects the properties of the gas. We breathe much more carbon dioxide—which composes 0.04% of air, and is a product of our own respiration as well—than carbon monoxide, yet it does not kill us. Why? Because the presence of the second oxygen atom prevents carbon dioxide from binding to the oxygen-carrying site in hemoglobin, making it far less toxic. Although high levels of carbon dioxide (greater than 10% of air) can be toxic for other reasons, lower levels can

Carbon monoxide molecule

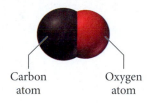

Carbon atom   Oxygen atom

Carbon dioxide molecule

Oxygen atom   Oxygen atom

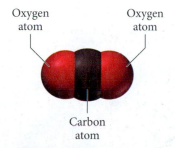

Carbon atom

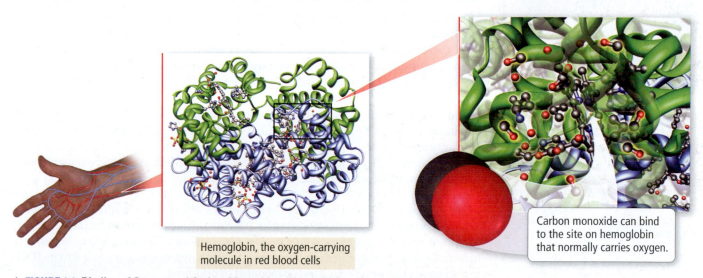

Hemoglobin, the oxygen-carrying molecule in red blood cells

Carbon monoxide can bind to the site on hemoglobin that normally carries oxygen.

▲ **FIGURE 1.1 Binding of Oxygen and Carbon Monoxide to Hemoglobin**   Hemoglobin, a large protein molecule, is the oxygen carrier in red blood cells. Each subunit of the hemoglobin molecule contains an iron atom to which oxygen binds. Carbon monoxide molecules can take the place of oxygen, thus reducing the amount of oxygen reaching the body's tissues.

In the study of chemistry, atoms are often portrayed as colored spheres, with each color representing a different kind of atom. For example, a black sphere represents a carbon atom, a red sphere represents an oxygen atom, and a white sphere represents a hydrogen atom. For a complete color code of atoms, see Appendix IIA.

enter the bloodstream with no adverse effects. Such is the molecular world. Any change in a molecule—such as the addition of an oxygen atom to carbon monoxide—will likely alter the properties of the substance composed by that molecule.

As another example, consider two other closely related molecules, water and hydrogen peroxide:

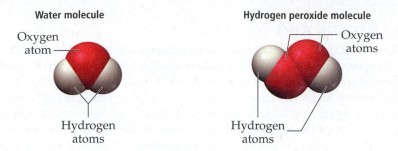

Water molecule

Oxygen atom

Hydrogen atoms

Hydrogen peroxide molecule

Oxygen atoms

Hydrogen atoms

A water molecule is composed of *one* oxygen atom and two hydrogen atoms. A hydrogen peroxide molecule is composed of *two* oxygen atoms and two hydrogen atoms. This seemingly small molecular difference results in a huge difference between water and hydrogen peroxide. Water is the familiar and stable liquid we all drink and bathe in. Hydrogen peroxide, in contrast, is an unstable liquid that, in its pure form, burns the skin on contact and is used in rocket fuel. When you pour water onto your hair, your hair simply becomes wet. However, if you put hydrogen peroxide in your hair—which you may have done if you have ever bleached your hair—a chemical reaction occurs that turns your hair blonde.

The hydrogen peroxide used as an antiseptic or bleaching agent is considerably diluted.

The details of how specific atoms bond to form a molecule—in a straight line, at a particular angle, in a ring, or in some other pattern—as well as the type of atoms in the molecule, determine everything about the substance that the molecule composes. If we want to understand the substances around us, we must understand the atoms and molecules that compose them—this is the central goal of chemistry. A good simple definition of **chemistry** is, therefore,

> **Chemistry—the science that seeks to understand the behavior of matter by studying the behavior of atoms and molecules**.

1.2 The Scientific Approach to Knowledge

Scientific knowledge is empirical—it is based on *observation* and *experiment*. Scientists observe and perform experiments on the physical world to learn about it. Some observations and experiments are qualitative (noting or describing how a process happens), but many are quantitative (measuring or quantifying something about the process). For example, Antoine Lavoisier (1743–1794), a French chemist who studied combustion, made careful measurements of the mass of objects before and after burning them in closed containers. He noticed that there was no change in the total mass of material within the container during combustion. Lavoisier made an important *observation* about the physical world.

▲ A painting of the French chemist Antoine Lavoisier with his wife, Marie, who helped him in his work by illustrating his experiments and translating scientific articles from English. Lavoisier, who also made significant contributions to agriculture, industry, education, and government administration, was executed during the French Revolution.

Observations often lead a scientist to formulate a **hypothesis**, a tentative interpretation or explanation of the observations. For example, Lavoisier explained his observations on combustion by hypothesizing that when a substance burns, it combines with a component of air. A good hypothesis is *falsifiable*, which means that it makes predictions that can be confirmed or refuted by further observations. Hypotheses are tested by **experiments**, highly controlled procedures designed to generate observations that can confirm or refute a hypothesis. The results of an experiment may support a hypothesis or prove it wrong. If it is proven wrong, the hypothesis must be modified or discarded.

In some cases, a series of similar observations can lead to the development of a **scientific law**, a brief statement that summarizes past observations and predicts future ones. For example, Lavoisier summarized his observations on combustion with the **law of conservation of mass**, which states, "In a chemical reaction, matter is neither created nor destroyed."

This statement summarized Lavoisier's observations on chemical reactions and predicted the outcome of future observations on reactions. Laws, like hypotheses, are also subject to experiments, which can add support to them or prove them wrong.

Scientific laws are not *laws* in the same sense as civil or governmental laws. Nature does not follow laws in the way that we obey the laws against speeding or running a red light. Rather, scientific laws *describe* how nature behaves—they are generalizations about what nature does. For that reason, some people find it more appropriate to refer to them as *principles* rather than *laws*.

One or more well-established hypotheses may form the basis for a scientific **theory**. A scientific theory is a model for the way nature is and tries to explain not merely what nature does, but why. As such, well-established theories are the pinnacle of scientific knowledge, often predicting behavior far beyond the observations or laws from which they were developed. A good example of a theory is the **atomic theory** proposed by English chemist John Dalton (1766–1844). Dalton explained the law of conservation of mass, as well as other laws and observations of the time, by proposing that matter is com-posed of small, indestructible particles called atoms. Since these particles merely rear-range in chemical changes (and do not form or vanish), the total amount of mass remains the same. Dalton's theory is a model for the physical world—it gives us insight into how nature works, and therefore *explains* our laws and observations.

Finally, the scientific approach returns to observations, usually in the form of experiments, to test theories. For example, scientists can test the atomic theory by try-ing to isolate single atoms, or by trying to image them (both of which, by the way, have already been accomplished). Theories are validated by experiments such as these; how-ever, theories can never be conclusively proven because some new observation or exper-iment always has the potential to reveal a flaw. Notice that the scientific approach to knowledge begins with observation and ends with observation, because an experiment is simply a highly controlled procedure for generating critical observations designed to test a theory or hypothesis. Each new set of observations has the potential to refine the original model. This approach, often called the **scientific method**, is summarized in **Figure 1.2▼**. Scientific laws, hypotheses, and theories are all subject to continued experi-mentation. If a law, hypothesis, or theory is proved wrong by an experiment, it must be revised and tested with new experiments. Over time, poor theories and laws are elimi-nated or corrected and good theories and laws—those consistent with experimental results—remain.

Established theories with strong experimental support are the most powerful pieces of scientific knowledge. You may have heard the phrase, "That is just a theory," as if theories are easily dismissible. Such a statement reveals a deep misunderstanding of the nature of a scientific theory. Well-established theories are as close to truth as we get in science. The idea that all matter is made of atoms is "just a theory," but it has over 200 years of experi-mental evidence to support it. It is a powerful piece of scientific knowledge on which many other scientific ideas have been built.

One last word about the scientific method: some people wrongly imagine science to be a strict set of rules and procedures that automatically lead to inarguable,

In Dalton's time, atoms were thought to be indestructible. Today, because of nuclear reactions, we know that atoms can be broken apart into their smaller components.

The Scientific Method

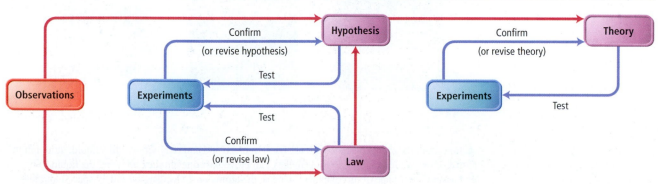

▲ **FIGURE 1.2 The Scientific Method**

objective facts. This is not the case. Even our diagram of the scientific method is only an idealization of real science, useful to help us see key distinctions. Doing real science requires hard work, care, creativity, and even a bit of luck. Scientific theories do not just fall out of data—they are crafted by men and women of great genius and creativity. A great theory is not unlike a master painting and many see a similar kind of beauty in both.

Conceptual Connection 1.1　Laws and Theories

Which statement best explains the difference between a law and a theory?

(a) A law is truth whereas a theory is mere speculation.

(b) A law summarizes a series of related observations, while a theory gives the underlying reasons for them.

(c) A theory describes *what* nature does; a law describes *why* nature does it.

1.3　The Classification of Matter

Matter is anything that occupies space and has mass. This book, your desk, your chair, and even your body are all composed of matter. Less obviously, the air around you is also matter—it too occupies space and has mass. We often call a specific instance of matter—such as air, water, or sand—a **substance**. We classify matter according to its state—solid, liquid, or gas—and according to its composition.

The States of Matter: Solid, Liquid, and Gas

The state of matter changes from solid to liquid to gas with increasing temperature.

Matter can exist in three different **states**: **solid**, **liquid**, and **gas**. In *solid matter*, atoms or molecules pack close to each other in fixed locations. Although the atoms and molecules in a solid vibrate, they do not move around or past each other. Consequently, a solid has a fixed volume and rigid shape. Ice, aluminum, and diamond are examples of solids. Solid matter may be **crystalline**, in which case its atoms or molecules are arranged in patterns

| Solid matter | Liquid matter | Gaseous matter |

▲ In a solid, the atoms or molecules are fixed in place and can only vibrate. In a liquid, although the atoms or molecules are closely packed, they can move past one another, allowing the liquid to flow and assume the shape of its container. In a gas, the atoms or molecules are widely spaced, making gases compressible as well as fluid.

Crystalline:
Regular 3-dimensional
pattern

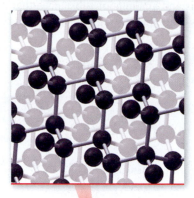

Diamond
C (*s*, diamond)

with long-range, repeating order (**Figure 1.3▲**), or it may be **amorphous**, in which case its atoms or molecules do not have any long-range order. Examples of *crystalline* solids include table salt and diamond; the well-ordered geometric shapes of salt and diamond crystals reflect the well-ordered geometric arrangement of their atoms. Examples of *amorphous* solids include glass and plastic. In *liquid matter*, atoms or molecules pack about as closely as they do in solid matter, but are free to move relative to each other, giving liquids a fixed volume but not a fixed shape. Liquids assume the shape of their container. Water, alcohol, and gasoline are all substances that are liquids at room temperature.

In *gaseous matter*, atoms or molecules have a lot of space between them and are free to move relative to one another, making gases *compressible* (**Figure 1.4▼**). When you squeeze a balloon or sit down on an air mattress, you force the atoms and molecules into a smaller space, so that they are closer together. Gases always assume the shape *and* volume of their container. Substances that are gases at room temperature include helium, nitrogen (the main component of air), and carbon dioxide.

◀ **FIGURE 1.4 The Compressibility of Gases**   Gases can be compressed—squeezed into a smaller volume—because there is so much empty space between atoms or molecules in the gaseous state.

Solid—not compressible   Gas—compressible

Classifying Matter According to Its Composition: Elements, Compounds, and Mixtures

In addition to classifying matter according to its state, we can classify it according to its **composition**, i.e., the kinds and amounts of substances that compose it. The following chart shows how to classify matter according to its composition:

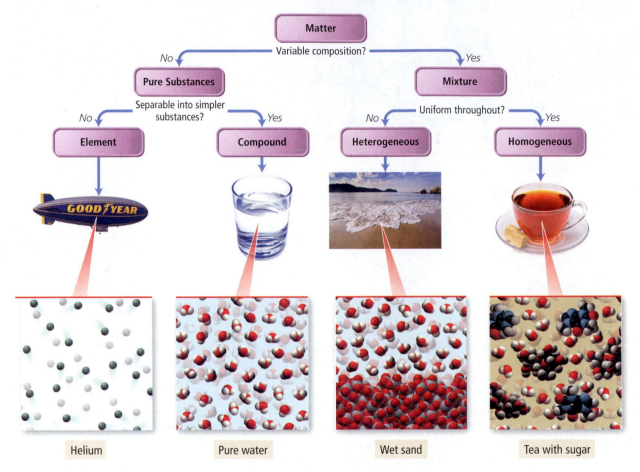

Helium   Pure water   Wet sand   Tea with sugar

The first division in the classification of matter depends on whether or not its composition can vary from one sample to another. For example, the composition of distilled (or pure) water never varies—it is always 100% water and is therefore a **pure substance**, a substance composed of only a single type of atom or molecule. In contrast, the composition of sweetened tea can vary considerably from one sample to another, depending, for instance, on the strength of the tea or how much sugar has been added. Sweetened tea is an example of a **mixture**, a substance composed of two or more different types of atoms or molecules that can be combined in continuously variable proportions.

Pure substances can be categorized into two types—elements and compounds—depending on whether or not they can be broken down into simpler substances. The helium in a blimp or party balloon is an example of an **element**, a substance that cannot be chemically broken down into simpler substances. Water is an example of a **compound**, a substance composed of two or more elements (hydrogen and oxygen) in fixed, definite proportions. On Earth, compounds are more common than pure elements because most elements combine with other elements to form compounds.

Mixtures can also be categorized into two types—heterogeneous and homogeneous—depending on how uniformly the substances within them mix. Wet sand is an example of a **heterogeneous mixture**, one in which the composition varies from one region to another. Sweetened tea is an example of a **homogeneous mixture**, one with the same composition throughout. Homogeneous mixtures have uniform compositions because the atoms or molecules that compose them mix uniformly. Heterogeneous mixtures are made up of distinct regions because the atoms or molecules that compose them separate. Here again we see that the properties of matter are determined by the atoms or molecules that compose it.

◯◯ Conceptual Connection 1.2 Pure Substances and Mixtures

Let a small circle represent an atom of one type of element and a small square represent at atom of a second type of element. Make a drawing of: (a) a pure substance composed of the two elements (in a one-to-one ratio); and (b) a homogenous mixture composed of the two elements; and (c) a heterogeneous mixture composed of the two elements.

1.4   Physical and Chemical Changes and Physical and Chemical Properties

Every day we witness changes in matter: ice melts, iron rusts, gasoline burns, fruit ripens, and water evaporates. What happens to the molecules that compose these samples of matter during such changes? The answer depends on the type of change. Changes that alter only state or appearance, but not composition, are **physical changes**. The atoms or molecules that compose a substance *do not change* their identity during a physical change. For example, when water boils, it changes its state from a liquid to a gas, but the gas remains composed of water molecules, which means this is a physical change (**Figure 1.5▼**).

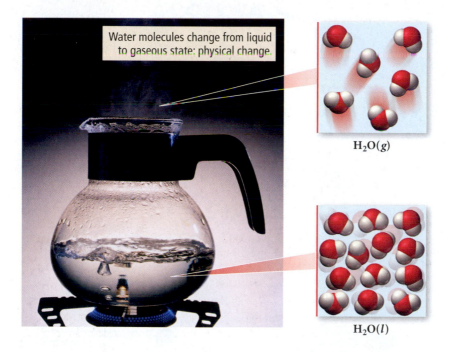

Water molecules change from liquid to gaseous state: physical change.

$H_2O(g)$

$H_2O(l)$

◀ **FIGURE 1.5 Boiling, a Physical Change**   When water boils, it turns into a gas but does not alter its chemical identity—the water molecules are the same in both the liquid and gaseous states. Boiling is a physical change, and the boiling point of water is a physical property.

In contrast, changes that alter the composition of matter are **chemical changes**. During a chemical change, atoms rearrange, transforming the original substances into different substances. For example, the rusting of iron is a chemical change. The atoms that compose iron (iron atoms) combine with oxygen molecules from air to form iron oxide, the orange substance we normally call rust (**Figure 1.6▶**). Some other examples of physical and chemical changes are shown in **Figure 1.7▶**.

Physical and chemical changes are manifestations of physical and chemical properties. A **physical property** is one that a substance displays without changing its composition, whereas a **chemical property** is one that a substance displays only by changing its composition via a chemical change. For example, the smell of gasoline is a physical property—gasoline does not change its composition when it exhibits its odor. The *flammability* of gasoline, however, is a chemical property—gasoline does change its composition when it burns, turning into completely new substances (primarily carbon dioxide and water). Physical properties include odor, taste, color, appearance, melting point, boiling point, and density. Chemical properties include corrosiveness, flammability, acidity, toxicity, and other such characteristics.

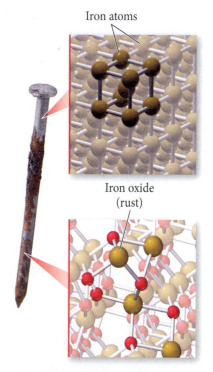

Iron atoms

Iron oxide (rust)

▲ **FIGURE 1.6 Rusting, a Chemical Change**   When iron rusts, the iron atoms combine with oxygen atoms to form a different chemical substance, the compound iron oxide. Rusting is a chemical change, and the tendency of iron to rust is a chemical property.

Physical Change and Chemical Change

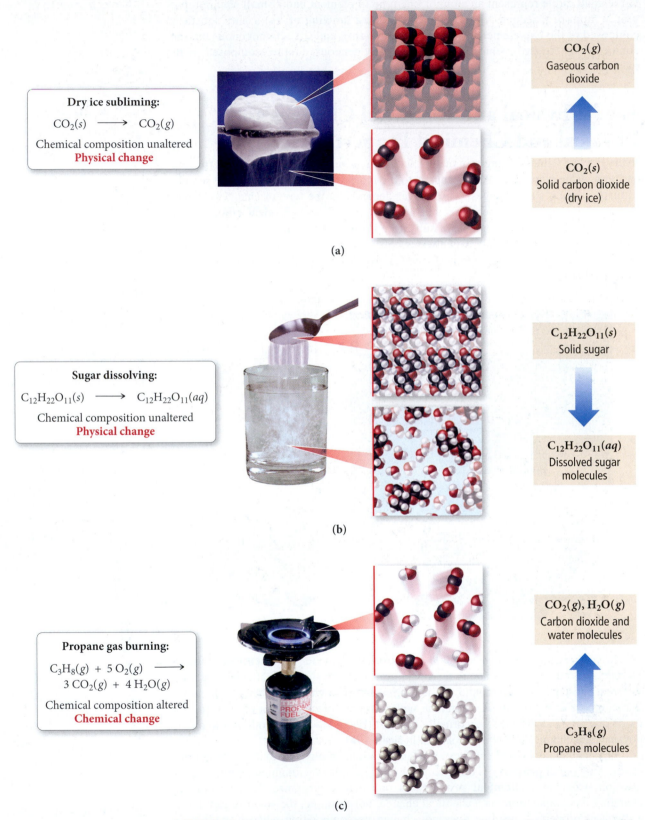

Dry ice subliming:

$$CO_2(s) \longrightarrow CO_2(g)$$

Chemical composition unaltered
Physical change

$CO_2(g)$
Gaseous carbon dioxide

$CO_2(s)$
Solid carbon dioxide (dry ice)

(a)

Sugar dissolving:

$$C_{12}H_{22}O_{11}(s) \longrightarrow C_{12}H_{22}O_{11}(aq)$$

Chemical composition unaltered
Physical change

$C_{12}H_{22}O_{11}(s)$
Solid sugar

$C_{12}H_{22}O_{11}(aq)$
Dissolved sugar molecules

(b)

Propane gas burning:

$$C_3H_8(g) + 5\,O_2(g) \longrightarrow$$
$$3\,CO_2(g) + 4\,H_2O(g)$$

Chemical composition altered
Chemical change

$CO_2(g)$, $H_2O(g)$
Carbon dioxide and water molecules

$C_3H_8(g)$
Propane molecules

(c)

▲ **FIGURE 1.7 Physical and Chemical Changes**   **(a)** The sublimation (the state change from solid to gas) of dry ice (solid CO_2) is a physical change. **(b)** The dissolution of sugar is a physical change. **(c)** The burning of propane is a chemical change.

The differences between physical and chemical changes are not always apparent. Only chemical examination can confirm whether any particular change is physical or chemical. In many cases, however, we can identify chemical and physical changes based on what we know about the changes. Changes in the state of matter, such as melting or boiling, or changes in the physical condition of matter, such as those that result from cutting or crushing, are typically physical changes. Changes involving chemical reactions—often evidenced by heat exchange or color changes—are chemical changes.

In Chapter 19 we will also learn about *nuclear changes*, which can involve atoms of one element changing into atoms of a different element.

A physical change results in a different form of the same substance, while a chemical change results in a completely different substance.

EXAMPLE 1.1 Physical and Chemical Changes and Properties

Determine whether each change is physical or chemical. What kind of property (chemical or physical) is demonstrated in each case?

(a) the evaporation of rubbing alcohol

(b) the burning of lamp oil

(c) the bleaching of hair with hydrogen peroxide

(d) the forming of frost on a cold night

SOLUTION

(a) When rubbing alcohol evaporates, it changes from liquid to gas, but it remains alcohol—this is a physical change. The volatility (the ability to evaporate easily) of alcohol is a physical property.

(b) Lamp oil burns because it reacts with oxygen in air to form carbon dioxide and water—this is a chemical change. The flammability of lamp oil is a chemical property.

(c) Applying hydrogen peroxide to hair changes pigment molecules in hair that give it color—this is a chemical change. The susceptibility of hair to bleaching is a chemical property.

(d) Frost forms on a cold night because water vapor in air changes its state to form solid ice—this is a physical change. The temperature at which water freezes is a physical property.

FOR PRACTICE 1.1

Determine whether each change is physical or chemical. What kind of property (chemical or physical) is demonstrated in each case?

(a) A copper wire is hammered flat.

(b) A nickel dissolves in acid to form a blue-green solution.

(c) Dry ice sublimes (changes into a gas) without melting.

(d) A match ignites when struck on a flint.

Answers to For Practice and For More Practice problems can be found in Appendix IV.

Conceptual Connection 1.3 Chemical and Physical Changes

The diagram to the right represents liquid water molecules in a pan.

Which diagram best represents the water molecules after they have been vaporized by the boiling of liquid water?

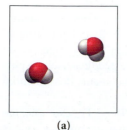

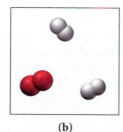

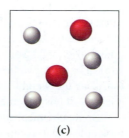

(a)   (b)   (c)

1.5 Energy: A Fundamental Part of Physical and Chemical Change

The physical and chemical changes that we have just discussed are usually accompanied by energy changes. For example, when water evaporates from your skin (a physical change), the water molecules absorb energy from your body, making you feel cooler. When you burn natural gas on the stove (a chemical change), energy is released, heating the food you are cooking. Understanding the physical and chemical changes of matter—that is, understanding chemistry—requires that we also understand energy changes and energy flow.

The scientific definition of **energy** is *the capacity to do work*. **Work** is the action of a force through a distance. For instance, when you push a box across the floor or pedal your bicycle down the street, you have done work.

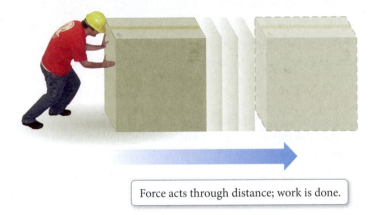

Force acts through distance; work is done.

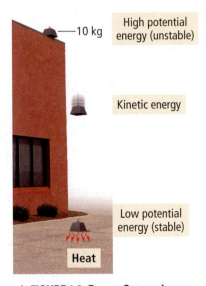

10 kg

High potential energy (unstable)

Kinetic energy

Low potential energy (stable)

Heat

▲ **FIGURE 1.8 Energy Conversions** Gravitational potential energy is converted into kinetic energy when the weight is released. The kinetic energy is converted mostly to thermal energy when the weight strikes the ground.

We will find in Chapter 19 that energy conservation is actually part of a more general law that allows for the interconvertibility of mass and energy.

The *total energy* of an object is a sum of its **kinetic energy**, the energy associated with its motion, and its **potential energy**, the energy associated with its position or composition. For example, a weight held several meters above the ground has potential energy due to its position within Earth's gravitational field (**Figure 1.8◄**). If the weight is dropped, it accelerates, and the potential energy is converted to kinetic energy. When the weight hits the ground, its kinetic energy is converted primarily to **thermal energy**, the energy associated with the temperature of an object. Thermal energy is actually a type of kinetic energy because it arises from the motion of the individual atoms or molecules that make up an object. When the weight hits the ground its kinetic energy is essentially transferred to the atoms and molecules that compose the ground, raising the temperature of the ground ever so slightly.

The first principle to note about the way that energy changes as the weight falls to the ground is that *energy is neither created nor destroyed*. The potential energy of the weight becomes kinetic energy as the weight accelerates toward the ground. The kinetic energy then becomes thermal energy when the weight hits the ground. The total amount of thermal energy that is released through the process is exactly equal to the difference between the initial and final potential energy of the weight. The idea that energy is neither created nor destroyed is known as the **law of conservation of energy**. Although energy can change from one type to another, and although it can flow from one object to another, the *total quantity* of energy does not change—it remains constant.

The second principle to note about the raised weight and its fall is that *systems with high potential energy tend to change in ways that lower their potential energy*. For this reason, objects or systems with high potential energy tend to be *unstable*. The weight lifted several meters from the ground is unstable because it contains a significant amount of localized potential energy. Unless restrained, the weight will naturally fall, lowering its potential energy. Some of the raised weight's potential energy can be harnessed to do work. For example, the weight can be attached to a rope that turns a paddle wheel or spins a drill as the weight falls. After it falls to the ground, the weight contains less potential energy—it has become more *stable*.

Some chemical substances are like the raised weight just described. For example, the molecules that compose gasoline have a relatively high potential energy—energy is

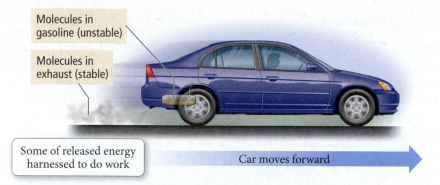

Molecules in
gasoline (unstable)

Molecules in
exhaust (stable)

Some of released energy
harnessed to do work

Car moves forward

▲ **FIGURE 1.9 Using Chemical Energy to Do Work**   The compounds produced when gasoline burns have less chemical potential energy than the gasoline molecules.

concentrated in them just as energy is concentrated in the raised weight. The molecules in the gasoline therefore tend to undergo chemical changes (specifically combustion) that lower their potential energy. As the energy of the gas molecules is released, some of it can be harnessed to do work, such as moving a car forward (**Figure 1.9▲**). The molecules that result from the chemical change have less potential energy than the original molecules in gasoline and are therefore more stable.

Chemical potential energy, such as that contained in the gasoline molecules, arises primarily from the forces between the electrically charged particles (protons and electrons) that compose atoms and molecules. We will learn more about those particles, as well as the properties of electrical charge, in Chapter 2, but for now, know that molecules contain specific, usually complex, arrangements of these charged particles. Some of these arrangements—such as the one within the molecules that compose gasoline—have a much higher potential energy than others. When gasoline combusts the arrangement of these particles changes, creating molecules with much lower potential energy and transferring a great deal of energy (mostly in the form of heat) to the surroundings.

Summarizing Energy:

▶ Energy is always conserved in a physical or chemical change; it is neither created nor destroyed.

▶ Systems with high potential energy tend to change in a direction that lowers their potential energy, releasing energy into the surroundings.

1.6   The Units of Measurement

In 1999, NASA lost the $125 million *Mars Climate Orbiter* because of confusion between English and metric units. The chairman of the commission that investigated the disaster concluded, "The root cause of the loss of the spacecraft was a failed translation of English units into metric units." As a result, the orbiter—which was supposed to monitor weather on Mars—descended too far into the Martian atmosphere and burned up. In chemistry, as in space exploration, **units**—standard quantities used to specify measurements—are critical. If you get them wrong, the consequences can be enormous.

The two most common unit systems are the **English system**, used in the United States, and the **metric system**, used in most of the rest of the world. The English system consists of units such as inches, yards, and pounds, while the metric system relies on centimeters, meters, and kilograms. The unit system used by scientists is based on the metric system and is called the **International System of Units (SI)**.

The Standard Units

The standard SI base units are shown in Table 1.1. For now, we will focus on the first four of these units including the *meter* as the standard unit of length, the *kilogram* as the standard unit of mass, the *second* as the standard unit of time, and the *kelvin* as the standard unit of temperature.

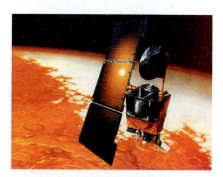

▲ The $125 million *Mars Climate Orbiter* was lost in the Martian atmosphere in 1999 because two groups of engineers failed to communicate to each other the units that they used in their calculations.

The abbreviation *SI* comes from the French, *Système International d'Unités.*

TABLE 1.1 SI Base Units

Quantity	Unit	Symbol
Length	Meter	m
Mass	Kilogram	kg
Time	Second	s
Temperature	Kelvin	K
Amount of substance	Mole	mol
Electric current	Ampere	A
Luminous intensity	Candela	cd

The Meter: A Measure of Length

The **meter (m)** is slightly longer than a yard (1 yard is 36 inches while 1 meter is 39.37 inches).

Yardstick

Meterstick

Thus, a 100-yard football field measures only 91.4 meters. The meter was originally defined as 1/10,000,000 of the distance from the equator to the north pole (through Paris). It is now defined more precisely as the distance light travels through a vacuum in a certain period of time, 1/299,792,458 second.

▲ A basketball player stands about 2 meters tall.

2 m

The Kilogram: A Measure of Mass

The **kilogram (kg)** is defined as the mass of a metal cylinder kept at the International Bureau of Weights and Measures at Sèvres, France. The kilogram is a measure of *mass*, a quantity different from *weight*. The **mass** of an object is a measure of the quantity of matter within it, while the weight of an object is a measure of the *gravitational pull* on the matter within it. If you weigh yourself on the moon, for example, its weaker gravity pulls on you with less force than does Earth's gravity, resulting in a lower weight. A 130-pound (lb) person on Earth only weighs 21.5 lb on the moon. However, the person's mass—the quantity of matter in his or her body—remains the same. One kilogram of mass is the equivalent of 2.205 lb of weight on Earth, so if we express mass in kilograms, a 130-lb person has a mass of approximately 59 kg and this book has a mass of about 2.0 kg. A second common unit of mass is the gram (g). One gram is 1/1000 kg.

▲ A nickel (5 cents) weighs about 5 grams.

 **Conceptual Connection 1.4 The Mass of a Gas**

A drop of water is put into a container and the container is sealed. The drop of water then vaporizes (turns from a liquid into a gas). Does the mass of the sealed container and its contents change upon vaporization?

The Second: A Measure of Time

For those of us who live in the United States, the **second (s)** is perhaps the most familiar SI unit. The second was originally defined in terms of the day and the year, but it is now defined more precisely as the duration of 9,192,631,770 periods of the radiation emitted from a certain transition in a cesium-133 atom.

The Kelvin: A Measure of Temperature

The **kelvin (K)** is the SI unit of **temperature**. The temperature of a sample of matter is a measure of the amount of average kinetic energy—the energy due to motion—of the atoms or molecules that compose the matter. For example, the molecules in a *hot* glass of

Temperature Scales

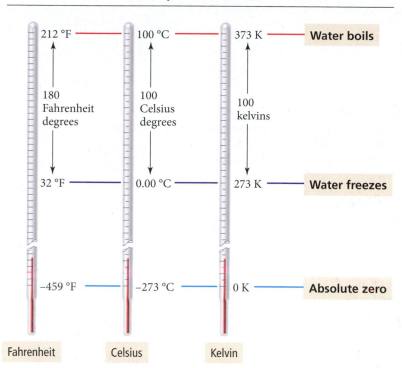

◀ **FIGURE 1.10 Comparison of the Fahrenheit, Celsius, and Kelvin Temperature Scales** The Fahrenheit degree is five-ninths the size of the Celsius degree and the kelvin. The zero point of the Kelvin scale is absolute zero (the lowest possible temperature), whereas the zero point of the Celsius scale is the freezing point of water.

water are, on average, moving faster than the molecules in a *cold* glass of water. Temperature is a measure of this molecular motion.

The three common temperature scales are shown in **Figure 1.10▲**. The most familiar in the United States is the **Fahrenheit (°F) scale**, shown on the left. On the Fahrenheit scale, water freezes at 32 °F and boils at 212 °F (at sea level). Room temperature is approximately 72 °F. The scale most often used by scientists and by most countries other than the United States is the **Celsius (°C) scale**, shown in the middle. On this scale, pure water freezes at 0 °C and boils at 100 °C (at sea level). Room temperature is approximately 22 °C. The Fahrenheit scale and the Celsius scale differ both in the size of their respective degrees and the temperature each designates as "zero." Both the Fahrenheit and Celsius scales allow for negative temperatures.

The SI unit for temperature, as we have seen, is the kelvin, shown on the right in Figure 1.10. The **Kelvin scale** (sometimes also called the *absolute scale*) avoids negative temperatures by assigning 0 K to the coldest temperature possible, absolute zero. Absolute zero (−273 °C or −459 °F) is the temperature at which molecular motion virtually stops. Lower temperatures do not exist. The size of the kelvin is identical to that of the Celsius degree—the only difference is the temperature that

Molecular motion does not *completely* stop at absolute zero because of the uncertainty principle in quantum mechanics, which we will discuss in Chapter 7.
Note that we give Kelvin temperatures in kelvins (*not* "degrees Kelvin") or K (*not* °K).

The Celsius Temperature Scale

0 °C – Water freezes	10 °C – Brisk fall day	22 °C – Room temperature	40 °C – Summer day in Death Valley

each designates as zero. You can convert between the temperature scales with the following formulas:

$$°C = \frac{(°F - 32)}{1.8}$$

$$K = °C + 273.15$$

Throughout this book you will see examples worked out in formats that are designed to help you develop problem-solving skills. The most common format uses two columns to guide you through the worked example. The left column describes the thought processes and steps used in solving the problem while the right column shows the implementation. The first example in this two-column format is Example 1.2.

EXAMPLE 1.2   Converting between Temperature Scales

A sick child has a temperature of 40.00 °C. What is the child's temperature in (a) K and (b) °F?

SOLUTION

(a) Begin by finding the equation that relates the quantity that is given (°C) and the quantity you are trying to find (K).	$K = °C + 273.15$
Since this equation gives the temperature in K directly, substitute in the correct value for the temperature in °C and calculate the answer.	$K = °C + 273.15$ $K = 40.00 + 273.15 = 313.15$ K
(b) To convert from °C to °F, first find the equation that relates these two quantities.	$°C = \dfrac{(°F - 32)}{1.8}$
Since this equation expresses °C in terms of °F, you must solve the equation for °F.	$°C = \dfrac{(°F - 32)}{1.8}$ $1.8(°C) = (°F - 32)$ $°F = 1.8(°C) + 32$
Now substitute °C into the equation and calculate the answer. *Note: The number of digits reported in this answer follows significant figure conventions, covered in Section 1.7.*	$°F = 1.8(°C) + 32$ $°F = 1.8(40.00 °C) + 32 = 104.00 °F$

FOR PRACTICE 1.2

Gallium is a solid metal at room temperature, but it will melt to a liquid in your hand. The melting point of gallium is 85.6 °F. What is this temperature on (a) the Celsius scale and (b) the Kelvin scale?

Prefix Multipliers

See Appendix IA for a brief discussion of scientific notation.

Scientific notation allows us to express very large or very small quantities in a compact way, using large positive or negative exponents to do so. For example, the diameter of a hydrogen atom can be written as 1.06×10^{-10} m. The International System of Units uses the **prefix multipliers** shown in Table 1.2 with the standard units. These multipliers change the value of the unit by powers of 10 (just like an exponent does in scientific notation). For example, the kilometer has the prefix "kilo" meaning 1000 or 10^3. Therefore,

$$1 \text{ kilometer} = 1000 \text{ meters} = 10^3 \text{ meters}$$

Similarly, the millimeter has the prefix "milli" meaning 0.001 or 10^{-3}.

$$1 \text{ millimeter} = 0.001 \text{ meters} = 10^{-3} \text{ meters}$$

When reporting a measurement, choose a prefix multiplier close to the size of the quantity being measured. For example, to state the diameter of a hydrogen atom, which is 1.06×10^{-10} m, use picometers (106 pm) or nanometers (0.106 nm) rather than micrometers or millimeters. Choose the prefix multiplier that is most convenient for a particular measurement.

Conceptual Connection 1.5 Prefix Multipliers

What prefix multiplier is appropriate for reporting a measurement of 5.57×10^{-5} m?

TABLE 1.2 SI Prefix Multipliers

Prefix	Symbol	Multiplier	
exa	E	1,000,000,000,000,000,000	(10^{18})
peta	P	1,000,000,000,000,000	(10^{15})
tera	T	1,000,000,000,000	(10^{12})
giga	G	1,000,000,000	(10^{9})
mega	M	1,000,000	(10^{6})
kilo	k	1000	(10^{3})
deci	d	0.1	(10^{-1})
centi	c	0.01	(10^{-2})
milli	m	0.001	(10^{-3})
micro	μ	0.000001	(10^{-6})
nano	n	0.000000001	(10^{-9})
pico	p	0.000000000001	(10^{-12})
femto	f	0.000000000000001	(10^{-15})
atto	a	0.000000000000000001	(10^{-18})

Derived Units: Volume and Density

A **derived unit** is a combination of other units. For example, the SI unit for speed is meters per second (m/s), a derived unit. Notice that this unit is formed from two other SI units—meters and seconds—put together. You are probably more familiar with speed in miles/hour or kilometers/hour—these are also examples of derived units. Two other common derived units are those for volume (SI base unit is m³) and density (SI base unit is kg/m³). We examine each of these individually.

Volume

Volume is a measure of space. Any unit of length, when cubed (raised to the third power), becomes a unit of volume. Thus, the cubic meter (m³), cubic centimeter (cm³), and cubic millimeter (mm³) are all units of volume. The cubic nature of volume is not always intuitive, and studies have shown that our brains are not naturally wired to think abstractly, as required to think about volume. For example, consider the following question: How many small cubes measuring 1 cm on each side are required to construct a large cube measuring 10 cm (or 1 dm) on a side?

The answer to this question, as you can see by carefully examining the unit cube in **Figure 1.11▶**, is 1000 small cubes. When you go from a linear, one-dimensional distance to three-dimensional volume, you must raise both the linear dimension *and* its unit to the third power (not multiply by 3). Thus the volume of a cube is equal to the length of its edge cubed:

$$\text{volume of cube} = (\text{edge length})^3$$

A cube with a 10-cm edge length has a volume of $(10 \text{ cm})^3$ or 1000 cm³, and a cube with a 100-cm edge length has a volume of $(100 \text{ cm})^3 = 1,000,000 \text{ cm}^3$.

Other common units of volume in chemistry are the **liter (L)** and the **milliliter (mL)**. One milliliter (10^{-3} L) is equal to 1 cm³. A gallon of gasoline contains 3.785 L. Table 1.3 lists some common units—for volume and other quantities—and their equivalents.

Relationship between Length and Volume

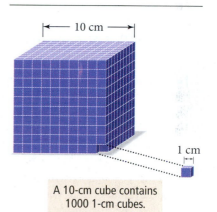

A 10-cm cube contains 1000 1-cm cubes.

▲ **FIGURE 1.11** The Relationship between Length and Volume

TABLE 1.3 Some Common Units and Their Equivalents

Length	Mass	Volume
1 kilometer (km) = 0.6214 mile (mi)	1 kilogram (kg) = 2.205 pounds (lb)	1 liter (L) = 1000 mL = 1000 cm³
1 meter (m) = 39.37 inches (in) = 1.094 yards (yd)	1 pound (lb) = 453.59 grams (g)	1 liter (L) = 1.057 quarts (qt)
1 foot (ft) = 30.48 centimeters (cm) (exact)	1 ounce (oz) = 28.35 grams (g)	1 U.S. gallon (gal) = 3.785 liters (L)
1 inch (in) = 2.54 centimeters (cm) (exact)		

Density

An old riddle asks, "Which weighs more, a ton of bricks or a ton of feathers?" The answer, of course, is neither—they both weigh the same (1 ton). If you answered bricks, you confused weight with density. The **density (*d*)** of a substance is the ratio of its mass (*m*) to its volume (*V*):

$$\text{Density} = \frac{\text{mass}}{\text{volume}} \quad \text{or} \quad d = \frac{m}{V}$$

Note that the *m* in this equation is in italic type, meaning that it stands for mass rather than for meters. In general, the symbols for units such as meters (m), seconds (s), or kelvins (K) appear in regular type while those for variables such as mass (*m*), volume (*V*), and time (*t*) appear in italics.

TABLE 1.4 **The Density of Some Common Substances at 20 °C**	
Substance	**Density (g/cm³)**
Charcoal (from oak)	0.57
Ethanol	0.789
Ice	0.917 (at 0 °C)
Water	1.00 (at 4 °C)
Sugar (sucrose)	1.58
Table salt (sodium chloride)	2.16
Glass	2.6
Aluminum	2.70
Titanium	4.51
Iron	7.86
Copper	8.96
Lead	11.4
Mercury	13.55
Gold	19.3
Platinum	21.4

Density is a characteristic physical property of materials and varies from one substance to another, as you can see in Table 1.4. The density of a substance also depends on its temperature. Density is an example of an **intensive property**, one that is *independent* of the amount of the substance. The density of aluminum, for example, is the same whether you have a gram or a kilogram. Intensive properties are often used to identify substances because these properties depend only on the type of substance, not on the amount of it. For example, one way to determine whether a substance is pure gold is to measure its density and compare it to the density of gold, 19.3 g/cm^3. Mass, in contrast, is an **extensive property**, one that depends on the amount of the substance. You could not use only the mass of a sample of gold to help identify it as gold.

The units of density are units of mass divided by volume. Although the SI derived unit for density is kg/m^3, the density of liquids and solids is most often expressed in g/cm^3 or g/mL. (Remember that cm^3 and mL are equivalent units.) Aluminum is among the least dense metals with a density of 2.70 g/cm^3, while platinum is one of the densest metals with a density of 21.4 g/cm^3.

Calculating Density

We can calculate the density of a substance by dividing the mass of a given amount of the substance by its volume. For example, suppose a small nugget suspected to be gold has a mass of 22.5 g and a volume of 2.38 cm^3. To find its density, we divide the mass by the volume:

$$d = \frac{m}{V} = \frac{22.5 \text{ g}}{2.38 \text{ cm}^3} = 9.45 \text{ g/cm}^3$$

In this case, the density reveals that the nugget is not pure gold.

EXAMPLE 1.3 Calculating Density

A man receives a ring he believes to be platinum from his fiancée. Before the wedding, he notices that the ring feels a little light for its size and decides to measure its density. He places the ring on a balance and finds that it has a mass of 3.15 grams. He then finds that the ring displaces 0.233 cm^3 of water. Is the ring made of platinum? (Note: The volume of irregularly shaped objects is often measured by the displacement of water. To use this method, the object is placed in water and the change in volume of the water is measured. This increase in the total volume represents the volume of water *displaced* by the object, and is equal to the volume of the object.)

Set up the problem by writing the important information that is *given* as well as the information that you are asked to *find*. In this case, we are to find the density of the ring and compare it to that of platinum. *Note: This standard way of setting up problems is discussed in detail in Section 1.7.*	**GIVEN** $m = 3.15 \text{ g}$ $\qquad V = 0.233 \text{ cm}^3$ **FIND** Density in g/cm^3
Next, write down the equation that defines density.	**EQUATION** $d = \dfrac{m}{V}$
Solve the problem by substituting the correct values of mass and volume into the expression for density.	**SOLUTION** $$d = \frac{m}{V} = \frac{3.15 \text{ g}}{0.233 \text{ cm}^3} = 13.5 \text{ g/cm}^3$$

The density of the ring is much too low to be platinum (platinum density is 21.4 g/cm^3), and the ring is therefore a fake.

FOR PRACTICE 1.3

The woman in the above example is shocked that the ring is fake and returns it. She buys a new ring that has a mass of 4.53 g and a volume of 0.212 cm^3. Is this ring genuine?

FOR MORE PRACTICE 1.3

A metal cube has an edge length of 11.4 mm and a mass of 6.67 g. Calculate the density of the metal and refer to Table 1.4 to determine the likely identity of the metal.

Conceptual Connection 1.6 Density

The density of copper decreases with increasing temperature (as does the density of most substances). Which statement about raising the temperature of a copper sample from room temperature to 95 °C is true?

(a) the copper sample becomes lighter

(b) the copper sample becomes heavier

(c) the copper sample expands

(d) the copper sample contracts

1.7 The Reliability of a Measurement

Recall from the opening example in this chapter (Section 1.1) that carbon monoxide is a colorless gas emitted by motor vehicles found in polluted air. The table below shows carbon monoxide concentrations in Los Angeles County (Long Beach) as reported by the U.S. Environmental Protection Agency (EPA) over the period 1998–2008.

Year	Carbon Monoxide Concentration (ppm)*
1998	11.5
2000	9.7
2002	8.5
2004	6.1
2006	5.5
2008	4.3

*Second maximum, 8 hour average; ppm = *parts per million*, defined as mL pollutant per million mL of air.

The first thing you should notice about these values is that they are decreasing over time. For this decrease, we can thank the Clean Air Act and its amendments, which have resulted in more efficient engines, in specially blended fuels, and consequently in cleaner air in all major U.S. cities over the last 30 years. The second thing you might notice is the number of digits to which the measurements are reported. The number of digits in a reported measurement indicates the certainty associated with that measurement. For example, a less certain set of measurements of carbon monoxide levels might be reported as follows:

Year	Carbon Monoxide Concentration (ppm)
1998	12
2000	10
2002	9
2004	6
2006	6
2008	4

Notice that the first set of data is reported to the nearest 0.1 ppm while the second set is reported to the nearest 1 ppm. Scientists agree on a standard way of reporting measured quantities in which the number of reported digits reflects the certainty in the measurement: more digits, more certainty; fewer digits, less certainty. Numbers are usually written so that the uncertainty is in the last reported digit. (That uncertainty is assumed to be ± 1 in the last digit unless otherwise indicated.) For example, by reporting the 2008 carbon monoxide concentration as 4.3 ppm (as in the first set of measurements), the scientists mean 4.3 ± 0.1 ppm. The carbon monoxide concentration is between 4.2 and 4.4 ppm. It might be 4.4 ppm, for example, but it could not be 5.0 ppm.

Estimation in Weighing

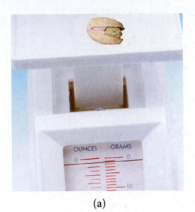

(a)

Markings every 1 g
Estimated reading 1.2 g

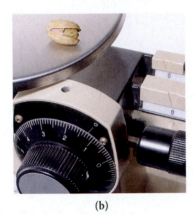

(b)

Markings every 0.1 g
Estimated reading 1.27 g

▲ **FIGURE 1.12 Estimation in Weighing** **(a)** This scale has markings every 1 g, so we estimate to the tenths place by mentally dividing the space into 10 equal spaces to estimate the last digit. This reading is 1.2 g. **(b)** Because this balance has markings every 0.1 g, we estimate to the hundredths place. This reading is 1.27 g.

In contrast, if the reported value was 4 ppm (as in the second set of measurements), this would mean 4 ± 1 ppm, or between 3 and 5 ppm. In general,

> **Scientific measurements are reported so that every digit is certain except the last, which is estimated.**

For example, consider the reported number:

certain   estimated

The first three digits are certain; the last digit is estimated.

The number of digits reported in a measurement depends on the measuring device. For example, consider weighing a pistachio nut on two different balances (**Figure 1.12◄**). The balance on the top has marks every 1 gram, while the balance on the bottom has marks every 0.1 gram. For the balance on the top, we mentally divide the space between the 1- and 2-gram marks into 10 equal spaces and estimate that the pointer is at about 1.2 grams. We then write the measurement as 1.2 grams indicating that we are sure of the "1" but have estimated the ".2." The balance on the bottom, with marks every *tenth* of a gram, allows us to write the result with more digits. The pointer is between the 1.2-gram mark and the 1.3-gram mark. We again divide the space between the two marks into 10 equal spaces and estimate the third digit. For the nut shown in figure (b), we report 1.27 g.

EXAMPLE 1.4 Reporting the Correct Number of Digits

The graduated cylinder shown has markings every 0.1 mL. Report the volume (which is read at the bottom of the meniscus) to the correct number of digits. (Note: The meniscus is the crescent-shaped surface at the top of a column of liquid.)

SOLUTION

Since the bottom of the meniscus is between the 4.5 and 4.6 mL markings, mentally divide the space between the markings into 10 equal spaces and estimate the next digit. In this case, you should report the result as 4.57 mL.

Meniscus

What if you estimated a little differently and wrote 4.56 mL? In general, one unit difference in the last digit is acceptable because the last digit is estimated and different people might estimate it slightly differently. However, if you wrote 4.63 mL, you would have misreported the measurement.

FOR PRACTICE 1.4

Record the temperature on the thermometer shown below to the correct number of digits.

Counting Significant Figures

The precision of a measurement—which depends on the instrument used to make the measurement—is key, not only when recording the measurement, but also when performing calculations that use the measurement. The preservation of this precision is conveniently accomplished by using *significant figures*. In any reported measurement, the non-place-holding digits—those that are not simply marking the decimal place—are called **significant figures** (or **significant digits**). *The greater the number of significant figures, the greater the certainty of the measurement.* The number 23.5 has three significant figures while the number 23.56 has four. To determine the number of significant figures in a number containing zeroes, we must distinguish between zeroes that are significant and those that simply mark the decimal place. For example, in the number 0.0008, the leading zeroes mark the decimal place but *do not* add to the certainty of the measurement and are therefore not significant; this number has only one significant figure. In contrast, the trailing zeroes in the number 0.000800 *do add* to the certainty of the measurement and are therefore counted as significant; this number has three significant figures.

To determine the number of significant figures in a number, follow these rules (examples are shown on the right).

Significant Figure Rules	Examples	
1. All nonzero digits are significant.	28.03	0.0540
2. Interior zeroes (zeroes between two digits) are significant.	408	7.0301
3. Leading zeroes (zeroes to the left of the first nonzero digit) are not significant. They only serve to locate the decimal point.	0.0032	0.00006
	not significant	
4. Trailing zeroes (zeroes at the end of a number) are categorized as follows:		
• Trailing zeroes after a decimal point are always significant.	45.000	3.5600
• Trailing zeroes before a decimal point (and after a nonzero number) are always significant.	140.00	25.0505
• Trailing zeroes before an implied decimal point are ambiguous and should be avoided by using scientific notation.	1200	ambiguous
	1.2×10^3	2 significant figures
	1.20×10^3	3 significant figures
	1.200×10^3	4 significant figures
• Some textbooks put a decimal point after one or more trailing zeroes if the zeroes are to be considered significant. We avoid that practice in this book, but you should be aware of it.	1200.	4 significant figures (common in some textbooks)

Exact Numbers

Exact numbers have no uncertainty, and thus do not limit the number of significant figures in any calculation. We can regard an exact number as having an unlimited number of significant figures. Exact numbers originate from three sources:

• From the accurate counting of discrete objects. For example, 3 atoms means 3.00000 . . . atoms.

• From defined quantities, such as the number of centimeters in 1 m. Because 100 cm is defined as 1 m,

$$100 \, \text{cm} = 1 \, \text{m} \qquad \text{means} \qquad 100.00000 \ldots \text{cm} = 1.0000000 \ldots \text{m}$$

• From integral numbers that are part of an equation. For example, in the equation, $radius = \frac{diameter}{2}$, the number 2 is exact and therefore has an unlimited number of significant figures.

EXAMPLE 1.5 Determining the Number of Significant Figures in a Number

How many significant figures are in each number?

(a) 0.04450 m   (b) 5.0003 km   (c) 10 dm = 1 m   (d) 1.000×10^5 s   (e) 0.00002 mm   (f) 10,000 m

SOLUTION

(a) 0.04450 m	*Four significant figures.* The two 4's and the 5 are significant (rule 1). The trailing zero is after a decimal point rule 4). The leading zeroes only mark the decimal place and are not significant (rule 3).
(b) 5.0003 km	*Five significant figures.* The 5 and 3 are significant (rule 1) as are the three interior zeroes (rule 2).
(c) 10 dm = 1 m	*Unlimited significant figures.* Defined quantities have an unlimited number of significant figures.
(d) 1.000×10^5 s	*Four significant figures.* The 1 is significant (rule 1). The trailing zeroes are after a decimal point and significant (rule 4).
(e) 0.00002 mm	*One significant figure.* The 2 is significant (rule 1). The leading zeroes only mark the decimal place and are not significant (rule 3).
(f) 10,000 m	*Ambiguous.* The 1 is significant (rule 1) but the trailing zeroes occur before an implied decimal point and are ambiguous (rule 4). Without more information, we would assume 1 significant figure. It is better to write this as 1×10^5 to indicate one significant figure or as 1.0000×10^5 to indicate five (rule 4).

FOR PRACTICE 1.5

How many significant figures are in each number?

(a) 554 km   (b) 7 pennies   (c) 1.01×10^5 m   (d) 0.00099 s   (e) 1.4500 km   (f) 21,000 m

Significant Figures in Calculations

When you use measured quantities in calculations, the results of the calculation must reflect the precision of the measured quantities. You should not lose or gain precision during mathematical operations. Follow these rules when carrying significant figures through calculations.

Rules for Calculations	Examples
1. In multiplication or division, the result carries the same number of significant figures as the factor with the fewest significant figures.	$1.052 \quad \times \quad 12.054 \quad \times \quad 0.53 \quad = \quad 6.7208 \quad = \quad 6.7$ (4 sig. figures) (5 sig. figures) (2 sig. figures) (2 sig. figures) $2.0035 \quad \div \quad 3.20 \quad = \quad 0.626094 \quad = \quad 0.626$ (5 sig. figures) (3 sig. figures) (3 sig. figures)
2. In addition or subtraction, the result carries the same number of decimal places as the quantity with the fewest decimal places.	$\begin{array}{r} 2.34\|5 \\ 0.07\| \\ +2.99\|75 \\ \hline 5.41\|25 = 5.41 \end{array} \qquad \begin{array}{r} 5.9\| \\ -0.2\|21 \\ \hline 5.6\|79 = 5.7 \end{array}$ In addition and subtraction, it is helpful to draw a line next to the number with the fewest decimal places. This line determines the number of decimal places in the answer.
3. When rounding to the correct number of significant figures, round down if the last (or leftmost) digit dropped is four or less; round up if the last (or leftmost) digit dropped is five or more.	To two significant figures: 5.372 rounds to 5.4 5.342 rounds to 5.3 5.352 rounds to 5.4 5.349 rounds to 5.3 Notice that only the *last (or leftmost) digit being dropped* determines in which direction to round—ignore all digits to the right of it.
4. To avoid rounding errors in multistep calculations round only the final answer—do not round intermediate steps. If you write down intermediate answers, keep track of significant figures by underlining the least-significant digit.	$6.78 \times 5.903 \times (5.489 - 5.01)$ $= 6.78 \times 5.903 \times 0.479$ $= 19.1707$ $= 19$ underline least significant digit

Notice that for multiplication or division, the quantity with the fewest *significant figures* determines the number of *significant figures* in the answer, but for addition and subtraction, the quantity with the fewest *decimal places* determines the number of *decimal places* in the answer. In multiplication and division, we focus on significant figures, but in addition and subtraction we focus on decimal places. When a problem involves addition or subtraction, the answer may have a different number of significant figures than the initial quantities. Keep this in mind in problems that involve both addition or subtraction and multiplication or division. For example,

$$\frac{1.002 - 0.999}{3.754} = \frac{0.003}{3.754} = 7.99 \times 10^{-4} = 8 \times 10^{-4}$$

The answer has only one significant figure, even though the initial numbers had three or four.

EXAMPLE 1.6 Significant Figures in Calculations

Perform the calculations to the correct number of significant figures.

(a) $1.10 \times 0.5120 \times 4.0015 \div 3.4555$

(b)
$$\begin{array}{r} 0.355 \\ +105.1 \\ -100.5820 \end{array}$$

(c) $4.562 \times 3.99870 \div (452.6755 - 452.33)$

(d) $(14.84 \times 0.55) - 8.02$

SOLUTION

(a) Round the intermediate result (in blue) to three significant figures to reflect the three significant figures in the least precisely known quantity (1.10).	$1.10 \times 0.5120 \times 4.0015 \div 3.4555$ $= 0.65219$ $= 0.652$
(b) Round the intermediate answer (in blue) to one decimal place to reflect the quantity with the fewest decimal places (105.1). Notice that 105.1 is *not* the quantity with the fewest significant figures, but it has the fewest decimal places and therefore determines the number of decimal places in the answer.	$\begin{array}{r} 0.355 \\ +105.1 \\ -100.5820 \\ \hline 4.8730 = 4.9 \end{array}$
(c) Mark the intermediate result to two decimal places to reflect the number of decimal places in the quantity within the parentheses having the fewest number of decimal places (452.33). Round the final answer to two significant figures to reflect the two significant figures in the least precisely known quantity (0.3 4 55).	$4.562 \times 3.99870 \div (452.6755 - 452.33)$ $= 4.562 \times 3.99870 \div 0.3455$ $= 52.79904$ $= 53$ 2 places of the decimal
(d) Mark the intermediate result to two significant figures to reflect the number of significant figures in the quantity within the parentheses having the fewest number of significant figures (0.55). Round the final answer to one decimal place to reflect the one decimal place in the least precisely known quantity (8.162).	$(14.84 \times 0.55) - 8.02$ $= 8.162 - 8.02$ $= 0.142$ $= 0.1$

FOR PRACTICE 1.6

Perform the calculations to the correct number of significant figures.

(a) $3.10007 \times 9.441 \times 0.0301 \div 2.31$

(b)
$$\begin{array}{r} 0.881 \\ +132.1 \\ -12.02 \end{array}$$

(c) $2.5110 \times 21.20 \div (44.11 + 1.223)$

(d) $(12.01 \times 0.3) + 4.811$

Precision and Accuracy

Scientific measurements are often repeated several times to increase confidence in the result. We can distinguish between two different kinds of certainty—called accuracy and precision—associated with such measurements. **Accuracy** refers to how close the measured value is to the actual value. **Precision** refers to how close a series of measurements are to one another or how reproducible they are. A series of measurements can be precise (close to one another in value and reproducible) but not accurate (not close to the true value). For example, consider the results of three students who repeatedly weighed a lead block known to have a true mass of 10.00 g (indicated by the solid horizontal blue line on the graphs).

	Student A	Student B	Student C
Trial 1	10.49 g	9.78 g	10.03 g
Trial 2	9.79 g	9.82 g	9.99 g
Trial 3	9.92 g	9.75 g	10.03 g
Trial 4	10.31 g	9.80 g	9.98 g
Average	**10.13 g**	**9.79 g**	**10.01 g**

- The results of student A are both inaccurate (not close to the true value) and imprecise (not consistent with one another). The inconsistency is the result of **random error**, error that has equal probability of being too high or too low. Almost all measurements have some degree of random error. Random error can, with enough trials, average itself out.

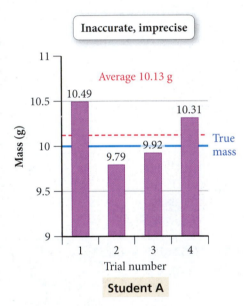

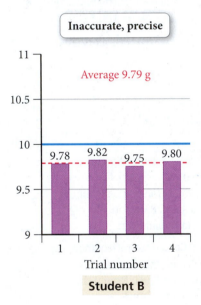

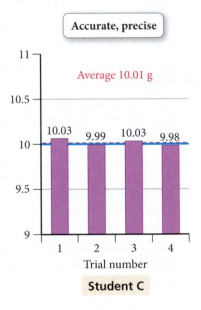

▲ Measurements are said to be precise if they are consistent with one another, but they are accurate only if they are close to the actual value.

- The results of student B are precise (close to one another in value) but inaccurate. The inaccuracy is the result of **systematic error**, error that tends toward being either too high or too low. Systematic error does not average out with repeated trials. For example, if a balance is not properly calibrated, it may systematically read too high or too low.

- The results of student C display little systematic error or random error—they are both accurate and precise.

1.8 Solving Chemical Problems

Learning to solve problems is one of the most important skills you will acquire in this course. No one succeeds in chemistry—or in life, really—without the ability to solve problems. Although no simple formula applies to every problem, there are some problem-solving strategies you can learn and you can begin to develop some chemical intuition. Many of the problems you will solve in this course can be thought of as *unit conversion problems*, where you are given one or more quantities and asked to convert them into different units. Other problems require the use of *specific equations* to get to the information you are trying to find. In the sections that follow, you will find strategies to help you solve both of these types of problems. Of course, many problems contain both conversions and equations, requiring the combination of these strategies, and some problems may require an altogether different approach.

Converting from One Unit to Another

In Section 1.6, we learned the SI unit system, the prefix multipliers, and a few other units. Knowing how to work with and manipulate these units in calculations is basic to solving chemical problems. In calculations, units help to determine correctness. Using units as a guide to solving problems is often called **dimensional analysis**. Units should always be included in calculations; they are multiplied, divided, and canceled like any other algebraic quantity.

Consider converting 12.5 inches (in) to centimeters (cm). We know from Table 1.3 that 1 in = 2.54 cm (exact), so we can use this quantity in the calculation as follows:

$$12.5 \; \cancel{in} \times \frac{2.54 \; cm}{1 \; \cancel{in}} = 31.8 \; cm$$

The unit, in, cancels and we are left with cm as our final unit. The quantity $\frac{2.54 \; cm}{1 \; in}$ is a **conversion factor**—a fractional quantity with the units we are *converting from* on the bottom and the units we are *converting to* on the top. Conversion factors are constructed from any two equivalent quantities. In this example, 2.54 cm = 1 in, so we construct the conversion factor by dividing both sides of the equality by 1 in and canceling the units.

$$2.54 \; cm = 1 \; in$$

$$\frac{2.54 \; cm}{1 \; in} = \frac{1 \; \cancel{in}}{1 \; \cancel{in}}$$

$$\frac{2.54 \; cm}{1 \; in} = 1$$

The quantity $\frac{2.54 \; cm}{1 \; in}$ is equivalent to 1, so multiplying by the conversion factor is mathematically equivalent to multiplying by 1. To convert the other way, from centimeters to inches, we must—using units as a guide—use a different form of the conversion factor. If you accidentally use the same form, you will get the wrong result, indicated by erroneous units. For example, suppose that you want to convert 31.8 cm to inches.

$$31.8 \; cm \times \frac{2.54 \; cm}{1 \; in} = \frac{80.8 \; cm^2}{in}$$

The units in the above answer (cm^2/in), as well as the value of the answer, are obviously wrong. When you solve a problem, always look at the final units. Are they the desired units? Always look at the magnitude of the numerical answer as well. Does it make sense? In this case, our mistake was the form of the conversion factor. It should have been inverted so that the units cancel as follows:

$$31.8 \; \cancel{cm} \times \frac{1 \; in}{2.54 \; \cancel{cm}} = 12.5 \; in$$

Conversion factors can be inverted because they are equal to 1 and the inverse of 1 is 1. Therefore,

$$\frac{2.54 \; cm}{1 \; in} = 1 = \frac{1 \; in}{2.54 \; cm}$$

Most unit conversion problems take the form:

Information given × conversion factor(s) = information sought

$$\cancel{\text{Given unit}} \quad \times \quad \frac{\text{desired unit}}{\cancel{\text{given unit}}} = \text{desired unit}$$

In this book, we diagram a problem solution using a *conceptual plan*. A conceptual plan is a visual outline that helps you to see the general flow of the problem. For unit conversions, the conceptual plan focuses on units and the conversion from one unit to another. The conceptual plan for converting in to cm is:

| in | → | cm |

$$\frac{2.54 \; cm}{1 \; in}$$

The conceptual plan for converting the other way, from cm to in, is just the reverse, with the reciprocal conversion factor:

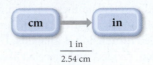

$$\frac{1 \text{ in}}{2.54 \text{ cm}}$$

Each arrow in a conceptual plan for a unit conversion has an associated conversion factor with the units of the previous step in the denominator and the units of the following step in the numerator. In the following section, we incorporate the idea of a conceptual plan into an overall approach to solving numerical chemical problems.

General Problem-Solving Strategy

In this book, we use a standard problem-solving procedure that can be adapted to many of the problems encountered in general chemistry and beyond. Solving any problem essentially requires you to assess the information given in the problem and devise a way to get to the information asked for. In other words, you must do the following:

- Identify the starting point (the *given* information).
- Identify the end point (what you must *find*).
- Devise a way to get from the starting point to the end point using what is given as well as what you already know or can look up. (This is the *conceptual plan*.)

In graphic form, we can represent this progression as

$$\textbf{Given} \longrightarrow \textbf{Conceptual Plan} \longrightarrow \textbf{Find}$$

One of the main difficulties beginning students encounter when trying to solve problems in general chemistry is not knowing where to begin. While no problem-solving procedure is applicable to all problems, the following four-step procedure can be helpful in working through many of the numerical problems you will encounter in this book.

1. **Sort.** Begin by sorting the information in the problem. *Given* information is the basic data provided by the problem—often one or more numbers with their associated units. *Find* indicates what information you will need for your answer.

2. **Strategize.** This is usually the most challenging part of solving a problem. In this process, you must develop a *conceptual plan*—a series of steps that will get you from the given information to the information you are trying to find. In the previous section, we presented a conceptual plans for simple unit conversion problems. Each arrow in a conceptual plan represents a computational step. On the left side of the arrow is the quantity you had before the step; on the right side of the arrow is the quantity you will have after the step; and below the arrow is the information you need to get from one to the other—the relationship between the quantities.

Most problems can be solved in more than one way. The solutions we derive in this book tend to be the most straightforward but certainly not the only way to solve the problem.

Often such relationships will take the form of conversion factors or equations. These may be given in the problem, in which case you will have written them down under "Given" in step 1. Usually, however, you will need other information—which may include physical constants, formulas, or conversion factors—to help get you from what you are given to what you must find. You may recall this information from what you have learned or you can look it up in the chapter or in tables within the book.

In some cases, you may get stuck at the strategize step. If you cannot figure out how to get from the given information to the information you are asked to find, you might try working backwards. For example, you may want to look at the units of the quantity you are trying to find and try to find conversion factors to get to the units of the given quantity. You may even try a combination of strategies; work forward, backward, or some of both. If you persist, you will develop a strategy to solve the problem.

3. **Solve.** This is the easiest part of solving a problem. Once you set up the problem properly and devise a conceptual plan, you simply follow the plan to solve the problem. Carry out any mathematical operations (paying attention to the rules for significant figures in calculations) and cancel units as needed.

4. **Check.** This is the step most often overlooked by beginning students. Experienced problem solvers always ask, does this answer make physical sense? Are the units correct? Is the number of significant figures correct? When solving multistep problems, errors can easily creep into the solution. You can catch most of these errors by simply checking the answer. For example, suppose you are calculating the number of atoms in a gold coin and end up with an answer of 1.1×10^{-6} atoms. Could the gold coin really be composed of one-millionth of one atom?

Here we apply this problem-solving procedure to unit conversion problems. The procedure is summarized in the left column and two examples of applying the procedure are shown in the middle and right columns. This three-column format is used in selected examples throughout this text. It allows you to see how a particular procedure can be applied to two different problems. Work through one problem first (from top to bottom) and then see how the same procedure is applied to the other problem. Being able to see the commonalities and differences between problems is a key part of developing problem-solving skills.

PROCEDURE FOR... **Solving Unit Conversion Problems**	EXAMPLE 1.7 **Unit Conversion** Convert 1.76 yards to centimeters.	EXAMPLE 1.8 **Unit Conversion** Convert 1.8 quarts to cubic centimeters.
SORT Begin by sorting the information in the problem into *Given* and *Find*.	**GIVEN** 1.76 yd **FIND** cm	**GIVEN** 1.8 qt **FIND** cm^3
STRATEGIZE Create a *conceptual plan* for the problem. Begin with the *given* quantity and symbolize each conversion step with an arrow. Below each arrow, write the appropriate conversion factor for that step. Focus on the units. The conceptual plan should end at the *find* quantity and its units. In these examples, the other information needed consists of relationships between the various units as shown.	**CONCEPTUAL PLAN** yd → m → cm $\dfrac{1\ m}{1.094\ yd}$ $\dfrac{100\ cm}{1\ m}$ **RELATIONSHIPS USED** 1.094 yd = 1 m $1\ cm = 1 \times 10^{-2}\ m$ (These conversion factors are from Tables 1.2 and 1.3.)	**CONCEPTUAL PLAN** qt → L → mL → cm^3 $\dfrac{1\ L}{1.057\ qt}$ $\dfrac{1000\ mL}{1\ L}$ $\dfrac{1\ cm^3}{1\ mL}$ **RELATIONSHIPS USED** 1.057 qt = 1 L $1\ mL = 1 \times 10^{-3}\ L$ $1\ mL = 1\ cm^3$ (These conversion factors are from Tables 1.2 and 1.3.)
SOLVE Follow the conceptual plan. Solve the equation(s) for the *find* quantity (if it is not already). Gather each of the quantities that must go into the equation in the correct units. (Convert to the correct units if necessary.) Substitute the numerical values and their units into the equation(s) and calculate the answer. Round the answer to the correct number of significant figures.	**SOLUTION** $1.76\ \text{yd} \times \dfrac{1\ \text{m}}{1.094\ \text{yd}} \times \dfrac{1\ cm}{1 \times 10^{-2}\ \text{m}}$ $= 160.8775\ cm$ $160.8775\ cm = 161\ cm$	**SOLUTION** $1.8\ \text{qt} \times \dfrac{1\ \text{L}}{1.057\ \text{qt}} \times \dfrac{1\ \text{mL}}{1 \times 10^{-3}\ \text{L}}$ $\times \dfrac{1\ cm^3}{1\ \text{mL}} = 1.70293 \times 10^3\ cm^3$ $1.70293 \times 10^3\ cm^3 = 1.7 \times 10^3\ cm^3$
CHECK Check your answer. Are the units correct? Does the answer make physical sense?	The units (cm) are correct. The magnitude of the answer (161) makes physical sense because a centimeter is a much smaller unit than a yard.	The units (cm^3) are correct. The magnitude of the answer (1700) makes physical sense because a cubic centimeter is a much smaller unit than a quart.
	FOR PRACTICE 1.7 Convert 288 cm to yards.	**FOR PRACTICE 1.8** Convert 9255 cm^3 to gallons.

Units Raised to a Power

When building conversion factors for units raised to a power, remember to raise both the number and the unit to the power. For example, to convert from in^2 to cm^2, we construct the conversion factor as follows:

$$2.54 \text{ cm} = 1 \text{ in}$$
$$(2.54 \text{ cm})^2 = (1 \text{ in})^2$$
$$(2.54)^2 \text{ cm}^2 = 1^2 \text{ in}^2$$
$$6.45 \text{ cm}^2 = 1 \text{ in}^2$$
$$\frac{6.45 \text{ cm}^2}{1 \text{ in}^2} = 1$$

Example 1.9 shows how to use conversion factors involving units raised to a power.

EXAMPLE 1.9 Unit Conversions Involving Units Raised to a Power

Calculate the displacement (the total volume of the cylinders through which the pistons move) of a 5.70-L automobile engine in cubic inches.

SORT Sort the information in the problem into *Given* and *Find*.	**GIVEN** 5.70 L **FIND** in^3
STRATEGIZE Write a conceptual plan. Begin with the given information and devise a path to the information that you are asked to find. Notice that for cubic units, the conversion factors must be cubed.	**CONCEPTUAL PLAN** 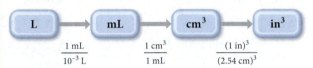 **RELATIONSHIPS USED** $1 \text{ mL} = 10^{-3} \text{ L}$ $1 \text{ mL} = 1 \text{ cm}^3$ $2.54 \text{ cm} = 1 \text{ in}$ (These conversion factors are from Tables 1.2 and 1.3.)
SOLVE Follow the conceptual plan to solve the problem. Round the answer to three significant figures to reflect the three significant figures in the least precisely known quantity (5.70 L). These conversion factors are all exact and therefore do not limit the number of significant figures.	**SOLUTION** $5.70 \text{ L} \times \dfrac{1 \text{ mL}}{10^{-3} \text{ L}} \times \dfrac{1 \text{ cm}^3}{1 \text{ mL}} \times \dfrac{(1 \text{ in})^3}{(2.54 \text{ cm})^3}$ $= 347.835 \text{ in}^3$ $= 348 \text{ in}^3$

CHECK The units of the answer are correct and the magnitude makes sense. The unit cubic inches is smaller than liters, so the volume in cubic inches should be larger than the volume in liters.

FOR PRACTICE 1.9

How many cubic centimeters are there in 2.11 yd^3?

FOR MORE PRACTICE 1.9

A vineyard has 145 acres of Chardonnay grapes. A particular soil supplement requires 5.50 grams for every square meter of vineyard. How many kilograms of the soil supplement are required for the entire vineyard? (1 km^2 = 247 acres)

EXAMPLE 1.10 Density as a Conversion Factor

The mass of fuel in a jet must be calculated before each flight to ensure that the jet is not too heavy to fly. A 747 is fueled with 173,231 L of jet fuel. If the density of the fuel is 0.768 g/cm³, what is the mass of the fuel in kilograms?

SORT Begin by sorting the information in the problem into *Given* and *Find*.	**GIVEN** fuel volume = 173,231 L density of fuel = 0. 768 g/cm³ **FIND** mass in kg
STRATEGIZE Draw a conceptual plan by beginning with the given quantity, in this case the volume in liters (L). The overall goal of this problem is to find the mass. You can convert between volume and mass using density (g/cm³). However, you must first convert the volume to cm³. Once you have converted the volume to cm³, use the density to convert to g. Finally convert g to kg.	**CONCEPTUAL PLAN** 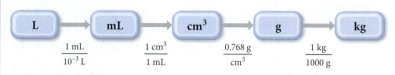 **RELATIONSHIPS USED** $1 \text{ mL} = 10^{-3} \text{ L}$ $1 \text{ mL} = 1 \text{ cm}^3$ $d = 0.768 \text{ g/cm}^3$ $1000 \text{ g} = 1 \text{ kg}$ (These conversion factors are from Tables 1.2 and 1.3.)
SOLVE Follow the conceptual plan to solve the problem. Round the answer to three significant figures to reflect the three significant figures in the density.	**SOLUTION** $173{,}231 \; \cancel{L} \times \dfrac{1 \; \cancel{mL}}{10^{-3} \; \cancel{L}} \times \dfrac{1 \; \cancel{cm^3}}{1 \; \cancel{mL}} \times \dfrac{0.768 \; \cancel{g}}{1 \; \cancel{cm^3}} \times \dfrac{1 \; kg}{1000 \; \cancel{g}} = 1.33 \times 10^5 \; kg$

CHECK The units of the answer (kg) are correct. The magnitude makes sense because the mass (1.33×10^5 kg) is similar in magnitude to the given volume (173,231 L or 1.73231×10^5 L), as expected for a density close to one (0.768 g/cm³).

FOR PRACTICE 1.10

Backpackers often use canisters of white gas to fuel a cooking stove's burner. If one canister contains 1.45 L of white gas, and the density of the gas is 0.710 g/cm³, what is the mass of the fuel in kilograms?

FOR MORE PRACTICE 1.10

A drop of gasoline has a mass of 22 mg and a density of 0.754 g/cm³. What is its volume in cubic centimeters?

Problems Involving an Equation

Problems involving equations can be solved in much the same way as problems involving conversions. Usually, in problems involving equations, you must find one of the variables in the equation, given the others. The *conceptual plan* concept outlined previously can be used for problems involving equations. For example, suppose you are given the mass (*m*) and volume (*V*) of a sample and asked to calculate its density. The conceptual plan shows how the *equation* takes you from the *given* quantities to the *find* quantity.

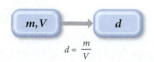

Instead of a conversion factor under the arrow, this conceptual plan has an equation. The equation shows the *relationship* between the quantities on the left of the arrow and the quantities on the right. Note that at this point, the equation need not be solved for the quantity on the right (although in this particular case it is). The procedure that follows, as well as the Examples 1.11 and 1.12, will guide you in developing a strategy to solve problems involving equations. We again use the three-column format here. Work through one problem from top to bottom and then see how the same general procedure is applied to the second problem.

PROCEDURE FOR... Solving Problems Involving Equations	EXAMPLE 1.11 Problems with Equations Find the radius (r), in centimeters, of a spherical water droplet with a volume (V) of 0.058 cm^3. For a sphere, $V = (4/3)\pi r^3$.	EXAMPLE 1.12 Problems with Equations Find the density (in g/cm^3) of a metal cylinder with a mass of 8.3 g, a length (l) of 1.94 cm, and a radius (r) of 0.55 cm. For a cylinder, $V = \pi r^2 l$.
SORT Begin by sorting the information in the problem into *Given* and *Find*.	**GIVEN** $V = 0.058$ cm^3 **FIND** r in cm	**GIVEN** $m = 8.3$ g $\qquad l = 1.94$ cm $\qquad\qquad r = 0.55$ cm **FIND** d in g/cm^3
STRATEGIZE Write a conceptual plan for the problem. Focus on the equation(s). The conceptual plan shows how the equation takes you from the given quantity (or quantities) to the find quantity. The conceptual plan may have several parts, involving other equations or required conversions. In these examples, you must use the geometrical relationships given in the problem statements as well as the definition of density, $d = m/V$, which you learned in this chapter.	**CONCEPTUAL PLAN** $$V = \frac{4}{3}\pi r^3$$ **RELATIONSHIPS USED** $$V = \frac{4}{3}\pi r^3$$	**CONCEPTUAL PLAN** $$V = \pi r^2 l$$ $$d = m/V$$ **RELATIONSHIPS USED** $$V = \pi r^2 l$$ $$d = \frac{m}{V}$$
SOLVE Follow the conceptual plan. Begin with the given quantity and its units. Multiply by the appropriate conversion factor(s), canceling units, to arrive at the find quantity. Round the answer to the correct number of significant figures by following the rules in Section 1.7. Remember that exact conversion factors do not limit significant figures.	**SOLUTION** $$V = \frac{4}{3}\pi r^3$$ $$r^3 = \frac{3}{4\pi}V$$ $$r = \left(\frac{3}{4\pi}V\right)^{1/3}$$ $$= \left(\frac{3}{4\pi}\, 0.058 \text{ cm}^3\right)^{1/3} = 0.24013 \text{ cm}$$ $0.24013 \text{ cm} = 0.24 \text{ cm}$	**SOLUTION** $$V = \pi r^2 l$$ $$= \pi(0.55 \text{ cm})^2\,(1.94 \text{ cm})$$ $$= 1.8436 \text{ cm}^3$$ $$d = \frac{m}{V}$$ $$= \frac{8.3 \text{ g}}{1.8436 \text{ cm}^3} = 4.50206 \text{ g/cm}^3$$ $4.50206 \text{ g/cm}^3 = 4.5 \text{ g/cm}^3$
CHECK Check your answer. Are the units correct? Does the answer make physical sense?	The units (cm) are correct and the magnitude seems right.	The units (g/cm^3) are correct. The magnitude of the answer seems correct for one of the lighter metals (see Table 1.4).
	FOR PRACTICE 1.11 Find the radius (r) of an aluminum cylinder that is 2.00 cm long and has a mass of 12.4 g. For a cylinder, $V = \pi r^2 l$.	**FOR PRACTICE 1.12** Find the density, in g/cm^3, of a metal cube with a mass of 50.3 g and an edge length (l) of 2.65 cm. For a cube, $V = l^3$.

CHAPTER IN REVIEW

Key Terms

Section 1.1
atoms (2)
molecules (2)
chemistry (3)

Section 1.2
hypothesis (3)
experiment (3)
scientific law (3)
law of conservation of
    mass (3)
theory (4)
atomic theory (4)
scientific method (4)

Section 1.3
matter (5)
substance (5)
state (5)
solid (5)
liquid (5)

gas (5)
crystalline (5)
amorphous (6)
composition (7)
pure substance (7)
mixture (7)
element (7)
compound (7)
heterogeneous mixture (7)
homogeneous mixture (7)

Section 1.4
physical change (8)
chemical change (8)
physical property (8)
chemical property (8)

Section 1.5
energy (11)
work (11)
kinetic energy (11)

potential energy (11)
thermal energy (11)
law of conservation of
    energy (11)

Section 1.6
units (12)
English system (12)
metric system (12)
International System of Units
    (SI) (12)
meter (m) (13)
kilogram (kg) (13)
mass (13)
second (s) (13)
kelvin (K) (13)
temperature (13)
Fahrenheit (°F) scale (14)
Celsius (°C) scale (14)
Kelvin scale (14)

prefix multipliers (15)
derived unit (16)
volume (16)
liter (L) (16)
milliliter (mL) (16)
density (d) (16)
intensive property (17)
extensive property (17)

Section 1.7
significant figures (significant
    digits) (20)
exact numbers (20)
accuracy (22)
precision (23)
random error (23)
systematic error (23)

Section 1.8
dimensional analysis (24)
conversion factor (24)

Key Concepts

Atoms and Molecules (1.1)

▶ All matter is composed of atoms and molecules.
▶ Chemistry is the science that investigates the properties and behavior of matter by examining atoms and molecules.

The Scientific Method (1.2)

▶ Science begins with the observation of the physical world. A number of related observations can often be subsumed in a summary statement or generalization called a scientific law.
▶ A hypothesis is a tentative interpretation or explanation of observations. One or more well-established hypotheses may prompt the development of a scientific theory, a model for nature that explains the underlying reasons for observations and laws.
▶ Laws, hypotheses, and theories all give rise to predictions that can be tested by experiments, carefully controlled procedures designed to produce critical new observations. If the predictions are not confirmed, the law, hypothesis, or theory must be modified or replaced.

The Classification of Matter (1.3)

▶ Matter can be classified according to its state (solid, liquid, or gas) or according to its composition (pure substance or mixture).
▶ A pure substance can either be an element, which is not decomposable into simpler substances, or a compound, which is composed of two or more elements in fixed proportions.
▶ A mixture can be either homogeneous, with the same composition throughout, or heterogeneous, with different compositions in different regions.

The Properties of Matter (1.4)

▶ There are two kinds of properties of matter—physical and chemical.
▶ Changes in matter in which its composition does not change are physical changes. Changes in matter in which its composition does change are chemical changes.

Energy (1.5)

▶ In chemical and physical changes, matter often exchanges energy with its surroundings. In these exchanges, the total energy is always conserved; energy is neither created nor destroyed.
▶ Systems with high potential energy tend to change in the direction of lower potential energy, releasing energy into the surroundings.

The Units of Measurement and Significant Figures (1.6, 1.7)

▶ Scientists use SI units, which are based on the metric system. The SI base units include the meter (m) for length, the kilogram (kg) for mass, the second (s) for time, and the kelvin (K) for temperature.
▶ Derived units are those formed from a combination of other units. Common derived units include volume (cm^3 or m^3) and density (g/cm^3).
▶ Measured quantities are reported so that the number of digits reflects the uncertainty in the measurement. The non-place-holding digits in a reported number are called significant figures.

Key Equations and Relationships

Relationship between Kelvin (K) and Celsius (°C) Temperature Scales (1.6)

$$K = °C + 273.15$$

Relationship between Celsius (°C) and Fahrenheit (°F) Temperature Scales (1.6)

$$°C = \frac{(°F - 32)}{1.8}$$

Relationship between Density (*d*), Mass (*m*), and Volume (*V*) (1.6)

$$d = \frac{m}{V}$$

Key Learning Objectives

Chapter Objectives	Assessment
Determining Physical and Chemical Changes and Properties (1.4)	Example 1.1 For Practice 1.1 Exercises 11–18
Converting between the Temperature Scales: Fahrenheit, Celsius, and Kelvin (1.6)	Example 1.2 For Practice 1.2 Exercises 19–22
Calculating the Density of a Substance (1.6)	Example 1.3 For Practice 1.3 For More Practice 1.3 Exercises 29–32
Reporting Scientific Measurements to the Correct Digit of Uncertainty (1.7)	Example 1.4 For Practice 1.4 Exercises 35, 36
Working with Significant Figures (1.7)	Examples 1.5, 1.6 For Practice 1.5, 1.6 Exercises 39, 40, 42, 45–50
Using Conversion Factors (1.8)	Examples 1.7, 1.8, 1.9, 1.10 For Practice 1.7, 1.8, 1.9, 1.10 For More Practice 1.9, 1.10 Exercises 53, 54, 58–61, 63, 64
Solving Problems Involving Equations (1.8)	Examples 1.11, 1.12 For Practice 1.11, 1.12 Exercises 77, 78

EXERCISES

Problems by Topic

Note: Answers to all odd-numbered Problems, numbered in blue, can be found in Appendix III. Exercises in the Problems by Topic section are paired, with each odd-numbered problem followed by a similar even-numbered problem. Exercises in the Cumulative Problems section are also paired, but somewhat more loosely. (Challenge Problems and Conceptual Problems, because of their nature, are unpaired.)

The Scientific Approach to Knowledge

1. Classify each statement as an observation, a law, or a theory.
   a. All matter is made of tiny, indestructible particles called atoms.
   b. When iron rusts in a closed container, the mass of the container and its contents does not change.
   c. In chemical reactions, matter is neither created nor destroyed.
   d. When a match burns, heat is evolved.

2. Classify each statement as an observation, a law, or a theory.
   a. Chlorine is a highly reactive gas.
   b. If elements are listed in order of increasing mass of their atoms, their chemical reactivity follows a repeating pattern.
   c. Neon is an inert (or nonreactive) gas.
   d. The reactivity of elements depends on the arrangement of their electrons.

3. A chemist decomposes several samples of carbon monoxide into carbon and oxygen and weighs the resultant elements. The results are shown below:

Sample	Mass of Carbon (g)	Mass of Oxygen (g)
1	6	8
2	12	16
3	18	24

a. Do you notice a pattern in these results?
Next, the chemist decomposes several samples of hydrogen peroxide into hydrogen and oxygen. The results are shown below:

Sample	Mass of Hydrogen (g)	Mass of Oxygen (g)
1	0.5	8
2	1	16
3	1.5	24

b. Do you notice a similarity between these results and those for carbon monoxide in part a?
c. Formulate a law from the observations in a and b.
d. Formulate a hypothesis that might explain your law in c.

4. When astronomers observe distant galaxies, they can tell that most of them are moving away from one another. In addition, the more distant the galaxies, the more rapidly they are likely to be moving away from each other. Devise a hypothesis to explain these observations.

The Classification and Properties of Matter

5. Classify each substance as a pure substance or a mixture. If it is a pure substance, classify it as an element or a compound. If it is a mixture, classify it as homogeneous or heterogeneous.
   a. sweat **b** carbon dioxide
   c. aluminum **d.** vegetable soup

6. Classify each substance as a pure substance or a mixture. If it is a pure substance, classify it as an element or a compound. If it is a mixture, classify it as homogeneous or heterogeneous.
   a. wine **b.** beef stew
   c. iron **d.** carbon monoxide

7. Complete the table.

Substance	Pure or mixture	Type
aluminum	pure	element
apple juice	_____	_____
hydrogen peroxide	_____	_____
chicken soup	_____	_____

8. Complete the table.

Substance	Pure or mixture	Type
water	pure	compound
coffee	_____	_____
ice	_____	_____
carbon	_____	_____

9. Determine whether each molecular diagrams represents a pure substance or a mixture. If it represents a pure substance, classify it as an element or a compound. If it represents a mixture, classify it as homogeneous or heterogeneous.

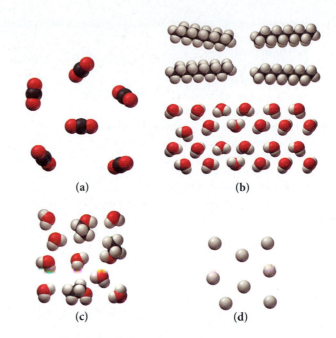

(a)   (b)

(c)   (d)

10. Determine whether each molecular diagram represents a pure substance or a mixture. If it represents a pure substance, classify it as an element or a compound. If it represents a mixture, classify it as homogeneous or heterogeneous.

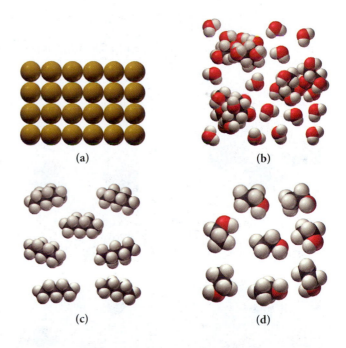

(a)   (b)

(c)   (d)

11. Several properties of isopropyl alcohol (also known as rubbing alcohol) are listed. Classify each property as physical or chemical.
   a. colorless **b.** flammable
   c. liquid at room temperature **d.** density = 0.79 g/mL
   e. mixes with water

12. Several properties of ozone (a pollutant in the lower atmosphere, but part of a protective shield against UV light in the

upper atmosphere) are listed. Which properties are physical and which are chemical?
a. bluish color
b. pungent odor
c. very reactive
d. decomposes on exposure to ultraviolet light
e. gas at room temperature

13. Classify each property as physical or chemical.
    a. the tendency of ethyl alcohol to burn
    b. the shine of silver
    c. the odor of paint thinner
    d. the flammability of propane gas

14. Classify each property as physical or chemical.
    a. the boiling point of ethyl alcohol
    b. the temperature at which dry ice evaporates
    c. the tendency of iron to rust
    d. the color of gold

15. Classify each change as physical or chemical.
    a. Natural gas burns in a stove.
    b. The liquid propane in a gas grill evaporates because the user left the valve open.
    c. The liquid propane in a gas grill burns in a flame.
    d. A bicycle frame rusts on repeated exposure to air and water.

16. Classify each change as physical or chemical.
    a. Sugar burns when heated on a skillet.
    b. Sugar dissolves in water.
    c. A platinum ring becomes dull because of continued abrasion.
    d. A silver surface becomes tarnished after exposure to air for a long period of time.

17. Based on the molecular diagram, classify each change as physical or chemical.

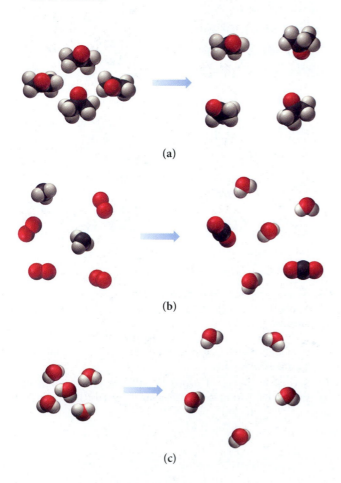

(a)

(b)

(c)

18. Based on the molecular diagram, classify each change as physical or chemical.

(a)

(b)

(c)

Units in Measurement

19. Perform each temperature conversion.
    a. 32 °F to °C (temperature at which water freezes)
    b. 77 K to °F (boiling point of liquid nitrogen)
    c. −109 °F to °C (sublimation point of dry ice)
    d. 98.6 °F to K (body temperature)

20. Perform each temperature conversion.
    a. 212 °F to °C (temperature of boiling water at sea level)
    b. 22 °C to K (approximate room temperature)
    c. 0.00 K to °F (coldest temperature possible, also known as absolute zero)
    d. 2.735 K to °C (average temperature of the universe as measured from background black body radiation)

21. The coldest temperature ever measured in the United States is −80 °F on January 23, 1971, in Prospect Creek, Alaska. Convert that temperature to °C and K. (Assume that −80 °F is accurate to two significant figures.)

22. The warmest temperature ever measured in the United States is 134 °F on July 10, 1913, in Death Valley, California. Convert that temperature to °C and K.

23. Use the prefix multipliers to express each measurement without any exponents.
  a. 3.8×10^{-8} s   **b.** 57×10^{-13} g
  c. 5.9×10^{7} L   **d.** 9.3×10^{8} m

24. Use scientific notation to express each quantity with only the base units (no prefix multipliers).
  a. 38 fs   **b.** 13.2 ns   **c.** 153 pm   **d.** 122 μm

25. Complete the table:
  a. 1245 kg 1.245×10^6 g 1.245×10^9 mg
  b. 515 km   _____ dm   _____ cm
  c. 122.355 s   _____ ms   _____ ks
  d. 3.345 kJ   _____ J   _____ mJ

26. Express the quantity 254,998 m in each unit.
  a. km   **b.** Mm   **c.** mm   **d.** cm

27. How many 1 cm squares would it take to construct a square that is 1 m on each side?

28. How many 1 cm cubes would it take to construct a cube that is 4 cm on edge?

Density

29. A new penny has a mass of 2.49 g and a volume of 0.349 cm³ Is the penny made of pure copper?

30. A titanium bicycle frame displaces 0.314 L of water and has a mass of 1.41 kg. What is the density of the titanium in g/cm³?

31. Glycerol is a syrupy liquid often used in cosmetics and soaps. A 3.25-L sample of pure glycerol has a mass of 4.10×10^3 g. What is the density of glycerol in g/cm³?

32. A supposedly gold nugget is tested to determine its density. It is found to displace 19.3 mL of water (which means it has a volume of 19.3 mL) and has a mass of 371 grams. Could the nugget be made of gold?

33. Ethylene glycol (antifreeze) has a density of 1.11 g/cm³.
  a. What is the mass in g of 417 mL of this liquid?
  b. What is the volume in L of 4.1 kg of this liquid?

34. Acetone (nail polish remover) has a density of 0.7857 g/cm³.
  a. What is the mass, in g, of 28.56 mL of acetone?
  b. What is the volume, in mL, of 6.54 g of acetone?

The Reliability of a Measurement and Significant Figures

35. Read each measurement to the correct number of significant figures. Laboratory glassware should always be read from the bottom of the meniscus.

(a)   (b)

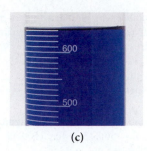

(c)

36. Read each measurment to the correct number of significant figures. Note: Laboratory glassware should always be read from the bottom of the meniscus. Digital balances normally display mass to the correct number of significant figures for that particular balance.

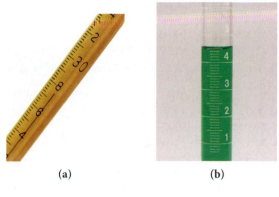

(a)   (b)

(c)

37. For each measurement, underline the zeroes that are significant and draw an x through the zeroes that are not.
  a. 1,050,501 km   **b.** 0.0020 m
  c. 0.000000000000002 s   **d.** 0.001090 cm

38. For each measurement, underline the zeroes that are significant and draw an x through the zeroes that are not.
  a. 180,701 mi   **b.** 0.001040 m
  c. 0.005710 km   **d.** 90,201 m

39. How many significant figures are in each measurement?
  a. 0.000312 m   **b.** 312,000 s
  c. 3.12×10^5 km   **d.** 13,127 s
  e. 2000

40. How many significant figures are in each measurement?
  a. 0.1111 s   **b.** 0.007 m
  c. 108,700 km   **d.** 1.563300×10^{11} m
  e. 30,800

41. Which quantities are exact numbers and therefore have an unlimited number of significant figures?
  a. $\pi = 3.14$
  b. 12 inches = 1 foot
  c. EPA gas mileage rating of 26 miles per gallon
  d. 1 gross = 144

42. Indicate the number of significant figures in each quantity. If the number is an exact number, indicate an unlimited number of significant figures.
  a. 284,796,887 (2001 U.S. population)
  b. 2.54 cm = 1 in
  c. 11.4 g/cm^3 (density of lead)
  d. 12 = 1 dozen

43. Round each number to four significant figures.
  a. 156.852   **b.** 156.842   **c.** 156.849   **d.** 156.899

44. Round each number to three significant figures.
  a. 79,845.82   **b.** 1.548937×10^7
  c. 2.3499999995   **d.** 0.000045389

Significant Figures in Calculations

45. Perform each calculation to the correct number of significant figures.
  a. $9.15 \div 4.970$
  b. $1.54 \times 0.03060 \times 0.69$
  c. $27.5 \times 1.82 \div 100.04$
  d. $(2.290 \times 10^6) \div (6.7 \times 10^4)$

46. Perform each calculation to the correct number of significant figures.
  a. $89.3 \times 77.0 \times 0.08$
  b. $(5.01 \times 10^5) \div (7.8 \times 10^2)$
  c. $4.005 \times 74 \times 0.007$
  d. $453 \div 2.031$

47. Perform each calculation to the correct number of significant figures.
  a. $43.7 - 2.341$ **b.** $17.6 + 2.838 + 2.3 + 110.77$
  c. $19.6 + 58.33 - 4.974$ **d.** $5.99 - 5.572$

48. Perform each calculation to the correct number of significant figures.
  a. $0.004 + 0.09879$ **b.** $1239.3 + 9.73 + 3.42$
  c. $2.4 - 1.777$ **d.** $532 + 7.3 - 48.523$

49. Perform each calculation to the correct number of significant figures.
  a. $(24.6681 \times 2.38) + 332.58$
  b. $(85.3 - 21.489) \div 0.0059$
  c. $(512 \div 986.7) + 5.44$
  d. $\left[(28.7 \times 10^5) \div 48.533\right] + 144.99$

50. Perform each calculation to the correct number of significant figures.
  a. $\left[(1.7 \times 10^6) \div (2.63 \times 10^5)\right] + 7.33$
  b. $(568.99 - 232.1) \div 5.3$
  c. $(9443 + 45 - 9.9) \times 8.1 \times 10^6$
  d. $(3.14 \times 2.4367) - 2.34$

Unit Conversions

51. Perform each unit conversion.
  a. 27.8 L to cm^3 **b.** 1898 mg to kg
  c. 198 km to cm

52. Perform each unit conversion.
  a. 28.9 nm to μm   **b.** 1432 cm^3 to L
  c. 1211 Tm to Gm

53. Perform each unit conversion between the English and metric systems.
  a. 228 cm to in   **b.** 2.55 kg to lb
  c. 2.41 L to qt   **d.** 157 mm to in

54. Perform each of unit conversion between the English and metric systems.
  a. 2.71 in to mm   **b.** 58 ft to cm
  c. 2169 kg to lb   **d.** 725 yd to km

55. A runner wants to run 10.0 km at a pace of 7.5 miles per hour. How many minutes must she run?

56. A cyclist rides at an average speed of 24 miles per hour. If she wants to bike 195 km, how long (in hours) must she ride?

57. A European automobile has a gas mileage of 14 km/L. What is the gas mileage in miles per gallon?

58. A gas can holds 5.0 gallons of gasoline. What is this quantity in cm^3?

59. A modest-sized house has an area of 195 m^2. What is its area in
  a. km^2 **b.** dm^2 **c.** cm^2

60. A bedroom has a volume of 115 m^3. What is its volume in
  a. km^3 **b.** dm^3 **c.** cm^3

61. The average U.S. farm occupies 435 acres. How many square miles is this? (1 acre = 43,560 ft^2, 1 mile = 5280 ft)

62. Total U.S. farmland occupies 954 million acres. How many square miles is this? (1 acre = 43,560 ft^2, 1 mile = 5280 ft). Total U.S. land area is 3.537 million square miles. What percentage of U.S. land is farmland?

63. An infant acetaminophen suspension contains 80 mg/0.80 mL suspension. The recommended dose is 15 mg/kg? body weight. How many mL of this suspension should be given to an infant weighing 14 lb? (Assume two significant figures.)

64. An infant ibuprofen suspension contains 100 mg/5.0 mL suspension. The recommended dose is 10 mg/kg body weight. How many mL of this suspension should be given to an infant weighing 18 lb? (Assume two significant figures.)

Cumulative Problems

65. There are exactly 60 seconds in a minute, there are exactly 60 minutes in an hour, there are exactly 24 hours in a mean solar day, and there are 365.24 solar days in a solar year. Find the number of seconds in a solar year. Be sure to give your answer with the correct number of significant figures.

66. Determine the number of picoseconds in 2.0 hours.

67. Classify each property as intensive or extensive.
  a. volume   **b.** boiling point
  c. temperature   **d.** electrical conductivity
  e. energy

68. At what temperatures will the readings on the Fahrenheit and Celsius thermometers be the same?

69. Suppose you have designed a new thermometer called the X thermometer. On the X scale the boiling point of water is 130 °X and the freezing point of water is 10 °X. At what temperature will the readings on the Fahrenheit and X thermometers be the same?

70. On a new Jekyll temperature scale, water freezes at 17 °J and boils at 97 °J. On another new temperature scale, the Hyde scale, water freezes at 0 °H and boils at 120 °H. If methyl alcohol boils at 84 °H, what is its boiling point on the Jekyll scale?

71. Do each calculation without using your calculator and give the answers to the correct number of significant figures.
    a. $1.76 \times 10^{-3}/8.0 \times 10^2$
    b. $1.87 \times 10^{-2} + 2 \times 10^{-4} - 3.0 \times 10^{-3}$
    c. $\left[(1.36 \times 10^5)(0.000322)/0.082\right](129.2)$

72. The value of the euro was recently $1.38 U.S. and the price of 1 liter of gasoline in France is 1.35 euro. What is the price of 1 gallon of gasoline in U.S. dollars in France?

73. A thief uses a can of sand to replace a solid gold cylinder that sits on a weight-sensitive, alarmed pedestal. The can of sand and the gold cylinder have exactly the same dimensions (length = 22 cm and radius = 3.8 cm).
    a. Calculate the mass of each cylinder (ignore the mass of the can itself). (density of gold = 19.3 g/cm^3, density of sand = 3.00 g/cm^3)
    b. Did the thief set off the alarm? Explain.

74. The proton has a radius of approximately 1.0×10^{-13} cm and a mass of 1.7×10^{-24} g. Determine the density of a proton. For a sphere $V = (4/3)\pi r^3$.

75. The density of titanium is 7.51 g/cm^3. What is the volume (in cubic inches) of 3.5 lb of titanium?

76. The density of iron is 7.86 g/cm^3. What is its density in pounds per cubic inch (lb/in^3)?

77. A steel cylinder has a length of 2.16 in, a radius of 0.22 in, and a mass of 41 g. What is the density of the steel in g/cm^3?

78. A solid aluminum sphere has a mass of 85 g. Use the density of aluminum to find the radius of the sphere in inches.

79. A backyard swimming pool holds 185 cubic yards (yd^3) of water. What is the mass of the water in pounds?

80. An iceberg has a volume of 7655 cubic feet. What is the mass of the ice (in kg) composing the iceberg?

81. The Toyota Prius, a hybrid electric vehicle, has an EPA gas mileage rating of 52 mi/gal in the city. How many kilometers can the Prius travel on 15 liters of gasoline?

82. The Honda Insight, a hybrid electric vehicle, has an EPA gas mileage rating of 57 mi/gal in the city. How many kilometers can the Insight travel on the amount of gasoline that would fit in a soda pop can? The volume of a soda pop can is 355 mL.

83. The single proton that forms the nucleus of the hydrogen atom has a radius of approximately 1.0×10^{-13} cm. The hydrogen atom itself has a radius of approximately 52.9 pm. What fraction of the space within the atom is occupied by the nucleus?

84. A sample of gaseous neon atoms at atmospheric pressure and 0 °C contains 2.69×10^{22} atoms per liter. The atomic radius of neon is 69 pm. What fraction of the space is occupied by the atoms themselves? What does this reveal about the separation between atoms in the gaseous phase?

85. A length of #8 copper wire (radius = 1.63 mm) has a mass of 24.0 kg and a resistance of 2.061 ohm per km (Ω/km). What is the overall resistance of the wire?

86. Rolls of aluminum foil are 304 mm wide and 0.016 mm thick. What maximum length of aluminum foil can be made from 1.10 kg of aluminum?

87. Liquid nitrogen has a density of 0.808 g/mL and boils at 77 K. Researchers often purchase liquid nitrogen in insulated 175-L tanks. The liquid vaporizes quickly to gaseous nitrogen (which has a density of 1.15 g/L at room temperature and atmospheric pressure) when the liquid is removed from the tank. Suppose that all 175 L of liquid nitrogen in a tank accidentally vaporized in lab that measured 10.00 m × 10.00 m × 2.50 m. What maximum fraction of the air in the room is displaced by the gaseous nitrogen?

88. Mercury is often used as an expansion medium in a thermometer. The mercury sits in a bulb on the bottom of the thermometer and rises up a thin capillary as the temperature rises. Suppose a mercury thermometer contains 3.380 g of mercury and has a capillary that is 0.200 mm in diameter. How far does the mercury rise in the capillary when the temperature changes from 0.0 °C to 25.0 °C. The density of mercury at these temperatures is 13.596 g/cm^3 and 13.534 g/cm^3, respectively.

Challenge Problems

89. In 1999, scientists discovered a new class of black holes with masses 100 to 10,000 times the mass of our sun, but occupying less space than our moon. Suppose that one of these black holes has a mass of 1×10^3 suns and a radius equal to one-half the radius of our moon. What is the density of the black hole in g/cm^3? The radius of our sun is 7.0×10^5 km is and it has an average density of 1.4×10^3 kg/m^3. The diameter of the moon is 2.16×10^3 miles.

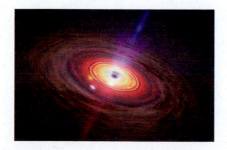

90. You learned in Section 1.7 that in 1997 Los Angeles County air had carbon monoxide (CO) levels of 15.0 ppm. An average human inhales about 0.50 L of air per breath and takes about 20 breaths per minute. How many milligrams of carbon monoxide does the average person inhale in an 8-hour period for this level of carbon monoxide pollution? Assume that the carbon monoxide has a density of 1.2 g/L. (Hint: 15.0 ppm CO means 15.0 L CO per 10^6 L air.)

91. Nanotechnology, the field of building ultrasmall structures one atom at a time, has progressed in recent years. One potential application of nanotechnology is the construction of artificial cells. The simplest cells would mimic red blood cells, the body's oxygen transporters. These nanocontainers, constructed of carbon, could be pumped full of oxygen and injected into a person's bloodstream. If the person needed additional oxygen—due to a heart attack perhaps, or for the purpose of space travel—these containers would slowly release oxygen into the blood, allowing tissues that would otherwise die to remain alive. Suppose that the cell-like nanocontainers were cubic and had an edge length of 25 nanometers.
    a. What is the volume of one nanocontainer? (Ignore the thickness of the nanocontainer's wall.)
    b. Suppose that each nanocontainer could contain pure oxygen pressurized to a density of 85 g/L. How many grams of oxygen could be contained by each nanocontainer?

c. Normal air contains about 0.28 g of oxygen per liter. An average human inhales about 0.50 L of air per breath and takes about 20 breaths per minute. How many grams of oxygen does a human inhale per hour? (Assume two significant figures.)

d. What is the minimum number of nanocontainers that a person would need in their bloodstream to provide 1 hour's worth of oxygen?

e. What is the minimum volume occupied by the number of nanocontainers calculated in part d? Is such a volume feasible, given that total blood volume in an adult is about 5 liters?

92. Approximate the percent increase in waist size that occurs when a 155 lb person gains 40.0 lbs of fat. Assume that the volume of the person can be modeled by a cylinder that is 4.0 feet tall. The average density of a human is about 1.0 g/cm^3 and the density of fat is 0.918 g/cm^3.

93. A box contains a mixture of small copper spheres and small lead spheres. The total volume of both metals is measured by the displacement of water to be 427 cm^3 and the total mass is 4.36 kg. What percent of the total volume of the spheres is copper?

Conceptual Problems

94. A volatile liquid (one that easily evaporates) is put into a jar and the jar is then sealed. Does the mass of the sealed jar and its contents change upon the vaporization of the liquid?

95. The following diagram represents solid carbon dioxide, also known as dry ice.

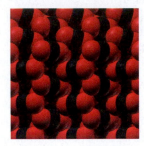

Which diagram best represents the dry ice after it has sublimed into a gas?

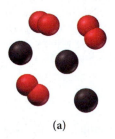

(a)

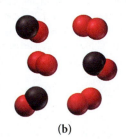

(b)

(c)

96. Let a triangle represent atoms of element A and a circle represent atoms of element B.
    a. Draw an atomic level view of a homogenous mixture of elements A and B.
    b. Draw an atomic view of the compound AB in a liquid state (molecules close together).

c. Draw an atomic view of the compound AB after it has undergone a physical change (such as evaporation).

d. Draw an atomic view of the compound after it has undergone a chemical change (such as decomposition of AB into A and B).

97. A cube has an edge length of 7 cm. If it is divided up into 1 cm cubes, how many 1 cm cubes are there?

98. Substance A has a density of 1.7 g/cm^3. Substance B has a density of 1.7 kg/m^3. Without doing any calculations, determine which substance is most dense.

99. For each box, examine the blocks attached to the balances. Based on their positions and sizes, determine which block is more dense (the dark block or the lighter-colored block), or if the relative densities cannot be determined. (Think carefully about the information being shown.)

(a)

(b)

(c)

100. Identify each statement as being most like an observation, a law, or a theory.

    a. All coastal areas experience two high tides and two low tides each day.

    b. The tides in Earth's oceans are caused mainly by the gravitational attraction of the moon.

    c. Yesterday, high tide in San Francisco Bay occurred at 2:43 a.m. and 3:07 p.m.

    d. Tides are higher at the full moon and new moon than at other times of the month.

Answers to Conceptual Connections

Laws and Theories

1.1 (b) A law simply summarizes a series of related observations, while a theory gives the underlying reasons for them.

Pure Substances and Mixtures

1.2

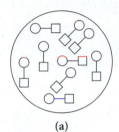

(a)

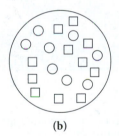

(b)

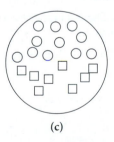

(c)

Chemical and Physical Changes

1.3 Diagram **(a)** best represents the water after vaporization. Vaporization is a physical change, so the molecules must remain the same before and after the change.

The Mass of a Gas

1.4 No. The water vaporizes and becomes a gas, but the water molecules are still present within the flask and have the same mass.

Prefix Multipliers

1.5 The prefix micro (10^{-6}) is appropriate. The measurement would be reported as 55.7 μm.

Density

1.6 (c) The sample expands. However, because its mass remains constant, its density decreases.

2 Atoms and Elements

These observations have tacitly led to the conclusion which seems universally adopted, that all bodies of sensible magnitude . . . are constituted of a vast number of extremely small particles, or atoms of matter . . .—John Dalton (1766–1844)

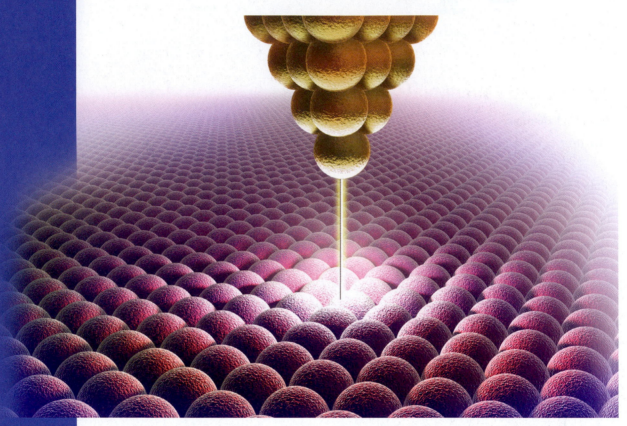

The tip of a scanning tunneling microscope (STM) moves across an atomic surface.

IF YOU CUT A PIECE OF GRAPHITE from the tip of a pencil into smaller and smaller pieces, how far could you go? Could you divide it forever? Would you eventually come upon some basic particles that were no longer divisible, not because of their sheer smallness, but because of the nature of matter? This fundamental question about matter has been asked by thinkers for over two millennia. The answer has varied over time. On the scale of everyday objects, matter appears continuous (or infinitely divisible) when viewed with the naked eye. Until about 200 years ago, many scientists thought that matter was indeed continuous—but they were wrong. Eventually, if you could divide the graphite from your pencil tip into smaller and smaller pieces (far smaller than the naked eye could see), you would end up with carbon atoms. The word atom comes from the Greek *atomos*, meaning "indivisible." You cannot divide a carbon atom into smaller pieces and still have carbon. Atoms compose all ordinary matter—if you want to understand matter, you must begin by understanding atoms.

2.1   Imaging and Moving Individual Atoms

On March 16, 1981, Gerd Binnig and Heinrich Rohrer worked late into the night in their laboratory at IBM in Zurich, Switzerland. They were measuring how an electrical current—flowing between a sharp metal tip and a flat metal surface—varied as the distance between the tip and the surface varied. The results of that night's experiment and subsequent results obtained over the next several months won Binnig and Rohrer a share of the 1986 Nobel Prize in Physics. It also led to the development of *scanning tunneling microscopy (STM)*, a technique that can image (make a visual representation of), and even move, individual atoms and molecules.

A scanning tunneling microscope works by moving an extremely sharp electrode (a conductor through which electricity can enter or leave) over a surface and measuring the resulting *tunneling current*, the electrical current that flows between the tip of the electrode and the surface even though the two are not in physical contact (**Figure 2.1▼**).

This tunneling current (as Binnig and Rohrer discovered that night in their laboratory at IBM) is extremely sensitive to distance, making it possible to maintain a precise separation between the electrode tip and the surface by moving the tip up or down to compensate for changes in current. As the tip goes over an atom, therefore, the tip moves up away from the surface to maintain constant current. As the tip goes over a gap between atoms, the tip moves down towards the surface to maintain constant current. By measuring the up-and-down movement of the tip as it is scanned horizontally across a surface, an image of the surface, showing the location of individual atoms, can be created, like the one shown in **Figure 2.2(a)▶**.

In other words, Binnig and Rohrer developed a type of microscope that could "see" atoms. Later work by other scientists showed that the STM could also be used to pick up and move individual atoms or molecules, allowing structures and patterns to be made one atom at a time. **Figure 2.2(b)▶**, for example, shows the Kanji characters for the word "atom" written with individual iron atoms on top of a copper surface. If all of the words

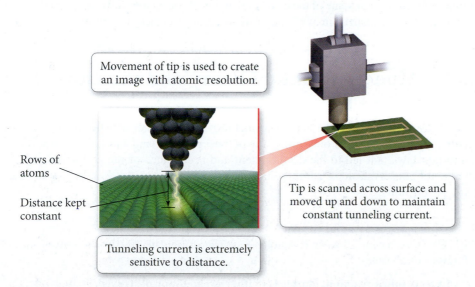

Movement of tip is used to create an image with atomic resolution.

Rows of atoms

Distance kept constant

Tip is scanned across surface and moved up and down to maintain constant tunneling current.

Tunneling current is extremely sensitive to distance.

▲ **FIGURE 2.1   Scanning Tunneling Microscopy**   In this technique, an atomically sharp tip is scanned across a surface. The tip is kept at a fixed distance from the surface by moving it up and down so the tunneling current remains constant. The motion of the tip is recorded to create an image of the surface with atomic resolution.

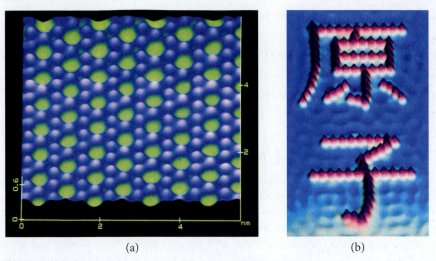

(a)   (b)

▲ **FIGURE 2.2 Imaging Atoms** (a) A scanning tunneling microscope image of iodine atoms (green) on a platinum surface (blue). (b) The Japanese Kanji characters for "atom" written with atoms.

in the books in the Library of Congress—29 million books occupying 530 miles of shelves—were written in letters the size of these Kanji characters, they would fit in an area of about 5 square millimeters.

As we discussed in Chapter 1, it was only 200 years ago that John Dalton proposed his atomic theory. Today scientists can not only image and move individual atoms, but are even beginning to build tiny machines out of just a few dozen atoms (an area of research called nanotechnology).

These atomic machines, and the atoms that compose them, are almost unimaginably small. To get an idea of the size of an atom, imagine picking up a grain of sand at your favorite beach. That grain contains more atoms than you could count in a lifetime. In fact, the number of atoms in one sand grain far exceeds the number of grains on the entire beach.

In spite of their small size, atoms are the key to connecting the macroscopic and microscopic worlds. An *atom* is the smallest identifiable unit of an *element*. There are about 91 different naturally occurring elements. In addition, scientists have succeeded in making over 20 synthetic elements (elements not found in nature). In this chapter, we introduce atoms: what they are made of, how they differ from one another, and how they are structured. We also introduce the elements composed of these different kinds of atoms, and learn about some of the characteristics of those elements. We will find that the elements can be organized in a way that reveals patterns in their properties.

> The exact number of naturally occurring elements is ambiguous, because some elements that were first discovered when they were synthesized are believed to be present in trace amounts in nature.

2.2 Modern Atomic Theory and the Laws That Led to It

Recall the discussion of the scientific method from Chapter 1. The theory that all matter is composed of atoms grew out of observations and laws. The three most important laws that led to the development and acceptance of the atomic theory are the law of conservation of mass, the law of definite proportions, and the law of multiple proportions.

The Law of Conservation of Mass

In 1789, as we saw in Chapter 1, Antoine Lavoisier formulated the **law of conservation of mass**, which states:

> **In a chemical reaction, matter is neither created nor destroyed.**

In other words, when you carry out any chemical reaction, the total mass of the substances involved in the reaction does not change. For example, consider the reaction between sodium and chlorine to form sodium chloride.

> We will see in Chapter 19 that this law is a slight oversimplification. However, the changes in mass in ordinary chemical processes are so minute that they can be ignored for all practical purposes.

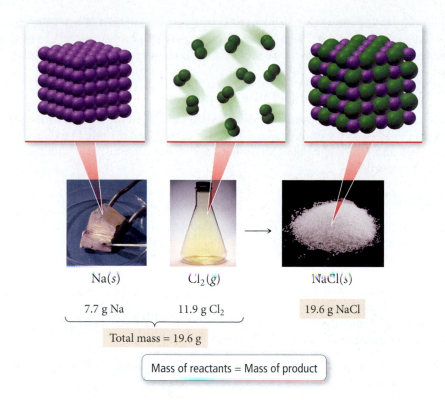

Na(s)   Cl$_2$(g)   NaCl(s)

7.7 g Na   11.9 g Cl$_2$ 19.6 g NaCl

Total mass = 19.6 g

Mass of reactants = Mass of product

The combined mass of the sodium and chlorine that react (the reactants) exactly equals the mass of the sodium chloride that results from the reaction (the product). This law is consistent with the idea that matter is composed of small, indestructible particles. Particles rearrange during a chemical reaction, but the amount of matter is conserved because the particles are indestructible (at least by chemical means).

 Conceptual Connection 2.1 **The Law of Conservation of Mass**

When a small log completely burns in a campfire, the mass of the ash is much less than the mass of the log. What happened to the matter that composed the log?

The Law of Definite Proportions

In 1797, a French chemist named Joseph Proust (1754–1826) made observations on the composition of compounds. He found that the elements composing a given compound always occur in fixed (or definite) proportions in all samples of the compound. In contrast, the components of a mixture can be present in any proportions whatsoever. He summarized his observations in the **law of definite proportions**, which states:

> **All samples of a given compound, regardless of their source or how they are prepared, have the same proportions of their constituent elements.**

For example, the decomposition of 18.0 g of water results in 16.0 g of oxygen and 2.0 g of hydrogen, or an oxygen-to-hydrogen mass ratio of

$$\text{Mass ratio} = \frac{16.0 \text{ g oxygen}}{2.0 \text{ g hydrogen}} = 8.0 \quad \text{or} \quad 8{:}1$$

This ratio holds for any sample of pure water, regardless of its origin. The law of definite proportions applies not only to water, but to every compound. Consider ammonia, a compound composed of nitrogen and hydrogen. Ammonia contains 14.0 g of nitrogen for every 3.0 g of hydrogen, resulting in a nitrogen-to-hydrogen mass ratio of:

$$\text{Mass ratio} = \frac{14.0 \text{ g nitrogen}}{3.0 \text{ g hydrogen}} = 4.7 \quad \text{or} \quad 4.7{:}1$$

Again, this ratio is the same for every sample of ammonia. The law of definite proportions also hints at the idea that matter is composed of atoms. Compounds have definite

The law of definite proportions is sometimes called the law of constant composition.

proportions of their constituent elements because they are composed of a definite ratio of atoms of each element, each with its own specific mass. Since the ratio of atoms is the same for all samples of a particular compound, the ratio of masses is also the same.

EXAMPLE 2.1 Law of Definite Proportions

Two samples of carbon dioxide are decomposed into their constituent elements. One sample produces 25.6 g of oxygen and 9.60 g of carbon, and the other produces 21.6 g of oxygen and 8.10 g of carbon. Show that these results are consistent with the law of definite proportions.

SOLUTION

To show this, calculate the mass ratio of one element to the other for both samples by dividing the mass of one element by the mass of the other. It is usually more convenient to divide the larger mass by the smaller one.

For the first sample:

$$\frac{\text{Mass oxygen}}{\text{Mass carbon}} = \frac{25.6 \text{ g}}{9.60 \text{ g}} = 2.67 \quad \text{or} \quad 2.67:1$$

For the second sample:

$$\frac{\text{Mass oxygen}}{\text{Mass carbon}} = \frac{21.6 \text{ g}}{8.10 \text{ g}} = 2.67 \quad \text{or} \quad 2.67:1$$

The ratios are the same for the two samples, so these results are consistent with the law of definite proportions.

FOR PRACTICE 2.1

Two samples of carbon monoxide are decomposed into their constituent elements. One sample produces 17.2 g of oxygen and 12.9 g of carbon, and the other sample produces 10.5 g of oxygen and 7.88 g of carbon. Show that these results are consistent with the law of definite proportions. (Answers to For Practice and For More Practice problems can be found in Appendix IV.)

The Law of Multiple Proportions

In 1804, John Dalton published his **law of multiple proportions**, which asserts the following principle:

> When two elements (call them A and B) form two different compounds, the masses of element B that combine with 1 g of element A can be expressed as a ratio of small whole numbers.

Dalton already suspected that matter was composed of atoms, so that when two elements, A and B, combine to form more than one compound, an atom of A combines with one, two, three, or more atoms of B (AB_1, AB_2, AB_3, etc.). Therefore the masses of B that react with a fixed mass of A are always related to one another as small whole-number ratios. For example, consider the compounds carbon monoxide and carbon dioxide, which we discussed in the opening section of Chapter 1 as well as in Example 2.1 and For Practice problem 2.1. Carbon monoxide and carbon dioxide are two compounds composed of the same two elements: carbon and oxygen. We saw in Example 2.1 that the mass ratio of oxygen to carbon in carbon dioxide is 2.67:1; therefore, 2.67 g of oxygen reacts with 1 g of carbon. In carbon monoxide, however, the mass ratio of oxygen to carbon is 1.33:1, or 1.33 g of oxygen to every 1 g of carbon.

The ratio of these two masses is itself a small whole number.

Carbon dioxide

Mass oxygen that combines with 1 g carbon = 2.67 g

Carbon monoxide

Mass oxygen that combines with 1 g carbon = 1.33 g

$$\frac{\text{Mass oxygen to 1 g carbon in carbon dioxide}}{\text{Mass oxygen to 1 g carbon in carbon monoxide}} = \frac{2.67\text{g}}{1.33\text{ g}} = 2.00$$

With the help of the molecular models, we can see why the ratio is 2:1—carbon dioxide contains two oxygen atoms to every carbon atom while carbon monoxide contains only one.

EXAMPLE 2.2 Law of Multiple Proportions

Nitrogen forms several compounds with oxygen, including nitrogen dioxide and dinitrogen monoxide. Nitrogen dioxide contains 2.28 g oxygen to every 1.00 g nitrogen while dinitrogen monoxide contains 0.570 g oxygen to every 1.00 g nitrogen. Show that these results are consistent with the law of multiple proportions.

SOLUTION

To show this, calculate the ratio of the mass of oxygen from one compound to the mass of oxygen in the other. Always divide the larger of the two masses by the smaller one.	$\dfrac{\text{Mass oxygen to 1 g nitrogen in nitrogen dioxide}}{\text{Mass oxygen to 1 g nitrogen in dinitrogen monoxide}} = \dfrac{2.28 \text{ g}}{0.570 \text{ g}} = 4.00$

The ratio is a small whole number (4); therefore these results are consistent with the law of multiple proportions.

FOR PRACTICE 2.2
Hydrogen and oxygen form both water and hydrogen peroxide. A sample of water is decomposed and forms 0.125 g hydrogen to every 1.00 g oxygen. A sample of hydrogen peroxide is decomposed and forms 0.250 g hydrogen to every 1.00 g oxygen. Show that these results are consistent with the law of multiple proportions.

 Conceptual Connection 2.2 **The Laws of Definite and Multiple Proportions**

Explain the difference between the law of definite proportions and the law of multiple proportions.

John Dalton and the Atomic Theory

In 1808, John Dalton explained the laws just discussed with his **atomic theory**, which includes the following concepts:

1. Each element is composed of tiny, indestructible particles called atoms.
2. All atoms of a given element have the same mass and other properties that distinguish them from the atoms of other elements.
3. Atoms combine in simple, whole-number ratios to form compounds.
4. Atoms of one element cannot change into atoms of another element. In a chemical reaction, atoms change the way that they are bound together with other atoms to form a new substance.

Today, as we have seen in the opening section of this chapter, the evidence for the atomic theory is overwhelming. Matter is indeed composed of atoms.

2.3   The Discovery of the Electron

By the end of the nineteenth century, scientists were convinced that matter was composed of atoms, the permanent, supposedly indestructible building blocks that composed everything. However, further experiments revealed that the atom itself is composed of even smaller, more fundamental particles.

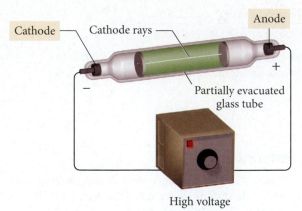

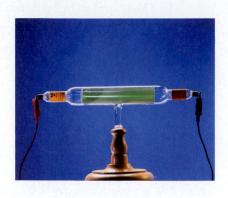

▲ FIGURE 2.3 **Cathode Ray Tube**

Cathode Rays

In the late 1800s an English physicist named J. J. Thomson (1856–1940), working at Cambridge University, performed experiments to probe the properties of **cathode rays**. Cathode rays are produced when a high electrical voltage is applied between two electrodes within a partially evacuated glass tube. This type of tube is called a **cathode ray tube** and is shown in **Figure 2.3▲**.

Cathode rays are emitted by the negatively charged electrode, the cathode, and travel to the positively charged electrode, the anode. The rays can be "seen" when they collide with the end of the tube, which is coated with fluorescent material that emits light when struck by the rays. Thomson found that these rays were actually streams of particles with the following properties: they travel in straight lines; they are independent of the composition of the material from which they originated (the cathode); and they carry a negative **electrical charge**. Electrical charge is a fundamental property of some of the particles that compose atoms. An electrically charged particle experiences attractive or repulsive forces—called *electrostatic forces*—with other electrically charged particles. The characteristics of electrical charge are summarized in the figure in the margin.

You have probably experienced the effects of electrical charge when brushing your hair on a dry day. The brushing action causes the accumulation of charged particles in your hair, which repel each other, making your hair stand on end.

J. J. Thomson measured the charge-to-mass ratio of the particles within cathode rays by deflecting them using electric and magnetic fields, as shown in **Figure 2.4▼**. The value he measured, -1.76×10^8 coulombs (C) per gram, implied that the cathode ray particle was about 2000 times lighter (less massive) than hydrogen, the lightest known atom. These results were incredible—the indestructible atom could apparently be chipped! J. J. Thomson had discovered the **electron**, a negatively charged, low mass particle present within all atoms.

Properties of Electrical Charge

Positive (red) and negative (yellow) electrical charges attract one another.

Positive charges repel one another. Negative charges repel one another.

+1 + (−1) = 0

Positive and negative charges of exactly the same magnitude sum to zero when combined.

▶ FIGURE 2.4 **Thomson's Measurement of the Charge-to-Mass Ratio of the Electron** J. J. Thomson used electric and magnetic fields to deflect the electron beam in a cathode ray tube.

Charge-to-Mass Ratio of the Electron

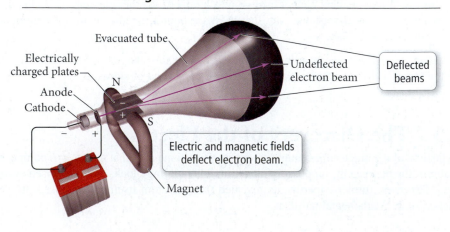

Evacuated tube

Electrically charged plates

Anode

Cathode

N

S

Undeflected electron beam

Deflected beams

Electric and magnetic fields deflect electron beam.

Magnet

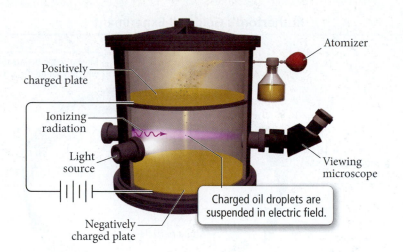

◀ **FIGURE 2.5 Millikan's Measurement of the Electron's Charge**   Millikan calculated the charge on oil droplets falling in an electric field. He found that it was always a whole-number multiple of -1.60×10^{-19} C, the charge of a single electron.

Millikan's Oil Drop Experiment: The Charge of the Electron

In 1909, American physicist Robert Millikan (1868–1953), working at the University of Chicago, performed his now famous oil drop experiment in which he deduced the charge of a single electron. The apparatus for the oil drop experiment is shown in **Figure 2.5▲**.

In Millikan's experiment, oil was sprayed into fine droplets using an atomizer. The droplets were allowed to fall through a small hole into the lower portion of the apparatus where they could be viewed with the help of a light source and a microscope. During their fall, the drops acquired electrons that had been produced when the air was bombarded with a type of energy (which we will learn more about later) called ionizing radiation. These negatively charged drops then interacted with the negatively charged plate at the bottom of the apparatus. (Remember that like charges repel each other.) By varying the amount of charge on the plates, the fall of the drops could be slowed, stopped, or even reversed.

By measuring the voltage required to halt the free fall of the drops, and figuring out the masses of the drops themselves (determined from their radii and density), Millikan calculated the charge of each drop. He then reasoned that, since each drop must contain an integral (whole) number of electrons, the charge of each drop must be a whole-number multiple of the electron's charge. Indeed, Millikan was correct; the measured charge on any drop was always a whole-number multiple of -1.60×10^{-19} C, the fundamental charge of a single electron. With this number in hand, and knowing Thomson's mass-to-charge ratio for electrons, we can deduce the mass of an electron as follows:

$$\text{Charge} \times \frac{\text{mass}}{\text{charge}} = \text{mass}$$

$$-1.60 \times 10^{-19}\ C \times \frac{g}{-1.76 \times 10^{8}\ C} = 9.10 \times 10^{-28}\ g$$

As Thomson had correctly deduced, this mass is about 2000 times lighter than hydrogen, the lightest atom.

 Conceptual Connection 2.3   The Millikan Oil Drop Experiment

Suppose that one of Millikan's oil drops has a charge of -4.8×10^{-19} C. How many excess electrons does the drop contain?

2.4   The Structure of the Atom

The discovery of negatively charged particles within atoms raised a new question. Since atoms are charge-neutral, they must contain positive charge that neutralizes the negative charge of the electrons—but how do the positive and negative charges within the atom fit together? Are atoms just a jumble of even more fundamental particles? Are they solid spheres? Do they have some internal structure? J. J. Thomson proposed that the negatively charged electrons were small particles held within a positively charged sphere, as shown in the margin at right. This model, the most popular of the time, became known as the plum-pudding model. The picture suggested by Thomson, to those of us not familiar

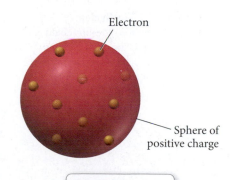

Electron

Sphere of positive charge

Plum-pudding model

▶ **FIGURE 2.6 Rutherford's Gold Foil Experiment** Rutherford directed alpha particles at a thin sheet of gold foil. Most of the particles passed through the foil, but a small fraction were deflected, and a few even bounced backward.

Rutherford's Gold Foil Experiment

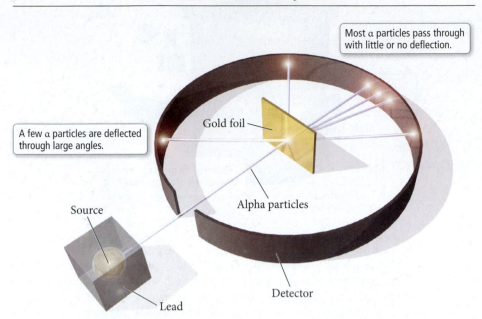

Most α particles pass through with little or no deflection.

A few α particles are deflected through large angles.

Gold foil

Alpha particles

Source

Detector

Lead

Alpha particles are about 7000 times more massive than electrons.

with plum pudding (an English dessert), is more like a blueberry muffin, where the blueberries are the electrons and the muffin is the positively charged sphere.

The discovery of **radioactivity**—the emission of small energetic particles from the core of certain unstable atoms—by scientists Henri Becquerel (1852–1908) and Marie Curie (1867–1934) at the end of the nineteenth century allowed the structure of the atom to be experimentally probed. At the time, three different types of radioactivity had been identified: alpha (α) particles, beta (β) particles, and gamma (γ) rays. We will discuss these and other types of radioactivity in more detail in Chapter 19. For now, just know that α particles are positively charged and that they are by far the most massive of the three.

In 1909, Ernest Rutherford (1871–1937), who had worked under Thomson and subscribed to his plum-pudding model, performed an experiment in an attempt to confirm it. His experiment, which employed α particles, proved it wrong instead. In the experiment, Rutherford directed the positively charged α particles at an ultrathin sheet of gold foil, as shown in **Figure 2.6▲**. These particles were to act as probes of the gold atoms' structure. If the gold atoms were indeed like blueberry muffins or plum pudding—with their mass and charge spread throughout the entire volume of the atom—these speeding probes should pass right through the gold foil with minimum deflection.

Rutherford performed the experiment, but the results were not as he expected. A majority of the particles did pass directly through the foil, but some particles were deflected, and some (1 in 20,000) even bounced back. The results puzzled Rutherford, who wrote that they were "about as credible as if you had fired a 15-inch shell at a piece of tissue paper and it came back and hit you." What must the structure of the atom be in order to explain this odd behavior?

Rutherford created a new model—a modern version of which is shown in **Figure 2.7▼** alongside the plum-pudding model—to explain his results.

▶ **FIGURE 2.7 The Nuclear Atom** Rutherford's results could not be explained by the plum-pudding model. Instead, they suggested that the atom has a small, dense nucleus.

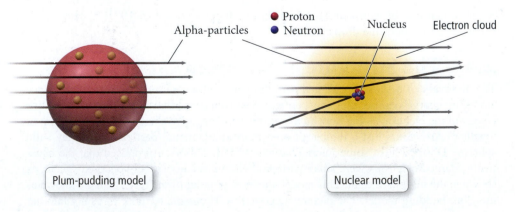

Alpha-particles

● Proton
● Neutron

Nucleus

Electron cloud

Plum-pudding model

Nuclear model

He realized that to account for the deflections he observed, the mass and positive charge of an atom must all be concentrated in a space much smaller than the size of the atom itself. He concluded that the structure of the atom must be far less uniform than that suggested by the plum-pudding model. It must contain large regions of empty space dotted with small regions of very dense matter. Using this idea, he proposed the **nuclear theory** of the atom, with three basic parts:

1. Most of the atom's mass and all of its positive charge are contained in a small core called the **nucleus**.

2. Most of the volume of the atom is empty space, throughout which tiny, negatively charged electrons are dispersed.

3. There are as many negatively charged electrons outside the nucleus as there are positively charged particles (named **protons**) within the nucleus, so that the atom is electrically neutral.

Although Rutherford's model was highly successful, scientists realized that it was incomplete. For example, hydrogen atoms contain one proton and helium atoms contain two, yet the helium-to-hydrogen mass ratio is 4:1. The helium atom must contain some additional mass. Later work by Rutherford and one of his students, British scientist James Chadwick (1891–1974), demonstrated that the previously unaccounted for mass was due to the presence of **neutrons**, neutral particles within the nucleus. The dense nucleus contains over 99.9% of the mass of the atom, but it occupies very little of its volume. For now, we can think of the electrons that surround the nucleus in analogy to the water droplets that make up a cloud—although their mass is almost negligibly small, they are dispersed over a very large volume. Consequently, the atom, like the cloud, is mostly empty space.

Rutherford's nuclear theory was a success and is still valid today. The revolutionary part of this theory is the idea that matter—at its core—is much less uniform than it appears. If the nucleus of the atom were the size of the period at the end of this sentence, the average electron would be about 10 meters away. Yet the period would contain almost the entire mass of the atom. Imagine what matter would be like if atomic structure broke down. What if matter were composed of atomic nuclei piled on top of each other like marbles? Such matter would be incredibly dense; a single grain of sand composed of solid atomic nuclei would have a mass of 5 million kilograms (or a weight of about 11 million pounds). Astronomers believe there are objects in the universe composed of such matter—they are called neutron stars.

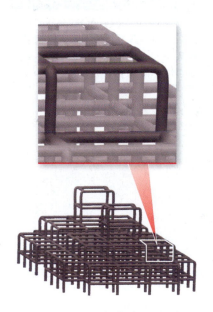

If matter really is mostly empty space, as Rutherford suggested, then why does it appear so solid? Why can we tap our knuckles on a table and feel a solid thump? Matter appears solid because the variation in its density is on such a small scale that our eyes cannot see it.

Imagine a scaffolding 100 stories high and the size of a football field as shown in the margin. It is mostly empty space. Yet if you viewed it from an airplane, it would appear as a solid mass. Matter is similar. When you tap your knuckle on the table, it is like one giant scaffolding (your finger) crashing into another (the table). Even though they are both primarily empty space, one does not fall into the other.

2.5 Subatomic Particles: Protons, Neutrons, and Electrons in Atoms

We have just seen that all atoms are composed of the same subatomic particles: protons, neutrons, and electrons. Protons and neutrons have nearly identical masses. In SI units, the mass of the proton is 1.67262×10^{-27} kg, and the mass of the neutron is 1.67493×10^{-27} kg. A more common unit to express these masses, however, is the **atomic mass unit (amu)**, defined as 1/12 the mass of a carbon atom containing six protons and six neutrons. The mass of a proton or neutron is approximately 1 amu. Electrons, in contrast, have an almost negligible mass of 0.00091×10^{-27} kg or 0.00055 amu. If a proton had the mass of a baseball, an electron would have the mass of a rice grain.

TABLE 2.1 Subatomic Particles

	Mass (kg)	Mass (amu)	Charge (relative)	Charge (C)
Proton	1.67262×10^{-27}	1.00727	+1	$+1.60218 \times 10^{-19}$
Neutron	1.67493×10^{-27}	1.00866	0	0
Electron	0.00091×10^{-27}	0.00055	−1	-1.60218×10^{-19}

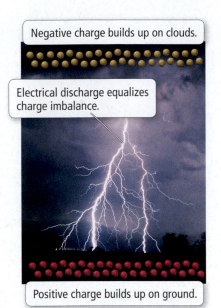

Negative charge builds up on clouds.

Electrical discharge equalizes charge imbalance.

Positive charge builds up on ground.

▲ When the normal charge balance of matter is disturbed, as happens during an electrical storm, it quickly equalizes, often in dramatic ways.

The proton and the electron both have electrical *charge*. We know from Millikan's oil drop experiment that the electron has a charge of -1.60×10^{-19} C. In atomic or relative units, the electron is assigned a charge of –1 and the proton is assigned a charge of +1. The charge of the proton and the electron are equal in magnitude but opposite in sign, so that when the two particles are paired, the charges sum to zero. The neutron has no charge.

Matter is usually charge-neutral (it has no overall charge) because protons and electrons are normally present in equal numbers. When matter does acquire charge imbalances, these imbalances usually equalize quickly, often in dramatic ways. For example, the shock you receive when touching a doorknob during dry weather is the equalization of a charge imbalance that developed as you walked across the carpet. Lightning, as shown in the margin, is an equalization of charge imbalances that develop during electrical storms.

If you had a sample of matter—even a tiny sample, such as a sand grain—that was composed only of protons or only of electrons, the repulsive forces inherent in that matter would be extraordinary, and the matter would be very unstable. Luckily, that is not how matter is. Table 2.1 summarizes the properties of protons, neutrons, and electrons.

Elements: Defined by Their Numbers of Protons

If all atoms are composed of the same subatomic particles, then what makes the atoms of one element different from those of another? The answer is the *number* of these particles. The number that is most important for the identity of an atom is the number of protons in its nucleus. In fact, the number of protons defines the element. For example, an atom with 2 protons in its nucleus is a helium atom; an atom with 6 protons in its nucleus is a carbon atom; and an atom with 92 protons in its nucleus is a uranium atom (**Figure 2.8▼**). The number of protons in an atom's nucleus is the **atomic number** and is given the symbol **Z**. The atomic numbers of known elements range from 1 to 117 (although additional elements may still be discovered). In the periodic table (**Figure 2.9▶**), described in more detail in Section 2.6, the elements are arranged so that those with similar properties occur in the same column. Each element, identified by a unique atomic number, is represented with a unique **chemical symbol**, a

▶ **FIGURE 2.8 How Elements Differ** Each element is defined by its unique atomic number (**Z**), the number of protons in the nucleus of every atom of that element.

The Number of Protons Defines the Element

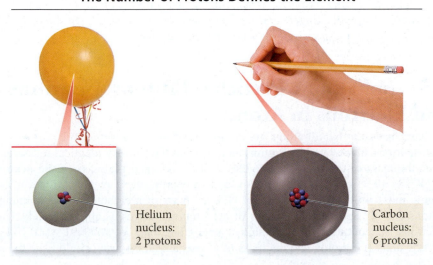

Helium nucleus: 2 protons

Carbon nucleus: 6 protons

The Periodic Table

Atomic number (Z)

4
Be
beryllium — Chemical symbol

— Name

[Periodic table follows, with elements arranged in standard layout]

1 H hydrogen	2 He helium

| 3 Li lithium | 4 Be beryllium | | | | | | | | | | | | | 5 B boron | 6 C carbon | 7 N nitrogen | 8 O oxygen | 9 F fluorine | 10 Ne neon |

| 11 Na sodium | 12 Mg magnesium | | | | | | | | | | | | | 13 Al aluminum | 14 Si silicon | 15 P phosphorus | 16 S sulfur | 17 Cl chlorine | 18 Ar argon |

| 19 K potassium | 20 Ca calcium | 21 Sc scandium | 22 Ti titanium | 23 V vanadium | 24 Cr chromium | 25 Mn manganese | 26 Fe iron | 27 Co cobalt | 28 Ni nickel | 29 Cu copper | 30 Zn zinc | 31 Ga gallium | 32 Ge germanium | 33 As arsenic | 34 Se selenium | 35 Br bromine | 36 Kr krypton |

| 37 Rb rubidium | 38 Sr strontium | 39 Y yttrium | 40 Zr zirconium | 41 Nb niobium | 42 Mo molybdenum | 43 Tc technetium | 44 Ru ruthenium | 45 Rh rhodium | 46 Pd palladium | 47 Ag silver | 48 Cd cadmium | 49 In indium | 50 Sn tin | 51 Sb antimony | 52 Te tellurium | 53 I iodine | 54 Xe xenon |

| 55 Cs cesium | 56 Ba barium | 57 La lanthanum | 72 Hf hafnium | 73 Ta tantalum | 74 W tungsten | 75 Re rhenium | 76 Os osmium | 77 Ir iridium | 78 Pt platinum | 79 Au gold | 80 Hg mercury | 81 Tl thallium | 82 Pb lead | 83 Bi bismuth | 84 Po polonium | 85 At astatine | 86 Rn radon |

| 87 Fr francium | 88 Ra radium | 89 Ac actinium | 104 Rf rutherfordium | 105 Db dubnium | 106 Sg seaborgium | 107 Bh bohrium | 108 Hs hassium | 109 Mt meitnerium | 110 Ds darmstadtium | 111 Rg roentgenium | 112 Cn copernicium | 113 ** | 114 ** | 115 ** | 116 ** | 117 ** | 118 ** |

| 58 Ce cerium | 59 Pr praseodymium | 60 Nd neodymium | 61 Pm promethium | 62 Sm samarium | 63 Eu europium | 64 Gd gadolinium | 65 Tb terbium | 66 Dy dysprosium | 67 Ho holmium | 68 Er erbium | 69 Tm thulium | 70 Yb ytterbium | 71 Lu lutetium |

| 90 Th thorium | 91 Pa protactinium | 92 U uranium | 93 Np neptunium | 94 Pu plutonium | 95 Am americium | 96 Cm curium | 97 Bk berkelium | 98 Cf californium | 99 Es einsteinium | 100 Fm fermium | 101 Md mendelevium | 102 No nobelium | 103 Lr lawrencium |

▲ **FIGURE 2.9 The Periodic Table** Each element is represented by its symbol and atomic number. Elements in the same column have similar properties.

one- or two-letter abbreviation listed directly below its atomic number on the periodic table. The chemical symbol for helium is He; for carbon, it is C; and for uranium, it is U. The chemical symbol and the atomic number always go together. If the atomic number is 2, the chemical symbol must be He. If the atomic number is 6, the chemical symbol must be C. This is just another way of saying that the number of protons defines the element.

Most chemical symbols are based on the name of the element. For example, the symbol for sulfur is S; for oxygen, O; and for chlorine, Cl. Several of the oldest known elements, however, have symbols based on their Latin names. Thus, the symbol for sodium is Na from the Latin *natrium*, and the symbol for tin is Sn from the Latin *stannum*. The names of elements sometimes describe their properties. For example, argon originates from the Greek word *argos* meaning inactive, referring to argon's chemical inertness (it does not react with other elements). Chlorine originates from the Greek word *chloros* meaning pale green, referring to chlorine's pale green color. Other elements, including helium, selenium, and mercury, are named after figures from Greek or Roman mythology or astronomical bodies. Still others (such as europium, polonium, and berkelium) are named for the places where they were discovered or where their discoverers were born. More recently, elements have been named after scientists; for example, curium for Marie Curie, einsteinium for Albert Einstein, and rutherfordium for Ernest Rutherford.

Isotopes: When the Number of Neutrons Varies

All atoms of a given element have the same number of protons; however, they do not necessarily have the same number of neutrons. Since neutrons have nearly the same mass as protons (1 amu), this means that—contrary to what John Dalton originally

96
Cm
curium

▲ Element 96 is named curium, after Marie Curie, co-discoverer of radioactivity.

proposed in his atomic theory—all atoms of a given element *do not* have the same mass. For example, all neon atoms contain 10 protons, but they may have 10, 11, or 12 neutrons. All three types of neon atoms exist, and each has a slightly different mass. Atoms with the same number of protons but different numbers of neutrons are **isotopes**. Some elements, such as beryllium (Be) and aluminum (Al), have only one naturally occurring isotope, while other elements, such as neon (Ne) and chlorine (Cl), have two or more.

Fortunately, the relative amount of each different isotope in a naturally occurring sample of a given element is usually the same. For example, in any natural sample of neon atoms, 90.48% of them are the isotope with 10 neutrons, 0.27% are the isotope with 11 neutrons, and 9.25% are the isotope with 12 neutrons. These percentages are called the **natural abundance** of the isotopes. Each element has its own characteristic natural abundance of isotopes.

The sum of the number of neutrons and protons in an atom is the **mass number** and is given the symbol A:

$$A = \text{number of protons (p)} + \text{number of neutrons (n)}$$

For neon, with 10 protons, the mass numbers of the three different naturally occurring isotopes are 20, 21, and 22, corresponding to 10, 11, and 12 neutrons, respectively.

Isotopes are often symbolized in the following way:

Mass number → $^{A}_{Z}\text{X}$ ← Chemical symbol
Atomic number →

where X is the chemical symbol, A is the mass number, and Z is the atomic number. Therefore, the symbols for the neon isotopes are

$$^{20}_{10}\text{Ne} \quad ^{21}_{10}\text{Ne} \quad ^{22}_{10}\text{Ne}$$

Notice that the chemical symbol, Ne, and the atomic number, 10, are redundant: if the atomic number is 10, the symbol must be Ne. The mass numbers, however, are different for different isotopes, reflecting the different number of neutrons in each one.

A second common notation for isotopes is the chemical symbol (or chemical name) followed by a dash and the mass number of the isotope.

Chemical symbol or name → X-A ← Mass number

In this notation, the neon isotopes are

Ne-20   Ne-21   Ne-22
neon-20   neon-21   neon-22

We summarize what we have learned about the neon isotopes in the following table:

Symbol	Number of Protons	Number of Neutrons	A (Mass Number)	Natural Abundance (%)
Ne-20 or $^{20}_{10}$Ne	10	10	20	90.48
Ne-21 or $^{21}_{10}$Ne	10	11	21	0.27
Ne-22 or $^{22}_{10}$Ne	10	12	22	9.25

Notice that all isotopes of a given element have the same number of protons (otherwise they would be different elements). Notice also that the mass number is the *sum* of the number of protons and the number of neutrons. The number of neutrons in an isotope is the difference between the mass number and the atomic number ($A - Z$). The different isotopes of an element generally exhibit the same chemical behavior—the three isotopes of neon, for example, all exhibit the same chemical inertness.

EXAMPLE 2.3   Atomic Numbers, Mass Numbers, and Isotope Symbols

(a) What are the atomic number (Z), mass number (A), and symbol of the chlorine isotope with 18 neutrons?

(b) How many protons, electrons, and neutrons are present in an atom of $^{52}_{24}Cr$?

SOLUTION

(a) From the periodic table, we find that the atomic number (Z) of chlorine is 17, so chlorine atoms have 17 protons.	$Z = 17$
The mass number (A) for the isotope with 18 neutrons is the sum of the number of protons (17) and the number of neutrons (18).	$A = 17 + 18 = 35$
The symbol for the chlorine isotope is its two-letter abbreviation with the atomic number (Z) in the lower left corner and the mass number (A) in the upper left corner.	$^{35}_{17}Cl$
(b) For $^{52}_{24}Cr$, the number of protons is the lower left number. Since this is a neutral atom, there are an equal number of electrons. The number of neutrons is equal to the upper left number minus the lower left number.	Number of protons $= Z = 24$ Number of electrons $= 24$ (neutral atom) Number of neutrons $= 52 - 24 = 28$

FOR PRACTICE 2.3

(a) What are the atomic number, mass number, and symbol for the carbon isotope with 7 neutrons?

(b) How many protons and neutrons are present in an atom of $^{39}_{19}K$?

Conceptual Connection 2.4   Isotopes

Carbon has two naturally occurring stable isotopes: C-12 (natural abundance is 98.93%) and C-13 (natural abundance is 1.07%). Using circles to represent protons and squares to represent neutrons, draw the nucleus of each isotope. How many C-13 atoms are present, on average, in a 10,000-atom sample of carbon?

Ions: Losing and Gaining Electrons

The number of electrons in a neutral atom is equal to the number of protons in its nucleus (given by the atomic number Z). During chemical changes, however, atoms often lose or gain electrons to form charged particles called **ions**. For example, neutral lithium (Li) atoms contain 3 protons and 3 electrons; however, in many chemical reactions lithium atoms lose 1 electron (e^-) to form Li^+ ions.

$$Li \rightarrow Li^+ + 1\ e^-$$

The charge of an ion is indicated in the upper right corner of the chemical symbol. The Li^+ ion contains 3 protons but only 2 electrons, resulting in a charge of 1+ (ion charges are written as the magnitude of the charge followed by the sign of the charge). The charge of an ion depends on how many electrons are gained or lost in forming the ion. Neutral fluorine (F) atoms contain 9 protons and 9 electrons; however, in many chemical reactions fluorine atoms gain 1 electron to form F^- ions.

$$F + 1\ e^- \rightarrow F^-$$

The F^- ion contains 9 protons and 10 electrons, resulting in a charge of 1–. For many elements, such as lithium and fluorine, the ion is much more common than the neutral atom. In fact, virtually all of the lithium and fluorine in nature are in the form of their ions.

Positively charged ions, such as Li^+, are **cations** and negatively charged ions, such as F^-, are **anions**. Ions behave quite differently than the atoms from which they are formed.

Neutral sodium atoms, for example, are chemically unstable, reacting violently with most things they contact. Sodium cations (Na^+), in contrast, are relatively inert—we eat them all the time in sodium chloride (NaCl or table salt). In ordinary matter, cations and anions always occur together so that matter is charge-neutral overall.

⊙⊙ Conceptual Connection 2.5 The Nuclear Atom and Isotopes

In light of the nuclear model for the atom, which statement is most likely to be true?

(a) For a given element, the isotope of an atom with a greater number of neutrons is larger than one with a smaller number of neutrons.

(b) For a given element, the size of an atom is the same for all of the element's isotopes.

2.6 Finding Patterns: The Periodic Law and the Periodic Table

The modern periodic table grew out of the work of Dmitri Mendeleev (1834–1907), a nineteenth-century Russian chemistry professor. In his time, about 65 different elements had been discovered. Through the work of a number of chemists, many of the properties of these elements—such as their relative masses, their chemical activity, and some of their physical properties—were known. However, there was no systematic way of organizing them.

In 1869, Mendeleev noticed that certain groups of elements had similar properties. Mendeleev found that when he listed elements in order of increasing mass, their properties recurred in a periodic pattern (**Figure 2.10▼**).

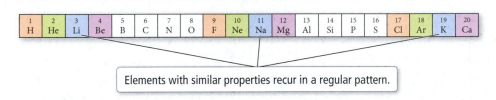

The Periodic Law

Elements with similar properties recur in a regular pattern.

Mendeleev summarized these observations in the **periodic law**, which states:

> **When the elements are arranged in order of increasing mass, certain sets of properties recur periodically.**

Mendeleev then organized all the known elements in a table consisting of a series of rows in which mass increases from left to right. The rows were arranged so that elements with similar properties aligned in the same vertical columns (**Figure 2.11◄**).

Since many elements had not yet been discovered, Mendeleev's table contained some gaps, which allowed him to predict the existence and even some properties of yet undiscovered elements. For example, Mendeleev predicted the existence of an element he called eka-silicon, which fell below silicon on the table. In 1886, eka-silicon was discovered by German chemist Clemens Winkler (1838–1904), who named it germanium, after his home country.

Mendeleev's original listing has evolved into the modern periodic table shown in **Figure 2.12▶**. In the modern table, elements are listed in order of increasing atomic number rather than increasing relative mass. The modern periodic table also contains more elements than Mendeleev's original table because more have been discovered since his time.

Mendeleev's periodic law was based on observation. Like all scientific laws, the periodic law summarized many observations but did not give the underlying reason for the observations—only theories do that. For now, we can simply accept the periodic law as it is, but in Chapters 7 and 8 we will examine a powerful theory—called quantum mechanics—that explains the law and gives the underlying reasons for it.

As shown in Figure 2.12, the elements in the periodic table can be broadly classified as metals, nonmetals, or metalloids. **Metals**, found on the lower left side and middle of the periodic table, have the following properties: they are good conductors of heat and electricity; they can be pounded into flat sheets (malleability); they can be drawn into wires

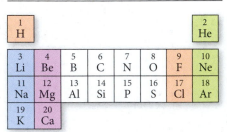

▲ Dmitri Mendeleev, a Russian chemistry professor who proposed the periodic law and arranged early versions of the periodic table, was honored on a Soviet postage stamp.

Periodic means exhibiting a repeating pattern.

▶ **FIGURE 2.10 Recurring Properties** These elements are listed in order of increasing atomic number. Elements with similar properties are shown in the same color. Notice that the colors form a repeating pattern, much like musical notes form a repeating pattern on a piano keyboard.

A Simple Periodic Table

1 H							2 He
3 Li	4 Be	5 B	6 C	7 N	8 O	9 F	10 Ne
11 Na	12 Mg	13 Al	14 Si	15 P	16 S	17 Cl	18 Ar
19 K	20 Ca						

Elements with similar properties fall into columns.

▲ **FIGURE 2.11 Making a Periodic Table** The elements in Figure 2.10 can be arranged in a table in which atomic number increases from left to right and elements with similar properties (as represented by the different colors) are aligned in columns.

Major Divisions of the Periodic Table

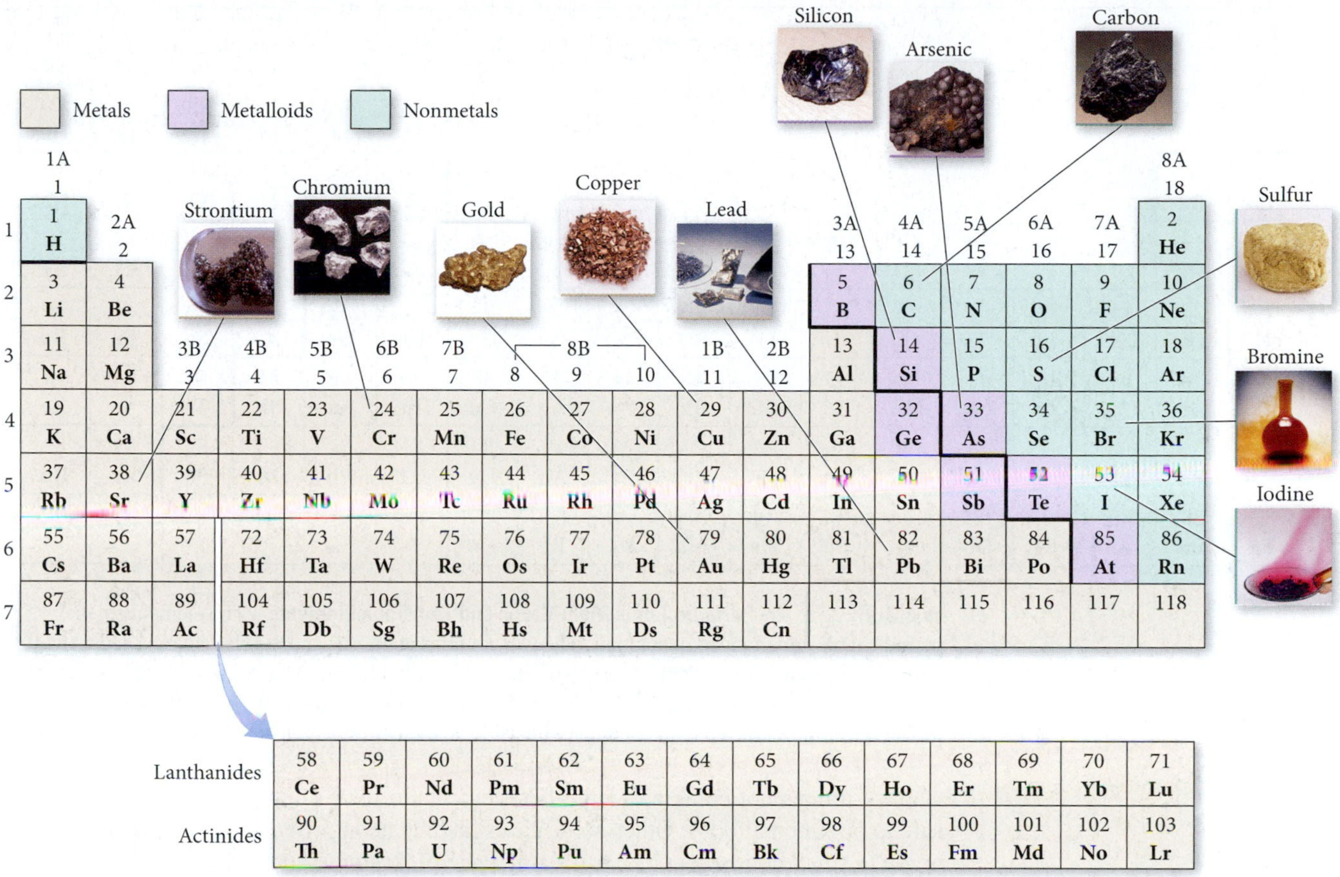

▲ **FIGURE 2.12 Metals, Nonmetals, and Metalloids**   The elements in the periodic table fall into these three broad classes.

(ductility); they are often shiny; and they tend to lose electrons when they undergo chemical changes. Good examples of metals include chromium, copper, strontium, and lead.

Nonmetals are found on the upper right side of the periodic table. The dividing line between metals and nonmetals is the zigzag diagonal line running from boron to astatine. Nonmetals have varied properties—some are solids at room temperature, others are liquids or gases—but as a whole they tend to be poor conductors of heat and electricity and they all tend to gain electrons when they undergo chemical changes. Good examples of nonmetals include oxygen, carbon, sulfur, bromine, and iodine.

Many of the elements that lie along the zigzag diagonal line that divides metals and nonmetals are **metalloids** that show mixed properties. Several metalloids are also classified as **semiconductors** because of their intermediate (and highly temperature-dependent) electrical conductivity. Our ability to change and control the conductivity of semiconductors makes them useful in the manufacture of the electronic chips and circuits central to computers, cellular telephones, and many other devices. Good examples of metalloids include silicon, arsenic, and antimony.

Metalloids are sometimes called semimetals.

The periodic table, as shown in **Figure 2.13▶**, can also be broadly divided into **main-group elements**, whose properties tend to be largely predictable based on their position in the periodic table, and **transition elements** or **transition metals**, whose properties tend to be less predictable based solely on their position in the periodic table. Main-group elements are in columns labeled with a number and the letter A. Transition elements are in columns labeled with a number and the letter B. An alternative labeling system does not use letters, but only the numbers 1–18. Both systems are shown in most of the periodic tables in this book. Each column within the main-group regions of the periodic table is a **family** or **group** of elements.

The elements within a group usually have similar properties. For example, the group 8A elements, referred to as the **noble gases**, are mostly unreactive. The most familiar noble gas is probably helium, used to fill buoyant balloons. Helium is chemically

Periods	Main-group elements		Transition elements										Main-group elements					
	1A 1 (Group number)	2A 2	3B 3	4B 4	5B 5	6B 6	7B 7	8B 8	8B 9	8B 10	1B 11	2B 12	3A 13	4A 14	5A 15	6A 16	7A 17	8A 18
1	1 H																	2 He
2	3 Li	4 Be											5 B	6 C	7 N	8 O	9 F	10 Ne
3	11 Na	12 Mg											13 Al	14 Si	15 P	16 S	17 Cl	18 Ar
4	19 K	20 Ca	21 Sc	22 Ti	23 V	24 Cr	25 Mn	26 Fe	27 Co	28 Ni	29 Cu	30 Zn	31 Ga	32 Ge	33 As	34 Se	35 Br	36 Kr
5	37 Rb	38 Sr	39 Y	40 Zr	41 Nb	42 Mo	43 Tc	44 Ru	45 Rh	46 Pd	47 Ag	48 Cd	49 In	50 Sn	51 Sb	52 Te	53 I	54 Xe
6	55 Cs	56 Ba	57 La	72 Hf	73 Ta	74 W	75 Re	76 Os	77 Ir	78 Pt	79 Au	80 Hg	81 Tl	82 Pb	83 Bi	84 Po	85 At	86 Rn
7	87 Fr	88 Ra	89 Ac	104 Rf	105 Db	106 Sg	107 Bh	108 Hs	109 Mt	110 Ds	111 Rg	112 Cn	113	114	115	116	117	118

▲ **FIGURE 2.13 The Periodic Table: Main-Group and Transition Elements** The elements in the periodic table are arranged in columns. The two columns at the left and the six columns at the right comprise the main-group elements. Each of these eight columns is a group or family. The properties of main-group elements can generally be predicted from their position in the periodic table. The properties of the elements in the middle of the table, known as transition elements, are less predictable.

stable—it does not combine with other elements to form compounds—and is therefore safe to put into balloons. Other noble gases include neon (often used in electronic signs), argon (a small component of Earth's atmosphere), krypton, and xenon.

The group 1A elements, the **alkali metals**, are all reactive metals. A marble-sized piece of sodium, for example, explodes violently when dropped into water. Other alkali metals include lithium, potassium, and rubidium.

The group 2A elements, the **alkaline earth metals**, are also fairly reactive, although not quite as reactive as the alkali metals. Calcium reacts fairly vigorously when dropped into water but will not explode as dramatically as sodium. Other alkaline earth metals include magnesium (a common low-density metal), strontium, and barium.

The group 7A elements, the **halogens**, are very reactive nonmetals. The most familiar halogen is probably chlorine, a greenish-yellow gas with a pungent odor. Chlorine is used as a sterilizing and disinfecting agent (because it reacts with important molecules in living organisms). Other halogens include bromine, a red-brown liquid that easily evaporates into a gas; iodine, a purple solid; and fluorine, a pale-yellow gas.

Ions and the Periodic Table

Recall that, in chemical reactions, metals tend to lose electrons (thus forming cations) and nonmetals tend to gain them (thus forming anions). The number of electrons lost or gained, and therefore the charge of the resulting ion, is often predictable for a given element, especially main-group elements. Main-group elements tend to form ions that have the same number of electrons as the nearest noble gas (the noble gas that has the number of electrons closest to that of the element).

- **A main-group metal tends to lose electrons, forming a cation with the same number of electrons as the nearest noble gas.**
- **A main-group nonmetal tends to gain electrons, forming an anion with the same number of electrons as the nearest noble gas.**

For example, lithium, a metal with three electrons, tends to lose one electron to form a 1+ cation having two electrons, the same number of electrons as helium. Chlorine, a nonmetal with 17 electrons, tends to gain one electron to form a 1– anion having 18 electrons, the same number of electrons as argon.

Elements That Form Ions with Predictable Charges

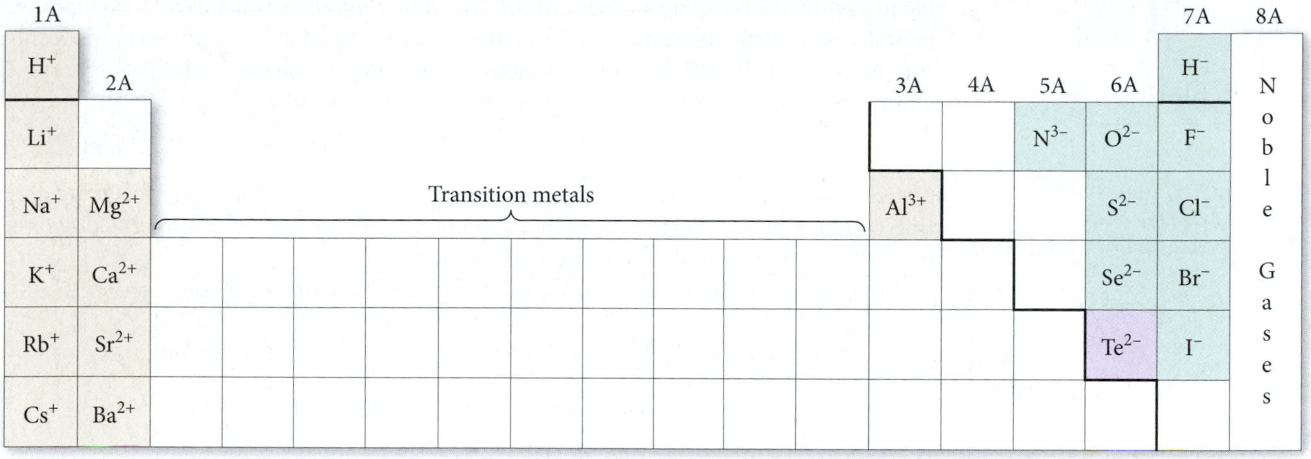

▲ **FIGURE 2.14 Elements That Form Ions with Predictable Charges**

In general, the alkali metals (group 1A) tend to lose one electron and therefore form 1+ ions. The alkaline earth metals (group 2A) tend to lose two electrons and therefore form 2+ ions. The halogens (group 7A) tend to gain one electron and therefore form 1– ions. The oxygen family nonmetals (group 6A) tend to gain two electrons and therefore form 2– ions. More generally, for main-group elements that form predictable cations, the charge of the cation is equal to the group number. For main-group elements that form predictable anions, the charge of the anion is equal to the group number minus eight. Transition elements form cations with different charges. The most common ions formed by main-group elements are shown in **Figure 2.14▲**. In Chapters 7 and 8, when we learn about quantum-mechanical theory, you will understand why these groups form ions as they do.

EXAMPLE 2.4 **Predicting the Charge of Ions**

Predict the charges of the monoatomic (single atom) ions formed by these main-group elements.

(a) Al   (b) S

SOLUTION

(a) Aluminum is a main-group metal and therefore will tend to lose electrons to form a cation with the same number of electrons as the nearest noble gas. Aluminum atoms have 13 electrons and the nearest noble gas is neon, which has 10 electrons. Aluminum will tend to lose 3 electrons to form a cation with a 3+ charge (Al^{3+}).

(b) Sulfur is a nonmetal and therefore will tend to gain electrons to form an anion with the same number of electrons as the nearest noble gas. Sulfur atoms have 16 electrons and the nearest noble gas is argon, which has 18 electrons. Sulfur will tend to gain 2 electrons to form an anion with a 2– charge (S^{2-}).

FOR PRACTICE 2.4

Predict the charges of the monoatomic ions formed by the main-group elements.

(a) N   (b) Rb

2.7 Atomic Mass: The Average Mass of an Element's Atoms

An important part of Dalton's atomic theory was that all atoms of a given element have the same mass. However, in Section 2.5, we learned that, because of isotopes, the atoms of a given element often have different masses, so Dalton was not completely correct. We can, however, calculate an average mass—called the **atomic mass**—for each element.

Atomic mass is sometimes called *atomic weight, average atomic mass,* or *average atomic weight.*

17
Cl
35.45
chlorine

When percentages are used in calculations, they are converted to their decimal value by dividing by 100.

The atomic mass of each element is listed directly beneath the element's symbol in the periodic table and represents the average mass of the isotopes that compose that element *weighted according to the natural abundance of each isotope*. For example, the periodic table lists the atomic mass of chlorine as 35.45 amu. Naturally occurring chlorine consists of 75.77% chlorine-35 atoms (mass 34.97 amu) and 24.23% chlorine-37 atoms (mass 36.97 amu). Its atomic mass is calculated as follows:

$$\text{Atomic mass} = 0.7577(34.97 \text{ amu}) + 0.2423(36.97 \text{ amu}) = 35.45 \text{ amu}$$

Notice that the atomic mass of chlorine is closer to 35 than 37. Naturally occurring chlorine contains more chlorine-35 atoms than chlorine-37 atoms, so the weighted average mass of chlorine is closer to 35 amu than to 37 amu.

In general, the atomic mass is calculated according to the equation:

$$\textbf{Atomic mass} = \sum_n (\textbf{fraction of isotope } n) \times (\textbf{mass of isotope } n)$$
$$= (\textbf{fraction of isotope 1} \times \textbf{mass of isotope 1})$$
$$+ \ (\textbf{fraction of isotope 2} \times \textbf{mass of isotope 2})$$
$$+ \ (\textbf{fraction of isotope 3} \times \textbf{mass of isotope 3}) + \cdots$$

where the fractions of each isotope are the percent natural abundances converted to their decimal values. The concept of atomic mass is useful because it allows us to assign a characteristic mass to each element, and as we will see shortly, it allows us to quantify the number of atoms in a sample of that element.

EXAMPLE 2.5 Atomic Mass

Copper has two naturally occurring isotopes: Cu-63 with mass 62.9396 amu and a natural abundance of 69.17%, and Cu-65 with mass 64.9278 amu and a natural abundance of 30.83%. Calculate the atomic mass of copper.

SOLUTION

Convert the percent natural abundances into decimal form by dividing by 100.	Fraction Cu-63 $= \dfrac{69.17}{100} = 0.6917$
	Fraction Cu-65 $= \dfrac{30.83}{100} = 0.3083$
Calculate the atomic mass using the equation given in the text.	Atomic mass $= 0.6917(62.9396 \text{ amu}) + 0.3083(64.9278 \text{ amu})$
	$= 43.5353 \text{ amu} + 20.0172 \text{ amu} = 63.5525 = 63.55 \text{ amu}$

FOR PRACTICE 2.5

Magnesium has three naturally occurring isotopes with masses of 23.99 amu, 24.99 amu, and 25.98 amu and natural abundances of 78.99%, 10.00%, and 11.01%, respectively. Calculate the atomic mass of magnesium.

FOR MORE PRACTICE 2.5

Gallium has two naturally occurring isotopes: Ga-69 with a mass of 68.9256 amu and a natural abundance of 60.11%, and Ga-71. Use the atomic mass of gallium listed in the periodic table to find the mass of Ga-71.

Conceptual Connection 2.6 Atomic Mass

Recall from Conceptual Connection 2.4 that carbon has two naturally occurring isotopes: C-12 (natural abundance is 98.93%; mass is 12.0000 amu) and C-13 (natural abundance is 1.07%; mass is 13.0034 amu). Without doing any calculations, determine which mass is closest to the atomic mass of carbon.

(a) 12.00 amu   (b) 12.50 amu   (c) 13.00 amu

2.8   Molar Mass: Counting Atoms by Weighing Them

Have you ever bought shrimp by *count*? Shrimp is normally sold by count, which tells you the number of shrimp per pound. For example, 41–50 count shrimp indicates that there are between 41 and 50 shrimp per pound. The smaller the count, the larger the shrimp. The big tiger prawns have counts as low as 10–15, which means that each shrimp can weigh up to 1/10 of a pound. The nice thing about categorizing shrimp this way is that you can count the shrimp by weighing them. For example, two pounds of 41–50 count shrimp contains between 82 and 100 shrimp.

A similar (but more precise) concept exists for atoms. Counting atoms is much more difficult than counting shrimp, yet we often need to know the number of atoms in a given mass of atoms. For example, intravenous fluids—fluids that are delivered to patients by directly dripping the fluid into their veins—are saline (salt) solutions that must have a specific number of sodium and chloride ions per liter of fluid in order to be effective. The result of using an intravenous fluid with the wrong number of sodium and chloride ions could be fatal.

Atoms are far too small to count by any ordinary means. As we saw earlier, even if you could somehow count atoms, and counted them 24 hours a day as long as you lived, you would barely begin to count the number of atoms in something as small as a sand grain. Therefore, if we want to know the number of atoms in anything of ordinary size, we count them by weighing.

The Mole: A Chemist's "Dozen"

When we count large numbers of objects, we often use units such as a dozen (12 objects) or a gross (144 objects) to organize our counting and to keep our numbers more manageable. With atoms, quadrillions of which may be in a speck of dust, we need a much larger number for this purpose. The chemist's "dozen" is called the **mole** (abbreviated mol) and is defined as the *amount* of material containing 6.0221421×10^{23} particles.

$$1 \text{ mol} = 6.0221421 \times 10^{23} \text{ particles}$$

This number is **Avogadro's number**, named after Italian physicist Amedeo Avogadro (1776–1856), and is a convenient number to use when working with atoms, molecules, and ions. In this book, we usually round Avogadro's number to four significant figures or 6.022×10^{23}. Notice that the definition of the mole is an *amount* of a substance. We will often refer to the number of moles of substance as the *amount* of the substance.

The first thing to understand about the mole is that it can specify Avogadro's number of anything. For example, 1 mol of marbles corresponds to 6.022×10^{23} marbles, and 1 mol of sand grains corresponds to 6.022×10^{23} sand grains. *One mole of anything is 6.022×10^{23} units of that thing.* One mole of atoms, ions, or molecules, however, makes up objects of everyday sizes. For example, 22 copper pennies contain approximately 1 mol of copper atoms and a tablespoon of water contains approximately 1 mol of water molecules.

The second, and more fundamental, thing to understand about the mole is how it gets its specific value.

> **The value of the mole is equal to the number of atoms in exactly 12 grams of pure carbon-12 (12 g C = 1 mol C atoms = 6.022×10^{23} C atoms).**

This definition of the mole gives us a relationship between mass (grams of carbon) and number of atoms (Avogadro's number). This relationship, as we will see shortly, allows us to count atoms by weighing them.

Converting between Number of Moles and Number of Atoms

Converting between number of moles and number of atoms is similar to converting between dozens of shrimp and number of shrimp. To convert between moles of atoms and number of atoms we use the conversion factors:

$$\frac{1 \text{ mol atoms}}{6.022 \times 10^{23} \text{ atoms}} \quad \text{or} \quad \frac{6.022 \times 10^{23} \text{ atoms}}{1 \text{ mol atoms}}$$

Example 2.6 demonstrates how to use these conversion factors.

Twenty-two copper pennies contain approximately 1 mol of copper atoms.

One tablespoon of water contains approximately one mole of water molecules.

Beginning in 1982, pennies became almost all zinc, with only a copper coating. Before 1982, however, pennies were mostly copper.

One tablespoon is approximately 15 mL; one mole of water occupies 18 mL.

EXAMPLE 2.6 Converting between Number of Moles and Number of Atoms

Calculate the number of copper atoms in 2.45 mol of copper (Cu).

SORT You are given the amount of copper in moles and asked to find the number of copper atoms.	**GIVEN** 2.45 mol Cu **FIND** Cu atoms
STRATEGIZE Convert between number of moles and number of atoms by using Avogadro's number as a conversion factor.	**CONCEPTUAL PLAN** mol Cu $\longrightarrow$ Cu atoms $$\frac{6.022 \times 10^{23} \text{ Cu atoms}}{1 \text{ mol Cu}}$$ **RELATIONSHIPS USED** $6.022 \times 10^{23} = 1$ mol (Avogadro's number)
SOLVE Follow the conceptual plan to solve the problem. Begin with 2.45 mol Cu and multiply by Avogadro's number to get to Cu atoms.	**SOLUTION** $2.45 \text{ mol Cu} \times \dfrac{6.022 \times 10^{23} \text{ Cu atoms}}{1 \text{ mol Cu}}$ $= 1.48 \times 10^{24}$ Cu atoms

CHECK Since atoms are small, it makes sense that the answer is large. The number of moles of copper is almost 2.5, so the number of atoms is almost 2.5 times Avogadro's number.

FOR PRACTICE 2.6

A pure silver ring contains 2.80×10^{22} silver atoms. How many moles of silver atoms does it contain?

Converting between Mass and Amount (Number of Moles)

To count atoms by weighing them, we need one other conversion factor—the mass of 1 mol of atoms. For the isotope carbon-12, we know that this mass is exactly 12 grams, which is numerically equivalent to carbon-12's atomic mass in atomic mass units. Since the masses of all other elements are defined relative to carbon-12, the same relationship holds for all elements.

The mass of 1 mol of atoms of an element is the **molar mass**.

> **An element's molar mass in grams per mole is numerically equal to the element's atomic mass in atomic mass units.**

For example, copper has an atomic mass of 63.55 amu and a molar mass of 63.55 g/mol. One mole of copper atoms therefore has a mass of 63.55 g. Just as the count for shrimp depends on the size of the shrimp, the mass of 1 mol of atoms depends on the element: 1 mol of aluminum atoms (which are lighter than copper atoms) has a mass of 26.98 g; 1 mol of carbon atoms (which are even lighter than aluminum atoms) has a mass of 12.01 g; and 1 mol of helium atoms (lighter still) has a mass of 4.003 g.

26.98 g aluminum = 1 mol aluminum = 6.022×10^{23} Al atoms   Al

12.01 g carbon = 1 mol carbon = 6.022×10^{23} C atoms   C

4.003 g helium = 1 mol helium = 6.022×10^{23} He atoms   He

The lighter the atom, the less mass it takes to make 1 mol.

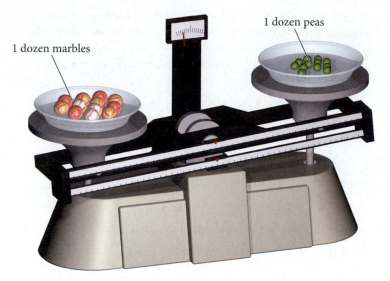

1 dozen marbles   1 dozen peas

▲ The two dishes contain the same number of objects (12), but the masses are different because peas are less massive than marbles. Similarly, a mole of light atoms will have less mass than a mole of heavier atoms.

Therefore, the molar mass of any element becomes a conversion factor between the mass (in grams) of that element and the amount (in moles) of that element. For carbon:

$$12.01 \text{ g C} = 1 \text{ mol C} \quad \text{or} \quad \frac{12.01 \text{ g C}}{\text{mol C}} \quad \text{or} \quad \frac{1 \text{ mol C}}{12.01 \text{ g C}}$$

Example 2.7 demonstrates how to use these conversion factors.

EXAMPLE 2.7   Converting between Mass and Amount (Number of Moles)

Calculate the amount of carbon (in moles) contained in a 0.0265 g pencil "lead." (Assume that the pencil lead is made of pure graphite, a form of carbon.)

SORT You are given the mass of carbon and asked to find the amount of carbon in moles.	**GIVEN** 0.0265 g C **FIND** mol C
STRATEGIZE Convert between mass and amount (in moles) of an element by using the molar mass of the element.	**CONCEPTUAL PLAN** g C $\longrightarrow$ mol C $\dfrac{1 \text{ mol}}{12.01 \text{ g}}$ **RELATIONSHIPS USED** 12.01 g C = 1 mol C (carbon molar mass)
SOLVE Follow the conceptual plan to solve the problem.	**SOLUTION** $0.0265 \text{ g C} \times \dfrac{1 \text{ mol C}}{12.01 \text{ g C}} = 2.21 \times 10^{-3} \text{ mol C}$

CHECK The given mass of carbon is much less than the molar mass of carbon. Therefore the answer (the amount in moles) is much less than 1 mol of carbon.

FOR PRACTICE 2.7

Calculate the amount of copper (in moles) in a 35.8 g pure copper sheet.

FOR MORE PRACTICE 2.7

Calculate the mass (in grams) of 0.473 mol of titanium.

We now have all the tools to count the number of atoms in a sample of an element by weighing it. To do this, first we obtain the mass of the sample. Then we convert it to amount in moles using the element's molar mass. Finally, we convert to number of atoms using Avogadro's number. The conceptual plan for these kinds of calculations takes the following form:

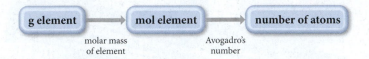

The examples that follow demonstrate these conversions.

EXAMPLE 2.8 The Mole Concept—Converting between Mass and Number of Atoms

How many copper atoms are in a copper penny with a mass of 3.10 g? (Assume that the penny is composed of pure copper.)

SORT You are given the mass of copper and asked to find the number of copper atoms.	**GIVEN** 3.10 g Cu **FIND** Cu atoms
STRATEGIZE Convert between the mass of an element in grams and the number of atoms of the element by first converting to moles (using the molar mass of the element) and then to number of atoms (using Avogadro's number).	**CONCEPTUAL PLAN** g Cu → mol Cu → number of Cu atoms $\dfrac{1 \text{ mol Cu}}{63.55 \text{ g Cu}}$ $\dfrac{6.022 \times 10^{23} \text{ Cu atoms}}{1 \text{ mol Cu}}$ **RELATIONSHIPS USED** 63.55 g Cu = 1 mol Cu (molar mass of copper) 6.022×10^{23} = 1 mol (Avogadro's number)
SOLVE Follow the conceptual plan to solve the problem. Begin with 3.10 g Cu and multiply by the appropriate conversion factors to arrive at the number of Cu atoms.	**SOLUTION** $3.10 \text{ g Cu} \times \dfrac{1 \text{ mol Cu}}{63.55 \text{ g Cu}} \times \dfrac{6.022 \times 10^{23} \text{ Cu atoms}}{1 \text{ mol Cu}}$ $= 2.94 \times 10^{22} \text{ Cu atoms}$

CHECK The answer (the number of copper atoms) is less than 6.022×10^{23} (one mole). This is consistent with the given mass of copper atoms, which is less than the molar mass of copper.

FOR PRACTICE 2.8

How many carbon atoms are there in a 1.3-carat diamond? Diamonds are a form of pure carbon. (1 carat = 0.20 grams)

FOR MORE PRACTICE 2.8

Calculate the mass of 2.25×10^{22} tungsten atoms.

Notice that numbers with large exponents, such as 6.022×10^{23}, are deceptively large. Twenty-two copper pennies contain 6.022×10^{23} or 1 mol of copper atoms, but 6.022×10^{23} pennies would cover Earth's entire surface to a depth of 300 m. Even objects small by everyday standards occupy a huge space when we have a mole of them. For example, a grain of sand has a mass of less than 1 mg and a diameter of less than 0.1 mm, yet 1 mol of sand grains would cover the state of Texas to a depth of several feet. For every increase of 1 in the exponent of a number, the number increases by a factor of 10, so 10^{23} is incredibly large. One mole has to be a large number, however, if it is to have practical value, because atoms are so small.

EXAMPLE 2.9 The Mole Concept

An aluminum sphere contains 8.55×10^{22} aluminum atoms. What is the radius of the sphere in centimeters? The density of aluminum is 2.70 g/cm³.

SORT You are given the number of aluminum atoms in a sphere and the density of aluminum. You are asked to find the radius of the sphere.

GIVEN 8.55×10^{22} Al atoms
$d = 2.70 \text{ g/cm}^3$
FIND radius (r) of sphere

STRATEGIZE The heart of this problem is density, which relates mass to volume, and though you aren't given the mass directly, you are given the number of atoms, which you can use to find mass.

1. Convert from number of atoms to number of moles using Avogadro's number as a conversion factor.

2. Convert from number of moles to mass using molar mass as a conversion factor.

3. Convert from mass to volume (in cm³) using density as a conversion factor.

4. Once you calculate the volume, find the radius from the volume using the formula for the volume of a sphere.

CONCEPTUAL PLAN

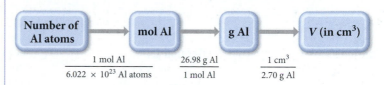

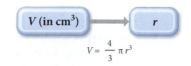

RELATIONSHIPS AND EQUATIONS USED

$6.022 \times 10^{23} = 1 \text{ mol}$ (Avogadro's number)
$26.98 \text{ g Al} = 1 \text{ mol Al}$ (molar mass of aluminum)
$2.70 \text{ g Al} = 1 \text{ cm}^3 \text{ Al}$ (density of aluminum)
$V = \dfrac{4}{3}\pi r^3$ (volume of a sphere)

SOLVE Follow the conceptual plan to solve the problem. Begin with 8.55×10^{22} Al atoms and multiply by the appropriate conversion factors to arrive at volume in cm³.

Then solve the equation for the volume of a sphere for r and substitute the volume to compute r.

SOLUTION

$8.55 \times 10^{22} \text{ Al atoms} \times \dfrac{1 \text{ mol Al}}{6.022 \times 10^{23} \text{ Al atoms}}$

$\times \dfrac{26.98 \text{ g Al}}{1 \text{ mol Al}} \times \dfrac{1 \text{ cm}^3}{2.70 \text{ g Al}} = 1.4187 \text{ cm}^3$

$V = \dfrac{4}{3}\pi r^3$

$r = \sqrt[3]{\dfrac{3v}{4\pi}} = \sqrt[3]{\dfrac{3(1.4187 \text{ cm}^3)}{4\pi}} = 0.697 \text{ cm}$

CHECK The units of the answer (cm) are correct. The magnitude cannot be estimated accurately, but a radius of about one-half of a centimeter is reasonable for just over one-tenth of a mole of aluminum atoms.

FOR PRACTICE 2.9
A titanium cube contains 2.86×10^{23} atoms. What is the edge length of the cube? The density of titanium is 4.50 g/cm³.

FOR MORE PRACTICE 2.9
Find the number of atoms in a copper rod with a length of 9.85 cm and a radius of 1.05 cm. The density of copper is 8.96 g/cm³.

⬤ Conceptual Connection 2.7 Avogadro's Number

Why is Avogadro's number defined as 6.022×10^{23} and not a simpler round number such as 1.00×10^{23}?

Conceptual Connection 2.8 **The Mole**

Without doing any calculations, determine which sample contains the most atoms.

(a) a 1 g sample of copper

(b) a 1 g sample of carbon

(c) a 10 g sample of uranium

CHAPTER IN REVIEW

Key Terms

Section 2.2
law of conservation of
    mass (41)
law of definite proportions (42)
law of multiple
    proportions (43)
atomic theory (44)

Section 2.3
cathode rays (45)
cathode ray tube (45)
electrical charge (45)
electron (45)

Section 2.4
radioactivity (47)
nuclear theory (48)
nucleus (48)
proton (48)
neutron (48)

Section 2.5
atomic mass unit (amu) (48)
atomic number (Z) (49)
chemical symbol (49)
isotope (51)
natural abundance (51)
mass number (A) (51)

ion (52)
cation (52)
anion (52)

Section 2.6
periodic law (53)
metal (53)
nonmetal (54)
metalloid (54)
semiconductor (54)
main-group elements (54)
transition elements (transition
    metals) (54)
family (group) (54)

noble gases (54)
alkali metals (55)
alkaline earth metals (55)
halogens (55)

Section 2.7
atomic mass (56)

Section 2.8
mole (mol) (58)
Avogadro's number (58)
molar mass (59)

Key Concepts

Imaging and Moving Individual Atoms (2.1)

▶ Although it was only 200 years ago that John Dalton proposed his atomic theory, technology has progressed to the level where individual atoms can be imaged and moved by techniques such as *scanning tunneling microscopy* (STM).

The Atomic Theory (2.2)

▶ Each element is composed of indestructible particles called atoms.
▶ All atoms of a given element have the same mass and other properties.
▶ Atoms combine in simple, whole-number ratios to form compounds.
▶ Atoms of one element cannot change into atoms of another element. In a chemical reaction, atoms change the way that they are bound together with other atoms to form a new substance.

The Electron (2.3)

▶ J. J. Thomson discovered the electron in the late 1800s through experiments examining the properties of cathode rays. He deduced that electrons are negatively charged, and then measured their charge-to-mass ratio.
▶ Robert Millikan measured the charge of the electron, which—in conjunction with Thomson's results—led to the calculation of the mass of an electron.

The Nuclear Atom (2.4)

▶ In 1909, Ernest Rutherford probed the inner structure of the atom by working with a form of radioactivity called alpha radiation and developed the nuclear theory of the atom.
▶ Nuclear theory states that the atom is mainly empty space, with most of its mass concentrated in a tiny region called the nucleus and most of its volume occupied by the relatively light electrons.

Subatomic Particles (2.5)

▶ Atoms are composed of three fundamental particles: the proton (1 amu, +1 charge), the neutron (1 amu, 0 charge), and the electron (~0 amu, −1 charge).
▶ The number of protons in the nucleus of the atom is the atomic number (Z) and defines the element.
▶ The sum of the number of protons and neutrons is the mass number (A).
▶ Atoms of an element that have different numbers of neutrons (and therefore different mass numbers) are called isotopes.
▶ Atoms that have lost or gained electrons become charged and are called ions. Cations are positively charged and anions are negatively charged.

The Periodic Table (2.6)

▶ The periodic table tabulates all known elements in order of increasing atomic number.

▶ The periodic table is arranged so that similar elements are grouped together in columns.

▶ Elements on the left side and in the center of the periodic table are metals and tend to lose electrons in their chemical changes.

▶ Elements on the upper right side of the periodic table are nonmetals and tend to gain electrons in their chemical changes.

▶ Elements located on the boundary between these two classes are metalloids.

Atomic Mass and the Mole (2.7, 2.8)

▶ The atomic mass of an element, listed directly below its symbol in the periodic table, is a weighted average of the masses of the naturally occurring isotopes of the element.

▶ One mole of an element is the amount of that element that contains Avogadro's number (6.022×10^{23}) of atoms.

▶ Any sample of an element with a mass (in grams) that equals its atomic mass contains one mole of the element. For example, the atomic mass of carbon is 12.01 amu, therefore 12.01 grams of carbon contains 1 mol of carbon atoms.

Key Equations and Relationships

Relationship between Mass Number (A), Number of Protons (p), and Number of Neutrons (n) (2.5)

$$A = \text{number of protons (p)} + \text{number of neutrons (n)}$$

Avogadro's Number (2.8)

$$1 \text{ mol} = 6.0221421 \times 10^{23} \text{ particles}$$

Atomic Mass (2.7)

$$\text{Atomic mass} = \sum_n (\text{fraction of isotope } n) \times (\text{mass of isotope } n)$$

Key Learning Objectives

Chapter Objectives	Assessment
Using the Law of Definite Proportions (2.2)	Example 2.1 For Practice 2.1 Exercises 3, 4
Using the Law of Multiple Proportions (2.2)	Example 2.2 For Practice 2.2 Exercises 7–10
Working with Atomic Numbers, Mass Numbers, and Isotope Symbols (2.5)	Example 2.3 For Practice 2.3 Exercises 21–28
Predicting the Charge of Ions (2.6)	Example 2.4 For Practice 2.4 Exercises 29–32
Calculating Atomic Mass (2.7)	Example 2.5 For Practice 2.5 For More Practice 2.5 Exercises 41–44
Converting between Moles and Number of Atoms (2.8)	Example 2.6 For Practice 2.6 Exercises 45, 46
Converting between Mass and Amount (in Moles) (2.8)	Example 2.7 For Practice 2.7 For More Practice 2.7 Exercises 47, 48
Using the Mole Concept (2.8)	Examples 2.8, 2.9 For Practice 2.8, 2.9 For More Practice 2.8, 2.9 Exercises 49–56, 72, 73

EXERCISES

Problems by Topic

Note: Answers to all odd-numbered Problems, numbered in blue, can be found in Appendix III. Exercises in the Problems by Topic section are paired, with each odd-numbered problem followed by a similar even-numbered problem. Exercises in the Cumulative Problems section are also paired, but somewhat more loosely. (Challenge Problems and Conceptual Problems, because of their nature, are unpaired.)

The Laws of Conservation of Mass, Definite Proportions, and Multiple Proportions

1. A hydrogen-filled balloon is ignited and 1.50 g of hydrogen react with 12.0 g of oxygen. How many grams of water vapor are formed? (Assume that water vapor is the only product.)

2. An automobile gasoline tank holds 21 kg of gasoline. When the gasoline burns, 84 kg of oxygen is consumed and carbon dioxide and water are produced. What is the total combined mass of carbon dioxide and water that is produced?

3. Two samples of carbon tetrachloride are decomposed into their constituent elements. One sample produces 38.9 g of carbon and 448 g of chlorine, and the other sample produces 14.8 g of carbon and 134 g of chlorine. Are these results consistent with the law of definite proportions? Explain why or why not.

4. Two samples of sodium chloride are decomposed into their constituent elements. One sample produces 6.98 g of sodium and 10.7 g of chlorine, and the other sample produces 11.2 g of sodium and 17.3 g of chlorine. Are these results consistent with the law of definite proportions? Explain why or why not.

5. The mass ratio of sodium to fluorine in sodium fluoride is 1.21:1. A sample of sodium fluoride produces 28.8 g of sodium upon decomposition. How much fluorine (in grams) forms?

6. Upon decomposition, one sample of magnesium fluoride produced 1.65 kg of magnesium and 2.57 kg of fluorine. A second sample produced 1.32 kg of magnesium. How much fluorine (in grams) did the second sample produce?

7. Two different compounds containing osmium and oxygen have the following masses of oxygen per gram of osmium: 0.168 and 0.3369 g. Show that these amounts are consistent with the law of multiple proportions.

8. Palladium forms three different compounds with sulfur. The mass of sulfur per gram of palladium in each compound is listed below:

Compound	Grams S per Gram Pd
A	0.603
B	0.301
C	0.151

Show that these masses are consistent with the law of multiple proportions.

9. Sulfur and oxygen form both sulfur dioxide and sulfur trioxide. When samples of these were decomposed the sulfur dioxide produced 3.49 g oxygen and 3.50 g sulfur, while the sulfur trioxide produced 6.75 g oxygen and 4.50 g sulfur. Calculate the mass of oxygen per gram of sulfur for each sample and show that these results are consistent with the law of multiple proportions.

10. Sulfur and fluorine form several different compounds including sulfur hexafluoride and sulfur tetrafluoride. Decomposition of a sample of sulfur hexafluoride produces 4.45 g of fluorine and 1.25 g of sulfur, while decomposition of a sample of sulfur tetrafluoride produces 4.43 g of fluorine and 1.87 g of sulfur. Calculate the mass of fluorine per gram of sulfur for each sample and show that these results are consistent with the law of multiple proportions.

Atomic Theory, Nuclear Theory, and Subatomic Particles

11. Which statements are consistent with Dalton's atomic theory as it was originally stated? Why?
   a. Sulfur and oxygen atoms have the same mass.
   b. All cobalt atoms are identical.
   c. Potassium and chlorine atoms combine in a 1:1 ratio to form potassium chloride.
   d. Lead atoms can be converted into gold.

12. Which statements are *inconsistent* with Dalton's atomic theory as it was originally stated? Why?
   a. All carbon atoms are identical.
   b. An oxygen atom combines with 1.5 hydrogen atoms to form a water molecule.
   c. Two oxygen atoms combine with a carbon atom to form a carbon dioxide molecule.
   d. The formation of a compound often involves the destruction of one or more atoms.

13. Which statements are consistent with Rutherford's nuclear theory as it was originally stated? Why?
   a. The volume of an atom is mostly empty space.
   b. The nucleus of an atom is small compared to the size of the atom.
   c. Neutral lithium atoms contain more neutrons than protons.
   d. Neutral lithium atoms contain more protons than electrons.

14. Which statements are inconsistent with Rutherford's nuclear theory as it was originally stated? Why?
   a. Since electrons are smaller than protons, and since a hydrogen atom contains only one proton and one electron, it must follow that the volume of a hydrogen atom is mostly due to the proton.
   b. A nitrogen atom has seven protons in its nucleus and seven electrons outside of its nucleus.
   c. A phosphorus atom has 15 protons in its nucleus and 150 electrons outside of its nucleus.
   d. The majority of the mass of a fluorine atom is due to its nine electrons.

15. A chemist in an imaginary universe, where electrons have a different charge than they do in our universe, performs the Millikan oil drop experiment to measure the electron's charge. The charges of several drops are recorded below. What is the charge of the electron in this imaginary universe?

Drop#	Charge
A	-6.9×10^{-19} C
B	-9.2×10^{-19} C
C	-11.5×10^{-19} C
D	-4.6×10^{-19} C

16. Imagine a unit of charge called the zorg. A chemist performs the oil drop experiment and measures the charge of each drop in zorgs. Based on the results below, what is the charge of the electron in zorgs (z)? How many electrons are in each drop?

Drop#	Charge
A	-4.8×10^{-9} z
B	-9.6×10^{-9} z
C	-6.4×10^{-9} z
D	-12.8×10^{-9} z

17. On a dry day, your body can accumulate static charge from walking across a carpet or from brushing your hair. If your body develops a charge of $-15 \ \mu C$ (microcoulombs), how many excess electrons has it acquired? What is their collective mass?

18. How many electrons does it take to the equal the mass of a proton?

19. Which statements about subatomic particles are true?
  a. If an atom has an equal number of protons and electrons, it is charge-neutral.
  b. Electrons are attracted to protons.
  c. Electrons are much lighter than neutrons.
  d. Protons have twice the mass of neutrons.

20. Which statements about subatomic particles are false?
  a. Protons and electrons have charges of the same magnitude but opposite sign.
  b. Protons have about the same mass as neutrons.
  c. Some atoms don't have any protons.
  d. Protons and neutrons have charges of the same magnitude but opposite sign.

Isotopes and Ions

21. Write isotopic symbols of the form $_{Z}^{A}X$ for each isotope.
  a. the sodium isotope with 12 neutrons
  b. the oxygen isotope with 8 neutrons
  c. the aluminum isotope with 14 neutrons
  d. the iodine isotope with 74 neutrons

22. Write isotopic symbols of the form X-A (e.g., C-13) for each isotope.
  a. the argon isotope with 22 neutrons
  b. the plutonium isotope with 145 neutrons
  c. the phosphorus isotope with 16 neutrons
  d. the fluorine isotope with 10 neutrons

23. Determine the number of protons and and the number of neutrons in each isotope.
  a. $_{7}^{14}N$ **b.** $_{11}^{23}Na$
  c. $_{86}^{222}Rn$ **d.** $_{82}^{208}Pb$

24. Determine the number of protons and the number of neutrons in each isotope.
  a. $_{19}^{40}K$ **b.** $_{88}^{226}Ra$
  c. $_{43}^{99}Tc$ **d.** $_{15}^{33}P$

25. The amount of carbon-14 in artifacts and fossils is often used to establish their age. Determine the number of protons and the number of neutrons in a carbon-14 isotope and write its symbol in the form $_{Z}^{A}X$.

26. Uranium-235 is used in nuclear fission. Determine the number of protons and the number of neutrons in uranium-235 and write its symbol in the form $_{Z}^{A}X$.

27. Determine the number of protons and the number of electrons in each ion.
  a. Ni^{2+} **b.** S^{2-}
  c. Br^{-} **d.** Cr^{3+}

28. Determine the number of protons and the number of electrons in each ion.
  a. Al^{3+} **b.** Se^{2-}
  c. Ga^{3+} **d.** Sr^{2+}

29. Predict the charge of the monoatomic (single atom) ion formed by each element.
  a. O   **b.** K
  c. Al   **d.** Rb

30. Predict the charge of the monoatomic (single atom) ion formed by each element.
  a. Mg   **b.** N
  c. F   **d.** Na

31. Fill in the blanks to complete the table.

Symbol	Ion Formed	Number of Electrons in Ion	Number of Protons in Ion
Ca	Ca^{2+}	_____	_____
_____	Be^{2+}	2	_____
Se	_____	_____	34
In	_____	_____	49

32. Fill in the blanks to complete the table.

Symbol	Ion Formed	Number of Electrons in Ion	Number of Protons in Ion
Cl	_____	_____	17
Te	_____	54	_____
Br	Br^{-}	_____	_____
_____	Sr^{2+}	_____	38

The Periodic Table and Atomic Mass

33. Write the name of each element and classify it as a metal, nonmetal, or metalloid.
  a. Na   **b.** Mg   **c.** Br
  d. N   **e.** As

34. Write the symbol for each element and classify it as a metal, nonmetal, or metalloid.
  a. lead   **b.** iodine   **c.** potassium
  d. silver   **e.** xenon

35. Determine whether or not each element is a main-group element.
  a. tellurium   **b.** potassium
  c. vanadium   **d.** manganese

36. Determine whether or not each element is a transition element.
  a. Cr   **b.** Br
  c. Mo   **d.** Cs

37. Classify each element as an alkali metal, alkaline earth metal, halogen, or noble gas.
  a. sodium   **b.** iodine
  c. calcium   **d.** barium
  e. krypton

38. Classify each element as an alkali metal, alkaline earth metal, halogen, or noble gas.
  a. F   **b.** Sr   **c.** K
  d. Ne   **e.** At

39. Which pair of elements do you expect to be most similar? Why?
  a. N and Ni   **b.** Mo and Sn
  c. Na and Mg   **d.** Cl and F
  e. Si and P

40. Which pair of elements do you expect to be most similar? Why?
  a. nitrogen and oxygen
  b. titanium and gallium
  c. lithium and sodium
  d. germanium and arsenic
  e. argon and bromine

41. Rubidium has two naturally occurring isotopes with the following masses and natural abundances:

Isotope	Mass (amu)	Abundance (%)
Rb-85	84.9118	72.15
Rb-87	86.9092	27.85

Calculate the atomic mass of rubidium.

42. Silicon has three naturally occurring isotopes with the following masses and natural abundances:

Isotope	Mass (amu)	Abundance (%)
Si-28	27.9769	92.2
Si-29	28.9765	4.67
Si-30	29.9737	3.10

Calculate the atomic mass of silicon.

43. An element has two naturally occurring isotopes. Isotope 1 has a mass of 120.9038 amu and a relative abundance of 57.4%, and isotope 2 has a mass of 122.9042 amu. Find the atomic mass of this element and, by comparison to the periodic table, identify it.

44. Bromine has only two naturally occurring isotopes (Br-79 and Br-81) and has an atomic mass of 79.904 amu. The mass of Br-81 is 80.9163 amu, and its natural abundance is 49.31%. Calculate the mass and natural abundance of Br-79.

The Mole Concept

45. Determine the number of sulfur atoms in 2.7 mol of sulfur.

46. How many moles of aluminum do 1.42×10^{24} aluminum atoms constitute?

47. What is the amount, in moles, in each elemental sample?
  a. 11.8 g Ar   **b.** 3.55 g Zn
  c. 26.1 g Ta   **d.** 0.211 g Li

48. What is the mass, in grams, of each elemental sample?
  a. 2.3×10^{-3} mol Sb   **b.** 0.0355 mol Ba
  c. 43.9 mol Xe   **d.** 1.3 mol W

49. How many silver atoms are there in 2.54 g of silver?

50. What is the mass of 9.71×10^{22} platinum atoms?

51. Determine the number of atoms in each sample.
  a. 5.18 g P   **b.** 2.26 g Hg
  c. 1.87 g Bi   **d.** 0.082 g Sr

52. Calculate the mass, in grams, of each sample.
  a. 1.1×10^{23} gold atoms
  b. 2.82×10^{22} helium atoms
  c. 1.8×10^{23} lead atoms
  d. 7.9×10^{21} uranium atoms

53. How many carbon atoms are there in a diamond (pure carbon) with a mass of 83 mg?

54. How many helium atoms are there in a helium blimp containing 427 kg of helium?

55. Calculate the average mass, in grams, of a platinum atom.

56. Using scanning tunneling microscopy, scientists at IBM wrote the initials of their company with 35 individual xenon atoms (as shown below). Calculate the total mass of these letters in grams.

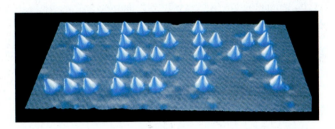

Cumulative Problems

57. A 7.83-g sample of HCN contains 0.290 g of H and 4.06 g of N. Find the mass of carbon in a sample of HCN with a mass of 3.37 g.

58. The ratio of sulfur to oxygen by mass in SO_2 is 1.0:1.0.
  a. Find the ratio of sulfur to oxygen by mass in SO_3.
  b. Find the ratio of sulfur to oxygen by mass in S_2O.

59. The ratio of oxygen to carbon by mass in carbon monoxide is 1.33:1.00. Find the formula of an oxide of carbon in which the ratio by mass of oxygen to carbon is 2.00:1.00.

60. The ratio of the mass of a nitrogen atom to the mass of an atom of ^{12}C is 7:6 and the ratio of the mass of nitrogen to oxygen in N_2O is 7:4. Find the mass of 1 mol of oxygen atoms.

61. An α particle, $^4He^{2+}$, has a mass of 4.00151 amu. Find the value of its charge-to-mass ratio in C/kg.

62. Naturally occurring iodine has an atomic mass of 126.9045. A 12.3849-g sample of naturally occurring iodine is accidentally contaminated with an additional 1.00070 g of ^{129}I, a synthetic radioisotope of iodine used in the treatment of certain diseases of the thyroid gland. The mass of ^{129}I is 128.9050 amu. Find the apparent "atomic mass" of the contaminated iodine.

63. Nuclei with the same number of *neutrons* but different mass numbers are called *isotones*. Write the symbols of four isotones of ^{236}Th.

64. Fill in the blanks to complete the table.

Symbol	Z	A	Number of p	Number of e⁻	Number of n	Charge
Si	14	_____	_____	14	14	_____
S^{2-}	_____	32	_____	_____	_____	2−
Cu^{2+}	_____	_____	_____	_____	24	2+
_____	15	_____	_____	15	16	_____

65. Fill in the blanks to complete the table.

Symbol	Z	A	Number of p	Number of e⁻	Number of n	Charge
_____	8	_____	_____	_____	8	2–
Ca^{2+}	20	_____	_____	_____	20	_____
Mg^{2+}	_____	25	_____	_____	13	2+
N^{3-}	_____	14	_____	10	_____	_____

66. Neutron stars are believed to be composed of solid nuclear matter, primarily neutrons. Assume the radius of a neutron to be approximattely 1.0×10^{-13} cm, and calculate its density. [Hint: For a sphere $V = (4/3)\pi r^3$.]

Assuming that a neutron star has the same density as a neutron, calculate the mass (in kg) of a small piece of a neutron star the size of a spherical pebble with a radius of 0.10 mm.

67. Carbon-12 contains 6 protons and 6 neutrons. The radius of the nucleus is approximately 2.7 fm (femtometers) and the radius of the atom is approximately 70 pm (picometers). Calculate the volume of the nucleus and the volume of the atom. What percentage of the carbon atom's volume is occupied by the nucleus? (Assume two significant figures.)

68. A penny has a thickness of approximately 1.0 mm. If you stacked Avogadro's number of pennies one on top of the other on Earth's surface, how far would the stack extend (in km)? [For comparison, the sun is about 150 million km from Earth and the nearest star (Proxima Centauri) is about 40 trillion km from Earth.]

69. Consider the stack of pennies in the previous problem. How much money (in dollars) would this represent? If this money were equally distributed among the world's population of 6.8 billion people, how much would each person receive? Would each person be a millionaire? Billionaire? Trillionaire?

70. The mass of an average blueberry is 0.75 g and the mass of an automobile is 2.0×10^3 kg. Find the number of automobiles whose total mass is the same as 1.0 mol blueberries.

71. Suppose that atomic masses were based on the assignment of a mass of 12.000 g to 1 mol of carbon, rather than 1 mol of ^{12}C. Find the atomic mass of oxygen.

72. A pure titanium cube has an edge length of 2.78 in. How many titanium atoms does it contain? Titanium has a density of 4.50 g/cm³.

73. A pure copper sphere has a radius 0.935 in. How many copper atoms does it contain? [The volume of a sphere is $(4/3)\pi r^3$ and the density of copper is 8.96 g/cm³.]

74. Boron has only two naturally occurring isotopes. The mass of boron-10 is 10.01294 amu and the mass of boron-11 is 11.00931 amu. Use the atomic mass of boron to calculate the relative abundances of the two isotopes.

75. Lithium has only two naturally occurring isotopes. The mass of lithium-6 is 6.01512 amu and the mass of lithium-7 is 7.01601 amu. Use the atomic mass of lithium to calculate the relative abundances of the two isotopes.

76. Common brass is a copper and zinc alloy containing 37.0 % zinc by mass and having a density of 8.48 g/cm³. A fitting composed of common brass has a total volume of 112.5 cm³. How many atoms (copper and zinc) does the fitting contain?

77. A 67.2 g sample of a gold and palladium alloy contains 2.49×10^{23} atoms. What is the composition (by mass) of the alloy?

78. The U.S. Environmental Protection Agency (EPA) sets limits on healthful levels of air pollutants. The maximum level that the EPA considers safe for lead air pollution is $1.5 \ \mu g/m^3$. If your lungs were filled with air containing this level of lead, how many lead atoms would be in your lungs? (Assume a total lung volume of 5.50 L.)

79. Pure gold is usually too soft to use in jewelry making, so it is often alloyed with other metals. How many gold atoms are in an 0.255 ounce, 18 K gold bracelet? (18 K gold is 75% gold by mass.)

Challenge Problems

80. In Section 2.8, it was stated that 1 mol of sand grains would cover the state of Texas to several feet. Estimate how many feet by assuming that the sand grains are cube-shaped, each one with an edge length of 0.10 mm. Texas has a land area of 268,601 square miles.

81. Use the concepts in this chapter to obtain an estimate for the number of atoms in the universe. Make the following assumptions: (a) Assume that all of the atoms in the universe are hydrogen atoms in stars. (This is not a ridiculous assumption because over three-fourths of the atoms in the universe are in fact hydrogen. Gas and dust between the stars represent only about 15% of the visible matter of our galaxy, and planets compose a far tinier fraction.) (b) Assume that the sun is a typical star composed of pure hydrogen with a density of 1.4 g/cm³ and a radius of 7×10^8 m. (c) Assume that each of the roughly 100 billion stars in the Milky Way galaxy contains the same number of atoms as our sun. (d) Assume that each of the 10 billion galaxies in the visible universe contains the same number of atoms as our Milky Way galaxy.

82. Consider a representation (shown on the next page) of 50 atoms of fictitious element westmontium (Wt). The red spheres represent Wt-296, the blue spheres Wt-297, and the green spheres Wt-298.

a. Assuming that the sample is statistically representative of a naturally occurring sample, calculate the percent natural abundance of each Wt isotope.

b. The mass of each Wt isotope is measured relative to C-12 and tabulated below. Use the mass of C-12 to convert each of the masses to amu and calculate the atomic mass of Wt.

Isotope	Mass
Wt-296	$24.6630 \times \text{Mass}(^{12}\text{C})$
Wt-297	$24.7490 \times \text{Mass}(^{12}\text{C})$
Wt-298	$24.8312 \times \text{Mass}(^{12}\text{C})$

83. Naturally occurring cobalt consists of only one isotope, ^{59}Co, whose relative atomic mass is 58.9332. A synthetic radioactive isotope of cobalt, ^{60}Co, relative atomic mass 59.9338, is used in radiation therapy for cancer. A 1.5886-g sample of cobalt has an apparent "atomic mass" of 58.9901. Find the mass of ^{60}Co in this sample.

84. A 7.36-g sample of copper is contaminated with 0.51 g of zinc. Suppose an atomic mass measurement was performed on this sample. What would be the measured atomic mass?

85. The ratio of the mass of O to the mass of N in N_2O_3 is 12:7. Another binary (two-element) compound of nitrogen has a ratio of O to N of 16:7 and contains two nitrogen atoms per molecule. What is the formula of the binary compound? What is the ratio of O to N in the next member of this series of compounds?

86. Naturally occurring magnesium has an atomic mass of 24.312 and consists of three isotopes. The major isotope is ^{24}Mg, natural abundance 78.99%, relative atomic mass 23.98504. The next most abundant isotope is ^{26}Mg, relative atomic mass 25.98259. The third isotope is ^{25}Mg whose natural abundance is in the ratio of 0.9083 to that of ^{26}Mg. Find the relative atomic mass of ^{25}Mg.

Conceptual Problems

87. Which statement is an example of the law of multiple proportions? Explain.
   a. Two different samples of water are found to have the same ratio of hydrogen to oxygen.
   b. When hydrogen and oxygen react to form water, the mass of water formed is exactly equal to the mass of hydrogen and oxygen that reacted.
   c. The mass ratio of oxygen to hydrogen in water is 8:1. The mass ratio of oxygen to hydrogen in hydrogen peroxide (a compound that only contains hydrogen and oxygen) is 16:1.

88. Suppose that one of Millikan's oil drops had a charge of -11.2×10^{-19} C. How many excess electrons does the drop contain?

89. Lithium has two naturally occurring isotopes: Li-6 (natural abundance is 7.5%) and Li-7 (natural abundance is 92.5%). Using circles to represent protons and squares to represent neutrons, draw the nucleus of each isotope. How many Li-6 atoms would be present, on average, in a 1000-atom sample of lithium?

90. As we saw in the previous problem, lithium has two naturally occurring isotopes: Li-6 (natural abundance is 7.5%; mass is 6.0151 amu) and Li-7 (natural abundance is 92.5 %; mass is 7.0160 amu). Without doing any calculations, determine which mass is closest to the atomic mass of Li.
   a. 6.00 amu   b. 6.50 amu   c. 7.00 amu

91. The mole is defined as the amount of a substance containing the same number of particles as exactly 12 grams of C-12. The amu is defined as 1/12 of the mass of an atom of C-12. Why is it important that both of these definitions reference the same isotope? What would be the result, for example, of defining the mole with respect to C-12 and the amu with respect to Ne-20?

92. Without doing any calculations, determine which sample contains the greatest amount of the element in moles. Which contains the greatest mass of the element?
   a. 55.0 g Cr   b. 45.0 g Ti   c. 60.0 g Zn

93. The atomic radii of the isotopes of an element are identical to one another. However, the atomic radii of the ions of an element are significantly different from the atomic radii of the neutral atom of the element. Explain.

Answers to Conceptual Connections

The Law of Conservation of Mass

2.1 Most of the matter that composed the log underwent a chemical change by reacting with oxygen molecules in the air and was released as gases (primarily carbon dioxide and water) into the air.

The Laws of Definite and Multiple Proportions

2.2 The law of definite proportions applies to two or more samples of the *same compound* and states that the ratio of one element to the other in the samples is always the same. The law of multiple proportions applies to two *different compounds* containing the same two elements (A and B) and states that the masses of B that combine with 1 g of A are related as a small whole-number ratio.

The Millikan Oil Drop Experiment

2.3 The drop contains three excess electrons ($3 \times (-1.6 \times 10^{-19}$ C) $= -4.8 \times 10^{-19}$ C).

Isotopes

2.4

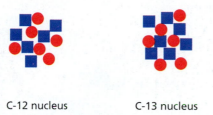

C-12 nucleus   C-13 nucleus

A 10,000-atom sample of carbon, on average, contains 107 C-13 atoms.

The Nuclear Atom and Isotopes

2.5 (b) The number of neutrons in the nucleus of an atom does not affect the atom's size because the nucleus is miniscule compared to the size of the atom itself.

Atomic Mass

2.6 (a) Since 98.93% of the atoms are C-12, we would expect the atomic mass to be very close to the mass of the C-12 isotope.

Avogadro's Number

2.7 Remember that Avogadro's number is defined with respect to carbon-12—it is the number equal to the number of atoms in exactly 12 g of carbon-12. If Avogadro's number were defined as 1.00×10^{23} (a nice round number), it would correspond to 1.99 g (an inconvenient number) of carbon-12 atoms. Avogadro's number is defined with respect to carbon-12 because, as you recall from Section 2.5, the amu (the basic mass unit used for all atoms) is defined relative to carbon-12. Therefore, the mass in grams of 1 mol of any element is equal to its atomic mass. As we have seen, these two definitions together make it possible to determine the number of atoms in a known mass of any element.

The Mole

2.8 (b) The carbon sample contains more atoms than the copper sample because carbon has a lower molar mass than copper. Carbon atoms are lighter than copper atoms, so a 1-g sample of carbon contains more atoms than a 1-g sample of copper. The carbon sample also contains more atoms than the uranium sample because, even though the uranium sample has 10 times the mass of the carbon sample, a uranium atom is more than 10 times as massive (238 g/mol for uranium versus 12 g/mol for carbon).

3 Molecules, Compounds, and Chemical Equations

Almost all aspects of life are engineered at the molecular level, and without understanding molecules we can only have a very sketchy understanding of life itself. —Francis Harry Compton Crick (1916–2004)

When a balloon filled with H_2 and O_2 is ignited, the two elements react violently to form H_2O.

HOW MANY DIFFERENT substances exist? Recall from Chapter 2 that there are about 91 different elements in nature, so there are at least 91 different substances. However, the world would be dull—not to mention lifeless—with only 91 different substances. Fortunately, elements combine with each other to form *compounds*. Just as combinations of only 26 letters in our English alphabet allow for an almost limitless number of words, each with its own specific meaning, so combinations of the 91 naturally occurring elements allow for an almost limitless number of compounds, each with its own specific properties. The great diversity of substances found in nature is a direct result of the ability of elements to form compounds. Life could not exist with just 91 different elements. It takes compounds, in all of their diversity, to make life possible.

3.1   Hydrogen, Oxygen, and Water

Hydrogen (H_2) is an explosive gas used as a fuel in space shuttles. Oxygen (O_2), also a gas, is a natural component of air. Oxygen is not itself flammable, but must be present for combustion (burning) to occur. Hydrogen and oxygen both have extremely low boiling points (see the following table). When hydrogen and oxygen combine to form the compound water (H_2O), however, a dramatically different substance results.

Selected Properties	Hydrogen	Oxygen	Water
Boiling Point	−253 °C	−183 °C	100 °C
State at Room Temperature	Gas	Gas	Liquid
Flammability	Explosive	Necessary for combustion	Used to extinguish flame

First of all, water is a liquid rather than a gas at room temperature, and its boiling point is hundreds of degrees higher than the boiling points of hydrogen and oxygen. Second, instead of being flammable (like hydrogen gas) or supporting combustion (like oxygen gas), water actually smothers flames. *Water is nothing like the hydrogen and oxygen from which it was formed.*

The dramatic difference between the elements hydrogen and oxygen and the compound water is typical of the differences between elements and the compounds that they form. When two elements combine to form a compound, an entirely new substance results. Common table salt, for example, is a highly stable compound composed of sodium and chlorine. Elemental sodium, by contrast, is a highly reactive, silvery metal that can explode on contact with water, and elemental chlorine is a corrosive, greenish-yellow gas that can be fatal if inhaled. Yet the compound that results from the combination of these two elements is sodium chloride (or table salt), a flavor enhancer that we sprinkle on our food.

Although some of the substances that we encounter in everyday life are elements, most are compounds. Free atoms are rare on Earth. As we discussed in Chapter 1, a compound is different from a mixture of elements. In a compound, elements combine in fixed, definite proportions; in a mixture, elements can mix in any proportions. For example, consider the difference between the hydrogen–oxygen mixture and water shown on the next page. A hydrogen–oxygen mixture can have any proportions of hydrogen and oxygen gas. Water, by contrast, is composed of water molecules that always contain two hydrogen atoms to every one oxygen atom. Water has a definite proportion of hydrogen to oxygen.

In this chapter you will learn about compounds: how to represent them, how to name them, how to distinguish among their different types, and how to write chemical equations showing how they form and change. You will also learn how to quantify the composition of a compound according to its constituent elements. This is important whenever we want to know how much of a particular element is contained within a particular compound. For example, patients with high blood pressure often have to reduce their sodium ion intake. Since the sodium ion is normally consumed in the form of sodium chloride, a high-blood-pressure patient needs to know how much sodium is in a given amount of sodium chloride. Similarly, an iron-mining company needs to know how much iron it can recover from a given amount of iron ore. This chapter provides the tools to understand and solve these kinds of problems.

Mixtures and Compounds

Hydrogen and Oxygen Mixture	Water (A Compound)
Can have any ratio of hydrogen to oxygen.	Water molecules have a fixed ratio of hydrogen (2 atoms) to oxygen (1 atom).

  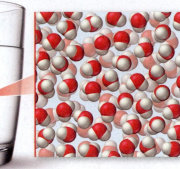

▲ The balloon in this illustration is filled with a mixture of hydrogen gas and oxygen gas. The proportions of hydrogen and oxygen are variable. The glass is filled with water, a compound of hydrogen and oxygen. The ratio of hydrogen to oxygen in water is fixed: water molecules always have two hydrogen atoms for each oxygen atom.

3.2   Chemical Bonds

Compounds are composed of atoms held together by *chemical bonds*. Chemical bonds are the result of the interactions between the charged particles—electrons and protons—that compose atoms. We can broadly classify most chemical bonds into two types: ionic and covalent. *Ionic bonds*—which occur between metals and nonmetals—involve the *transfer* of electrons from one atom to another. *Covalent bonds*—which occur between two or more nonmetals—involve the *sharing* of electrons between two atoms.

Ionic Bonds

Recall from Chapter 2 that metals have a tendency to lose electrons and that nonmetals have a tendency to gain them. Therefore, when a metal interacts with a nonmetal, it can transfer one or more of its electrons to the nonmetal. The metal atom then becomes a *cation* (a positively charged ion) and the nonmetal atom becomes an *anion* (a negatively charged ion) as shown in **Figure 3.1▶**. These oppositely charged ions are attracted to one another by electrostatic forces—they form an **ionic bond**. The result is an ionic compound, which in the solid phase is composed of a lattice—a regular three-dimensional array—of alternating cations and anions.

Covalent Bonds

When a nonmetal bonds with another nonmetal, neither atom transfers its electron to the other. Instead some electrons are *shared* between the two bonding atoms. The shared electrons interact with the nuclei of both atoms, lowering their potential energy through electrostatic interactions with the nuclei. The resulting bond is called a **covalent bond**. We can begin to understand the stability of a covalent bond by considering the most stable (or lowest potential energy) configuration of a negative charge interacting with two positive charges (which are separated by some small distance). As you can see in **Figure 3.2▶**, the arrangement in which the negative charge lies *between* the two positive charges has the lowest potential energy because in this arrangement the negative charge can interact with *both positive charges*. Similarly, shared electrons in a covalent chemical bond hold the bonding atoms together by attracting the positively charged nuclei of both bonding atoms.

The Formation of an Ionic Compound

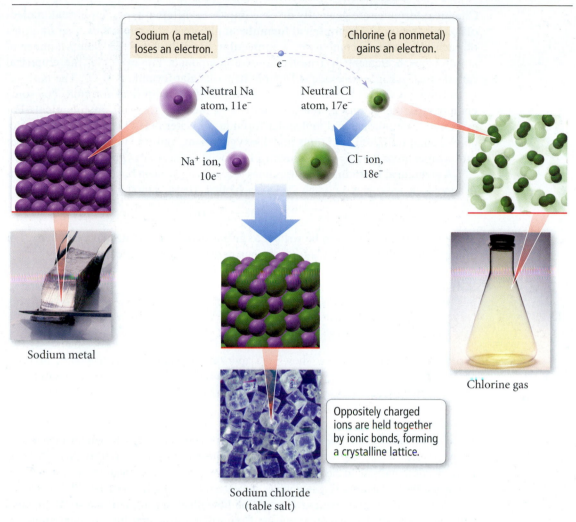

Sodium (a metal) loses an electron.

Chlorine (a nonmetal) gains an electron.

e^-

Neutral Na atom, $11e^-$

Neutral Cl atom, $17e^-$

Na^+ ion, $10e^-$

Cl^- ion, $18e^-$

Sodium metal

Chlorine gas

Oppositely charged ions are held together by ionic bonds, forming a crystalline lattice.

Sodium chloride (table salt)

▲ **FIGURE 3.1 The Formation of an Ionic Compound**   An atom of sodium (a metal) loses an electron to an atom of chlorine (a nonmetal), creating a pair of oppositely charged ions. The sodium cation is then attracted to the chloride anion and the two are held together as part of a crystalline lattice.

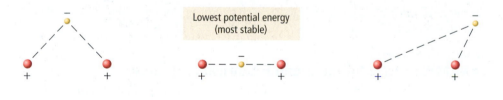

Lowest potential energy (most stable)

◀ **FIGURE 3.2 The Stability of a Covalent Bond**   The potential energy of a negative charge interacting with two positive charges is lowest when the negative charge is directly between the two positive charges.

3.3   Representing Compounds: Chemical Formulas and Molecular Models

The quickest and easiest way to represent a compound is with its **chemical formula**, which indicates which elements are present in the compound and the relative number of atoms or ions of each. For example, H_2O is the chemical formula for water—it indicates that water consists of hydrogen and oxygen atoms in a two-to-one ratio. The formula contains the symbol for each element and a subscript indicating the relative number of atoms of the element. A subscript of 1 is typically omitted. Chemical formulas normally list the more metallic (or more positively charged) elements first, followed by the less metallic (or more negatively charged) elements. Other examples of common chemical formulas include NaCl for sodium chloride, indicating sodium and chloride ions in a one-to-one ratio; CO_2 for carbon dioxide, indicating carbon and oxygen atoms in a one-to-two ratio; and CCl_4 for carbon tetrachloride, indicating carbon and chlorine in a one-to-four ratio.

Types of Chemical Formulas

Chemical formulas are generally divided into three different types: empirical, molecular, and structural. An **empirical formula** simply gives the *relative* number of atoms of each element in a compound. A **molecular formula** gives the *actual* number of atoms of each element in a molecule of a compound. For example, the empirical fomula for hydrogen peroxide is HO, but its molecular formula is H_2O_2. The molecular formula is always a whole-number multiple of the empirical formula. For some compounds, the empirical formula and the molecular formula are identical. For example, the empirical and molecular formula for water is H_2O because water molecules contain 2 hydrogen atoms and 1 oxygen atom, and no simpler whole-number ratio can express the relative number of hydrogen atoms to oxygen atoms.

A **structural formula**, which uses lines to represent covalent bonds, shows how atoms in a molecule are connected or bonded to each other. The structural formula for H_2O_2 is

$$H-O-O-H$$

Structural formulas may also be written to give a sense of the molecule's geometry. For example, the structural formula for hydrogen peroxide can be written

$$\begin{array}{c} H \\ \diagdown \\ O-O \\ \diagdown \\ H \end{array}$$

Writing the formula this way shows the approximate angles between bonds, giving a sense of the molecule's shape. Structural formulas can also show different types of bonds that occur between molecules. For example, the structural formula for carbon dioxide is

$$O=C=O$$

The two lines between the carbon and oxygen atoms represent a double bond, which is generally stronger and shorter than a single bond (represented by a single line). A single bond corresponds to one shared electron pair while a double bond corresponds to two shared electron pairs. We will discuss single, double, and even triple bonds in more detail in Chapter 9.

The type of formula we use depends on how much we know about the compound and how much we want to communicate. Notice that a structural formula communicates the most information, while an empirical formula communicates the least.

You can find answers to Conceptual Connections at the end of the chapter.

 Conceptual Connection 3.1   Structural Formulas

In water, the oxygen atom is connected to each hydrogen atom by a single bond. Write a structural formula for water.

EXAMPLE 3.1   Molecular and Empirical Formulas

Write empirical formulas for the compounds represented by the molecular formulas.

(a) C_4H_8 **(b)** B_2H_6 **(c)** CCl_4

SOLUTION

To determine the empirical formula from a molecular formula, divide the subscripts by the greatest common factor (the largest number that divides exactly into all of the subscripts).

(a) For C_4H_8, the greatest common factor is 4. The empirical formula is therefore CH_2.

(b) For B_2H_6, the greatest common factor is 2. The empirical formula is therefore BH_3.

(c) For CCl_4, the only common factor is 1, so the empirical formula and the molecular formula are identical.

Answers to For Practice and For More Practice problems can be found in Appendix IV.

FOR PRACTICE 3.1

Write the empirical formula for the compounds represented by the molecular formulas.

(a) C_5H_{12} **(b)** Hg_2Cl_2 **(c)** $C_2H_4O_2$

Molecular Models

Molecular models are a more accurate and complete way to specify a compound. **Ball-and-stick models** represent atoms as balls and chemical bonds as sticks; how the two connect reflects a molecule's shape. The balls are normally color-coded to specific elements. For example, carbon is customarily black, hydrogen is white, nitrogen is blue, and oxygen is red. (For a complete list of colors of elements in the molecular models used in this book see Appendix IIA.)

In **space-filling molecular models**, atoms fill the space between each other to more closely represent our best estimates for how a molecule might appear if scaled to a visible size. For example, consider the following ways to represent a molecule of methane, the main component of natural gas:

CH_4	H–C–H (with H above and below)		
Molecular formula	Structural formula	Ball-and-stick model	Space-filling model

Hydrogen
Carbon
Nitrogen
Oxygen
Fluorine
Phosphorus
Sulfur
Chlorine

The molecular formula of methane shows the number and type of each atom in the molecule: one carbon atom and four hydrogen atoms. The structural formula indicates how the atoms are connected: the carbon atom is bonded to the four hydrogen atoms. The ball-and-stick model clearly shows the geometry of the molecule: the carbon atom sits in the center of a *tetrahedron* formed by the four hydrogen atoms. The space-filling model gives the best sense of the relative sizes of the atoms and how they merge together in bonding.

Throughout this book, you will see molecules represented in all of these ways. As you look at these representations, keep in mind what you learned in Chapter 1: the details about a molecule—the atoms that compose it, the lengths of the bonds between atoms, the angles of the bonds between atoms, and its overall shape—determine the properties of the substance that the molecule composes. If any of these details were to change, the properties of the substance would change. Table 3.1 (on p. 78) shows various compounds represented in the different ways we have just discussed.

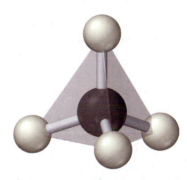

▲ A tetrahedron is a three-dimensional geometrical shape characterized by four equivalent triangular faces.

Conceptual Connection 3.2 Representing Molecules

Based on what you learned in Chapter 2 about atoms, what part of the atom do you think the spheres in the molecular models just shown represent? If you were to superimpose a nucleus on one of these spheres, how big would you draw it?

3.4 An Atomic-Level View of Elements and Compounds

Recall from Chapter 1 that we can categorize pure substances as either elements or compounds. We can further subcategorize elements and compounds according to the basic units that compose them, as shown in **Figure 3.3▶**. Elements may be either atomic or molecular. Compounds may be either molecular or ionic.

Atomic elements are elements that exist in nature with single atoms as their basic units. Most elements fall into this category. For example, helium is composed of helium atoms, aluminum is composed of aluminum atoms, and iron is composed of iron atoms. **Molecular elements** do not normally exist in nature with single atoms as their basic units. Instead, these elements exist as molecules—two or more atoms of the element bonded together. Most molecular elements exist as *diatomic* molecules. For example, hydrogen is composed of H_2 molecules, nitrogen is composed of N_2 molecules, and chlorine is composed of Cl_2 molecules. A few molecular elements exist as *polyatomic*

Diatomic chlorine molecules

▲ The basic units that compose chlorine gas are diatomic chlorine molecules.

TABLE 3.1 Benzene, Acetylene, Glucose, and Ammonia

Name of Compound	Empirical Formula	Molecular Formula	Structural Formula	Ball-and-Stick Model	Space-Filling Model
Benzene	CH	C_6H_6			
Acetylene	CH	C_2H_2	$H-C\equiv C-H$		
Glucose	CH_2O	$C_6H_{12}O_6$			
Ammonia	NH_3	NH_3			

Classification of Elements and Compounds

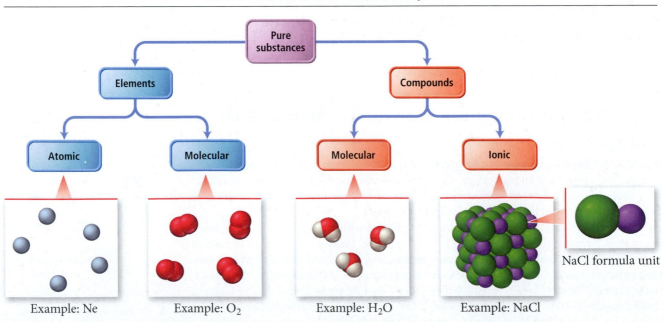

▲ **FIGURE 3.3   A Molecular View of Elements and Compounds**

Molecular Elements

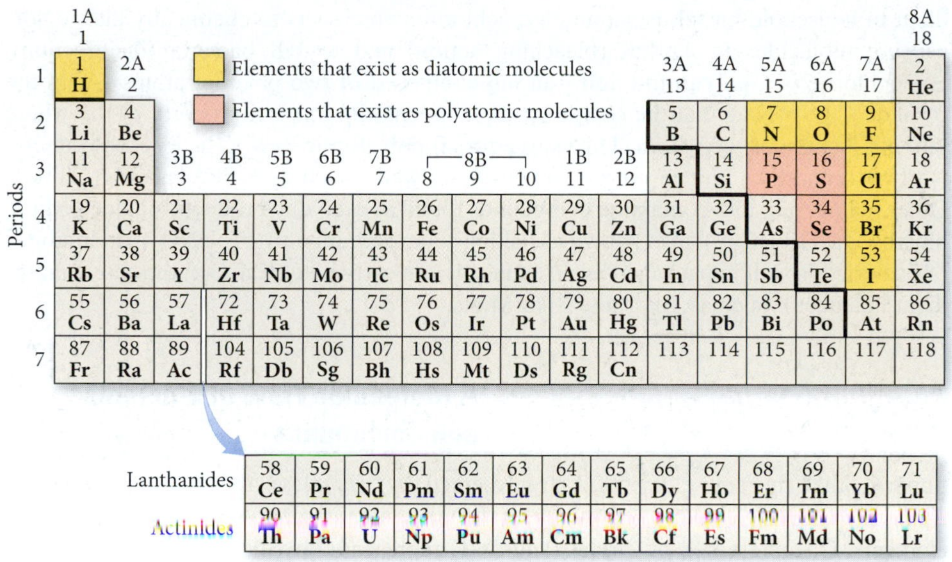

▲ **FIGURE 3.4 Molecular Elements** The highlighted elements exist primarily as diatomic molecules (yellow) or polyatomic molecules (red).

molecules. Phosphorus, for example, exists as P_4 and sulfur exists as S_8. The elements that exist primarily as diatomic or polyatomic molecules are shown in the periodic table in **Figure 3.4▲**.

   Molecular compounds are usually composed of two or more covalently bonded nonmetals. The basic units of molecular compounds are molecules composed of the constituent atoms. For example, water is composed of H_2O molecules, dry ice is composed of CO_2 molecules, and propane (often used as a fuel for grills) is composed of C_3H_8 molecules as shown in **Figure 3.5(a)▼**.

   Ionic compounds are composed of cations (usually one type of metal) and anions (usually one or more nonmetals) bound together by ionic bonds. The basic unit of an ionic compound is the **formula unit**, the smallest electrically neutral collection of ions. Formula units are different from molecules in that they do not exist as discrete entities, but rather as part of a larger lattice. For example, the ionic compound table salt, with the formula unit NaCl, is composed of Na^+ and Cl^- ions in a one-to-one ratio. In table salt, Na^+ and Cl^- ions exist in a three-dimensional array. However, because ionic bonds are not directional, no one Na^+ ion pairs with a specific Cl^- ion. Rather, as you can see in **Figure 3.5(b)▼**, any one Na^+ cation is surrounded by Cl^- anions and vice versa.

Some ionic compounds, such as K_2NaPO_4, for example, contain more than one type of metal ion.

People occasionally refer to formula units as molecules, but this is *not* correct since ionic compounds do not contain distinct molecules.

A Molecular Compound

An Ionic Compound

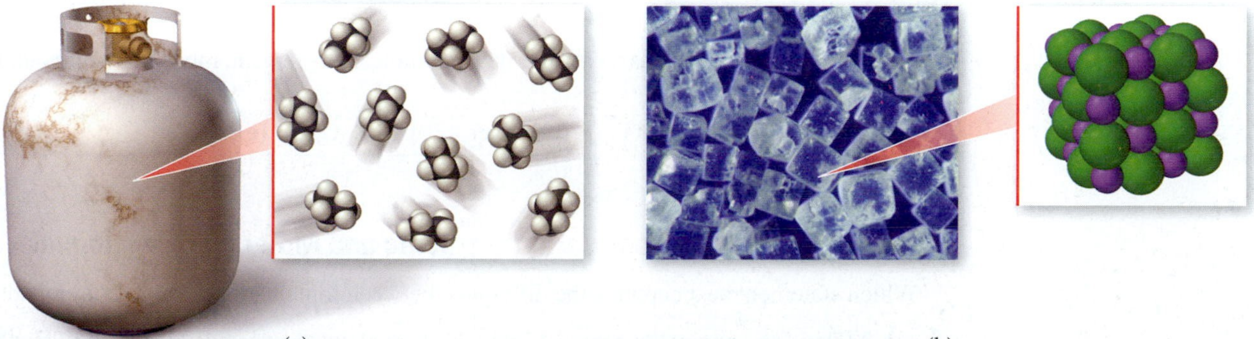

(a)

(b)

▲ **FIGURE 3.5 Molecular and Ionic Compounds** (a) Propane is an example of a molecular compound. The basic units that compose propane gas are propane (C_3H_8) molecules. (b) Table salt (NaCl) is an ionic compound. Its formula unit is the simplest charge-neutral collection of ions: one Na^+ ion and one Cl^- ion.

Active Ingredient:
Sodium Hypochlorite.....6.0%
Other Ingredients:.......94.0%
Total:....................100.0%
(Yields 5.7% available chlorine)

KEEP OUT OF REACH OF CHILDREN

DANGER: CORROSIVE.

FIRST AID: IF IN EYES: Hold eye open and rinse
slowly and gently with water for 15–20 minutes.
Remove contact lenses, if present, after the first
5 minutes, then continue rinsing eye. IF ON SKIN OR
CLOTHING: Take off contaminated clothing. Rinse skin
immediately with plenty of water for 15–20 minutes.
IN EITHER CASE, CALL A POISON CONTROL CENTER
OR DOCTOR IMMEDIATELY FOR TREATMENT ADVICE.
See back panel for additional precautionary labeling.

▲ Polyatomic ions are common in household products such as bleach, which contains sodium hypochlorite (NaClO).

Many common ionic compounds contain ions that are themselves composed of a group of covalently bonded atoms with an overall charge. For example, the active ingredient in household bleach is sodium hypochlorite, which acts to chemically alter color-causing molecules in clothes (bleaching action) and to kill bacteria (disinfection). Hypochlorite is a **polyatomic ion**—an ion composed of two or more atoms—with the formula ClO^-. (Note that the charge on the hypochlorite ion is a property of the whole ion, not just the oxygen atom. This is true for all polyatomic ions.) The hypochlorite ion is often found as a unit in other compounds as well [such as KClO and $Mg(ClO)_2$]. Other polyatomic ion–containing compounds found in everyday products include sodium bicarbonate ($NaHCO_3$), also known as baking soda; sodium nitrite ($NaNO_2$), an inhibitor of bacterial growth in packaged meats; and calcium carbonate ($CaCO_3$), the active ingredient in antacids such as Tums and Alka-Mints.

 Conceptual Connection 3.3 **A Molecular View of Elements and Compounds**

Suppose that the two elements A (represented by triangles) and B (represented by squares) form a molecular compound with the molecular formula A_2B, and that two other elements, C (represented by circles) and D (D represented by diamonds) form an ionic compound with the formula CD. Draw a molecular level view of each compound.

EXAMPLE 3.2 Classifying Substances as Atomic Elements, Molecular Elements, Molecular Compounds, or Ionic Compounds

Classify each substance as an atomic element, molecular element, molecular compound, or ionic compound.

(a) xenon   (b) $NiCl_2$ (c) bromine   (d) NO_2 (e) $NaNO_3$

SOLUTION

(a) Xenon is an element and it is not one of the elements that exist as diatomic or polyatomic molecules (Figure 3.4); therefore, it is an atomic element.

(b) $NiCl_2$ is a compound composed of a metal (left side of the periodic table) and nonmetal (right side of the periodic table); therefore, it is an ionic compound.

(c) Bromine is one of the elements that exist as diatomic molecules; therefore, it is a molecular element.

(d) NO_2 is a compound composed of a nonmetal and a nonmetal; therefore, it is a molecular compound.

(e) $NaNO_3$ is a compound composed of a metal and a polyatomic ion; therefore, it is an ionic compound.

FOR PRACTICE 3.2

Classify each substance as an atomic element, molecular element, molecular compound, or ionic compound.

(a) fluorine   (b) N_2O (c) silver   (d) K_2O (e) Fe_2O_3

 Conceptual Connection 3.4 **Ionic and Molecular Compounds**

Which statement best captures the difference between ionic and molecular compounds?

(a) Molecular compounds contain highly directional covalent bonds, which results in the formation of molecules—discrete particles that do not covalently bond to each other. Ionic compounds contain nondirectional ionic bonds, which results (in the solid phase) in the formation of ionic lattices—extended networks of alternating cations and anions.

(b) Molecular compounds contain covalent bonds in which one of the atoms shares an electron with the other one, resulting in a new force that holds the atoms together in a covalent molecule. Ionic compounds contain ionic bonds in which one atom donates an electron to the other, resulting in a new force that holds the ions together in pairs (in the solid phase).

(c) The main difference between ionic and covalent compounds is the types of elements that compose them, not the way that the atoms bond together.

(d) A molecular compound is composed of covalently bonded molecules. An ionic compound is composed of ionically bonded molecules (in the solid phase).

3.5 Ionic Compounds: Formulas and Names

Ionic compounds occur throughout Earth's crust as minerals. Examples include limestone ($CaCO_3$), a type of sedimentary rock; gibbsite $\left[Al(OH)_3\right]$, an aluminum-containing mineral; and soda ash (Na_2CO_3), a natural deposit.

◀ Calcite (left) is the main component of limestone, marble, and other forms of calcium carbonate ($CaCO_3$) commonly found in Earth's crust. Trona (right) is a crystalline form of hydrated sodium carbonate ($Na_3H(CO_3)_2 \cdot 2H_2O$).

Ionic compounds are also in the foods that we eat. Examples include table salt (NaCl), the most common flavor enhancer, calcium carbonate ($CaCO_3$), a source of calcium necessary for bone health, and potassium chloride (KCl), a source of potassium necessary for fluid balance and muscle function. Ionic compounds are generally very stable because the attractions between cations and anions within ionic compounds are strong, and because each ion interacts with several oppositely charged ions in the crystalline lattice.

Writing Formulas for Ionic Compounds

Since ionic compounds are charge neutral, and since many elements form only one type of ion with a predictable charge, the formulas for many ionic compounds can be deduced from their constituent elements. For example, the formula for the ionic compound composed of sodium and chlorine must be NaCl because, in compounds, Na always forms 1+ cations and Cl always forms 1− anions. In order for the compound to be charge-neutral, it must contain one Na^+ cation to every one Cl^- anion. The formula for the ionic compound composed of calcium and chlorine can only be $CaCl_2$ because Ca always forms 2+ cations and Cl always forms 1− anions. In order for this compound to be charge neutral, it must contain one Ca^{2+} cation to every two Cl^- anions.

▲ Ionic compounds are common in food and consumer products such as light salt (a mixture of NaCl and KCl) and Tums™ ($CaCO_3$).

See Figure 2.14 to review the elements that form ions with a predictable charge.

Summarizing Ionic Compound Formulas:

▶ Ionic compounds always contain positive and negative ions.

▶ In a chemical formula, the sum of the charges of the positive ions (cations) must always equal the sum of the charges of the negative ions (anions).

▶ The formula reflects the smallest whole-number ratio of ions.

To write the formula for an ionic compound, follow the procedure in the left column of the following page. Examples 3.3 and 3.4, provided in the center and right columns, show how to apply the procedure.

PROCEDURE FOR... Writing Formulas for Ionic Compounds	**EXAMPLE 3.3** Writing Formulas for Ionic Compounds	**EXAMPLE 3.4** Writing Formulas for Ionic Compounds
	Write the formula for the ionic compound that forms between aluminum and oxygen.	Write the formula for the ionic compound that forms between calcium and oxygen.
1. Write the symbol for the metal cation and its charge followed by the symbol for the nonmetal anion and its charge. Obtain charges from the element's group number in the periodic table (refer to Figure 2.14).	$Al^{3+} \quad O^{2-}$	$Ca^{2+} \quad O^{2-}$
2. Adjust the subscript on each cation and anion to balance the overall charge.	$Al^{3+} \qquad\qquad O^{2-}$ $\downarrow$ Al_2O_3	$Ca^{2+} \qquad\qquad O^{2-}$ $\downarrow$ CaO
3. Check that the sum of the charges of the cations equals the sum of the charges of the anions.	cations: $2(3+) = 6+$ anions: $3(2-) = 6-$ The charges cancel.	cations: $2+$ anions: $2-$ The charges cancel.
	FOR PRACTICE 3.3 Write the formula for the compound formed between potassium and sulfur.	**FOR PRACTICE 3.4** Write the formula for the compound formed between aluminum and nitrogen.

Naming Ionic Compounds

Some ionic compounds—such as NaCl (table salt) and $NaHCO_3$ (baking soda)—have **common names**, which are nicknames of sorts that can be learned only through familiarity. However, chemists have developed **systematic names** for different types of compounds including ionic ones. Systematic names can be determined by looking at the chemical formula of a compound. Conversely, the formula of a compound can be deduced from its systematic name.

The first step in naming an ionic compound is identifying it as one. Remember, *ionic compounds are usually formed between metals and nonmetals*; any time you see a metal and one or more nonmetal together in a chemical formula, you can assume you have an ionic compound. Ionic compounds can be categorized into two types, depending on the metal in the compound. The first type contains metals whose charge is invariant from one compound to another. In other words, whenever one of these metals forms an ion, the ion always has the same charge.

Since the charge of the metal in this first type of ionic compound is always the same, it need not be specified in the name of the compound. Sodium, for instance, has a 1+ charge in all of its compounds. Some examples of these types of metals are listed in Table 3.2, and their charges can be inferred from their group number in the periodic table.

The second type of ionic compound contains metals whose charges can be different in different compounds—each of these metals can form more than one kind of cation and the charge must therefore be specified for a given compound. Iron, for instance, has a 2+ charge in some of its compounds and a 3+ charge in others. Metals of this type are often in the section of the periodic table known as the *transition metals* (**Figure 3.6◄**).

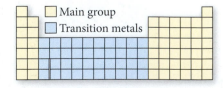

▲ **FIGURE 3.6 Transition Metals**
Metals that can have different charges in different compounds are usually, but not always, transition metals.

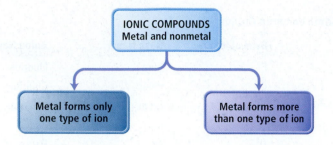

However, some transition metals, such as Zn and Ag, have the same charge in all of their compounds (as shown in Table 3.2), and some main group metals, such as lead and tin, have charges that can vary from one compound to another.

TABLE 3.2 Metals Whose Charge Is Invariant from One Compound to Another

Metal	Ion	Name	Group Number
Li	Li^+	Lithium	1A
Na	Na^+	Sodium	1A
K	K^+	Potassium	1A
Rb	Rb^+	Rubidium	1A
Cs	Cs^+	Cesium	1A
Be	Be^{2+}	Beryllium	2A
Mg	Mg^{2+}	Magnesium	2A
Ca	Ca^{2+}	Calcium	2A
Sr	Sr^{2+}	Strontium	2A
Ba	Ba^{2+}	Barium	2A
Al	Al^{3+}	Aluminum	3A
Zn	Zn^{2+}	Zinc	*
Sc	Sc^{3+}	Scandium	*
Ag**	Ag^+	Silver	*

*The charge of these metals cannot be inferred from their group number.
**Silver does sometimes form compounds with other charges, but these are rare.

Naming Binary Ionic Compounds Containing a Metal That Forms Only One Type of Cation

Binary compounds contain only two different elements. The names for binary ionic compounds take the form

For example, the name for KCl consists of the name of the cation, *potassium*, followed by the base name of the anion, *chlor*, with the ending *-ide*. Its full name is *potassium chloride*.

KCl   potassium chloride

The name for CaO consists of the name of the cation, *calcium*, followed by the base name of the anion, *ox*, with the ending *-ide*. Its full name is *calcium oxide*.

CaO   calcium oxide

The base names for various nonmetals and their most common charges in ionic compounds are shown in Table 3.3.

TABLE 3.3 Some Common Anions

Nonmetal	Symbol for Ion	Base Name	Anion Name
Fluorine	F^-	fluor	Fluoride
Chlorine	Cl^-	chlor	Chloride
Bromine	Br^-	brom	Bromide
Iodine	I^-	iod	Iodide
Oxygen	O^{2-}	ox	Oxide
Sulfur	S^{2-}	sulf	Sulfide
Nitrogen	N^{3-}	nitr	Nitride
Phosphorus	P^{3-}	phosph	Phosphide

EXAMPLE 3.5 Naming Ionic Compounds Containing a Metal That Forms Only One Type of Cation

Determine the name for the compound $CaBr_2$.

SOLUTION

The cation is *calcium*. The anion is from bromine, which becomes *bromide*. The correct name is *calcium bromide*.

FOR PRACTICE 3.5
Determine the name for the compound Ag_3N.

FOR MORE PRACTICE 3.5
Write the formula for rubidium sulfide.

Naming Binary Ionic Compounds Containing a Metal That Forms More than One Kind of Cation

For these types of metals, the name of the cation is followed by a Roman numeral (in parentheses) indicating its charge in that particular compound. For example, we distinguish between Fe^{2+} and Fe^{3+} as follows:

$$Fe^{2+} \quad iron(II)$$
$$Fe^{3+} \quad iron(III)$$

The full names therefore have the form

Note that there is no space between the name of the cation and the parenthetical number indicating its charge.

name of cation (metal)	charge of cation (metal) in Roman numerals in parentheses	base name of anion (nonmetal) + *-ide*

The charge of the metal cation is obtained by inference from the sum of the charges of the nonmetal anions—remember that the sum of all the charges must be zero. Table 3.4 shows some of the metals that form more than one cation and the values of their most common charges. For example, in $CrBr_3$, the charge of chromium must be 3+ in order for the compound to be charge-neutral with three Br^- anions. The cation is therefore named as

$$Cr^{3+} \quad chromium(III)$$

The full name of the compound is

$$CrBr_3 \quad \text{chromium(III) bromide}$$

Similarly, in CuO, the charge of copper must be 2+ in order for the compound to be charge neutral with one O^{2-} anion. The cation is therefore named as follows:

$$Cu^{2+} \quad \text{copper(II)}$$

The full name of the compound is

$$CuO \quad \text{copper(II) oxide}$$

TABLE 3.4 Some Metals That Form Cations with Different Charges

Metal	Ion	Name	Older Name*
Chromium	Cr^{2+}	Chromium(II)	Chromous
	Cr^{3+}	Chromium(III)	Chromic
Iron	Fe^{2+}	Iron(II)	Ferrous
	Fe^{3+}	Iron(III)	Ferric
Cobalt	Co^{2+}	Cobalt(II)	Cobaltous
	Co^{3+}	Cobalt(III)	Cobaltic
Copper	Cu^+	Copper(I)	Cuprous
	Cu^{2+}	Copper(II)	Cupric
Tin	Sn^{2+}	Tin(II)	Stannous
	Sn^{4+}	Tin(IV)	Stannic
Mercury	Hg_2^{2+}	Mercury(I)	Mercurous
	Hg^{2+}	Mercury(II)	Mercuric
Lead	Pb^{2+}	Lead(II)	Plumbous
	Pb^{4+}	Lead(IV)	Plumbic

*An older naming system substitutes the names found in this column for the name of the metal and its charge. Under this system, chromium(II) oxide is named chromous oxide. In this system, the suffix *-ous* indicates the ion with the lesser charge and *-ic* indicates the ion with the greater charge. We will *not* use the older system in this text.

EXAMPLE 3.6 Naming Ionic Compounds Containing a Metal That Forms More than One Kind of Cation

Determine the name for the compound $PbCl_4$.

SOLUTION

The charge on Pb must be 4+ for the compound to be charge neutral with 4 Cl^- anions. The name for $PbCl_4$ consists of the name of the cation, *lead*, followed by the charge of the cation in parentheses *(IV)*, followed by the base name of the anion, *chlor*, with the ending *-ide*. The full name is *lead(IV) chloride*.

$$PbCl_4 \quad \text{lead(IV) chloride}$$

FOR PRACTICE 3.6

Determine the name for the compound FeS.

FOR MORE PRACTICE 3.6

Write the formula for ruthenium(IV) oxide.

Naming Ionic Compounds Containing Polyatomic Ions

We name ionic compounds containing polyatomic ions in the same way as other ionic compounds, except that we use the name of the polyatomic ion whenever it occurs. Table 3.5 lists common polyatomic ions and their formulas. For example, $NaNO_2$ is named according to its cation, Na^+, *sodium,* and its polyatomic anion, NO_2^-, *nitrite.* Its full name is *sodium nitrite.*

$$NaNO_2 \quad \text{sodium nitrite}$$

$FeSO_4$ is named according to its cation, *iron,* its charge *(II)*, and its polyatomic ion *sulfate.* The full name is *iron(II) sulfate.*

$$FeSO_4 \quad \text{iron(II) sulfate}$$

If the compound contains both a polyatomic cation and a polyatomic anion, use the names of both polyatomic ions. For example, NH_4NO_3 is *ammonium nitrate.*

$$NH_4NO_3 \quad \text{ammonium nitrate}$$

You must be able to recognize polyatomic ions in a chemical formula, so become familiar with Table 3.5. Most polyatomic ions are **oxyanions**, anions containing oxygen and another element. Notice that when a series of oxyanions contains different numbers of oxygen atoms, they are named systematically according to the number of oxygen atoms in the ion. If there are only two ions in the series, the one with more oxygen atoms is given the ending *-ate* and the one with fewer is given the ending *-ite.* For example, NO_3^- is *nitrate* and NO_2^- is *nitrite.*

$$NO_3^- \quad \text{nit}rate$$

$$NO_2^- \quad \text{nit}rite$$

If there are more than two ions in the series then the prefixes *hypo-*, meaning *less than,* and *per-*, meaning *more than,* are used. So ClO^- is hypochlorite, meaning less oxygen than chlorite, and ClO_4^- is perchlorate, meaning more oxygen than chlorate.

The other halides (halogen ions) form similar series with similar names. Thus, IO_3^- is iodate and BrO_3^- is bromate.

$$ClO^- \quad \textit{hypo}\text{chlor}\textit{ite}$$

$$ClO_2^- \quad \text{chlor}\textit{ite}$$

$$ClO_3^- \quad \text{chlor}\textit{ate}$$

$$ClO_4^- \quad \textit{per}\text{chlor}\textit{ate}$$

TABLE 3.5 Some Common Polyatomic Ions

Name	Formula	Name	Formula
Acetate	$C_2H_3O_2^-$	Hypochlorite	ClO^-
Carbonate	CO_3^{2-}	Chlorite	ClO_2^-
Hydrogen carbonate (or bicarbonate)	HCO_3^-	Chlorate	ClO_3^-
Hydroxide	OH^-	Perchlorate	ClO_4^-
Nitrite	NO_2^-	Permanganate	MnO_4^-
Nitrate	NO_3^-	Sulfite	SO_3^{2-}
Chromate	CrO_4^{2-}	Hydrogen sulfite (or bisulfite)	HSO_3^-
Dichromate	$Cr_2O_7^{2-}$	Sulfate	SO_4^{2-}
Phosphate	PO_4^{3-}	Hydrogen sulfate (or bisulfate)	HSO_4^-
Hydrogen phosphate	HPO_4^{2-}	Cyanide	CN^-
Dihydrogen phosphate	$H_2PO_4^-$	Peroxide	O_2^{2-}
Ammonium	NH_4^+		

EXAMPLE 3.7 Naming Ionic Compounds That Contain a Polyatomic Ion

Determine the name for the compound $Li_2Cr_2O_7$.

SOLUTION

The name for $Li_2Cr_2O_7$ consists of the name of the cation, *lithium*, followed by the name of the polyatomic ion, *dichromate*. Its full name is *lithium dichromate*.

$$Li_2Cr_2O_7 \quad \text{lithium dichromate}$$

FOR PRACTICE 3.7

Determine the name for the compound $Sn(ClO_3)_2$.

FOR MORE PRACTICE 3.7

Write a formula for cobalt(II) phosphate.

Hydrated Ionic Compounds

Some ionic compounds—called **hydrates**—contain a specific number of water molecules associated with each formula unit. For example, Epsom salts has the formula $MgSO_4 \cdot 7H_2O$ and the systematic name magnesium sulfate heptahydrate. The seven H_2O molecules associated with the formula unit are called *waters of hydration*. Waters of hydration can usually be removed by heating the compound. **Figure 3.7▼** shows a sample of copper(II) sulfate pentahydrate being heated. The hydrate is blue and the anhydrous salt (the salt without the associated water molecules) is white. Hydrates are named just as other ionic compounds, but they are given the additional name "*prefix*hydrate," where the prefix indicates the number of water molecules associated with each formula unit.

Some other common examples of hydrated ionic compounds and their names are as follows:

$$CaSO_4 \cdot \frac{1}{2}H_2O \quad \text{calcium sulfate hemihydrate}$$

$$BaCl_2 \cdot 2H_2O \quad \text{barium chloride dihydrate}$$

$$CuSO_4 \cdot 5H_2O \quad \text{copper(II) sulfate pentahydrate}$$

Common hydrate prefixes
hemi = 1/2
mono = 1
di = 2
tri = 3
tetra = 4
penta = 5
hexa = 6
hepta = 7
octa = 8

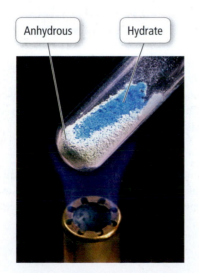

Anhydrous   Hydrate

◄ **FIGURE 3.7 Hydrates** Copper(II) sulfate pentahydrate is blue, but heating the compound removes the waters of hydration leaving the pale anhydrous copper(II) sulfate.

3.6 Molecular Compounds: Formulas and Names

In contrast to ionic compounds, the formula for a molecular compound *cannot* easily be determined based on its constituent elements because the same group of elements may form many different molecular compounds, each with a different formula. For example, we learned in Chapter 1 that carbon and oxygen form both CO and CO_2, and that hydrogen and oxygen form both H_2O and H_2O_2. Nitrogen and oxygen form all of the following unique molecular compounds: NO, NO_2, N_2O, N_2O_3, N_2O_4, and N_2O_5. In Chapter 9, you will learn how to understand the stability of these various combinations of the same elements. For now, we focus on naming a molecular compound based on its formula or writing its formula based on its name.

Naming Molecular Compounds

Like ionic compounds, many molecular compounds have common names. For example, H_2O and NH_3 have the common names *water* and *ammonia*. However, the sheer number of existing molecular compounds—numbering in the millions—requires a systematic approach to naming them.

The first step in naming a molecular compound is identifying it as one. Remember, *molecular compounds form between two or more nonmetals*. In this section, we learn how to name binary (two-element) molecular compounds. Their names have the form

When writing the name of a molecular compound, as when writing the formula, the first element is the more metal-like one (toward the left and bottom of the periodic table). Always write the name of the element with the smallest group number first. If the two elements lie in the same group, then write the element with the greatest row number first. The prefixes given to each element indicate the number of atoms present:

These prefixes are the same as those used in naming hydrates.

mono = 1	hexa = 6
di = 2	hepta = 7
tri = 3	octa = 8
tetra = 4	nona = 9
penta = 5	deca = 10

If there is only one atom of the *first element* in the formula, the prefix *mono-* is normally omitted. For example, NO_2 is named according to the first element, *nitrogen*, with no prefix because *mono-* is omitted for the first element, followed by the prefix *di*, to indicate two oxygen atoms, followed by the base name of the second element, *ox*, with the ending *-ide*. Its full name is *nitrogen dioxide*.

$$NO_2 \quad \text{nitrogen dioxide}$$

When a prefix ends with "o" and the base name begins with "o," the first "o" is often dropped. So mono-oxide becomes *monoxide*.

The compound N_2O, sometimes called laughing gas, is named similarly except that we use the prefix *di-* before nitrogen to indicate two nitrogen atoms and the prefix *mono-* before oxide to indicate one oxygen atom. Its entire name is *dinitrogen monoxide*.

$$N_2O \quad \text{dinitrogen monoxide}$$

EXAMPLE 3.8 Naming Molecular Compounds

Name each compound.

(a) NI_3 **(b)** PCl_5 **(c)** P_4S_{10}

SOLUTION

(a) The name of the compound is the name of the first element, *nitrogen,* followed by the base name of the second element, *iod*, prefixed by *tri-* to indicate three and given the suffix *-ide*.

NI_3 nitrogen triiodide

(b) The name of the compound is the name of the first element, *phosphorus,* followed by the base name of the second element, *chlor*, prefixed by *penta-* to indicate five and given the suffix *-ide*.

PCl_5 phosphorus pentachloride

(c) The name of the compound is the name of the first element, *phosphorus,* prefixed by *tetra-* to indicate four, followed by the base name of the second element, *sulf*, prefixed by *deca-* to indicate ten and given the suffix *-ide*.

P_4S_{10} tetraphosphorus decasulfide

FOR PRACTICE 3.8

Name the compound N_2O_5.

FOR MORE PRACTICE 3.8

Write a formula for phosphorus tribromide.

 Conceptual Connection 3.5 Nomenclature

The compound NCl_3 is nitrogen trichloride, but $AlCl_3$ is simply aluminum chloride. Why?

Naming Acids

We can define acids in a number of ways, as we will see in Chapter 15. For now, we define **acids** as molecular compounds that release hydrogen ions (H^+) when dissolved in water. Acids are composed of hydrogen, usually written first in their formula, and one or more nonmetals, written second. For example, HCl is a molecular compound that, when dissolved in water, forms $H^+(aq)$ and $Cl^-(aq)$ ions, where *aqueous (aq)* means *dissolved in water*. Therefore, HCl is an acid when dissolved in water. To distinguish

▲ Many fruits are acidic and have the characteristically sour taste of acids.

between gaseous HCl (which is named hydrogen monochloride because it is a molecular compound) and HCl in solution (which is named as an acid), we write the former as HCl(g) and the latter as HCl(aq).

Acids are characterized by their sour taste and their ability to dissolve many metals. Since the acid HCl(aq) is present in stomach fluids, its sour taste becomes painfully obvious when you vomit. HCl(aq), hydrochloric acid, also dissolves some metals. If you put a strip of zinc into a test tube of HCl(aq), it slowly dissolves as the H$^+$(aq) ions convert the zinc metal into Zn^{2+}(aq) cations.

Acids are present in many foods, such as lemons and limes, and are used in household products, such as bathroom cleaner and Lime-Away. In this section, we discuss how to name them; in Chapter 15 you will learn more about their properties. We can categorize acids into two types: binary acids and oxyacids.

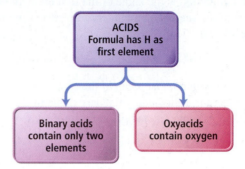

Naming Binary Acids

Binary acids are composed of hydrogen and a nonmetal. The names for binary acids have the form

For example, HCl(aq) is hydro*chlor*ic acid and HBr(aq) is hydro*brom*ic acid.

    HCl(aq)   hydrochloric acid   HBr(aq)   hydrobromic acid

EXAMPLE 3.9 Naming Binary Acids

Determine the name of HI(aq).

SOLUTION

The base name of I is *iod* so the name is hydroiodic acid.

        HI(aq)   hydroiodic acid

FOR PRACTICE 3.9

Determine the name of HF(aq).

Naming Oxyacids

Oxyacids contain hydrogen and an oxyanion (an anion containing a nonmetal and oxygen). The common oxyanions are listed in the table of polyatomic ions (Table 3.5). For example, HNO$_3$(aq) contains the nitrate (NO$_3{}^-$) ion, H$_2$SO$_3$(aq) contains the sulfite

(SO_3^{2-}) ion, and H_2SO_4(*aq*) contains the sulfate (SO_4^{2-}) ion. Notice that these acids are simply a combination of one or more H^+ ions with an oxyanion. The number of H^+ ions depends on the charge of the oxyanion so that the formula is always charge neutral. The names of oxyacids depend on the ending of the oxyanion and have the following forms:

oxyanions ending with *-ate*

base name of oxyanion + *-ic* | **acid**

oxyanions ending with *-ate*

oxyanions ending with *-ite*

base name of oxyanion + *-ous* | **acid**

oxyanions ending with *-ite*

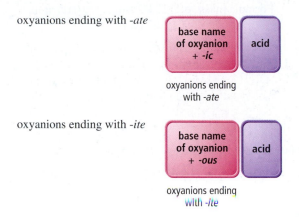

So HNO_3(*aq*) is nitric acid (oxyanion is nitrate), and H_2SO_3(*aq*) is sulfurous acid (oxyanion is sulfite).

HNO_3(*aq*) nitric acid $\qquad$ H_2SO_3(*aq*) sulfurous acid

EXAMPLE 3.10 Naming Oxyacids

Determine the name of $HC_2H_3O_2$(*aq*).

SOLUTION

The oxyanion is acetate, which ends in *-ate*; therefore, the name of the acid is *acetic acid.*

$HC_2H_3O_2$(*aq*) acetic acid

FOR PRACTICE 3.10

Determine the name of HNO_2(*aq*).

FOR MORE PRACTICE 3.10

Write the formula for perchloric acid.

3.7 Formula Mass and the Mole Concept for Compounds

In Chapter 2, we defined the average mass of an atom of an element as the *atomic mass* for that element. Similarly, we now define the average mass of a molecule (or a formula unit) of a compound as the **formula mass** for that compound. The terms *molecular mass* or *molecular weight* have the same meaning as formula mass. For any compound, the formula mass is the sum of the atomic masses of all the atoms in its chemical formula.

Formula mass = (Number of atoms of 1st element in chemical formula × Atomic mass of 1st element) + (Number of atoms of 2nd element in chemical formula × Atomic mass of 2nd element) + …

For example, the formula mass of carbon dioxide, CO_2, is

$$\text{Formula mass} = 12.01 \text{ amu} + 2(16.00 \text{ amu})$$
$$= 44.01 \text{ amu}$$

and that of sodium oxide, Na_2O, is

$$\text{Formula mass} = 2(22.99 \text{ amu}) + 16.00 \text{ amu}$$
$$= 61.98 \text{ amu}$$

EXAMPLE 3.11 Calculating Formula Mass

Calculate the formula mass of glucose, $C_6H_{12}O_6$.

SOLUTION

To find the formula mass, we sum the atomic masses of each atom in the chemical formula:

$$\text{Formula mass} = 6 \times (\text{atomic mass C}) + 12 \times (\text{atomic mass H}) + 6 \times (\text{atomic mass O})$$

$$= 6(12.01 \text{ amu}) \qquad + 12(1.008 \text{ amu}) \qquad + 6(16.00 \text{ amu})$$

$$= 180.16 \text{ amu}$$

FOR PRACTICE 3.11
Calculate the formula mass of calcium nitrate.

Molar Mass of a Compound

Remember, ionic compounds do not contain individual molecules. In casual language, the smallest electrically neutral collection of ions is sometimes called a molecule but is more correctly called a formula unit.

In Chapter 2 (Section 2.8), we saw that an element's molar mass—the mass in grams of one mole of its atoms—is numerically equivalent to its atomic mass. We then used the molar mass in combination with Avogadro's number to determine the number of atoms in a given mass of the element. The same concept applies to compounds. The *molar mass of a compound*—the mass in grams of 1 mol of its molecules or formula units—is numerically equivalent to its formula mass. For example, we just calculated the formula mass of CO_2 to be 44.01 amu. The molar mass is, therefore,

$$CO_2 \text{ molar mass} = 44.01 \text{ g/mol}$$

Using Molar Mass to Count Molecules by Weighing

The molar mass of CO_2 provides us with a conversion factor between mass (in grams) and amount (in moles) of CO_2. Suppose we want to find the number of CO_2 molecules in a sample of dry ice (solid CO_2) with a mass of 10.8 g. This calculation is analogous to Example 2.8, where we found the number of atoms in a sample of copper of a given mass. We begin with the mass of 10.8 g and use the molar mass to convert to the amount in moles. Then we use Avogadro's number to convert to number of molecules. The conceptual plan is as follows:

Conceptual Plan

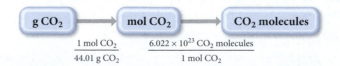

To solve the problem, we follow the conceptual plan, beginning with 10.8 g CO_2, converting to moles, and then converting to molecules.

Solution

$$10.8 \text{ g } CO_2 \times \frac{1 \text{ mol } CO_2}{44.01 \text{ g } CO_2} \times \frac{6.022 \times 10^{23} \text{ CO}_2 \text{ molecules}}{1 \text{ mol } CO_2}$$

$$= 1.48 \times 10^{23} \text{ CO}_2 \text{ molecules}$$

EXAMPLE 3.12 The Mole Concept—Converting between Mass and Number of Molecules

An aspirin tablet contains 325 mg of acetylsalicylic acid ($C_9H_8O_4$). How many acetylsalicylic acid molecules does it contain?

SORT You are given the mass of acetylsalicylic acid and asked to find the number of molecules.	**GIVEN** 325 mg $C_9H_8O_4$ **FIND** number of $C_9H_8O_4$ molecules
STRATEGIZE Convert between mass and the number of molecules of a compound by first converting to moles (using the molar mass of the compound) and then to the number of molecules (using Avogadro's number). You need both the molar mass of acetylsalicylic acid and Avogadro's number as conversion factors. You also need the conversion factor between g and mg.	**CONCEPTUAL PLAN** 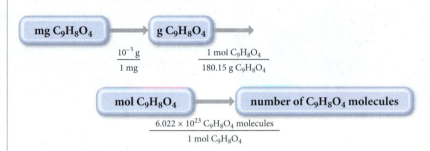 **RELATIONSHIPS USED** $1 \text{ mg} = 10^{-3} \text{ g}$ $C_9H_8O_4$ molar mass $= 9(12.01) + 8(1.008) + 4(16.00)$ $\qquad\qquad = 180.15 \text{ g/mol}$ $6.022 \times 10^{23} = 1 \text{ mol}$
SOLVE Follow the conceptual plan to solve the problem.	**SOLUTION** $325 \text{ mg } C_9H_8O_4 \times \dfrac{10^{-3} \text{ g}}{1 \text{ mg}} \times \dfrac{1 \text{ mol } C_9H_8O_4}{180.15 \text{ g } C_9H_8O_4} \times$ $\dfrac{6.022 \times 10^{23} \text{ C}_9H_8O_4 \text{ molecules}}{1 \text{ mol } C_9H_8O_4} = 1.09 \times 10^{21} \text{ C}_9H_8O_4 \text{ molecules}$

CHECK The units of the answer, $C_9H_8O_4$ molecules, are correct. The magnitude seems appropriate because it is smaller than Avogadro's number, as expected, since we have less than one mole of acetylsalicylic acid.

FOR PRACTICE 3.12

Find the number of ibuprofen molecules in a tablet containing 200.0 mg of ibuprofen ($C_{13}H_{18}O_2$).

FOR MORE PRACTICE 3.12

What is the mass of a drop of water containing 3.55×10^{22} H_2O molecules?

Throughout this book, we use space-filling molecular models to represent molecules. Which number is a good estimate for the scaling factor used in these models? For example, by approximately what number would you have to multiply the radius of an actual oxygen atom to get the radius of the sphere used to represent the oxygen atom in the water molecule shown in the margin?

(a) 10   **(b)** 10^4 **(c)** 10^8 **(d)** 10^{16}

3.8   Composition of Compounds

A chemical formula, in combination with the molar masses of its constituent elements, indicates the relative quantities of each element in a compound, which is extremely useful information. For example, about 30 years ago, scientists began to suspect that synthetic compounds known as chlorofluorocarbons (or CFCs) were destroying ozone (O_3) in Earth's upper atmosphere.

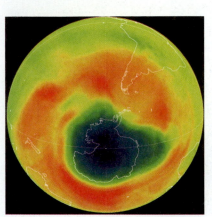

▲ The ozone hole over Antarctica is caused by the chlorine in chlorofluorocarbons. The dark blue color indicates depressed ozone levels.

Upper atmospheric ozone is important because it shields life on Earth from the sun's harmful ultraviolet light. CFCs are chemically inert compounds that were used primarily as refrigerants and industrial solvents. Over time, however, CFCs began to accumulate in the atmosphere. In the upper atmosphere sunlight breaks bonds within CFCs, resulting in the release of chlorine atoms. The chlorine atoms then react with ozone, converting it into O_2. Therefore, the harmful part of CFCs is the chlorine atoms that they carry. How can we determine the mass of chlorine in a given mass of a CFC?

One way to express how much of an element is in a given compound is to use the element's mass percent composition for that compound. The **mass percent composition** or more simply **mass percent** of an element is that element's percentage of the compound's total mass. We can calculate the mass percent of element X in a compound from the chemical formula as follows:

$$\text{mass percent of element X} = \frac{\text{mass of element X in 1 mol of compound}}{\text{mass of 1 mol of the compound}} \times 100\%$$

Suppose, for example, that we want to calculate the mass percent composition of Cl in the chlorofluorocarbon CCl_2F_2. The mass percent Cl is

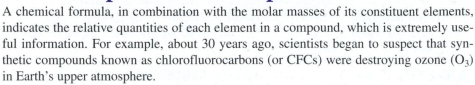

$$\text{Mass percent Cl} = \frac{2 \times \text{Molar mass Cl}}{\text{Molar mass } CCl_2F_2} \times 100\%$$

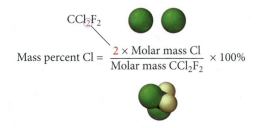

We must multiply the molar mass of Cl by two because the chemical formula has a subscript of 2 for Cl, indicating that 1 mol of CCl_2F_2 contains 2 mol of Cl atoms. The molar mass of CCl_2F_2 is computed as follows:

$$\text{Molar mass} = 12.01 \text{ g/mol} + 2(35.45 \text{ g/mol}) + 2(19.00 \text{ g/mol})$$

$$= 120.91 \text{ g/mol}$$

So the mass percent of Cl in CCl_2F_2 is

$$\begin{aligned}
\text{Mass percent Cl} &= \frac{2 \times \text{molar mass Cl}}{\text{molar mass } CCl_2F_2} \times 100\% \\
&= \frac{2 \times 35.45 \text{ g/mol}}{120.91 \text{ g/mol}} \times 100\% \\
&= 58.64\%
\end{aligned}$$

EXAMPLE 3.13 Mass Percent Composition

Calculate the mass percent of Cl in Freon-112 ($C_2Cl_4F_2$), a CFC refrigerant.

SORT You are given the molecular formula of Freon-112 and asked to find the mass percent of Cl.	**GIVEN** $C_2Cl_4F_2$ **FIND** mass percent Cl
STRATEGIZE The molecular formula indicates that there are 4 mol of Cl in each mole of Freon-112. Find the mass percent composition from the chemical formula by using the equation that defines mass percent. The conceptual plan shows how you can use the mass of Cl in 1 mol of $C_2Cl_4F_2$ and the molar mass of $C_2Cl_4F_2$ to determine the mass percent of Cl.	**CONCEPTUAL PLAN** Mass % Cl $= \dfrac{4 \times \text{molar mass Cl}}{\text{molar mass } C_2Cl_4F_2} \times 100\%$ **RELATIONSHIPS USED** Mass percent of element X $= \dfrac{\text{mass of element X in 1 mol of compound}}{\text{mass of 1 mol of compound}} \times 100\%$
SOLVE Calculate the necessary parts of the equation and substitute the values into the equation to find mass percent Cl.	**SOLUTION** $4 \times \text{molar mass Cl} = 4(35.45 \text{ g/mol}) = 141.8 \text{ g/mol}$ Molar mass $C_2Cl_4F_2 = 2(12.01 \text{ g/mol}) + 4(35.45 \text{ g/mol}) + 2(19.00 \text{ g/mol})$ $= 24.02 \text{ g/mol} + 141.8 \text{ g/mol} + 38.00 \text{ g/mol} = 203.8 \text{ g/mol}$ Mass % Cl $= \dfrac{4 \times \text{molar mass Cl}}{\text{molar mass } C_2Cl_4F_2} \times 100\%$ $= \dfrac{141.8 \text{ g/mol}}{203.8 \text{ g/mol}} \times 100\%$ $= 69.58\%$

CHECK The units of the answer (%) are correct and the magnitude is reasonable because (a) it is between 0 and 100% and (b) chlorine is the heaviest atom in the molecule and there are four of them.

FOR PRACTICE 3.13

Acetic acid ($HC_2H_3O_2$) is the active ingredient in vinegar. Calculate the mass percent composition of oxygen in acetic acid.

FOR MORE PRACTICE 3.13

Calculate the mass percent composition of sodium in sodium oxide.

 Conceptual Connection 3.7 Mass Percent Composition

In For Practice 3.13 you calculated the mass percent of oxygen in acetic acid ($HC_2H_3O_2$). Without doing any calculations, predict whether the mass percent of *carbon* in acetic acid would be greater or smaller. Explain.

 Conceptual Connection 3.8 Chemical Formula and Mass Percent Composition

Without doing any calculations, order the elements in this compound in order of decreasing mass percent composition.

$$C_6H_6O$$

Conversion Factors from Chemical Formulas

Mass percent composition is one way to understand how much chlorine is in a particular chlorofluorocarbon or, more generally, how much of a constituent element is present in a given mass of any compound. However, we can also approach this question another way. Chemical formulas indicate the inherent relationships between atoms (or moles of atoms) and molecules (or moles of molecules). For example, the formula for CCl_2F_2 tells us that 1 mol of CCl_2F_2 contains 2 mol of Cl atoms. We write the ratio as follows:

$$1 \text{ mol } CCl_2F_2 : 2 \text{ mol Cl}$$

With ratios such as these—that come from the chemical formula—we can directly determine the amounts of the constituent elements present in a given amount of a compound without having to calculate mass percent composition. For example, we calculate the number of moles of Cl in 38.5 mol of CCl_2F_2 as follows:

Conceptual Plan

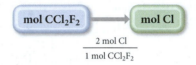

Solution

$$38.5 \text{ mol } CCl_2F_2 \times \frac{2 \text{ mol Cl}}{1 \text{ mol } CCl_2F_2} = 77.0 \text{ mol Cl}$$

We often want to know, however, not the *amount in moles* of an element in a certain number of moles of compound, but the *mass in grams* (or other units) of a constituent element in a given *mass* of the compound. For example, suppose we want to know the mass (in grams) of Cl contained in 25.0 g CCl_2F_2. *The relationship inherent in the chemical formula (2 mol Cl:1 mol CCl_2F_2) applies to amount in moles, not to mass.* Therefore, we must first convert the mass of CCl_2F_2 to moles CCl_2F_2. Then we use the conversion factor from the chemical formula to convert to moles Cl. Finally, we use the molar mass of Cl to convert to grams Cl. The calculation proceeds as follows:

Conceptual Plan

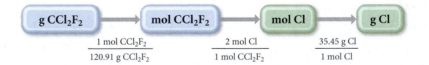

Solution

$$25.0 \text{ g } CCl_2F_2 \times \frac{1 \text{ mol } CCl_2F_2}{120.91 \text{ g } CCl_2F_2} \times \frac{2 \text{ mol Cl}}{1 \text{ mol } CCl_2F_2} \times \frac{35.45 \text{ g Cl}}{1 \text{ mol Cl}} = 14.7 \text{ g Cl}$$

Notice that we must convert from g CCl_2F_2 to mol CCl_2F_2 *before* we can use the chemical formula as a conversion factor.

The general form for solving problems where you are asked to find the mass of an element present in a given mass of a compound is

$$\text{Mass compound} \rightarrow \text{moles compound} \rightarrow \text{moles element} \rightarrow \text{mass element}$$

We use the molar mass to convert between mass and moles, and we use relationships inherent in the chemical to convert between moles and moles.

EXAMPLE 3.14 Chemical Formulas as Conversion Factors

Hydrogen may potentially be used in the future as a fuel to replace gasoline. Most major automobile companies are developing vehicles that run on hydrogen. These cars are environmentally friendly because their only emission is water vapor. One way to obtain hydrogen for fuel is to use an emission-free energy source such as wind power to split hydrogen from water. What mass of hydrogen (in grams) does 1.00 gallon of water contain? (The density of water is 1.00 g/mL.)

SORT You are given a volume of water and asked to find the mass of hydrogen it contains. You are also given the density of water.

GIVEN 1.00 gal H_2O
$d_{H_2O} = 1.00$ g/mL
FIND g H

STRATEGIZE The first part of the conceptual plan shows how you can convert the units of volume from gallons to liters and then to mL. It also shows how you can use the density to convert mL to g.
The second part of the conceptual plan is the basic sequence of mass → moles → moles → mass. Convert between moles and mass using the appropriate molar masses, and convert from mol H_2O to mol H using the conversion factor derived from the molecular formula.

CONCEPTUAL PLAN

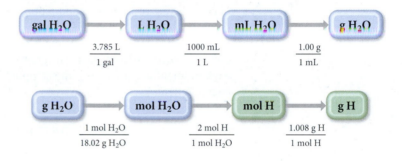

RELATIONSHIPS USED

3.785 L = 1 gal (Table 1.3)

1000 mL = 1 L

1.00 g H_2O = 1 mL H_2O (density of H_2O)

Molar mass H_2O = 2(1.008) + 16.00 = 18.02 g/mol

2 mol H : 1 mol H_2O

1.008 g H = 1 mol H

SOLVE Follow the conceptual plan to solve the problem.

SOLUTION

$$1.00 \, \cancel{gal \, H_2O} \times \frac{3.785 \, \cancel{L}}{1 \, \cancel{gal}} \times \frac{1000 \, \cancel{mL}}{1 \, \cancel{L}} \times \frac{1.0 \, g}{\cancel{mL}} = 3.785 \times 10^3 \, g \, H_2O$$

$$3.785 \times 10^3 \, \cancel{g \, H_2O} \times \frac{1 \, \cancel{mol \, H_2O}}{18.02 \, \cancel{g \, H_2O}} \times \frac{2 \, \cancel{mol \, H}}{1 \, \cancel{mol \, H_2O}} \times \frac{1.008 \, g \, H}{1 \, \cancel{mol \, H}} = 4.23 \times 10^2 \, g \, H$$

CHECK The units of the answer (g H) are correct. Since a gallon of water is about 3.8 L, its mass is about 3.8 kg. H is a light atom, so its mass should be significantly less than 3.8 kg.

FOR PRACTICE 3.14
Determine the mass of oxygen in a 7.2 g sample of $Al_2(SO_4)_3$.

FOR MORE PRACTICE 3.14
Butane (C_4H_{10}) is a liquid fuel used in lighters. How many grams of carbon are in a lighter containing 7.25 mL of butane? (The density of liquid butane is 0.601 g/mL.)

3.9   Determining a Chemical Formula from Experimental Data

In Section 3.8, we calculated mass percent composition from a chemical formula. But can we also do the reverse? Can we calculate a chemical formula from mass percent composition? This question is important because laboratory analyses of compounds do not often give chemical formulas directly, but only the relative masses of each element present in a compound. For example, if we decompose water into hydrogen and oxygen in the laboratory, we can measure the masses of hydrogen and oxygen produced. Can we get a chemical formula from this kind of data? The answer is a qualified yes. We can get a chemical formula, but it is an empirical formula (not a molecular formula). To get a molecular formula, we need additional information, such as the molar mass of the compound.

Suppose we decompose a sample of water in the laboratory and find that it produces 0.857 g of hydrogen and 6.86 g of oxygen. How do we determine an empirical formula from these data? We know that an empirical formula represents a ratio of atoms or moles of atoms, but not a ratio of masses. So the first thing we must do is convert our data from mass (in grams) to amount (in moles). How many moles of each element are present in the sample? To convert to moles, use the molar mass of each element:

$$\text{Moles H} = 0.857 \; \cancel{\text{g H}} \times \frac{1 \; \text{mol H}}{1.008 \; \cancel{\text{g H}}} = 0.850 \; \text{mol H}$$

$$\text{Moles O} = 6.86 \; \cancel{\text{g O}} \times \frac{1 \; \text{mol O}}{16.00 \; \cancel{\text{g O}}} = 0.429 \; \text{mol O}$$

From these data, we know there are 0.850 mol H for every 0.429 mol O. We can now write a pseudoformula for water:

$$H_{0.850}O_{0.429}$$

To get the smallest whole-number subscripts in our formula, we divide all the subscripts by the smallest one, in this case 0.429:

$$H_{\frac{0.850}{0.429}} O_{\frac{0.429}{0.429}} = H_{1.98}O = H_2O$$

Our empirical formula for water, which also happens to be the molecular formula, is H_2O. Use the following procedure to obtain the empirical formula of any compound from experimental data giving the relative masses of the constituent elements. The left column outlines the procedure, and the examples in the center and right columns show how to apply the procedure.

PROCEDURE FOR...	EXAMPLE 3.15	EXAMPLE 3.16
Obtaining an Empirical Formula from Experimental Data	**Obtaining an Empirical Formula from Experimental Data**	**Obtaining an Empirical Formula from Experimental Data**
	A compound containing nitrogen and oxygen is decomposed in the laboratory and produces 24.5 g nitrogen and 70.0 g oxygen. Calculate the empirical formula of the compound.	A laboratory analysis of aspirin determined the following mass percent composition: C 60.00% H 4.48% O 35.52% Find the empirical formula.

1. Write down (or compute) as *given* the masses of each element present in a sample of the compound. If you are given mass percent composition, assume a 100-g sample and calculate the masses of each element from the given percentages.	**GIVEN** 24.5 g N, 70.0 g O **FIND** empirical formula	**GIVEN** In a 100-g sample: 60.00 g C, 4.48 g H, 35.52 g O **FIND** empirical formula
2. Convert each of the masses in step 1 to moles by using the appropriate molar mass for each element as a conversion factor.	$24.5 \ \text{g N} \times \dfrac{1 \ \text{mol N}}{14.01 \ \text{g N}} = 1.75 \ \text{mol N}$ $70.0 \ \text{g O} \times \dfrac{1 \ \text{mol O}}{16.00 \ \text{g O}} = 4.38 \ \text{mol O}$	$60.00 \ \text{g C} \times \dfrac{1 \ \text{mol C}}{12.01 \ \text{g C}} = 4.996 \ \text{mol C}$ $4.48 \ \text{g H} \times \dfrac{1 \ \text{mol H}}{1.008 \ \text{g H}} = 4.44 \ \text{mol H}$ $35.52 \ \text{g O} \times \dfrac{1 \ \text{mol O}}{16.00 \ \text{g O}} = 2.220 \ \text{mol O}$
3. Write down a pseudoformula for the compound using the number of moles of each element (from step 2) as subscripts.	$N_{1.75}O_{4.38}$	$C_{4.996}H_{4.44}O_{2.220}$
4. Divide all the subscripts in the formula by the smallest subscript.	$N_{\frac{1.75}{1.75}}O_{\frac{4.38}{1.75}} \rightarrow N_1O_{2.5}$	$C_{\frac{4.996}{2.220}}H_{\frac{4.44}{2.220}}O_{\frac{2.220}{2.220}} \rightarrow C_{2.25}H_2O_1$
5. If the subscripts are not whole numbers, multiply all the subscripts by a small whole number (see table) to get whole-number subscripts.	$N_1O_{2.5} \times 2 \rightarrow N_2O_5$ The correct empirical formula is N_2O_5.	$C_{2.25}H_2O_1 \times 4 \rightarrow C_9H_8O_4$ The correct empirical formula is $C_9H_8O_4$.

Fractional Subscript	Multiply by This
0.20	5
0.25	4
0.33	3
0.40	5
0.50	2
0.66	3
0.75	4
0.80	5

FOR PRACTICE 3.15

A sample of a compound is decomposed in the laboratory and produces 165 g carbon, 27.8 g hydrogen, and 220.2 g oxygen. Calculate the empirical formula of the compound.

FOR PRACTICE 3.16

Ibuprofen, an aspirin substitute, has the following mass percent composition:

C 75.69%, H 8.80%, O 15.51%.

What is the empirical formula of ibuprofen?

Calculating Molecular Formulas for Compounds

We can find the molecular formula of a compound from the empirical formula if we also know the molar mass of the compound. Recall from Section 3.3 that the molecular formula is always a whole-number multiple of the empirical formula:

$$\text{Molecular formula} = \text{empirical formula} \times n, \text{ where } n = 1, 2, 3, \ldots$$

Suppose we want to find the molecular formula for fructose (a sugar found in fruit) from its empirical formula, CH_2O, and its molar mass, 180.2 g/mol. We know that the molecular formula is a whole-number multiple of CH_2O:

$$\text{Molecular formula} = (CH_2O) \times n$$
$$= C_nH_{2n}O_n$$

We also know that the molar mass is a whole-number multiple of the **empirical formula molar mass**, the sum of the masses of all the atoms in the empirical formula.

$$\text{Molar mass} = \text{empirical formula molar mass} \times n$$

For a particular compound, the value of n in both cases is the same. Therefore, we can find n by calculating the ratio of the molar mass to the empirical formula molar mass:

$$n = \frac{\text{molar mass}}{\text{empirical formula molar mass}}$$

For fructose, the empirical formula molar mass is:

empirical formula molar mass

$$= 12.01 \text{ g/mol} + 2(1.01 \text{ g/mol}) + 16.00 \text{ g/mol} = 30.03 \text{ g/mol}$$

Therefore, n is

$$n = \frac{180.2 \text{ g/mol}}{30.03 \text{ g/mol}} = 6$$

We can use this value of n to find the molecular formula:

$$\text{Molecular formula} = (CH_2O) \times 6 = C_6H_{12}O_6$$

EXAMPLE 3.17 Calculating a Molecular Formula from an Empirical Formula and Molar Mass

Butanedione—a main component in the smell and taste of butter and cheese—contains the elements carbon, hydrogen, and oxygen. The empirical formula of butanedione is C_2H_3O, and its molar mass is 86.09 g/mol. Find its molecular formula.

SORT You are given the empirical formula and molar mass of butanedione and asked to find the molecular formula.	**GIVEN** Empirical formula = C_2H_3O molar mass = 86.09 g/mol **FIND** molecular formula
STRATEGIZE A molecular formula is always a whole-number multiple of the empirical formula. Divide the molar mass by the empirical formula molar mass to get the whole number.	Molecular formula = empirical formula $\times n$ $n = \dfrac{\text{molar mass}}{\text{empirical formula molar mass}}$
SOLVE Calculate the empirical formula molar mass.	Empirical formula molar mass $= 2(12.01 \text{ g/mol}) + 3(1.008 \text{ g/mol})$ $+16.00 \text{ g/mol} = 43.04 \text{ g/mol}$
Divide the molar mass by the empirical formula molar mass to find n.	$n = \dfrac{\text{molar mass}}{\text{empirical formula molar mass}}$ $= \dfrac{86.09 \text{ g/mol}}{43.04 \text{ g/mol}} = 2$
Multiply the empirical formula by n to obtain the molecular formula.	Molecular formula = $C_2H_3O \times 2$ $= C_4H_6O_2$

CHECK Check the answer by calculating the molar mass of the formula as follows:
$4(12.01 \text{ g/mol}) + 6(1.008 \text{ g/mol}) + 2(16.00 \text{ g/mol}) = 86.09 \text{ g/mol}$
The calculated molar mass is in agreement with the given molar mass. The answer is correct.

FOR PRACTICE 3.17

A compound has the empirical formula CH and a molar mass of 78.11 g/mol. Find its molecular formula.

FOR MORE PRACTICE 3.17

A compound with the percent composition shown below has a molar mass of 60.10 g/mol. Find its molecular formula.
C, 39.97%
H, 13.41%
N, 46.62%

Combustion Analysis

In the previous section, you learned how to calculate the empirical formula of a compound from the relative masses of its constituent elements. Another common (and related) way to obtain empirical formulas for unknown compounds, especially those containing carbon and hydrogen, is **combustion analysis**. In combustion analysis, the unknown compound undergoes combustion (or burning) in the presence of pure oxygen, as shown in **Figure 3.8▼**. All of the carbon in the sample is converted to CO_2, and all of the hydrogen is converted to H_2O. The CO_2 and H_2O produced are weighed, and the numerical relationships between moles inherent in the formulas for CO_2 and H_2O (1 mol CO_2 : 1 mol C and 1 mol H_2O : 2 mol H) are used to determine the amounts of C and H in the original sample. Any other elemental constituents, such as O, Cl, or N, can be determined by subtracting the original mass of the sample from the sum of the masses of C and H. The examples that follow show how to perform these calculations for a sample containing only C and H and for a sample containing C, H, and O.

Combustion is a type of *chemical reaction*. We discuss chemical reactions and how to represent them more thoroughly in Section 3.10.

Combustion Analysis

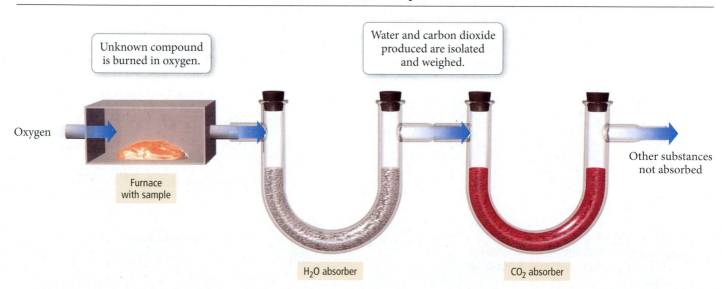

Unknown compound is burned in oxygen.

Water and carbon dioxide produced are isolated and weighed.

Oxygen

Furnace with sample

H_2O absorber

CO_2 absorber

Other substances not absorbed

▲ **FIGURE 3.8 Combustion Analysis Apparatus** The sample to be analyzed is placed in a furnace and burned in oxygen. The water and carbon dioxide produced are absorbed into separate containers and weighed.

PROCEDURE FOR... Determining an Empirical Formula from Combustion Analysis	**EXAMPLE 3.18** Determining an Empirical Formula from Combustion Analysis	**EXAMPLE 3.19** Determining an Empirical Formula from Combustion Analysis
	Upon combustion, a compound containing only carbon and hydrogen produced 1.83 g CO_2 and 0.901 g H_2O. Find the empirical formula of the compound.	Upon combustion, a 0.8233 g sample of a compound containing only carbon, hydrogen, and oxygen produced 2.445 g CO_2 and 0.6003 g H_2O. Find the empirical formula of the compound.
1. Write down as *given* the masses of each combustion product and the mass of the sample (if given).	**GIVEN** 1.83 g CO_2, 0.901 g H_2O **FIND** empirical formula	**GIVEN** 0.8233 g sample, 2.445 g CO_2, 0.6003 g H_2O **FIND** empirical formula
2. Convert the masses of CO_2 and H_2O from step 1 to moles by using the appropriate molar mass for each compound as a conversion factor.	$1.83 \text{ g } CO_2 \times \dfrac{1 \text{ mol } CO_2}{44.01 \text{ g } CO_2}$ $= 0.0416 \text{ mol } CO_2$ $0.901 \text{ g } H_2O \times \dfrac{1 \text{ mol } H_2O}{18.02 \text{ g } H_2O}$ $= 0.0500 \text{ mol } H_2O$	$2.445 \text{ g } CO_2 \times \dfrac{1 \text{ mol } CO_2}{44.01 \text{ g } CO_2}$ $= 0.05556 \text{ mol } CO_2$ $0.6003 \text{ g } H_2O \times \dfrac{1 \text{ mol } H_2O}{18.01 \text{ g } H_2O}$ $= 0.03331 \text{ mol } H_2O$
3. Convert the moles of CO_2 and moles of H_2O from step 2 to moles of C and moles of H using the conversion factors inherent in the chemical formulas of CO_2 and H_2O.	$0.0416 \text{ mol } CO_2 \times \dfrac{1 \text{ mol C}}{1 \text{ mol } CO_2}$ $= 0.0416 \text{ mol C}$ $0.0500 \text{ mol } H_2O \times \dfrac{2 \text{ mol H}}{1 \text{ mol } H_2O}$ $= 0.100 \text{ mol H}$	$0.05556 \text{ mol } CO_2 \times \dfrac{1 \text{ mol C}}{1 \text{ mol } CO_2}$ $= 0.05556 \text{ mol C}$ $0.03331 \text{ mol } H_2O \times \dfrac{2 \text{ mol H}}{1 \text{ mol } H_2O}$ $= 0.06662 \text{ mol H}$
4. If the compound contains an element other than C and H, find the mass of the other element by subtracting the sum of the masses of C and H (obtained in step 3) from the mass of the sample. Finally, convert the mass of the other element to moles.	This sample contains no elements besides C and H, so proceed to next step.	$\text{Mass C} = 0.05556 \text{ mol C} \times \dfrac{12.01 \text{ g C}}{\text{mol C}}$ $= 0.6673 \text{ g C}$ $\text{Mass H} = 0.06662 \text{ mol H} \times \dfrac{1.008 \text{ g H}}{\text{mol H}}$ $= 0.06715 \text{ g H}$ $\text{Mass O} = 0.8233 \text{ g } -$ $(0.6673 \text{ g} + 0.06715 \text{ g}) = 0.0889 \text{ g}$ $\text{Mol O} = 0.0889 \text{ g O} \times \dfrac{\text{mol O}}{16.00 \text{ g O}}$ $= 0.00556 \text{ mol O}$
5. Write down a pseudoformula for the compound using the number of moles of each element (from steps 3 and 4) as subscripts.	$C_{0.0416}H_{0.100}$	$C_{0.05556}H_{0.06662}O_{0.00556}$
6. Divide all the subscripts in the formula by the smallest subscript. (Round all subscripts that are within 0.1 of a whole number.)	$C_{\frac{0.0416}{0.0416}} H_{\frac{0.100}{0.0416}} \rightarrow C_1H_{2.4}$	$C_{\frac{0.05556}{0.00556}} H_{\frac{0.06662}{0.00556}} O_{\frac{0.00556}{0.00556}} \rightarrow C_{10}H_{12}O_1$

| 7. If the subscripts are not whole numbers, multiply all the subscripts by a small whole number to get whole-number subscripts. | $C_1H_{2.4} \times 5 \rightarrow C_5H_{12}$
The correct empirical formula is C_5H_{12}. | The subscripts are whole numbers; no additional multiplication is needed. The correct empirical formula is $C_{10}H_{12}O$. |

FOR PRACTICE 3.18

Upon combustion, a compound containing only carbon and hydrogen produced 1.60 g CO_2 and 0.819 g H_2O. Find the empirical formula of the compound.

FOR PRACTICE 3.19

Upon combustion, a 0.8009 g sample of a compound containing only carbon, hydrogen, and oxygen produced 1.6004 g CO_2 and 0.6551 g H_2O. Find the empirical formula of the compound.

3.10 Writing and Balancing Chemical Equations

The method of combustion analysis in Section 3.9 involves a **chemical reaction**, a process in which one or more substances are converted into one or more different substances. Compounds form and change through chemical reactions. As we have seen, water can be produced by the reaction of hydrogen with oxygen. A **combustion reaction** is a particular type of chemical reaction in which a substance combines with oxygen to form one or more oxygen-containing compounds. Combustion reactions also emit heat, which provides the energy that our society depends on. The heat produced in the combustion of gasoline, for example, helps expand the gaseous combustion products in a car engine's cylinders, which push the pistons and propel the car. We use the heat released by the combustion of *natural gas* to cook food and to heat our homes.

We represent a chemical reaction with a **chemical equation**. For example, we represent the combustion of natural gas by the equation

$$CH_4 + O_2 \rightarrow CO_2 + H_2O$$
reactants   products

The substances on the left side of the equation are the **reactants**, and the substances on the right side are the **products**. We often specify the states of each reactant or product in parentheses next to the formula as follows:

$$CH_4(g) + O_2(g) \rightarrow CO_2(g) + H_2O(g)$$

The (g) indicates that these substances are gases in the reaction. The common states of reactants and products and their symbols used in chemical equations are summarized in Table 3.6.

If you look more closely at our equation for the combustion of natural gas, you should immediately notice a problem.

$$CH_4(g) + O_2(g) \longrightarrow CO_2(g) + H_2O(g)$$

2 O atoms   **2 O atoms + 1 O atom = 3 O atoms**

The left side of the equation has two oxygen atoms while the right side has three. The reaction as written violates the law of conservation of mass because an oxygen atom formed out of nothing. Notice also that there are four hydrogen atoms on the left and only two on the right.

$$CH_4(g) + O_2(g) \longrightarrow CO_2(g) + H_2O(g)$$

4 H atoms   **2 H atoms**

Two hydrogen atoms have vanished, again violating mass conservation. To correct these problems—that is, to write an equation that more closely represents *what actually happens*—we must **balance** the equation. We must change the coefficients (the numbers *in front of* the chemical formulas), not the subscripts (the numbers *within* the chemical formulas), to ensure that the number of each type of atom on the left side of the equation is equal to the number on the right side. New atoms do not form during a reaction, nor do atoms vanish—matter is always conserved.

The reason that you cannot change the subscripts when balancing a chemical equation is that changing the subscripts changes the substance itself, while changing the coefficients simply changes the number of molecules of the substance. For example, 2 H_2O is two water molecules, but H_2O_2 is hydrogen peroxide, a drastically different compound.

TABLE 3.6 States of Reactants and Products in Chemical Equations

Abbreviation	State
(g)	Gas
(l)	Liquid
(s)	Solid
(aq)	Aqueous (water solution)

When we add coefficients to the reactants and products to balance an equation, we change the number of molecules in the equation but not the *kind of* molecules. To balance the equation for the combustion of methane, we put the coefficient 2 before O_2 in the reactants, and the coefficient 2 before H_2O in the products.

$$CH_4(g) \ + \ 2\,O_2(g) \ \longrightarrow \ CO_2(g) \ + \ 2\,H_2O(g)$$

The equation is now balanced; the numbers of each type of atom on either side of the equation are equal. The balanced equation tells us that one CH_4 molecule reacts with 2 O_2 molecules to form 1 CO_2 molecule and 2 H_2O molecules. We can verify that the equation is balanced by summing the number of each type of atom on each side of the equation.

$$CH_4(g) + 2\,O_2(g) \ \rightarrow \ CO_2(g) + 2\,H_2O(g)$$

Reactants	Products
1 C atom (1 × $\underline{C}H_4$)	1 C atom (1 × $\underline{C}O_2$)
4 H atoms (1 × $C\underline{H}_4$)	4 H atoms (2 × $\underline{H}_2O$)
4 O atoms (2 × $\underline{O}_2$)	4 O atoms (1 × $C\underline{O}_2$ + 2 × $H_2\underline{O}$)

The number of each type of atom on both sides of the equation is now equal—the equation is balanced.

Writing Balanced Chemical Equations

We balance many chemical equations simply by trial and error. However, some guidelines can be useful. For example, balancing the atoms in the most complex substances first and the atoms in the simplest substances (such as pure elements) last often makes the process shorter. The following illustrations of how to balance chemical equations are presented in a three-column format. The general guidelines are shown on the left, with two examples of how to apply them on the right. This procedure is meant only as a flexible guide, not a rigid set of steps.

PROCEDURE FOR... Balancing Chemical Equations	**EXAMPLE 3.20** **Balancing Chemical Equations**	**EXAMPLE 3.21** **Balancing Chemical Equations**
	Write a balanced equation for the reaction between solid cobalt(III) oxide and solid carbon to produce solid cobalt and carbon dioxide gas.	Write a balanced equation for the combustion of gaseous butane (C_4H_{10}), a fuel used in portable stoves and grills, in which it combines with gaseous oxygen to form gaseous carbon dioxide and gaseous water.
1. Write an unbalanced equation by writing chemical formulas for each of the reactants and products. Review Sections 3.5 and 3.6 for nomenclature rules. (If an unbalanced equation is provided, go to step 2.)	$Co_2O_3(s) + C(s) \rightarrow Co(s) + CO_2(g)$	$C_4H_{10}(g) + O_2(g) \rightarrow CO_2(g) + H_2O(g)$

2. Balance atoms that occur in more complex substances first. Always balance atoms in compounds before atoms in pure elements.	**Begin with O:** $$Co_2O_3(s) + C(s) \rightarrow Co(s) + CO_2(g)$$ 3 O atoms → 2 O atoms To balance O, put a 2 before $Co_2O_3(s)$ and a 3 before $CO_2(g)$. $$2\,Co_2O_3(s) + C(s) \rightarrow$$ $$Co(s) + 3\,CO_2(g)$$ 6 O atoms → 6 O atoms	**Begin with C:** $$C_4H_{10}(g) + O_2(g) \rightarrow CO_2(g) + H_2O(g)$$ 4 C atoms → 1 C atom To balance C, put a 4 before $CO_2(g)$. $$C_4H_{10}(g) + O_2(g) \rightarrow 4\,CO_2(g) + H_2O(g)$$ 4 C atoms → 4 C atoms **Balance H:** $$C_4H_{10}(g) + O_2(g) \rightarrow 4\,CO_2(g) + H_2O(g)$$ 10 H atoms → 2 H atoms To balance H, put a 5 before $H_2O(g)$: $$C_4H_{10}(g) + O_2(g) \rightarrow 4\,CO_2(g) + 5\,H_2O(g)$$ 10 H atoms → 10 H atoms
3. Balance atoms that occur as free elements on either side of the equation last. Always balance free elements by adjusting their coefficients.	**Balance Co:** $$2\,Co_2O_3(s) + C(s) \rightarrow$$ $$Co(s) + 3\,CO_2(g)$$ 4 Co atoms → 1 Co atom To balance Co, put a 4 before $Co(s)$. $$2\,Co_2O_3(s) + C(s) \rightarrow$$ $$4\,Co(s) + 3\,CO_2(g)$$ 4 Co atoms → 4 Co atoms **Balance C:** $$2\,Co_2O_3(s) + C(s) \rightarrow$$ $$4\,Co(s) + 3\,CO_2(g)$$ 1 C atoms → 3 C atoms To balance C, put a 3 before C(s). $$2\,Co_2O_3(s) + 3\,C(s) \rightarrow$$ $$4\,Co(s) + 3\,CO_2(g)$$	**Balance O:** $$C_4H_{10}(g) + O_2(g) \rightarrow 4\,CO_2(g) + 5H_2O(g)$$ 2 O atoms → 8 O + 5 O = 13 O atoms To balance O, put a 13/2 before $O_2(g)$: $$C_4H_{10}(g) + 13/2\,O_2(g) \rightarrow 4\,CO_2(g) + 5\,H_2O(g)$$ 13 O atoms → 13 O atoms
4. If the balanced equation contains coefficient fractions, clear these by multiplying the entire equation by the denominator of the fraction.	This step is not necessary in this example. Proceed to step 5.	$$\left[C_4H_{10}(g) + 13/2\,O_2(g) \rightarrow 4\,CO_2(g) + 5\,H_2O(g) \right] \times 2$$ $$2\,C_4H_{10}(g) + 13\,O_2(g) \rightarrow 8\,CO_2(g) + 10\,H_2O(g)$$
5. Check to make certain the equation is balanced by summing the total number of each type of atom on both sides of the equation.	$$2\,Co_2O_3(s) + 3\,C(s) \rightarrow 4\,Co(s) + 3\,CO_2(g)$$ <table><tr><td>Left</td><td>Right</td></tr><tr><td>4 Co atoms</td><td>4 Co atoms</td></tr><tr><td>6 O atoms</td><td>6 O atoms</td></tr><tr><td>3 C atoms</td><td>3 C atoms</td></tr></table> The equation is balanced.	$$2\,C_4H_{10}(g) + 13\,O_2(g) \rightarrow 8\,CO_2(g) + 10\,H_2O(g)$$ <table><tr><td>Left</td><td>Right</td></tr><tr><td>8 C atoms</td><td>8 C atoms</td></tr><tr><td>20 H atoms</td><td>20 H atoms</td></tr><tr><td>26 O atoms</td><td>26 O atoms</td></tr></table> The equation is balanced.
	FOR PRACTICE 3.20 Write a balanced equation for the reaction between solid silicon dioxide and solid carbon that produces solid silicon monocarbide and carbon monoxide gas.	**FOR PRACTICE 3.21** Write a balanced equation for the combustion of gaseous ethane (C_2H_6), a minority component of natural gas, in which it combines with gaseous oxygen to form gaseous carbon dioxide and gaseous water.

Conceptual Connection 3.9   Balanced Chemical Equations

Which quantities must always be the same on both sides of a chemical equation?

(a) the number of atoms of each type   **(b)** the number of molecules of each type

(c) the number of moles of each type of molecule   **(d)** the sum of the masses of all substances involved

PROCEDURE FOR...	**EXAMPLE 3.22**
Balancing Chemical Equations Containing Ionic Compounds with Polyatomic Ions	**Balancing Chemical Equations Containing Ionic Compounds with Polyatomic Ions**
	Write a balanced equation for the reaction between aqueous strontium chloride and aqueous lithium phosphate to form solid strontium phosphate and aqueous lithium chloride.
1. Write an unbalanced equation by writing chemical formulas for each of the reactants and products. Review Sections 3.5 and 3.6 for naming rules. (If an unbalanced equation is provided, go to step 2.)	$SrCl_2(aq) + Li_3PO_4(aq) \rightarrow Sr_3(PO_4)_2(s) + LiCl(aq)$
2. Balance metal ions (cations) first. If a polyatomic cation exists on both sides of the equation, balance it as a unit.	**Begin with Sr^{2+}:** $SrCl_2(aq) + Li_3PO_4(aq) \rightarrow Sr_3(PO_4)_2(s) + LiCl(aq)$ 1 Sr^{2+} ion $\rightarrow$ 3 Sr^{2+} ions To balance Sr^{2+}, put a 3 before $SrCl_2(aq)$. 3 $SrCl_2(aq) + Li_3PO_4(aq) \rightarrow Sr_3(PO_4)_2(s) + LiCl(aq)$ 3 Sr^{2+} ion $\rightarrow$ 3 Sr^{2+} ions **Balance Li^+:** 3 $SrCl_2(aq) + Li_3PO_4(aq) \rightarrow Sr_3(PO_4)_2(s) + LiCl(aq)$ 3 Li^+ ions $\rightarrow$ 1 Li^+ ion To balance Li+, put a 3 before $LiCl(aq)$. 3 $SrCl_2(aq) + Li_3PO_4(aq) \rightarrow Sr_3(PO_4)_2(s) + 3 LiCl(aq)$ 3 Li^+ ions $\rightarrow$ 3 Li^+ ions
3. Balance nonmetal ions (anions) second. If a polyatomic anion exists on both sides of the equation, balance it as a unit.	**Balance PO_4^{3-}:** 3 $SrCl_2(aq) + Li_3PO_4(aq) \rightarrow Sr_3(PO_4)_2(s) + 3 LiCl(aq)$ 1 PO_4^{3-} ion $\rightarrow$ 2 PO_4^{3-} ions To balance PO_4^{3-}, put a 2 before $Li_3PO_4(aq)$. 3 $SrCl_2(aq) + 2 Li_3PO_4(aq) \rightarrow Sr_3(PO_4)_2(s) + 3 LiCl(aq)$ 2 PO_4^{3-} ion $\rightarrow$ 2 PO_4^{3-} ions **Balance Cl^-:** 3 $SrCl_2(aq) + 2 Li_3PO_4(aq) \rightarrow Sr_3(PO_4)_2(s) + 3 LiCl(aq)$ 6 Cl^- ions $\rightarrow$ 3 Cl^- ion To balance Cl^-, replace the 3 before $LiCl(aq)$ with a 6. This also corrects the balance for Li^+, which was thrown off in the previous step. 3 $SrCl_2(aq) + 2 Li_3PO_4(aq) \rightarrow Sr_3(PO_4)_2(s) + 6 LiCl(aq)$ 6 Cl^- atoms $\rightarrow$ 6 Cl^- atoms
4. Check to make certain the equation is balanced by summing the total number of each type of ion on both sides of the equation.	3 $SrCl_2(aq) + 2 Li_3PO_4(aq) \rightarrow Sr_3(PO_4)_2(s) + 6 LiCl(aq)$ Left Right 3 Sr^{2+} ions 3 Sr^{2+} ions 6 Li^+ ions 6 Li^+ ions 2 PO_4^{3-} ions 2 PO_4^{3-} ions 6 Cl^- ions 6 Cl^- ions The equation is balanced.
	FOR PRACTICE 3.22 Write a balanced equation for the reaction between aqueous lead(II) nitrate and aqueous potassium chloride to form solid lead(II) chloride and aqueous potassium nitrate.

Organic compounds are discussed in more detail in Chapter 20.

3.11 Organic Compounds

Early chemists divided compounds into two types: organic and inorganic. Organic compounds came from living things. Sugar—obtained from sugarcane or the sugar beet—is a common example of an organic compound. Inorganic compounds, on the other hand, came from the Earth. Salt—mined from the ground or from the ocean—is a common example of an inorganic compound.

Eighteenth-century chemists could synthesize inorganic compounds in the laboratory, but not organic compounds, so a clear division existed between the two different types of compounds. Today, chemists can synthesize both organic and inorganic compounds, and even though organic chemistry is a subfield of chemistry, the differences between organic and inorganic compounds are primarily organizational (not fundamental).

Organic compounds are composed of carbon and hydrogen and a few other elements, including nitrogen, oxygen, and sulfur. The key element to organic chemistry, however, is carbon. Carbon always forms four bonds in compounds. For example, the simplest organic compound is methane, CH_4. The chemistry of carbon is unique and complex because carbon frequently bonds to itself to form chain, branched, and ring structures. This versatility allows carbon to be the backbone of millions of different chemical compounds, which is why even a survey of organic chemistry requires a year-long course. For now, all you really need to know is that the simplest organic compounds are called **hydrocarbons** and that they are composed of only carbon and hydrogen. Hydrocarbons compose common fuels such as oil, gasoline, liquid propane gas, and natural gas. Table 3.7 lists some common hydrocarbons and their names.

Structural formula | Space-filling model

Methane, CH_4

TABLE 3.7 Common Hydrocarbons

Name	Molecular Formula	Structural Formula	Space-Filling Model	Common Uses
Methane	CH_4			Primary component of natural gas
Propane	C_3H_8			LP gas for grills and outdoor stoves
n-Butane*	C_4H_{10}			Common fuel for lighters
n-Pentane*	C_5H_{12}			Component of gasoline
Ethene	C_2H_4			Ripening agent in fruit
Ethyne	C_2H_2	$H-C\equiv C-H$		Fuel for welding torches

*The "n" in the names of these hydrocarbons stands for normal, which indicates a straight chain.

CHAPTER IN REVIEW

Key Terms

Section 3.2
ionic bond (75)
covalent bond (75)

Section 3.3
chemical formula (75)
empirical formula (75)
molecular formula (75)
structural formula (76)
ball-and-stick model (77)
space-filling molecular
  model (77)

Section 3.4
atomic element (77)

molecular element (77)
molecular compound (79)
ionic compound (79)
formula unit (79)
polyatomic ion (80)

Section 3.5
common name (82)
systematic name (82)
binary compound (83)
oxyanion (86)
hydrate (87)

Section 3.6
acid (89)

binary acid (90)
oxyacid (90)

Section 3.7
formula mass (91)

Section 3.8
mass percent composition
  (mass percent) (94)

Section 3.9
empirical formula molar
  mass (100)
combustion analysis (101)

Section 3.10
chemical reaction (103)
combustion reaction (103)
chemical equation (103)
reactants (103)
products (103)
balanced chemical
  equation (103)

Section 3.11
organic compound (107)
hydrocarbon (107)

Key Concepts

Chemical Bonds (3.2)

▶ Chemical bonds, the forces that hold atoms together in compounds, arise from the interactions between nuclei and electrons in atoms.

▶ In an ionic bond, one or more electrons have been *transferred* from one atom to another, forming a cation (positively charged) and an anion (negatively charged). The two ions are drawn together by the attraction between the opposite charges.

▶ In a covalent bond, one or more electrons are *shared* between two atoms. The atoms are held together by the attraction between their nuclei and the shared electrons.

Representing Molecules and Compounds (3.3, 3.4)

▶ A compound is represented with a chemical formula, which indicates the elements present and the number of atoms of each.

▶ An empirical formula gives only the *relative* number of atoms, while a molecular formula gives the *actual* number present in the molecule.

▶ Structural formulas show how the atoms are bonded together, while molecular models show the geometry of the molecule.

▶ Compounds can be divided into two types: molecular compounds, formed between two or more covalently bonded nonmetals; and ionic compounds, usually formed between a metal ionically bonded to one or more nonmetals. The smallest identifiable unit of a molecular compound is a molecule, and the smallest identifiable unit of an ionic compound is a formula unit: the smallest electrically neutral collection of ions.

▶ Elements can also be divided into two types: molecular elements—which occur as (mostly diatomic) molecules—and atomic elements—which occur as individual atoms.

Naming Inorganic Ionic and Molecular Compounds and Acids (3.5, 3.6)

▶ A flowchart for naming simple inorganic compounds is provided at the end of this section. Use this chart to name inorganic compounds.

Formula Mass and the Mole Concept for Compounds (3.7)

▶ The formula mass of a compound is the sum of the atomic masses of all the atoms in the chemical formula. Like the atomic masses of elements, the formula mass characterizes the average mass of a molecule (or a formula unit).

▶ The mass of one mole of a compound (in grams) is the molar mass and is numerically equal to its formula mass (in amu).

Chemical Composition (3.8, 3.9)

▶ The mass percent composition of a compound is each element's percentage of the total compound's mass. The mass percent composition can be determined from the compound's chemical formula and the molar masses of its elements.

▶ The chemical formula of a compound provides the relative number of atoms (or moles) of each element in a compound, and can therefore be used to determine numerical relationships between moles of the compound and moles of its constituent elements.

▶ If the mass percent composition and molar mass of a compound are known, its empirical and molecular formulas can be determined.

Writing and Balancing Chemical Equations (3.10)

▶ In chemistry, we represent chemical reactions with chemical equations. The substances on the left side of a chemical equation are the reactants and the substances on the right side are the products.

▶ Chemical equations are balanced when the number of each type of atom on the left side of the equation is equal to the number on the right side.

Organic Compounds (3.11)

▶ Organic compounds—originally derived only from living organisms but now readily synthesized in the laboratory—are composed of carbon, hydrogen, and a few other elements such as nitrogen, oxygen, and sulfur.

▶ The simplest organic compounds are hydrocarbons, composed of only carbon and hydrogen.

Key Equations and Relationships

Formula Mass (3.7)

$$\left(\begin{array}{c}\text{\# atoms of 1st element} \\ \text{in chemical formula}\end{array} \times \begin{array}{c}\text{atomic mass} \\ \text{of 1st element}\end{array}\right) + \left(\begin{array}{c}\text{\# atoms of 2nd element} \\ \text{in chemical formula}\end{array} \times \begin{array}{c}\text{atomic mass} \\ \text{of 2nd element}\end{array}\right) + \; \cdots$$

Mass Percent Composition (3.8)

$$\text{Mass \% of element X} = \frac{\text{mass of X in 1 mol compound}}{\text{mass of 1 mol compound}} \times 100\%$$

Empirical Formula Molar Mass (3.9)

$$\text{Molecular formula} = n \times (\text{empirical formula})$$

$$n = \frac{\text{molar mass}}{\text{empirical formula molar mass}}$$

Inorganic Nomenclature Summary Chart

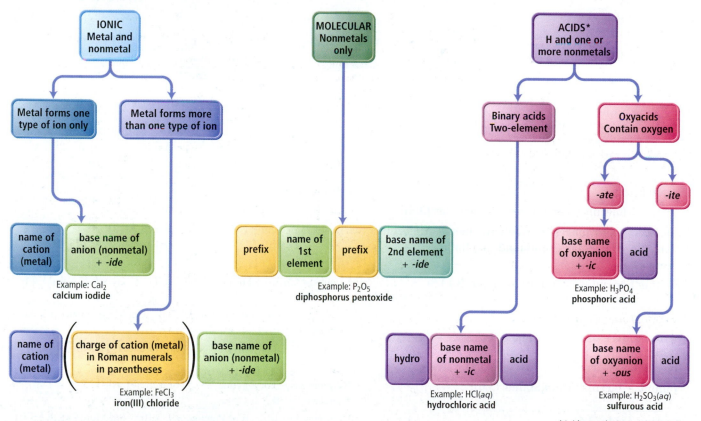

*Acids must be in aqueous solution.

Using the Flowchart

The examples below demonstrate how to name compounds using the flowchart. The path through the flowchart is shown below each compound and is followed by the correct name for the compound.

(a) $MgCl_2$

Magnesium chloride

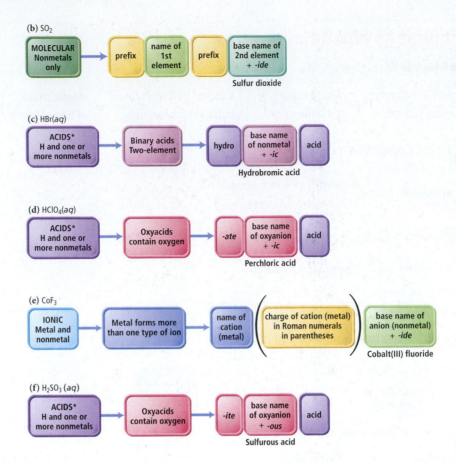

Key Learning Objectives

Chapter Objectives	Assessment
Writing Molecular and Empirical Formulas (3.3)	Example 3.1 For Practice 3.1 Exercises 1–4
Classifying Substances as Atomic Elements, Molecular Elements, Molecular Compounds, or Ionic Compounds (3.4)	Example 3.2 For Practice 3.2 Exercises 5–10
Writing Formulas for Ionic Compounds (3.5)	Examples 3.3, 3.4 For Practice 3.3, 3.4 Exercises 11–14, 23, 24
Naming Ionic Compounds (3.5)	Examples 3.5, 3.6 For Practice 3.5, 3.6 For More Practice 3.5, 3.6 Exercises 15–20
Naming Ionic Compounds Containing Polyatomic Ions (3.5)	Example 3.7 For Practice 3.7 For More Practice 3.7 Exercises 21–24
Naming Molecular Compounds (3.6)	Example 3.8 For Practice 3.8 For More Practice 3.8 Exercises 27–30
Naming Acids (3.6)	Examples 3.9, 3.10 For Practice 3.9, 3.10 For More Practice 3.10 Exercises 31–34
Calculating Formula Mass (3.7)	Example 3.11 For Practice 3.11 Exercises 35, 36
Using Formula Mass to Count Molecules by Weighing (3.7)	Example 3.12 For Practice 3.12 For More Practice 3.12 Exercises 39–44
Calculating Mass Percent Composition (3.8)	Example 3.13 For Practice 3.13 For More Practice 3.13 Exercises 45–50
Using Chemical Formulas as Conversion Factors (3.8)	Example 3.14 For Practice 3.14 For More Practice 3.14 Exercises 57, 58
Obtaining an Empirical Formula from Experimental Data (3.9)	Examples 3.15, 3.16 For Practice 3.15, 3.16 Exercises 59–64

Calculating a Molecular Formula from an Empirical Formula and Molar Mass (3.9)	Example 3.17 For Practice 3.17 For More Practice 3.17 Exercises 65–66
Determining an Empirical Formula from Combustion Analysis (3.9)	Examples 3.18, 3.19 For Practice 3.18, 3.19 Exercises 67–70
Balancing Chemical Equations (3.10)	Examples 3.20, 3.21, 3.22 For Practice 3.20, 3.21, 3.22 Exercises 71–82

EXERCISES

Problems by Topic

Note: Answers to all odd-numbered Problems, numbered in blue, can be found in Appendix III. Exercises in the Problems by Topic section are paired, with each odd-numbered problem followed by a similar even-numbered problem. Exercises in the Cumulative Problems section are also paired, but somewhat more loosely. (Challenge Problems and Conceptual Problems, because of their nature, are unpaired.)

Chemical Formulas and Molecular View of Elements and Compounds

1. Determine the number of each type of atom in each formula:
   a. $Ca_3(PO_4)_2$ **b.** $SrCl_2$ **c.** KNO_3 **d.** $Mg(NO_2)_2$

2. Determine the number of each type of atom in each formula:
   a. $Ba(OH)_2$ **b.** NH_4Cl **c.** $NaCN$ **d.** $Ba(HCO_3)_2$

3. Write a chemical formula for each molecular model. (See Appendix IIA for color codes.)

(a)   (b)   (c)

4. Write a chemical formula for each molecular model. (See Appendix IIA for color codes.)

(a)   (b)   (c)

5. Classify each element as atomic or molecular.
   a. neon   **b.** fluorine
   c. potassium   **d.** nitrogen

6. Identify the elements that have molecules as their basic units.
   a. hydrogen   **b.** iodine
   c. lead   **d.** oxygen

7. Classify each compound as ionic or molecular.
   a. CO_2 **b.** $NiCl_2$
   c. NaI **d.** PCl_3

8. Classify each compound as ionic or molecular.
   a. CF_2Cl_2 **b.** CCl_4
   c. PtO_2 **d.** SO_3

9. Based on the molecular views, classify each substance as an atomic element, a molecular element, an ionic compound, or a molecular compound.

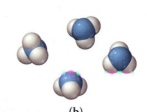

(a)   (b)

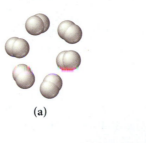

(c)

10. Based on the molecular views, classify each substance as an atomic element, a molecular element, an ionic compound, or a molecular compound.

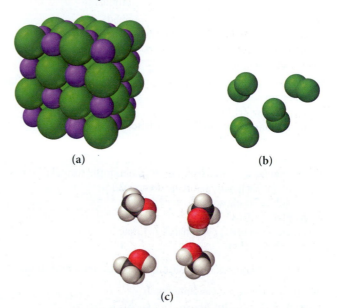

(a)   (b)

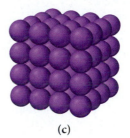

(c)

Formulas and Names for Ionic Compounds

11. Write a formula for the ionic compound that forms between each pair of elements.
    a. magnesium and sulfur   **b.** barium and oxygen
    c. strontium and bromine **d.** beryllium and chlorine

12. Write a formula for the ionic compound that forms between each pair of elements.
    a. aluminum and sulfur   b. aluminum and oxygen
    c. sodium and oxygen   d. strontium and iodine

13. Write a formula for the compound that forms between barium and each polyatomic ion:
    a. hydroxide   b. chromate   c. phosphate   d. cyanide

14. Write a formula for the compound that forms between sodium and each polyatomic ion:
    a. carbonate   b. phosphate
    c. hydrogen phosphate   d. acetate

15. Name each ionic compound.
    a. Mg_3N_2 b. KF   c. Na_2O d. Li_2S

16. Name each ionic compound.
    a. CsF   b. KI   c. $SrCl_2$ d. $BaCl_2$

17. Name each ionic compound.
    a. $SnCl_4$ b. PbI_2 c. Fe_2O_3 d. CuI_2

18. Name each ionic compound.
    a. SnO_2 b. $HgBr_2$ c. $CrCl_2$ d. $CrCl_3$

19. Name each ionic compound.
    a. SnO   b. Cr_2S_3 c. RbI   d. $BaBr_2$

20. Name each ionic compound.
    a. BaS   b. $FeCl_3$ c. $PbCl_4$ d. $SrBr_2$

21. Name each ionic compound.
    a. $CuNO_2$ b. $Mg(C_2H_3O_2)_2$
    c. $Ba(NO_3)_2$ d. $Pb(C_2H_3O_2)_2$
    e. $KClO_3$ f. $PbSO_4$

22. Name each ionic compound.
    a. $Ba(OH)_2$ b. NH_4I c. $NaBrO_4$
    d. $Fe(OH)_3$ e. $CoSO_4$ f. KClO

23. Write a formula for each ionic compound.
    a. sodium hydrogen sulfite   b. lithium permanganate
    c. silver nitrate   d. potassium sulfate
    e. rubidium hydrogen sulfate   f. potassium hydrogen carbonate

24. Write a formula for each ionic compound.
    a. copper(II) chloride   b. copper(I) iodate
    c. lead(II) chromate   d. calcium fluoride
    e. potassium hydroxide   f. iron(II) phosphate

25. Determine the name from the formula or the formula from the name for each hydrated ionic compound.
    a. $CoSO_4 \cdot 7H_2O$
    b. iridium(III) bromide tetrahydrate
    c. $Mg(BrO_3)_2 \cdot 6H_2O$
    d. potassium carbonate dihydrate

26. Determine the name from the formula or the formula from the name for each hydrated ionic compound.
    a. cobalt(II) phosphate octahydrate
    b. $BeCl_2 \cdot 2H_2O$
    c. chromium(III) phosphate trihydrate
    d. $LiNO_2 \cdot H_2O$

Formulas and Names for Molecular Compounds and Acids

27. Name each molecular compound.
    a. CO   b. NI_3 c. $SiCl_4$
    d. N_4Se_4 e. I_2O_5

28. Name each molecular compound.
    a. SO_3 b. SO_2 c. BrF_5
    d. NO   e. XeO_3

29. Write a formula for each molecular compound.
    a. phosphorus trichloride   b. chlorine monoxide

    c. disulfur tetrafluoride   d. phosphorus pentafluoride
    e. diphosphorus pentasulfide

30. Write a formula for each molecular compound.
    a. boron tribromide   b. dichlorine monoxide
    c. xenon tetrafluoride   d. carbon tetrabromide
    e. diboron tetrachloride

31. Name each acid.
    a. HI   b. HNO_3 c. H_2CO_3 d. $HC_2H_3O_2$

32. Name each acid.
    a. HCl   b. $HClO_2$ c. H_2SO_4 d. HNO_2

33. Write formulas for each acid.
    a. hydrofluoric acid
    b. hydrobromic acid
    c. sulfurous acid

34. Write formulas for each acid.
    a. phosphoric acid
    b. hydrocyanic acid
    c. chlorous acid

Formula Mass and the Mole Concept for Compounds

35. Calculate the formula mass for each compound.
    a. NO_2 b. C_4H_{10} c. $C_6H_{12}O_6$ d. $Cr(NO_3)_3$

36. Calculate the formula mass for each compound.
    a. $MgBr_2$ b. HNO_2 c. CBr_4 d. $Ca(NO_3)_2$

37. Calculate the number of moles in each sample.
    a. 72.5 g CCl_4 b. 12.4 g $C_{12}H_{22}O_{11}$
    c. 25.2 kg C_2H_2
    d. 12.3 g of dinitrogen monoxide

38. Calculate the mass of each sample.
    a. 15.7 mol HNO_3
    b. 1.04×10^{-3} mol H_2O_2
    c. 72.1 mmol SO_2
    d. 1.23 mol xenon difluoride

39. How many molecules are in each sample?
    a. 3.5 g H_2O b. 254 g CBr_4
    c. 18.3 g O_2 d. 26.9 g C_8H_{10}

40. Calculate the mass (in g) of each sample.
    a. 3.87×10^{21} SO_3 molecules
    b. 1.3×10^{24} H_2O molecules
    c. 2.55×10^{24} O_3 molecules
    d. 1.54×10^{21} CCl_2F_2 molecules

41. Calculate the mass (in g) of a single water molecule.

42. Calculate the mass (in g) of a single glucose molecule ($C_6H_{12}O_6$).

43. A sugar crystal contains approximately 1.8×10^{17} sucrose ($C_{12}H_{22}O_{11}$) molecules. How many moles are present in the sugar crystal? What is the mass of the crystal?

44. A salt crystal has a mass of 0.12 mg. How many moles of Na^+ are present in the crystal? How many Na^+ ions are present?

Composition of Compounds

45. Calculate the mass percent composition of carbon in each carbon compound.
    a. CH_4 b. C_2H_6 c. C_2H_2 d. C_2H_5Cl

46. Calculate the mass percent composition of nitrogen in each nitrogen compound.
    a. N_2O b. NO   c. NO_2 d. HNO_3

47. Most fertilizers consist of nitrogen-containing compounds. Examples include NH_3, $CO(NH_2)_2$, NH_4NO_3, and $(NH_4)_2SO_4$. The nitrogen content in these compounds is needed for protein synthesis in plants. Calculate the mass percent composition of

nitrogen in each of the fertilizers named above. Which fertilizer has the highest nitrogen content?

48. Iron is mined from the Earth as iron ore. Common ores include Fe_2O_3 (hematite), Fe_3O_4 (magnetite), and $FeCO_3$ (siderite). Calculate the mass percent composition of iron for each of these iron ores. Which ore has the highest iron content?

49. Copper(II) fluoride contains 37.42% F by mass. Use this percentage to calculate the mass of fluorine (in grams) contained in 72.4 g of copper(II) fluoride.

50. Silver chloride, often used in silver plating, contains 75.27% Ag. Calculate the mass of silver chloride required to plate 112 mg of pure silver.

51. The iodide ion is a dietary mineral essential to good nutrition. In countries where potassium iodide is added to salt, iodine deficiency or goiter has been almost completely eliminated. The recommended daily allowance (RDA) for iodine is 150 μg/day. How much potassium iodide (76.45% I) should be consumed to meet the RDA?

52. The American Dental Association recommends that an adult female should consume 3.0 mg of fluoride (F^-) per day to prevent tooth decay. If the fluoride is consumed as sodium fluoride (45.24% F), what amount of sodium fluoride contains the recommended amount of fluoride?

53. Write a ratio showing the relationship between the molar amounts of each element for each molecular image.

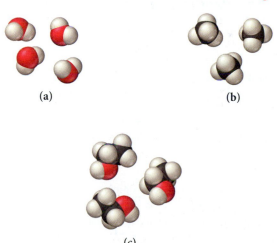

(a)   (b)

(c)

54. Write a ratio showing the relationship between the molar amounts of each element for each molecular image.

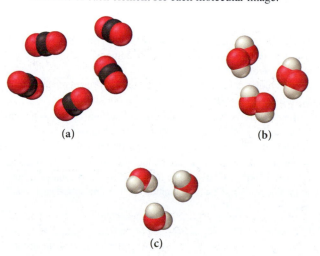

(a)   (b)

(c)

55. Determine the number of moles of hydrogen atoms in each sample.
   a. 0.0885 mol C_4H_{10} b. 1.3 mol CH_4
   c. 2.4 mol C_6H_{12} d. 1.87 mol C_8H_{18}

56. Determine the number of moles of oxygen atoms in each sample.
   a. 4.88 mol H_2O_2 b. 2.15 mol N_2O
   c. 0.0237 mol H_2CO_3 d. 24.1 mol CO_2

57. Calculate the number of grams of sodium in 8.5 g of each sodium-containing food additive.
   a. NaCl (table salt)
   b. Na_3PO_4 (sodium phosphate)
   c. $NaC_7H_5O_2$ (sodium benzoate)
   d. $Na_2C_6H_6O_7$ (sodium hydrogen citrate)

58. How many kilograms of chlorine are in 25 kg of each chlorofluorocarbon (CFC)?
   a. CF_2Cl_2 b. $CFCl_3$ c. $C_2F_3Cl_3$ d. CF_3Cl

Chemical Formulas from Experimental Data

59. Samples of several compounds are decomposed and the masses of their constituent elements are shown below. Calculate the empirical formula for each compound.
   a. 1.651 g Ag, 0.1224 g O
   b. 0.672 g Co, 0.569 g As, 0.486 g O
   c. 1.443 g Se, 5.841 g Br

60. Samples of several compounds are decomposed and the masses of their constituent elements are shown below. Calculate the empirical formula for each compound.
   a. 1.245 g Ni, 5.381 g I   b. 2.677 g Ba, 3.115 g Br
   c. 2.128 g Be, 7.557 g S, 15.107 g O

61. Calculate the empirical formula for each stimulant based on its elemental mass percent composition:
   a. nicotine (found in tobacco leaves): C 74.03%, H 8.70%, N 17.27%
   b. caffeine (found in coffee beans): C 49.48%, H 5.19%, N 28.85%, O 16.48%

62. Calculate the empirical formula for each natural flavor based on its elemental mass percent composition:
   a. methyl butyrate (component of apple taste and smell): C 58.80%, H 9.87%, O 31.33%
   b. vanillin (responsible for the taste and smell of vanilla): C 63.15%, H 5.30%, O 31.55%

63. A 0.77 mg sample of nitrogen reacts with chlorine to form 6.61 mg of the chloride. What is the empirical formula of the nitrogen chloride?

64. A 45.2 mg sample of phosphorus reacts with selenium to form 131.6 mg of the selenide. What is the empirical formula of the phosphorus selenide?

65. The empirical formula and molar mass of several compounds are listed below. Find the molecular formula of each compound.
   a. C_6H_7N, 186.24 g/mol   b. C_2HCl, 181.44 g/mol
   c. $C_5H_{10}NS_2$, 296.54 g/mol

66. The molar mass and empirical formula of several compounds are listed below. Find the molecular formula of each compound.
   a. C_4H_9, 114.22 g/mol   b. CCl, 284.77 g/mol
   c. C_3H_2N, 312.29 g/mol

67. Combustion analysis of a hydrocarbon produced 33.01 g CO_2 and 13.51 g H_2O. Calculate the empirical formula of the hydrocarbon.

68. Combustion analysis of naphthalene, a hydrocarbon used in mothballs, produced 8.80 g CO_2 and 1.44 g H_2O. Calculate the empirical formula for naphthalene.

69. The foul odor of rancid butter is due largely to butyric acid, a compound containing carbon, hydrogen, and oxygen. Combustion analysis of a 4.30 g sample of butyric acid produced 8.59 g CO_2 and 3.52 g H_2O. Find the empirical formula for butyric acid.

70. Tartaric acid is the white, powdery substance that coats sour candies such as Sour Patch Kids. Combustion analysis of a 12.01 g sample of tartaric acid—which contains only carbon, hydrogen, and oxygen—produced 14.08 g CO_2 and 4.32 g H_2O. Find the empirical formula for tartaric acid.

Writing and Balancing Chemical Equations

71. Sulfuric acid is a component of acid rain that is formed when gaseous sulfur dioxide pollutant reacts with gaseous oxygen and liquid water. Write a balanced chemical equation for this reaction. (Note: this is a simplified representation of this reaction.)

72. Nitric acid is a component of acid rain that forms when gaseous nitrogen dioxide pollutant reacts with gaseous oxygen and liquid water to form aqueous nitric acid. Write a balanced chemical equation for this reaction. (Note: this is a simplified representation of this reaction.)

73. In a popular classroom demonstration, solid sodium is added to liquid water and reacts to produce hydrogen gas and aqueous sodium hydroxide. Write a balanced chemical equation for this reaction.

74. When iron rusts, solid iron reacts with gaseous oxygen to form solid iron(III) oxide. Write a balanced chemical equation for this reaction.

75. Write a balanced chemical equation for the fermentation of sucrose ($C_{12}H_{22}O_{11}$) by yeasts in which the aqueous sugar reacts with water to form aqueous ethyl alcohol (C_2H_5OH) and carbon dioxide gas.

76. Write a balanced equation for the photosynthesis reaction in which gaseous carbon dioxide and liquid water react in the presence of chlorophyll to produce aqueous glucose ($C_6H_{12}O_6$) and oxygen gas.

77. Write a balanced chemical equation for each reaction.
   a. Solid lead(II) sulfide reacts with aqueous hydrobromic acid to form solid lead(II) bromide and dihydrogen monosulfide gas.

   b. Gaseous carbon monoxide reacts with hydrogen gas to form gaseous methane (CH_4) and liquid water.

   c. Aqueous hydrochloric acid reacts with solid manganese(IV) oxide to form aqueous manganese(II) chloride, liquid water, and chlorine gas.

   d. Liquid pentane (C_5H_{12}) reacts with gaseous oxygen to form carbon dioxide and liquid water.

78. Write a balanced chemical equation for each reaction.
   a. Solid copper reacts with solid sulfur to form solid copper(I) sulfide.

   b. Solid iron(III) oxide reacts with hydrogen gas to form solid iron and liquid water.

   c. Sulfur dioxide gas reacts with oxygen gas to form sulfur trioxide gas.

   d. Gaseous ammonia (NH_3) reacts with gaseous oxygen to form gaseous nitrogen monoxide and gaseous water.

79. Write a balanced chemical equation for the reaction of aqueous sodium carbonate with aqueous copper(II) chloride to form solid copper(II) carbonate and aqueous sodium chloride.

80. Write a balanced chemical equation for the reaction of aqueous potassium hydroxide with aqueous iron(III) chloride to form solid iron(III) hydroxide and aqueous potassium chloride.

81. Balance each chemical equation.
   a. $CO_2(g) + CaSiO_3(s) + H_2O(l) \rightarrow$
       $SiO_2(s) + Ca(HCO_3)_2(aq)$
   b. $Co(NO_3)_3(aq) + (NH_4)_2S(aq) \rightarrow Co_2S_3(s) + NH_4NO_3(aq)$
   c. $Cu_2O(s) + C(s) \rightarrow Cu(s) + CO(g)$
   d. $H_2(g) + Cl_2(g) \rightarrow HCl(g)$

82. Balance each chemical equation.
   a. $Na_2S(aq) + Cu(NO_3)_2(aq) \rightarrow NaNO_3(aq) + CuS(s)$
   b. $N_2H_4(l) \rightarrow NH_3(g) + N_2(g)$
   c. $HCl(aq) + O_2(g) \rightarrow H_2O(l) + Cl_2(g)$
   d. $FeS(s) + HCl(aq) \rightarrow FeCl_2(aq) + H_2S(g)$

Organic Compounds

83. Classify each compound as organic or inorganic:
   a. $CaCO_3$ **b.** C_4H_8 **c.** $C_4H_6O_6$ **d.** LiF

84. Classify each compound as organic or inorganic:
   a. C_8H_{18} **b.** CH_3NH_2 **c.** CaO   **d.** $FeCO_3$

Cumulative Problems

85. How many molecules of ethanol (C_2H_5OH) (the alcohol in alcoholic beverages) are present in 165 mL of ethanol? The density of ethanol is 0.789 g/cm^3.

86. A drop of water has a volume of approximately 0.05 mL. How many water molecules does it contain? The density of water is 1.0 g/cm^3.

87. Determine the chemical formula of each compound and then use it to calculate the mass percent composition of each constituent element.
   a. potassium chromate   **b.** lead(II) phosphate
   c. sulfurous acid   **d.** cobalt(II) bromide

88. Determine the chemical formula of each compound and then use it to calculate the mass percent composition of each constituent element.
   a. perchloric acid   **b.** phosphorus pentachloride
   c. nitrogen triiodide   **d.** carbon dioxide

89. A Freon leak in the air conditioning system of an older car releases 32 g of CF_2Cl_2 per month. What mass of chlorine is emitted into the atmosphere each year by this car?

90. A Freon leak in the air-conditioning system of a large building releases 17 kg of CHF_2Cl per month. If the leak continues, how many kilograms of Cl will be emitted into the atmosphere each year?

91. A metal (M) forms a compound with the formula MCl_3. If the compound contains 65.57% Cl by mass, what is the identity of the metal?

92. A metal (M) forms an oxide with the formula M_2O. If the oxide contains 16.99% O by mass, what is the identity of the metal?

93. Estradiol is a female sexual hormone that causes maturation and maintenance of the female reproductive system. Elemental analysis of estradiol gives this mass percent composition: C 79.37%, H 8.88%, O 11.75%. The molar mass of estradiol is 272.37 g/mol. Find the molecular formula of estradiol.

94. Fructose is a common sugar found in fruit. Elemental analysis of fructose gives this mass percent composition: C 40.00%, H 6.72%, O 53.28%. The molar mass of fructose is 180.16 g/mol. Find the molecular formula of fructose.

95. Combustion analysis of a 13.42 g sample of equilin (which contains only carbon, hydrogen, and oxygen) produces 39.61 g CO_2 and 9.01 g H_2O. The molar mass of equilin is 268.34 g/mol. Find the molecular formula for equilin.

96. Estrone, which contains only carbon, hydrogen, and oxygen, is a female sexual hormone that occurs in the urine of pregnant women. Combustion analysis of a 1.893 g sample of estrone produces 5.545 g of CO_2 and 1.388 g H_2O. The molar mass of estrone is 270.36 g/mol. Find the molecular formula for estrone.

97. Epsom salts is a hydrated ionic compound with the following formula: $MgSO_4 \cdot xH_2O$. A sample of Epsom salts with a mass of 4.93 g is heated to drive off the water of hydration. The mass of the sample after complete dehydration is 2.41 g. Find the number of waters of hydration (x) in Epsom salts.

98. A hydrate of copper(II) chloride has the following formula: $CuCl_2 \cdot xH_2O$. The water in a 3.41 g sample of the hydrate is driven off by heating. The remaining sample has a mass of 2.69 g. Find the number of waters of hydration (x) in the hydrate.

99. A compound of molar mass 177 g/mol contains only carbon, hydrogen, bromine, and oxygen. Analysis reveals that the compound contains eight times as much carbon as hydrogen by mass. Find the molecular formula.

100. The following data were obtained from experiments to find the molecular formula of benzocaine, a local anesthetic, which contains only carbon, hydrogen, nitrogen, and oxygen. Complete combustion of a 3.54-g sample of benzocaine with excess O_2 formed 8.49 g of CO_2 and 2.14 g H_2O. Another sample of mass 2.35 g was found to contain 0.199 g of N. The molar mass of benzocaine is 165 g/mol. Find the molar formula of benzocaine.

101. Find the total number of atoms in a sample of cocaine hydrochloride, $C_{17}H_{22}ClNO_4$, of mass 23.5 mg.

102. Vanadium forms four different oxides in which the percent by mass of vanadium is respectively 76%, 68%, 61%, and 56%. List the formula and the name of each of these oxides.

103. The chloride of an unknown metal is believed to have the formula MCl_3. A 2.395 g sample of the compound is found to contain 3.606×10^{-2} mol Cl. Find the atomic mass of M.

104. A chromium-containing compound has the formula $Fe_xCr_yO_4$ and is 28.59 % oxygen by mass. Find x and y.

105. A phosphorus compound that contains 34.00% phosphorus by mass has the formula X_3P_2. Identify the element X.

106. A particular brand of beef jerky contains 0.0552% sodium nitrite by mass and is sold in an 8.00 oz bag. What mass of sodium does the sodium nitrite contribute to the bag of beef jerky?

107. Phosphorus is obtained primarily from ores containing calcium phosphate. If a particular ore contains 57.8% calcium phosphate, what minimum mass of the ore must be processed to obtain 1.00 kg of phosphorus?

Challenge Problems

108. A mixture of NaCl and NaBr has a mass of 2.00 g and contains 0.75 g of Na. What is the mass of NaBr in the mixture?

109. Three pure compounds form when 1.00 g samples of element X combine with, respectively, 0.472 g, 0.630 g, and 0.789 g of element Z. The first compound has the formula X_2Z_3. Find the empirical formulas of the other two compounds.

110. A mixture of $CaCO_3$ and $(NH_4)_2CO_3$ is 61.9% CO_3 by mass. Find the mass percent of $CaCO_3$ in the mixture.

111. A mixture of 50.0 g of S and 1.00×10^2 g of Cl_2 reacts completely to form S_2Cl_2 and SCl_2. Find the mass of S_2Cl_2 formed.

112. Because of increasing evidence of damage to the ozone layer, chlorofluorocarbon (CFC) production was banned in 1996. However, there are about 100 million auto air conditioners that still use CFC-12 (CF_2Cl_2). These air conditioners are recharged from stockpiled supplies of CFC-12. If each of the 100 million automobiles contains 1.1 kg of CFC-12 and leaks 25% of its CFC-12 into the atmosphere per year, how much chlorine, in kg, enters the atmosphere each year due to auto air conditioners? (Assume two significant figures in your calculations.)

113. A particular coal contains 2.55% sulfur by mass. When the coal is burned, it produces SO_2 emissions which combine with rainwater to produce sulfuric acid. Use the formula of sulfuric acid to calculate the mass percent of S in sulfuric acid. Then determine how much sulfuric acid (in metric tons) is produced by the combustion of 1.0 metric ton of this coal. (A metric ton is 1000 kg.)

114. Lead is found in Earth's crust in several different lead ores. Suppose a certain rock is composed of 38.0% PbS (galena), 25.0% $PbCO_3$ (cerussite), and 17.4% $PbSO_4$ (anglesite). The remainder of the rock is composed of substances containing no lead. How much of this rock (in kg) must be processed to obtain 5.0 metric tons of lead? (A metric ton is 1000 kg.)

115. A 2.52 g sample of a compound containing only carbon, hydrogen, nitrogen, oxygen, and sulfur was burned in excess O_2 to yield 4.23 g of CO_2 and 1.01 g of H_2O. Another sample of the same compound, of mass 4.14 g, yielded 2.11 g of SO_3. A reaction to determine the nitrogen content was carried out on a third sample with a mass of 5.66 g and yielded

2.27 g HNO_3 (assume that all of the nitrogen in HNO_3 came from the compound). Calculate the empirical formula of the compound.

116. A compound of molar mass 229 g/mol contains only carbon, hydrogen, iodine, and sulfur. Analysis shows that a sample of the compound contains six times as much carbon as hydrogen, by mass. Calculate the molecular formula of the compound.

117. The elements X and Y form a compound that is 40% X and 60% Y by mass. The atomic mass of X is twice that of Y. What is the empirical formula of the compound?

118. A compound composed of elements X and Y is ⅓ X by mass. The atomic mass of element X is ¾ the atomic mass of element Y. Find the empirical formula of the compound.

Conceptual Problems

119. When molecules are represented by molecular models, what does each sphere represent? How big is the nucleus of an atom in comparison to the sphere used to represent an atom in a molecular model?

120. Without doing any calculations, determine which element in each compound will have the highest mass percent composition.

a. CO   **b.** N_2O
c. $C_6H_{12}O_6$ **d.** NH_3

121. Explain the problem with the following statement and correct it. "The chemical formula for ammonia (NH_3) indicates that ammonia contains three grams of hydrogen to each gram of nitrogen."

122. Explain the problem with the following statement and correct it. "When a chemical equation is balanced, the number of molecules of each type on both sides of the equation will be equal."

123. Without doing any calculations, arrange the elements in H_2SO_4 in order of decreasing mass percent composition.

124. Element A is an atomic element and element B is a diatomic molecular element. Using circles to represent atoms of A and squares to represent atoms of B, draw molecular-level views of each element.

Answers to Conceptual Connections

Structural Formulas

3.1 H—O—H

Representing Molecules

3.2 The spheres represent the electron cloud of the atom. It would be nearly impossible to draw a nucleus to scale on any of the space-filling molecular models—on this scale, the nucleus would be too small to see.

A Molecular View of Elements and Compounds

3.3

A_2B

CD

Ionic and Molecular Compounds

3.4 Choice (a) best describes the difference between ionic and molecular compounds. The (b) answer is incorrect because

there are no "new" forces in bonding (just rearrangements that result in lower potential energy), and because ions do not group together in pairs in the solid phase. The (c) answer is incorrect because the main difference between ionic and molecular compounds is the way that the atoms bond. The (d) answer is incorrect because ionic compounds do not contain molecules.

Nomenclature

3.5 This conceptual connection addresses one of the main errors you can make in nomenclature: the failure to correctly categorize the compound. Remember that you must first determine whether the compound is an ionic compound, a molecular compound, or an acid, and then name it accordingly. NCl_3 is a molecular compound (two or more nonmetals), and therefore prefixes indicate the number of each type of atom—so NCl_3 is nitrogen trichloride. The compound $AlCl_3$, however, is an ionic compound (metal and nonmetal), and therefore does not require prefixes—so $AlCl_3$ is aluminum chloride.

Molecular Models and the Size of Molecules

3.6 Answer: (c) Atomic radii range in the hundreds of picometers, while the spheres in these models have radii of less than a centimeter. The scaling factor is therefore about 10^8 (100 million).

Mass Percent Composition

3.7 The mass percent of carbon in acetic acid will be smaller than the mass percent of oxygen because even though the formula ($HC_2H_3O_2$) contains the same molar amounts of the two elements, carbon is lighter (it has a lower molar mass) than oxygen.

Chemical Formula and Mass Percent Composition

3.8 C > O > H. Since carbon and oxygen differ in atomic mass by only 4 amu, and since there are 6 carbon atoms in the formula, we can conclude that carbon constitutes the greatest fraction of the mass. Oxygen is next because its mass is 16 times that of hydrogen and there are only 6 hydrogen atoms to every 1 oxygen atom.

Balanced Chemical Equations

3.9 (a) and (d) are correct. When the number of atoms of each type is balanced, the sum of the masses of the substances involved will be the same on both sides of the equation. Since molecules change during a chemical reaction, their number is not the same on both sides, nor is the number of moles necessarily the same.

4 Chemical Quantities and Aqueous Reactions

I feel sorry for people who don't understand anything about chemistry. They are missing an important source of happiness. —Linus Pauling (1901–1994)

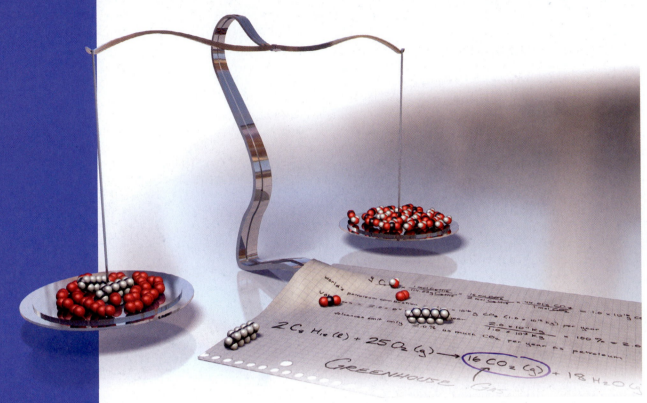

The molecular models on this balance represent the reactants and products in the combustion of octane, a component of petroleum. One of the products, carbon dioxide, is the main greenhouse gas implicated in climate change.

WHAT ARE THE RELATIONSHIPS between the amounts of reactants in a chemical reaction and the amounts of products that are formed? How do we best describe and understand these relationships? The first half of this chapter focuses on chemical stoichiometry—the numerical relationships between the amounts of reactants and products in chemical reactions. In Chapter 3, you learned how to write balanced chemical equations for chemical reactions. Now we examine more closely the meaning of those balanced equations. In the second half of this chapter, we turn to describing chemical reactions that occur in water. You have probably witnessed many of these types of reactions in your daily life because they are so common. Have you ever mixed baking soda with vinegar and observed the subsequent bubbling? Or have you ever noticed the hard water deposits that form on plumbing fixtures? These reactions—and many others, including those that occur within the watery environment of living cells—are aqueous chemical reactions, the subject of the second half of this chapter.

4.1 Climate Change and the Combustion of Fossil Fuels

The temperature outside my office today is a cool 48 °F, lower than normal for this time of year on the West Coast. However, today's "chill" pales in comparison with how cold it would be without the presence of *greenhouse gases* in the atmosphere. These gases act like the glass of a greenhouse, allowing sunlight to enter the atmosphere and warm Earth's surface, but preventing some of the heat generated by the sunlight from escaping, as shown in **Figure 4.1▼**. The balance between incoming and outgoing energy from the sun determines Earth's average temperature.

Without greenhouse gases in the atmosphere, more heat energy would escape, and Earth's average temperature would be about 60 °F colder than it is now. The temperature outside of my office today would be below 0 °F, and even the sunniest U.S. cities would most likely be covered with snow. However, if the concentration of greenhouse gases in the atmosphere were to increase, Earth's average temperature would rise.

In recent years scientists have become increasingly concerned because the amount of atmospheric carbon dioxide (CO_2)—Earth's most significant greenhouse gas in terms of climate change—is rising. More CO_2 enhances the atmosphere's ability to hold heat and may therefore lead to *global warming*, an increase in Earth's average temperature. Since 1860, atmospheric CO_2 levels have risen by 35% (**Figure 4.2▶**), and Earth's average temperature has risen by 0.6 °C (about 1.1 °F), as shown in **Figure 4.3▶**.

Most scientists now believe that the primary cause of rising atmospheric CO_2 concentration is the burning of fossil fuels (natural gas, petroleum, and coal), which provide 90% of our society's energy. Some people, however, have suggested that fossil fuel combustion does not significantly contribute to global warming and climate change. They argue, for example, that the amount of carbon dioxide emitted into the atmosphere by volcanic eruptions far exceeds that from fossil fuel combustion. Which group is right?

The Greenhouse Effect

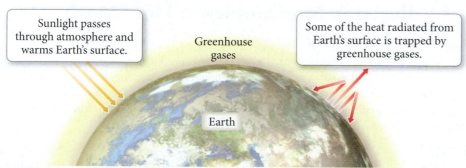

▲ FIGURE 4.1 The Greenhouse Effect Greenhouse gases in the atmosphere act as a one-way filter. They allow solar energy to pass through and warm Earth's surface, but they prevent heat from being radiated back out into space.

Atmospheric Carbon Dioxide

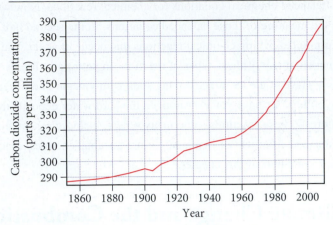

▲ **FIGURE 4.2 Carbon Dioxide Concentrations in the Atmosphere**   The rise in carbon dioxide levels shown in this graph is due largely to fossil fuel combustion.

Global Temperature

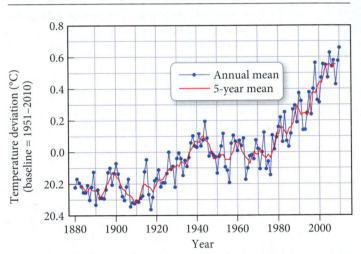

▲ **FIGURE 4.3 Global Temperature**   Average temperatures worldwide have risen by about 0.6 °C since 1880.

We could judge the validity of the naysayers' argument if we could calculate the amount of carbon dioxide emitted by fossil fuel combustion and compare it to that released by volcanic eruptions. As you will see in the next section of the chapter, at this point in your study of chemistry, you have enough knowledge to do just that.

4.2   Reaction Stoichiometry: How Much Carbon Dioxide?

The amount of carbon dioxide emitted by fossil fuel combustion is related to the amount of fossil fuel that is burned—the balanced chemical equations for the combustion reactions give the exact relationships between these amounts. In this discussion, we use octane (a component of gasoline) as a representative fossil fuel. The balanced equation for the combustion of octane is:

$$2 \, C_8H_{18}(l) + 25 \, O_2(g) \longrightarrow 16 \, CO_2(g) + 18 \, H_2O(g)$$

The balanced chemical equation shows that 16 CO_2 molecules are produced for every 2 molecules of octane burned. This numerical relationship between molecules can be extended to the amounts in moles as follows:

The coefficients in a chemical reaction specify the relative amounts in moles of each of the substances involved in the reaction.

In other words, from the equation, we know that 16 moles of CO_2 are produced for every 2 moles of octane burned. The numerical relationship between chemical amounts in a balanced chemical equation is called reaction **stoichiometry**. Stoichiometry allows us to predict the amounts of products that will form in a chemical reaction based on the amounts of reactants that undergo the reaction. Stoichiometry also allows us to determine the amount of reactants necessary to form a given amount of product. These calculations are central to chemistry, allowing chemists to plan and carry out chemical reactions to obtain products in the desired quantities.

> Stoichiometry is pronounced stoy-kee-AHM-e-tree.

Making Pizza: The Relationships among Ingredients

The concepts of stoichiometry are similar to those in a cooking recipe. Calculating the amount of carbon dioxide produced by the combustion of a given amount of a fossil fuel is analogous to calculating the number of pizzas that can be made from a given amount of cheese. For example, suppose we use the following pizza recipe:

$$1 \text{ crust } + 5 \text{ ounces tomato sauce } + 2 \text{ cups cheese } \longrightarrow 1 \text{ pizza}$$

The recipe shows the numerical relationships between the pizza ingredients. It says that if we have 2 cups of cheese—and enough of everything else—we can make 1 pizza. We can write this relationship as a ratio between the cheese and the pizza:

$$2 \text{ cups cheese : 1 pizza}$$

What if we have 6 cups of cheese? Assuming that we have enough of everything else, we can use the above ratio as a conversion factor to calculate the number of pizzas:

$$6 \text{ cups cheese} \times \frac{1 \text{ pizza}}{2 \text{ cups cheese}} = 3 \text{ pizzas}$$

Six cups of cheese are sufficient to make 3 pizzas. The pizza recipe contains numerical ratios between other ingredients as well, including the following:

$$1 \text{ crust : 1 pizza}$$
$$5 \text{ ounces tomato sauce : 1 pizza}$$

Making Molecules: Mole-to-Mole Conversions

In a balanced chemical equation, we have a "recipe" for how reactants combine to form products. From our balanced equation for the combustion of octane, for example, we can write the following stoichiometric ratio:

$$2 \text{ mol } C_8H_{18}(l) : 16 \text{ mol } CO_2(g)$$

We can use this ratio to determine how many moles of CO_2 are produced for a given number of moles of C_8H_{18} burned. Suppose that we burn 22.0 moles of C_8H_{18}; how many moles of CO_2 are produced? We can use the ratio from the balanced chemical equation in the same way that we used the ratio from the pizza recipe. This ratio acts as a conversion factor allowing us to convert from the amount in moles of the reactant (C_8H_{18}) to the amount in moles of the product (CO_2):

$$22.0 \; \text{mol} \; C_8H_{18} \times \frac{16 \; \text{mol} \; CO_2}{2 \; \text{mol} \; C_8H_{18}} = 176 \; \text{mol} \; CO_2$$

The combustion of 22 moles of C_8H_{18} adds 176 moles of CO_2 to the atmosphere.

Making Molecules: Mass-to-Mass Conversions

According to the U.S. Department of Energy, the world burned 3.1×10^{10} barrels of petroleum in 2010, the equivalent of approximately 3.5×10^{15} g of gasoline. Let's estimate the mass of CO_2 emitted into the atmosphere from burning this much gasoline by using the combustion of 3.5×10^{15} g octane as the representative reaction. This calculation is similar to the one we just did, except that we are given the *mass* of octane instead of the *amount* of octane in moles. Consequently, we must first convert the mass (in grams) to the amount (in moles). The general conceptual plan for calculations in which you are given the mass of a reactant or product in a chemical reaction and asked to find the mass of a different reactant or product is:

where A and B are two different substances involved in the reaction. We use the molar mass of A to convert from the mass of A to the amount of A (in moles). We use the appropriate ratio from the balanced chemical equation to convert from the amount of A (in moles) to the amount of B (in moles). And finally, we use the molar mass of B to convert from the amount of B (in moles) to the mass of B. To calculate the mass of CO_2 emitted upon the combustion of 3.5×10^{15} g of octane, therefore, we use the following conceptual plan:

Conceptual Plan

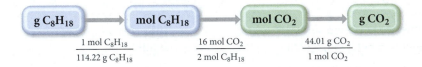

Relationships Used

    2 mol C_8H_{18} : 16 mol CO_2 (from the chemical equation)
    molar mass C_8H_{18} = 114.22 g/mol
    molar mass CO_2 = 44.01 g/mol

Solution

We then follow the conceptual plan to solve the problem, beginning with g C_8H_{18} and canceling units to arrive at g CO_2:

$$3.5 \times 10^{15} \; \text{g} \; C_8H_{18} \times \frac{1 \; \text{mol} \; C_8H_{18}}{114.22 \; \text{g} \; C_8H_{18}} \times \frac{16 \; \text{mol} \; CO_2}{2 \; \text{mol} \; C_8H_{18}} \times \frac{44.01 \; \text{g} \; CO_2}{1 \; \text{mol} \; CO_2} = 1.1 \times 10^{16} \; \text{g} \; CO_2$$

The world's petroleum combustion produces 1.1×10^{16} g CO_2 (1.1×10^{13} kg) per year. In comparison, volcanoes produce about 2.0×10^{11} kg CO_2 per year.[*] In other words, volcanoes emit only $\dfrac{2.0 \times 10^{11} \text{ kg}}{1.1 \times 10^{13} \text{ kg}} \times 100\% = 1.8\%$ as much CO_2 per year as petroleum combustion. The argument that volcanoes emit more carbon dioxide than fossil fuel combustion is blatantly incorrect. Additional examples of stoichiometric calculations follow.

> The percentage of CO_2 emitted by volcanoes relative to all fossil fuels is even less than 2% because CO_2 is also emitted by the combustion of coal and natural gas.

EXAMPLE 4.1 Stoichiometry

In photosynthesis, plants convert carbon dioxide and water into glucose ($C_6H_{12}O_6$) according to the reaction:

$$6\ CO_2(g) + 6\ H_2O(l) \xrightarrow{\text{sunlight}} 6\ O_2(g) + C_6H_{12}O_6(aq)$$

Suppose you determine that a particular plant consumes 37.8 g CO_2 in one week. Assuming that there is more than enough water present to react with all of the CO_2, what mass of glucose (in grams) can the plant synthesize from the CO_2?

SORT The problem gives the mass of carbon dioxide and asks you to find the mass of glucose that can be produced.	**GIVEN** 37.8 g CO_2 **FIND** g $C_6H_{12}O_6$

STRATEGIZE The conceptual plan follows the general pattern of mass A → amount A (in moles) → amount B (in moles) → mass B. From the chemical equation, you can deduce the relationship between moles of carbon dioxide and moles of glucose. Use the molar masses to convert between grams and moles.

CONCEPTUAL PLAN

$$\begin{array}{cccc}
\boxed{\text{g } CO_2} & \boxed{\text{mol } CO_2} & \boxed{\text{mol } C_6H_{12}O_6} & \boxed{\text{g } C_6H_{12}O_6} \\
\dfrac{1 \text{ mol } CO_2}{44.01 \text{ g } CO_2} & \dfrac{1 \text{ mol } C_6H_{12}O_6}{6 \text{ mol } CO_2} & \dfrac{180.2 \text{ g } C_6H_{12}O_6}{1 \text{ mol } C_6H_{12}O_6} &
\end{array}$$

RELATIONSHIPS USED

molar mass CO_2 = 44.01 g/mol

6 mol CO_2 : 1 mol $C_6H_{12}O_6$

molar mass $C_6H_{12}O_6$ = 180.16 g/mol

SOLVE Follow the conceptual plan to solve the problem. Begin with g CO_2 and use the conversion factors to arrive at g $C_6H_{12}O_6$.

SOLUTION

$$37.8 \text{ g } \cancel{CO_2} \times \frac{1 \text{ mol } \cancel{CO_2}}{44.01 \text{ g } \cancel{CO_2}} \times \frac{1 \text{ mol } \cancel{C_6H_{12}O_6}}{6 \text{ mol } \cancel{CO_2}} \times \frac{180.16 \text{ g } C_6H_{12}O_6}{1 \text{ mol } \cancel{C_6H_{12}O_6}} = 25.8 \text{ g } C_6H_{12}O_6$$

CHECK The units of the answer are correct. The magnitude of the answer (25.8 g) is less than the initial mass of CO_2 (37.8 g). This is reasonable because each carbon in CO_2 has two oxygen atoms associated with it, while in $C_6H_{12}O_6$ each carbon has only one oxygen atom associated with it and two hydrogen atoms, which are much lighter than oxygen. Therefore the mass of glucose produced should be less than the mass of carbon dioxide for this reaction.

FOR PRACTICE 4.1

Magnesium hydroxide, the active ingredient in milk of magnesia, neutralizes stomach acid, primarily HCl, according to the reaction:

$$Mg(OH)_2(aq) + 2\ HCl(aq) \longrightarrow 2H_2O(l) + MgCl_2(aq)$$

What mass of HCl, in grams, is neutralized by a dose of milk of magnesia containing 3.26 g $Mg(OH)_2$?

[*]Gerlach, T. M., Present-day CO_2 emissions from volcanoes; *Eos, Transactions, American Geophysical Union*, Vol. 72, No. 23, June 4, 1991, pp. 249 and 254–255.

EXAMPLE 4.2 Stoichiometry

Sulfuric acid (H_2SO_4) is a component of acid rain that forms when SO_2, a pollutant, reacts with oxygen and water according to the simplified reaction:

$$2 \, SO_2(g) + O_2(g) + 2 \, H_2O(l) \longrightarrow 2 \, H_2SO_4(aq)$$

The generation of the electricity used in a typical medium-sized home produces about 25 kg of SO_2 per year. Assuming that there is more than enough O_2 and H_2O, what mass of H_2SO_4, in kg, can form from this much SO_2?

SORT The problem gives the mass of sulfur dioxide and asks you to find the mass of sulfuric acid.	**GIVEN** 25 kg SO_2 **FIND** kg H_2SO_4
STRATEGIZE The conceptual plan follows the standard format of mass → amount (in moles) → amount (in moles) → mass. Since the original quantity of SO_2 is given in kg, you must first convert to grams. You can deduce the relationship between moles of sulfur dioxide and moles of sulfuric acid from the chemical equation. Since the final quantity is requested in kg, convert to kg at the end.	**CONCEPTUAL PLAN** 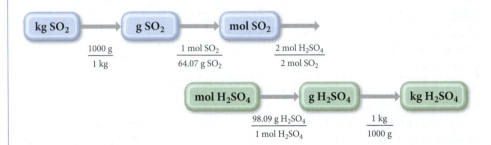 **RELATIONSHIPS USED** 1 kg = 1000 g molar mass SO_2 = 64.07 g/mol 2 mol SO_2 : 2 mol H_2SO_4 molar mass H_2SO_4 = 98.09 g/mol
SOLVE Follow the conceptual plan to solve the problem. Begin with the given amount of SO_2 in kilograms and use the conversion factors to arrive at kg H_2SO_4.	**SOLUTION** $25 \text{ kg SO}_2 \times \dfrac{1000 \text{ g}}{1 \text{ kg}} \times \dfrac{1 \text{ mol SO}_2}{64.07 \text{ g SO}_2} \times \dfrac{2 \text{ mol H}_2\text{SO}_4}{2 \text{ mol SO}_2}$ $\times \dfrac{98.09 \text{ g H}_2\text{SO}_4}{1 \text{ mol H}_2\text{SO}_4} \times \dfrac{1 \text{ kg}}{1000 \text{ g}} = 38 \text{ kg H}_2\text{SO}_4$

CHECK The units of the final answer are correct. The magnitude of the final answer (38 kg H_2SO_4) is larger than the amount of SO_2 given (25 kg). This is reasonable because in the reaction each SO_2 molecule "gains weight" by reacting with O_2 and H_2O.

FOR PRACTICE 4.2

Another component of acid rain is nitric acid, which forms when NO_2, also a pollutant, reacts with oxygen and water according to the simplified equation:

$$4 \, NO_2(g) + O_2(g) + 2 \, H_2O(l) \longrightarrow 4 \, HNO_3(aq)$$

The generation of the electricity used in a medium-sized home produces about 16 kg of NO_2 per year. Assuming that there is plenty of O_2 and H_2O, what mass of HNO_3, in kg, can form from this amount of NO_2 pollutant?

Conceptual Connection 4.1 Stoichiometry

Under certain conditions, sodium can react with oxygen to form sodium oxide according to the reaction:

$$4\,Na(s) + O_2(g) \longrightarrow 2\,Na_2O(s)$$

A flask contains the amount of oxygen represented by the diagram on the right.

Which image best represents the amount of sodium required to completely react with all of the oxygen in the flask?

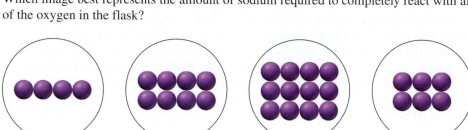

(a) (b) (c) (d)

Conceptual Connection 4.2 Stoichiometry

Consider the generic chemical equation A + 3B $\longrightarrow$ 2C. Let circles represent molecules of A, squares represent molecules of B, and triangles represent molecules of C, If the diagram below represents the amount of B available for reaction, draw similar diagrams showing (a) the amount of A necessary to completely react with B, and (b) the amount of C that forms if A completely reacts.

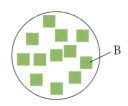

B

4.3   Limiting Reactant, Theoretical Yield, and Percent Yield

Let's return to our pizza analogy to understand three more concepts important in reaction stoichiometry: *limiting reactant*, *theoretical yield*, and *percent yield*. Recall our pizza recipe from Section 4.2:

1 crust + 5 ounces tomato sauce + 2 cups cheese $\longrightarrow$ 1 pizza

Suppose that we have 4 crusts, 10 cups of cheese, and 15 ounces of tomato sauce. How many pizzas can we make?

We have enough crusts to make:

$$4\ \text{crusts} \times \frac{1\ \text{pizza}}{1\ \text{crust}} = 4\ \text{pizzas}$$

We have enough cheese to make:

$$10\ \text{cups cheese} \times \frac{1\ \text{pizza}}{2\ \text{cups cheese}} = 5\ \text{pizzas}$$

We have enough tomato sauce to make:

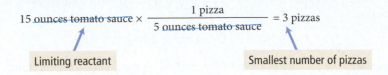

$$15 \text{ ounces tomato sauce} \times \frac{1 \text{ pizza}}{5 \text{ ounces tomato sauce}} = 3 \text{ pizzas}$$

Limiting reactant   Smallest number of pizzas

The term limiting reagent *is sometimes used in place of* limiting reactant.

We have enough crusts for 4 pizzas, enough cheese for 5 pizzas, but enough tomato sauce for only 3 pizzas. Consequently, unless we get more ingredients, we can make only 3 pizzas. The tomato sauce *limits* how many pizzas we can make. If the pizza recipe were a chemical reaction, the tomato sauce would be the **limiting reactant**, the reactant that limits the amount of product in a chemical reaction. Notice that the limiting reactant is simply the reactant that makes *the least amount of product*. The reactants that *do not* limit the amount of product—such as the crusts and the cheese in this example—are said to be *in excess*. If this were a chemical reaction, 3 pizzas would be the **theoretical yield**, the amount of product that can be made in a chemical reaction based on the amount of limiting reactant.

Let us carry this analogy one step further. Suppose we go on to cook our pizzas and accidentally burn one of them. So even though we theoretically have enough ingredients for 3 pizzas, we end up with only 2. If this were a chemical reaction, the 2 pizzas would be our **actual yield**, the amount of product actually produced by a chemical reaction. (The actual yield is always equal to or less than the theoretical yield because at least a small amount of product is usually lost to other reactions or does not form during a reaction.) Finally, our **percent yield**, the percentage of the theoretical yield that was actually attained, is calculated as follows:

Actual yield

$$\% \text{ yield } = \frac{2 \text{ pizzas}}{3 \text{ pizzas}} \times 100\% = 67\%$$

Theoretical yield

Since one of our pizzas burned, we obtained only 67% of our theoretical yield.

Summarizing: Limiting Reactant and Yield

▶ **The limiting reactant** (or **limiting reagent**) is the reactant that is completely consumed in a chemical reaction and limits the amount of product.

▶ **The reactant in excess** is any reactant that occurs in a quantity greater than is required to completely react with the limiting reactant.

▶ **The theoretical yield** is the amount of product that can be made in a chemical reaction based on the amount of limiting reactant.

▶ **The actual yield** is the amount of product actually produced by a chemical reaction.

▶ **The percent yield** is calculated as $\dfrac{\text{actual yield}}{\text{theoretical yield}} \times 100\%$.

Now let's apply these concepts to a chemical reaction. Recall from Section 3.10 our balanced equation for the combustion of methane:

$$CH_4(g) + 2\,O_2(g) \longrightarrow CO_2(g) + 2\,H_2O(l)$$

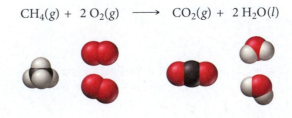

If we start out with 5 CH_4 molecules and 8 O_2 molecules, what is our limiting reactant? What is our theoretical yield of carbon dioxide molecules? We first calculate the number of CO_2 molecules that can be made from 5 CH_4 molecules:

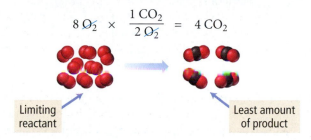

$$5 \; CH_4 \; \times \; \frac{1 \; CO_2}{1 \; CH_4} \; = \; 5 \; CO_2$$

We then calculate the number of CO_2 molecules that can be made from 8 O_2 molecules:

$$8 \; O_2 \; \times \; \frac{1 \; CO_2}{2 \; O_2} \; = \; 4 \; CO_2$$

Limiting reactant

Least amount of product

We have enough CH_4 to make 5 CO_2 molecules and enough O_2 to make 4 CO_2 molecules; therefore O_2 is the limiting reactant and 4 CO_2 molecules is the theoretical yield. The CH_4 is in excess.

⬭ Conceptual Connection 4.3 Limiting Reactant and Theoretical Yield

Nitrogen and hydrogen gas react to form ammonia according to the reaction:

$$N_2(g) + 3 \; H_2(g) \longrightarrow 2 \; NH_3(g)$$

If a flask contains a mixture of reactants represented by the image on the right, which of the following images best represents the mixture after the reactants have reacted as completely as possible? What is the limiting reactant? Which reactant is in excess?

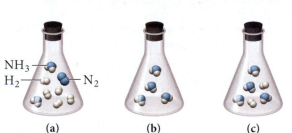

(a)   (b)   (c)

Limiting Reactant, Theoretical Yield, and Percent Yield from Initial Reactant Masses

When working in the laboratory, we normally measure the initial quantities of reactants in grams, not in number of molecules. To find the limiting reactant and theoretical yield from initial masses, we must first convert the masses to amounts in moles. Consider, for example, the following reaction:

$$2 \; Mg(s) + O_2(g) \longrightarrow 2 \; MgO(s)$$

If we have 42.5 g Mg and 33.8 g O_2, what is the limiting reactant and theoretical yield?

To solve this problem, we must determine which of the reactants makes the least amount of product.

Conceptual Plan

We find the limiting reactant by calculating how much product can be made from each reactant. Since we are given the initial quantities in grams, and stoichiometric relationships are between moles, we must first convert to moles. We then convert from moles of the reactant to moles of product. The reactant that makes the *least amount of product* is the limiting reactant. The conceptual plan is:

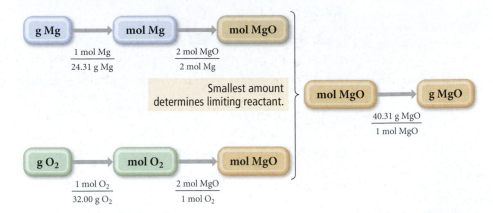

In this conceptual plan, we compare the number of moles of MgO made by each reactant, and convert only the smaller amount to grams. (Alternatively, we can convert both quantities to grams and determine the limiting reactant based on the mass of the product.)

Relationships Used

molar mass Mg = 24.31 g Mg

molar mass O_2 = 32.00 g O_2

2 mol Mg : 2 mol MgO

1 mol O_2 : 2 mol MgO

molar mass MgO = 40.31 g MgO

Solution

Beginning with the masses of each reactant, we follow the conceptual plan to calculate how much product can be made from each:

$$42.5 \text{ g Mg} \times \frac{1 \text{ mol Mg}}{24.31 \text{ g Mg}} \times \frac{2 \text{ mol MgO}}{2 \text{ mol Mg}} = 1.7483 \text{ mol MgO}$$

Limiting reactant

Least amount of product

$$1.7483 \text{ mol MgO} \times \frac{40.31 \text{ g MgO}}{1 \text{ mol MgO}} = 70.5 \text{ g MgO}$$

$$33.8 \text{ g O}_2 \times \frac{1 \text{ mol O}_2}{32.00 \text{ g O}_2} \times \frac{2 \text{ mol MgO}}{1 \text{ mol O}_2} = 2.1125 \text{ mol MgO}$$

Since Mg makes the least amount of product, it is the limiting reactant and O_2 is in excess. Notice that the limiting reactant is not necessarily the reactant with the least mass. In this case, the mass of O_2 is less than the mass of Mg, yet Mg is the limiting reactant because it makes the least amount of MgO. The theoretical yield is therefore 70.5 g of MgO, the mass of product possible based on the limiting reactant.

Now suppose that when the synthesis is carried out, the actual yield of MgO is 55.9 g. What is the percent yield? The percent yield is calculated as follows:

$$\% \text{ yield} = \frac{\text{actual yield}}{\text{theoretical yield}} \times 100\% = \frac{55.9 \text{ g}}{70.5 \text{ g}} \times 100\% = 79.3\%$$

EXAMPLE 4.3 Limiting Reactant and Theoretical Yield

Ammonia, NH_3, can be synthesized by the reaction:

$$2\, NO(g) + 5\, H_2(g) \longrightarrow 2\, NH_3(g) + 2\, H_2O(g)$$

Starting with 86.3 g NO and 25.6 g H_2, find the theoretical yield of ammonia in grams.

SORT You are given the mass of each reactant in grams and asked to find the theoretical yield of a product.	**GIVEN** 86.3 g NO, 25.6 g H_2 **FIND** theoretical yield of NH_3

STRATEGIZE Determine which reactant makes the least amount of product by converting from grams of each reactant to moles of the reactant to moles of the product. Use molar masses to convert between grams and moles and use the stoichiometric relationships (deduced from the chemical equation) to convert between moles of reactant and moles of product. The reactant that makes *the least amount of product* is the limiting reactant. Convert the number of moles of product obtained using the limiting reactant to grams of product.

CONCEPTUAL PLAN

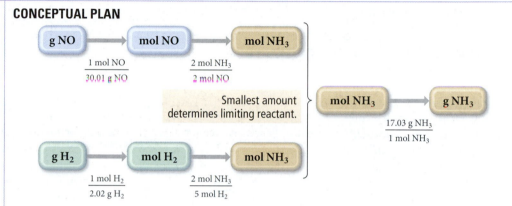

RELATIONSHIPS USED

molar mass NO = 30.01 g/mol
molar mass H_2 = 2.02 g/mol
2 mol NO : 2 mol NH_3 (from chemical equation)
5 mol H_2 : 2 mol NH_3 (from chemical equation)
molar mass NH_3 = 17.03 g/mol

SOLVE Beginning with the given mass of each reactant, calculate the amount of product that can be made in moles. Convert the amount of product made by the limiting reactant to grams—this is the theoretical yield.

SOLUTION

$$86.3\ \text{g NO} \times \frac{1\ \text{mol NO}}{30.01\ \text{g NO}} \times \frac{2\ \text{mol NH}_3}{2\ \text{mol NO}} = 2.8\underline{7}57\ \text{mol NH}_3$$

Limiting reactant   Least amount of product

$$2.8\underline{7}57\ \text{mol NH}_3 \times \frac{17.03\ \text{g NH}_3}{\text{mol NH}_3} = 49.0\ \text{g NH}_3$$

$$25.6\ \text{g H}_2 \times \frac{1\ \text{mol H}_2}{2.02\ \text{g H}_2} \times \frac{2\ \text{mol NH}_3}{5\ \text{mol H}_2} = 5.0\underline{6}93\ \text{mol NH}_3$$

Since NO makes the least amount of product, it is the limiting reactant, and the theoretical yield of ammonia is 49.0 g.

CHECK The units of the answer (g NH_3) are correct. The magnitude (49.0 g) seems reasonable given that 86.3 g NO is the limiting reactant. NO contains one oxygen atom per nitrogen atom and NH_3 contains three hydrogen atoms per nitrogen atom. Since three hydrogen atoms have less mass than one oxygen atom, it is reasonable that the mass of NH_3 obtained is less than the mass of NO.

FOR PRACTICE 4.3

Ammonia can also be synthesized by the reaction:

$$3\, H_2(g) + N_2(g) \longrightarrow 2\, NH_3(g)$$

What is the theoretical yield of ammonia, in kg, that can be synthesized from 5.22 kg of H_2 and 31.5 kg of N_2?

EXAMPLE 4.4 Limiting Reactant and Theoretical Yield

Titanium metal can be obtained from its oxide according to the balanced equation:

$$TiO_2(s) + 2\,C(s) \longrightarrow Ti(s) + 2\,CO(g)$$

When 28.6 kg of C reacts with 88.2 kg of TiO$_2$, 42.8 kg of Ti is produced. Find the limiting reactant, theoretical yield (in kg), and percent yield.

SORT You are given the mass of each reactant and the mass of product formed. You are asked to find the limiting reactant, theoretical yield, and percent yield.	**GIVEN** 28.6 kg C, 88.2 kg TiO$_2$, 42.8 kg Ti produced **FIND** limiting reactant, theoretical yield, % yield

STRATEGIZE Determine which of the reactants makes the least amount of product by converting from kilograms of each reactant to moles of product. Convert between grams and moles using molar mass. Convert between moles of reactant and moles of product using the stoichiometric relationships derived from the chemical equation. The reactant that makes the *least amount of product* is the limiting reactant. Determine the theoretical yield (in kg) by converting the number of moles of product obtained with the limiting reactant to kilograms of product.	**CONCEPTUAL PLAN** 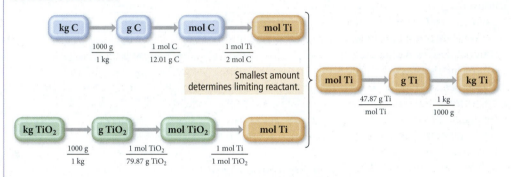 **RELATIONSHIPS USED** 1000 g = 1 kg molar mass of C = 12.01 g/mol molar mass of TiO$_2$ = 79.87 g/mol 1 mol TiO$_2$: 1 mol Ti 2 mol C : 1 mol Ti molar mass of Ti = 47.87 g/mol

SOLVE Beginning with the actual amount of each reactant, calculate the amount of product that can be made in moles. Convert the amount of product made by the limiting reactant to kilograms—this is the theoretical yield.	**SOLUTION** $$28.6\ \text{kg C} \times \frac{1000\ \text{g}}{1\ \text{kg}} \times \frac{1\ \text{mol C}}{12.01\ \text{g C}} \times \frac{1\ \text{mol Ti}}{2\ \text{mol C}} = 1.1907 \times 10^3\ \text{mol Ti}$$ Limiting reactant Least amount of product $$88.2\ \text{kg TiO}_2 \times \frac{1000\ \text{g}}{1\ \text{kg}} \times \frac{1\ \text{mol TiO}_2}{79.87\ \text{g TiO}_2} \times \frac{1\ \text{mol Ti}}{1\ \text{mol TiO}_2} = 1.1043 \times 10^3\ \text{mol Ti}$$ $$1.1043 \times 10^3\ \text{mol Ti} \times \frac{47.87\ \text{g Ti}}{1\ \text{mol Ti}} \times \frac{1\ \text{kg}}{1000\ \text{g}} = 52.9\ \text{kg Ti}$$ Since TiO$_2$ makes the least amount of product, it is the limiting reactant and 52.9 kg Ti is the theoretical yield.
Calculate the percent yield by dividing the actual yield (42.8 kg Ti) by the theoretical yield.	$$\% \text{ yield} = \frac{\text{actual yield}}{\text{theoretical yield}} \times 100\% = \frac{42.8\ \text{kg}}{52.9\ \text{kg}} \times 100\% = 80.9\%$$

CHECK The theoretical yield has the correct units (kg Ti) and has a reasonable magnitude compared to the mass of TiO_2. Since Ti has a lower molar mass than TiO_2, the amount of Ti made from TiO_2 should have a lower mass. The percent yield is reasonable (under 100% as it should be).

FOR PRACTICE 4.4

The following reaction is used to obtain iron from iron ore:

$$Fe_2O_3(s) + 3\,CO(g) \longrightarrow 2\,Fe(s) + 3\,CO_2(g)$$

The reaction of 167 g Fe_2O_3 with 85.8 g CO produces 72.3 g Fe. Find the limiting reactant, theoretical yield, and percent yield.

4.4   Solution Concentration and Solution Stoichiometry

Chemical reactions in which the reactants are dissolved in water are among the most common and important. The reactions that occur in lakes, streams, and oceans, as well as the reactions that occur in every cell of our bodies, take place in water. A homogeneous mixture of two or more substances—such as salt and water—is called a **solution**. The majority component of a solution is the **solvent**, and the minority component is the **solute**. A solution in which water acts as the solvent is an **aqueous solution**. In this section, we first examine how to quantify the concentration of a solution (the amount of solute relative to solvent) and then turn to applying the principles of stoichiometry, which you learned in the previous section, to reactions occurring in solution.

Solution Concentration

The amount of solute in a solution is variable. For example, we can add just a little salt to water to make a **dilute solution**, one that contains a small amount of solute relative to the solvent, or we can add a lot of salt to water to make a **concentrated solution**, one that contains a large amount of solute relative to the solvent. A common way to express solution concentration is **molarity (M)**, the amount of solute (in moles) divided by the volume of solution (in liters).

$$\text{Molarity (M)} = \frac{\text{amount of solute (in mol)}}{\text{volume of solution (in L)}}$$

Note that molarity is a ratio of the amount of solute per liter of *solution*, not per liter of solvent. To make an aqueous solution of a specified molarity, we usually put the solute into a flask and then add water until we have the desired volume of solution. For example, to make 1 L of a 1 M NaCl solution, we add 1 mol of NaCl to a flask and then add water to make 1 L of solution (**Figure 4.4▶**). We *do not* combine 1 mol of NaCl with 1 L of water because the resulting solution would have a total volume exceeding 1 L and therefore a molarity of less than 1 M. To calculate molarity, divide the amount of the solute in moles by the volume of the solution (solute *and* solvent) in liters, as shown in Example 4.5.

▶ **FIGURE 4.4 Preparing a 1.00 Molar NaCl Solution**

Preparing a Solution of Specified Concentration

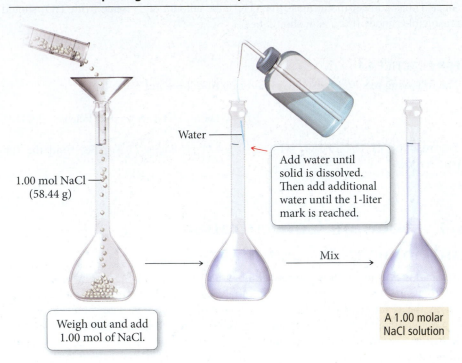

1.00 mol NaCl
(58.44 g)

Water

Add water until
solid is dissolved.
Then add additional
water until the 1-liter
mark is reached.

Mix

Weigh out and add
1.00 mol of NaCl.

A 1.00 molar
NaCl solution

EXAMPLE 4.5 Calculating Solution Concentration

If 25.5 g KBr is dissolved in enough water to make 1.75 L of solution, what is the molarity of the solution?

SORT You are given the mass of KBr and the volume of a solution and asked to find its molarity.	**GIVEN** 25.5 g KBr, 1.75 L of solution **FIND** molarity (M)
STRATEGIZE When formulating the conceptual plan, think about the definition of molarity, the amount of solute *in moles* per liter of solution. You are given the mass of KBr, so first use the molar mass of KBr to convert from g KBr to mol KBr. Then use the number of moles of KBr and liters of solution to find the molarity.	**CONCEPTUAL PLAN** 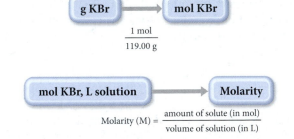 **g KBr** → **mol KBr** $\dfrac{1\ \text{mol}}{119.00\ \text{g}}$ **mol KBr, L solution** → **Molarity** $\text{Molarity (M)} = \dfrac{\text{amount of solute (in mol)}}{\text{volume of solution (in L)}}$ **RELATIONSHIP USED** molar mass of KBr $= 119.00\ \text{g/mol}$
SOLVE Follow the conceptual plan. Begin with g KBr and convert to mol KBr, then use mol KBr and L solution to calculate molarity.	**SOLUTION** $25.5\ \cancel{\text{g KBr}} \times \dfrac{1\ \text{mol KBr}}{119.00\ \cancel{\text{g KBr}}} = 0.21429\ \text{mol KBr}$ $\text{molarity (M)} = \dfrac{\text{amount of solute (in mol)}}{\text{volume of solution (in L)}}$ $= \dfrac{0.21429\ \text{mol KBr}}{1.75\ \text{L solution}} = 0.122\ \text{M}$

CHECK The units of the answer (M) are correct. The magnitude is reasonable. Common solutions range in concentration from 0 to about 18 M. Concentrations significantly above 18 M are suspect and should be double-checked.

Using Molarity in Calculations

The molarity of a solution can be used as a conversion factor between moles of the solute and liters of the solution. For example, a 0.500 M NaCl solution contains 0.500 mol NaCl for every liter of solution:

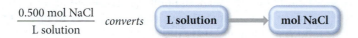

$$\frac{0.500 \text{ mol NaCl}}{\text{L solution}} \quad converts \quad \boxed{\text{L solution}} \longrightarrow \boxed{\text{mol NaCl}}$$

This conversion factor converts from L solution to mol NaCl. If you want to go the other way, simply invert the conversion factor:

$$\frac{\text{L solution}}{0.500 \text{ mol NaCl}} \quad converts \quad \boxed{\text{mol NaCl}} \longrightarrow \boxed{\text{L solution}}$$

Example 4.6 shows how to use molarity in this way.

EXAMPLE 4.6 Using Molarity in Calculations

How many liters of a 0.125 M NaOH solution contain 0.255 mol of NaOH?

SORT You are given the concentration of a NaOH solution. You are asked to find the volume of the solution that contains a given amount (in moles) of NaOH.	**GIVEN** 0.125 M NaOH solution, 0.255 mol NaOH **FIND** volume of NaOH solution (in L)
STRATEGIZE The conceptual plan begins with mol NaOH and shows the conversion to L of solution using the molarity as a conversion factor.	**CONCEPTUAL PLAN** 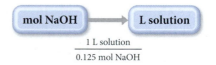 $\boxed{\text{mol NaOH}} \longrightarrow \boxed{\text{L solution}}$ $\dfrac{1 \text{ L solution}}{0.125 \text{ mol NaOH}}$ **RELATIONSHIPS USED** $0.125 \text{ M NaOH} = \dfrac{0.125 \text{ mol NaOH}}{1 \text{ L solution}}$
SOLVE Follow the conceptual plan. Begin with mol NaOH and convert to L solution.	**SOLUTION** $0.255 \ \cancel{\text{mol NaOH}} \times \dfrac{1 \text{ L solution}}{0.125 \ \cancel{\text{mol NaOH}}} = 2.04 \text{ L solution}$

CHECK The units of the answer (L) are correct. The magnitude seems reasonable because the solution contains 0.125 mol per liter. Therefore, roughly 2 L contains the given amount of moles (0.255 mol).

Conceptual Connection 4.4 **Solutions**

If 25 grams of salt is dissolved in 251 grams of water, what is the mass of the resulting solution?

(a) 251 g

(b) 276 g

(c) 226 g

Solution Dilution

> When diluting acids, always add the concentrated acid to the water. Never add water to concentrated acid solutions, as the heat generated may cause the concentrated acid to splatter and burn your skin.

To save space in laboratory storerooms, solutions are often stored in concentrated forms called **stock solutions**. For example, hydrochloric acid is often stored as a 12 M stock solution. However, many lab procedures call for much less concentrated hydrochloric acid solutions, so we must dilute the stock solution to the required concentration. How do we know how much of the stock solution to use? The easiest way to solve dilution problems is to use the following dilution equation:

$$M_1V_1 = M_2V_2 \qquad\qquad [4.1]$$

where M_1 and V_1 are the molarity and volume of the initial concentrated solution, and M_2 and V_2 are the molarity and volume of the final diluted solution. This equation works because the molarity multiplied by the volume gives the number of moles of solute, which is the same in both solutions.

$$M_1V_1 = M_2V_2$$
$$\text{mol}_1 = \text{mol}_2$$

In other words, the number of moles of solute does not change when we dilute a solution.

For example, suppose a laboratory procedure calls for 3.00 L of a 0.500 M $CaCl_2$ solution. How should we prepare this solution from a 10.0 M stock solution? We can solve Equation 4.1 for V_1, the volume of the stock solution required for the dilution, and then substitute in the correct values to calculate it.

$$M_1V_1 = M_2V_2$$

$$V_1 = \frac{M_2V_2}{M_1}$$

$$= \frac{0.500 \;\cancel{\text{mol}/\text{L}} \times 3.00 \text{ L}}{10.0 \;\cancel{\text{mol}/\text{L}}}$$

$$= 0.150 \text{ L}$$

Consequently, we make the solution by adding enough water to 0.150 L of the stock solution to create a total volume of 3.00 L (V_2). The resulting solution will be 0.500 M in $CaCl_2$ (**Figure 4.5▶**).

EXAMPLE 4.7 Solution Dilution

To what volume should you dilute 0.200 L of a 15.0 M NaOH solution to obtain a 3.00 M NaOH solution?

SORT You are given the initial volume, initial concentration, and final concentration of a solution, and you need to find the final volume.	**GIVEN** $V_1 = 0.200$ L $M_1 = 15.0$ M $M_2 = 3.00$ M **FIND** V_2

STRATEGIZE Equation 4.1 relates the initial and final volumes and concentrations for solution dilution problems. You are asked to find V_2. The other quantities (V_1, M_1, and M_2) are all given in the problem.

CONCEPTUAL PLAN

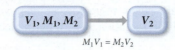

$$M_1V_1 = M_2V_2$$

RELATIONSHIPS USED

$$M_1V_1 = M_2V_2$$

SOLVE Begin with the solution dilution equation and solve it for V_2. Substitute in the required quantities and calculate V_2.

Make the solution by diluting 0.200 L of the stock solution to a total volume of 1.00 L (V_2). The resulting solution will have a concentration of 3.00 M.

SOLUTION $M_1V_1 = M_2V_2$

$$V_2 = \frac{M_1V_1}{M_2}$$

$$= \frac{15.0 \text{ mol/L} \times 0.200 \text{ L}}{3.00 \text{ mol/L}}$$

$$= 1.00 \text{ L}$$

CHECK The final units (L) are correct. The magnitude of the answer is reasonable because the solution is diluted from 15.0 M to 3.00 M, a factor of five. Therefore the volume should increase by a factor of five.

Diluting a Solution

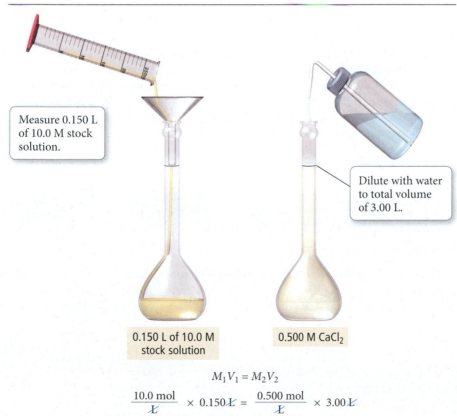

Measure 0.150 L of 10.0 M stock solution.

Dilute with water to total volume of 3.00 L.

0.150 L of 10.0 M stock solution

0.500 M $CaCl_2$

$$M_1V_1 = M_2V_2$$

$$\frac{10.0 \text{ mol}}{L} \times 0.150 \text{ L} = \frac{0.500 \text{ mol}}{L} \times 3.00 \text{ L}$$

$$1.50 \text{ mol} = 1.50 \text{ mol}$$

◀ **FIGURE 4.5 Preparing 3.00 L of 0.500 M $CaCl_2$ from a 10.0 M Stock Solution**

Conceptual Connection 4.5 Solution Dilution

The figure below represents a small volume within 500 mL of an aqueous ethanol (CH_3CH_2OH) solution. (The water molecules have been omitted for clarity.)

Which of the following images best represents the same volume of the solution after adding an additional 500 mL of water?

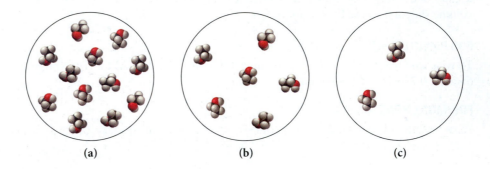

| (a) | (b) | (c) |

Solution Stoichiometry

In Section 4.2 we discussed how the coefficients in chemical equations are used as conversion factors between the amounts of reactants (in moles) and the amounts of products (in moles). In reactions involving aqueous reactants and products, it is often convenient to specify their quantities in terms of volumes and concentrations. We can then use these quantities to calculate the amounts in moles of reactants or products and use the stoichiometric coefficients to convert these to amounts of other reactants or products. The general conceptual plan for these kinds of calculations is:

We use molarity to convert between solution volumes and amount of solute, and we use the stoichiometric coefficients from the balanced chemical equation to convert between amounts of A and B. Example 4.8 demonstrates solution stoichiometry.

EXAMPLE 4.8 Solution Stoichiometry

What volume (in L) of 0.150 M KCl solution is required to completely react with 0.150 L of a 0.175 M $Pb(NO_3)_2$ solution according to the following balanced chemical equation?

$$2 KCl(aq) + Pb(NO_3)_2(aq) \longrightarrow PbCl_2(s) + 2 KNO_3(aq)$$

SORT You are given the volume and concentration of a $Pb(NO_3)_2$ solution.
You are asked to find the volume of KCl solution (of a given concentration) required to react with it.

GIVEN 0.150 L of $Pb(NO_3)_2$ solution, 0.175 M $Pb(NO_3)_2$ solution, 0.150 M KCl solution
FIND volume KCl solution (in L)

STRATEGIZE The conceptual plan has the form: volume A → amount A (in moles) → amount B (in moles) → volume B. The molar concentrations of the KCl and $Pb(NO_3)_2$ solutions can be used as conversion factors between the number of moles of reactants in these solutions and their volumes. The stoichiometric coefficients from the balanced equation are used to convert between number of moles of $Pb(NO_3)_2$ and number of moles of KCl.

CONCEPTUAL PLAN

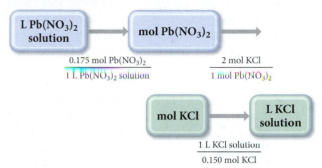

RELATIONSHIPS USED

$$M\left[Pb(NO_3)_2\right] = \frac{0.175 \text{ mol } Pb(NO_3)_2}{1 \text{ L } Pb(NO_3)_2 \text{ solution}}$$

$$2 \text{ mol KCl : } 1 \text{ mol } Pb(NO_3)_2$$

$$M(KCl) = \frac{0.150 \text{ mol KCl}}{1 \text{ L KCl solution}}$$

SOLVE Begin with L $Pb(NO_3)_2$ solution and follow the conceptual plan to arrive at L KCl solution.

SOLUTION

$$0.150 \text{ L } Pb(NO_3)_2 \text{ solution} \times \frac{0.175 \text{ mol } Pb(NO_3)_2}{1 \text{ L } Pb(NO_3)_2 \text{ solution}}$$

$$\times \frac{2 \text{ mol KCl}}{1 \text{ mol } Pb(NO_3)_2} \times \frac{1 \text{ L KCl solution}}{0.150 \text{ mol KCl}}$$

$$= 0.350 \text{ L KCl solution}$$

CHECK The final units (L KCl solution) are correct. The magnitude (0.350 L) seems reasonable because the reaction stoichiometry requires 2 mol of KCl per mole of $Pb(NO_3)_2$. Since the concentrations of the two solutions are not very different (0.150 M compared to 0.175 M), the volume of KCl required should be roughly two times the 0.150 L of $Pb(NO_3)_2$ given in the problem.

FOR PRACTICE 4.8

What volume (in mL) of a 0.150 M HNO_3 solution is required to completely react with 35.7 mL of a 0.108 M Na_2CO_3 solution according to the following balanced chemical equation?

$$Na_2CO_3(aq) + 2 HNO_3(aq) \longrightarrow 2 NaNO_3(aq) + CO_2(g) + H_2O(l)$$

FOR MORE PRACTICE 4.8

In the reaction from For Practice 4.8, what mass (in grams) of carbon dioxide is formed?

Solute and Solvent Interactions

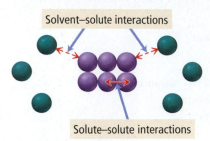

▲ **FIGURE 4.6 Solute and Solvent Interactions** When a solid is put into a solvent, the interactions between solvent and solute particles compete with the interactions among the solute particles themselves.

▲ **FIGURE 4.7 Charge Distribution in a Water Molecule** An uneven distribution of electrons within the water molecule causes the oxygen side of the molecule to have a partial negative charge and the hydrogen side to have a partial positive charge.

4.5 Types of Aqueous Solutions and Solubility

Consider two familiar aqueous solutions: salt water and sugar water. Salt water is a homogeneous mixture of NaCl and H_2O, and sugar water is a homogeneous mixture of $C_{12}H_{22}O_{11}$ and H_2O. You may have made these solutions yourself by adding solid table salt or solid sugar to water. As you stir either of these two substances into the water, it seems to disappear. However, you know that the original substance is still present because you can taste saltiness or sweetness in the water. How do solids such as salt and sugar dissolve in water?

When a solid is put into a liquid solvent, the attractive forces that hold the solid together (the solute–solute interactions) come into competition with the attractive forces between the solvent molecules and the particles that compose the solid (the solvent–solute interactions), as shown in **Figure 4.6◄**. For example, when sodium chloride is put into water, there is a competition between the attraction of Na^+ cations and Cl^- anions to each other (due to their opposite charges) and the attraction of Na^+ and Cl^- to water molecules. The attraction of Na^+ and Cl^- to water is based on the *polar nature* of the water molecule. For reasons we discuss later in this book (Section 9.6), the oxygen atom in water is electron-rich, giving it a partial negative charge (δ^-), as shown in **Figure 4.7◄**. The hydrogen atoms, in contrast, are electron-poor, giving them a partial positive charge (δ^+). As a result, the positively charged sodium ions are strongly attracted to the oxygen side of the water molecule (which has a partial negative charge), and the negatively charged chloride ions are attracted to the hydrogen side of the water molecule (which has a partial positive charge), as shown in **Figure 4.8▼**. In the case of NaCl, the attraction between the separated ions and the water molecules overcomes the attraction of sodium and chloride ions to each other, and the sodium chloride dissolves in the water (**Figure 4.9▼**).

Interactions in a Sodium Chloride Solution

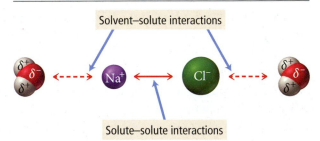

▲ **FIGURE 4.8 Solute and Solvent Interactions in a Sodium Chloride Solution** When sodium chloride is put into water, the attraction of Na^+ and Cl^- ions to water molecules competes with the attraction among the oppositely charged ions themselves.

Dissolution of an Ionic Compound

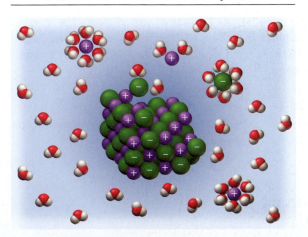

▲ **FIGURE 4.9 Sodium Chloride Dissolving in Water** The attraction between water molecules and the ions of sodium chloride allows NaCl to dissolve in the water.

Electrolyte and Nonelectrolyte Solutions

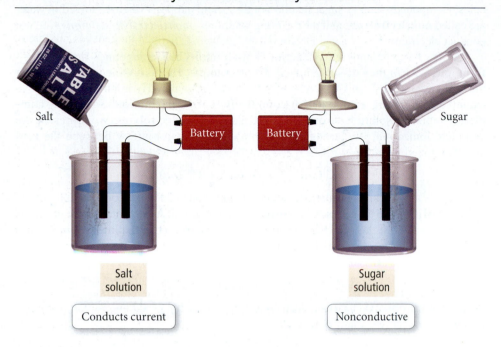

Salt
Battery Battery Sugar

Salt solution Sugar solution

Conducts current Nonconductive

◀ **FIGURE 4.10 Electrolyte and Nonelectrolyte Solutions** A solution of salt (an electrolyte) conducts electrical current. A solution of sugar (a nonelectrolyte) does not.

Electrolyte and Nonelectrolyte Solutions

The difference in the way that salt (an ionic compound) and sugar (a molecular compound) dissolve in water illustrates a fundamental difference between types of solutions. As **Figure 4.10▲** shows, a salt solution conducts electricity while a sugar solution does not. As we have just seen with sodium chloride, ionic compounds dissociate into their component ions when they dissolve in water. An NaCl solution, represented as NaCl(aq), does not contain NaCl units, only Na$^+$ ions and Cl$^-$ ions. The dissolved ions act as charge carriers, allowing the solution to conduct electricity. Substances that dissolve in water to form solutions that conduct electricity are called **electrolytes**. Substances such as sodium chloride that completely dissociate into ions when they dissolve in water are **strong electrolytes**, and the resulting solutions are strong electrolyte solutions.

In contrast to sodium chloride, sugar is a molecular compound. Most molecular compounds—with the important exception of acids, which we discuss in the next paragraph—dissolve in water as intact molecules. Sugar dissolves because the attraction between sugar molecules and water molecules—both of which contain a distribution of electrons that results in partial positive and partial negative charges—overcomes the attraction of sugar molecules to each other (**Figure 4.11▼**). However, in contrast to a sodium chloride solution (which is composed of dissociated ions), a sugar solution is

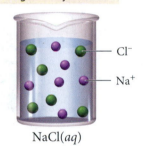

Strong electrolyte solution

Cl$^-$

Na$^+$

NaCl(aq)

Unlike soluble ionic compounds, which contain ions and therefore *dissociate* in water, acids are molecular compounds that *ionize* in water.

Interactions between Sugar and Water Molecules

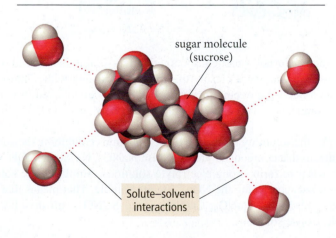

sugar molecule (sucrose)

Solute–solvent interactions

◀ **FIGURE 4.11 Sugar and Water Interactions** Partial charges on sugar molecules and water molecules (discussed more fully in Chapter 11) result in attractions between the sugar molecules and the water molecules.

Nonelectrolyte solution

$C_{12}H_{22}O_{11}(aq)$

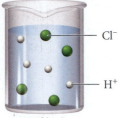

Strong acid

Cl⁻

H⁺

$HCl(aq)$

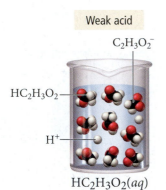

Weak acid

$C_2H_3O_2^-$

$HC_2H_3O_2$

H^+

$HC_2H_3O_2(aq)$

▲ AgCl does not significantly dissolve in water, but remains as a white powder at the bottom of the beaker.

composed of intact $C_{12}H_{22}O_{11}$ molecules homogeneously mixed with the water molecules. Compounds such as sugar that do not dissociate into ions when dissolved in water are called **nonelectrolytes**, and the resulting solutions—*nonelectrolyte solutions*—do not conduct electricity.

Acids, first encountered in Section 3.6, are molecular compounds that ionize—form ions—when they dissolve in water. For example, HCl is a molecular compound that ionizes into H^+ and Cl^- when it dissolves in water. Hydrochloric acid is an example of a **strong acid**, one that completely ionizes in solution. Since they completely ionize in solution, strong acids are also strong electrolytes. We represent the complete ionization of a strong acid with a single reaction arrow between the acid and its ionized form:

$$HCl(aq) \longrightarrow H^+(aq) + Cl^-(aq)$$

Many acids are **weak acids**; they do not completely ionize in water. For example, acetic acid ($HC_2H_3O_2$), the acid present in vinegar, is a weak acid. A solution of a weak acid is composed mostly of the nonionized acid—only a small percentage of the acid molecules ionize. We represent the partial ionization of a weak acid with opposing half arrows between the reactants and products:

$$HC_2H_3O_2(aq) \rightleftharpoons H^+(aq) + C_2H_3O_2^-(aq)$$

Weak acids are classified as **weak electrolytes** and the resulting solutions—*weak electrolyte solutions*—conduct electricity only weakly.

The Solubility of Ionic Compounds

You have just seen that, when an ionic compound dissolves in water, the resulting solution contains—not the intact ionic compound itself—but its component ions dissolved in water. However, not all ionic compounds dissolve in water. If we add AgCl to water, for example, it remains solid and appears as a white powder at the bottom of the water.

In general, a compound is termed **soluble** if it dissolves in water and **insoluble** if it does not. However, these classifications are a bit of an oversimplification. In reality, solubility is a continuum. Compounds exhibit a very wide range of solubilities, and even "insoluble" compounds dissolve to some extent, though usually orders of magnitude less than soluble compounds. For example, silver nitrate is soluble. If we mix solid $AgNO_3$ with water, it dissolves and forms a strong electrolyte solution. Silver chloride, on the other hand, is almost completely insoluble. If we mix solid AgCl with water, virtually all of it remains as a solid within the liquid water.

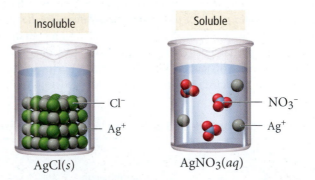

Insoluble

Cl⁻

Ag⁺

$AgCl(s)$

Soluble

NO_3^-

Ag⁺

$AgNO_3(aq)$

Determining whether or not a particular compound is soluble is a complex process. In Section 12.3, we examine more closely the energy associated with solution formation. For now, we can follow a set of empirical rules that have been inferred from observations of many ionic compounds. These are called the *solubility rules* and are summarized in Table 4.1.

For example, the solubility rules state that compounds containing the sodium ion are soluble. That means that compounds such as NaBr, $NaNO_3$, Na_2SO_4, NaOH, and Na_2CO_3 all dissolve in water to form strong electrolyte solutions. Similarly, the solubility rules state that compounds containing the NO_3^- ion are soluble. That means that compounds such as $AgNO_3$, $Pb(NO_3)_2$, $NaNO_3$, $Ca(NO_3)_2$, and $Sr(NO_3)_2$ all dissolve in water to form strong electrolyte solutions.

TABLE 4.1 Solubility Rules for Ionic Compounds in Water

Compounds Containing the Following Ions Are Generally Soluble	Exceptions
Li^+, Na^+, K^+, and NH_4^+	None
NO_3^- and $C_2H_3O_2^-$	None
Cl^-, Br^-, and I^-	When these ions pair with Ag^+, Hg_2^{2+}, or Pb^{2+}, the resulting compounds are insoluble.
SO_4^{2-}	When SO_4^{2-} pairs with Sr^{2+}, Ba^{2+}, Pb^{2+}, Ag^+, or Ca^{2+}, the resulting compound is insoluble.

Compounds Containing the Following Ions Are Generally Insoluble	Exceptions
OH^- and S^{2-}	When these ions pair with Li^+, Na^+, K^+, or NH_4^+, the resulting compounds are soluble.
	When S^{2-} pairs with Ca^{2+}, Sr^{2+}, or Ba^{2+}, the resulting compound is soluble.
	When OH^- pairs with Ca^{2+}, Sr^{2+}, or Ba^{2+}, the resulting compound is soluble.
CO_3^{2-} and PO_4^{3-}	When these ions pair with Li^+, Na^+, K^+, or NH_4^+, the resulting compounds are soluble.

Notice that when compounds containing polyatomic ions such as NO_3^- dissolve, the polyatomic ions remain as intact units when they dissolve.

The solubility rules also state that, with some exceptions, compounds containing the CO_3^{2-} ion are insoluble. Therefore, compounds such as $CuCO_3$, $CaCO_3$, $SrCO_3$, and $FeCO_3$ do not dissolve in water. Note that the solubility rules include many exceptions.

EXAMPLE 4.9 Predicting Whether an Ionic Compound Is Soluble

Predict whether each compound is soluble or insoluble.

(a) $PbCl_2$ (b) $CuCl_2$ (c) $Ca(NO_3)_2$ (d) $BaSO_4$

SOLUTION

(a) Insoluble. Compounds containing Cl^- are normally soluble, but Pb^{2+} is an exception.

(b) Soluble. Compounds containing Cl^- are normally soluble and Cu^{2+} is not an exception.

(c) Soluble. Compounds containing NO_3^- are always soluble.

(d) Insoluble. Compounds containing SO_4^{2-} are normally soluble, but Ba^{2+} is an exception.

FOR PRACTICE 4.9

Predict whether each of the following compounds is soluble or insoluble.

(a) NiS (b) $Mg_3(PO_4)_2$ (c) Li_2CO_3 (d) NH_4Cl

4.6 Precipitation Reactions

Have you ever taken a bath in hard water? Hard water contains dissolved ions such as Ca^{2+} and Mg^{2+} that diminish the effectiveness of soap. These ions react with soap to form a gray curd that may appear as "bathtub ring" after you drain the tub. Hard water is particularly troublesome when washing clothes. Imagine how a white shirt would look covered with the gray curd from the bathtub. Consequently, most laundry detergents include substances designed to remove Ca^{2+} and Mg^{2+} from the laundry mixture. The

▲ The reaction of ions in hard water with soap produces a gray curd that can be seen after bathwater is drained.

most common substance used for this purpose is sodium carbonate, which dissolves in water to form sodium cations (Na^+) and carbonate (CO_3^{2-}) anions:

$$Na_2CO_3(aq) \longrightarrow 2\,Na^+(aq) + CO_3^{2-}(aq)$$

Sodium carbonate is soluble, but calcium carbonate and magnesium carbonate are not (see the solubility rules in Table 4.1). Consequently, the carbonate anions react with dissolved Mg^{2+} and Ca^{2+} ions in hard water to form solids that *precipitate* from (or come out of) solution:

$$Mg^{2+}(aq) + CO_3^{2-}(aq) \longrightarrow MgCO_3(s)$$

$$Ca^{2+}(aq) + CO_3^{2-}(aq) \longrightarrow CaCO_3(s)$$

The precipitation of these ions prevents their reaction with the soap, eliminating curd and keeping shirts white instead of gray.

The reactions between CO_3^{2-} and Mg^{2+} and Ca^{2+} are examples of **precipitation reactions**, reactons in which a solid or **precipitate** forms upon mixing two solutions. Precipitation reactions are common in chemistry. As another example, consider potassium iodide and lead(II) nitrate, which both form colorless, strong electrolyte solutions when dissolved in water. When the two solutions are combined, however, a brilliant yellow precipitate forms (**Figure 4.12▼**). This precipitation reaction can be described with the chemical equation:

$$2\,KI(aq) + Pb(NO_3)_2(aq) \longrightarrow 2\,KNO_3(aq) + PbI_2(s)$$

▶ **FIGURE 4.12 Precipitation of Lead(II) Iodide**   When a potassium iodide solution is mixed with a lead(II) nitrate solution, a yellow lead(II) iodide precipitate forms.

Precipitation Reaction

$2\,KI(aq)$	$+$	$Pb(NO_3)_2(aq)$	$\longrightarrow$	$2\,KNO_3(aq)$	$+$	$PbI_2(s)$
(soluble)		(soluble)		(soluble)		(insoluble)

When a potassium iodide solution is mixed with a lead(II) nitrate solution, a yellow lead(II) iodide precipitate forms.

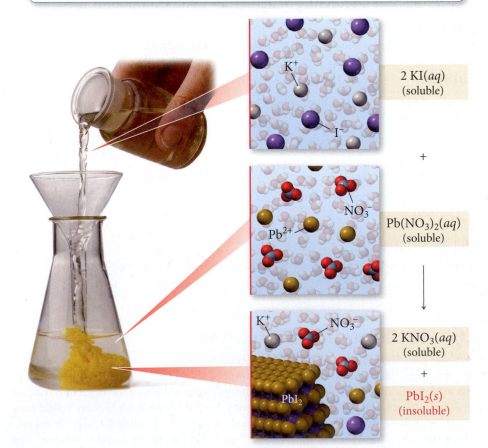

$2\,KI(aq)$ (soluble)

$+$

$Pb(NO_3)_2(aq)$ (soluble)

$2\,KNO_3(aq)$ (soluble)

$+$

$PbI_2(s)$ (insoluble)

Precipitation reactions do not always occur when two aqueous solutions are mixed. For example, if solutions of KI(*aq*) and NaCl(*aq*) are combined, nothing happens:

$$\text{KI}(aq) + \text{NaCl}(aq) \longrightarrow \text{NO REACTION}$$

The key to predicting precipitation reactions is to understand that only *insoluble* compounds form precipitates. In a precipitation reaction, two solutions containing soluble compounds combine and an insoluble compound precipitates. For example, consider the precipitation reaction described previously:

$$2 \text{ KI}(aq) + \text{Pb(NO}_3)_2(aq) \rightarrow \text{PbI}_2(s) + 2 \text{ KNO}_3(aq)$$
$$\text{soluble} \qquad \text{soluble} \qquad \text{insoluble} \qquad \text{soluble}$$

KI and $\text{Pb(NO}_3)_2$ are both soluble, but the precipitate, PbI_2, is insoluble. Before mixing, KI(*aq*) and $\text{Pb(NO}_3)_2(aq)$ are both dissociated in their respective solutions:

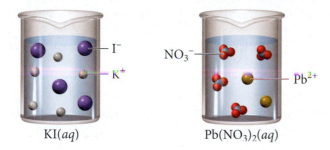

The instant that the solutions are mixed, all four ions are present:

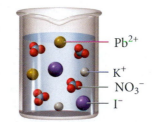

KI(*aq*) and $\text{Pb(NO}_3)_2(aq)$

However, two new compounds—one or both of which might be insoluble—are now possible. The cation from each compound can pair with the anion from the other compound to form possibly insoluble products (we will learn more about why this happens in the chapter on solutions):

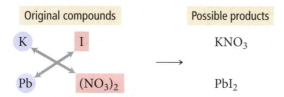

If the possible products are both soluble, then no reaction occurs. If one or both of the possible products are insoluble, a precipitation reaction occurs. In this case, KNO_3 is soluble, but PbI_2 is insoluble. Consequently, PbI_2 precipitates as shown.

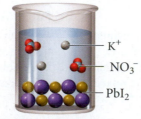

$\text{PbI}_2(s)$ and $\text{KNO}_3(aq)$

To predict whether a precipitation reaction will occur when two solutions are mixed and to write an equation for the reaction, use the procedure that follows. The steps are outlined in the left column, and two examples of applying the procedure are shown in the center and right columns.

PROCEDURE FOR... **Writing Equations for Precipitation Reactions**	**EXAMPLE 4.10** **Writing Equations for Precipitation Reactions** Write an equation for the precipitation reaction that occurs (if any) when solutions of potassium carbonate and nickel(II) chloride are mixed.	**EXAMPLE 4.11** **Writing Equations for Precipitation Reactions** Write an equation for the precipitation reaction that occurs (if any) when solutions of sodium nitrate and lithium sulfate are mixed.
1. Write the formulas of the two compounds being mixed as reactants in a chemical equation.	$K_2CO_3(aq) + NiCl_2(aq) \longrightarrow$	$NaNO_3(aq) + Li_2SO_4(aq) \longrightarrow$
2. Below the equation, write the formulas of the products that could form from the reactants. Obtain these by combining the cation from each reactant with the anion from the other. Make sure to write correct formulas for these ionic compounds, as described in Section 3.5.	$K_2CO_3(aq) + NiCl_2(aq) \longrightarrow$ Possible products KCl $NiCO_3$	$NaNO_3(aq) + Li_2SO_4(aq) \longrightarrow$ Possible products $LiNO_3$ Na_2SO_4
3. Consult the solubility rules to determine whether any of the possible products are insoluble.	KCl is soluble. (Compounds containing Cl^- are usually soluble and K^+ is not an exception.) $NiCO_3$ is insoluble. (Compounds containing CO_3^{2-} are usually insoluble and Ni^{2+} is not an exception.)	$LiNO_3$ is soluble. (Compounds containing NO_3^- are soluble and Li^+ is not an exception.) Na_2SO_4 is soluble. (Compounds containing SO_4^{2-} are generally soluble and Na^+ is not an exception.)
4. If all of the possible products are soluble, there will be no precipitate. Write NO REACTION after the arrow.	Since this example has an insoluble product, we proceed to the next step.	Since this example has no insoluble product, there is no reaction. $NaNO_3(aq) + Li_2SO_4(aq) \longrightarrow$ NO REACTION
5. If any of the possible products are insoluble, write their formulas as the products of the reaction using (*s*) to indicate solid. Write any soluble products with (*aq*) to indicate aqueous.	$K_2CO_3(aq) + NiCl_2(aq) \longrightarrow$ $NiCO_3(s) + KCl(aq)$	
6. Balance the equation. Remember to adjust only coefficients, not subscripts.	$K_2CO_3(aq) + NiCl_2(aq) \longrightarrow$ $NiCO_3(s) + 2\,KCl(aq)$	
	FOR PRACTICE 4.10 Write an equation for the precipitation reaction that occurs (if any) when solutions of ammonium chloride and iron(III) nitrate are mixed.	**FOR PRACTICE 4.11** Write an equation for the precipitation reaction that occurs (if any) when solutions of sodium hydroxide and copper(II) bromide are mixed.

Conceptual Connection 4.6 **Precipitation Reactions**

Consider the generic ionic compounds with the formulas AX and BY and the following solubility rules:

AX soluble; BY soluble; AY soluble; BX insoluble.

Let A^+ ions be represented by circles, B^+ ions be represented by squares, X^- ions be represented by triangles, Y^- ions be represented by diamonds. Solutions of the two compounds (AX and BY) can be represented as follows:

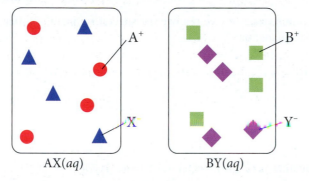

Draw a molecular level representation showing the result of mixing the two solutions above and write an equation to represent the reaction.

4.7   Representing Aqueous Reactions: Molecular, Ionic, and Complete Ionic Equations

Consider the following equation for a precipitation reaction.

$$Pb(NO_3)_2(aq) + 2\ KCl(aq) \longrightarrow PbCl_2(s) + 2\ KNO_3(aq)$$

This equation is a **molecular equation**, an equation showing the complete neutral formulas for each compound in the reaction as if they existed as molecules. However, in actual solutions of soluble ionic compounds, dissociated substances are present as ions. Equations for reactions occurring in aqueous solution can be written to better show the dissociated nature of dissolved ionic compounds. For example, we can rewrite the above equation as:

$$Pb^{2+}(aq) + 2\ NO_3^-(aq) + 2\ K^+(aq) + 2\ Cl^-(aq) \longrightarrow$$
$$PbCl_2(s) + 2\ K^+(aq) + 2\ NO_3^-(aq)$$

Equations such as this, which list individually all of the ions present as either reactants or products in a chemical reaction, are **complete ionic equations**.

Notice that in the complete ionic equation, some of the ions in solution appear unchanged on both sides of the equation. These ions are called **spectator ions** because they do not participate in the reaction.

$$Pb^{2+}(aq) + 2\ NO_3^-(aq) + 2\ K^+(aq) + 2\ Cl^-(aq) \longrightarrow$$
$$PbCl_2(s) + 2\ K^+(aq) + 2\ NO_3^-(aq)$$

Spectator ions

To simplify the equation, and to show more clearly what is happening, we can omit spectator ions:

$$Pb^{2+}(aq) + 2\ Cl^-(aq) \longrightarrow PbCl_2(s)$$

Equations such as this one, which show only the species that actually change during the reaction, are called **net ionic equations**.

As another example, consider the reaction between HCl(aq) and KOH(aq):

$$\text{HCl}(aq) + \text{KOH}(aq) \longrightarrow \text{H}_2\text{O}(l) + \text{KCl}(aq)$$

Since HCl, KOH, and KCl all exist in solution primarily as independent ions, the complete ionic equation is:

$$\text{H}^+(aq) + \text{Cl}^-(aq) + \text{K}^+(aq) + \text{OH}^-(aq) \longrightarrow$$
$$\text{H}_2\text{O}(l) + \text{K}^+(aq) + \text{Cl}^-(aq)$$

To write the net ionic equation, we remove the spectator ions, those that are unchanged on both sides of the equation:

$$\text{H}^+(aq) + \cancel{\text{Cl}^-}(aq) + \cancel{\text{K}^+}(aq) + \text{OH}^-(aq) \longrightarrow \text{H}_2\text{O}(l) + \cancel{\text{K}^+}(aq) + \cancel{\text{Cl}^-}(aq)$$

Spectator ions

The net ionic equation is $\text{H}^+(aq) + \text{OH}^-(aq) \longrightarrow \text{H}_2\text{O}(l)$.

Summarizing: Aqueous Equations

▶ **A molecular equation** is a chemical equation showing the complete, neutral formulas for every compound in a reaction.

▶ **A complete ionic equation** is a chemical equation showing all of the species as they are actually present in solution.

▶ **A net ionic equation** is an equation showing only the species that actually change during the reaction.

EXAMPLE 4.12 Writing Complete Ionic and Net Ionic Equations

Consider the following precipitation reaction occurring in aqueous solution:

$$3\,\text{SrCl}_2(aq) + 2\,\text{Li}_3\text{PO}_4(aq) \longrightarrow \text{Sr}_3(\text{PO}_4)_2(s) + 6\,\text{LiCl}(aq)$$

Write the complete ionic equation and net ionic equation for this reaction.

SOLUTION Write the complete ionic equation by separating aqueous ionic compounds into their constituent ions. The $\text{Sr}_3(\text{PO}_4)_2(s)$, since it precipitates as a solid, remains as one unit.	**Complete ionic equation:** $3\,\text{Sr}^{2+}(aq) + 6\,\text{Cl}^-(aq) + 6\,\text{Li}^+(aq) + 2\,\text{PO}_4{}^{3-}(aq) \longrightarrow$ $\text{Sr}_3(\text{PO}_4)_2(s) + 6\,\text{Li}^+(aq) + 6\,\text{Cl}^-(aq)$
Write the net ionic equation by eliminating the spectator ions, those that do not change from one side of the reaction to the other.	**Net ionic equation:** $3\,\text{Sr}^{2+}(aq) + 2\,\text{PO}_4{}^{3-}(aq) \longrightarrow \text{Sr}_3(\text{PO}_4)_2(s)$

FOR PRACTICE 4.12

Consider the following reaction occurring in aqueous solution:

$$2\,\text{HI}(aq) + \text{Ba(OH)}_2(aq) \longrightarrow 2\,\text{H}_2\text{O}(l) + \text{BaI}_2(aq)$$

Write the complete ionic equation and net ionic equation for this reaction.

FOR MORE PRACTICE 4.12

Write complete ionic and net ionic equations for the following reaction occurring in aqueous solution:

$$2\,\text{AgNO}_3(aq) + \text{MgCl}_2(aq) \longrightarrow 2\,\text{AgCl}(s) + \text{Mg(NO}_3)_2(aq)$$

4.8 Acid–Base and Gas-Evolution Reactions

Two other important classes of reactions that occur in aqueous solution are acid–base reactions and gas-evolution reactions. In an **acid–base reaction** (also called a **neutralization reaction**), an acid reacts with a base and the two neutralize each other, producing water (or in some cases a weak electrolyte). In a **gas-evolution reaction**, a gas forms, resulting in bubbling. In both cases, as in precipitation reactions, the reactions occur when the anion from one reactant combines with the cation of the other. In addition, many gas-evolution reactions are also acid–base reactions.

Acid–Base Reactions

Our stomachs contain hydrochloric acid, which acts in the digestion of food. Certain foods or stress, however, can increase the stomach's acidity to uncomfortable levels, causing acid stomach or heartburn. Antacids are over-the-counter medicines that work by reacting with, and neutralizing, stomach acid. Antacids employ different *bases*—substances that produce hydroxide (OH^-) ions in water—as neutralizing agents. Milk of magnesia, for example, contains $Mg(OH)_2$ and Mylanta contains $Al(OH)_3$. All antacids, however, have the same effect of neutralizing stomach acid through *acid–base reactions* and relieving heartburn.

Recall from Chapter 3 that an acid forms H^+ ions in solution, and we just saw that a base is a substance that produces OH^- ions in solution:

▶ **Acid Substance that produces H^+ ions in aqueous solution**
▶ **Base Substance that produces OH^- ions in aqueous solution**

These definitions of acids and bases are called the **Arrhenius definitions**, after Swedish chemist Svante Arrhenius (1859–1927). In Chapter 15, we will learn more general definitions of acid–base behavior, but these are sufficient to describe neutralization reactions.

According to the Arrhenius definition, HCl is an acid because it produces H^+ ions in solution:

$$HCl(aq) \longrightarrow H^+(aq) + Cl^-(aq)$$

An H^+ ion is a bare proton. Protons associate with water molecules in solution to form **hydronium ions (Figure 4.13▶)**:

$$H^+(aq) + H_2O(l) \longrightarrow H_3O^+(aq)$$

Chemists use $H^+(aq)$ and $H_3O^+(aq)$ interchangeably to mean the same thing—an H^+ ion dissolved in water. The ionization of HCl and other acids is often written to show the association of the proton with a water molecule to form the hydronium ion:

$$HCl(aq) + H_2O(l) \longrightarrow H_3O^+(aq) + Cl^-(aq)$$

Some acids—called **polyprotic acids**—contain more than one ionizable proton and release them sequentially. For example, sulfuric acid, H_2SO_4, is a **diprotic acid**. It is strong in its first ionizable proton, but weak in its second:

$$H_2SO_4(aq) \longrightarrow H^+(aq) + HSO_4^-(aq)$$
$$HSO_4^-(aq) \rightleftharpoons H^+(aq) + SO_4^{2-}(aq)$$

According to the the Arrhenius definition, NaOH is a base because it produces OH^- ions in solution:

$$NaOH(aq) \longrightarrow Na^+(aq) + OH^-(aq)$$

In analogy to diprotic acids, some bases, such as $Sr(OH)_2$, for example, produce two moles of OH^- per mole of the base.

$$Sr(OH)_2(aq) \longrightarrow Sr^{2+}(aq) + 2\,OH^-(aq)$$

Common acids and bases are listed in Table 4.2. Acids and bases are in many everyday substances.

▲ In a gas-evolution reaction, such as the reaction of hydrochloric acid with limestone ($CaCO_3$) to produce CO_2, bubbling typically occurs as the gas is released.

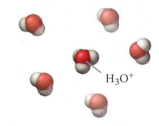

▲ **FIGURE 4.13 The Hydronium Ion** Protons normally associate with water molecules in solution to form H_3O^+ ions, which in turn interact with other water molecules.

▲ Acids are found in lemons, limes, and vinegar. Vitamin C and aspirin are also acids.

▲ Many common household products contain bases.

TABLE 4.2 Some Common Acids and Bases

Name of Acid	Formula	Name of Base	Formula
Hydrochloric acid	HCl	Sodium hydroxide	NaOH
Hydrobromic acid	HBr	Lithium hydroxide	LiOH
Hydroiodic acid	HI	Potassium hydroxide	KOH
Nitric acid	HNO_3	Calcium hydroxide	$Ca(OH)_2$
Sulfuric acid	H_2SO_4	Barium hydroxide	$Ba(OH)_2$
Perchloric acid	$HClO_4$	Ammonia*	NH_3 (weak base)
Acetic acid	$HC_2H_3O_2$ (weak acid)		
Hydrofluoric acid	HF (weak acid)		

*Ammonia does not contain OH^-, but it produces OH^- in a reaction with water that occurs only to a small extent:
$NH_3(aq) + H_2O(l) \rightleftharpoons NH_4^+(aq) + OH^-(aq).$

When an acid and a base are mixed, the $H^+(aq)$ from the acid—whether it is weak or strong—combines with the $OH^-(aq)$ from the base to form $H_2O(l)$ (**Figure 4.14▼**). For example, consider the reaction between hydrochloric acid and sodium hydroxide:

$$HCl(aq) + NaOH(aq) \longrightarrow H_2O(l) + NaCl(aq)$$

Acid   Base   Water   Salt

▶ **FIGURE 4.14 Acid–Base Reaction**   The reaction between hydrochloric acid and sodium hydroxide forms water and a salt, sodium chloride, which remains dissolved in the solution.

Acid–Base Reaction

$$HCl(aq) + NaOH(aq) \longrightarrow H_2O(l) + NaCl(aq)$$

The reaction between hydrochloric acid and sodium hydroxide forms water and a salt, sodium chloride, which remains dissolved in the solution.

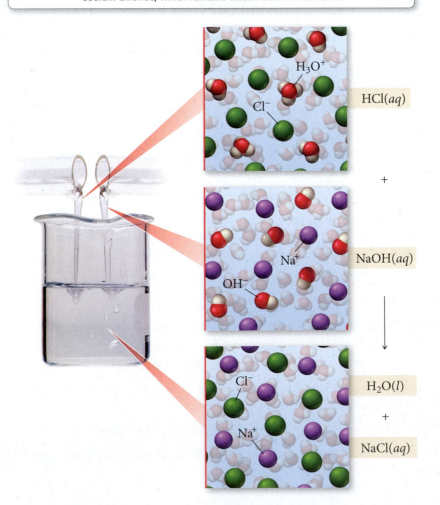

H_3O^+

Cl^-

HCl(aq)

+

Na^+

OH^-

NaOH(aq)

Cl^-

$H_2O(l)$

+

Na^+

NaCl(aq)

Acid–base reactions generally form water and an ionic compound—called a **salt**—that usually remains dissolved in the solution. The net ionic equation for many acid–base reactions is

$$H^+(aq) + OH^-(aq) \longrightarrow H_2O(l)$$

> The word *salt* in this sense applies to any ionic compound and is more general than the common usage, which refers only to table salt (NaCl).

Another example of an acid–base reaction is the reaction between sulfuric acid and potassium hydroxide:

$$\underset{\text{acid}}{H_2SO_4(aq)} + \underset{\text{base}}{2\ KOH(aq)} \longrightarrow \underset{\text{water}}{2\ H_2O(l)} + \underset{\text{salt}}{K_2SO_4(aq)}$$

Again, notice the pattern of acid and base reacting to form water and a salt.

Acid + Base $\longrightarrow$ Water + Salt (acid–base reactions)

When writing equations for acid–base reactions, write the formula of the salt using the procedure for writing formulas of ionic compounds from Section 3.5.

EXAMPLE 4.13 Writing Equations for Acid–Base Reactions

Write a molecular and net ionic equation for the reaction between aqueous HI and aqueous $Ba(OH)_2$.

SOLUTION You must first recognize these substances as an acid and a base. Begin by writing the unbalanced equation in which the acid and the base combine to form water and a salt.	$\underset{\text{acid}}{HI(aq)} + \underset{\text{base}}{Ba(OH)_2(aq)} \longrightarrow$ $\qquad\qquad \underset{\text{water}}{H_2O(l)} + \underset{\text{salt}}{BaI_2(aq)}$
Next, balance the equation; this is the molecular equation.	$2\ HI(aq) + Ba(OH)_2(aq) \longrightarrow$ $\qquad\qquad 2H_2O(l) + BaI_2(aq)$
Write the net ionic equation by removing the spectator ions.	$2\ H^+(aq) + 2\ OH^-(aq) \longrightarrow 2\ H_2O(l)$ or simply $H^+(aq) + OH^-(aq) \longrightarrow H_2O(l)$

FOR PRACTICE 4.13

Write a molecular and a net ionic equation for the reaction that occurs between aqueous H_2SO_4 and aqueous LiOH.

Gas-Evolution Reactions

Aqueous reactions that form a gas when two solutions are mixed are *gas-evolution reactions*. Some gas-evolution reactions form a gaseous product directly when the cation of one reactant combines with the anion of the other. For example, when sulfuric acid reacts with lithium sulfide, dihydrogen sulfide gas forms:

$$H_2SO_4(aq) + Li_2S(aq) \longrightarrow \underset{\text{gas}}{H_2S(g)} + Li_2SO_4(aq)$$

Other gas-evolution reactions often form an intermediate product that then decomposes (breaks down into component elements) into a gas. For example, when aqueous hydrochloric acid is mixed with aqueous sodium bicarbonate the following reaction occurs (**Figure 4.15▶**):

$$HCl(aq) + NaHCO_3(aq) \rightarrow H_2CO_3(aq) + NaCl(aq) \longrightarrow$$
$$\qquad\qquad\qquad H_2O(l) + \underset{\text{gas}}{CO_2(g)} + NaCl(aq)$$

> Many gas-evolution reactions such as this one are also acid–base reactions. In Chapter 15 we will learn how ions such as CO_3^{2-} act as bases in aqueous solution.

Gas-Evolution Reaction

$$NaHCO_3(aq) + HCl(aq) \longrightarrow H_2O(l) + NaCl(aq) + CO_2(g)$$

When aqueous sodium bicarbonate is mixed with aqueous hydrochloric acid, gaseous CO_2 bubbles are the result of the reaction.

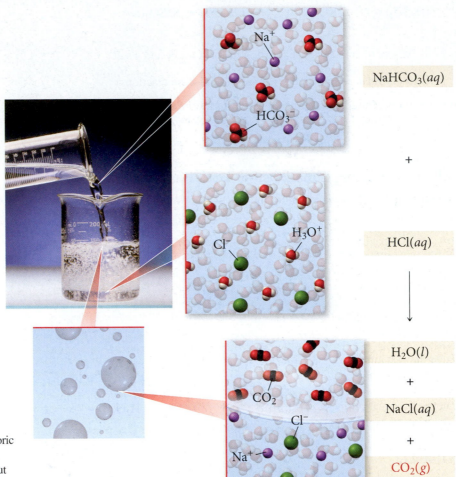

▶ **FIGURE 4.15 Gas-Evolution Reaction** When aqueous hydrochloric acid is mixed with aqueous sodium bicarbonate, gaseous CO_2 bubbles out of the reaction mixture.

The intermediate product, H_2CO_3, is not stable and decomposes into H_2O and gaseous CO_2. Other important gas-evolution reactions form H_2SO_3 or NH_4OH as intermediate products:

The intermediate product NH_4OH provides a convenient way to think about this reaction, but the extent to which it actually forms is debatable.

$$HCl(aq) + NaHSO_3(aq) \longrightarrow H_2SO_3(aq) + NaCl(aq) \longrightarrow$$
$$H_2O(l) + SO_2(g) + NaCl(aq)$$
$$NH_4Cl(aq) + NaOH(aq) \longrightarrow NH_4OH(aq) + NaCl(aq) \longrightarrow$$
$$H_2O(l) + NH_3(g) + NaCl(aq)$$

Table 4.3 list the main types of compounds that form gases in aqueous reactions, as well as the gases that form or evolve.

TABLE 4.3 Types of Compounds That Undergo Gas-Evolution Reactions

Reactant Type	Intermediate Product	Gas Evolved	Example
Sulfides	None	H_2S	$2 HCl(aq) + K_2S(aq) \rightarrow H_2S(g) + 2 KCl(aq)$
Carbonates and bicarbonates	H_2CO_3	CO_2	$2 HCl(aq) + K_2CO_3(aq) \rightarrow H_2O(l) + CO_2(g) + 2 KCl(aq)$
Sulfites and bisulfites	H_2SO_3	SO_2	$2 HCl(aq) + K_2SO_3(aq) \rightarrow H_2O(l) + SO_2(g) + 2 KCl(aq)$
Ammonium	NH_4OH	NH_3	$NH_4Cl(aq) + KOH(aq) \rightarrow H_2O(l) + NH_3(g) + KCl(aq)$

EXAMPLE 4.14 Writing Equations for Gas-Evolution Reactions

Write a molecular equation for the gas-evolution reaction that occurs when aqueous nitric acid and aqueous sodium carbonate are mixed.

Begin by writing an unbalanced equation in which the cation of each reactant combines with the anion of the other.	$HNO_3(aq) + Na_2CO_3(aq) \longrightarrow$ $\qquad\qquad\qquad H_2CO_3(aq) + NaNO_3(aq)$
You must then recognize that $H_2CO_3(aq)$ decomposes into $H_2O(l)$ and $CO_2(g)$ and write these products into the equation.	$HNO_3(aq) + Na_2CO_3(aq) \longrightarrow H_2O(l) + CO_2(g) + NaNO_3(aq)$
Finally, balance the equation.	$2\,HNO_3(aq) + Na_2CO_3(aq) \longrightarrow H_2O(l) + CO_2(g) + 2\,NaNO_3(aq)$

FOR PRACTICE 4.14

Write a molecular equation for the gas-evolution reaction that occurs when aqueous hydrobromic acid and aqueous potassium sulfite are mixed.

FOR MORE PRACTICE 4.14

Write a net ionic equation for the reaction that occurs when hydroiodic acid and calcium sulfide are mixed.

4.9 Oxidation–Reduction Reactions

Oxidation–reduction reactions or redox reactions are reactions in which electrons are transferred from one reactant to the other. The rusting of iron, the bleaching of hair, and the production of electricity in batteries involve redox reactions. Many redox reactions involve the reaction of a substance with oxygen (**Figure 4.16▼**):

$$4\,Fe(s) + 3\,O_2(g) \longrightarrow 2\,Fe_2O_3(s) \qquad \text{(rusting of iron)}$$
$$2\,C_8H_{18}(l) + 25\,O_2(g) \longrightarrow 16\,CO_2(g) + 18\,H_2O(g) \quad \text{(combustion of octane)}$$
$$2\,H_2(g) + O_2(g) \longrightarrow 2\,H_2O(g) \qquad \text{(combustion of hydrogen)}$$

Oxidation–reduction reactions are covered in more detail in Chapter 18.

Oxidation–Reduction Reaction

$$2\,H_2(g) + O_2(g) \longrightarrow 2\,H_2O(g)$$

Hydrogen and oxygen in the balloon react to form gaseous water.

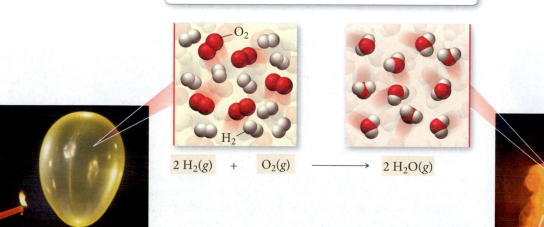

$$2\,H_2(g) + O_2(g) \longrightarrow 2\,H_2O(g)$$

▲ **FIGURE 4.16 Oxidation–Reduction Reaction** When heat is applied, the hydrogen in the balloon reacts explosively with oxygen to form gaseous water.

Oxidation–Reduction Reaction without Oxygen

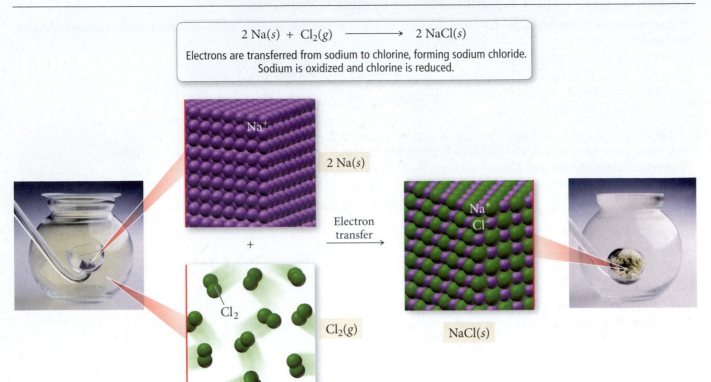

$$2\,Na(s) + Cl_2(g) \longrightarrow 2\,NaCl(s)$$

Electrons are transferred from sodium to chlorine, forming sodium chloride.
Sodium is oxidized and chlorine is reduced.

Na⁺

2 Na(s)

+

Cl₂

Electron
transfer

Na⁺
Cl⁻

Cl₂(g)

NaCl(s)

▲ **FIGURE 4.17 Oxidation–Reduction without Oxygen**   When sodium reacts with chlorine, electrons are transferred from the sodium to the chlorine, resulting in the formation of sodium chloride. In this redox reaction, sodium is oxidized and chlorine is reduced.

The reaction between sodium and oxygen forms other oxides as well.

Helpful Mnemonic O I L R I G—
Oxidation **I**s **L**oss; **R**eduction **I**s **G**ain.

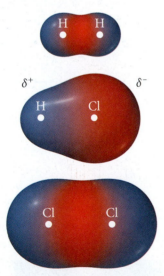

▲ **FIGURE 4.18 Redox with Partial Electron Transfer**   When hydrogen bonds to chlorine, the electrons are unevenly shared, resulting in an increase of electron density (reduction) for chlorine and a decrease in electron density (oxidation) for hydrogen.

However, redox reactions need not involve oxygen. Consider, for example, the reaction between sodium and chlorine to form sodium chloride (NaCl), depicted in **Figure 4.17▲**:

$$2\,Na(s) + Cl_2(g) \longrightarrow 2\,NaCl(s)$$

This reaction is similar to the reaction between sodium and oxygen to form sodium oxide:

$$4\,Na(s) + O_2(g) \longrightarrow 2\,Na_2O(s)$$

In both cases, a metal (which has a tendency to lose electrons) reacts with a nonmetal (which has a tendency to gain electrons). In both cases, metal atoms lose electrons to nonmetal atoms. A fundamental definition of **oxidation** is the loss of electrons, and a fundamental definition of **reduction** is the gain of electrons.

The transfer of electrons, however, need not be a *complete* transfer (as occurs in the formation of an ionic compound) for the reaction to qualify as oxidation–reduction. For example, consider the reaction between hydrogen gas and chlorine gas:

$$H_2(g) + Cl_2(g) \longrightarrow 2\,HCl(g)$$

Even though hydrogen monochloride is a molecular compound with a covalent bond, and even though the hydrogen has not completely transferred its electron to chlorine during the reaction, you can see from the electron density diagrams (**Figure 4.18◄**) that hydrogen has lost some of its electron density—it has *partially* transferred its electron to chlorine. Therefore, in the above reaction, hydrogen is oxidized and chlorine is reduced and the reaction is a redox reaction.

Oxidation States

Identifying a reaction between a metal and a nonmetal as a redox reaction is fairly straightforward because the metal becomes a cation and the nonmetal becomes an anion (electron transfer is obvious). However, how do we identify redox reactions that occur between nonmetals? Chemists have devised a scheme to track electrons before and after a chemical reaction. In this scheme—which is like bookkeeping for electrons—all shared electrons are assigned to the atom that attracts the electrons most strongly. Then a number, called the **oxidation state** or **oxidation number**, is given to each atom based on the electron assignments. In other words, the oxidation number of an atom in a compound is the "charge" it would have if all shared electrons were assigned to the atom with a greater attraction for those electrons.

For example, consider HCl. Since chlorine attracts electrons more strongly than hydrogen, we assign the two shared electrons in the bond to chlorine; then H (which has lost an electron in our assignment) has an oxidation state of +1 and Cl (which has gained one electron in our assignment) has an oxidation state of −1 Notice that, in contrast to ionic charges, which are usually written with the sign of the charge after the magnitude (1+ and 1−, for example), oxidation states are written with the sign of the charge before the magnitude (+1 and −1, for example). Use the following rules to assign oxidation states to atoms in elements and compounds.

> The ability of an element to attract electrons in a chemical bond is called electronegativity. Electronegativity is covered in more detail in Section 9.6.

Rules for Assigning Oxidation States	Examples
(These rules are hierarchical. If any two rules conflict, follow the rule that is higher on the list.)	

1. The oxidation state of an atom in a free element is 0.

$$Cu \qquad Cl_2$$
$$\text{0 ox state} \qquad \text{0 ox state}$$

2. The oxidation state of a monoatomic ion is equal to its charge.

$$Ca^{2+} \qquad Cl^-$$
$$\text{+2 ox state} \qquad \text{−1 ox state}$$

3. The sum of the oxidation states of all atoms in:
   - A neutral molecule or formula unit is 0.

$$H_2O$$
$$2(\text{H ox state}) + 1(\text{O ox state}) = 0$$

   - An ion is equal to the charge of the ion.

$$NO_3^-$$
$$1(\text{N ox state}) + 3(\text{O ox state}) = -1$$

4. In their compounds, metals have positive oxidation states.
   - Group 1A metals *always* have an oxidation state of +1.

$$NaCl$$
$$\text{+1 ox state}$$

   - Group 2A metals *always* have an oxidation state of +2.

$$CaF_2$$
$$\text{+2 ox state}$$

5. In their compounds, nonmetals are assigned oxidation states according to the table at right. Entries at the top of the table take precedence over entries at the bottom of the table.

> Do not confuse oxidation state with ionic charge. Unlike ionic charge—which is a real property of an ion—the oxidation state of an atom is merely a theoretical (but useful) construct.

When assigning oxidation states, keep these points in mind:

- The oxidation state of any given element generally depends on what other elements are present in the compound. (The exceptions are the group 1A and 2A metals, which are *always* +1 and +2 respectively.)
- Rule 3 must always be followed. Therefore, when following the hierarchy shown in rule 5, give priority to the element(s) highest on the list and then assign the oxidation state of the element lowest on the list using rule 3.
- When assigning oxidation states to elements that are not covered by rules 4 and 5 (such as carbon) use rule 3 to deduce their oxidation state once all other oxidation states have been assigned.

Oxidation States of Nonmetals

Nonmetal	Oxidation State	Example
Fluorine	−1	MgF_2 −1 ox state
Hydrogen	+1	H_2O +1 ox state
Oxygen	−2	CO_2 −2 ox state
Group 7A	−1	CCl_4 −1 ox state
Group 6A	−2	H_2S −2 ox state
Group 5A	−3	NH_3 −3 ox state

EXAMPLE 4.15 Assigning Oxidation States

Assign an oxidation state to each atom in each compound.

(a) Cl_2 **(b)** Na^+ **(c)** KF   **(d)** CO_2 **(e)** SO_4^{2-} **(f)** K_2O_2

SOLUTION

Since Cl_2 is a free element, the oxidation state of both Cl atoms is 0 (rule 1).	**(a)** Cl_2 ClCl 0 0
Since Na^+ is a monoatomic ion, the oxidation state of the Na^+ ion is +1 (rule 2).	**(b)** Na^+ Na^+ +1
The oxidation state of K is +1 (rule 4). The oxidation state of F is –1 (rule 5). Since this is a neutral compound, the sum of the oxidation states is 0.	**(c)** KF KF + 1 –1 sum: + 1 –1 = 0
The oxidation state of oxygen is –2 (rule 5). Deduce the oxidation state of carbon by using rule 3, which states that the sum of the oxidation states of all the atoms must be 0.	**(d)** CO_2 (C ox state) + 2(O ox state) = 0 (C ox state) + 2(−2) = 0 C ox state = + 4 CO_2 + 4 −2 sum: + 4 + 2(−2) = 0
The oxidation state of oxygen is –2 (rule 5). You would ordinarily expect the oxidation state of S to be –2 (rule 5). However, if that were the case, the sum of the oxidation states would not equal the charge of the ion. Since O is higher on the list than S, it takes priority. Deduce the oxidation state of sulfur by setting the *sum* of all of the oxidation states equal to –2 (the charge of the ion).	**(e)** SO_4^{2-} (S ox state) + 4(O ox state) = −2 (S ox state) + 4(−2) = −2 S ox state = +6 SO_4^{2-} +6 −2 sum: +6 + 4(−2) = −2
The oxidation state of potassium is +1 (rule 4). You would ordinarily expect the oxidation state of O to be –2 (rule 5), but rule 4 takes priority. Deduce the oxidation state of O by setting the sum of all of the oxidation states equal to 0.	**(f)** K_2O_2 2(K ox state) + 2(O ox state) = 0 2(+1) + 2(O ox state) = 0 O ox state = −1 K_2O_2 + 1 −1 sum: 2(+ 4) + 2(−1) = 0

FOR PRACTICE 4.15

Assign an oxidation state to each atom in the following species.

(a) Cr   **(b)** Cr^{3+} **(c)** CCl_4 **(d)** $SrBr_2$ **(e)** SO_3 **(f)** NO_3^-

In most cases, oxidation states are positive or negative integers; however, on occasion an atom within a compound can have a fractional oxidation state. For example, consider KO_2. The oxidation state is assigned as follows:

$$KO_2$$
$$+1 \quad -\tfrac{1}{2}$$
$$\text{sum:} +1 + 2\left(-\frac{1}{2}\right) = 0$$

In KO_2, oxygen has a $-\frac{1}{2}$ oxidation state. Although this seems unusual, it is accepted because oxidation states are merely an imposed electron bookkeeping scheme, not an actual physical quantity.

Identifying Redox Reactions

Oxidation states can be used to identify redox reactions, even between nonmetals. For example, is the following reaction between carbon and sulfur a redox reaction?

$$C + 2\,S \longrightarrow CS_2$$

If so, what element is oxidized? What element is reduced? We can use the oxidation state rules to assign oxidation states to all elements on both sides of the equation.

Carbon changed from an oxidation state of 0 to an oxidation state of +4. In terms of our electron bookkeeping scheme (the assigned oxidation state), carbon *lost electrons* and was *oxidized*. Sulfur changed from an oxidation state of 0 to an oxidation state of –2. In terms of our electron bookkeeping scheme, sulfur *gained electrons* and was *reduced*. In terms of oxidation states, oxidation and reduction are defined as follows.

▶ **Oxidation   An increase in oxidation state**

▶ **Reduction   A decrease in oxidation state**

| Remember that a reduction is a *reduction* in oxidation state.

EXAMPLE 4.16 Using Oxidation States to Identify Oxidation and Reduction

Use oxidation states to identify the element that is being oxidized and the element that is being reduced in the following redox reaction.

$$Mg(s) + 2\,H_2O(l) \longrightarrow Mg(OH)_2(aq) + H_2(g)$$

SOLUTION

Begin by assigning oxidation states to each atom in the reaction.

$$
\begin{array}{ccccc}
Mg(s) + & 2\,H_2O(l) & \longrightarrow & Mg(OH)_2(aq) + & H_2(g) \\
\text{Oxidation states:} \quad 0 & +1 \ -2 & & +2 \ -2 \ +1 & 0
\end{array}
$$

Since Mg increased in oxidation state, it was oxidized. Since H decreased in oxidation state, it was reduced.

FOR PRACTICE 4.16

Use oxidation states to identify the element that is being oxidized and the element that is being reduced in the following redox reaction.

$$Sn(s) + 4\,HNO_3(aq) \longrightarrow SnO_2(s) + 4\,NO_2(g) + 2\,H_2O(g)$$

FOR MORE PRACTICE 4.16

Determine if each reaction is a redox reaction. If the reaction is a redox reaction, identify which element is oxidized and which is reduced.

(a) $Hg_2(NO_3)_2(aq) + 2\,KBr(aq) \longrightarrow Hg_2Br_2(s) + 2\,KNO_3(aq)$

(b) $4\,Al(s) + 3\,O_2(g) \longrightarrow 2\,Al_2O_3(s)$

(c) $CaO(s) + CO_2(g) \longrightarrow CaCO_3(s)$

Notice that *oxidation and reduction must occur together.* If one substance loses electrons (oxidation), then another substance must gain electrons (reduction). A substance that causes the oxidation of another substance is an **oxidizing agent**. Oxygen, for example, is an excellent oxidizing agent because it causes the oxidation of many other substances. In a redox reaction, *the oxidizing agent is always reduced.* A substance that causes the reduction of another substance is a **reducing agent**. Hydrogen, for example, as well as the group 1A and group 2A metals (because of their tendency to lose electrons) are excellent reducing agents. In a redox reaction, *the reducing agent is always oxidized.*

In Section 18.2 you will learn more about redox reactions, including how to balance them. For now, be able to identify redox reactions, as well as oxidizing and reducing agents, according to these guidelines.

Redox reactions include:

- **Any reaction in which there is a change in the oxidation states of atoms between the reactants and the products.**

In a redox reaction:

- **The oxidizing agent oxidizes another substance (and is itself reduced).**
- **The reducing agent reduces another substance (and is itself oxidized).**

EXAMPLE 4.17 Identifying Redox Reactions, Oxidizing Agents, and Reducing Agents

Determine whether each reaction is an oxidation–reduction reaction. If the reaction is an oxidation–reduction, identify the oxidizing agent and the reducing agent.

(a) $2\,Mg(s) + O_2(g) \longrightarrow 2\,MgO(s)$

(b) $2\,HBr(aq) + Ca(OH)_2(aq) \longrightarrow 2\,H_2O(l) + CaBr_2(aq)$

(c) $Zn(s) + Fe^{2+}(aq) \longrightarrow Zn^{2+}(aq) + Fe(s)$

SOLUTION

This is a redox reaction because magnesium increases in oxidation number (oxidation) and oxygen decreases in oxidation number (reduction).	**(a)** $2\,\underset{0}{Mg}(s) + \underset{0}{O_2}(g) \longrightarrow 2\,\underset{+2\;-2}{MgO}(s)$ Oxidation (Mg: 0 → +2); Reduction (O: 0 → −2) Oxidizing agent: O_2 Reducing agent: Mg
This is not a redox reaction because none of the atoms undergoes a change in oxidation number.	**(b)** $2\,\underset{+1\;-1}{HBr}(aq) + \underset{+2\;-2+1}{Ca(OH)_2}(aq) \longrightarrow 2\,\underset{+1\;-2}{H_2O}(l) + \underset{+2\;-1}{CaBr_2}(aq)$
This is a redox reaction because zinc increases in oxidation number (oxidation) and iron decreases in oxidation number (reduction).	**(c)** $\underset{0}{Zn}(s) + \underset{+2}{Fe^{2+}}(aq) \longrightarrow \underset{+2}{Zn^{2+}}(aq) + \underset{0}{Fe}(s)$ Oxidation (Zn: 0 → +2); Reduction (Fe: +2 → 0) Oxidizing agent: Fe^{2+} Reducing agent: Zn

FOR PRACTICE 4.17

Determine whether each reaction is a redox reaction. For all redox reactions, identify the oxidizing agent and the reducing agent.

(a) $2\,Li(s) + Cl_2(g) \longrightarrow 2\,LiCl(s)$

(b) $2\,Al(s) + 3\,Sn^{2+}(aq) \longrightarrow 2\,Al^{3+}(aq) + 3\,Sn(s)$

(c) $Pb(NO_3)_2(aq) + 2\,LiCl(aq) \longrightarrow PbCl_2(s) + 2\,LiNO_3(aq)$

(d) $C(s) + O_2(g) \longrightarrow CO_2(g)$

 **Conceptual Connection 4.7 Oxidation and Reduction**

Which statement is true regarding redox reactions?

(a) A redox reaction can occur without any changes in the oxidation states of the elements within the reactants and products of a reaction.

(b) If any of the reactants or products in a reaction contains oxygen, the reaction is a redox reaction.

(c) In a reaction, oxidation can occur independently of reduction.

(d) In a redox reaction, any increase in the oxidation state of a reactant must be accompanied by a decrease in the oxidation state of a reactant.

Combustion Reactions

We encountered combustion reactions, a type of redox reaction, in the opening section of this chapter. Combustion reactions are important because most of our society's energy is derived from them (**Figure 4.19▼**).

Combustion reactions are characterized by the reaction of a substance with O_2 to form one or more oxygen-containing compounds, often including water. Combustion reactions also emit heat. For example, as we saw earlier in this chapter, natural gas (CH_4) reacts with oxygen to form carbon dioxide and water:

$$CH_4(g) + 2\,O_2(g) \longrightarrow CO_2(g) + 2\,H_2O(g)$$

Oxidation state: $-4\ +1$ 0 $+4\ -2$ $+1\ -2$

In this reaction, carbon is oxidized and oxygen is reduced. Ethanol, the alcohol in alcoholic beverages, also reacts with oxygen in a combustion reaction to form carbon dioxide and water:

$$C_2H_5OH(l) + 3\,O_2(g) \longrightarrow 2\,CO_2(g) + 3\,H_2O(g)$$

Compounds containing carbon and hydrogen—or carbon, hydrogen, and oxygen—always form carbon dioxide and water upon complete combustion. Other combustion reactions include the reaction of carbon with oxygen to form carbon dioxide:

$$C(s) + O_2(g) \longrightarrow CO_2(g)$$

and the reaction of hydrogen with oxygen to form water:

$$2\,H_2(g) + O_2(g) \longrightarrow 2\,H_2O(g)$$

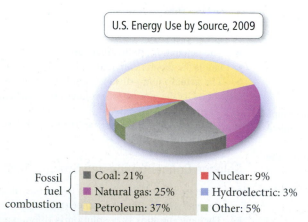

U.S. Energy Use by Source, 2009

Fossil fuel combustion
- Coal: 21%
- Natural gas: 25%
- Petroleum: 37%
- Nuclear: 9%
- Hydroelectric: 3%
- Other: 5%

▲ **FIGURE 4.19 U.S. Energy Consumption**   About 83% of the energy used in the United States in 2009 was produced by combustion reactions.

Source: U.S. Energy Information Administration, *Annual Energy Review*, 2009

EXAMPLE 4.18 Writing Equations for Combustion Reactions

Write a balanced equation for the combustion of liquid methyl alcohol (CH_3OH).

SOLUTION

Begin by writing an unbalanced equation showing the reaction of CH_3OH with O_2 to form CO_2 and H_2O.	$CH_3OH(l) + O_2(g) \longrightarrow$ $CO_2(g) + H_2O(g)$
Balance the equation using the guidelines in Section 3.10.	$2\,CH_3OH(l) + 3\,O_2(g) \longrightarrow$ $2\,CO_2(g) + 4\,H_2O(g)$

FOR PRACTICE 4.18

Write a balanced equation for the complete combustion of liquid C_2H_5SH.

CHAPTER IN REVIEW

Key Terms

Section 4.2
stoichiometry (121)

Section 4.3
limiting reactant (126)
theoretical yield (126)
actual yield (126)
percent yield (126)

Section 4.4
solution (131)
solvent (131)
solute (131)
aqueous solution (131)
dilute solution (131)
concentrated solution (131)

molarity (M) (131)
stock solution (134)

Section 4.5
electrolyte (139)
strong electrolyte (139)
nonelectrolyte (140)
strong acid (140)
weak acid (140)
weak electrolyte (140)
soluble (140)
insoluble (140)

Section 4.6
precipitation reaction (142)
precipitate (142)

Section 4.7
molecular equation (145)
complete ionic equation (145)
spectator ion (145)
net ionic equation (146)

Section 4.8
acid–base reaction
   (neutralization
   reaction) (147)
gas-evolution reaction (147)
Arrhenius definitions (147)
hydronium ion (147)

polyprotic acid (147)
diprotic acid (147)
salt (149)

Section 4.9
oxidation–reduction (redox)
   reaction (151)
oxidation (152)
reduction (152)
oxidation state (oxidation
   number) (153)
oxidizing agent (156)
reducing agent (156)

Key Concepts

Climate Change and the Combustion of Fossil Fuels (4.1)

▶ Climate change, caused by rising atmospheric carbon dioxide levels, is potentially harmful. The largest carbon dioxide source is the burning of fossil fuels, which can be verified by reaction stoichiometry.

Reaction Stoichiometry (4.2)

▶ Reaction stoichiometry refers to the numerical relationships between the reactants and products in a balanced chemical equation.

▶ Reaction stoichiometry allows us to predict, for example, the amount of product that can be formed for a given amount of reactant, or how much of one reactant is required to react with a given amount of another.

Limiting Reactant, Theoretical Yield, and Percent Yield (4.3)

▶ The limiting reactant in a chemical reaction is the reactant that is present in the smallest stoichiometric quantity; it will be completely

consumed in the reaction and it limits the amount of product that can be made.

▶ Any reactant that does not limit the amount of product is said to be in excess.

▶ The amount of product that can be made from the limiting reactant is the theoretical yield.

▶ The actual yield—always equal to or less than the theoretical yield—is the amount of product that is actually made when the reaction is carried out.

▶ The percentage of the theoretical yield that is actually produced is the percent yield.

Solution Concentration and Stoichiometry (4.4)

▶ An aqueous solution is a homogeneous mixture of water (the solvent) with another substance (the solute).

▶ The concentration of a solution is often expressed in molarity, the number of moles of solute per liter of solution.

▶ We can use the molarities and volumes of reactant solutions to predict the amount of product that will form in an aqueous reaction.

Aqueous Solutions and Precipitation Reactions (4.5, 4.6)

▶ Solutes that completely dissociate (or completely ionize in the case of the acids) to ions in solution are strong electrolytes and are good conductors of electricity.

▶ Solutes that only partially dissociate (or partially ionize) are weak electrolytes, and solutes that do not dissociate (or ionize) at all are nonelectrolytes.

▶ A substance that dissolves in water to form a solution is said to be soluble.

▶ In a precipitation reaction, two aqueous solutions are mixed and a solid—or precipitate—forms.

▶ The solubility rules are an empirical set of guidelines that help predict the solubilities of ionic compounds; these rules are especially useful when determining whether or not a precipitate will form.

Equations for Aqueous Reactions (4.7)

▶ We can represent an aqueous reaction with a molecular equation, which shows the complete neutral formula for each compound in the reaction.

▶ An aqueous reaction can also be represented with a complete ionic equation, which shows the dissociated nature of the aqueous ionic compounds.

▶ A third representation of an aqueous reaction is a net ionic equation, in which the spectator ions—those that do not change in the course of the reaction—are left out of the equation.

Acid–Base and Gas-Evolution Reactions (4.8)

▶ In an acid–base reaction, an acid, a substance which produces H^+ in solution, reacts with a base, a substance which produces OH^- in solution, and the two neutralize each other, producing water (or in some cases a weak electrolyte).

▶ In gas-evolution reactions, two aqueous solutions are combined and a gas is produced.

Oxidation–Reduction Reactions (4.9)

▶ In oxidation–reduction reactions, one substance transfers electrons to another substance.

▶ The substance that loses electrons is oxidized and the one that gains them is reduced.

▶ An oxidation state is a charge given to each atom in a redox reaction by assigning all shared electrons to the atom with the greater attraction for those electrons. Oxidation states are an imposed electronic bookkeeping scheme, not an actual physical state.

▶ The oxidation state of an atom increases upon oxidation and decreases upon reduction.

▶ A combustion reaction is a specific type of oxidation—a reduction reaction in which a substance reacts with oxygen—emitting heat and forming one or more oxygen-containing products.

Key Equations and Relationships

Mass-to-Mass Conversion: Stoichiometry (4.2)

mass A ⟶ amount A (in moles) ⟶

amount B (in moles) ⟶ mass B

Percent Yield (4.3)

$$\% \text{ yield} = \frac{\text{actual yield}}{\text{theoretical yield}} \times 100\%$$

Molarity (M): Solution Concentration (4.4)

$$M = \frac{\text{amount of solute (in mol)}}{\text{volume of solution (in L)}}$$

Solution Dilution (4.4)

$$M_1V_1 = M_2V_2$$

Solution Stoichiometry (4.4)

volume A ⟶ amount A (in moles) ⟶

amount B (in moles) ⟶ volume B

Key Learning Objectives

Chapter Objectives	Assessment
Calculating Stoichiometric Quantities (4.2)	Examples 4.1, 4.2 For Practice 4.1, 4.2 Exercises 7–12
Determining the Limiting Reactant and Calculating Theoretical and Percent Yield (4.3)	Examples 4.3, 4.4 For Practice 4.3, 4.4 Exercises 19–23
Calculating and Using Molarity as a Conversion Factor (4.4)	Examples 4.5, 4.6 For Practice 4.5, 4.6 For More Practice 4.5, 4.6 Exercises 25–30
Determining Solution Dilutions (4.4)	Example 4.7 For Practice 4.7 For More Practice 4.7 Exercises 33, 34
Using Solution Stoichiometry to Find Volumes and Amounts (4.4)	Example 4.8 For Practice 4.8 For More Practice 4.8 Exercises 35–37
Predicting whether a Compound Is Soluble (4.5)	Example 4.9 For Practice 4.9 Exercises 43, 44
Writing Equations for Precipitation Reactions (4.6)	Examples 4.10, 4.11 For Practice 4.10, 4.11 Exercises 45–48

Chapter Objectives	Assessment
Writing Complete Ionic and Net Ionic Equations (4.7)	Example 4.12 For Practice 4.12 For More Practice 4.12 Exercises 49, 50
Writing Equations for Acid–Base Reactions (4.8)	Example 4.13 For Practice 4.13 Exercises 53, 54
Writing Equations for Gas-Evolution Reactions (4.8)	Example 4.14 For Practice 4.14 For More Practice 4.14 Exercises 57, 58
Assigning Oxidation States (4.9)	Example 4.15 For Practice 4.15 Exercises 59–62
Identifying Redox Reactions, Oxidizing Agents, and Reducing Agents Using Oxidation States (4.9)	Examples 4.16, 4.17 For Practice 4.16, 4.17 For More Practice 4.16 Exercises 63, 64
Writing Equations for Combustion Reactions (4.9)	Example 4.18 For Practice 4.18 Exercises 65, 66

EXERCISES

Problems by Topic

Reaction Stoichiometry

1. Consider the unbalanced equation for the combustion of hexane:

$$C_6H_{14}(g) + O_2(g) \rightarrow CO_2(g) + H_2O(g)$$

Balance the equation and determine how many moles of O_2 are required to react completely with 4.9 moles C_6H_{14}.

2. Consider the unbalanced equation for the neutralization of acetic acid:

$$HC_2H_3O_2(aq) + Ba(OH)_2(aq) \rightarrow$$
$$H_2O(l) + Ba(C_2H_3O_2)_2(aq)$$

Balance the equation and determine how many moles of $Ba(OH)_2$ are required to completely neutralize 0.107 mole of $HC_2H_3O_2$.

3. For the reaction shown, calculate how many moles of NO_2 form when each amount of reactant completely reacts.

$$2 N_2O_5(g) \rightarrow 4 NO_2(g) + O_2(g)$$

  a. 1.3 mol N_2O_5 b. 5.8 mol N_2O_5
  c. 10.5 g N_2O_5 d. 1.55 kg N_2O_5

4. For the reaction shown, calculate how many moles of NH_3 form when each amount of reactant completely reacts.

$$3 N_2H_4(l) \rightarrow 4 NH_3(g) + N_2(g)$$

  a. 5.3 mol N_2H_4 b. 2.28 mol N_2H_4
  c. 32.5 g N_2H_4 d. 14.7 kg N_2H_4

5. Consider the balanced equation:

$$SiO_2(s) + 3 C(s) \rightarrow SiC(s) + 2 CO(g)$$

Complete the table showing the appropriate number of moles of reactants and products. If the number of moles of a *reactant* is provided, fill in the required amount of the other reactant, as well as the moles of each product formed. If the number of moles of a *product* is provided, fill in the required amount of each reactant to make that amount of product, as well as the amount of the other product that is made.

Mol SiO₂	Mol C	Mol SiC	Mol CO
3			
	6		

Mol SiO₂	Mol C	Mol SiC	Mol CO
			10
2.8			
	1.55		

6. Consider the balanced equation:

$$2 N_2H_4(g) + N_2O_4(g) \rightarrow 3 N_2(g) + 4 H_2O(g)$$

Complete the table showing the appropriate number of moles of reactants and products. If the number of moles of a reactant is provided, fill in the required amount of the other reactant, as well as the moles of each product formed. If the number of moles of a product is provided, fill in the required amount of each reactant to make that amount of product, as well as the amount of the other product that is made.

Mol N₂H₄	Mol N₂O₄	Mol N₂	Mol H₂O
2			
	5		
			10
2.5			
	4.2		
		11.8	

7. Hydrobromic acid dissolves solid iron according to the reaction:

$$Fe(s) + 2 HBr(aq) \rightarrow FeBr_2(aq) + H_2(g)$$

What mass of HBr (in g) would dissolve a 4.8-g pure iron bar on a padlock? What mass of H_2 would be produced by the complete reaction of the iron bar?

8. Sulfuric acid dissolves aluminum metal according to the reaction:

$$2 Al(s) + 3 H_2SO_4(aq) \rightarrow Al_2(SO_4)_3(aq) + 3 H_2(g)$$

Suppose you wanted to dissolve an aluminum block with a mass of 12.7 g. What minimum mass of H_2SO_4 (in g) would you need? What mass of H_2 gas (in g) would be produced by the complete reaction of the aluminum block?

9. For each reaction, calculate the mass (in grams) of the product formed when 2.5 g of the underlined reactant completely reacts. Assume that there is more than enough of the other reactant.
   a. $\underline{Ba}(s) + Cl_2(g) \rightarrow BaCl_2(s)$
   b. $\underline{CaO}(s) + CO_2(g) \rightarrow CaCO_3(s)$
   c. $2\,\underline{Mg}(s) + O_2(g) \rightarrow 2\,MgO(s)$
   d. $4\,\underline{Al}(s) + 3\,O_2(g) \rightarrow 2\,Al_2O_3(s)$

10. For each reaction, calculate the mass (in grams) of the product formed when 10.4 g of the underlined reactant completely reacts. Assume that there is more than enough of the other reactant.
    a. $2\,\underline{K}(s) + Cl_2(g) \rightarrow 2\,KCl(s)$
    b. $2\,\underline{K}(s) + Br_2(l) \rightarrow 2\,KBr(s)$
    c. $4\,\underline{Cr}(s) + 3\,O_2(g) \rightarrow 2\,Cr_2O_3(s)$
    d. $2\,\underline{Sr}(s) + O_2(g) \rightarrow 2\,SrO(s)$

11. For each acid–base reaction, calculate the mass (in grams) of each acid necessary to completely react with and neutralize 4.85 g of the base.
    a. $HCl(aq) + NaOH(aq) \rightarrow H_2O(l) + NaCl(aq)$
    b. $2\,HNO_3(aq) + Ca(OH)_2(aq) \rightarrow$
    $$2\,H_2O(l) + Ca(NO_3)_2(aq)$$
    c. $H_2SO_4(aq) + 2\,KOH(aq) \rightarrow 2\,H_2O(l) + K_2SO_4(aq)$

12. For each precipitation reaction, calculate how many grams of the first reactant are necessary to completely react with 55.8 g of the second reactant.
    a. $2\,KI(aq) + Pb(NO_3)_2(aq) \rightarrow PbI_2(s) + 2\,KNO_3(aq)$
    b. $Na_2CO_3(aq) + CuCl_2(aq) \rightarrow$
    $$CuCO_3(s) + 2\,NaCl(aq)$$
    c. $K_2SO_4(aq) + Sr(NO_3)_2(aq) \rightarrow$
    $$SrSO_4(s) + 2\,KNO_3(aq)$$

Limiting Reactant, Theoretical Yield, and Percent Yield

13. For the reaction shown, find the limiting reactant for each of the initial amounts of reactants.

    $$2\,Na(s) + Br_2(g) \rightarrow 2\,NaBr(s)$$

    a. 2 mol Na, 2 mol Br_2 b. 1.8 mol Na, 1.4 mol Br_2
    c. 2.5 mol Na, 1 mol Br_2 d. 12.6 mol Na, 6.9 mol Br_2

14. For the reaction shown, find the limiting reactant for each of the initial amounts of reactants.

    $$4\,Al(s) + 3\,O_2(g) \rightarrow 2\,Al_2O_3(s)$$

    a. 1 mol Al, 1 mol O_2 b. 4 mol Al, 2.6 mol O_2
    c. 16 mol Al, 13 mol O_2 d. 7.4 mol Al, 6.5 mol O_2

15. Consider the reaction:

    $$4\,HCl(g) + O_2(g) \rightarrow 2\,H_2O(g) + 2\,Cl_2(g)$$

    Each molecular diagram represents an initial mixture of the reactants. How many molecules of Cl_2 are formed from the reaction mixture that produces the greatest amount of products?

(a)   (b)   (c)

16. Consider the reaction:

    $$2\,CH_3OH(g) + 3\,O_2(g) \rightarrow 2\,CO_2(g) + 4\,H_2O(g)$$

    Each molecular diagram represents an initial mixture of the reactants. How many CO_2 molecules are formed from the reaction mixture that produces the greatest amount of products?

 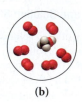 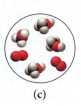

(a)   (b)   (c)

17. For the reaction shown, calculate the theoretical yield of the product (in moles) for each set of initial amounts of reactants.

    $$Ti(s) + 2\,Cl_2(g) \rightarrow TiCl_4(l)$$

    a. 4 mol Ti, 4 mol Cl_2 b. 7 mol Ti, 17 mol Cl_2
    c. 12.4 mol Ti, 18.8 mol Cl_2

18. For the reaction shown, calculate the theoretical yield of product (in moles) for each set of initial amounts of reactants.

    $$2\,Mn(s) + 2\,O_2(g) \rightarrow 2\,MnO_2(s)$$

    a. 3 mol Mn, 3 mol O_2 b. 4 mol Mn, 7 mol O_2
    c. 27.5 mol Mn, 43.8 mol O_2

19. For the reaction shown, calculate the theoretical yield of product (in grams) for each set of the initial amounts of reactants.

    $$2\,Al(s) + 3\,Cl_2(g) \rightarrow 2\,AlCl_3(s)$$

    a. 2.0 g Al, 2.0 g Cl_2 b. 7.5 g Al, 24.8 g Cl_2
    c. 0.235 g Al, 1.15 g Cl_2

20. For the reaction shown, calculate the theoretical yield of the product (in grams) for each set of the initial amounts of reactants.

    $$Ti(s) + 2\,F_2(g) \rightarrow TiF_4(s)$$

    a. 5.0 g Ti, 5.0 g F_2 b. 2.4 g Ti, 1.6 g F_2
    c. 0.233 g Ti, 0.288 g F_2

21. Lead ions can be precipitated from solution with KCl according to the reaction:

    $$Pb^{2+}(aq) + 2\,KCl(aq) \rightarrow PbCl_2(s) + 2\,K^+(aq)$$

    When 28.5 g KCl is added to a solution containing 25.7 g Pb^{2+}, a $PbCl_2$ precipitate forms. The precipitate is filtered, and dried, and found to have a mass of 29.4 g. Determine the limiting reactant, theoretical yield of $PbCl_2$, and percent yield for the reaction.

22. Magnesium oxide can be made by heating magnesium metal in the presence of oxygen. The balanced equation for the reaction is

    $$2\,Mg(s) + O_2(g) \rightarrow 2\,MgO(s)$$

    When 10.1 g of Mg reacts with 10.5 g O_2, 11.9 g MgO is collected. Determine the limiting reactant, theoretical yield, and percent yield for the reaction.

23. Urea (CH_4N_2O) is a common fertilizer that can be synthesized by the reaction of ammonia (NH_3) with carbon dioxide as follows:

    $$2\,NH_3(aq) + CO_2(aq) \rightarrow CH_4N_2O(aq) + H_2O(l)$$

    In an industrial synthesis of urea, a chemist combines 136.4 kg of ammonia with 211.4 kg of carbon dioxide and obtains 168.4 kg of urea. Determine the limiting reactant, theoretical yield of urea, and percent yield for the reaction.

24. Many computer chips are manufactured from silicon, which occurs in nature as SiO_2. When SiO_2 is heated to melting, it reacts with solid carbon to form liquid silicon and carbon monoxide gas. In an industrial preparation of silicon, 155.8 kg of SiO_2 reacts with 78.3 kg of carbon to produce 66.1 kg of silicon. Determine the limiting reactant, theoretical yield, and percent yield for the reaction.

Solution Concentration and Solution Stoichiometry

25. Calculate the molarity of each solution.
  a. 4.3 mol of LiCl in 2.8 L solution
  b. 22.6 g $C_6H_{12}O_6$ in 1.08 L of solution
  c. 45.5 mg NaCl in 154.4 mL of solution

26. Calculate the molarity of each solution.
  a. 0.11 mol of $LiNO_3$ in 5.2 L of solution
  b. 61.3 g C_2H_6O in 2.44 L of solution
  c. 15.2 mg KI in 102 mL of solution

27. How many moles of KCl are in each solution?
  a. 0.556 L of a 2.3 M KCl solution
  b. 1.8 L of a 0.85 M KCl solution
  c. 114 mL of a 1.85 M KCl solution

28. What volume of 0.200 M ethanol solution contains each of the number of moles of ethanol?
  a. 0.45 mol ethanol
  b. 1.22 mol ethanol
  c. 1.2×10^{-2} mol ethanol

29. A laboratory procedure calls for making 500.0 mL of a 1.3 M $NaNO_3$ solution. What mass of $NaNO_3$ (in g) is needed?

30. A chemist wants to make 7.2 L of a 0.350 M $CaCl_2$ solution. What mass of $CaCl_2$ (in g) should the chemist use?

31. If 123 mL of a 1.1 M glucose solution is diluted to 500.0 mL, what is the molarity of the diluted solution?

32. If 3.5 L of a 4.8 M $SrCl_2$ solution is diluted to 45 L, what is the molarity of the diluted solution?

33. To what volume should you dilute 50.0 mL of a 12 M stock HNO_3 solution to obtain a 0.100 M HNO_3 solution?

34. To what volume should you dilute 25 mL of a 10.0 M H_2SO_4 solution to obtain a 0.150 M H_2SO_4 solution?

35. Consider the precipitation reaction:

$$2\,Na_3PO_4(aq) + 3\,CuCl_2(aq) \longrightarrow$$
$$Cu_3(PO_4)_2(s) + 6\,NaCl(aq)$$

What volume of 0.175 M Na_3PO_4 solution is necessary to completely react with 95.4 mL of 0.102 M $CuCl_2$?

36. Consider the precipitation reaction:

$$Li_2S(aq) + Co(NO_3)_2(aq) \longrightarrow 2\,LiNO_3(aq) + CoS(s)$$

What volume of 0.150 M Li_2S solution is required to completely react with 125 mL of 0.150 M $Co(NO_3)_2$?

37. What is the minimum amount of 6.0 M H_2SO_4 necessary to produce 25.0 g of H_2 (g) according to the following reaction?

$$2\,Al(s) + 3\,H_2SO_4(aq) \longrightarrow Al_2(SO_4)_3(aq) + 3\,H_2(g)$$

38. What is the molarity of $ZnCl_2$ that forms when 25.0 g of zinc completely reacts with $CuCl_2$ according to the following reaction? Assume a final volume of 275 mL.

$$Zn(s) + CuCl_2(aq) \longrightarrow ZnCl_2(aq) + Cu(s)$$

39. A 25.0 mL sample of a 1.20 M potassium chloride solution is mixed with 15.0 mL of a 0.900 M barium nitrate solution and this precipitation reaction occurs:

$$2\,KCl(aq) + Ba(NO_3)_2(aq) \longrightarrow BaCl_2(s) + 2\,KNO_3(aq)$$

The solid $BaCl_2$ is collected, dried, and found to have a mass of 2.45 g. Determine the limiting reactant, the theoretical yield, and the percent yield.

40. A 55.0 mL sample of a 0.102 M potassium sulfate solution is mixed with 35.0 mL of a 0.114 M lead acetate solution and this precipitation reaction occurs:

$$K_2SO_4(aq) + Pb(C_2H_3O_2)_2(aq) \longrightarrow 2\,KC_2H_3O_2(aq) + PbSO_4(s)$$

The solid $PbSO_4$ is collected, dried, and found to have a mass of 1.01 g. Determine the limiting reactant, the theoretical yield, and the percent yield.

Types of Aqueous Solutions and Solubility

41. Each compound listed is soluble in water. For each compound, do you expect the resulting aqueous solution to conduct electrical current?
  a. CsCl
  b. CH_3OH
  c. $Ca(NO_2)_2$
  d. $C_6H_{12}O_6$

42. Classify each compound as a strong electrolyte or nonelectrolyte.
  a. $MgBr_2$
  b. $C_{12}H_{22}O_{11}$
  c. Na_2CO_3
  d. KOH

43. Determine whether each compound is soluble or insoluble. If the compound is soluble, write the ions present in solution.
  a. $AgNO_3$
  b. $Pb(C_2H_3O_2)_2$
  c. KNO_3
  d. $(NH_4)_2S$

44. Determine whether each compound is soluble or insoluble. For the soluble compounds, write the ions present in solution.
  a. AgI
  b. $Cu_3(PO_4)_2$
  c. $CoCO_3$
  d. K_3PO_4

Precipitation Reactions

45. Complete and balance each equation. If no reaction occurs, write NO REACTION.
  a. $LiI(aq) + BaS(aq) \longrightarrow$
  b. $KCl(aq) + CaS(aq) \longrightarrow$
  c. $CrBr_2(aq) + Na_2CO_3(aq) \longrightarrow$
  d. $NaOH(aq) + FeCl_3(aq) \longrightarrow$

46. Complete and balance each equation. If no reaction occurs, write NO REACTION.
  a. $NaNO_3(aq) + KCl(aq) \longrightarrow$
  b. $NaCl(aq) + Hg_2(C_2H_3O_2)_2(aq) \longrightarrow$
  c. $(NH_4)_2SO_4(aq) + SrCl_2(aq) \longrightarrow$
  d. $NH_4Cl(aq) + AgNO_3(aq) \longrightarrow$

47. Write a molecular equation for the precipitation reaction that occurs (if any) when each set of solutions is mixed. If no reaction occurs, write NO REACTION.
  a. potassium carbonate and lead(II) nitrate
  b. lithium sulfate and lead(II) acetate
  c. copper(II) nitrate and magnesium sulfide
  d. strontium nitrate and potassium iodide

48. Write a molecular equation for the precipitation reaction that occurs (if any) when each set of solutions is mixed. If no reaction occurs, write NO REACTION.
  a. sodium chloride and lead(II) acetate
  b. potassium sulfate and strontium iodide

c. cesium chloride and calcium sulfide
d. chromium(III) nitrate and sodium phosphate

Ionic and Net Ionic Equations

49. Write balanced complete ionic and net ionic equations for each reaction.
   a. $HCl(aq) + LiOH(aq) \rightarrow H_2O(l) + LiCl(aq)$
   b. $MgS(aq) + CuCl_2(aq) \rightarrow CuS(s) + MgCl_2(aq)$
   c. $NaOH(aq) + HNO_3(aq) \rightarrow H_2O(l) + NaNO_3(aq)$
   d. $Na_3PO_4(aq) + NiCl_2(aq) \rightarrow Ni_3(PO_4)_2(s) + NaCl(aq)$

50. Write balanced complete ionic and net ionic equations for each reaction.
   a. $K_2SO_4(aq) + CaI_2(aq) \rightarrow CaSO_4(s) + KI(aq)$
   b. $NH_4Cl(aq) + NaOH(aq) \rightarrow$
$$H_2O(l) + NH_3(g) + NaCl(aq)$$
   c. $AgNO_3(aq) + NaCl(aq) \rightarrow AgCl(s) + NaNO_3(aq)$
   d. $HC_2H_3O_2(aq) + K_2CO_3(aq) \rightarrow$
$$H_2O(l) + CO_2(g) + KC_2H_3O_2(aq)$$

51. Mercury ions (Hg_2^{2+}) can be removed from solution by precipitation with Cl^-. Suppose that a solution contains aqueous $Hg_2(NO_3)_2$. Write complete ionic and net ionic equations to show the reaction of aqueous $Hg_2(NO_3)_2$ with aqueous sodium chloride to form solid Hg_2Cl_2 and aqueous sodium nitrate.

52. Lead ions can be removed from solution by precipitation with sulfate ions. Suppose that a solution contains lead(II) nitrate. Write complete ionic and net ionic equations to show the reaction of aqueous lead(II) nitrate with aqueous potassium sulfate to form solid lead(II) sulfate and aqueous potassium nitrate.

Acid–Base and Gas-Evolution Reactions

53. Write balanced molecular and net ionic equations for the reaction between hydrobromic acid and potassium hydroxide.

54. Write balanced molecular and net ionic equations for the reaction between nitric acid and calcium hydroxide.

55. Complete and balance each acid–base reaction equation.
   a. $H_2SO_4(aq) + Ca(OH)_2(aq) \rightarrow$
   b. $HClO_4(aq) + KOH(aq) \rightarrow$
   c. $H_2SO_4(aq) + NaOH(aq) \rightarrow$

56. Complete and balance each acid–base reaction equation.
   a. $HI(aq) + LiOH(aq) \rightarrow$
   b. $HC_2H_3O_2(aq) + Ca(OH)_2(aq) \rightarrow$
   c. $HCl(aq) + Ba(OH)_2(aq) \rightarrow$

57. Complete and balance each gas-evolution reaction equation.
   a. $HBr(aq) + NiS(s) \rightarrow$
   b. $NH_4I(aq) + NaOH(aq) \rightarrow$
   c. $HBr(aq) + Na_2S(aq) \rightarrow$
   d. $HClO_4(aq) + Li_2CO_3(aq) \rightarrow$

58. Complete and balance each gas-evolution reaction equation.
   a. $HNO_3(aq) + Na_2SO_3(aq) \rightarrow$
   b. $HCl(aq) + KHCO_3(aq) \rightarrow$
   c. $HC_2H_3O_2(aq) + NaHSO_3(aq) \rightarrow$
   d. $(NH_4)_2SO_4(aq) + Ca(OH)_2(aq) \rightarrow$

Oxidation–Reduction and Combustion

59. Assign oxidation states to each atom in each ion or compound.
   a. Ag b. Ag^+
   c. CaF_2 d. H_2S
   e. CO_3^{2-} f. CrO_4^{2-}

60. Assign oxidation states to each atom in each ion or compound.
   a. Cl_2 b. Fe^{3+}
   c. $CuCl_2$ d. CH_4
   e. $Cr_2O_7^{2-}$ f. HSO_4^-

61. What is the oxidation state of Cr in each compound?
   a. CrO
   b. CrO_3
   c. Cr_2O_3

62. What is the oxidation state of Cl in each ion?
   a. ClO^-
   b. ClO_2^-
   c. ClO_3^-
   d. ClO_4^-

63. Which reactions are redox reactions? For each redox reaction, identify the oxidizing agent and the reducing agent.
   a. $4\,Li(s) + O_2(g) \rightarrow 2\,Li_2O(s)$
   b. $Mg(s) + Fe^{2+}(aq) \rightarrow Mg^{2+}(aq) + Fe(s)$
   c. $Pb(NO_3)_2(aq) + Na_2SO_4(aq) \rightarrow PbSO_4(s) + 2\,NaNO_3(aq)$
   d. $HBr(aq) + KOH(aq) \rightarrow H_2O(l) + KBr(aq)$

64. Which reactions are redox reactions? For each redox reaction, identify the oxidizing agent and the reducing agent.
   a. $Al(s) + 3\,Ag^+(aq) \rightarrow Al^{3+}(aq) + 3\,Ag(s)$
   b. $SO_3(g) + H_2O(l) \rightarrow H_2SO_4(aq)$
   c. $Ba(s) + Cl_2(g) \rightarrow BaCl_2(s)$
   d. $Mg(s) + Br_2(l) \rightarrow MgBr_2(s)$

65. Complete and balance each combustion reaction equation.
   a. $S(s) + O_2(g) \rightarrow$
   b. $C_3H_6(g) + O_2(g) \rightarrow$
   c. $Ca(s) + O_2(g) \rightarrow$
   d. $C_5H_{12}S(l) + O_2(g) \rightarrow$

66. Complete and balance each combustion reaction equation.
   a. $C_4H_6(g) + O_2(g) \rightarrow$
   b. $C(s) + O_2(g) \rightarrow$
   c. $CS_2(s) + O_2(g) \rightarrow$
   d. $C_3H_8O(l) + O_2(g) \rightarrow$

Cumulative Problems

67. The density of a 20.0% by mass ethylene glycol ($C_2H_6O_2$) solution in water is 1.03 g/mL. Find the molarity of the solution.

68. Find the percent by mass of sodium chloride in a 1.35 M NaCl solution. The density of the solution is 1.05 g/mL.

69. Sodium bicarbonate is often used as an antacid to neutralize excess hydrochloric acid in an upset stomach. What mass of hydrochloric acid (in grams) can be neutralized by 2.5 g of sodium bicarbonate? (Hint: Begin by writing a balanced equation for the reaction between aqueous sodium bicarbonate and aqueous hydrochloric acid.)

70. Toilet bowl cleaners often contain hydrochloric acid to dissolve the calcium carbonate deposits that accumulate within a toilet bowl. What mass of calcium carbonate (in grams) can be dissolved by 3.8 g of HCl? (Hint: Begin by writing a balanced equation for the reaction between hydrochloric acid and calcium carbonate.)

71. The combustion of gasoline produces carbon dioxide and water. Assume gasoline to be pure octane (C_8H_{18}) and calculate the mass (in kg) of carbon dioxide that is added to the atmosphere per 1.0 kg of octane burned. (Hint: Begin by writing a balanced equation for the combustion reaction.)

72. Many home barbeques are fueled with propane gas (C_3H_8). What mass of carbon dioxide (in kg) is produced upon the complete combustion of 18.9 L of propane (approximate contents of one 5-gallon tank)? Assume that the density of the liquid propane in the tank is 0.621 g/mL. (Hint: Begin by writing a balanced equation for the combustion reaction.)

73. Aspirin can be made in the laboratory by reacting acetic anhydride ($C_4H_6O_3$) with salicylic acid ($C_7H_6O_3$) to form aspirin ($C_9H_8O_4$) and acetic acid ($C_2H_4O_2$). The balanced equation is

$$C_4H_6O_3 + C_7H_6O_3 \rightarrow C_9H_8O_4 + HC_2H_3O_2$$

In a laboratory synthesis, a student begins with 3.00 mL of acetic anhydride (density = 1.08 g/mL) and 1.25 g of salicylic acid. Once the reaction is complete, the student collects 1.22 g of aspirin. Determine the limiting reactant, theoretical yield of aspirin, and percent yield for the reaction.

74. The combustion of liquid ethanol (C_2H_5OH) produces carbon dioxide and water. After 4.62 mL of ethanol (density = 0.789 g/mL) is allowed to burn in the presence of 15.55 g of oxygen gas, 3.72 mL of water (density = 1.00 g/mL) is collected. Determine the limiting reactant, theoretical yield of H_2O, and percent yield for the reaction. (Hint: Write a balanced equation for the combustion of ethanol.)

75. A loud classroom demonstration involves igniting a hydrogen-filled balloon. The hydrogen within the balloon reacts explosively with oxygen in the air to form water. If the balloon is filled with a mixture of hydrogen and oxygen, the explosion is even louder than if the balloon is filled only with hydrogen; the intensity of the explosion depends on the relative amounts of oxygen and hydrogen within the balloon. Look at the molecular views representing different amounts of hydrogen and oxygen in four different balloons. Based on the balanced chemical equation, which balloon will make the loudest explosion?

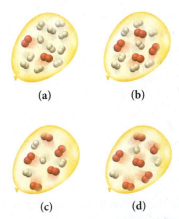

(a)   (b)

(c)   (d)

76. A hydrochloric acid solution will neutralize a sodium hydroxide solution. Look at the molecular views showing one beaker of HCl and four beakers of NaOH. Which NaOH beaker will just neutralize the HCl beaker? Begin by writing a balanced chemical equation for the neutralization reaction.

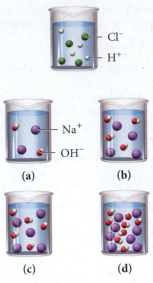

(a)   (b)

(c)   (d)

77. Predict the products of each reaction and write balanced molecular equations for each. If no reaction occurs, write NO REACTION.
   a. $HCl(aq) + Hg_2(NO_3)_2(aq) \rightarrow$
   b. $KHSO_3(aq) + HNO_3(aq) \rightarrow$
   c. aqueous ammonium chloride and aqueous lead(II) nitrate
   d. aqueous ammonium chloride and aqueous calcium hydroxide

78. Predict the products of each of these reactions and write balanced molecular equations for each. If no reaction occurs, write NO REACTION.
   a. $H_2SO_4(aq) + HNO_3(aq) \rightarrow$
   b. $Cr(NO_3)_3(aq) + LiOH(aq) \rightarrow$
   c. liquid pentanol ($C_5H_{12}O$) and gaseous oxygen
   d. aqueous strontium sulfide and aqueous copper(II) sulfate

79. Hard water often contains dissolved Ca^{2+} and Mg^{2+} ions. One way to soften water is to add phosphates. The phosphate ion forms insoluble precipitates with calcium and magnesium ions, removing them from solution. Suppose that a solution is 0.050 M in calcium chloride and 0.085 M in magnesium nitrate. What mass of sodium phosphate must be added to 1.5 L of this solution to completely eliminate the hard water ions? Assume complete reaction.

80. An acid solution is 0.100 M in HCl and 0.200 M in H_2SO_4. What volume of a 0.150 M KOH solution must be added to 500.0 mL of the acidic solution to completely neutralize all of the acid?

81. Find the mass of barium metal (in grams) that must react with O_2 to produce enough barium oxide to prepare 1.0 L of a 0.10 M solution of OH^-.

82. A solution contains Cr^{3+} ion and Mg^{2+} ion. The addition of 1.00 L of 1.51 M NaF solution is required to cause the complete precipitation of these ions as $CrF_3(s)$ and $MgF_2(s)$. The total mass of the precipitate is 49.6 g. Find the mass of Cr^{3+} in the original solution.

83. The nitrogen in sodium nitrate and in ammonium sulfate is available to plants as fertilizer. Which is the more economical source of nitrogen, a fertilizer containing 30.0% sodium nitrate by weight and costing $9.00 per 100 lb or one containing 20.0% ammonium sulfate by weight and costing $8.10 per 100 lb.

84. Find the volume of 0.110 M hydrochloric acid necessary to react completely with 1.52 g $Al(OH)_3$.

85. Treatment of gold metal with BrF_3 and KF produces Br_2 and $KAuF_4$, a salt of gold. Identify the oxidizing agent and the reducing agent in this reaction. Find the mass of the gold salt that forms when a 73.5-g mixture of equal masses of all three reactants is prepared.

86. A solution is prepared by mixing 0.10 L of 0.12 M sodium chloride with 0.23 L of a 0.18 M $MgCl_2$ solution. What volume of a 0.20 M silver nitrate solution is required to precipitate all the Cl^- ion in the solution as AgCl?

87. A solution contains one or more of the following ions: Ag^+, Ca^{2+}, and Cu^{2+}. When sodium chloride is added to the solution, no precipitate forms. When sodium sulfate is added to the solution, a white precipitate forms. The precipitate is filtered off and sodium carbonate is added to the remaining solution, producing a precipitate. Which ions were present in the original solution? Write net ionic equations for the formation of each of the precipitates observed.

88. A solution contains one or more of the following ions Hg_2^{2+}, Ba^{2+}, and Fe^{2+}. When potassium chloride is added to the solution, a precipitate forms. The precipitate is filtered off and potassium sulfate is added to the remaining solution, producing no precipitate. When potassium carbonate is added to the remaining solution, a precipitate forms. Which ions were present in the original solution? Write net ionic equations for the formation of each of the precipitates observed.

89. A liquid level mixture contains 30.35% hexane (C_6H_{14}), 15.85% heptane (C_7H_{16}), and the rest is octane (C_8H_{18}). What maximum mass of carbon dioxide is produced by the complete combustion of 10.0 kg of this fuel mixture?

90. An ilmenite–sand mixture contains 22.8% ilmenite by mass and the first reaction is carried out with a 90.8% yield. If the second reaction is carried out with an 85.9% yield, what mass of titanium can be obtained from 1.00 kg of the ilmenite–sand mixture?

Challenge Problems

91. Lakes that have been acidified by acid rain (HNO_3 and H_2SO_4) can be neutralized by a process called liming, in which limestone ($CaCO_3$) is added to the acidified water. What mass of limestone (in kg) would be required to completely neutralize a 15.2 billion-liter lake that is 1.8×10^{-5} M in H_2SO_4 and 8.7×10^{-6} M in HNO_3?

92. Recall from Section 4.6 that sodium carbonate is often added to laundry detergents to soften hard water and make the detergent more effective. Suppose that a particular detergent mixture is designed to soften hard water that is 3.5×10^{-3} M in Ca^{2+} and 1.1×10^{-3} M in Mg^{2+} and that the average capacity of a washing machine is 19.5 gallons of water. If the detergent requires using 0.65 kg detergent per load of laundry, determine what percentage (by mass) of the detergent must be sodium carbonate in order to completely precipitate all of the calcium and magnesium ions in an average load of laundry water.

93. Lead poisoning is a serious condition resulting from the ingestion of lead in food, water, or other environmental sources. It affects the central nervous system, leading to a variety of symptoms such as distractibility, lethargy, and loss of motor coordination. Lead poisoning is treated with chelating agents, substances that bind to metal ions, allowing the lead to be eliminated in the urine. A modern chelating agent used for this purpose is succimer ($C_4H_6O_4S_2$). Suppose you are trying to determine the appropriate dose for succimer treatment of lead poisoning. What minimum mass of succimer (in mg) is needed to bind all of the lead in a patient's bloodstream? Assume that

patient blood lead levels are 45 $\mu g/dL$, that total blood volume is 5.0 L, and that one mole of succimer binds one mole of lead.

94. A particular kind of emergency breathing apparatus—often placed in mines, caves, or other places where oxygen might become depleted or where the air might become poisoned—works via the chemical reaction:

$$4 KO_2(s) + 2 CO_2(g) \longrightarrow 2 K_2CO_3(s) + 3 O_2(g)$$

Notice that the reaction produces O_2, which can be breathed, and absorbs CO_2, a product of respiration. Suppose you work for a company interested in producing a self-rescue breathing apparatus (based on the above reaction) which would allow the user to survive for 10 minutes in an emergency situation. What are the important chemical considerations in designing such a unit? Estimate how much KO_2 would be required for the apparatus. (Find any necessary additional information—such as human breathing rates—from appropriate sources. Assume that normal air is 20% oxygen.)

95. Metallic aluminum reacts with MnO_2 at elevated temperatures to form manganese metal and aluminum oxide. A mixture of the two reactants is 67.2% mole percent Al. Find the theoretical yield (in grams) of manganese from the reaction of 250 g of this mixture.

96. Hydrolysis of the compound B_5H_9 forms boric acid, H_3BO_3. Fusion of boric acid with sodium oxide forms a borate salt, $Na_2B_4O_7$. Without writing complete equations, find the mass (in grams) of B_5H_9 required to form 151 g of the borate salt by this reaction sequence.

Conceptual Problems

97. Consider the reaction:

$$4 K(s) + O_2(g) \longrightarrow 2 K_2O(s)$$

The molar mass of K is 39.09 g/mol and that of O_2 is 32.00 g/mol. Without doing any calculations, pick the conditions under which potassium is the limiting reactant and explain your reasoning.
   a. 170 g K, 31 g O_2 b. 16 g K, 2.5 g O_2
   c. 165 kg K, 28 kg O_2 d. 1.5 g K, 0.38 g O_2

98. Consider the reaction:

$$2 NO(g) + 5 H_2(g) \longrightarrow 2 NH_3(g) + 2 H_2O(g)$$

A reaction mixture initially contains 5 moles of NO and 10 moles of H_2. Without doing any calculations, determine which of the following best represents the mixture after the reactants have reacted as completely as possible. Explain your reasoning.
   a. 1 mol NO, 0 mol H_2, 4 mol NH_3, 4 mol H_2O
   b. 0 mol NO, 1 mol H_2, 5 mol NH_3, 5 mol H_2O

c. 3 mol NO, 5 mol H$_2$, 2 mol NH$_3$, 2 mol H$_2$O
d. 0 mol NO, 0 mol H$_2$, 4 mol NH$_3$, 4 mol H$_2$O

99. The circle below represents 1.0 liter of a solution with a solute concentration of 1 M:

Explain what you would add (the amount of solute or volume of solvent) to the solution above so as to obtain a solution represented by each of the following:

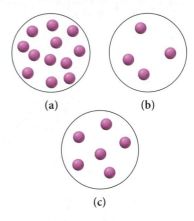

(a)   (b)

(c)

100. Consider the reaction:

$$2 \, N_2H_4(g) + N_2O_4(g) \longrightarrow 3 \, N_2(g) + 4 \, H_2O(g)$$

Consider the following representation of an initial mixture of N$_2$H$_4$ and N$_2$O$_4$:

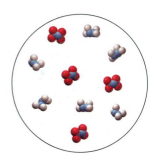

Which of the following best represents the reaction mixture after the reactants have reacted as completely as possible?

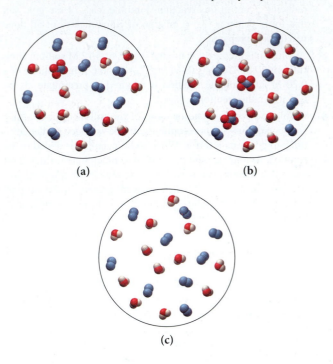

(a)   (b)

(c)

101. Consider the generic ionic compounds with the formulas A$_2$X and BY$_2$ and the following solubility rules:

A$_2$X soluble; BY$_2$ soluble; AY insoluble; BX soluble.

Let A$^+$ ions be represented by circles, B^{2+} ions be represented by squares, X^{2-} ions be represented by triangles, Y$^-$ ions be represented by diamonds. Solutions of the two compounds (A$_2$X and BY$_2$) can be represented as follows:

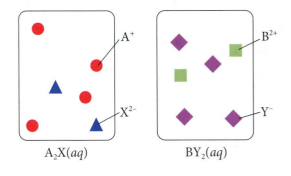

A$_2$X(aq)   BY$_2$(aq)

Draw a molecular-level representation showing the result of mixing the two solutions above and write an equation to represent the reaction.

Answers to Conceptual Connections

Stoichiometry

4.1 (c) Since each O$_2$ molecule reacts with 4 Na atoms, 12 Na atoms are required to react with 3 O$_2$ molecules.

Stoichiometry
4.2

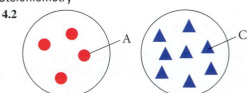

Limiting Reactant and Theoretical Yield

4.3 (c) Nitrogen is the limiting reactant and there is enough nitrogen to make 4 NH_3 molecules. Hydrogen is in excess and two hydrogen molecules remain after the reactants have reacted as completely as possible.

Solutions

4.4 (b) The mass of a solution is equal to the mass of the solute plus the mass of the solvent. Although the solute seems to disappear, it really does not, and its mass becomes part of the mass of the solution, in accordance with the law of mass conservation.

Solution Dilution

4.5 (c) Since the volume has doubled, the concentration is halved, so the same volume should contain half as many solute molecules.

Precipitation Reactions

4.6

$$AX(aq) + BY(aq) \longrightarrow BX(s) + AY(aq)$$

Oxidation and Reduction

4.7 (d) Since oxidation and reduction must occur together, an increase in the oxidation state of a reactant will always be accompanied by a decrease in the oxidation state of a reactant.

5 Gases

So many of the properties of matter, especially when in the gaseous form, can be deduced from the hypothesis that their minute parts are in rapid motion, the velocity increasing with the temperature, that the precise nature of this motion becomes a subject of rational curiosity.—James Clerk Maxwell (1831–1879)

The cork in this champagne bottle is expelled by the buildup of pressure, which results from the constant collisions of gas molecules with the surfaces around them.

WE CAN SURVIVE FOR WEEKS without food, days without water, but only minutes without air. Fortunately, we live at the bottom of a vast ocean of air held to Earth's surface by gravity. We inhale a lungful of air every few seconds, keep some of the molecules for our own use, add some of the molecules that our bodies no longer need, and exhale the mixture back into the surrounding air. The air around us is a *gas*, matter in its gaseous state. What are the fundamental properties of a gas? What laws describe its behavior? What theories explain these properties and laws? The gaseous state is the simplest and best-understood state of matter. In this chapter, we examine that state.

5.1   Breathing: Putting Pressure to Work

Every day, without even thinking about it, you move approximately 8500 liters of air into and out of your lungs. The total mass of this air is about 11 kg (or 24 pounds). How do you do it? The simple answer is *pressure*. You rely on your body's ability to create pressure differences to move air into and out of your lungs. **Pressure** is the force exerted per unit area by gas particles (molecules or atoms) as they strike the surfaces around them (**Figure 5.1▶**). Just as a ball exerts a force when it bounces against a wall, so a gaseous molecule exerts a force when it collides with a surface. The result of all these collisions is pressure—a constant force on the surfaces exposed to any gas. The total pressure exerted by a gas depends on several factors, including the concentration of gas particles in the sample; the higher the concentration, the greater the pressure.

When you inhale, the muscles that surround your chest cavity expand the volume of your lungs. The expanded volume results in a lower concentration of gas molecules (the number of molecules does not change, but since the volume increases, the *concentration* goes down). This in turn results in fewer molecular collisions, which results in lower pressure. The external pressure (the pressure outside of your lungs) remains relatively constant and is now higher than the pressure within your lungs. As a result, gaseous molecules flow into your lungs (from higher pressure to lower pressure). When you exhale, the process is reversed. The chest cavity muscles relax, which *decreases* the lung volume, increasing the pressure within the lungs and forcing air back out. In this way, over your lifetime, you will take about half a billion breaths and move about 250 million liters of air through your lungs. With each breath you create pressure differences that allow you to obtain the oxygen that you need to live.

5.2   Pressure: The Result of Molecular Collisions

Air can hold up a jumbo jet or knock down a building. How? Air contains gas molecules in constant motion that collide with each other and with the surfaces around them. Each collision exerts only a small force, but when these forces are summed over the many molecules in air, they can add up to a substantial force. As we have just seen, the result of the constant collisions between the atoms or molecules in a gas and the surfaces around them is called pressure. Because of pressure, we can drink from straws, inflate basketballs, and breathe. Variation in pressure in Earth's atmosphere creates wind, and changes in pressure help us to predict weather. Pressure is all around us and even inside us. The pressure exerted by a gas sample, as defined previously, is the force per unit area that results from the collisions of gas particles with the surrounding surfaces:

$$\text{Pressure} = \frac{\text{force}}{\text{area}} = \frac{F}{A} \qquad [5.1]$$

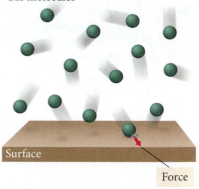

Gas molecules

Surface

Force

▲ **FIGURE 5.1 Gas Pressure** Pressure is the force per unit area exerted by gas particles colliding with the surfaces around them.

◀ Pressure variations in Earth's atmosphere create wind and weather. The Hs in this map indicate regions of high pressure, usually associated with clear weather. The Ls indicate regions of low pressure, usually associated with unstable weather. The map shows a typhoon off the northeast coast of Japan. The isobars, or lines of constant pressure, are labeled in hectopascals (100 Pa).

▶ **FIGURE 5.2 Pressure and Particle Density** A low density of gas particles results in low pressure. A high density of gas particles results in high pressure.

Pressure and Density

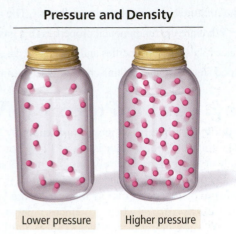

Lower pressure   Higher pressure

The pressure exerted by a gas depends on several factors, including, as we just discussed, the number of gas particles in a given volume (**Figure 5.2▲**). Since the number of gas particles in a given volume generally decreases with increasing altitude, *pressure decreases with increasing altitude.* Above 30,000 ft, for example, where most commercial airplanes fly, the pressure is so low that you could pass out for lack of oxygen. For this reason, most airplane cabins are artificially pressurized.

You may sometimes feel the effect of a drop in pressure as a brief pain in your ears. This pain arises from the air-containing cavities within your ear (**Figure 5.3▼**). When you ascend a mountain, the external pressure (the pressure that surrounds you) drops, while the pressure within your ear cavities (the internal pressure) remains the same. This creates an imbalance—the greater internal pressure forces your eardrum to bulge outward, causing pain. With time, and with the help of a yawn or two, the excess air within your ear's cavities escapes, equalizing the internal and external pressure and relieving the pain.

Pressure Units

Pressure is measured in a number of different units. A common unit of pressure, the **millimeter of mercury (mmHg)**, originates from how pressure is measured with a **barometer** (**Figure 5.4▶**). A barometer is an evacuated glass tube, the tip of which is

▶ **FIGURE 5.3 Pressure Imbalance** The pain you feel in your ears upon ascending a mountain is caused by a pressure imbalance between the cavities in your ears and the outside air.

Pressure Imbalance

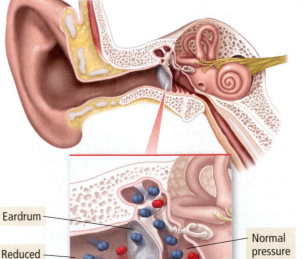

Eardrum

Reduced external pressure

Normal pressure

Eardrum bulges outward, causing pain.

The Mercury Barometer

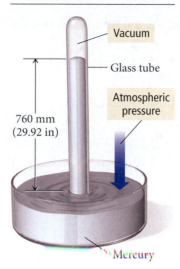

Vacuum

Glass tube

Atmospheric pressure

760 mm
(29.92 in)

Mercury

▲ **FIGURE 5.4 The Mercury Barometer** Average atmospheric pressure at sea level can support a column of mercury 760 mm in height.

submerged in a pool of mercury. The liquid mercury is forced upward into the evacuated tube by atmospheric pressure on the liquid's surface. Because mercury is so dense (13.5 times more dense than water), atmospheric pressure can support a column of Hg that is only about 0.760 m or 760 mm (about 30 in) tall. (By contrast, atmospheric pressure can support a column of water that is about 10.3 m tall.) The mercury column rises with increasing atmospheric pressure or falls with decreasing atmospheric pressure. The unit *millimeter of mercury* is often called a **torr**, after the Italian physicist Evangelista Torricelli (1608–1647) who invented the barometer.

$$1 \text{ mmHg} = 1 \text{ torr}$$

A second unit of pressure is the **atmosphere (atm)**, the average pressure at sea level. Since one atmosphere of pressure pushes a column of mercury to a height of 760 mm, 1 atm and 760 mmHg are equal:

$$1 \text{ atm} = 760 \text{ mmHg}$$

A fully inflated mountain bike tire has a pressure of about 6 atm, and the pressure at the top of Mt. Everest is about 0.31 atm.

The SI unit of pressure is the **pascal (Pa)**, defined as 1 newton (N) per square meter.

$$1 \text{ Pa} = 1 \text{ N/m}^2$$

The pascal is a much smaller unit of pressure than the atmosphere:

$$1 \text{ atm} = 101{,}325 \text{ Pa}$$

Other common units of pressure include inches of mercury (in Hg) and pounds per square inch (psi).

$$1 \text{ atm} = 29.92 \text{ in Hg} \quad 1 \text{ atm} = 14.7 \text{ psi}$$

Table 5.1 summarizes these units.

TABLE 5.1 Common Units of Pressure

Unit	Abbreviation	Average Air Pressure at Sea Level
Pascal (1 N/m^2)	Pa	101,325 Pa
Pounds per square inch	psi	14.7 psi
Torr (1 mmHg)	torr	760 torr (exact)
Inches of mercury	in Hg	29.92 in Hg
Atmosphere	atm	1 atm

EXAMPLE 5.1 Converting between Pressure Units

A high-performance road bicycle tire is inflated to a total pressure of 132 psi. What is this pressure in mmHg?

SORT The problem gives a pressure in psi and asks you to convert the units to mmHg.	**GIVEN** 132 psi **FIND** mmHg
STRATEGIZE Begin with the given value in psi and convert to atm and then to mmHg.	**CONCEPTUAL PLAN** 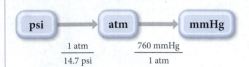 **RELATIONSHIPS USED** 1 atm = 14.7 psi 760 mmHg = 1 atm (both from Table 5.1)
SOLVE Follow the conceptual plan to solve the problem. Begin with 132 psi and use the conversion factors to arrive at the pressure in mmHg.	**SOLUTION** $132 \text{ psi} \times \dfrac{1 \text{ atm}}{14.7 \text{ psi}} \times \dfrac{760 \text{ mmHg}}{1 \text{ atm}} = 6.82 \times 10^3 \text{ mmHg}$

CHECK The units of the answer are correct. The magnitude of the answer (6.82×10^3 mmHg) is greater than the given pressure in psi. This is reasonable because mmHg is a much smaller unit than psi.

FOR PRACTICE 5.1
The Weather Channel reports the barometric pressure as 30.44 in Hg. Convert this pressure to psi.

FOR MORE PRACTICE 5.1
Convert a pressure of 23.8 in Hg to kPa.

5.3 The Simple Gas Laws: Boyle's Law, Charles's Law, and Avogadro's Law

A sample of gas has four basic physical properties: pressure (P), volume (V), temperature (T), and amount in moles (n). These properties are interrelated—when one changes, it affects one or more of the others. The simple gas laws describe the relationships between pairs of these properties. For example, one simple gas law describes how *volume* varies with *pressure* at constant temperature and amount of gas; another law describes how volume varies with *temperature* at constant pressure and amount of gas. These laws were deduced from observations in which two of the four basic properties were held constant in order to elucidate the relationship between the other two.

Boyle's Law: Volume and Pressure

In the early 1660s, the pioneering English scientist Robert Boyle (1627–1691) and his assistant Robert Hooke (1635–1703) used a J-tube (**Figure 5.5▶**) to measure the volume of a sample of gas at different pressures. After trapping a sample of air in the J-tube, they added mercury to increase the pressure on the gas. They found an *inverse relationship* between volume and pressure—an increase in one results in a decrease in the other—as shown in **Figure 5.6▶**. This relationship is now known as **Boyle's law**.

$$\text{Boyle's law: } V \propto \frac{1}{P} \quad (\text{constant T and } n)$$

Boyle's law assumes constant temperature and constant amount of gas.

Boyle's law follows from the idea that pressure results from the collisions of the gas particles with the walls of their container. If the volume of a gas sample is decreased, the

The J-Tube

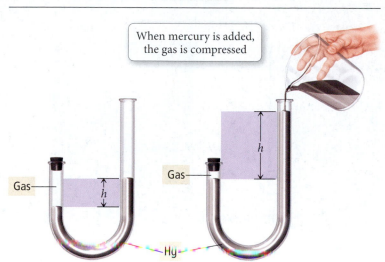

When mercury is added, the gas is compressed

Gas

h

Gas

h

Hg

◀ **FIGURE 5.5 The J-Tube** In a J-tube, a sample of gas is trapped by a column of mercury. The pressure on the gas can be increased by increasing the height (h) of mercury in the column.

same number of gas particles is crowded into a smaller volume, resulting in more collisions with the walls and therefore an increase in the pressure (**Figure 5.7▶**).

Scuba divers learn about Boyle's law during certification because it explains why a diver should not ascend toward the surface without continuous breathing. For every 10 m of depth that a diver descends in water, she experiences an additional 1 atm of pressure due to the weight of the water above her (**Figure 5.8▶**). The pressure regulator used in scuba diving delivers air at a pressure that matches the external pressure; otherwise the diver could not inhale the air because the muscles that surround the chest cavity are not strong enough to expand the volume against the greatly increased external pressure. When a diver is at a depth of 20 m below the surface, the regulator delivers air at a pressure of 3 atm to match the 3 atm of pressure around the diver (1 atm due to normal atmospheric pressure and 2 additional atmospheres due to the weight of the water at 20 m). Suppose that a diver inhaled a lungful of air at a pressure of 3 atm and swam quickly to the surface (where the pressure drops to 1 atm) while holding her breath. What would happen to the volume of air in her lungs? Since the pressure decreases by a factor of 3, the volume of the air in her lungs would increase by a factor of 3, severely damaging her lungs and possibly killing her. Boyle's law can be used to calculate the volume of a gas following a pressure

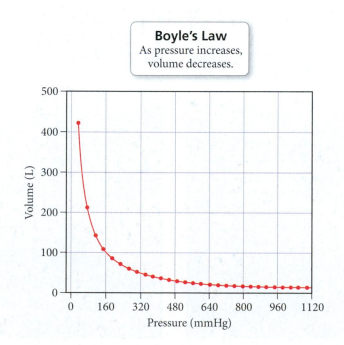

Boyle's Law
As pressure increases, volume decreases.

Volume (L)

Pressure (mmHg)

◀ **FIGURE 5.6 Volume versus Pressure** A plot of the volume of a gas sample—as measured in a J-tube—versus pressure. The plot shows that volume and pressure are inversely related.

▶ **FIGURE 5.7 Molecular Interpretation of Boyle's Law** As the volume of a gas sample decreases, gas particles collide with surrounding surfaces more frequently, resulting in greater pressure.

Volume versus Pressure: A Molecular View

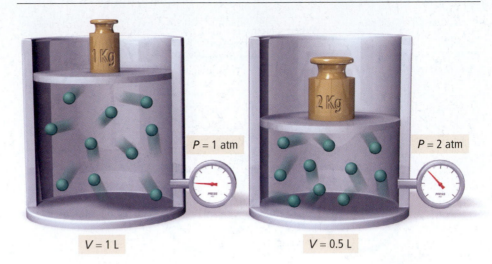

change or the pressure of a gas following a volume change *as long as the temperature and the amount of gas remain constant.* For these types of calculations, we write Boyle's law in a slightly different way.

| If two quantities are proportional, one is equal to the other multiplied by a constant.

$$\text{Since } V \propto \frac{1}{P}, \quad \text{then } V = (\text{constant}) \times \frac{1}{P} \quad \text{or} \quad V = \frac{(\text{constant})}{P}$$

If we multiply both sides by P, we get

$$PV = \text{constant}$$

This relationship shows that if the pressure increases, the volume decreases, but the product $P \times V$ is always equal to the same constant. For two different sets of conditions, we can say that

$$P_1 V_1 = \text{constant} = P_2 V_2$$

or

$$P_1 V_1 = P_2 V_2 \qquad\qquad [5.2]$$

where P_1 and V_1 are the initial pressure and volume of the gas and P_1 and V_2 are the final pressure and volume.

▶ **FIGURE 5.8 Increase in Pressure with Depth** For every 10 m of depth, a diver experiences approximately one additional atmosphere of pressure due to the weight of the surrounding water. At 20 m, for example, the diver experiences approximately 3 atm of pressure (1 atm of normal atmospheric pressure plus an additional 2 atm due to the weight of the water).

EXAMPLE 5.2 Boyle's Law

As discussed in the opening section of this chapter, you inhale by increasing your lung volume. A woman has an initial lung volume of 2.75 L, which is filled with air at an atmospheric pressure of 1.02 atm. If she increases her lung volume to 3.25 L without inhaling any additional air, what is the pressure in her lungs?

| To solve the problem, first solve Boyle's law (Equation 5.2) for P_2 and then substitute the given quantities to calculate P_2. | **SOLUTION**

$P_1V_1 = P_2V_2$

$P_2 = \dfrac{V_1}{V_2}P_1$

$= \dfrac{2.75 \text{ L}}{3.25 \text{ L}} 1.02 \text{ atm}$

$= 0.863 \text{ atm}$ |

FOR PRACTICE 5.2

A snorkeler takes a syringe filled with 16 mL of air from the surface, where the pressure is 1.0 atm, to an unknown depth. The volume of the air in the syringe at this depth is 7.5 mL. What is the pressure at this depth? If the pressure increases by an additional 1 atm for every 10 m of depth, how deep is the snorkeler?

Charles's Law: Volume and Temperature

Suppose we keep the pressure of a gas sample constant and measure its volume at a number of different temperatures. The results of several such measurements are shown in **Figure 5.9▼**. From the plot we can see the relationship between volume and temperature: the volume of a gas increases with increasing temperature. Looking at the plot more closely, however, reveals more—volume and temperature are *linearly related*. If two variables are linearly related, then plotting one against the other produces a straight line.

Another interesting feature emerges if we extend or *extrapolate* the line in the plot backward from the lowest measured temperature. The extrapolated line shows that the gas should have a zero volume at –273.15 °C. Recall from Chapter 1 that –273.15 °C corresponds to 0 K (zero on the Kelvin scale), the coldest possible temperature. The extrapolated line indicates that below –273.15 °C, the gas would have a negative volume, which is physically impossible. For this reason, we refer to 0 K as *absolute zero*—colder temperatures do not exist.

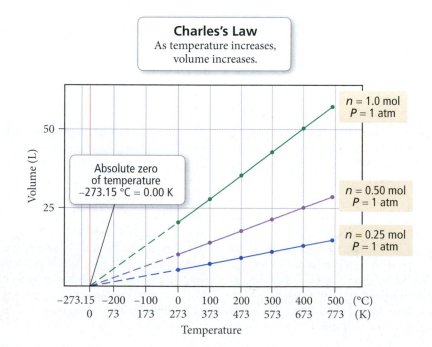

◄ FIGURE 5.9 Volume versus Temperature   The volume of a fixed amount of gas at a constant pressure increases linearly with increasing temperature in kelvins. (The extrapolated lines cannot be measured experimentally because all gases condense into liquids before –273.15 °C is reached.)

Volume versus Temperature: A Molecular View

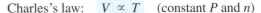

▲ **FIGURE 5.10 Molecular Interpretation of Charles's Law**   If a balloon is moved from an ice water bath to a boiling water bath, its volume will expand as the gas particles within the balloon move faster (due to the increased temperature) and collectively occupy more space.

The first person to carefully quantify the relationship between the volume of a gas and its temperature was J. A. C. Charles (1746–1823), a French mathematician and physicist. Charles was interested in gases and was among the first people to ascend in a hydrogen-filled balloon. The direct proportionality between volume and temperature is named **Charles's law** after him.

$$\text{Charles's law:} \quad V \propto T \quad \text{(constant } P \text{ and } n)$$

When the temperature of a gas sample is increased, the gas particles move faster; collisions with the walls are more frequent, and the force exerted with each collision is greater. The only way for the pressure (the force per unit area) to remain constant is for the gas to occupy a larger volume so that collisions become less frequent and occur over a larger area (**Figure 5.10▲**).

Charles's law explains why the second floor of a house is usually a bit warmer than the ground floor. According to Charles's law, when air is heated, its volume increases, resulting in a lower density. The warm, less dense air tends to rise in a room filled with colder, denser air. Similarly, Charles's law explains why a hot-air balloon can take flight. The gas that fills a hot-air balloon is warmed with a burner, increasing its volume and lowering its density, and causing it to float in the colder, denser surrounding air.

You can experience Charles's law directly by holding a partially inflated balloon over a warm toaster. As the air in the balloon warms, you can feel the balloon expanding. Alternatively, you can put an inflated balloon into liquid nitrogen and see that it becomes smaller as it cools (**Figure 5.11▼**).

Charles's law can be used to calculate the volume of a gas following a temperature change or the temperature of a gas following a volume change *as long as the pressure and the amount of gas are constant.* For these types of calculations, we rearrange Charles's law as follows:

$$\text{Since } V \propto T, \text{ then } V = \text{constant} \times T$$

If we divide both sides by T, we get

$$V/T = \text{constant}$$

If the temperature increases, the volume increases in direct proportion so that the quotient, V/T, is always equal to the same constant. So, for two different measurements, we can say that

$$V_1/T_1 = \text{constant} = V_2/T_2,$$

or

$$\frac{V_1}{T_1} = \frac{V_2}{T_2} \qquad [5.3]$$

where V_1 and T_1 are the initial volume and temperature of the gas and V_2 and T_2 are the final volume and temperature. *The temperatures must always be expressed in kelvins (K)*, because, as we can see from Figure 5.9, the volume of a gas is directly proportional to its absolute temperature, not its temperature in °C For example, doubling the temperature of a gas sample from 1 °C to 2 °C does not double its volume, but doubling the temperature from 200 K to 400 K does.

▲ A hot-air balloon floats because the hot air is less dense than the surrounding cold air.

▲ **FIGURE 5.11 The Effect of Temperature on Volume**   If a balloon is placed into liquid nitrogen (77 K), it shrivels up as the air within it cools and occupies less volume at the same external pressure.

EXAMPLE 5.3 Charles's Law

A sample of gas has a volume of 2.80 L at an unknown temperature. When the sample is submerged in ice water at $T = 0.00\,°C$, its volume decreases to 2.57 L. What was its initial temperature (in K and in °C)?

To solve the problem, first solve Charles's law for T_1.	**SOLUTION** $$\frac{V_1}{T_1} = \frac{V_2}{T_2}$$ $$T_1 = \frac{V_1}{V_2}\, T_2$$
Before you substitute the numerical values to calculate T_1, you must convert the temperature to kelvins (K). *Remember, gas law problems must always be worked with Kelvin temperatures.*	$T_2(K) = 0.00 + 273.15 = 273.15\text{ K}$
Substitute T_2 and the other given quantities to calculate T_1.	$$T_1 = \frac{V_1}{V_2}\, T_2$$ $$= \frac{2.80\ \cancel{L}}{2.57\ \cancel{L}}\, 273.15\text{ K}$$ $$= 297.6\text{ K}$$
Calculate T_1 in °C by subtracting 273.15 from the value in kelvins.	$T_1(°C) = 297.6 - 273.15 = 24\,°C$

FOR PRACTICE 5.3

A gas in a cylinder with a moveable piston has an initial volume of 88.2 mL. If the gas is heated from 35 °C to 155 °C, what is its final volume (in mL)?

Conceptual Connection 5.1   Boyle's Law and Charles's Law

The pressure on a sample of a fixed amount of gas is doubled, and then the temperature of the gas in kelvins is doubled. What is the final volume of the gas?

(a) The final volume is twice the initial volume.

(b) The final volume of the gas is four times the initial volume.

(c) The final volume of the gas is one-half the initial volume.

(d) The final volume of the gas is one-fourth the initial volume.

(e) The final volume of the gas is the same as the initial volume.

Avogadro's Law: Volume and Amount (in Moles)

So far, we have discussed the relationships between volume and pressure, and volume and temperature, but we have considered only a constant amount of a gas. What happens when the amount of gas changes? The volume of a gas sample (at constant temperature and pressure) as a function of the amount of gas (in moles) in the sample is shown in **Figure 5.12▶**. We can see that the relationship between volume and amount is linear. As we might expect, extrapolation to zero moles shows zero volume. This relationship, first stated formally by Amedeo Avogadro, is called **Avogadro's law**:

> Avogadro's law assumes constant temperature and pressure and is independent of the nature of the gas.

$$\text{Avogadro's law}\quad V \propto n \quad \text{(constant } T \text{ and } P\text{)}$$

When the amount of gas in a sample is increased at constant temperature and pressure, its volume increases in direct proportion because the greater number of gas particles fill more space.

You experience Avogadro's law when you inflate a balloon. With each exhaled breath, you add more gas particles to the inside of the balloon, increasing its volume. Avogadro's law can be used to calculate the volume of a gas following a change in the

▶ **FIGURE 5.12 Volume versus Number of Moles** The volume of a gas sample increases linearly with the number of moles of gas in the sample.

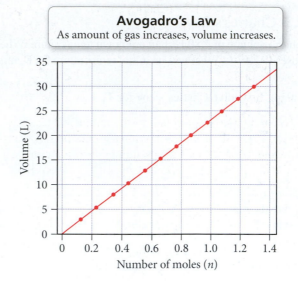

Avogadro's Law
As amount of gas increases, volume increases.

amount of the gas *as long as the pressure and temperature of the gas are constant.* For these types of calculations, we express Avogadro's law as

$$\frac{V_1}{n_1} = \frac{V_2}{n_2}$$ [5.4]

where V_1 and n_1 are the initial volume and number of moles of the gas and V_2 and n_2 are the final volume and number of moles. In calculations, we use Avogadro's law in a manner similar to the other gas laws, as shown Example 5.4.

EXAMPLE 5.4 Avogadro's Law

A male athlete in a kinesiology research study has a lung volume of 6.15 L during a deep inhalation. At this volume, his lungs contain 0.254 moles of air. During exhalation, his lung volume decreases to 2.55 L. How many moles of gas did the athlete exhale? Assume constant temperature and pressure.

To solve the problem, first solve Avogadro's law for n_2. Then substitute the given quantities to calculate n_2.	**SOLUTION** $$\frac{V_1}{n_1} = \frac{V_2}{n_2}$$ $$n_2 = \frac{V_2}{V_1}n_1$$
Since the lungs already contain 0.254 mol of air, compute the amount of air exhaled by subtracting the result from 0.254 mol. (In Chapter 1, we introduced the practice of underlining the least (rightmost) significant digit of intermediate answers but not rounding the final answer until the very end of the calculation. We continue that practice in this chapter. However, in order to avoid unneccessary notation, we will not carry additional digits in cases, such as this one, where doing so would not affect the final answer.)	$$= \frac{2.55\ \cancel{L}}{6.15\ \cancel{L}}\,0.254\ \text{mol}$$ $$= 0.10\underline{5}\ \text{mol}$$ moles exhaled $= 0.254\ \text{mol} - 0.105\ \text{mol}$ $$= 0.149\ \text{mol}$$

FOR PRACTICE 5.4
A chemical reaction occurring in a cylinder equipped with a moveable piston produces 0.621 mol of a gaseous product. If the cylinder contained 0.120 mol of gas before the reaction and had an initial volume of 2.18 L, what was its volume after the reaction? (Assume constant pressure and temperature and that the initial amount of gas completely reacts.)

5.4 The Ideal Gas Law

The relationships that we have covered so far can be combined into a single law that encompasses all of them. So far, we know that

$$V \propto \frac{1}{P} \quad \text{(Boyle's law)}$$

$$V \propto T \quad \text{(Charles's law)}$$

$$V \propto n \quad \text{(Avogadro's law)}$$

Combining these three expressions, we get

$$V \propto \frac{nT}{P}$$

The volume of a gas is directly proportional to the number of moles of gas and to the temperature of the gas but is inversely proportional to the pressure of the gas. We can replace the proportionality sign with an equals sign by incorporating R, a proportionality constant called the *ideal gas constant*:

$$V = \frac{RnT}{P}$$

Rearranging, we get

$$PV = nRT \qquad\qquad [5.5]$$

This equation is called the **ideal gas law**, and a hypothetical gas that exactly follows this law is called an **ideal gas**. The value of R, the **ideal gas constant**, is the same for all gases and has the value:

$$R = 0.08206 \ \frac{\text{L} \cdot \text{atm}}{\text{mol} \cdot \text{K}}$$

The ideal gas law contains within it the simple gas laws that we have learned as summarized in **Figure 5.13▼**. The ideal gas law also shows how other pairs of variables are related. For example, from Charles's law we know that $V \propto T$ at constant pressure and constant number of moles. But what if we heat a sample of gas at constant *volume* and constant number of moles? This question applies to the warning labels on aerosol cans such as hair spray or deodorants. These labels warn against excessive heating or incineration of the can, even after the contents are used up. Why? An "empty" aerosol can is not really empty but contains a fixed amount of gas trapped in a fixed volume. What would happen if you heated the can? We can rearrange the ideal gas law by dividing both sides by V to clearly see the relationship between pressure and temperature at constant volume and constant number of moles:

$$PV = nRT$$

$$P = \frac{nRT}{V} = \left(\frac{nR}{V} \right) T$$

Since n and V are constant and since R is always a constant

$$P = (\text{constant}) \times T$$

This relationship between pressure and temperature is also known as *Gay-Lussac's law*. As the temperature of a fixed amount of gas in a fixed volume increases, the pressure increases. In an aerosol can, this pressure increase can blow the can apart, which is why aerosol cans should not be heated or incinerated. They might explode.

◀ FIGURE 5.13 The Ideal Gas Law and Simple Gas Laws The ideal gas law contains the simple gas laws within it.

Ideal Gas Law

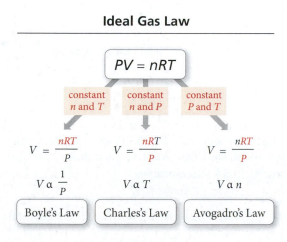

$$PV = nRT$$

constant n and T	constant n and P	constant P and T
$V = \dfrac{nRT}{P}$	$V = \dfrac{nRT}{P}$	$V = \dfrac{nRT}{P}$
$V \propto \dfrac{1}{P}$	$V \propto T$	$V \propto n$
Boyle's Law	Charles's Law	Avogadro's Law

▶ The labels on most aerosol cans warn against incineration. Since the volume of the can is constant, an increase in temperature causes an increase in pressure and can cause the can to explode.

> or persist. **INGESTION:** Have victim rinse mouth thoroughly with water. Do not induce vomiting without medical advice. If vomiting occurs naturally, have victim lean forward to reduce risk of aspiration. **WARNING:** Product is not required to be labeled as flammable as described in 16 CFR 1500.3 and 1500.45. Keep away from heat, sparks, and flames, Do not puncture or incinerate container. Do not expose to heat or store at temperatures above 120°F (49°C) as container may burst. Avoid prolonged exposure to sunlight. Use with adequate ventilation. KEEP OUT OF REACH OF CHILDREN.
> Contains: Liquefied Petroleum Gas.

We can use the ideal gas law to determine the value of any one of the four variables (P, V, n, or T) given the other three. However, each of the quantities in the ideal gas law *must be expressed* in the units within R:

L = liters
atm = atmospheres
mol = moles
K = kelvins

- pressure (P) in atm
- volume (V) in L
- moles (n) in mol
- temperature (T) in K

EXAMPLE 5.5 Ideal Gas Law I

Calculate the volume occupied by 0.845 mol of nitrogen gas at a pressure of 1.37 atm and a temperature of 315 K.

SORT The problem gives you the number of moles of nitrogen gas, the pressure, and the temperature. You are asked to find the volume.	**GIVEN** $n = 0.845$ mol, $P = 1.37$ atm, $T = 315$ K **FIND** V
STRATEGIZE You are given three of the four variables (P, T, and n) in the ideal gas law and asked to find the fourth (V). The conceptual plan shows how the ideal gas law provides the relationship between the given quantities and the quantity to be found.	**CONCEPTUAL PLAN** $n, P, T \longrightarrow V$ $PV = nRT$ **RELATIONSHIPS USED** $PV = nRT$ (ideal gas law)
SOLVE To solve the problem, first solve the ideal gas law for V. Then substitute the given quantities to calculate V.	**SOLUTION** $PV = nRT$ $V = \dfrac{nRT}{P}$ $V = \dfrac{0.845 \text{ mol} \times 0.08206 \dfrac{\text{L} \cdot \text{atm}}{\text{mol} \cdot \text{K}} \times 315 \text{ K}}{1.37 \text{ atm}}$ $= 15.9$ L

CHECK The units of the answer are correct. The magnitude of the answer (15.9 L) makes sense because, as you will see in the next section, 1 mole of an ideal gas under standard temperature and pressure (273 K and 1 atm) occupies 22.4 L. Although the temperature and pressure in this problem are not standard, they are close enough for a ballpark check of the answer. Since this gas sample contains 0.845 mol, a volume of 15.9 L is reasonable.

FOR PRACTICE 5.5

An 8.50 L tire is filled with 0.552 mol of gas at a temperature of 305 K. What is the pressure (in atm and psi) of the gas in the tire?

EXAMPLE 5.6 Ideal Gas Law II

Calculate the number of moles of gas in a 3.24 L basketball inflated to a *total pressure* of 24.3 psi at 25 °C. (Note: The *total pressure* is not the same as the pressure read on the type of pressure gauge used for checking a car or bicycle tire. That pressure, called the *gauge pressure*, is the *difference* between the total pressure and atmospheric pressure. In this case, if atmospheric pressure is 14.7 psi, the gauge pressure would be 9.6 psi. However, for calculations involving the ideal gas law, you must use the *total pressure* of 24.3 psi.)

SORT The problem gives you the pressure, the volume, and the temperature. You are asked to find the number of moles of gas.	**GIVEN** $P = 24.3$ psi, $V = 3.24$ L, $T(°C) = 25$ °C **FIND** n

STRATEGIZE The conceptual plan shows how the ideal gas law provides the relationship between the given quantities and the quantity to be found.	**CONCEPTUAL PLAN** 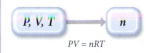 $PV = nRT$ **RELATIONSHIP USED** $$PV = nRT \quad \text{(ideal gas law)}$$

SOLVE To solve the problem, first solve the ideal gas law for n.

SOLUTION

$$PV = nRT$$

$$n = \frac{PV}{RT}$$

Before substituting into the equation, convert P and T into the correct units.

$$P = 24.3 \text{ psi} \times \frac{1 \text{ atm}}{14.7 \text{ psi}} = 1.6531 \text{ atm}$$

(Since rounding the intermediate answer would result in a slightly different final answer, we mark the least significant digit in the intermediate answer, but don't round until the end.)

$$T(K) = 25 + 273 = 298 \text{ K}$$

Finally, substitute into the equation and calculate n.

$$n = \frac{1.6531 \text{ atm} \times 3.24 \text{ L}}{0.08206 \frac{\text{L} \cdot \text{atm}}{\text{mol} \cdot \text{K}} \times 298 \text{ K}}$$

$$= 0.219 \text{ mol}$$

CHECK The units of the answer are correct. The magnitude of the answer (0.219 mol) makes sense because, as you will see in the next section, one mole of an ideal gas under standard temperature and pressure (273 K and 1 atm) occupies 22.4 L. At a pressure that is 65% higher than this, the volume of 1 mol of gas would be proportionally lower. Since this gas sample occupies 3.24 L, the answer of 0.219 mol is reasonable.

FOR PRACTICE 5.6
What volume does 0.556 mol of gas occupy at a pressure of 715 mmHg and a temperature of 58 °C?

FOR MORE PRACTICE 5.6
Find the pressure in mmHg of a 0.133-g sample of helium gas in a 648-mL container at a temperature of 32 °C.

5.5 Applications of the Ideal Gas Law: Molar Volume, Density, and Molar Mass of a Gas

We just examined how the ideal gas law can be used to calculate one of the variables (*P, V, T*, or *n*) given the other three. We now turn to three other applications of the ideal gas law: molar volume, density, and molar mass.

Molar Volume at Standard Temperature and Pressure

The volume occupied by one mole of a substance is its **molar volume**. For gases, we often specify the molar volume under conditions known as **standard temperature** (*T* = 0 °C or 273 K) **and pressure** (*P* = 1.00 atm), abbreviated as **STP**. Using the ideal gas law, we can determine that the molar volume of an ideal gas at STP is

$$V = \frac{nRT}{P}$$

$$= \frac{1.00 \text{ mol} \times 0.08206 \dfrac{\text{L} \cdot \text{atm}}{\text{mol} \cdot \text{K}} \times 273 \text{ K}}{1.00 \text{ atm}}$$

$$= 22.4 \text{ L}$$

The molar volume of an ideal gas at STP is useful because—as we saw in the *Check* steps of Examples 5.5 and 5.6—it gives us a way to approximate the volume of an ideal gas under conditions that are close to STP.

 **Conceptual Connection 5.2 Molar Volume**

Assuming ideal behavior, which gas sample will have the greatest volume at STP?

(a) 1 g of H_2 **(b)** 1 g of O_2 **(c)** 1 g of Ar

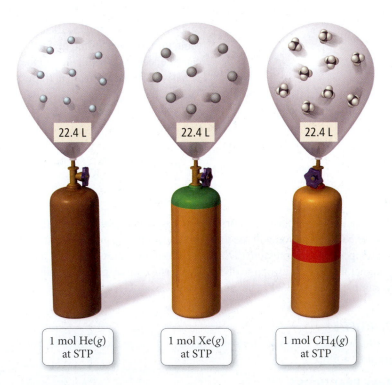

22.4 L   22.4 L   22.4 L

1 mol He(*g*)   1 mol Xe(*g*)   1 mol CH_4(*g*)
at STP   at STP   at STP

▲ One mole of any gas occupies approximately 22.4 L at standard temperature (273 K) and pressure (1.0 atm).

Density of a Gas

Since one mole of an ideal gas occupies 22.4 L under standard temperature and pressure, the density of an ideal gas can be easily calculated under standard temperature and pressure. Since density is mass/volume, and since the mass of one mole of a gas is its molar mass, the *density of a gas under standard temperature and pressure* is given by the relationship:

$$\text{Density} = \frac{\text{molar mass}}{\text{molar volume}}$$

For example, the densities of helium and nitrogen gas at STP are calculated as follows:

$$d_{\text{He}} = \frac{4.00 \text{ g/mol}}{22.4 \text{ L/mol}} = 0.179 \text{ g/L} \quad d_{\text{N}_2} = \frac{28.02 \text{ g/mol}}{22.4 \text{ L/mol}} = 1.25 \text{ g/L}$$

Notice that *the density of a gas is directly proportional to its molar mass*. The greater the molar mass of a gas, the more dense the gas. For this reason, a gas with a molar mass lower than that of air tends to rise in air. For example, both helium and hydrogen gas (molar masses of 4.00 and 2.02 g/mol, respectively) have molar masses that are lower than the average molar mass of air (approximately 28.8 g/mol). Therefore a balloon filled with either helium or hydrogen gas floats in air.

> The detailed composition of air is covered in Section 5.6. The main components of air are nitrogen (about four-fifths) and oxygen (about one-fifth).

We can calculate the density of a gas more generally (under any conditions) by using the ideal gas law. For example, we can arrange the ideal gas law as follows:

$$PV = nRT$$

$$\frac{n}{V} = \frac{P}{RT}$$

Since the left-hand side of this equation has units of moles/liter, it represents the molar density. We can obtain the density in grams/liter from molar density by multiplying by the molar mass ($\mathcal{M}$):

$$\underset{\text{Molar density}}{\frac{\text{moles}}{\text{liter}}} \times \underset{\text{Molar mass}}{\frac{\text{grams}}{\text{mole}}} = \underset{\substack{\text{Density in} \\ \text{grams/liter}}}{\frac{\text{grams}}{\text{liter}}}$$

Therefore

$$d = \frac{P\mathcal{M}}{RT} \qquad\qquad [5.6]$$

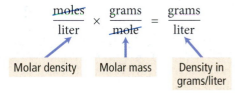

$$\overset{\text{Density}}{\overbrace{\underset{\text{Molar density}}{\frac{n}{V}} \; \underset{\text{Molar mass}}{\mathcal{M}}}} = \frac{P\mathcal{M}}{RT}$$

$$d = \frac{P\mathcal{M}}{RT}$$

Notice that, as expected, density increases with increasing molar mass. Notice also that as we learned in Section 5.3, density decreases with increasing temperature.

EXAMPLE 5.7 Density

Calculate the density of nitrogen gas at 125 °C and a pressure of 755 mmHg.

SORT The problem gives you the temperature and pressure of a gas and asks you to find its density. The problem also states that the gas is nitrogen.	**GIVEN** $T = (°C) = 125 °C$, $P = 755$ mmHg **FIND** d
STRATEGIZE Equation 5.6 provides the relationship between the density of a gas and its temperature, pressure, and molar mass. The temperature and pressure are given and you can compute the molar mass from the formula of the gas, which we know is N_2.	**CONCEPTUAL PLAN** 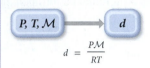 $$d = \frac{P\mathcal{M}}{RT}$$ **RELATIONSHIPS USED** $$d = \frac{P\mathcal{M}}{RT} \quad \text{(density of a gas)}$$ Molar mass $N_2 = 28.02$ g/mol
SOLVE To solve the problem, you must gather each of the required quantities in the correct units. Convert the temperature to kelvins and the pressure to atmospheres. Now substitute the quantities into the equation to calculate density.	**SOLUTION** $T(K) = 125 + 273 = 398$ K $P = 755 \text{ mmHg} \times \dfrac{1 \text{ atm}}{760 \text{ mmHg}} = 0.99342 \text{ atm}$ $d = \dfrac{P\mathcal{M}}{RT}$ $= \dfrac{0.99342 \text{ atm} \left(28.02 \dfrac{\text{g}}{\text{mol}}\right)}{0.08206 \dfrac{\text{L} \cdot \text{atm}}{\text{mol} \cdot \text{K}}(398 \text{ K})}$ $= 0.852 \text{ g/L}$

CHECK The units of the answer are correct. The magnitude of the answer (0.852 g/L) makes sense because earlier we calculated the density of nitrogen gas at STP as 1.25 g/L. Since the temperature is higher than standard temperature and pressure, the density should be lower.

FOR PRACTICE 5.7
Calculate the density of xenon gas at a pressure of 742 mmHg and a temperature of 45 °C.

FOR MORE PRACTICE 5.7
A gas has a density of 1.43 g/L at a temperature of 23 °C and a pressure of 0.789 atm. Calculate the molar mass of the gas.

 Conceptual Connection 5.3 Density of a Gas

Arrange the following gases in order of increasing density at STP: Ne, Cl_2, F_2, and O_2.

Molar Mass of a Gas

The ideal gas law can be used in combination with mass measurements to calculate the molar mass of an unknown gas. Usually, the mass and volume of an unknown gas are measured under conditions of known pressure and temperature. Then, the amount of the gas in moles is determined from the ideal gas law. Finally, the molar mass is computed by dividing the mass (in grams) by the amount (in moles) as shown in the following example.

EXAMPLE 5.8 Molar Mass of a Gas

A sample of gas has a mass of 0.311 g. Its volume is 0.225 L at a temperature of 55 °C and a pressure of 886 mmHg. Find its molar mass.

SORT The problem gives you the mass of a gas sample, along with its volume, temperature, and pressure. You are asked to find the molar mass.	**GIVEN** $m = 0.311$ g, $V = 0.225$ L, $T(°C) = 55$ °C, $P = 886$ mmHg **FIND** molar mass (g/mol)
STRATEGIZE The conceptual plan has two parts. In the first part, use the ideal gas law to find the number of moles of gas. In the second part, use the definition of molar mass to find the molar mass.	**CONCEPTUAL PLAN** 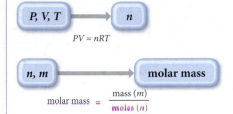 **RELATIONSHIPS USED** $PV = nRT$ $\text{Molar mass} = \dfrac{\text{mass }(m)}{\text{moles }(n)}$
SOLVE To find the number of moles, first solve the ideal gas law for n.	**SOLUTION** $PV = nRT$ $n = \dfrac{PV}{RT}$
Before substituting into the equation for n, convert the pressure to atm and the temperature to K.	$P = 886 \ \text{mmHg} \times \dfrac{1 \ \text{atm}}{760 \ \text{mmHg}} = 1.1658 \ \text{atm}$ $T(\text{K}) = 55 + 273 = 328 \ \text{K}$
Now, substitute into the equation and calculate n, the number of moles.	$n = \dfrac{1.1658 \ \text{atm} \times 0.225 \ \text{L}}{0.08206 \ \dfrac{\text{L} \cdot \text{atm}}{\text{mol} \cdot \text{K}} \times 328 \ \text{K}}$ $= 9.7454 \times 10^{-3} \ \text{mol}$
Finally, use the number of moles (n) and the given mass (m) to find the molar mass.	$\text{molar mass} = \dfrac{\text{mass }(m)}{\text{moles }(n)}$ $= \dfrac{0.311 \ \text{g}}{9.7454 \times 10^{-3} \ \text{mol}}$ $= 31.9 \ \text{g/mol}$

CHECK The units of the answer are correct. The magnitude of the answer (31.9 g/mol) is a reasonable number for a molar mass. If you calculated some very small number (for example, any number smaller than 1) or a very large number, you probably made some mistake. Most common gases have molar masses between two and several hundred grams per mole.

FOR PRACTICE 5.8
A sample of gas has a mass of 827 mg. Its volume is 0.270 L at a temperature of 88 °C and a pressure of 975 mmHg. Find its molar mass.

5.6   Mixtures of Gases and Partial Pressures

Many gas samples are not pure, but are mixtures of gases. Dry air, for example, is a mixture containing nitrogen, oxygen, argon, carbon dioxide, and a few other gases in smaller amounts (Table 5.2).

TABLE 5.2 Composition of Dry Air	
Gas	**Percent by Volume (%)**
Nitrogen (N_2)	78
Oxygen (O_2)	21
Argon (Ar)	0.9
Carbon dioxide (CO_2)	0.04

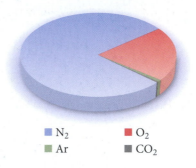

■ N_2 ■ O_2
■ Ar   ■ CO_2

Because the particles in an ideal gas do not interact (as we discuss in more detail in Section 5.8), each component in an ideal gas mixture acts independently of the others. For example, the nitrogen molecules in air exert a certain pressure—78% of the total pressure—that is independent of the presence of the other gases in the mixture. Likewise, the oxygen molecules in air exert a certain pressure—21% of the total pressure—that is also independent of the presence of the other gases in the mixture. The pressure due to any individual component in a gas mixture is the **partial pressure** (P_n) of that component and can be calculated from the ideal gas law by assuming that each gas component acts independently.

$$P_n = n_n \frac{RT}{V}$$

Notice that this equation is the ideal gas law applied to one individual component gas. For a multicomponent gas mixture, the partial pressure of each component can be calculated from the ideal gas law as follows:

$$P_a = n_a \frac{RT}{V}; \ P_b = n_b \frac{RT}{V}; \ P_c = n_c \frac{RT}{V}; \ ... \qquad [5.7]$$

The sum of the partial pressures of the components in a gas mixture must equal the total pressure:

$$P_{total} = P_a + P_b + P_c + ... \qquad [5.8]$$

where P_{total} is the total pressure and P_a, P_b, P_c, ..., are the partial pressures of the components. This is known as **Dalton's law of partial pressures**.

Combining Equations 5.7 and 5.8, we get

$$P_{total} = P_a + P_b + P_c + ...$$

$$= n_a \frac{RT}{V} + n_b \frac{RT}{V} + n_c \frac{RT}{V} + ...$$

$$= (n_a + n_b + n_c + ...) \frac{RT}{V} \qquad [5.9]$$

$$= (n_{total}) \frac{RT}{V}$$

The total number of moles in the mixture, when substituted into the ideal gas law, gives the total pressure of the sample, as we might expect.

If we divide Equation 5.7 by Equation 5.9, we get the following result:

$$\frac{P_a}{P_{total}} = \frac{n_a \ (\cancel{RT/V})}{n_{total} \ (\cancel{RT/V})} = \frac{n_a}{n_{total}} \qquad [5.10]$$

The quantity n_a/n_{total}, the number of moles of a component in a mixture divided by the total number of moles in the mixture, is the **mole fraction** (χ_a):

$$\chi_a = \frac{n_a}{n_{total}} \qquad [5.11]$$

Rearranging Equation 5.10 and substituting the definition of mole fraction gives

$$\frac{P_a}{P_{total}} = \frac{n_a}{n_{total}}$$

$$P_a = \frac{n_a}{n_{total}} P_{total} = \chi_a P_{total}$$

or simply

$$P_a = \chi_a P_{total} \qquad [5.12]$$

The partial pressure of a component in a gaseous mixture is its mole fraction multiplied by the total pressure. For gases, the mole fraction of a component is equivalent to its percent by volume divided by 100%. Therefore, based on Table 5.2, we calculate the partial pressure of nitrogen (P_{N_2}) in air at 1.00 atm as follows:

$$P_{N_2} = 0.78 \times 1.00 \text{ atm}$$

$$= 0.78 \text{ atm}$$

Similarly, the partial pressure of oxygen in air at 1.00 atm is 0.21 atm and the partial pressure of Ar in air is 0.01 atm. Applying Dalton's law of partial pressures to air at 1.00 atm:

We can ignore the contribution of the CO_2 and other trace gases because they are so small.

$$P_{total} = P_{N_2} + P_{O_2} + P_{Ar}$$

$$P_{total} = 0.78 \text{ atm} + 0.21 \text{ atm} + 0.01 \text{ atm}$$

$$= 1.00 \text{ atm}$$

Conceptual Connection 5.4 Partial Pressures

A gas mixture contains an equal number of moles of He and Ne. The total pressure of the mixture is 3.0 atm. What are the partial pressures of He and Ne?

EXAMPLE 5.9 Total Pressure and Partial Pressures

A 1.00 L mixture of helium, neon, and argon has a total pressure of 662 mmHg at 298 K. If the partial pressure of helium is 341 mmHg and the partial pressure of neon is 112 mmHg, what mass of argon is present in the mixture?

SORT The problem gives you the partial pressures of two of the three components in a gas mixture, along with the total pressure, the volume, and the temperature, and asks you find the mass of the third component.

GIVEN $P_{He} = 341$ mmHg, $P_{Ne} = 112$ mmHg,
$\quad\quad P_{total} = 662$ mmHg, $V = 1.00$ L, $T = 298$ K
FIND m_{Ar}

STRATEGIZE You can find the mass of argon from the number of moles of argon, which you can calculate from the partial pressure of argon and the ideal gas law. Begin by finding the partial pressure of argon from Dalton's law of partial pressures.

Then use the partial pressure of argon together with the volume of the sample and the temperature to find the number of moles of argon.

Finally, use the molar mass of argon to calculate the mass of argon from the number of moles of argon.

CONCEPTUAL PLAN

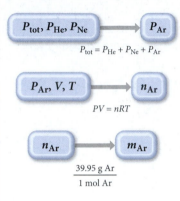

$P_{tot} = P_{He} + P_{Ne} + P_{Ar}$

$PV = nRT$

$$\frac{39.95 \text{ g Ar}}{1 \text{ mol Ar}}$$

RELATIONSHIPS USED

$P_{total} = P_{He} + P_{Ne} + P_{Ar}$ (Dalton's law)
$PV = nRT$ (ideal gas law)
molar mass Ar = 39.95 g/mol

SOLVE Follow the conceptual plan. To find the partial pressure of argon, solve the equation for P_{Ar} and substitute the values of the other partial pressures to calculate P_{Ar}.

SOLUTION

$P_{total} = P_{He} + P_{Ne} + P_{Ar}$

$P_{Ar} = P_{total} - P_{He} - P_{Ne}$

$\quad\quad = 662 \text{ mmHg} - 341 \text{ mmHg} - 112 \text{ mmHg}$

$\quad\quad = 209 \text{ mmHg}$

Convert the partial pressure from mmHg to atm and use it in the ideal gas law to calculate the amount of argon in moles.

$209 \text{ mmHg} \times \dfrac{1 \text{ atm}}{760 \text{ mmHg}} = 0.275 \text{ atm}$

$n = \dfrac{PV}{RT} = \dfrac{0.275 \text{ atm} (1.00 \text{ L})}{0.08206 \dfrac{\text{L} \cdot \text{atm}}{\text{mol} \cdot \text{K}} (298 \text{ K})} = 1.125 \times 10^{-2} \text{ mol Ar}$

Use the molar mass of argon to convert from amount of argon in moles to mass of argon.

$1.125 \times 10^{-2} \text{ mol Ar} \times \dfrac{39.95 \text{ g Ar}}{1 \text{ mol Ar}} = 0.449 \text{ g Ar}$

CHECK The units of the answer are correct. The magnitude of the answer makes sense because the volume is 1.0 L, which at STP would contain about 1/22 mol. Since the partial pressure of argon in the mixture is about 1/3 of the total pressure, we roughly estimate about 1/66 of one molar mass of argon, which is fairly close to the answer we got.

FOR PRACTICE 5.9
A sample of hydrogen gas is mixed with water vapor. The mixture has a total pressure of 755 torr and the water vapor has a partial pressure of 24 torr. What amount (in moles) of hydrogen gas is contained in 1.55 L of this mixture at 298 K?

EXAMPLE 5.10 Partial Pressures and Mole Fractions

A 12.5 L scuba diving tank is filled with a helium–oxygen (heliox) mixture containing 24.2 g of He and 4.32 g of O_2 at 298 K. Calculate the mole fraction and partial pressure of each component in the mixture and the total pressure of the mixture.

SORT The problem gives the masses of two gases in a mixture and the volume and temperature of the mixture. You are asked to find the mole fraction and partial pressure of each component, as well as the total pressure.	**GIVEN** $m_{He} = 24.2$ g, $m_{O_2} = 4.32$ g, $V = 12.5$ L, $T = 298$ K **FIND** $\chi_{He}, \chi_{O_2}, P_{He}, P_{O_2}, P_{total}$

STRATEGIZE The conceptual plan has several parts. To calculate the mole fraction of each component, you must first find the number of moles of each component. Therefore, in the first part of the conceptual plan, convert the masses to moles using the molar masses.

In the second part, calculate the mole fraction of each component using the mole fraction definition.

To calculate *partial pressures* you must calculate the *total pressure* and then use the mole fractions from the previous part to calculate the partial pressures. Calculate the total pressure from the sum of the moles of both components. (Alternatively, you could calculate the partial pressures of the components individually, using the number of moles of each component. Then you could sum them to obtain the total pressure.)

Last, use the mole fractions of each component and the total pressure to calculate the partial pressure of each component.

CONCEPTUAL PLAN

$$\chi_{He} = \frac{n_{He}}{n_{He} + n_{O_2}} ; \chi_{O_2} = \frac{n_{O_2}}{n_{He} + n_{O_2}}$$

$$P_{total} = \frac{(n_{He} + n_{O_2})RT}{V}$$

$$P_{He} = \chi_{He} P_{total} ; P_{O_2} = \chi_{O_2} P_{total}$$

RELATIONSHIPS USED

$\chi_a = n_a/n_{total}$ (mole fraction definition)

$P_{total}V = n_{total} RT$ (ideal gas law)

$P_a = \chi_a P_{total}$

SOLVE Follow the plan to solve the problem. Begin by converting each of the masses to amounts in moles.

SOLUTION

$$24.2 \text{ g He} \times \frac{1 \text{ mol He}}{4.00 \text{ g He}} = 6.05 \text{ mol He}$$

$$4.32 \text{ g } O_2 \times \frac{1 \text{ mol } O_2}{32.00 \text{ g } O_2} = 0.135 \text{ mol } O_2$$

Calculate each of the mole fractions.

$$\chi_{He} = \frac{n_{He}}{n_{He} + n_{O_2}} = \frac{6.05 \text{ mol}}{6.05 \text{ mol} + 0.135 \text{ mol}} = 0.97817$$

$$\chi_{O_2} = \frac{n_{O_2}}{n_{He} + n_{O_2}} = \frac{0.135 \text{ mol}}{6.05 \text{ mol} + 0.135 \text{ mol}} = 0.021827$$

Calculate the total pressure.

$$P_{total} = \frac{(n_{He} + n_{O_2})RT}{V}$$

$$= \frac{(6.05 \text{ mol} + 0.135 \text{ mol})\left(0.08206 \frac{L \cdot atm}{mol \cdot K}\right)(298 \text{ K})}{12.5 \text{ L}}$$

$$= 12.099 \text{ atm}$$

Finally, calculate the partial pressure of each component.

$$P_{He} = \chi_{He} P_{total} = 0.97817 \times 12.099 \text{ atm}$$
$$= 11.8 \text{ atm}$$

$$P_{O_2} = \chi_{O_2} P_{total} = 0.021827 \times 12.099 \text{ atm}$$
$$= 0.264 \text{ atm}$$

CHECK The units of the answers are correct and the magnitudes are reasonable.

FOR PRACTICE 5.10

A diver breathes a heliox mixture with an oxygen mole fraction of 0.050. What must the total pressure be for the partial pressure of oxygen to be 0.21 atm?

Vapor pressure is covered in detail in Chapter 11.

Collecting Gases Over Water

When the product of a chemical reaction is gaseous, it is often collected by the displacement of water. For example, suppose we want to use the following reaction as a source of hydrogen gas:

$$Zn(s) + 2\,HCl(aq) \longrightarrow ZnCl_2(aq) + H_2(g)$$

As the hydrogen gas forms, it bubbles through the water and gathers in the collection flask (**Figure 5.14▼**). The hydrogen gas collected in this way is not pure, however. It is mixed with water vapor because some water molecules evaporate and mix with the hydrogen molecules.

The partial pressure of water in the resulting mixture is its **vapor pressure**, which depends on the temperature (Table 5.3). Vapor pressure increases with increasing temperature because higher temperatures cause more water molecules to evaporate.

Suppose we collect hydrogen gas over water at a total pressure of 758.2 mmHg and a temperature of 25 °C. What is the partial pressure of the hydrogen gas? We know that the total pressure is 758.2 mmHg and that the partial pressure of water is 23.78 mmHg (its vapor pressure at 25 °C):

TABLE 5.3 Vapor Pressure of Water versus Temperature

Temperature (°C)	Pressure (mmHg)	Temperature (°C)	Pressure (mmHg)
0	4.58	55	118.2
5	6.54	60	149.6
10	9.21	65	187.5
15	12.79	70	233.7
20	17.55	75	289.1
25	23.78	80	355.1
30	31.86	85	433.6
35	42.23	90	525.8
40	55.40	95	633.9
45	71.97	100	760.0
50	92.6		

$$P_{total} = P_{H_2} + P_{H_2O}$$

$$758.2\text{ mmHg} = P_{H_2} + 23.78\text{ mmHg}$$

Appendix IIE contains a more complete table of the vapor pressure of water versus temperature.

Therefore,

$$P_{H_2} = 758.2\text{ mmHg} - 23.78\text{ mmHg}$$

$$= 734.4\text{ mmHg}$$

The partial pressure of the hydrogen in the mixture will be 734.4 mmHg.

▶ **FIGURE 5.14 Collecting a Gas Over Water**   When the gaseous product of a chemical reaction is collected over water, the product molecules become mixed with water molecules. The pressure of water in the final mixture is equal to the vapor pressure of water at the temperature at which the gas is collected.

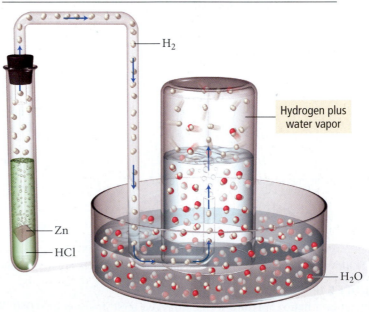

Collecting a Gas Over Water

H₂

Hydrogen plus water vapor

Zn

HCl

H₂O

EXAMPLE 5.11 Collecting Gases Over Water

In order to determine the rate of photosynthesis, the oxygen gas emitted by an aquatic plant is collected over water at a temperature of 293 K and a total pressure of 755.2 mmHg. Over a specific time period, a total of 1.02 L of gas is collected. What mass of oxygen gas (in grams) formed?

SORT The problem gives the volume of gas collected over water as well as the temperature and the pressure. You are asked to find the mass in grams of oxygen formed.	**GIVEN** $V = 1.02$ L, $P_{total} = 755.2$ mmHg, $T = 293$ K **FIND** g O_2

STRATEGIZE You can determine the mass of oxygen from the amount of oxygen in moles, which you can calculate from the ideal gas law if you know the partial pressure of oxygen. Since the oxygen is mixed with water vapor, you find the partial pressure of oxygen in the mixture by subtracting the partial pressure of water at 293 K (20 °C) from the total pressure.

Next, use the ideal gas law to find the number of moles of oxygen from its partial pressure, volume, and temperature.

Finally, use the molar mass of oxygen to convert the number of moles to grams.

CONCEPTUAL PLAN

$$P_{O_2} = P_{total} - P_{H_2O}(20\,°C)$$

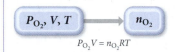

$$P_{O_2}V = n_{O_2}RT$$

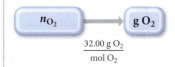

$$\frac{32.00 \text{ g } O_2}{\text{mol } O_2}$$

RELATIONSHIP USED

$$P_{total} = P_a + P_b + P_c + ... \quad \text{(Dalton's law)}$$

$$PV = nRT \quad \text{(ideal gas law)}$$

SOLVE Follow the conceptual plan to solve the problem. Begin by calculating the partial pressure of oxygen in the oxygen/water mixture. You can find the partial pressure of water at 20 °C in Table 5.3.

SOLUTION

$$P_{O_2} = P_{total} - P_{H_2O}(20\,°C)$$
$$= 755.2 \text{ mmHg} - 17.55 \text{ mmHg}$$
$$= 737.65 \text{ mmHg}$$

Next, solve the ideal gas law for number of moles.

$$n_{O_2} = \frac{P_{O_2}V}{RT}$$

Before substituting into the ideal gas law, you need to convert the partial pressure of oxygen from mmHg to atm.

$$737.65 \text{ mmHg} \times \frac{1 \text{ atm}}{760 \text{ mmHg}} = 0.97059 \text{ atm}$$

Next, substitute into the ideal gas law to find the number of moles of oxygen.

$$n_{O_2} = \frac{P_{O_2}V}{RT} = \frac{0.97059 \text{ atm}(1.02 \text{ L})}{0.08206 \frac{\text{L} \cdot \text{atm}}{\text{mol} \cdot \text{K}}(293 \text{ K})}$$
$$= 4.1175 \times 10^{-2} \text{ mol}$$

Finally, use the molar mass of oxygen to convert to grams of oxygen.

$$4.1175 \times 10^{-2} \text{ mol } O_2 \times \frac{32.00 \text{ g } O_2}{1 \text{ mol } O_2} = 1.32 \text{ g } O_2$$

CHECK The answer is in the correct units. You can quickly check the magnitude of the answer by using molar volume. Under STP one liter is about 1/22 of 1 mole. Therefore the answer should be about 1/22 the molar mass of oxygen (1/22 × 32 = 1.45). The magnitude of the answer seems reasonable.

FOR PRACTICE 5.11

A common way to make hydrogen gas in the laboratory is to place a metal such as zinc in hydrochloric acid (see Figure 5.14). The hydrochloric acid reacts with the metal to produce hydrogen gas, which is then collected over water. Suppose a student carries out this reaction and collects a total of 154.4 mL of gas at a pressure of 742 mmHg and a temperature of 25 °C. What mass of hydrogen gas (in mg) did the student collect?

5.7   Gases in Chemical Reactions: Stoichiometry Revisited

In Chapter 4, you learned how the coefficients in chemical equations can be used as conversion factors between number of moles of reactants and number of moles of products in a chemical reaction. These conversion factors can be used to determine, for example, the mass of product obtained in a chemical reaction based on a given mass of reactant, or the mass of one reactant needed to react completely with a given mass of another reactant. The general conceptual plan for these kinds of calculations is

where A and B are two different substances involved in the reaction and the conversion factor between amounts (in moles) of each comes from the stoichiometric coefficients in the balanced chemical equation.

In reactions involving *gaseous* reactants or products, we often specify the quantity of a gas in terms of its volume at a given temperature and pressure. As we have seen, stoichiometric relationships are always between amounts in moles. However, we can use the ideal gas law to find the amounts in moles from the volumes, or find the volumes from the amounts in moles.

$$n = \frac{PV}{RT} \quad V = \frac{nRT}{P}$$

The general conceptual plan for these kinds of calculations is

The following examples show this kind of calculation.

> The pressures here could also be partial pressures.

EXAMPLE 5.12   Gases in Chemical Reactions

Methanol (CH_3OH) can be synthesized by the reaction

$$CO(g) + 2\,H_2(g) \longrightarrow CH_3OH(g)$$

What volume (in liters) of hydrogen gas, measured at a temperature of 355 K and a pressure of 738 mmHg, is required to synthesize 35.7 g of methanol?

SORT You are given the mass of methanol, the product of a chemical reaction. You are asked to find the required volume of one of the reactants (hydrogen gas) at a specified temperature and pressure.	**GIVEN** 35.7 g CH_3OH, $T = 355$ K, $P = 738$ mmHg **FIND** V_{H_2}
STRATEGIZE The required volume of hydrogen gas can be calculated from the number of moles of hydrogen gas, which can be obtained from the number of moles of methanol via the stoichiometry of the reaction. First, find the number of moles of methanol from its mass by using the molar mass.	**CONCEPTUAL PLAN** g CH_3OH → mol CH_3OH $\dfrac{1\ mol\ CH_3OH}{32.04\ g\ CH_3OH}$

Then use the stoichiometric relationship from the balanced chemical equation to find the number of moles of hydrogen needed to form that quantity of methanol.

Finally, substitute the number of moles of hydrogen together with the pressure and temperature into the ideal gas law to find the volume of hydrogen.

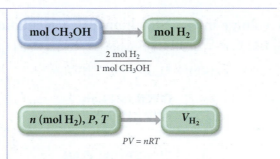

RELATIONSHIPS USED

$PV = nRT$ (ideal gas law)

2 mol H_2 : 1 mol CH_3OH
  (from balanced chemical equation)

molar mass $CH_3OH = 32.04$ g/mol

SOLVE Follow the conceptual plan to solve the problem. Begin by using the mass of methanol to get the number of moles of methanol.

Next, convert the number of moles of methanol to moles of hydrogen.

Finally, use the ideal gas law to find the volume of hydrogen. Before substituting into the equation, you must convert the pressure to atmospheres.

SOLUTION

$$35.7 \text{ g } CH_3OH \times \frac{1 \text{ mol } CH_3OH}{32.04 \text{ g } CH_3OH}$$

$$= 1.1142 \text{ mol } CH_3OH$$

$$1.1142 \text{ mol } CH_3OH \times \frac{2 \text{ mol } H_2}{1 \text{ mol } CH_3OH}$$

$$= 2.2284 \text{ mol } H_2$$

$$V_{H_2} = \frac{n_{H_2} RT}{P}$$

$$P = 738 \text{ mmHg} \times \frac{1 \text{ atm}}{760 \text{ mmHg}}$$

$$= 0.97105 \text{ atm}$$

$$V_{H_2} = \frac{(2.2284 \text{ mol})\left(0.08206 \dfrac{L \cdot atm}{mol \cdot K}\right)(355 \text{ K})}{0.97105 \text{ atm}}$$

$$= 66.9 \text{ L}$$

CHECK The units of the answer are correct. The magnitude of the answer (66.9 L) seems reasonable since you are given slightly more than 1 molar mass of methanol, which is therefore slightly more than 1 mol of methanol. From the equation you can see that 2 mol hydrogen is required to make 1 mol methanol. Therefore, the answer must be slightly greater than 2 mol hydrogen. Under standard temperature and pressure, slightly more than 2 mol hydrogen occupies slightly more than 2×22.4 L = 44.8 L. At a temperature greater than standard temperature, the volume would be even greater; therefore, the magnitude of the answer is reasonable.

FOR PRACTICE 5.12
In the following reaction, 4.58 L of O_2 was formed at $P = 745$ mmHg and $T = 308$ K. How many grams of Ag_2O decomposed?

$$2 \, Ag_2O(s) \longrightarrow 4 \, Ag(s) + O_2(g)$$

FOR MORE PRACTICE 5.12
In the above reaction, what mass of $Ag_2O(s)$ (in grams) is required to form 388 mL of oxygen gas at $P = 734$ mmHg and 25.0 °C?

Molar Volume and Stoichiometry

In Section 5.5, we saw that, under standard temperature and pressure, 1 mol of an ideal gas occupies 22.4 L. Consequently, if a reaction is occurring at or near standard temperature and pressure, we can use 1 mol = 22.4 L as a conversion factor in stoichiometric calculations, as shown Example 5.13.

EXAMPLE 5.13 Using Molar Volume in Gas Stoichiometric Calculations

How many grams of water form when 1.24 L of H_2 gas at STP completely reacts with O_2?

$$2\ H_2(g) + O_2(g) \longrightarrow 2\ H_2O(g)$$

SORT You are given the volume of hydrogen gas (a reactant) at STP and asked to determine the mass of water that forms upon complete reaction.	**GIVEN** 1.24 L H_2 **FIND** g H_2O
STRATEGIZE Since the reaction occurs under standard temperature and pressure, you can convert directly from the volume (in L) of hydrogen gas to the amount in moles. Then use the stoichiometric relationship from the balanced equation to find the number of moles of water that forms. Finally, use the molar mass of water to obtain the mass of water.	**CONCEPTUAL PLAN** 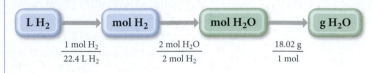 **RELATIONSHIPS USED** 1 mol = 22.4 L (at STP) 2 mol H_2 : 2 mol H_2O (from balanced equation) molar mass H_2O = 18.02 g/mol
SOLVE Follow the conceptual plan to solve the problem.	**SOLUTION** $1.24\ \cancel{L\ H_2} \times \dfrac{1\ \cancel{mol\ H_2}}{22.4\ \cancel{L\ H_2}} \times \dfrac{2\ \cancel{mol\ H_2O}}{2\ \cancel{mol\ H_2}} \times \dfrac{18.02\ g\ H_2O}{1\ \cancel{mol\ H_2O}}$ $= 0.998\ g\ H_2O$

CHECK The units of the answer are correct. The magnitude of the answer (0.998 g) is about 1/18 of the molar mass of water, roughly equivalent to the approximately 1/22 of a mole of hydrogen gas given, as expected for a stoichiometric relationship between number of moles of hydrogen and number of moles of water that is 1:1.

FOR PRACTICE 5.13

How many liters of oxygen (at STP) are required to form 10.5 g of H_2O?

$$2\ H_2(g) + O_2(g) \longrightarrow 2\ H_2O(g)$$

 Conceptual Connection 5.5 Pressure and Number of Moles

Nitrogen and hydrogen react to form ammonia according to the equation:

$$N_2(g) + 3\ H_2(g) \rightleftharpoons 2\ NH_3(g)$$

Consider the representations of the initial mixture of reactants and the resulting mixture after the reaction has been allowed to react for some time:

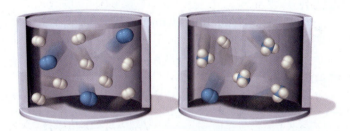

If the volume is kept constant, and nothing is added to the reaction mixture, what happens to the total pressure during the course of the reaction?

(a) the pressure increases

(b) the pressure decreases

(c) the pressure does not change

5.8   Kinetic Molecular Theory: A Model for Gases

In Chapter 1, you learned how the scientific method proceeds from observations to laws and eventually to theories. Remember that laws summarize behavior—for example, Charles's law summarizes *how* the volume of a gas depends on temperature—while theories give the underlying reasons for the behavior. A theory of gas behavior explains, for example, *why* the volume of a gas increases with increasing temperature.

The simplest model for the behavior of gases is the **kinetic molecular theory**. In this theory, a gas is modeled as a collection of particles (either molecules or atoms, depending on the gas) in constant motion (**Figure 5.15▶**). A single particle moves in a straight line until it collides with another particle (or with the wall of the container). The basic postulates (or assumptions) of kinetic molecular theory are the following:

1. **The size of a particle is negligibly small.** Kinetic molecular theory assumes that the particles themselves occupy no volume, even though they have mass. This postulate is justified because, under normal pressures, the space between atoms or molecules in a gas is very large compared to the size of the atoms or molecules themselves. For example, in a sample of argon gas under STP conditions, only about 0.01% of the volume is occupied by atoms and the average distance from one argon atom to another is 3.3 nm. In comparison, the atomic radius of argon is 97 pm. If an argon atom were the size of a golf ball, its nearest neighbor would be, on average, just over 4 ft away at STP.

2. **The average kinetic energy of a particle is proportional to the temperature in kelvins.** The motion of atoms or molecules in a gas is due to thermal energy, which distributes itself among the particles in the gas. At any given moment, some particles are moving faster than others—there is a distribution of velocities—but the higher the temperature, the faster the overall motion, and the greater the average kinetic energy. Notice that *kinetic energy* $\left(\frac{1}{2} mv^2\right)$—not *velocity*—is proportional to temperature. The atoms in a sample of helium and a sample of argon at the same temperature have the same average kinetic energy but not the same average velocity. Since the helium atoms are lighter, they must move faster to have the same kinetic energy as argon atoms.

3. **The collision of one particle with another (or with the walls) is completely elastic.** This means that when two particles collide, they may *exchange energy*, but there is no overall *loss of energy*. Any kinetic energy lost by one particle is completely gained by the other. This is the case because the particles have no "stickiness," and they are not deformed by the collision. In other words, an encounter between two particles in kinetic molecular theory is more like the collision between two billiard balls than between two lumps of clay (**Figure 5.16▼**). Between collisions, the particles do not exert any forces on one another.

Kinetic Molecular Theory

▲ **FIGURE 5.15   A Model for Gas Behavior**   In the kinetic molecular theory of gases, a gas sample is modeled as a collection of particles in constant straight-line motion. The size of each particle is negligibly small and their collisions are elastic.

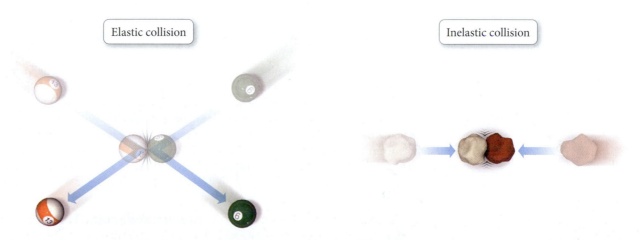

Elastic collision

Inelastic collision

▲ **FIGURE 5.16   Elastic versus Inelastic Collisions**   When two billiard balls collide, the collision is elastic—the total kinetic energy of the colliding bodies is the same before and after the collision. When two lumps of clay collide, the collision is inelastic—the kinetic energy of the colliding bodies is dissipated as heat during the collision.

If you start with the postulates of kinetic molecular theory, you can mathematically derive the ideal gas law. In other words, the ideal gas law follows directly from kinetic molecular theory, which gives us confidence that the assumptions of the theory are valid, at least under conditions where the ideal gas law works. Let's see how the concept of pressure and each of the gas laws we have examined follow conceptually from kinetic molecular theory.

The Nature of Pressure

In Section 5.2, we defined pressure as force divided by area:

$$P = \frac{F}{A}$$

The force (F) associated with an individual collision is given by $F = ma$, where m is the mass of the particle and a is its acceleration as it changes its direction of travel due to the collision.

According to kinetic molecular theory, a gas is a collection of particles in constant motion. The motion results in collisions between the particles and the surfaces around them. As each particle collides with a surface, it exerts a force upon that surface. The result of many particles in a gas sample exerting forces on the surfaces around them is a constant pressure.

Boyle's Law

Boyle's law states that, for a constant number of particles at constant temperature, the volume of a gas is inversely proportional to its pressure. If you decrease the volume of a gas, you force the gas particles to occupy a smaller space. It follows from kinetic molecular theory that, as long as the temperature remains the same, the result is a greater number of collisions with the surrounding surfaces and therefore a greater pressure.

Charles's Law

Charles's law states that, for a constant number of particles at constant pressure, the volume of a gas is proportional to its temperature. According to kinetic molecular theory, when you increase the temperature of a gas, the average speed, and thus the average kinetic energy, of the particles increases. Since this greater kinetic energy results in more frequent collisions and more force per collision, the pressure of the gas would increase if its volume were held constant (Gay-Lussac's law). The only way for the pressure to remain constant is for the volume to increase. The greater volume spreads the collisions out over a greater area, so that the pressure (defined as force per unit area) is unchanged.

Avogadro's Law

Avogadro's law states that, at constant temperature and pressure, the volume of a gas is proportional to the number of particles. According to kinetic molecular theory, when you increase the number of particles in a gas sample, the number of collisions with the surrounding surfaces increases. Since the greater number of collisions would result in a greater overall force on surrounding surfaces, the only way for the pressure to remain constant is for the volume to increase so that the number of particles per unit volume (and thus the number of collisions) remains constant.

Dalton's Law

Dalton's law states that the total pressure of a gas mixture is the sum of the partial pressures of its components. In other words, according to Dalton's law, the components in a gas mixture act identically to, and independently of, one another. According to kinetic molecular theory, the particles have negligible size and they do not interact. Consequently, the only property that would distinguish one type of particle from another is its mass. However, even particles of different masses have the same average kinetic energy at a given temperature, so they exert the same force upon collision with a surface. Consequently, adding components to a gas mixture—even different *kinds* of gases—has the same effect as simply adding more particles. The partial pressures of all the components sum to the overall pressure.

 Conceptual Connection 5.6 **Kinetic Molecular Theory I**

Draw a depiction, as described by kinetic molecular theory, of a gas sample containing equal molar amounts of argon and xenon. Use red dots to represent argon atoms and blue dots to represent xenon atoms. Give each atom a "tail" to represent its velocity relative to the others in the mixture.

Temperature and Molecular Velocities

According to kinetic molecular theory, particles of different masses have the same average kinetic energy at a given temperature. The kinetic energy of a particle depends on its mass and velocity according to the equation:

$$KE = \frac{1}{2}mv^2$$

The only way for particles of different masses to have the same kinetic energy is if they are traveling at different velocities, as we saw in Conceptual Connection 5.6.

In a gas at a given temperature, lighter particles travel faster (on average) than heavier ones.

In kinetic molecular theory, we define the root mean square velocity (u_{rms}) of a particle as follows:

$$u_{rms} = \sqrt{\overline{u^2}} \qquad [5.13]$$

where $\overline{u^2}$ is the average of the squares of the particle velocities. Even though the root mean square velocity of a collection of particles is not identical to the average velocity, the two are close in value and conceptually similar. Root mean square velocity is simply a special type of average. The average kinetic energy of one mole of gas particles is then given by

$$KE_{avg} = \frac{1}{2}N_A m\overline{u^2} \qquad [5.14]$$

where N_A is Avogadro's number.

Postulate 2 of the kinetic molecular theory states that the average kinetic energy is proportional to the temperature in kelvins. The constant of proportionality in this relationship is $(3/2)R$:

$$KE_{avg} = (3/2)RT \qquad [5.15]$$

where R is the gas constant, but in different units ($R = 8.314 \text{ J/mol} \cdot \text{K}$) from those we saw in Section 5.4. If we combine Equations 5.14 and 5.15, and solve for $\overline{u^2}$, we get

$$(1/2)N_A m\overline{u^2} = (3/2)RT$$

$$\overline{u^2} = \frac{(3/2)RT}{(1/2)N_A m} = \frac{3RT}{N_A m}$$

Taking the square root of both sides we get

$$\sqrt{\overline{u^2}} = u_{rms} = \sqrt{\frac{3RT}{N_A m}} \qquad [5.16]$$

In Equation 5.16, m is the mass of a particle in kg and N_A is Avogadro's number. The product $N_A m$, then, is the molar mass in kg/mol. If we call this quantity $\mathcal{M}$, then the expression for root mean square velocity as a function of temperature becomes the following important result:

$$u_{rms} = \sqrt{\frac{3RT}{\mathcal{M}}} \qquad [5.17]$$

The root mean square velocity of a collection of gas particles is proportional to the square root of the temperature in kelvins and inversely proportional to the square root of the molar mass of the particles in kilograms per mole. The root mean square velocity of nitrogen molecules at 25 °C, for example, is 515 m/s (1152 mi/hr). The root mean square velocity of hydrogen molecules at room temperature is 1920 m/s (4295 mi/hr). Notice that the lighter molecules move much faster at a given temperature.

The root mean square velocity, as we have seen, is a kind of average velocity. Some particles are moving faster and some are moving slower than this average. The velocities of all the particles in a gas sample form a distribution like the ones shown in **Figure 5.17▶**. We can see from these distributions that some particles are indeed traveling at the root mean square velocity. However, many particles are traveling

The $(3/2)R$ proportionality constant comes from a derivation that is beyond our current scope.

The joule (J) is a unit of energy that is covered in more detail in Section 6.1.
$$\left(1 \text{ J} = 1 \text{ kg}\frac{m^2}{s^2}\right)$$

Variation of Velocity Distribution with Molar Mass

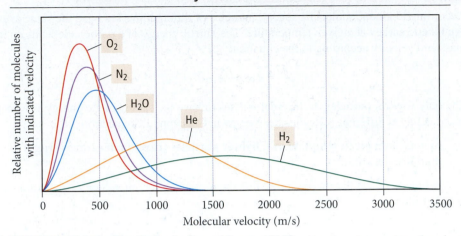

▲ **FIGURE 5.17** **Velocity Distribution for Several Gases at 25 °C** At a given temperature, there is a distribution of velocities among the particles in a sample of gas. The exact shape and peak of the distribution varies with the molar mass of the gas.

Variation of Velocity Distribution with Temperature

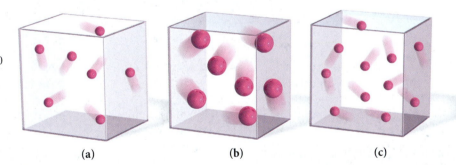

▲ **FIGURE 5.18** **Velocity Distribution for Nitrogen at Several Temperatures** As the temperature of a gas sample is increased, the velocity distribution of the molecules shifts toward higher velocities and becomes less sharply peaked.

faster and many slower than the root mean square velocity. For lighter molecules, the velocity distribution is shifted toward higher velocities and the curve becomes broader, indicating a wider range of velocities. The velocity distribution for nitrogen at different temperatures is shown in **Figure 5.18◄**. As the temperature increases, the root mean square velocity increases and the distribution becomes broader.

Conceptual Connection 5.7 Kinetic Molecular Theory II

Which of the following samples of an ideal gas, all at the same temperature, will have the greatest pressure? Assume that the mass of the particle is proportional to its size.

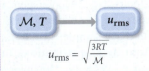

(a) (b) (c)

EXAMPLE 5.14 Root Mean Square Velocity

Calculate the root mean square velocity of oxygen molecules at 25 °C.

SORT You are given the kind of molecule and the temperature and asked to find the root mean square velocity.	**GIVEN** O_2, $t = 25\ °C$ **FIND** u_{rms}
STRATEGIZE The conceptual plan for this problem shows how the molar mass of oxygen and the temperature (in kelvins) can be used with the equation that defines the root mean square velocity to compute the root mean square velocity.	**CONCEPTUAL PLAN** $\mathcal{M}, T \longrightarrow u_{rms}$ $u_{rms} = \sqrt{\dfrac{3RT}{\mathcal{M}}}$

SOLVE First gather the required quantities in the correct units. Note that molar mass must be in kg/mol and temperature must be in kelvins.

SOLUTION

$T = 25 + 273 = 298$ K

$$\mathcal{M} = \frac{32.00 \text{ g } O_2}{1 \text{ mol } O_2} \times \frac{1 \text{ kg}}{1000 \text{ g}} = \frac{32.00 \times 10^{-3} \text{ kg } O_2}{1 \text{ mol } O_2}$$

Substitute the quantities into the equation to calculate root mean square velocity. Note that $1 \text{ J} = 1 \text{ kg} \cdot \text{m}^2/\text{s}^2$.

$$u_{rms} = \sqrt{\frac{3RT}{\mathcal{M}}}$$

$$= \sqrt{\frac{3\left(8.314 \dfrac{\text{J}}{\text{mol} \cdot \text{K}}\right)(298 \text{ K})}{\dfrac{32.00 \times 10^{-3} \text{ kg } O_2}{1 \text{ mol } O_2}}}$$

$$= \sqrt{2.32 \times 10^5 \frac{\text{J}}{\text{kg}}}$$

$$= \sqrt{2.32 \times 10^5 \frac{\dfrac{\text{kg} \cdot \text{m}^2}{\text{s}^2}}{\text{kg}}} = 482 \text{ m/s}$$

CHECK The units of the answer are correct. The magnitude of the answer seems reasonable because oxygen is slightly heavier than nitrogen and should therefore have a slightly lower root mean square velocity at the same temperature. Recall that earlier we stated the root mean square velocity of nitrogen to be 515 m/s at 25 °C.

FOR PRACTICE 5.14
Calculate the root mean square velocity of gaseous xenon atoms at 25 °C.

5.9 Mean Free Path, Diffusion, and Effusion of Gases

We have just learned that the root mean square velocity of gas molecules at room temperature is in the range of hundreds of meters per second. But suppose that your sister just put on too much perfume in the bathroom only 2 m away. Why does it take a minute or two before you can smell the fragrance? Although most molecules in a perfume bottle have higher molar masses than nitrogen, their velocities are still hundreds of meters per second. Why the delay? The answer is that even though gaseous particles travel at tremendous speeds, they also travel in very haphazard paths (**Figure 5.19▶**). To a perfume molecule, the path from the perfume bottle in the bathroom to your nose 2 m away is much like the path you would take through a busy shopping mall during a clearance sale. The molecule travels only a short distance before it collides with another molecule and changes direction, only to collide again with another molecule, and so on. In fact, at room temperature and atmospheric pressure, a gas molecule in the air experiences several billion collisions per second. The average distance that a molecule travels between collisions is its **mean free path**. At room temperature and atmospheric pressure, the mean free path of a nitrogen molecule with a molecular diameter of 300 pm (four times the covalent radius) is 93 nm, or about 310 molecular diameters. If the nitrogen molecule was the size of a golf ball, it would travel about 40 ft between collisions. Mean free path increases with *decreasing* pressure. Under conditions of ultrahigh vacuum (10^{-10} torr), the mean free path of a nitrogen molecule is hundreds of kilometers.

The process by which gas molecules spread out in response to a concentration gradient is called **diffusion**, and even though the particles undergo many collisions, the root mean square velocity still influences the rate of diffusion. Heavier molecules diffuse

The transport of molecules within a ventilated room is also enhanced by air currents.

Typical Gas Molecule Path

The average distance between collisions is the mean free path.

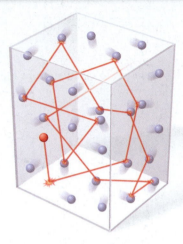

▲ **FIGURE 5.19** **Mean Free Path** A molecule in a volume of gas follows a haphazard path, experiencing many collisions with other molecules.

more slowly than lighter ones, so the first molecules you smell from a perfume mixture (in a room with no air currents) are the lighter ones.

A process related to diffusion is **effusion**, the process by which a gas escapes from a container into a vacuum through a small hole (**Figure 5.20▼**). The rate of effusion is also related to root mean square velocity—heavier molecules effuse more slowly than lighter ones. The rate of effusion—the amount of gas that effuses in a given time—is inversely proportional to the square root of the molar mass of the gas as follows:

$$\text{rate} \propto \frac{1}{\sqrt{\mathcal{M}}}$$

The ratio of effusion rates of two different gases is given by **Graham's law of effusion**, named after Thomas Graham (1805–1869):

$$\frac{\text{rate}_A}{\text{rate}_B} = \sqrt{\frac{\mathcal{M}_B}{\mathcal{M}_A}} \qquad [5.18]$$

In this expression, rate_A and rate_B are the effusion rates of gases A and B, and $\mathcal{M}_A$ and $\mathcal{M}_B$ are their molar masses.

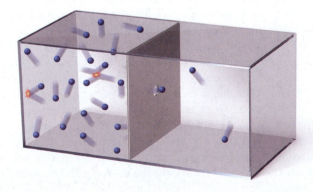

▲ **FIGURE 5.20** **Effusion** Effusion is the escape of a gas from a container into a vacuum through a small hole.

EXAMPLE 5.15 Graham's Law of Effusion

An unknown gas effuses at a rate that is 0.462 times that of nitrogen gas (at the same temperature). Calculate the molar mass of the unknown gas in g/mol.

SORT You are given the ratio of effusion rates for the unknown gas and nitrogen and asked to find the molar mass of the unknown gas.	**GIVEN** $\dfrac{\text{Rate}_{\text{unk}}}{\text{Rate}_{N_2}} = 0.462$ **FIND** $\mathcal{M}_{\text{unk}}$
STRATEGIZE The conceptual plan uses Graham's law of effusion. You are given the ratio of rates and you know the molar mass of the nitrogen. You can use Graham's law to find the molar mass of the unknown gas.	**CONCEPTUAL PLAN** 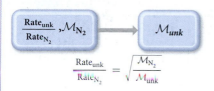 $\dfrac{\text{Rate}_{\text{unk}}}{\text{Rate}_{N_2}} = \sqrt{\dfrac{\mathcal{M}_{N_2}}{\mathcal{M}_{\text{unk}}}}$ **RELATIONSHIP USED** $\dfrac{\text{rate}_A}{\text{rate}_B} = \sqrt{\dfrac{\mathcal{M}_B}{\mathcal{M}_A}}$ (Graham's law)
SOLVE Solve the equation for $\mathcal{M}_{\text{unk}}$ and substitute the correct values to calculate it.	**SOLUTION** $\dfrac{\text{rate}_{\text{unk}}}{\text{rate}_{N_2}} = \sqrt{\dfrac{\mathcal{M}_{N_2}}{\mathcal{M}_{\text{unk}}}}$ $\mathcal{M}_{\text{unk}} = \dfrac{\mathcal{M}_{N_2}}{\left(\dfrac{\text{rate}_{\text{unk}}}{\text{rate}_{N_2}}\right)^2} = \dfrac{28.02 \text{ g/mol}}{(0.462)^2} = 131 \text{ g/mol}$

CHECK The units of the answer are correct. The magnitude of the answer seems reasonable for the molar mass of a gas. In fact, from the answer, we can even conclude that the gas is probably xenon, which has a molar mass of 131.29 g/mol.

FOR PRACTICE 5.15
Find the ratio of effusion rates of hydrogen gas and krypton gas.

5.10   Real Gases: The Effects of Size and Intermolecular Forces

One mole of an ideal gas has a volume of 22.41 L at STP. **Figure 5.21▶** shows the molar volume of several real gases at STP. As you can see, most of these gases have a volume that is very close to 22.41 L, meaning that they are acting very nearly as ideal gases. Gases behave ideally when both of the following are true: (a) the volume of the gas particles is negligible compared to the space between them; and (b) the forces between the gas particles are not significant. At STP, these assumptions are valid for most common gases. However, these assumptions break down at higher pressures or lower temperatures.

The Effect of the Finite Volume of Gas Particles

The finite volume of gas particles becomes important at high pressure, when the particles themselves occupy a significant portion of the total gas volume (**Figure 5.22▶**). We can see the effect of particle volume by comparing the molar volume of argon to the molar volume of an

▶ **FIGURE 5.21 Molar Volumes of Real Gases** The molar volumes of several gases at STP are all close to 22.41 L, indicating that their departures from ideal behavior are small.

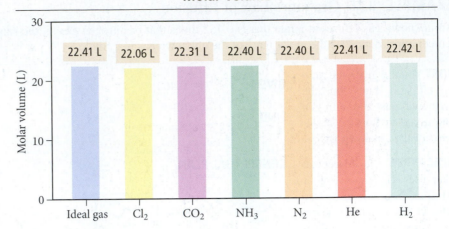

Molar Volume

| 22.41 L | 22.06 L | 22.31 L | 22.40 L | 22.40 L | 22.41 L | 22.42 L |

Ideal gas   Cl₂   CO₂   NH₃   N₂   He   H₂

ideal gas as a function of pressure at 500 K as shown in **Figure 5.23▼**. At low pressures, the molar volume of argon is nearly identical to that of an ideal gas. But as the pressure increases, the molar volume of argon becomes *greater than* that of an ideal gas. At the higher pressures, the argon atoms themselves occupy a significant portion of the gas volume, making the actual volume greater than that predicted by the ideal gas law.

In 1873, Johannes van der Waals (1837–1923) modified the ideal gas equation to fit the behavior of real gases. From the graph for argon, we can see that the ideal gas law predicts a volume that is too small. Van der Waals suggested a small correction factor that accounts for the volume of the gas particles themselves:

Ideal behavior $$V = \frac{nRT}{P}$$

Corrected for volume of gas particles $V = \dfrac{nRT}{P} + nb$ [5.19]

▶ **FIGURE 5.22 Particle Volume and Ideal Behavior** As a gas is compressed, the gas particles themselves begin to occupy a significant portion of the total gas volume, leading to deviations from ideal behavior.

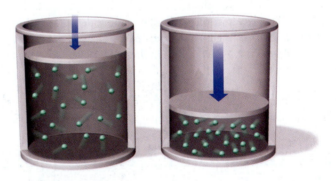

▶ **FIGURE 5.23 The Effect of Particle Volume on Molar Volume** At high pressures, 1 mol of argon occupies a larger volume than would 1 mol of an ideal gas because of the volume of the argon atoms themselves. (This example was chosen to minimize the effects of intermolecular forces, which are very small in argon at 500 K, thereby isolating the effect of particle volume.)

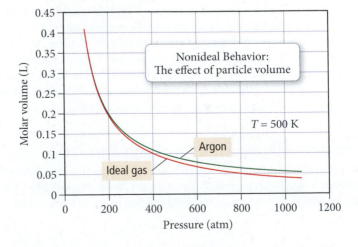

Nonideal Behavior: The effect of particle volume

$T = 500$ K

Argon

Ideal gas

The correction adds the quantity nb to the volume, where n is the number of moles and b is a constant that depends on the gas (see Table 5.4). We can rearrange the corrected equation as follows:

$$(V - nb) = \frac{nRT}{P} \qquad [5.20]$$

The Effect of Intermolecular Forces

Intermolecular forces are attractions between the atoms or molecules in a gas. These attractions are small and therefore do not matter much at low pressure, when the molecules are too far apart to "feel" the attractions. They also do not matter too much at high temperatures because the molecules have a lot of kinetic energy. When two particles with high kinetic energies collide, a weak attraction between them does not affect the collision very much. At lower temperatures, however, the collisions occur with less kinetic energy, and weak attractions do affect the collisions. We can understand this difference with an analogy to billiard balls. Imagine two billiard balls that are coated with a substance that makes them slightly sticky. If they collide when moving at high velocities, the stickiness will not have much of an effect—the balls bounce off one another as if the sticky substance was not even present. However, if the two billiard balls collide when moving very slowly (say barely rolling) the sticky substance would have an effect—the billiard balls might even stick together and not bounce off one another.

The effect of these weak attractions between particles is a lower number of collisions with the surfaces of the container, thereby lowering the pressure compared to that of an ideal gas. We can see the effect of intermolecular forces by comparing the pressure of 1.0 mol of xenon gas to the pressure of 1.0 mol of an ideal gas as a function of temperature and at a fixed volume of 1.0 L, as shown in **Figure 5.24▼**. At high temperature, the pressure of the xenon gas is nearly identical to that of an ideal gas. But at lower temperatures, the pressure of xenon is *less than* that of an ideal gas. At the lower temperatures, the xenon atoms spend more time interacting with each other and less time colliding with the walls, making the actual pressure less than that predicted by the ideal gas law.

Ideal behavior $\qquad\qquad P = \dfrac{nRT}{V}$

Corrected for intermolecular forces $\quad P = \dfrac{nRT}{V} - a\left(\dfrac{n}{V}\right)^2 \qquad [5.21]$

The correction subtracts the quantity $a(n/V)^2$ from the pressure, where n is the number of moles, V is the volume, and a is a constant that depends on the gas (see Table 5.4). Notice that the correction factor increases as n/V (the number of moles of particles per unit volume) increases because a greater concentration of particles makes it more likely that they will interact with one another. We can rearrange the corrected equation as follows:

$$P + a\left(\frac{n}{V}\right)^2 = \frac{nRT}{V} \qquad [5.22]$$

TABLE 5.4   Van der Waals Constants for Common Gases

Gas	$a\,(L^2 \cdot atm/mol^2)$	$b\,(L/mol)$
He	0.0342	0.02370
Ne	0.211	0.0171
Ar	1.35	0.0322
Kr	2.32	0.0398
Xe	4.19	0.0511
H_2	0.244	0.0266
N_2	1.39	0.0391
O_2	1.36	0.0318
Cl_2	6.49	0.0562
H_2O	5.46	0.0305
CH_4	2.25	0.0428
CO_2	3.59	0.0427
CCl_4	20.4	0.1383

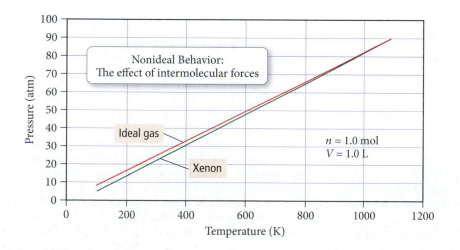

◀ FIGURE 5.24 The Effect of Intermolecular Forces on Pressure   At low temperatures, the pressure of xenon is less than an ideal gas would exert because interactions among xenon molecules reduce the number of collisions with the walls of the container.

Van der Waals Equation

We can now combine the effects of particle volume (Equation 5.20) and particle intermolecular forces (Equation 5.22) into one equation that describes nonideal gas behavior:

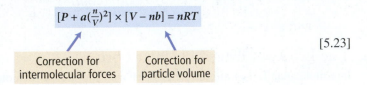

$$[P + a(\tfrac{n}{V})^2] \times [V - nb] = nRT$$

Correction for intermolecular forces

Correction for particle volume

[5.23]

This equation is the **van der Waals equation** and can be used to calculate the properties of a gas under nonideal conditions.

CHAPTER IN REVIEW

Key Terms

Section 5.1
pressure (169)

Section 5.2
millimeter of mercury (mmHg) (170)
barometer (170)
torr (171)
atmosphere (atm) (171)
pascal (Pa) (171)

Section 5.3
Boyle's law (172)
Charles's law (176)
Avogadro's law (177)

Section 5.4
ideal gas law (179)
ideal gas (179)
ideal gas constant (179)

Section 5.5
standard temperature and pressure (STP) (182)
molar volume (183)

Section 5.6
partial pressure (P_n) (186)
Dalton's law of partial pressures (186)
mole fraction (χ_a) (187)
vapor pressure (190)

Section 5.8
kinetic molecular theory (195)

Section 5.9
mean free path (199)
diffusion (199)
effusion (200)
Graham's law of effusion (200)

Section 5.10
van der Waals equation (204)

Key Concepts

Pressure (5.1, 5.2)

▶ Gas pressure is the force per unit area that results from gas particles colliding with the surfaces around them. Pressure is measured in a number of units including mmHg, torr, Pa, psi, in Hg, and atm.

The Simple Gas Laws (5.3)

▶ The simple gas laws express relationships between pairs of variables when the other variables are held constant.
▶ Boyle's law states that the volume of a gas is inversely proportional to its pressure.
▶ Charles's law states that the volume of a gas is directly proportional to its temperature.
▶ Avogadro's law states that the volume of a gas is directly proportional to the amount (in moles).

The Ideal Gas Law and Its Applications (5.4, 5.5)

▶ The ideal gas law, $PV = nRT$, gives the relationship among all four gas variables and contains the simple gas laws within it. The ideal gas law can be used to find one of the four variables given the other three.
▶ The ideal gas law can also be used to calculate the molar volume of an ideal gas, which is 22.4 L at STP, and to calculate the density and molar mass of a gas.

Mixtures of Gases and Partial Pressures (5.6)

▶ In a mixture of gases, each gas acts independently of the others so that any overall property of the mixture is the sum of the properties of the individual components.
▶ The pressure of any individual component is its partial pressure.

Gas Stoichiometry (5.7)

▶ In reactions involving gaseous reactants and products, quantities are often reported in volumes at specified pressures and temperatures. These quantities can be converted to amounts (in moles) using the ideal gas law. Then the stoichiometric coefficients from the balanced equation can be used to determine the stoichiometric amounts of other reactants or products. The general form for these types of calculations is often as follows: volume A → amount A (in moles) → amount B (in moles) → quantity of B (in desired units).
▶ In cases where the reaction is carried out at STP, the molar volume at STP (22.4 L = 1 mol) can be used to convert between volume in liters and amount in moles.

Kinetic Molecular Theory and Its Applications (5.8, 5.9)

▶ Kinetic molecular theory is a quantitative model for gases with three main assumptions: (1) the gas particles are negligibly small; (2) the average kinetic energy of a gas particle is proportional to the temperature in kelvins; and (3) the collision of one gas particle with another is completely elastic (the particles do not stick together). The gas laws all follow from the kinetic molecular theory.
▶ Kinetic molecular theory can be used to derive the expression for the root mean square velocity of gas particles. This velocity is inversely proportional to the square root of the molar mass of the gas, and therefore—at a given temperature—smaller gas particles are (on average) moving more quickly than larger ones.
▶ The kinetic molecular theory also allows us to predict the mean free path of a gas particle (the distance it travels between collisions) and relative rates of diffusion or effusion.

Real Gases (5.10)

▶ Real gases differ from ideal gases to the extent that they do not fit the assumptions of kinetic molecular theory. These assumptions tend to break down at high pressures, where the volume is higher than predicted because the particles are no longer negligibly small compared to the space between them.

▶ The assumptions also break down at low temperatures where the pressure is lower than predicted because the attraction between molecules combined with low kinetic energies causes partially inelastic collisions.

▶ Van der Waals equation can be used to predict gas properties under nonideal conditions.

Key Equations and Relationships

Relationship between Pressure (P), Force (F), and Area (A) (5.2)

$$P = \frac{F}{A}$$

Boyle's Law: Relationship between Pressure (P) and Volume (V) (5.3)

$$V \propto \frac{1}{P}$$
$$P_1V_1 = P_2V_2$$

Charles's Law: Relationship between Volume (V) and Temperature (T) (5.3)

$$V \propto T \text{ (in K)}$$
$$\frac{V_1}{T_1} = \frac{V_2}{T_2}$$

Avogadro's Law: Relationship between Volume (V) and Amount in Moles (n) (5.3)

$$V \propto n$$
$$\frac{V_1}{n_1} = \frac{V_2}{n_2}$$

Ideal Gas Law: Relationship between Volume (V), Pressure (P), Temperature (T), and Amount (n) (5.4)

$$PV = nRT$$

Dalton's Law: Relationship between Partial Pressures (P_n) in Mixture of Gases and Total Pressure (P_{total}) (5.6)

$$P_{total} = P_a + P_b + P_c + \ldots$$
$$P_a = \frac{n_aRT}{V} \quad P_b = \frac{n_bRT}{V} \quad P_c = \frac{n_cRT}{V}$$

Mole Fraction (χ_a) (5.6)

$$\chi_a = \frac{n_a}{n_{total}}$$
$$P_a = \chi_a P_{total}$$

Average Kinetic Energy (KE_{avg}) (5.8)

$$KE_{avg} = \frac{3}{2}RT$$

Relationship between Root Mean Square Velocity (u_{rms}) and Temperature (T) (5.8)

$$u_{rms} = \sqrt{\frac{3RT}{\mathcal{M}}}$$

Relationship of Effusion Rates of Two Different Gases (5.9)

$$\frac{\text{rate A}}{\text{rate B}} = \sqrt{\frac{\mathcal{M}_B}{\mathcal{M}_A}}$$

Van der Waals Equation: The Effects of Volume and Intermolecular Forces on Nonideal Gas Behavior (5.10)

$$[P + a(n/V)^2] \times (V - nb) = nRT$$

Key Learning Objectives

Chapter Objectives	Assessment
Converting between Pressure Units (5.2)	Example 5.1 For Practice 5.1 For More Practice 5.1 Exercises 1–4
Relating Volume and Pressure: Boyle's Law (5.3)	Example 5.2 For Practice 5.2 Exercises 5, 6
Relating Volume and Temperature: Charles's Law (5.3)	Example 5.3 For Practice 5.3 Exercises 7, 8
Relating Volume and Moles: Avogadro's Law (5.3)	Example 5.4 For Practice 5.4 Exercises 9, 10
Determining P, V, n, or T using the Ideal Gas Law (5.4)	Examples 5.5, 5.6 For Practice 5.5, 5.6 For More Practice 5.6 Exercises 11–18, 21, 22
Relating the Density of a Gas to Its Molar Mass (5.5)	Example 5.7 For Practice 5.7 For More Practice 5.7 Exercises 25, 26
Calculating the Molar Mass of a Gas with the Ideal Gas Law (5.5)	Example 5.8 For Practice 5.8 Exercises 27–30
Calculating Total Pressure, Partial Pressures, and Mole Fractions of Gases in a Mixture (5.6)	Examples 5.9, 5.10, 5.11 For Practice 5.9, 5.10, 5.11 Exercises 31, 32, 35, 37, 38, 40
Relating the Amounts of Reactants and Products in Gaseous Reactions: Stoichiometry (5.7)	Examples 5.12, 5.13 For Practice 5.12, 5.13 For More Practice 5.12 Exercises 41–47
Calculating the Root Mean Square Velocity of a Gas (5.8)	Example 5.14 For Practice 5.14 Exercises 53, 54
Calculating the Effusion Rate or the Ratio of Effusion Rates of Two Gases (5.9)	Example 5.15 For Practice 5.15 Exercises 55–58

EXERCISES

Problems by Topic

Converting between Pressure Units

1. The pressure in Denver, Colorado (elevation 5280 ft), averages about 24.9 in Hg. Convert this pressure to
   a. atm   **b.** mmHg   **c.** psi   **d.** Pa

2. The pressure on top of Mt. Everest averages about 235 mmHg. Convert this pressure to
   a. torr   **b.** psi   **c.** in Hg   **d.** atm

3. The North American record for highest recorded barometric pressure is 31.85 in Hg, set in 1989 in Northway, Alaska. Convert this pressure to
   a. mmHg   **b.** atm
   c. torr   **d.** kPa (kilopascals)

4. The world record for lowest barometric pressure (at sea level) was 652.5 mmHg recorded inside Typhoon Tip on October 12, 1979, in the Western Pacific Ocean. Convert this pressure to
   a. torr   **b.** atm   **c.** in Hg   **d.** psi

Simple Gas Laws

5. A sample of gas has an initial volume of 2.8 L at a pressure of 755 mmHg. If the volume of the gas is increased to 3.7 L, what will the pressure be?

6. A sample of gas has an initial volume of 32.6 L at a pressure of 1.3 atm. If the sample is compressed to a volume of 13.8 L, what will its pressure be?

7. A 37.2 mL sample of gas in a cylinder is warmed from 22 °C to 81 °C. What is its volume at the final temperature?

8. A syringe containing 1.25 mL of oxygen gas is cooled from 91.3 °C to 0.0 °C. What is the final volume of oxygen gas?

9. A balloon contains 0.128 mol of gas and has a volume of 2.76 L. If an additional 0.073 mol of gas is added to the balloon (at the same temperature and pressure), what will its final volume be?

10. A cylinder with a moveable piston contains 0.87 mol of gas and has a volume of 334 mL. What will its volume be if an additional 0.22 mol of gas is added to the cylinder? (Assume constant temperature and pressure.)

Ideal Gas Law

11. What is the volume occupied by 0.128 mol of helium gas at a pressure of 0.97 atm and a temperature of 325 K?

12. What is the pressure in a 15.0 L cylinder filled with 0.448 mol of nitrogen gas at a temperature of 305 K?

13. A cylinder contains 28.5 L of oxygen gas at a pressure of 1.8 atm and a temperature of 298 K. How much gas (in moles) is in the cylinder?

14. What is the temperature of 0.52 mol of gas at a pressure of 1.3 atm and a volume of 11.8 L?

15. An automobile tire has a maximum rating of 38.0 psi (gauge pressure). The tire is inflated (while cold) to a volume of 11.8 L and a gauge pressure of 36.0 psi at a temperature of 12.0 °C. Driving on a hot day, the tire warms to 65.0 °C and its volume expands to 12.2 L. Does the pressure in the tire exceed its maximum rating? (Note: The *gauge pressure* is the *difference* between the total pressure and atmospheric pressure. In this case, assume that atmospheric pressure is 14.7 psi.)

16. A weather balloon is inflated to a volume of 28.5 L at a pressure of 748 mmHg and a temperature of 28.0 °C. The balloon rises in the atmosphere to an altitude of approximately 25,000 feet, where the pressure is 385 mmHg and the temperature is −15.0 °C. Assuming the balloon can freely expand, calculate the volume of the balloon at this altitude.

17. A piece of dry ice (solid carbon dioxide) with a mass of 28.8 g is allowed to sublime (convert from solid to gas) into a large balloon. Assuming that all of the carbon dioxide ends up in the balloon, what will be the volume of the balloon at a temperature of 22 °C and a pressure of 742 mmHg?

18. A 1.0 L container of liquid nitrogen is kept in a closet measuring 1.0 m by 1.0 m by 2.0 m. Assuming that the container is completely full, that the temperature is 25.0 °C, and that the atmospheric pressure is 1.0 atm, calculate the percent (by volume) of air that would be displaced if all of the liquid nitrogen evaporated. (Liquid nitrogen has a density of 0.807 g/mL.)

19. Which gas sample will have the greatest pressure? Explain. (Assume that the temperature of all samples is the same.)

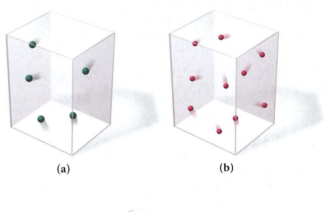

(a)   (b)

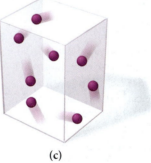

(c)

20. This picture represents a sample of gas at a pressure of 1 atm, a volume of 1 L, and a temperature of 25 °C. Draw a similar picture showing what would happen if the volume were reduced to 0.5 L and the temperature increased to 250 °C. What would happen to the pressure?

21. Aerosol cans carry clear warnings against incineration because of the high pressures that can develop upon heating. Suppose that a can contains a residual amount of gas at a pressure of 755 mmHg

and a temperature of 25 °C. What would be the pressure if the can were heated to 1155 °C?

22. A sample of nitrogen gas in a 1.75 L container exerts a pressure of 1.35 atm at 25 °C. What is the pressure if the volume of the container is maintained constant and the temperature is raised to 355 °C?

Molar Volume, Density, and Molar Mass of a Gas

23. Use the molar volume of a gas at STP to determine the volume (in L) occupied by 15.0 g of neon at STP.

24. Use the molar volume of a gas at STP to calculate the density (in g/L) of carbon dioxide gas at STP.

25. What is the density (in g/L) of hydrogen gas at 20.0 °C and a pressure of 1655 psi?

26. A sample of N_2O gas has a density of 2.85 g/L at 298 K. What is the pressure of the gas (in mmHg)?

27. An experiment shows that a 248 mL gas sample has a mass of 0.433 g at a pressure of 745 mmHg and a temperature of 28 °C. What is the molar mass of the gas?

28. An experiment shows that a 113 mL gas sample has a mass of 0.171 g at a pressure of 721 mmHg and a temperature of 32 °C. What is the molar mass of the gas?

29. A sample of gas has a mass of 38.8 mg. Its volume is 224 mL at a temperature of 55 °C and a pressure of 886 torr. Find the molar mass of the gas.

30. A sample of gas has a mass of 0.555 g. Its volume is 117 mL at a temperature of 85 °C and a pressure of 753 mmHg. Find the molar mass of the gas.

Partial Pressure

31. A gas mixture contains each of the following gases at the indicated partial pressures: N_2, 315 torr; O_2, 134 torr; and He, 219 torr. What is the total pressure of the mixture? What mass of each gas is present in a 2.15 L sample of this mixture at 25.0 °C?

32. A gas mixture with a total pressure of 765 mmHg contains each of the following gases at the indicated partial pressures: CO_2, 235 mmHg; Ar, 112 mmHg; and O_2, 124 mmHg. The mixture also contains helium gas. What is the partial pressure of the helium gas? What mass of helium gas is present in a 20.0 L sample of this mixture at 273 K?

33. A 1.20 g sample of dry ice is added to a 755 mL flask containing nitrogen gas at a temperature of 25.0 °C and a pressure of 725 mmHg. The dry ice is allowed to sublime (convert from solid to gas) and the mixture is allowed to return to 25.0 °C. What is the total pressure in the flask?

34. A 275 mL flask contains pure helium at a pressure of 752 torr. A second flask with a volume of 475 mL contains pure argon at a pressure of 722 torr. If the two flasks are connected through a stopcock and the stopcock is opened, what are the partial pressures of each gas and the total pressure?

35. A gas mixture contains 1.25 g N_2 and 0.85 g O_2 in a 1.55 L container at 18 °C. Calculate the mole fraction and partial pressure of each component in the gas mixture.

36. What is the mole fraction of oxygen gas in air (see Table 5.2)? What volume of air contains 10.0 g of oxygen gas at 273 K and 1.00 atm?

37. The hydrogen gas formed in a chemical reaction is collected over water at 30.0 °C at a total pressure of 732 mmHg. What is the partial pressure of the hydrogen gas? If the total volume of gas collected is 722 mL, what mass of hydrogen gas is collected?

38. The air in a bicycle tire is bubbled through water and collected at 25 °C. If the total volume of gas collected is 5.45 L at a

temperature of 25 °C and a pressure of 745 torr, how many moles of gas were in the bicycle tire?

39. The zinc within a copper-plated penny will dissolve in hydrochloric acid if the copper coating is filed down in several spots (so that the hydrochloric acid can reach the zinc). The reaction between the acid and the zinc is as follows: $2H^+(aq) + Zn(s) \rightarrow H_2(g) + Zn^{2+}(aq)$. When the zinc in a certain penny dissolves, the total volume of gas collected over water at 25 °C is 0.951 L at a total pressure of 748 mmHg. What mass of hydrogen gas is collected?

40. A heliox deep-sea diving mixture contains 2.0 g of oxygen to every 98.0 g of helium. What is the partial pressure of oxygen when this mixture is delivered at a total pressure of 8.5 atm?

Reaction Stoichiometry Involving Gases

41. Consider the chemical reaction

$$C(s) + H_2O(g) \longrightarrow CO(g) + H_2(g)$$

How many liters of hydrogen gas is formed from the complete reaction of 15.7 g C? Assume that the hydrogen gas is collected at a pressure of 1.0 atm and a temperature of 355 K.

42. Consider the chemical reaction

$$2 H_2O(l) \longrightarrow 2 H_2(g) + O_2(g)$$

What mass of H_2O is required to form 1.4 L of O_2 at a temperature of 315 K and a pressure of 0.957 atm?

43. CH_3OH can be synthesized by the reaction

$$CO(g) + 2 H_2(g) \longrightarrow CH_3OH(g)$$

What volume of H_2 gas (in L), measured at 748 mmHg and 86 °C, is required to synthesize 25.8 g CH_3OH? How many liters of CO gas, measured under the same conditions, is required?

44. Oxygen gas reacts with powdered aluminum according to the reaction

$$4 Al(s) + 3 O_2(g) \longrightarrow 2 Al_2O_3(s)$$

What volume of O_2 gas (in L), measured at 782 mmHg and 25 °C, is required to completely react with 53.2 g Al?

45. Automobile air bags inflate following a serious impact. The impact triggers the chemical reaction

$$2 NaN_3(s) \longrightarrow 2 Na(s) + 3 N_2(g)$$

If an automobile air bag has a volume of 11.8 L, what mass of NaN_3 (in g) is required to fully inflate the air bag upon impact? Assume STP conditions.

46. Lithium reacts with nitrogen gas according to the reaction

$$6 Li(s) + N_2(g) \longrightarrow 2 Li_3N(s)$$

What mass of lithium (in g) is required to react completely with 58.5 mL of N_2 gas at STP?

47. Hydrogen gas (a potential future fuel) can be formed by the reaction of methane with water according to the equation

$$CH_4(g) + H_2O(g) \longrightarrow CO(g) + 3 H_2(g)$$

In a particular reaction, 25.5 L of methane gas (measured at a pressure of 732 torr and a temperature of 25 °C) is mixed with 22.8 L of water vapor (measured at a pressure of 702 torr and a temperature of 125 °C). The reaction produces 26.2 L of hydrogen gas measured at STP. What is the percent yield of the reaction?

48. Ozone is depleted in the stratosphere by chlorine from CF_3Cl according to the following set of equations:

$$CF_3Cl + UV\ light \longrightarrow CF_3 + Cl$$

$$Cl + O_3 \longrightarrow ClO + O_2$$

$$O_3 + UV \text{ light} \longrightarrow O_2 + O$$

$$ClO + O \longrightarrow Cl + O_2$$

What total volume of ozone measured at a pressure of 25.0 mmHg and a temperature of 225 K can be destroyed when all of the chlorine from 15.0 g of CF_3Cl goes through ten cycles of the reactions shown?

49. Chlorine gas reacts with fluorine gas to form chlorine trifluoride.

$$Cl_2(g) + 3 F_2(g) \longrightarrow 2 ClF_3(g)$$

A 2.00 L reaction vessel, initially at 298 K, contains chlorine gas at a partial pressure of 337 mmHg and fluorine gas at a partial pressure of 729 mmHg. Identify the limiting reactant and determine the theoretical yield of ClF_3 in grams.

50. Carbon monoxide gas reacts with hydrogen gas to form methanol.

$$CO(g) + 2 H_2(g) \longrightarrow CH_3OH(g)$$

A 1.50 L reaction vessel, initially at 305 K, contains carbon monoxide gas at a partial pressure of 232 mmHg and hydrogen gas at a partial pressure of 397 mmHg. Identify the limiting reactant and determine the theoretical yield of methanol in grams.

Kinetic Molecular Theory

51. Consider a 1.0 L sample of helium gas and a 1.0 L sample of argon gas, both at room temperature and atmospheric pressure.
 a. Do the atoms in the helium sample have the same *average kinetic energy* as the atoms in the argon sample?
 b. Do the atoms in the helium sample have the same *average velocity* as the atoms in the argon sample?
 c. Do the argon atoms, since they are more massive, exert a greater pressure on the walls of the container? Explain.
 d. Which gas sample would have the fastest rate of effusion?

52. A flask at room temperature contains exactly equal amounts (in moles) of nitrogen and xenon.
 a. Which gas exerts the greater partial pressure?
 b. The molecules or atoms of which gas have the greater average velocity?
 c. The molecules of which gas have the greater average kinetic energy?
 d. If a small hole were opened in the flask, which gas would effuse more quickly?

53. Calculate the root mean square velocity and kinetic energy of F_2, Cl_2, and Br_2 at 298 K. Rank the three halogens with respect to their rate of effusion.

54. Calculate the root mean square velocity and kinetic energy of CO, CO_2, and SO_3 at 298 K. Which gas has the greatest velocity? The greatest kinetic energy? The greatest effusion rate?

55. Uranium-235 can be separated from U-238 by fluorinating the uranium to form UF_6 (which is a gas) and then taking advantage of the different rates of effusion and diffusion for compounds containing the two isotopes. Calculate the ratio of effusion rates for $^{238}UF_6$ and $^{235}UF_6$. The atomic mass of U-235 is 235.054 amu and that of U-238 is 238.051 amu.

56. Calculate the ratio of effusion rates for Ar and Kr.

57. A sample of neon effuses from a container in 76 seconds. The same amount of an unknown noble gas requires 155 seconds. Identify the gas.

58. A sample of N_2O effuses from a container in 42 seconds. How long would it take the same amount of gaseous I_2 to effuse from the same container under identical conditions?

59. This graph shows the distribution of molecular velocities for two different molecules (A and B) at the same temperature. Which molecule has the higher molar mass? Which molecule would have the higher rate of effusion?

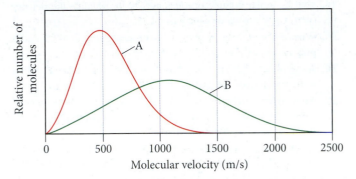

60. This graph shows the distribution of molecular velocities for the same molecule at two different temperatures (T_1 and T_2). Which temperature is greater? Explain.

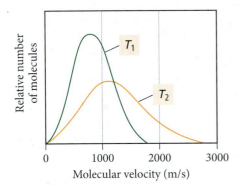

Real Gases

61. Which postulate of the kinetic molecular theory is no longer valid under conditions of high pressure? Explain.

62. Which postulate of the kinetic molecular theory is no longer valid under conditions of low temperature? Explain.

63. Use the van der Waals equation and the ideal gas equation to calculate the volume of 1.000 mol of neon at a pressure of 500.0 atm and a temperature of 355.0 K. Explain why the two values are different.

64. Use the van der Waals equation and the ideal gas equation to calculate the pressure exerted by 1.000 mol of Cl_2 in a volume of 5.000 L at a temperature of 273.0 K. Explain why the two values are different.

Cumulative Problems

65. Modern pennies are composed of zinc coated with copper. A student determines the mass of a penny to be 2.482 g and then makes several scratches in the copper coating (to expose the underlying zinc). The student puts the scratched penny in hydrochloric acid, where the following reaction occurs between the zinc and the HCl (the copper remains undissolved):

$$Zn(s) + 2 HCl(aq) \longrightarrow H_2(g) + ZnCl_2(aq)$$

The student collects the hydrogen produced over water at 25 °C. The collected gas occupies a volume of 0.899 L at a total pressure of 791 mmHg. Calculate the percent zinc in the penny. (Assume that all the Zn in the penny dissolves.)

66. A 2.85 g sample of an unknown chlorofluorocarbon is decomposed and produces 564 mL of chlorine gas at a pressure of 752 mmHg and a temperature of 298 K. What is the percent chlorine (by mass) in the unknown chlorofluorocarbon?

67. The mass of an evacuated 255 mL flask is 143.187 g. The mass of the flask filled with 267 torr of an unknown gas at 25 °C is 143.289 g. Calculate the molar mass of the unknown gas.

68. A 118 mL flask is evacuated and found to have a mass of 97.129 g. When the flask is filled with 768 torr of helium gas at 35 °C, it is found to have a mass of 97.171 g. Was the helium gas pure?

69. A gaseous hydrogen and carbon containing compound is decomposed and found to contain 82.66% carbon and 17.34% hydrogen by mass. The mass of 158 mL of the gas, measured at 556 mmHg and 25 °C, was found to be 0.275 g. What is the molecular formula of the compound?

70. A gaseous hydrogen and carbon containing compound is decomposed and found to contain 85.63% C and 14.37% H by mass. The mass of 258 mL of the gas, measured at STP, was found to be 0.646 g. What is the molecular formula of the compound?

71. Consider the following reaction:

$$2\,NiO(s) \longrightarrow 2\,Ni(s) + O_2(g)$$

If O_2 is collected over water at 40.0 °C and a total pressure of 745 mmHg, what volume of gas will be collected for the complete reaction of 24.78 g of NiO?

72. The following reaction forms 15.8 g of $Ag(s)$:

$$2\,Ag_2O(s) \longrightarrow 4\,Ag(s) + O_2(g)$$

What total volume of gas forms if it is collected over water at a temperature of 25 °C and a total pressure of 752 mmHg?

73. When hydrochloric acid is poured over potassium sulfide, 42.9 mL of hydrogen sulfide gas is produced at a pressure of 752 torr and 25.8 °C. Write an equation for the gas-evolution reaction and determine how much potassium sulfide (in grams) reacted.

74. Consider the following reaction:

$$2\,SO_2(g) + O_2(g) \longrightarrow 2\,SO_3(g)$$

a. If 285.5 mL of SO_2 is allowed to react with 158.9 mL of O_2 (both measured at 315 K and 50.0 mmHg), what is the limiting reactant and the theoretical yield of SO_3?

b. If 187.2 mL of SO_3 is collected (measured at 315 K and 50.0 mmHg), what is the percent yield for the reaction?

75. Ammonium carbonate decomposes upon heating according to the balanced equation

$$(NH_4)_2CO_3(s) \longrightarrow 2\,NH_3(g) + CO_2(g) + H_2O(g)$$

Calculate the total volume of gas produced at 22 °C and 1.02 atm by the complete decomposition of 11.83 g of ammonium carbonate.

76. Ammonium nitrate decomposes explosively upon heating according to the balanced equation

$$2\,NH_4NO_3(s) \longrightarrow 2\,N_2(g) + O_2(g) + 4\,H_2O(g)$$

Calculate the total volume of gas (at 125 °C and 748 mmHg) produced by the complete decomposition of 1.55 kg of ammonium nitrate.

77. Olympic cyclists fill their tires with helium to make them lighter. Calculate the mass of air in an air-filled tire and the mass of helium in a helium-filled tire. What is the mass

difference between the two? Assume that the volume of the tire is 855 mL, that it is filled to a total pressure of 125 psi, and that the temperature is 25 °C. Also, assume an average molar mass for air of 28.8 g/mol.

78. In a common classroom demonstration, a balloon is filled with air and submerged in liquid nitrogen. The balloon contracts as the gases within the balloon cool. Suppose the balloon initially contains 2.95 L of air at a temperature of 25.0 °C and a pressure of 0.998 atm. Calculate the expected volume of the balloon upon cooling to –196 °C (the boiling point of liquid nitrogen). When the demonstration is carried out, the actual volume of the balloon decreases to 0.61 L. How well does the observed volume of the balloon compare to your calculated value? Explain the difference.

79. Gaseous ammonia can be injected into the exhaust stream of a coal-burning power plant to reduce the pollutant NO to N_2 according to the reaction

$$4\,NH_3(g) + 4\,NO(g) + O_2(g) \longrightarrow 4\,N_2(g) + 6\,H_2O(g)$$

Suppose that the exhaust stream of a power plant has a flow rate of 335 L/s at a temperature of 955 K, and that the exhaust contains a partial pressure of NO of 22.4 torr. What should be the flow rate of ammonia delivered at 755 torr and 298 K into the stream to react completely with the NO if the ammonia is 65.2% pure (by volume)?

80. The emission of NO_2 by fossil fuel combustion can be prevented by injecting gaseous urea into the combustion mixture. The urea reduces NO (which oxidizes in air to form NO_2) according to the reaction

$$2\,CO(NH_2)_2(g) + 4\,NO(g) + O_2(g) \longrightarrow$$
$$4\,N_2(g) + 2\,CO_2(g) + 4\,H_2O(g)$$

Suppose that the exhaust stream of an automobile has a flow rate of 2.55 L/s at 655 K and contains a partial pressure of NO of 12.4 torr. What total mass of urea is necessary to react completely with the NO formed during 8.0 hours of driving?

81. An ordinary gasoline can measuring 30.0 cm by 20.0 cm by 15.0 cm is evacuated with a vacuum pump. Assuming that virtually all of the air can be removed from inside the can, and that atmospheric pressure is 14.7 psi, what is the total force (in pounds) on the surface of the can? Will the can withstand the force?

82. Twenty-five milliliters of liquid nitrogen (density = 0.807 g/mL) is poured into a cylindrical container with a radius of 10.0 cm and a length of 20.0 cm. The container initially contains only air at a pressure of 760.0 mmHg (atmospheric pressure) and a temperature of 298 K. If the liquid nitrogen completely vaporizes, what is the total force (in lb) on the interior of the container at 298 K?

83. A 160.0 L helium tank contains pure helium at a pressure of 1855 psi and a temperature of 298 K. How many 3.5 L helium balloons can be filled from the helium in the tank?

(Assume an atmospheric pressure of 1.0 atm and a temperature of 298 K.)

84. A 11.5 mL sample of liquid butane (density = 0.573 g/mL) is evaporated in an otherwise empty container at a temperature of 28.5 °C. The pressure in the container following evaporation is 892 torr. What is the volume of the container?

85. A scuba diver creates a spherical bubble with a radius of 2.5 cm at a depth of 30.0 m where the total pressure (including atmospheric pressure) is 4.00 atm. What is the radius of the bubble when it reaches the surface of the water? (Assume atmospheric pressure to be 1.00 atm and the temperature to be 298 K.)

86. A particular balloon can be stretched to a maximum surface area of 1257 cm². The balloon is filled with 3.0 L of helium gas at a pressure of 755 torr and a temperature of 298 K. The balloon is then allowed to rise in the atmosphere. Assuming an atmospheric temperature of 273 K, at what pressure will the balloon burst? (Assume the balloon to be in the shape of a sphere.)

87. A catalytic converter in an automobile uses a palladium or platinum catalyst (a substance that increases the rate of a reaction without being consumed by the reaction) to convert carbon monoxide gas to carbon dioxide according to the following reaction:

$$2 CO(g) + O_2(g) \longrightarrow 2 CO_2(g)$$

A chemist researching the effectiveness of a new catalyst combines a 2.0 : 1.0 mole ratio mixture of carbon monoxide and oxygen gas (respectively) over the catalyst in a 2.45 L flask at a total pressure of 745 torr and a temperature of 552 °C. When the reaction is complete, the pressure in the flask has dropped to 552 torr. What percentage of the carbon monoxide was converted to carbon dioxide?

88. A quantity of N_2 occupies a volume of 1.0 L at 300 K and 1.0 atm. The gas expands to a volume of 3.0 L as the result of a change in both temperature and pressure. Find the density of the gas at these new conditions.

89. A mixture of $CO(g)$ and $O_2(g)$ in a 1.0 L container at 1.0×10^3 K has a total pressure of 2.2 atm. After some time the total pressure falls to 1.9 atm as the result of the formation of CO_2. Find the amount of CO_2 that forms.

90. The radius of a xenon atom is 1.3×10^{-8} cm. A 100 mL flask is filled with Xe at a pressure of 1.0 atm and a temperature of 273 K. Calculate the fraction of the volume that is occupied by Xe atoms. (Hint: The atoms are spheres.)

91. A natural gas storage tank is a cylinder with a moveable top whose volume can change only as its height changes. Its radius remains fixed. The height of the cylinder is 22.6 m on a day when the temperature is 22 °C. The next day the height of the cylinder increases to 23.8 m as the gas expands because of a heat wave. Find the temperature, assuming that the pressure and amount of gas in the storage tank have not changed.

92. A mixture of 8.0 g CH_4 and 8.0 g Xe is placed in a container and the total pressure is found to be 0.44 atm. Find the partial pressure of CH_4.

93. A steel container of volume 0.35 L can withstand pressures up to 88 atm before exploding. Find the mass of helium that can be stored in this container at 299 K.

94. Binary compounds of alkali metals and hydrogen react with water to liberate $H_2(g)$. The H_2 from the reaction of a sample of NaH that has an excess of water fills a volume of 0.490 L above the water. The temperature of the gas is 35 °C and the total pressure is 758 mmHg. Find the mass of H_2 liberated and the mass of NaH that reacted.

95. In a given diffusion apparatus, 15.0 mL of HBr gas diffused in 1.0 min. In the same apparatus and under the same conditions, 20.3 mL of an unknown gas diffused in 1.0 min. The unknown gas is a hydrocarbon. Determine its molecular formula.

96. A gas mixture composed of helium and argon has a density of 0.670 g/L at 755 mmHg and 298 K. What is the composition of the mixture by volume?

97. A gas mixture contains 75.2% nitrogen and 24.8% krypton by mass. What is the partial pressure of krypton in the mixture if the total pressure is 745 mmHg?

Challenge Problems

98. The world burns approximately 9.0×10^{12} kg of fossil fuel per year. Use the combustion of octane as the representative reaction and determine the mass of carbon dioxide (the most significant greenhouse gas) formed per year by this combustion. The current concentration of carbon dioxide in the atmosphere is approximately 387 ppm (by volume). By what percentage does the concentration increase in one year due to fossil fuel combustion? Approximate the average properties of the entire atmosphere by assuming that the atmosphere extends from sea level to 15 km and that it has an average pressure of 381 torr and average temperature of 275 K. Assume Earth is a perfect sphere with a radius of 6371 km.

99. The atmosphere slowly oxidizes hydrocarbons in a number of steps that eventually convert the hydrocarbon into carbon dioxide and water. The overall reactions of a number of such steps for methane gas is

$$CH_4(g) + 5 O_2(g) + 5 NO(g) \longrightarrow CO_2(g) + H_2O(g)$$
$$+ 5 NO_2(g) + 2 OH(g)$$

Suppose that an atmospheric chemist combines 155 mL of methane at STP, 885 mL of oxygen at STP, and 55.5 mL of NO at STP in a 2.0 L flask. The reaction is allowed to stand for several weeks at 275 K. If the reaction reaches 90.0% of completion (90.0% of the limiting reactant is consumed), what are the partial pressures of each of the reactants and products in the flask at 275 K? What is the total pressure in the flask?

100. Two identical balloons are filled to the same volume, one with air and one with helium. The next day, the volume of the air-filled balloon has decreased by 5.0%. By what percent has the volume of the helium-filled balloon decreased? (Assume that the air is four-fifths nitrogen and one-fifth oxygen, and that the temperature did not change.)

101. A mixture of $CH_4(g)$ and $C_2H_6(g)$ has a total pressure of 0.53 atm. Just enough $O_2(g)$ is added to the mixture to bring about its complete combustion to $CO_2(g)$ and $H_2O(g)$. The total pressure of the two product gases is found to be 2.2 atm. Assuming constant volume and temperature, find the mole fraction of CH_4 in the mixture.

102. A sample of $C_2H_2(g)$ has a pressure of 7.8 kPa. After some time, a portion of it reacts to form $C_6H_6(g)$. The total pressure of the mixture of gases is then 3.9 kPa. Assume the volume and the temperature do not change. Find the fraction of $C_2H_2(g)$ that has undergone reaction.

103. A 10 L container is filled with 0.10 mol of $H_2(g)$ and heated to 3000 K causing some of the $H_2(g)$ to decompose into H(g).

The pressure is found to be 3.0 atm. Find the partial pressure of the H(g) that forms from H_2 at this temperature. (Assume two significant figures for the temperature.)

104. A mixture of $NH_3(g)$ and $N_2H_4(g)$ is placed in a sealed container at 300 K. The total pressure is 0.50 atm. The container is heated to 1200 K at which time both substances decompose completely according to the equations $2 NH_3(g) \rightarrow N_2(g) + 3 H_2(g)$ and $N_2H_4(g) \rightarrow N_2(g) + 2 H_2(g)$ After decomposition is complete, the total pressure at 1200 K is

found to be 4.5 atm. Find the percent of $N_2H_4(g)$ in the original mixture. (Assume two significant figures for the temperature.)

105. A quantity of CO gas occupies a volume of 0.48 L at 1.0 atm and 275 K. The pressure of the gas is lowered and its temperature is raised until its volume is 1.3 L. Find the density of the CO under the new conditions.

106. When $CO_2(g)$ is put in a sealed container at 701 K and a pressure of 10.0 atm and is heated to 1401 K, the pressure rises to 22.5 atm. Some of the CO_2 decomposes to CO and O_2. Calculate the mole percent of CO_2 that decomposes.

Conceptual Problems

107. When the driver of an automobile applies the brakes, the passengers are pushed toward the front of the car, but a helium balloon is pushed toward the back of the car. Upon forward acceleration, the passengers are pushed toward the back of the car, but the helium balloon is pushed toward the front of the car. Why?

108. The following reaction occurs in a closed container:

$$A(g) + 2 B(g) \longrightarrow 2 C(g)$$

A reaction mixture initially contains 1.5 L of A and 2.0 L of B. Assuming that the volume and temperature of the reaction mixture remain constant, what is the percent change in pressure if the reaction goes to completion?

109. One mole of nitrogen and one mole of neon are combined in a closed container at STP. How big is the container?

110. Exactly equal amounts (in moles) of gas A and gas B are combined in a 1 L container at room temperature. Gas B has a molar mass that is twice that of gas A. Which of the following is true for the mixture of gases and why?
   a. The molecules of gas B have greater kinetic energy than those of gas A.

   b. Gas B has a greater partial pressure than gas A.
   c. The molecules of gas B have a greater average velocity than those of gas A.
   d. Gas B makes a greater contribution to the average density of the mixture than gas A.

111. Which gas would you expect to deviate most from ideal behavior under conditions of low temperature: F_2, Cl_2, or Br_2? Explain.

112. The volume of a sample of a fixed amount of gas is decreased from 2.0 L to 1.0 L. The temperature of the gas in Kelvins is then doubled. What is the final pressure of the gas in terms of the initial pressure?

113. Which gas sample has the greatest volume at STP?
   a. 10.0 g Kr
   b. 10.0 g Xe
   c. 10.0 g He

114. Draw a depiction of a gas sample, as described by kinetic molecular theory, containing equal molar amounts of helium, neon, and krypton. Use different color dots to represent each element. Give each atom a "tail" to represent its velocity relative to the others in the mixture.

Answers to Conceptual Connections

Boyle's Law and Charles's Law

5.1 (e) The final volume of the gas is the same as the initial volume because doubling the pressure *decreases* the volume by a factor of two, but doubling the temperature *increases* the volume by a factor of two. The two changes in volume are equal in magnitude but opposite in sign, resulting in a final volume that is equal to the initial volume.

Molar Volume

5.2 (a) Since 1 g of H_2 contains the greatest number of particles (due to H_2 having the lowest molar mass of the set), it will occupy the greatest volume.

Density of a Gas

5.3 $Ne < O_2 < F_2 < Cl_2$

Partial Pressures

5.4 $P_{He} = 1.5$ atm; $P_{Ne} = 1.5$ atm. Since the number of moles of each gas are equal, the mole fraction of each gas is 0.50 and the partial pressure of each gas is simply $0.50 \times P_{total}$.

Pressure and Number of Moles

5.5 (b) Since the total number of gas molecules decreases, the total pressure—the sum of all the partial pressures—must also decrease.

Kinetic Molecular Theory I

5.6 Although the velocity "tails" have different lengths, the average length of the tails on the argon atoms in your drawing should be longer than the average length of the tails on the xenon atoms. Since the argon atoms are lighter, they must on average move faster than the xenon atoms to have the same kinetic energy.

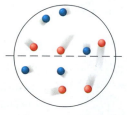

Kinetic Molecular Theory II

5.7 (c) Since the temperature and the volume are both constant, the ideal gas law tells us that the pressure will depend solely on the number of particles. Sample (c) has the greatest number of particles per unit volume, and so it will have the greatest pressure. The pressures of (a) and (b) at a given temperature will be identical. Even though the particles in (b) are more massive than those in (a), they have the same average kinetic energy at a given temperature. The particles in (b) therefore move more slowly, and so exert the same pressure as the particles in (a).

6 Thermochemistry

There is a fact, or if you wish, a law, governing all natural phenomena that are known to date. There is no exception to this law—it is exact as far as we know. The law is called the conservation of energy. It states that there is a certain quantity, which we call energy, that does not change in the manifold changes which nature undergoes. —Richard P. Feynman (1918–1988)

A chemical hand warmer contains substances that react with oxygen to emit heat.

WE HAVE SPENT THE FIRST FEW CHAPTERS of this book examining one of the two major components of our universe—matter. We now turn our attention to the other major component—energy. As far as we know, matter and energy make up the physical universe. Unlike matter, energy is not something we can touch with our fingers or hold in our hand, but we experience it in many ways. The warmth of sunlight, the feel of wind on our faces, and the force that presses us back when a car accelerates are all manifestations of energy and its interconversions (the changes of energy from one form to another). And of course energy and its uses are critical to society and to the world's economy. The standard of living around the globe is strongly correlated with access to and use of energy resources. Most of those resources are chemical ones, and their advantages as well as their drawbacks can be understood in terms of chemistry.

6.1 Chemical Hand Warmers

My family loves to snowboard. However, my wife hates being cold (with a passion), especially in her hands and toes. Her solution is the chemical hand warmer, a small pouch that comes sealed in a plastic package. She opens the package and places the pouch in her glove or boot. The pouch slowly warms up and keeps her hand (or foot) warm all day long.

Warming your hands with chemical hand warmers involves many of the principles of **thermochemistry**, the study of the relationships between chemistry and energy. When you open the package that contains the hand warmer, the contents are exposed to air, and a *reaction* that gives off heat to its surroundings occurs. The most commonly available hand warmers use the oxidation of iron as the reaction:

$$4\ Fe(s)\ +\ 3\ O_2(g)\ \longrightarrow\ 2\ Fe_2O_3(s)$$

The most important product of this reaction is not a substance—it is *heat*. We'll define heat more carefully later, but heat is what you feel when you touch something that is warmer than your hand (in this case, the hot hand warmer). Although some of the heat is lost through the minute openings in your gloves (which is why my wife prefers mittens) most of it is transferred to your hands and to the pocket of air surrounding your hands, resulting in a temperature increase. The magnitude of the temperature increase depends on the size of the hand warmer and the size of your glove (as well as some other details). But in general, the size of the temperature increase is proportional to the amount of heat released by the reaction.

In this chapter, we examine the relationship between chemical reactions and energy. Specifically, we look at how chemical reactions can *exchange* energy with their surroundings and how to quantify the magnitude of those exchanges. These kinds of calculations are important, not only for chemical hand warmers, but also to many other important processes such as the heating of homes and the production of energy for our society.

6.2 The Nature of Energy: Key Definitions

Recall that we briefly examined energy in Section 1.5. At that point, we defined **energy** as the capacity to do work and defined **work** as the result of a force acting through a distance. When you push a box across the floor, you have done work. Consider another example of work: a billiard ball rolling across a billiard table and colliding straight on with a second, stationary billiard ball. The rolling ball has *energy* due to its motion. When it collides with another ball it does *work*, resulting in the *transfer* of energy from one ball to the other. The second billiard ball absorbs the energy and begins to roll across the table.

As we just saw with chemical hand warmers, energy can also be transferred through **heat**, the flow of energy caused by a temperature difference. For example, if you hold a cup of coffee in your hand, energy is transferred, in the form of heat, from the hot coffee to your cooler hand. Think of *energy* as something that an object or set of objects possesses. Think of *heat* and *work* as ways that objects or sets of objects *exchange* energy.

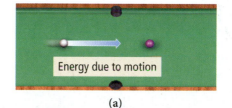

Energy due to motion

(a)

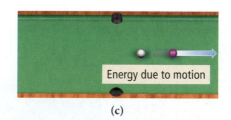

Work

Energy transfer

(b)

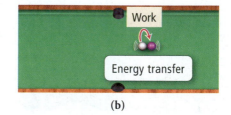

Energy due to motion

(c)

▲ **(a)** A rolling billiard ball has energy due to its motion. **(b)** When the ball collides with a second ball it does work, transferring energy to the second ball. **(c)** The second ball now has energy as it rolls away from the collision.

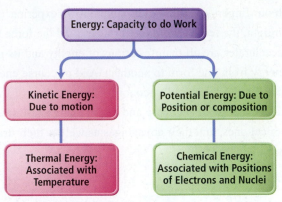

▲ FIGURE 6.1 The Different Manifestations of Energy

The energy contained in a rolling billiard ball is an example of **kinetic energy**, the energy associated with the *motion* of an object. The energy contained in a hot cup of coffee is **thermal energy**, the energy associated with the *temperature* of an object. Thermal energy is actually a type of kinetic energy because it arises from the motions of atoms or molecules within a substance.

If you raise a billiard ball off the table, you increase its **potential energy**, the energy associated with the *position* or *composition* of an object. The potential energy of the billiard ball, for example, is a result of its position in Earth's gravitational field. Raising the ball off the table, against Earth's gravitational pull, gives it more potential energy. Another example of potential energy is the energy contained in a compressed spring. When you compress a spring, you push against the forces that maintain the spring's uncompressed shape, and you store energy as potential energy. **Chemical energy**, the energy associated with the relative positions of electrons and nuclei in atoms and molecules, is also a form of potential energy. Some chemical compounds, such as the methane in natural gas or the iron in a chemical hand warmer, are like a compressed spring—they contain potential energy and a chemical reaction can release that potential energy. These different kinds of energy are summarized in **Figure 6.1▲**.

The **law of conservation of energy** states that *energy can be neither created nor destroyed*. However, energy can be transferred from one object to another, and it can assume different forms. For example, if you drop a raised billiard ball, some of its potential energy becomes kinetic energy as the ball falls toward the table, as shown in **Figure 6.2▼**. If you release a compressed spring, the potential energy becomes kinetic energy as the spring expands outward, as shown in **Figure 6.3▶**. When iron reacts with

Energy Transformation I

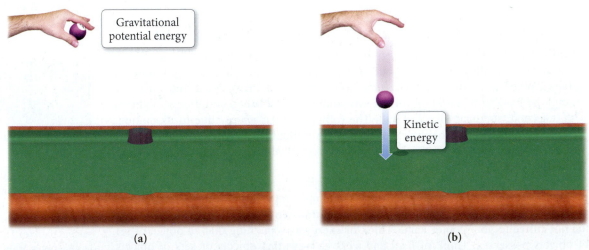

(a)   (b)

▲ FIGURE 6.2 **Energy Transformation: Potential and Kinetic Energy**   (a) A billiard ball held above the table has gravitational potential energy. (b) When the ball is released, the potential energy is transformed into kinetic energy, the energy of motion.

Energy Transformation II

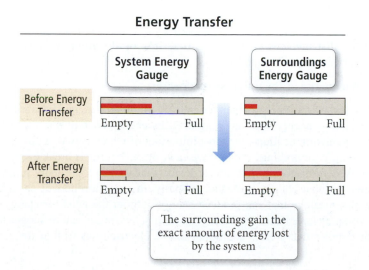

(a) **(b)**

◄ **FIGURE 6.3 Energy Transformation: Potential and Kinetic Energy** **(a)** A compressed spring has potential energy. **(b)** When the spring is released, the potential energy is transformed into kinetic energy.

oxygen within your chemical hand warmer, the chemical energy of the iron and oxygen becomes thermal energy that increases the temperature of your hand and glove.

A good way to understand and track energy changes is to define the **system** under investigation. The system may be a beaker of chemicals in the lab, or it may be the iron reacting in a hand warmer. The system's **surroundings** are everything with which the system can exchange energy. If we define the chemicals in a beaker as the system, the surroundings may include the water that the chemicals are dissolved in (for aqueous solutions), the beaker itself, the lab bench on which the beaker sits, the air in the room, and so on. For the iron in the hand warmer, the surroundings include your hand, your glove, the air in the glove, and even the air outside of the glove.

In an energy exchange, energy is transferred between the system and the surroundings, as shown in **Figure 6.4▼**. If the system loses energy, the surroundings gain the same exact amount of energy, and vice versa. When the iron within the chemical hand warmer reacts, the system loses energy to the surroundings, producing the desired temperature increase within your gloves.

Units of Energy

We can deduce the units of energy from the definition of kinetic energy. An object of mass m, moving at velocity v, has a kinetic energy KE given by

$$KE = \frac{1}{2}mv^2$$

[6.1]

Energy Transfer

	System Energy Gauge	Surroundings Energy Gauge
Before Energy Transfer	Empty ─────── Full	Empty ─────── Full
After Energy Transfer	Empty ─────── Full	Empty ─────── Full

The surroundings gain the exact amount of energy lost by the system

◄ **FIGURE 6.4 Energy Transfer** If a system and surroundings had energy gauges (which would measure energy content in the way a fuel gauge measures fuel content), an energy transfer in which the system transfers energy to the surroundings would result in a decrease in the energy content of the system and an increase in the energy content of the surroundings. The total amount of energy, however, must be conserved.

TABLE 6.1 Energy Conversion Factors*

1 calorie (cal)	= 4.184 joules (J)
1 Calorie (Cal) or kilocalorie (kcal)	= 1000 cal = 4184 J
1 kilowatt-hour (kWh)	= 3.60 × 10⁶ J

*All conversion factors in this table are exact.

TABLE 6.2 Energy Uses in Various Units

Unit	Amount Required to Raise Temperature of 1 g of Water by 1 °C	Amount Required to Light 100-W Bulb for 1 Hour	Amount Used by Human Body in Running 1 Mile (Approximate)	Amount Used by Average U.S. Citizen in 1 Day
joule (J)	4.18	3.60×10^5	4.2×10^5	9.0×10^8
calorie (cal)	1.00	8.60×10^4	1.0×10^5	2.2×10^8
Calorie (Cal)	0.00100	86.0	100.	2.2×10^5
kilowatt-hour (kWh)	1.16×10^{-6}	0.100	0.12	2.5×10^2

The SI unit of mass is the kg and the unit of velocity is m/s; the SI unit of energy is therefore $kg \cdot m^2/s^2$, defined as the **joule (J)**, named after the English scientist James Joule (1818–1889).

$$1 \, kg\frac{m^2}{s^2} = 1 \, J$$

One joule is a relatively small amount of energy—for example, a 100-watt light bulb uses 3.6×10^5 J in 1 hour. Therefore, we often use the kilojoule (kJ) in our energy discussions and calculations (1 kJ = 1000 J). A second commonly used unit of energy is the **calorie (cal)**, originally defined as the amount of energy required to raise the temperature of 1 g of water by 1 °C. The current definition is 1 cal = 4.184 J (exact); a calorie is a larger unit than a joule. A related energy unit is the nutritional, or uppercase "C" **Calorie (Cal)**, equivalent to 1000 lowercase "c" calories. The Calorie is the same as a kilocalorie (kcal): 1 Cal = 1 kcal = 1000 cal. Electricity bills typically are based on another, even larger, energy unit, the **kilowatt-hour (kWh)**: 1 kWh = 3.60×10^6 J. Electricity costs $0.08–$0.15 per kWh. Table 6.1 shows various energy units and their conversion factors. Table 6.2 shows the amount of energy required for various processes in these units.

3.6 × 10⁵ J or 0.10 kWh used in 1 hour

▲ A watt (W) is 1 J/s, so a 100-W lightbulb uses 100 J every second or 3.6×10^5 J every hour.

The "calorie" referred to on all nutritional labels (regardless of the capitalization) is always the uppercase C Calorie.

6.3 The First Law of Thermodynamics: There Is No Free Lunch

Thermodynamics is the general study of energy and its interconversions. The laws of thermodynamics are among the most fundamental in all of science, governing virtually every process that involves change. The **first law of thermodynamics** is the law of energy conservation, stated as follows:

The total energy of the universe is constant.

In other words, because energy is neither created nor destroyed, and because the universe does not exchange energy with anything else, its energy content does not change. The first law has many implications, the most important of which is that, with energy, you do not get something for nothing. The best we can do with energy is break even—there is no free lunch. According to the first law, a device that would continually produce energy with no energy input cannot exist. Occasionally, the media report or speculate on the discovery of a machine that can produce energy without the need for energy input. For example, you may have heard someone propose an electric car that recharges itself while driving, or a motor that creates additional usable electricity as well as the electricity to power itself. Although hybrid (electric and gasoline powered) vehicles can capture energy from braking and use that energy to recharge their batteries, they could never run indefinitely unless you add fuel. As for the motor that powers an external load as well as itself—no such thing exists. Our society has a continual need for energy, and as our current energy resources dwindle, new energy sources will be required. However, those sources must also follow the first law of thermodynamics—energy must be conserved.

Internal Energy

The **internal energy** (E) of a system is the sum of the kinetic and potential energies of all of the particles that compose the system. Internal energy is a **state function**, which means that its value depends *only on the state of the system*, not on how the system arrived at that state. The state of a chemical system is specified by parameters such as temperature, pressure, concentration, and phase (solid, liquid, or gas). We can explain state functions with a mountain-climbing analogy as shown in **Figure 6.5▼**. The elevation at any point during a climb is a state function. For example, when you reach 10,000 ft, your elevation is 10,000 ft, no matter how you got there. The distance you traveled to get there, by contrast, is not a state function. You could have climbed the mountain by any number of routes, each requiring you to cover a different distance.

Since state functions depend only on the state of the system, the value of a change in a state function is always the difference between its final and initial values. For example, if you start climbing a mountain at an elevation of 3000 ft, and reach the summit at 10,000 feet, then your elevation change is 7000 ft, regardless of what path you took.

Like an altitude change, an internal energy change (ΔE) is the difference in internal energy between the final and initial states:

$$\Delta E = E_{final} - E_{initial}$$

In a chemical system, the reactants constitute the initial state and the products constitute the final state. So ΔE is the difference in internal energy between the products and the reactants.

$$\Delta E = E_{products} - E_{reactants} \qquad [6.2]$$

For example, consider the reaction between carbon and oxygen to form carbon dioxide:

$$C(s) + O_2(g) \longrightarrow CO_2(g)$$

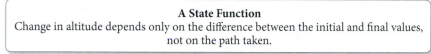

A State Function
Change in altitude depends only on the difference between the initial and final values, not on the path taken.

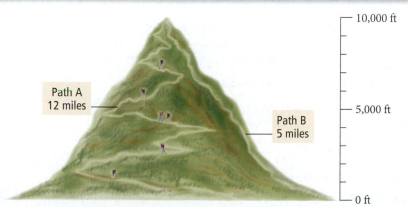

Path A
12 miles

Path B
5 miles

10,000 ft

5,000 ft

0 ft

▲ **FIGURE 6.5 Altitude is a state function.** The change in altitude during a climb depends only on the difference between the final and initial altitudes.

Just as we can portray the changes that occur when climbing a mountain with an *altitude* diagram which depicts the *altitude* before and after the climb (see Figure 6.5), we can portray the energy changes that occur during a reaction with an *energy* diagram, which compares the *internal energy* of the reactants and the products:

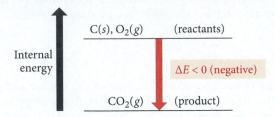

The vertical axis of the diagram is internal *energy*, which increases as you move up on the diagram. The reactants are *higher* on the diagram than the products because they have higher internal energy. As the reaction occurs, the reactants become products, which have lower internal energy. Therefore, energy is given off by the reaction and ΔE (that is, $E_{products} - E_{reactants}$) is *negative*.

Where does the energy lost by the reactants (as they transform to products) go? If we define the thermodynamic *system* as the reactants and products of the reaction, then energy flows *out of the system* and *into the surroundings*.

According to the first law, energy must be conserved. Therefore, the amount of energy lost by the system must exactly equal the amount gained by the surroundings.

$$\Delta E_{sys} = -\Delta E_{surr} \qquad [6.3]$$

Now, suppose the reaction is reversed:

$$CO_2(g) \longrightarrow C(s) + O_2(g)$$

The energy level diagram is nearly identical for this reversed reaction, with one important difference: $CO_2(g)$ is now the reactant and $C(s)$ and $O_2(g)$ are the products. Instead of decreasing in energy as the reaction occurs, the system increases in energy:

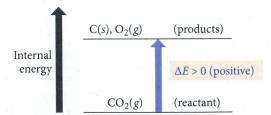

The difference, ΔE, is *positive* and energy flows *into the system* and *out of the surroundings*.

Summarizing Energy Flow:

▶ If the reactants have a higher internal energy than the products, ΔE_{sys} is negative and energy flows out of the system into the surroundings.

▶ If the reactants have a lower internal energy than the products, ΔE_{sys} is positive and energy flows into the system from the surroundings.

You can think of the internal energy of the system in the same way you think about the balance in a checking account. Energy flowing *out of* the system is like a withdrawal and therefore carries a negative sign. Energy flowing *into* the system is like a deposit and carries a positive sign.

As we saw in Section 6.2, a system can exchange energy with its surroundings through *heat* and *work*:

According to the first law of thermodynamics, the change in the internal energy of the system (ΔE) must be the sum of the heat transferred (q) and the work done (w):

$$\Delta E = q + w \qquad \text{[6.4]}$$

In the above equation, and from this point forward, we follow the standard convention that ΔE (with no subscript) refers to the internal energy change of the *system*. As shown in Table 6.3, energy entering the system through heat or work carries a positive sign, and energy leaving the system through heat or work carries a negative sign. Again, the system is like a checking account—withdrawals are negative and deposits are positive.

Let's define our system as the previously discussed billiard ball rolling across a pool table. The rolling ball has a certain initial amount of kinetic energy. When it reaches the other end of the table, the rolling ball collides head-on with a second ball. Let's assume that the first ball loses all of its kinetic energy, so that it remains completely still (it has no kinetic energy) at the point of collision. The total change in internal energy (ΔE) for the first ball is the difference between its initial kinetic energy and its final kinetic energy (which is zero); the first billiard ball lost all of its energy. What happened to that energy? According to the first law, it must have been transferred to the surroundings. In fact, the energy lost by the system must *exactly equal* the amount gained by the surroundings.

$$\Delta E_{sys} = -\Delta E_{surr}$$

| Energy lost by first ball | Energy gained by surroundings |

The surroundings include both the pool table and the second ball. The pool table absorbs some of the ball's kinetic energy as the ball rolls down the table. Minute bumps on the table surface cause friction, which slows the ball down by converting kinetic energy to heat (q). Upon collision, the second ball absorbs some of the ball's kinetic energy in the form of work (w).

Although it is always the case that $\Delta E_{sys} = -\Delta E_{surr}$, the exact amount of *work* done on the second ball depends on the quality of the billiard table. On a smooth, high-quality billiard table, the amount of energy lost to friction is relatively small, as shown in **Figure 6.6(a)▶**. The speed of the first ball is not reduced by much as it travels across the table and a great deal of its original kinetic energy is available to perform work when it collides with the second ball. In contrast, on a rough, poor-quality table, the ball loses much of its initial kinetic energy as heat, leaving only a relatively small amount available for work, as shown in **Figure 6.6(b)▶**.

TABLE 6.3 Sign Conventions for q, w, and ΔE		
q (heat)	+ system *gains* thermal energy	− system *loses* thermal energy
w (work)	+ work done *on* the system	− work done *by* the system
ΔE (change in internal energy)	+ energy flows *into* the system	− energy flows *out* of the system

▶ **FIGURE 6.6 Energy, Work, and Heat** (a) On a smooth table, most of the first billiard ball's initial kinetic energy is transferred to the second ball as work. Only a small amount is lost to heat. (b) On a rough table, most of the first billiard ball's initial kinetic energy is lost to heat. Only a small amount is left to do work on the second billiard ball.

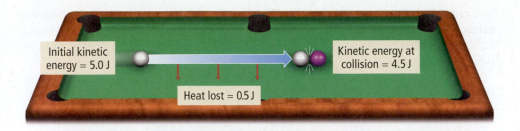

Initial kinetic energy = 5.0 J
Kinetic energy at collision = 4.5 J
Heat lost = 0.5 J

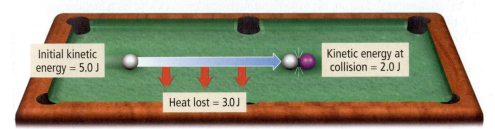

$w = -4.5$ J
$q = -0.5$ J
$\Delta E = -5.0$ J
Kinetic energy after collision = 0 J

(a) Smooth table

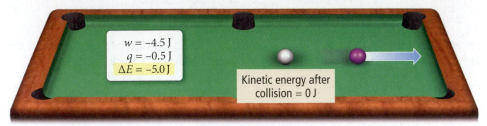

Initial kinetic energy = 5.0 J
Kinetic energy at collision = 2.0 J
Heat lost = 3.0 J

$w = -2.0$ J
$q = -3.0$ J
$\Delta E = -5.0$ J
Kinetic energy after collision = 0 J

(b) Rough table

Notice that the respective amounts of energy converted to heat and work depend on the details of the pool table and the path taken, while the change in internal energy of the rolling ball does not. In other words, since internal energy is a state function, the value of ΔE for the process in which the ball moves across the table and collides with another ball depends only on the ball's initial and final velocities. Work and heat, however, are *not* state functions; therefore, the values of q and w depend on the details of the ball's journey across the table. On the smooth table, w is greater in magnitude than q; on the rough table, q is greater in magnitude than w. However, ΔE (the sum of q and w) is constant.

Conceptual Connection 6.1 System and Surroundings

The following are fictitious internal energy gauges for a chemical system and its surroundings:

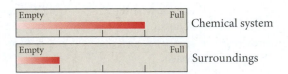

Empty   Full   Chemical system

Empty   Full   Surroundings

Which of the energy gauges best represents the same system and surroundings following an energy exchange in which ΔE_{sys} is negative?

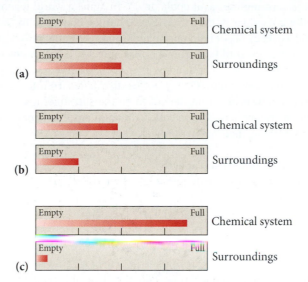

Conceptual Connection 6.2 Heat and Work

Identify each energy exchange as heat or work and determine whether the sign of heat or work (relative to the system) is positive or negative.

(a) An ice cube melts and cools the surrounding beverage. (The ice cube is the system.)

(b) A metal cylinder is rolled up a ramp. (The metal cylinder is the system; assume no friction.)

(c) Steam condenses on skin, causing a burn. (The condensing steam is the system.)

EXAMPLE 6.1 Internal Energy, Heat, and Work

A potato cannon provides a good example of the heat and work associated with a chemical reaction. In a potato cannon, a potato is stuffed into a long cylinder that is capped on one end and open at the other. Some kind of fuel is introduced under the potato at the capped end—usually through a small hole—and ignited. The potato then shoots out of the cannon, sometimes flying hundreds of feet, and heat is given off to the surroundings. If the burning of the fuel performs 855 J of work on the potato and produces 1422 J of heat, what is ΔE for the burning of the fuel? (Note: A potato cannon can be dangerous and should not be constructed without proper training and experience.)

SOLUTION To solve the problem, substitute the values of q and w into the equation for ΔE. Since work is done by the system on the surroundings, w is negative. Similarly, since heat is released by the system to the surroundings, q is also negative.	$\begin{aligned} \Delta E &= q + w \\ &= -1422 \text{ J} - 855 \text{ J} \\ &= -2277 \text{ J} \end{aligned}$

FOR PRACTICE 6.1

A cylinder and piston assembly (defined as the system) is warmed by an external flame. The contents of the cylinder expand, doing work on the surroundings by pushing the piston outward against the external pressure. If the system absorbs 559 J of heat and does 488 J of work during the expansion, what is the value of ΔE?

6.4 Quantifying Heat and Work

In the previous section, we calculated ΔE based on *given values of q and w*. We now turn to *calculating q (heat) and w (work)* based on changes in temperature and volume.

Heat

Recall from Section 6.2, that *heat* is the exchange of thermal energy between a system and its surroundings caused by a temperature difference. Notice the distinction between heat and temperature. Temperature is a *measure* of the thermal energy of a sample of matter. Heat is the *transfer* of thermal energy.

The reason for this one-way transfer is related to the second law of thermodynamics, which states that energy tends to distribute itself among the greatest number of particles possible. We will cover the second law of thermodynamics in more detail in Chapter 17.

Thermal energy always flows from matter at higher temperatures to matter at lower temperatures. For example, a hot cup of coffee transfers thermal energy—as heat—to the lower temperature surroundings as it cools down. Imagine a world where the cooler surroundings actually got colder as they transferred thermal energy to the hot coffee, which got hotter. Such a world exists only in our imaginations (or in the minds of science fiction writers), because the transfer of heat from a hotter object to a colder one is a fundamental principle of our universe—no exception has ever been observed. Consequently, the thermal energy in the molecules within the hot coffee distributes itself to the molecules in the surroundings. The heat transfer from the coffee to the surroundings stops when the two reach the same temperature, a condition called **thermal equilibrium**. At thermal equilibrium, there is no additional net transfer of heat.

Temperature Changes and Heat Capacity

When a system absorbs heat (q) its temperature changes by ΔT:

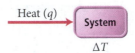

Experimental measurements demonstrate that the heat absorbed by a system and its corresponding temperature change are directly proportional: $q \propto \Delta T$. The constant of proportionality between q and ΔT is *heat capacity* (C), a measure of the system's ability to absorb thermal energy without undergoing a large change in temperature.

$$q = C \times \Delta T$$

Heat capacity

[6.5]

Notice that the higher the heat capacity of a system, the smaller the change in temperature for a given amount of absorbed heat. The **heat capacity (C)** of a system is usually defined as the quantity of heat required to change the temperature of the system by 1 °C. As we can see by solving Equation 6.5 for heat capacity, the units of heat capacity are those of heat (usually J) divided by those of temperature (usually °C).

$$C = \frac{q}{\Delta T} = \frac{J}{°C}$$

▲ The high heat capacity of the water surrounding San Francisco results in relatively cool summer temperatures.

In order to understand two important properties of heat capacity, consider putting a steel saucepan on a kitchen flame. The saucepan's temperature rises rapidly as it absorbs heat from the flame. However, if you add some water to the saucepan, the temperature rises more slowly. Why? The first reason is that, when you add the water, the same amount of heat must now warm more matter, so the temperature rises more slowly. In other words, heat capacity is an extensive property—it depends on the amount of matter being heated (see Section 1.6). The second (and more fundamental) reason is that water is more resistant to temperature change than steel—water has an intrinsically higher capacity to absorb heat without undergoing a large temperature change. The measure of the *intrinsic capacity* of a substance to absorb heat is called its **specific heat capacity (C_s)**, the amount of heat required to raise the temperature of *1 gram* of the substance by 1 °C. The units of specific heat capacity (also called *specific heat*) are J/g·°C. Table 6.4 lists the values of the specific heat capacity for several substances. Heat capacity is sometimes reported as **molar heat capacity**, the amount of heat required to raise the temperature of *1 mole* of a substance by 1 °C. The units of molar heat capacity are J/mol·°C. You can see from these definitions that *specific* heat capacity and *molar* heat capacity are intensive properties—they depend on the *kind* of substance being heated, not on the amount.

Notice that water has the highest specific heat capacity of all the substances in Table 6.4—changing its temperature requires a lot of heat. If you have ever experienced the drop in temperature that occurs when traveling from an inland region to the coast during the summer, you have experienced the effects of water's high specific heat capacity. On a summer's day in California, for example, the temperature difference between Sacramento (an inland city) and San Francisco (a coastal city) can be 18 °C (30 °F)—San Francisco

TABLE 6.4 Specific Heat Capacities of Some Common Substances

Substance	Specific Heat Capacity, C_s (J/g·°C)*
Elements	
Lead	0.128
Gold	0.128
Silver	0.235
Copper	0.385
Iron	0.449
Aluminum	0.903
Compounds	
Ethanol	2.42
Water	4.18
Materials	
Glass (Pyrex)	0.75
Granite	0.79
Sand	0.84

*At 298 K.

enjoys a cool 20 °C (68 °F), while Sacramento bakes at nearly 38 °C (100 °F). Yet the intensity of sunlight falling on these two cities is the same. Why the large temperature difference? San Francisco sits on a peninsula, surrounded by the Pacific Ocean. Water, with its high heat capacity, absorbs much of the sun's heat without undergoing a large increase in temperature, keeping San Francisco cool. Sacramento, by contrast, is about 160 km (100 mi) inland. The land surrounding Sacramento, with its low heat capacity, undergoes a large increase in temperature as it absorbs a similar amount of heat.

The specific heat capacity of a substance can be used to quantify the relationship between the amount of heat added to a given amount of the substance and the corresponding temperature increase. The equation that relates these quantities is

$$\text{heat} = \text{mass} \times \text{specific heat capacity} \times \text{temperature change} \qquad [6.6]$$

$$q = m \times C_s \times \Delta T$$

where q is the amount of heat in J, m is the mass of the substance in g, C_s is the specific heat capacity in J/g · °C, and ΔT is the temperature change in °C. Example 6.2 demonstrates the use of this equation.

ΔT in °C is equal to ΔT in K, but not equal to ΔT in °F (Section 1.6).

EXAMPLE 6.2 Temperature Changes and Heat Capacity

Suppose you find a copper penny (minted pre-1982 when pennies were almost entirely copper) in the snow. How much heat is absorbed by the penny as it warms from the temperature of the snow, which is −8.0 °C, to the temperature of your body, 37.0 °C? Assume the penny is pure copper and has a mass of 3.10 g.

SORT You are given the mass of copper as well as its initial and final temperature. You are asked to find the heat required for the given temperature change.	**GIVEN** $m = 3.10$ g copper $\quad\quad T_i = -8.0\,°C$ $\quad\quad T_f = 37.0\,°C$ **FIND** q
STRATEGIZE The equation $q = m \times C_s \times \Delta T$ gives the relationship between the amount of heat (q) and the temperature change (ΔT).	**CONCEPTUAL PLAN** 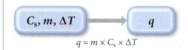 **RELATIONSHIPS USED** $q = m \times C_s \times \Delta T$ (Equation 6.6) $C_s = 0.385$ J/g · °C (Table 6.4)
SOLVE Gather the necessary quantities for the equation in the correct units and substitute these into the equation to calculate q.	**SOLUTION** $\Delta T = T_f - T_i$ $\quad\quad = 37.0\,°C - (-8.0\,°C) = 45.0\,°C$ $q = m \times C_s \times \Delta T$ $\quad\quad = 3.10\ \cancel{g} \times 0.385\ \dfrac{J}{\cancel{g} \cdot \cancel{°C}} \times 45.0\,\cancel{°C} = 53.7\text{J}$

CHECK The units (J) are correct for heat. The sign of q is *positive*, as it should be since the penny *absorbed* heat from the surroundings.

FOR PRACTICE 6.2

To determine whether a shiny gold-colored rock is actually gold, a chemistry student decides to measure its heat capacity. She first weighs the rock and finds it has a mass of 4.7 g. She then finds that upon absorption of 57.2 J of heat, the temperature of the rock rises from 25 °C to 57 °C. Find the specific heat capacity of the substance composing the rock and determine whether the value is consistent with the rock being pure gold.

FOR MORE PRACTICE 6.2

A 55.0-g aluminum block initially at 27.5 °C absorbs 725 J of heat. What is the final temperature of the aluminum?

Conceptual Connection 6.3 The Heat Capacity of Water

Suppose you are cold-weather camping and decide to heat some objects to bring into your sleeping bag for added warmth. You place a large water jug and a rock of equal mass near the fire. Over time, both the rock and the water jug warm to about 38 °C (100 °F). If you could bring only one into your sleeping bag, which one should you choose to keep you the warmest? Why?

Thermal Energy Transfer

As we noted earlier, when two substances of different temperature are combined, thermal energy flows as heat from the hotter substance to the cooler one. If we assume that the two substances are thermally isolated from everything else, then the heat lost by one substance exactly equals the heat gained by the other (according the law of energy conservation). If we define one substance as the system and the other as the surroundings, we can quantify the heat exchange as follows:

$$q_{sys} = -q_{surr}$$

Suppose a block of metal initially at 55 °C is submerged into water initially at 25 °C. Thermal energy transfers as heat from the metal to the water:

$$q_{metal} = -q_{water}$$

The metal will get colder and the water will get warmer until the two substances reach the same temperature (thermal equilibrium). The exact temperature change that occurs depends on the masses of the metal and the water and on their specific heat capacities. Since $q = m \times C_s \times \Delta T$ we can arrive at the following relationship:

$$q_{metal} = -q_{water}$$
$$m_{metal} \times C_{s,metal} \times \Delta T_{metal} = -m_{water} \times C_{s,water} \times \Delta T_{water}$$

Example 6.3 shows how to work with thermal energy transfer.

EXAMPLE 6.3 Thermal Energy Transfer

A 32.5 g cube of aluminum initially at 45.8 °C is submerged into 105.3 g of water at 15.4 °C. What is the final temperature of both substances at thermal equilibrium? (Assume that the aluminum and the water are thermally isolated from everything else.)

SORT You are given the masses of aluminum and water and their initial temperatures. You are asked to find the final temperature.	**GIVEN** $m_{Al} = 32.5$ g; $m_{H_2O} = 105.3$ g; $T_{i,Al} = 45.8$ °C; $T_{i,H_2O} = 15.4$ °C **FIND** T_f
STRATEGIZE The heat lost by the aluminum (q_{Al}) equals the heat gained by the water (q_{H_2O}). Use the relationship between q and ΔT and the given variables to find a relationship between ΔT_{Al} and ΔT_{H_2O}. Use the relationship between ΔT_{Al} and ΔT_{H_2O} (that you just found in the previous step) along with the initial temperatures of the aluminum and the water to determine the final temperature. Note that at thermal equilibrium, the final temperature of the aluminum and the water is the same, that is $T_{f,Al} = T_{f,H_2O} = T_f$	**CONCEPTUAL PLAN** $q_{Al} = -q_{water}$ $\boxed{m_{Al}, C_{s,Al} m_{H_2O}, C_{s,H_2O}} \longrightarrow \boxed{\Delta T_{Al} = \text{constant} \times \Delta T_{H_2O}}$ $m_{Al} \times C_{s,Al} \times \Delta T_{Al} = -m_{H_2O} \times C_{s,H_2O} \times \Delta T_{H_2O}$ $\boxed{T_{i,Al}; T_{i,H_2O}} \longrightarrow \boxed{T_f}$ $\Delta T_{Al} = \text{constant} \times \Delta T_{H_2O}$ **RELATIONSHIPS USED** $C_{s,H_2O} = 4.18$ J/g·°C; $C_{s,Al} = 0.903$ J/g·°C (Table 6.4) $q = m \times C_s \times \Delta T$ (Equation 6.6)

SOLVE Write the equation for the relationship between the heat lost by the aluminum (q_{Al}) and the heat gained by the water (q_{H_2O}) and substitute $q = m \times C_s \times \Delta T$ for each substance. Substitute the values of m (given) and C_s (from Table 6.4) for each substance and solve the equation for ΔT_{Al}. (Alternatively, you can solve the equation for ΔT_{H_2O}.)	**SOLUTION** $$q_{Al} = -q_{water}$$ $$m_{Al} \times C_{s,Al} \times \Delta T_{Al} = -m_{water} \times C_{s,water} \times \Delta T_{water}$$ $$32.5\ \cancel{g} \times \frac{0.903\ J}{\cancel{g} \cdot °\cancel{C}} \cdot \Delta T_{Al} = -105.3\ \cancel{g} \times \frac{4.18\ J}{\cancel{g} \cdot °\cancel{C}} \cdot \Delta T_{H_2O}$$ $$29.\underline{3}47 \cdot \Delta T_{Al} = -440.\underline{1}5 \cdot \Delta T_{H_2O}$$ $$\Delta T_{Al} = -14.\underline{9}98 \cdot \Delta T_{H_2O}$$
Substitute $T_f - T_i$ for each ΔT and solve for T_f.	$$T_f - T_{i,Al} = -14.\underline{9}98\,(T_f - T_{i,H_2O})$$ $$T_f = -14.\underline{9}98 \cdot T_f + 14.\underline{9}98 \cdot T_{i,H_2O} + T_{i,Al}$$ $$15.\underline{9}98 \cdot T_f = 14.\underline{9}98 \cdot T_{i,H_2O} + T_{i,Al}$$
Substitute the initial temperatures of aluminum and water into the relationship from the previous step and solve the expression for the final temperature (T_f).	$$T_f = \frac{14.\underline{9}98 \cdot T_{i,H_2O} + T_{i,Al}}{15.\underline{9}98} = \frac{14.\underline{9}98 \cdot 15.4°C + 45.8\ °C}{15.\underline{9}98}$$ $$= 17.3\ °C$$

CHECK The units °C are correct. The final temperature of the mixture is closer to the initial temperature of the *water* than the *aluminum*. This makes sense for two reasons: (1) Water has a higher specific heat capacity than aluminum; and (2) There is more water than aluminum. Since the aluminum loses the same amount of heat that is gained by the water, the greater mass and specific heat capacity of the water make the temperature change in the water *less than* the temperature change in the aluminum.

FOR PRACTICE 6.3

A block of copper of unknown mass has an initial temperature of 65.4 °C. The copper is immersed in a beaker containing 95.7 g of water at 22.7 °C. When the two substances reach thermal equilibrium, the final temperature is 24.2 °C. What is the mass of the copper block?

 Conceptual Connection 6.4 Thermal Energy Transfer

Substances A and B, initially at different temperatures, come in contact with each other and reach thermal equilibrium. The mass of substance A is twice the mass of substance B. The specific heat capacity of substance B is twice the specific heat capacity of substance A. Which statement is true about the final temperature of the two substances once thermal equilibrium is reached?

(a) The final temperature will be closer to the initial temperature of substance A than substance B.

(b) The final temperature will be closer to the initial temperature of substance B than substance A.

(c) The final temperature will be exactly midway between the initial temperatures of substances A and B.

Work: Pressure–Volume Work

We have seen that energy transfer between a system and its surroundings occurs via heat (q) and work (w). We just saw how to calculate the *heat* associated with an observed *temperature* change. We now turn to calculating the *work* associated with an observed *volume* change. Although there are several types of work that a chemical reaction can do, for now we will limit ourselves to what is called **pressure–volume work**. We have already defined work as a force acting through a distance. Pressure–volume work occurs

▲ The combustion of gasoline within an engine's cylinders does pressure–volume work that ultimately results in the car's motion.

when the force is the result of a volume change against an external pressure. Pressure–volume work occurs, for example, in the cylinder of an automobile engine. The combustion of gasoline causes gases within the cylinders to expand, pushing the piston outward and ultimately moving the wheels of the car.

The relationship between a volume change (ΔV) and work (w) is given by the equation:

$$w = -P\Delta V$$

where P is the external pressure. When the volume of a cylinder *increases* against a constant external pressure, ΔV is positive and w is negative. This makes sense because, as the volume of the cylinder expands, work is done *on* the surroundings *by* the system. When the volume of a cylinder *decreases* under a constant external pressure, ΔV is negative and w is positive. This also makes sense because, as the volume of the cylinder contracts, work is done *on* the system *by* the surroundings. The units of the work obtained by using this equation are those of pressure (usually atm) multiplied by those of volume (usually L). To convert between $L \cdot atm$ and J, use the conversion factor $101.3 \text{ J} = 1 \text{ L} \cdot atm$.

EXAMPLE 6.4 Pressure–Volume Work

Inflating a balloon requires the inflator to do pressure–volume work on the surroundings. If a balloon is inflated from a volume of 0.100 L to 1.85 L against an external pressure of 1.00 atm, how much work is done (in joules)?

SORT You are given the initial and final volumes of the balloon and the pressure against which it expands. The balloon and its contents are the system.	**GIVEN** $V_1 = 0.100$ L, $V_2 = 1.85$ L, $P = 1.00$ atm **FIND** w
STRATEGIZE The equation $w = -P\,\Delta V$ specifies the amount of work done during a volume change against an external pressure.	**CONCEPTUAL PLAN**  $w = -P\Delta V$
SOLVE To solve the problem, calculate the value of ΔV and substitute it, together with P, into the equation.	**SOLUTION** $\Delta V = V_2 - V_1$ $\quad = 1.85 \text{ L} - 0.100 \text{ L}$ $\quad = 1.75 \text{ L}$ $w = -P\,\Delta V$ $\quad = -1.00 \text{ atm} \times 1.75 \text{ L}$ $\quad = -1.75 \text{ L} \cdot atm$
The units of the answer ($L \cdot atm$) can be converted to joules using $101.3 \text{ J} = 1 \text{ L} \cdot atm$.	$-1.75 \, \cancel{L \cdot atm} \times \dfrac{101.3 \text{ J}}{1 \, \cancel{L \cdot atm}} = -177 \text{ J}$

CHECK The units (J) are correct for work. The sign of the work is negative, as it should be for an expansion: work is done on the surroundings by the expanding balloon.

FOR PRACTICE 6.4

A cylinder equipped with a piston expands against an external pressure of 1.58 atm. If the initial volume is 0.485 L and the final volume is 1.245 L, how much work (in J) is done?

FOR MORE PRACTICE 6.4

When fuel is burned in a cylinder equipped with a piston, the volume expands from 0.255 L to 1.45 L against an external pressure of 1.02 atm. In addition, 875 J is emitted as heat. What is ΔE for the burning of the fuel?

6.5 Measuring ΔE for Chemical Reactions: Constant-Volume Calorimetry

We now have a complete picture of how a system exchanges energy with its surroundings via heat and pressure–volume work:

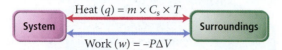

From Section 6.3, we know that the change in internal energy that occurs during a chemical reaction (ΔE) is a measure of *all of the energy* (heat and work) exchanged with the surroundings ($\Delta E = q + w$). Therefore, we can measure the changes in temperature (to calculate heat) and the changes in volume (to calculate work) that occur during a chemical reaction, and then sum them together to calculate ΔE. However, an easier way to obtain the value of ΔE for a chemical reaction is to force all of the energy change associated with a reaction to appear as heat rather than work. Then, we only have to measure the temperature change caused by the heat flow.

Recall that $\Delta E = q + w$ and that $w = -P\Delta V$. If a reaction is carried out at constant volume, $\Delta V = 0$ and $w = 0$. The heat evolved (given off), called the *heat at constant volume* (q_v), is equivalent to ΔE_{rxn}.

$$\Delta E_{rxn} = q_v + w \quad \text{(Equals zero at constant volume)}$$

$$\Delta E_{rxn} = q_v \qquad [6.7]$$

We can measure the heat evolved in a chemical reaction through *calorimetry*. In **calorimetry**, the thermal energy exchanged between the reaction (defined as the system) and the surroundings is measured by observing the change in temperature of the surroundings.

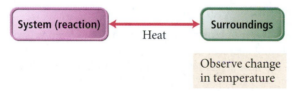

Observe change in temperature

The magnitude of the temperature change in the surroundings depends on the magnitude of ΔE for the reaction and on the heat capacity of the surroundings.

Figure 6.7▶ shows a **bomb calorimeter**, a piece of equipment designed to measure ΔE for combustion reactions. In a bomb calorimeter, a tight-fitting, sealed lid forces the reaction to occur at constant volume. The sample to be burned (of known mass) is placed into a cup equipped with an ignition wire. The cup is sealed into a stainless steel container, called a *bomb*, that is filled with oxygen gas. The bomb is then placed in a water-filled, insulated container equipped with a stirrer and a thermometer. The sample is ignited using a wire coil, and the temperature is monitored with the thermometer. The temperature change (ΔT) is then related to the heat absorbed by the entire calorimeter assembly (q_{cal}) by the following equation:

$$q_{cal} = C_{cal} \times \Delta T \qquad [6.8]$$

where C_{cal} is the heat capacity of the entire calorimeter assembly (which is usually determined in a separate measurement involving the burning of a substance that gives off a known amount of heat). If no heat escapes from the calorimeter, the amount of heat *gained by* the calorimeter must exactly equal that *released by* the reaction (the two are equal in magnitude but opposite in sign):

$$q_{cal} = -q_{rxn} \qquad [6.9]$$

Since the reaction is occurring under conditions of constant volume, $q_{rxn} = q_v = \Delta E_{rxn}$. This measured quantity is the change in the internal energy of the reaction for the specific

The heat capacity of the calorimeter, C_{cal}, has units of energy over temperature; its value accounts for all of the heat absorbed by all of the components within the calorimeter (including the water).

Calorimeter A bomb calorimeter is used to measure changes in internal energy for combustion reactions.

The Bomb Calorimeter

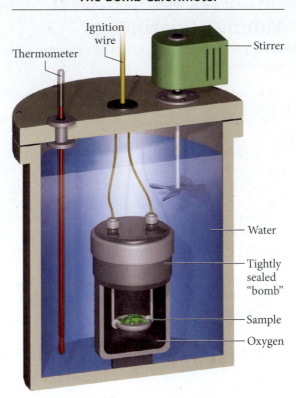

Thermometer

Ignition wire

Stirrer

Water

Tightly sealed "bomb"

Sample

Oxygen

amount of reactant burned. To get ΔE_{rxn} per mole of a particular reactant—a more general quantity—we divide by the number of moles that actually reacted, as shown in Example 6.5.

EXAMPLE 6.5 Measuring ΔE_{rxn} in a Bomb Calorimeter

When 1.010 g of sucrose ($C_{12}H_{22}O_{11}$) undergoes combustion in a bomb calorimeter, the temperature rises from 24.92 °C to 28.33 °C. Find ΔE_{rxn} for the combustion of sucrose in kJ/mol sucrose. The heat capacity of the bomb calorimeter, determined in a separate experiment, is 4.90 kJ/°C. (You can ignore the heat capacity of the small sample of sucrose because it is negligible compared to the heat capacity of the calorimeter.)

SORT You are given the mass of sucrose, the heat capacity of the calorimeter, and the initial and final temperatures. You are asked to find the change in internal energy for the reaction.	**GIVEN** 1.010 g $C_{12}H_{22}O_{11}$, $T_i = 24.92$ °C, $T_f = 28.33$ °C, $C_{cal} = 4.90$ kJ/°C **FIND** ΔE_{rxn}
STRATEGIZE The conceptual plan has three parts. In the first part, use the temperature change and the heat capacity of the calorimeter to find q_{cal}.	**CONCEPTUAL PLAN** 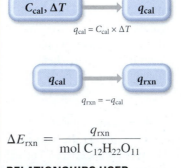 $$C_{cal}, \Delta T \longrightarrow q_{cal}$$ $q_{cal} = C_{cal} \times \Delta T$
In the second part, use q_{cal} to get q_{rxn} (which just involves changing the sign). Since the bomb calorimeter ensures constant volume, q_{rxn} is equivalent to ΔE_{rxn} for the amount of sucrose burned.	$$q_{cal} \longrightarrow q_{rxn}$$ $q_{rxn} = -q_{cal}$
In the third part, divide q_{rxn} by the number of moles of sucrose to get ΔE_{rxn} per mole of sucrose.	$$\Delta E_{rxn} = \frac{q_{rxn}}{\text{mol } C_{12}H_{22}O_{11}}$$ **RELATIONSHIPS USED** $q_{cal} = C_{cal} \times \Delta T = -q_{rxn}$ molar mass $C_{12}H_{22}O_{11} = 342.3$ g/mol

SOLVE Determine the necessary quantities in the correct units and substitute these into the equation to calculate q_{cal}.	**SOLUTION** $$\Delta T = T_f - T_i$$ $$= 28.33\ °C - 24.92\ °C$$ $$= 3.41\ °C$$ $$q_{cal} = C_{cal} \times \Delta T$$ $$q_{cal} = 4.90\ \frac{kJ}{°C} \times 3.41\ °C$$ $$= 16.7\ kJ$$
Find q_{rxn} by taking the negative of q_{cal}.	$$q_{rxn} = -q_{cal} = -16.7\ kJ$$
Find ΔE_{rxn} per mole of sucrose by dividing q_{rxn} by the number of moles of sucrose (calculated from the given mass of sucrose and its molar mass).	$$\Delta E_{rxn} = \frac{q_{rxn}}{\text{mol } C_{12}H_{22}O_{11}}$$ $$= \frac{-16.7\ kJ}{1.010\ \text{g } C_{12}H_{22}O_{11} \times \frac{1\ \text{mol } C_{12}H_{22}O_{11}}{342.3\ \text{g } C_{12}H_{22}O_{11}}}$$ $$= -5.66 \times 10^3\ kJ/\text{mol } C_{12}H_{22}O_{11}$$

CHECK The units of the answer (kJ) are correct for a change in internal energy. The sign of ΔE_{rxn} is negative, as it should be for a combustion reaction that gives off energy.

FOR PRACTICE 6.5

When 1.550 g of liquid hexane (C_6H_{14}) undergoes combustion in a bomb calorimeter, the temperature rises from 25.87 °C to 38.13 °C. Find ΔE_{rxn} for the reaction in kJ/mol hexane. The heat capacity of the bomb calorimeter, determined in a separate experiment, is 5.73 kJ/°C.

FOR MORE PRACTICE 6.5

The combustion of toluene has a ΔE_{rxn} of -3.91×10^3 kJ/mol. When 1.55 g of toluene (C_7H_8) undergoes combustion in a bomb calorimeter, the temperature rises from 23.12 °C to 37.57 °C. Find the heat capacity of the bomb calorimeter.

6.6 Enthalpy: The Heat Evolved in a Chemical Reaction at Constant Pressure

We have just seen that when a chemical reaction occurs in a sealed container under conditions of constant volume, the energy evolved in the reaction appears only as heat. However, when a chemical reaction occurs open to the atmosphere under conditions of constant pressure—for example, a reaction occurring in an open beaker or the burning of natural gas in a furnace—the energy evolved may appear as both heat and work. As we have also seen, ΔE_{rxn} is a measure of the *total energy change* (both heat and work) that occurs during the reaction. However, in many cases, we are interested only in the heat exchanged, not the work done. For example, when we burn natural gas in the furnace to heat our homes, we do not really care how much work the combustion reaction does on the atmosphere by expanding against it—we just want to know how much heat is given off to warm the home. Under conditions of constant pressure, a thermodynamic quantity called *enthalpy* provides exactly this.

The **enthalpy (*H*)** of a system is defined as the sum of its internal energy and the product of its pressure and volume:

$$H = E + PV \qquad [6.10]$$

Since internal energy, pressure, and volume are all state functions, enthalpy is also a state function. The *change in enthalpy* (ΔH) for any process occurring under constant pressure is given by the following expression:

$$\Delta H = \Delta E + P\Delta V \qquad [6.11]$$

To better understand this expression, we can interpret the two terms on the right with the help of relationships already familiar to us. We saw previously that $\Delta E = q + w$. Thus, if we represent the heat at constant pressure as q_p, then the change in internal energy at constant pressure is $\Delta E = q_p + w$. In addition, from our definition of pressure–volume work, we know that $P\Delta V = -w$. Substituting these expressions into the expression for ΔH gives us

$$\Delta H = \Delta E + P\Delta V$$

$$= (q_p + w) + P\Delta V$$

$$= q_p + w - w$$

$$\Delta H = q_p \qquad [6.12]$$

So we can see that ΔH is equal to q_p, the heat at constant pressure.

Conceptually (and often numerically), ΔH and ΔE are similar: they both represent changes in a state function for the system. However, ΔE is a measure of *all of the energy* (heat and work) exchanged with the surroundings, while ΔH is a measure of only the heat exchanged under conditions of constant pressure. For chemical reactions that do not exchange much work with the surroundings—that is, those that do not cause a large change in the reaction volume as they occur— ΔH and ΔE are nearly identical in value. For chemical reactions that produce or consume large amounts of gas, and therefore produce large volume changes, ΔH and ΔE can be slightly different in value.

Conceptual Connection 6.5 The Difference between ΔH and ΔE

Lighters are usually fueled by butane (C_4H_{10}). When 1 mole of butane burns at constant pressure, it produces 2658 kJ of heat and does 3 kJ of work. What are the values of ΔH and ΔE for the combustion of 1 mole of butane?

The signs of ΔH and ΔE follow the same conventions. A positive ΔH means that heat flows into the system as the reaction occurs. A chemical reaction with a positive ΔH, called an **endothermic reaction**, absorbs heat from its surroundings. A chemical cold pack, often used to ice athletic injuries, is a good example of an endothermic reaction. When a barrier separating the reactants in a chemical cold pack is broken, the substances mix, react, and absorb heat from the surroundings. The surroundings—possibly including your bruised wrist—get *colder* because they *lose* energy as the cold pack absorbs it.

A chemical reaction with a negative ΔH, called an **exothermic reaction**, gives off heat to its surroundings. The reaction occurring in the chemical hand warmer discussed in Section 6.1 is a good example of an exothermic reaction. As the reaction occurs, heat is given off into the surroundings (including your hand and glove) making them warmer. The burning of natural gas is another good example of an exothermic reaction. As the gas burns, it gives off heat, raising the temperature of its surroundings.

The term exothermic and endothermic also apply to *physical* changes. The evaporation of water, for example, is endothermic.

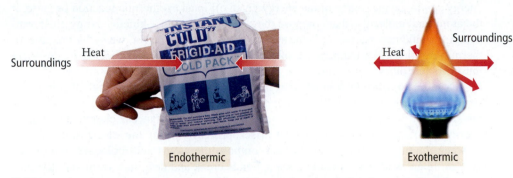

Surroundings — Heat → Endothermic

Heat ← Surroundings → Exothermic

▲ The reaction that occurs in a chemical cold pack is endothermic—it absorbs energy from the surroundings (in this case your wrist). The combustion of natural gas is an exothermic reaction—it gives off energy to the surroundings.

Summarizing Enthalpy:

▶ The value of ΔH for a chemical reaction is the amount of heat absorbed or evolved in the reaction under conditions of constant pressure.

▶ An endothermic reaction has a *positive* ΔH and absorbs heat from the surroundings. An endothermic reaction feels cold to the touch.

▶ An exothermic reaction has a *negative* ΔH and gives off heat to the surroundings. An exothermic reaction feels warm to the touch.

EXAMPLE 6.6 Exothermic and Endothermic Processes

Identify each process as endothermic or exothermic and indicate the sign of ΔH.

(a) sweat evaporating from skin

(b) water freezing in a freezer

(c) wood burning in a fire

SOLUTION

(a) Sweat evaporating from skin cools the skin and is therefore endothermic, with a positive ΔH. The skin must supply heat to the water in order for it to continue to evaporate.

(b) Water freezing in a freezer releases heat and is therefore exothermic, with a negative ΔH. The refrigeration system in the freezer must remove this heat for the water to continue to freeze.

(c) Wood burning in a fire releases heat and is therefore exothermic, with a negative ΔH.

FOR PRACTICE 6.6

Identify each process as endothermic or exothermic and indicate the sign of ΔH.

(a) an ice cube melting

(b) nail polish remover evaporating after it is accidentally spilled on the skin

(c) gasoline burning within the cylinder of an automobile engine

Exothermic and Endothermic Processes: A Molecular View

When a chemical system undergoes a change in enthalpy, where does the energy come from or go to? For example, we just saw that an exothermic chemical reaction gives off *thermal energy*—what is the source of that energy?

First, we know that the emitted thermal energy *does not* come from the original thermal energy of the system. Recall from Section 6.2 that the thermal energy of a system is the total kinetic energy of the atoms and molecules that compose the system. This kinetic energy *cannot* be the source of the energy given off in an exothermic reaction because, if the atoms and molecules that compose the system were to lose kinetic energy, their temperature would necessarily fall—the system would get colder. Yet, we know that in exothermic reactions, the temperature of the system and the surroundings rises. So there must be some other source of energy.

Recall also from Section 6.1 that the internal energy of a chemical system is the sum of its kinetic energy and its *potential energy. This potential energy is the source in an exothermic chemical reaction.* Under normal circumstances, chemical potential energy (or simply chemical energy) arises primarily from the electrostatic forces between the protons and electrons that compose the atoms and molecules within the system. In an exothermic reaction, some bonds break and new ones form, and the protons and electrons go from an arrangement of higher potential energy to one of lower potential energy. As the molecules rearrange, their potential energy is converted into kinetic energy, the heat emitted in the reaction. In an endothermic reaction, the opposite happens: as some bonds break and others form, the protons and electrons go from an arrangement of lower potential energy to one of higher potential energy, absorbing thermal energy in the process.

Conceptual Connection 6.6   Exothermic and Endothermic Reactions

If an endothermic reaction absorbs heat, then why does it feel cold to the touch?

Stoichiometry Involving ΔH: Thermochemical Equations

The enthalpy change for a chemical reaction, abbreviated ΔH_{rxn}, is also called the **enthalpy of reaction** or **heat of reaction** and is an extensive property, one that depends on the amount of material. In other words, the amount of heat generated or absorbed when a chemical reaction occurs depends on the *amounts* of reactants that actually react. We usually specify ΔH_{rxn} in combination with the balanced chemical equation for the reaction. The magnitude of ΔH_{rxn} is for the stoichiometric amounts of reactants and products for the reaction *as written.* For example, the balanced equation and ΔH_{rxn} for the combustion of propane (the main component of liquid propane or LP gas) are as follows:

$$C_3H_8(g) + 5\,O_2(g) \longrightarrow 3\,CO_2(g) + 4\,H_2O(g) \quad \Delta H_{rxn} = -2044\ \text{kJ}$$

This means that when 1 mol of C_3H_8 reacts with 5 mol of O_2 to form 3 moles of CO_2 and 4 mol of H_2O, 2044 kJ of heat is emitted. We can write these relationships in the same way that we expressed stoichiometric relationships in Chapter 4, as ratios between two quantities. For example, for the reactants, we write the following:

$$1\ \text{mol}\ C_3H_8: -2044\ \text{kJ} \quad \text{or} \quad 5\ \text{mol}\ O_2: -2044\ \text{kJ}$$

The ratios mean that 2044 kJ of heat is evolved when 1 mol of C_3H_8 and 5 mol of O_2 completely react. These ratios can then be used to construct conversion factors between amounts of reactants or products and the quantity of heat emitted (for exothermic reactions) or absorbed (for endothermic reactions). To find out how much heat is emitted upon the combustion of a certain mass in grams of C_3H_8, we use the following conceptual plan:

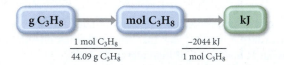

We use the molar mass to convert between grams and moles, and the stoichiometric relationship between moles of C_3H_8 and the heat of reaction to convert between moles and kilojoules, as shown in Example 6.7.

EXAMPLE 6.7 Stoichiometry Involving ΔH

An LP gas tank in a home barbeque contains 13.2 kg of propane, C_3H_8. Calculate the heat (in kJ) associated with the complete combustion of all of the propane in the tank.

$$C_3H_8(g) + 5\,O_2(g) \longrightarrow 3\,CO_2(g) + 4\,H_2O(g) \qquad \Delta H_{rxn} = -2044\text{ kJ}$$

SORT You are given the mass of propane and asked to find the heat evolved in its combustion.	**GIVEN** 13.2 kg C_3H_8 **FIND** q

STRATEGIZE Starting with kg C_3H_8, convert to g C_3H_8 and then use the molar mass of C_3H_8 to find the number of moles. Next, use the stoichiometric relationship between mol C_3H_8 and kJ to find the heat evolved.

CONCEPTUAL PLAN

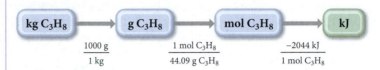

$$\boxed{\text{kg } C_3H_8} \longrightarrow \boxed{\text{g } C_3H_8} \longrightarrow \boxed{\text{mol } C_3H_8} \longrightarrow \boxed{\text{kJ}}$$

$$\frac{1000\text{ g}}{1\text{ kg}} \qquad \frac{1\text{ mol } C_3H_8}{44.09\text{ g } C_3H_8} \qquad \frac{-2044\text{ kJ}}{1\text{ mol } C_3H_8}$$

RELATIONSHIPS USED
1000 g = 1 kg
molar mass C_3H_8 = 44.09 g/mol
1 mol C_3H_8: −2044 kJ (from balanced equation)

SOLVE Follow the conceptual plan to solve the problem. Begin with 13.2 kg C_3H_8 and multiply by the appropriate conversion factors to arrive at kJ.

SOLUTION

$$13.2\text{ kg } C_3H_8 \times \frac{1000\text{ g}}{1\text{ kg}} \times \frac{1\text{ mol } C_3H_8}{44.09\text{ g } C_3H_8} \times \frac{-2044\text{ kJ}}{1\text{ mol } C_3H_8} = -6.12 \times 10^5\text{ kJ}$$

CHECK The units of the answer (kJ) are correct for energy. The answer is negative, as it should be for heat evolved by the reaction.

FOR PRACTICE 6.7
Ammonia reacts with oxygen according to the equation:

$$4\,NH_3(g) + 5\,O_2(g) \longrightarrow 4\,NO(g) + 6\,H_2O(g) \qquad \Delta H_{rxn} = -906\text{ kJ}$$

Calculate the heat (in kJ) associated with the complete reaction of 155 g of NH_3.

FOR MORE PRACTICE 6.7
What mass of butane in grams is necessary to produce 1.5×10^3 kJ of heat? What mass of CO_2 is produced?

$$C_4H_{10}(g) + 13/2\,O_2(g) \longrightarrow 4\,CO_2(g) + 5\,H_2O(g) \qquad \Delta H_{rxn} = -2658\text{ kJ}$$

6.7 Constant-Pressure Calorimetry: Measuring ΔH_{rxn}

For many aqueous reactions, ΔH_{rxn} can be measured fairly simply using a **coffee-cup calorimeter** (**Figure 6.8▶**). The calorimeter consists of two Styrofoam coffee cups, one inserted into the other, to provide insulation from the laboratory environment. The calorimeter is equipped with a thermometer and a stirrer. The reaction is carried out in a specifically measured quantity of solution within the calorimeter, so that the mass of the solution is known. During the reaction, the heat evolved (or absorbed) causes a temperature change in the solution, which is measured by the thermometer. If we know the specific heat capacity of the solution, normally assumed to be that of water, we can calculate q_{soln}, the heat absorbed by or lost from the solution (which is acting as the surroundings) using the following equation:

$$q_{soln} = m_{soln} \times C_{s,soln} \times \Delta T$$

Since the insulated calorimeter prevents heat from escaping, we can assume that the heat gained by the solution equals that lost by the reaction (or vice versa):

$$q_{rxn} = -q_{soln}$$

Since the reaction happens under conditions of constant pressure (open to the atmosphere), $q_{rxn} = q_p = \Delta H_{rxn}$. This measured quantity is the heat of reaction for the

The Coffee-Cup Calorimeter

Thermometer

Glass stirrer

Cork lid (loose fitting)

Two nested Styrofoam® cups containing reactants in solution

▲ **FIGURE 6.8 The Coffee-Cup Calorimeter** A coffee-cup calorimeter measures enthalpy changes for chemical reactions in solution.

This equation assumes that no heat is lost to the calorimeter itself. If heat absorbed by the calorimeter is accounted for, the equation becomes $q_{rxn} = -(q_{soln} + q_{cal})$.

specific amount (measured ahead of time) of reactants that reacted. To get ΔH_{rxn} per mole of a particular reactant—a more general quantity—we divide by the number of moles that actually reacted, as shown in Example 6.8.

EXAMPLE 6.8 Measuring ΔH_{rxn} in a Coffee-Cup Calorimeter

Magnesium metal reacts with hydrochloric acid according to this balanced equation:

$$Mg(s) + 2\,HCl(aq) \longrightarrow MgCl_2(aq) + H_2(g)$$

In an experiment to determine the enthalpy change for this reaction, 0.158 g of Mg metal is combined with enough HCl to make 100.0 mL of solution in a coffee-cup calorimeter. The HCl is sufficiently concentrated so that the Mg completely reacts. The temperature of the solution rises from 25.6 °C to 32.8 °C as a result of the reaction. Find ΔH_{rxn} for the reaction as written. Use 1.00 g/mL as the density of the solution and $C_{s,\,soln} = 4.18\ J/g\cdot°C$ as the specific heat capacity of the solution.

SORT You are given the mass of magnesium, the volume of solution, the initial and final temperatures, the density of the solution, and the heat capacity of the solution. You are asked to find the change in enthalpy for the reaction.

GIVEN 0.158 g Mg
   100.0 mL soln
   $T_i = 25.6$ °C
   $T_f = 32.8$ °C
   $d = 1.00$ g/mL
   $C_{s,\,soln} = 4.18\ J/g\cdot°C$
FIND ΔH_{rxn}

STRATEGIZE The conceptual plan has three parts. In the first part, use the temperature change and the other given quantities, together with the equation $q = m \times C_s \times \Delta T$, to find q_{soln}.

CONCEPTUAL PLAN

$q = m \times C_s \times \Delta T$

In the second part, use q_{soln} to get q_{rxn} (which simply involves changing the sign). Because the pressure is constant, q_{rxn} is equivalent to ΔH_{rxn} for the amount of magnesium that reacts.

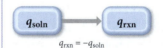

$q_{rxn} = -q_{soln}$

In the third part, divide q_{rxn} by the number of moles of magnesium to get ΔH_{rxn} per mole of magnesium.

$$\Delta H_{rxn} = \frac{q_{rxn}}{mol\ Mg}$$

RELATIONSHIPS USED $q = m \times C_s \times \Delta T$
   $q_{rxn} = -q_{soln}$

SOLVE Determine the necessary quantities in the correct units for the equation $q = m \times C_s \times \Delta T$ and substitute these into the equation to calculate q_{soln}. Notice that the sign of q_{soln} is *positive*, meaning that the solution *absorbs heat* from the reaction.

SOLUTION
$C_{s,\,soln} = 4.18\ J/g\cdot\ °C$

$m_{soln} = 100.0\ \text{mL soln} \times \dfrac{1.00\ g}{1\ \text{mL soln}} = 1.00 \times 10^2\ g$

$\Delta T = T_f - T_i = 32.8\ °C - 25.6\ °C = 7.2\ °C$

$q_{soln} = m_{soln} \times C_{s,\,soln} \times \Delta T$

$= 1.00 \times 10^2\ g \times 4.18\dfrac{J}{g\cdot °C} \times 7.2\ °C = 3.0 \times 10^3\ J$

Find q_{rxn} by simply taking the negative of q_{soln}. Notice that q_{rxn} is negative, as expected for an *exothermic* reaction.
Finally, find ΔH_{rxn} per mole of magnesium by dividing q_{rxn} by the number of moles of magnesium that react. Find the number of moles of magnesium from the given mass of magnesium and its molar mass.

$q_{rxn} = -q_{soln} = -3.0 \times 10^3\ J$

$\Delta H_{rxn} = \dfrac{q_{rxn}}{mol\ Mg}$

$= \dfrac{-3.0 \times 10^3\ J}{0.158\ \text{g Mg} \times \dfrac{1\ mol\ Mg}{24.31\text{g Mg}}}$

$= -4.6 \times 10^5\ J/mol\ Mg$

Since the stoichiometric coefficient for magnesium in the balanced chemical equation is 1, the calculated value represents ΔH_{rxn} for the reaction as written.

$Mg(s) + 2\,HCl(aq) \longrightarrow MgCl_2(aq) + H_2(g)$

$\Delta H_{rxn} = -4.6 \times 10^5\ J$

CHECK The units of the answer (J) are correct for the change in enthalpy of a reaction. The sign is negative, as expected for an exothermic reaction.

FOR PRACTICE 6.8

The addition of hydrochloric acid to a silver nitrate solution precipitates silver chloride according to the reaction:

$$AgNO_3(aq) + HCl(aq) \longrightarrow AgCl(s) + HNO_3(aq)$$

When 50.0 mL of 0.100 M $AgNO_3$ is combined with 50.0 mL of 0.100 M HCl in a coffee-cup calorimeter, the temperature changes from 23.40 °C to 24.21 °C. Calculate ΔH_{rxn} for the reaction as written. Use 1.00 g/mL as the density of the solution and $C = 4.18$ J/g·°C as the specific heat capacity.

 Conceptual Connection 6.7 Constant-Pressure versus Constant-Volume Calorimetry

The same reaction, with exactly the same amount of reactant, is conducted in a bomb calorimeter and in a coffee-cup calorimeter. In one of the measurements, $q_{rxn} = -12.5$ kJ and in the other $q_{rxn} = -11.8$ kJ. Which value was obtained in the bomb calorimeter? (Assume that the reaction has a positive ΔV in the coffee-cup calorimeter.)

6.8 Relationships Involving ΔH_{rxn}

We now turn our attention to three quantitative relationships between a chemical equation and ΔH_{rxn}.

1. If a chemical equation is multiplied by some factor, ΔH_{rxn} is also multiplied by the same factor.

Recall from Section 6.6 that ΔH_{rxn} is an extensive property; therefore, it depends on the quantity of reactants undergoing reaction. We also discussed that ΔH_{rxn} is usually reported for a reaction involving stoichiometric amounts of reactants. For example, for a reaction A + 2 B $\longrightarrow$ C, ΔH_{rxn} is usually reported as the amount of heat emitted or absorbed when 1 mol A reacts with 2 mol B to form 1 mol C. Thus, if a chemical equation is multiplied by a factor, then ΔH_{rxn} is also multiplied by the same factor. For example,

$$A + 2B \longrightarrow C \qquad \Delta H_1$$
$$2A + 4B \longrightarrow 2C \qquad \Delta H_2 = 2 \times \Delta H_1$$

2. If a chemical equation is reversed, ΔH_{rxn} changes sign.

Recall from Section 6.6 that ΔH_{rxn} is a state function, which means that its value depends only on the initial and final states of the system.

$$\Delta H = H_{final} - H_{initial}$$

When a reaction is reversed, the final state becomes the initial state and vice versa. Consequently, ΔH_{rxn} changes sign, as exemplified by the following:

$$A + 2B \longrightarrow C \qquad \Delta H_1$$
$$C \longrightarrow A + 2B \qquad \Delta H_2 = -\Delta H_1$$

3. If a chemical equation can be expressed as the sum of a series of steps, ΔH_{rxn} for the overall equation is the sum of the heats of reactions for each step.

This last relationship, known as **Hess's law** also follows from the enthalpy of reaction being a state function. Since ΔH_{rxn} is dependent only on the initial and final states, and not on the pathway the reaction follows, ΔH obtained from summing the individual steps that lead to an overall reaction must be the same as ΔH for that overall reaction. For example,

$$\begin{array}{ll} A + 2B \longrightarrow C & \Delta H_1 \\ \underline{C \longrightarrow 2D} & \underline{\Delta H_2} \\ A + 2B \longrightarrow 2D & \Delta H_3 = \Delta H_1 + \Delta H_2 \end{array}$$

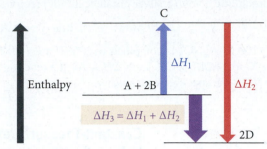

FIGURE 6.9 Hess's Law The change in enthalpy for a stepwise process is the sum of the enthalpy changes of the steps.

We illustrate Hess's law with the energy level diagram shown in **Figure 6.9▲**.

These three quantitative relationships make it possible to determine ΔH for a reaction without directly measuring it in the laboratory. (For some reactions, direct measurement can be difficult.) If we can find related reactions (with known ΔH's) that sum to the reaction of interest, we can find ΔH for the reaction of interest. For example, the following reaction between $C(s)$ and $H_2O(g)$ is an industrially important way to generate hydrogen gas:

$$C(s) + H_2O(g) \longrightarrow CO(g) + H_2(g) \qquad \Delta H_{rxn} = \; ?$$

We can find ΔH_{rxn} from the following reactions with known ΔH's:

$$C(s) + O_2(g) \longrightarrow CO_2(g) \qquad \Delta H = -393.5 \text{ kJ}$$
$$2\, CO(g) + O_2(g) \longrightarrow 2\, CO_2(g) \qquad \Delta H = -566.0 \text{ kJ}$$
$$2\, H_2(g) + O_2(g) \longrightarrow 2\, H_2O(g) \qquad \Delta H = -483.6 \text{ kJ}$$

We just have to determine how to sum these reactions to get the overall reaction of interest. We do this by manipulating the reactions with known ΔH's in such a way as to get the reactants of interest on the left, the products of interest on the right, and other species to cancel.

Since the first reaction has $C(s)$ as a reactant, and the reaction of interest also has $C(s)$ as a reactant, we write the first reaction unchanged.

$$C(s) + O_2(g) \longrightarrow CO_2(g) \qquad\qquad \Delta H = -393.5 \text{ kJ}$$

The second reaction has 2 mol of $CO(g)$ as a reactant. However, the reaction of interest has 1 mol of $CO(g)$ as a product. Therefore, we reverse the second reaction, change the sign of ΔH, and multiply the reaction and ΔH by $^1/_2$.

$$^1/_2 \times [2\, CO_2(g) \longrightarrow 2\, CO(g) + O_2(g)] \qquad \Delta H = {}^1/_2 \times (+566.0 \text{ kJ})$$

The third reaction has $H_2(g)$ as a reactant. In the reaction of interest, however, $H_2(g)$ is a product. Therefore, we reverse the equation and change the sign of ΔH. In addition, to obtain coefficients that match the reaction of interest, and to cancel O_2, we must multiply the reaction and ΔH by $^1/_2$.

$$^1/_2 \times [2\, H_2O(g) \longrightarrow 2\, H_2(g) + O_2(g)] \qquad \Delta H = {}^1/_2 \times (+483.6 \text{ kJ})$$

Lastly, we rewrite the three reactions after multiplying through by the indicated factors and show how they sum to the reaction of interest. ΔH for the reaction of interest is simply the sum of the ΔH's for the steps.

$$C(s) + O_2(g) \longrightarrow CO_2(g) \qquad\qquad \Delta H = -393.5 \text{ kJ}$$
$$CO_2(g) \longrightarrow CO(g) + {}^1/_2 O_2(g) \qquad \Delta H = +283.0 \text{ kJ}$$
$$H_2O(g) \longrightarrow H_2(g) + {}^1/_2\, O_2(g) \qquad \Delta H = +241.8 \text{ kJ}$$
$$\overline{C(s) + H_2O(g) \longrightarrow CO(g) + H_2(g) \qquad \Delta H_{rxn} = +131.3 \text{ kJ}}$$

EXAMPLE 6.9 Hess's Law

Find ΔH_{rxn} for the reaction:

$$3 \, C(s) + 4 \, H_2(g) \longrightarrow C_3H_8(g)$$

Use the following reactions with known ΔH's:

$$C_3H_8(g) + 5 \, O_2(g) \longrightarrow 3 \, CO_2(g) + 4 \, H_2O(g) \qquad \Delta H = -2043 \text{ kJ}$$

$$C(s) + O_2(g) \longrightarrow CO_2(g) \qquad \Delta H = -393.5 \text{ kJ}$$

$$2 \, H_2(g) + O_2(g) \longrightarrow 2 \, H_2O(g) \qquad \Delta H = -483.6 \text{ kJ}$$

SOLUTION

To work this and other Hess's law problems, manipulate the reactions with known ΔH's in such a way as to get the reactants of interest on the left, the products of interest on the right, and other species to cancel.

Since the first reaction has C_3H_8 as a reactant, and the reaction of interest has C_3H_8 as a product, reverse the first reaction and change the sign of ΔH.	$3 \, CO_2(g) + 4 \, H_2O(g) \longrightarrow C_3H_8(g) + 5 \, O_2(g) \quad \Delta H = +2043 \text{ kJ}$
The second reaction has C as a reactant and CO_2 as a product, just as required in the reaction of interest. However, the coefficient for C is 1, and in the reaction of interest, the coefficient for C is 3. Therefore multiply this equation and its ΔH by 3.	$3 \times \left[C(s) + O_2(g) \longrightarrow CO_2(g) \right] \quad \Delta H = 3 \times (-393.5 \text{ kJ})$
The third reaction has H_2 as a reactant, as required. However, the coefficient for H_2 is 2, and in the reaction of interest, the coefficient for H_2 is 4. Therefore multiply this reaction and its ΔH by 2.	$2 \times \left[2 \, H_2(g) + O_2(g) \longrightarrow 2 \, H_2O(g) \right] \quad \Delta H = 2 \times (-483.6 \text{ kJ})$
Lastly, rewrite the three reactions after multiplying through by the indicated factors and show how they sum to the reaction of interest. Then ΔH for the reaction of interest is the sum of the ΔH's for the steps.	$3 \, CO_2(g) + 4 \, H_2O(g) \longrightarrow C_3H_8(g) + 5 \, O_2(g) \quad \Delta H = +2043 \text{ kJ}$ $3 \, C(s) + 3 \, O_2(g) \longrightarrow 3 \, CO_2(g) \qquad\qquad\qquad \Delta H = -1181 \text{ kJ}$ $\underline{4 \, H_2(g) + 2 \, O_2(g) \longrightarrow 4 \, H_2O(g) \qquad\qquad\quad \Delta H = -967.2 \text{ kJ}}$ $3 \, C(s) + 4 \, H_2(g) \longrightarrow C_3H_8(g) \qquad\qquad\quad \Delta H_{rxn} = -105 \text{ kJ}$

FOR PRACTICE 6.9

Find ΔH_{rxn} for the reaction:

$$N_2O(g) + NO_2(g) \longrightarrow 3 \, NO(g)$$

Use the following reactions with known ΔH's:

$$2 \, NO(g) + O_2(g) \longrightarrow 2 \, NO_2(g) \qquad \Delta H = -113.1 \text{ kJ}$$

$$N_2(g) + O_2(g) \longrightarrow 2 \, NO(g) \qquad \Delta H = +182.6 \text{ kJ}$$

$$2 \, N_2O(g) \longrightarrow 2 \, N_2(g) + O_2(g) \qquad \Delta H = -163.2 \text{ kJ}$$

FOR MORE PRACTICE 6.9

Find ΔH_{rxn} for the reaction:

$$3 \, H_2(g) + O_3(g) \longrightarrow 3 \, H_2O(g)$$

Use the following reactions with known ΔH's:

$$2 \, H_2(g) + O_2(g) \longrightarrow 2 \, H_2O(g) \qquad \Delta H = -483.6 \text{ kJ}$$

$$3 \, O_2(g) \longrightarrow 2 \, O_3(g) \qquad \Delta H = +285.4 \text{ kJ}$$

6.9 Enthalpies of Reaction from Standard Heats of Formation

We have studied two ways to determine ΔH for a chemical reaction: experimentally through calorimetry and inferentially through Hess's law. We now turn to a third and more convenient way to determine ΔH for a large number of chemical reactions: from tabulated *standard enthalpies of formation*.

Standard States and Standard Enthalpy Changes

Recall that ΔH is the *change* in enthalpy for a chemical reaction—the difference in enthalpy between the products and the reactants. Since we are interested in *changes* in enthalpy (and not in absolute values of enthalpy itself), we are free to define the *zero* of enthalpy as conveniently as possible. Returning to our mountain-climbing analogy, a change in altitude (like a change in enthalpy) is an absolute quantity. The altitude itself (like enthalpy itself), however, is a relative quantity, defined relative to some standard. In the case of altitude the standard is sea level. We must define a similar, albeit slightly more complex, standard for enthalpy. This standard has three parts: the **standard state**, the **standard enthalpy change ($\Delta H°$)**, and the **standard enthalpy of formation ($\Delta H_f°$)**.

The standard state was changed in 1997 to a pressure of 1 bar, which is very close to 1 atm (1 atm = 1.013 bar). Both standards are now in common use.

1. **Standard State**

   • *For a Gas:* The standard state for a gas is the pure gas at a pressure of exactly 1 atmosphere.

   • *For a Liquid or Solid:* The standard state for a liquid or solid is the pure substance in its most stable form at a pressure of 1 atm and at the temperature of interest (often taken to be 25 °C).

   • *For a Substance in Solution:* The standard state for a substance in solution is a concentration of exactly 1 M.

2. **Standard Enthalpy Change ($\Delta H°$)**

   • The change in enthalpy for a process when all reactants and products are in their standard states. The degree sign indicates standard states.

3. **Standard Enthalpy of Formation ($\Delta H_f°$)**

   • *For a Pure Compound:* The change in enthalpy when 1 mole of the compound forms from its constituent elements in their standard states.

   • *For a Pure Element in Its Standard State:* $\Delta H_f° = 0$.

Assigning the value of zero to the standard enthalpy of formation for an element in its standard state is the equivalent of assigning an altitude of zero to sea level—we can then measure all subsequent changes in altitude relative to sea level. Similarly, we can measure all changes in enthalpy relative to those of pure elements in their standard states. For example, consider the standard enthalpy of formation of methane gas at 25 °C:

The carbon in this equation must be graphite (the most stable form of carbon at 1 atm and 25 °C).

$$C(s, \text{graphite}) + 2\,H_2(g) \longrightarrow CH_4(g) \qquad \Delta H_f° = -74.6 \text{ kJ/mol}$$

For methane, as with most compounds, $\Delta H_f°$ is negative. If we think of pure elements in their standard states as *sea level*, then most compounds lie *below sea level*. The chemical equation for the enthalpy of formation of a compound is always written to form 1 mole of the compound, so $\Delta H_f°$ has the units of kJ/mol. Table 6.5 lists $\Delta H_f°$ values for some selected compounds. A more complete list can be found in Appendix IIB.

TABLE 6.5 Standard Enthalpies of Formation, ΔH_f°, at 298 K

Formula	ΔH_f° (kJ/mol)	Formula	ΔH_f° (kJ/mol)
Bromine		$F_2(g)$	0
$Br(g)$	111.9	$HF(g)$	−273.3
$Br_2(l)$	0	*Hydrogen*	
$HBr(g)$	−36.3	$H(g)$	218.0
Calcium		$H_2(g)$	0
$Ca(s)$	0	*Nitrogen*	
$CaO(s)$	−634.9	$N_2(g)$	0
$CaCO_3(s)$	−1207.6	$NH_3(g)$	−45.9
Carbon		$NH_4NO_3(s)$	−365.6
$C(s, \text{graphite})$	0	$NO(g)$	91.3
$C(s, \text{diamond})$	1.88	$N_2O(g)$	81.6
$CO(g)$	−110.5	*Oxygen*	
$CO_2(g)$	−393.5	$O_2(g)$	0
$CH_4(g)$	−74.6	$O_3(g)$	142.7
$CH_3OH(l)$	−238.6	$H_2O(g)$	−241.8
$C_2H_2(g)$	227.4	$H_2O(l)$	2285.8
$C_2H_4(g)$	52.4	*Silver*	
$C_2H_6(g)$	−84.68	$Ag(s)$	0
$C_2H_5OH(l)$	−277.6	$AgCl(s)$	−127.0
$C_3H_8(g)$	−103.85	*Sodium*	
$C_3H_6O(l, \text{acetone})$	−248.4	$Na(s)$	0
$C_3H_8O(l, \text{isopropanol})$	−318.1	$Na(g)$	107.5
$C_6H_6(l)$	49.1	$NaCl(s)$	−411.2
$C_6H_{12}O_6(s, \text{glucose})$	−1273.3	$Na_2CO_3(s)$	−1130.7
$C_{12}H_{22}O_{11}(s, \text{sucrose})$	−2226.1	$NaHCO_3(s)$	−950.8
Chlorine		*Sulfur*	
$Cl(g)$	121.3	$S_8(s, \text{rhombic})$	0
$Cl_2(g)$	0	$S_8(s, \text{monoclinic})$	0.3
$HCl(g)$	−92.3	$SO_2(g)$	−296.8
Fluorine		$SO_3(g)$	−395.7
$F(g)$	79.38	$H_2SO_4(l)$	−814.0

EXAMPLE 6.10 Standard Enthalpies of Formation

Write an equation for the formation of **(a)** $MgCO_3(s)$ and **(b)** $C_6H_{12}O_6(s)$ from their elements in their standard states. Include the value of ΔH_f° for each equation.

SOLUTION

(a) $MgCO_3(s)$

Write the equation with the elements in $MgCO_3$ in their standard states as the reactants and 1 mol of $MgCO_3$ as the product.

$$Mg(s) + C(s, \text{graphite}) + O_2(g) \longrightarrow MgCO_3(s)$$

Balance the equation and look up ΔH_f° in Appendix IIB. (Use fractional coefficients so that the product of the reaction is 1 mol of $MgCO_3$.)

$$Mg(s) + C(s, \text{graphite}) + \tfrac{3}{2}O_2(g) \longrightarrow MgCO_3(s)$$
$$\Delta H_f^\circ = -1095.8 \text{ kJ/mol}$$

(b) $C_6H_{12}O_6(s)$

Write the equation with the elements in $C_6H_{12}O_6$ in their standard states as the reactants and 1 mol of $C_6H_{12}O_6$ as the product.

$$C(s, \text{graphite}) + H_2(g) + O_2(g) \longrightarrow C_6H_{12}O_6(s)$$

Balance the equation and look up ΔH_f° in Appendix IIB.

$$6\,C(s, \text{graphite}) + 6\,H_2(g) + 3\,O_2(g) \longrightarrow C_6H_{12}O_6(s)$$
$$\Delta H_f^\circ = -1273.3 \text{ kJ/mol}$$

FOR PRACTICE 6.10

Write an equation for the formation of **(a)** $NaCl(s)$ and **(b)** $Pb(NO_3)_2(s)$ from their elements in their standard states. Include the value of ΔH_f° for each equation.

Calculating the Standard Enthalpy Change for a Reaction

We have just seen that the standard heat of formation corresponds to the *formation* of a compound from its constituent elements in their standard states.

$$\text{elements} \longrightarrow \text{compound} \quad \Delta H_f^\circ$$

Therefore, the *negative* of the standard heat of formation corresponds to the *decomposition* of a compound into its constituent elements in their standard states.

$$\text{compound} \longrightarrow \text{elements} \quad -\Delta H_f^\circ$$

We can use these two concepts—the decomposition of a compound into its elements and the formation of a compound from its elements—to calculate the enthalpy change of any reaction by mentally taking the reactants through two steps. In the first step we *decompose the reactants* into their constituent elements in their standard states; in the second step we *form the products* from the constituent elements in their standard states.

$$
\begin{array}{ll}
\text{reactants} \longrightarrow \text{elements} & \Delta H_1 = -\sum \Delta H_f^\circ(\text{reactants}) \\
\text{elements} \longrightarrow \text{products} & \Delta H_2 = +\sum \Delta H_f^\circ(\text{products}) \\
\hline
\text{reactants} \longrightarrow \text{products} & \Delta H_{rxn}^\circ = \Delta H_1 + \Delta H_2
\end{array}
$$

In these equations, $\sum$ means "the sum of" so that ΔH_1 is the sum of the negatives of the heats of formation of the reactants and ΔH_2 is the sum of the heats of formation of the products.

We can generalize this process as follows:

To calculate ΔH_{rxn}°, subtract the heats of formations of the reactants multiplied by their stoichiometric coefficients from the heats of formation of the products multiplied by their stoichiometric coefficients.

In the form of an equation,

$$\Delta H^\circ_{rxn} = \sum n_p \Delta H^\circ_f (\text{products}) - \sum n_r \Delta H^\circ_f (\text{reactants}) \qquad [6.13]$$

In this equation, n_p represents the stoichiometric coefficients of the products, n_r represents the stoichiometric coefficients of the reactants, and ΔH°_f represents the standard enthalpies of formation. Keep in mind when using this equation that elements in their standard states have $\Delta H^\circ_f = 0$. The following examples demonstrate this process.

EXAMPLE 6.11 ΔH°_{rxn} and Standard Enthalpies of Formation

Use the standard enthalpies of formation to determine ΔH°_{rxn} for the reaction:

$$4\,NH_3(g) + 5\,O_2(g) \longrightarrow 4\,NO(g) + 6\,H_2O(g)$$

SORT You are given the balanced equation and asked to find the enthalpy of reaction.

GIVEN $4\,NH_3(g) + 5\,O_2(g) \longrightarrow 4\,NO(g) + 6\,H_2O(g)$
FIND ΔH°_{rxn}

STRATEGIZE To calculate ΔH°_{rxn} from standard enthalpies of formation, subtract the heats of formation of the reactants multiplied by their stoichiometric coefficients from the heats of formation of the products multiplied by their stoichiometric coefficients.

CONCEPTUAL PLAN

$$\Delta H^\circ_{rxn} = \sum n_p \Delta H^\circ_f (\text{products}) -$$
$$\sum n_r \Delta H^\circ_f (\text{reactants})$$

SOLVE Begin by looking up (in Appendix IIB) the standard enthalpy of formation for each reactant and product. Remember that the standard enthalpy of formation of pure elements in their standard state is zero. Calculate ΔH°_{rxn} by substituting into the equation.

SOLUTION

Reactant or product	ΔH°_f (kJ/mol, from Appendix IIB)
$NH_3(g)$	−45.9
$O_2(g)$	0.0
$NO(g)$	+91.3
$H_2O(g)$	−241.8

$$
\begin{aligned}
\Delta H^\circ_{rxn} &= \sum n_p \Delta H^\circ_f(\text{products}) - \sum n_r \Delta H^\circ_f(\text{reactants}) \\
&= \left[4(\Delta H^\circ_{f,\,NO(g)}) + 6(\Delta H^\circ_{f,\,H_2O(g)})\right] - \\
&\quad\quad \left[4(\Delta H^\circ_{f,\,NH_3(g)}) + 5(\Delta H^\circ_{f,\,O_2(g)})\right] \\
&= \left[4(+91.3\,kJ) + 6(-241.8\,kJ)\right] - \\
&\quad\quad \left[4(-45.9\,kJ) + 5(0.0\,kJ)\right] \\
&= -1085.6\,kJ - (-183.6\,kJ) \\
&= -902.0\,kJ
\end{aligned}
$$

CHECK The units of the answer (kJ) are correct. The answer is negative, which means that the reaction is exothermic.

FOR PRACTICE 6.11

The thermite reaction, in which powdered aluminum reacts with iron oxide, is highly exothermic.

$$2\,Al(s) + Fe_2O_3(s) \longrightarrow Al_2O_3(s) + 2\,Fe(s)$$

Use standard enthalpies of formation to find ΔH°_{rxn} for the thermite reaction.

▲ The reaction of powdered aluminum with iron oxide, known as the thermite reaction, releases a large amount of heat.

EXAMPLE 6.12 ΔH_{rxn}° and Standard Enthalpies of Formation

A city of 100,000 people uses approximately 1.0×10^{11} kJ of energy per day. Suppose all of that energy comes from the combustion of liquid octane (C_8H_{18}) to form gaseous water and gaseous carbon dioxide. Use standard enthalpies of formation to calculate ΔH_{rxn}° for the combustion of octane and then determine how many kilograms of octane would be necessary to provide this amount of energy.

SORT You are given the amount of energy used and asked to find the mass of octane required to produce the energy.	**GIVEN** 1.0×10^{11} kJ **FIND** kg C_8H_{18}

STRATEGIZE The conceptual plan has three parts. In the first part, write a balanced equation for the combustion of octane. In the second part, calculate ΔH_{rxn}° from the ΔH_f°'s of the reactants and products. In the third part, convert from kilojoules of energy to moles of octane using the conversion factor found in step 2, and then convert from moles of octane to mass of octane using the molar mass.	**CONCEPTUAL PLAN** (1) Write balanced equation. (2)  $\Delta H_{rxn}^{\circ} = \Sigma n_p \Delta H_f^{\circ} (\text{products}) - \Sigma n_r \Delta H_f^{\circ} (\text{reactants})$ (3) 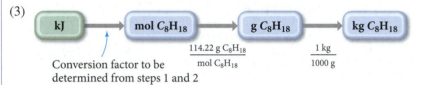 Conversion factor to be determined from steps 1 and 2 **RELATIONSHIPS USED** molar mass $C_8H_{18} = 114.22$ g/mol 1 kg = 1000 g

SOLVE Begin by writing the balanced equation for the combustion of octane. For convenience, do not clear the $\dfrac{25}{2}$ fraction in order to keep the coefficient on octane as 1.	**SOLUTION STEP 1** $C_8H_{18}(l) + \dfrac{25}{2} O_2(g) \longrightarrow 8\,CO_2(g) + 9\,H_2O(g)$

Look up (in Appendix IIB) the standard enthalpy of formation for each reactant and product and then calculate ΔH_{rxn}°.	**SOLUTION STEP 2** <table><tr><td>Reactant or product</td><td>ΔH_f° (kJ/mol from Appendix IIB)</td></tr><tr><td>$C_8H_{18}(l)$</td><td>−250.1</td></tr><tr><td>$O_2(g)$</td><td>0.0</td></tr><tr><td>$CO_2(g)$</td><td>−393.5</td></tr><tr><td>$H_2O(g)$</td><td>−241.8</td></tr></table>

$$\Delta H_{rxn}^{\circ} = \sum n_p \Delta H_f^{\circ}(\text{products}) - \sum n_r \Delta H_f^{\circ}(\text{reactants})$$

$$= \left[8(\Delta H_{f,\,CO_2(g)}^{\circ}) + 9(\Delta H_{f,\,H_2O(g)}^{\circ}) \right] - \left[1(\Delta H_{f,\,C_8H_{18}(l)}^{\circ}) + \frac{25}{2}(\Delta H_{f,\,O_2(g)}^{\circ}) \right]$$

$$= \left[8(-393.5\ \text{kJ}) + 9(-241.8\ \text{kJ}) \right] - \left[1(-250.1\ \text{kJ}) + \frac{25}{2}(0.0\ \text{kJ}) \right]$$

$$= -5324.2\ \text{kJ} - (-250.1\ \text{kJ})$$

$$= -5074.1\ \text{kJ}$$

| From steps 1 and 2 build a conversion factor between mol C_8H_{18} and kJ.

 Follow step 3 of the conceptual plan. Begin with -1.0×10^{11} kJ (since the city uses this much energy, the reaction must emit it, and therefore the sign is negative) and follow the steps to arrive at kg octane. | **SOLUTION STEP 3**
 1 mol C_8H_{18}: -5074.1 kJ

 $$-1.0 \times 10^{11} \text{ kJ} \times \frac{1 \text{ mol } C_8H_{18}}{-5074.1 \text{ kJ}} \times \frac{114.22 \text{ g } C_8H_{18}}{1 \text{ mol } C_8H_{18}} \times \frac{1 \text{ kg}}{1000 \text{ g}} = 2.3 \times 10^6 \text{ kg } C_8H_{18}$$ |

CHECK The units of the answer (kg C_8H_{18}) are correct. The answer is positive, as it should be for mass. The magnitude is fairly large, but it is expected to be so because this amount of octane is intended to provide the energy for an entire city.

FOR PRACTICE 6.12

Dry chemical hand warmers utilize the oxidation of iron to form iron oxide according to the following reaction: $4 \text{ Fe}(s) + 3 \text{ O}_2(g) \longrightarrow 2 \text{ Fe}_2\text{O}_3(s)$. Calculate ΔH°_{rxn} for this reaction and calculate how much heat is produced from a hand warmer containing 15.0 g of iron powder.

CHAPTER IN REVIEW

Key Terms

Section 6.1
thermochemistry (213)

Section 6.2
energy (213)
work (213)
heat (213)
kinetic energy (214)
thermal energy (214)
potential energy (214)
chemical energy (214)
law of conservation of
  energy (214)
system (215)
surroundings (215)

joule (J) (216)
calorie (cal) (216)
Calorie (Cal) (216)
kilowatt-hour (kWh) (216)

Section 6.3
thermodynamics (216)
first law of thermodynamics
  (216)
internal energy (E) (217)
state function (217)

Section 6.4
thermal equilibrium (222)
heat capacity (C) (222)

specific heat capacity (C_s)
  (222)
molar heat capacity (222)
pressure–volume work (225)

Section 6.5
calorimetry (227)
bomb calorimeter (227)

Section 6.6
enthalpy (H) (230)
endothermic reaction (230)
exothermic reaction (230)
enthalpy (heat) of reaction
  (ΔH_{rxn}) (232)

Section 6.7
coffee-cup calorimeter (233)

Section 6.8
Hess's law (235)

Section 6.9
standard state (238)
standard enthalpy change
  (ΔH°) (238)
standard enthalpy of formation
  (ΔH°_f) (238)

Key Concepts

The Nature of Energy and Thermodynamics (6.1, 6.2, 6.3)

▶ Energy, which is measured with the SI unit of joules (J), is the capacity to do work.
▶ Work is the result of a force acting through a distance.
▶ There are many different kinds of energy, including kinetic energy, thermal energy, potential energy, and chemical energy, a type of potential energy associated with the relative positions of electrons and nuclei in atoms and molecules.
▶ According to the first law of thermodynamics, energy can be converted from one form to another, but the total amount of energy is always conserved.
▶ The internal energy (E) of a system is the sum of all of its kinetic and potential energy. Internal energy is a state function, which

means that it depends only on the state of the system and not on the pathway by which that state is achieved.
▶ A chemical system exchanges energy with its surroundings through heat (the transfer of thermal energy caused by a temperature difference) or work. The total change in internal energy is the sum of these two quantities.

Heat and Work (6.4)

▶ We quantify heat using the equation $q = m \times C_s \times \Delta T$. In this expression, C_s is the specific heat capacity, the amount of heat required to change the temperature of 1 g of the substance by 1 °C. Compared to most substances, water has a very high heat capacity; it takes a lot of heat to change its temperature.

▶ The type of work most characteristic of chemical reactions is pressure–volume work, which occurs when a gas expands against an external pressure. Pressure–volume work can be quantified with the equation $w = -P\Delta V$.

▶ The change in internal energy (ΔE) that occurs during a chemical reaction is the sum of the heat (q) exchanged and the work (w) done: $\Delta E = q + w$.

Enthalpy (6.6)

▶ The heat evolved in a chemical reaction occurring at constant pressure is the change in enthalpy (ΔH) for the reaction. Like internal energy, enthalpy is a state function.

▶ An endothermic reaction has a positive enthalpy of reaction, whereas an exothermic reaction has a negative enthalpy of reaction.

▶ We can use the enthalpy of reaction to determine stoichiometrically the heat evolved when a specific amount of reactant reacts.

Calorimetry (6.5, 6.7)

▶ Calorimetry is a method of measuring ΔE or ΔH for a reaction.

▶ In bomb calorimetry, the reaction is carried out under conditions of constant volume, so $\Delta E = q_v$. The temperature change of the calorimeter can therefore be used to calculate ΔE for the reaction.

▶ When a reaction takes place at constant pressure, energy may be released both as heat and as work. In coffee-cup calorimetry, a reaction is carried out under atmospheric pressure in a solution, so $\Delta E = \Delta H$. We use the temperature change of the solution to calculate ΔH for the reaction.

Calculating ΔH_{rxn} (6.8, 6.9)

▶ The enthalpy of reaction (ΔH_{rxn}) can be calculated from known thermochemical data in two ways.

▶ The first way is by using the following relationships: (a) when a reaction is multiplied by a factor, ΔH_{rxn} is multiplied by the same factor; (b) when a reaction is reversed, ΔH_{rxn} changes sign; and (c) if a chemical reaction can be expressed as a sum of two or more steps, ΔH_{rxn} is the sum of the ΔH's for the individual steps (Hess's law). Together, these relationships can be used to determine the enthalpy change of an unknown reaction from reactions with known enthalpy changes.

▶ The second way to calculate ΔH_{rxn} from known thermochemical data is by using tabulated standard enthalpies of formation for the reactants and products of the reaction. These are usually tabulated for substances in their standard states, and the enthalpy of reaction is called the standard enthalpy of reaction (ΔH_{rxn}°). For any reaction, ΔH_{rxn}° is obtained by subtracting the sum of the enthalpies of formation of the reactants multiplied by their stoichiometric coefficients from the sum of the enthalpies of formation of the products multiplied by their stoichiometric coefficients.

Key Equations and Relationships

Kinetic Energy (6.1)

$$KE = \frac{1}{2}mv^2$$

Change in Internal Energy (ΔE) of a Chemical System (6.3)

$$\Delta E = E_{products} - E_{reactants}$$

Energy Flow between System and Surroundings (6.3)

$$\Delta E_{system} = -\Delta E_{surroundings}$$

Relationship between Internal Energy (ΔE), Heat (q), and Work (w) (6.3)

$$\Delta E = q + w$$

Relationship between Heat (q), Temperature (T), and Heat Capacity (C) (6.4)

$$q = C \times \Delta T$$

Relationship between Heat (q), Mass (m), Temperature (T), and Specific Heat Capacity of a Substance (C_s) (6.4)

$$q = m \times C_s \times \Delta T$$

Relationship between Work (w), Pressure (P), and Change in Volume (ΔV) (6.4)

$$w = -P\Delta V$$

Change in Internal Energy (ΔE) of System at Constant Volume (6.5)

$$\Delta E = q_v$$

Heat of a Bomb Calorimeter (q_{cal}) (6.5)

$$q_{cal} = C_{cal} \times \Delta T$$

Heat Exchange between a Calorimeter and a Reaction (6.5)

$$q_{cal} = -q_{rxn}$$

Relationship between Enthalpy (ΔH), Internal Energy (ΔE), Pressure (P), and Volume (V) (6.6)

$$\Delta H = \Delta E + P\Delta V$$

Relationship between Enthalpy of a Reaction (ΔH_{rxn}°) and the Heats of Formation (ΔH_f°) (6.9)

$$\Delta H_{rxn}^\circ = \sum n_p \Delta H_f^\circ(\text{products}) - \sum n_r \Delta H_f^\circ(\text{reactants})$$

Key Learning Objectives

EXERCISES

Problems by Topic

Note: Answers to all odd-numbered Problems, numbered in blue, can be found in Appendix III. Exercises in the Problems by Topic section are paired, with each odd-numbered problem followed by a similar even-numbered problem. Exercises in the Cumulative Problems section are also paired, but somewhat more loosely. (Challenge Problems and Conceptual Problems, because of their nature, are unpaired.)

Energy Units

1. Perform each conversion between energy units.
   a. 3.55×10^4 J to cal   b. 1025 Cal to J
   c. 355 kJ to cal   d. 125 kWh to J

2. Perform each conversion between energy units.
   a. 1.58×10^3 kJ to kcal   b. 865 cal to kJ
   c. 1.93×10^4 J to Cal   d. 1.8×10^4 kJ to kWh

3. Suppose that a person eats a diet of 2285 Calories per day. Convert this energy into each alternative unit.
   a. J   b. kJ   c. kWh

4. A frost-free refrigerator uses 685 kWh of electrical energy per year. Express this amount of energy in each alternative unit.
   a. J   b. kJ   c. Cal

Internal Energy, Heat, and Work

5. Which statement is true of the internal energy of a system and its surroundings during an energy exchange with a negative ΔE_{sys}?
   a. The internal energy of the system increases and the internal energy of the surroundings decreases.
   b. The internal energy of both the system and the surroundings increases.
   c. The internal energy of both the system and the surroundings decreases.
   d. The internal energy of the system decreases and the internal energy of the surroundings increases.

6. During an energy exchange, a chemical system absorbs energy from its surroundings. What is the sign of ΔE_{sys} for this process? Explain.

7. Identify each energy exchange as primarily heat or work and determine whether the sign of ΔE is positive or negative for the system.
   a. Sweat evaporates from skin, cooling the skin. (The evaporating sweat is the system.)
   b. A balloon expands against an external pressure. (The contents of the balloon is the system.)
   c. An aqueous chemical reaction mixture is warmed with an external flame. (The reaction mixture is the system.)

8. Identify each energy exchange as heat or work and determine whether the sign of ΔE is positive or negative for the system.
   a. A rolling billiard ball collides with another billiard ball. The first billiard ball (defined as the system) stops rolling after the collision.
   b. A book is dropped to the floor (the book is the system).
   c. A father pushes his daughter on a swing (the daughter and the swing are the system).

9. A system releases 625 kJ of heat and does 105 kJ of work on the surroundings. What is the change in internal energy of the system?

10. A system absorbs 272 kJ of heat and the surroundings do 125 kJ of work on the system. What is the change in internal energy of the system?

11. The gas in a piston (defined as the system) is warmed and absorbs 655 J of heat. The expansion performs 344 J of work on the surroundings. What is the change in internal energy for the system?

12. The air in an inflated balloon (defined as the system) is warmed over a toaster and absorbs 115 J of heat. As it expands, it does 77 kJ of work. What is the change in internal energy for the system?

Heat, Heat Capacity, and Work

13. Two identical coolers are packed for a picnic. Each cooler is packed with twenty-four 12-ounce soft drinks and 5 pounds of ice. However, the drinks that went into cooler A were refrigerated for several hours before they were packed in the cooler, while the drinks that went into cooler B were packed at room temperature. When the two coolers are opened 3 hours later, most of the ice in cooler A is still ice, while nearly all of the ice in cooler B has melted. Explain this difference.

14. A kilogram of aluminum metal and a kilogram of water are each warmed to 75 °C and placed in two identical insulated containers. One hour later, the two containers are opened and the temperature of each substance is measured. The aluminum has cooled to 35 °C while the water has cooled only to 66 °C. Explain this difference.

15. How much heat is required to warm 1.75 L of water from 25.0 °C to 100.0 °C? (Assume a density of 1.0 g/mL for the water.)

16. How much heat is required to warm 1.75 kg of sand from 25.0 °C to 100.0 °C?

17. Suppose that 25 g of each substance is initially at 27.0 °C. What is the final temperature of each substance upon absorbing 2.35 kJ of heat?
    a. gold   b. silver
    c. aluminum   d. water

18. An unknown mass of each substance, initially at 23.0 °C, absorbs 1.95×10^3 J of heat. The final temperature is recorded as indicated. Find the mass of each substance.
    a. Pyrex glass ($T_f = 55.4$ °C)   b. sand ($T_f = 62.1$ °C)
    c. ethanol ($T_f = 44.2$ °C)   d. water ($T_f = 32.4$ °C)

19. How much work (in J) is required to expand the volume of a pump from 0.0 L to 2.5 L against an external pressure of 1.1 atm?

20. During a breath, the average human lung expands by about 0.50 L. If this expansion occurs against an external pressure of 1.0 atm, how much work (in J) is done during the expansion?

21. The air within a piston equipped with a cylinder absorbs 565 J of heat and expands from an initial volume of 0.10 L to a final volume of 0.85 L against an external pressure of 1.0 atm. What is the change in internal energy of the air within the piston?

22. A gas is compressed from an initial volume of 5.55 L to a final volume of 1.22 L by an external pressure of 1.00 atm. During the compression the gas releases 124 J of heat. What is the change in internal energy of the gas?

Enthalpy and Thermochemical Stoichiometry

23. When 1 mol of a fuel is burned at constant pressure, it produces 3452 kJ of heat and does 11 kJ of work. What are the values of ΔE and ΔH for the combustion of the fuel?

24. The change in internal energy for the combustion of 1.0 mol of octane at a pressure of 1.0 atm is 5084.3 kJ. If the change in enthalpy is 5074.1 kJ, how much work is done during the combustion?

25. Determine whether each process is exothermic or endothermic and indicate the sign of ΔH.
    a. natural gas burning on a stove
    b. isopropyl alcohol evaporating from skin
    c. water condensing from steam

26. Determine whether each process is exothermic or endothermic and indicate the sign of ΔH.
    a. dry ice evaporating
    b. a sparkler burning
    c. the reaction that occurs in a chemical cold pack often used to ice athletic injuries

27. Consider the thermochemical equation for the combustion of acetone (C_3H_6O), the main ingredient in nail polish remover.

$$C_3H_6O(l) + 4\,O_2(g) \longrightarrow 3\,CO_2(g) + 3\,H_2O(g)$$
$$\Delta H^\circ_{rxn} = -1790 \text{ kJ}$$

If a bottle of nail polish remover contains 177 mL of acetone, how much heat is released by its complete combustion? The density of acetone is 0.788 g/mL.

28. What mass of natural gas (CH_4) must you burn to emit 267 kJ of heat?

$$CH_4(g) + 2\,O_2(g) \longrightarrow CO_2(g) + 2\,H_2O(g)$$
$$\Delta H^\circ_{rxn} = -802.3 \text{ kJ}$$

29. Nitromethane (CH_3NO_2) burns in air to produce significant amounts of heat.

$$2\,CH_3NO_2(l) + 3/2\,O_2(g) \longrightarrow 2\,CO_2(g) + 3\,H_2O(l) + N_2(g)$$
$$\Delta H^\circ_{rxn} = -1418 \text{ kJ}$$

How much heat is produced by the complete reaction of 5.56 kg of nitromethane?

30. Titanium reacts with iodine to form titanium(III) iodide, emitting heat.

$$2\,Ti(s) + 3\,I_2(g) \longrightarrow 2\,TiI_3(s)$$
$$\Delta H^\circ_{rxn} = -839 \text{ kJ}$$

Determine the masses of titanium and iodine that reacts if 1.55×10^3 kJ of heat is emitted by the reaction.

31. The propane fuel (C_3H_8) used in gas barbeques burns according to this thermochemical equation:

$$C_3H_8(g) + 5\,O_2(g) \longrightarrow 3\,CO_2(g) + 4\,H_2O(g)$$
$$\Delta H^\circ_{rxn} = -2217 \text{ kJ}$$

If a pork roast must absorb 1.6×10^3 kJ to fully cook, and if only 10% of the heat produced by the barbeque is actually absorbed by the roast, what mass of CO_2 is emitted into the atmosphere during the grilling of the pork roast?

32. Charcoal is primarily carbon. Determine the mass of CO_2 produced by burning enough carbon (in the form of charcoal) to produce 5.00×10^2 kJ of heat.

$$C(s) + O_2(g) \longrightarrow CO_2(g) \qquad \Delta H^\circ_{rxn} = -393.5 \text{ kJ}$$

33. A silver block, initially at 58.5 °C, is submerged into 100.0 g of water at 24.8 °C in an insulated container. The final temperature of the mixture upon reaching thermal equilibrium is 26.2 °C. What is the mass of the silver block?

34. A 32.5 g iron rod, initially at 22.7 °C, is submerged into an unknown mass of water at 63.2 °C, in an insulated container. The final temperature of the mixture upon reaching thermal equilibrium is 59.5 °C. What is the mass of the water?

35. A 31.1 g wafer of pure gold initially at 69.3 °C is submerged into 64.2 g of water at 27.8 °C in an insulated container. What is the final temperature of both substances at thermal equilibrium?

36. A 2.85 g lead weight, initially at 10.3 °C, is submerged in 7.55 g of water at 52.3 °C in an insulated container. What is the final temperature of both substances at thermal equilibrium?

37. Two substances, A and B, initially at different temperatures, come into contact and reach thermal equilibrium. The mass of substance A is 6.15 g and its initial temperature is 20.5 °C. The mass of substance B is 25.2 g and its initial temperature is 52.7 °C. The final temperature of both substances at thermal equilibrium is 46.7 °C. If the specific heat capacity of substance B is 1.17 J/g·°C, what is the specific heat capacity of substance A?

38. A 2.74 g sample of a substance suspected of being pure gold is warmed to 72.1 °C and submerged into 15.2 g of water initially at 24.7 °C. The final temperature of the mixture is 26.3 °C. What is the heat capacity of the unknown substance? Could the substance be pure gold?

Calorimetry

39. Exactly 1.5 g of a fuel is burned under conditions of constant pressure and then again under conditions of constant volume. In measurement A the reaction produces 25.9 kJ of heat, and in measurement B the reaction produces 23.3 kJ of heat. Which measurement (A or B) corresponds to conditions of constant pressure? Which one corresponds to conditions of constant volume? Explain.

40. To obtain the largest possible amount of heat from a chemical reaction in which there is a large increase in the number of moles of gas, should you carry out the reaction under conditions of constant volume or constant pressure? Explain.

41. When 0.514 g of biphenyl ($C_{12}H_{10}$) undergoes combustion in a bomb calorimeter, the temperature rises from 25.8 °C to 29.4 °C. Find ΔE_{rxn} for the combustion of biphenyl in kJ/mol biphenyl. The heat capacity of the bomb calorimeter, determined in a separate experiment, is 5.86 kJ/°C.

42. Mothballs are composed primarily of the hydrocarbon naphthalene ($C_{10}H_8$). When 1.025 g of naphthalene is burned in a bomb calorimeter, the temperature rises from 24.25 °C to 32.33 °C. Find ΔE_{rxn} for the combustion of naphthalene. The heat capacity of the calorimeter, determined in a separate experiment, is 5.11 kJ/°C.

43. Zinc metal reacts with hydrochloric acid according to this balanced equation.

$$Zn(s) + 2\,HCl(aq) \longrightarrow ZnCl_2(aq) + H_2(g)$$

When 0.103 g of $Zn(s)$ is combined with enough HCl to make 50.0 mL of solution in a coffee-cup calorimeter, all of the zinc reacts, raising the temperature of the solution from 22.5 °C to 23.7 °C. Find ΔH_{rxn} for this reaction as written. (Use 1.0 g/mL for the density of the solution and 4.18 J/g·°C as the specific heat capacity.)

44. Instant cold packs, often used to ice athletic injuries on the field, contain ammonium nitrate and water separated by a thin plastic divider. When the divider is broken, the ammonium nitrate dissolves according to the following endothermic reaction.

$$NH_4NO_3(s) \longrightarrow NH_4^+(aq) + NO_3^-(aq)$$

In order to measure the enthalpy change for this reaction, 1.25 g of NH_4NO_3 is dissolved in enough water to make 25.0 mL of solution. The initial temperature is 25.8 °C, and the final temperature (after the solid dissolves) is 21.9 °C. Calculate the change in enthalpy for the reaction in kJ. (Use 1.0 g/mL as the density of the solution and 4.18 J/g·C as the specific heat capacity.)

Quantitative Relationships Involving ΔH and Hess's Law

45. For each set of reactions, determine the value of ΔH_2 in terms of ΔH_1.

   a. $A + B \longrightarrow 2\,C \qquad \Delta H_1$
       $2\,C \longrightarrow A + B \qquad \Delta H_2 = ?$
   b. $A + \frac{1}{2}B \longrightarrow C \qquad \Delta H_1$
       $2\,A + B \longrightarrow 2\,C \qquad \Delta H_2 = ?$
   c. $A \longrightarrow B + 2\,C \qquad \Delta H_1$
       $\frac{1}{2}B + C \longrightarrow \frac{1}{2}A \qquad \Delta H_2 = ?$

46. Consider the generic reaction

$$A + 2\,B \longrightarrow C + 3\,D \quad \Delta H = 155\,kJ$$

Determine the value of ΔH for each related reaction:

   a. $3\,A + 6\,B \longrightarrow 3\,C + 9\,D$
   b. $C + 3\,D \longrightarrow A + 2\,B$
   c. $\frac{1}{2}C + \frac{3}{2}D \longrightarrow \frac{1}{2}A + B$

47. Calculate ΔH_{rxn} for the reaction

$$Fe_2O_3(s) + 3\,CO(g) \longrightarrow 2\,Fe(s) + 3\,CO_2(g)$$

Given these reactions and their ΔH's.

$$2\,Fe(s) + \tfrac{3}{2}O_2(g) \longrightarrow Fe_2O_3(s) \quad \Delta H = -824.2\,kJ$$
$$CO(g) + \tfrac{1}{2}O_2(g) \longrightarrow CO_2(g) \quad \Delta H = -282.7\,kJ$$

48. Calculate ΔH_{rxn} for the reaction

$$CaO(s) + CO_2(g) \longrightarrow CaCO_3(s)$$

Given these reactions and their ΔH's.

$$Ca(s) + CO_2(g) + \tfrac{1}{2}O_2(g) \longrightarrow CaCO_3(s) \quad \Delta H = -812.8\,kJ$$
$$2\,Ca(s) + O_2(g) \longrightarrow 2\,CaO(s) \quad \Delta H = -1269.8\,kJ$$

49. Calculate ΔH_{rxn} for the reaction

$$5\,C(s) + 6\,H_2(g) \longrightarrow C_5H_{12}(l)$$

Given these reactions and their ΔH's.

$$C_5H_{12}(l) + 8\,O_2(g) \longrightarrow 5\,CO_2(g) + 6\,H_2O(g)$$
$$\Delta H = -3505.8\,kJ$$
$$C(s) + O_2(g) \longrightarrow CO_2(g) \quad \Delta H = -393.5\,kJ$$
$$2\,H_2(g) + O_2(g) \longrightarrow 2\,H_2O(g) \quad \Delta H = -483.5\,kJ$$

50. Calculate ΔH_{rxn} for the reaction

$$CH_4(g) + 4\,Cl_2(g) \longrightarrow CCl_4(g) + 4\,HCl(g)$$

Given these reactions and and their ΔH's.

$$C(s) + 2\,H_2(g) \longrightarrow CH_4(g) \quad \Delta H = -74.6\,kJ$$
$$C(s) + 2\,Cl_2(g) \longrightarrow CCl_4(g) \quad \Delta H = -95.7\,kJ$$
$$H_2(g) + Cl_2(g) \longrightarrow 2\,HCl(g) \quad \Delta H = -184.6\,kJ$$

Enthalpies of Formation and ΔH

51. Write an equation for the formation of each compound from its elements in their standard states, and find ΔH_f° for each from Appendix IIB.

   a. $NH_3(g)$ **b.** $CO_2(g)$ **c.** $Fe_2O_3(s)$ **d.** $CH_4(g)$

52. Write an equation for the formation of each compound from its elements in their standard states, and find ΔH_f° for each from Appendix IIB.

   a. $NO_2(g)$ **b.** $MgCO_3(s)$ **c.** $C_2H_4(g)$ **d.** $CH_3OH(l)$

53. Hydrazine (N_2H_4) is a fuel used by some spacecraft. It is normally oxidized by N_2O_4 according to the equation

$$N_2H_4(l) + N_2O_4(g) \longrightarrow 2 N_2O(g) + 2 H_2O(g)$$

Calculate ΔH°_{rxn} for this reaction using standard enthalpies of formation.

54. Pentane (C_5H_{12}) is a component of gasoline that burns according to the balanced equation

$$C_5H_{12}(l) + 8 O_2(g) \longrightarrow 5 CO_2(g) + 6 H_2O(g)$$

Calculate ΔH°_{rxn} for this reaction using standard enthalpies of formation. (The standard enthalpy of formation of liquid pentane is -146.8 kJ/mol.)

55. Use standard enthalpies of formation to calculate ΔH°_{rxn} for each reaction:
a. $C_2H_4(g) + H_2(g) \longrightarrow C_2H_6(g)$
b. $CO(g) + H_2O(g) \longrightarrow H_2(g) + CO_2(g)$
c. $3 NO_2(g) + H_2O(l) \longrightarrow 2 HNO_3(aq) + NO(g)$
d. $Cr_2O_3(s) + 3 CO(g) \longrightarrow 2 Cr(s) + 3 CO_2(g)$

56. Use standard enthalpies of formation to calculate ΔH°_{rxn} for each reaction:
a. $2 H_2S(g) + 3 O_2(g) \longrightarrow 2 H_2O(l) + 2 SO_2(g)$
b. $SO_2(g) + \frac{1}{2} O_2(g) \longrightarrow SO_3(g)$
c. $C(s) + H_2O(g) \longrightarrow CO(g) + H_2(g)$
d. $N_2O_4(g) + 4 H_2(g) \longrightarrow N_2(g) + 4 H_2O(g)$

57. During photosynthesis, plants use energy from sunlight to form glucose ($C_6H_{12}O_6$) and oxygen from carbon dioxide and water. Write a balanced equation for photosynthesis and calculate ΔH°_{rxn}.

58. Ethanol can be made from the fermentation of crops and is used as a fuel additive to gasoline. Write a balanced equation for the combustion of ethanol and calculate ΔH°_{rxn}.

59. Top fuel dragsters and funny cars burn nitromethane as fuel according to the balanced combustion equation

$$2 CH_3NO_2(l) + \tfrac{3}{2} O_2(g) \longrightarrow 2 CO_2(g) + 3 H_2O(l) + N_2(g)$$

The standard enthalpy of combustion for nitromethane is -709.2 kJ/mol (-1418 kJ for the reaction as written above). Calculate the standard enthalpy of formation (ΔH°_f) for nitromethane.

60. The explosive nitroglycerin ($C_3H_5N_3O_9$) decomposes rapidly upon igntition or sudden impact according to the balanced equation

$$4 C_3H_5N_3O_9(l) \longrightarrow 12 CO_2(g) + 10 H_2O(g) + 6 N_2(g) + O_2(g)$$
$$\Delta H^\circ_{rxn} = -5678 \text{ kJ}$$

Calculate the standard enthalpy of formation (ΔH°_f) for nitroglycerin.

Cumulative Problems

61. The kinetic energy of a rolling billiard ball is given by $KE = \frac{1}{2}mv^2$. Suppose a 0.17 kg billiard ball is rolling down a pool table with an initial speed of 4.5 m/s. As it travels, it loses some of its energy as heat. The ball slows down to 3.8 m/s and then collides straight-on with a second billiard ball of equal mass. The first billiard ball completely stops and the second one rolls away with a velocity of 3.8 m/s. Assume the first billiard ball is the system and calculate w, q, and ΔE for the process.

62. A 100 W lightbulb is placed in a cylinder equipped with a moveable piston. The lightbulb is turned on for 0.015 hour, and the assembly expands from an initial volume of 0.85 L to a final volume of 5.88 L against an external pressure of 1.0 atm. Use the wattage of the lightbulb and the time it is on to calculate ΔE in joules (assume that the cylinder and lightbulb assembly is the system and assume two significant figures). Calculate w and q.

63. Sweat cools the body because evaporation is an endothermic process

$$H_2O(l) \longrightarrow H_2O(g) \quad \Delta H^\circ_{rxn} = +44.01 \text{ kJ}$$

Estimate the mass of water that must evaporate from the skin to cool the body by 0.50 °C. Assume a body mass of 95 kg and assume that the specific heat capacity of the body is 4.0 J/g · °C.

64. LP gas burns according to the exothermic reaction.

$$C_3H_8(g) + 5 O_2(g) \longrightarrow 3 CO_2(g) + 4 H_2O(g)$$
$$\Delta H^\circ_{rxn} = -2044 \text{ kJ}$$

What mass of LP gas is necessary to heat 1.5 L of water from room temperature (25.0 °C) to boiling (100.0 °C)? Assume that during heating, 15% of the heat emitted by the LP gas combustion goes to heat the water. The rest is lost as heat to the surroundings.

65. Use standard enthalpies of formation to calculate the standard change in enthalpy for the melting of ice. (The ΔH°_f for $H_2O(s)$ is -291.8 kJ/mol.) Use this value to calculate the mass of ice required to cool 355 mL of a beverage from room temperature (25.0 °C) to 0.0 °C. Assume that the specific heat capacity and density of the beverage are the same as those of water.

66. Dry ice is solid carbon dioxide. Instead of melting, solid carbon dioxide sublimes according to the equation

$$CO_2(s) \longrightarrow CO_2(g)$$

When dry ice is added to warm water, heat from the water causes the dry ice to sublime more quickly. The evaporating carbon dioxide produces a dense fog often used to create special effects. In a simple dry ice fog machine, dry ice is added to warm water in a Styrofoam cooler. The dry ice produces fog until it evaporates away, or until the water gets too cold to sublime the dry ice quickly enough. Suppose that a small Styrofoam cooler holds 15.0 liters of water heated to 85 °C. Use standard enthalpies of formation to calculate the change in enthalpy for dry ice sublimation, and calculate the mass of dry ice that should be added to the water so that the dry ice completely sublimes away when the water reaches 25 °C. Assume no heat loss to the surroundings. (The ΔH°_f for $CO_2(s)$ is -427.4 kJ/mol.)

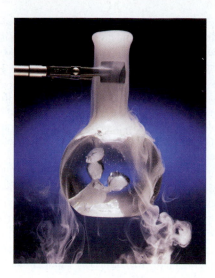

When carbon dioxide sublimes, the gaseous CO_2 is cold enough to cause water vapor in the air to condense, forming fog.

67. A 25.5 g aluminum block is warmed to 65.4 °C and plunged into an insulated beaker containing water initially at 22.2 °C. The aluminum and the water are allowed to come to thermal equilibrium. Assuming that no heat is lost, what is the final temperature of the water and aluminum?

68. If 50.0 mL of ethanol (density = 0.789 g/mL) initially at 7.0 °C is mixed with 50.0 mL of water (density = 1.0 g/mL) initially at 28.4 °C in an insulated beaker, and assuming that no heat is lost, what is the final temperature of the mixture?

69. Palmitic acid ($C_{16}H_{32}O_2$) is a dietary fat found in beef and butter. The caloric content of palmitic acid is typical of fats in general. Write a balanced equation for the complete combustion of palmitic acid and calculate the standard enthalpy of combustion. What is the caloric content of palmitic acid in Cal/g? Do the same calculation for table sugar (sucrose, $C_{12}H_{22}O_{11}$). Which dietary substance (sugar or fat) contains more Calories per gram? The standard enthalpy of formation of palmitic acid is −208 kJ/mol and that of sucrose is −2226.1 kJ/mol. (Use $H_2O(l)$ in the balanced chemical equations because the metabolism of these compounds produces liquid water.)

70. Hydrogen and methanol have both been proposed as alternatives to hydrocarbon fuels. Write balanced reactions for the complete combustion of hydrogen and methanol and use standard enthalpies of formation to calculate the amount of heat released per kilogram of the fuel. Which fuel contains the most energy in the least mass? How does the energy of these fuels compare to that of octane (C_8H_{18})?

71. Derive a relationship between ΔH and ΔE for a process in which the temperature of a fixed amount of an ideal gas changes.

72. Under certain nonstandard conditions, oxidation by $O_2(g)$ of 1 mol of $SO_2(g)$ to $SO_3(g)$ absorbs 89.5 kJ. The heat of formation of $SO_3(g)$ is −204.2 kJ under these conditions. Find the heat of formation of $SO_2(g)$.

73. One tablespoon of peanut butter has a mass of 16 g. It is combusted in a calorimeter whose heat capacity is 120.0 kJ/°C. The temperature of the calorimeter rises from 22.2 °C to 25.4 °C. Find the food caloric content of peanut butter.

74. A mixture of 2.0 mol of $H_2(g)$ and 1.0 mol of $O_2(g)$ is placed in a sealed evacuated container made of a perfect insulating material at 25 °C. The mixture is ignited with a spark and it reacts to form liquid water. Find the temperature of the water.

75. A 20.0 L volume of an ideal gas in a cylinder with a piston is at a pressure of 3.0 atm. Enough weight is suddenly removed from the piston to lower the external pressure to 1.5 atm. The gas then expands at constant temperature until its pressure is 1.5 atm. Find ΔE, ΔH, q, and w for this change in state.

76. When 10.00 g of phosphorus is burned in $O_2(g)$ to form $P_4O_{10}(s)$, enough heat is generated to raise the temperature of 2950 g of water from 18.0 °C to 38.0 °C. Calculate the heat of formation of $P_4O_{10}(s)$ under these conditions.

77. A gaseous fuel mixture contains 25.3% methane (CH_4), 38.2% ethane (C_2H_6), and the rest propane (C_3H_8) by volume. When the fuel mixture contained in a 1.55 L tank, stored at 755 mm Hg and 298 K, undergoes complete combustion, how much heat is emitted? (Assume that the water produced by the combustion is in the gaseous state.)

78. A gaseous fuel mixture stored at 745 mm Hg and 298 K contains only methane (CH_4) and propane (C_3H_8). When 11.7 L of this fuel mixture is burned, it produces 769 kJ of heat. What is the mole fraction of methane in the mixture? (Assume that the water produced by the combustion is in the gaseous state.)

Challenge Problems

79. A typical frostless refrigerator uses 655 kWh of energy per year in the form of electricity. Suppose that all of this electricity is generated at a power plant that burns coal containing 3.2% sulfur by mass and that all of the sulfur is emitted as SO_2 when the coal is burned. If all of the SO_2 goes on to react with rainwater to form H_2SO_4, what mass of H_2SO_4 is produced by the annual operation of the refrigerator? (Hint: Assume that the remaining percentage of the coal is carbon and begin by calculating $\Delta H°_{rxn}$ for the combustion of carbon.)

80. A large sport utility vehicle has a mass of 2.5×10^3 kg. Calculate the mass of CO_2 emitted into the atmosphere upon accelerating the SUV from 0.0 mph to 65.0 mph. Assume that the required energy comes from the combustion of octane with 300% efficiency. (Hint: Use $KE = \frac{1}{2}mv^2$ to calculate the kinetic energy required for the acceleration.)

81. Combustion of natural gas (primarily methane) occurs in most household heaters. The heat given off in this reaction raises the temperature of the air in the house. Assuming that all the energy given off in the reaction goes to heating up only the air in the house, determine the mass of methane required to heat the air in a house by 10.0 °C. Assume each of the following: house dimensions are 30.0 m × 30.0 m × 3.0 m; specific heat capacity of air is 30 J/K·mol; 1.00 mol of air occupies 22.4 L for all temperatures concerned.

82. When backpacking in the wilderness, hikers often boil water to sterilize it for drinking. Suppose that you are planning a backpacking trip and will need to boil 35 L of water for your group. What volume of fuel should you bring? Assume that the fuel has an average formula of C_7H_{16}; 15% of the heat generated from combustion goes to heat the water (the rest is lost to the surroundings); the density of the fuel is 0.78 g/mL; the initial temperature of the water is 25.0 °C; and the standard enthalpy of formation of C_7H_{16} is -224.4 kJ/mol.

83. An ice cube of mass 9.0 g is added to a cup of coffee, whose temperature is 90.0 °C which contains 120.0 g of liquid. Assume the specific heat capacity of the coffee is the same as that of water. The heat of fusion of ice (the heat associated with ice melting) is 6.0 kJ/mol. Find the temperature of the coffee after the ice melts.

84. Find ΔH, ΔE, q, and w for the freezing of water at -10.0 °C. The specific heat capacity of ice is 2.04 J/g · °C and its heat of fusion is -332 J/g.

85. Starting from the relationship between temperature and kinetic energy for an ideal gas, find the value of the molar heat capacity of an ideal gas when its temperature is changed at constant volume. Find its molar heat capacity when its temperature is changed at constant pressure.

86. An amount of an ideal gas expands from 12.0 L to 24.0 L at a constant pressure of 1.0 atm. Then the gas is cooled at a constant volume of 24.0 L back to its original temperature. Then it contracts back to its original volume. Find the total heat flow for the entire process.

87. The heat of vaporization of water at 373 K is 40.7 kJ/mol. Find q, w, ΔE, and ΔH for the evaporation of 454 g of water at this temperature.

88. Find ΔE, ΔH, q, and w for the change in state of 1.0 mol $H_2O(l)$ at 80.0 °C to $H_2O(g)$ at 110.0 °C. The heat capacity of $H_2O(l) = 75.3$ J/mol K, the heat capacity of $H_2O(g) = 25.0$ J/mol K, and the heat of vaporization of H_2O is 40.7×10^3 J/mol at 100 °C.

Conceptual Problems

89. Which statement is true of the internal energy of the system and its surroundings following a process in which $\Delta E_{sys} = +65$ kJ. Explain.
   a. The system and the surroundings both lose 65 kJ of energy.
   b. The system and the surroundings both gain 65 kJ of energy.
   c. The system loses 65 kJ of energy and the surroundings gain 65 kJ of energy.
   d. The system gains 65 kJ of energy and the surroundings lose 65 kJ of energy.

90. The internal energy of an ideal gas depends only on its temperature. Which expression is true of an isothermal (constant-temperature) expansion of an ideal gas against a constant external pressure? Explain.
   a. ΔE is positive   b. w is positive
   c. q is positive   d. ΔE is negative

91. Which expression describes the heat evolved in a chemical reaction when the reaction is carried out at constant pressure? Explain.
   a. $\Delta E - w$ b. ΔE c. $\Delta E - q$

Answers to Conceptual Connections

System and Surroundings

6.1 The correct answer is (a). When ΔE_{sys} is negative, energy flows out of the system and into the surroundings. The energy increase in the surroundings must exactly match the decrease in the system.

Heat and Work

6.2 (a) heat, sign is positive (b) work, sign is positive (c) heat, sign is negative

The Heat Capacity of Water

6.3 Bring the water because it has the higher heat capacity and will therefore release more heat as it cools.

Thermal Energy Transfer

6.4 (c) The specific heat capacity of substance B is twice that of A, but since the mass of B is half that of A, the quantity m × C_s will be identical for both substances so that the final temperature is exactly midway between the two initial temperatures.

The Difference between ΔH and ΔE

6.5 ΔH represents only the heat exchanged; therefore $\Delta H = -2658$ kJ. ΔE represents the heat *and work* exchanged; therefore $\Delta E = -2661$ kJ. The signs of both ΔH and ΔE are negative because heat and work are flowing out of the system and into the surroundings. Notice that the values of ΔH and ΔE are similar in magnitude, as is the case for many chemical reactions.

Exothermic and Endothermic Reactions

6.6 An endothermic reaction feels cold to the touch because the reaction (acting here as the system) absorbs heat from the surroundings. When you touch the vessel in which the reaction occurs, you, being part of the surroundings, lose heat to the system (the reaction), which results in the feeling of cold. The heat absorbed by the reaction is not used to increase its temperature, but rather becomes potential energy stored in chemical bonds.

Constant-Pressure versus Constant-Volume Calorimetry

6.7 The value of q_{rxn} with the greater magnitude (-12.5 kJ) must have come from the bomb calorimeter. Recall that $\Delta E_{rxn} = q_{rxn} + w_{rxn}$. In a bomb calorimeter, the energy change that occurs in the course of the reaction all takes the form of heat (q). In a coffee-cup calorimeter, the amount of energy released as heat may be smaller because some of the energy may be used to do work (w).

The Quantum-Mechanical Model of the Atom

Anyone who is not shocked by quantum mechanics has not understood it.
—Neils Bohr (1885–1962)

The thought experiment known as Schrödinger's cat was intended to show that the strangeness of the quantum world does not transfer to the macroscopic world.

THE EARLY PART OF THE TWENTIETH century brought changes that revolutionized how we think about physical reality, especially in the atomic realm. Before that time, all descriptions of the behavior of matter had been deterministic—the present set of conditions completely determining the future. Quantum mechanics changed that. This new theory suggested that for subatomic particles—electrons, neutrons, and protons—the present does NOT completely determine the future. For example, if you shoot one electron down a path and measure where it lands, a second electron shot down the same path under the same conditions will not necessarily follow the same course but instead will most likely land in a different place!

Quantum-mechanical theory was developed by several unusually gifted scientists including Albert Einstein, Neils Bohr, Louis de Broglie, Max Planck, Werner Heisenberg, P. A. M. Dirac, and Erwin Schrödinger. These scientists did not necessarily feel comfortable with their own theory. Bohr said, "Anyone who is not shocked by quantum mechanics has not understood it." Schrödinger wrote, "I don't like it, and I'm sorry I ever had anything to do with it." Albert Einstein disbelieved the very theory he helped create, stating, "God does not play dice with the universe." In fact, Einstein attempted to disprove quantum mechanics—without success—until he died. However, quantum mechanics was able to account for fundamental observations, including the very stability of atoms, which could not be understood within the framework of classical physics. Today, quantum mechanics forms the foundation of chemistry—explaining, for example, the periodic table and the behavior of the elements in chemical bonding—as well as providing the practical basis for lasers, computers, and countless other applications.

7.1   Schrödinger's Cat

Atoms and the particles that compose them are unimaginably small. Electrons have a mass of less than a trillionth of a trillionth of a gram, and a size so small that it is immeasurable. Electrons are *small* in the absolute sense of the word—they are among the smallest particles that make up matter. And yet, as we have seen, an atom's electrons determine many of its chemical and physical properties. If we are to understand these properties, we must try to understand electrons.

In the early 20th century, scientists discovered that the *absolutely small* (or *quantum*) world of the electron behaves differently than the *large* (or *macroscopic*) world that we are used to observing. Chief among these differences is the idea that, when unobserved, *absolutely small particles like electrons can simultaneously be in two different states at the same time*. For example, through a process called radioactive decay (see Chapter 19) an atom can emit small (that is, *absolutely* small) energetic particles from its nucleus. In the macroscopic world, something either emits an energetic particle or it doesn't. In the quantum world, however, the unobserved atom can be in a state in which it is doing both—emitting the particle and not emitting the particle—simultaneously. At first, this seems absurd. The absurdity resolves itself, however, upon observation. When we set out to measure the emitted particle, the act of measurement actually forces the atom into one state or other.

Early 20th century physicists struggled with this idea. Austrian physicist Erwin Schrödinger, in an attempt to demonstrate that this quantum strangeness could never transfer itself to the macroscopic world, published a paper in 1935 that contained a thought experiment about a cat, now known as Schrödinger's cat. In the thought experiment, the cat is put into a steel chamber that contains radioactive atoms such as the one described in the previous paragraph. The chamber is equipped with a mechanism that, upon the emission of an energetic particle by one of the radioactive atoms, causes a hammer to break a flask of hydrocyanic acid, a poison. If the flask breaks, the poison is released and the cat dies.

Now here comes the absurdity: if the steel chamber is closed, the whole system remains unobserved, and the radioactive atom is in a state in which it has emitted the particle and not emitted the particle (with equal probability). Therefore the cat is both dead and undead. Schrödinger put it this way: "[the steel chamber would have] *in it the living and dead cat (pardon the expression) mixed or smeared out in equal parts*." When the chamber is opened, the act of observation forces the entire system into one state or the other: the cat is either dead or alive, not both. However, while unobserved, the cat is both dead and alive. The absurdity of the both dead and not dead cat in Schrödinger's thought experiment was meant to demonstrate how quantum strangeness does not transfer to the macroscopic world.

In this chapter, we examine the **quantum-mechanical model** of the atom, a model that explains the strange behavior of electrons. In particular, we focus on how the model describes electrons as they exist within atoms, and how those electrons determine the chemical and physical properties of elements. We have already learned much about those properties. We know, for example, that some elements are metals and that others are nonmetals. We know that the noble gases are chemically inert and that the alkali metals are chemically reactive. We know that sodium tends to form 1+ ions and that fluorine tends to form 1− ions. But we have not explored *why*. The quantum-mechanical model explains why. In doing so, it explains the modern periodic table and provides the basis for our understanding of chemical bonding.

7.2   The Nature of Light

Before we explore electrons and their behavior within the atom, we must understand a few things about light. As quantum mechanics developed, light was (surprisingly) found to have many characteristics in common with electrons. Chief among these is the *wave–particle duality* of light. Certain properties of light are best described by thinking of it as a wave, while other properties are best described by thinking of it as a particle. In this chapter, we first explore the wave behavior of light, and then its particle behavior. We then turn to electrons to see how they display the same wave–particle duality.

The Wave Nature of Light

Light is **electromagnetic radiation**, a type of energy embodied in oscillating electric and magnetic fields. A *magnetic field* is a region of space where a magnetic particle experiences a force (think of the space around a magnet). An *electric field* is a region of space where an electrically charged particle experiences a force. Electromagnetic radiation can be described as a wave composed of oscillating, mutually perpendicular electric and magnetic fields propagating through space, as shown in **Figure 7.1▼**. In a vacuum, these waves move at a constant speed of 3.00×10^8 m/s (186,000 mi/s)—fast enough to circle Earth in one-seventh of a second. This great speed explains the delay between the moment when you see a firework in the sky and the moment when you hear the sound of its explosion. The light from the exploding firework reaches your eye almost instantaneously. The sound, traveling much more slowly (340 m/s), takes longer. The same thing happens in a thunderstorm—you see the flash of the lightning immediately, but the sound of the thunder takes a few seconds to reach you.

An electromagnetic wave, like all waves, can be characterized by its *amplitude* and its *wavelength*. In the graphical representation shown below, the **amplitude** of the wave is the vertical height of a crest (or depth of a trough). The amplitude of the electric and magnetic field waves in light is related to the *intensity* or brightness of the light—the greater the amplitude, the greater the intensity. The **wavelength (λ)** of the wave is the distance in space between adjacent crests (or any two analogous points) and is measured in units of distance such as the meter, micrometer, or nanometer.

The symbol λ is the Greek letter lambda, pronounced "lamb-duh."

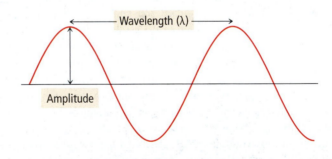

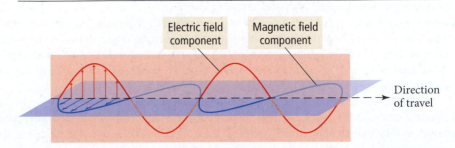

► FIGURE 7.1 Electromagnetic Radiation   Electromagnetic radiation can be described as a wave composed of oscillating electric and magnetic fields. The fields oscillate in perpendicular planes.

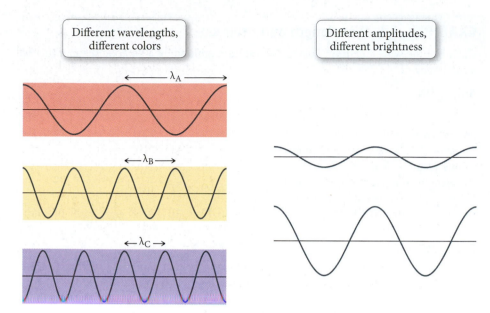

Different wavelengths, different colors

Different amplitudes, different brightness

▲ **FIGURE 7.2 Wavelength and Amplitude** Wavelength and amplitude are independent properties. The wavelength of light determines its color. The amplitude, or intensity, determines its brightness.

Wavelength and amplitude are both related to the amount of energy carried by a wave. Imagine trying to swim out from a shore that is being pounded by waves. Greater amplitude (higher waves) or shorter wavelength (more closely spaced, and thus steeper, waves) make the swim more difficult. Notice also that amplitude and wavelength can vary independently of one another, as shown in **Figure 7.2▲**. A wave can have a large amplitude and a long wavelength, or a small amplitude and a short wavelength. The most energetic waves have large amplitudes and short wavelengths.

Like all waves, light is also characterized by its **frequency (ν)**, the number of cycles (or wave crests) that pass through a stationary point in a given period of time. The units of frequency are cycles per second (cycle/s) or simply s^{-1}. An equivalent unit of frequency is the hertz (Hz), defined as 1 cycle/s. The frequency of a wave is directly proportional to the speed at which the wave is traveling—the faster the wave, the more crests will pass a fixed location per unit time. Frequency is also *inversely* proportional to the wavelength (λ)—the farther apart the crests, the fewer that pass a fixed location per unit time. For light, therefore, we write

$$\nu = \frac{c}{\lambda}$$

[7.1]

where the speed of light, c, and the wavelength, λ, are expressed using the same unit of distance. Therefore, wavelength and frequency represent different ways of specifying the same information—if we know one, we can readily calculate the other.

For *visible light*—light that can be seen by the human eye—wavelength (or, alternatively, frequency) determines color. White light, as produced by the sun or by a lightbulb, contains a spectrum of wavelengths and therefore a spectrum of colors. We see these colors—red, orange, yellow, green, blue, indigo, and violet—in a rainbow or when white light is passed through a prism (**Figure 7.3▶**). Red light, with a wavelength of about 750 nanometers (nm), has the longest wavelength of visible light; violet light, with a wavelength of about 400 nm, has the shortest. The presence of a variety of wavelengths in white light is responsible for the colors that we perceive. When a substance absorbs some colors while reflecting others, it appears colored. For example, a red shirt appears red because it reflects predominantly red light while absorbing most other colors (**Figure 7.4▶**). Our eyes see only the reflected light, making the shirt appear red.

▲ **FIGURE 7.3 Components of White Light** White light can be decomposed into its constituent colors, each with a different wavelength, by passing it through a prism. The array of colors makes up the spectrum of visible light.

The symbol ν is the Greek letter nu, pronounced "noo."

$nano = 10^{-9}$

▲ **FIGURE 7.4 The Color of an Object** A red shirt is red is because it reflects predominantly red light while absorbing most other colors.

EXAMPLE 7.1 Wavelength and Frequency

Calculate the wavelength (in nm) of the red light emitted by a barcode scanner that has a frequency of $4.62 \times 10^{14}\,s^{-1}$.

SOLUTION

You are given the frequency of the light and asked to find its wavelength. Use Equation 7.1, which relates frequency to wavelength. You can convert the wavelength from meters to nanometers by using the conversion factor between the two (1 nm = 10^{-9} m).	$\nu = \dfrac{c}{\lambda}$ $\lambda = \dfrac{c}{\nu} = \dfrac{3.00 \times 10^8\,\text{m}/\cancel{s}}{4.62 \times 10^{14}\,1/\cancel{s}}$ $= 6.49 \times 10^{-7}\,\text{m}$ $= 6.49 \times 10^{-7}\,\cancel{\text{m}} \times \dfrac{1\,\text{nm}}{10^{-9}\,\cancel{\text{m}}} = 649\,\text{nm}$

FOR PRACTICE 7.1

A laser used to dazzle the audience in a rock concert emits green light with a wavelength of 515 nm. Calculate the frequency of the light.

The Electromagnetic Spectrum

Visible light makes up only a tiny portion of the entire **electromagnetic spectrum**, which includes all known wavelengths of electromagnetic radiation. **Figure 7.5▼** shows the main regions of the electromagnetic spectrum, ranging in wavelength from 10^{-15} m (gamma rays) to 10^5 m (radio waves).

As we noted previously, short-wavelength light inherently has greater energy than long-wavelength light. Therefore, the most energetic forms of electromagnetic radiation have the shortest wavelengths. The form of electromagnetic radiation with the shortest wavelength is the **gamma (γ) ray**. Gamma rays are produced by the sun, other stars, and certain unstable atomic nuclei on Earth. Human exposure to gamma rays is dangerous because the high energy of gamma rays can damage biological molecules.

Next on the electromagnetic spectrum, with longer wavelengths than gamma rays, are **X-rays**, familiar to us from their medical use. X-rays pass through many substances that block visible light and are therefore used to image bones and internal organs. Like gamma rays, X-rays are sufficiently energetic to damage biological molecules. While several yearly exposures to X-rays are relatively harmless, excessive exposure to X-rays increases cancer risk.

Sandwiched between X-rays and visible light in the electromagnetic spectrum is **ultraviolet (UV) radiation**, most familiar to us as the component of sunlight that produces a

The Electromagnetic Spectrum

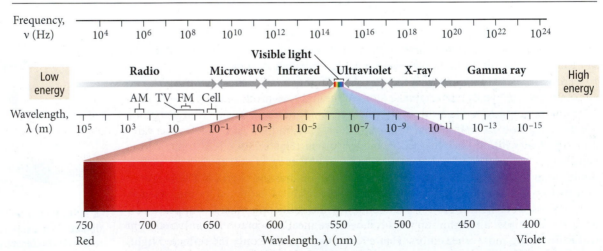

▲ **FIGURE 7.5 The Electromagnetic Spectrum** The right side of the spectrum consists of high-energy, high-frequency, short-wavelength radiation. The left side consists of low-energy, low-frequency, long-wavelength radiation. Visible light constitutes a small segment in the middle.

sunburn or suntan. While not as energetic as gamma rays or X-rays, ultraviolet light still carries enough energy to damage biological molecules. Excessive exposure to ultraviolet light increases the risk of skin cancer and cataracts and causes premature wrinkling of the skin.

Next on the spectrum is **visible light**, ranging from violet (shorter wavelength, higher energy) to red (longer wavelength, lower energy). Visible light—as long as the intensity is not too high—does not carry enough energy to damage biological molecules. It does, however, cause certain molecules in our eyes to change their shape, sending a signal to our brains that results in vision. Beyond visible light lies **infrared (IR) radiation**. The heat you feel when you place your hand near a hot object is infrared radiation. All warm objects, including human bodies, emit infrared light. Although infrared light is invisible to our eyes, infrared sensors can detect it and are used in night vision technology to "see" in the dark.

At longer wavelengths still, are **microwaves**, used for radar and in microwave ovens. Although microwave radiation has longer wavelengths and therefore lower energies than visible or infrared light, it is efficiently absorbed by water and can therefore heat substances that contain water. The longest wavelengths are those of **radio waves**, which are used to transmit the signals responsible for AM and FM radio, cellular telephones, television, and other forms of communication.

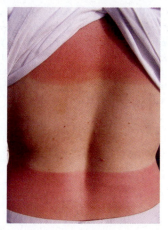

▲ Suntans and sunburns are produced by ultraviolet light from the sun.

Interference and Diffraction

Waves, including electromagnetic waves, interact with each other in a characteristic way called **interference**: they can cancel each other out or build each other up, depending on their alignment upon interaction. For example, if waves of equal amplitude from two sources are *in phase* when they interact—that is, they align with overlapping crests—a wave with twice the amplitude results. This is called **constructive interference**.

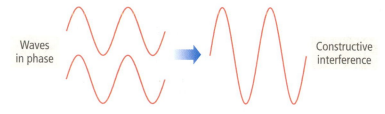

Waves in phase → Constructive interference

On the other hand, if the waves are completely *out of phase*—that is, they align so that the crest from one source overlaps the trough from the other source—the waves cancel by **destructive interference**.

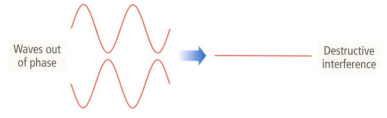

Waves out of phase → Destructive interference

▲ Warm objects emit infrared light, which is invisible to the eye but can be captured on film or by detectors to produce an infrared photograph.
(© Sierra Pacific Innovations. All rights reserved. SPI CORP, www.x20.org.)

▲ When a reflected wave meets an incoming wave near the shore, the two waves interfere constructively for an instant, producing a large amplitude spike.

Understanding interference in waves is critical to understanding the wave nature of the electron, as we will soon see.

When a wave encounters an obstacle or a slit that is comparable in size to its wavelength, it bends around it—a phenomenon called **diffraction** (**Figure 7.6▶**). The diffraction of light through two slits separated by a distance comparable to the wavelength of the light results in an *interference pattern*, as shown in **Figure 7.7▶**. Each slit acts as a new wave source, and the two new waves interfere with each other. The resulting pattern consists of a series of bright and dark lines that can be viewed on a screen (or recorded on a film) placed at a short distance behind the slits. At the center of the screen, the two waves travel equal distances and interfere constructively to produce a bright line. However, a small distance away from the center in either direction, the two waves travel slightly different distances, so that they are out of phase. At the point where the difference in distance is one-half of a wavelength, the interference is destructive and a dark line appears on the screen. Moving a bit further away from the center produces constructive interference again because the difference between the paths is one whole wavelength. The end result is the interference pattern shown. Notice that interference results from the ability of a wave to diffract through the two slits—this is an inherent property of waves.

▶ **FIGURE 7.6 Diffraction** This view of waves from above shows how they are bent, or diffracted, when they encounter an obstacle or slit with a size comparable to their wavelength. When a wave passes through a small opening, it spreads out. Particles, by contrast, do not diffract; they simply pass through the opening.

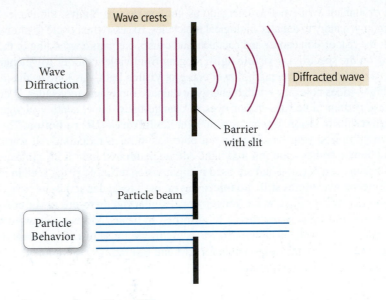

Interference from Two Slits

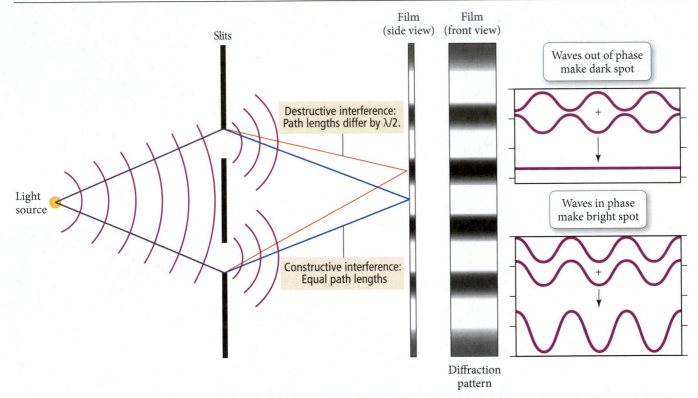

▲ **FIGURE 7.7 Interference from Two Slits** When a beam of light passes through two small slits, the two resulting waves interfere with each other. Whether the interference is constructive or destructive at any given point depends on the difference in the path lengths traveled by the waves. The resulting interference pattern can be viewed as a series of bright and dark lines on a screen.

The Particle Nature of Light

The term *classical*, as in classical electromagnetic theory or classical mechanics, refers to descriptions of matter and energy before the advent of quantum mechanics.

Prior to the early 1900s, and especially after the discovery of the diffraction of light, light was thought to be purely a wave phenomenon. Its behavior was described adequately by classical electromagnetic theory, which treated the electric and magnetic fields that constitute light as waves propagating through space. However, a number of discoveries brought the classical view into question. Chief among those for light was the *photoelectric effect*.

The **photoelectric effect** was the observation that many metals eject electrons when light shines upon them, as shown in **Figure 7.8▶**. The light dislodges an electron from the metal when it shines on the metal, much like an ocean wave might dislodge a rock from a cliff when it breaks on a cliff. Classical electromagnetic theory attributed this effect to the

The Photoelectric Effect

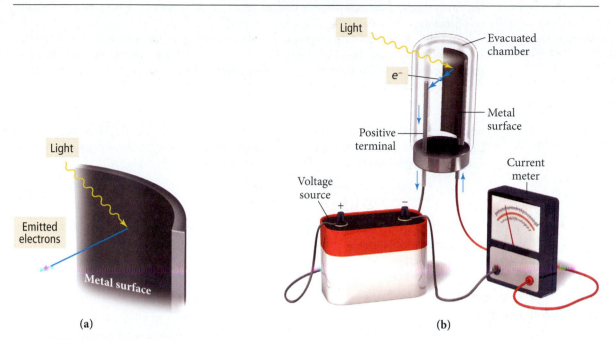

(a)   (b)

▲ **FIGURE 7.8 The Photoelectric Effect**   (a) When sufficiently energetic light shines on a metal surface, electrons are emitted. (b) The emitted electrons can be measured as an electrical current.

transfer of energy from the light to the electron in the metal, dislodging the electron. In this description, changing either the wavelength (color) or the amplitude (intensity) of the light should affect the ejection of electrons (just as changing the wavelength or intensity of the ocean wave would affect the dislodging of rocks from the cliff). In other words, according to the classical description, the rate at which electrons were ejected from a metal due to the photoelectric effect could be increased by using either light of shorter wavelength or light of higher intensity (brighter light). If a dim light were used, the classical description predicted that there would be a *lag time* between the initial shining of the light and the subsequent ejection of an electron. The lag time was the minimum amount of time required for the dim light to transfer sufficient energy to the electron to dislodge it (much as there would be a lag time for small waves to finally dislodge a rock from a cliff).

However, when observed in the laboratory, it was found that high-frequency, low-intensity light produced electrons without the predicted lag time. Furthermore, experiments showed that the light used to eject electrons in the photoelectric effect had a *threshold frequency*, below which no electrons were ejected from the metal, no matter how long or how brightly the light shone on the metal. In other words, low-frequency (long-wavelength) light would not eject electrons from a metal regardless of its intensity or its duration. But high-frequency (short-wavelength) light would eject electrons, even if its intensity were low. *This is like observing that long wavelength waves crashing on a cliff would not dislodge rocks even if their amplitude (wave height) was large, but that short wavelength waves crashing on the same cliff would dislodge rocks even if their amplitude was small.* **Figure 7.9▼** is a graph of the

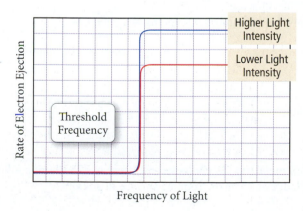

◀ **FIGURE 7.9 The Photoelectric Effect**   A plot of the electron ejection rate versus frequency of light for the photoelectric effect. Electrons are only ejected when the energy of a photon exceeds the energy with which an electron is held to the metal. The frequency at which this occurs is called the *threshold frequency*.

rate of electron ejection from the metal versus the frequency of light used. Notice that increasing the intensity of the light does not change the threshold frequency. What could explain this odd behavior?

In 1905, Albert Einstein proposed a bold explanation of this observation: *light energy must come in packets*. In other words, light was *not* like ocean waves, but more like particles. According to Einstein, the amount of energy (E) in a light packet depends on its frequency (ν) according to the equation:

$$E = h\nu \qquad [7.2]$$

where h, called *Planck's constant*, has the value $h = 6.626 \times 10^{-34}$ J·s. A *packet* of light is called a **photon** or a **quantum** of light. Since $\nu = c/\lambda$, the energy of a photon can also be expressed in terms of wavelength as follows:

$$E = \frac{hc}{\lambda} \qquad [7.3]$$

> Einstein was not the first to suggest that energy was quantized. Max Planck used the idea in 1900 to account for certain characteristics of radiation from hot bodies. However, he did not suggest that light actually traveled in discrete packets.

> The energy of a photon is directly proportional to its frequency.

> The energy of a photon is inversely proportional to its wavelength.

Unlike classical electromagnetic theory, in which light was viewed purely as a wave whose intensity was *continuously variable*, Einstein suggested that light was *lumpy*. From this perspective, a beam of light is *not* a wave propagating through space, but a shower of particles, each with energy $h\nu$.

EXAMPLE 7.2 Photon Energy

A nitrogen gas laser pulse with a wavelength of 337 nm contains 3.83 mJ of energy. How many photons does it contain?

SORT You are given the wavelength and total energy of a light pulse and asked to find the number of photons it contains.	**GIVEN** $E_{pulse} = 3.83$ mJ $\lambda = 337$ nm **FIND** number of photons

STRATEGIZE In the first part of the conceptual plan, calculate the energy of an individual photon from its wavelength.

CONCEPTUAL PLAN

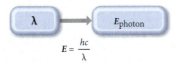

$$E = \frac{hc}{\lambda}$$

In the second part, divide the total energy of the pulse by the energy of a photon to determine the number of photons in the pulse.

$$\frac{E_{pulse}}{E_{photon}} = \text{number of photons}$$

RELATIONSHIPS USED $E = hc/\lambda$ (Equation 7.3)

SOLVE To execute the first part of the conceptual plan, convert the wavelength to meters and substitute it into the equation to calculate the energy of a 337-nm photon.

SOLUTION

$$\lambda = 337 \text{ nm} \times \frac{10^{-9} \text{ m}}{1 \text{ nm}} = 3.37 \times 10^{-7} \text{ m}$$

$$E_{photon} = \frac{hc}{\lambda} = \frac{(6.626 \times 10^{-34} \text{ J·s})\left(3.00 \times 10^{8} \frac{\text{m}}{\text{s}}\right)}{3.37 \times 10^{-7} \text{ m}}$$
$$= 5.8985 \times 10^{-19} \text{ J}$$

To execute the second part of the conceptual plan, convert the energy of the pulse from mJ to J. Then divide the energy of the pulse by the energy of a photon to obtain the number of photons.

$$3.83 \text{ mJ} \times \frac{10^{-3} \text{ J}}{1 \text{ mJ}} = 3.83 \times 10^{-3} \text{ J}$$

$$\text{number of photons} = \frac{E_{pulse}}{E_{photon}} = \frac{3.83 \times 10^{-3} \text{ J}}{5.8985 \times 10^{-19} \text{ J}}$$
$$= 6.49 \times 10^{15} \text{ photons}$$

FOR PRACTICE 7.2

A 100-watt lightbulb radiates energy at a rate of 100 J/s. (The watt, a unit of power, or energy over time, is defined as 1 J/s.) If all of the light emitted has a wavelength of 525 nm, how many photons are emitted per second? (Assume three significant figures in this calculation.)

FOR MORE PRACTICE 7.2

The energy required to dislodge electrons from sodium metal via the photoelectric effect is 275 kJ/mol. What wavelength (in nm) of light has sufficient energy per photon to dislodge an electron from the surface of sodium?

EXAMPLE 7.3 Wavelength, Energy, and Frequency

Arrange these three types of electromagnetic radiation—visible light, X-rays, and microwaves—in order of increasing:

(a) wavelength **(b)** frequency **(c)** energy per photon

SOLUTION

Examine Figure 7.5 and note that X-rays have the shortest wavelength, followed by visible light and then microwaves.	**(a)** wavelength X-rays < visible < microwaves
Since frequency and wavelength are inversely proportional—the longer the wavelength the shorter the frequency—the ordering with respect to frequency is the reverse order with respect to wavelength.	**(b)** frequency microwaves < visible < X-rays
Energy per photon decreases with increasing wavelength, but increases with increasing frequency; therefore, the ordering with respect to energy per photon is the same as for frequency.	**(c)** energy per photon microwaves < visible < X-rays

FOR PRACTICE 7.3

Arrange these colors of visible light—green, red, and blue—in order of increasing:

(a) wavelength **(b)** frequency **(c)** energy per photon

Einstein's idea that light was *quantized* elegantly explains the photoelectric effect. The emission of electrons from the metal depends on whether or not a single photon has sufficient energy (as given by $h\nu$) to dislodge a single electron. For an electron bound to the metal with binding energy ϕ, the threshold frequency is reached when the energy of the photon is equal to ϕ.

The symbol ϕ is the Greek letter phi, pronounced "fee."

Threshold frequency condition

$$h\nu = \phi$$

Energy of   Binding energy of
photon   emitted electron

Low-frequency light will not eject electrons because no single photon has the minimum energy necessary to dislodge the electron. Increasing the *intensity* of low-frequency light simply increases the number of low-energy photons, but does not produce any single photon with greater energy. In contrast, increasing the *frequency* of the light, even at low intensity, increases the energy of each photon, allowing the photons to dislodge electrons with no lag time.

As the frequency of the light is increased past the threshold frequency, the excess energy of the photon (beyond what is needed to dislodge the electron) is transferred to the electron in the form of kinetic energy. The kinetic energy (KE) of the ejected electron, therefore, is the difference between the energy of the photon ($h\nu$) and the binding energy of the electron, as given by the equation

$$KE = h\nu - \phi$$

Although the quantization of light explained the photoelectric effect, the wave explanation of light continued to have explanatory power as well, depending on the circumstances of the particular observation. So the principle that slowly emerged (albeit with some measure of resistance) is what we now call the *wave–particle duality of light*. Sometimes light appears to behave like a wave, at other times like a particle. Which behavior you observe depends on the particular experiment performed.

Conceptual Connection 7.1 The Photoelectric Effect

Light of three different wavelengths—325 nm, 455 nm, and 632 nm—was shone on a metal surface. The observations for each wavelength, labeled A, B, and C, were as follows:

Observation A: No photoelectrons were observed.

Observation B: Photoelectrons with a kinetic energy of 155 kJ/mol were observed.

Observation C: Photoelectrons with a kinetic energy of 51 kJ/mol were observed.

Which observation corresponds to which wavelength of light?

7.3   Atomic Spectroscopy and the Bohr Model

The discovery of the particle nature of light began to break down the division that existed in nineteenth-century physics between electromagnetic radiation, which was thought of as a wave phenomenon, and the small particles (protons, neutrons, and electrons) that compose atoms, which were thought to follow Newton's laws of motion (see Section 7.4). Just as the photoelectric effect suggested the particle nature of light, so certain observations of atoms began to suggest a wave nature for particles. The most important of these came from *atomic spectroscopy*, the study of the electromagnetic radiation absorbed and emitted by atoms.

When an atom absorbs energy—in the form of heat, light, or electricity—it often reemits that energy as light. For example, a neon sign is composed of one or more glass tubes filled with neon gas. When an electric current is passed through the tube, the neon atoms absorb some of the electrical energy and reemit it as the familiar red light of a neon sign. If the atoms in the tube are not neon atoms but those of a different gas, the emitted light is a different color. Atoms of each element emit light of a characteristic color. Mercury atoms, for example, emit light that appears blue, helium atoms emit light that appears violet, and hydrogen atoms emit light that appears reddish (**Figure 7.10◄**).

Closer investigation of the light emitted by atoms reveals that it contains several distinct wavelengths. Just as the white light from a lightbulb can be separated into its constituent wavelengths by passing it through a prism, so can the light emitted by an element when it is heated, as shown in **Figure 7.11▶**. The result is a series of bright lines called an **emission spectrum**. The emission spectrum of a particular element is always the same—it consists of the same bright lines at the same characteristic wavelengths—and can be used to identify the element. For example, light arriving from a distant star contains the emission spectra of the elements that compose the star. Analysis of the light allows us to identify the elements present in the star.

Notice the differences between a white light spectrum and the emission spectra of hydrogen, helium, and barium. The white light spectrum is *continuous*; there are no sudden interruptions in the intensity of the light as a function of wavelength—it consists of light of all wavelengths. The emission spectra of hydrogen, helium, and barium, however, are not continuous—they consist of bright lines at specific wavelengths, with complete darkness in between. That is, only certain discrete wavelengths of light are present. Classical physics could not explain why these spectra consisted of discrete lines. In fact, according to classical physics, an atom composed of an electron orbiting a nucleus should emit a continuous white light spectrum. Even more problematic, the electron should lose energy as it emits the light, and spiral into the nucleus.

Johannes Rydberg, a Swedish mathematician, analyzed many atomic spectra and developed an equation (shown in the margin) that predicted the wavelengths of the hydrogen emission spectrum. However, his equation gave little insight into *why* atomic spectra were discrete, *why* atoms were stable, or *why* his equation worked.

The Danish physicist Neils Bohr (1885–1962) attempted to develop a model for the atom that explained atomic spectra. In his model, electrons travel around the nucleus in circular orbits (similar to those of the planets around the sun). However, in

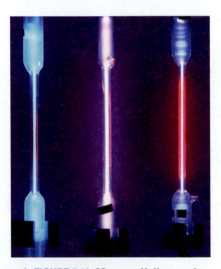

▲ The familiar red light from a neon sign is emitted by neon atoms that have absorbed electrical energy, which they reemit as visible radiation.

Remember that the color of visible light is determined by its wavelength.

▲ **FIGURE 7.10 Mercury, Helium, and Hydrogen**   Each element emits a characteristic color.

The Rydberg equation is $1/\lambda = R(1/m^2 - 1/n^2)$, where R is the Rydberg constant ($1.097 \times 10^7 \text{ m}^{-1}$) and m and n are integers.

Emission Spectra

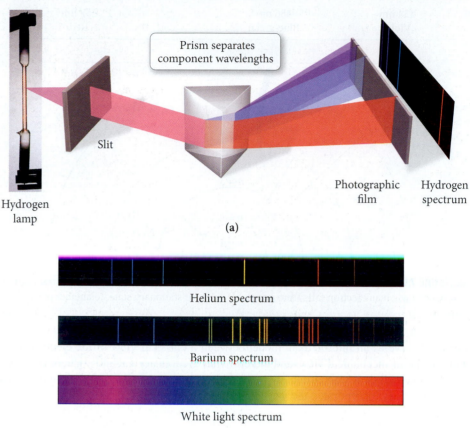

(a)

Helium spectrum

Barium spectrum

White light spectrum

(b)

▲ **FIGURE 7.11 Emission Spectra**   (a) The light emitted from a hydrogen, helium, or barium lamp consists of specific wavelengths, which can be separated by passing the light through a prism. (b) The resulting bright lines constitute an emission spectrum characteristic of the element that produced it.

contrast to planetary orbits—which can theoretically exist at any distance from the sun—Bohr's orbits could exist only at specific, fixed distances from the nucleus. The energy of each Bohr orbit was also fixed, or *quantized*. Bohr called these orbits *stationary states* and suggested that, although they obeyed the laws of classical mechanics, they also possessed "a peculiar, mechanically unexplainable, stability." We now know that the stationary states were really manifestations of the wave nature of the electron, which we expand upon shortly. Bohr further proposed that, in contradiction to classical electromagnetic theory, no radiation was emitted by an electron orbiting the nucleus in a stationary state. It was only when an electron jumped, or made a *transition*, from one stationary state to another that radiation was emitted or absorbed (**Figure 7.12▶**).

The transitions between stationary states in a hydrogen atom are quite unlike any transitions that you might imagine in the macroscopic world. The electron is *never observed between states*, only in one state or the next—the transition between states is instantaneous. The emission spectrum of an atom consists of discrete lines because the states exist only at specific, fixed energies. The energy of the photon created when an electron makes a transition from one stationary state to another is the energy difference between the two stationary states. Transitions between stationary states that are closer together, therefore, produce light of lower energy (longer wavelength) than transitions between stationary states that are farther apart.

In spite of its initial success in explaining the line spectrum of hydrogen (including the correct wavelengths), the Bohr model left many unanswered questions. It did,

The Bohr Model and Emission Spectra

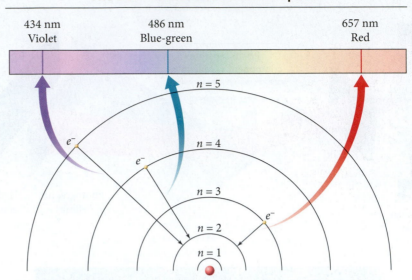

▲ **FIGURE 7.12 The Bohr Model and Emission Spectra** In the Bohr model, each spectral line is produced when an electron falls from one stable orbit, or stationary state, to another of lower energy.

however, serve as an intermediate model between a classical view of the electron and a fully quantum-mechanical view, and therefore has great historical and conceptual importance. Nonetheless, it was ultimately replaced by a more complete quantum-mechanical theory that fully incorporated the wave nature of the electron.

7.4 The Wave Nature of Matter: the de Broglie Wavelength, the Uncertainty Principle, and Indeterminacy

The heart of the quantum-mechanical theory that replaced Bohr's model is the wave nature of the electron, first proposed by Louis de Broglie (1892–1987) in 1924 and confirmed by experiments in 1927. It seemed incredible at the time, but electrons—which were thought of as particles and known to have mass—also have a wave nature. The wave nature of the electron is seen most clearly in its diffraction. If an electron beam is aimed at two closely spaced slits, and a series (or array) of detectors is arranged to detect the electrons after they pass through the slits, an interference pattern similar to that observed for light is recorded behind the slits (**Figure 7.13a▶**). The detectors at the center of the array (midway between the two slits) detect a large number of electrons—exactly the opposite of what you would expect for particles (**Figure 7.13b▶**). Moving outward from this center spot, the detectors alternately detect small numbers of electrons and then large numbers again and so on, forming an interference pattern characteristic of waves.

The first evidence of electron wave properties was provided by the Davisson-Germer experiment of 1927, in which electrons were observed to undergo diffraction by a metal crystal.

For interference to occur, the spacing of the slits has to be on the order of atomic dimensions.

It is critical to understand that the interference pattern described here is *not caused by pairs of electrons interfering with each other, but rather by single electrons interfering with themselves.* If the electron source is turned down to a very low level, so that electrons come out only one at a time, *the interference pattern remains.* In other words, we can design an experiment in which electrons come out of the source singly. We can then record where each electron strikes the detector after it has passed through the slits. If we record the positions of thousands of electrons over a long period of time, we find the same interference pattern shown in Figure 7.13(a). This leads us to an important conclusion: *The wave nature of the electron is an inherent property of individual electrons.* Recall from Section 7.1 that unobserved electrons can simultaneously occupy two different states. In this case, the unobserved electron goes through both slits—it exists in two states simultaneously, just like Schrödinger's cat—and interferes with itself. As it turns out, this wave nature is what explains the existence of stationary states (in the Bohr model)

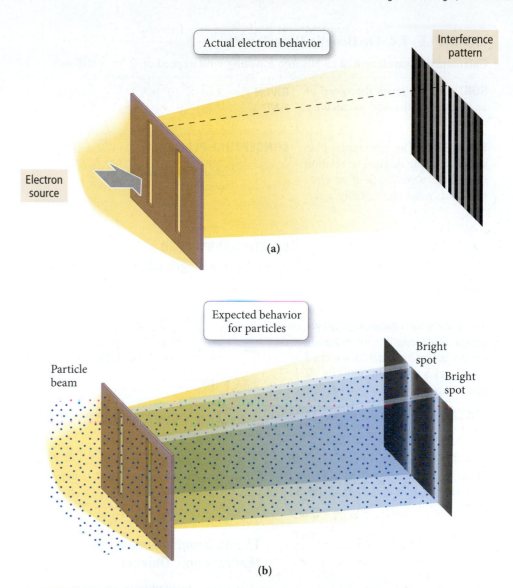

▲ **FIGURE 7.13 Electron Diffraction** When a beam of electrons goes through two closely spaced slits (**a**), an interference pattern is created, as if the electrons were waves. By contrast, a beam of particles passing through two slits (**b**) should simply produce two smaller beams of particles. Notice that for particle beams, there is a dark line directly behind the center of the two slits, in contrast to wave behavior, which produces a bright line.

and prevents the electrons in an atom from crashing into the nucleus as they are predicted to do according to classical physics. We now turn to three important manifestations of the electron's wave nature: the de Broglie wavelength, the uncertainty principle, and indeterminacy.

The de Broglie Wavelength

As we have seen, a single electron traveling through space has a wave nature; its wavelength is related to its kinetic energy (the energy associated with its motion). The faster the electron is moving, the higher its kinetic energy and the shorter its wavelength. The wavelength (λ) of an electron of mass m moving at velocity v is given by the **de Broglie relation**:

$$\lambda = \frac{h}{mv} \qquad \text{de Broglie relation} \qquad [7.4]$$

where h is Planck's constant. Notice that the velocity of a moving electron is related to its wavelength—knowing one is equivalent to knowing the other.

The mass of an object (m) times its velocity (v) is its momentum. Therefore, the wavelength of an electron is inversely proportional to its momentum.

EXAMPLE 7.4 De Broglie Wavelength

Calculate the wavelength of an electron traveling with a speed of 2.65×10^6 m/s.

SORT You are given the speed of an electron and asked to calculate its wavelength.	**GIVEN** $v = 2.65 \times 10^6$ m/s **FIND** λ
STRATEGIZE The conceptual plan shows how the de Broglie relation relates the wavelength of an electron to its mass and velocity.	**CONCEPTUAL PLAN** $\lambda = \dfrac{h}{mv}$ **RELATIONSHIPS USED** $\lambda = h/mv$ (de Broglie relation, Equation 7.4)
SOLVE Substitute the velocity, Planck's constant, and the mass of an electron to calculate the electron's wavelength. To correctly cancel the units, break down the J in Planck's constant into its SI base units $(1 \text{ J} = 1 \text{ kg} \cdot \text{m}^2/\text{s}^2)$.	**SOLUTION** $\lambda = \dfrac{h}{mv} = \dfrac{6.626 \times 10^{-34} \dfrac{\text{kg} \cdot \text{m}^2}{\text{s}^2} \text{s}}{(9.11 \times 10^{-31} \text{ kg})\left(2.65 \times 10^6 \dfrac{\text{m}}{\text{s}}\right)}$ $= 2.74 \times 10^{-10}$ m

CHECK The units of the answer (m) are correct. The magnitude of the answer is very small, as expected for the wavelength of an electron.

FOR PRACTICE 7.4

What is the velocity of an electron having a de Broglie wavelength that is approximately the length of a chemical bond? Assume this length to be 1.2×10^{-10} m.

Conceptual Connection 7.2 The de Broglie Wavelength of Macroscopic Objects

Since quantum-mechanical theory is universal, it applies to all objects, regardless of size. Therefore, according to the de Broglie relation, a thrown baseball should also exhibit wave properties. Why do we not observe such properties at the ballpark?

The Uncertainty Principle

The wave nature of the electron is difficult to reconcile with its particle nature. How can a single entity behave as both a wave and a particle? We can begin to answer this question by returning to the single-electron diffraction experiment. Specifically, we can ask the question: how does a single electron aimed at a double slit produce an interference pattern? We saw previously that the electron travels through both slits and interferes with itself. This idea is testable. We simply have to observe the single electron as it travels through both of the slits. If it travels through both slits simultaneously, our hypothesis is correct. But here is where nature gets tricky.

Any experiment designed to observe the electron as it travels through the slits results in the detection of an electron "particle" traveling through a single slit and no interference pattern. Recall from Section 7.1 that an *unobserved* electron can occupy two different states; however, the act of observation forces it into one state or the other. Similarly, the act of observing the electron as it travels through both slits forces it go through only one slit. The following electron diffraction experiment is designed to "watch" which slit the electron travels through by using a laser beam placed directly behind the slits.

An electron that crosses a laser beam produces a tiny "flash"—a single photon is scattered at the point of crossing. A flash behind a particular slit indicates an electron

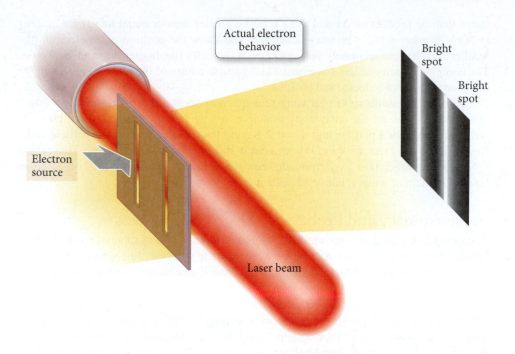

passing through that slit. However, when the experiment is performed, the flash always originates either from one slit *or* the other, but *never* from both at once. Futhermore, the interference pattern, which was present without the laser, is now absent. With the laser on, the electrons hit positions directly behind each slit, as if they were ordinary particles.

As it turns out, no matter how hard we try, or whatever method we set up, *we can never see the interference pattern and simultaneously determine which hole the electron goes through.* It has never been done, and most scientists agree that it never will. In the words of P. A. M. Dirac (1902–1984),

> There is a limit to the fineness of our powers of observation and the smallness of the accompanying disturbance—a limit which is inherent in the nature of things and can never be surpassed by improved technique or increased skill on the part of the observer.

The single electron diffraction experiment demonstrates that you cannot simultaneously observe both the wave nature and the particle nature of the electron. When you try to observe which hole the electron goes through (associated with the particle nature of the electron) you lose the interference pattern (associated with the wave nature of the electron). When you try to observe the interference pattern, you cannot determine which hole the electron goes through. The wave nature and particle nature of the electron are said to be **complementary properties**. Complementary properties exclude one another—the more you know about one, the less you know about the other. Which of two complementary properties you observe depends on the experiment you perform—in quantum mechanics, the observation of an event affects its outcome.

As we just saw in the de Broglie relation, the *velocity* of an electron is related to its *wave nature*. The *position* of an electron, however, is related to its *particle nature*. (Particles have well-defined positions, but waves do not.) Consequently, our inability to observe the electron simultaneously as both a particle and a wave means that we cannot simultaneously measure its position and its velocity. Werner Heisenberg formalized this idea with the equation:

$$\Delta x \times m\Delta v \geq \frac{h}{4\pi} \qquad \text{Heisenberg's uncertainty principle} \qquad [7.5]$$

where Δx is the uncertainty in the position, Δv is the uncertainty in the velocity, m is the mass of the particle, and h is Planck's constant. **Heisenberg's uncertainty principle**

▲ Werner Heisenberg (1901–1976)

states that the product of Δx and $m\Delta v$ must be greater than or equal to a finite number ($h/4\pi$). In other words, the more accurately you know the position of an electron (the smaller Δx) the less accurately you can know its velocity (the bigger Δv) and vice versa. The complementarity of the wave nature and particle nature of the electron results in the complementarity of velocity and position.

Although Heisenberg's uncertainty principle may seem puzzling, it actually solves a great puzzle. Without the uncertainty principle, we are left with the question: how can something be *both* a particle and a wave? Saying that an object is both a particle and a wave is like saying that an object is both a circle and a square, a contradiction. Heisenberg solved the contradiction by introducing complementarity—an electron is observed as *either* a particle or a wave, but never both at once.

Indeterminacy and Probability Distribution Maps

According to classical physics, and in particular Newton's laws of motion, particles move in a *trajectory* (or path) that is determined by the particle's velocity (the speed and direction of travel), its position, and the forces acting on it. Even if you are not familiar with Newton's laws, you probably have an intuitive sense of them. For example, when you chase a baseball in the outfield, you visually predict where the ball will land by observing its path. You do this by noting its initial position and velocity, watching how these are affected by the forces acting on it (gravity, air resistance, wind), and then inferring its trajectory, as shown in **Figure 7.14▼**. If you knew only the ball's velocity, or only its position (imagine a still photo of the baseball in the air), you could not predict its landing spot. In classical mechanics, both position and velocity are required to predict a trajectory.

Newton's laws of motion are **deterministic**—the present *determines* the future. This means that if two baseballs are hit consecutively with the same velocity from the same position under identical conditions, they will land in exactly the same place. The same is not true of electrons. We have just seen that we cannot simultaneously know the position and velocity of an electron; therefore, we cannot know its trajectory. In quantum mechanics, trajectories are replaced with *probability distribution maps*, as shown in **Figure 7.15▼**.

> Remember that velocity includes speed as well as direction of travel.

▶ **FIGURE 7.14 The Concept of Trajectory** In classical mechanics, the position and velocity of a particle determine its future trajectory, or path. Thus, an outfielder can catch a baseball by observing its position and velocity, allowing for the effects of forces acting on it, such as gravity, and estimating its trajectory. (For simplicity, air resistance and wind are not shown.)

The Classical Concept of Trajectory

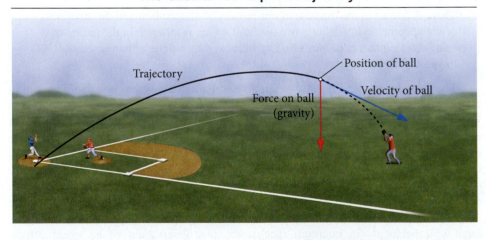

Trajectory

Position of ball

Velocity of ball

Force on ball
(gravity)

Classical
trajectory

Quantum-mechanical
probability distribution map

▲ **FIGURE 7.15 Trajectory versus Probability** In quantum mechanics, we cannot calculate deterministic trajectories. Instead, it is necessary to think in terms of probability maps: statistical pictures of where a quantum-mechanical particle, such as an electron, is most likely to be found. In this hypothetical map, darker shading indicates greater probability.

A probability distribution map is a statistical map that shows where an electron is likely to be found under a given set of conditions.

To understand the concept of a probability distribution map, let us return to baseball. Imagine a baseball thrown from the pitcher's mound to a catcher behind home plate (**Figure 7.16▶**). The catcher can watch the baseball's path, predict exactly where it will cross home plate, and place his mitt in the correct place to catch it. As we have seen, this would be impossible for an electron. If an electron were thrown from the pitcher's mound to home plate, it would generally land in a different place every time, even if it were thrown in exactly the same way. This behavior is called **indeterminacy**. Unlike a baseball, whose future path is *determined* by its position and velocity when it leaves the pitcher's hand, the future path of an electron is indeterminate, and can only be described statistically.

In the quantum-mechanical world of the electron, the catcher could not know exactly where the electron will cross the plate for any given throw. However, if he kept track of hundreds of identical electron throws, the catcher could observe a reproducible *statistical pattern* of where the electron crosses the plate. He could even draw a map of the strike zone showing the probability of an electron crossing a certain area, as shown in **Figure 7.17▼**. This would be a probability distribution map. In the sections that follow, we discuss quantum-mechanical electron *orbitals*, which are essentially probability distribution maps for electrons as they exist within atoms.

▲ **FIGURE 7.16 Trajectory of a Macroscopic Object** A baseball follows a well-defined trajectory from the hand of the pitcher to the mitt of the catcher.

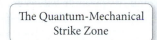
The Quantum-Mechanical Strike Zone

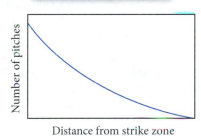

Distance from strike zone

▲ **FIGURE 7.17 The Quantum-Mechanical Strike Zone** An electron does not have a well-defined trajectory. However, we can construct a probability distribution map to show the relative probability of it crossing home plate at different points.

7.5 Quantum Mechanics and the Atom

As we have seen, the position and velocity of the electron are complementary properties—if we know one accurately, the other becomes indeterminate. Since velocity is directly related to energy (we have seen that kinetic energy equals $\frac{1}{2} mv^2$), position and *energy* are also complementary properties—the more you know about one, the less you know about the other. Many of the properties of an element, however, depend on the energies of its electrons. For example, whether an electron is transferred from one atom to another to form an ionic bond depends in part on the relative energies of the electron in the two atoms. In the following paragraphs, we describe the probability distribution maps for electron states in which the electron has well-defined energy, but not well-defined position. In other words, for each state, we can specify the *energy* of the electron precisely, but not its location at a given instant. Instead, the electron's position is described in terms of an **orbital**, a probability distribution map showing where the electron is likely to be found. Since chemical bonding often involves the sharing of electrons between atoms to form covalent bonds, the spatial distribution of atomic electrons is important to bonding.

These states are known as energy *eigenstates*.

The mathematical derivation of energies and orbitals for electrons in atoms comes from solving the Schrödinger equation for the atom of interest. The general form of the Schrödinger equation is:

$$H\psi = E\psi \qquad [7.6]$$

An operator is different from a normal algebraic entity. In general, an operator transforms a mathematical function into another mathematical function. For example, d/dx is an operator that means "take the derivative of." When d/dx operates on a function (such as x^2) it returns another function ($2x$).

The symbol ψ is the Greek letter psi, pronounced "sigh."

The symbol H stands for the Hamiltonian operator, a set of mathematical operations that represent the total energy (kinetic and potential) of the electron within the atom. The symbol E is the actual energy of the electron. The symbol ψ is the **wave function**, a mathematical function that describes the wavelike nature of the electron. A plot of the wave function squared (ψ^2) represents an orbital, a position probability distribution map of the electron.

Solutions to the Schrödinger Equation for the Hydrogen Atom

When the Schrödinger equation is solved, it yields many solutions—many possible wave functions. The wave functions themselves are fairly complicated mathematical functions, and we do not examine them in detail in this book. Instead, we introduce graphical representations (or plots) of the orbitals that correspond to the wave functions. Each orbital is specified by three interrelated **quantum numbers**: n, the **principal quantum number**; l, the **angular momentum quantum number** (sometimes called the *azimuthal quantum number*); and m_l the **magnetic quantum number**. These quantum numbers all have integer values, as had been hinted at by both the Rydberg equation and Bohr's model. A fourth quantum number, m_s, the **spin quantum number**, specifies the orientation of the spin of the electron. We examine each of these quantum numbers individually.

The Principal Quantum Number (n)

The principal quantum number is an integer that determines the overall size and energy of an orbital. Its possible values are $n = 1, 2, 3, \ldots$ and so on. For the hydrogen atom, the energy of an electron in an orbital with quantum number n is given by

$$E_n = -2.18 \times 10^{-18} \text{ J}\left(\frac{1}{n^2}\right) \qquad (n = 1, 2, 3, \ldots) \qquad [7.7]$$

The energy is negative because the energy of the electron in the atom is less than the energy of the electron when it is very far away from the atom (which is taken to be zero). Notice that orbitals with higher values of n have greater (less negative) energies, as shown in the energy level diagram below. Notice also that, as n increases, the spacing between the energy levels becomes smaller.

$$n = 4 \underline{\qquad} E_4 = -1.36 \times 10^{-19} \text{ J}$$
$$n = 3 \underline{\qquad} E_3 = -2.42 \times 10^{-19} \text{ J}$$

$$n = 2 \underline{\qquad} E_2 = -5.45 \times 10^{-19} \text{ J}$$

Energy

$$n = 1 \underline{\qquad} E_1 = -2.18 \times 10^{-18} \text{ J}$$

The Angular Momentum Quantum Number (l)

The angular momentum quantum number is an integer that determines the shape of the orbital. We consider these shapes in Section 7.6. The possible values of l are $0, 1, 2, \ldots, (n-1)$. In other words, for a given value of n, l can be any integer (including 0) up to $n - 1$. For example, if $n = 1$, then the only possible value of l is 0; if $n = 2$, the possible values of l are 0 and 1. In order to avoid confusion between n and l, values of l are often assigned letters as follows:

The values of l beyond 3 are designated with letters in alphabetical order so that $l = 4$ is designated g, $l = 5$ is designated h, and so on.

Value of l	Letter Designation
$l = 0$	s
$l = 1$	p
$l = 2$	d
$l = 3$	f

Conceptual Connection 7.3 The Relationship Between *n* and *l*

What is the full range of possible values of *l* for *n* = 3?

(a) 0 (or *s*)　　**(b)** 0 and 1 (or *s* and *p*)　　**(c)** 0, 1, and 2 (or *s*, *p*, and *d*)　　**(d)** 0, 1, 2, and 3 (or *s*, *p*, *d*, and *f*)

The Magnetic Quantum Number (m_l)

The magnetic quantum number is an integer that specifies the orientation of the orbital. We consider these orientations in Section 7.6. The possible values of m_l are the integer values (including zero) ranging from $-l$ to $+l$. For example, if $l = 0$, then the only possible value of m_l is 0; if $l = 1$, the possible values of m_l are -1, 0, and $+1$.

 Conceptual Connection 7.4 The Relationship between *l* and m_l

What is the full range of possible values of m_l for $l = 2$?

(a) 0, 1, and 2　　**(b)** 0　　**(c)** -1, 0 and $+1$　　**(d)** -2, -1, 0, $+1$, and $+2$

The Spin Quantum Number (m_s)

The spin quantum number specifies the orientation of the *spin* of the electron. **Electron spin** is a fundamental property of an electron (like its negative charge). One electron does not have more or less spin than another—all electrons have the same amount of spin. The orientation of the electron's spin is quantized, with only two possibilities that we can call spin up ($m_s = +1/2$) and spin down ($m_s = -1/2$). The spin quantum number becomes important when we begin to consider how electrons occupy orbitals (Section 8.3). For now, we will focus on the first three quantum numbers.

> The idea of a "spinning" electron is something of a metaphor. A more correct way to express the same idea is to say that an electron has inherent angular momentum.

The Hydrogen Atom Orbitals

Each specific combination of the first three quantum numbers (n, l, and m_l) specifies one atomic orbital. For example, the orbital with $n = 1$, $l = 0$, and $m_l = 0$ is known as the 1*s* orbital. The 1 in 1*s* is the value of *n* and the *s* specifies that $l = 0$. There is only one 1*s* orbital in an atom, and its m_l value is zero. Orbitals with the same value of *n* are said to be in the same **principal level** (or **principal shell**). Orbitals with the same value of *n* and *l* are said to be in the same **sublevel** (or **subshell**). The following diagram shows all of the orbitals, each represented by a small square, in the first three principal levels.

Principal level (specified by *n*)	*n* = 1	*n* = 2		*n* = 3		
Sublevel (specified by *n* and *l*)	*l* = 0	*l* = 0	*l* = 1	*l* = 0	*l* = 1	*l* = 2
	1*s* sublevel	2*s* sublevel	2*p* sublevels	3*s* sublevel	3*p* sublevels	3*d* sublevels
Orbital (specified by *n*, *l*, and m_l)	□ $m_l = 0$	□ $m_l = 0$	□□□ $m_l = -1, 0, +1$	□ $m_l = 0$	□□□ $m_l = -1, 0, +1$	□□□□□ $m_l = -2, -1, 0, +1, +2$

For example, the $n = 2$ level contains the $l = 0$ and $l = 1$ sublevels. Within the $n = 2$ level, the $l = 0$ sublevel—called the 2*s* sublevel—contains only one orbital (the 2*s* orbital), with $m_l = 0$. The $l = 1$ sublevel—called the 2*p* sublevel—contains three 2*p* orbitals, with $m_l = -1, 0, +1$.

In general, notice the following:

- The number of sublevels in any level is equal to *n*, the principal quantum number. Therefore, the $n = 1$ level has one sublevel, the $n = 2$ level has two sublevels, etc.
- The number of orbitals in any sublevel is equal to $2l + 1$. Therefore, the *s* sublevel ($l = 0$) has one orbital, the *p* sublevel ($l = 1$) has three orbitals, the *d* sublevel ($l = 2$) has five orbitals, etc.
- The number of orbitals in a level is equal to n^2. Therefore, the $n = 1$ level has one orbital, the $n = 2$ level has four orbitals, the $n = 3$ level has nine orbitals, etc.

EXAMPLE 7.5 Quantum Numbers I

What are the quantum numbers and names (for example, $2s$, $2p$) of the orbitals in the $n = 4$ principal level? How many $n = 4$ orbitals exist?

SOLUTION

You first determine the possible values of l (from the given value of n). You then determine the possible values of m_l for each possible value of l. For a given value of n, the possible values of l are $0, 1, 2, \ldots , (n - 1)$.	$n = 4$; therefore $l = 0, 1, 2,$ and 3

For a given value of l, the possible values of m_l are the integer values including zero ranging from $-l$ to $+l$. The name of an orbital is its principal quantum number (n) followed by the letter corresponding to the value l.
The total number of orbitals is given by n^2.

l	Possible m_l Values	Orbital Name(s)
0	0	$4s$ (1 orbital)
1	$-1, 0, +1$	$4p$ (3 orbitals)
2	$-2, -1, 0, +1, +2$	$4d$ (5 orbitals)
3	$-3, -2, -1, 0, +1, +2, +3$	$4f$ (7 orbitals)

Total number of orbitals $= 4^2 = 16$

FOR PRACTICE 7.5

List the quantum numbers associated with all of the $5d$ orbitals. How many $5d$ orbitals exist?

EXAMPLE 7.6 Quantum Numbers II

These sets of quantum numbers are each supposed to specify an orbital. One set, however, is erroneous. Which one and why?

(a) $n = 3$; $l = 0$; $m_l = 0$

(b) $n = 2$; $l = 1$; $m_l = -1$

(c) $n = 1$; $l = 0$; $m_l = 0$

(d) $n = 4$; $l = 1$; $m_l = -2$

SOLUTION

Choice **(d)** is erroneous because, for $l = 1$, the possible values of m_l are only -1, 0, and $+1$.

FOR PRACTICE 7.6

Each of the following sets of quantum numbers is supposed to specify an orbital. However, each set contains one quantum number that is not allowed. Replace the quantum number that is not allowed with one that is allowed.

(a) $n = 3$; $l = 3$; $m_l = +2$

(b) $n = 2$; $l = 1$; $m_l = -2$

(c) $n = 1$; $l = 1$; $m_l = 0$

Atomic Spectroscopy Explained

Quantum theory explains the atomic spectra of atoms discussed in Section 7.3. Each wavelength in the emission spectrum of an atom corresponds to an electron *transition* between quantum-mechanical orbitals. When an atom absorbs energy, an electron in a lower energy level orbital is *excited* or promoted to a higher energy level orbital, as shown in **Figure 7.18▼**. In this new configuration, however, the atom is unstable, and the electron quickly falls back or *relaxes* to a lower energy orbital. As it does so, it releases a photon of light containing an amount of energy precisely equal to the energy difference between the two energy levels. We saw previously (see Equation 7.7) that the energy of an orbital with principal quantum number n is given by $E_n = -2.18 \times 10^{-18} \, \text{J}(1/n^2)$, where $n = 1, 2, 3, \ldots$. Therefore, the *difference* in energy between two levels n_{initial} and n_{final} is given by $\Delta E = E_{\text{final}} - E_{\text{initial}}$. If we substitute the expression for E_n into the expression for ΔE, we get the following important expression for the change in energy that occurs in an atom when an electron changes energy levels:

$$\Delta E = E_{\text{final}} - E_{\text{initial}}$$

$$= -2.18 \times 10^{-18} \, \text{J}\left(\frac{1}{n_{\text{f}}^2}\right) - \left[-2.18 \times 10^{-18} \, \text{J}\left(\frac{1}{n_{\text{i}}^2}\right)\right]$$

$$\Delta E = -2.18 \times 10^{-18} \, \text{J}\left(\frac{1}{n_{\text{f}}^2} - \frac{1}{n_{\text{i}}^2}\right) \qquad [7.8]$$

For example, suppose that an electron in a hydrogen atom relaxes from an orbital in the $n = 3$ level to an orbital in the $n = 2$ level. Recall that the energy of an orbital in the hydrogen atom depends only on n and is given by $E_n = -2.18 \times 10^{-18} \, \text{J}(1/n^2)$, where $n = 1, 2, 3, \ldots$. Therefore, ΔE, the energy difference corresponding to the transition from $n = 3$ to $n = 2$, is determined as follows:

$$\Delta E_{\text{atom}} = E_2 - E_3$$

$$= -2.18 \times 10^{-18} \, \text{J}\left(\frac{1}{2^2}\right) - \left[-2.18 \times 10^{-18} \, \text{J}\left(\frac{1}{3^2}\right)\right]$$

$$= -2.18 \times 10^{-18} \, \text{J}\left(\frac{1}{2^2} - \frac{1}{3^2}\right)$$

$$= -3.03 \times 10^{-19} \, \text{J}$$

The energy carries a negative sign because the atom *emits* the energy as it relaxes from $n = 3$ to $n = 2$. Since energy must be conserved, the exact amount of energy emitted by the atom is carried away by the photon:

$$\Delta E_{\text{atom}} = -E_{\text{photon}}$$

Excitation and Radiation

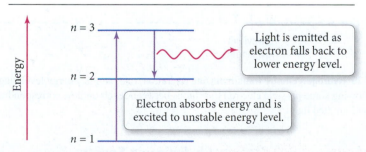

▲ **FIGURE 7.18 Excitation and Radiation** When an atom absorbs energy, an electron can be excited from an orbital in a lower energy level to an orbital in a higher energy level. The electron in this "excited state" is unstable, however, and relaxes to a lower energy level, releasing energy in the form of electromagnetic radiation.

This energy then determines the frequency and wavelength of the photon. Since the wavelength of the photon is related to its energy as $E = hc/\lambda$, we calculate the wavelength of the photon as follows:

$$\lambda = \frac{hc}{E}$$

$$= \frac{(6.626 \times 10^{-34} \text{ J} \cdot \text{s})(3.00 \times 10^8 \text{ m/s})}{3.03 \times 10^{-19} \text{ J}}$$

$$= 6.56 \times 10^{-7} \text{ m or } 656 \text{ nm}$$

The Rydberg equation, $1/\lambda = R(1/m^2 - 1/n^2)$, can be derived from the relationships just covered. We leave this derivation to an exercise (see Problem 7.62).

Consequently, the light emitted by an excited hydrogen atom as it relaxes from an orbital in the $n = 3$ level to an orbital in the $n = 2$ level has a wavelength of 656 nm (red). We can similarly calculate the light emitted due to a transition from $n = 4$ to $n = 2$ to be 486 nm (green). Notice that transitions between orbitals that are further apart in energy produce light that is higher in energy, and therefore shorter in wavelength, than transitions between orbitals that are closer together. **Figure 7.19▼** shows several of the transitions in the hydrogen atom and their corresponding wavelengths.

Hydrogen Energy Transitions and Radiation

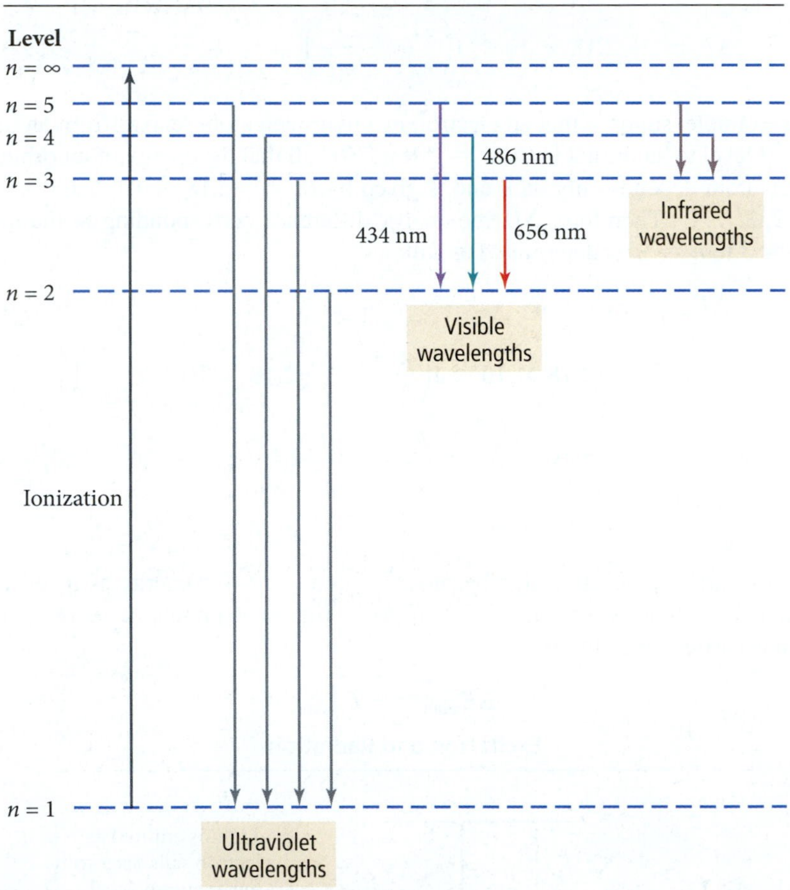

▲ **FIGURE 7.19 Hydrogen Energy Transitions and Radiation**   An atomic energy level diagram for hydrogen, showing some possible electron transitions between levels and the corresponding wavelengths of emitted light.

Conceptual Connection 7.5 Emission Spectra

Which transition will result in emitted light with the shortest wavelength?

(a) $n = 5 \rightarrow n = 4$

(b) $n = 4 \rightarrow n = 3$

(c) $n = 3 \rightarrow n = 2$

EXAMPLE 7.7 Wavelength of Light for a Transition in the Hydrogen Atom

Determine the wavelength of light emitted when an electron in a hydrogen atom makes a transition from an orbital in $n = 6$ to an orbital in $n = 5$.

SORT You are given the energy levels of an atomic transition and asked to find the wavelength of emitted light.

GIVEN $n = 6 \rightarrow n = 5$

FIND λ

STRATEGIZE In the first part of the conceptual plan, calculate the energy of the electron in the $n = 6$ and $n = 5$ orbitals using Equation 7.7 and subtract to find ΔE_{atom}.

In the second part, find E_{photon} by taking the negative of ΔE_{atom} and then calculate the wavelength corresponding to a photon of this energy using Equation 7.3. (The difference in sign between E_{photon} and ΔE_{atom} applies only to emission. *The energy of a photon must always be positive.*)

CONCEPTUAL PLAN

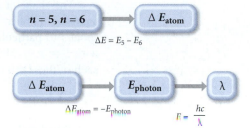

RELATIONSHIPS USED

$$E_n = -2.18 \times 10^{-18} \text{ J}(1/n^2)$$

$$E = hc/\lambda$$

SOLVE Follow the conceptual plan. Begin by calculating ΔE_{atom}.

SOLUTION

$$\Delta E_{atom} = E_5 - E_6$$

$$= -2.18 \times 10^{-18} \text{ J}\left(\frac{1}{5^2}\right) - \left[-2.18 \times 10^{-18} \text{ J}\left(\frac{1}{6^2}\right)\right]$$

$$= -2.18 \times 10^{-18} \text{ J}\left(\frac{1}{5^2} - \frac{1}{6^2}\right)$$

$$= -2.6644 \times 10^{-20} \text{ J}$$

Calculate E_{photon} by changing the sign of ΔE_{atom}.

$$E_{photon} = -\Delta E_{atom} = +2.6644 \times 10^{-20} \text{ J}$$

Solve the equation relating the energy of a photon to its wavelength for λ. Substitute the energy of the photon and calculate λ.

$$E = \frac{hc}{\lambda}$$

$$\lambda = \frac{hc}{E}$$

$$= \frac{(6.626 \times 10^{-34} \text{ J} \cdot \text{s})(3.00 \times 10^8 \text{ m/s})}{2.6644 \times 10^{-20} \text{ J}}$$

$$= 7.46 \times 10^{-6} \text{ m}$$

CHECK The units of the answer (m) are correct for wavelength. The magnitude seems reasonable because 10^{-6} m is in the infrared region of the electromagnetic spectrum. We know that transitions from $n = 3$ or $n = 4$ to $n = 2$ lie in the visible region, so it makes sense that a transition between levels of higher n value (which are energetically closer to one another) would result in light of longer wavelength.

FOR PRACTICE 7.7

Determine the wavelength of the light absorbed when an electron in a hydrogen atom makes a transition from an orbital in $n = 2$ to an orbital in $n = 7$.

FOR MORE PRACTICE 7.7

An electron in the $n = 6$ level of the hydrogen atom relaxes to a lower energy level, emitting light of $\lambda = 93.8$ nm. Find the principal level to which the electron relaxed.

7.6 The Shapes of Atomic Orbitals

As we noted previously, the shapes of atomic orbitals are important because covalent chemical bonds depend on the sharing of the electrons that occupy these orbitals. In one model of chemical bonding, for example, a bond consists of the overlap of atomic orbitals on adjacent atoms. Therefore the shapes of the overlapping orbitals determine the shape of the molecule. Although we limit ourselves in this chapter to the orbitals of the hydrogen atom, we will see in Chapter 8 that the orbitals of all atoms can be approximated as being hydrogen-like and therefore have very similar shapes to those of hydrogen.

The shape of an atomic orbital is determined primarily by l, the angular momentum quantum number. Recall that each value of l is assigned a letter that therefore corresponds to particular orbitals. For example, the orbitals with $l = 0$ are s orbitals; those with $l = 1$, p orbitals; those with $l = 2$, d orbitals, etc. We now examine the shape of each of these orbitals.

s Orbitals ($l = 0$)

The lowest energy orbital is the spherically symmetrical $1s$ orbital shown in **Figure 7.20a▼**. This image is actually a three-dimensional plot of the wave function squared (ψ^2), which represents **probability density**, the probability (per unit volume) of finding the electron at a point in space.

$$\psi^2 = \text{probability density} = \frac{\text{probability}}{\text{unit volume}}$$

The magnitude of ψ^2 in this plot is proportional to the density of the dots shown in the image. The high dot density near the nucleus indicates a higher probability density for the electron there. As you move away from the nucleus, the probability density decreases. **Figure 7.20(b)▼** shows a plot of probability density (ψ^2) versus r, the distance from the nucleus. This is essentially a slice through the three-dimensional plot of ψ^2 and shows how the probability density decreases as r increases.

We can understand probability density with the help of a thought experiment. Imagine an electron in the $1s$ orbital located within the volume surrounding the nucleus. Imagine also taking a photograph of the electron every second for 10 or 15 minutes. In one photograph, the electron is very close to the nucleus, in another it is farther away, and so on. Each photo has a dot showing the electron's position relative to the nucleus when the photo was taken. Remember that you can never predict where the electron will be for any one photo. However, if you took hundreds of photos and superimposed all of them, you would have a plot similar to Figure 7.20(a)—a statistical representation of how likely the electron is to be found at each point.

An atomic orbital can also be represented by a geometrical shape that encompasses the volume where the electron is likely to be found most frequently—typically, 90% of

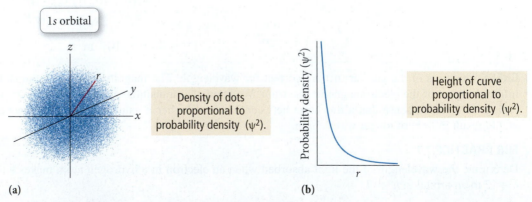

(a)   (b)

▲ **FIGURE 7.20 The 1s Orbital: Two Representations**   In **(a)** the dot density is proportional to the electron probability density. In **(b)**, the height of the curve is proportional to the electron probability density. The x-axis is r, the distance from the nucleus.

the time. For example, the $1s$ orbital can be represented as the three-dimensional sphere shown in **Figure 7.21▶**. If we were to superimpose the dot-density representation of the $1s$ orbital on the shape representation, 90% of the dots would be within the sphere, meaning that when the electron is in the $1s$ orbital it has a 90% chance of being found within the sphere.

The plots we have just seen represent probability *density*. However, they are a bit misleading because they seem to imply that the electron is most likely to be found *at the nucleus*. To get a better idea of where the electron is most likely to be found, we can use a plot called the **radial distribution function**, shown in **Figure 7.22▶** for the $1s$ orbital. The radial distribution function represents the *total probability of finding the electron within a thin spherical shell at a distance r from the nucleus*.

$$\text{Total radial probability (at a given } r) = \frac{\text{probability}}{\text{unit volume}} \times \text{volume of shell at } r$$

The radial distribution function represents, not *probability density at a point r*, but *total probability at a radius r*. In contrast to probability density, which has a maximum at the nucleus, the radial distribution function has a value of *zero* at the nucleus. It increases to a maximum at 52.9 pm and then decreases again with increasing r.

The shape of the radial distribution function is the result of multiplying together two functions with opposite trends in r:

1. the probability density function (ψ^2), which is the probability per unit volume, has a maximum at the nucleus, and decreases with increasing r
2. the volume of the thin shell, which is zero at the nucleus and increases with increasing r.

At the nucleus ($r = 0$) the probability *density* is at a maximum; however, the volume of a thin spherical shell is zero, so the radial distribution function is zero. As r increases, the volume of the thin spherical shell increases. We can see this by analogy to an onion. A spherical shell at a distance r from the nucleus is like a layer in an onion at a distance r from its center. If the layers of the onion all have the same thickness, then the volume of any one layer—think of this as the total amount of onion in the layer—is greater as r increases. Similarly, the volume of any one spherical shell in the radial distribution function increases with increasing distance from the nucleus, resulting in a greater total probability of finding the electron within that shell. Close to the nucleus, this increase in volume with increasing r outpaces the decrease in probability density, producing a maximum at 52.9 pm. Farther out, however, the density falls off faster than the volume increases.

The maximum in the radial distribution function, 52.9 pm, turns out to be the very same radius that Bohr had predicted for the innermost orbit of the hydrogen atom. However, there is a significant conceptual difference between the two radii. In the Bohr model, every time you probe the atom (in its lowest energy state), you would find the electron at a radius of 52.9 pm. In the quantum-mechanical model, you would generally find the electron at various radii, with 52.9 pm having the greatest probability.

The probability densities and radial distribution functions for the $2s$ and $3s$ orbitals are shown in **Figure 7.23▶**. Like the $1s$ orbital, these orbitals are spherically symmetric. These orbitals are larger in size, however, than the $1s$ orbital, and, unlike the $1s$ orbital, they contain *nodes*. A **node** is a point where the wave function (ψ), and therefore the probability density (ψ^2) and radial distribution function, all go through zero. A node in a wave function is much like a node in a standing wave on a vibrating string. We can see nodes in an orbital most clearly by actually looking at a slice through the orbital. Plots of probability density and the radial distribution function as a function of r both reveal the presence of nodes. The probability of finding the electron at a node is zero.

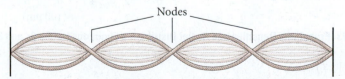

Nodes

▲ The nodes in quantum-mechanical atomic orbitals are three-dimensional analogs of the nodes we find on a vibrating string.

When an orbital is represented as shown below, the surface shown is one of constant probability. The probability of finding the electron at any point on the surface is the same.

$1s$ orbital surface

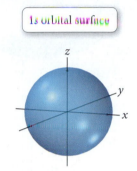

▲ **FIGURE 7.21 The 1s Orbital Surface**   In this representation, the surface of the sphere encompasses the volume where the electron is found 90% of the time when the electron is in the $1s$ orbital.

$1s$ Radial Distribution Function

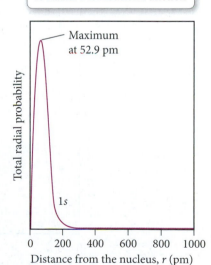

Maximum at 52.9 pm

$1s$

Total radial probability

Distance from the nucleus, r (pm)

▲ **FIGURE 7.22 The Radial Distribution Function for the 1s Orbital**   The curve shows the total probability of finding the electron within a thin shell at a distance r from the nucleus.

The 2s and 3s Orbitals

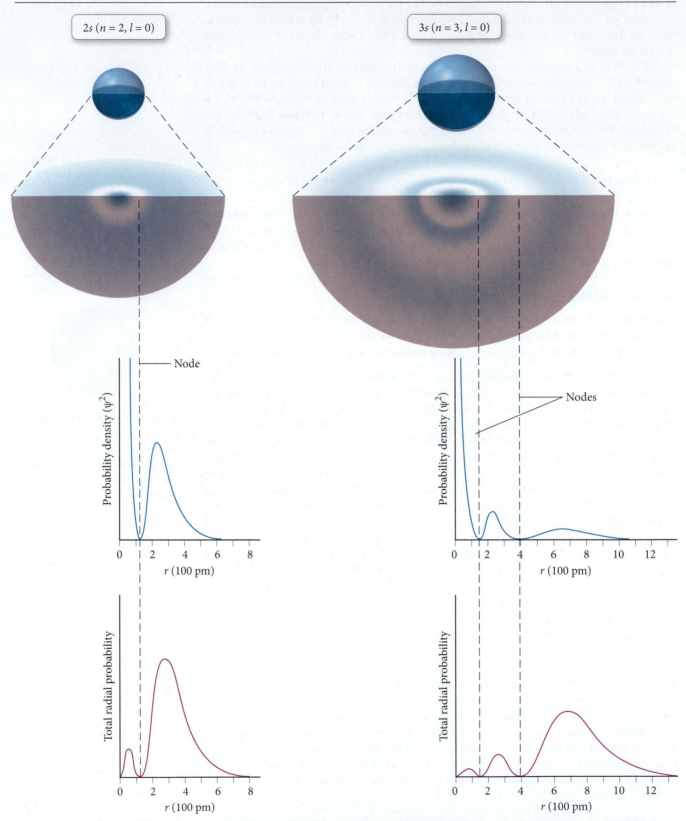

▲ **FIGURE 7.23** **Probability Densities and Radial Distribution Functions for the 2s and 3s Orbitals**

A nodal plane is a plane where the electron probability density is zero. For example, in the d_{xy} orbitals, the nodal planes lie in the xz and yz planes.

p Orbitals (l = 1)

Each principal level with $n = 2$ or greater contains three p orbitals ($m_l = -1, 0, +1$). The three $2p$ orbitals and their radial distribution functions are shown in **Figure 7.24▶**. The p orbitals are not spherically symmetric like the s orbitals, but have two *lobes* of electron density on

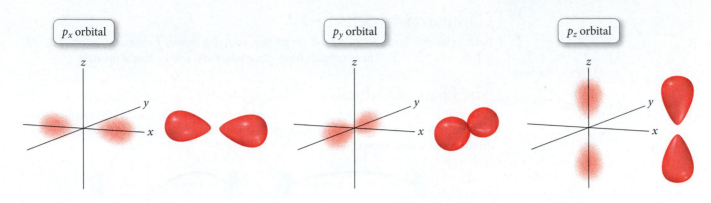

either side of the nucleus and a node located at the nucleus. The three p orbitals differ only in their orientation and are orthogonal (mutually perpendicular) to one another. It is convenient to define an x-, y-, and z-axis system and then label each p orbital as p_x, p_y, and p_z. The $3p$, $4p$, $5p$, and higher p orbitals are all similar in shape to the $2p$ orbitals, but they contain additional nodes (like the higher s orbitals) and are progressively larger in size.

d Orbitals ($l = 2$)

Each principal level with $n = 3$ or greater contains five d orbitals ($m_l = -2, -1, 0, +1, +2$). The five $3d$ orbitals are shown in **Figure 7.25▼**. Four of these orbitals have a cloverleaf shape, with four lobes of electron density around the nucleus and two perpendicular nodal planes. The d_{xy}, d_{xz}, and d_{yz} orbitals are oriented along the xy, xz, and yz planes, respectively, and their lobes are oriented *between* the corresponding axes. The four lobes of the $d_{x^2-y^2}$ orbital are oriented along the x- and y-axes. The d_{z^2} orbital is different in shape from the other four, having two lobes oriented along the z-axis and a donut-shaped ring along the xy plane. The $4d$, $5d$, $6d$, etc., orbitals are all similar in shape to the $3d$ orbitals, but they contain additional nodes and are progressively larger in size.

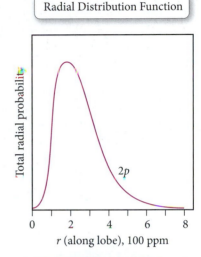

▲ FIGURE 7.24 The 2p Orbitals and Their Radial Distribution Function
The radial distribution function is the same for all three $2p$ orbitals when the x-axis of the graph is taken as the axis containing the lobes of the orbital.

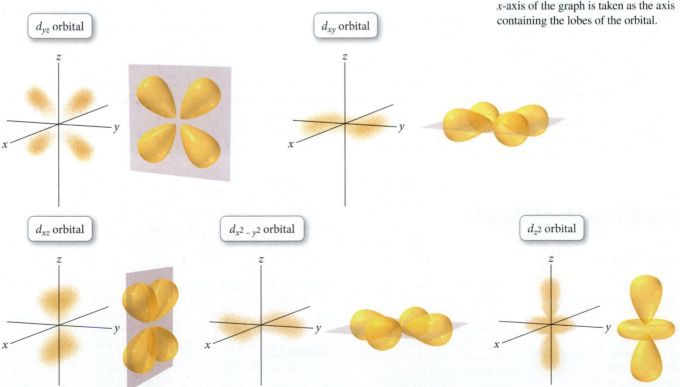

▲ FIGURE 7.25 The 3d Orbitals

f Orbitals (*l* = 3)

Each principal level with *n* = 4 or greater contains seven *f* orbitals (m_l = −3, −2, −1, 0, +1, +2, +3). These orbitals have more lobes and nodes than *d* orbitals.

The Phase of Orbitals

The orbitals we have just seen are three-dimensional waves. We can understand an important property of these orbitals by analogy to one-dimensional waves. Consider the one-dimensional waves shown here:

The wave on the left has a positive amplitude over its entire length, while the wave on the right has a positive amplitude over half of its length and a negative amplitude over the other half. The sign of the amplitude of a wave—positive or negative—is known as its **phase**. In these images, blue indicates positive phase and red indicates negative phase. The phase of a wave determines how it interferes with another wave as we saw in Section 7.2.

Just as a one-dimensional wave has a phase, so does a three-dimensional wave. We often represent the phase of a quantum mechanical orbital with color. For example, we can represent the phase of a 1s and 2p orbital as follows:

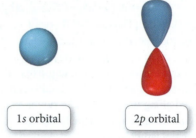

1s orbital   2p orbital

In these depictions, blue represents positive phase and red represents negative phase. The 1s orbital is all one phase, while the 2p orbital exhibits two different phases. The phase of quantum mechanical orbitals is important in bonding, as we shall see in Chapter 10.

The Shapes of Atoms

If some orbitals are shaped like dumbbells and three-dimensional cloverleafs, and if most of the volume of an atom is empty space diffusely occupied by electrons in these orbitals, then why do we often depict atoms as spheres?

Atoms are usually drawn as spheres because most atoms contain many electrons occupying a number of different orbitals. Therefore, the shape of an atom is obtained by superimposing all of its orbitals. If we superimpose the *s*, *p*, and *d* orbitals we get a spherical shape, as shown in **Figure 7.26◄**.

▲ **FIGURE 7.26 Why Atoms Are Spherical** Atoms are depicted as roughly spherical because all the orbitals together make up roughly spherical shape.

CHAPTER IN REVIEW

Key Terms

Section 7.1
quantum-mechanical model (253)

Section 7.2
electromagnetic radiation (254)
amplitude (254)
wavelength (λ) (254)
frequency (ν) (255)
electromagnetic spectrum (256)

gamma rays (256)
X-rays (256)
ultraviolet (UV) radiation (256)
visible light (257)
infrared (IR) radiation (257)
microwaves (257)
radio waves (257)
interference (257)
constructive interference (257)

destructive interference (257)
diffraction (257)
photoelectric effect (258)
photon (quantum) (260)

Section 7.3
emission spectrum (262)

Section 7.4
de Broglie relation (265)

complementary properties (267)
Heisenberg's uncertainty principle (267)
deterministic (268)
indeterminacy (269)

Section 7.5
orbital (269)
wave function (270)
quantum number (270)

principal quantum number (n) (270)

angular momentum quantum number (l) (270)

magnetic quantum number (m_l) (270)

spin quantum number (m_s) (271)

electron spin (271)

principal level (shell) (271)

sublevel (subshell) (271)

Section 7.6

probability density (276)

radial distribution function (277)

node (277)

phase (280)

Key Concepts

The Realm of Quantum Mechanics (7.1)

▶ The theory of quantum mechanics explains the behavior of particles in the atomic and subatomic realms. These particles include photons (particles of light) and electrons.

▶ Because the electrons of an atom determine many of its chemical and physical properties, quantum mechanics is foundational to understanding chemistry.

The Nature of Light (7.2)

▶ Light is a type of electromagnetic radiation—a form of energy embodied in oscillating electric and magnetic fields that travels though space at 3.00×10^8 m/s. Light has both a wave nature and a particle nature.

▶ The wave nature of light is characterized by its wavelength—the distance between wave crests—and the ability of light to experience interference (constructive or destructive) and diffraction. Its particle nature is characterized by the energy carried in each photon.

▶ The electromagnetic spectrum includes all wavelengths of electromagnetic radiation from gamma rays (high energy per photon, short wavelength) to radio waves (low energy per photon, long wavelength). Visible light is a tiny sliver in the middle of the electromagnetic spectrum.

Atomic Spectroscopy (7.3)

▶ Atomic spectroscopy is the study of the light absorbed and emitted by atoms when an electron makes a transition from one energy level to another.

▶ The wavelengths absorbed or emitted depend on the energy differences between the levels involved in the transition; large energy differences result in short wavelengths and small energy differences result in long wavelengths.

The Wave Nature of Matter (7.4)

▶ Electrons have a wave nature with an associated wavelength, as quantified by the de Broglie relation.

▶ The wave nature and particle nature of matter are complementary, which means that the more you know of one, the less you know of the other. The wave–particle duality of electrons is quantified in Heisenberg's uncertainty principle, which states that there is a limit to how well we can know both the position of an electron (associated with the electron's particle nature) and the velocity of an electron (associated with the electron's wave nature)—the more accurately one is measured, the greater the uncertainty in the other.

▶ The inability to simultaneously know both the position and the velocity of an electron results in indeterminacy, the inability to predict a trajectory for an electron. Consequently electron behavior is described differently than the behavior of everyday-sized particles.

▶ The trajectory we normally associate with macroscopic objects is replaced, for electrons, with statistical descriptions that show, not the electron's path, but the region where it is most likely to be found.

The Quantum-Mechanical Model of the Atom (7.5, 7.6)

▶ The most common way to describe electrons in atoms according to quantum mechanics is to solve the Schrödinger equation for the energy states of the electrons within the atom. When the electron is in these states, its energy is well-defined but its position is not. The position of an electron is described by a probability distribution map called an orbital.

▶ The solutions to the Schrödinger equation (including the energies and orbitals) are characterized by three quantum numbers: n, l, and m_l.

▶ The principal quantum number (n) determines the energy of the electron and the size of the orbital; the angular momentum quantum number (l) determines the shape of the orbital; and the magnetic quantum number (m_l) determines the orientation of the orbital.

Key Equations and Relationships

Relationship between Frequency (ν), Wavelength (λ), and the Speed of Light (c) (7.2)

$$\nu = \frac{c}{\lambda}$$

Relationship between Energy (E), Frequency (ν), Wavelength (λ), and Planck's Constant (h) (7.2)

$$E = h\nu$$

$$E = \frac{hc}{\lambda}$$

de Broglie Relation: Relationship between Wavelength (λ), Mass (m), and Velocity (v) of a Particle (7.4)

$$\lambda = \frac{h}{mv}$$

Heisenberg's Uncertainty Principle: Relationship between a Particle's Uncertainty in Position (Δx) and Uncertainty in Velocity (Δv) (7.4)

$$\Delta x \times m\Delta v \geq \frac{h}{4\pi}$$

Energy of an Electron in an Orbital with Quantum Number n in a Hydrogen Atom (7.5)

$$E_n = -2.18 \times 10^{-18} \text{ J} \left(\frac{1}{n^2}\right) \quad (n = 1, 2, 3, \ldots)$$

Key Learning Objectives

Chapter Objectives	Assessment
Calculating the Wavelength and Frequency of Light (7.2)	Example 7.1 For Practice 7.1 Exercises 5, 6
Calculating the Energy of a Photon (7.2)	Example 7.2 For Practice 7.2 For More Practice 7.2 Exercises 7–12
Relating Wavelength, Energy, and Frequency to the Electromagnetic Spectrum (7.2)	Example 7.3 For Practice 7.3 Exercises 3, 4
Using the de Broglie Relation to Calculate Wavelength (7.4)	Example 7.4 For Practice 7.4 Exercises 15–18
Relating Quantum Numbers to One Another and to Their Corresponding Orbitals (7.5)	Examples 7.5, 7.6 For Practice 7.5, 7.6 Exercises 21–28
Relating the Wavelength of Light to Transitions in the Hydrogen Atom (7.5)	Example 7.7 For Practice 7.7 For More Practice 7.7 Exercises 35–38

EXERCISES

Problems by Topic

Electromagnetic Radiation

1. The distance from the sun to Earth is 1.496×10^8 km. How long does it take light to travel from the sun to Earth?

2. The star nearest to our sun is Proxima Centauri, at a distance of 4.3 light-years from the sun. A light-year is the distance that light travels in one year (365 days). How far away, in km, is Proxima Centauri from the sun?

3. List these types of electromagnetic radiation in order of (i) increasing wavelength and (ii) increasing energy per photon:
   a. radio waves   b. microwaves
   c. infrared radiation   d. ultraviolet radiation

4. List these types of electromagnetic radiation in order of (i) increasing frequency and (ii) decreasing energy per photon:
   a. gamma rays   b. radio waves
   c. microwaves   d. visible light

5. Calculate the frequency of each wavelength of electromagnetic radiation:
   a. 632.8 nm (wavelength of red light from helium–neon laser)
   b. 503 nm (wavelength of maximum solar radiation)
   c. 0.052 nm (a wavelength contained in medical X-rays)

6. Calculate the wavelength of each frequency of electromagnetic radiation:
   a. 100.2 MHz (typical frequency for FM radio broadcasting)
   b. 1070 kHz (typical frequency for AM radio broadcasting) (assume four significant figures)
   c. 835.6 MHz (common frequency used for cell phone communication)

7. Calculate the energy of a photon of electromagnetic radiation at each of the wavelengths indicated in Problem 5.

8. Calculate the energy of a photon of electromagnetic radiation at each of the frequencies indicated in Problem 6.

9. A laser pulse with wavelength 532 nm contains 4.88 mJ of energy. How many photons are in the laser pulse?

10. A heat lamp produces 41.7 watts of power at a wavelength of 6.5 μm. How many photons are emitted per second? (1 watt = 1 J/s)

11. Determine the energy of 1 mol of photons for each kind of light. (Assume three significant figures.)
    a. infrared radiation (1500 nm)
    b. visible light (500 nm)
    c. ultraviolet radiation (150 nm)

12. How much energy is contained in 1 mol of each type of photon?
    a. X-ray photons with a wavelength of 0.155 nm
    b. γ-ray photons with a wavelength of 2.55×10^{-5} nm

The Wave Nature of Matter and the Uncertainty Principle

13. Make a sketch of the interference pattern that results from the diffraction of electrons passing through two closely spaced slits.

14. What happens to the interference pattern described in Problem 13 if the rate of electrons going through the slits is decreased to one electron per hour? What happens to the pattern if we try to determine which slit the electron goes through by using a laser placed directly behind the slits?

15. Calculate the wavelength of an electron traveling at 1.15×10^5 m/s.

16. An electron has a de Broglie wavelength of 225 nm. What is the speed of the electron?

17. Calculate the de Broglie wavelength of a 143-g baseball traveling at 95 mph. Why is the wave nature of matter not important for a baseball?

18. A 0.22-caliber handgun fires a 2.7-g bullet at a velocity of 765 m/s. Calculate the de Broglie wavelength of the bullet. Is the wave nature of matter significant for bullets?

Orbitals and Quantum Numbers

19. Which electron is, on average, closer to the nucleus: an electron in a 2s orbital or an electron in a 3s orbital?

20. Which electron is, on average, further from the nucleus: an electron in a 3p orbital or an electron in a 4p orbital?

21. What are the possible values of l for each value of n?
    a. 1   b. 2   c. 3   d. 4

22. What are the possible values of m_l for each value of l?

    a. 0   **b.** 1   **c.** 2   **d.** 3

23. For the $n = 3$ level, list all the possible values of l and m_l. How many orbitals does the $n = 3$ level contain?

24. For the $n = 4$ level, list all the possible values of l and m_l. How many orbitals does the $n = 4$ level contain?

25. What are the possible values of m_s?

26. What do each of the possible values of m_s in the previous problem specify?

27. Which set of quantum numbers *cannot* occur together to specify an orbital?

    a. $n = 2, l = 1, m_l = -1$

    b. $n = 3, l = 2, m_l = 0$

    c. $n = 3, l = 3, m_l = 2$

    d. $n = 4, l = 3, m_l = 0$

28. Which combinations of n and l represent real orbitals and which are impossible?

    a. $1s$ **b.** $2p$ **c.** $4s$ **d.** $2d$

29. Make a sketch of the $1s$ and $2p$ orbitals. How would the $2s$ and $3p$ orbitals differ from the $1s$ and $2p$ orbitals?

30. Make a sketch of the $3d$ orbitals. How would the $4d$ orbitals differ from the $3d$ orbitals?

Atomic Spectroscopy

31. An electron in a hydrogen atom is excited with electrical energy to an excited state with $n = 2$. The atom then emits a photon. What is the value of n for the electron after the emission?

32. Determine whether each transition in the hydrogen atom corresponds to absorption or emission of energy.

    a. $n = 3 \rightarrow n = 1$

    b. $n = 2 \rightarrow n = 4$

    c. $n = 4 \rightarrow n = 3$

33. According to the quantum-mechanical model for the hydrogen atom, which electron transition produces light with the longer wavelength: $2p \rightarrow 1s$ or $3p \rightarrow 1s$?

34. According to the quantum-mechanical model for the hydrogen atom, which transition produces light with the longer wavelength: $3p \rightarrow 2s$ or $4p \rightarrow 3p$?

35. Calculate the wavelength of the light emitted when an electron in a hydrogen atom makes each transition and indicate the region of the electromagnetic spectrum (infrared, visible, ultraviolet, etc.) where the light is found.

    a. $n = 2 \rightarrow n = 1$ **b.** $n = 3 \rightarrow n = 1$

    c. $n = 4 \rightarrow n = 2$ **d.** $n = 5 \rightarrow n = 2$

36. Calculate the frequency of the light emitted when an electron in a hydrogen atom makes each transition:

    a. $n = 4 \rightarrow n = 3$ **b.** $n = 5 \rightarrow n = 1$

    c. $n = 5 \rightarrow n = 4$ **d.** $n = 6 \rightarrow n = 5$

37. An electron in the $n = 7$ level of the hydrogen atom relaxes to a lower energy level, emitting light of 397 nm. What is the value of n for the level to which the electron relaxed?

38. An electron in a hydrogen atom relaxes to the $n = 4$ level, emitting light of 114 THz. What is the value of n for the level in which the electron originated?

Cumulative Problems

39. Ultraviolet radiation and radiation of shorter wavelengths can damage biological molecules because they carry enough energy to break bonds within the molecules. A carbon–carbon bond requires 348 kJ/mol to break. What is the longest wavelength of radiation with enough energy to break carbon–carbon bonds?

40. The human eye contains a molecule called 11-*cis*-retinal that changes conformation when struck with light of sufficient energy. The change in conformation triggers a series of events that results in an electrical signal being sent to the brain. The minimum energy required to change the conformation of 11-*cis*-retinal within the eye is about 164 kJ/mol. Calculate the longest wavelength visible to the human eye.

41. An argon ion laser puts out 5.0 W of continuous power at a wavelength of 532 nm. The diameter of the laser beam is 5.5 mm. If the laser is pointed toward a pinhole with a diameter of 1.2 mm, how many photons will travel through the pinhole per second? Assume that the light intensity is equally distributed throughout the entire cross-divisional area of the beam. (1 W = 1 J/s)

42. A green leaf has a surface area of 2.50 cm². If solar radiation is 1000 W/m², how many photons strike the leaf every second? Assume three significant figures and an average wavelength of 504 nm for solar radiation.

43. In a technique used for surface analysis called Auger electron spectroscopy (AES), electrons are accelerated toward a metal surface. These electrons cause the emissions of secondary electrons—called Auger electrons—from the metal surface. The kinetic energy of the auger electrons depends on the composition of the surface. The presence of oxygen atoms on the surface results in auger electrons with a kinetic energy of approximately 506 eV. What is the de Broglie wavelength of this electron?

$$\left[KE = \tfrac{1}{2}mv^2; 1 \text{ electron volt (eV)} = 1.602 \times 10^{-19} \text{ J} \right]$$

44. An X-ray photon of wavelength 0.989 nm strikes a surface. The emitted electron has a kinetic energy of 969 eV. What is the binding energy of the electron in kJ/mol?

$$\left[KE = \tfrac{1}{2}mv^2; 1 \text{ electron volt (eV)} = 1.602 \times 10^{-19} \text{ J} \right]$$

45. Ionization involves completely removing an electron from an atom. How much energy is required to ionize a hydrogen atom in its ground (or lowest energy) state? What wavelength of light contains enough energy in a single photon to ionize a hydrogen atom?

46. The energy required to ionize sodium is 496 kJ/mol. What minimum frequency of light is required to ionize sodium?

47. Suppose that in an alternate universe, the possible values of l were the integer values from 0 to n (instead of 0 to $n - 1$). Assuming no other differences from this universe, how many orbitals would exist in each level?

    a. $n = 1$ **b.** $n = 2$ **c.** $n = 3$

48. Suppose that, in an alternate universe, the possible values of m_l were the integer values including 0 ranging from $-l - 1$ to $l + 1$ (instead of simply $-l$ to $+l$). How many orbitals would exist in each sublevel?

    a. s sublevel   **b.** p sublevel   **c.** d sublevel

49. An atomic emission spectrum of hydrogen shows three wavelengths: 1875 nm, 1282 nm, and 1093 nm. Assign these wavelengths to transitions in the hydrogen atom.

50. An atomic emission spectrum of hydrogen shows three wavelengths: 121.5 nm, 102.6 nm, and 97.23 nm. Assign these wavelengths to transitions in the hydrogen atom.

51. The binding energy of electrons in a metal is 193 kJ/mol. Find the threshold frequency of the metal.

52. In order for a thermonuclear fusion reaction of two deuterons ($^{2}_{1}H^{+}$) to occur, the deuterons must collide each with a velocity of about 1×10^{6} m/s. Find the wavelength of such a deuteron.

53. The speed of sound in air is 344 m/s at room temperature. The lowest frequency of a large organ pipe is 30 s^{-1} and the highest frequency of a piccolo is 1.5×10^{4} s^{-1}. Find the difference in wavelength between these two sounds.

54. The distance from Earth to the sun is 1.5×10^{8} km. Find the number of crests in a light wave of frequency 1.0×10^{14} s^{-1} traveling from the sun to the Earth.

55. The iodine molecule can be photodissociated into iodine atoms in the gas phase with light of wavelengths shorter than about 792 nm. A 100.0 mL glass tube contains 55.7 mtorr of gaseous iodine at 25.0 °C. What minimum amount of light energy must be absorbed by the iodine in the tube to dissociate 15.0% of the molecules?

56. A 5.00 mL ampule of a 0.100 M solution of naphthalene in hexane is excited with a flash of light. The naphthalene emits 15.5 J of energy at an average wavelength of 349 nm. What percentage of the naphthalene molecules emitted a photon?

57. A laser produces 20.0 mW of red light. In 1.00 hr, the laser emits 2.29×10^{20} photons. What is the wavelength of the laser?

58. A 1064 nm laser consumes 150.0 watts of electrical power and produces a stream of 1.33×10^{19} photons per second. What is the percent efficiency of the laser in converting electrical power to light?

Challenge Problems

59. An electron confined to a one-dimensional box has energy levels given by the equation

$$E_n = n^2h^2/8\,mL^2$$

where n is a quantum number with possible values of $1, 2, 3, \ldots$, m is the mass of the particle, and L is the length of the box.
   a. Calculate the energies of the $n = 1$, $n = 2$, and $n = 3$ levels for an electron in a box with a length of 155 pm.
   b. Calculate the wavelength of light required to make a transition from $n = 1 \rightarrow n = 2$ and from $n = 2 \rightarrow n = 3$. In what region of the electromagnetic spectrum do these wavelengths lie?

60. The energy of a vibrating molecule is quantized much like the energy of an electron in the hydrogen atom. The energy levels of a vibrating molecule are given by the equation

$$E_n = (n + \tfrac{1}{2})h\nu$$

where n is a quantum number with possible values of $1, 2, \ldots$, and ν is the frequency of vibration. The vibration frequency of HCl is approximately 8.85×10^{13} s^{-1}. What minimum energy is required to excite a vibration in HCl? What wavelength of light is required to excite this vibration?

61. The wave functions for the 1s and 2s orbitals are specified by these equations:

   1s $\psi = (1/\pi)^{1/2}(1/a_0^{3/2})\exp(-r/a_0)$

   2s $\psi = (1/32\pi)^{1/2}(1/a_0^{3/2})(2 - r/a_0)\exp(-r/a_0)$

where a_0 is a constant ($a_0 = 53$ pm) and r is the distance from the nucleus. Make a plot of each of these wave functions for values of

r ranging from 0 pm to 200 pm. Describe the differences in the plots and identify the node in the 2s wave function.

62. Before quantum mechanics was developed, Johannes Rydberg developed the following equation that predicted the wavelengths (λ) in the atomic spectrum of hydrogen:

$$1/\lambda = R(1/m^2 - 1/n^2)$$

In this equation R is a constant and m and n are integers. Use the quantum-mechanical model for the hydrogen atom to derive the Rydberg equation.

63. Find the velocity of an electron emitted by a metal whose threshold frequency is 2.25×10^{14} s^{-1} when it is exposed to visible light of wavelength 5.00×10^{-7} m.

64. Water is exposed to infrared radiation of wavelength 2.8×10^{-4} cm. Assume that all the radiation is absorbed and converted to heat. How many photons will be required to raise the temperature of 2.0 g of water by 2.0 K?

65. The 2005 Nobel Prize in Physics was given, in part, to scientists who had made ultrashort pulses of light. These pulses are important in making measurements involving very short time periods. One challenge in making such pulses is the uncertainty principle, which can be stated with respect to energy and time as $\Delta E \cdot \Delta t > h/4\pi$. What is the energy uncertainty (ΔE) associated with a short pulse of laser light that lasts for only 5.0 femtoseconds (fs)? Suppose the low-energy end of the pulse had a wavelength of 722 nm. What is the wavelength of the high-energy end of the pulse that is limited only by the uncertainty principle?

66. A metal whose threshold frequency is 6.71×10^{14} s^{-1} emits an electron with a velocity of 6.95×10^{5} m/s when radiation of 1.01×10^{15} s^{-1} strikes the metal. Use these data to calculate the mass of the electron.

Conceptual Problems

67. Explain the difference between the Bohr model for the hydrogen atom and the quantum-mechanical model. Is the Bohr model consistent with Heisenberg's uncertainty principle?

68. The light emitted from one of these electronic transitions ($n = 4 \rightarrow n = 3$ or $n = 3 \rightarrow n = 2$) in the hydrogen atom caused the photoelectric effect in a particular metal while light

from the other transition did not. Which transition was able to cause the photoelectric effect and why?

69. Which transition in the hydrogen atom will result in emitted light with the longest wavelength?
   a. $n = 4 \rightarrow n = 3$ b. $n = 2 \rightarrow n = 1$
   c. $n = 3 \rightarrow n = 2$

70. Determine whether an interference pattern is observed on the other side of the slits in each experiment.

   a. An electron beam is aimed at two closely spaced slits. The beam is attenuated to produce only 1 electron per minute.

   b. An electron beam is aimed at two closely spaced slits. A light beam is placed at each slit to determine when an electron goes through the slit.

   c. A high-intensity light beam is aimed at two closely spaced slits.

   d. A gun is fired at a solid wall containing two closely spaced slits. (Will the bullets that pass through the slits form an interference pattern on the other side of the solid wall?)

Answers to Conceptual Connections

The Photoelectric Effect

7.1 Observation A corresponds to 632 nm; observation B corresponds to 325 nm; and observation C corresponds to 455 nm. The shortest wavelength of light (highest energy per photon) must correspond to the photoelectrons with the greatest kinetic energy. The longest wavelength of light (lowest energy per photon) must correspond to the observation where no photoelectrons were observed.

The de Broglie Wavelength of Macroscopic Objects

7.2 Because of the baseball's large mass, its de Broglie wavelength is minuscule. (For a 150-g baseball, λ is on the order of 10^{-34} m.) This minuscule *wavelength* is insignificant compared to the size of the baseball itself, and therefore its effects are not measurable.

The Relationship between n and l

7.3 **(c)** Since l can have a maximum value of $n - 1$, and since $n = 3$, then l can have a maximum value of 2.

The Relationship between l and m_l

7.4 **(d)** Since m_l can have the integer values (including 0) between $-l$ and $+l$, and since $l = 2$, the possible values of m_l are $-2, -1, 0, +1,$ and $+2$.

Emission Spectra

7.5 **(c)** The energy difference between $n = 3$ and $n = 2$ is greatest because the energy spacings get closer together with increasing n. The greater energy difference results in an emitted photon of greater energy and therefore shorter wavelength.

Periodic Properties of the Elements

Beginning students of chemistry often think of the science as a mere collection of disconnected data to be memorized by brute force. Not at all! Just look at it properly and everything hangs together and makes sense. —Isaac Asimov (1920–1992)

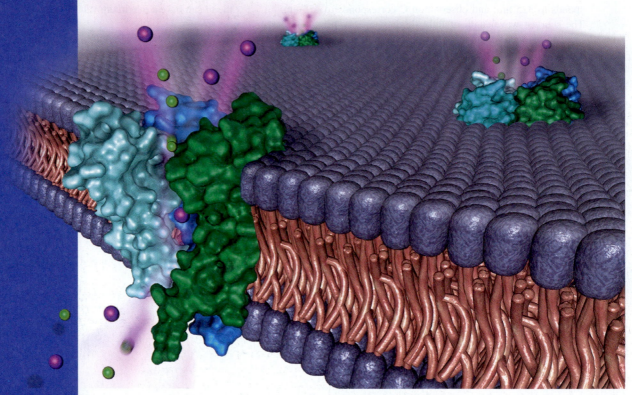

In order for a nerve cell to transmit a signal, sodium and potassium ions must flow in opposite directions through specific ion channels in the cell membrane.

GREAT ADVANCES IN SCIENCE come not only when a scientist sees something new, but also when a scientist sees what everyone else has seen in a new way. In other words, great scientists often see patterns where others have seen only disjointed facts. Such was the case in 1869 when Dmitri Mendeleev, a Russian chemistry professor, saw a pattern in the properties of elements. Mendeleev's insight led to the periodic law and the periodic table, arguably the single most important tool for the chemist. Mendeleev's periodic law and the periodic table summarize a large number of observations. Remember, however, that science proceeds by devising theories that explain the underlying reasons for laws and observations. Quantum mechanics, which we covered in Chapter 7, is the theory that explains the underlying reasons for the periodic table. Quantum mechanics explains the arrangement of the periodic table by relating chemical behavior to how electrons fill quantum-mechanical orbitals. In this chapter, we see a continuation of the theme we have been developing since page one of this book—the properties of macroscopic substances (in this case, the elements in the periodic table) are explained by the properties of the particles that compose them (in this case, atoms and their electrons).

8.1   Nerve Signal Transmission

As you sit reading this book, tiny pumps in the membranes of your cells are working hard to transport ions—especially sodium (Na^+) and potassium (K^+)—through those membranes. Amazingly, the two ions are pumped in opposite directions. Sodium ions are pumped *out of cells*, while potassium ions are pumped *into cells*. The result is a *chemical gradient* for each ion: the concentration of sodium is higher outside the cell than within, while just the opposite is true for potassium. These ion pumps are analogous to the water pumps in a high-rise building that pump water against the force of gravity to a tank on the roof. Structures within the membrane, called ion channels, are like the building's faucets. When these open, sodium and potassium ions flow back down their gradients—sodium flowing in and potassium flowing out. This movement of ions is the basis for the transmission of nerve signals in the brain and throughout the body. Every move you make, every thought you have, and every sensation you experience is mediated by these ion movements.

How do the pumps and channels differentiate between sodium and potassium ions so as to selectively move one out of the cell and the other into the cell? To answer this question, we must examine the ions more closely. Both ions are cations of group 1A metals. All group 1A metals tend to lose one electron to form cations with a 1+ charge, so that cannot be the decisive factor. However, potassium (atomic number 19) lies directly below sodium in the periodic table (atomic number 11), indicating that it has more protons, neutrons, and electrons than sodium. How do these additional subatomic particles affect the properties of potassium (relative to sodium)? As we will see in this chapter, although a higher atomic number does not always result in a larger ion (or atom), it does in the cases of sodium and potassium. The potassium ion has a radius of 133 pm while the sodium ion has a radius of 95 pm. (Recall that 1 pm = 10^{-12} m.) The pumps and channels within cell membranes are so sensitive that they can distinguish between the sizes of these two ions and selectively allow only one or the other to pass.

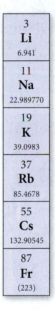

| 3 |
| **Li** |
| 6.941 |
| 11 |
| **Na** |
| 22.989770 |
| 19 |
| **K** |
| 39.0983 |
| 37 |
| **Rb** |
| 85.4678 |
| 55 |
| **Cs** |
| 132.90545 |
| 87 |
| **Fr** |
| (223) |

▲The group 1A metals. Potassium lies directly beneath sodium in the periodic table.

The relationship between the size of the ion and its ability to move through an ion channel is complex, involving a number of factors that are beyond the scope of this text.

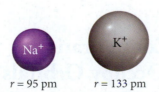

Na^+ K^+

$r = 95$ pm $r = 133$ pm

The relative size of sodium and potassium ions is an example of a **periodic property**: one that is predictable based on an element's position within the periodic table. In this chapter, we examine several periodic properties of elements, including atomic radius, ionization energy, and electron affinity. We will see that these properties, as well as the overall arrangement of the periodic table, are explained by quantum-mechanical theory, which we examined in Chapter 7. The arrangement of elements in the periodic table— originally based on similarities in their properties—reflects how electrons fill quantum-mechanical orbitals.

8.2 The Development of the Periodic Table

The modern periodic table is credited primarily to the Russian chemist Dmitri Mendeleev (1834–1907), even though a similar organization had been suggested by the German chemist Julius Lothar Meyer (1830–1895). Recall from Chapter 2 that Mendeleev's table is based on the periodic law, which states that when elements are arranged in order of increasing mass, their properties recur periodically. Mendeleev arranged the elements in a table in which mass increased from left to right and elements with similar properties fell in the same columns.

Mendeleev's arrangement was a huge success, allowing him to predict the existence and properties of yet undiscovered elements such as eka-aluminum (later discovered and named gallium) and eka-silicon (later discovered and named germanium), whose properties are summarized in **Figure 8.1▼**. (*Eka* means "the one beyond" or "the next one.") However, Mendeleev did have some difficulties. For example, according to accepted values of atomic masses, tellurium (with higher mass) should come *after* iodine. But based on their properties, Mendeleev placed tellurium *before* iodine and suggested that the mass of tellurium was erroneous. The mass was correct; later work by the English physicist Henry Moseley (1887–1915) showed that listing elements according to *atomic number*, rather than atomic mass, resolved this problem and resulted in even better correlation with elemental properties.

The theory that explains the existence of the periodic law (Mendeleev's table is really just an expression of the periodic law) is quantum-mechanical theory. We now turn to exploring the connection between the periodic table and quantum-mechanical theory.

▲ Dmitri Mendeleev is credited with the arrangement of the periodic table.

Gallium (eka-aluminum)

	Mendeleev's predicted properties	Actual properties
Atomic mass	About 68 amu	69.72 amu
Melting point	Low	29.8 °C
Density	5.9 g/cm^3	5.90 g/cm^3
Formula of oxide	X$_2$O$_3$	Ga$_2$O$_3$
Formula of chloride	XCl$_3$	GaCl$_3$

Germanium (eka-silicon)

	Mendeleev's predicted properties	Actual properties
Atomic mass	About 72 amu	72.64 amu
Density	5.5 g/cm^3	5.35 g/cm^3
Formula of oxide	XO$_2$	GeO$_2$
Formula of chloride	XCl$_4$	GeCl$_4$

▲ **FIGURE 8.1 Eka-aluminum and eka-silicon** Mendeleev's arrangement of elements in the periodic table allowed him to predict the existence of these elements, now known as gallium and germanium, and anticipate their properties.

8.3 Electron Configurations: How Electrons Occupy Orbitals

Quantum-mechanical theory describes the behavior of electrons in atoms. Since chemical bonding involves the transfer or sharing of electrons, quantum-mechanical theory helps us understand and describe chemical behavior. As we saw in Chapter 7, electrons in atoms exist within orbitals. An **electron configuration** for an atom shows the particular orbitals that electrons occupy for that atom. For example, consider the **ground state**—or lowest energy state—electron configuration for a hydrogen atom:

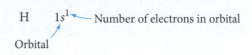

H   1s^1 ← Number of electrons in orbital

Orbital

The electron configuration indicates that hydrogen's one electron is in the $1s$ orbital. Electrons generally occupy the lowest energy orbitals available. Since the $1s$ orbital is the lowest energy orbital in hydrogen (see Section 7.5), the electron occupies that orbital. The solutions to the Schrödinger equation (the atomic orbitals and their energies) that we described in Chapter 7 were for the hydrogen atom. What do the atomic orbitals of other atoms look like? What are their relative energies?

Unfortunately, the Schrödinger equation for multielectron atoms is so complicated—because of new terms introduced into the equation by the interactions of the electrons with one another—that it cannot be solved exactly. However, approximate solutions indicate that the orbitals in multielectron atoms are hydrogen-like—they are similar to the s, p, d, and f orbitals that we discussed in Chapter 7. In order to see how the electrons in multielectron atoms occupy these hydrogen-like orbitals, we must examine two additional concepts: *the effects of electron spin*, a fundamental property of all electrons (first discussed in Section 7.5) that affects the number of electrons that are allowed in one orbital; and *sublevel energy splitting*, which applies to multielectron atoms and determines the order of orbital filling within a level.

Electron Spin and the Pauli Exclusion Principle

The electron configuration of hydrogen ($1s^1$) can be represented in a slightly different way by an **orbital diagram**, which gives similar information, but symbolizes the electron as an arrow in a box that represents the orbital. The orbital diagram for a hydrogen atom is

H   ↑
    $1s$

In an orbital diagram, the direction of the arrow (pointing up or pointing down) represents the orientation of the *electron's spin*. Recall from Section 7.5 that the orientation of the electron's spin is quantized, with only two possibilities—spin up ($m_s = +\frac{1}{2}$) and spin down ($m_s = -\frac{1}{2}$). In an orbital diagram, $m_s = +\frac{1}{2}$ is represented with a half-arrow pointing up (↑) and $m_s = -\frac{1}{2}$ is represented with a half-arrow pointing down (↓). In a collection of hydrogen atoms, the electrons in about half of the atoms are spin up and the other half are spin down. Since no other electrons are present within the atom, we conventionally represent the hydrogen atom electron configuration with its one electron as spin up.

Helium is the first element on the periodic table that contains two electrons. Its two electrons occupy the $1s$ orbital.

He $1s^2$

How do the spins of the two electrons in helium align relative to each other? The answer to this question is described by the **Pauli exclusion principle**, formulated by Wolfgang Pauli (1900–1958) in 1925:

> **Pauli exclusion principle: No two electrons in an atom can have the same four quantum numbers**.

Since two electrons occupying the same orbital have three identical quantum numbers (n, l, and m_l), they must have different spin quantum numbers. Since there are only two possible spin quantum numbers ($+\frac{1}{2}$ and $-\frac{1}{2}$), the Pauli exclusion principle implies that *each orbital can have a maximum of only two electrons, with opposing spins*. By applying the exclusion principle, we can write an electron configuration and orbital diagram for helium as follows:

Electron configuration   Orbital diagram
He $1s^2$ ↑↓
                                 $1s$

The table below shows the four quantum numbers for each of the two electrons in helium.

n	l	m_l	m_s
1	0	0	$+\frac{1}{2}$
1	0	0	$-\frac{1}{2}$

The two electrons have three quantum numbers in common (because they are in the same orbital) but have different spin quantum numbers (as indicated by the opposing half-arrows in the orbital diagram).

Sublevel Energy Splitting in Multielectron Atoms

A major difference in the (approximate) solutions to the Schrödinger equation for multi-electron atoms compared to the solutions for the hydrogen atom is the energy ordering of the orbitals. In the hydrogen atom, the energy of an orbital depends only on n, the principal quantum number. For example, the $3s$, $3p$, and $3d$ orbitals (which are all empty for hydrogen in its lowest energy state) all have the same energy—they are **degenerate**. The orbitals within a principal level of a *multielectron atom*, by contrast, are not degenerate—their energy depends on the value of l. We say that the energies of the sublevels are *split*. In general, the lower the value of l within a principal level, the lower the energy of the corresponding orbital. Thus, for a given value of n:

$$E(s \text{ orbital}) < E(p \text{ orbital}) < E(d \text{ orbital}) < E(f \text{ orbital})$$

In order to understand sublevel energy splitting, we must examine three key concepts associated with the energy of an electron in the vicinity of a nucleus: 1) Coulomb's law, which describes the interactions between charged particles; 2) shielding, which describes how one electron can shield another electron from the full charge of the nucleus; and 3) penetration, which describes how one atomic orbital can overlap spatially with another, thus penetrating into a region that is close to the nucleus (and therefore less shielded from nuclear charge).

Coulomb's Law

The attractions and repulsions between charged particles, first introduced in Section 2.4, are described by **Coulomb's law**, which states that the potential energy (E) of two charged particles depends on their charges (q_1 and q_2) and their separation (r):

$$E = \frac{1}{4\pi\varepsilon_0}\frac{q_1 q_2}{r} \tag{8.1}$$

In this equation, ε_0 is a constant ($\varepsilon_0 = 8.85 \times 10^{-12}\ \text{C}^2/\text{J}\cdot\text{m}$). The potential energy is positive for charges of the same sign (plus $\times$ plus, or minus $\times$ minus), and negative for charges of opposite sign (plus $\times$ minus, or minus $\times$ plus). The *magnitude* of the potential energy depends inversely on the separation between the charged particles. We can draw three important conclusions from Coulomb's law:

- For like charges, the potential energy (E) is positive and decreases as the particles get *farther apart* (as r increases). Since systems tend toward lower potential energy, like charges repel each other (in much the same way that the like poles of two magnets repel each other).

- For opposite charges, the potential energy is negative and becomes more negative as the particles get *closer together* (as r decreases). Therefore opposite charges (like opposite poles on a magnet) *attract each other*.

- The *magnitude* of the interaction between charged particles increases as the charges of the particles increases. Consequently, an electron with a charge of 1− is more strongly attracted to a nucleus with a charge of 2+ than it would be to a nucleus with a charge of 1+.

Conceptual Connection 8.1 Coulomb's Law

According to Coulomb's law, what happens to the potential energy of two oppositely charged particles as they get closer together?

(a) Their potential energy decreases.

(b) Their potential energy increases.

(c) Their potential energy does not change.

Shielding

For multielectron atoms, any one electron experiences both the positive charge of the nucleus (which is attractive) and the negative charges of the other electrons (which are repulsive). Within an atom or ion, we can think of the repulsion of one electron by other electrons as *screening* or **shielding** that electron from the full effects of the nuclear charge. Consider a lithium ion (Li^+). Since the lithium ion contains two electrons, its electron configuration is identical to that of helium:

$$Li^+ \; 1s^2$$

Now imagine bringing a third electron toward the lithium ion. When the third electron is far from the nucleus, it experiences the 3+ charge of the nucleus through the *screen* or *shield* of the 2− charge of the two 1s electrons, as shown in **Figure 8.2(a)▼**. We can think of the third electron as experiencing an **effective nuclear charge** (Z_{eff}) of approximately 1+ (3+ from the nucleus and 2− from the electrons, for a net charge of 1+). The inner electrons in effect *shield* the outer electron from the full nuclear charge.

Penetration

Now imagine allowing this third electron to come closer to the nucleus. As the electron *penetrates* the electron cloud of the 1s electrons, it begins to experience the 3+ charge of the nucleus more fully because it is less shielded by the intervening electrons. If the electron could somehow get closer to the nucleus than the 1s electrons, it would experience the full 3+ charge, as shown in **Figure 8.2(b)▼**. In other words, as the outer electron *penetrates* into the region occupied by the inner electrons, it experiences a greater nuclear charge and therefore (according to Coulomb's law) a lower energy.

Electron Spatial Distributions and Sublevel Splitting

We now have examined the concepts we need to understand the energy splitting of the sublevels within a principal level. The splitting is a result of the spatial distributions of

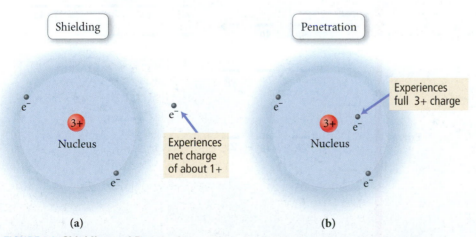

▲ **FIGURE 8.2 Shielding and Penetration**   **(a)** An electron far from the nucleus is partly shielded by the electrons in the 1s orbital, reducing the effective net nuclear charge that it experiences. **(b)** An electron that penetrates the electron cloud of the 1s orbital experiences more of the nuclear charge.

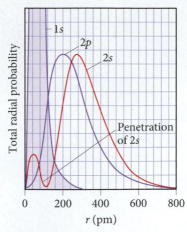

▲ FIGURE 8.3 Radial Distribution Functions for the 1s, 2s, and 2p Orbitals The bump near $r = 0$ (near the nucleus) for the 2s orbital represents a significant probability of the electron being found very close to the nucleus.

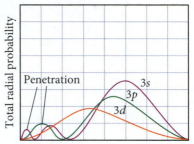

▲ FIGURE 8.4 Radial Distribution Functions for the 3s, 3p, and 3d Orbitals The 3s electrons penetrate most deeply into the inner orbitals, are least shielded, and experience the greatest effective nuclear charge. The 3d electrons penetrate least. This accounts for the energy ordering of the sublevels: $d > p > s$.

electrons within a sublevel. Recall from Section 7.6 that the radial distribution function for an atomic orbital shows the total probability of finding the electron within a thin spherical shell at a distance r from the nucleus. **Figure 8.3◄** shows the radial distribution functions of the 2s and 2p orbitals superimposed on one another (the radial distribution function of the 1s orbital is also shown).

Notice that, in general, an electron in a 2p orbital has a greater probability of being found closer to the nucleus than an electron in a 2s orbital. We might initially expect, therefore, that the 2p orbital would be lower in energy. However, exactly the opposite is the case—the 2s orbital is actually lower in energy, *but only when the 1s orbital is occupied*. (When the 1s orbital is empty, the 2s and 2p orbitals are degenerate.) Why? The reason is the bump near $r = 0$ (near the nucleus) for the 2s orbital. This bump represents a significant probability of the electron being found very close to the nucleus. Even more importantly, this area of the probability penetrates into the 1s orbital—it gets into the region where shielding by the 1s electrons is less effective. In contrast, most of the probability in the radial distribution function of the 2p orbital lies *outside* the radial distribution function of the 1s orbital. Consequently, almost all of the 2p orbital is shielded from nuclear charge by the 1s orbital. The end result is that the 2s orbital—since it experiences more of the nuclear charge due to its greater *penetration*—is lower in energy than the 2p orbital. The results are similar when we compare the 3s, 3p, and 3d orbitals. The s orbitals penetrate more fully than the p orbitals, which in turn penetrate more fully than the d orbitals, as shown in **Figure 8.4◄**.

Figure 8.5▼ shows the energy ordering of a number of orbitals in multielectron atoms. Notice these points about the diagram in Figure 8.5:

- Because of penetration, the sublevels of each principal level are *not* degenerate for multielectron atoms.

- In the fourth and fifth principal levels, the effects of penetration become so significant that the 4s orbital lies lower in energy than the 3d orbitals and the 5s orbital lies lower in energy than the 4d orbitals.

- The energy separations between one set of orbitals and the next become smaller for 4s orbitals and beyond so that the relative energy ordering of these orbitals can actually vary among elements. These variations result in anomalies in the electron configurations of the transition metals and their ions (as we shall see later).

General Energy Ordering of Orbitals for Multielectron Atoms

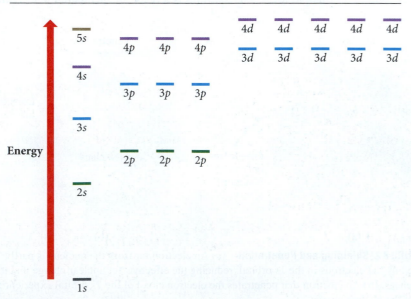

▲ FIGURE 8.5 General Energy Ordering of Orbitals for Multielectron Atoms

Conceptual Connection 8.2 Penetration and Shielding

Which statement is true?

(a) An orbital that penetrates into the region occupied by core electrons is more shielded from nuclear charge than an orbital that does not penetrate and will therefore have a higher energy.

(b) An orbital that penetrates into the region occupied by core electrons is less shielded from nuclear charge than an orbital that does not penetrate and will therefore have a higher energy.

(c) An orbital that penetrates into the region occupied by core electrons is less shielded from nuclear charge than an orbital that does not penetrate and will therefore have a lower energy.

(d) An orbital that penetrates into the region occupied by core electrons is more shielded from nuclear charge than an orbital that does not penetrate and will therefore have a lower energy.

Electron Configurations for Multielectron Atoms

Now that we know the energy ordering of orbitals in multielectron atoms, we can build ground state electron configurations for the rest of the elements. Since we know that electrons occupy the lowest energy orbitals available when the atom is in its ground state, and that only two electrons (with opposing spins) are allowed in each orbital, we can systematically build up the electron configurations for the elements. The pattern of orbital filling that reflects what we have just learned is known as the **aufbau principle** (the German word *aufbau* means "build up"). For lithium, which has three electrons, the electron configuration and orbital diagram are

> Unless otherwise specified, we will use the term "electron configuration" to mean the ground state (or lowest energy) configuration.

Electron configuration | Orbital diagram

Li $1s^2 2s^1$

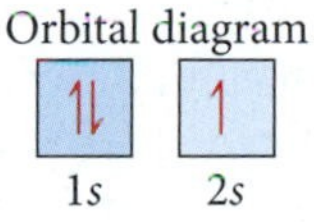

$\begin{array}{cc} \uparrow\downarrow & \uparrow \\ 1s & 2s \end{array}$

For carbon, with atomic number 6 and therefore 6 electrons, the electron configuration and orbital diagram are

> Remember that the number of electrons in an atom is equal to its atomic number.

Electron configuration | Orbital diagram

C $1s^2 2s^2 2p^2$

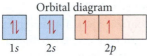

$\begin{array}{ccc} 1s & 2s & 2p \end{array}$

Notice that the $2p$ electrons occupy the p orbitals (of equal energy) singly, rather than pairing in one orbital. This way of filling orbitals is known as **Hund's rule**, which states that *when filling degenerate orbitals, electrons fill them singly first, with parallel spins*. Hund's rule is a result of an atom's tendency to find the lowest energy state possible. When two electrons occupy separate orbitals of equal energy, the repulsive interaction between them is lower than when they occupy the same orbital because the electrons are spread out over a larger region of space.

Summarizing Orbital Filling:

▶ Electrons occupy orbitals so as to minimize the energy of the atom; therefore, lower energy orbitals fill before higher energy orbitals. Orbitals fill in the following order: $1s\ 2s\ 2p\ 3s\ 3p\ 4s\ 3d\ 4p\ 5s\ 4d\ 5p\ 6s$.

▶ Orbitals can hold no more than two electrons each. When two electrons occupy the same orbital, their spins are opposite. This is another way of expressing the Pauli exclusion principle (no two electrons in one atom can have the same four quantum numbers).

▶ When orbitals of identical energy are available, electrons first occupy these orbitals singly with parallel spins rather than in pairs. Once the orbitals of equal energy are half-full, the electrons start to pair (Hund's rule).

Consider the electron configurations and orbital diagrams for elements with atomic numbers 3–10.

Symbol	Number of electrons	Electron configuration	Orbital diagram
Li	3	$1s^2 2s^1$	$\uparrow\downarrow$ (1s) $\uparrow$ (2s)
Be	4	$1s^2 2s^2$	$\uparrow\downarrow$ (1s) $\uparrow\downarrow$ (2s)
B	5	$1s^2 2s^2 2p^1$	$\uparrow\downarrow$ (1s) $\uparrow\downarrow$ (2s) $\uparrow$ □ □ (2p)
C	6	$1s^2 2s^2 2p^2$	$\uparrow\downarrow$ (1s) $\uparrow\downarrow$ (2s) $\uparrow$ $\uparrow$ □ (2p)
N	7	$1s^2 2s^2 2p^3$	$\uparrow\downarrow$ (1s) $\uparrow\downarrow$ (2s) $\uparrow$ $\uparrow$ $\uparrow$ (2p)
O	8	$1s^2 2s^2 2p^4$	$\uparrow\downarrow$ (1s) $\uparrow\downarrow$ (2s) $\uparrow\downarrow$ $\uparrow$ $\uparrow$ (2p)
F	9	$1s^2 2s^2 2p^5$	$\uparrow\downarrow$ (1s) $\uparrow\downarrow$ (2s) $\uparrow\downarrow$ $\uparrow\downarrow$ $\uparrow$ (2p)
Ne	10	$1s^2 2s^2 2p^6$	$\uparrow\downarrow$ (1s) $\uparrow\downarrow$ (2s) $\uparrow\downarrow$ $\uparrow\downarrow$ $\uparrow\downarrow$ (2p)

Notice that, as a result of Hund's rule, the p orbitals fill with single electrons before the electrons pair.

The electron configuration of neon represents the complete filling of the $n = 2$ principal level. When writing electron configurations for elements beyond neon, or beyond any other noble gas, the electron configuration of the previous noble gas—sometimes called the *inner electron configuration*—is often abbreviated by the symbol for the noble gas in square brackets. For example, the electron configuration of sodium is

$$\text{Na} \quad 1s^2 2s^2 2p^6 3s^1$$

This configuration can also be written using $[\text{Ne}]$ to represent the inner electrons:

$$\text{Na} \quad [\text{Ne}] \, 3s^1$$

$[\text{Ne}]$ represents $1s^2 2s^2 2p^6$, the electron configuration for neon.

To write an electron configuration for an element, first find its atomic number from the periodic table—this number equals the number of electrons. Then use the order of filling to distribute the electrons in the appropriate orbitals. Remember that each orbital can hold a maximum of 2 electrons. Consequently,

- The s sublevel has only one orbital and can therefore hold only 2 electrons.
- The p sublevel has three orbitals and can therefore hold 6 electrons.
- The d sublevel has five orbitals and can therefore hold 10 electrons.
- The f sublevel has seven orbitals and can therefore hold 14 electrons.

EXAMPLE 8.1   Electron Configurations

Write electron configurations for each element:

(a) Mg   **(b)** P   **(c)** Br   **(d)** Al

SOLUTION

(a) Mg Magnesium has 12 electrons. Distribute two of these into the $1s$ orbital, two into the $2s$ orbital, six into the $2p$ orbitals, and two into the $3s$ orbital.	Mg $1s^2 2s^2 2p^6 3s^2$ or $[\text{Ne}]\, 3s^2$
(b) P Phosphorus has 15 electrons. Distribute two of these into the $1s$ orbital, two into the $2s$ orbital, six into the $2p$ orbitals, two into the $3s$ orbital, and three into the $3p$ orbitals.	P $1s^2 2s^2 2p^6 3s^2 3p^3$ or $[\text{Ne}]\, 3s^2 3p^3$
(c) Br Bromine has 35 electrons. Distribute two of these into the $1s$ orbital, two into the $2s$ orbital, six into the $2p$ orbitals, two into the $3s$ orbital, six into the $3p$ orbitals, two into the $4s$ orbital, ten into the $3d$ orbitals, and five into the $4p$ orbitals.	Br $1s^2 2s^2 2p^6 3s^2 3p^6 4s^2 3d^{10} 4p^5$ or $[\text{Ar}]\, 4s^2 3d^{10} 4p^5$
(d) Al Aluminum has 13 electrons. Distribute two of these into the $1s$ orbital, two into the $2s$ orbital, six into the $2p$ orbitals, two into the $3s$ orbital, and one into the $3p$ orbital.	Al $1s^2 2s^2 2p^6 3s^2 3p^1$ or $[\text{Ne}]\, 3s^2 3p^1$

FOR PRACTICE 8.1

Write electron configurations for each element:

(a) Cl   **(b)** Si   **(c)** Sr   **(d)** O

EXAMPLE 8.2   Writing Orbital Diagrams

Write an orbital diagram for sulfur and determine the number of unpaired electrons.

SOLUTION

Since sulfur is atomic number 16 it has 16 electrons and the electron configuration is $1s^2 2s^2 2p^6 3s^2 3p^4$. Draw a box for each orbital putting the lowest energy orbital ($1s$) on the far left and proceeding to orbitals of higher energy to the right.	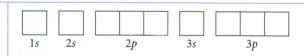
Distribute the 16 electrons into the boxes representing the orbitals allowing a maximum of two electrons per orbital and remembering Hund's rule. You can see from the diagram that sulfur has two unpaired electrons.	

FOR PRACTICE 8.2

Write an orbital diagram for Ar and determine the number of unpaired electrons.

Conceptual Connection 8.3 **Electron Configurations and Quantum Numbers**

What are the four quantum numbers for each of the two electrons in a $4s$ orbital?

8.4   Electron Configurations, Valence Electrons, and the Periodic Table

Recall that Mendeleev arranged the periodic table so that elements with similar chemical properties lie in the same column. We can begin to make the connection between an element's properties and its electron configuration by superimposing the electron configurations of the first 18 elements onto a partial periodic table, as shown in **Figure 8.6▼**. As we move to the right across a row, the orbitals are simply filling in the correct order. With each subsequent row, the highest principal quantum number increases by one. Notice that as we move down a column, *the number of electrons in the outermost principal energy level (highest n value) remains the same.* The key connection between the macroscopic world (an element's chemical properties) and the microscopic world (an atom's electronic structure) lies in these outermost electrons.

An atom's **valence electrons** are those that are important in chemical bonding. *For main-group elements, the valence electrons are those in the outermost principal energy level* (Figure 8.6). For transition elements, we also count the outermost d electrons among the valence electrons (even though they are not in an outermost principal energy level). The chemical properties of an element depend on its valence electrons, which are important in bonding because they are held most loosely (and are therefore the easiest to lose or share). We can now see *why* the elements in a column of the periodic table have similar chemical properties: *they have the same number of valence electrons.*

Valence electrons are distinguished from all of the other electrons in an atom, which are called **core electrons**. Core electrons are those in *complete* principal energy levels and those in *complete* d and f sublevels. For example, silicon, with the electron configuration $1s^2 2s^2 2p^6 3s^2 3p^2$ has 4 valence electrons (those in the $n = 3$ principal level) and 10 core electrons.

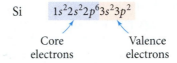

Outer Electron Configurations of Elements 1–18

1A							8A
1 **H** $1s^1$	2A	3A	4A	5A	6A	7A	2 **He** $1s^2$
3 **Li** $2s^1$	4 **Be** $2s^2$	5 **B** $2s^2 2p^1$	6 **C** $2s^2 2p^2$	7 **N** $2s^2 2p^3$	8 **O** $2s^2 2p^4$	9 **F** $2s^2 2p^5$	10 **Ne** $2s^2 2p^6$
11 **Na** $3s^1$	12 **Mg** $3s^2$	13 **Al** $3s^2 3p^1$	14 **Si** $3s^2 3p^2$	15 **P** $3s^2 3p^3$	16 **S** $3s^2 3p^4$	17 **Cl** $3s^2 3p^5$	18 **Ar** $3s^2 3p^6$

▲ **FIGURE 8.6** **Outer Electron Configurations of the First 18 Elements in the Periodic Table**

EXAMPLE 8.3 Valence Electrons and Core Electrons

Write the electron configuration for Ge and identify the valence electrons and the core electrons.

SOLUTION

Write the electron configuration for Ge by determining the total number of electrons from germanium's atomic number (32) and distributing them into the appropriate orbitals as described previously.	Ge $\quad 1s^2 2s^2 2p^6 3s^2 3p^6 4s^2 3d^{10} 4p^2$
Since germanium is a main-group element, its valence electrons are those in the outermost principal energy level. For germanium, the $n = 1, 2$, and 3 principal levels are complete (or full) and the $n = 4$ principal level is outermost. Consequently, the $n = 4$ electrons are valence electrons and the rest are core electrons. *Note: In this book, electron configurations are always written with the orbitals in the order of filling. However, writing electron configurations in order of increasing principal quantum number is also common. The electron configuration of germanium written in order of increasing principal quantum number is* Ge $\quad 1s^2 2s^2 2p^6 3s^2 3p^6 3d^{10} 4s^2 4p^2$	4 valence electrons Ge $\quad 1s^2 2s^2 2p^6 3s^2 3p^6 4s^2 3d^{10} 4p^2$ 28 core electrons

FOR PRACTICE 8.3

Write the electron configuration for phosphorus and identify the valence electrons and core electrons.

Orbital Blocks in the Periodic Table

A pattern similar to what we just saw for the first 18 elements exists for the entire periodic table, as shown in **Figure 8.7▶**. Note that, because of the filling order of orbitals, the periodic table can be divided into blocks representing the filling of particular sublevels. The first two columns on the left side of the periodic table comprise the *s* block, with outer electron configurations of ns^1 (the alkali metals) and ns^2 (the alkaline earth metals). The six columns on the right side of the periodic table comprise the *p* block, with outer electron configurations of $ns^2 np^1$, $ns^2 np^2$, $ns^2 np^3$, $ns^2 np^4$, $ns^2 np^5$ (halogens), and $ns^2 np^6$ (noble gases). The transition elements comprise the *d* block, and the lanthanides and actinides (also called the inner transition elements) comprise the *f* block. (For compactness, the *f* block is normally printed below the *d* block instead of being imbedded within it.)

You can see that *the number of columns in a block corresponds to the maximum number of electrons that can occupy the particular sublevel of that block.* The *s* block has 2 columns (corresponding to one *s* orbital holding a maximum of two electrons); the *p* block has 6 columns (corresponding to three *p* orbitals with two electrons each); the *d* block has 10 columns (corresponding to five *d* orbitals with two electrons each); and the *f* block has 14 columns (corresponding to seven *f* orbitals with two electrons each).

Notice also that, except for helium, *the number of valence electrons for any main-group element is equal to its lettered group number.* For example, we can tell that chlorine has 7 valence electrons because it is in group number 7A.

Lastly, note that, for main-group elements, *the row number in the periodic table is equal to the number (or n value) of the highest principal level.* For example, chlorine is in row 3, and its highest principal level is the $n = 3$ level.

Helium is an exception. Even though it lies in the column with an outer electron configuration of $ns^2 np^6$, its electron configuration is simply $1s^2$.

Recall from Chapter 2 that main-group elements are those in the two far left columns (groups 1A, 2A) and the six far right columns (groups 3A–8A) of the periodic table.

Summarizing Periodic Table Organization:

▶ The periodic table is divisible into four blocks corresponding to the filling of the four quantum sublevels (*s*, *p*, *d*, and *f*).

▶ The group number of a main-group element is equal to the number of valence electrons for that element.

▶ The row number of a main-group element is equal to the highest principal quantum number of that element.

Orbital Blocks of the Periodic Table

Groups

Period	1A 1	2A 2	3B 3	4B 4	5B 5	6B 6	7B 7	8B 8	8B 9	8B 10	1B 11	2B 12	3A 13	4A 14	5A 15	6A 16	7A 17	8A 18
1	1 H $1s^1$																	2 He $1s^2$
2	3 Li $2s^1$	4 Be $2s^2$											5 B $2s^22p^1$	6 C $2s^22p^2$	7 N $2s^22p^3$	8 O $2s^22p^4$	9 F $2s^22p^5$	10 Ne $2s^22p^6$
3	11 Na $3s^1$	12 Mg $3s^2$											13 Al $3s^23p^1$	14 Si $3s^23p^2$	15 P $3s^23p^3$	16 S $3s^23p^4$	17 Cl $3s^23p^5$	18 Ar $3s^23p^6$
4	19 K $4s^1$	20 Ca $4s^2$	21 Sc $4s^23d^1$	22 Ti $4s^23d^2$	23 V $4s^23d^3$	24 Cr $4s^13d^5$	25 Mn $4s^23d^5$	26 Fe $4s^23d^6$	27 Co $4s^23d^7$	28 Ni $4s^23d^8$	29 Cu $4s^13d^{10}$	30 Zn $4s^23d^{10}$	31 Ga $4s^24p^1$	32 Ge $4s^24p^2$	33 As $4s^24p^3$	34 Se $4s^24p^4$	35 Br $4s^24p^5$	36 Kr $4s^24p^6$
5	37 Rb $5s^1$	38 Sr $5s^2$	39 Y $5s^24d^1$	40 Zr $5s^24d^2$	41 Nb $5s^14d^4$	42 Mo $5s^14d^5$	43 Tc $5s^24d^5$	44 Ru $5s^14d^7$	45 Rh $5s^14d^8$	46 Pd $4d^{10}$	47 Ag $5s^14d^{10}$	48 Cd $5s^24d^{10}$	49 In $5s^25p^1$	50 Sn $5s^25p^2$	51 Sb $5s^25p^3$	52 Te $5s^25p^4$	53 I $5s^25p^5$	54 Xe $5s^25p^6$
6	55 Cs $6s^1$	56 Ba $6s^2$	57 La $6s^25d^1$	72 Hf $6s^25d^2$	73 Ta $6s^25d^3$	74 W $6s^25d^4$	75 Re $6s^25d^5$	76 Os $6s^25d^6$	77 Ir $6s^25d^7$	78 Pt $6s^15d^9$	79 Au $6s^15d^{10}$	80 Hg $6s^25d^{10}$	81 Tl $6s^26p^1$	82 Pb $6s^26p^2$	83 Bi $6s^26p^3$	84 Po $6s^26p^4$	85 At $6s^26p^5$	86 Rn $6s^26p^6$
7	87 Fr $7s^1$	88 Ra $7s^2$	89 Ac $7s^26d^1$	104 Rf $7s^26d^2$	105 Db $7s^26d^3$	106 Sg $7s^26d^4$	107 Bh	108 Hs	109 Mt	110 Ds	111 Rg	112 Cn	113 Uut	114 Uuq	115 Uup	116 Uuh	117 Uus	118 Uuo

■ *s*-block elements ■ *p*-block elements
■ *d*-block elements ■ *f*-block elements

Lanthanides	58 Ce $6s^24f^15d^1$	59 Pr $6s^24f^3$	60 Nd $6s^24f^4$	61 Pm $6s^24f^5$	62 Sm $6s^24f^6$	63 Eu $6s^24f^7$	64 Gd $6s^24f^75d^1$	65 Tb $6s^24f^9$	66 Dy $6s^24f^{10}$	67 Ho $6s^24f^{11}$	68 Er $6s^24f^{12}$	69 Tm $6s^24f^{13}$	70 Yb $6s^24f^{14}$	71 Lu $6s^24f^{14}6d^1$
Actinides	90 Th $7s^26d^2$	91 Pa $7s^25f^26d^1$	92 U $7s^25f^36d^1$	93 Np $7s^25f^46d^1$	94 Pu $7s^25f^6$	95 Am $7s^25f^7$	96 Cm $7s^25f^76d^1$	97 Bk $7s^25f^9$	98 Cf $7s^25f^{10}$	99 Es $7s^25f^{11}$	100 Fm $7s^25f^{12}$	101 Md $7s^25f^{13}$	102 No $7s^25f^{14}$	103 Lr $7s^25f^{14}6d^1$

▲ **FIGURE 8.7** The *s*, *p*, *d*, and *f* Blocks of the Periodic Table

Writing an Electron Configuration for an Element from Its Position in the Periodic Table

The organization of the periodic table allows us to write the electron configuration for any element based on its position in the periodic table. For example, suppose we want to write an electron configuration for Cl. The *inner electron configuration* of Cl is that of the noble gas that precedes it in the periodic table, Ne. So we can represent the inner electron configuration with [Ne]. The *outer electron configuration*—the configuration of the electrons beyond the previous noble gas—is obtained by tracing the elements between Ne and Cl and assigning electrons to the appropriate orbitals, as shown here. Remember that the highest *n* value is given by the row number (3 for chlorine).

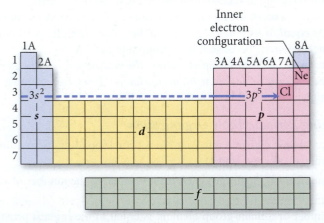

So, we begin with [Ne], then add in the two 3*s* electrons as we trace across the *s* block, followed by five 3*p* electrons as we trace across the *p* block to Cl, which is in the fifth column of the *p* block. The electron configuration is

$$Cl \quad [Ne] 3s^2 3p^5$$

Notice that Cl is in column 7A and therefore has 7 valence electrons and an outer electron configuration of $ns^2 np^5$.

EXAMPLE 8.4 Writing Electron Configurations from the Periodic Table

Use the periodic table to write an electron configuration for selenium (Se).

SOLUTION

The atomic number of Se is 34. The noble gas that precedes Se in the periodic table is argon, so the inner electron configuration is $[Ar]$. Obtain the outer electron configuration by tracing the elements between Ar and Se and assigning electrons to the appropriate orbitals. Begin with $[Ar]$. Because Se is in row 4, add two $4s$ electrons as you trace across the s block (n = row number). Next, add ten $3d$ electrons as you trace across the d block (n = row number − 1). Lastly, add four $4p$ electrons as you trace across the p block to Se, which is in the fourth column of the p block (n = row number).

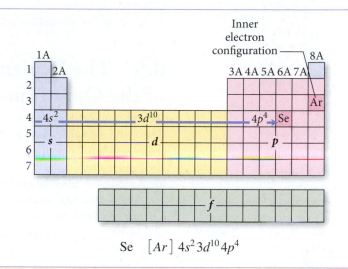

$$Se \quad [Ar]\, 4s^2 3d^{10} 4p^4$$

FOR PRACTICE 8.4

Use the periodic table to determine the electron configuration for bismuth (Bi).

FOR MORE PRACTICE 8.4

Use the periodic table to write an electron configuration for iodine (I).

The Transition and Inner Transition Elements

The electron configurations of the transition elements (d block) and inner transition elements (f block) exhibit trends that differ somewhat from those of the main-group elements. As we move to the right across a row in the d block, the d orbitals fill as shown here:

21 **Sc** $4s^2 3d^1$	22 **Ti** $4s^2 3d^2$	23 **V** $4s^2 3d^3$	24 **Cr** $4s^1 3d^5$	25 **Mn** $4s^2 3d^5$	26 **Fe** $4s^2 3d^6$	27 **Co** $4s^2 3d^7$	28 **Ni** $4s^2 3d^8$	29 **Cu** $4s^1 3d^{10}$	30 **Zn** $4s^2 3d^{10}$
39 **Y** $5s^2 4d^1$	40 **Zr** $5s^2 4d^2$	41 **Nb** $5s^1 4d^4$	42 **Mo** $5s^1 4d^5$	43 **Tc** $5s^2 4d^5$	44 **Ru** $5s^1 4d^7$	45 **Rh** $5s^1 4d^8$	46 **Pd** $4d^{10}$	47 **Ag** $5s^1 4d^{10}$	48 **Cd** $5s^2 4d^{10}$

However, *the principal quantum number of the d orbitals that fill across each row in the transition series is equal to the row number minus one*. In the fourth row, the $3d$ orbitals fill, and in the fifth row, the $4d$ orbitals fill, and so on. This happens because, as we discussed in Section 8.3, the $4s$ orbital is generally lower in energy than the $3d$ orbital (because it more efficiently penetrates into the region occupied by the core electrons). The result is that the $4s$ orbital fills before the $3d$ orbital, even though its principal quantum number (n = 4) is higher.

Keep in mind, however, that the $4s$ and the $3d$ orbitals are extremely close to each other in energy and their relative energy ordering depends on the exact species under consideration and the exact electron configuration (first discussed in Section 8.3); this causes some irregular behavior in the transition metals. For example, notice that, in the first transition series of the d block, the outer configuration is $4s^2 3d^x$ with two exceptions: Cr is $4s^1 3d^5$ and Cu is $4s^1 3d^{10}$. This behavior is related to the closely spaced $3d$ and $4s$ energy levels and the additional stability associated with a half-filled or completely filled sublevel. Actual electron configurations are always determined experimentally (through spectroscopy) and

8A

2 He $1s^2$
10 Ne $2s^2 2p^6$
18 Ar $3s^2 3p^6$
36 Kr $4s^2 4p^6$
54 Xe $5s^2 5p^6$
86 Rn $6s^2 6p^6$

Noble gases

▲ The noble gases all have eight valence electrons except for helium, which has two. They have full outer energy levels and are particularly stable and unreactive.

1A	**2A**
3 Li $2s^1$	4 Be $2s^2$
11 Na $3s^1$	12 Mg $3s^2$
19 K $4s^1$	20 Ca $4s^2$
37 Rb $5s^1$	38 Sr $5s^2$
55 Cs $6s^1$	56 Ba $6s^2$
87 Fr $7s^1$	88 Ra $7s^2$

| **Alkali metals** | **Alkaline earth metals** |

▲ The alkali metals all have one valence electron. They are one electron beyond a stable electron configuration and tend to lose that electron in their reactions.

▲ The alkaline earth metals all have two valence electrons. They are two electrons beyond a stable electron configuration and tend to lose those electrons in their reactions.

do not always conform to simple patterns, as is the case for Cr and Cu. Nonetheless, the patterns we have described allow us to accurately predict electron configurations for most of the elements in the periodic table.

As we move across the *f* block, the *f* orbitals fill. However, the principal quantum number of the *f* orbitals that fill across each row in the inner transition series is the row number *minus two*. (In the sixth row, the 4*f* orbitals fill, and in the seventh row, the 5*f* orbitals fill.) In addition, within the inner transition series, the close energy spacing of the 5*d* and 4*f* orbitals sometimes causes an electron to enter a 5*d* orbital instead of the expected 4*f* orbital. For example, the electron configuration of gadolinium is $[Xe]\,6s^2 4f^7 5d^1$ (instead of the expected $[Xe]\,6s^2 4f^8$).

8.5 The Explanatory Power of the Quantum-Mechanical Model

We can now see how the quantum-mechanical model accounts for the chemical properties of the elements, such as the inertness of helium or the reactivity of hydrogen, and (more generally) how it accounts for the periodic law. *The chemical properties of elements are largely determined by the number of valence electrons they contain.* Their properties are periodic because the number of valence electrons is periodic.

Since elements within a column in the periodic table have the same number of valence electrons, they also have similar chemical properties. The noble gases, for example, all have eight valence electrons, except for helium, which has two. Although we do not cover the quantitative (or numerical) aspects of the quantum-mechanical model in this book, calculations of the overall energy of atoms with eight valence electrons (or two for helium) show that they are particularly stable. In other words, when a quantum level is completely full, the overall energy of the electrons that occupy that level is particularly low. Those electrons *cannot* lower their energy by reacting with other atoms or molecules, so the corresponding atom is unreactive or inert. Consequently, the noble gases are the most chemically stable or relatively unreactive family in the periodic table.

Elements with electron configurations *close* to those of the noble gases are the most reactive because they can attain noble gas electron configurations by losing or gaining a small number of electrons. For example, alkali metals (group 1A) are the most reactive metals because their outer electron configuration (ns^1) is one electron beyond a noble gas configuration. They react to lose the ns^1 electron, obtaining a noble gas configuration. This explains why—as we saw in Chapter 2—the group 1A metals tend to form 1+ cations. Similarly, alkaline earth metals, with an outer electron configuration of ns^2, also tend to be reactive metals, losing their ns^2 electrons to form 2+ cations. This does not mean that forming an ion with a noble gas configuration is in itself energetically favorable. In fact, forming cations always *requires energy*. But when the cation formed has a noble gas configuration, the energy cost of forming the cation is often less than the energy payback that occurs when that cation forms ionic bonds with anions, as we shall see in Chapter 9.

Elements That Form Ions with Predictable Charges

	1A 1	2A 2											3A 13	4A 14	5A 15	6A 16	7A 17	8A 18
1	Li⁺														N³⁻	O²⁻	F⁻	
2	Na⁺	Mg²⁺	3B 3	4B 4	5B 5	6B 6	7B 7	8	8B 9	10	1B 11	2B 12	Al³⁺			S²⁻	Cl⁻	
3	K⁺	Ca²⁺														Se²⁻	Br⁻	
4	Rb⁺	Sr²⁺														Te²⁻	I⁻	
5	Cs⁺	Ba²⁺																

▲ **FIGURE 8.8 Elements That Form Ions with Predictable Charges** Notice that each ion has a noble gas electron configuration.

On the right side of the periodic table, halogens are among the most reactive nonmetals because of their $ns^2 np^5$ electron configurations. They are only one electron short of a noble gas configuration and tend to react to gain that one electron, forming $1-$ ions. **Figure 8.8◄**, first introduced in Chapter 2, shows the elements that form predictable ions. Notice how the charges of these ions reflect their electron configurations—in their reactions, these elements form ions with noble gas electron configurations.

8.6   Periodic Trends in the Size of Atoms and Effective Nuclear Charge

In previous chapters, we saw that the volume of an atom is taken up primarily by its electrons (Chapter 2) occupying quantum-mechanical orbitals (Chapter 7). We also saw that these orbitals do not have a definite boundary, but represent only a statistical probability distribution for where the electron is found. So how do we define the size of an atom? One way to define atomic radii is to consider the distance between *nonbonding* atoms in molecules or atoms that are in direct contact. For example, krypton can be frozen into a solid in which the krypton atoms are touching each other but are not bonded together. The distance between the centers of adjacent krypton atoms—which can be determined from the solid's density—is twice the radius of a krypton atom. An atomic radius determined in this way is called the **nonbonding atomic radius** or the **van der Waals radius**. The van der Waals radius represents the radius of an atom when it is not bonded to another atom.

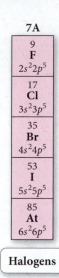

7A
| 9 |
| **F** |
| $2s^2 2p^5$ |
| 17 |
| **Cl** |
| $3s^2 3p^5$ |
| 35 |
| **Br** |
| $4s^2 4p^5$ |
| 53 |
| **I** |
| $5s^2 5p^5$ |
| 85 |
| **At** |
| $6s^2 6p^5$ |

Halogens

▲ The halogens all have seven valence electrons. They are one electron short of a stable electron configuration and tend to gain one electron in their reactions.

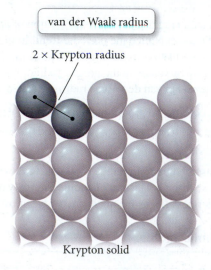

van der Waals radius

2 × Krypton radius

Krypton solid

The **bonding atomic radius** or **covalent radius**, another way to indicate the size of an atom, is defined differently for nonmetals and metals, as follows:

Nonmetals: one-half the distance between two of the atoms bonded together

Metals: one-half the distance between two of the atoms next to each other in a crystal of the metal

For example, the distance between Br atoms in Br_2 is 228 pm; therefore, the Br covalent radius is assigned to be one-half of 228 pm or 114 pm.

Similar radii can be assigned to all elements in the periodic table that form chemical bonds or form metallic crystals. A more general term, the **atomic radius**, refers to a set of average bonding radii determined from measurements on a large number of elements and compounds. The atomic radius represents the radius of an atom when it is bonded to another atom and is always smaller than the van der Waals radius. The approximate bond length of any two covalently bonded atoms is simply the sum of their atomic radii. For example, the approximate bond length for ICl is iodine's atomic radius (133 pm) plus chlorine's atomic radius (99 pm), for a bond length of 232 pm. (The actual experimentally measured bond length in ICl is 232.07 pm.)

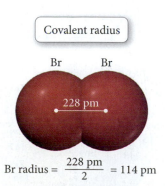

Covalent radius

Br   Br

228 pm

Br radius = $\dfrac{228 \text{ pm}}{2}$ = 114 pm

▲ The covalent radius of bromine is one-half the distance between two bonded bromine atoms.

Atomic Radii

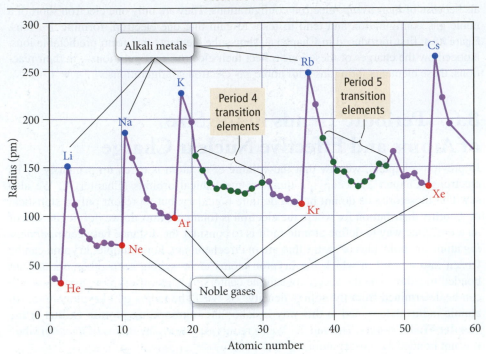

▶ **FIGURE 8.9 Atomic Radius versus Atomic Number**   Notice the periodic trend in atomic radius, starting at a peak with each alkali metal and falling to a minimum with each noble gas.

The bonding radii of some elements, such as helium and neon, must be approximated since they do not form chemical bonds or metallic crystals.

Figure 8.9▲ shows the atomic radius plotted as a function of atomic number for the first 57 elements in the periodic table. Notice the periodic trend in the radii. Atomic radii peak with each alkali metal. **Figure 8.10▼** is a relief map of atomic radii for most of the elements in the periodic table. The general trends in the atomic radii of main-group elements, which are the same as trends observed in van der Waals radii, are stated as follows:

1. As you move down a column (or group) in the periodic table, atomic radius increases.
2. As you move to the right across a row (or period) in the periodic table, atomic radius decreases.

We can understand the observed trend in radius as we move down a column in light of the trends in the sizes of atomic orbitals. The atomic radius is largely determined by the valence electrons, the electrons farthest from the nucleus. As we move down a column in the

Trends in Atomic Radius

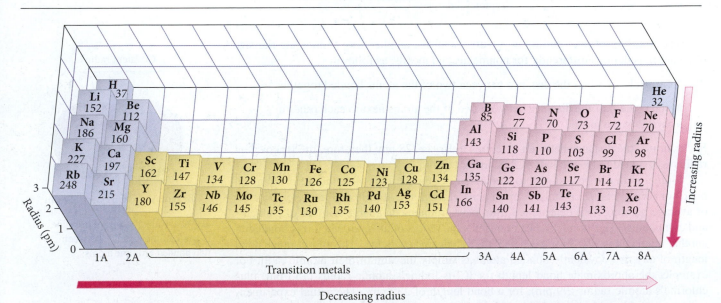

▲ **FIGURE 8.10 Trends in Atomic Radius**   In general, atomic radii increase as you move down a column and decrease as you move to the right across a row in the periodic table.

periodic table, the highest principal quantum number (n) of the valence electrons increases. Consequently, the valence electrons occupy larger orbitals, resulting in larger atoms.

The observed trend in atomic radius as we move to the right across a row, however, is a bit more complex. To understand this trend, we revisit some concepts from Section 8.3, including effective nuclear charge and shielding.

Effective Nuclear Charge

The trend in atomic radius as we move to the right across a row in the periodic table is determined by the inward pull of the nucleus on the electrons in the outermost principal energy level (highest n value). According to Coulomb's law the attraction between a nucleus and an electron increases with increasing magnitude of nuclear charge. For example, compare the H atom to the He$^+$ ion.

$$H \quad 1s^1$$

$$He^+ \quad 1s^1$$

It takes 1312 kJ/mol of energy to remove the $1s$ electron from hydrogen, but 5251 kJ/mol of energy to remove it from He$^+$. Why? Although each electron is in a $1s$ orbital, the electron in the helium ion is attracted to the nucleus with a 2+ charge, while the electron in the hydrogen atom is attracted to the nucleus by only a 1+ charge. Therefore, the electron in the helium ion is held more tightly, making it more difficult to remove and making the helium ion smaller than the hydrogen atom.

As we saw in Section 8.3, any one electron in a multielectron atom experiences both the positive charge of the nucleus (which is attractive) and the negative charges of the other electrons (which are repulsive). Consider again the outermost electron in the lithium atom:

$$Li \quad 1s^2 2s^1$$

As shown in **Figure 8.11▼**, even though the $2s$ orbital penetrates into the $1s$ orbital to some degree, the majority of the $2s$ orbital is outside of the $1s$ orbital. Therefore, the electron in the $2s$ orbital is partially *screened* or *shielded* from the 3+ charge of the nucleus by the 2− charge of the $1s$ (or core) electrons, reducing the net charge experienced by the $2s$ electron.

As we have seen, we can define the average or net charge experienced by an electron as the *effective nuclear charge*. The effective nuclear charge experienced by a particular electron in an atom is the *actual nuclear charge (Z)* minus *the charge shielded by other electrons (S)*:

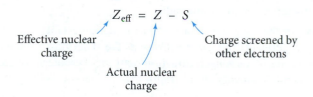

Effective nuclear charge

Actual nuclear charge

Charge screened by other electrons

Screening and Effective Nuclear Charge

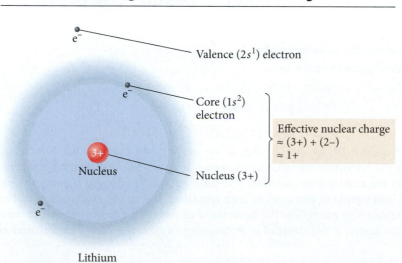

e^- — Valence ($2s^1$) electron

e^- — Core ($1s^2$) electron

Effective nuclear charge
$\approx (3+) + (2-)$
$\approx 1+$

3+

Nucleus — Nucleus (3+)

e^-

Lithium

◀ **FIGURE 8.11 Screening and Effective Nuclear Charge**   The valence electron in lithium experiences the 3+ charge of the nucleus through the screen of the 2− charge of the core electrons. The effective nuclear charge acting on the valence electron is approximately 1+.

For lithium, we can estimate that the two core electrons shield the valence electron from the nuclear charge with high efficiency (S is nearly 2). The effective nuclear charge experienced by lithium's valence electron is therefore slightly greater than 1+.

Now consider the valence electrons in beryllium (Be), with atomic number 4. Its electron configuration is

$$\text{Be} \quad 1s^2 2s^2$$

To estimate the effective nuclear charge experienced by the $2s$ electrons, we must distinguish between two different types of shielding: (1) the shielding of the outermost electrons by the core electrons and (2) the shielding of the outermost electrons by *each other*. The key to understanding the trend in atomic radius is understanding the difference between these two types of shielding. In general:

> **Core electrons efficiently shield electrons in the outermost principal energy level from nuclear charge, but outermost electrons do not efficiently shield one another from nuclear charge.**

In other words, the two outermost electrons in beryllium experience the 4+ charge of the nucleus through the shield of the two $1s$ core electrons without shielding each other from that charge very much. We can estimate that the shielding (S) experienced by any one of the outermost electrons due to the core electrons is nearly 2, but that the shielding due to the other outermost electron is nearly zero. The effective nuclear charge experienced by beryllium's outermost electrons is therefore slightly greater than 2+.

Notice that the effective nuclear charge experienced by beryllium's outermost electrons is greater than that experienced by lithium's outermost electron. Consequently, beryllium's outermost electrons are held more tightly than lithium's, resulting in a smaller atomic radius for beryllium. The effective nuclear charge experienced by an atom's outermost electrons continues to become more positive as you move to the right across the rest of the second row in the periodic table, resulting in successively smaller atomic radii. The same trend is generally observed in all main-group elements.

Summarizing Atomic Radii for Main-Group Elements:

▶ As you move down a column in the periodic table, the principal quantum number (n) of the electrons in the outermost principal energy level increases, resulting in larger orbitals and therefore larger atomic radii.

▶ As you move to the right across a row in the periodic table, the effective nuclear charge (Z_{eff}) experienced by the electrons in the outermost principal energy level increases, resulting in a stronger attraction between the outermost electrons and the nucleus and therefore smaller atomic radii.

 Conceptual Connection 8.4 **Effective Nuclear Charge**

Which electrons experience the greatest effective nuclear charge?

(a) The valence electrons in Mg

(b) The valence electrons in Al

(c) The valence electrons in S

Atomic Radii and the Transition Elements

From Figure 8.10, we can see that as we move down the first two rows of a column within the transition metals, the elements follow the same general trend in atomic radii as the main-group elements (the radii get larger). However, with the exception of the first couple of elements in each transition series, the atomic radii of the transition elements *do not* follow the same trend as the main-group elements as we move to the right across a row. Instead of decreasing in size, *the radii of transition elements stay roughly constant across each row.* Why? The difference is that, across a row of transition elements, the number of electrons in the outermost principal energy level (highest n value) is nearly constant (recall from Section 8.3, for example, that the 4s orbital fills before the 3d). As another proton is added to the nucleus with each successive element, another electron is added as well, but the electron goes into an $n_{highest}-1$ orbital. The number of outermost electrons stays constant and they experience a roughly constant effective nuclear charge, keeping the radius approximately constant.

EXAMPLE 8.5   Atomic Size

On the basis of periodic trends, choose the larger atom from each pair (if possible). Explain your choices.

(a) N or F   (b) C or Ge   (c) N or Al   (d) Al or Ge

SOLUTION

(a) N atoms are larger than F atoms because, as you trace the path between N and F on the periodic table, you move to the right within the same period (row). As you move to the right across a row, the effective nuclear charge experienced by the outermost electrons increases, resulting in a smaller radius.

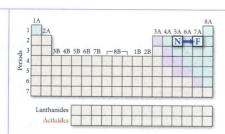

(b) Ge atoms are larger than C atoms because, as you trace the path between C and Ge on the periodic table, you move down a column. Atomic size increases as you move down a column because the outermost electrons occupy orbitals with a higher principal quantum number that are therefore larger, resulting in a larger atom.

(c) Al atoms are larger than N atoms because, as you trace the path between N and Al on the periodic table, you move down a column (atomic size increases) and then to the left across a row (atomic size increases). These effects add together for an overall increase.

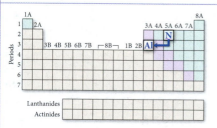

(d) Based on periodic trends alone, you cannot tell which atom is larger, because as you trace the path between Al and Ge you go to the right across a row (atomic size decreases) and then down a column (atomic size increases). These effects tend to oppose each other, and it is not easy to tell which will predominate.

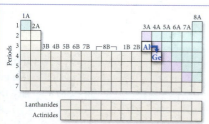

FOR PRACTICE 8.5

On the basis of periodic trends, choose the larger atom from each pair (if possible):

(a) Sn or I   (b) Ge or Po   (c) Cr or W   (d) F or Se

FOR MORE PRACTICE 8.5

Arrange these elements in order of decreasing radius: S, Ca, F, Rb, Si.

8.7   Ions: Electron Configurations, Magnetic Properties, Ionic Radii, and Ionization Energy

As we have seen, ions are simply atoms (or groups of atoms) that have lost or gained electrons. In this section, we examine periodic trends in ionic electron configurations, magnetic properties, ionic radii, and ionization energies.

Electron Configurations and Magnetic Properties of Ions

We can deduce the electron configuration of a main-group monoatomic ion from the electron configuration of the neutral atom and the charge of the ion. For anions, we *add* the number of electrons indicated by the magnitude of the charge of the anion. For example, the electron configuration of fluorine (F) is $1s^2 2s^2 2p^5$ and that of the fluoride ion (F⁻) is $1s^2 2s^2 2p^6$.

We determine the electron configuration of cations by *subtracting* the number of electrons indicated by the magnitude of the charge. For example, the electron configuration of lithium (Li) is $1s^2 2s^1$ and that of the lithium ion (Li⁺) is $1s^2 2s^0$ (or simply $1s^2$). For main-group cations, we remove the required number of electrons in the reverse order of filling. However, an important exception occurs for transition metal cations. When writing the electron configuration of a transition metal cation, we *remove the electrons in the highest n-value orbitals first, even if this does not correspond to the order of filling.* For example, the electron configuration of vanadium is:

$$V \quad [Ar] \, 4s^2 3d^3$$

The V^{2+} ion, however, has the following electron configuration:

$$V^{2+} \quad [Ar] \, 4s^0 3d^3$$

In other words, for transition metal cations, the order in which electrons are removed upon ionization is *not* the reverse of the filling order. During filling, we normally fill the $4s$ orbital before the $3d$ orbital. When a fourth period transition metal ionizes, however, it normally loses its $4s$ electrons before its $3d$ electrons. Why this odd behavior? The full answer to this question is beyond the scope of this text, but the following two factors contribute to this behavior.

- As discussed previously, the ns and $(n-1)d$ orbitals are extremely close in energy and, depending on the exact configuration, can vary in relative energy ordering.
- As the $(n-1)d$ orbitals begin to fill in the first transition series, the increasing nuclear charge stabilizes the $(n-1)d$ orbitals relative to the ns orbitals. This happens because the $(n-1)d$ orbitals are not outermost (or highest n) orbitals and are therefore not effectively shielded from the increasing nuclear charge by the ns orbitals.

The experimental observation is that an $ns^0(n-1)d^x$ configuration is lower in energy than an $ns^2(n-1)d^{x-2}$ configuration for transition metal ions. Therefore, when writing electron configurations for transition metals, we remove the ns electrons before the $(n-1)d$ electrons.

The magnetic properties of transition metal ions support these assignments. An unpaired electron generates a magnetic field due to its spin. Consequently, if an atom or ion contains unpaired electrons, it will be attracted by an external magnetic field, and we say that the atom or ion is **paramagnetic**.

An atom or ion in which all electrons are paired is not attracted to an external magnetic field—it is in fact slightly repelled—and we say that the atom or ion is **diamagnetic**. The zinc atom, for example, is diamagnetic.

$$Zn \quad [Ar] \, 4s^2 3d^{10}$$

The magnetic properties of the zinc ion provide confirmation that the $4s$ electrons are indeed lost before $3d$ electrons in the ionization of zinc. If zinc lost two $3d$ electrons upon ionization, then the Zn^{2+} would become paramagnetic (because the two electrons would come out of two different filled d orbitals, leaving each of them with one unpaired electron). However, the zinc ion, like the zinc atom, is diamagnetic because the $4s$ electrons are lost instead.

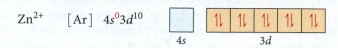

$$Zn^{2+} \quad [Ar] \ 4s^0 3d^{10}$$

4s   3d

Similar observations in other transition metals confirm that the ns electrons are lost before the $(n-1)d$ electrons upon ionization.

EXAMPLE 8.6 Electron Configurations and Magnetic Properties for Ions

Write the electron configuration and orbital diagram for each ion and determine whether the ion is diamagnetic or paramagnetic.

(a) Al^{3+} **(b)** S^{2-} **(c)** Fe^{3+}

SOLUTION

(a) Al^{3+}

Begin by writing the electron configuration of the neutral atom. Since this ion has a 3+ charge, remove three electrons to write the electron configuration of the ion. Write the orbital diagram by drawing half-arrows to represent each electron in boxes representing the orbitals. Because there are no unpaired electrons, Al^{3+} is diamagnetic.

Al $[Ne] \ 3s^2 3p^1$
Al^{3+} $[Ne]$ or $[He] \ 2s^2 2p^6$

Al^{3+} $[He]$
                       2s   2p
Diamagnetic

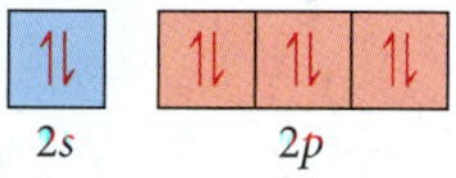

(b) S^{2-}

Begin by writing the electron configuration of the neutral atom. Since this ion has a 2− charge, add two electrons to write the electron configuration of the ion. Write the orbital diagram by drawing half-arrows to represent each electron in boxes representing the orbitals. Because there are no unpaired electrons, S^{2-} is diamagnetic.

S $[Ne] \ 3s^2 3p^4$
S^{2-} $[Ne] \ 3s^2 3p^6$

S^{2-} $[Ne]$
                    3s   3p
Diamagnetic

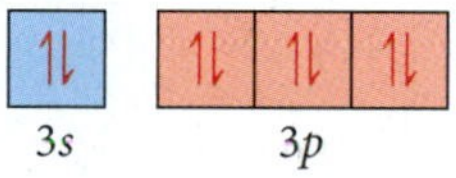

(c) Fe^{3+}

Begin by writing the electron configuration of the neutral atom. Since this ion has a 3+ charge, remove three electrons to write the electron configuration of the ion. Since it is a transition metal, remove the electrons from the $4s$ orbital before removing electrons from the $3d$ orbitals. Write the orbital diagram by drawing half-arrows to represent each electron in boxes representing the orbitals. Because there are unpaired electrons, Fe^{3+} is paramagnetic.

Fe $[Ar] \ 4s^2 3d^6$
Fe^{3+} $[Ar] \ 4s^0 3d^5$

Fe^{3+} $[Ar]$
                         4s   3d
Paramagnetic

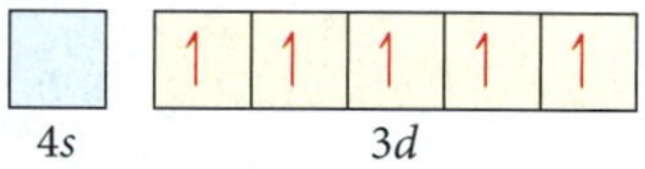

FOR PRACTICE 8.6

Write the electron configuration and orbital diagram for each ion and predict whether the ion will be paramagnetic or diamagnetic.

(a) Co^{2+} **(b)** N^{3-} **(c)** Ca^{2+}

Ionic Radii

What happens to the radius of an atom when it becomes a cation? An anion? Consider, for example, the difference between the Na atom and the Na^+ ion. Their electron configurations are:

Na $[Ne] \ 3s^1$

Na^+ $[Ne]$

The sodium atom has an outer $3s$ electron and a neon core. Since the $3s$ electron is the outermost electron, and since it is shielded from the nuclear charge by the core electrons, it contributes greatly to the size of the sodium atom. The sodium cation, having lost the outermost $3s$ electron, has only the neon core and carries a charge of $1+$. Without the $3s$ electron, the sodium cation (ionic radius = 95 pm) becomes much smaller than the sodium atom (covalent radius = 186 pm). The trend is the same with all cations and their atoms, as shown in **Figure 8.12▼**. In general,

Cations are much smaller than their corresponding atoms.

What about anions? Consider, for example, the difference between Cl and Cl⁻. Their electron configurations are as follows:

$$Cl \quad [Ne]3s^2 3p^5$$
$$Cl^- \quad [Ne]3s^2 3p^6$$

The chlorine anion has one additional outermost electron, but no additional proton to increase the nuclear charge. The extra electron increases the repulsions among the

Radii of Atoms and Their Cations (pm)

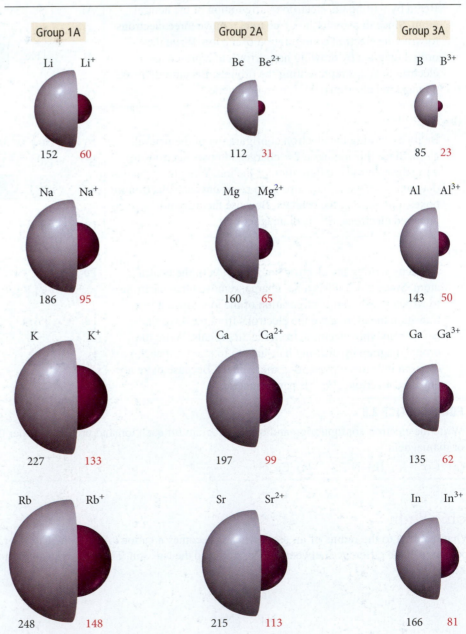

▶ **FIGURE 8.12 Sizes of Atoms and Their Cations** Atomic and ionic radii (pm) for the first three columns of main-group elements.

outermost electrons, resulting in a chloride anion that is larger than the chlorine atom. The trend is the same with all anions and their atoms, as shown in **Figure 8.13▼**. In general,

Anions are much larger than their corresponding atoms.

We can observe an interesting trend in ionic size by examining the radii of an *isoelectronic* series of ions—ions with the same number of electrons. For example, consider the following ions and their radii:

$$S^{2-} \text{ (184 pm)} \quad Cl^{-} \text{ (181 pm)} \quad K^{+} \text{ (133 pm)} \quad Ca^{2+} \text{ (99 pm)}$$

18 electrons	18 electrons	18 electrons	18 electrons
16 protons	17 protons	19 protons	20 protons

Each of these ions has 18 electrons in exactly the same orbitals, but the radius of the ions gets successively smaller. Why? The reason is the progressively greater number of protons. The S^{2-} ion has 16 protons, and therefore a charge of 16+ pulling on 18 electrons. The Ca^{2+} ion, however, has 20 protons, and therefore a charge of 20+ pulling on the same 18 electrons. The result is a much smaller radius. For a given number of electrons, a greater nuclear charge results in a smaller atom or ion.

Radii of Atoms and Their Anions (pm)

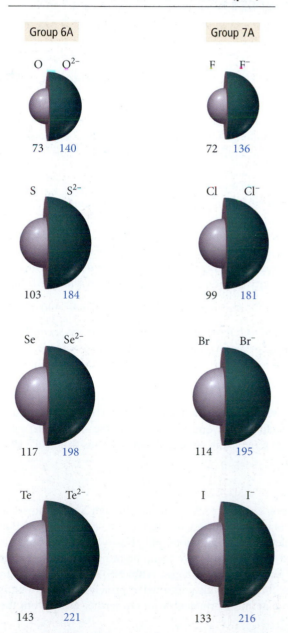

▲ **FIGURE 8.13 Sizes of Atoms and Their Anions** Atomic and ionic radii for groups 6A and 7A in the periodic table.

EXAMPLE 8.7 Ion Size

Choose the larger atom or ion from each pair:

(a) S or S^{2-} (b) Ca or Ca^{2+} (c) Br^- or Kr

SOLUTION

(a) The S^{2-} ion is larger than an S atom because anions are larger than the atoms from which they are formed.

(b) A Ca atom is larger than Ca^{2+} because cations are smaller than the atoms from which they are formed.

(c) A Br^- ion is larger than a Kr atom because, although they are isoelectronic, Br^- has one fewer proton than Kr, resulting in a lesser pull on the electrons and therefore a larger radius.

FOR PRACTICE 8.7

Choose the larger atom or ion from each pair:

(a) K or K^+ (b) F or F^- (c) Ca^{2+} or Cl^-

FOR MORE PRACTICE 8.7

Arrange the following in order of decreasing radius: Ca^{2+}, Ar, Cl^-.

 Conceptual Connection 8.5 Ions, Isotopes, and Atomic Size

In the previous sections, we have seen how the number of electrons and the number of protons affects the size of an atom or ion. However, we have not considered how the number of neutrons affects the size of an atom. Why not? Would you expect isotopes—for example, C-12 and C-13—to have different atomic radii?

Ionization Energy

The **ionization energy (IE)** of an atom or ion is the energy required to remove an electron from the atom or ion in the gaseous state (see Figure 7.19). Ionization energy is always positive because removing an electron always takes energy. (The process is similar to an endothermic reaction, which absorbs heat and therefore has a positive ΔH.) The energy required to remove the first electron is the *first ionization energy* (IE_1). The first ionization energy of sodium can be represented with the following equation:

$$Na(g) \longrightarrow Na^+(g) + 1\ e^- \qquad IE_1 = 496\ kJ/mol$$

The energy required to remove the second electron is the *second ionization energy* (IE_2), the energy required to remove the third electron is the *third ionization energy* (IE_3), and so on. The second ionization energy of sodium can be represented as follows:

$$Na^+(g) \longrightarrow Na^{2+}(g) + 1\ e^- \qquad IE_2 = 4560\ kJ/mol$$

Notice that the second ionization energy is not the energy required to remove *two* electrons from sodium (that quantity would be the sum of IE_1 and IE_2), but rather the energy required to remove the second of two electrons from Na^+. We look at trends in IE_1 and IE_2 separately.

Trends in First Ionization Energy

The first ionization energies of the elements through Xe are shown in **Figure 8.14▶**. Notice the periodic trend in ionization energy, peaking at each noble gas. Based on what we have learned about electron configurations and effective nuclear charge, how can we account for the observed trend? As we have seen, the principal quantum number, n, increases as we move down a column. Within a given sublevel, orbitals with higher principal quantum numbers are larger than orbitals with smaller principal quantum numbers. Consequently, electrons in the outermost principal level are farther away from the positively charged

First Ionization Energies

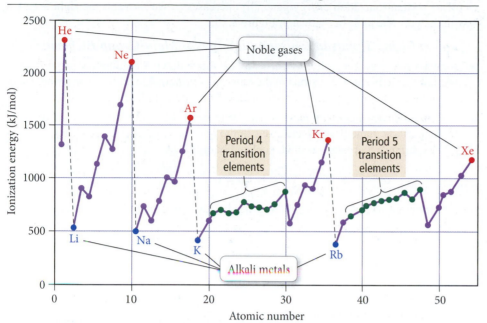

◀ **FIGURE 8.14 First Ionization Energy versus Atomic Number for the Elements through Xenon** Ionization energy starts at a minimum with each alkali metal and rises to a peak with each noble gas. (Compare with Figure 8.9.)

nucleus—and are therefore held less tightly—as we move down a column. This results in a lower ionization energy as we move down a column, as we see in **Figure 8.15▼**.

What about the trend as we move to the right across a row? For example, would it take more energy to remove an electron from Na or from Cl, two elements on either end of the third row in the periodic table? We know that Na has an outer electron configuration of $3s^1$ and Cl has an outer electron configuration of $3s^2 3p^5$. As discussed previously, the outermost electrons in chlorine experience a higher effective nuclear charge than the outermost electrons in sodium (which is why chlorine has a smaller atomic radius than sodium). Consequently, we would expect chlorine to have a higher ionization energy than

Trends in First Ionization Energy

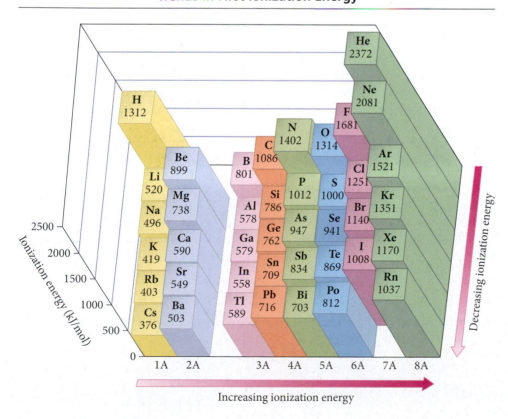

◀ **FIGURE 8.15 Trends in Ionization Energy** Ionization energy increases as we move to the right across a row and decreases as we move down a column in the periodic table.

sodium, which is in fact the case. A similar argument can be made for other main-group elements so that ionization energy generally increases as we move to the right across a row in the periodic table, as shown in Figure 8.15.

Summarizing Trends in Ionization Energy for Main-Group Elements:

▶ Ionization energy generally *decreases* as we move down a column (or group) in the periodic table because electrons in the outermost principal level become farther away from the positively charged nucleus and are therefore held less tightly.

▶ Ionization energy generally *increases* as we move to the right across a period (or row) in the periodic table because electrons in the outermost principal energy level generally experience a greater effective nuclear charge (Z_{eff}).

EXAMPLE 8.8 Ionization Energy

On the basis of periodic trends, determine the element with the higher first ionization energy in each pair (if possible):

(a) Al or S **(b)** As or Sb **(c)** N or Si **(d)** O or Cl

SOLUTION

(a) Al or S

S has a higher ionization energy than Al because, as you trace the path between Al and S on the periodic table, you move to the right within the same row. Ionization energy increases as you go to the right because of increasing effective nuclear charge.

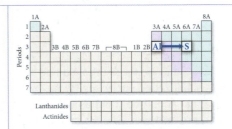

(b) As or Sb

As has a higher ionization energy than Sb because, as you trace the path between As and Sb on the periodic table, you move down a column. Ionization energy decreases as you go down a column because of the increasing size of orbitals with increasing *n*.

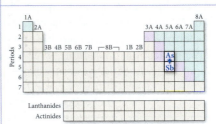

(c) N or Si

N has a higher ionization energy than Si because, as you trace the path between N and Si on the periodic table, you move down a column (ionization energy decreases) and then to the left across a row (ionization energy decreases). These effects sum together for an overall decrease.

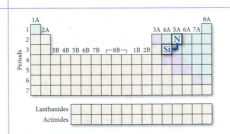

(d) O or Cl

Based on periodic trends alone, it is impossible to tell which has a higher ionization energy because, as you trace the path between O and Cl, you go to the right across a row (ionization energy increases) and then down a column (ionization energy decreases). These effects tend to oppose each other, and it is not obvious which will dominate.

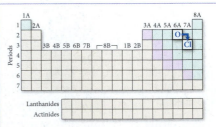

FOR PRACTICE 8.8

On the basis of periodic trends, determine the element with the higher first ionization energy in each pair (if possible):

(a) Sn or I **(b)** Ca or Sr **(c)** C or P **(d)** F or S

FOR MORE PRACTICE 8.8

Arrange the following elements in order of decreasing first ionization energy: S, Ca, F, Rb, Si.

Exceptions to Trends in First Ionization Energy

If we carefully examine Figure 8.15, we can see some exceptions to the trends in first ionization energies. For example, boron has a smaller ionization energy than beryllium, even though it lies to the right of beryllium in the same row. This exception is caused by the move from the *s* block to the *p* block. Recall from Section 8.3 that the $2p$ orbital penetrates into nuclear region *less than* the $2s$ orbital. Consequently, the $1s$ electrons shield the electron in the $2p$ orbital from nuclear charge more than they shield the electrons in the $2s$ orbital. The result, as we saw in Section 8.3, is that the $2p$ orbitals are higher in energy, and therefore the electron is easier to remove (it has a lower ionization energy). Similar exceptions occur for aluminum and gallium, both directly below boron in group 3A.

Another exception occurs between nitrogen and oxygen. Although oxygen is to the right of nitrogen in the same row, it has a lower ionization energy. This exception is caused by the repulsion between electrons when they must occupy the same orbital. Examine the electron configurations and orbital diagrams of nitrogen and oxygen:

N $\quad 1s^2 2s^2 2p^3$ $\quad$ ↑↓ $\quad$ ↑↓ $\quad$ ↑ $\quad$ ↑ $\quad$ ↑
$\qquad\qquad\qquad\qquad\quad$ 1s $\quad\;$ 2s $\qquad\;$ 2p

O $\quad 1s^2 2s^2 2p^4$ $\quad$ ↑↓ $\quad$ ↑↓ $\quad$ ↑↓ $\quad$ ↑ $\quad$ ↑
$\qquad\qquad\qquad\qquad\quad$ 1s $\quad\;$ 2s $\qquad\;$ 2p

Nitrogen has three electrons in three *p* orbitals, while oxygen has four. In nitrogen, the $2p$ orbitals are half-filled (which makes the configuration particularly stable). Oxygen's fourth electron must pair with another electron, making it easier to remove. Similar exceptions occur for S and Se, directly below oxygen in group 6A.

Trends in Second and Successive Ionization Energies

Notice the trends in the first, second, and third ionization energies of sodium (group 1A) and magnesium (group 2A), as shown in the margin below.

For sodium, there is a huge jump between the first and second ionization energies. For magnesium, the ionization energy roughly doubles from the first to the second, but then a huge jump occurs between the second and third ionization energies. What is the reason for these jumps?

We can understand these trends by examining the electron configurations of sodium and magnesium:

$$\text{Na} \quad [\text{Ne}]\, 3s^1$$

$$\text{Mg} \quad [\text{Ne}]\, 3s^2$$

The first ionization of sodium involves removing the valence electron in the $3s$ orbital. Recall that these valence electrons are held more loosely than the core electrons, and that the resulting ion has a noble gas configuration, which is particularly stable. Consequently, the first ionization energy is fairly low. Conversely, the second ionization of sodium involves removing a core electron from an ion with a noble gas configuration. This requires a tremendous amount of energy, making the value of IE$_2$ very high.

As with sodium, the first ionization of magnesium involves removing a valence electron in the $3s$ orbital. However, this requires a bit more energy than the corresponding ionization of sodium because of the trends in Z_{eff} that we discussed earlier (Z_{eff} increases as we move to the right across a row). The second ionization of magnesium also involves removing an outer electron in the $3s$ orbital, but this time from an ion with a 1+ charge (instead of from a neutral atom). This requires roughly twice the energy as removing the electron from the neutral atom. The third ionization of magnesium is analogous to

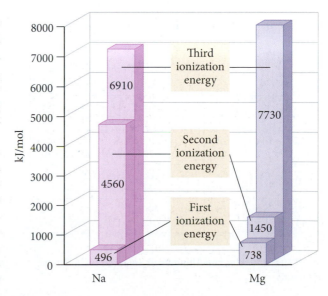

TABLE 8.1 Successive Values of Ionization Energies for the Elements Sodium through Argon (kJ/mol)

Element	IE$_1$	IE$_2$	IE$_3$	IE$_4$	IE$_5$	IE$_6$	IE$_7$
Na	496	4560					
Mg	738	1450	7730		*Core electrons*		
Al	578	1820	2750	11,600			
Si	786	1580	3230	4360	16,100		
P	1012	1900	2910	4960	6270	22,200	
S	1000	2250	3360	4560	7010	8500	27,100
Cl	1251	2300	3820	5160	6540	9460	11,000
Ar	1521	2670	3930	5770	7240	8780	12,000

the second ionization of sodium—it requires removing a core electron from an ion with a noble gas configuration. This requires a tremendous amount of energy, making the value of IE$_3$ very high.

As shown in Table 8.1, similar trends exist for the successive ionization energies of many elements. The ionization energy increases fairly uniformly with each successive removal of an outermost electron, but then it takes a large jump with the removal of the first core electron.

Conceptual Connection 8.6 Ionization Energies and Chemical Bonding

Based on what you just learned about ionization energies, explain why valence electrons are more important than core electrons in determining the reactivity and bonding in atoms.

8.8 Electron Affinities and Metallic Character

The type of periodic behavior we have just seen is also exhibited by an element's electron affinity and its metallic character. Electron affinity is a measure of how easily an atom will accept an additional electron. Since chemical bonding involves the transfer or sharing of electrons, electron affinity is crucial to chemical bonding. Metallic character is important because of the high proportion of metals in the periodic table and the crucial role they play in our lives. Of the roughly 110 elements, 87 are metals. We examine each of these periodic properties individually.

Electron Affinity

The **electron affinity (EA)** of an atom or ion is the energy change associated with the gaining of an electron by the atom in the gaseous state. The electron affinity is usually—though not always—negative because an atom or ion usually releases energy when it gains an electron. (The process is analogous to an exothermic reaction, which releases heat and therefore has a negative ΔH.) In other words, the coulombic attraction between the nucleus of an atom and the incoming electron usually results in the release of energy as the electron is gained. For example, the electron affinity of chlorine can be represented with the equation:

$$Cl(g) + 1\ e^- \longrightarrow Cl^-(g) \qquad EA = -349\ \text{kJ/mol}$$

Electron affinities for a number of main-group elements are shown in **Figure 8.16◄**. As you can see, the trends in electron affinity

Electron Affinities (kJ/mol)

1A		3A	4A	5A	6A	7A	8A
H −73	2A						**He** >0
Li −60	**Be** >0	**B** −27	**C** −122	**N** >0	**O** −141	**F** −328	**Ne** >0
Na −53	**Mg** >0	**Al** −43	**Si** −134	**P** −72	**S** −200	**Cl** −349	**Ar** >0
K −48	**Ca** −2	**Ga** −30	**Ge** −119	**As** −78	**Se** −195	**Br** −325	**Kr** >0
Rb −47	**Sr** −5	**In** −30	**Sn** −107	**Sb** −103	**Te** −190	**I** −295	**Xe** >0

▲ **FIGURE 8.16 Electron Affinities of Selected Main-Group Elements**

are not as regular as trends in other properties we have examined. For example, we might expect electron affinities to become relatively more positive (so that the addition of an electron is less exothermic) as we move down a column because the electron is entering orbitals that are successively farther from the nucleus. This trend applies to the group 1A metals but does not hold for the other columns in the periodic table.

There is a more regular trend in electron affinity as we move to the right across a row, however. Based on the periodic properties we have learned so far, would you expect more energy to be released when an electron is gained by Na or Cl? We know that Na has an outer electron configuration of $3s^1$ and Cl has an outer electron configuration of $3s^2 3p^5$. Since adding an electron to chlorine gives it a noble gas configuration and adding an electron to sodium does not, and since the outermost electrons in chlorine experience a higher Z_{eff} than the outermost electrons in sodium, we would expect chlorine to have a more negative electron affinity—the process should be more exothermic for chlorine. This is in fact the case. For main-group elements, electron affinity generally becomes more negative (more exothermic) as we move to the right across a row in the periodic table. The halogens (group 7A) therefore have the most negative electron affinities. However, exceptions do occur. For example, notice that nitrogen and the other group 5A elements do not follow the general trend. These elements have $ns^2 np^3$ outer electron configurations. When an electron is added to this configuration, it must pair with another electron in an already occupied p orbital. The repulsion between two electrons occupying the same orbital causes the electron affinity to be more positive for these elements than for elements in the previous column.

Summarizing Electron Affinity for Main-Group Elements:

▶ Most groups of the periodic table do not exhibit any definite trend in electron affinity. Among the group 1A metals, however, electron affinity becomes more positive as we move down the column (adding an electron becomes less exothermic).

▶ Electron affinity generally becomes more negative (adding an electron becomes more exothermic) as we move to the right across a period (or row) in the periodic table.

Metallic Character

As we discussed in Chapter 2, metals are good conductors of heat and electricity; they can be pounded into flat sheets (malleability); they can be drawn into wires (ductility); they are often shiny; and they tend to lose electrons in chemical reactions. Nonmetals, by contrast, have more varied physical properties; some are solids at room temperature, others are gases, but in general they tend to be poor conductors of heat and electricity, and they all tend to gain electrons in chemical reactions. As we move to the right across a row in the periodic table, ionization energy increases and electron affinity becomes more negative, which means that elements on the left side of the periodic table are more likely to lose electrons than elements on the right side of the periodic table, which are more likely to gain them. The other properties associated with metals follow the same general trend (even though we do not quantify them here). Consequently, as shown in **Figure 8.17▶**,

> **As we move to the right across a period (or row) in the periodic table, metallic character decreases.**

As we move down a column in the periodic table, ionization energy decreases, making electrons more likely to be lost in chemical reactions. Consequently,

> **As we move down a column (or family) in the periodic table, metallic character increases.**

These trends, based on the quantum-mechanical model, explain the distribution of metals and nonmetals that we discussed in Chapter 2. Metals are found on the left side and toward the center of the periodic table and nonmetals on the upper right side. The change in chemical behavior from metallic to nonmetallic can be seen most clearly as we proceed to the right across period 3, or down along group 5A, of the periodic table, as can be seen in **Figure 8.18▶**.

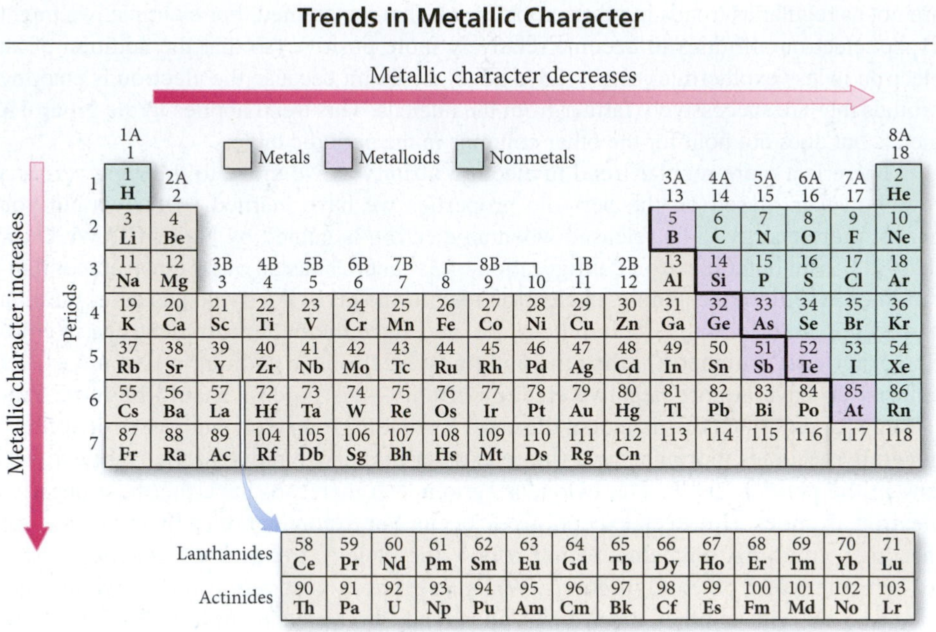

Trends in Metallic Character

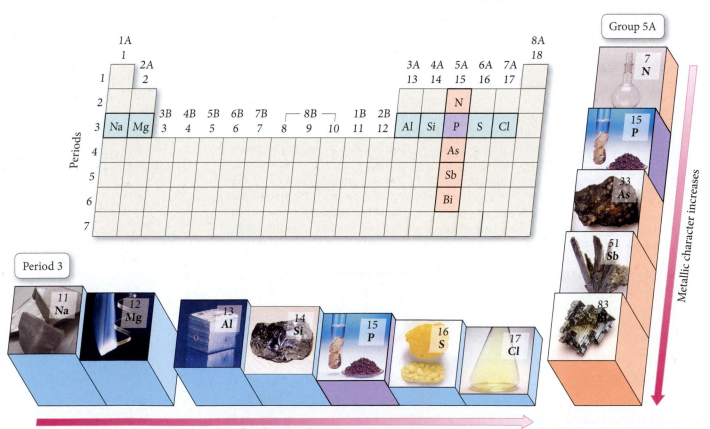

▲ **FIGURE 8.17 Trends in Metallic Character** Metallic character decreases as we move to the right across a row and increases as we move down a column in the periodic table.

Trends in Metallic Character

▲ **FIGURE 8.18 Trends in Metallic Character** As we move down group 5A in the periodic table, metallic character increases. As we move across row 3 in the periodic table, metallic character decreases.

EXAMPLE 8.9 Metallic Character

On the basis of periodic trends, choose the more metallic element from each pair (if possible):

(a) Sn or Te   **(b)** P or Sb   **(c)** Ge or In   **(d)** S or Br

SOLUTION

(a) Sn or Te
Sn is more metallic than Te because, as you trace the path between Sn and Te on the periodic table, you move to the right within the same period. Metallic character decreases as you go to the right.

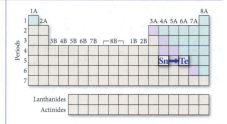

(b) P or Sb
Sb is more metallic than P because, as you trace the path between P and Sb on the periodic table, you move down a column. Metallic character increases as you go down a column.

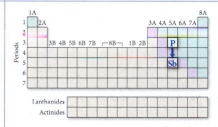

(c) Ge or In
In is more metallic than Ge because, as you trace the path between Ge and In on the periodic table, you move down a column (metallic character increases) and then to the left across a row (metallic character increases). These effects add together for an overall increase.

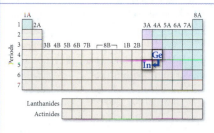

(d) S or Br
Based on periodic trends alone, you cannot tell which is more metallic because as you trace the path between S and Br, you go to the right across a row (metallic character decreases) and then down a column (metallic character increases). These effects tend to oppose each other, and it is not easy to tell which will predominate.

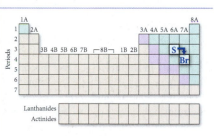

FOR PRACTICE 8.9

On the basis of periodic trends, choose the more metallic element from each pair (if possible):

(a) Ge or Sn   **(b)** Ga or Sn   **(c)** P or Bi   **(d)** B or N

FOR MORE PRACTICE 8.9

Arrange the following elements in order of increasing metallic character: Si, Cl, Na, Rb.

 Conceptual Connection 8.7 Periodic Trends

Use the trends in ionization energy and electron affinity to explain why sodium chloride has the formula NaCl and not Na_2Cl or $NaCl_2$.

CHAPTER IN REVIEW

Key Terms

Section 8.1
periodic property (287)

Section 8.3
electron configuration (288)
ground state (288)
orbital diagram (289)
Pauli exclusion principle (289)
degenerate (290)
Coulomb's law (290)

shielding (291)
effective nuclear charge
  (Z_{eff}) (291)
penetration (291)
aufbau principle (293)
Hund's rule (293)

Section 8.4
valence electrons (296)
core electrons (296)

Section 8.6
van der Waals radius (nonbonding atomic radius) (301)
covalent radius (bonding
  atomic radius) (301)
atomic radius (301)

Section 8.7
paramagnetic (306)

diamagnetic (306)
ionization energy (IE) (310)

Section 8.8
electron affinity (EA) (314)

Key Concepts

Periodic Properties and the Development of the Periodic Table (8.1, 8.2)

▶ The periodic table was primarily developed by Dmitri Mendeleev in the nineteenth century. Mendeleev arranged the elements in a table so that atomic mass increased from left to right in a row and elements with similar properties fell in the same columns.

▶ Periodic properties are those that are predictable based on an element's position within the periodic table. Periodic properties include atomic radius, ionization energy, electron affinity, density, and metallic character.

▶ Quantum mechanics explains the periodic table by explaining how electrons fill the quantum-mechanical orbitals within the atoms that compose the elements.

Electron Configurations (8.3)

▶ An electron configuration for an atom shows which quantum-mechanical orbitals the atom's electrons occupy. For example, the electron configuration of helium ($1s^2$) shows that helium's two electrons exist within the $1s$ orbital.

▶ The order of filling quantum-mechanical orbitals in multielectron atoms is $1s\ 2s\ 2p\ 3s\ 3p\ 4s\ 3d\ 4p\ 5s\ 4d\ 5p\ 6s$.

▶ According to the Pauli exclusion principle, each orbital can hold a maximum of two electrons with opposing spins.

▶ According to Hund's rule, orbitals of the same energy first fill singly with electrons with parallel spins, before pairing.

Electron Configurations and the Periodic Table (8.4, 8.5)

▶ Because quantum-mechanical orbitals fill sequentially with increasing atomic number, we can infer the electron configuration of an element from its position in the periodic table.

▶ The most stable configurations are those with completely full principal energy levels. Therefore, the most stable and unreactive elements are the noble gases.

▶ Elements with one or two valence electrons are among the most active metals, readily losing their valence electrons to attain noble gas configurations.

▶ Elements with six or seven valence electrons are among the most active nonmetals, readily gaining enough electrons to attain a noble gas configuration.

Effective Nuclear Charge and Periodic Trends in Atomic Size (8.6)

▶ The size of an atom is largely determined by its outermost electrons. As we move down a column in the periodic table, the principal quantum number (n) of the outermost electrons increases, resulting in successively larger orbitals and therefore larger atomic radii.

▶ As we move across a row in the periodic table, atomic radii decrease because the effective nuclear charge—the net or average charge experienced by the atom's outermost electrons—increases.

▶ The atomic radii of the transition elements stay roughly constant across each row because, as we move across a row, electrons are added to the $n_{highest}-1$ orbitals while the number of highest n electrons stays roughly constant.

Ion Properties (8.7)

▶ The electron configuration of an ion can be determined by adding or subtracting the number of electrons corresponding to the charge of the ion to the electron configuration of the neutral atom.

▶ For main-group cations, the order of removing electrons is the same as the order in which they are added in building up the electron configuration.

▶ For transition metal cations, the ns electrons are removed before the $(n-1)d$ electrons.

▶ The radius of a cation is much *smaller* than that of the corresponding atom, and the radius of an anion is much *larger* than that of the corresponding atom.

▶ The ionization energy—the energy required to remove an electron from an atom in the gaseous state—generally decreases as we move down a column in the periodic table and increases when moving to the right across a row.

▶ Successive ionization energies for valence electrons increase smoothly from one to the next, but the ionization energy increases dramatically for the first core electron.

Electron Affinities and Metallic Character (8.8)

▶ Electron affinity—the energy associated with an element in its gaseous state gaining an electron—does not show a general trend as we move down a column in the periodic table, but it generally becomes more negative (more exothermic) to the right across a row.

▶ Metallic character—the tendency to lose electrons in a chemical reaction—generally increases down a column in the periodic table and decreases to the right across a row.

Key Equations and Relationships

Order of Filling Quantum-Mechanical Orbitals (8.3)

$1s\ 2s\ 2p\ 3s\ 3p\ 4s\ 3d\ 4p\ 5s\ 4d\ 5p\ 6s$

Key Learning Objectives

Chapter Objectives	Assessment		
Writing Electron Configurations (8.3)	Example 8.1	For Practice 8.1	Exercises 1, 2
Writing Orbital Diagrams (8.3)	Example 8.2	For Practice 8.2	Exercises 3, 4
Valence Electrons and Core Electrons (8.4)	Example 8.3	For Practice 8.3	Exercises 11, 12
Electron Configurations from the Periodic Table (8.4)	Example 8.4 Exercises 5, 6	For Practice 8.4	For More Practice 8.4
Using Periodic Trends to Predict Atomic Size (8.6)	Example 8.5 Exercises 21–24	For Practice 8.5	For More Practice 8.5
Writing Electron Configurations for Ions (8.7)	Example 8.6	For Practice 8.6	Exercises 25, 26
Using Periodic Trends to Predict Ion Size (8.7)	Example 8.7 Exercises 29–32	For Practice 8.7	For More Practice 8.7
Using Periodic Trends to Predict Relative Ionization Energies (8.7)	Example 8.8 Exercises 33–36	For Practice 8.8	For More Practice 8.8
Predicting Metallic Character Based on Periodic Trends (8.8)	Example 8.9 Exercises 41–44	For Practice 8.9	For More Practice 8.9

EXERCISES

Problems by Topic

Electron Configurations

1. Write full electron configurations for each element:
   a. P b. C c. Na d. Ar

2. Write full electron configurations for each element:
   a. O b. Si c. Ne d. K

3. Write full orbital diagrams for each element:
   a. N b. F c. Mg d. Al

4. Write full orbital diagrams for each element:
   a. S b. Ca c. Ne d. He

5. Use the periodic table to write electron configurations for each element. Represent core electrons with the symbol of the previous noble gas in brackets.
   a. P b. Ge c. Zr d. I

6. Use the periodic table to determine the element corresponding to each electron configuration.
   a. $[Ar]\ 4s^2 3d^{10} 4p^6$ b. $[Ar]\ 4s^2 3d^2$
   c. $[Kr]\ 5s^2 4d^{10} 5p^2$ d. $[Kr]\ 5s^2$

7. Use the periodic table to determine each quantity:
   a. The number of $2s$ electrons in Li
   b. The number of $3d$ electrons in Cu
   c. The number of $4p$ electrons in Br
   d. The number of $4d$ electrons in Zr

8. Use the periodic table to determine each quantity:
   a. The number of $3s$ electrons in Mg
   b. The number of $3d$ electrons in Cr
   c. The number of $4d$ electrons in Y
   d. The number of $6p$ electrons in Pb

9. Name an element in the fourth period (row) of the periodic table with:
   a. five valence electrons
   b. four $4p$ electrons
   c. three $3d$ electrons
   d. a complete outer shell

10. Name an element in the third period (row) of the periodic table with:
    a. three valence electrons
    b. four $3p$ electrons
    c. six $3p$ electrons
    d. two $3s$ electrons and no $3p$ electrons

Valence Electrons and Simple Chemical Behavior from the Periodic Table

11. Determine the number of valence electrons in each element.
    a. Ba b. Cs c. Ni d. S

12. Determine the number of valence electrons for each element. Which elements do you expect to lose electrons in their chemical reactions? Which do you expect to gain electrons?
    a. Al b. Sn c. Br d. Se

13. Which outer electron configuration would you expect to belong to a reactive metal? To a reactive nonmetal?
    a. ns^2 b. $ns^2 np^6$ c. $ns^2 np^5$ d. $ns^2 np^2$

14. Which outer electron configuration would you expect to belong to a noble gas? To a metalloid?
    a. ns^2 b. $ns^2 np^6$ c. $ns^2 np^5$ d. $ns^2 np^2$

Coulomb's Law and Effective Nuclear Charge

15. According to Coulomb's law, which pair of charged particles has the lowest potential energy?
   a. A particle with a 1− charge separated by 150 pm from a particle with a 2+ charge.
   b. A particle with a 1− charge separated by 150 pm from a particle with a 1+ charge.
   c. A particle with a 1− charge separated by 100 pm from a particle with a 3+ charge.

16. According to Coulomb's law, rank the interactions between charged particles from lowest potential energy to highest potential energy.
   a. A 1+ charge and a 1− charge separated by 100 pm.
   b. A 2+ charge and a 1− charge separated by 100 pm.
   c. A 1+ charge and a 1− charge separated by 100 pm.
   d. A 1+ charge and a 1− charge separated by 200 pm.

17. Which electrons experience a greater effective nuclear charge, the valence electrons in beryllium, or the valence electrons in nitrogen? Why?

18. Arrange the atoms according to decreasing effective nuclear charge experienced by their valence electrons: S, Mg, Al, Si.

19. If core electrons completely shielded valence electrons from nuclear charge (i.e., if each core electron reduced nuclear charge by 1 unit) and if valence electrons did not shield one another from nuclear charge at all, what would be the effective nuclear charge experienced by the valence electrons of these atoms?
   a. K   b. Ca   c. O   d. C

20. In Section 8.6, we estimated the effective nuclear charge on beryllium's valence electrons to be slightly greater than 2+. What would a similar treatment predict for the effective nuclear charge on boron's valence electrons? Would you expect the effective nuclear charge to be different for boron's $2s$ electrons compared to its $2p$ electron? How so? (Hint: Consider the shape of the $2p$ orbital compared to that of the $2s$ orbital.)

Atomic Radius

21. Choose the larger atom from each pair:
   a. Al or In   b. Si or N   c. P or Pb   d. C or F

22. Choose the larger atom from each pair:
   a. Sn or Si   b. Br or Ga   c. Sn or Bi   d. Se or Sn

23. Arrange the elements in order of increasing atomic radius: Ca, Rb, S, Si, Ge, F.

24. Arrange the elements in order of decreasing atomic radius: Cs, Sb, S, Pb, Se.

Ionic Electron Configurations, Ionic Radii, Magnetic Properties, and Ionization Energy

25. Write an electron configuration for each ion:
   a. O^{2-} b. Br^- c. Sr^{2+} d. Co^{3+}
   e. Cu^{2+}

26. Write an electron configuration for each ion:
   a. Cl^- b. P^{3-} c. K^+ d. Mo^{3+}
   e. V^{3+}

27. Write an orbital diagram for each ion and determine if the ion is diamagnetic or paramagnetic.
   a. V^{5+} b. Cr^{3+} c. Ni^{2+} d. Fe^{3+}

28. Write an orbital diagram for each ion and determine if the ion is diamagnetic or paramagnetic.
   a. Cd^{2+} b. Au^+ c. Mo^{3+} d. Zr^{2+}

29. Pick the larger species from each pair:
   a. Li or Li^+ b. I^- or Cs^+
   c. Cr or Cr^{3+} d. O or O^{2-}

30. Pick the larger species from each pair:
   a. Sr or Sr^{2+} b. N or N^{3-}
   c. Ni or Ni^{2+} d. S^{2-} or Ca^{2+}

31. Arrange the isoelectronic series in order of decreasing radius: F^-, Ne, O^{2-}, Mg^{2+}, Na^+.

32. Arrange the isoelectronic series in order of increasing atomic radius: Se^{2-}, Kr, Sr^{2+}, Rb^+, Br^-.

33. Choose the element with the highest first ionization energy from each pair:
   a. Br or Bi   b. Na or Rb
   c. As or At   d. P or Sn

34. Choose the element with the highest first ionization energy from each pair:
   a. P or I   b. Si or Cl
   c. P or Sb   d. Ga or Ge

35. Arrange the elements in order of increasing first ionization energy: Si, F, In, N.

36. Arrange the elements in order of decreasing first ionization energy: Cl, S, Sn, Pb.

37. For each element, predict where the "jump" occurs for successive ionization energies. (For example, does the jump occur between the first and second ionization energies, the second and third, or the third and fourth?)
   a. Be   b. N   c. O   d. Li

38. Consider this set of successive ionization energies:
   $IE_1 = 578$ kJ/mol
   $IE_2 = 1820$ kJ/mol
   $IE_3 = 2750$ kJ/mol
   $IE_4 = 11,600$ kJ/mol
   To which third period element do these ionization values belong?

Electron Affinities and Metallic Character

39. Choose the element with the more negative (more exothermic) electron affinity from each pair:
   a. Na or Rb   b. B or S
   c. C or N   d. Li or F

40. Choose the element with the more negative (more exothermic) electron affinity from each pair:
   a. Mg or S   b. K or Cs
   c. Si or P   d. Ga or Br

41. Choose the more metallic element from each pair:
   a. Sr or Sb   b. As or Bi
   c. Cl or O   d. S or As

42. Choose the more metallic element from each pair:
   a. Sb or Pb   b. K or Ge
   c. Ge or Sb   d. As or Sn

43. Arrange the elements in order of increasing metallic character: Fr, Sb, In, S, Ba, Se.

44. Arrange the elements in order of decreasing metallic character: Sr, N, Si, P, Ga, Al.

Cumulative Problems

45. Bromine is a highly reactive liquid while krypton is an inert gas. Explain the difference based on their electron configurations.

46. Potassium is a highly reactive metal while argon is an inert gas. Explain the difference based on their electron configurations.

47. Both vanadium and its 3+ ion are paramagnetic. Use electron configurations to explain this.

48. Use electron configurations to explain why copper is paramagnetic while its 1+ ion is not.

49. If you were trying to find a substitute for K^+ in nerve signal transmission, where would you begin your search? What ions would be most like K^+? For each ion you propose, explain the ways in which it would be similar to K^+ and the ways it would be different. Refer to periodic trends in your explanation.

50. Where would you begin a search for a substitute for Na^+ in nerve signal transmission? What ions would be most like Na^+? For each ion you propose, explain the ways in which it would be similar to Na^+ and the ways it would be different. Refer to periodic trends in your discussion.

51. Life on Earth evolved around the element carbon. Based on periodic properties, what two or three elements would you expect to be most like carbon?

52. Which pair of elements would you expect to have the most similar atomic radii, and why?
  a. Si and Ga   **b.** Si and Ge   **c.** Si and As

53. Consider these elements: N, Mg, O, F, Al.
  a. Write an electron configuration for each element.
  b. Arrange the elements in order of decreasing atomic radius.
  c. Arrange the elements in order of increasing ionization energy.
  d. Use the electron configurations in part a to explain the differences between your answers to parts b and c.

54. Consider these elements: P, Ca, Si, S, Ga.
  a. Write an electron configuration for each element.
  b. Arrange the elements in order of decreasing atomic radius.
  c. Arrange the elements in order of increasing ionization energy.
  d. Use the electron configurations in part a to explain the differences between your answers to parts b and c.

55. Explain why atomic radius decreases as we move to the right across a row for main-group elements but not for transition elements.

56. Explain why vanadium (radius = 134 pm) and copper (radius = 128 pm) have nearly identical atomic radii, even

though the atomic number of copper is about 25% higher than that of vanadium. What would you predict about the relative densities of these two metals? Look up the densities in a reference book, periodic table, or on the Web. Are your predictions correct?

57. The lightest noble gases, such as helium and neon, are completely inert—they do not form any chemical compounds whatsoever. The heavier noble gases, in contrast, do form a limited number of compounds. Explain this difference in terms of trends in fundamental periodic properties.

58. The lightest halogen is also the most chemically reactive, and reactivity generally decreases as we move down the column of halogens in the periodic table. Explain this trend in terms of periodic properties.

59. Write general outer electron configurations ($ns^x np^y$) for groups 6A and 7A in the periodic table. The electron affinity of each group 7A element is more negative than that of each corresponding group 6A element. Use the electron configurations to explain this observation.

60. The electron affinity of each group 5A element is more positive than that of each corresponding group 4A element. Use the outer electron configurations to suggest a reason for this behavior.

61. Elements 35 and 53 have similar chemical properties. Based on their electronic configurations predict the atomic number of a heavier element that should share these chemical properties.

62. Write the electronic configurations of the six cations that form from sulfur by the loss of one to six electrons. For those cations that have unpaired electrons, write orbital diagrams.

63. Use Coulomb's law to calculate the ionization energy in kJ/mol of an atom composed of a proton and an electron separated by 100.0 pm. What wavelength of light would have sufficient energy to ionize the atom?

64. The first ionization energy of sodium is 496 kJ/mol. Use Coulomb's law to estimate the average distance between the sodium nucleus and the 3s electron. How does this distance compare to the atomic radius of sodium? Explain the difference.

65. The heaviest known alkaline earth metal is radium, atomic number 88. Find the atomic numbers of the as yet undiscovered next two members of the series.

66. From its electronic configuration, predict which of the first 10 elements would be most similar in chemical behavior to the as yet undiscovered element 165.

Challenge Problems

67. Consider the densities and atomic radii of the noble gases at 25 °C:

Element	Atomic Radius (pm)	Density (g/L)
He	32	0.18
Ne	70	0.90
Ar	98	—
Kr	112	3.75
Xe	130	—
Rn	—	9.73

  a. Estimate the densities of argon and xenon by interpolation from the data.

  b. Estimate of the density of the yet undiscovered element with atomic number 118 by extrapolation from the data.

  c. Use the molar mass of neon to estimate the mass of a neon atom. Then use the atomic radius of neon to calculate the average density of a neon atom. How does this density compare to the density of neon gas? What does this comparison suggest about the nature of neon gas?

  d. Use the densities and molar masses of krypton and neon to calculate the number of atoms of each found in a volume of 1.0 L. Use these values to estimate the number of atoms that occur in 1.0 L of Ar. Now use the molar mass of argon to estimate the density of Ar. How does this estimate compare to that in part a?

68. As we have seen, the periodic table is a result of empirical observation (i.e., the periodic law), but quantum-mechanical theory explains *why* the table is so arranged. Suppose that, in another universe, quantum theory was such that there were one *s* orbital but only two *p* orbitals (instead of three) and only three *d* orbitals (instead of five). Draw out the first four periods of the periodic table in this alternative universe. Which elements would be the equivalent of the noble gases? Halogens? Alkali metals?

69. Consider the metals in the first transition series. Use periodic trends to predict a trend in density as you move to the right across the series.

70. Imagine a universe in which the value of m_s can be $+\frac{1}{2}$, 0, and $-\frac{1}{2}$. Assuming that all the other quantum numbers can take only the values possible in our world and that the Pauli exclusion principle applies, give the following:
   a. the new electronic configuration of neon
   b. the atomic number of the element with a completed $n = 2$ shell
   c. the number of unpaired electrons in fluorine

71. A carbon atom can absorb radiation of various wavelengths with resulting changes in its electronic configuration. Write orbital diagrams for the electronic configuration of carbon that would result from absorption of the three longest wavelengths of radiation it can absorb.

72. Only trace amounts of the synthetic element darmstadtium, atomic number 110, have been obtained. The element is so highly unstable that no observations of its properties have been possible. Based on its position in the periodic table, propose three different reasonable valence electron configurations for this element.

73. What is the atomic number of the as yet undiscovered element in which the 8*s* and 8*p* electron energy levels fill? Predict the chemical behavior of this element.

74. The trend in second ionization energy for the elements from lithium to fluorine is not a smooth one. Predict which of these elements has the highest second ionization energy and which has the lowest and explain. Of the elements N, O, and F, O has the highest and N the lowest second ionization energy. Explain.

75. Unlike the elements in groups 1A and 2A, those in group 3A do not show a smooth decrease in first ionization energy in going down the column. Explain the irregularities.

76. Using the data in Figures 8.15 and 8.16, calculate ΔE for the reaction

$$Na(g) + Cl(g) \longrightarrow Na^+(g) + Cl^-(g)$$

77. Even though adding two electrons to O or S forms an ion with a noble gas electron configuration, the second electron affinity of both of these elements is positive. Explain.

78. In Section 2.6 we discussed the metalloids, which form a diagonal band separating the metals from the nonmetals. There are other instances in which elements such as lithium and magnesium that are diagonal to each other have comparable metallic character. Suggest an explanation for this observation.

Conceptual Problems

79. Imagine that in another universe, atoms and elements are identical to ours, except that atoms with six valence electrons have particular stability (in contrast to our universe where atoms with eight valence electrons have particular stability). Give an example of an element in the alternative universe that corresponds to each:
   a. a noble gas
   b. a reactive nonmetal
   c. a reactive metal

80. The outermost valence electron in atom A experiences an effective nuclear charge of 2+ and is on average 225 pm from the nucleus. The outermost valence electron in atom B experiences and effective nuclear charge of 1+ and is on average 175 pm from the nucleus. Which atom (A or B) has the highest first ionization energy? Explain.

81. Determine whether each statement is true or false regarding penetration and shielding. (Assume that all lower energy orbitals are fully occupied.)

   a. An electron in a 3*s* orbital is more shielded than an electron in a 2*s* orbital.
   b. An electron in a 3*s* orbital penetrates into the region occupied by core electrons more than electrons in a 3*p* orbital.
   c. An electron in an orbital that penetrates closer to the nucleus will always experience more shielding than an electron in an orbital that does not penetrate as far.
   d. An electron in an orbital that penetrates close to the nucleus will tend to experience a higher effective nuclear charge than one that does not.

82. Give a combination of four quantum numbers that could be assigned to an electron occupying a 5*p* orbital. Do the same for an electron occupying a 6*d* orbital.

83. Use the trends in ionization energy and electron affinity to explain why calcium fluoride has the formula CaF_2 and not Ca_2F or CaF.

Answers to Conceptual Connections

Coulomb's Law

8.1 (a) Since the charges are opposite, the potential energy of the interaction is negative. As the charges gets closer together, *r* becomes smaller and the potential energy decreases.

Penetration and Shielding

8.2 (c) Penetration results in less shielding from nuclear charge and therefore lower energy.

Electron Configurations and Quantum Numbers

8.3 $n = 4, l = 0, m_l = 0, m_s = +\frac{1}{2}; n = 4, l = 0, m_l = 0, m_s = -\frac{1}{2}$

Effective Nuclear Charge

8.4 (c) Since Z_{eff} increases from left to right across a row in the periodic table, the valence electrons in S experience a greater effective nuclear charge than the valence electrons in Al or in Mg.

Ions, Isotopes, and Atomic Size

8.5 The isotopes of an element all have the same radii for two reasons: (1) neutrons are negligibly small compared to the size of an atom and therefore extra neutrons do not increase atomic size; (2) neutrons have no charge and therefore do not attract electrons in the way that protons do.

Ionization Energies and Chemical Bonding

8.6 As you can see from the successive ionization energies of any element, valence electrons are held most loosely and can therefore be transferred or shared most easily. Core electrons, on the other hand, are held tightly and are not easily transferred or shared. Consequently, valence electrons are most important to chemical bonding.

Periodic Trends

8.7 The $3s$ electron in sodium has a relatively low ionization energy (496 kJ/mol) because it is a valence electron. The energetic cost for sodium to lose a second electron is extraordinarily high (4560 kJ/mol) because the next electron to be lost is a core electron ($2p$). Similarly, the electron affinity of chlorine to gain one electron (-349 kJ/mol) is highly exothermic because the added electron completes chlorine's valence shell. The gain of a second electron by the negatively charged chlorine anion would not be so favorable. Therefore, we would expect sodium and chlorine to combine in a 1:1 ratio.

9 Chemical Bonding I: The Lewis Model

Theories are nets cast to catch what we call 'the world': to rationalize, to explain, and to master it. We endeavor to make the mesh ever finer and finer.
—Karl Popper (1902–1994)

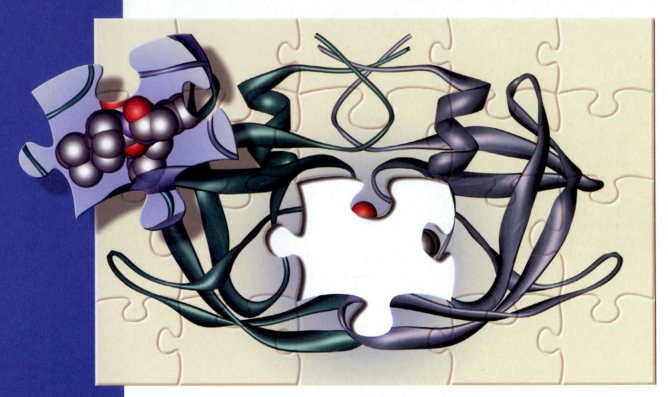

The AIDS drug Indinavir—*shown here as the missing piece in a puzzle depicting the protein HIV-protease—was developed with the help of chemical bonding theories.*

$\mathbf{C}$HEMICAL BONDING IS AT THE HEART of chemistry. The bonding theories that we are about to examine are—as Karl Popper, a philosopher of science, eloquently states above—nets cast to understand the world. In the next two chapters, we will examine three theories, with successively finer "meshes." The first is the Lewis model, which can be practiced on the back of an envelope. With just a few dots, dashes, and chemical symbols, we can understand and predict a myriad of chemical observations. The second is valence bond theory, which treats electrons in a more quantum-mechanical manner, but stops short of viewing them as belonging to the entire molecule. The third is molecular orbital theory, essentially a full quantum-mechanical treatment of the molecule and its electrons as a whole. Molecular orbital theory has great predictive power, but that power comes at the expense of great complexity and intensive computational requirements. Which theory is "correct"? Remember that theories are models that help us understand and predict behavior. All three of these theories are extremely useful, depending on exactly what aspect of chemical bonding we want to predict or understand.

9.1   Bonding Models and AIDS Drugs

In 1989, researchers used X-ray crystallography—a technique in which X-rays are scattered from crystals of the molecule of interest—to determine the structure of a molecule called HIV-protease. HIV-protease is a protein (a class of large biological molecules) synthesized by the human immunodeficiency virus (HIV). This particular protein is crucial to the virus's ability to multiply and cause acquired immune deficiency syndrome, or AIDS. Without HIV-protease, HIV cannot spread in the human body because the virus cannot replicate. In other words, without HIV-protease, AIDS can't develop.

With knowledge of the HIV-protease structure, drug companies set out to create a molecule that would disable HIV-protease by adhering to the working part of the molecule, called the active site. To design such a molecule, researchers used *bonding theories*—models that predict how atoms bond together to form molecules—to simulate the shape of potential drug molecules and determine how they would interact with the protease molecule. By the early 1990s, these companies had developed several drug molecules that worked. Since these molecules inhibit the action of HIV-protease, they are called *protease inhibitors*. In human trials, protease inhibitors, when given in combination with other drugs, have decreased the viral count in HIV-infected individuals to undetectable levels. Although protease inhibitors are not a cure for AIDS, many AIDS patients are still alive today because of these drugs.

Bonding theories are central to chemistry because they explain how atoms form molecules. They explain why some combinations of atoms are stable and others are not. Bonding theories explain why table salt is NaCl and not $NaCl_2$ and why water is H_2O and not H_3O. Bonding theories also predict the shapes of molecules—a topic in our next chapter—which in turn determine many of their physical and chemical properties. The bonding theory we will cover in this chapter is called the **Lewis model**, named after the American chemist who developed it, G. N. Lewis (1875–1946). In the Lewis model, we represent valence electrons as dots and draw **Lewis electron-dot structures** (or simply **Lewis structures**) to represent molecules. These structures, which are fairly simple to draw, have tremendous predictive power. We can use the Lewis model to predict whether a particular set of atoms will form a stable molecule and what that molecule might look like. Although we will also learn about more advanced theories in the following chapter, the Lewis model remains the simplest method for making quick, everyday predictions about most molecules.

9.2   Types of Chemical Bonds

We begin our discussion of chemical bonding by asking why bonds form in the first place. This seemingly simple question is vitally important. Imagine a universe without chemical bonding. Such a universe would contain only 91 different kinds of substances

▲ G.N. Lewis

(the 91 naturally occurring elements). With such a poor diversity of substances, life would be impossible, and we would not be around to wonder why. The answer to this question is not simple and involves not only quantum mechanics but also some thermodynamics that we will not introduce until Chapter 17. Nonetheless, we can address an important *part* of the answer now: *chemical bonds form because they lower the potential energy between the charged particles that compose atoms.*

As we already know, atoms are composed of particles with positive charges (the protons in the nucleus) and negative charges (the electrons). When two atoms approach each other, the electrons of one atom are attracted to the nucleus of the other according to Coulomb's law (see Section 8.3) and vice versa. However, at the same time, the electrons of each atom repel the electrons of the other, and the nucleus of each atom repels the nucleus of the other. The result is a complex set of interactions among a potentially large number of charged particles. If these interactions lead to an overall net reduction of potential energy between the charged particles, a chemical bond forms. Bonding theories help us to predict the circumstances under which bonds form and also the properties of the resultant molecules.

We can broadly classify chemical bonds into three types depending on the kind of atoms involved in the bonding (**Figure 9.1▼**).

We learned in Chapter 8 that metals tend to have low ionization energies (it is relatively easy to remove electrons from them) and that nonmetals tend to have negative electron affinities (it is energetically favorable for them to gain electrons). When a metal bonds with a nonmetal, it transfers one or more electrons to the nonmetal. The metal atom thus becomes a cation and the nonmetal atom an anion. These oppositely charged ions are then attracted to one another, lowering their overall potential energy as indicated by Coulomb's law. The resulting bond is an **ionic bond**.

We also learned in Chapter 8 that nonmetals tend to have high ionization energies (it is difficult to remove electrons from them). Therefore, when a nonmetal bonds with another nonmetal, neither atom transfers electrons to the other. Instead, some electrons are *shared* between the two bonding atoms. The shared electrons interact with the nuclei

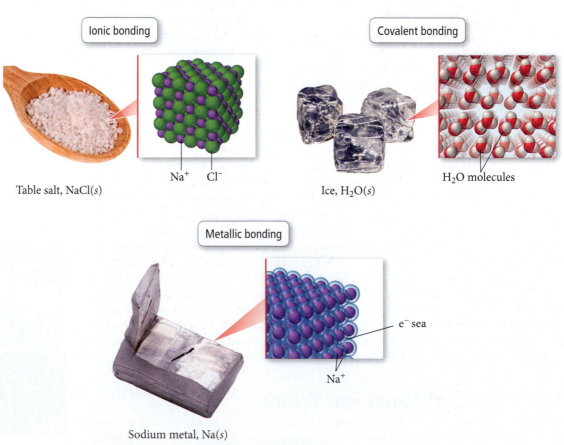

▲ **FIGURE 9.1 Ionic, Covalent, and Metallic Bonding**

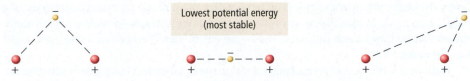

▲ **FIGURE 9.2 Some Possible Configurations of One Electron and Two Protons**

of both of the bonding atoms, lowering their potential energy in accordance with Coulomb's law. The resulting bond is a **covalent bond**. Recall from Section 3.2 that we can understand the stability of a covalent bond by considering the most stable arrangement (the one with the lowest potential energy) of a negatively charged particle and two positively charged particles separated by a small distance. As you can see in **Figure 9.2▲**, the arrangement in which the negatively charged particle lies *between* the two positively charged ones has the lowest potential energy. This is because in this arrangement, the negatively charged particle interacts most strongly with *both of the positively charged ones*. In a sense, the negatively charged particle holds the two positively charged ones together. Similarly, shared electrons in a covalent chemical bond *hold* the bonding atoms together by attracting the positive charges of their nuclei.

A third type of bonding, called **metallic bonding**, occurs in metals. Since metals have low ionization energies, they tend to lose electrons easily. In the simplest model for metallic bonding—called the *electron sea* model—all of the atoms in a metal lattice pool their valence electrons. These pooled electrons are no longer localized on a single atom, but delocalized over the entire metal. The positively charged metal atoms are then attracted to the sea of electrons, holding the metal together. We discuss metallic bonding in more detail in Section 9.11. The table shown here summarizes the three different types of bonding.

Types of Atoms	Type of Bond	Characteristic of Bond
Metal and nonmetal	Ionic	Electrons transferred
Nonmetal and nonmetal	Covalent	Electrons shared
Metal and metal	Metallic	Electrons pooled

9.3 Representing Valence Electrons with Dots

Recall from Chapter 8, that, for main-group elements, valence electrons are those electrons in the outermost principal energy level. Since valence electrons are held most loosely, and since chemical bonding involves the transfer or sharing of electrons between two or more atoms, valence electrons are most important in bonding, so the Lewis model focuses on these. In a Lewis structure, the valence electrons of main-group elements are represented as dots surrounding the symbol for the element. For example, the electron configuration of O is

Remember, the number of valence electrons for any main group element is equal to the group number of the element (except for helium, which is in group 8A but has only two valence electrons).

$1s^2 2s^2 2p^4$ — 6 valence electrons

And the Lewis structure is

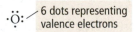
$\cdot \ddot{\text{O}} :$ — 6 dots representing valence electrons

Each dot represents a valence electron. The dots are placed around the element's symbol with a maximum of two dots per side. The Lewis structures for all of the row 2 elements are

While the exact location of dots is not critical, in this book we first place dots singly before pairing (except for helium which always has two paired dots).

Li·   ·Be·   ·Ḃ·   ·Ċ·   ·N̈:   ·Ö: :F̈:   :N̈e:

Lewis structures provide a simple way to visualize the number of valence electrons in a main-group atom. Notice that atoms with eight valence electrons—which are particularly stable because they have a full outer level—are easily identified because they have eight dots, an **octet**.

Helium is somewhat of an exception. Its electron configuration and Lewis structure are

$$1s^2 \qquad \text{He:}$$

The two dots in the Lewis structure of helium are always paired; the paired dots are referred to as a **duet**. For helium, a duet represents a stable electron configuration because the $n = 1$ quantum level fills with only two electrons.

In the Lewis model, a **chemical bond** is the sharing or transfer of electrons to attain stable electron configurations for the bonding atoms. If electrons are transferred, as occurs between a metal and a nonmetal, the bond is an *ionic bond*. If the electrons are shared, as occurs between two nonmetals, the bond is a *covalent bond*. In either case, the bonding atoms obtain stable electron configurations; since the stable configuration is usually eight electrons in the outermost level, this is known as the **octet rule**. Notice that the Lewis model does not attempt to address attractions and repulsions between electrons and nuclei on neighboring atoms. The energy changes that occur because of these interactions are central to chemical bonding (as we saw in Section 9.2), yet the Lewis model ignores them because calculating these energy changes is extremely complicated. Instead the Lewis model uses the simple octet rule, a practical approach which predicts what we see in nature for a large number of compounds—hence the success and longevity of the Lewis model.

9.4 Ionic Bonding: Lewis Structures and Lattice Energies

Although the Lewis model's strength is in modeling covalent bonding (which we discuss in detail in the next section of the chapter), it can also be applied to ionic bonding. In the Lewis model, we represent ionic bonding by moving electron dots from the metal to the nonmetal and allowing the resultant ions to form a crystalline lattice composed of alternating cations and anions.

Ionic Bonding and Electron Transfer

To see how ionic bonding is formulated in the Lewis model, consider potassium and chlorine, which have the following Lewis structures:

$$\text{K}\cdot \qquad :\ddot{\text{Cl}}:$$

When potassium and chlorine bond, potassium transfers its valence electron to chlorine:

$$\text{K}\cdot \; + \; :\ddot{\text{Cl}}: \; \longrightarrow \; \text{K}^+\left[:\ddot{\text{Cl}}:\right]^-$$

The transfer of the electron gives chlorine an octet (shown as eight dots around chlorine) and leaves potassium without any valence electrons but with an octet in the *previous* principal energy level (which becomes its outermost level).

$$\text{K} \qquad 1s^2 2s^2 2p^6 3s^2 3p^6 4s^1$$
$$\text{K}^+ \qquad 1s^2 2s^2 2p^6 \underbrace{3s^2 3p^6}\ 4s^0$$

Octet in previous level

Potassium, because it has lost an electron, becomes positively charged, while chlorine, which has gained an electron, becomes negatively charged. The Lewis structure of an anion is usually written within brackets with the charge in the upper right-hand corner, outside the brackets. The positive and negative charges attract one another, resulting in the compound KCl.

The Lewis model predicts the correct chemical formulas for ionic compounds. For the compound that forms between K and Cl, for example, the Lewis model

Recall that solid ionic compounds do not contain distinct molecules, but rather are composed of alternating positive and negative ions in a three-dimensional crystalline array.

predicts one potassium cation to every one chlorine anion, KCl. In nature, when we examine the compound formed between potassium and chlorine, we indeed find one potassium ion to every chloride ion. As another example, consider the ionic compound formed between sodium and sulfur. The Lewis structures for sodium and sulfur are

$$Na\cdot \quad \cdot \ddot{S}:$$

Notice that sodium must lose its one valence electron in order to have an octet (in the previous principal energy level), while sulfur must gain two electrons to get an octet. Consequently, the compound that forms between sodium and sulfur requires two sodium atoms for every one sulfur atom. The Lewis structure is

$$2Na^+ \left[:\ddot{S}:\right]^{2-}$$

The two sodium atoms each lose their one valence electron while the sulfur atom gains two electrons and gets an octet. The Lewis model predicts that the correct chemical formula is Na_2S. When we look at the compound formed between sodium and sulfur in nature, the formula predicted by the Lewis model is exactly what we see.

EXAMPLE 9.1 Using the Lewis Model to Predict the Chemical Formula of an Ionic Compound

Use the Lewis model to predict the formula for the compound that forms between calcium and chlorine.

SOLUTION

Draw Lewis structures for calcium and chlorine based on their respective number of valence electrons, determined from their group number in the periodic table.	Ca: :$\ddot{C}$l:
Calcium must lose its two valence electrons (to be left with an octet in its previous principal level), while chlorine only needs to gain one electron to get an octet. Draw two chloride anions, each with an octet and a 1− charge, and one calcium cation with a 2+ charge. Draw the chloride anions in brackets and indicate the charges on each ion.	Ca^{2+} $2\left[:\ddot{C}l:\right]^-$
Finally, write the formula with subscripts that indicate the number of atoms.	$CaCl_2$

FOR PRACTICE 9.1

Use the Lewis model to predict the formula for the compound that forms between magnesium and nitrogen.

Lattice Energy: The Rest of the Story

The formation of an ionic compound from its constituent elements is usually quite exothermic. For example, when sodium chloride (table salt) forms from elemental sodium and chlorine, 411 kJ of heat is evolved in the following violent reaction:

$$Na(s) + {}^1/_2\,Cl_2(g) \longrightarrow NaCl(s) \quad \Delta H^\circ_f = -411 \text{ kJ/mol}$$

Where does this energy come from? You might think that it comes solely from the tendency of metals to lose electrons and nonmetals to gain electrons—but it does not. In fact, the transfer of an electron from sodium to chlorine—by itself—actually absorbs energy. The first ionization energy of sodium is +496 kJ/mol, and the electron affinity of Cl is only −349 kJ/mol. So based only on these energies, the reaction should be *endothermic* by +147 kJ/mol. So why is the reaction so exothermic?

The answer lies in the **lattice energy** ($\Delta H^\circ_{lattice}$)—the energy associated with the formation of a crystalline lattice of alternating cations and anions from the gaseous ions. Since the sodium ions are positively charged and the chlorine ions negatively charged, there is a lowering of potential energy—as prescribed by Coulomb's law—when these ions come together to form a lattice. That energy is emitted as heat when the lattice forms, as shown in **Figure 9.3▶**. Since the formation of the lattice is exothermic, lattice energy is always negative.

The exothermicity of the formation of the crystalline NaCl lattice from sodium cations and chloride anions more than compensates for the endothermicity of the electron transfer process. In other words, the formation of ionic compounds is not exothermic because

▶ **FIGURE 9.3 Lattice Energy**
The lattice energy of an ionic compound is the energy associated with the formation of a crystalline lattice of the compound from the gaseous ions.

Lattice Energy of an Ionic Compound

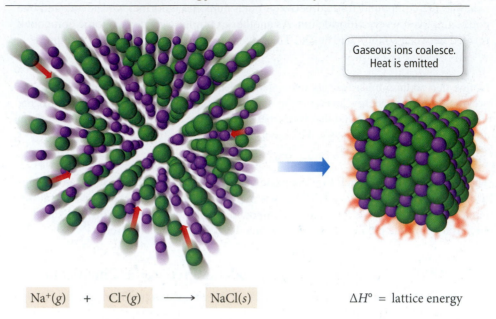

Gaseous ions coalesce. Heat is emitted

$$Na^+(g) \quad + \quad Cl^-(g) \quad \longrightarrow \quad NaCl(s) \qquad\qquad \Delta H° = \text{lattice energy}$$

sodium "wants" to lose electrons and chlorine "wants" to gain them; rather, it is exothermic because of the large amount of heat released when sodium and chloride ions coalesce to form a crystalline lattice.

Trends in Lattice Energies: Ion Size

Consider the lattice energies of the alkali metal chlorides:

Metal Chloride	Lattice Energy (kJ/mol)
LiCl	−834
NaCl	−787
KCl	−701
CsCl	−657

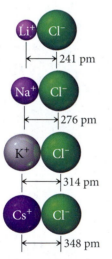

▲ Bond lengths of the group 1A metal chlorides.

Why do you suppose that the magnitude of the lattice energy decreases as we move down the column? We know from the periodic trends discussed in Chapter 8 that ionic radius increases as we move down a column in the periodic table (see Section 8.7). We also know, from our discussion of Coulomb's law in Section 8.3, that the potential energy of oppositely charged ions becomes less negative (or more positive) as the distance between ions increases. As the size of the alkali metal ions increases down the column, so does the distance between the metal cations and the chloride anions. Therefore, the magnitude of the lattice energy decreases accordingly, making the formation of the chlorides less exothermic and the compounds less stable. In other words, as the ionic radii increase as we move down the column, the ions cannot get as close to each other and therefore do not release as much energy when the lattice forms.

Trends in Lattice Energies: Ion Charge

Consider the lattice energies of the following two compounds:

Compound	Lattice Energy (kJ/mol)
NaF	−910
CaO	−3414

Why is the magnitude of the lattice energy of CaO so much greater than the lattice energy of NaF? Na^+ has a radius of 95 pm and F^- has a radius of 136 pm, resulting in a distance between ions of 231 pm. Ca^{2+} has a radius of 99 pm and O^{2-} has a radius of 140 pm, resulting in a distance between ions of 239 pm. Even though the separation between the calcium and oxygen is slightly greater (which would tend to lower the lattice energy), the lattice energy is almost four times greater. The explanation lies in the charges of the ions. Recall from our discussion of Coulomb's law that the magnitude of the potential energy of two interacting charges depends not only on the distance between the charges, but also on the product of the charges:

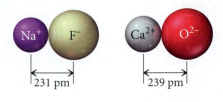

$$E = \frac{1}{4\pi\varepsilon_0} \frac{q_1 q_2}{r}$$

For NaF, E is proportional to $(1+)(1-) = 1-$, while for CaO, E is proportional to $(2+)(2-) = 4-$, so the relative stabilization for CaO relative to NaF should be roughly four times greater, as observed in the lattice energy.

Summarizing Trends in Lattice Energies:

▶ *Lattice energies become less exothermic (less negative) with increasing ionic radius.*

▶ *Lattice energies become more exothermic (more negative) with increasing magnitude of ionic charge.*

EXAMPLE 9.2 Predicting Relative Lattice Energies

Arrange the ionic compounds in order of increasing *magnitude* of lattice energy: CaO, KBr, KCl, SrO.

SOLUTION

KBr and KCl should have lattice energies of smaller magnitude than CaO and SrO because of their lower ionic charges (1+, 1− compared to 2+, 2−). Between KBr and KCl, we expect KBr to have a lattice energy of lower magnitude due to the larger ionic radius of the bromide ion relative to the chloride ion. Between CaO and SrO, we expect SrO to have a lattice energy of lower magnitude due to the larger ionic radius of the strontium ion relative to the calcium ion.

Order of increasing *magnitude* of lattice energy:

KBr < KCl < SrO < CaO

Actual lattice energy values:

Compound	Lattice Energy (kJ/mol)
KBr	−671
KCl	−701
SrO	−3217
CaO	−3414

FOR PRACTICE 9.2

Arrange the compounds in order of increasing magnitude of lattice energy: LiBr, KI, and CaO.

FOR MORE PRACTICE 9.2

Which compound has the greater lattice energy, NaCl or $MgCl_2$?

Ionic Bonding: Models and Reality

Earlier in this section, we developed a model for ionic bonding based on the Lewis model. The value of a model is in how well it accounts for what we see in the real world (through observations). Can the ionic bonding model explain the properties of ionic compounds, including their high melting and boiling points, their tendency *not to conduct* electricity as solids, and their tendency *to conduct* electricity when dissolved in water?

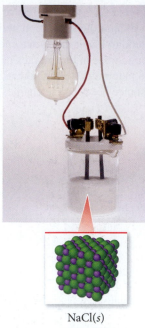

NaCl(*s*)

NaCl(*aq*)

▲ Solid sodium chloride does not conduct electricity.

▲ When sodium chloride dissolves in water, the resulting solution contains mobile ions that can create an electric current.

We modeled ionic solids as a lattice of individual ions held together by coulombic forces that are *nondirectional* (which means that, as you move away from the center of an ion, the forces are equally strong in all directions). To melt the solid, these forces must be overcome, which requires a significant amount of heat. Therefore, our model accounts for the high melting points of ionic solids. In our model, electrons are transferred from the metal to the nonmetal, but the transferred electrons remain localized on one atom. In other words, our model does not include any free electrons that might conduct electricity (the movement or flow of electrons in response to an electric potential (or voltage) is electrical current). In addition, the ions themselves are fixed in place; therefore, our model accounts for the nonconductivity of ionic solids. When our idealized ionic solid dissolves in water, however, the cations and anions dissociate, forming free ions in solution. These ions can move in response to electrical forces, creating an electrical current. Thus, our model predicts that solutions of ionic compounds conduct electricity.

Conceptual Connection 9.1 **Melting Points of Ionic Solids**

Use the ionic bonding model to determine which has the higher melting point, NaCl or MgO. Explain the relative ordering.

9.5 Covalent Bonding: Lewis Structures

The Lewis model provides us with a very simple and useful model for covalent bonding. In the Lewis model, we represent covalent bonding by depicting neighboring atoms as sharing some (or all) of their valence electrons in order to attain octets (or duets for hydrogen).

Single Covalent Bonds

To see how covalent bonding is conceived in terms of the Lewis model, consider hydrogen and oxygen, which have the following Lewis structures:

In water, hydrogen and oxygen share their unpaired valence electrons so that each hydrogen atom gets a duet and the oxygen atom gets an octet:

H:Ö:H

The shared electrons—those that appear in the space between the two atoms—count towards the octets (or duets) of *both of the atoms*.

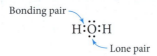

Duet   Octet   Duet

A pair of electrons that is shared between two atoms is a **bonding pair**, while a pair that is associated with only one atom—and therefore not involved in bonding—is a **lone pair**.

> Sometimes lone pair electrons are also called nonbonding electrons.

Bonding pair
H:Ö:H
Lone pair

We often represent bonding pair electrons by dashes to emphasize that they constitute a chemical bond.

> Keep in mind that *one* dash always stands for *two* electrons (a bonding pair).

H—Ö—H

The Lewis model explains why the halogens form diatomic molecules. Consider the Lewis structure of chlorine:

:Cl.

If two Cl atoms pair together, they can each get an octet:

:Cl:Cl: or   :Cl—:Cl:

When we examine elemental chlorine, it indeed exists as a diatomic molecule, just as the Lewis model predicts. The same is true for the other halogens.

Similarly, the Lewis model predicts that hydrogen, which has the Lewis structure

H·

should exist as H_2. When two hydrogen atoms share their valence electrons, each ends up with a duet, a stable configuration for hydrogen.

H:H   *or*   H—H

Again, the Lewis model is correct. In nature, elemental hydrogen exists as H_2 molecules.

Double and Triple Covalent Bonds

In the Lewis model, two atoms may share more than one electron pair to get octets. For example, if two oxygen atoms join together, they must share two electron pairs in order for each oxygen atom to have an octet. Each oxygen atom now has an octet because the additional bonding pair counts toward the octet of both oxygen atoms.

·Ö: + ·Ö:
↓
:Ö::Ö: *or* :Ö=Ö:

Octet   Octet

When two electron pairs are shared between two atoms, the resulting bond is a **double bond**. In general, double bonds are shorter and stronger than single bonds. Atoms can also share three electron pairs. Consider the Lewis structure of N_2. Since each N atom has

We explore the characteristics of
multiple bonds more fully in Section 9.10.

five valence electrons, the Lewis structure for N_2 has 10 electrons. Both nitrogen atoms can attain octets only by sharing three electron pairs:

$$:N:::N: \quad or \quad :N\equiv N:$$

The bond is a **triple bond**. Triple bonds are even shorter and stronger than double bonds. When we examine nitrogen in nature, we find that it indeed exists as a diatomic molecule with a very strong bond between the two nitrogen atoms. The bond is so strong that it is difficult to break, making N_2 a relatively unreactive molecule.

Covalent Bonding: Models and Reality

The Lewis model predicts the properties of molecular compounds in many ways. First, it accounts for why particular combinations of atoms form molecules and others do not. For example, why is water H_2O and not H_3O? We can write a good Lewis structure for H_2O, but not for H_3O.

$$H-\ddot{O}-H \qquad \overset{\displaystyle H}{\underset{\displaystyle \bullet\bullet}{\mid}}H-O-H$$

Oxygen has nine electrons
(one electron beyond an octet)

So the Lewis model predicts that H_2O should be stable, while H_3O should not be, and that is in fact the case. However, if we remove an electron from H_3O, we get H_3O^+, which should be stable (according to the Lewis model) because, by removing the extra electron, oxygen gets an octet.

$$\left[\overset{\displaystyle H}{\underset{\displaystyle \bullet\bullet}{\mid}}H-O-H\right]^+$$

This ion, called the hydronium ion, is in fact stable in aqueous solutions (see Section 4.8). The Lewis model also predicts other possible combinations for hydrogen and oxygen. For example, we can write a Lewis structure for H_2O_2 as follows:

$$H-\ddot{O}-\ddot{O}-H$$

Indeed, H_2O_2, or hydrogen peroxide, exists and is often used as a disinfectant and a bleach.

The Lewis model also accounts for why covalent bonds are highly *directional*. The attraction between two covalently bonded atoms is due to the sharing of one or more electron pairs in the space between them. Thus, each bond links just one specific pair of atoms—*in contrast to ionic bonds, which are nondirectional and hold together an entire array of ions.* As a result, the fundamental units of covalently bonded compounds are individual molecules. These molecules can interact with one another in a number of different ways that we will cover in Chapter 11. However, the interactions *between* molecules (intermolecular forces) are generally much weaker than the bonding interactions within a molecule (intramolecular forces), as shown in **Figure 9.4▶**. When a molecular compound melts or boils, the molecules themselves remain intact—only the relatively weak interactions between molecules must be overcome. Consequently, molecular compounds tend to have lower melting and boiling points than ionic compounds.

Conceptual Connection 9.2 Energy and the Octet Rule

In what way is this statement incomplete (or misleading)? *Atoms form bonds in order to satisfy the octet rule.*

Molecular Compound

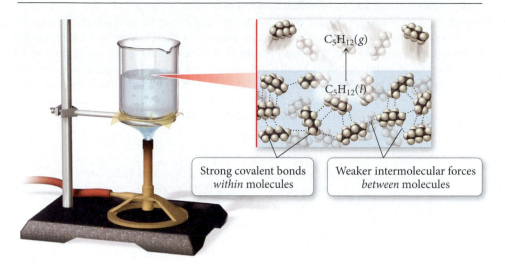

◀ **FIGURE 9.4 Intermolecular and Intramolecular Forces**   The covalent bonds between atoms of a molecule are much stronger than the interactions between molecules. To boil a molecular substance, we only have to overcome the relatively weak intermolecular forces, so molecular compounds generally have low boiling points.

9.6   Electronegativity and Bond Polarity

We know from Chapter 7 that representing electrons with dots, as we do in the Lewis model, is a drastic oversimplification. As we have already discussed, this does not invalidate the Lewis model—which is an extremely useful model—but we must recognize and compensate for its inherent limitations. One limitation of representing electrons as dots, and covalent bonds as two dots shared between two atoms, is that the shared electrons always appear to be *equally* shared. That is not the case. For example, consider the Lewis structure of hydrogen fluoride.

$$H : \ddot{\underset{\cdot\cdot}{F}} :$$

The two shared electron dots sitting between the H and the F atoms appear to be equally shared between hydrogen and fluorine. However, based on laboratory measurements, we know they are not. When HF is put in an electric field, the molecules orient as shown in **Figure 9.5▼**. From this observation, we know that the hydrogen side of the molecule must have a slight positive charge and the fluorine side of the molecule must have a slight negative charge. We represent this as:

$$\overset{+\longrightarrow}{H-F} \quad or \quad \overset{\delta^+ \quad \delta^-}{H-F}$$

The red arrow on the left, with a positive sign on the tail, shows that the left side of the molecule has a partial positive charge and that the right side of the molecule (the side the arrow is pointing *toward*) has a partial negative charge. Similarly, the $\delta+$ (delta plus) represents a partial positive charge and the $\delta-$ (delta minus) represents a partial negative

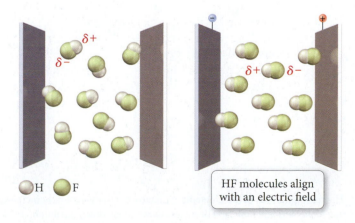

◀ **FIGURE 9.5 Orientation of Gaseous Hydrogen Fluoride in an Electric Field**   Because one side of the HF molecule has a slight positive charge and the other side a slight negative charge, the molecules will align themselves with an external electric field.

▲ **FIGURE 9.6 Electron Density Plot for the HF Molecule**   The F end of the molecule, with its partial negative charge, is pink; the H end, with its partial positive charge, is blue.

charge. Does this make the bond ionic? No. In an ionic bond, the electron is essentially *transferred* from one atom to another. In HF, it is *unequally shared*. In other words, even though the Lewis structure of HF portrays the bonding electrons as residing *between* the two atoms, in reality the electron density is greater on the fluorine atom than on the hydrogen atom (**Figure 9.6◄**). The bond is said to be *polar*—having a positive pole and a negative pole. A **polar covalent bond** is intermediate in nature between a pure covalent bond and an ionic bond. In fact, the categories of pure covalent and ionic are really two extremes within a broad continuum. Most covalent bonds between dissimilar atoms are actually *polar covalent*, somewhere between the two extremes.

Electronegativity

The ability of an atom to attract electrons to itself in a chemical bond (which results in polar bonds) is called **electronegativity**. We say that fluorine is more *electronegative* than hydrogen because it takes a greater share of the electron density in HF.

Electronegativity was quantified by the American chemist Linus Pauling in his book, *The Nature of the Chemical Bond*. Pauling assigned an electronegativity of 4.0 to fluorine (the most electronegative element on the periodic table), and developed the electronegativity values shown in **Figure 9.7▼**. For main-group elements, notice the periodic trends in electronegativity from Figure 9.7:

Notice the difference between *electronegativity* (the ability of an atom to attract electrons to itself in a covalent bond) and *electron affinity* (the energy associated with the addition of an electron to a gas phase atom). The two terms are related but not identical.

- Electronegativity generally increases across a period in the periodic table.
- Electronegativity generally decreases down a column in the periodic table.
- Fluorine is the most electronegative element.
- Francium is the least electronegative element (sometimes called the most *electropositive*).

Trends in Electronegativity

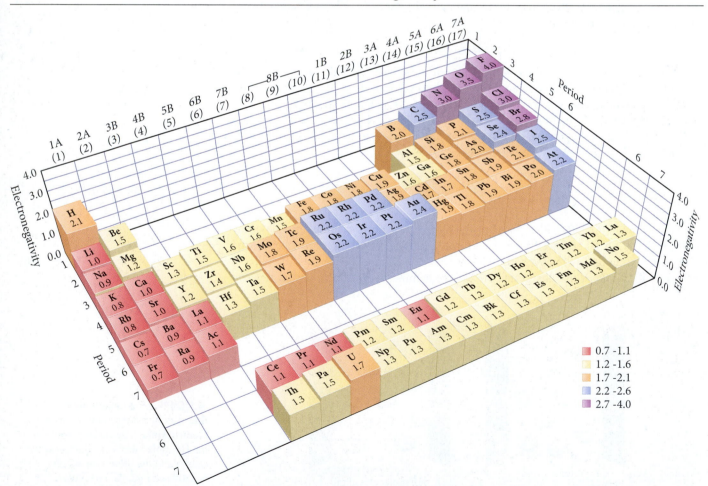

▲ **FIGURE 9.7 Electronegativities of the Elements**   Electronegativity generally increases as we move across a row in the periodic table and decreases as we move down a column.

The periodic trends in electronegativity are consistent with other periodic trends we have seen. In general, electronegativity is inversely related to atomic size—the larger the atom, the less ability it has to attract electrons to itself in a chemical bond.

 Conceptual Connection 9.3 Periodic Trends in Electronegativity

Arrange these elements in order of decreasing electronegativy: P, Na, N, Al

Bond Polarity, Dipole Moment, and Percent Ionic Character

The degree of polarity in a chemical bond depends on the electronegativity difference (sometimes abbreviated ΔEN) between the two bonding atoms. The greater the electronegativity difference, the more polar the bond. If two atoms with identical electronegativities form a covalent bond, they share the electrons equally, and the bond is purely covalent or *nonpolar*. For example, the chlorine molecule, composed of two chlorine atoms (which have identical electronegativities), has a covalent bond in which electrons are evenly shared.

If there is a large electronegativity difference between the two atoms in a bond, such as typically occurs between a metal and a nonmetal, the electron from the metal is almost completely transferred to the nonmetal, and the bond is ionic. For example, sodium and chlorine form an ionic bond.

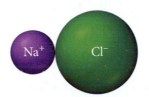

If there is an intermediate electronegativity difference between the two atoms, such as between two different nonmetals, then the bond is polar covalent. For example, HCl has a polar covalent bond.

While all attempts to divide the bond polarity continuum into specific regions are necessarily arbitrary, it is helpful to classify bonds as covalent, polar covalent, and ionic, based on the electronegativity difference between the bonding atoms as shown in Table 9.1 and **Figure 9.8▶**.

TABLE 9.1 The Effect of Electronegativity Difference on Bond Type

Electronegativity Difference (ΔEN)	Bond Type	Example
Small (0–0.4)	Covalent	Cl_2
Intermediate (0.4–2.0)	Polar covalent	HCl
Large (2.0+)	Ionic	NaCl

We can quantify the polarity of a bond with a quantity referred to as the bond's *dipole moment*. A **dipole moment** (μ) is a measure of the separation of positive and negative charge. The magnitude of a dipole moment created by separating two particles of equal but opposite charges of magnitude q by a distance r is given by the equation:

$$\mu = qr$$

▶ **FIGURE 9.8** Electronegativity Difference (ΔEN) and Bond Type

The Continuum of Bond Types

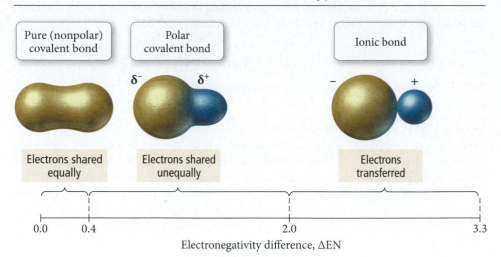

We can get a sense for the dipole moment of a completely ionic bond by calculating the dipole moment that results from separating a proton and an electron ($q = 1.6 \times 10^{-19}$ C) by a distance of $r = 130$ pm (the approximate length of a short chemical bond)

$$\mu = qr$$
$$= (1.6 \times 10^{-19} \text{ C}) (130 \times 10^{-12} \text{ m})$$
$$= 2.1 \times 10^{-29} \text{ C} \cdot \text{m}$$
$$= 6.2 \text{ D}$$

TABLE 9.2	Dipole Moments of Several Molecules in the Gas Phase	
Molecule	ΔEN	Dipole Moment (D)
Cl_2	0	0
ClF	1.0	0.88
HF	1.9	1.82
LiF	3.0	6.33

The debye (D) is a common unit used for reporting dipole moments (1 D = 3.34×10^{-30} C $\cdot$ m). We would therefore expect the dipole moment of completely ionic bonds with bond lengths close to 130 pm to be about 6 D. The smaller the magnitude of the charge separation, and the smaller the distance between the charges, the smaller the dipole moment. Table 9.2 shows the dipole moments of several molecules along with the electronegativity differences of their atoms.

By comparing the *actual* dipole moment of a bond to what the dipole moment would be if the electron were completely transferred from one atom to the other, we can get a sense of the degree to which the electron is transferred (or the degree to which the bond is ionic). A quantity called the **percent ionic character** is the ratio of a bond's actual dipole moment to the dipole moment it would have if the electron were completely transferred from one atom to the other, multiplied by 100%:

$$\text{Percent ionic character} = \frac{\text{measured dipole moment of bond}}{\text{dipole moment if electron were completely transferred}} \times 100\%$$

For example, suppose a diatomic molecule with a bond length of 130 pm has a dipole moment of 3.5 D. We previously calculated that separating a proton and an electron by 130 pm results in a dipole moment of 6.2 D. Therefore, the percent ionic character of the bond is:

$$\text{Percent ionic character} = \frac{3.5 \text{ D}}{6.2 \text{ D}} \times 100\%$$
$$= 56\%$$

A bond in which an electron is completely transferred from one atom to another would have 100% ionic character (although even the most ionic bonds do not reach this ideal). Figure 9.9▶ shows the percent ionic character of a number of diatomic gas-phase molecules plotted against the electronegativity difference between the bonding atoms. As expected, the percent ionic character generally increases as the electronegativity difference increases. However, as you can see and as we just mentioned, no bond is 100% ionic. In general, bonds with greater than 50% ionic character are referred to as ionic bonds.

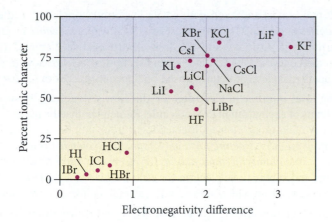

EXAMPLE 9.3 Classifying Bonds as Pure Covalent, Polar Covalent, or Ionic

Classify the bond formed between each pair of atoms as covalent, polar covalent, or ionic.

(a) Sr and F   **(b)** N and Cl   **(c)** N and O

SOLUTION

(a) From Figure 9.7, find the electronegativities of Sr (1.0) and of F (4.0). The electronegativity difference (ΔEN) is ΔEN $= 4.0 - 1.0 = 3.0$. Using Table 9.1, classify this bond as ionic.

(b) From Figure 9.7, find the electronegativities of N (3.0) and of Cl (3.0). The electronegativity difference (ΔEN) is ΔEN $= 3.0 - 3.0 = 0$. Using Table 9.1, classify this bond as covalent.

(c) From Figure 9.7, find the electronegativities of N (3.0) and of O (3.5). The electronegativity difference (ΔEN) is ΔEN $= 3.5 - 3.0 = 0.5$. Using Table 9.1, classify this bond as polar covalent.

FOR PRACTICE 9.3

Classify the bond formed between each pair of atoms as pure covalent, polar covalent, or ionic.

(a) I and I   **(b)** Cs and Br   **(c)** P and O

9.7 Lewis Structures of Molecular Compounds and Polyatomic Ions

We now turn to the basic sequence of steps involved in actually writing Lewis structures for given combinations of atoms.

Writing Lewis Structures for Molecular Compounds

To write the Lewis structure for a molecular compound, follow these steps:

1. **Write the correct skeletal structure for the molecule.** The Lewis structure of a molecule must have the atoms in the correct positions. For example, you could not write a Lewis structure for water if you started with the hydrogen atoms next to each other and the oxygen atom at the end (H H O). In nature, oxygen is the central atom and the hydrogen atoms are *terminal* (at the ends). The correct skeletal structure is H O H. The only way to determine the skeletal structure of a molecule with absolute certainty is to examine its structure experimentally. However, you can write likely skeletal structures by remembering two guidelines. First, *hydrogen atoms are always terminal*. Hydrogen does not ordinarily occur as a central atom because central atoms must

There are a few exceptions to this rule, such as diborane (B_2H_6) which contains *bridging hydrogens*, but these are rare and cannot be addressed by the Lewis model.

Often, chemical formulas are written in a way that provides clues to how the atoms are bonded together. For example, CH_3OH indicates that three hydrogen atoms and the oxygen atom are bonded to the carbon atom, but the fourth hydrogen atom is bonded to the oxygen atom.

Sometimes distributing all the remaining electrons to the central atom results in more than an octet. This is called an expanded octet and is covered in Section 9.9.

form at least two bonds, and hydrogen, which has only a single valence electron to share and requires only a duet, can form just one. Second, *put the more electronegative elements in terminal positions* and the less electronegative elements (other than hydrogen) in the central position. Later in this section, you learn how to distinguish between competing skeletal structures based on the concept of formal charge.

2. **Calculate the total number of electrons for the Lewis structure by summing the valence electrons of each atom in the molecule.** Remember that the number of valence electrons for any main-group element is equal to its group number in the periodic table. *If you are writing a Lewis structure for a polyatomic ion, you must consider the charge of the ion when calculating the total number of electrons.* Add one electron for each negative charge and subtract one electron for each positive charge. Don't worry about which electron comes from which atom—only the total number is important.

3. **Distribute the electrons among the atoms, giving octets (or duets in the case of hydrogen) to as many atoms as possible.** Begin by placing two electrons between every two atoms. These represent the minimum number of bonding electrons. Then distribute the remaining electrons as lone pairs, first to terminal atoms, and then to the central atom, giving octets (or duets for hydrogen) to as many atoms as possible.

4. **If any atoms lack an octet, form double or triple bonds as necessary to give them octets.** Do this by moving lone electron pairs from terminal atoms into the bonding region with the central atom.

A brief version of this procedure is shown below in the left column. Examples 9.4 and 9.5 demonstrate the procedure.

PROCEDURE FOR... **Writing Lewis Structures for Covalent Compounds**	**EXAMPLE 9.4** **Writing Lewis Structures** Write the Lewis Structure for CO_2.	**EXAMPLE 9.5** **Writing Lewis Structures** Write the Lewis Structure for NH_3.
	SOLUTION	**SOLUTION**
1. Write the correct skeletal structure for the molecule.	Because carbon is the less electronegative atom, put it in the central position. O C O	Since hydrogen is always terminal, put nitrogen in the central position. H N H H
2. Calculate the total number of electrons for the Lewis structure by summing the valence electrons of each atom in the molecule.	Total number of electrons for Lewis structure = $\left(\begin{array}{c}\text{number of}\\\text{valence}\\\text{e}^-\text{ for C}\end{array}\right) + 2\left(\begin{array}{c}\text{number of}\\\text{valence}\\\text{e}^-\text{ for O}\end{array}\right)$ $= 4 + 2(6) = 16$	Total number of electrons for Lewis structure = $\left(\begin{array}{c}\text{number of}\\\text{valence}\\\text{e}^-\text{ for N}\end{array}\right) + 3\left(\begin{array}{c}\text{number of}\\\text{valence}\\\text{e}^-\text{ for H}\end{array}\right)$ $= 5 + 3(1) = 8$
3. Distribute the electrons among the atoms, giving octets (or duets for hydrogen) to as many atoms as possible. Begin with the bonding electrons, and then proceed to lone pairs on terminal atoms, and finally to lone pairs on the central atom.	Bonding electrons are first. O:C:O (4 of 16 electrons used) Lone pairs on terminal atoms are next. :Ö:C:Ö: (16 of 16 electrons used)	Bonding electrons are first. H:N:H H (6 of 8 electrons used) Lone pairs on terminal atoms are next, but none are needed on hydrogen. Lone pairs on central atom are last. H—N̈—H ‖ H (8 of 8 electrons used)

| 4. If any atom lacks an octet, form double or triple bonds as necessary to give them octets. | Carbon lacks an octet in this case. Move lone pairs from the oxygen atoms to bonding regions to form double bonds. 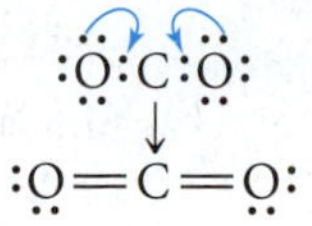

$$\ddot{\text{:O:C:O:}}$$
$$\downarrow$$
$$\text{:O}=\text{C}=\text{O:}$$ | Since all of the atoms have octets (or duets for hydrogen), the Lewis structure for NH_3 is complete as shown above. |
| | **FOR PRACTICE 9.4**
Write the Lewis structure for CO. | **FOR PRACTICE 9.5**
Write the Lewis structure for H_2CO. |

Writing Lewis Structures for Polyatomic Ions

We write Lewis structures for polyatomic ions by following the same procedure, but we pay special attention to the charge of the ion when calculating the number of electrons. We add one electron for each negative charge and subtract one electron for each positive charge. The Lewis structure for a polyatomic ion is usually written within brackets with the charge of the ion in the upper right-hand corner, outside the bracket.

EXAMPLE 9.6 Writing Lewis Structures for Polyatomic Ions

Write the Lewis structure for the NH_4^+ ion.

SOLUTION

Begin by writing the skeletal structure. Since hydrogen is always terminal, put the nitrogen atom in the central position.	H H N H H
Calculate the total number of electrons for the Lewis structure by summing the number of valence electrons for each atom and subtracting 1 for the 1+ charge.	Total number of electrons for Lewis structure = (number of valence e$^-$ in N) + 4 (number of valence e$^-$ in H) − 1 = 5 + 4(1) − 1 = 8 Subtract 1 e$^-$ to account for 1$^+$ charge of ion.
Place two bonding electrons between every two atoms. Since all of the atoms have complete octets, no double bonds are necessary.	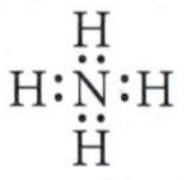 H $$\text{H:}\ddot{\text{N}}\text{:H}$$ H (8 of 8 electrons used)
Lastly, write the Lewis structure in brackets with the charge of the ion in the upper right-hand corner.	$$\left[\begin{array}{c} \text{H} \\ | \\ \text{H}-\text{N}-\text{H} \\ | \\ \text{H} \end{array}\right]^+$$

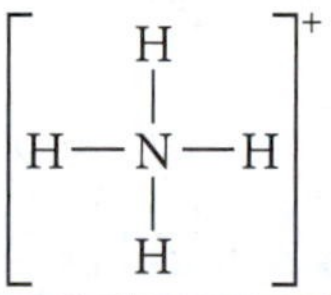

FOR PRACTICE 9.6
Write the Lewis structure for the hypochlorite ion, ClO^-.

9.8 Resonance and Formal Charge

We need two additional concepts to write the best possible Lewis structure for a large number of compounds. The concepts are *resonance*, used when two or more valid Lewis structures can be drawn for the same compound, and *formal charge*, an electron bookkeeping system that allows us to discriminate between alternative Lewis structures.

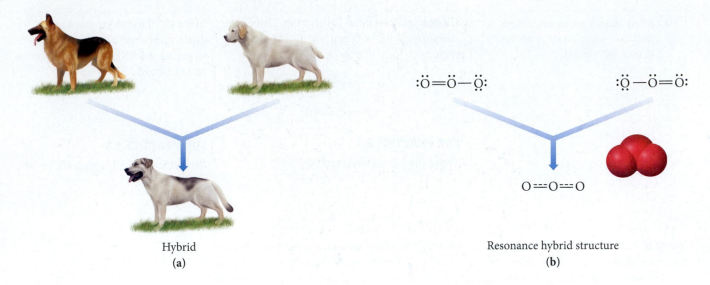

Hybrid
(a)

Resonance hybrid structure
(b)

▲ **FIGURE 9.10 Hybridization** Just as the offspring of two different dog breeds is a hybrid that is intermediate between the two breeds **(a)**, the structure of a resonance hybrid is intermediate between that of the contributing resonance structures **(b)**.

Resonance

When writing Lewis structures, we may find that, for some molecules, we can write more than one valid Lewis structure. For example, consider writing a Lewis structure for O_3. The following two Lewis structures, with the double bond on alternate sides, are equally correct:

$$:\ddot{O}=\ddot{O}-\ddot{O}: \quad :\ddot{O}-\ddot{O}=\ddot{O}:$$

In cases such as this—where there are two or more Lewis structures for the same molecule—we find that, in nature, the molecule exists as an *average* of the two Lewis structures. Any *one* of the two Lewis structures for O_3 would predict that O_3 should contain two different kinds of bonds (one double bond and one single bond). However, when we experimentally examine the structure of O_3, we find that both bonds are equivalent and intermediate in strength and length between a double bond and single bond. We account for this in the Lewis model by representing the molecule with both structures, called **resonance structures**, with a double-headed arrow between them:

$$:\ddot{O}=\ddot{O}-\ddot{O}: \quad \longleftrightarrow \quad :\ddot{O}-\ddot{O}=\ddot{O}:$$

The actual structure of the molecule is intermediate between the two resonance structures and is called a **resonance hybrid**. The term *hybrid* comes from breeding and means the offspring of two animals or plants of different varieties or breeds. If we breed a Labrador retriever with a German shepherd, we get a *hybrid* that is intermediate between the two breeds (**Figure 9.10(a)▲**). Similarly, the structure of a resonance hybrid is intermediate between the two resonance structures (**Figure 9.10(b)▲**). However, the only structure that actually exists is the hybrid structure—the individual resonance structures do not exist and are merely a convenient way to describe the real structure.

EXAMPLE 9.7 Writing Resonance Structures

Write the Lewis structure for the NO_3^- ion. Include resonance structures.

SOLUTION

Begin by writing the skeletal structure. Since nitrogen is the least electronegative atom, put it in the central position.	O O N O
Calculate the total number of electrons for the Lewis structure by summing the number of valence electrons for each atom and adding 1 for the 1− charge.	Total number of electrons for Lewis structure = (number of valence e⁻ in N) + 3 (number of valence e⁻ in O) + 1 = 5 + 3(6) + 1 = 24 Add 1 e⁻ to account for 1− charge of ion.

Place two bonding electrons between each pair of atoms.	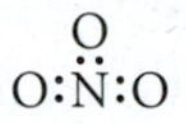 O O:N:O (6 of 24 electrons used)
Distribute the remaining electrons, first to terminal atoms. There are not enough electrons to complete the octet on the central atom.	 :Ö: :Ö:N:Ö: (24 of 24 electrons used)
Form a double bond by moving a lone pair from one of the oxygen atoms into the bonding region with nitrogen. Enclose the structure in brackets and include the charge.	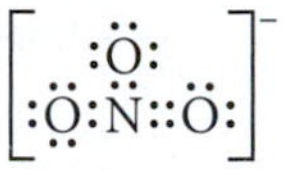 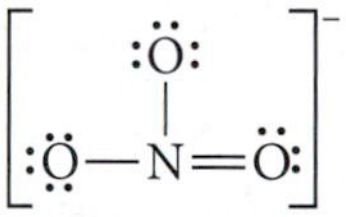 $$\left[\begin{array}{c} :\ddot{O}: \\ :\ddot{O}:N::\ddot{O}: \end{array} \right]^{-}$$ *or* $$\left[\begin{array}{c} :\ddot{O}: \\ \mid \\ :\ddot{O}-N=\ddot{O}: \end{array} \right]^{-}$$
Since the double bond can equally well form with any of the three oxygen atoms, write all three structures as resonance structures. (The actual space filling model of NO_3^- is shown here for comparison. Note that all three bonds are equal in length.)	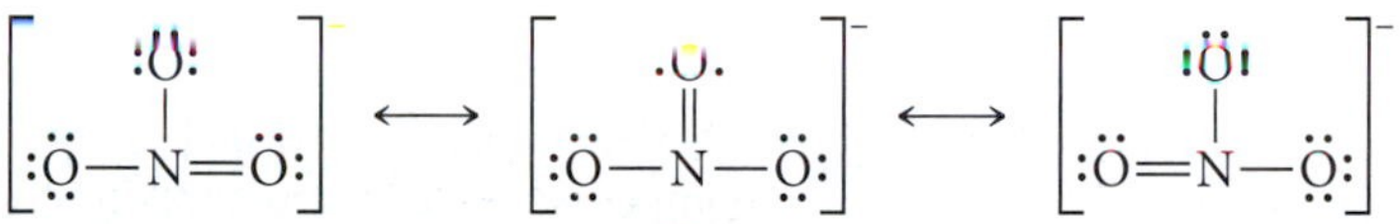  $$\left[\begin{array}{c} :\ddot{O}: \\ \mid \\ :\ddot{O}-N=\ddot{O}: \end{array} \right]^{-} \longleftrightarrow \left[\begin{array}{c} \cdot\ddot{O}\cdot \\ \parallel \\ :\ddot{O}-N-\ddot{O}: \end{array} \right]^{-} \longleftrightarrow \left[\begin{array}{c} :\ddot{O}: \\ \mid \\ :\ddot{O}=N-\ddot{O}: \end{array} \right]^{-}$$ NO_3^-

In the examples of resonance hybrids that we have examined so far, the contributing structures have been equivalent (or equally valid) Lewis structures. In these cases, the true structure is an equally weighted average of the resonance structures. In some cases, however, we can write resonance structures that are not equivalent. For reasons we cover in the following sections of this chapter—such as formal charge, for example—one possible resonance structure may be somewhat better than another. In such cases, the true structure may still be an average of the resonance structures, but with the better resonance structure contributing more to the true structure. In other words, multiple nonequivalent resonance structures may be weighted differently in their contributions to the true overall structure of a molecule (see Example 9.8).

Formal Charge

Formal charge is a fictitious charge assigned to each atom in a Lewis structure that helps us to distinguish among competing Lewis structures. The **formal charge** of an atom in a Lewis structure is *the charge it would have if all bonding electrons were shared equally between the bonded atoms.* In other words, formal charge is the calculated charge for an atom if we completely ignore the effects of electronegativity. For example, we know that because fluorine is more electronegative than hydrogen, HF has a dipole moment—the hydrogen atom has a slight positive charge and the fluorine atom has a slight negative charge. However, the *formal charges* of hydrogen and fluorine in HF (the charges computed if we ignore their differences in electronegativity) are both zero.

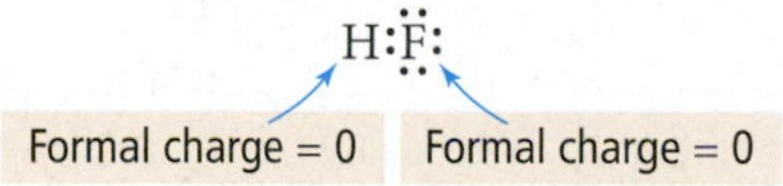

$$H:\ddot{F}:$$

Formal charge = 0   Formal charge = 0

The formal charge on an atom in a Lewis structure is the difference between the number of the atom's valence electrons and the number of electrons that it "owns" in the Lewis structure. An atom in a Lewis structure "owns" all of its lone pair electrons and one-half of its bonding electrons.

$$\text{Formal charge} = \text{number of valence electrons} -$$

$$(\text{number of lone pair electrons} + {}^1/_2 \text{ number of bonding electrons})$$

So the formal charge of hydrogen in HF is calculated as follows:

$$\text{Formal charge} = 1 - \left[0 + \tfrac{1}{2}(2) \right] = 0$$

Number of valence electrons for H

Number of electrons that H "owns" in the Lewis structure

Similarly, the formal charge of fluorine in HF is calculated as follows:

$$\text{Formal charge} = 7 - \left[6 + \tfrac{1}{2}(2) \right] = 0$$

Number of valence electrons for F

Number of electrons that F "owns" in the Lewis structure

The concept of formal charge is useful because it can help us distinguish between competing skeletal structures or competing resonance structures. In general, the following rules apply:

1. The sum of all formal charges in a neutral molecule must be zero.
2. The sum of all formal charges in an ion must equal the charge of the ion.
3. Small (or zero) formal charges on individual atoms are better than large ones.
4. When formal charge cannot be avoided, negative formal charge should reside on the most electronegative atom.

Let's use formal charge to choose between the competing skeletal structures for hydrogen cyanide shown below. Notice that both skeletal structures equally satisfy the octet rule. The formal charge of each atom in the structure is calculated below it.

	Structure A			Structure B		
	H —	C ≡	N:	H —	N ≡	C:
number of valence e⁻	1	4	5	1	5	4
−number of lone pair e⁻	−0	−0	−2	−0	−0	−2
−¹/₂ (number of bonding e⁻)	−¹/₂ (2)	−¹/₂ (8)	−¹/₂ (6)	−¹/₂ (2)	−¹/₂ (8)	−¹/₂ (6)
Formal charge	**0**	**0**	**0**	**0**	**+1**	**−1**

As required, the sum of the formal charges for each of these structures is zero (as it should be for neutral molecules). However, structure B has formal charges on both the N atom and the C atom, while structure A has no formal charges on any atom. Furthermore, in structure B, the negative formal charge is not on the most electronegative element (nitrogen is more electronegative than carbon). Consequently, structure A is the better Lewis structure. Since atoms in the middle of a molecule tend to have more bonding electrons and fewer lone pairs, they will also tend to have more positive formal charges. Consequently, the best skeletal structures usually have the least electronegative atom in the central position, as we learned in step 1 of our procedure for writing Lewis structures.

Both HCN and HNC exist, but—as predicted by formal charge—HCN is more stable than HNC.

EXAMPLE 9.8 Assigning Formal Charges

Assign formal charges to each atom in the resonance forms of the cyanate ion (OCN⁻). Which resonance form is likely to contribute most to the correct structure of OCN⁻?

A   B   C

$$[:\ddot{O}\!-\!C\!\equiv\!N:]^{-} \quad [:\ddot{O}\!=\!C\!=\!\dot{\ddot{N}}:]^{-} \quad [:O\!\equiv\!C\!-\!\dot{\ddot{N}}:]^{-}$$

SOLUTION

Calculate the formal charge on each atom by finding the number of valence electrons and subtracting the number of lone pair electrons and one-half the number of bonding electrons.		A $[:\ddot{O}-C\equiv N:]^-$			B $[:\ddot{o}=C=\ddot{N}:]^-$			C $[:O\equiv C-\ddot{N}:]^-$		
	Number of valence e⁻	6	4	5	6	4	5	6	4	5
	−number of lone pair e⁻	−6	−0	−2	−4	−0	−4	−2	−0	−6
	$-\frac{1}{2}$(number of bond e⁻)	−1	−4	−3	−2	−4	−2	−3	−4	−1
	Formal charge	**−1**	**0**	**0**	**0**	**0**	**−1**	**+1**	**0**	**−2**

The sum of all formal charges for each structure is −1, as it should be for a 1− ion. Structures A and B have the least amount of formal charge and are therefore preferred over structure C. Structure A is preferable to B because it has the negative formal charge on the more electronegative atom. We thus expect structure A to make the biggest contribution to the resonance forms of the cyanate ion.

FOR PRACTICE 9.8

Assign formal charges to each atom in the resonance forms of N_2O. Which resonance form is likely to contribute most to the correct structure of N_2O?

A	B	C
$:\ddot{N}=N=\ddot{O}:$	$:N\equiv N-\ddot{O}:$	$:\ddot{N}-N\equiv O:$

FOR MORE PRACTICE 9.8

Assign formal charges to each of the atoms in the nitrate ion (NO_3^-). The Lewis structure for the nitrate ion is shown in Example 9.7.

9.9   Exceptions to the Octet Rule: Odd-Electron Species, Incomplete Octets, and Expanded Octets

The octet rule in the Lewis model has some exceptions that we must accommodate. They include (1) *odd-electron species*, molecules or ions with an odd number of electrons; (2) *incomplete octets*, molecules or ions with *fewer than eight electrons* around an atom; and (3) *expanded octets*, molecules or ions with *more than eight electrons* around an atom. Let's examine each of these exceptions individually.

Odd-Electron Species

Molecules and ions with an odd number of electrons in their Lewis structures are called **free radicals** (or simply *radicals*). For example, nitrogen monoxide—a pollutant found in motor vehicle exhaust—has 11 electrons. If we try to write a Lewis structure for nitrogen monoxide, we can't achieve octets for both atoms:

$$:N::\ddot{O}: \quad \text{or} \quad :N=\ddot{O}:$$

The nitrogen atom does not have an octet, so this Lewis structure does not satisfy the octet rule. Yet, nitrogen monoxide exists, especially in polluted air. Why? As with any simple theory, the Lewis model is not sophisticated enough to handle every single case. It is impossible to write good Lewis structures for free radicals, yet some of these molecules exist in nature. Perhaps it is a testament to the Lewis model, however, that *relatively few* such molecules exist and that, in general, they tend to be somewhat unstable and reactive.

The unpaired electron in nitrogen monoxide is put on the nitrogen rather than the oxygen in order to minimize formal charges.

⬡ Conceptual Connection 9.4   Odd-Electron Species

Which molecule would you expect to be a free radical?

(a) CO   **(b)** CO_2 **(c)** N_2O **(d)** NO_2

Incomplete Octets

Another significant exception to the octet rule involves elements that tend to form *incomplete octets*. The most important of these elements is boron, which forms compounds with only six electrons around B, rather than eight. For example, BF_3 and BH_3 lack an octet for B.

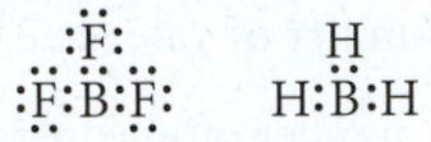

Beryllium compounds, such as BeH,
also have incomplete octets.

$$:\ddot{F}:$$
$$:\ddot{F}:B:\ddot{F}: \qquad H:B:H$$

You might be wondering why we don't simply form double bonds to increase the number of electrons around B. For BH_3, of course, we can't, because there are no additional electrons to move into the bonding region. For BF_3, however, we could attempt to give B an octet by moving a lone pair from an F atom into the bonding region with B.

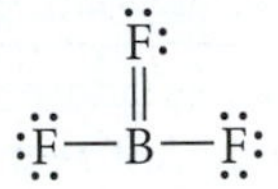

$$\ddot{F}:$$
$$\|$$
$$:\ddot{F}-B-\ddot{F}:$$

This Lewis structure has octets for all atoms, including boron. However, when we assign formal charges to this structure, we get a negative formal charge on B and a positive formal charge on F:

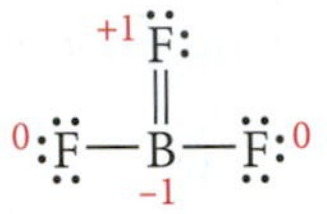

$$^{+1}\ddot{F}:$$
$$\|$$
$$^{0}:\ddot{F}-\underset{-1}{B}-\ddot{F}:^{0}$$

The positive formal charge on fluorine—the most electronegative element in the periodic table—makes this an unfavorable structure. This leaves us with some questions. Do we complete the octet on B at the expense of giving fluorine a positive formal charge? Or do we leave B without an octet in order to avoid the positive formal charge on fluorine? The answers to these kinds of questions are not always clear because we are pushing the limits of the Lewis model. In the case of boron, we usually accept the incomplete octet as the better Lewis structure. However, doing so does not rule out the possibility that the doubly bonded Lewis structure might be a minor contributing resonance structure. The ultimate answer to these kinds of questions must come from experiment. Experimental measurements of the B—F bond length in BF_3 suggest that the bond may be slightly shorter than expected for a single B—F bond, indicating that it may indeed have a small amount of double-bond character.

BF_3 can complete its octet in another way—via a chemical reaction. The Lewis model predicts that BF_3 might react in ways to complete its octet, and indeed it does. BF_3 reacts with NH_3 as follows:

When nitrogen bonds to boron, the
nitrogen atom provides both of the
electrons. This kind of bond is called
a *coordinate covalent bond*.

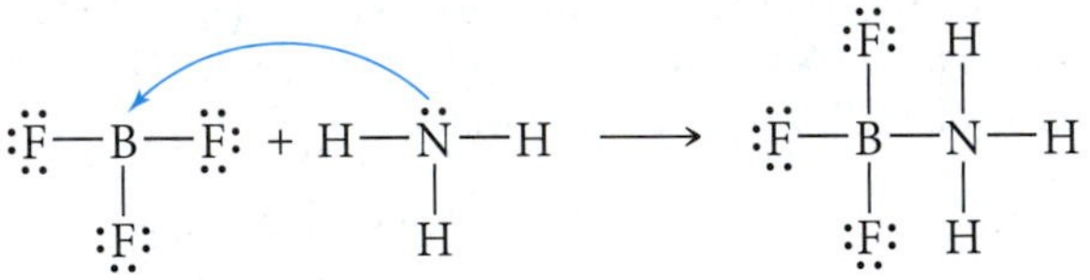

The product has complete octets (or duets for H) for all atoms in the structure.

Expanded Octets

Elements in the third row of the periodic table and beyond often exhibit *expanded octets* of up to 12 (and occasionally 14) electrons. Consider the Lewis structures of arsenic pentafluoride and sulfur hexafluoride.

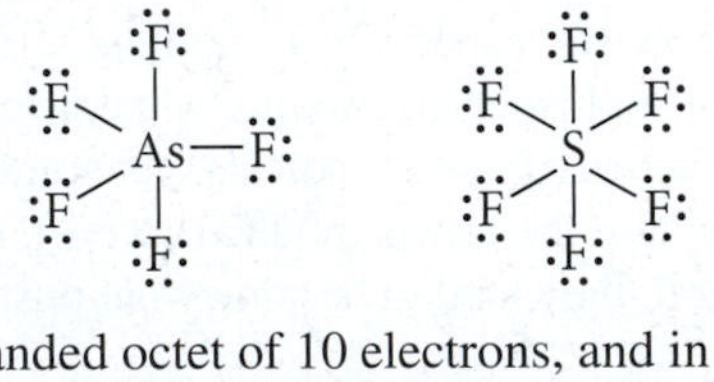

In AsF_5 arsenic has an expanded octet of 10 electrons, and in SF_6 sulfur has an expanded octet of 12 electrons. Both of these compounds exist and are stable. Ten- and twelve-electron expanded octets are common in third-period elements and beyond because the *d* orbitals in these elements are energetically accessible (they are not much higher in energy than the orbitals occupied by the valence electrons) and can accommodate the extra electrons (see Section 8.3). Expanded octets *never* occur in second-period elements.

In some Lewis structures, we must make the decision whether or not to expand an octet in order to lower formal charge. Consider the Lewis structure of H_2SO_4.

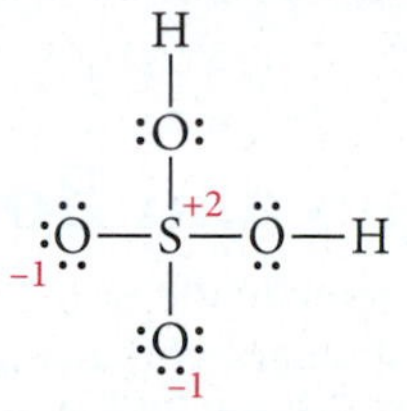

Notice that two of the oxygen atoms have a -1 formal charge and that sulfur has a $+2$ formal charge. While this amount of formal charge is acceptable, especially since the negative formal charge resides on the more electronegative atom, it is possible to eliminate the formal charge by expanding the octet on sulfur:

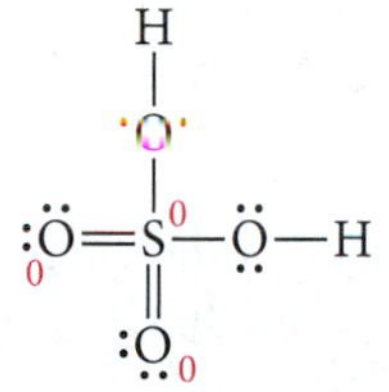

Which of these two Lewis structures for H_2SO_4 is better? Again, the answer is not straight-forward. Experiments show that the sulfur–oxygen bond lengths in the two sulfur–oxygen bonds without the hydrogen atoms are shorter than expected for sulfur–oxygen single bonds, indicating that the double-bonded Lewis structure plays an important role in the bonding in H_2SO_4. In general, we can expand octets in third-row (or beyond) elements to lower formal charge. However, we must *never* expand the octets of second-row elements. Second-row elements do not have energetically accessible *d* orbitals and never exhibit expanded octets.

EXAMPLE 9.9 Writing Lewis Structures for Compounds Having Expanded Octets

Write the Lewis structure for XeF_2.

SOLUTION

Begin by writing the skeletal structure. Since xenon is the less electronegative atom, put it in the central position.	F Xe F
Calculate the total number of electrons for the Lewis structure by summing the number of valence electrons for each atom.	Total number of electrons for Lewis structure = (number of valence e⁻ in Xe) + 2(number of valence e⁻ in F) = 8 + 2(7) = 22
Place two bonding electrons between each pair of atoms.	F:Xe:F (4 of 22 electrons used)
Distribute the remaining electrons to give octets to as many atoms as possible, beginning with terminal atoms and finishing with the central atom. Arrange additional electrons (beyond an octet) around the central atom, giving it an expanded octet of up to 12 electrons.	:Ḟ:Xe:Ḟ: (16 of 22 electrons used) :Ḟ:Xe:Ḟ: or :Ḟ—Xe—Ḟ: (22 of 22 electrons used)

FOR PRACTICE 9.9
Write the Lewis structure for XeF_4.

FOR MORE PRACTICE 9.9
Write the Lewis structure for H_3PO_4. If necessary, expand the octet on appropriate atoms to lower formal charge.

Conceptual Connection 9.5 **Expanded Octets**

Which molecule could have an expanded octet?

(a) H_2CO_3 **(b)** H_3PO_4 **(c)** HNO_2

9.10 Bond Energies and Bond Lengths

In Chapter 6, we learned how to calculate the standard enthalpy change for a chemical reaction ($\Delta H°_{rxn}$) from tabulated standard enthalpies of formation. However, at times it may not be possible to find standard enthalpies of formation for all of the reactants and products of a reaction. In such cases, we can use individual *bond energies* to estimate enthalpy changes of reaction. In this section, we introduce the concept of bond energy and then show how bond energies can be used to calculate enthalpy changes of reaction. We also look at average bond lengths for a number of commonly encountered bonds.

Bond Energy

Bond energy is also called bond enthalpy or bond dissociation energy.

The **bond energy** of a chemical bond is the energy required to break 1 mole of the bond in the gas phase. For example, the bond energy of the Cl—Cl bond in Cl_2 is 243 kJ/mol.

$$Cl_2(g) \longrightarrow 2\,Cl(g) \quad \Delta H = 243\ kJ$$

The bond energy of HCl is 431 kJ/mol.

$$HCl(g) \longrightarrow H(g) + Cl(g) \quad \Delta H = 431\ kJ$$

Bond energies are always positive, because it always takes energy to break a bond. We say that the HCl bond is *stronger* than the Cl_2 bond because it requires more energy to break it. In general, compounds with stronger bonds tend to be more chemically stable, and therefore less chemically reactive, than compounds with weaker bonds. The triple bond in N_2 has a bond energy of 946 kJ/mol.

$$N_2(g) \longrightarrow N(g) + N(g) \quad \Delta H = 946\ kJ$$

It is a very strong and stable bond, which explains nitrogen's relative inertness.

The bond energy of a particular bond in a polyatomic molecule is a little more difficult to determine because a particular type of bond can have different bond energies in different molecules. For example, consider the C—H bond. In CH_4, the energy required to break one C—H bond is 438 kJ/mol.

$$H_3C\!-\!H(g) \longrightarrow H_3C(g) + H(g) \quad \Delta H = 438\ kJ$$

However, the energy required to break a C—H bond in other molecules varies slightly, as shown here.

$$
\begin{array}{ll}
F_3C\!-\!H(g) \longrightarrow F_3C(g) + H(g) & \Delta H = 446\ kJ \\
Br_3C\!-\!H(g) \longrightarrow Br_3C(g) + H(g) & \Delta H = 402\ kJ \\
Cl_3C\!-\!H(g) \longrightarrow Cl_3C(g) + H(g) & \Delta H = 401\ kJ
\end{array}
$$

It is useful to calculate an *average bond energy* for a chemical bond, which is an average of the bond energies for that bond in a large number of compounds. For our limited number of compounds listed above, we calculate an average C—H bond energy of 422 kJ/mol. Table 9.3 lists average bond energies for a number of common chemical bonds averaged over a large number of compounds. Notice that the C—H bond energy is listed as 414 kJ/mol, which is not too different from the value we calculated from a limited number of compounds. Notice also that bond energies depend not only on the kind of atoms involved in the bond, but also on the type of bond: single, double, or triple. In general, for a particular pair of atoms, triple bonds are stronger than double bonds, which are in turn stronger than single bonds. For example, consider the bond energies of carbon–carbon triple, double, and single bonds listed at left.

Bond	Bond Energy (kJ/mol)
C≡C	837 kJ/mol
C=C	611 kJ/mol
C—C	347 kJ/mol

TABLE 9.3   Average Bond Energies

Bond	Bond Energy (kJ/mol)	Bond	Bond Energy (kJ/mol)	Bond	Bond Energy (kJ/mol)
H—H	436	N—N	163	Br—F	237
H—C	414	N=N	418	Br—Cl	218
H—N	389	N≡N	946	Br—Br	193
H—O	464	N—O	222	I—Cl	208
H—S	368	N=O	590	I—Br	175
H—F	565	N—F	272	I—I	151
H—Cl	431	N—Cl	200	Si—H	323
H—Br	364	N—Br	243	Si—Si	226
H—I	297	N—I	159	Si—C	301
C—C	347	O—O	142	Si=O	368
C=C	611	O=O	498	Si=Cl	464
C≡C	837	O—F	190	S—O	265
C—N	305	O—Cl	203	S=O	523
C=N	615	O—I	234	S=S	418
C≡N	891	F—F	159	S—F	327
C—O	360	Cl—F	253	S—Cl	253
C=O	736*	Cl—Cl	243	S—Br	218
C≡O	1072	S—S	266	C—Cl	339

*799 in CO_2

Using Average Bond Energies to Estimate Enthalpy Changes for Reactions

Average bond energies are useful in *estimating* the enthalpy change of a reaction. Consider the following reaction:

$$H_3C—H(g) + Cl—Cl(g) \longrightarrow H_3C—Cl(g) + H—Cl(g)$$

We can imagine this reaction occurring by the breaking of a C—H bond and a Cl—Cl bond and the forming of a C—Cl bond and an H—Cl bond. Since breaking bonds is endothermic (positive bond energy) and forming bonds is exothermic (negative bond energy), we can calculate the overall enthalpy change as a sum of the enthalpy changes associated with breaking the required bonds in the reactants and forming the required bonds in the products, as shown in **Figure 9.11►** (on p. 350).

$$H_3C—H(g) + Cl—Cl(g) \longrightarrow H_3C—Cl(g) + H—Cl(g)$$

Bonds Broken		Bonds Formed	
C—H break	+414 kJ	C—Cl form	−339 kJ
Cl—Cl break	+243 kJ	H—Cl form	−431 kJ

Sum (Σ) ΔH's *bonds broken*: +657 kJ   *Sum* (Σ) ΔH's *bonds formed*: −770 kJ

$$\Delta H_{rxn} = \Sigma(\Delta H\text{'s bonds broken}) + \Sigma(\Delta H\text{'s bonds formed})$$

$$= +657 \text{ kJ} - 770 \text{ kJ}$$

$$= -113 \text{ kJ}$$

We find that $\Delta H_{rxn} = -113$ kJ. Calculating $\Delta H°_{rxn}$ from tabulated enthalpies of formation—as we learned in Chapter 6—gives $\Delta H°_{rxn} = -101$ kJ fairly close to the value we obtained from average bond energies. In general, we can calculate ΔH_{rxn} from average bond energies by summing the changes in enthalpy for all of the bonds that are

▶ **FIGURE 9.11 Estimating ΔH_{rxn} from Bond Energies** The enthalpy change of a reaction can be approximated by summing the enthalpy changes involved in breaking old bonds and those involved in forming new ones.

Estimating the Enthalpy Change of a Reaction from Bond Energies

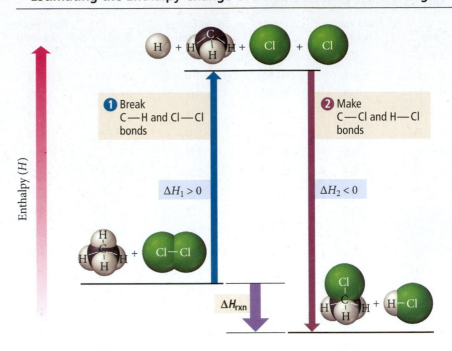

broken and adding the sum of the enthalpy changes for all of the bonds that are formed. Remember that ΔH is positive for breaking bonds and negative for forming them:

$$\Delta H_{rxn} = \underbrace{\Sigma(\Delta H\text{'s bonds broken})}_{\text{Positive}} + \underbrace{\Sigma(\Delta H\text{'s bonds formed})}_{\text{Negative}}$$

As we can see from the above equation:

- A reaction is *exothermic* when weak bonds break and strong bonds form.
- A reaction is *endothermic* when strong bonds break and weak bonds form.

Scientists often say that "energy is stored in chemical bonds or in a chemical compound," which may sound as if breaking the bonds in the compound releases energy. For example, we often hear in biology that energy is stored in glucose or in ATP. However, *breaking a chemical bond always requires energy*. When we say that energy is stored in a compound, or that a compound is energy rich, we mean that the compound can undergo a reaction in which weak bonds break and strong bonds form, thereby releasing net energy in the overall process. However, *it is always the forming of chemical bonds that releases energy*.

Conceptual Connection 9.6 Bond Energies and ΔH_{rxn}

The reaction between hydrogen and oxygen to form water is highly exothermic. Which statement about the formation of water is true?

(a) The energy needed to break the required bonds is greater than the energy released when the new bonds form.

(b) The energy needed to break the required bonds is less than the energy released when the new bonds form.

(c) The energy needed to break the required bonds is about the same as the energy released when the new bonds form.

EXAMPLE 9.10 Calculating ΔH_{rxn} from Bond Energies

Hydrogen gas, a potential future fuel, can be made by the reaction of methane gas and steam.

$$CH_4(g) + 2\,H_2O(g) \longrightarrow 4\,H_2(g) + CO_2(g)$$

Use bond energies to calculate ΔH_{rxn} for this reaction.

SOLUTION

Begin by rewriting the reaction using the Lewis structures of the molecules involved.	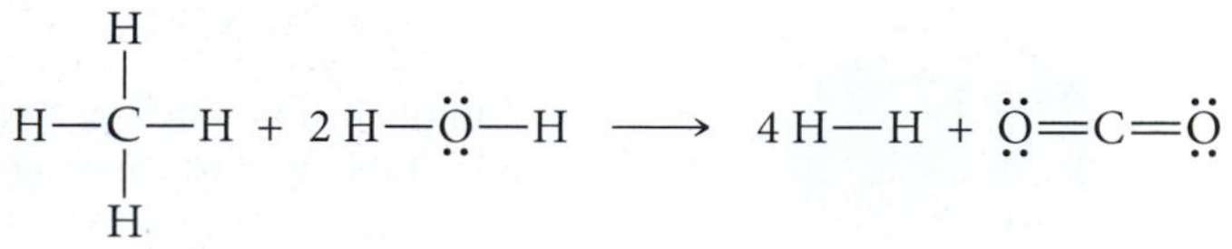 H—C—H + 2 H—Ö—H ⟶ 4 H—H + Ö=C=Ö (with H above and below the C)
Determine which bonds are broken in the reaction and sum the bond energies of these.	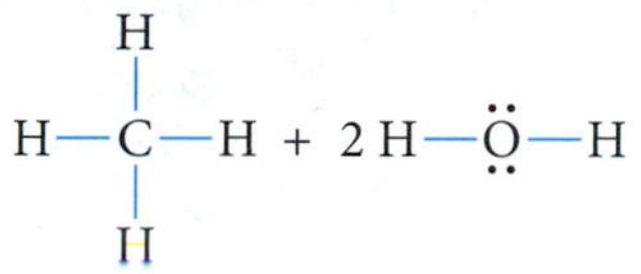 H—C—H + 2 H—Ö—H (with H above and below C) $\Sigma(\Delta H's \text{ bonds broken})$ $= 4(C-H) + 4(O-H)$ $= 4(414\text{ kJ}) + 4(464\text{ kJ})$ $= 3512\text{ kJ}$
Determine which bonds are formed in the reaction and sum the negatives of the bond energies of these.	 4 H—H + Ö=C=Ö $\Sigma(\Delta H's \text{ bonds formed})$ $= -4(H-H) - 2(C=O)$ $= -4(436\text{ kJ}) - 2(799\text{ kJ})$ $= -3342\text{ kJ}$
Find ΔH_{rxn} by summing the results of the previous two steps.	$\Delta H_{rxn} = \Sigma(\Delta H's \text{ bonds broken}) + \Sigma(\Delta H's \text{ bonds formed})$ $= 3512 - 3342$ $= 1.70 \times 10^2\text{ kJ}$

FOR PRACTICE 9.10

One potential future fuel is methanol (CH_3OH). Write a balanced equation for the combustion of gaseous methanol and use bond energies to calculate the enthalpy of combustion of methanol in kJ/mol.

FOR MORE PRACTICE 9.10

Use bond energies to calculate ΔH_{rxn} for the reaction. $N_2(g) + 3\,H_2(g) \longrightarrow 2\,NH_3(g)$

Bond Lengths

Just as we can tabulate average bond energies, which represent the average energy of a bond between two particular atoms in a large number of compounds, we can tabulate average **bond lengths**, which represent the average length of a bond between two particular atoms in a large number of compounds. Like bond energies, bond lengths depend not only on the kind of atoms involved in the bond, but also on the type of bond: single, double, or triple. In general, for a particular pair of atoms, triple bonds are shorter than double bonds, which are in turn shorter than single bonds. For example, consider the bond lengths (shown here with bond strengths, repeated from earlier in this section) of carbon–carbon triple, double, and single bonds.

Bond	Bond Length (pm)	Bond Strength (kJ/mol)
C≡C	120 pm	837 kJ/mol
C=C	134 pm	611 kJ/mol
C—C	154 pm	347 kJ/mol

Bond Lengths

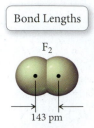

F₂

143 pm

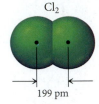

Cl₂

199 pm

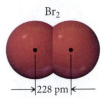

Br₂

228 pm

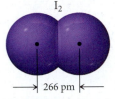

I₂

266 pm

▲ Bond lengths in the diatomic halogen molecules.

Notice that, as the bond gets longer, it also becomes weaker. This relationship between the length of a bond and the strength of a bond does not necessarily hold true for all bonds. For example, consider the following series of nitrogen–halogen single bonds:

Bond	Bond Length (pm)	Bond Strength (kJ/mol)
N—F	139	272
N—Cl	191	200
N—Br	214	243
N—I	222	159

Although the bonds generally get weaker as they get longer, the trend is not a smooth one. Table 9.4 lists average bond lengths for a number of common bonds.

TABLE 9.4 Average Bond Lengths

Bond	Bond Length (pm)	Bond	Bond Length (pm)	Bond	Bond Length (pm)
H—H	74	C—C	154	N—N	145
H—C	110	C=C	134	N=N	123
H—N	100	C≡C	120	N≡N	110
H—O	97	C—N	147	N—O	136
H—S	132	C=N	128	N=O	120
H—F	92	C≡N	116	O—O	145
H—Cl	127	C—O	143	O=O	121
H—Br	141	C=O	120	F—F	143
H—I	161	C—Cl	178	Cl—Cl	199
				Br—Br	228
				I—I	266

9.11   Bonding in Metals: The Electron Sea Model

So far, we have explored simple Lewis models for bonding between a metal and a non-metal (ionic bonding) and for bonding between two nonmetals (covalent bonding). We have seen how these models account for and predict the properties of ionic and molecular compounds. The last type of bonding that we examine in this chapter is metallic bonding, which occurs between metals.

As we have already discussed, metals have a tendency to lose electrons, which means that they have relatively low ionization energies. When metal atoms bond together to form a solid, each metal atom donates one or more electrons to an *electron sea*. For example, we can think of sodium metal as an array of positively charged Na^+ ions immersed in a sea of negatively charged electrons (e^-), as shown in **Figure 9.12◄**. Each sodium atom donates its one valence electron to the "sea" and becomes a sodium ion. The sodium cations are then held together by their attraction to the sea of electrons.

Although this model is simple, it accounts for many of the properties of metals. For example, metals conduct electricity because—in contrast to ionic solids where electrons are localized on an ion—the electrons in a metal are free to move in response to an electric potential (or voltage) and generate an electric current. Metals are also excellent conductors of heat, again because of the highly mobile electrons, which help to disperse thermal energy throughout the metal.

The *malleability* of metals (their ability to be pounded into sheets) and the *ductility* of metals (their ability to be drawn into wires) are also accounted for by this model. Since there are no localized or specific "bonds" in a metal, it can be deformed relatively easily by forcing the metal ions to slide past one another. The electron sea easily accommodates these deformations by flowing into the new shape.

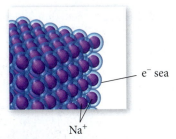

e⁻ sea

Na⁺

▲ **FIGURE 9.12 The Electron Sea Model for Sodium**   In this model of metallic bonding, Na^+ ions are immersed in a "sea" of electrons.

CHAPTER IN REVIEW

Key Terms

Section 9.1
Lewis model (325)
Lewis electron-dot
    structures (Lewis
    structures) (325)

Section 9.2
ionic bond (326)
covalent bond (327)
metallic bonding (327)

Section 9.3
octet (328)
duet (328)
chemical bond (328)
octet rule (328)

Section 9.4
lattice energy ($\Delta H^\circ_{lattice}$) (329)

Section 9.5
bonding pair (333)

lone pair (333)
double bond (333)
triple bond (334)

Section 9.6
polar covalent bond (336)
electronegativity (336)
dipole moment (μ) (337)
percent ionic
    character (338)

Section 9.8
resonance structures (342)
resonance hybrid (342)
formal charge (343)

Section 9.9
free radical (345)

Section 9.10
bond energy (348)
bond length (351)

Key Concepts

Bonding Models (9.1)

▶ Theories that predict how and why atoms bond together are central to chemistry, because they explain compound stability and molecular shape.

Types of Chemical Bonds (9.2)

▶ Chemical bonds can be divided into three general types: ionic bonds, which occur between a metal and a nonmetal; covalent bonds, which occur between two nonmetals; and metallic bonds, which occur within metals.

▶ In an ionic bond, an electron is transferred from the metal to the nonmetal and the resultant ions are attracted to each other by coulombic forces.

▶ In a covalent bond, nonmetals share electrons that interact with the nuclei of both atoms via coulombic forces, holding the atoms together.

▶ In a metallic bond, the atoms form a lattice in which each metal loses electrons to an "electron sea." The attraction of the positively charged metal ions to the electron sea holds the metal together.

The Lewis Model and Electron Dots (9.3)

▶ In the Lewis model, chemical bonds are formed when atoms transfer (ionic bonding) or share (covalent bonding) valence electrons to attain noble gas electron configurations.

▶ The Lewis model represents valence electrons as dots surrounding the symbol for an element. When two or more elements bond together, the dots are transferred or shared so that every atom gets eight dots, an octet (or two dots, a duet, in the case of hydrogen).

Ionic Lewis Structures and Lattice Energy (9.4)

▶ In an ionic Lewis structure involving main-group metals, the metal transfers its valence electrons (dots) to the nonmetal.

▶ The formation of most ionic compounds from their elements is exothermic because of lattice energy, the energy released when metal cations and nonmetal anions coalesce to form the solid; the smaller the radius of the ions and the greater their charge, the more exothermic the lattice energy.

Covalent Lewis Structures, Electronegativity, and Polarity (9.5, 9.6, 9.7)

▶ In a covalent Lewis structure, neighboring atoms share valence electrons to attain octets (or duets).

▶ A single shared electron pair constitutes a single bond, while two or three shared pairs constitute double or triple bonds, respectively.

▶ The shared electrons in a covalent bond are not always *equally* shared; when two dissimilar nonmetals form a covalent bond, the electron density is greater on the more electronegative element. The result is a polar bond, with one element carrying a partial positive charge and the other a partial negative charge.

▶ Electronegativity—the ability of an atom to attract electrons to itself in a chemical bond—increases as we move to the right across a row in the periodic table and decreases as we move down a column.

▶ Elements with very different electronegativities form ionic bonds; those with very similar electronegativities form nonpolar covalent bonds; and those with intermediate electronegativity differences form polar covalent bonds.

Resonance and Formal Charge (9.8)

▶ Some molecules are best represented not by a single Lewis structure, but by two or more resonance structures. The actual structure of the molecule is then a resonance hybrid: a combination or average of the contributing structures.

▶ The formal charge of an atom in a Lewis structure is the charge the atom would have if all bonding electrons were shared equally between bonding atoms.

▶ In general, the best Lewis structures will have the fewest atoms with formal charge and any negative formal charge will be on the most electronegative atom.

Exceptions to the Octet Rule (9.9)

▶ Although the octet rule is normally used in drawing Lewis structures, some exceptions occur.

▶ These exceptions include odd-electron species, which necessarily have Lewis structures with only 7 electrons around an atom. Such molecules, called free radicals, tend to be unstable and chemically reactive.

▶ Other exceptions to the octet rule include incomplete octets—usually totaling 6 electrons (especially important in compounds containing boron)—and expanded octets—usually 10 or 12 electrons (important in compounds containing elements from the third row of the periodic table and below). Expanded octets never occur in second-period elements.

Bond Energies and Bond Lengths (9.10)

▶ The bond energy of a chemical bond is the energy required to break 1 mole of the bond in the gas phase.

▶ Average bond energies for a number of different bonds are tabulated and can be used to estimate enthalpies of reaction.

▶ Average bond lengths are also tabulated.

▶ In general, triple bonds are shorter and stronger than double bonds, which are in turn shorter and stronger than single bonds.

Bonding in Metals (9.11)

▶ When metal atoms bond together to form a solid, each metal atom donates one or more electrons to an *electron sea*. The metal cations are then held together by their attraction to the sea of electrons.

▶ This simple model accounts for the electrical conductivity, thermal conductivity, malleability, and ductility of metals.

Key Equations and Relationships

Coulomb's Law: Potential Energy (*E*) of Two Charged Particles with Charges q_1 and q_2 Separated by a Distance *r* (9.2)

$$E = \frac{1}{4\pi\varepsilon_0}\frac{q_1 q_2}{r} \qquad \varepsilon_0 = 8.85 \times 10^{-12}\,\mathrm{C^2/\,J\cdot m}$$

Dipole Moment (*μ*): Separation of Two Particles of Equal but Opposite Charges of Magnitude *q* by a Distance *r* (9.6)

$$\mu = qr$$

Percent Ionic Character (9.6)

$$\text{Percent ionic character} = \frac{\text{measured dipole moment of bond}}{\text{dipole moment if electron were completely transferred}} \times 100\%$$

Formal Charge (9.8)

$$\text{Formal charge} = \text{number of valance electrons} - (\text{number of lone pair electrons} + {}^1\!/_2 \text{ number of shared electrons})$$

Enthalpy Change of a Reaction ΔH_{rxn}: Relationship of Bond Energies (9.10)

$$\Delta H_{rxn} = \sum(\Delta H's \text{ bonds broken}) + \sum(\Delta H's \text{ bonds formed})$$

Key Learning Objectives

Chapter Objectives	Assessment
Predicting Chemical Formulas of an Ionic Compound (9.4)	Example 9.1 For Practice 9.1 Exercises 7, 8
Predicting Relative Lattice Energies (9.4)	Example 9.2 For Practice 9.2 For More Practice 9.2 Exercise 12
Classifying Bonds: Pure Covalent, Polar Covalent, or Ionic (9.6)	Example 9.3 For Practice 9.3 Exercises 17, 18
Writing Lewis Structures for Covalent Compounds (9.7)	Examples 9.4, 9.5 For Practice 9.4, 9.5 Exercises 15, 16
Writing Lewis Structures for Polyatomic Ions (9.7)	Example 9.6 For Practice 9.6 Exercises 21–24
Writing Resonance Lewis Structures (9.8)	Example 9.7 For Practice 9.7 Exercises 23, 24
Assigning Formal Charges to Assess Competing Resonance Structures (9.8)	Example 9.8 For Practice 9.8 For More Practice 9.8 Exercises 25, 26
Writing Lewis Structures for Compounds Having Expanded Octets (9.9)	Example 9.9 For Practice 9.9 For More Practice 9.9 Exercises 33, 34
Calculating ΔH_{rxn} from Bond Energies (9.10)	Example 9.10 For Practice 9.10 For More Practice 9.10 Exercises 37, 38

EXERCISES

Problems by Topic

Valence Electrons and Dot Structures

1. Write an electron configuration for N. Then write the Lewis structure for N and show which electrons from the electron configuration are included in the Lewis structure.

2. Write an electron configuration for Ne. Then write the Lewis structure for Ne and show which electrons from the electron configuration are included in the Lewis structure.

3. Write the Lewis structure for each atom or ion:
   a. Al
   b. Na^+
   c. Cl
   d. Cl^-

4. Write the Lewis structure for each atom or ion:
   a. S^{2-}
   b. Mg
   c. Mg^{2+}
   d. P

Ionic Lewis Structures and Lattice Energy

5. Write the Lewis structure for each ionic compound.
   a. NaF
   b. CaO
   c. $SrBr_2$
   d. K_2O

6. Write the Lewis structure for each ionic compound.
   a. SrO
   b. Li_2S
   c. CaI_2
   d. RbF

7. Use the Lewis model to determine the formula for the compound that forms between:
   a. Sr and Se
   b. Ba and Cl
   c. Na and S
   d. Al and O

8. Use the Lewis model to determine the formula for the compound that forms between:
   a. Ca and N
   b. Mg and I
   c. Ca and S
   d. Cs and F

9. Explain this trend in the lattice energies of the alkaline earth metal oxides:

Metal Oxide	Lattice Energy (kJ/mol)
MgO	-3795
CaO	-3414
SrO	-3217
BaO	-3029

10. Rubidium iodide has a lattice energy of -617 kJ/mol, while potassium bromide has a lattice energy of -671 kJ/mol. Why is the lattice energy of potassium bromide more exothermic than the lattice energy of rubidium iodide?

11. The lattice energy of CsF is -744 kJ/mol, whereas that of BaO is -3029 kJ/mol. Explain this large difference in lattice energy.

12. Arrange these substances in order of increasing magnitude of lattice energy: KCl, SrO, RbBr, CaO.

Simple Covalent Lewis Structures, Electronegativity, and Bond Polarity

13. Use covalent Lewis structures to explain why each element (or family of elements) occurs as diatomic molecules:
   a. hydrogen
   b. the halogens
   c. oxygen
   d. nitrogen

14. Use covalent Lewis structures to explain why the compound that forms between nitrogen and hydrogen has the formula NH_3. Show why NH_2 and NH_4 are not stable.

15. Write the Lewis structure for each molecule:
   a. PH_3
   b. SCl_2
   c. HI
   d. CH_4

16. Write the Lewis structure for each molecule:
   a. NF_3
   b. HBr
   c. SBr_2
   d. CCl_4

17. Determine whether a bond between each pair of atoms would be pure covalent, polar covalent, or ionic.
   a. Br and Br
   b. C and Cl
   c. C and S
   d. Sr and O

18. Determine whether a bond between each pair of atoms would be pure covalent, polar covalent, or ionic.
   a. C and N
   b. N and S
   c. K and F
   d. N and N

19. Draw the Lewis structure for CO with an arrow representing the dipole moment. Refer to Figure 9.9 to estimate the percent ionic character of the CO bond.

20. Draw the Lewis structure for BrF with an arrow representing the dipole moment. Refer to Figure 9.9 to estimate the percent ionic character of the BrF bond.

Covalent Lewis Structures, Resonance, and Formal Charge

21. Write the Lewis structure for each molecule or ion:
   a. CI_4
   b. N_2O
   c. SiH_4
   d. Cl_2CO
   e. H_3COH
   f. OH^-
   g. BrO^-

22. Write a Lewis structure for each molecule or ion:
    a. N_2H_2
    b. N_2H_4
    c. C_2H_2
    d. C_2H_4
    e. H_3COCH_3
    f. CN^-
    g. NO_2^-

23. Write the Lewis structure that obeys the octet rule for each molecule or ion. Include resonance structures if necessary and assign formal charges to each atom.
    a. SeO_2
    b. CO_3^{2-}
    c. ClO^-
    d. NO_2^-

24. Write the Lewis structure that obeys the octet rule for each ion. Include resonance structures if necessary and assign formal charges to each atom.
    a. ClO_3^-
    b. ClO_4^-
    c. NO_3^-
    d. NH_4^+

25. Use formal charge to determine which of the two Lewis structures is better:

$$H-\overset{\overset{\displaystyle H}{|}}{C}=\ddot{\underset{\displaystyle ..}{S}} \qquad H-\overset{\overset{\displaystyle H}{|}}{S}=\ddot{\underset{\displaystyle ..}{C}}$$

26. Use formal charge to determine which of the two Lewis structures is better:

$$H-\overset{\overset{\displaystyle H}{|}}{\underset{\underset{\displaystyle H}{|}}{S}}-\ddot{C}-H \qquad H-\overset{\overset{\displaystyle H}{|}}{\underset{\underset{\displaystyle H}{|}}{C}}-\ddot{S}-H$$

27. How important is this resonance structure to the overall structure of carbon dioxide? Explain.

$$:O\equiv C-\ddot{O}:$$

28. In N_2O, nitrogen is the central atom and the oxygen atom is terminal. In OF_2, however, oxygen is the central atom. Use formal charges to explain this.

Odd-Electron Species, Incomplete Octets, and Expanded Octets

29. Write the Lewis structure for each molecule.
    a. BCl_3
    b. NO_2
    c. BH_3

30. Write the Lewis structure for each molecule.
    a. BBr_3
    b. NO
    c. ClO_2

31. Write the Lewis structure for each ion. Include resonance structures if necessary and assign formal charges to all atoms. If necessary, expand the octet on the central atom to lower formal charge.
    a. PO_4^{3-}
    b. CN^-
    c. SO_3^{2-}
    d. ClO_2^-

32. Write the Lewis structure for each molecule or ion. Include resonance structures if necessary and assign formal charges to all atoms. If necessary, expand the octet on the central atom to lower formal charge.
    a. SO_4^{2-}
    b. HSO_4^-
    c. SO_3
    d. BrO_2^-

33. Write a Lewis structure for each molecule or ion. Use expanded octets as necessary.
    a. PF_5
    b. I_3^-
    c. SF_4
    d. GeF_4

34. Write the Lewis structure for each molecule or ion. Use expanded octets as necessary.
    a. ClF_5
    b. AsF_6^-
    c. Cl_3PO
    d. IF_5

Bond Energies and Bond Lengths

35. Consider these three compounds: HCCH, H_2CCH_2, H_3CCH_3. Order these compounds in order of increasing carbon–carbon bond strength and in order of decreasing carbon–carbon bond length.

36. Which of the two compounds has the strongest nitrogen–nitrogen bond? The shortest nitrogen–nitrogen bond?

$$H_2NNH_2, HNNH$$

37. Hydrogenation reactions add hydrogen across double bonds in hydrocarbons and other organic compounds. Use average bond energies to calculate ΔH_{rxn} for this hydrogenation reaction:

$$H_2C=CH_2(g) + H_2(g) \longrightarrow H_3C-CH_3(g)$$

38. Ethanol is a fuel we may use as a possible energy source in the future. Use average bond energies to calculate ΔH_{rxn} for the combustion of ethanol.

$$CH_3CH_2OH(g) + 3\,O_2(g) \longrightarrow 2\,CO_2(g) + 3\,H_2O(g)$$

Cumulative Problems

39. Write an appropriate Lewis structure for each compound. Make certain to distinguish between ionic and covalent compounds.
    a. BI_3
    b. K_2S
    c. HCFO
    d. PBr_3

40. Write an appropriate Lewis structure for each compound. Make certain to distinguish between ionic and covalent compounds.
    a. Al_2O_3
    b. ClF_5
    c. MgI_2
    d. XeO_4

41. Each compound contains both ionic and covalent bonds. Write ionic Lewis structures for each of them, including the covalent structure for the ion in brackets. Write resonance structures if necessary.
    a. $BaCO_3$
    b. $Ca(OH)_2$
    c. KNO_3
    d. $LiIO$

42. Each compound contains both ionic and covalent bonds. Write ionic Lewis structures for each of them, including the covalent structure for the ion in brackets. Write resonance structures if necessary.
    a. $RbIO_2$
    b. NH_4Cl
    c. KOH
    d. $Sr(CN)_2$

43. Carbon ring structures are common in organic chemistry. Draw the Lewis structure for each carbon ring structure, including any necessary resonance structures.
    a. C_4H_8
    b. C_4H_4
    c. C_6H_{12}
    d. C_6H_6

44. Amino acids are the building blocks of proteins. The simplest amino acid is glycine (H_2NCH_2COOH). Draw the Lewis structure for glycine. (Hint: The central atoms in the skeletal structure are nitrogen bonded to carbon which is bonded to another carbon. The two oxygen atoms are bonded directly to the rightmost carbon atom.)

45. Formic acid is responsible for the sting in biting ants. By mass, formic acid is 26.10% C, 4.38% H, and 69.52% O. The molar mass of formic acid is 46.02 g/mol. Find the molecular formula of formic acid and draw its Lewis structure.

46. Diazomethane is a highly poisonous, explosive compound because it readily reacts to form gaseous nitrogen as a product. Diazomethane has the following composition by mass: 28.57% C; 4.80% H; and 66.64% N. The molar mass of diazomethane is 42.04 g/mol. Find the molecular formula of diazomethane, draw its Lewis structure, and assign formal charges to each atom. Explain why diazomethane is not very stable.

47. The reaction of $Fe_2O_3(s)$ with $Al(s)$ to form $Al_2O_3(s)$ and $Fe(s)$ is called the thermite reaction and is highly exothermic. What role does lattice energy play in the exothermicity of the reaction?

48. NaCl has a lattice energy -787 kJ/mol. Consider a hypothetical salt XY. X^{3+} has the same radius of Na^+ and Y^{3-} has the same radius as Cl^-. Estimate the lattice energy of XY.

49. Draw the Lewis structure for nitric acid (the hydrogen atom is attached to one of the oxygen atoms). Include all three resonance structures by alternating the double bond among the three oxygen atoms. Use formal charge to determine which of the resonance structures is most important to the structure of nitric acid.

50. Phosgene (Cl_2CO) is a poisonous gas that was used as a chemical weapon during World War I and is a potential agent for chemical terrorism. Draw the Lewis structure of phosgene. Include all three resonance forms by alternating the double bond among the three terminal atoms. Which resonance structure is the best?

51. The cyanate ion (OCN^-) and the fulminate ion (CNO^-) share the same three atoms, but have vastly different properties. The cyanate ion is stable, while the fulminate ion is unstable and forms explosive compounds. The resonance structures of the cyanate ion were explored in Example 9.8. Draw the Lewis structure for the fulminate ion—including possible resonance forms—and use formal charge to explain why the fulminate ion is less stable (and therefore more reactive) than the cyanate ion.

52. Use Lewis structures to explain why Br_3^- and I_3^- are stable, while F_3^- is not.

53. Draw the Lewis structure for each organic compound from its condensed structural formula.
    a. C_3H_8
    b. CH_3OCH_3
    c. CH_3COCH_3
    d. CH_3COOH
    e. CH_3CHO

54. Draw the Lewis structure for each organic compound from its condensed structural formula.
    a. C_2H_4
    b. CH_3NH_2
    c. $HCHO$
    d. CH_3CH_2OH
    e. $HCOOH$

55. Draw the Lewis structure for $HCSNH_2$. (The carbon and nitrogen atoms are bonded together and the sulfur atom is bonded to the carbon atom.) Label each bond in the molecule as polar or nonpolar.

56. Draw the Lewis structure for urea, H_2NCONH_2, the compound primarily responsible for the smell of urine. (The central carbon atom is bonded to both nitrogen atoms and to the oxygen atom.) Does urea contain polar bonds? Which bond in urea is most polar?

57. Some theories of aging suggest that free radicals cause certain diseases and perhaps aging in general. According to the Lewis model, such odd-electron species molecules are not chemically stable and will quickly react with other molecules. Free radicals may react with molecules in the cell, such as DNA, attacking them and causing cancer or other diseases. Free radicals may also attack molecules on the surfaces of cells, making them appear foreign to the body's immune system. The immune system then attacks the cells and destroys them, weakening the body. Draw the Lewis structures for each free radical implicated in this theory of aging.
    a. O_2^-
    b. O^-
    c. OH
    d. CH_3OO (unpaired electron on terminal oxygen)

58. Free radicals are important in many environmentally significant reactions. For example, photochemical smog—smog that forms as a result of the action of sunlight on air pollutants—is formed in part by two steps:

$$NO_2 \xrightarrow{\text{UV light}} NO + O$$

$$O + O_2 \longrightarrow O_3$$

The product of this reaction, ozone, is a pollutant in the lower atmosphere. (Upper atmospheric ozone is a natural part of the atmosphere that protects life on Earth from ultraviolet light.) Ozone is an eye and lung irritant and also accelerates the weathering of rubber products. Rewrite the above reactions using the Lewis structure of each reactant and product. Identify the free radicals.

59. If hydrogen were used as a fuel, it could be burned according to the reaction:

$$H_2(g) + \text{\small{1/2}}\, O_2(g) \longrightarrow H_2O(g)$$

Use average bond energies to calculate ΔH_{rxn} for this reaction and also for the combustion of methane (CH_4). Which fuel yields more energy per mole? Per gram?

60. Calculate ΔH_{rxn} for the combustion of octane (C_8H_{18}), a component of gasoline, by using average bond energies and then calculate it again using enthalpies of formation from Appendix IIA. What is the percent difference between the two results? Which result would you expect to be more accurate?

61. Draw the Lewis structure for each compound.
    a. Cl_2O_7 (no $Cl-Cl$ bond)
    b. H_3PO_3 (two OH bonds)
    c. H_3AsO_4

62. The azide ion, N_3^-, is a symmetrical ion, all of whose contributing structures have formal charges. Draw three important contributing Lewis structures for this ion.

63. List the gas-phase ion pairs in order of the quantity of energy released when they form from separated gas-phase ions. Start with the pair that releases the least energy.
$$Na^+F^-, Mg^{2+}F^-, Na^+O^{2-}, Mg^{2+}O^{2-}, Al^{3+}O^{2-}.$$

64. Calculate $\Delta H°$ for the reaction $H_2(g) + Br_2(g) \longrightarrow 2\ HBr(g)$ using the bond energy values. The $\Delta H°_f$ of $HBr(g)$ is not equal to one-half of the value calculated. Account for the difference.

65. A compound composed of only carbon and hydrogen is 7.743% hydrogen by mass. Propose the Lewis structure for the compound.

66. A compound composed of only carbon and chlorine is 85.5% chlorine by mass. Propose the Lewis structure for the compound.

Challenge Problems

67. The main component of acid rain (H_2SO_4) forms from SO_2 pollutant in the atmosphere via the following series of steps:
$$SO_2 + OH\cdot \longrightarrow HSO_3\cdot$$
$$HSO_3\cdot + O_2 \longrightarrow SO_3 + HOO\cdot$$
$$SO_3 + H_2O \longrightarrow H_2SO_4$$
Draw the Lewis structure for each of the species in these steps and use bond energies and Hess's law to estimate ΔH_{rxn} for the overall process. (Use 265 kJ/mol for the $S-O$ single bond energy.)

68. A 0.167-g sample of an unknown acid requires 27.8 mL of 0.100 M NaOH to titrate to the equivalence point. Elemental analysis of the acid gives the following percentages by mass: 40.00% C; 6.71% H; 53.29% O. Determine the molecular formula, molar mass, and Lewis structure of the unknown acid.

69. Use the dipole moments of HF and HCl (listed at the end of this problem) together with the percent ionic character of each bond (Figure 9.9) to estimate the bond length in each molecule. How well does your estimated bond length agree with the bond length given in Table 9.4?
$$HCl \quad \mu = 1.08\ D$$
$$HF \quad \mu = 1.82\ D$$

70. Use average bond energies together with the standard enthalpy of formation of $C(g)$ (718.4 kJ/mol) to estimate the standard enthalpy of formation of gaseous benzene, $C_6H_6(g)$. (Remember that average bond energies apply to the gas phase only.) Compare the value you obtain using average bond energies to the actual standard enthalpy of formation of gaseous benzene, 82.9 kJ/mol. What does the difference between these two values tell you about the stability of benzene?

71. The standard state of phosphorus at 25 °C is P_4. This molecule has four equivalent P atoms, no double or triple bonds, and no expanded octets. Draw its Lewis structure.

72. A compound with the formula C_8H_8 does not contain any double or triple bonds. All the carbon atoms are chemically identical and all the hydrogen atoms are chemically identical. Draw the Lewis structure for this molecule.

73. Find the oxidation number of each sulfur in the molecule H_2S_4, which has a linear arrangement of its atoms.

74. Ionic solids of the O^- and O^{3-} anions do not exist, while ionic solids of the O^{2-} anion are common. Explain.

75. The standard state of sulfur is solid rhombic sulfur. Use the appropriate standard heats of formation given in Appendix II to find the average bond energy of the $S=O$ in SO_2.

Conceptual Problems

76. Which statement is true of an endothermic reaction?
    a. Strong bonds break and weak bonds form.
    b. Weak bonds break and strong bonds form.
    c. The bonds that break and those that form are of approximately the same strength.

77. When a firecracker explodes, energy is obviously released. The compounds in the firecracker can be viewed as being "energy rich." What does this mean? Explain the source of the energy in terms of chemical bonds.

78. A fundamental difference between compounds containing ionic bonds and those containing covalent bonds is the existence of molecules. Explain why molecules exist in solid covalent compounds but do not exist in solid ionic compounds.

79. In the first chapter of this book, we described the scientific method and put a special emphasis on scientific models or theories. In this chapter, we looked carefully at a model for chemical bonding (the Lewis model). Why is this theory successful? Can you name some of the limitations of the theory?

Answers to Conceptual Connections

Melting Points of Ionic Solids

9.1 We would expect MgO to have the higher melting point because, in our bonding model, the magnesium and oxygen ions are held together in a crystalline lattice by charges of 2+ for magnesium and 2− for oxygen. In contrast, the NaCl lattice is held together by charges of 1+ for sodium and 1− for chlorine. The experimentally measured melting points of these compounds are 801 °C for NaCl and 2852 °C for MgO, in accordance with our model.

Energy and the Octet Rule

9.2 The reasons that atoms form bonds are complex. One contributing factor is the lowering of their potential energy. The octet rule is just a handy way to predict the combinations of atoms that will have a lower potential energy when they bond together.

Periodic Trends in Electronegativity

9.3 $N > P > Al > Na$

Odd-Electron Species

9.4 (d) NO_2 because the sum of the valence electrons of its atoms is an odd number.

Expanded Octets

9.5 (b) The only molecule in this group that could have an expanded octet is H_3PO_4 because phosphorus is a third period element. Expanded octets never occur in second period elements (such as C and N).

Bond Energies and ΔH_{rxn}

9.6 (b) In a highly exothermic reaction, the energy needed to break bonds is less than the energy released when the new bonds form, resulting in a net release of energy.

10 Chemical Bonding II: Molecular Shapes, Valence Bond Theory, and Molecular Orbital Theory

No theory ever solves all the puzzles with which it is confronted at a given time; nor are the solutions already achieved often perfect. —Thomas Kuhn (1922–1996)

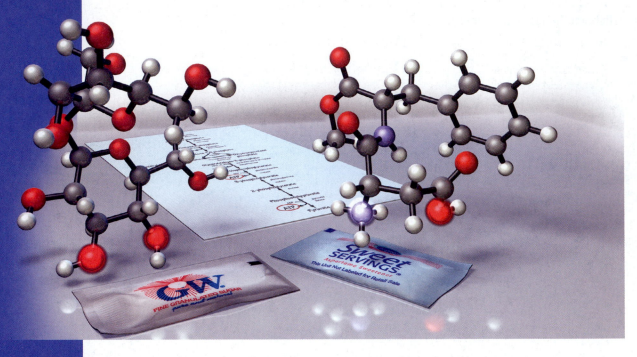

Similarities between the shape of sugar and aspartame give both molecules the ability to stimulate a sweet taste sensation.

I N CHAPTER 9, WE STUDIED a simple model for chemical bonding called the Lewis model. This model helps us explain and predict the combinations of atoms that form stable molecules. When we combine the Lewis model with the idea that valence electron groups repel one another—the basis of an approach known as VSEPR theory—we can predict the general shape of a molecule from its Lewis structure. We address molecular shape and its importance in the first part of this chapter. We then move on to discuss two additional bonding theories—valence bond theory and molecular orbital theory—that are progressively more sophisticated, at the cost of being more complex, than the Lewis model. As you work through this chapter, our second on chemical bonding, keep in mind the importance of this topic. In our universe, elements join together to form compounds—that makes many things possible, including our own existence.

10.1   Artificial Sweeteners: Fooled by Molecular Shape

Artificial sweeteners, such as aspartame (Nutrasweet), taste sweet but have few or no calories. Why? *Because taste and caloric value are independent properties of foods*. The caloric value of a food depends on the amount of energy released when the food is metabolized. For example, sucrose (table sugar) is metabolized by oxidation to carbon dioxide and water:

$$C_{12}H_{22}O_{11} + 12\ O_2 \longrightarrow 12\ CO_2 + 11\ H_2O \quad \Delta H^{\circ}_{rxn} = -5644\ \text{kJ}$$

When your body metabolizes a mole of sucrose, it obtains 5644 kJ of energy. Some artificial sweeteners, such as saccharin, for example, are not metabolized at all—they just pass through the body unchanged—and therefore have no caloric value. Other artificial sweeteners, such as aspartame, are metabolized but have a much lower caloric content (for a given amount of sweetness) than sucrose.

The *taste* of a food, in contrast, is independent of its metabolism. The sensation of taste originates in the tongue, where specialized cells called taste cells act as highly sensitive and specific molecular detectors. These cells can discern sugar molecules from the thousands of different types of molecules present in a mouthful of food. The main factors for this discrimination are the sugar molecule's shape and charge distribution.

The surface of a taste cell contains specialized protein molecules called taste receptors. A particular *tastant*—a molecule that we can taste—fits snugly (just as a key fits into a lock) into a special pocket on the taste receptor protein of the taste cell called the *active site*. A sugar molecule just fits into the active site of the sugar receptor protein called T1r3. When the sugar molecule (the key) enters the active site (the lock), the different subunits of the T1r3 protein split apart. This split causes ion channels in the cell membrane to open, resulting in nerve signal transmission (see Section 8.1). The nerve signal reaches the brain and registers a sweet taste.

Artificial sweeteners taste sweet because they fit into the receptor pocket that normally binds sucrose. In fact, both aspartame and saccharin actually bind to the active site in the T1r3 protein more strongly than does sugar! For this reason, artificial sweeteners are "sweeter than sugar." Aspartame, for example, is 200 times sweeter than sugar; it takes 200 times as much sugar as aspartame to trigger the same amount of nerve signal transmission from taste cells.

The lock-and-key fit between the active site of a protein and a particular molecule is important not only to taste but to many other biological functions as well. Immune response, the sense of smell, and many types of drug action all depend on shape-specific interactions between molecules and proteins. In fact, our ability to determine the shapes of key biological molecules is largely responsible for the revolution in biology that has occurred over the last 50 years.

In this chapter, we look at ways to predict and account for the shapes of molecules. The molecules we examine are much smaller than the T1r3 protein molecules we just discussed, but the same principles apply to both. The simple model we develop to account for molecular shape is called *valence shell electron pair repulsion* (VSEPR) theory, and it is used in conjunction with the Lewis model. We then proceed to introduce two additional bonding theories: valence bond theory and molecular orbital theory. These bonding theories also predict and account for molecular shape as well as other properties of molecules.

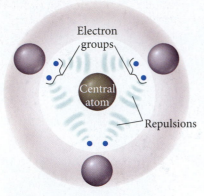

▲**FIGURE 10.1 Repulsion between Electron Groups** The basic idea of VSEPR theory is that repulsions between electron groups determine molecular geometry.

| Beryllium often forms incomplete octets, as it does in this structure.

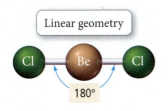

| A double bond counts as one electron group.

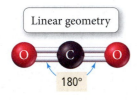

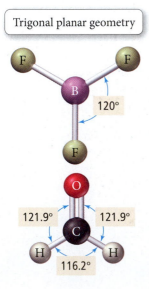

10.2 VSEPR Theory: The Five Basic Shapes

Valence shell electron pair repulsion (VSEPR) theory, is based on the idea that **electron groups**—lone pairs, single bonds, multiple bonds, and even single electrons—repel one another through coulombic forces. The repulsion between electron groups on interior atoms of a molecule, therefore, determines the geometry of the molecule (**Figure 10.1◄**). The preferred geometry is the one in which the electron groups have the maximum separation (and therefore the minimum energy) possible. Consequently, for molecules having just one interior atom—the central atom—molecular geometry depends on (a) the number of electron groups around the central atom and (b) how many of those electron groups are bonding groups and how many are lone pairs. We first look at the molecular geometries associated with two to six electron groups around the central atom when all of those groups are bonding groups (single or multiple bonds). The resulting geometries constitute the five basic shapes of molecules. We will then consider how these basic shapes are modified if one or more of the electron groups are lone pairs.

Two Electron Groups: Linear Geometry

Consider the Lewis structure of $BeCl_2$, which has two electron groups (two single bonds) about the central atom:

$$:\ddot{Cl}:Be:\ddot{Cl}:$$

According to VSEPR theory, the geometry of $BeCl_2$ is determined by the repulsion between these two electron groups, which can maximize their separation by assuming a 180° bond angle or a **linear geometry**. Experimental measurements of the geometry of $BeCl_2$ indicate that the molecule is indeed linear, as predicted by the theory.

Molecules that form only two single bonds, with no lone pairs, are rare because they do not follow the octet rule. However, the same geometry is observed in all molecules that have two electron groups (and no lone pairs). Consider the Lewis structure of CO_2, which has two electron groups (the two double bonds) around the central carbon atom:

$$:\ddot{O}=C=\ddot{O}:$$

According to VSEPR theory, the two double bonds repel each other (just as the two single bonds in $BeCl_2$ repel each other), resulting in a linear geometry for CO_2. Experimental observations confirm that CO_2 is indeed a linear molecule, as predicted.

Three Electron Groups: Trigonal Planar Geometry

The Lewis structure of BF_3 (another molecule with an incomplete octet) has three electron groups around the central atom.

$$:\ddot{F}:\ddot{B}:\ddot{F}:$$

These three electron groups maximize their separation by assuming 120° bond angles in a plane—a **trigonal planar geometry**. Experimental observations of the structure of BF_3 again confirm the predictions of VSEPR theory.

Another molecule with three electron groups, formaldehyde, has one double bond and two single bonds around the central atom.

$$H-\overset{\overset{\displaystyle :O:}{\|}}{C}-H$$

Since formaldehyde has three electron groups around the central atom, we initially predict that the bond angles should also be 120°. However, experimental observations show that the HCO bond angles are 121.9° and that the HCH bond angle is 116.2°. These bond angles are close to the idealized 120° that we originally predicted, but the HCO bond angles are slightly greater than the HCH bond angle because the double bond contains more electron density than the single bond and therefore exerts a slightly

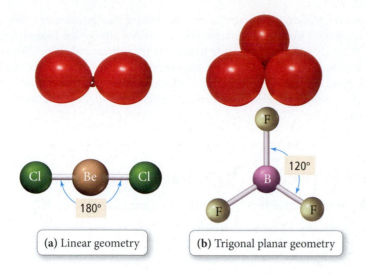

(a) Linear geometry

(b) Trigonal planar geometry

greater repulsion on the single bonds. In general, *different types of electron groups exert slightly different repulsions—the resulting bond angles reflect these differences.*

 Conceptual Connection 10.1 Electron Groups and Molecular Geometry

In determining electron geometry, why do we consider only the electron groups on the central atom? Why don't we consider electron groups on terminal atoms?

Four Electron Groups: Tetrahedral Geometry

The VSEPR geometries of molecules with two or three electron groups around the central atom are two-dimensional and can therefore easily be visualized and represented on paper. For molecules with four or more electron groups around the central atom, the geometries are three-dimensional and are therefore more difficult to imagine and to draw. One common way to help visualize these shapes is by visualizing balloons tied together. In this analogy, each electron group around a central atom is like a balloon tied to a central point. The bulkiness of the balloons causes them to spread out as much as possible, much as the repulsion between electron groups causes them to position themselves as far apart as possible. For example, if you tie two balloons together, they assume a roughly linear arrangement, as shown in **Figure 10.2(a)▲**, analogous to the linear geometry of BeCl₂ that we just examined. Note that the balloons do not represent atoms, but *electron groups*. Similarly, if you tie three balloons together—in analogy to three electron groups—they assume a trigonal planar geometry, as shown in **Figure 10.2(b)▲**, much like our BF₃ molecule. If you tie *four* balloons together, they assume a three-dimensional **tetrahedral geometry** with 109.5° angles between the balloons. That is, the balloons point toward the vertices of a *tetrahedron*—a geometrical shape with four identical faces, each an equilateral triangle—as shown at right.

Methane is an example of a molecule with four electron groups around the central atom.

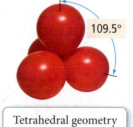

Tetrahedral geometry

Tetrahedron

Tetrahedral geometry

For four electron groups, the tetrahedron is the three-dimensional shape that allows the maximum separation among the groups. The repulsions among the four electron groups in the C—H bonds cause the molecule to assume the tetrahedral shape. When we write the Lewis structure of CH_4 on paper, it may seem that the molecule should be square planar, with bond angles of 90°. However, in three dimensions, the electron groups can get farther away from each other by forming the tetrahedral geometry, as shown by our balloon analogy.

Conceptual Connection 10.2 Molecular Geometry

What is the geometry of the HCN molecule? The Lewis structure of HCN is:

H—C≡N:

(a) linear

(b) trigonal planar

(c) tetrahedral

Five Electron Groups: Trigonal Bipyramidal Geometry

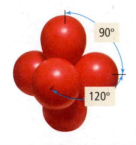

Five electron groups around a central atom assume a **trigonal bipyramidal geometry**, like that of five balloons tied together. In this structure, three of the groups lie in a single plane, as in the trigonal planar configuration, while the other two are positioned above and below this plane. The angles in the trigonal bipyramidal structure are not all the same. The angles between the *equatorial positions* (the three bonds in the trigonal plane) are 120°, while the angle between the *axial positions* (the two bonds on either side of the trigonal plane) and the trigonal plane is 90°. As an example of a molecule with five electron groups around the central atom, consider PCl_5:

Trigonal bipyramidal geometry   Trigonal bipyramid

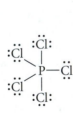

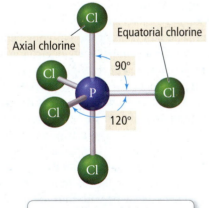

Trigonal bipyramidal geometry

The three equatorial chlorine atoms are separated by 120° bond angles and the two axial chlorine atoms are separated from the equatorial atoms by 90° bond angles.

Six Electron Groups: Octahedral Geometry

Octahedral geometry   Octahedron

Six electron groups around a central atom assume an **octahedral geometry**, like that of six balloons tied together. In this structure—named after the eight-sided geometrical shape called the octahedron—four of the groups lie in a single plane, with one group above the plane and another below it.

The angles in this geometry are all 90°. As an example of a molecule with six electron groups around the central atom, consider SF_6:

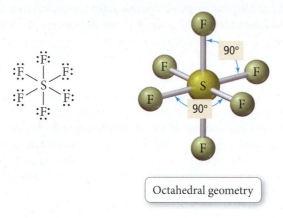

Octahedral geometry

You can see that the structure of this molecule is highly symmetrical. All six bonds are equivalent.

EXAMPLE 10.1 VSEPR Theory and the Basic Shapes

Determine the molecular geometry of NO_3^-.

SOLUTION

The molecular geometry of NO_3^- is determined by the number of electron groups around the central atom (N). Begin by drawing a Lewis structure of NO_3^-.	NO_3^- has $5 + 3(6) + 1 = 24$ valence electrons. The Lewis structure is as follows: 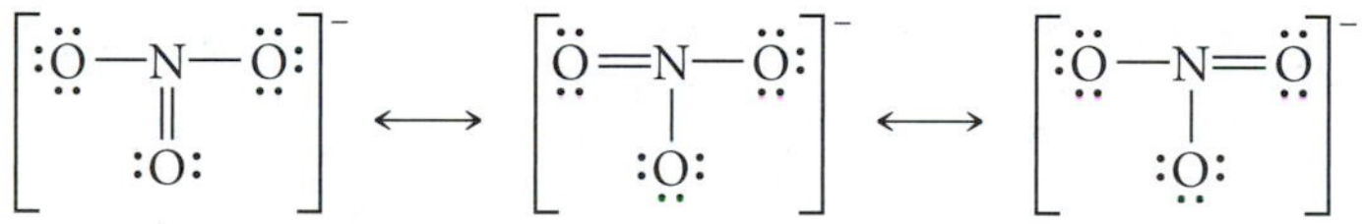 $$\left[\ddot{\underset{..}{O}}{-}N{-}\ddot{\underset{..}{O}}\underset{\|}{}\atop{:O:}\right]^{-} \longleftrightarrow \left[\ddot{O}{=}N{-}\ddot{\underset{..}{O}}\atop{:O:}\right]^{-} \longleftrightarrow \left[\ddot{\underset{..}{O}}{-}N{=}\ddot{O}\atop{:O:}\right]^{-}$$ The hybrid structure is intermediate between these three and has three equivalent bonds.
Use the Lewis structure, or any one of the resonance structures, to determine the number of electron groups around the central atom.	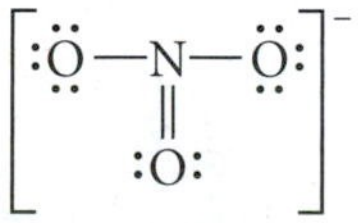 $$\left[\ddot{\underset{..}{O}}{-}N{-}\ddot{\underset{..}{O}}\atop{:O:}\right]^{-}$$ The nitrogen atom has three electron groups.
Based on the number of electron groups, determine the geometry that minimizes the repulsions between the groups.	The electron geometry that minimizes the repulsions between three electron groups is trigonal planar. Because there are no lone pairs on the central atom, the molecular geometry is also trigonal planar. 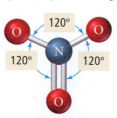 Since the three bonds are equivalent, they each exert the same repulsion on the other two and the molecule has three equal bond angles of 120°.

FOR PRACTICE 10.1
Determine the molecular geometry of CCl_4.

10.3 VSEPR Theory: The Effect of Lone Pairs

Each of the examples we have just examined has only bonding electron groups around the central atom. What happens in molecules that have lone pairs around the central atom as well? These lone pairs also repel other electron groups, as we see in the examples that follow.

Four Electron Groups with Lone Pairs

In the Lewis structure of ammonia, shown at left, the central nitrogen atom has four electron groups (one lone pair and three bonding pairs) that repel one another. If we do not distinguish between bonding electron groups and lone pairs, we find that the **electron geometry**—the geometrical arrangement of the *electron groups*—is still tetrahedral, as we expect for four electron groups. However, the **molecular geometry**—the geometrical arrangement of the atoms—is **trigonal pyramidal**, as shown at left. Notice that although the electron geometry and the molecular geometry are different, *the electron geometry is relevant to the molecular geometry*. The lone pair exerts its influence on the bonding pairs.

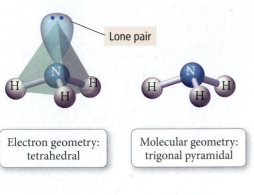

Electron geometry: tetrahedral

Molecular geometry: trigonal pyramidal

As we saw previously, different kinds of electron groups generally result in different amounts of repulsion. Lone pair electrons generally exert slightly greater repulsions than bonding electrons. If all four electron groups in NH_3 exerted equal repulsions on one another, the bond angles in the molecule would all be the ideal tetrahedral angle, 109.5°. However, the actual angle between N—H bonds in ammonia is slightly smaller, 107°. A lone electron pair is more spread out in space than a bonding electron pair because the lone pair is attracted to only one nucleus while the bonding pair is attracted to two (**Figure 10.3▼**). The lone pair occupies more of the angular space around a nucleus, exerting a greater repulsive force on neighboring electrons and compressing the N—H bond angles.

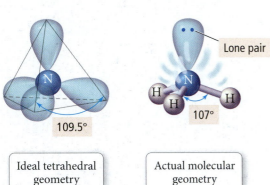

Lone pair

109.5°

Ideal tetrahedral geometry

107°

Actual molecular geometry

Consider also H_2O. Its Lewis structure is

$$H - \ddot{O} - H$$

Since it has four electron groups (two bonding pairs and two lone pairs), its *electron geometry* is also tetrahedral, but its *molecular geometry* is **bent**, as shown at left. As in NH_3, the bond angles in H_2O are smaller (104.5°) than the ideal tetrahedral bond angles because of the greater repulsion exerted by the lone pair electrons. The bond angle in H_2O is even smaller than in NH_3 because H_2O has *two* lone pairs of electrons on the central oxygen atom. These lone pairs compress the H_2O bond angle to an even greater extent than in NH_3.

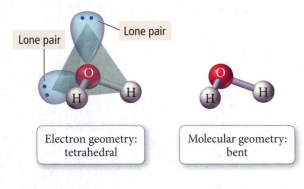

Lone pair   Lone pair

Electron geometry: tetrahedral

Molecular geometry: bent

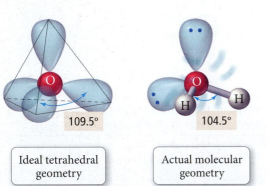

109.5°

104.5°

Ideal tetrahedral geometry

Actual molecular geometry

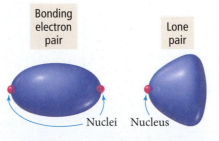

Bonding electron pair

Lone pair

Nuclei   Nucleus

▲ **FIGURE 10.3 Nonbonding versus Bonding Electron Pairs** A lone electron pair occupies more space than a bonding pair.

Effect of Lone Pairs on Molecular Geometry

No lone pairs	One lone pair	Two lone pairs

109.5°
CH_4

107°
NH_3

104.5°
H_2O

◄ **FIGURE 10.4 The Effect of Lone Pairs on Molecular Geometry** The bond angles get progressively smaller as the number of lone pairs on the central atom increases from zero in CH_4 to one in NH_3 to two in H_2O.

In general, electron group repulsions vary as follows:

Lone pair–lone pair > Lone pair–bonding pair > Bonding pair– bonding pair

Most repulsive   Least repulsive

We see the effects of this ordering in the progressively smaller bond angles of CH_4, NH_3, and H_2O, shown in **Figure 10.4▲**. The relative ordering of repulsions also plays a role in determining the geometry of molecules with five and six electron groups when one or more of those groups are lone pairs.

Five Electron Groups with Lone Pairs

The Lewis structure of SF_4, is shown at right. The central sulfur atom has five electron groups (one lone pair and four bonding pairs). The *electron geometry*, due to the five electron groups, is trigonal bipyramidal. In determining the molecular geometry, notice that the lone pair can occupy either an equatorial position or an axial position within the trigonal bipyramidal electron geometry. Which position is most favorable? To answer this question, we must consider that, as we have just seen, lone pair–bonding pair repulsions are greater than bonding pair–bonding pair repulsions. Therefore, the lone pair will occupy the position that minimizes its interaction with the bonding pairs. If the lone pair were in an axial position, it would have three 90° interactions with bonding pairs. In an equatorial position, however, it has only two 90° interactions. Consequently, the lone pair occupies an equatorial position and the resulting molecular geometry is called **seesaw**, because it resembles a seesaw (or teeter-totter).

When two of the five electron groups around the central atom are lone pairs, as in BrF_3, the lone pairs occupy two of the three equatorial positions—again minimizing 90°

The seesaw molecular geometry is sometimes called an *irregular tetrahedron*.

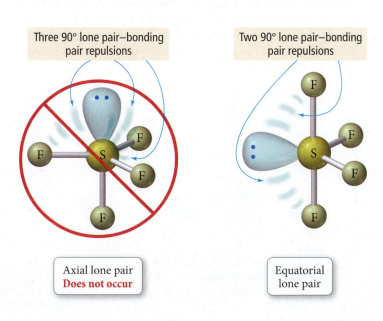

Three 90° lone pair–bonding pair repulsions

Two 90° lone pair–bonding pair repulsions

Axial lone pair
Does not occur

Equatorial lone pair

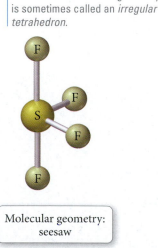

Molecular geometry: seesaw

interactions with bonding pairs and also avoiding a lone pair–lone pair 90° repulsion. The resulting molecular geometry is **T-shaped**.

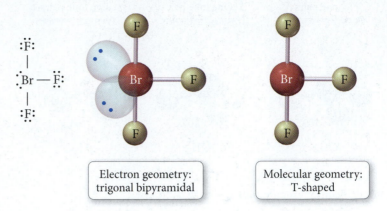

| Electron geometry: | Molecular geometry: |
| trigonal bipyramidal | T-shaped |

When three of the five electron groups around the central atom are lone pairs, as in XeF_2, the lone pairs occupy all three of the equatorial positions and the resulting molecular geometry is linear.

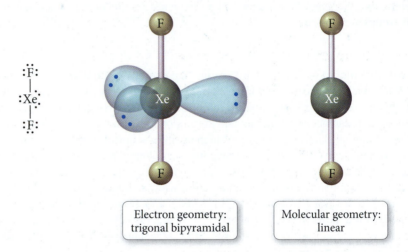

| Electron geometry: | Molecular geometry: |
| trigonal bipyramidal | linear |

Six Electron Groups with Lone Pairs

The Lewis structure of BrF_5 is shown below. The central bromine atom has six electron groups (one lone pair and five bonding pairs). The electron geometry, due to the six electron groups, is octahedral. Since all six positions in the octahedral geometry are equivalent, the lone pair can be situated in any one of these positions. The resulting molecular geometry is called **square pyramidal**.

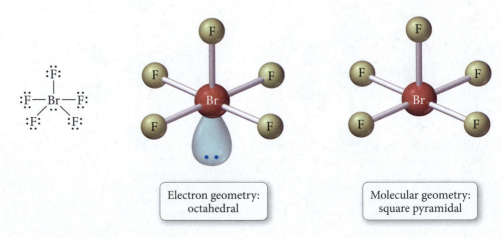

| Electron geometry: | Molecular geometry: |
| octahedral | square pyramidal |

When two of the six electron groups around the central atom are lone pairs, as in XeF_4, the lone pairs occupy positions across from one another (to minimize lone pair–lone pair repulsions), and the resulting molecular geometry is **square planar**.

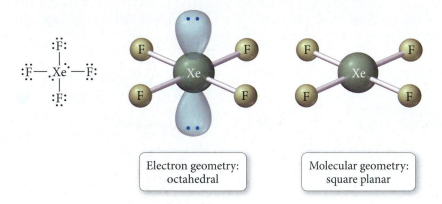

Electron geometry:
octahedral

Molecular geometry:
square planar

Summarizing VSEPR Theory:

▶ The geometry of a molecule is determined by the number of electron groups on the central atom (or on all interior atoms, if there is more than one).

▶ The number of electron groups is determined from the Lewis structure of the molecule. If the Lewis structure contains resonance structures, use any one of the resonance structures to determine the number of electron groups.

▶ Each of the following counts as a single electron group: a lone pair, a single bond, a double bond, a triple bond, or a single electron.

▶ The geometry of the electron groups is determined by minimizing their repulsions as summarized in Table 10.1. In general, electron group repulsions vary as follows:

Lone pair–lone pair > lone pair–bonding pair > bonding pair–bonding pair

▶ Bond angles can vary from the idealized angles because double and triple bonds occupy more space than single bonds, and lone pairs occupy more space than bonding groups. The presence of lone pairs usually makes bond angles smaller than the ideal angle for the particular geometry.

Conceptual Connection 10.3 Molecular Geometry and Electron Group Repulsions

Which statement is always true according to VSEPR theory?

(a) The shape of a molecule is determined by repulsions among bonding electron groups.

(b) The shape of a molecule is determined by repulsions among nonbonding electron groups.

(c) The shape of a molecule is determined by the polarity of its bonds.

(d) The shape of a molecule is determined by repulsions among all electron groups on the central atom.

10.4 VSEPR Theory: Predicting Molecular Geometries

To determine the geometry of a molecule, follow the procedure shown in examples 10.2 and 10.3. As in many previous examples, we give the steps in the left column and provide two examples of applying the steps in the center and right columns.

TABLE 10.1 Electron and Molecular Geometries

Electron Groups*	Bonding Groups	Lone Pairs	Electron Geometry	Molecular Geometry	Approximate Bond Angles	Example
2	2	0	Linear	Linear	180°	$:\ddot{O}=C=\ddot{O}:$
3	3	0	Trigonal planar	Trigonal planar	120°	$:\ddot{F}-B(-\ddot{F}:)\ddot{F}:$
3	2	1	Trigonal planar	Bent	<120°	$:\ddot{O}=\ddot{S}-\ddot{O}:$
4	4	0	Tetrahedral	Tetrahedral	109.5°	$H-C(-H)(-H)H$
4	3	1	Tetrahedral	Trigonal pyramidal	<109.5°	$H-N(-H)H$
4	2	2	Tetrahedral	Bent	<109.5°	$H-\ddot{O}-H$
5	5	0	Trigonal bipyramidal	Trigonal bipyramidal	120° (equatorial) 90° (axial)	$:\ddot{Cl}-P(:\ddot{Cl})(:\ddot{Cl})(\ddot{Cl}:)\ddot{Cl}:$
5	4	1	Trigonal bipyramidal	Seesaw	<120° (equatorial) <90° (axial)	$\ddot{F}-\ddot{S}(-\ddot{F})(-\ddot{F})\ddot{F}$
5	3	2	Trigonal bipyramidal	T-shaped	<90°	$:\ddot{F}-\ddot{Br}(-\ddot{F}:)\ddot{F}:$
5	2	3	Trigonal bipyramidal	Linear	180°	$:\ddot{F}-\ddot{Xe}-\ddot{F}:$
6	6	0	Octahedral	Octahedral	90°	$\ddot{F}-S(-\ddot{F})(-\ddot{F})(-\ddot{F})(-\ddot{F})\ddot{F}$
6	5	1	Octahedral	Square pyramidal	<90°	$:\ddot{F}-Br(-\ddot{F})(-\ddot{F})(-\ddot{F})\ddot{F}:$
6	4	2	Octahedral	Square planar	90°	$:\ddot{F}-Xe(-\ddot{F})(-\ddot{F})\ddot{F}:$

*Count only electron groups around the central atom. Each of the following is considered one electron group: a lone pair, a single bond, a double bond, a triple bond, or a single electron.

PROCEDURE FOR... Predicting Molecular Geometries	**EXAMPLE 10.2** **Predicting Molecular Geometries** Predict the geometry and bond angles of PCl_3.	**EXAMPLE 10.3** **Predicting Molecular Geometries** Predict the geometry and bond angles of ICl_4^-.
1. Draw the Lewis structure for the molecule.	PCl_3 has 26 valence electrons. :C̈l: │ :C̈l—P—C̈l:	ICl_4^- has 36 valence electrons. ⎡ :C̈l: ⎤⁻ │ │ :C̈l—I—C̈l: │ :C̈l:
2. Determine the total number of electron groups around the central atom. Lone pairs, single bonds, double bonds, triple bonds, and single electrons each count as one group.	The central atom (P) has four electron groups.	The central atom (I) has six electron groups.
3. Determine the number of bonding groups and the number of lone pairs around the central atom. These should sum to your result from step 2. Bonding groups include single bonds, double bonds, and triple bonds.	:C̈l: │ :C̈l—P—C̈l: Lone pair Three of the four electron groups around P are bonding groups and one is a lone pair.	Lone pairs ⎡ :C̈l: ⎤ :C̈l—I—C̈l: :C̈l: Four of the six electron groups around I are bonding groups and two are lone pairs.
4. Use Table 10.1 to determine the electron geometry and molecular geometry. If no lone pairs are present around the central atom, the bond angles will be that of the ideal geometry. If lone pairs are present, the bond angles may be smaller than the ideal geometry.	The electron geometry is tetrahedral (four electron groups) and the molecular geometry—the shape of the molecule—is *trigonal pyramidal* (three bonding groups and one lone pair). Because of the presence of a lone pair, the bond angles are less than 109.5°. 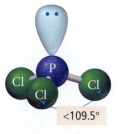 Trigonal pyramidal	The electron geometry is octahedral (six electron groups) and the molecular geometry—the shape of the molecule—is *square planar* (four bonding groups and two lone pairs). Even though lone pairs are present, the bond angles are 90° because the lone pairs are symmetrically arranged and do not compress the I—Cl bond angles. 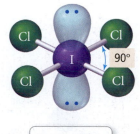 Square planar
	FOR PRACTICE 10.2 Predict the molecular geometry and bond angle of ClNO.	**FOR PRACTICE 10.3** Predict the molecular geometry of I_3^-.

Representing Molecular Geometries on Paper

Because molecular geometries are three-dimensional, they are often difficult to represent on two-dimensional paper. Many chemists use the following notation for bonds to indicate three-dimensional structures on two-dimensional paper.

Straight line Bond in plane of paper	*Hatched wedge* Bond going into the page	*Solid wedge* Bond coming out of the page

Some examples of the molecular geometries used in this book are shown here using this notation.

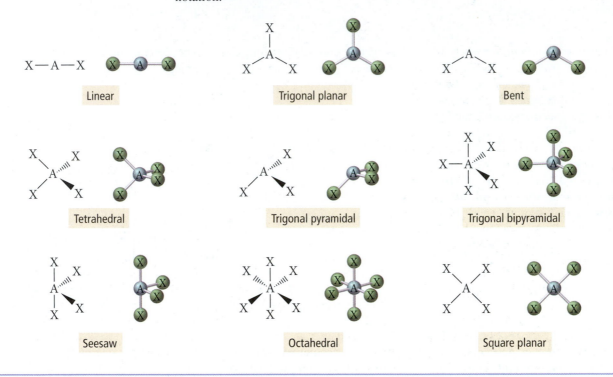

Linear   Trigonal planar   Bent

Tetrahedral   Trigonal pyramidal   Trigonal bipyramidal

Seesaw   Octahedral   Square planar

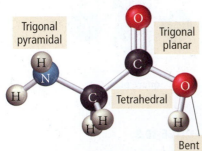

Four interior atoms

Glycine

Trigonal pyramidal   Trigonal planar   Tetrahedral   Bent

Predicting the Shapes of Larger Molecules

Larger molecules may have two or more *interior* atoms. When predicting the shapes of these molecules, we apply the principles we just covered to each interior atom. For example, glycine, an amino acid found in many proteins such as those involved in taste, contains four interior atoms: one nitrogen atom, two carbon atoms, and an oxygen atom. To determine the shape of glycine, we determine the geometry about each interior atom as follows:

Atom	Number of Electron Groups	Number of Lone Pairs	Molecular Geometry
Nitrogen	4	1	Trigonal pyramidal
Leftmost carbon	4	0	Tetrahedral
Rightmost carbon	3	0	Trigonal planar
Oxygen	4	2	Bent

Using the geometries of each atom, we can determine the entire three-dimensional shape of the molecule as shown at left.

EXAMPLE 10.4 **Predicting the Shape of Larger Molecules**

Predict the geometry of each interior atom in methanol (CH_3OH) and make a sketch of the molecule.

SOLUTION

Begin by drawing the Lewis structure of CH_3OH. CH_3OH contains two interior atoms: one carbon atom and one oxygen atom. To determine the shape of methanol, determine the geometry of each interior atom.

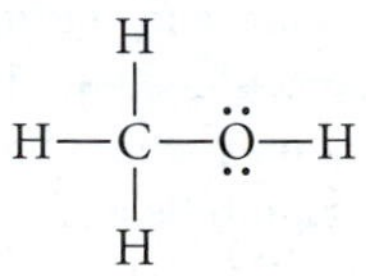

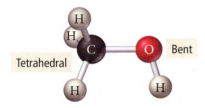

Atom	Number of Electron Groups	Number of Lone Pairs	Molecular Geometry
Carbon	4	0	Tetrahedral
Oxygen	4	2	Bent

Using the geometries of each of these, draw a three-dimensional sketch of the molecule as shown here.

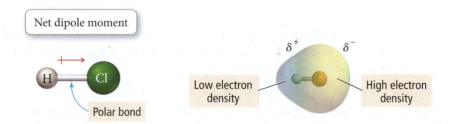

FOR PRACTICE 10.4

$$\overset{\displaystyle O}{\overset{\displaystyle \|}{H_3C-C-OH}}$$

Predict the geometry of each interior atom in acetic acid ($H_3C-C-OH$) and make a sketch of the molecule.

10.5 Molecular Shape and Polarity

In Chapter 9, we discussed polar bonds. Entire molecules can also be polar, depending on their shape and the nature of their bonds. For example, if a diatomic molecule has a polar bond, the molecule as a whole will be polar.

The figure at right above is an electron density model of HCl. Yellow indicates moderately high electron density, red indicates very high electron density, and blue indicates low electron density. Notice that the electron density is greater around the more electronegative atom (chlorine). Thus the molecule itself is polar. If the bond in a diatomic molecule is nonpolar, the molecule as a whole will be nonpolar.

In polyatomic molecules, the presence of polar bonds may or may not result in a polar molecule, depending on the molecular geometry. If the molecular geometry is such that the dipole moments of individual polar bonds sum together to a net dipole moment, then the molecule will be polar. However, if the molecular geometry is such that the dipole moments of the individual polar bonds cancel each other (that is, sum to zero), then the molecule will be nonpolar. It all depends on the geometry of the molecule. For example, consider carbon dioxide:

$$:\ddot{O}=C=\ddot{O}:$$

Each $C=O$ bond in CO_2 is polar because oxygen and carbon have significantly different electronegativities (3.5 and 2.5, respectively). But since CO_2 is a linear molecule, the polar bonds directly oppose one another and the dipole moment of one bond exactly opposes the dipole moment of the other—the two dipole moments sum to zero and the *molecule* is nonpolar. Dipole moments can cancel each other because they are *vector quantities*; they have both a magnitude and a direction. Think of each polar bond as a vector, pointing in the direction of the more electronegative atom. The length of the vector is proportional to the electronegativity difference between the bonds. In CO_2, we have two identical vectors pointing in exactly opposite directions—the vectors sum to zero, much as $+1$ and -1 sum to zero:

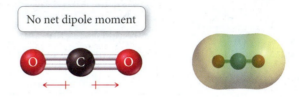

Notice that the electron density model shows regions of moderately high electron density (yellow) positioned symmetrically on either end of the molecule with a region of low electron density (blue) located in the middle.

In contrast, consider water:

$$H-\ddot{O}-H$$

The $O-H$ bonds in water are also polar; oxygen and hydrogen have electronegativities of 3.5 and 2.1, respectively. However, the water molecule is not linear but bent, so the two dipole moments do not sum to zero. If we imagine each bond as a vector pointing toward oxygen (the more electronegative atom) we see that, because of the angle between the vectors, they do not cancel, but sum to an overall vector or a net dipole moment.

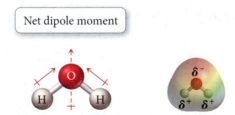

The electron density model shows a region of very high electron density at the oxygen end of the molecule. Consequently, water is a polar molecule. Table 10.2 summarizes common geometries and molecular polarity.

Summarizing How to Determine if a Molecule is Polar:

▶ *Draw a Lewis structure for the molecule and determine its molecular geometry.*

▶ *Determine if the molecule contains polar bonds.* A bond is polar if the two bonding atoms have sufficiently different electronegativities (see Figure 9.9). If the molecule contains polar bonds, superimpose a vector, pointing toward the more electronegative atom, on each bond. Make the length of the vector proportional to the electronegativity difference between the bonding atoms.

▶ *Determine if the polar bonds add together to form a net dipole moment.* Sum the vectors corresponding to the polar bonds. If the vectors sum to zero, the molecule is nonpolar. If the vectors sum to a net vector, the molecule is polar.

TABLE 10.2 Common Cases of Adding Dipole Moments to Determine whether a Molecule Is Polar

Nonpolar

The dipole moments of two identical polar bonds pointing in opposite directions will cancel. The molecule is nonpolar.

Polar

The dipole moments of two polar bonds with an angle of less than 180° between them will not cancel. The resultant dipole moment vector is shown in red. The molecule is polar.

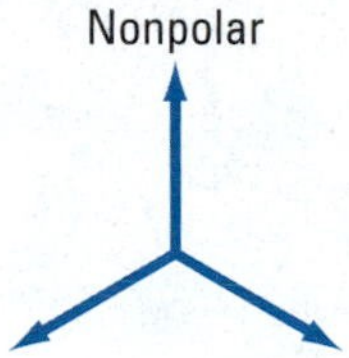

Nonpolar

The dipole moments of three identical polar bonds at 120° from each other will cancel. The molecule is nonpolar.

Nonpolar

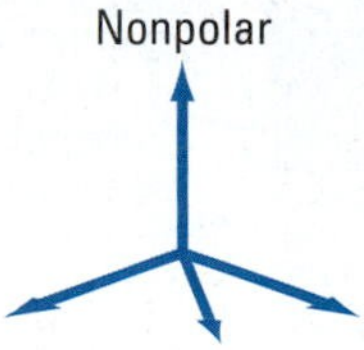

The dipole moments of four identical polar bonds in a tetrahedral arrangement (109.5° from each other) will cancel. The molecule is nonpolar.

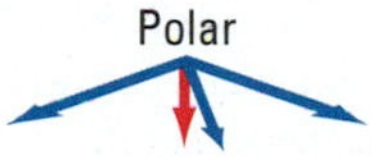

Polar

The dipole moments of three polar bonds in a trigonal pyramidal arrangement (109.5° from each other) will not cancel. The resultant dipole moment vector is shown in red. The molecule is polar.

Note: In all cases where the dipoles of two or more polar bonds cancel, the bonds are assumed to be identical. If one or more of the bonds are different from the other(s), the dipoles will not cancel and the molecule will be polar.

EXAMPLE 10.5 Determining Molecule Polarity

Determine if NH_3 is polar.

SOLUTION

Draw the Lewis structure for the molecule and determine its molecular geometry.	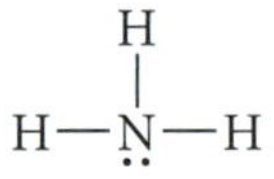 The Lewis structure has three bonding groups and one lone pair about the central atom. Therefore the molecular geometry is trigonal pyramidal.
Determine whether the molecule contains polar bonds. Sketch the molecule and superimpose a vector for each polar bond. The relative length of each vector should be proportional to the electronegativity difference between the atoms forming each bond. The vector should point in the direction of the more electronegative atom.	The electronegativities of nitrogen and hydrogen are 3.0 and 2.1, respectively. Therefore the bonds are polar. 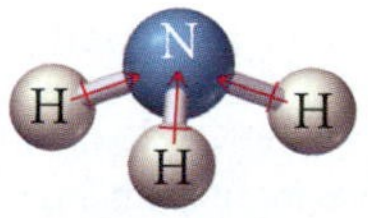
Determine whether the polar bonds add together to form a net dipole moment. Examine the symmetry of the vectors (representing dipole moments) and determine whether they cancel each other or sum to a net dipole moment.	The three dipole moments sum to a net dipole moment. The molecule is polar. 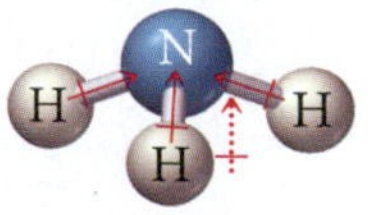

FOR PRACTICE 10.5

Determine if CF_4 is polar.

Opposite partial charges on molecules
attract one another.

▲ **FIGURE 10.5 Interaction of Polar Molecules** The north pole of one magnet attracts the south pole of another magnet. In an analogous way (although the forces involved are different), the positively charged end of one molecule attracts the negatively charged end of another. As a result of this electrical attraction, polar molecules interact strongly with one another.

Polar and nonpolar molecules have different properties. Water and oil do not mix, for example, because water molecules are polar and the molecules that compose oil are generally nonpolar. Polar molecules interact strongly with other polar molecules because the positive end of one molecule is attracted to the negative end of another, just as the south pole of a magnet is attracted to the north pole of another magnet (**Figure 10.5◄**). A mixture of polar and nonpolar molecules is similar to a mixture of small magnetic particles and nonmagnetic ones. The magnetic particles (which are like polar molecules) clump together, excluding the nonmagnetic particles (which are like nonpolar molecules) and separating into distinct regions.

Oil is nonpolar.

Water is polar.

▲ Oil and water do not mix because water molecules are polar and the molecules that compose oil are nonpolar.

▲ A mixture of polar and nonpolar molecules is analogous to a mixture of magnetic marbles (opaque) and nonmagnetic marbles (transparent). As with the magnetic marbles, mutual attraction causes polar molecules to clump together, excluding the nonpolar molecules.

10.6 Valence Bond Theory: Orbital Overlap as a Chemical Bond

In the Lewis model, we use "dots" to represent electrons in bonding atoms. We know from quantum-mechanical theory, however, that such a treatment is oversimplified. More advanced bonding theories treat electrons in a quantum-mechanical manner. These more advanced theories are actually extensions of quantum mechanics applied to molecules. Although a detailed quantitative treatment of these theories is beyond the scope of this book, we introduce them in a *qualitative* manner in the sections that follow. Keep in mind, however, that modern *quantitative* approaches to chemical bonding using these theories accurately predict many of the properties of molecules—such as bond lengths, bond strengths, molecular geometries, and dipole moments—that we have been discussing in this book.

The simpler of the two more advanced bonding theories is called **valence bond theory**. According to valence bond theory, electrons reside in quantum-mechanical orbitals localized on individual atoms. In many cases, these orbitals are the standard *s*, *p*, *d*, and *f* atomic orbitals that we learned about in Chapter 7. In other cases, these orbitals are *hybridized atomic orbitals* that are a kind of blend or combination of two or more standard atomic orbitals.

When two atoms approach each other, the electrons and nucleus of one atom interact with the electrons and nucleus of the other atom. In valence bond theory, we calculate the effect of these interactions on the energies of the electrons in the atomic orbitals. If the energy of the system is lowered because of the interactions, then a chemical bond forms. If the energy of the system is raised by the interactions, then a chemical bond does not form.

The interaction energy is usually calculated as a function of the internuclear distance between the two bonding atoms. For example, **Figure 10.6▶** shows the calculated interaction energy between two hydrogen atoms as a function of the distance between them. The *y*-axis of the graph is the potential energy of the interaction between the electron and nucleus of one hydrogen atom and the electron and nucleus of the other. The *x*-axis is the

Valence bond theory is an application of a more general quantum-mechanical approximation method called *perturbation theory*. In perturbation theory, a more complex system (such as a molecule) is viewed as a simpler system (such as two atoms) that is slightly altered or perturbed by some additional force or interaction (such as the interaction between the two atoms).

Interaction Energy of Two Hydrogen Atoms

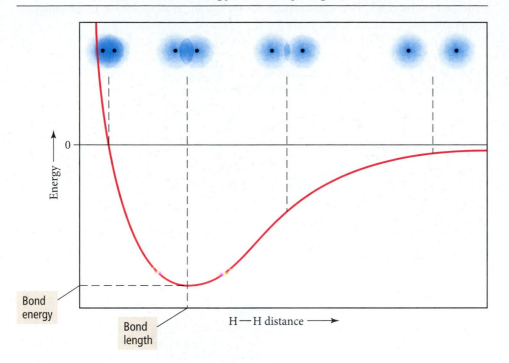

separation (or internuclear distance) between the two atoms. As you can see from the graph, when the atoms are far apart (right side of the graph), the interaction energy is nearly zero because the two atoms do not interact to any significant extent. As the atoms get closer, the interaction energy becomes negative. This is a net stabilization that attracts one hydrogen atom to the other. If the atoms get too close, however, the interaction energy begins to rise, primarily because of the mutual repulsion of the two positively charged nuclei. The most stable point on the curve occurs at the minimum of the interaction energy—this is the equilibrium bond length. At this distance, the two atomic $1s$ orbitals have a significant amount of overlap and the electrons spend time in the internuclear region where they can interact with both nuclei. The value of the interaction energy at the equilibrium bond distance is the bond energy.

When we apply valence bond theory to a number of atoms and their corresponding molecules, we can make the following general observation: *the interaction energy is usually negative (or stabilizing) when the interacting atomic orbitals contain a total of two electrons that can spin-pair.* Most commonly, the two electrons come from the two half-filled orbitals, but in some cases, the two electrons can come from one filled orbital overlapping with a completely empty orbital (this is called a coordinate covalent bond). In other words, when two atoms with half-filled orbitals approach each other, the half-filled orbitals *overlap*—parts of the orbitals occupy the same space—and the electrons occupying them align with opposite spins. This results in a net energy stabilization that constitutes a covalent chemical bond. The resulting geometry of the molecule emerges from the geometry of the overlapping orbitals.

When *completely filled* orbitals overlap, the interaction energy is positive (or destabilizing) and no bond forms.

Summarizing Valence Bond Theory:

▶ The valence electrons of the atoms in a molecule reside in quantum-mechanical atomic orbitals. The orbitals can be the standard s, p, d, and f orbitals or they may be hybrid combinations of these.

▶ A chemical bond results from the overlap of two half-filled orbitals with spin-pairing of the two valence electrons (or less commonly the overlap of a completely filled orbital with an empty orbital).

▶ The shape of the molecule is determined by the geometry of the overlapping orbitals.

Let's apply the general concepts of valence bond theory to explain bonding in hydrogen sulfide, H_2S. The valence electron configurations of the atoms in the molecule are as follows:

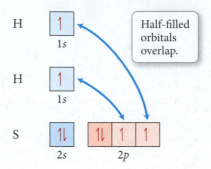

The hydrogen atoms each have one half-filled orbital, and the sulfur atom has two half-filled orbitals. The half-filled orbitals on each hydrogen atom overlap with the two half-filled orbitals on the sulfur atom, forming two chemical bonds:

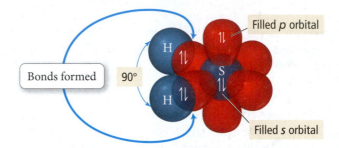

To show the spin-pairing of the electrons in the overlapping orbitals, we superimpose a half-arrow for each electron in each half-filled orbital and show that, within a bond, the electrons are spin-paired (one half-arrow pointing up and the other pointing down). We also superimpose paired half-arrows in the filled sulfur *s* and *p* orbitals to represent the lone pair electrons in those orbitals. (Since those orbitals are full, they are not involved in bonding.)

A quantitative calculation of H_2S using valence bond theory yields bond energies, bond lengths, and bond angles. In our more qualitative treatment, we simply show how orbital overlap leads to bonding and make a rough sketch of the molecule based on the overlapping orbitals. Notice that, because the overlapping orbitals on the central atom (sulfur) are *p* orbitals, and because *p* orbitals are oriented at 90° to one another, the predicted bond angle is 90°. The actual bond angle in H_2S is 92°. In the case of H_2S, a simple valence bond treatment matches well with the experimentally measured bond angle (in contrast to VSEPR theory, which predicts a bond angle of less than 109.5°).

◯◯ Conceptual Connection 10.4 What Is a Chemical Bond?

How you answer the question, *what is a chemical bond*, depends on the bonding model. Answer the following questions:

(a) What is a covalent chemical bond according to the Lewis model?

(b) What is a covalent chemical bond according to valence bond theory?

(c) Why are the answers different?

10.7   Valence Bond Theory: Hybridization of Atomic Orbitals

Although the overlap of half-filled *standard* atomic orbitals adequately explains the bonding in H_2S, it cannot adequately explain the bonding in many other molecules. For example, suppose we try to explain the bonding between hydrogen and carbon using the same approach. The valence electron configurations of H and C are as follows:

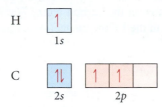

H $\quad$ $1s$

C $\quad$ $2s$ $\quad$ $2p$

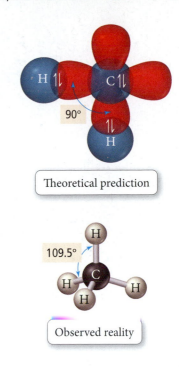

Theoretical prediction

Observed reality

Carbon has only two half-filled orbitals and should therefore form only two bonds with two hydrogen atoms. We would therefore predict that carbon and hydrogen should form a molecule with the formula CH_2 and with a bond angle of 90° (corresponding to the angle between any two p orbitals).

However, from experiment, we know that the stable compound formed from carbon and hydrogen is CH_4 (methane), with bond angles of 109.5°. The experimental reality is different from our prediction in two ways. The first is that carbon forms bonds to four hydrogen atoms, not two. The second is that the bond angles are much larger than the angle between two p orbitals. Valence bond theory accounts for the bonding in CH_4 and many other polyatomic molecules by incorporating a concept called *orbital hybridization*.

So far, we have assumed that the overlapping orbitals that form chemical bonds are simply the standard s, p, or d atomic orbitals. Valence bond theory treats the electrons in a molecule as if they occupied these standard atomic orbitals, but this is a major oversimplification. The concept of hybridization in valence bond theory is essentially a step toward recognizing that *the orbitals in a molecule are not necessarily the same as the orbitals in an atom*. **Hybridization** is a mathematical procedure in which the standard atomic orbitals are combined to form new atomic orbitals called **hybrid orbitals** that correspond more closely to the actual distribution of electrons in chemically bonded atoms. Hybrid orbitals are still localized on individual atoms, but they have different shapes and energies from those of standard atomic orbitals.

Why do we hypothesize that electrons in some molecules occupy hybrid orbitals? In valence bond theory, a chemical bond is the overlap of two orbitals that together contain two electrons. The greater the overlap, the stronger the bond and the lower the energy. In hybrid orbitals, the electron probability density is more concentrated in a single directional lobe, allowing greater overlap with the orbitals of other atoms. In other words, hybrid orbitals *minimize* the energy of the molecule by *maximizing* the orbital overlap in a bond.

Hybridization, however, is not a free lunch—in most cases it actually costs some energy. So hybridization occurs only to the degree that the energy payback through bond formation is large. In general, therefore, the more bonds that an atom forms, the greater the tendency of its orbitals to hybridize. Central or interior atoms, which form the most bonds, have the greatest tendency to hybridize. Terminal atoms, which form the fewest bonds, have the least tendency to hybridize. *In this book, we will focus on the hybridization of interior atoms and assume that all terminal atoms—those bonding to only one other atom—are unhybridized.* Hybridization is particularly important in carbon, which tends to form four bonds in its compounds and therefore always hybridizes.

Although we cannot show the procedure for obtaining hybrid orbitals in mathematical detail, we can make the following general statements regarding hybridization:

- The *number of standard atomic orbitals* added together always equals the *number of hybrid orbitals* formed. The total number of orbitals is conserved.
- The *particular combinations* of standard atomic orbitals added together determine the *shapes and energies* of the hybrid orbitals formed.
- The *type of hybridization that occurs* is the one that yields the *lowest overall energy for the molecule*. Since actual energy calculations are beyond the scope of this book, we use electron geometries as determined by VSEPR theory to predict the type of hybridization.

In Section 10.8, we examine *molecular orbital theory*, which treats electrons in a molecule as occupying orbitals that belong to the molecule as a whole.

As we saw in Chapter 9, the word *hybrid* comes from breeding. A *hybrid* is an offspring of two animals or plants of different standard races or breeds. Similarly, a hybrid orbital is a product of mixing two or more standard atomic orbitals.

In a more detailed treatment, hybridization is not an all-or-nothing process—it can occur to varying degrees that are not always easy to predict. We saw earlier, for example, that sulfur does not hybridize very much in forming H_2S.

sp^3 Hybridization

We can account for the tetrahedral geometry in CH_4 by the hybridization of the one $2s$ orbital and the three $2p$ orbitals on the carbon atom. The four new orbitals that result, called sp^3 hybrids, are shown in the following energy diagram.

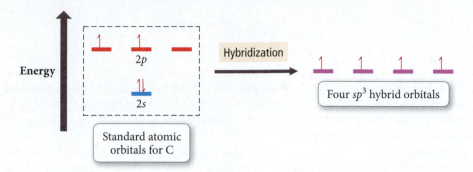

The notation "sp^3" indicates that the hybrid orbitals are mixtures of one s orbital and three p orbitals. Notice that the hybrid orbitals all have the same energy—they are degenerate. The shapes of the sp^3 hybrid orbitals are shown in **Figure 10.7▼**. Note also

Formation of sp^3 Hybrid Orbitals

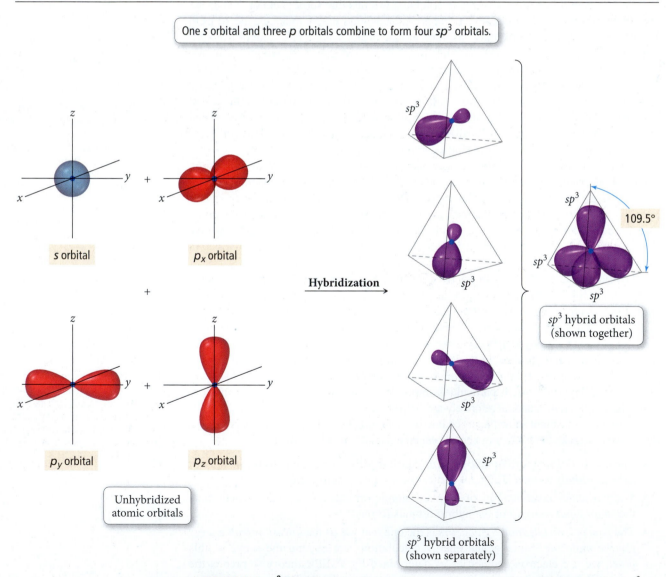

▲ **FIGURE 10.7 sp^3 Hybridization**   One s orbital and three p orbitals combine to form four sp^3 hybrid orbitals.

that the four hybrid orbitals are arranged in a tetrahedral geometry with 109.5° angles between them.

The orbital diagram for carbon using these hybrid orbitals is:

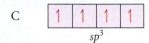

Carbon's four valence electrons occupy these orbitals singly with parallel spins as dictated by Hund's rule. With this electron configuration, carbon has four half-filled orbitals and can form four bonds with four hydrogen atoms as follows:

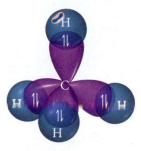

The geometry of the *overlapping orbitals* (the hybrids) is tetrahedral, with angles of 109.5° between the orbitals, so the *resulting geometry of the molecule* is tetrahedral, with 109.5° bond angles, in agreement with the experimentally measured geometry of and with the predicted VSEPR geometry.

Hybrid orbitals are good for forming chemical bonds because they tend to maximize overlap with other orbitals. However, if the central atom of a molecule contains lone pairs, hybrid orbitals can also accommodate them. For example, the nitrogen orbitals in ammonia are sp^3 hybrids. Three of the hybrids are involved in bonding with three hydrogen atoms, but the fourth hybrid contains a lone pair. The presence of the lone pair, however, does lower the tendency of nitrogen's orbitals to hybridize. Therefore the bond angle in NH_3 is 107°, a bit closer to the unhybridized p orbital bond angle of 90°.

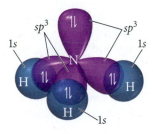

sp^2 Hybridization and Double Bonds

Hybridization of one s and two p orbitals results in three sp^2 hybrids and one leftover unhybridized p orbital.

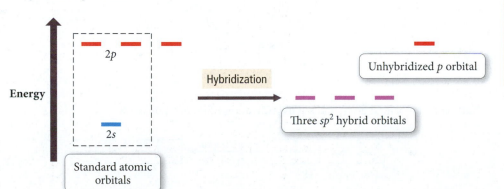

In valence bond theory, the particular hybridization scheme to follow (sp^2 versus sp^3 for example) for a given molecule is determined computationally, which is beyond the scope of this book. Here we will determine the particular hybridization scheme from the VSEPR geometry of the molecule, as shown later in this section.

Formation of *sp²* Hybrid Orbitals

One *s* orbital and two *p* orbitals combine to form three *sp²* orbitals.

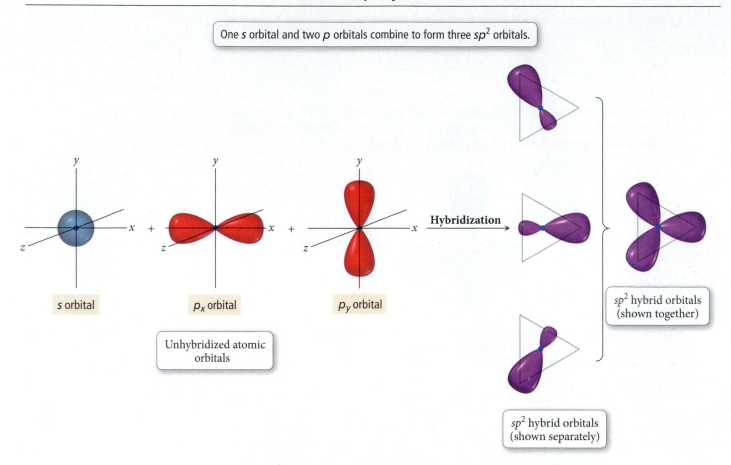

s orbital

p_x orbital

p_y orbital

Unhybridized atomic orbitals

Hybridization

sp² hybrid orbitals (shown together)

sp² hybrid orbitals (shown separately)

▲ **FIGURE 10.8 *sp²* Hybridization**   One *s* orbital and two *p* orbitals combine to form three *sp²* hybrid orbitals. One *p* orbital remains unhybridized.

The notation "*sp²*" indicates that the hybrids are mixtures of one *s* orbital and two *p* orbitals. The shapes of the *sp²* hybrid orbitals are shown in **Figure 10.8▲**. Notice that the three hybrid orbitals are arranged in a trigonal planar geometry with 120° angles between them. The unhybridized *p* orbital is oriented perpendicular to the three hybridized orbitals.

Consider H_2CO, an example of a molecule with *sp²* hybrid orbitals. The unhybridized valence electron configurations of each of the atoms are:

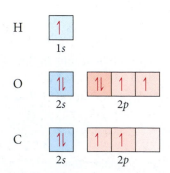

Carbon is the central atom and the hybridization of its orbitals is *sp²*:

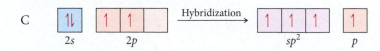

Each of the sp^2 orbitals is half-filled. The remaining electron occupies the leftover p orbital, even though it is slightly higher in energy. We can now see that the carbon atom has four half-filled orbitals and can therefore form four bonds: two with two hydrogen atoms and two (a double bond) with the oxygen atom. We draw the molecule and the overlapping orbitals as follows:

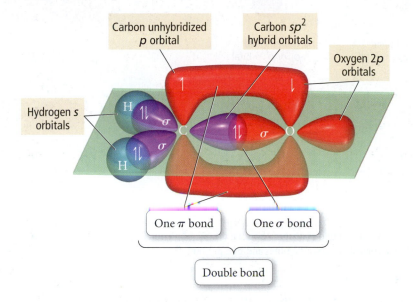

Notice the overlap between the half-filled p orbitals on the carbon and oxygen atoms. When p orbitals overlap this way (side by side) the resulting bond is a **pi (π) bond**, and the electron density is above and below the internuclear axis. When orbitals overlap end to end, as in all of the rest of the bonds in the molecule, the resulting bond is a **sigma (σ) bond (Figure 10.9▶)**. We can therefore label all the bonds in the molecule using a notation that specifies the type of bond (σ or π) as well as the type of overlapping orbitals. We have included this notation, as well as the Lewis structure of H_2CO for comparison, in the following bonding diagram for H_2CO:

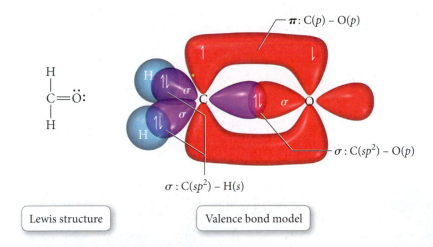

Notice the correspondence between valence bond theory and the Lewis model. In both cases, the central carbon atom is forming four bonds: two single bonds and one double bond. However, valence bond theory gives us more insight into the bonds. The double bond between carbon and oxygen according to valence bond theory consists of two different *kinds* of bonds—one σ and one π—while in the Lewis model the two bonds within the double bond appear identical. *Double bonds in the Lewis model always correspond to one σ and one π bond in valence bond theory.* In this sense, valence bond theory gives us more insight into the nature of a double bond than the Lewis model.

One—and only one—σ bond forms between any two atoms. Additional bonds must be π bonds.

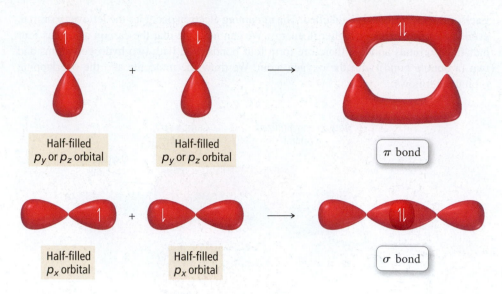

▲ **FIGURE 10.9 Sigma and Pi Bonding** When orbitals overlap end to end, they form a sigma (σ) bond. When orbitals overlap side by side, the result is a pi (π) bond. Two atoms can form only one sigma bond. A single bond is a sigma bond; a double bond consists of a sigma bond and a pi bond; a triple bond consists of a sigma bond and two pi bonds.

Valence bond theory allows us to see that rotation about a double bond is severely restricted. Because of the side-by-side overlap of the p orbitals, the π bond must essentially break for rotation to occur. Valence bond theory also shows us the types of orbitals involved in the bonding and their shapes. In H_2CO, the sp^2 hybrid orbitals on the central atom are trigonal planar with 120° angles between them, so the resulting predicted geometry of the molecule is trigonal planar with 120° bond angles. The experimentally measured bond angles in H_2CO, as discussed previously, are 121.9° for the HCO bond and 116.2° for the HCH bond angle, close to the predicted values.

Although rotation about a double bond is highly restricted, rotation about a single bond is relatively unrestricted. Consider, for example, the structures of two chlorinated hydrocarbons, 1,2-dichloroethane and 1,2-dichloroethene.

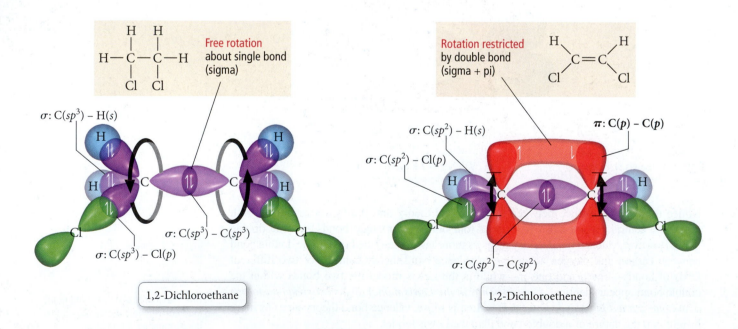

The hybridization of the carbon atoms in 1,2-dichloroethane is sp^3, resulting in relatively free rotation about the sigma single bond. Consequently, there is no difference between the following two structures at room temperature because they quickly interconvert:

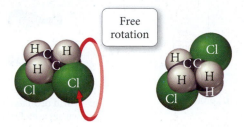

In contrast, rotation about the double bond (sigma + pi) in 1,2-dichloroethene is restricted, so that, at room temperature, 1,2-dichloroethene can exist in two forms:

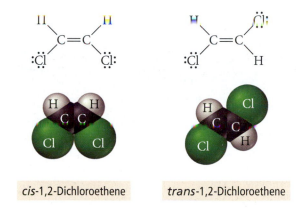

| *cis*-1,2-Dichloroethene | *trans*-1,2-Dichloroethene |

These two forms of 1,2-dichloroethene are indeed different compounds with different properties. We distinguish between them with the designations *cis* (meaning "same side") and *trans* (meaning "opposite sides"). Compounds such as these, with the same molecular formula but different structures or different spatial arrangement of atoms, are called *isomers*. In other words, nature can—and does—make different compounds out of the same atoms by arranging the atoms in different ways. Isomerism is common throughout chemistry and especially important in organic chemistry, as we shall see in Chapter 20.

Conceptual Connection 10.5   Single and Double Bonds

In Section 9.10 we learned that double bonds were stronger and shorter than single bonds. For example, a C—C single bond has an average bond energy of 347 kJ/mole while a C=C double bond has an average bond energy of 611 kJ/mole. Use valence bond theory to explain why a double bond is *not* simply twice as strong as a single bond.

sp Hybridization and Triple Bonds

Hybridization of one *s* and one *p* orbital results in two *sp* hybrid orbitals and two leftover unhybridized *p* orbitals.

The shapes of the *sp* hybrid orbitals are shown in **Figure 10.10▶**. Notice that the two *sp* hybrid orbitals are arranged in a linear geometry with a 180° angle between them. The unhybridized *p* orbitals are oriented in the plane that is perpendicular to the hybridized *sp* orbitals.

Formation of *sp* Hybrid Orbitals

One *s* orbital and one *p* orbital combine to form two *sp* orbitals.

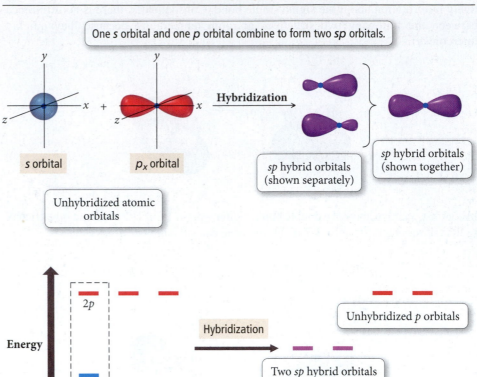

▶ **FIGURE 10.10** *sp* **Hybridization**
One *s* orbital and one *p* orbital combine to form two *sp* hybrid orbitals. Two *p* orbitals (not shown) remain unhybridized.

Acetylene, HC≡CH, is a molecule with *sp* hybrid orbitals. The valence electron configurations (showing hybridization) of the atoms are as follows:

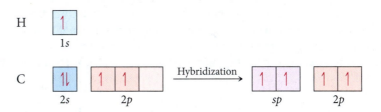

The two interior carbon atoms have *sp* hybridized orbitals, leaving two unhybridized 2*p* orbitals on each carbon atom. Each carbon atom then has four half-filled orbitals and can form four bonds: one with a hydrogen atom and three (a triple bond) with the other carbon atom. We draw the molecule and the overlapping orbitals as follows:

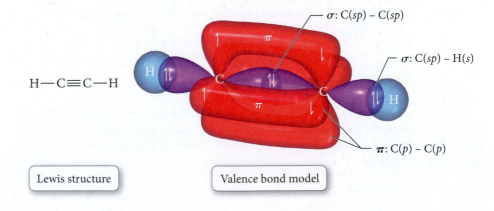

Notice that the triple bond between the two carbon atoms consists of two π bonds (overlapping p orbitals) and one σ bond (overlapping sp orbitals). The sp orbitals on the carbon atoms are linear with $180°$ between them, so the resulting geometry of the molecule is linear with $180°$ bond angles, in agreement with the experimentally measured geometry of HC$\equiv$CH, and also in agreement with the prediction of VSEPR theory.

sp^3d and sp^3d^2 Hybridization

We know from the Lewis model that elements occurring in the third row of the periodic table (or below) can exhibit expanded octets. The equivalent concept in valence bond theory is hybridization involving the d orbitals. For third-period elements, the $3d$ orbitals become involved in hybridization because their energies are close to the energies of the $3s$ and $3p$ orbitals. The hybridization of one s orbital, three p orbitals, and one d orbital results in sp^3d hybrid orbitals, as shown in **Figure 10.11(a)▼**. The five sp^3d hybrid orbitals have a trigonal bipyramidal arrangement, as in **Figure 10.11(b)▼**. As an example of sp^3d hybridization, consider arsenic pentafluoride, AsF_5. The arsenic atom bonds to five fluorine atoms by overlap between the sp^3d hybrid orbitals on arsenic and p orbitals on the fluorine atoms, as shown here:

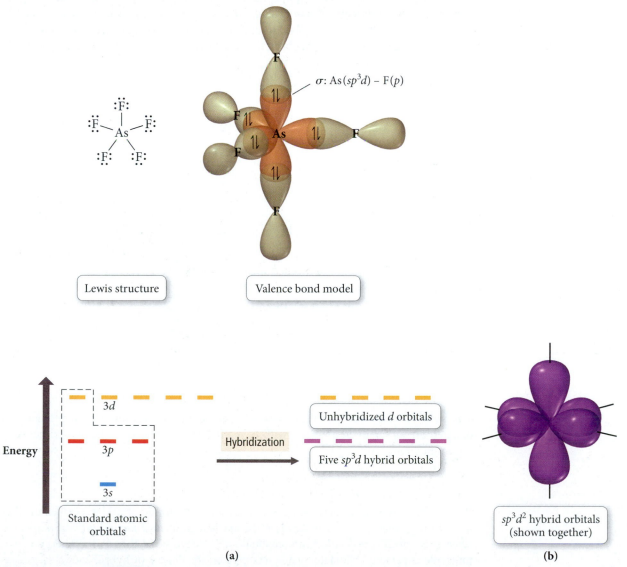

Lewis structure

Valence bond model

σ: As(sp^3d) – F(p)

Energy

Standard atomic orbitals

Hybridization

Unhybridized d orbitals

Five sp^3d hybrid orbitals

(a)

sp^3d^2 hybrid orbitals (shown together)

(b)

▲ **FIGURE 10.11** sp^3d **Hybridization** One s orbital, three p orbitals, and one d orbital combine to form five sp^3d hybrid orbitals.

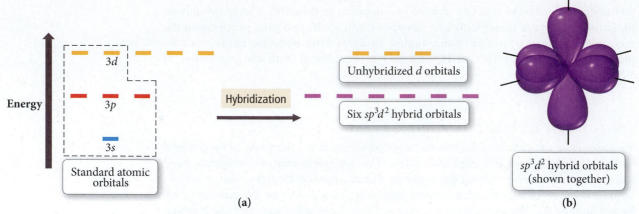

(a)   (b)

▲ **FIGURE 10.12** sp^3d^2 **Hybridization**   One s orbital, three p orbitals, and two d orbitals combine to form six sp^3d^2 hybrid orbitals.

The sp^3d orbitals on the arsenic atom are trigonal bipyramidal, so the molecular geometry is trigonal bipyramidal.

The hybridization of one s orbital, three p orbitals, and *two d orbitals* results in sp^3d^2 hybrid orbitals, as shown in **Figure 10.12(a)▲**. The six sp^3d^2 hybrid orbitals have an octahedral geometry, as in **Figure 10.12(b)▲**. In sulfur hexafluoride, SF_6, the sulfur atom bonds to six fluorine atoms by overlap between the sp^3d^2 hybrid orbitals on sulfur and p orbitals on the fluorine atoms, as shown here:

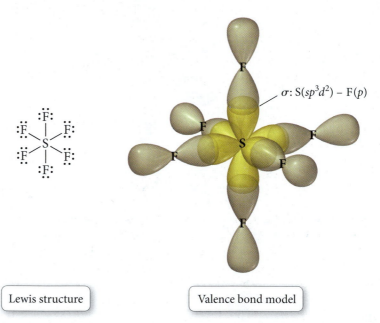

Lewis structure   Valence bond model

The sp^3d^2 orbitals on the sulfur atom are octahedral, so the molecular geometry is octahedral, again in agreement with VSEPR theory and with the experimentally observed geometry.

Writing Hybridization and Bonding Schemes

We have now seen examples of the five main types of atomic orbital hybridization. *But how do we know which hybridization scheme best describes the orbitals of a specific atom in a specific molecule?* In computational valence bond theory, the energy of the molecule is actually calculated using a computer; the degree of hybridization as well as

TABLE 10.3 Hybridization Scheme from Electron Geometry

Number of Electron Groups	Electron Geometry (from VSEPR Theory)	Hybridization Scheme	
2	Linear	sp	
3	Trigonal planar	sp^2	120°
4	Tetrahedral	sp^2	109.5
5	Trigonal bipyramidal	sp^3d	90° 120°
6	Octahedral	sp^3d^2	90° 90°

the type of hybridization are varied to find the combination that gives the molecule the lowest overall energy. For our purposes, we will assign a hybridization scheme from the electron geometry—determined using VSEPR theory—of the central atom (or interior atoms) of the molecule. The five VSEPR electron geometries and the corresponding hybridization schemes are shown in Table 10.3. For example, if the electron geometry of the central atom is tetrahedral, then the hybridization is sp^3; if the electron geometry is octahedral, then the hybridization is sp^3d^2, and so on.

We are now ready to put the Lewis model and valence bond theory together to describe bonding in molecules. In the procedure and examples that follow, you will learn how to write a *hybridization and bonding scheme* for a molecule. This scheme involves drawing a Lewis structure for the molecule, determining its geometry using VSEPR theory, determining the correct hybridization of the interior atoms, drawing the molecule with its overlapping orbitals, and labeling each bond with the σ and π notation followed by the type of overlapping orbitals. As you can see, this procedure involves virtually everything you have learned about bonding in this chapter and the previous one. The procedure for writing a hybridization and bonding scheme is in the left column on the following page with two examples of how to apply the procedure in the columns to the right of it.

PROCEDURE FOR... Hybridization and Bonding Scheme Procedure	EXAMPLE 10.6 Hybridization and Bonding Scheme Write a hybridization and bonding scheme for bromine trifluoride, BrF$_3$.	EXAMPLE 10.7 Hybridization and Bonding Scheme Write a hybridization and bonding scheme for acetaldehyde, H$_3$C—C(=O)—H
1. Write the Lewis structure for the molecule.	SOLUTION BrF$_3$ has 28 valence electrons and the following Lewis structure:	SOLUTION Acetaldehyde has 18 valence electrons and the following Lewis structure:
2. Use VSEPR theory to predict the electron geometry about the central atom (or interior atoms).	The bromine atom has five electron groups and therefore has a trigonal bipyramidal electron geometry.	The leftmost carbon atom has four electron groups and a tetrahedral electron geometry. The rightmost carbon atom has three electron groups and trigonal planar geometry.
3. Refer to Table 10.3 to select the correct hybridization for the central atom (or interior atoms) based on the electron geometry.	A trigonal bipyramidal electron geometry corresponds to sp^3d hybridization.	The leftmost carbon atom is sp^3 hybridized, and the rightmost carbon atom is sp^2 hybridized.
4. Sketch the molecule, beginning with the central atom and its orbitals. Show overlap with the appropriate orbitals on the terminal atoms.	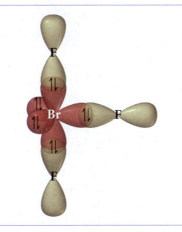	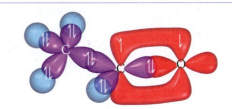
5. Label all bonds using the σ and π notation followed by the type of overlapping orbitals.	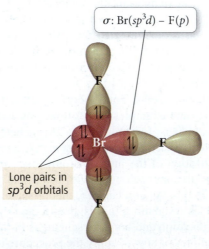 σ: Br(sp^3d) – F(p) Lone pairs in sp^3d orbitals	σ: C(sp^3) – H(s) π: C(p) – O(p) σ: C(sp^2) – H(s) σ: C(sp^3) – C(sp^2) σ: C(sp^2) – O(p)

FOR PRACTICE 10.6
Write a hybridization and bonding scheme for XeF$_4$.

FOR PRACTICE 10.7
Write a hybridization and bonding scheme for HCN.

EXAMPLE 10.8 **Hybridization and Bonding Scheme**

Use valence bond theory to write a hybridization and bonding scheme for ethene, $H_2C{=}CH_2$.

SOLUTION

1. Write the Lewis structure for the molecule.	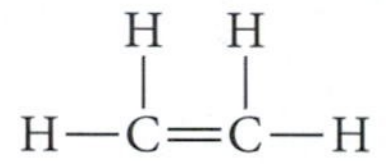 <div align="center">H H | | H—C=C—H</div>
2. Use VSEPR theory to predict the electron geometry about the central atom (or interior atoms).	The molecule has two interior atoms. Since each atom has three electron groups (one double bond and two single bonds), the electron geometry about each atom is trigonal planar.
3. Use Table 10.3 to select the correct hybridization for the central atom (or interior atoms) based on the electron geometry.	A trigonal planar geometry corresponds to sp^2 hybridization.
4. Sketch the molecule, beginning with the central atom and its orbitals. Show overlap with the appropriate orbitals on the terminal atoms.	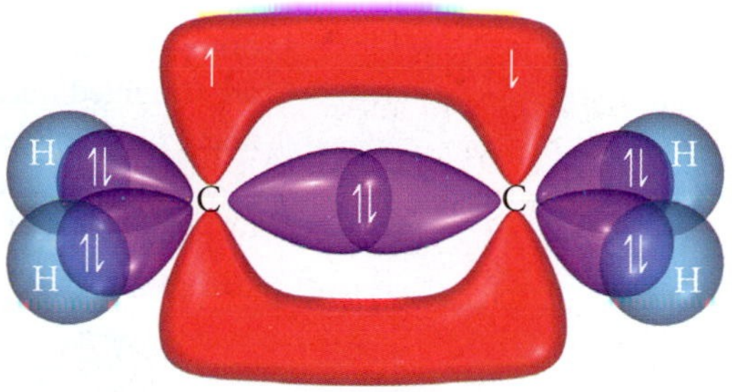
5. Label all bonds using the σ and π notation followed by the type of overlapping orbitals.	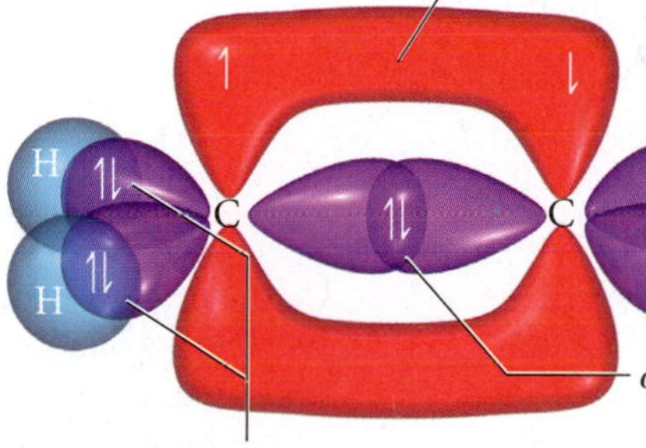

FOR PRACTICE 10.8

Use valence bond theory to write a hybridization and bonding scheme for CO_2.

FOR MORE PRACTICE 10.8

What is the hybridization of the central iodine atom in I_3^-?

10.8   Molecular Orbital Theory: Electron Delocalization

Valence bond theory can explain a number of aspects of chemical bonding—such as the rigidity of a double bond—but it also has limitations. Recall that in valence bond theory, we treat electrons as if they reside in the quantum-mechanical orbitals that we calculated *for atoms*. This is a significant oversimplification that we partially compensate for by hybridization. Nevertheless, we can do even better.

Recall from Chapter 7 that the mathematical derivation of energies and orbitals for electrons *in atoms* comes from solving the Schrödinger equation for the atom of interest. For a molecule, you can theoretically do the same thing. The resulting orbitals would be the actual *molecular* orbitals of the molecule as a whole (in contrast to valence bond theory, in which the orbitals are those of individual atoms). As it turns out, however, solving the Schrödinger equation exactly for even the simplest molecules is impossible without making some approximations.

Molecular orbital theory is a specific application of a more general quantum-mechanical approximation technique called the *variational method*. In this method, the energy of a trial function within the Schrödinger equation is minimized.

In **molecular orbital (MO) theory**, we do not actually solve the Schrödinger equation for a molecule directly. Instead, we use a trial function, an "educated guess" as to what the solution might be. In other words, instead of mathematically solving the Schrödinger equation, which would give us a mathematical function describing an orbital, we start with a trial mathematical function for the orbital. We then test the trial function to see how well it works.

Linear Combination of Atomic Orbitals (LCAO)

When molecular orbitals are computed mathematically, it is actually the *wave functions* corresponding to the orbitals that are combined.

The simplest trial functions that work reasonably well in molecular orbital theory turn out to be linear combinations of atomic orbitals (LCAOs). An LCAO molecular orbital is a *weighted linear sum—analogous to a weighted average—of the valence atomic orbitals* of the atoms in the molecule. At first glance, this concept might seem very similar to that of hybridization in valence bond theory. However, in valence bond theory, hybrid orbitals are weighted linear sums of the valence atomic orbitals of a *particular atom*, and the hybrid orbitals remain *localized* on that atom. In molecular orbital theory, the molecular orbitals are weighted linear sums of the valence atomic orbitals of *all the atoms* in a molecule, and many of the molecular orbitals are *delocalized* over the entire molecule.

Consider the H_2 molecule. One of the molecular orbitals for H_2 is an equally weighted sum of the $1s$ orbital from one atom and the $1s$ orbital from the other. We can represent this pictorially and energetically as follows:

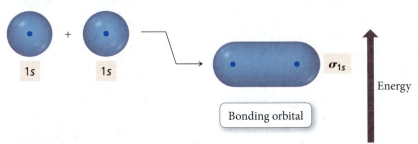

The name of this molecular orbital is σ_{1s}. The σ comes from the shape of the orbital, which looks like a σ bond in valence bond theory, and the $1s$ comes from its formation by a linear sum of $1s$ orbitals. The σ_{1s} orbital is lower in energy than either of the two $1s$ atomic orbitals from which it was formed. For this reason, this orbital is called a **bonding orbital**. When electrons occupy bonding molecular orbitals, the energy of the electrons is lower than it would be if they were occupying atomic orbitals.

You can think of a molecular orbital for a molecule in much the same way that you think about an atomic orbital in an atom. Electrons will seek the lowest energy molecular orbital available, but just as an atom has more than one atomic orbital (and some may be empty), so a molecule has more than one molecular orbital (and some may be empty). The next molecular orbital of H_2 is approximated by summing the $1s$ orbital on one hydrogen atom with the *negative* (opposite phase) of the $1s$ orbital on the other hydrogen atom. The different phases of the orbitals result in *destructive* interference between them. The resulting molecular orbital therefore has a node between the two atoms. The different colors (red and blue) on either side of the node represent the different phases of the orbital (see Section 7.6).

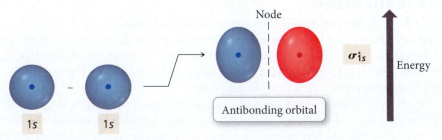

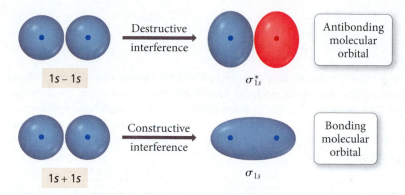

▲ **FIGURE 10.13 Formation of Bonding and Antibonding Orbitals**
Constructive interference between two atomic orbitals gives rise to a molecular orbital that is lower in energy than the atomic orbitals. This is the bonding orbital. Destructive interference between two atomic orbitals gives rise to a molecular orbital that is higher in energy than the atomic orbitals. This is the antibonding orbital.

The name of this molecular orbital is σ_{1s}^*. The star indicates that this orbital is an **antibonding orbital**. Electrons in antibonding orbitals have higher energies than they did in their respective atomic orbitals and therefore tend to raise the energy of the system (relative to the unbonded atoms).

In general, when two atomic orbitals are added together to form molecular orbitals, one of the resultant molecular orbitals is lower in energy (the bonding orbital) than the atomic orbitals and the other is higher in energy (the antibonding orbital). Remember that electrons in orbitals behave like waves. The bonding molecular orbital arises out of constructive interference between the atomic orbitals because the two orbitals have the same phase. The antibonding orbital arises out of destructive interference between the atomic orbitals because *subtracting* one from the other means the two interacting orbitals have opposite phases (**Figure 10.13▲**).

For this reason, the bonding orbital has an *increased* electron density in the internuclear region while the antibonding orbital has a *node* (zero electron density) in the internuclear region. The increased electron density in the internuclear region lowers the energy of bonding orbitals compared to the orbitals in nonbonded atoms. The lower electron density of antibonding orbitals in the internuclear region raises their energy compared to the orbitals in nonbonded atoms.

We put all of this together in the molecular orbital energy diagram for H_2 as follows:

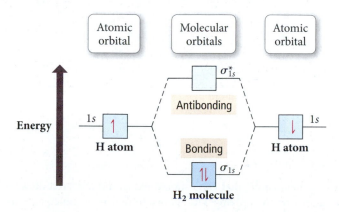

The molecular orbital diagram shows that two hydrogen atoms can lower their overall energy by forming H_2 because the electrons can move from higher energy atomic orbitals into the lower energy σ_{1s} bonding molecular orbital. In molecular orbital theory, we define the **bond order** of a diatomic molecule such as H_2 as follows:

$$\text{Bond order} = \frac{(\text{number of electrons in bonding MOs}) - (\text{number of electrons in antibonding MOs})}{2}$$

For H_2, the bond order is

$$H_2 \text{ bond order} = \frac{2-0}{2} = 1$$

A positive bond order means that there are more electrons in bonding molecular orbitals than in antibonding molecular orbitals. The electrons will therefore have lower energy than they did in the orbitals of the isolated atoms and a chemical bond will form. In general, the higher bond order, the stronger the bond. A negative or zero bond order indicates that a bond will *not* form between the atoms. For example, consider the MO diagram for He_2:

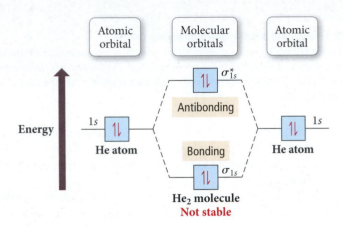

Notice that the two additional electrons must go into the higher energy antibonding orbital. There is no net stabilization by joining two helium atoms to form a helium molecule, as indicated by the bond order:

$$He_2 \text{ bond order} = \frac{2-2}{2} = 0$$

So according to MO theory, He_2 should not exist as a stable molecule, and indeed it does not. An interesting case is the helium–helium ion, He_2^+, with the following MO diagram:

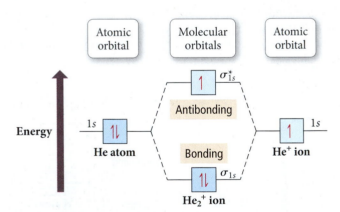

The bond order is $\frac{1}{2}$, indicating that He_2^+ should exist, and indeed it does.

Before we move on to applying MO theory to diatomic molecules from the second row of the periodic table, let's summarize the main ideas in MO theory as follows.

Summarizing LCAO–MO Theory

▶ Molecular orbitals (MOs) can be approximated by a linear combination of atomic orbitals (AOs). The total number of MOs formed from a particular set of AOs will always equal the number of AOs in the set.

▶ When two AOs combine to form two MOs, one MO will be lower in energy (the bonding MO) and the other will be higher energy (the antibonding MO).

▶ When assigning the electrons of a molecule to MOs, fill the lowest energy MOs first with a maximum of two spin-paired electrons per orbital.

▶ When assigning electrons to two MOs of the same energy, follow Hund's rule—fill the orbitals singly first, with parallel spins, before pairing.

▶ The bond order in a diatomic molecule is the number of electrons in bonding MOs minus the number in antibonding MOs divided by two. Stable bonds require a positive bond order (more electrons in bonding MOs than in antibonding MOs).

Notice the power of the molecular orbital approach. Every electron that enters a bonding molecular orbital stabilizes the molecule or polyatomic ion and every electron that enters an antibonding molecular orbital destabilizes it. The emphasis on electron pairs has been removed. One electron in a bonding molecular orbital stabilizes half as much as two, so a bond order of one-half is nothing mysterious.

EXAMPLE 10.9 Bond Order

Use molecular orbital theory to predict the bond order in H_2^-. Is the H_2^- bond stronger or weaker than the H_2 bond?

SOLUTION

The H_2^- ion has three electrons. Assign the three electrons to the molecular orbitals, filling lower energy orbitals first and proceeding to higher energy orbitals.

Calculate the bond order by subtracting the number of electrons in antibonding orbitals from the number in bonding orbitals and dividing the result by two.

$$H_2^- \text{ bond order} = \frac{2-1}{2} = +\tfrac{1}{2}$$

Since the bond order is positive, H_2^- should be stable. However, the bond order of H_2^- is lower than the bond order of H_2 (which is 1); therefore, the bond in H_2^- is weaker than in H_2.

FOR PRACTICE 10.9

Use molecular orbital theory to predict the bond order in H_2^+. Is the H_2^+ bond a stronger or weaker bond than the H_2 bond?

Period Two Homonuclear Diatomic Molecules

Homonuclear diatomic molecules (molecules made up of two atoms of the same kind) formed from second-period elements have between 2 and 16 valence electrons. To explain bonding in these molecules, we must consider the next set of higher energy molecular orbitals, which can be approximated by linear combinations of the valence atomic orbitals of the period 2 elements.

The core electrons can be ignored, as they are in other models for bonding, because these electrons do not contribute significantly to chemical bonding.

We begin with Li_2. Even though lithium is normally a metal, we can use MO theory to predict whether or not the Li_2 molecule should exist in the gas phase. The molecular orbitals in Li_2 are approximated as linear combinations of the $2s$ atomic orbitals and look much like the H_2 molecular orbitals formed from linear combinations of the $1s$ orbitals. The MO diagram for Li_2 is therefore as follows:

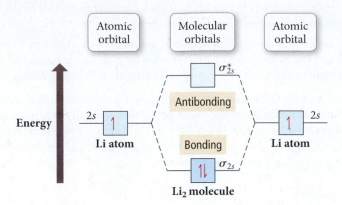

The two valence electrons of Li_2 occupy a bonding molecular orbital. We would predict that the Li_2 molecule is stable with a bond order of 1. Experiments confirm this prediction. In contrast, consider the MO diagram for Be_2:

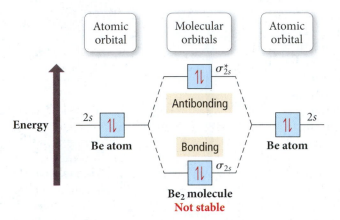

The four valence electrons of Be_2 occupy one bonding MO and one antibonding MO. The bond order is 0 and we predict that Be_2 should not be stable. This prediction is also consistent with experimental findings.

The next homonuclear molecule composed of second-row elements is B_2, which has six total valence electrons to accommodate. We can approximate the next higher energy molecular orbitals for B_2 and the rest of the period 2 diatomic molecules as linear combinations of the $2p$ orbitals taken pairwise. Since the three $2p$ orbitals orient along three orthogonal axes, we must assign similar axes to the molecule. In this book, we assign the internuclear axis to be the x direction. Then the LCAO–MOs that result from combining the $2p_x$ orbitals—the ones that lie along the internuclear axis—from each atom are represented pictorially:

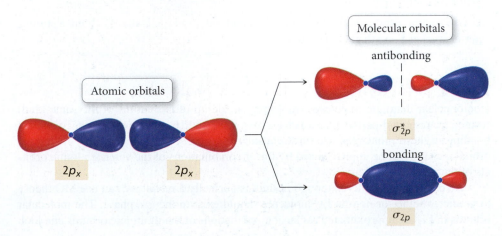

The bonding MO in this pair looks something like a candy in a wrapper, with increased electron density in the internuclear region due to constructive interference between the two $2p$ atomic orbitals. It has the characteristic σ shape (it is cylindrically symmetrical about the bond axis) and is therefore called the σ_{2p} bonding orbital. The antibonding orbital, called σ_{2p}^*, has a node between the two nuclei (due to destructive interference between the two $2p$ orbitals) and is higher in energy than either of the $2p_x$ orbitals.

The LCAO–MOs that result from combining the $2p_y$ orbitals from each atom are represented pictorially as:

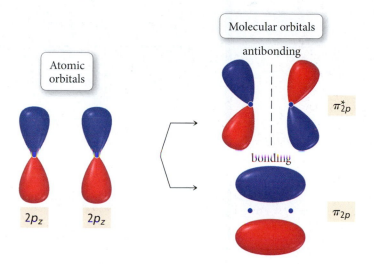

Notice that in this case the p orbitals are added together in a side-by-side orientation (in contrast to the $2p_x$ orbitals which were oriented end to end). The resultant molecular orbitals consequently have a different shape. The electron density in the bonding molecular orbital is above and below the internuclear axis with a nodal plane that includes the internuclear axis. This orbital resembles the electron density distribution of a π bond in valence bond theory. We call this orbital the π_{2p} orbital. The corresponding antibonding orbital has an additional node *between* the nuclei (perpendicular to the internuclear axis) and is called the π_{2p}^* orbital.

The LCAO–MOs that result from combining the $2p_z$ orbitals from each atom are represented pictorially as:

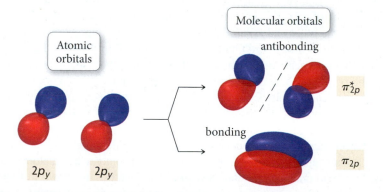

The only difference between the $2p_z$ and the $2p_y$ atomic orbitals is a 90° rotation about the internuclear axis. Consequently, the only difference between the resulting MOs is a 90° rotation about the internuclear axis. The energies and the names of the bonding and antibonding MOs obtained from the combination of the $2p_z$ AOs are identical to those obtained from the combination of the $2p_y$ AOs.

Before we can draw MO diagrams for B_2 and the other second-period diatomic molecules, we need to determine the relative energy ordering of the MOs obtained from the $2p$ AO combinations. This is not a simple issue. The relative ordering of MOs obtained from LCAO–MO theory is usually determined computationally. There is no single order that will work for all molecules. For second-period diatomic molecules, calculations reveal that the energy ordering for B_2, C_2, and N_2 is slightly different than that for O_2, F_2, and Ne_2 as follows:

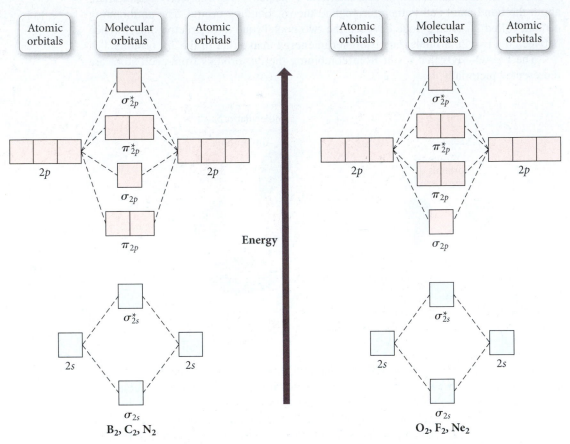

▲ Molecular orbital energy diagrams for second-period diatomic molecules show that the energy ordering of the π_{2p} and σ_{2p} molecular orbitals can vary.

The reason for the difference in energy ordering can only be explained by going back to our LCAO–MO model. In our simplified treatment, we assumed that the MOs that result from the second-period AOs could be calculated pairwise. In other words, we took the linear combination of a $2s$ from one atom with the $2s$ from another, a $2p_x$ from one atom with a $2p_x$ from the other and so on. However, in a more comprehensive treatment, the MOs are formed from linear combinations that include all of the AOs that are relatively close to each other in energy and of the correct symmetry. Specifically, in a more detailed treatment, the two $2s$ orbitals and the two $2p_x$ orbitals should all be combined to form a total of four molecular orbitals. The extent to which you include this type of mixing affects the energy levels of the corresponding MOs, as shown in **Figure 10.14▶**. The bottom line is that s–p mixing is significant in B_2, C_2, and N_2 but not in O_2, F_2, and Ne_2. The result is different energy ordering, depending on the specific molecule.

The MO energy diagrams for all of the second-period homonuclear diatomic molecules, as well as their bond orders, bond energies, and bond lengths, are shown in **Figure 10.15▶**. Notice that as bond order increases, the bond gets stronger (greater bond energy) and shorter (smaller bond length). For B_2, with six electrons, the bond order is 1. For C_2, the bond order is 2, and for N_2, the bond order reaches a maximum with a value of 3. Recall that the Lewis structure of N_2 has a triple bond, so both the Lewis model and molecular orbital theory predict a strong bond for N_2, which is confirmed by experimental observation.

In O_2, the two additional electrons occupy antibonding orbitals and the bond order is 2. Notice in the figure that these two electrons are unpaired—they occupy the π_{2p}^* orbitals *singly with parallel spins*, as indicated by Hund's rule. The presence of unpaired electrons in the molecular orbital diagram of oxygen is significant because oxygen is known from experiment to be **paramagnetic**, which means that it is attracted to a magnetic field. The paramagnetism of oxygen can be demonstrated by suspending liquid oxygen between

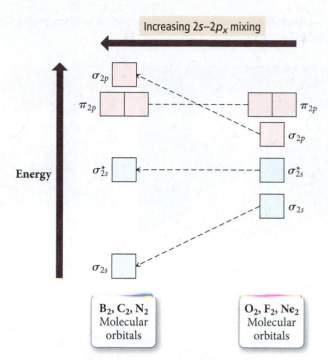

Increasing $2s$–$2p_x$ mixing

Energy

σ_{2p}
π_{2p}
σ_{2p}
σ^*_{2s} σ^*_{2s}
σ_{2s}
σ_{2s}

B$_2$, C$_2$, N$_2$
Molecular
orbitals

O$_2$, F$_2$, Ne$_2$
Molecular
orbitals

▲ **FIGURE 10.14   The Effects of $2s$–$2p$ Mixing**   The degree of mixing between two orbitals decreases with increasing energy difference between them. Mixing of the $2s$ and $2p_x$ orbitals is therefore greater in B$_2$, C$_2$, and N$_2$ than in O$_2$, F$_2$, and Ne$_2$ because in B, C, and N the energy levels of the atomic orbitals are more closely spaced than in O, F, and Ne. This mixing produces a change in energy ordering for the π_{2p} and σ_{2p} molecular orbitals.

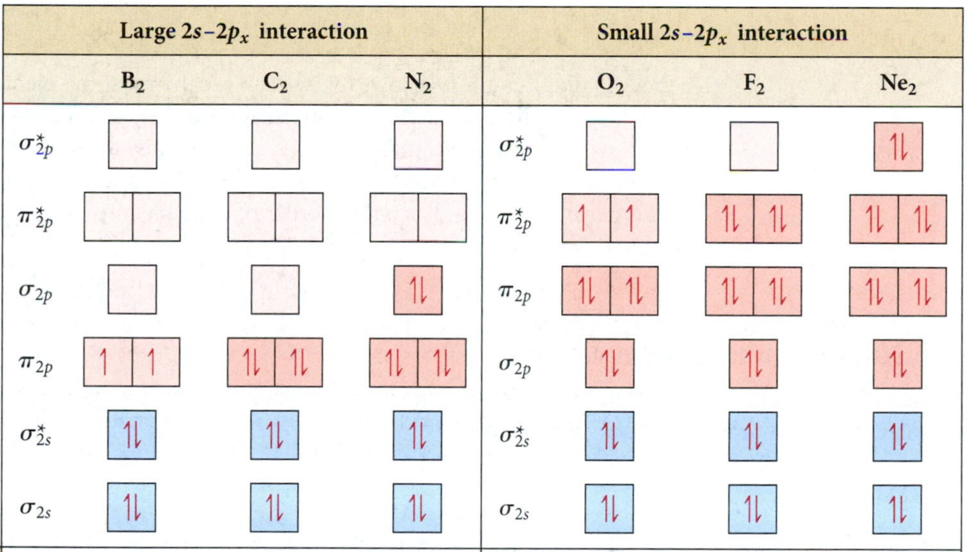

	Large $2s$–$2p_x$ interaction				Small $2s$–$2p_x$ interaction		
	B$_2$	**C$_2$**	**N$_2$**		**O$_2$**	**F$_2$**	**Ne$_2$**
σ^*_{2p}				σ^*_{2p}			⇅
π^*_{2p}				π^*_{2p}	↑ ↑	⇅ ⇅	⇅ ⇅
σ_{2p}			⇅	π_{2p}	⇅ ⇅	⇅ ⇅	⇅ ⇅
π_{2p}	↑ ↑	⇅ ⇅	⇅ ⇅	σ_{2p}	⇅	⇅	⇅
σ^*_{2s}	⇅	⇅	⇅	σ^*_{2s}	⇅	⇅	⇅
σ_{2s}	⇅	⇅	⇅	σ_{2s}	⇅	⇅	⇅
Bond order	1	2	3		2	1	0
Bond energy (kJ/mol)	290	620	946		498	159	—
Bond length (pm)	159	131	110		121	143	—

▲ **FIGURE 10.15   Molecular Orbital Energy Diagrams for Second-Row Homonuclear Diatomic Molecules**

the poles of a magnet. This magnetic property is the direct result of *unpaired electrons*, whose spin and movement around the nucleus (more accurately known as orbital angular momentum) generate tiny magnetic fields. When a paramagnetic substance is placed in an external magnetic field, the magnetic fields of each atom align with the external field, creating the attraction (much as two magnets attract each other when properly oriented). In contrast, when the electrons in an atom are all *paired*, the magnetic fields caused by electron spin and orbital angular momentum tend to cancel each other, resulting in diamagnetism. A **diamagnetic** substance is not attracted to a magnetic field (and is, in fact, slightly repelled).

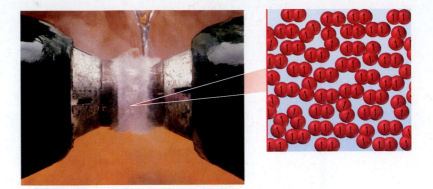

▶ Liquid oxygen can be suspended between the poles of a magnet because it is paramagnetic. It contains unpaired electrons that generate tiny magnetic fields, which align with and interact with the external field.

In the Lewis structure of O_2, as well as in the valence bond model of O_2, all of its electrons seem to be paired:

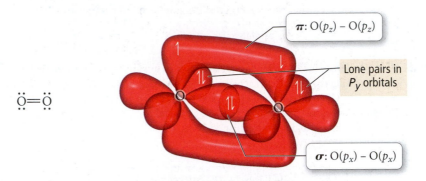

$\overset{..}{O}=\overset{..}{O}$

$\boldsymbol{\pi}: O(p_z) - O(p_z)$

Lone pairs in P_y orbitals

$\boldsymbol{\sigma}: O(p_x) - O(p_x)$

In the MO diagram for O_2, however, we can see the unpaired electrons. Molecular orbital theory is more powerful in that it can account for the paramagnetism of O_2—it gives us a picture of bonding that more closely corresponds to what we see in experiment. Continuing along the second-row homonuclear diatomic molecules we see that F_2 has a bond order of 1 and Ne_2 has a bond order of 0, again consistent with experiment since F_2 exists and Ne_2 does not.

EXAMPLE 10.10   Molecular Orbital Theory

Draw an MO energy diagram and determine the bond order for the N_2^- ion. Would you expect the bond to be stronger or weaker than in the N_2 molecule? Is N_2^- diamagnetic or paramagnetic?

SOLUTION

Write an energy level diagram for the molecular orbitals in N_2^-. Use the energy ordering for N_2.	σ_{2p}^*
	π_{2p}^*
	σ_{2p}
	π_{2p}
	σ_{2s}^*
	σ_{2s}

The N_2^- ion has 11 valence electrons (5 for each nitrogen atom plus 1 for the negative charge). Assign the electrons to the molecular orbitals beginning with the lowest energy orbitals and following Hund's rule.

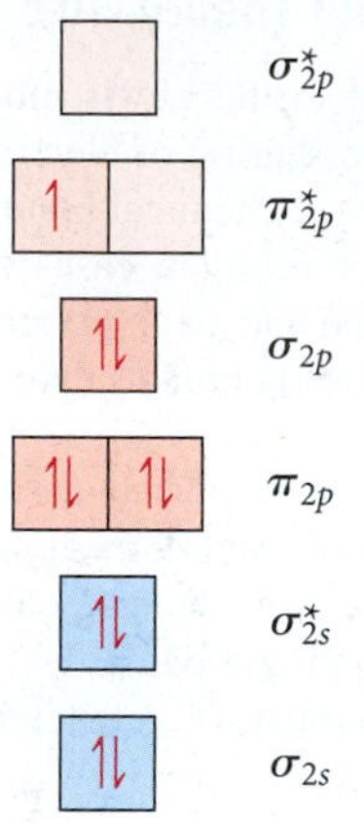

☐	σ_{2p}^*
↑ |	π_{2p}^*
↑↓	σ_{2p}
↑↓ | ↑↓	π_{2p}
↑↓	σ_{2s}^*
↑↓	σ_{2s}

Calculate the bond order by subtracting the number of electrons in antibonding orbitals from the number in bonding orbitals and dividing the result by two.

$$N_2^- \text{ bond order} = \frac{8 - 3}{2} = +2.5$$

The bond order is 2.5, which is a lower bond order than in the N_2 molecule (bond order $= 3$); therefore, the bond is weaker. The MO diagram shows that the N_2^- ion has one unpaired electron and is therefore paramagnetic.

FOR PRACTICE 10.10

Draw an MO energy diagram and determine the bond order for the N_2^+ ion. Do you expect the bond to be stronger or weaker than in the N_2 molecule? Is N_2^+ diamagnetic or paramagnetic?

FOR MORE PRACTICE 10.10

Use molecular orbital theory to determine the bond order of Ne_2.

CHAPTER IN REVIEW

Key Terms

Section 10.2
valence shell electron pair repulsion (VSEPR) theory (362)
electron groups (362)
linear geometry (362)
trigonal planar geometry (362)
tetrahedral geometry (363)
trigonal bipyramidal geometry (364)

octahedral geometry (364)

Section 10.3
electron geometry (366)
molecular geometry (366)
trigonal pyramidal geometry (366)
bent geometry (366)
seesaw geometry (367)
T-shaped geometry (368)

square pyramidal geometry (368)
square planar geometry (369)

Section 10.6
valence bond theory (376)

Section 10.7
hybridization (379)
hybrid orbitals (379)
pi (π) bond (383)
sigma (σ) bond (383)

Section 10.8
molecular orbital (MO) theory (392)
bonding orbital (392)
antibonding orbital (393)
bond order (393)
paramagnetic (398)
diamagnetic (400)

Key Concepts

Molecular Shape and VSEPR Theory (10.1–10.4)

▶ The properties of molecules are directly related to their shapes. In VSEPR theory, molecular shapes are determined by the repulsions between electron groups on the central atom. An electron group can be a single bond, double bond, triple bond, lone pair, or even a single electron.

▶ The five basic molecular shapes are linear (two electron groups), trigonal planar (three electron groups), tetrahedral (four electron groups), trigonal bipyramidal (five electron groups), and octahedral (six electron groups).

▶ When lone pairs are present on the central atom, the *electron geometry* is still one of the five basic shapes, but one or more positions are occupied by lone pairs. The *molecular geometry* is therefore different from the electron geometry. Lone pairs are positioned so as to minimize repulsions with other lone pairs and with bonding pairs.

Polarity (10.5)

▶ The polarity of a polyatomic molecule containing polar bonds depends on its geometry. If the dipole moments of the polar bonds are aligned in such a way that they cancel one another, the molecule will not be polar. If they are aligned in such a way as to sum, the molecule will be polar.

▶ Highly symmetric molecules tend to be nonpolar, while asymmetric molecules containing polar bonds tend to be polar. The polarity of a molecule dramatically affects its properties. For example, water (polar) and oil (nonpolar) do not mix.

Valence Bond Theory (10.6, 10.7)

▶ In contrast to the Lewis model, in which a covalent chemical bond is the sharing of electrons represented by dots, in valence bond theory a chemical bond is the overlap of half-filled atomic orbitals (or in some cases the overlap between a completely filled orbital and an empty one).

▶ The overlapping orbitals may be the standard atomic orbitals, such as $1s$ or $2p$ or they may be hybridized atomic orbitals, which are mathematical combinations of the standard orbitals on a single atom. The basic hybridized orbitals are sp, sp^2, sp^3, sp^3d, and sp^3d^2.

▶ The geometry of the molecule is determined by the geometry of the overlapping orbitals.

▶ In our treatment of valence bond theory, we use the molecular geometry determined by VSEPR theory to determine the correct hybridization scheme.

▶ In valence bond theory, we distinguish between two types of bonds, σ (sigma) and π (pi).

▶ In a σ bond, the orbital overlap occurs in the region that lies directly between the two bonding atoms.

▶ In a π bond, formed from the side-by-side overlap of p orbitals, the overlap occurs above and below the region that lies directly between the two bonding atoms.

▶ Rotation about a σ bond is relatively free, while rotation about a π bond is restricted.

Molecular Orbital Theory (10.8)

▶ The simplest molecular orbitals are linear combinations of atomic orbitals (LCAOs), weighted averages of the atomic orbitals of the different atoms in the molecule.

▶ When two atomic orbitals combine to form molecular orbitals, they will form one molecular orbital of lower energy (the bonding orbital) and one of higher energy (the antibonding orbitals).

▶ A set of molecular orbitals fills in much the same way as atomic orbitals.

▶ The stability of the molecule and the strength of the bond depend on the number of electrons in bonding orbitals compared to the number in antibonding orbitals.

Key Equations and Relationships

Bond Order of a Diatomic Molecule (10.8)

$$\text{Bond order} = \frac{(\text{number of electrons in bonding MOs}) - (\text{number of electrons in antibonding MOs})}{2}$$

Key Learning Objectives

Chapter Objectives	Assessment
Using VSEPR Theory to Predict the Basic Shapes of Molecules (10.2)	Example 10.1 For Practice 10.1 Exercises 1, 2
Predicting Molecular Geometries Using VSEPR Theory and the Effects of Lone Pairs (10.4)	Examples 10.2, 10.3 For Practice 10.2, 10.3 Exercises 5, 6
Predicting the Shapes of Larger Molecules (10.4)	Example 10.4 For Practice 10.4 Exercises 11, 12, 15, 16
Using Molecular Shape to Determine Polarity of a Molecule (10.5)	Example 10.5 For Practice 10.5 Exercises 19–22
Writing Hybridization and Bonding Schemes Using Valence Bond Theory (10.7)	Examples 10.6, 10.7, 10.8 For Practice 10.6, 10.7, 10.8 For More Practice 10.8 Exercises 31–36
Drawing Molecular Orbital Diagrams to Predict Bond Order and Magnetism of a Diatomic Molecule (10.8)	Examples 10.9, 10.10 For Practice 10.9, 10.10 For More Practice 10.10 Exercises 41, 42, 45–48

EXERCISES

Problems by Topic

VSEPR Theory and Molecular Geometry

1. A molecule with the formula AB_3 has a trigonal pyramidal geometry. How many electron groups are on the central atom (A)?

2. A molecule with the formula AB_3 has a trigonal planar geometry. How many electron groups are on the central atom?

3. For each molecular geometry, give the number of total electron groups, the number of bonding groups, and the number of lone pairs on the central atom.

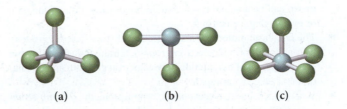

(a)   (b)   (c)

4. For each molecular geometry, give the number of total electron groups, the number of bonding groups, and the number of lone pairs on the central atom.

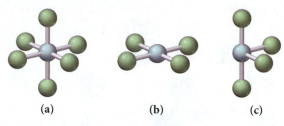

(a)   (b)   (c)

5. Determine the electron geometry, molecular geometry, and idealized bond angles for each molecule. In which cases do you expect deviations from the idealized bond angle?
   a. PF_3 **b.** SBr_2 **c.** $CHCl_3$ **d.** CS_2

6. Determine the electron geometry, molecular geometry, and idealized bond angles for each molecule. In which cases do you expect deviations from the idealized bond angle?
   a. CF_4 **b.** NF_3 **c.** OF_2 **d.** H_2S

7. Which species has the smaller bond angle, H_3O^+ or H_2O? Explain.

8. Which species has the smaller bond angle, ClO_4^- or ClO_3^-? Explain.

9. Determine the molecular geometry and make a sketch of each molecule or ion using the bond conventions shown in the "Representing Molecular Geometries on Paper" Box in Section 10.4.
   a. SF_4 **b.** ClF_3 **c.** IF_2^- **d.** IBr_4^-

10. Determine the molecular geometry and make a sketch of each molecule or ion using the bond conventions shown in the "Representing Molecular Geometries on Paper" Box in Section 10.4.
   a. BrF_5 **b.** SCl_6 **c.** PF_5 **d.** IF_4^+

11. Determine the molecular geometry of each interior atom and make a sketch of molecule.
   a. C_2H_2 (skeletal structure HCCH)
   b. C_2H_4 (skeletal structure H_2CCH_2)
   c. C_2H_6 (skeletal structure H_3CCH_3)

12. Determine the molecular geometry of each interior atom and make a sketch of each molecule.
   a. N_2
   b. N_2H_2 (skeletal structure HNNH)
   c. N_2H_4 (skeletal structure H_2NNH_2)

13. Each ball-and-stick model shows the electron and molecular geometry of a generic molecule. Explain what is wrong with each molecular geometry and provide the correct molecular geometry, given the number of lone pairs and bonding groups on the central atom.

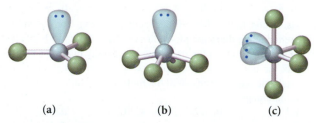

(a)   (b)   (c)

14. Each ball-and-stick model shows the electron and molecular geometry of a generic molecule. Explain what is wrong with each molecular geometry and provide the correct molecular geometry, given the number of lone pairs and bonding groups on the central atom.

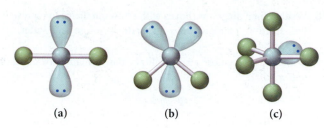

(a)   (b)   (c)

15. Determine the geometry about each interior atom in each molecule and sketch the molecule. (Skeletal structure is indicated in parentheses.)
   a. CH_3OH (H_3COH)   **b.** CH_3OCH_3 (H_3COCH_3)
   c. H_2O_2 (HOOH)

16. Determine the geometry about each interior atom in each molecule and sketch the molecule. (Skeletal structure is indicated in parentheses.)
   a. CH_3NH_2 (H_3CNH_2)
   b. $CH_3CO_2CH_3$ ($H_3CCOOCH_3$ both O atoms attached to second C)
   c. NH_2CO_2H (H_2NCOOH both O atoms attached to C)

Molecular Shape and Polarity

17. Explain why CO_2 and CCl_4 are both nonpolar even though they contain polar bonds.

18. CH_3F is a polar molecule, even though the tetrahedral geometry often leads to nonpolar molecules. Explain.

19. Determine whether each molecule in Problem 5 is polar or nonpolar.

20. Determine whether each molecule in Problem 6 is polar or nonpolar.

21. Determine whether each molecule is polar or nonpolar.
   a. SCl_2 **b.** SCl_4 **c.** $BrCl_5$

22. Determine whether each molecule is polar or nonpolar.
   a. $SiCl_4$ **b.** CF_2Cl_2 **c.** SeF_6 **d.** IF_5

Valence Bond Theory

23. The valence electron configurations of several atoms are shown below. How many bonds can each atom make without hybridization?
   a. Be $2s^2$ **b.** P $3s^23p^3$ **c.** F $2s^22p^5$

24. The valence electron configurations of several atoms are shown below. How many bonds can each atom make without hybridization?
   a. B $2s^22p^1$ **b.** N $2s^22p^3$ **c.** O $2s^22p^4$

25. Write orbital diagrams (boxes with arrows in them) to represent the electron configurations—without hybridization—for all the atoms in PH_3. Circle the electrons involved in bonding. Draw a three-dimensional sketch of the molecule and show orbital overlap. What bond angle do you expect from the unhybridized orbitals? Does valence bond theory agree with the experimentally measured bond angle of 93.3°?

26. Write orbital diagrams (boxes with arrows in them) to represent the electron configurations—without hybridization—for all the atoms in SF_2. Circle the electrons involved in bonding. Draw a three-dimensional sketch of the molecule and show orbital overlap. What bond angle do you expect from the unhybridized orbitals? Does valence bond theory agree with the experimentally measured bond angle of 98.2°?

27. Write orbital diagrams (boxes with arrows in them) to represent the electron configuration of carbon before and after sp^3 hybridization.

28. Write orbital diagrams (boxes with arrows in them) to represent the electron configurations of carbon before and after *sp* hybridization.

29. Which hybridization scheme allows the formation of at least one π bond?

$$sp^3, sp^2, sp^3d^2$$

30. Which hybridization scheme allows the central atom to form more than four bonds?

$$sp^3, sp^3d, sp^2$$

31. Write a hybridization and bonding scheme for each molecule. Sketch the molecule, including overlapping orbitals, and label all bonds using the notation shown in Examples 10.6 and 10.7.
    a. CCl_4 b. NH_3 c. OF_2 d. CO_2

32. Write a hybridization and bonding scheme for each molecule. Sketch the molecule, including overlapping orbitals, and label all bonds using the notation shown in Examples 10.6 and 10.7.
    a. CH_2Br_2 b. SO_2 c. NF_3 d. BF_3

33. Write a hybridization and bonding scheme for each molecule or ion. Sketch the structure, including overlapping orbitals, and label all bonds using the notation shown in Examples 10.6 and 10.7.
    a. $COCl_2$ (carbon is the central atom)   b. BrF_5
    c. XeF_2 d. I_3^-

34. Write a hybridization and bonding scheme for each molecule or ion. Sketch the structure, including overlapping orbitals, and label all bonds using the notation shown in Examples 10.6 and 10.7.
    a. SO_3^{2-} b. PF_6^- c. BrF_3 d. HCN

35. Write a hybridization and bonding scheme for each molecule containing more than one interior atom. Indicate the hybridization about each interior atom. Sketch the structure, including overlapping orbitals, and label all bonds using the notation shown in Examples 10.6 and 10.7.
    a. N_2H_2 (skeletal structure HNNH)
    b. N_2H_4 (skeletal structure H_2NNH_2)
    c. CH_3NH_2 (skeletal structure H_3CNH_2)

36. Write a hybridization and bonding scheme for each molecule containing more than one interior atom. Indicate the hybridization about each interior atom. Sketch the structure, including overlapping orbitals, and label all bonds using the notation shown in Examples 10.6 and 10.7.
    a. C_2H_2 (skeletal structure HCCH)
    b. C_2H_4 (skeletal structure H_2CCH_2)
    c. C_2H_6 (skeletal structure H_3CCH_3)

37. Consider the structure of the amino acid alanine. Indicate the hybridization about each interior atom.

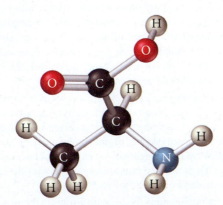

38. Consider the structure of the amino acid aspartic acid. Indicate the hybridization about each interior atom.

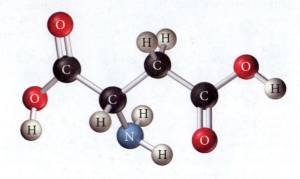

Molecular Orbital Theory

39. Sketch the bonding molecular orbital that results from the linear combination of two $1s$ orbitals. Indicate the region where interference occurs and state the kind of interference (constructive or destructive).

40. Sketch the antibonding molecular orbital that results from the linear combination of two $1s$ orbitals. Indicate the region where interference occurs and state the kind of interference (constructive or destructive).

41. Draw an MO energy diagram and predict the bond order of Be_2^+ and Be_2^-. Do you expect these molecules to exist in the gas phase?

42. Draw an MO energy diagram and predict the bond order of Li_2^+ and Li_2^-. Do you expect these molecules to exist in the gas phase?

43. Sketch the bonding and antibonding molecular orbitals that result from linear combinations of the $2p_x$ atomic orbitals in a homonuclear diatomic molecule. (The $2p_x$ orbitals' lobes are oriented along the bonding axis.)

44. Sketch the bonding and antibonding molecular orbitals that result from linear combinations of the $2p_z$ atomic orbitals in a homonuclear diatomic molecule. (The $2p_z$ orbitals' lobes are oriented perpendicular to the bonding axis.) How do these molecular orbitals differ from those obtained from linear combinations of the $2p_y$ atomic orbitals? (The $2p_y$ orbitals are oriented perpendicular to the bonding axis, but also perpendicular to the $2p_z$ orbitals.)

45. Using the molecular orbital energy ordering for second-row homonuclear diatomic molecules in which the π_{2p} orbitals lie at *lower* energy than the σ_{2p}, draw MO energy diagrams and predict the bond order in a molecule or ion with each number of total valence electrons. Will the molecule or ion be diamagnetic or paramagnetic?
    a. 4   b. 6   c. 8   d. 9

46. Using the molecular orbital energy ordering for second-row homonuclear diatomic molecules in which the π_{2p} orbitals lie at *higher* energy than the σ_{2p}, draw MO energy diagrams and predict the bond order in a molecule or ion with each number of total valence electrons. Will the molecule or ion be diamagnetic or paramagnetic?
    a. 10   b. 12   c. 13   d. 14

47. Use molecular orbital theory to predict whether or not each molecule or ion should exist in a relatively stable form.
    a. H_2^{2-} b. Ne_2 c. He_2^{2+} d. F_2^{2-}

48. Use molecular orbital theory to predict whether or not each molecule or ion should exist in a relatively stable form.
  a. C_2^{2+} **b.** Li_2 **c.** Be_2^{2+} **d.** Li_2^{2-}

49. According to MO theory, which molecule or ion has the highest bond order? Highest bond energy? Shortest bond length?

$$C_2, C_2^+, C_2^-$$

50. According to MO theory, which molecule or ion has the highest bond order? Highest bond energy? Shortest bond length?

$$O_2, O_2^-, O_2^{2-}$$

Cumulative Problems

51. For each compound, draw an appropriate Lewis structure, determine the geometry using VSEPR theory, determine whether the molecule is polar, identify the hybridization of all interior atoms, and make a sketch of the molecule, according to valence bond theory, showing orbital overlap.
  a. COF_2 (carbon is the central atom)
  b. S_2Cl_2 (ClSSCl)
  c. SF_4

52. For each compound, draw an appropriate Lewis structure, determine the geometry using VSEPR theory, determine whether the molecule is polar, identify the hybridization of all interior atoms, and make a sketch of the molecule, according to valence bond theory, showing orbital overlap.
  a. IF_5
  b. CH_2CHCH_3
  c. CH_3SH

53. Amino acids are biological compounds that link together to form proteins, the workhorse molecules in living organisms. The skeletal structures of several simple amino acids are shown below. For each skeletal structure, complete the Lewis structure, determine the geometry and hybridization about each interior atom, and make a sketch of the molecule, using the bond conventions of Section 10.4.

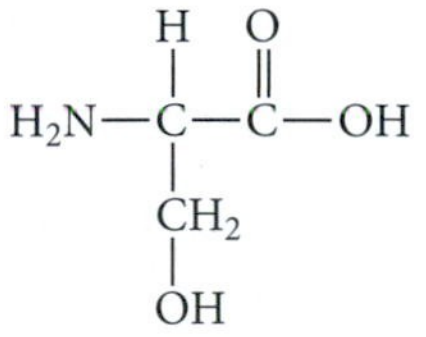

(a) serine

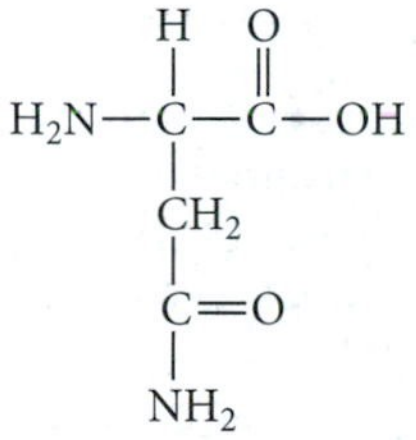

(b) asparagine

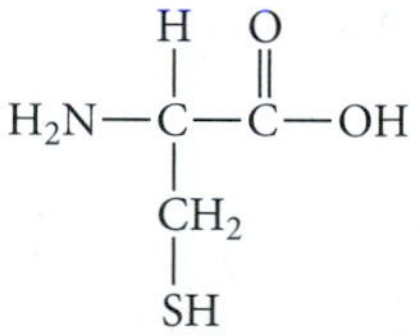

(c) cysteine

54. The genetic code is based on four different bases with the structures shown above and to the right. Assign a geometry and hybridization to each interior atom in these four bases.
  a. cytosine
  b. adenine
  c. thymine
  d. guanine

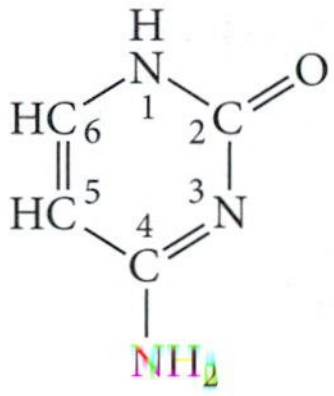

(a)

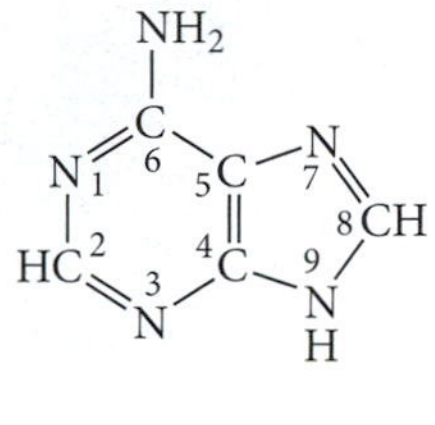

(b)

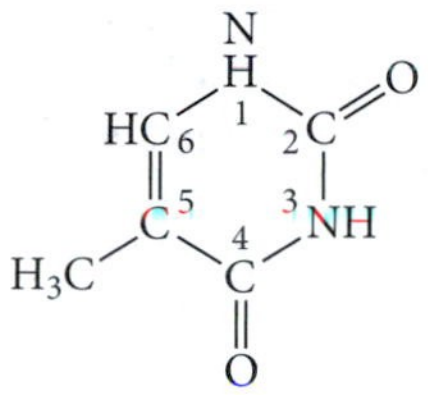

(c)

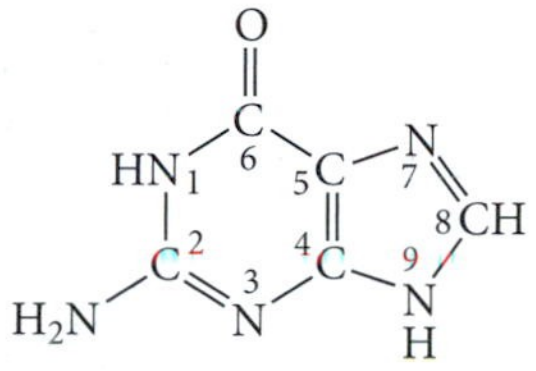

(d)

55. The structure of caffeine, present in coffee and many soft drinks, is shown below. How many pi bonds are present in caffeine? How many sigma bonds? Insert the lone pairs in the molecule. What kinds of orbitals do the lone pairs occupy?

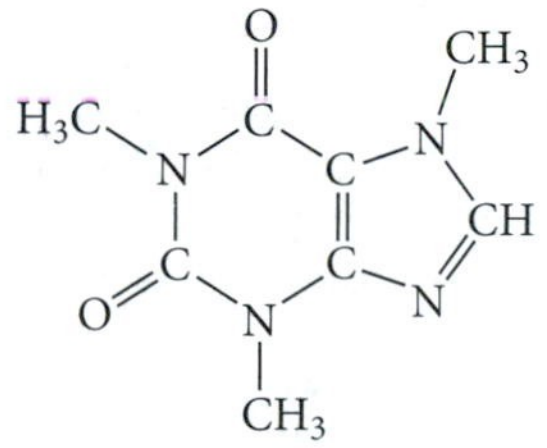

56. The structure of acetylsalicylic acid (aspirin) is shown below. How many pi bonds are present in acetylsalicylic acid? How many sigma bonds? What parts of the molecule are free to rotate? What parts are rigid?

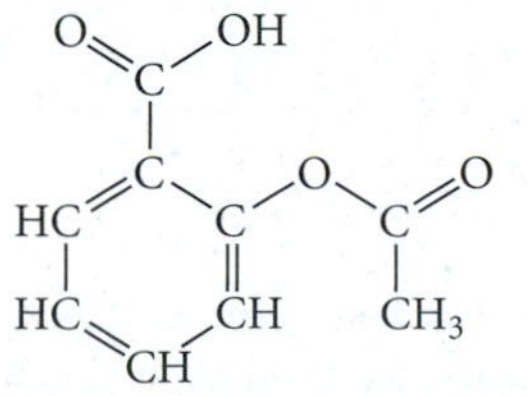

57. Most vitamins can be classified either as fat soluble, which means that they tend to accumulate in the body (so that taking too much can be harmful), or water soluble, which means that they tend to be quickly eliminated from the body in urine. Examine the structural formula and space-filling model for each vitamin and determine whether each one is fat soluble (mostly nonpolar) or water soluble (mostly polar).

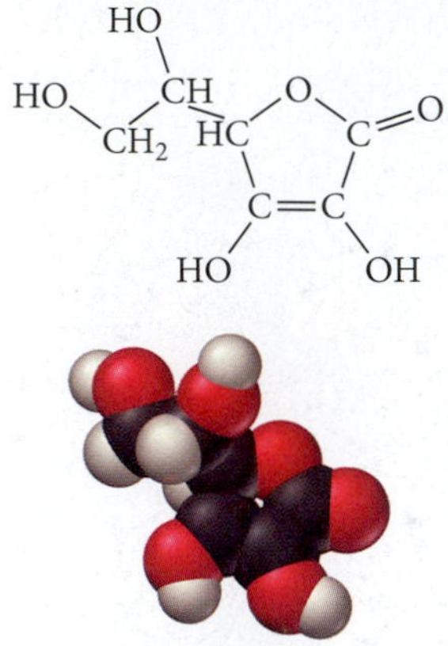

(a) vitamin C

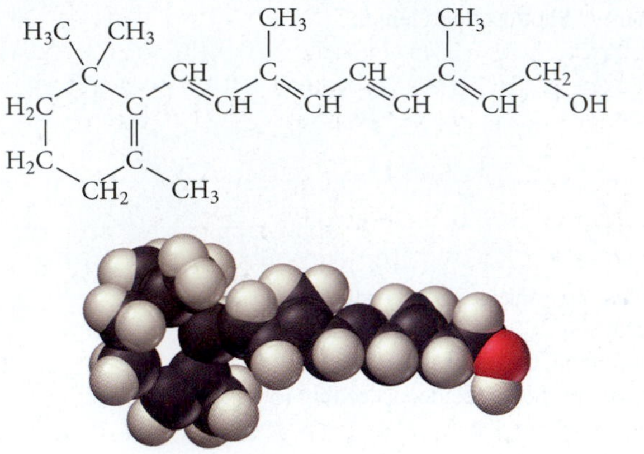

(b) vitamin A

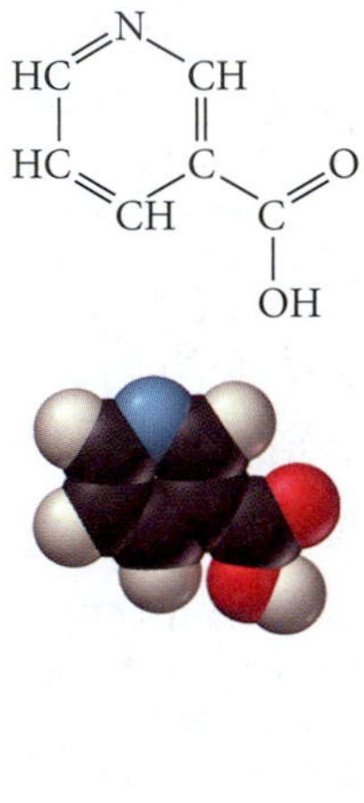

(c) niacin (vitamin B$_3$)

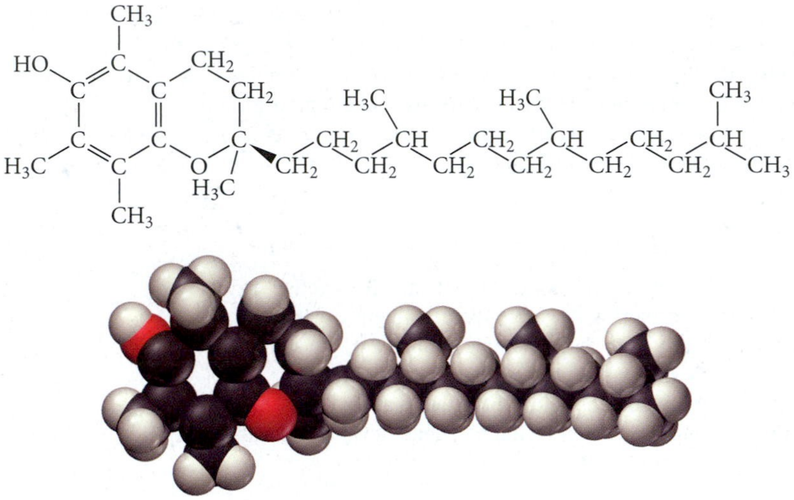

(d) vitamin E

58. Water does not easily remove grease from dishes or hands because grease is nonpolar and water is polar. The addition of soap to water, however, allows the grease to dissolve. Look at the structure of sodium stearate (a soap) below and suggest how it works.

$$CH_3(CH_2)_{16}\overset{\displaystyle O}{\overset{\displaystyle \|}{C}}\text{—}O^-Na^+$$

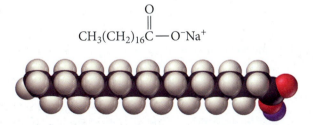

59. Bromine can form compounds or ions with any number of fluorine atoms from one to five. Write the formulas of all five of these species, assign a hybridization, and describe their electron and molecular geometries.

60. The compound C$_3$H$_4$ has two double bonds. Describe its bonding and geometry, using a valence bond approach.

61. Draw the Lewis structure of a molecule with the formula C$_4$H$_6$Cl$_2$ that has a dipole moment of 0.

62. Indicate which orbitals overlap to form the σ bonds in: (a) BeBr$_2$, (b) HgCl$_2$, (c) ICN

Challenge Problems

63. In VSEPR theory, which uses the Lewis model to determine molecular geometry, the trend of decreasing bond angle in CH$_4$, NH$_3$, and H$_2$O is accounted for by the greater repulsion of lone pair electrons compared to bonding pair electrons. How would this trend be accounted for in valence bond theory?

64. *cis*-2-Butene isomerizes to *trans*-2-butene via the following reaction:

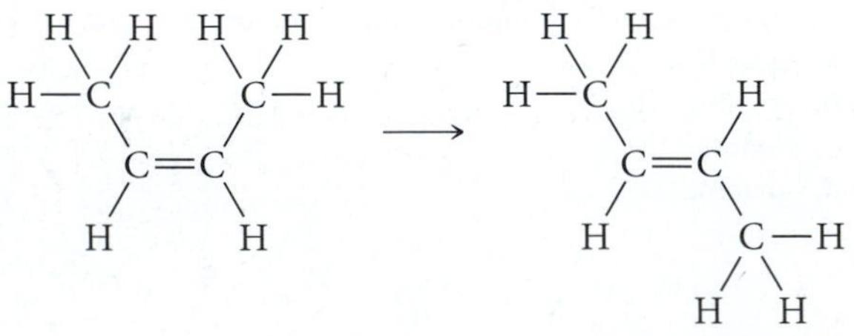

a. If isomerization requires breaking the pi bond, what minimum energy is required for isomerization in J/mol? In J/molecule?

b. If the energy for isomerization comes from light, what minimum frequency of light is required? In what portion of the electromagnetic spectrum does this frequency lie?

65. The species NO_2, NO_2^+, and NO_2^-, in which N is the central atom, have very different bond angles. Predict how these bond angles might differ in relation to the ideal angles and justify your prediction.

66. The bond angles increase steadily in the series PF_3, PCl_3, PBr_3, and PI_3. After consulting the data on atomic radii in Chapter 8, provide an explanation for this observation.

67. Draw the Lewis structure for acetamide (CH_3CONH_2), an organic compound, and determine the geometry about each interior atom. Experiments show that the geometry about the nitrogen atom in acetamide is nearly planar. What resonance structure can account for the planar geometry about the nitrogen atom?

68. Draw the Lewis structures for two possible compounds with the formula CH_3NO_2. Which of the two compounds do you expect to have the larger ONO bond angle?

Conceptual Problems

69. Pick the statement below that best captures the fundamental idea behind VSEPR theory. Explain what is wrong with each of the other statements.

a. The angle between two or more bonds is determined primarily by the repulsions between the electrons within those bonds and other (lone pair) electrons on the central atom of a molecule. Each of these electron groups (bonding electrons or lone pair electrons) will lower its potential energy by maximizing its separation from other electron groups, thus determining the geometry of the molecule.

b. The angle between two or more bonds is determined primarily by the repulsions between the electrons within those bonds. Each of these bonding electrons will lower its potential energy by maximizing its separation from other electron groups, thus determining the geometry of the molecule.

c. The geometry of a molecule is determined by the shapes of the overlapping orbitals that form the chemical bonds. Therefore, to determine the geometry of a molecule, you must determine the shapes of the orbitals involved in bonding.

70. Suppose that a molecule has four bonding groups and one lone pair on the central atom. Suppose further that the molecule is confined to two dimensions (this is a purely hypothetical assumption for the sake of understanding the principles behind VSEPR theory). Make a sketch of the molecule and estimate the bond angles.

71. How does each of the three major bonding theories (the Lewis model, valence bond theory, and molecular orbital theory) define a single chemical bond? A double bond? A triple bond? How are these definitions similar? How are they different?

72. The most stable forms of the nonmetals in groups 4A, 5A, and 6A of the first period are molecules with multiple bonds. Beginning with the second period, the most stable forms of the nonmetals of these groups are molecules without multiple bonds. Propose an explanation based on valence bond theory for this observation.

Answers to Conceptual Connections

Electron Groups and Molecular Geometry

10.1 The geometry of a molecule is determined by how the terminal atoms are arranged around the central atom, which is in turn determined by how the electron groups are arranged around the central atom. The electron groups on the terminal atoms do not affect this arrangement.

Molecular Geometry

10.2 Linear. HCN has two electron groups (the single bond and the triple bond) resulting in a linear geometry.

Molecular Geometry and Electron Group Repulsions

10.3 (d) All electron groups on the central atom determine the shape of a molecule according to VSEPR theory.

What Is a Chemical Bond?

10.4 (a) In the Lewis model, a covalent chemical bond is the sharing of electrons (represented by dots). **(b)** In valence bond theory, a covalent chemical bond is the overlap of half-filled atomic orbitals. **(c)** The answers are different because the Lewis model and valence bond theory are different *models* for chemical bonding. They both make useful and often similar predictions, but the assumptions of each model are different, and so are their respective descriptions of a chemical bond.

Single and Double Bonds

10.5 Applying valence bond theory, we see that a double bond is actually composed of two different kinds of bonds, one σ and one π. The orbital overlap in the π bond is side-to-side between two p orbitals and consequently not as efficient as the end-to-end overlap in a σ bond. Since the strength of the bond depends in part on the degree of overlap between the orbitals and the π bond is weaker than a σ bond, the bond energy of a double bond is less than twice the bond energy of a single σ bond.

Liquids, Solids, and Intermolecular Forces

It's a wild dance floor there at the molecular level. —Roald Hoffmann (1937–)

Recent studies suggest that the gecko's remarkable ability to climb walls and adhere to surfaces depends on intermolecular forces.

WE LEARNED IN Chapter 1 that matter exists primarily in three states: solid, liquid, and gas. In Chapter 5, we examined the gas state. In this chapter we turn to the solid and liquid states, known collectively as the condensed states (or condensed phases). The solid and liquid states are more similar to each other than they are to the gas state. In the gas state, the constituent particles—atoms or molecules—are separated by large distances and do not interact with each other very much. In the condensed states, the constituent particles are close together and exert moderate to strong attractive forces on one another. Unlike the gas state, for which we have a good, simple quantitative model (kinetic molecular theory) to describe and predict behavior, we have no such model for the condensed states. In fact, modeling the condensed states is an active area of research today. In this chapter, we focus primarily on describing the condensed states and their properties and on providing some qualitative guidelines to help us understand those properties.

11.1 Climbing Geckos and Intermolecular Forces

The gecko can run up a polished glass window in seconds or even walk across a ceiling. It can support its entire weight by a single toe in contact with a surface. How? Recent work by several scientists points to *intermolecular forces*—attractive forces that exist *between* all molecules and atoms—as the reason that the gecko can perform its gravity-defying feats. Intermolecular forces are the forces that hold many liquids and solids—such as water and ice, for example—together.

The key to the gecko's sticky feet lies in the millions of microhairs, called *setae*, that line its toes. Each seta is between 30 and 130 μm long and branches out to end in several hundred flattened tips called *spatulae*, as you can see in the photo. This unique structure results in many contact points, affording the gecko's toes unusually close contact with the surfaces it climbs. The close contact allows intermolecular forces—which are significant only at short distances—to hold the gecko to the wall.

▲ Each of the millions of microhairs on a gecko's feet branches out to end in flattened tips called *spatulae*.

More generally, intermolecular forces are responsible for the very existence of the condensed states. The state of a sample of matter—solid, liquid, or gas—depends on the magnitude of intermolecular forces between the constituent particles relative to the amount of thermal energy in the sample. Recall from Chapter 6 that the molecules and atoms composing matter are in constant random motion that increases with increasing temperature. The energy associated with this motion is called *thermal energy*. When thermal energy is high relative to intermolecular forces, matter tends to be gaseous. When thermal energy is low relative to intermolecular forces, matter tends to be liquid or solid.

11.2 Solids, Liquids, and Gases: A Molecular Comparison

To begin to understand the differences between the three common states of matter, consider Table 11.1, which shows the density and molar volume of water in its three different states, along with molecular representations of each state. Notice that the densities of the solid and liquid states are much greater than the density of the gas state. Notice also that the solid and liquid states are more similar in density and molar volume to one another than they are to the gas state. The molecular representations show the reason for these differences. The molecules in liquid water and ice are in close contact with one another—essentially touching—while those in gaseous water are separated by large distances. The molecular representation of gaseous water in Table 11.1 is out of proportion—the water molecules in the figure should be much farther apart for their size. (Only a fraction of a molecule could be included in the figure if it were drawn to scale.) From the molar volumes, we know that 18.0 mL of liquid water (slightly more than a tablespoon) would occupy 30.5 L when converted to gas at 100 °C. The low density of gaseous water results from this large separation between molecules.

Notice also that, for water, the solid is slightly less dense than the liquid. This is *atypical* behavior. Most solids are slightly denser than their corresponding liquids because the molecules move closer together upon freezing. As we will see in Section 11.9, ice is less dense than liquid water because the unique crystal structure of ice results in water molecules moving slightly farther apart upon freezing.

From the molecular perspective, one major difference between liquids and solids is the freedom of movement of the constituent molecules or atoms. Even though the atoms or molecules in a liquid are in close contact, thermal energy can partially overcome the attractions between them, allowing them to move around one another. Not so in solids,

TABLE 11.1 The Three States of Water				
Phase	Temperature (°C)	Density (g/cm³, at 1 atm)	Molar Volume	Molecular View
Gas (steam)	100	5.90×10^{-4}	30.5 L	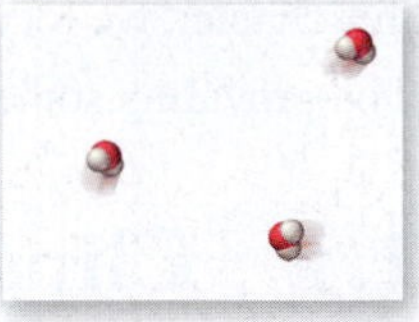
Liquid (water)	20	0.998	18.0 mL	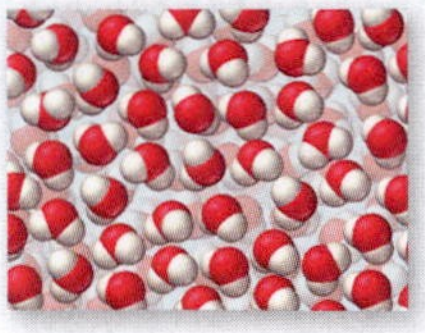
Solid (ice)	0	0.917	19.6 mL	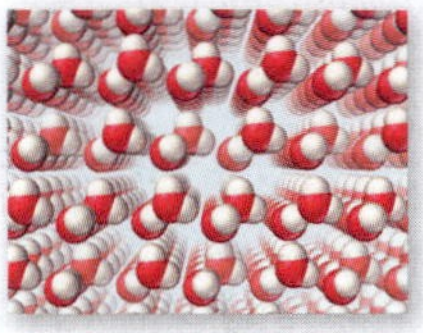

TABLE 11.2 Properties of the States of Matter				
State	Density	Shape	Volume	Strength of Intermolecular Forces[*]
Gas	Low	Indefinite	Indefinite	Weak
Liquid	High	Indefinite	Definite	Moderate
Solid	High	Definite	Definite	Strong

*Relative to thermal energy.

where the atoms or molecules are virtually locked in their positions, only vibrating back and forth about a fixed point. Table 11.2 summarizes the properties of liquids and solids, as well as the properties of gases for comparison.

Liquids assume the shape of their containers because the atoms or molecules that compose them are free to flow (or move around one another) **Figure 11.1◄**. Liquids are not easily compressed because the molecules or atoms that compose them are in close contact—they cannot be pushed much closer together. The molecules in a gas, by contrast, have a great deal of space between them and are easily forced into a smaller volume by an increase in external pressure **Figure 11.2▼**.

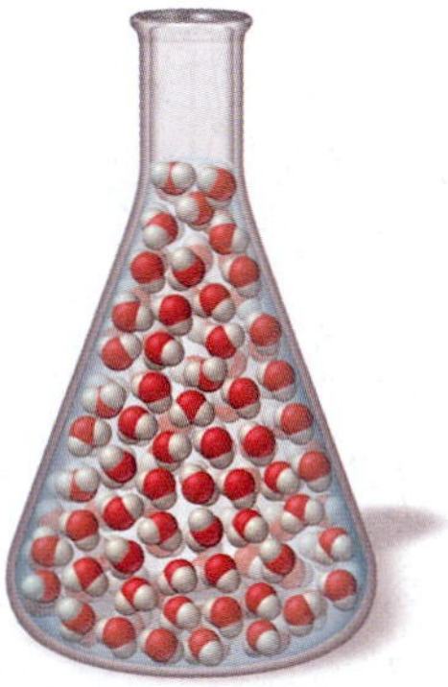

▲ **FIGURE 11.1 Liquids Assume the Shapes of Their Containers** When you pour water into a flask, it assumes the shape of the flask because water molecules are free to flow.

▶ **FIGURE 11.2 Gases Are Compressible** Molecules in a liquid are closely spaced and are not easily compressed. Molecules in a gas have a great deal of space between them, making gases compressible.

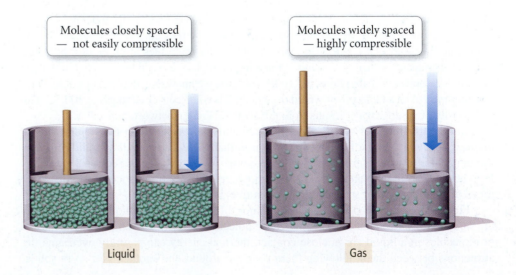

Molecules closely spaced — not easily compressible

Molecules widely spaced — highly compressible

Liquid

Gas

Solids have a definite shape because, in contrast to liquids and gases, the molecules or atoms that compose solids are fixed in place. Like liquids, solids have a definite volume and generally cannot be compressed because the molecules or atoms composing them are already in close contact. Solids may be **crystalline,** in which case the atoms or molecules that compose them are arranged in a well-ordered three-dimensional array, or they may be **amorphous,** in which case the atoms or molecules that compose them have no long-range order.

By some definitions, an amorphous solid is considered a unique state, different from the normal solid state because it lacks any long-range order.

Changes between States

One state of matter can be transformed to another by changing the temperature, pressure, or both. For example, solid ice can be converted to liquid water by heating, and liquid water can be converted to solid ice by cooling. The following diagram shows the three states of matter and the changes in conditions that commonly induce transitions between them.

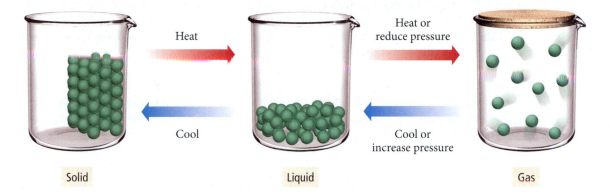

Solid   Liquid   Gas

Notice that transitions between the liquid and gas state can be achieved not only by heating and cooling, but also through changes in pressure. In general, increases in pressure favor the denser state, so increasing the pressure of a gas sample results in a transition to the liquid state. The most familiar example of this phenomenon is the LP (liquid propane) gas used as a fuel for outdoor grills and lanterns. Propane is a gas at room temperature and atmospheric pressure. However, it liquefies at pressures exceeding about 2.7 atm. The propane you buy in a tank is under pressure and therefore in the liquid form. When you open the tank, some of the propane escapes as a gas, lowering the pressure in the tank for a brief moment. Immediately, however, some of the liquid propane evaporates, replacing the gas that escaped. Storing gases like propane as liquids is efficient because, in their liquid form, they occupy much less space.

 Conceptual Connection 11.1 State Changes

The molecular diagram below shows a sample of liquid water.

Which diagram best depicts the vapor emitted from a pot of boiling of water?

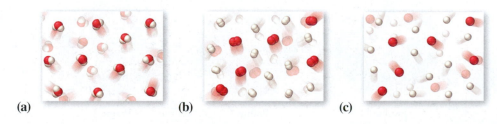

(a)   **(b)**   **(c)**

11.3 Intermolecular Forces: The Forces That Hold Condensed States Together

Intermolecular forces originate from the interactions between charges, partial charges, and temporary charges on molecules (or atoms and ions), much as bonding forces originate from interactions between charged particles in atoms. Recall from Section 8.3 that according to Coulomb's law the potential energy (E) of two oppositely charged particles (with charges q_1 and q_2) decreases (becomes more negative) with increasing magnitude of charge and with decreasing separation (r) between them:

$$E = \frac{1}{4\pi\varepsilon_0}\frac{q_1 q_2}{r}$$ (When q_1 and q_2 are opposite in sign, E is negative.)

Therefore, as we have seen, protons and electrons are attracted to each other because their potential energy decreases as they get closer together. Similarly, molecules with partial or temporary charges are attracted to each other because *their* potential energy decreases as they get closer together. However, intermolecular forces, even the strongest ones, are generally *much weaker* than bonding forces.

The reason for the relative weakness of intermolecular forces compared to bonding forces is also related to Coulomb's law. Bonding forces are the result of large charges (the charges of protons and electrons) interacting at very close distances. Intermolecular forces are the result of smaller charges (as we shall see in the following discussion) interacting at greater distances. For example, consider the interaction between two water molecules in liquid water:

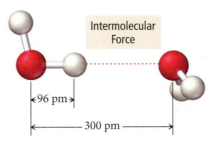

The length of an O—H bond in liquid water is 96 pm; however, the average distance between water molecules in liquid water is about 300 pm. The larger distances between molecules, as well as the smaller charges involved (partial charges on the hydrogen and oxygen atoms), result in weaker forces. To break the O—H bonds in water, you have to heat the water to thousands of degrees Celsius. However, to completely overcome the intermolecular forces *between* water molecules, you have to heat water only to its boiling point, 100 °C. We will examine several different types of intermolecular forces, including dispersion forces, dipole–dipole forces, hydrogen bonding, and ion–dipole forces. The first three of these can potentially occur in pure substances *and* mixtures; the last one is found only in mixtures.

Dispersion Force

The nature of dispersion forces was first recognized by Fritz W. London (1900–1954), a German-American physicist.

The one intermolecular force present in all molecules and atoms is the **dispersion force** (also called the London force). Dispersion forces are the result of fluctuations in the electron distribution within molecules or atoms. Since all atoms and molecules have electrons, they all exhibit dispersion forces. The electrons in an atom or molecule may, *at any one instant*, be unevenly distributed. For example, imagine a frame-by-frame movie of a helium atom in which each "frame" captures the position of the helium atom's two electrons.

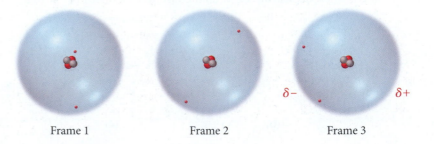

Frame 1   Frame 2   Frame 3

In any one frame, the electrons may not be symmetrically arranged around the nucleus. In frame 3, for example, helium's two electrons are on the left side of the helium atom. At that instant, the left side will have a slightly negative charge ($\delta -$). The right side of the atom, which temporarily has no electrons, will have a slightly positive charge ($\delta +$) because of the the nucleus. This fleeting charge separation is called an *instantaneous dipole* or a *temporary dipole*. As shown in **Figure 11.3▼**, an instantaneous dipole on one helium atom induces an instantaneous dipole on its neighboring atoms because the positive end of the instantaneous dipole attracts electrons in the neighboring atoms. The neighboring atoms then attract one another—the positive end of one instantaneous dipole attracting the negative end of another. This attraction is the dispersion force.

Dispersion Force

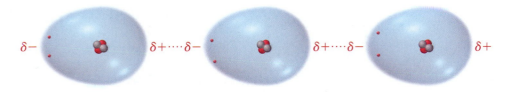

An instantaneous dipole on any one helium atom induces instantaneous dipoles on neighboring atoms, which then attract one another.

$\delta -$ $\delta + \cdots \delta -$ $\delta + \cdots \delta -$ $\delta +$

◄ **FIGURE 11.3 Dispersion Interactions** The temporary dipole in one helium atom induces a temporary dipole in its neighbor. The resulting attraction between the positive and negative charges creates the dispersion force.

The *magnitude* of the dispersion force depends on how easily the electrons in the atom or molecule can move or *polarize* in response to an instantaneous dipole, which in turn depends on the size (or volume) of the electron cloud. A larger electron cloud results in a greater dispersion force because the electrons are held less tightly by the nucleus and can therefore polarize more easily. If all other variables are constant, the dispersion force increases with increasing molar mass because molecules or atoms of higher molar mass generally have more electrons dispersed over a greater volume. Consider the boiling points of the noble gases displayed in Table 11.3. As the molar masses and electron cloud volumes of the noble gases increase, the greater dispersion forces result in increasing boiling points.

Molar mass alone, however, does not determine the magnitude of the dispersion force. For example, compare the molar masses and boiling points of *n*-pentane and neopentane:

To polarize means to form a dipole moment (see Section 9.6).

n-Pentane
molar mass = 72.15 g/mol
boiling point = 36.1 °C

Neopentane
molar mass = 72.15 g/mol
boiling point = 9.5 °C

These molecules have identical molar masses, but *n*-pentane has a higher boiling point than neopentane. Why? Because the two molecules have different shapes. The *n*-pentane molecules are long and can interact with one another along their entire length, as shown in **Figure 11.4(a)▶**. In contrast, the bulky, round shape of neopentane molecules results in a smaller area of interaction between neighboring molecules, as shown in **Figure 11.4(b)▶**, and thus a lower boiling point.

Although molecular shape and other factors must always be considered in determining the magnitude of dispersion forces, molar mass can act as a guide when comparing dispersion forces within a family of similar elements or compounds, as shown in **Figure 11.5▶** for some selected *n*-alkanes.

TABLE 11.3 Boiling Points of the Noble Gases			
Noble Gas		**Molar Mass (g/mol)**	**Boiling Point (K)**
He	●	4.00	4.2
Ne	●	20.18	27
Ar	●	39.95	87
Kr	●	83.80	120
Xe	●	131.30	165

▶ **FIGURE 11.4 Dispersion Force and Molecular Shape** (a) The straight shape of *n*-pentane molecules allows them to interact with one another along the entire length of the molecule. (b) The nearly spherical shape of neopentane molecules allows for only a small area of interaction. Thus, dispersion forces are weaker in neopentane than in *n*-pentane, resulting in a lower boiling point.

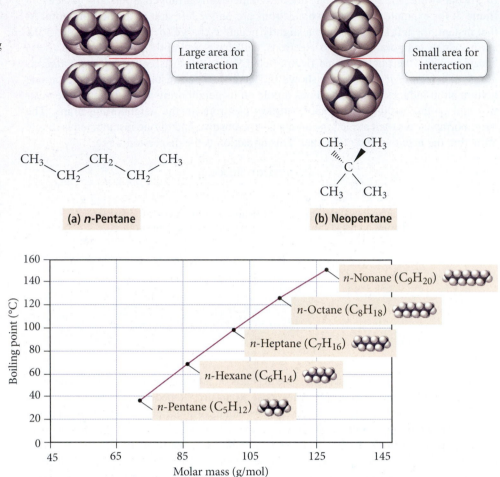

(a) *n*-Pentane

(b) Neopentane

▲ **FIGURE 11.5 Boiling Points of the *n*-Alkanes** The boiling points of the *n*-alkanes rise with increasing molar mass and the consequent stronger dispersion forces.

Dipole–Dipole Interaction

The positive end of a polar molecule is attracted to the negative end of its neighbor.

$\delta+$ $\delta-$ ······ $\delta+$ $\delta-$

▲ **FIGURE 11.6 Dipole–Dipole Interaction**

(Q) **Conceptual Connection 11.2 Dispersion Forces**

Which halogen has the highest boiling point?

(a) Cl_2 **(b)** Br_2 **(c)** I_2

Dipole–Dipole Force

The **dipole–dipole force** exists in all molecules that are polar. Polar molecules have **permanent dipoles** that interact with the permanent dipoles of neighboring molecules, as shown in **Figure 11.6◀**. The positive end of one permanent dipole is attracted to the negative end of another; this attraction is the dipole–dipole force. Example 11.1 shows how to determine if a compound exhibits dipole–dipole forces.

EXAMPLE 11.1 Dipole–Dipole Forces

Determine whether each molecule has dipole–dipole forces.

(a) CO_2 **(b)** CH_2Cl_2 **(c)** CH_4

SOLUTION

A molecule will have dipole–dipole forces if it is polar. To determine whether a molecule is polar, (1) *determine whether the molecule contains polar bonds* and (2) *determine whether the polar bonds add together to form a net dipole moment* (Section 9.6).

(a) CO_2	(a) O=C=O
(1) Since the electronegativity of carbon is 2.5 and that of oxygen is 3.5 (Figure 9.9), CO_2 has polar bonds. (2) The geometry of CO_2 is linear. Consequently, the dipoles of the polar bonds cancel, so the molecule is *not polar* and does not have dipole–dipole forces.	No dipole forces present
(b) CH_2Cl_2	CH_2Cl_2
(1) The electronegativity of C is 2.5, that of H is 2.1, and that of Cl is 3.5. Consequently, CH_2Cl_2 has two polar bonds (C—Cl) and two bonds that are nearly nonpolar (C—H). (2) The geometry of CH_2Cl_2 is tetrahedral. Since the C—Cl bonds and the C—H bonds are different, their dipoles do not cancel but sum to a net dipole moment. Therefore the molecule is polar and has dipole–dipole forces.	Dipole forces present
(c) CH_4	CH_4
(1) Since the electronegativity of C is 2.5 and that of hydrogen is 2.1 the C—H bonds are nearly nonpolar. (2) In addition, since the geometry of the molecule is tetrahedral, any slight polarities that the bonds might have will cancel. CH_4 is therefore nonpolar and does not have dipole–dipole forces.	No dipole forces present

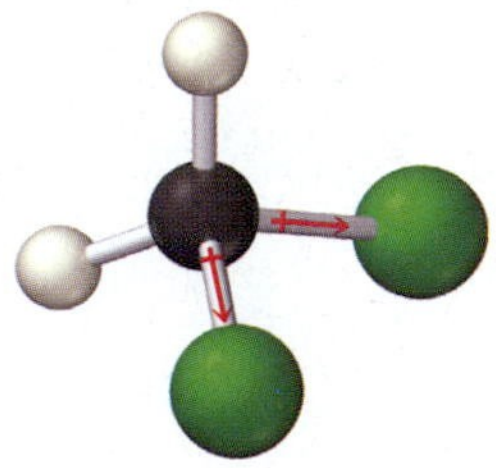

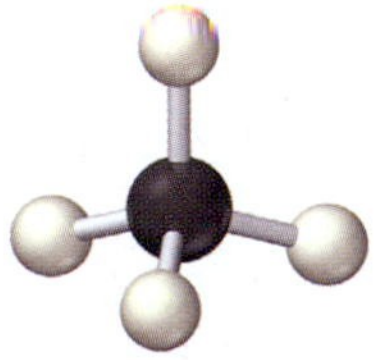

FOR PRACTICE 11.1

Which molecules have dipole–dipole forces?

(a) CI_4 (b) CH_3Cl (c) HCl

Polar molecules have higher melting and boiling points than nonpolar molecules of similar molar mass. Remember that all molecules (including polar ones) have dispersion forces. Polar molecules have, *in addition,* dipole–dipole forces. This additional attractive force raises their melting and boiling points relative to nonpolar molecules of similar molar mass. For example, consider the following two compounds:

Name	Formula	Molar Mass (amu)	Structure	bp (°C)	mp (°C)
Formaldehyde	CH_2O	30.03	H—C—H with =O	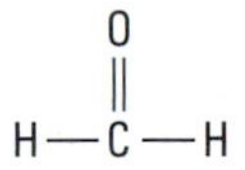  −19.5	−92
Ethane	C_2H_6	30.07	H—C—C—H (with H's)	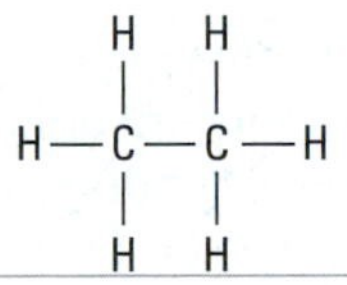  −88	−172

Formaldehyde is polar, and therefore has a higher melting point and boiling point than nonpolar ethane even though the two compounds have the same molar mass. **Figure 11.7▶** shows the boiling point of a series of molecules with similar molar mass but progressively greater dipole moments. Notice that the boiling points increase with increasing dipole moment.

The polarity of molecules composing liquids is also important in determining the **miscibility**—the ability to mix without separating into two phases—of liquids. In general, polar liquids are miscible with other polar liquids but are not miscible with nonpolar liquids. For example, water, a polar liquid, is not miscible with pentane (C_5H_{12}), a

▶ **FIGURE 11.7 Dipole Moment and Boiling Point** These molecules all have similar molar masses but different dipole moments. The boiling points increase with increasing dipole moment.

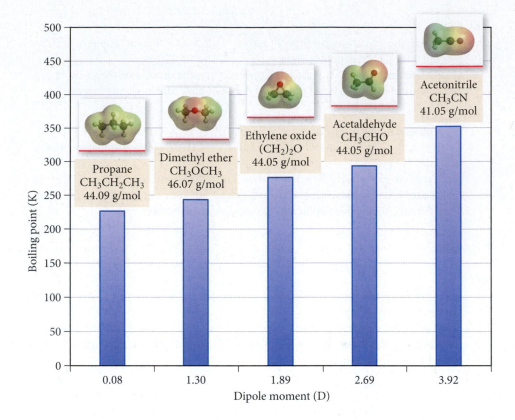

nonpolar liquid **Figure 11.8▼**. Similarly, water and oil (also nonpolar) do not mix. Consequently, oily hands or oily stains on clothes cannot be washed with plain water. The water will not mix with the oil.

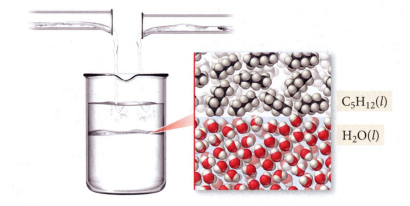

▶ **FIGURE 11.8 Polar and Nonpolar Compounds** Water and pentane do not mix because water molecules are polar and pentane molecules are nonpolar.

Hydrogen Bonding

Polar molecules containing hydrogen atoms bonded directly to small electronegative atoms—most importantly fluorine, oxygen, or nitrogen—exhibit an intermolecular force called **hydrogen bonding.** HF, NH_3, and H_2O all undergo hydrogen bonding. The hydrogen bond is a sort of *super* dipole–dipole force. The large electronegativity difference between hydrogen and these electronegative elements means that the H atom will have a fairly large partial positive charge ($\delta+$) when bonded to F, O, or N, while the F, O, or N atom will have a fairly large partial negative charge ($\delta-$). In addition, since these atoms are all quite small, they can approach one another very closely. The result is a strong attraction between the hydrogen in each of these molecules and the F, O, or N *on its neighbors*, an attraction called a **hydrogen bond.** For example, in HF, the hydrogen is strongly attracted to the fluorine on neighboring molecules **Figure 11.9▶**.

Hydrogen bonds should not be confused with chemical bonds. Chemical bonds occur *between individual atoms within a molecule*, whereas hydrogen bonds—like dispersion

Hydrogen Bonding

> When H bonds directly to F, O, or N, the bonding atoms acquire relatively large partial charges, giving rise to strong dipole–dipole attractions between neighboring molecules.

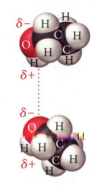

◄ FIGURE 11.9 Hydrogen Bonding in HF   The hydrogen of one HF molecule, with its partial positive charge, is attracted to the fluorine of its neighbor, with its partial negative charge. This dipole–dipole interaction is an example of a hydrogen bond.

forces and dipole–dipole forces—are intermolecular forces that occur *between molecules*. A typical hydrogen bond is only 2–5% as strong as a typical covalent chemical bond. Hydrogen bonds are, however, the strongest of the three *intermolecular* forces we have discussed so far. Substances composed of molecules that form hydrogen bonds have higher melting and boiling points than substances composed of molecules that do not form hydrogen bonds. For example, consider the following two compounds:

Name	Formula		Molar Mass (amu)	Structure	bp (°C)	mp (°C)
Ethanol	C_2H_6O		46.07	CH_3CH_2OH	78.3	−114.1
Dimethyl Ether	C_2H_6O		46.07	CH_3OCH_3	−22.0	−138.5

▲ FIGURE 11.10 Hydrogen Bonding in Ethanol

Since ethanol contains hydrogen bonded directly to oxygen, ethanol molecules form hydrogen bonds with each other as shown in **Figure 11.10►**. Consequently, ethanol is a liquid at room temperature. Dimethyl ether, in contrast, has an identical molar mass but does not exhibit hydrogen bonding because the oxygen atom is not bonded directly to hydrogen, resulting in lower boiling and melting points. Dimethyl ether is a gas at room temperature.

Water is another good example of a molecule with hydrogen bonding **Figure 11.11►**. **Figure 11.12▼** shows the boiling points of the simple hydrogen compounds of the group 4A and group 6A elements. Notice that, in general, boiling points increase with increasing molar mass, as expected based on increasing dispersion forces. However, because of hydrogen bonding, the boiling point of water (100 °C) is much higher than expected based on its molar mass (18.0 g/mol). Without hydrogen bonding, all the water on our planet would be gaseous.

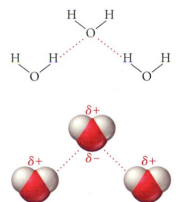

▲ FIGURE 11.11 Hydrogen Bonding in Water

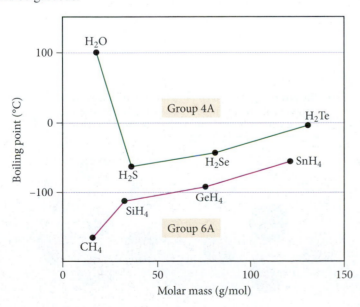

◄ FIGURE 11.12 Boiling Points of Group 4A and 6A Compounds Because of hydrogen bonding, the boiling point of water is anomalous compared to the boiling points of other hydrogen-containing compounds.

EXAMPLE 11.2 Hydrogen Bonding

One of the following compounds is a liquid at room temperature. Which one and why?

$$O$$
$$\|$$
$$H-C-H$$
Formaldehyde

$$H$$
$$|$$
$$H-C-F$$
$$|$$
$$H$$
Fluoromethane

$$H-O-O-H$$
Hydrogen peroxide

SOLUTION

The three compounds have similar molar masses:

Formaldehyde	30.03 g/mol
Fluoromethane	34.03 g/mol
Hydrogen peroxide	34.02 g/mol

So the strengths of their dispersion forces are similar. All three compounds are also polar, so they have dipole–dipole forces. Hydrogen peroxide, however, is the only compound that also contains H bonded directly to F, O, or N. Therefore it also has hydrogen bonding and is likely to have the highest boiling point of the three. Since the example stated that only one of the compounds was a liquid, we can safely assume that hydrogen peroxide is the liquid. Note that, although fluoromethane *contains* both H and F, H is not *directly bonded* to F, so fluoromethane does not have hydrogen bonding as an intermolecular force. Similarly, although formaldehyde *contains* both H and O, H is not *directly bonded* to O, so formaldehyde does not have hydrogen bonding either.

FOR PRACTICE 11.2

Which has the higher boiling point, HF or HCl? Why?

Ion–Dipole Force

The **ion–dipole force** occurs when an ionic compound is mixed with a polar compound and is especially important in aqueous solutions of ionic compounds. For example, when sodium chloride is mixed with water, the sodium and chloride ions interact with water molecules via ion–dipole forces, as shown in **Figure 11.13▼**. Notice that the positive sodium ions interact with the negative poles of water molecules, while the negative chloride ions interact with the positive poles. Ion–dipole forces are the strongest of the types of intermolecular forces we have discussed and are responsible for the ability of ionic substances to form solutions with water. We discuss aqueous solutions more thoroughly in Chapter 12. Table 11.4 summarizes the intermolecular forces we have discussed.

Ion–Dipole Forces

The positively charged end of a polar molecule such as H₂O is attracted to negative ions and the negatively charged end of the molecule is attracted to positive ions.

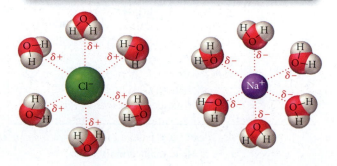

▶ **FIGURE 11.13 Ion–Dipole Forces** Ion–dipole forces exist between Na⁺ and the negative ends of H₂O molecules and between Cl⁻ and the positive ends of H₂O molecules.

TABLE 11.4 Types of Intermolecular Forces

Type	Present in	Molecular perspective	Strength
Dispersion	All molecules and atoms	$\delta-$ $\delta+$ ···· $\delta-$ $\delta+$	
Dipole–dipole	Polar molecules	$\delta+$ $\delta-$ ····· $\delta+$ $\delta-$	
Hydrogen bonding	Molecules containing H bonded to F, O, or N	$\delta+$ $\delta+$ $\delta-$ $\delta+$ $\delta-$ $\delta-$	
Ion–dipole	Mixtures of ionic compounds and polar compounds	$\delta-$ $\delta-$ $\delta-$ $+$ $\delta-$ $\delta-$ $\delta-$	

Conceptual Connection 11.3 Intermolecular Forces and Boiling Point

Which substance has the highest boiling point?

(a) CH_3OH **(b)** CO **(c)** N_2

11.4 Intermolecular Forces in Action: Surface Tension, Viscosity, and Capillary Action

The most important manifestation of intermolecular forces is the very existence of liquids and solids. In liquids, we also observe several other manifestations of intermolecular forces including surface tension, viscosity, and capillary action.

Surface Tension

A fly fisherman delicately casts a small metal fly (a hook with a few feathers and strings attached to make it look like a fly) onto the surface of a moving stream. The fly floats on the surface of the water—even though the metal composing the hook is denser than water—and attracts trout. Why? The hook floats because of *surface tension*, which results from the tendency of liquids to minimize their surface area.

We can understand surface tension by carefully examining **Figure 11.14▶**, which depicts the intermolecular forces experienced by a molecule at the surface of the liquid compared to those experienced by a molecule in the interior. Notice that a molecule at the surface has relatively fewer neighbors with which to interact—and is therefore inherently less stable because it has higher potential energy—than those in the interior. (Remember that the attractive interactions with other molecules *lower* potential energy.) In order to increase the surface area of the liquid, molecules from the interior have to be moved to the surface, and, since molecules at the surface have a higher potential energy than those in the interior, this movement requires energy. Therefore, liquids tend to minimize their

▲ A trout fly can float on water because of surface tension.

Recall from Section 11.3 that the interactions between molecules lower their potential energy in much the same way that the interaction between protons and electrons lowers their potential energy, in accordance with Coulomb's law.

► **FIGURE 11.14 The Origin of Surface Tension** Molecules at the liquid surface have a higher potential energy than those in the interior. As a result, liquids tend to minimize their surface area, and the surface behaves like a membrane or "skin."

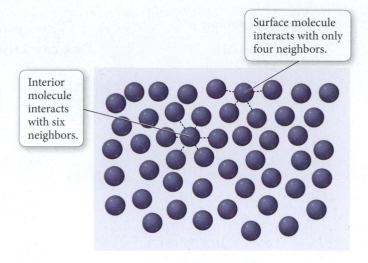

Interior molecule interacts with six neighbors.

Surface molecule interacts with only four neighbors.

▲ **FIGURE 11.15 Spherical Water** On the Space Shuttle in orbit, under weightless conditions, water coalesces into nearly perfect spheres held together by intermolecular forces between water molecules.

surface area. The **surface tension** of a liquid is the energy required to increase the surface area by a unit amount. For example, at room temperature, water has a surface tension of $72.8 \ mJ/m^2$—it takes 72.8 mJ to increase the surface area of water by one square meter.

Why does surface tension allow the fly fisherman's hook to float on water? The tendency for liquids to minimize their surface creates a kind of skin at the surface that resists penetration. For the hook to sink into the water, the water's surface area must increase slightly—an increase that is resisted by the surface tension. A fisherman's fly, even though it is denser than water, will float on the surface of the water. A slight tap on the fly will overcome the surface tension and cause it to sink.

Surface tension decreases with decreasing intermolecular forces. You could not float a fisherman's fly on benzene (C_6H_6), for example, because the dispersion forces among the molecules composing benzene are significantly weaker than the hydrogen bonds among water molecules. The surface tension of benzene is only $28 \ mJ/m^2$—just 40% that of water.

Surface tension is also the reason that small water droplets (those not large enough to be distorted by gravity) form nearly perfect spheres. On the Space Shuttle, the complete absence of gravity allows even large samples of water to form nearly perfect spheres **Figure 11.15◄**. Why? Just as gravity pulls the matter of a planet or star inward to form a sphere, so intermolecular forces among collections of water molecules pull the water into a sphere. A sphere is the geometrical shape with the smallest ratio of surface area to volume; therefore, the formation of a sphere minimizes the number of molecules at the surface, minimizing the potential energy of the system.

Viscosity

Another manifestation of intermolecular forces is **viscosity,** the resistance of a liquid to flow. Motor oil, for example, is more viscous than gasoline, and maple syrup is more viscous than water. Viscosity is measured in a unit called the poise (P), defined as $1 \ g/cm \cdot s$. The viscosity of water at room temperature is approximately one centipoise (cP). Viscosity is greater in substances with stronger intermolecular forces because if molecules are more strongly attracted to each other they do not flow around each other as freely. Viscosity also depends on molecular shape, increasing in longer molecules that can interact over a greater area and possibly become entangled. Table 11.5 lists the viscosity of several hydrocarbons.

TABLE 11.5 Viscosity of Several Hydrocarbons at 20 °C			
Hydrocarbon	**Molar Mass (g/mol)**	**Formula**	**Viscosity (cP)**
n-Pentane	72.15	$CH_3CH_2CH_2CH_2CH_3$	0.240
n-Hexane	86.17	$CH_3CH_2CH_2CH_2CH_2CH_3$	0.326
n-Heptane	100.2	$CH_3CH_2CH_2CH_2CH_2CH_3$	0.409
n-Octane	114.2	$CH_3CH_2CH_2CH_2CH_2CH_2CH_3$	0.542
n-Nonane	128.3	$CH_3CH_2CH_2CH_2CH_2CH_2CH_2CH_3$	0.711

Viscosity also depends on temperature because thermal energy partially overcomes the intermolecular forces, allowing molecules to flow past each other more easily. Table 11.6 lists the viscosity of water as a function of temperature. Nearly all liquids become less viscous as temperature increases.

Capillary Action

Medical technicians often take advantage of **capillary action**—the ability of a liquid to flow against gravity up a narrow tube—when taking a blood sample. The technician pokes the patient's finger with a pin, squeezes some blood out of the puncture, and then collects the blood with a thin tube. When the tube's tip comes into contact with the blood, the blood is drawn into the tube by capillary action. The same force plays a role when trees and plants draw water from the soil.

Capillary action results from a combination of two forces: the attraction between molecules in a liquid, called *cohesive forces*, and the attraction between these molecules and the surface of the tube, called *adhesive forces*. The adhesive forces cause the liquid to spread out over the interior surface of the tube, while the cohesive forces cause the liquid to stay together. If the adhesive forces are greater than the cohesive forces (as is the case for the blood in a glass tube), the attraction to the surface draws the liquid up the tube while the cohesive forces pull along those molecules not in direct contact with the tube walls. The blood rises up the tube until the force of gravity balances the capillary action—the thinner the tube, the higher the rise. If the adhesive forces are smaller than the cohesive forces (as is the case for liquid mercury), the liquid does not rise up the tube at all.

We can see the result of the differences in the relative magnitudes of cohesive and adhesive forces by comparing the meniscus of water to the meniscus of mercury **Figure 11.16▶**. (The meniscus is the curved shape of a liquid surface within a tube.) The meniscus of water is concave because the *adhesive forces* are greater than the cohesive forces, causing the edges of the water to creep up the sides of the tube a bit, forming the familiar cupped shape. The meniscus of mercury is convex because the *cohesive forces*—due to metallic bonding between the atoms—are greater than the adhesive forces. The mercury atoms crowd toward the interior of the liquid to maximize their interactions with each other, resulting in the upward bulge at the center of the liquid surface.

11.5 Vaporization and Vapor Pressure

We now turn our attention to vaporization, the process by which thermal energy can overcome intermolecular forces and produce a state change from liquid to gas. We will first discuss the process of vaporization itself, then the energetics of vaporization, and finally the concepts of vapor pressure, dynamic equilibrium, and critical point. Vaporization is a common occurrence that we experience every day and even depend on to maintain proper body temperature.

The Process of Vaporization

Imagine the water molecules within a beaker of water sitting on a table at room temperature and open to the atmosphere **Figure 11.17▶**. The molecules are in constant motion due to thermal energy. If we could actually see the molecules at the surface, we would witness Roald Hoffmann's "wild dance floor" (see the chapter-opening quote) because of all the vibrating, jostling, and molecular movement. *The higher the temperature, the greater the average energy of the collection of molecules.* However, at any one time, some molecules will have more thermal energy than the average and some will have less.

The distributions of thermal energies for the molecules in a sample of water at two different temperatures are shown in **Figure 11.18▶**. The molecules at the high end of the distribution curve have enough energy to break free from the surface—where molecules are held less tightly than in the interior due to fewer neighbor–neighbor interactions—and into the gas state. This process is called **vaporization,** the state transition from liquid to gas. Some of the water molecules in the *gas state*, at the low end of the energy distribution curve for the gaseous molecules, can plunge back into the water and be captured by

TABLE 11.6 Viscosity of Liquid Water at Several Temperatures

Temperature (°C)	Viscosity (cP)
20	1.002
40	0.653
60	0.467
80	0.355
100	0.282

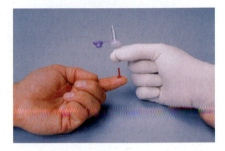

▲ Blood is drawn into a capillary tube by capillary action.

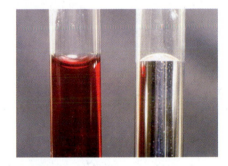

▲ **FIGURE 11.16 Meniscuses of Water and Mercury** The meniscus of water (dyed red for visibility at left) is concave because water molecules are more strongly attracted to the glass wall than to one another. The meniscus of mercury is convex because mercury atoms are more strongly attracted to one another than to the glass walls.

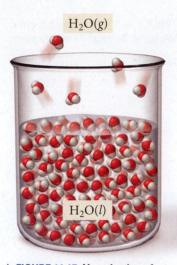

$H_2O(g)$

$H_2O(l)$

▲ **FIGURE 11.17 Vaporization of Water**   Some molecules in an open beaker have enough kinetic energy to vaporize from the surface of the liquid.

intermolecular forces. This process—the opposite of vaporization—is called **condensation,** the transition from gas to liquid.

Although both evaporation and condensation occur in a beaker open to the atmosphere, under normal conditions evaporation takes place at a greater rate because most of the newly evaporated molecules escape into the surrounding atmosphere and never come back. The result is a noticeable decrease in the water level within the beaker over time (usually several days).

What happens if we increase the temperature of the water within the beaker? Because of the shift in the energy distribution to higher energies (see Figure 11.18), more molecules now have enough energy to break free and evaporate, so vaporization occurs more quickly. What happens if we spill the water on the table or floor? The same amount of water is now spread over a wider area, resulting in more molecules at the surface of the liquid. Since molecules at the surface have the greatest tendency to evaporate—because they are held less tightly—the vaporization also occurs more quickly.

What happens if the liquid in the beaker is not water, but some other substance with weaker intermolecular forces, such as acetone? The weaker intermolecular forces allow more molecules to evaporate at a given temperature, again increasing the rate of vaporization. Liquids that vaporize easily are termed **volatile,** while those that do not vaporize easily are termed **nonvolatile.** Acetone is more volatile than water. Motor oil is virtually nonvolatile at room temperature.

Summarizing the Process of Vaporization:

▶ The rate of vaporization increases with increasing temperature.

▶ The rate of vaporization increases with increasing surface area.

▶ The rate of vaporization increases with decreasing strength of intermolecular forces.

The Energetics of Vaporization

To understand the energetics of vaporization, consider again a beaker of water from the molecular point of view, except now the beaker is thermally insulated so that heat from the surroundings cannot enter the beaker. What happens to the temperature of the water left in the beaker as molecules evaporate? To answer this question, think about the energy distribution curve again (see Figure 11.18). The molecules that leave the beaker are the ones at the high end of the energy curve—the most energetic. If no additional heat enters the beaker, the average energy of the entire collection of molecules goes down—much as the class average on an exam goes down if you eliminate the highest-scoring students— and as a result, vaporization slows. So vaporization is an *endothermic* process; it takes energy to vaporize the molecules in a liquid. Another way to understand the endothermicity

See Chapter 6 to review endothermic and exothermic processes.

▶ **FIGURE 11.18 Distribution of Thermal Energy**   The thermal energies of the molecules in a liquid are distributed over a range. The peak energy increases with increasing temperature.

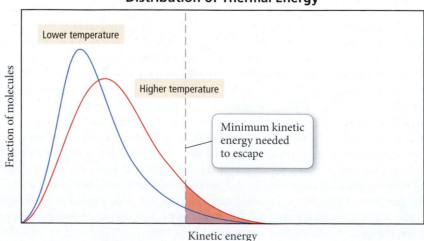

Distribution of Thermal Energy

Lower temperature

Higher temperature

Minimum kinetic energy needed to escape

Fraction of molecules

Kinetic energy

of vaporization is to remember that vaporization requires overcoming the intermolecular forces that hold liquids together. Since energy must be absorbed to pull the molecules apart, the process is endothermic.

Our bodies use the endothermic nature of vaporization for cooling. When you overheat, you sweat, causing your skin to be covered with liquid water. As this water evaporates, it absorbs heat from your body, cooling your skin as the water and the heat leave your body. A fan makes you feel cooler because it blows newly vaporized water away from your skin, allowing more sweat to vaporize and causing even more cooling. High humidity, on the other hand, slows down the net rate of evaporation, preventing cooling. When the air already contains large amounts of water vapor, the sweat evaporates more slowly, making the body's cooling system less efficient.

Condensation, the opposite of vaporization, is exothermic—heat is released when a gas condenses to a liquid. If you have ever accidentally put your hand above a steaming kettle, or opened a bag of microwaved popcorn too soon, you may have experienced a *steam burn*. As the steam condenses to a liquid on your skin, it releases a lot of heat, causing the burn. The condensation of water vapor is also the reason that winter overnight temperatures in coastal regions, which tend to have water vapor in the air, do not get as low as in deserts, which tend to have dry air. As the air temperature in a coastal area drops, water condenses out of the air, releasing heat and preventing the temperature from dropping further. In deserts, there is little moisture in the air to condense, so the temperature drop is greater.

Heat of Vaporization

The amount of heat required to vaporize one mole of a liquid to gas is called the **heat (or enthalpy) of vaporization (ΔH_{vap}).** The heat of vaporization of water at its normal boiling point of 100 °C is + 40.7 kJ/mol:

$$H_2O(l) \longrightarrow H_2O(g) \qquad \Delta H_{vap} = +40.7 \text{ kJ/mol}$$

The sign conventions of ΔH were introduced in Chapter 6.

The heat of vaporization is always positive because the process is endothermic—energy must be absorbed to vaporize a substance. The heat of vaporization is somewhat temperature dependent. For example, at 25 °C the heat of vaporization of water is + 44.0 kJ/mol, slightly more than at 100 °C because the water contains less thermal energy. Table 11.7 lists the heats of vaporization of several liquids at their boiling points and at 25 °C.

The heat of vaporization is also called the enthalpy of vaporization.

When a substance condenses from a gas to a liquid, the same amount of heat is involved, but the heat is emitted rather than absorbed.

$$H_2O(g) \longrightarrow H_2O(l) \qquad \Delta H = -\Delta H_{vap} = -40.7 \text{ kJ} \qquad (\text{at } 100 \text{ °C})$$

When one mole of water condenses, it releases 40.7 kJ of heat. The sign of ΔH in this case is negative because the process is exothermic.

We can use the heat of vaporization of a liquid to calculate the amount of heat energy required to vaporize a given mass of the liquid (or the amount of heat given off by the condensation of a given mass of liquid), using concepts similar to those covered in Section 6.5 (stoichiometry of ΔH). We essentially use the heat of vaporization as a conversion factor between number of moles of a liquid and the amount of heat required to vaporize it (or the amount of heat emitted when it condenses), as shown in Example 11.3.

TABLE 11.7 Heats of Vaporization of Several Liquids at Their Boiling Points and at 25 °C

Liquid	Chemical Formula	Normal Boiling Point (°C)	ΔH_{vap} (kJ/mol) at Boiling Point	ΔH_{vap} (kJ/mol) at 25 °C
Water	H_2O	100	40.7	44.0
Rubbing alcohol (isopropyl alcohol)	C_3H_8O	82.3	39.9	45.4
Acetone	C_3H_6O	56.1	29.1	31.0
Diethyl ether	$C_4H_{10}O$	34.6	26.5	27.1

EXAMPLE 11.3 Using the Heat of Vaporization in Calculations

Calculate the mass of water (in g) that can be vaporized at its boiling point with 155 kJ of heat.

SORT You are given a certain amount of heat in kilojoules and asked to find the mass of water that can be vaporized.	**GIVEN** 155 kJ **FIND** g H_2O
STRATEGIZE The heat of vaporization gives the relationship between heat absorbed and moles of water vaporized. Begin with the given amount of heat (in kJ) and convert to moles of water that can be vaporized. Then use the molar mass as a conversion factor to convert from moles of water to mass of water.	**CONCEPTUAL PLAN** 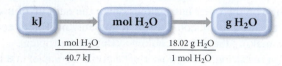 **RELATIONSHIPS USED** $\Delta H_{vap} = 40.7 \text{ kJ/mol}$ (at 100 °C) 18.02 g H_2O = 1 mol H_2O
SOLVE Follow the conceptual plan to solve the problem.	$155 \text{ kJ} \times \dfrac{1 \text{ mol } H_2O}{40.7 \text{ kJ}} \times \dfrac{18.02 \text{ g } H_2O}{1 \text{ mol } H_2O} = 68.6 \text{ g } H_2O$

FOR PRACTICE 11.3

Calculate the amount of heat (in kJ) required to vaporize 2.58 kg of water at its boiling point.

FOR MORE PRACTICE 11.3

Suppose that 0.48 g of water at 25 °C condenses on the surface of a 55-g block of aluminum that is initially at 25 °C. If the heat released during condensation goes only toward heating the metal, what is the final temperature (in °C) of the metal block? (The specific heat capacity of aluminum is 0.903 J/g °C.)

Vapor Pressure and Dynamic Equilibrium

We have already seen that if a container of water is left uncovered at room temperature, the water slowly evaporates away. But what happens if the container is sealed? Imagine a sealed evacuated flask—one from which the air has been removed—containing liquid water, as shown in **Figure 11.19▼**. Initially, the water molecules evaporate, as they did in the open beaker. However, because of the seal, the evaporated molecules cannot escape into the atmosphere. As water molecules enter the gas state, some start condensing back into the liquid. When the concentration (or partial pressure) of gaseous water molecules

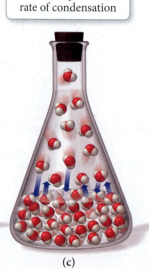

Dynamic equilibrium:
Rate of evaporation =
rate of condensation

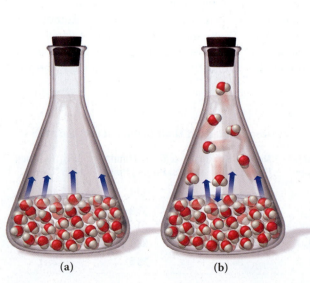

▶ **FIGURE 11.19 Vaporization in a Sealed Flask** **(a)** When water is placed into a sealed container, water molecules begin to vaporize. **(b)** As water molecules build up in the gas state, they begin to recondense into the liquid. **(c)** When the rate of evaporation equals the rate of condensation, dynamic equilibrium is reached.

(a)   (b)   (c)

increases, the rate of condensation also increases. However, as long as the water remains at a constant temperature, the rate of evaporation remains constant. Eventually the rate of condensation and the rate of vaporization become equal—**dynamic equilibrium** has been reached **Figure 11.20▶**. Although condensation and vaporization continue at equal rates, the concentration of water vapor above the liquid is constant.

The pressure of a gas in dynamic equilibrium with its liquid is called its **vapor pressure.** The vapor pressure of a particular liquid depends on the intermolecular forces present in the liquid and the temperature. Weak intermolecular forces result in volatile substances with high vapor pressures because the intermolecular forces are easily overcome by thermal energy. Strong intermolecular forces result in nonvolatile substances with low vapor pressures.

A liquid in dynamic equilibrium with its vapor is a balanced system that tends to return to equilibrium if disturbed. For example, consider a sample of *n*-pentane (a component of gasoline) at 25 °C in a cylinder equipped with a moveable piston **Figure 11.21(a)▼**. The cylinder contains no other gases except *n*-pentane vapor in dynamic equilibrium with the liquid. Since the vapor pressure of *n*-pentane at 25 °C is 510 mmHg, the pressure in the cylinder is 510 mmHg. Now, what happens when the piston is moved upward to expand the volume within the cylinder? Initially, the pressure in the cylinder drops below 510 mmHg in accordance with Boyle's law. Then, however, more liquid vaporizes until equilibrium is reached once again **Figure 11.21(b)▼**. If the volume of the cylinder is expanded again, the same thing happens—the pressure initially drops and more *n*-pentane vaporizes to bring the system back into equilibrium. Further expansion will cause the same result *as long as some liquid n-pentane remains in the cylinder.*

Conversely, what happens if the piston is lowered, decreasing the volume in the cylinder? Initially, the pressure in the cylinder rises above 510 mmHg, but then some of the gas condenses into liquid until equilibrium is reached again **Figure 11.21(c)▼**.

We can describe the tendency of a system in dynamic equilibrium to return to equilibrium with the following general statement:

> **When a system in dynamic equilibrium is disturbed, the system responds so as to minimize the disturbance and return to a state of equilibrium.**

If the pressure above the system is decreased, the pressure increases (some of the liquid evaporates) so that the system returns to equilibrium. If the pressure above a liquid–vapor system in equilibrium is increased, the pressure of the system drops (some of the gas condenses) so that the system returns to equilibrium. This basic principle—often called Le Châtelier's principle—is applicable to any chemical system in equilibrium, as we will see in Chapter 14.

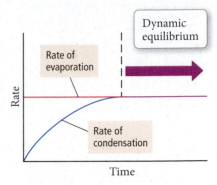

▲ FIGURE 11.20 Dynamic Equilibrium Dynamic equilibrium occurs when the rate of condensation is equal to the rate of evaporation.

Boyle's law is discussed in Section 5.3.

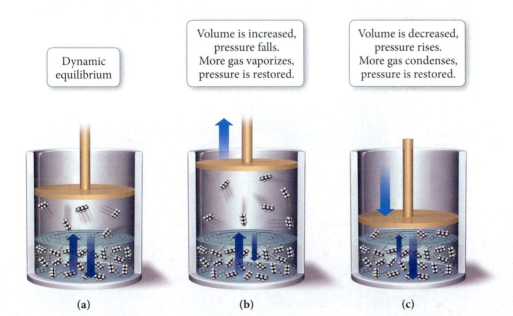

◀ FIGURE 11.21 Dynamic Equilibrium in *n*-Pentane (a) Liquid *n*-pentane is in dynamic equilibrium with its vapor. (b) When the volume is increased, the pressure drops and some liquid is converted to gas to bring the pressure back up. (c) When the volume is decreased, the pressure increases and some gas is converted to liquid to bring the pressure back down.

Conceptual Connection 11.4 Vapor Pressure

What happens to the vapor pressure of a substance when its surface area is increased at constant temperature?

(a) The vapor pressure increases.

(b) The vapor pressure remains the same.

(c) The vapor pressure decreases.

Temperature Dependence of Vapor Pressure and Boiling Point

When the temperature of a liquid increases, its vapor pressure rises because the higher thermal energy increases the number of molecules that have enough energy to vaporize (see Figure 11.18). Because of the shape of the energy distribution curve, however, a small change in temperature makes a large difference in the number of molecules that have enough energy to vaporize, which results in a large increase in vapor pressure. For example, the vapor pressure of water at 25 °C is 23.3 torr, while at 60 °C the vapor pressure is 149.4 torr. **Figure 11.22▼** shows the vapor pressure of water and several other liquids as a function of temperature.

The **boiling point** of a liquid is *the temperature at which its vapor pressure equals the external pressure.* When a liquid reaches its boiling point, the thermal energy is enough for molecules in the interior of the liquid (not just those at the surface) to break free of their neighbors and enter the gas state **Figure 11.23▼**. The bubbles you see in boiling water are pockets of gaseous water that have formed within the liquid water. The bubbles float to the surface and leave as gaseous water or steam.

The **normal boiling point** of a liquid is *the temperature at which its vapor pressure equals 1 atm.* The normal boiling point of pure water is 100 °C. However, at a lower pressure, water boils at a lower temperature. In Denver, Colorado, where the altitude is around 1600 meters (5200 feet) above sea level, for example, the average atmospheric pressure is about 83% of what it is at sea level, and water boils at approximately 94 °C. For this reason, it takes slightly longer to cook food in boiling water in Denver than in San Francisco (which is at sea level). Once the boiling point of a liquid is reached, additional heating only causes more rapid boiling; it does not raise the temperature of the liquid above its boiling point, as shown in the *heating curve* in **Figure 11.24▶**. Therefore, boiling water at 1 atm will always have a temperature of 100 °C. *As long as liquid water is present, its temperature cannot rise above its boiling point.* However, after all the water has been converted to steam, the temperature of the steam can continue to rise beyond 100 °C.

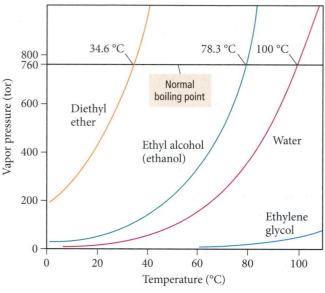

▲ **FIGURE 11.22 Vapor Pressure of Several Liquids at Different Temperatures**   At higher temperatures, more molecules have enough thermal energy to escape into the gas state, so vapor pressure increases with increasing temperature.

▶ **FIGURE 11.23 Boiling**   A liquid boils when thermal energy is high enough to cause molecules in the interior of the liquid to become gaseous, forming bubbles that rise to the surface.

The Clausius–Clapeyron Equation

Now, let's return our attention to Figure 11.22. As you can see from the graph, the vapor pressure of a liquid increases with increasing temperature. However, *the relationship is not linear*. In other words, doubling the temperature results in more than a doubling of the vapor pressure. The relationship between vapor pressure and temperature is exponential, and can be expressed as follows:

$$P_{vap} = \beta \exp\left(\frac{-\Delta H_{vap}}{RT}\right) \qquad [11.1]$$

In this expression P_{vap} is the vapor pressure, β is a constant that depends on the gas, ΔH_{vap} is the heat of vaporization, R is the gas constant (8.314 J/mol·K), and T is the temperature in kelvins. Equation 11.1 can be rearranged by taking the natural logarithm of both sides as follows:

$$\ln P_{vap} = \ln\left[\beta \exp\left(\frac{-\Delta H_{vap}}{RT}\right)\right] \qquad [11.2]$$

Since $\ln AB = \ln A + \ln B$, we can rearrange the right side of Equation 11.2:

$$\ln P_{vap} = \ln \beta + \ln\left[\exp\left(\frac{-\Delta H_{vap}}{RT}\right)\right] \qquad [11.3]$$

Since $\ln e^x = x$ (see Appendix IB), we can simplify Equation 11.3:

$$\ln P_{vap} = \ln \beta + \frac{-\Delta H_{vap}}{RT} \qquad [11.4]$$

A slight additional rearrangement gives us the following important result:

$$\ln P_{vap} = \frac{-\Delta H_{vap}}{R}\left(\frac{1}{T}\right) + \ln \beta \qquad \text{Clausius–Clapeyron equation}$$

$$y = m(x) + b \qquad \text{(equation for a line)}$$

Notice the parallel relationship between the **Clausius–Clapeyron equation** and the equation for a straight line. Just as a plot of y versus x yields a straight line with slope m and intercept b, so a plot of $\ln P_{vap}$ (equivalent to y) versus $1/T$ (equivalent to x) gives a straight line with slope $-\Delta H_{vap}/R$ (equivalent to m) and y-intercept $\ln \beta$ (equivalent to b), as shown in **Figure 11.25▼**. The Clausius–Clapeyron equation gives a linear relationship—not between the vapor pressure and the temperature (which have an exponential relationship)—but between the *natural log* of the vapor pressure and the *inverse* of temperature. This is a common technique in the analysis of chemical

▲ **FIGURE 11.24 Temperature during Boiling**   The temperature of water during boiling remains at 100 °C.

Using the Clausius–Clapeyron equation in this way ignores the relatively small temperature dependence of ΔH_{vap}.

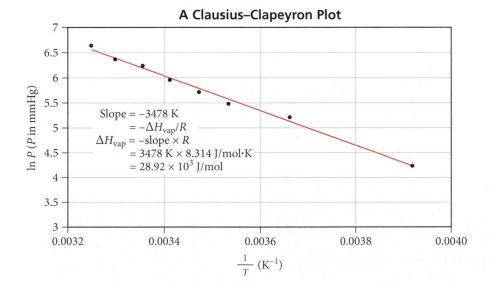

A Clausius–Clapeyron Plot

Slope = −3478 K
= −ΔH_{vap}/R
ΔH_{vap} = −slope × R
= 3478 K × 8.314 J/mol·K
= 28.92 × 10³ J/mol

◀ **FIGURE 11.25 A Clausius–Clapeyron Plot for Diethyl Ether (CH₃CH₂OCH₂CH₃)**   A plot of the natural log of the vapor pressure versus the inverse of the temperature in K yields a straight line with slope $-\Delta H_{vap}/R$.

data. If two variables are not linearly related, it is often convenient to find ways to graph *functions of those variables* that are linearly related.

The Clausius–Clapeyron equation leads to a convenient way to measure the heat of vaporization in the laboratory. We measure the vapor pressure of a liquid as a function of temperature and create a plot of the natural log of the vapor pressure versus the inverse of the temperature. We can then determine the slope of the line to find the heat of vaporization, as shown in Example 11.4.

EXAMPLE 11.4 Using the Clausius–Clapeyron Equation to Determine Heat of Vaporization from Experimental Measurements of Vapor Pressure

The vapor pressure of dichloromethane was measured as a function of temperature, and the following results were obtained:

Temperature (K)	Vapor Pressure (torr)
200	0.8
220	4.5
240	21
260	71
280	197
300	391

Determine the heat of vaporization of dichloromethane.

SOLUTION

To find the heat of vaporization, use an Excel spreadsheet or a graphing calculator to make a plot of the natural log of vapor pressure ($\ln P$) as a function of the inverse of the temperature in kelvins ($1/T$). Then fit the points to a line and determine the slope of the line. The slope of the best fitting line is -3805 K. Since the slope equals $-\Delta H_{vap}/R$, we find the heat of vaporization as follows:

$$\text{slope} = -\Delta H_{vap}/R$$

$$\Delta H_{vap} = -slope \times R$$

$$= -(-3805 \text{ K})(8.314 \text{ J}/\text{mol} \cdot \text{K})$$

$$= 3.16 \times 10^4 \text{ J}/\text{mol}$$

$$= 31.6 \text{ kJ}/\text{mol}$$

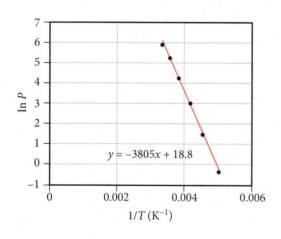

$y = -3805x + 18.8$

FOR PRACTICE 11.4

The vapor pressure of carbon tetrachloride was measured as a function of the temperature and the following results were obtained:

Temperature (K)	Vapor Pressure (torr)
255	11.3
265	21.0
275	36.8
285	61.5
295	99.0
300	123.8

Determine the heat of vaporization of carbon tetrachloride.

The Clausius–Clapeyron equation can also be expressed in a two-point form that can be used with just two measurements of vapor pressure and temperature to determine the heat of vaporization.

$$\ln \frac{P_2}{P_1} = \frac{-\Delta H_{vap}}{R}\left(\frac{1}{T_2} - \frac{1}{T_1}\right)$$ Clausius–Clapeyron equation (two-point form)

This form of the equation can also be used to predict the vapor pressure of a liquid at any temperature if you know the enthalpy of vaporization and the normal boiling point (or the vapor pressure at some other temperature), as shown in Example 11.5.

> The two-point method is generally inferior to plotting multiple points because fewer data points result in greater possible error.

EXAMPLE 11.5 Using the Two-Point Form of the Clausius–Clapeyron Equation to Predict the Vapor Pressure at a Given Temperature

Methanol has a normal boiling point of 64.6 °C and a heat of vaporization (ΔH_{vap}) of 35.2 kJ/mol. What is the vapor pressure of methanol at 12.0 °C?

SORT The problem gives you the normal boiling point of methanol (the temperature at which the vapor pressure is 760 mmHg) and the heat of vaporization. You are asked to find the vapor pressure at a specified temperature, which is also given.	**GIVEN** $T_1(°C) = 64.6 °C$ $P_1 = 760$ torr $\Delta H_{vap} = 35.2$ kJ/mol $T_2(°C) = 12.0 °C$ **FIND** P_2
STRATEGIZE The conceptual plan is essentially the Clausius-Clapeyron equation, which relates the given and find quantities.	**CONCEPTUAL PLAN** $$\ln \frac{P_2}{P_1} = \frac{-\Delta H_{vap}}{R}\left(\frac{1}{T_2} - \frac{1}{T_1}\right)$$ (Clausius–Clapeyron equation, two-point form)
SOLVE First, convert T_1 and T_2 from °C to K.	**SOLUTION** $T_1(K) = T_1(°C) + 273.15$ $= 64.6 + 273.15$ $= 337.8$ K $T_2(K) = T_2(°C) + 273.15$ $= 12.0 + 273.15$ $= 285.2$ K
Then, substitute the required values into the Clausius–Clapeyron equation and solve for P_2.	$$\ln \frac{P_2}{P_1} = \frac{-\Delta H_{vap}}{R}\left(\frac{1}{T_2} - \frac{1}{T_1}\right)$$ $$\ln \frac{P_2}{P_1} = \frac{-35.2 \times 10^3 \frac{J}{mol}}{8.314 \frac{J}{mol \cdot K}}\left(\frac{1}{285.2K} - \frac{1}{337.8K}\right)$$ $= -2.31$ $\frac{P_2}{P_1} = e^{-2.31}$ $P_2 = P_1(e^{-2.31})$ $= 760$ torr(0.0993) $= 75.4$ torr

CHECK The units of the answer are correct. The magnitude of the answer makes sense because vapor pressure should be significantly lower at the lower temperature.

FOR PRACTICE 11.5

Propane has a normal boiling point of −42.0 °C and a heat of vaporization (ΔH_{vap}) of 19.04 kJ/mol. What is the vapor pressure of propane at 25.0 °C?

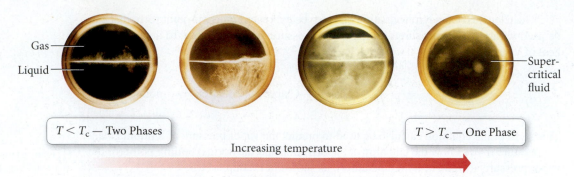

Gas

Liquid

Super-
critical
fluid

$T < T_c$ — Two Phases

$T > T_c$ — One Phase

Increasing temperature

▲ **FIGURE 11.26 Critical Point Transition**   As *n*-pentane is heated in a sealed container, it undergoes a transition to a supercritical fluid. At the critical point, the meniscus separating the liquid and gas disappears, and the fluid becomes supercritical—neither a liquid nor a gas.

The Critical Point: The Transition to an Unusual State of Matter

We have considered the vaporization of a liquid in a container open to the atmosphere with and without heating, and the vaporization of a liquid in a *sealed* container without heating. We now examine the vaporization of a liquid in a *sealed* container *while heating*. Consider liquid *n*-pentane in equilibrium with its vapor in a sealed container initially at 25 °C. At this temperature, the vapor pressure of *n*-pentane is 0.67 atm. What happens if the liquid is heated? As the temperature rises, more *n*-pentane vaporizes and the pressure within the container increases. At 100 °C, the pressure is 5.5 atm, and at 190 °C the pressure is 29 atm. As more and more gaseous *n*-pentane is forced into the same amount of space, the density of the *gas* becomes higher and higher. At the same time, the increasing temperature causes the density of the *liquid* to become lower and lower. At 197 °C, the meniscus between the liquid and gaseous *n*-pentane disappears and the gas and liquid states commingle to form a *supercritical fluid* **Figure 11.26▲**. For any substance, the *temperature* at which this transition occurs is called the **critical temperature (T_c)**—it represents the temperature above which the liquid cannot exist (regardless of pressure). The *pressure* at which this transition occurs is called the **critical pressure (P_c)**—it represents the pressure required to bring about a transition to a liquid at the critical temperature.

11.6   Sublimation and Fusion

In Section 11.5, we examined a beaker of liquid water at room temperature from the molecular viewpoint. Now, let's examine a block of ice at −10 °C from the same molecular perspective, paying close attention to two common processes: sublimation and fusion.

Sublimation

Even though a block of ice is solid, the water molecules have thermal energy which causes each one to vibrate about a fixed point. The motion is much less than in a liquid, but significant nonetheless. Like the molecules in liquids, at any one time some molecules in a solid block of ice will have more thermal energy than the average and some will have less. The molecules with high enough thermal energy can break free from the ice surface—where molecules are held less tightly than in the interior due to fewer neighbor–neighbor interactions—and go directly into the gas state **Figure 11.27◀**. This process is called **sublimation,** the transition from solid to gas. Some of the water molecules in the gas state (those at the low end of the energy distribution curve for the gaseous molecules) can collide with the surface of the ice and be captured by the intermolecular forces with other molecules. This process—the opposite of sublimation—is called **deposition,** the transition from gas to solid. As is the case with liquids, the pressure of a gas in dynamic equilibrium with its solid is the vapor pressure of the solid.

Although both sublimation and deposition are happening on the surface of an ice block open to the atmosphere at −10 °C, sublimation is usually happening at a greater

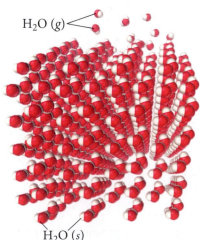

H_2O (*g*)

H_2O (*s*)

▲ **FIGURE 11.27 The Sublimation of Ice**   The water molecules at the surface of an ice cube can sublime directly into the gas state.

rate because most of the newly sublimed molecules escape into the surrounding atmosphere and never come back. The result is a noticeable decrease in the size of the ice block over time (even though the temperature is below the melting point).

If you live in a cold climate, you may have noticed the disappearance of ice and snow from the ground even though the temperature remains below 0 °C. Similarly, ice cubes left in the freezer for a long time slowly shrink, even though the freezer is always set below 0 °C. In both cases, the ice is subliming, turning directly into water vapor. Ice also sublimes out of frozen foods. You may have noticed, for example, the gradual growth of ice crystals on the *inside* of airtight plastic food-storage bags in your freezer. The ice crystals are composed of water that has sublimed out of the food and redeposited on the surface of the bag or on the surface of the food. For this reason, food that remains frozen for too long becomes dried out. Such dehydration can be avoided to some degree by freezing foods to colder temperatures, a process called deep-freezing. The colder temperature lowers the vapor pressure of ice and preserves the food longer. Freezer burn on meats is another common manifestation of sublimation. When meat is improperly stored (that is, when its container is not airtight) sublimation continues unabated. The result is the dehydration of the surface of the meat, which becomes discolored and loses flavor and texture.

A substance commonly associated with sublimation is solid carbon dioxide or dry ice, which does not melt under atmospheric pressure no matter what the temperature. However, at −78 °C the CO_2 molecules have enough energy to leave the surface of the dry ice and become gaseous through sublimation.

▲ The ice crystals that form on frozen food are due to sublimation of water from the food and redeposition on its surface.

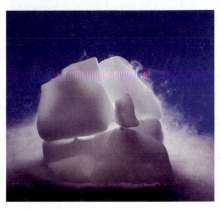

▲ Dry ice (solid CO_2) sublimes but does not melt at atmospheric pressure.

Fusion

Let's return to our ice block and examine what happens at the molecular level as we increase its temperature. The increasing thermal energy causes the water molecules to vibrate faster and faster. At the **melting point** (0 °C for water), the molecules have enough thermal energy to overcome the intermolecular forces that hold them at their stationary points, and the solid turns into a liquid. This process is called **melting or fusion,** the transition from solid to liquid. The opposite of melting is **freezing,** the transition from liquid to solid. Once the melting point of a solid is reached, additional heating only causes more rapid melting; it does not raise the temperature of the solid above its melting point **Figure 11.28▶**. Only after all of the ice has melted will additional heating raise the temperature of the liquid water past 0 °C. A mixture of water *and* ice will always have a temperature of 0 °C (at 1 atm pressure).

The term fusion is used for melting because if you heat several crystals of a solid, they will *fuse* into a continuous liquid upon melting.

Energetics of Melting and Freezing

The most common way to cool a beverage quickly is to drop several ice cubes into it. As the ice melts, the drink cools because melting is endothermic—the melting ice absorbs heat from the liquid. The amount of heat required to melt 1 mol of a solid is called the **heat of fusion** (ΔH_{fus}). The heat of fusion for water is 6.02 kJ/mol:

$$H_2O(s) \longrightarrow H_2O(l) \qquad \Delta H_{fus} = 6.02 \text{ kJ/mol}$$

The heat of fusion is positive because melting is endothermic.

Freezing, the opposite of melting, is exothermic—heat is released when a liquid freezes into a solid. For example, as water in your freezer turns into ice, it releases heat, which must be removed by the refrigeration system of the freezer. If the refrigeration system did not remove the heat, the water would not completely freeze into ice. The heat released as the water began to freeze would warm the freezer, preventing further freezing. The change in enthalpy for freezing has the same magnitude as the heat of fusion but the opposite sign.

$$H_2O(l) \longrightarrow H_2O(s) \qquad \Delta H = -\Delta H_{fus} = -6.02 \text{ kJ/mol}$$

Different substances have different heats of fusion, as shown in Table 11.8.

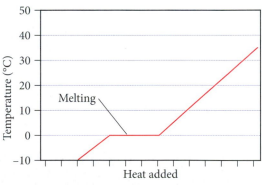

▲ **FIGURE 11.28 Temperature during Melting** The temperature of water during melting remains at 0.0 °C as long as both solid and liquid water remain.

TABLE 11.8 Heats of Fusion of Several Substances			
Liquid	Chemical Formula	Melting Point (°C)	ΔH_{vap} (kJ/mol)
Water	H_2O	0.00	6.02
Rubbing alcohol (isopropyl alcohol)	C_3H_8O	−89.5	5.37
Acetone	C_3H_6O	−94.8	5.69
Diethyl ether	$C_4H_{10}O$	−116.3	7.27

In general, the heat of fusion is significantly less than the heat of vaporization, as shown in **Figure 11.29▼**. We have already mentioned that the solid and liquid states are closer to each other in many ways than they are to the gas state. It takes less energy to melt 1 mol of ice into liquid than it does to vaporize 1 mol of liquid water into gas because vaporization requires complete separation of molecules from one another, so the intermolecular forces must be completely overcome. Melting, however, requires that intermolecular forces be only partially overcome, allowing molecules to move around one another while still remaining in contact.

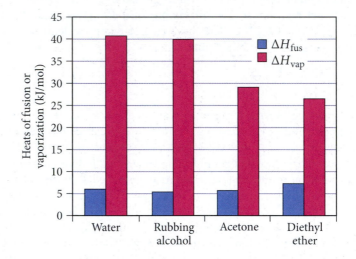

▶ **FIGURE 11.29 Heat of Fusion and Heat of Vaporization**   Typical heats of fusion are significantly less than heats of vaporization.

11.7   Heating Curve for Water

We can combine and build on the concepts from the previous two sections by examining the *heating curve* for 1.00 mol of water at 1.00 atm pressure shown in **Figure 11.30▶**. The *y*-axis of the heating curve represents the temperature of the water sample. The *x*-axis represents the amount of heat added (in kilojoules) during heating. As you can see from the diagram, the process can be divided into five segments: (1) ice warming; (2) ice melting into liquid water; (3) liquid water warming; (4) liquid water vaporizing into steam; and (5) steam warming.

In two of these segments (2 and 4) the temperature is constant as heat is added because the added heat goes into producing the transition, not increasing the temperature. The two states are in equilibrium during the transition and the temperature remains constant. The amount of heat required to achieve the state change is given by $q = n\,\Delta H$.

In the other three segments (1, 3, and 5), temperature increases linearly. These segments represent the heating of a single state in which the deposited heat raises the temperature in accordance with the substance's specific heat capacity ($q = mC_s\,\Delta T$).

Conceptual Connection 11.5   Cooling Water with Ice

The specific heat capacity of ice is $C_{s,\,ice} = 2.09\ \text{J/g}\cdot{}^\circ\text{C}$ and the heat of fusion of ice is 6.02 kJ/mol. When a small ice cube at −10 °C is put into a cup of water at room temperature, which plays a larger role in cooling the liquid water: the warming of the ice from −10 °C to 0 °C, or the melting of the ice?

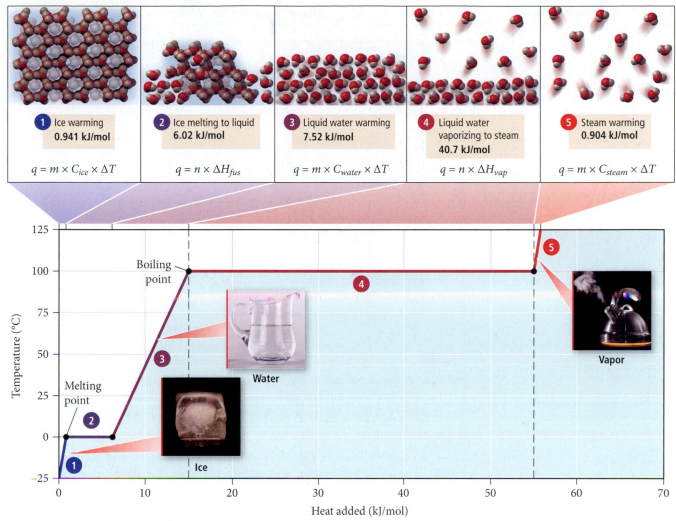

① Ice warming
0.941 kJ/mol

$q = m \times C_{ice} \times \Delta T$

② Ice melting to liquid
6.02 kJ/mol

$q = n \times \Delta H_{fus}$

③ Liquid water warming
7.52 kJ/mol

$q = m \times C_{water} \times \Delta T$

④ Liquid water vaporizing to steam
40.7 kJ/mol

$q = n \times \Delta H_{vap}$

⑤ Steam warming
0.904 kJ/mol

$q = m \times C_{steam} \times \Delta T$

▲ **FIGURE 11.30 Heating Curve for Water**   The graph shows the temperature as a function of heat added while heating one mole of water.

11.8 Phase Diagrams

Throughout most of this chapter, we have examined how the state of a substance changes with temperature and pressure. We can combine both the temperature dependence and pressure dependence of the state of a particular substance in a graph called a *phase diagram*. A **phase diagram** is a map of the state of a substance as a function of pressure (on the *y*-axis) and temperature (on the *x*-axis). Let's first examine the major features of a phase diagram, then turn to navigating within a phase diagram.

The Major Features of a Phase Diagram

Consider the phase diagram of water as an example **Figure 11.31▶**. The *y*-axis displays the pressure in torr and the *x*-axis shows the temperature in degrees Celsius. The main features of the phase diagram can be categorized as regions, lines, and points. We examine each of these individually.

Regions

Any of the three main regions—solid, liquid, and gas—in the phase diagram represent conditions where that particular state is stable. For example, under any of the temperatures and pressures within the liquid region in the phase diagram of water, the liquid is the stable state. Notice that the point 25 °C and 760 torr falls within the liquid region, as we know from everyday experience. In general, low temperature and high pressure favor the solid state; high temperature and low pressure favor the gas state; and intermediate conditions favor the

▶ **FIGURE 11.31 Phase Diagram for Water**

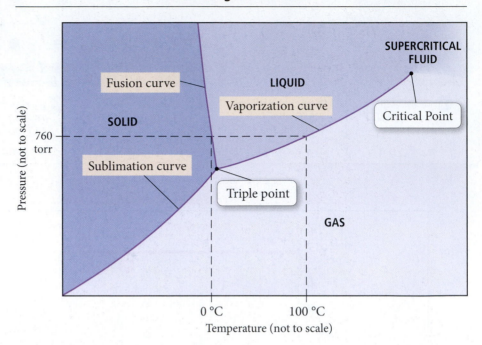

Phase Diagram for Water

The triple point of a substance such as water can be reproduced anywhere to calibrate a thermometer or pressure gauge with a known temperature and pressure.

liquid state. A sample of matter that is not in the state indicated by its phase diagram for a given set of conditions will convert to that state when those conditions are imposed. For example, steam that is cooled to room temperature at 1 atm will condense to liquid.

Lines
Each of the lines (curves) in the phase diagram represents a set of temperatures and pressures at which the substance is in equilibrium between the two states on either side of the line. For example, in the phase diagram for water, consider the curved line beginning just beyond 0 °C separating the liquid from the gas. This line is the vaporization curve (also called the vapor pressure curve) for water that we examined in Section 11.5. At any of the temperatures and pressures that fall along this line, the liquid and gas states of water are equally stable and in equilibrium. For example, at 100 °C and 760 torr pressure, water and its vapor are in equilibrium—they are equally stable and will coexist. The other two major lines in a phase diagram are the sublimation curve (separating the solid and the gas) and the fusion curve (separating the solid and the liquid).

The Triple Point
The **triple point** *in a phase diagram represents the unique set of conditions at which three states are equally stable and in equilibrium.* In the phase diagram for water, the triple point occurs at 0.0098 °C and 4.58 torr. Under these unique conditions (and only under these conditions), the solid, liquid, and gas states of water are equally stable and will coexist in equilibrium.

The Critical Point
As we learned in Section 11.5, at the critical temperature and pressure, the liquid and gas states coalesce into a *supercritical fluid*. *The* **critical point** *in a phase diagram represents the temperature and pressure above which a supercritical fluid exists.*

Navigation within a Phase Diagram
Changes in the temperature or pressure of a sample can be represented by movement within the phase diagram. Consider the phase diagram of carbon dioxide (dry ice) shown in **Figure 11.32▶**. An increase in the temperature of a block of solid carbon dioxide at 1 atm can be indicated by horizontal movement in the phase diagram as shown by line A, which crosses the sublimation curve at −78.5 °C. At this temperature, the solid sublimes to a gas, as you may have observed with dry ice. A change in pressure can be indicated by vertical movement in the phase diagram. For example, a sample of gaseous carbon dioxide at 0 °C

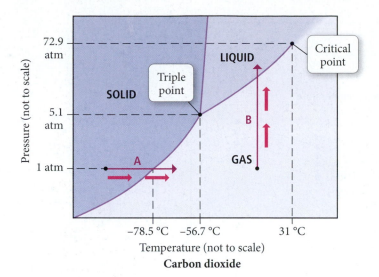

and 1 atm could be converted to a liquid by increasing the pressure, as shown by line B.
Notice that the fusion curve for carbon dioxide has a positive slope—as the temperature
increases the pressure also increases—in contrast to the fusion curve for water, which has
a negative slope. The behavior of carbon dioxide is more typical than that of water. The
fusion curve within the phase diagrams for most substances has a positive slope because
increasing pressure favors the denser state, which for most substances is the solid state.

 Conceptual Connection 11.6 Phase Diagrams

A substance has a triple point at −24.5 °C and 225 mm Hg. What is most likely to happen
to a solid sample of the substance as it is warmed from −35 °C to 0 °C at a pressure of
220 mm Hg?

(a) The solid will melt into a liquid.

(b) The solid will sublime into a gas.

(c) Nothing (the solid will remain as a solid).

11.9 Water: An Extraordinary Substance

Water is the most common and important liquid on Earth. It fills our oceans, lakes, and
streams. In its solid form, it caps our mountains, and in its gaseous form, it humidifies our
air. We drink water, we sweat water, and we excrete bodily wastes dissolved in water.
Indeed, the majority of our body mass *is* water. Life is impossible without water, and in
most places on Earth where liquid water exists, life exists. Recent evidence for water on
Mars in the past has fueled hopes of finding life or evidence of past life there. And though
it may not always be obvious to us (because we take water for granted), this familiar substance turns out to have many remarkable properties.

Among liquids, water is unique. It has a low molar mass (18.02 g/mol), yet it is a liquid
at room temperature. Other main-group hydrides have higher molar masses but lower boiling
points, as shown in **Figure 11.33▶**. No other substance of similar molar mass (except for HF)
comes close to being a liquid at room temperature. The high boiling point of water (in spite of
its low molar mass) can be understood by examining its molecular structure. The bent geometry of the water molecule and the highly polar nature of the O—H bonds result in a molecule with a significant dipole moment. Water's two O—H bonds (hydrogen directly bonded
to oxygen) allow a water molecule to form strong hydrogen bonds with four other water molecules **Figure 11.34▶**, resulting in a relatively high boiling point. Water's high polarity also
allows it to dissolve many other polar and ionic compounds, and even a number of nonpolar
gases such as oxygen and carbon dioxide (by inducing a dipole moment in their molecules).
Consequently, water is the main solvent within living organisms, transporting nutrients and
other important compounds throughout the body. Water is also the main solvent in our environment, allowing aquatic animals to survive by breathing dissolved oxygen and allowing
aquatic plants to survive by using dissolved carbon dioxide for photosynthesis.

▲ The *Phoenix Mars Lander* looked
for evidence of life in frozen water that
lies below the surface of Mars's north
polar region.

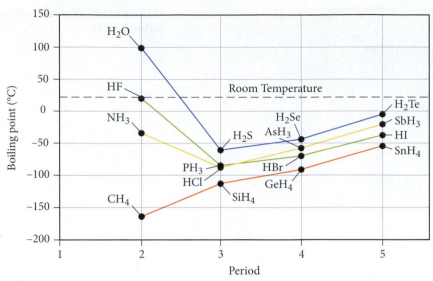

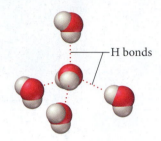

▲ **FIGURE 11.34 Hydrogen Bonding in Water** A water molecule can form four strong hydrogen bonds with four other water molecules.

▲ **FIGURE 11.33 Boiling Points of Main Group Hydrides** Water is the only common main-group hydride that is a liquid at room temperature.

▲ When lettuce freezes, the water within its cells expands, rupturing the cells.

We have already seen in Section 6.3 that water has an exceptionally high specific heat capacity, which has a moderating effect on the climate of coastal cities. In some cities, such as San Francisco, for example, the daily fluctuation in temperature can be less than 10 °C. This same moderating effect occurs over the entire planet, two-thirds of which is covered by water. Without water, the daily temperature fluctuations on our planet might be more like those on Mars, where temperature fluctuations of 63 °C (113 °F) have been measured between midday and early morning. Imagine awakening to below freezing temperatures, only to bake at summer desert temperatures in the afternoon! The presence of water on Earth and its uniquely high specific heat capacity is largely responsible for our much smaller daily fluctuations.

The way water freezes is also unique. Unlike other substances, which contract upon freezing, water expands upon freezing. Consequently, ice is less dense than liquid water, and it floats. This seemingly trivial property has significant consequences. The frozen layer of ice at the surface of a winter lake insulates the water in the lake from further freezing. If this ice layer sank, it would kill bottom-dwelling aquatic life and possibly allow the lake to freeze solid, eliminating virtually all life in the lake.

The expansion of water upon freezing, however, is one reason that most organisms do not survive freezing. When the water within a cell freezes, it expands and often ruptures the cell, just as water freezing within a pipe bursts the pipe. Many foods, especially those with high water content, do not survive freezing very well either. Have you ever tried, for example, to freeze your own vegetables? If you put lettuce or spinach in the freezer, it will be limp and damaged when you defrost it.

11.10 Crystalline Solids: Unit Cells and Basic Structures

Solids may be crystalline (comprising a well-ordered array of atoms or molecules) or amorphous (having no long-range order). Crystalline solids are composed of atoms or molecules arranged in structures with long-range order (see Section 11.2). If you have ever visited the mineral section of a natural history museum and seen crystals with smooth faces and well-defined angles between them, or if you have carefully observed the hexagonal shapes of snowflakes, you have witnessed some of the effects of the underlying order in crystalline solids. The often beautiful geometric shape that you see on the macroscopic scale is the result of a specific structural arrangement—called the **crystalline lattice**—on the molecular and atomic scale.

The crystalline lattice of any solid is nature's way of aggregating the particles to minimize their energy. The crystalline lattice can be represented by a small collection of atoms, ions, or molecules called the **unit cell.** When the unit cell is repeated over and

▲ The well-defined angles and smooth faces of crystalline solids reflect the underlying order of the atoms composing them.

◀ The hexagonal shape of a snowflake derives from the hexagonal arrangement of water molecules in crystalline ice.

over—like the tiles of a floor or the pattern in a wallpaper design, but in three dimensions—the entire lattice can be reproduced. For example, consider the two-dimensional crystalline lattice shown below. The unit cell for this lattice is the dark-colored square. Each circle represents a *lattice point*, a point in space occupied by an atom, ion, or molecule. Notice that by repeating and moving the pattern in the square throughout the two-dimensional space, we can generate the entire lattice.

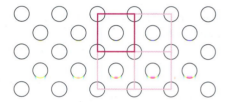

Unit cells are often classified by their symmetry, and many different unit cells exist. In this book, we focus primarily on *cubic unit cells* (although we look at one hexagonal unit cell). Cubic unit cells are characterized by equal edge lengths and 90° angles at their corners. The three cubic unit cells—simple cubic, body-centered cubic, and face-centered cubic—along with some of their basic characteristics, are shown in **Figure 11.35▼**.

Simple cubic

$l = 2r$

▲ In the simple cubic lattice, the atoms touch along each edge so that the edge length is $2r$.

Cubic Cell Name	Atoms per Unit Cell	Structure	Coordination Number	Edge Length in terms of r	Packing Efficiency (fraction of volume occupied)
Simple Cubic	1		6	$2r$	52%
Body-centered Cubic	2		8	$\dfrac{4r}{\sqrt{3}}$	68%
Face-centered Cubic	4		12	$2\sqrt{2}r$	74%

▲ **FIGURE 11.35 The Cubic Crystalline Lattices**   The different atom colors in this figure are for clarity only. All the atoms within each structure are identical.

Simple Cubic Unit Cell

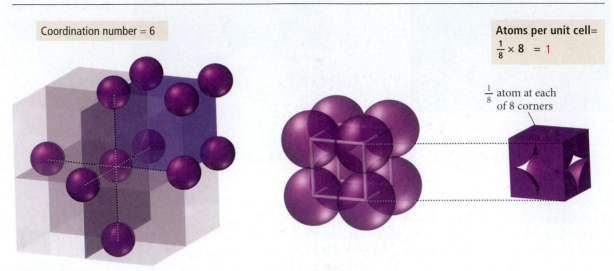

Coordination number = 6

Atoms per unit cell=
$$\frac{1}{8} \times 8 = 1$$

$\frac{1}{8}$ atom at each
of 8 corners

▲ **FIGURE 11.36** Simple Cubic Crystal Structure

Unit cells, such as the cubic ones shown here, are customarily portrayed with "whole" atoms, even though only a part of the whole atom may actually be in the unit cell.

The **simple cubic** unit cell **Figure 11.36**▲ consists of a cube with one atom at each corner. The atoms touch along each edge of the cube, so the edge length is twice the radius of the atoms ($l = 2r$). Note that even though the unit cell may seem as though it contains eight atoms, it actually contains only one. Each corner atom is shared by eight other unit cells. Any one unit cell actually contains only one-eighth of each of the eight atoms at its corners, for a total of only one atom per unit cell.

A characteristic feature of any unit cell is the **coordination number,** the number of atoms with which each atom is in *direct contact*. The coordination number represents the number of atoms with which a particular atom can have a strong interaction. The coordination number for the simple cubic unit cell is 6, because any one atom touches only six others, as you can see in Figure 11.36. A quantity closely related to the coordination number is the **packing efficiency,** the percentage of the volume of the unit cell occupied by the spheres. The higher the coordination number, the greater the packing efficiency. The simple cubic unit cell has a packing efficiency of 52%—there is a lot of empty space in the simple cubic unit cell.

The **body-centered cubic** unit cell **Figure 11.37**▼ consists of a cube with one atom at each corner and one atom of the same kind in the very center of the cube. Note that in the body-centered unit cell, the atoms *do not* touch along each edge of the cube, but rather

Body-Centered Cubic Unit Cell

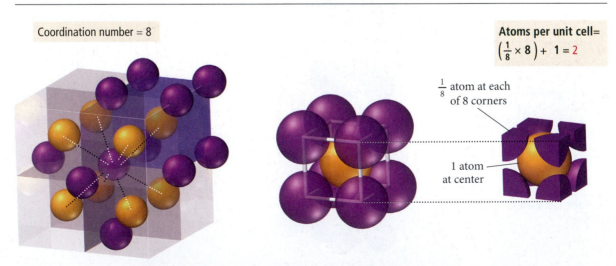

Coordination number = 8

Atoms per unit cell=
$$\left(\frac{1}{8} \times 8\right) + 1 = 2$$

$\frac{1}{8}$ atom at each
of 8 corners

1 atom
at center

▲ **FIGURE 11.37** Body-Centered Cubic Crystal Structure   The different atom colors in this figure are for clarity only. All the atoms within each structure are identical.

Body-centered cubic

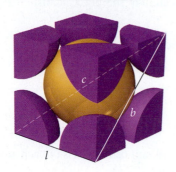

$$c^2 = b^2 + l^2 \qquad b^2 = l^2 + l^2$$
$$c = 4r \qquad b^2 = 2l^2$$
$$(4r)^2 = 2l^2 + l^2$$
$$(4r)^2 = 3l^2$$
$$l^2 = \frac{(4r)^2}{3}$$
$$l = \frac{4r}{\sqrt{3}}$$

◄ In the body-centered cubic lattice, the atoms touch only along the cube diagonal. The edge length is $4r/\sqrt{3}$.

Face-centered cubic

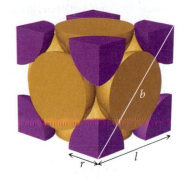

touch along the diagonal line that runs from one corner, through the middle of the cube, to the opposite corner. The edge length in terms of the atomic radius is therefore $l = 4r/\sqrt{3}$ as shown in the diagram above. The body-centered unit cell contains two atoms per unit cell because the center atom is not shared with any other neighboring cells. The coordination number of the body-centered cubic unit cell is 8, which you can see by observing the atom in the very center of the cube, which touches the eight atoms at the corners. The packing efficiency is 68%, significantly higher than for the simple cubic unit cell. In this structure, any one atom strongly interacts with more atoms than in the simple cubic unit cell.

The **face-centered cubic** unit cell **Figure 11.38▼** is characterized by a cube with one atom at each corner and one atom of the same kind in the center of each cube face. Note that in the face-centered unit cell (like the body-centered unit cell), the atoms *do not* touch along each edge of the cube. Instead, the atoms touch *along the face diagonal*. The edge length in terms of the atomic radius is therefore $l = 2\sqrt{2}r$, as shown in the figure at right. The face-centered unit cell contains four atoms per unit cell because the center atoms on each of the six faces are shared between two unit cells. So there are $\frac{1}{2} \times 6 = 3$ face-centered atoms plus $1/8 \times 8 = 1$ corner atoms, for a total of four atoms per unit cell. The coordination number of the face-centered cubic unit cell is 12 and its packing efficiency is 74%. In this structure, any one atom strongly interacts with more atoms than in either the simple cubic unit cell or the body-centered cubic unit cell.

$$b^2 = l^2 + l^2 = 2l^2$$
$$b = 4r$$
$$(4r)^2 = 2l^2$$
$$l^2 = \frac{(4r)^2}{2}$$
$$l = \frac{4r}{\sqrt{2}}$$
$$= 2\sqrt{2}r$$

▲ In the face-centered cubic lattice, the atoms touch along a face diagonal. The edge length is $2\sqrt{2}r$.

Face-Centered Cubic Unit Cell

Face-centered cubic:
extended structure
Coordination number = 12

Face-centered cubic: unit cell
Atoms/unit $= \left(\frac{1}{8} \times 8\right) + \left(\frac{1}{2} \times 6\right) = 4$

$\frac{1}{8}$ atom
at 8 corners

$\frac{1}{2}$ atom
at 6 faces

▲ **FIGURE 11.38 Face-Centered Cubic Crystal Structure** The different atom colors in this figure are for clarity only. All the atoms within each structure are identical.

EXAMPLE 11.6 Relating Density to Crystal Structure

Aluminum crystallizes with a face-centered cubic unit cell. The radius of an aluminum atom is 143 pm. Calculate the density of solid crystalline aluminum in g/cm^3.

SORT You are given the radius of an aluminum atom and its crystal structure. You are asked to find the density of solid aluminum.	**GIVEN** $r = 143$ pm, face-centered cubic **FIND** d

STRATEGIZE The conceptual plan is based on the definition of density.

Since the unit cell has the physical properties of the entire crystal, find the mass and volume of the unit cell and use these to calculate its density.

CONCEPTUAL PLAN

$d = m/V$

m = mass of unit cell
 = number of atoms in unit cell × mass of each atom

V = volume of unit cell
 = (edge length)^3

SOLVE Begin by finding the mass of the unit cell. Obtain the mass of an aluminum atom from its molar mass. Since the face-centered cubic unit cell contains four atoms per unit cell, multiply the mass of aluminum by 4 to get the mass of a unit cell.

SOLUTION

$$m(\text{Al atm}) = 26.98 \frac{g}{mol} \times \frac{1 mol}{6.022 \times 10^{23} \text{ atoms}}$$
$$= 4.481 \times 10^{-23} \text{ g}$$
$$m(\text{unit cell}) = 4(4.481 \times 10^{-23} \text{ g})$$
$$= 1.792 \times 10^{-22} \text{ g}$$

Next, calculate the edge length (l) of the unit cell (in m) from the atomic radius of aluminum. For the face-centered cubic structure, $l = 2\sqrt{2}r$.

$$l = 2\sqrt{2}r$$
$$= 2\sqrt{2}(143 \text{ pm})$$
$$= 2\sqrt{2} (143 \times 10^{-12} \text{ m})$$
$$= 4.045 \times 10^{-10} \text{ m}$$

Calculate the volume of the unit cell (in cm) by converting the edge length to cm and cubing the edge length. (We use centimeters because we want to report the density in units of g/cm^3.)

$$V = l^3$$
$$= \left(4.045 \times 10^{-10} \text{ m} \times \frac{1 \text{ cm}}{10^{-2} \text{ m}} \right)^3$$
$$= 6.618 \times 10^{-23} \text{ cm}^3$$

Finally, calculate the density by dividing the mass of the unit cell by the volume of the unit cell.

$$d = \frac{m}{V} = \frac{1.792 \times 10^{-22} \text{ g}}{6.618 \times 10^{23} \text{ cm}^3}$$
$$= 2.71 \text{ g/cm}^3$$

CHECK The units of the answer are correct. The magnitude of the answer is reasonable because the density is greater than 1 g/cm^3 (as we would expect for metals), but still not too high (aluminum is a low-density metal).

FOR PRACTICE 11.6

Chromium crystallizes with a body-centered cubic unit cell. The radius of a chromium atom is 125 pm. Calculate the density of solid crystalline chromium in g/cm^3.

Closest-Packed Structures

Another way to envision crystal structures, especially useful in metals where bonds are not usually directional, is to think of the atoms as stacking in layers, much as fruit is stacked at the grocery store. For example, the simple cubic structure can be envisioned as one layer of atoms arranged in a square pattern with the next layer stacking directly over the first, so that the atoms in one layer align exactly on top of the atoms in the layer beneath it, as shown at left.

As we saw previously, this crystal structure has a great deal of empty space—only 52% of the volume is occupied by the spheres, and the coordination number is 6.

More space-efficient packing can be achieved by aligning neighboring rows of atoms within a layer not in a square pattern, but in a pattern with one row offset from the next by one-half a sphere, as shown at the top right of the next page.

In this way, the atoms pack more closely to each other in any one layer. We can further increase the packing efficiency by placing the next layer *not directly on top of the first*, but again offset so that any one atom actually sits in the indentation formed by three atoms in the layer beneath it, as shown here.

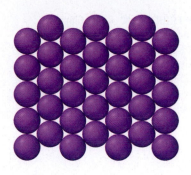

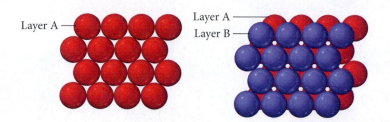

Layer A

Layer A
Layer B

This kind of packing leads to two different crystal structures called *closest-packed structures*, both of which have packing efficiencies of 74% and coordination numbers of 12. In the first of these two closest-packed structures—called **hexagonal closest packing**—the third layer of atoms aligns exactly on top of the first, as shown below.

The pattern from one layer to the next is ABAB . . . with alternating layers aligning exactly on top of one another. Notice that the central atom in layer B of this structure is touching 6 atoms in its own layer, 3 atoms in the layer above it, and 3 atoms in the layer below, for a coordination number of 12. The unit cell for this crystal structure is not a cubic unit cell, but a hexagonal one, as shown in **Figure 11.39▼**.

In the second of the two closest-packed structures—called **cubic closest packing**—the third layer of atoms is offset from the first, as shown here.

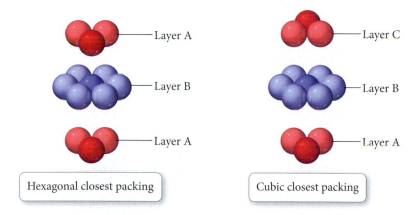

Layer A

Layer B

Layer A

Hexagonal closest packing

Layer C

Layer B

Layer A

Cubic closest packing

The different atom colors in this figure are for clarity only. All the atoms within each structure are identical.

Hexagonal Closest Packing

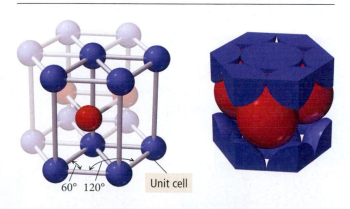

60° 120°

Unit cell

◀ **FIGURE 11.39 Hexagonal Closest-Packing Crystal Structure**   The unit cell is outlined in bold. The different atom colors in this figure are for clarity only. All the atoms within the structure are identical.

The pattern from one layer to the next is ABCABC . . . with every fourth layer aligning with the first. Although not easy to visualize, the unit cell for cubic closest packing is the

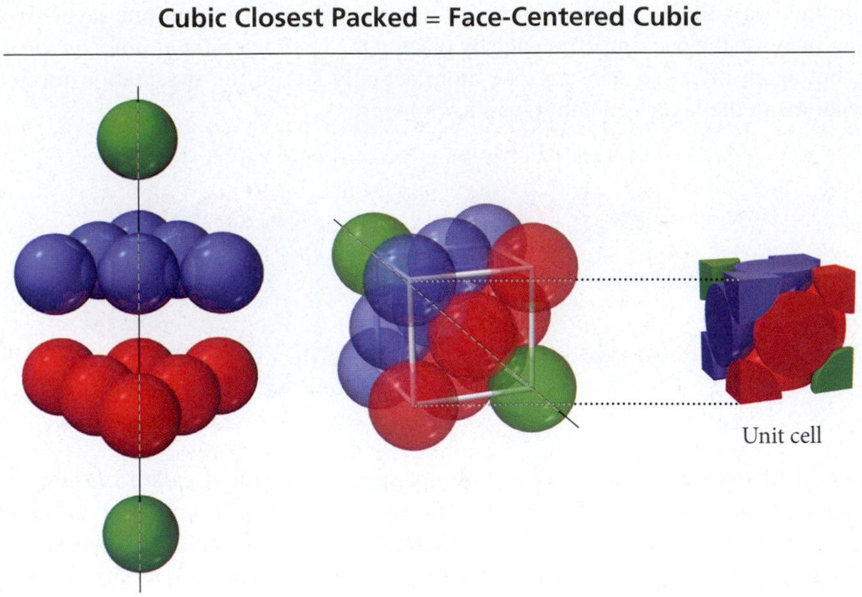

▶ **FIGURE 11.40 Cubic Closest-Packing Crystal Structure**   The unit cell of the cubic closest-packed structure is face-centered cubic. The different atom colors in this figure are for clarity only. All the atoms within each structure are identical.

Cubic Closest Packed = Face-Centered Cubic

Unit cell

face-centered cubic unit cell, as shown in **Figure 11.40▲**. Therefore the cubic closest-packed structure is identical to the face-centered cubic unit cell structure.

11.11   Crystalline Solids: The Fundamental Types

Crystalline solids can be divided into three categories—molecular, ionic, and atomic—based on the individual units that compose the solid. Atomic solids can themselves be divided into three categories—nonbonded, metallic, and network covalent—depending on the types of interactions between atoms within the solid. **Figure 11.41▼** shows the different categories of crystalline solids.

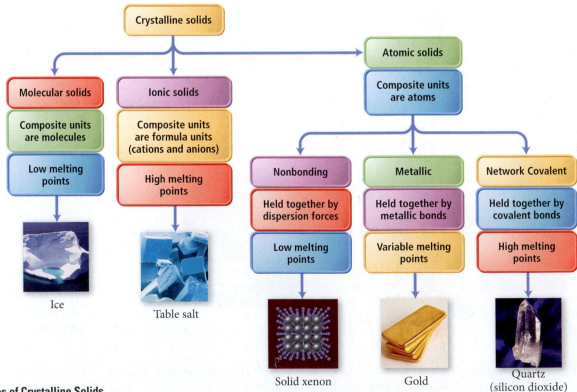

▶ **FIGURE 11.41 Types of Crystalline Solids**

Molecular Solids

Molecular solids are those solids whose composite units are *molecules*. The lattice sites in a crystalline molecular solid are therefore occupied by molecules. Ice (solid H_2O) and dry ice (solid CO_2) are examples of molecular solids. Molecular solids are held together by the kinds of intermolecular forces—dispersion forces, dipole–dipole forces, and hydrogen bonding—that we have discussed in this chapter. Molecular solids as a whole tend to have low to moderately low melting points. However, strong intermolecular forces (such as the hydrogen bonds in water) can increase the melting points of some molecular solids.

Ionic Solids

Ionic solids are those solids whose composite units are ions. Table salt (NaCl) and calcium fluoride (CaF_2) are examples of ionic solids. Ionic solids are held together by the coulombic interactions that occur between the cations and anions occupying the lattice sites in the crystal. The coordination number of the unit cell for an ionic compound, therefore, represents the number of close cation–anion interactions. Since these interactions lower potential energy, the crystal structure of a particular ionic compound will be the one that maximizes the coordination number, while accommodating both charge neutrality (each unit cell must be charge neutral) and the different sizes of the cations and anions that compose the particular compound. In general, the more similar the radii of the cation and the anion, the higher the coordination number.

Cesium chloride (CsCl) is a good example of an ionic compound containing cations and anions of similar size (Cs^+ radius $= 167$ pm; Cl^- radius $= 181$ pm). In the cesium chloride structure, the chloride ions occupy the lattice sites of a simple cubic cell and one cesium ion lies in the very center of the cell, as shown in **Figure 11.42▶**. The coordination number is 8; each cesium ion is in direct contact with eight chloride ions (and vice versa). Notice that the cesium chloride unit cell contains one chloride anion $\left[(8 \times \frac{1}{8}) = 1\right]$ and one cesium cation (the cesium ion in the middle belongs entirely to the unit cell and complete chloride atoms are shown even though only a fraction of each is part of the single unit cell) for a ratio of Cs to Cl of 1:1, just as in the formula for the compound. Calcium sulfide (CaS) adopts the same structure as cesium chloride.

The crystal structure of sodium chloride must accommodate the more disproportionate sizes of Na^+ (radius $= 97$ pm) and Cl^- (radius $= 181$ pm). If ion size were the only consideration, the larger chloride anion could theoretically fit many of the smaller sodium cations around it, but charge neutrality requires that each sodium cation be surrounded by an equal number of chloride anions. Therefore, the coordination number is limited by the number of chloride anions that can fit around the relatively small sodium cation. The structure that minimizes the energy is shown in **Figure 11.43▶** and has a coordination number of 6 (each chloride anion is surrounded by six sodium cations and vice versa). You can visualize this structure, called the *rock salt* structure, as chloride anions occupying the lattice sites of a face-centered cubic structure with the smaller sodium cations occupying the holes between the anions. (Alternatively, you can visualize this structure as the *sodium cations* occupying the lattice sites of a face-centered cubic structure with the *larger chloride anions* occupying the spaces between the cations.) Each unit cell contains four chloride anions $\left[(8 \times \frac{1}{8}) + (6 \times \frac{1}{2}) = 4\right]$ and four sodium cations $\left[(12 \times \frac{1}{4}) + 1 = 4\right]$, resulting in a ratio of 1:1, just as in the formula of the compound. Other compounds exhibiting the sodium chloride structure include LiF, KCl, KBr, AgCl, MgO, and CaO.

An even greater disproportion between the sizes of the cations and anions in a compound makes a coordination number of even 6 physically impossible. For example, in ZnS (Zn^{2+} radius $= 74$ pm; S^{2-} radius $= 184$ pm) the crystal structure, shown in **Figure 11.44▶**, has a coordination number of only 4. You can visualize this structure, called the *zinc blende* structure, as sulfide anions occupying the lattice sites of a face-centered cubic structure with the smaller zinc cations occupying four of the eight tetrahedral holes located directly beneath each corner atom. A tetrahedral hole is the empty space that lies in the center of a tetrahedral arrangement of four atoms, as shown in the margin on the top of the next page. Each unit cell contains four sulfide anions $\left[(8 \times \frac{1}{8}) + (6 \times \frac{1}{2}) = 4\right]$ and four zinc cations (each of the four zinc cations is completely contained within the unit cell), resulting in a ratio of 1:1, just as in the formula of the compound. Other compounds exhibiting the zinc blende structure include CuCl, AgI, and CdS.

Cesium chloride (CsCl)

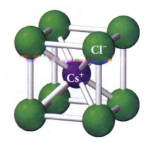

▲ **FIGURE 11.42 Cesium Chloride Unit Cell**

Sodium chloride (NaCl)

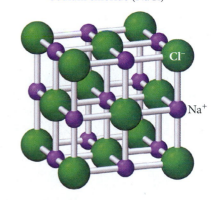

▲ **FIGURE 11.43 Sodium Chloride Unit Cell**

Zinc blende (ZnS)

▲ **FIGURE 11.44 Zinc Sulfide (Zinc Blende) Unit Cell**

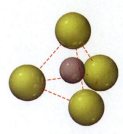

▲ A tetrahedral hole

Calcium fluoride (CaF₂)

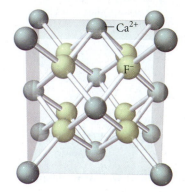

▲ **FIGURE 11.45 Calcium Fluoride Unit Cell**

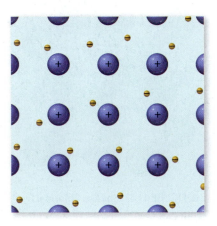

▲ **FIGURE 11.46 The Electron Sea Model**   In the electron sea model for metals, the metal cations exist in a "sea" of electrons.

When the ratio of cations to anions is not 1:1, the crystal structure must accommodate the unequal number of cations and anions. Many compounds that contain a cation to anion ratio of 1:2 adopt the *fluorite (CaF₂) structure* shown in **Figure 11.45◄**. You can visualize this structure as calcium cations occupying the lattice sites of a face-centered cubic structure with the larger fluoride anions occupying all eight of the tetrahedral holes located directly beneath each corner atom. Each unit cell contains four calcium cations $\left[(8 \times \frac{1}{8}) + (6 \times \frac{1}{2}) = 4\right]$ and eight fluoride anions (each of the eight fluoride anions is completely contained within the unit cell), resulting in a cation to anion ratio of 1:2, just as in the formula of the compound. Other compounds exhibiting the fluorite structure include PbF_2, SrF_2, and $BaCl_2$. Compounds with a cation to anion ratio of 2:1 often exhibit the *antifluorite structure*, in which the anions occupy the lattice sites of a face-centered cubic structure and the cations occupy the tetrahedral holes beneath each corner atom.

Since the forces holding ionic solids together are strong coulombic forces (or ionic bonds), and since these forces are much stronger than the intermolecular forces discussed previously, ionic solids tend to have much higher melting points than molecular solids. For example, sodium chloride melts at 801 °C, while carbon disulfide (CS_2)—a molecular solid with a higher molar mass—melts at −110 °C.

Atomic Solids

Solids whose composite units are individual atoms are called **atomic solids.** Solid xenon (Xe), iron (Fe), and silicon dioxide (SiO_2) are examples of atomic solids. As we saw in Figure 11.41 on page 442, atomic solids can themselves be divided into three categories—*nonbonding atomic solids, metallic atomic solids, and network covalent atomic solids*—each held together by a different kind of force.

Nonbonding atomic solids, a group that consists of only the noble gases in their solid form, are held together by relatively weak dispersion forces. In order to maximize these interactions, nonbonding atomic solids form closest-packed structures, maximizing their coordination numbers and minimizing the distance between them. Nonbonding atomic solids have very low melting points, which increase uniformly with molar mass. Argon, for example, has a melting point of −189 °C and xenon has a melting point of −112 °C.

Metallic atomic solids, such as iron or gold, are held together by *metallic bonds*, which in the simplest model are represented by the interaction of metal cations with the sea of electrons that surrounds them, as described in Section 9.11 **Figure 11.46◄**.

Since metallic bonds are not directional, metals also tend to form closest-packed crystal structures. For example, nickel crystallizes in the cubic closest-packed structure and zinc crystallizes in the hexagonal closest-packed structure **Figure 11.47▼**. Metallic bonds are of varying strengths. Some metals, such as mercury, have melting points below room temperature, whereas other metals, such as iron, have relatively high melting points (iron melts at 1809 °C).

Network covalent atomic solids, such as diamond, graphite, and silicon dioxide, are held together by covalent bonds. The crystal structures of these solids are restricted by the geometrical constraints of the covalent bonds (which tend to be more directional than intermolecular forces, ionic bonds, or metallic bonds) so they *do not* tend to form closest-packed structures.

▶ **FIGURE 11.47 Closest-Packed Crystal Structures in Metals**   Nickel crystallizes in the cubic closest-packed structure. Zinc crystallizes in the hexagonal closest-packed structure.

Nickel (Ni)   Zinc (Zn)

Covalent bonds   Dispersion forces

(a) Diamond   (b) Graphite

◀ **FIGURE 11.48** **Network Covalent Atomic Solids** **(a)** In diamond, each carbon atom forms four covalent bonds to four other carbon atoms in a tetrahedral geometry. **(b)** In graphite, carbon atoms are arranged in sheets. Within each sheet, the atoms are covalently bonded to one another by a network of sigma and pi bonds. Neighboring sheets are held together by dispersion forces.

In diamond **Figure 11.48(a)▲**, each carbon atom forms four covalent bonds to four other carbon atoms in a tetrahedral geometry. This structure extends throughout the entire crystal, so that a diamond crystal can be thought of as a giant molecule, held together by these covalent bonds. Since covalent bonds are very strong, covalent atomic solids have high melting points. Diamond is estimated to melt at about 3800 °C. The electrons in diamond are confined to the covalent bonds and are not free to flow. Therefore diamond does not conduct electricity.

In graphite **Figure 11.48(b)▲**, carbon atoms are arranged in sheets. Within each sheet, carbon atoms are covalently bonded to each other by a network of sigma and pi bonds. The electrons within the pi bonds are delocalized over the entire sheet, making graphite a good electrical conductor along the sheets. The bond length between carbon atoms *within a sheet* is 142 pm. However, the bonding *between* sheets is much different. The separation between sheets is 341 pm. There are no covalent bonds between sheets, only relatively weak dispersion forces. Consequently, the sheets slide past each other relatively easily, which explains the slippery feel of graphite and its extensive use as a lubricant.

The silicates (extended arrays of silicon and oxygen) are the most common network covalent atomic solids. Geologists estimate that 90% of Earth's crust is composed of silicates. The basic silicon oxygen compound is silica (SiO_2), which in its most common crystalline form is called quartz. The structure of quartz consists of an array of SiO_4 tetrahedra with shared oxygen atoms, as shown in **Figure 11.49(a)▼**. The strong silicon–oxygen covalent bonds that hold quartz together result in its high melting point of about 1600 °C. Common glass is also composed of SiO_2, but in its amorphous form **Figure 11.49(b)▼**.

Sigma and pi bonds are discussed in Section 10.7.

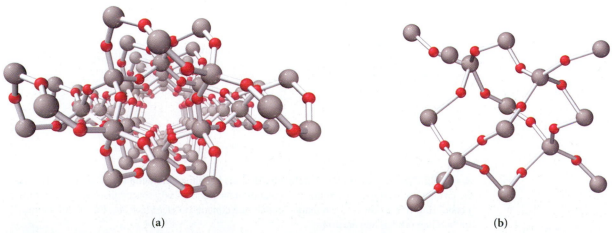

(a)   (b)

▲ **FIGURE 11.49** **The Structure of Quartz** **(a)** Quartz consists of an array of SiO_4 tetrahedra with shared oxygen atoms. **(b)** Glass is amorphous SiO_2.

11.12   Crystalline Solids: Band Theory

In Section 9.11, we explored a model for bonding in metals called the *electron sea model*. We now turn to a model for bonding in solids that is both more sophisticated and more broadly applicable—it applies to both metallic solids and covalent solids. The model is called **band theory** and it grows out of molecular orbital theory, first covered in Section 10.8.

Recall that in molecular orbital theory, we combined the atomic orbitals of the atoms within a molecule to form molecular orbitals. These molecular orbitals were not localized on individual atoms, but *delocalized over the entire molecule*. Similarly, in band theory, we combine the atomic orbitals of the atoms within a solid crystal to form orbitals that are not localized on individual atoms, but delocalized over the entire *crystal*. In some sense then, the crystal is like a very large molecule and its valence electrons occupy the molecular orbitals formed from the atomic orbitals of each atom in the crystal.

We begin our discussion of band theory by considering a series of molecules constructed from individual lithium atoms. The energy levels of the atomic orbitals and resulting molecular orbitals for Li, Li_2, Li_3, Li_4, and Li_N (where N is a large number on the order of 10^{23}) are shown in **Figure 11.50▼**. The lithium atom has a single electron in a single 2*s* atomic orbital. The Li_2 molecule contains two electrons and two molecular orbitals. The electrons occupy the lower energy bonding orbital—the higher energy, or antibonding, molecular orbital is empty. The Li_4 molecule contains four electrons and four molecular orbitals. The electrons occupy the two bonding molecular orbitals—the two antibonding orbitals are empty.

The Li_N molecule contains N electrons and N molecular orbitals. However, because there are so many molecular orbitals, the energy spacings between them are infinitesimally small; they are no longer discrete energy levels, but rather form a *band* of energy levels. One half of the orbitals in the band ($N/2$) are bonding molecular orbitals and (at 0 K) contain the N valence electrons. The other $N/2$ molecular orbitals are antibonding and (at 0 K) are completely empty. If the atoms composing a solid have *p* orbitals available, then the same process leads to another band of orbitals at higher energies.

In band theory, electrons become mobile when they make a transition from the highest occupied molecular orbital into higher energy empty molecular orbitals. For this reason, the occupied molecular orbitals are often called the *valence band* and the unoccupied orbitals are called the *conduction band*. In lithium metal, the highest occupied molecular orbital lies in the middle of a band of orbitals, and the energy difference between it and the next higher energy orbital is infinitesimally small. Therefore, above 0 K, electrons

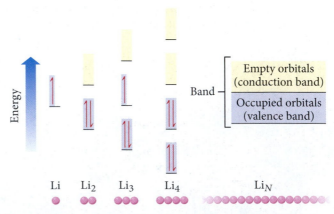

▲ **FIGURE 11.50 Energy Levels of Molecular Orbitals in Lithium Molecules**   When many Li atoms are present, the energy levels of the molecular orbitals are so closely spaced that they fuse to form a band. Half of the orbitals are bonding orbitals and contain valence electrons; the other half are antibonding orbitals and are empty.

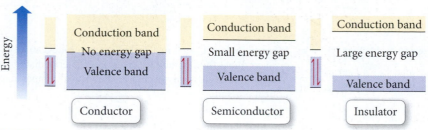

▲ **FIGURE 11.51 Band Gap** In a conductor, there is no energy gap between the valence band and the conduction band. In semiconductors there is a small energy gap, and in insulators there is a large energy gap.

can easily make the transition from the valence band to the conduction band. Since electrons in the conduction band are mobile, lithium, like all metals, is a good electrical conductor. Mobile electrons in the conduction band are also responsible for the thermal conductivity of metals. When a metal is heated, electrons are excited to higher energy molecular orbitals. These electrons can then quickly transport the thermal energy throughout the crystal lattice.

In metals, the valence band and conduction band are always energetically continuous—the energy difference between the top of the valence band and the bottom of the conduction band is infinitesimally small. In semiconductors and insulators, however, an energy gap, called the **band gap,** exists between the valence band and conduction band as shown in **Figure 11.51**▲. In insulators, the band gap is large, and electrons are not promoted into the conduction band at ordinary temperatures, resulting in no electrical conductivity. In semiconductors, the band gap is small, allowing some electrons to be promoted at ordinary temperatures and resulting in limited conductivity. However, the conductivity of semiconductors can be controlled by adding minute amounts of other substances to the semiconductor. These substances, called *dopants*, are minute impurities that result in additional electrons in the conduction band or electron holes in the valence band. The addition or subtraction of electrons affects the conductivity.

CHAPTER IN REVIEW

Key Terms

Section 11.2
crystalline (411)
amorphous (411)

Section 11.3
dispersion force (412)
dipole–dipole force
(414)
permanent dipole (414)
miscibility (415)
hydrogen bonding (416)
hydrogen bond (416)
ion–dipole force (418)

Section 11.4
surface tension (420)
viscosity (420)
capillary action (421)

Section 11.5
vaporization (421)

condensation (422)
volatile (422)
nonvolatile (422)
heat of vaporization
(ΔH_{vap}) (423)
dynamic equilibrium (425)
vapor pressure (425)
boiling point (426)
normal boiling point (426)
Clausius–Clapeyron
equation (427)
critical temperature
(T_c) (430)
critical pressure
(P_c) (430)

Section 11.6
sublimation (430)
deposition (430)
melting point (431)

melting (fusion) (431)
freezing (431)
heat of fusion (ΔH_{fus}) (431)

Section 11.8
phase diagram (433)
triple point (434)
critical point (434)

Section 11.10
crystalline lattice (436)
unit cell (436)
simple cubic (438)
coordination
number (438)
packing efficiency (438)
body-centered cubic (438)
face-centered cubic (439)
hexagonal closest packing (441)
cubic closest packing (441)

Section 11.11
molecular solids (443)
ionic solids (443)
atomic solids (444)
nonbonding atomic solids (444)
metallic atomic solids (444)
network covalent atomic
solids (444)

Section 11.12
band theory (446)
band gap (447)

Key Concepts

Solids, Liquids, and Intermolecular Forces (11.1, 11.2, 11.3)

▶ The forces that hold molecules or atoms together in a liquid or solid are intermolecular forces. The strength of the intermolecular forces in a substance determines its state.

▶ Dispersion forces are present in all elements and compounds because they result from the fluctuations in electron distribution within atoms and molecules. These are the weakest intermolecular forces, but they are significant in molecules with high molar masses.

▶ Dipole–dipole forces, generally stronger than dispersion forces, are present in all polar molecules.

▶ Hydrogen bonding occurs among polar molecules that contain hydrogen atoms bonded directly to fluorine, oxygen, or nitrogen.

▶ Ion–dipole forces occur in ionic compounds mixed with polar compounds, and are especially important in aqueous solutions.

Surface Tension, Viscosity, and Capillary Action (11.4)

▶ Surface tension results from the tendency of liquids to minimize their surface area in order to maximize the interactions between their constituent particles, thus lowering their potential energy. Surface tension causes water droplets to form spheres and allows insects and fishing flies to "float" on the surface of water.

▶ Viscosity is the resistance of a liquid to flow. Viscosity increases with increasing strength of intermolecular forces and decreases with increasing temperature.

▶ Capillary action is the ability of a liquid to flow against gravity up a narrow tube. It is the result of adhesive forces, the attraction between the molecules and the surface of the tube, and cohesive forces, the attraction between the molecules in the liquid.

Vaporization and Vapor Pressure (11.5, 11.7)

▶ Vaporization, the transition from liquid to gas, occurs when thermal energy overcomes the intermolecular forces present in a liquid. The opposite process is condensation. Vaporization is endothermic and condensation is exothermic.

▶ The rate of vaporization increases with increasing temperature, increasing surface area, and decreasing strength of intermolecular forces.

▶ The heat of vaporization (ΔH_{vap}) is the heat required to vaporize one mole of a liquid.

▶ In a sealed container, a solution and its vapor will come into dynamic equilibrium, at which point the rate of vaporization equals the rate of condensation. The pressure of a gas that is in dynamic equilibrium with its liquid is called its vapor pressure.

▶ The vapor pressure of a substance increases with increasing temperature and with decreasing strength of its intermolecular forces.

▶ The boiling point of a liquid is the temperature at which its vapor pressure equals the external pressure.

▶ The Clausius–Clapeyron equation expresses the relationship between the vapor pressure of a substance and its temperature and can be used to calculate the heat of vaporization from experimental measurements.

▶ When a liquid is heated in a sealed container it eventually forms a supercritical fluid, which has properties intermediate between a liquid and a gas. This occurs at the critical temperature and critical pressure.

Fusion and Sublimation (11.6, 11.7)

▶ Sublimation is the transition from solid to gas. The opposite process is deposition.

▶ Fusion, or melting, is the transition from solid to liquid. The opposite process is freezing.

▶ The heat of fusion (ΔH_{fus}) is the amount of heat required to melt one mole of a solid. Fusion is endothermic.

▶ The heat of fusion is generally less than the heat of vaporization because intermolecular forces do not have to be completely overcome for melting to occur.

Phase Diagrams (11.8)

▶ A phase diagram is a map of the states of a substance as a function of its pressure (y-axis) and temperature (x-axis).

▶ The regions in a phase diagram represent conditions under which a single stable state (solid, liquid, gas) exists.

▶ The lines represent conditions under which two states are in equilibrium.

▶ The triple point represents the conditions under which all three states coexist.

▶ The critical point is the temperature and pressure above which a supercritical fluid exists.

The Uniqueness of Water (11.9)

▶ Water is a liquid at room temperature despite its low molar mass. Water forms strong hydrogen bonds, resulting in its high boiling point.

▶ The high polarity of water enables it to dissolve many polar and ionic compounds, and even nonpolar gases.

▶ Water expands upon freezing, so that ice is less dense than liquid water.

Crystalline Structures (11.10–11.12)

▶ The crystal lattice in crystalline solids is represented by a unit cell, a structure that can reproduce the entire lattice when repeated in all three dimensions.

▶ Three basic cubic unit cells are the simple cubic, the body-centered cubic, and the face-centered cubic.

▶ Some crystal lattices can also be depicted as closest-packed structures, including the hexagonal closest-packing structure (not cubic) and the cubic closest-packing structure (which has a face-centered cubic unit cell).

▶ The basic types of crystal solids are molecular, ionic, and atomic solids. Atomic solids can themselves be divided into three different types: nonbonded, metallic, and covalent.

▶ Band theory is a model for bonding in solids in which the atomic orbitals of the atoms are combined and delocalized over the entire crystal solid.

Key Equations and Relationships

Clausius–Clapeyron Equation: Relationship between Vapor Pressure (P_{vap}), the Heat of Vaporization (H_{vap}), and Temperature (T) (11.5)

$$\ln P_{vap} = \frac{-\Delta H_{vap}}{RT} + \ln \beta \quad (\beta \text{ is a constant})$$

$$\ln \frac{P_2}{P_1} = \frac{-\Delta H_{vap}}{R}\left[\frac{1}{T_2} - \frac{1}{T_1}\right]$$

Key Learning Objectives

Chapter Objectives	Assessment
Determining Whether a Molecule Has Dipole–Dipole Forces (11.3)	Example 11.1 For Practice 11.1 Exercises 1–12
Determining Whether a Molecule Displays Hydrogen Bonding (11.3)	Example 11.2 For Practice 11.2 Exercises 1–12
Using the Heat of Vaporization in Calculations (11.5)	Example 11.3 For Practice 11.3 Exercise 23–26 For More Practice 11.3
Using the Clausius–Clapeyron Equation (11.5)	Examples 11.4, 11.5 For Practice 11.4, 11.5 Exercises 27–30
Relating Density to Crystal Structure (11.10)	Example 11.6 For Practice 11.6 Exercises 49–52

EXERCISES

Problems by Topic

Intermolecular Forces

1. Determine the kinds of intermolecular forces present in each element or compound:
   a. N_2 b. NH_3 c. CO   d. CCl_4

2. Determine the kinds of intermolecular forces present in each element or compound:
   a. Kr   b. NCl_3 c. SiH_4 d. HF

3. Determine the kinds of intermolecular forces present in each element or compound:
   a. HCl   b. H_2O c. Br_2 d. He

4. Determine the kinds of intermolecular forces present in each element or compound:
   a. PH_3 b. HBr   c. CH_3OH d. I_2

5. Arrange these compounds in order of increasing boiling point. Explain your reasoning.
   a. CH_4 b. CH_3CH_3
   c. CH_3CH_2Cl d. CH_3CH_2OH

6. Arrange these compounds in order of increasing boiling point. Explain your reasoning.
   a. H_2S b. H_2Se c. H_2O

7. For each pair of compounds, pick the one with the highest boiling point. Explain your reasoning.
   a. CH_3OH or CH_3SH
   b. CH_3OCH_3 or CH_3CH_2OH
   c. CH_4 or CH_3CH_3

8. For each pair of compounds, pick the one with the higher boiling point. Explain your reasoning.
   a. NH_3 or CH_4 b. CS_2 or CO_2 c. CO_2 or NO_2

9. For each pair of compounds, pick the one with the higher vapor pressure at a given temperature. Explain your reasoning.
   a. Br_2 or I_2 b. H_2S or H_2O c. NH_3 or PH_3

10. For each pair of compounds, pick the one with the higher vapor pressure at a given temperature. Explain your reasoning.
    a. CH_4 or CH_3Cl
    b. $CH_3CH_2CH_2OH$ or CH_3OH
    c. CH_3OH or H_2CO

11. Determine whether each pair of substances forms a homogeneous solution when combined. For those that form homogeneous solutions, indicate the type of forces that are involved.
    a. CCl_4 and H_2O b. KCl and H_2O
    c. Br_2 and CCl_4 d. CH_3CH_2OH and H_2O

12. Determine whether each pair of compounds forms a homogeneous solution when combined. For those that form homogeneous solutions, indicate the type of forces that are involved.
    a. $CH_3CH_2CH_2CH_2CH_3$ and $CH_3CH_2CH_2CH_2CH_2CH_3$
    b. CBr_4 and H_2O
    c. $LiNO_3$ and H_2O
    d. CH_3OH and $CH_3CH_2CH_2CH_2CH_3$

Surface Tension, Viscosity, and Capillary Action

13. Which compound would you expect to have greater surface tension, acetone $[(CH_3)_2CO]$ or water (H_2O)? Explain.

14. Water (a) "wets" some surfaces and beads up on others. Mercury (b), in contrast, beads up on almost all surfaces. Explain this difference.

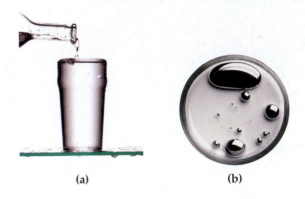

(a)   (b)

15. The structures of two isomers of heptane are shown here. Which of these two compounds would you expect to have the greater viscosity?

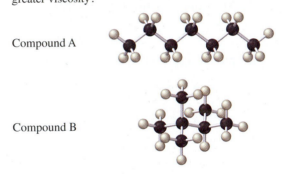

Compound A

Compound B

16. The viscosity of motor oil is important to the correct functioning of an engine. Multigrade motor oils contain polymers (long molecules made up of repeating structural units) that coil at low temperatures but unwind at high temperatures. Explain why this helps the viscosity of the motor oil depend less on temperature than it would otherwise.

17. Water in a glass tube that contains grease or oil residue displays a flat meniscus (left), whereas water in a clean glass tube displays a concave meniscus (right). Explain.

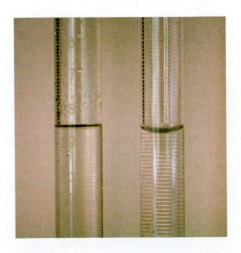

18. When a thin glass tube is put into water, the water rises 1.4 cm. When the same tube is put into hexane, the hexane rises only 0.4 cm. Explain.

Vaporization and Vapor Pressure

19. Which will evaporate more quickly: 55 mL of water in a beaker with a diameter of 4.5 cm, or 55 mL of water in a dish with a diameter of 12 cm? Will the vapor pressure of the water be different in the two containers? Explain.

20. Which will evaporate more quickly: 55 mL of water (H_2O) in a beaker or 55 mL of acetone $[(CH_3)_2CO]$ in an identical beaker under identical conditions? Is the vapor pressure of the two substances different? Explain.

21. Spilling room-temperature water over your skin on a hot day will cool you down. Spilling vegetable oil (of the same temperature as the water) over your skin on a hot day will not. Explain.

22. Why is the heat of vaporization of water greater at room temperature than it is at its boiling point?

23. The human body obtains 1065 kJ of energy from a candy bar. If this energy were used to vaporize water at 100.0 °C, how much water (in liters) could be vaporized? (Assume the density of water is 1.00 g/mL.)

24. A 45.0 mL sample of water is heated to its boiling point. How much heat (in kJ) is required to vaporize it? (Assume a density of 1.00 g/mL.)

25. Suppose that 0.88 g of water condenses on a 75.0-g block of iron that is initially at 22 °C. If the heat released during condensation goes only to warming the iron block, what is the final temperature (in °C) of the iron block? (Assume a constant enthalpy of vaporization for water of 44.0 kJ/mol.)

26. Suppose that 1.02 g of rubbing alcohol (C_3H_8O) evaporates from a 55.0-g aluminum block. If the aluminum block is initially at 25 °C, what is the final temperature of the block after the evaporation of the alcohol? Assume that the heat required for the vaporization of the alcohol comes only from the aluminum block and that the alcohol vaporizes at 25 °C.

27. The vapor pressure of ammonia at several different temperatures is shown in the table. Use the data to determine the heat of vaporization and normal boiling point of ammonia.

Temperature (K)	Pressure (torr)
200	65.3
210	134.3
220	255.7
230	456.0
235	597.0

28. The vapor pressure of nitrogen at several different temperatures is shown in the table. Use the data to determine the heat of vaporization and normal boiling point of nitrogen.

Temperature (K)	Pressure (torr)
65	130.5
70	289.5
75	570.8
80	1028
85	1718

29. Ethanol has a heat of vaporization of 38.56 kJ/mol and a normal boiling point of 78.4 °C. What is the vapor pressure of ethanol at 15 °C?

30. Benzene has a heat of vaporization of 30.72 kJ/mol and a normal boiling point of 80.1 °C. At what temperature does benzene boil when the external pressure is 445 torr?

Sublimation and Fusion

31. How much energy is released when 47.5 g of water freezes?

32. Calculate the amount of heat required to completely sublime 25.0 g of solid dry ice (CO_2) at its sublimation temperature. The heat of sublimation for carbon dioxide is 32.3 kJ/mol.

33. An 8.5-g ice cube is placed into 255 g of water. Calculate the temperature change in the water upon the complete melting of the ice. Assume that all of the energy required to melt the ice comes from the water.

34. How much ice (in grams) would have to melt to lower the temperature of 352 mL of water from 25 °C to 5 °C? (Assume the density of water is 1.0 g/mL.)

35. How much heat (in kJ) is required to warm 10.0 g of ice, initially at −10.0 °C, to steam at 110.0 °C? The heat capacity of ice is 2.09 J/g·°C and that of steam is 2.09 J/g·°C.

36. How much heat (in kJ) is evolved in converting 1.00 mol of steam at 145.0 °C to ice at −50.0 °C? The heat capacity of steam is 2.09 J/g·°C and of ice is 2.09 J/g·°C.

Phase Diagrams

37. Consider the phase diagram shown here. Identify the states present at points *a* through *g*.

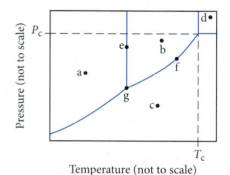

Temperature (not to scale)

38. Consider the phase diagram for iodine and answer each question.
  a. What is the normal boiling point for iodine?
  b. What is the melting point for iodine at 1 atm?
  c. What state is present at room temperature and normal atmospheric pressure?
  d. What state is present at 186 °C and 1.0 atm?

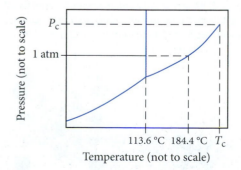

Temperature (not to scale)

39. Nitrogen has a normal boiling point of 77.3 K and a melting point (at 1 atm) of 63.1 K. Its critical temperature is 126.2 K and critical pressure is 2.55×10^4 torr. It has a triple point at 63.1 K and 94.0 torr. Sketch the phase diagram for nitrogen. Does nitrogen have a stable liquid state at 1 atm?

40. Argon has a normal boiling point of 87.2 K and a melting point (at 1 atm) of 84.1 K. Its critical temperature is 150.8 K and critical pressure is 48.3 atm. It has a triple point at 83.7 K and 0.68 atm. Sketch the phase diagram for argon. Which has the greater density, solid argon or liquid argon?

41. The phase diagram for sulfur is shown here. The rhombic and monoclinic phases are two solid states with different structures.

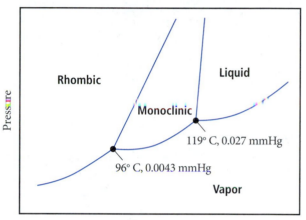

Temperature

  a. Below what pressure will solid sulfur sublime?
  b. Which of the two solid states of sulfur is most dense?

42. The high-pressure phase diagram of ice is shown here. Notice that, under high pressure, ice can exist in several different solid forms. What three forms of ice are present at the triple point marked O? What is the density of ice II compared to ice I (the familiar form of ice)? Would ice III sink or float in liquid water?

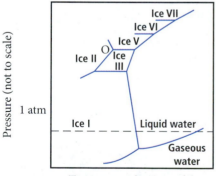

Temperature (not to scale)

The Uniqueness of Water

43. Water has a high boiling point for its relatively low molar mass. Why?

44. Water is a good solvent for many substances. What is the molecular basis for this property and why is it significant?

45. Explain the role of water in moderating Earth's climate.

46. How is the density of solid water compared to that of liquid water atypical among substances? Why is this significant?

Types of Solids and Their Structures

47. Determine the number of atoms per unit cell for each metal.

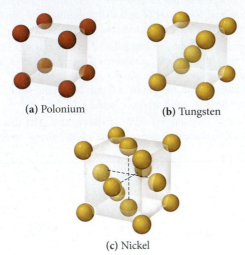

(a) Polonium   (b) Tungsten

(c) Nickel

48. Determine the coordination number for each structure.

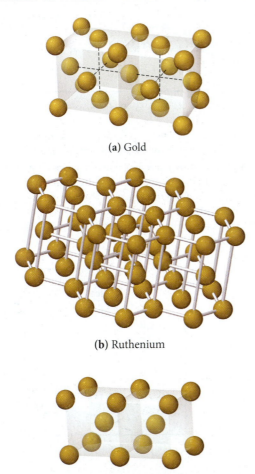

(a) Gold

(b) Ruthenium

(c) Chromium

49. Platinum crystallizes with the face-centered cubic unit cell. The radius of a platinum atom is 139 pm. Calculate the edge length of the unit cell and the density of platinum in g/cm³.

50. Molybdenum crystallizes with the body-centered unit cell. The radius of a molybdenum atom is 136 pm. Calculate the edge length of the unit cell and the density of molybdenum.

51. Rhodium has a density of 12.41 g/cm³ and crystallizes with the face-centered cubic unit cell. Calculate the radius of a rhodium atom.

52. Barium has a density of 3.59 g/cm³ and crystallizes with the body-centered cubic unit cell. Calculate the radius of a barium atom.

53. Polonium crystallizes with a simple cubic structure. It has a density of 9.3 g/cm³ a radius of 167 pm, and a molar mass of 209 g/mol. Use this data to estimate Avogadro's number (the number of atoms in one mole).

54. Palladium crystallizes with a face-centered cubic structure. It has a density of 12.0 g/cm³ a radius of 138 pm, and a molar mass of 106.42 g/mol. Use this data to estimate Avogadro's number.

55. Identify each solid as molecular, ionic, or atomic.
  a. $Ar(s)$ **b.** $H_2O(s)$ **c.** $K_2O(s)$ **d.** $Fe(s)$

56. Identify each solid as molecular, ionic, or atomic.
  a. $CaCl_2(s)$ **b.** $CO_2(s)$ **c.** $Ni(s)$ **d.** $I_2(s)$

57. Which solid has the highest melting point? Why?

$$Ar(s), CCl_4(s), LiCl(s), CH_3OH(s)$$

58. Which solid has the highest melting point? Why?

$$C(s \text{ diamond}), Kr(s), NaCl(s), H_2O(s)$$

59. Which solid in each pair has the higher melting point and why?
  a. $TiO_2(s)$ or $HOOH(s)$ **b.** $CCl_4(s)$ or $SiCl_4(s)$
  c. $Kr(s)$ or $Xe(s)$ **d.** $NaCl(s)$ or $CaO(s)$

60. Which solid in each pair has the higher melting point and why?
  a. $Fe(s)$ or $CCl_4(s)$ **b.** $KCl(s)$ or $HCl(s)$
  c. $Ti(s)$ or $Ne(s)$ **d.** $H_2O(s)$ or $H_2S(s)$

61. An oxide of titanium crystallizes with the following unit cell (titanium = gray; oxygen = red). What is the formula of the oxide?

62. An oxide of rhenium crystallizes with the following unit cell (rhenium = gray; oxygen = red). What is the formula of the oxide?

63. The unit cells for cesium chloride and barium(II) chloride are pictured here. Show that the ratio of cations to anions in each unit cell corresponds to the ratio of cations to anions in the formula of each compound.

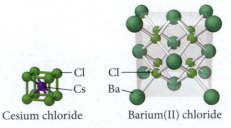

Cesium chloride   Barium(II) chloride

64. The unit cells for lithium oxide and silver iodide are pictured here. Show that the ratio of cations to anions in each unit cell corresponds to the ratio of cations to anions in the formula of each compound.

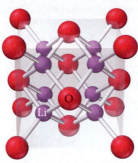

Lithium oxide

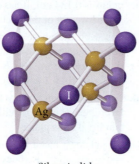

Silver iodide

Band Theory

65. Which substance would you expect to have little or no band gap?
    a. Zn(s)   **b.** Si(s)   **c.** As(s)

66. How many molecular orbitals are present in the valence band of a sodium crystal with a mass of 5.45 g?

Cumulative Problems

67. Explain the trend in the melting points of the hydrogen halides.

HI	$-50.8\,°C$
HBr	$-88.5\,°C$
HCl	$-114.8\,°C$
HF	$-83.1\,°C$

68. Explain the trend in the boiling points of the compounds.

H_2Te	$-2\,°C$
H_2Se	$-41.5\,°C$
H_2S	$-60.7\,°C$
H_2O	$+100\,°C$

69. The vapor pressure of water at 25 °C is 23.76 torr. If 1.25 g of water is enclosed in a 1.5-L container, will any liquid be present? If so, what mass of liquid?

70. The vapor pressure of CCl_3F at 300 K is 856 torr. If 11.5 g of CCl_3F is enclosed in a 1.0-L container, will any liquid be present? If so, what mass of liquid?

71. Examine the phase diagram for carbon dioxide shown in Figure 11.32. What transitions occur as you uniformly increase the pressure on a gaseous sample of carbon dioxide from 5.0 atm at −56 °C to 75 atm at −56 °C?

72. Carbon tetrachloride displays a triple point at 249.0 K and a melting point (at 1 atm) of 250.3 K. Which state of carbon tetrachloride is more dense, the solid or the liquid? Explain.

73. Four ice cubes at exactly 0 °C having a total mass of 53.5 g are combined with 115 g of water at 75 °C in an insulated container. If no heat is lost to the surroundings, what will be the final temperature of the mixture?

74. A sample of steam with a mass of 0.552 g and at a temperature of 100 °C condenses into an insulated container holding 4.25 g of water at 5.0 °C. Assuming that no heat is lost to the surroundings, what will be the final temperature of the mixture?

75. Air conditioners not only cool air, but dry it as well. Suppose that a room in a home measures 6.0 m × 10.0 m × 2.2 m. If the outdoor temperature is 30 °C and the vapor pressure of water in the air is 85% of the vapor pressure of water at this temperature, what mass of water must be removed from the air each time the volume of air in the room is cycled through the air conditioner? The vapor pressure for water at 30 °C is 31.8 torr.

76. A sealed flask contains 0.55 g of water at 28 °C. The vapor pressure of water at this temperature is 28.36 mmHg. What does the minimum volume of the flask have to be for no liquid water to be present in the flask?

77. Silver iodide crystallizes in the zinc blende structure. The separation between nearest neighbor cations and anions is approximately 325 pm and the melting point is 558 °C. Cesium chloride, by contrast, crystallizes in the cesium chloride structure shown in Figure 11.42. Even though the separation between nearest neighbor cations and anions is greater (348 pm), the melting point is still higher (645 °C). Explain why the melting point of cesium chloride is higher than that of silver iodide.

78. Copper iodide crystallizes in the zinc blende structure. The separation between nearest neighbor cations and anions is approximately 311 pm and the melting point is 606 °C. Potassium chloride, by contrast, crystallizes in the rock salt structure. Even though the separation between nearest neighbor cations and anions is greater (319 pm), the melting point is still higher (776 °C). Explain why the melting point of potassium chloride is higher than that of copper iodide.

79. Consider the face-centered cubic structure shown here:

a. What is the length of the line (labeled c) that runs diagonally across one of the faces of the cube in terms of r (the atomic radius)?

b. Use the answer to part a and the Pythagorean theorem to derive the expression for the edge length (l) in terms of r.

80. Consider the body-centered cubic structure shown here:

a. What is the length of the line (labeled c) that runs from one corner of the cube diagonally through the center of the cube to the other corner in terms of r (the atomic radius)?

b. Use the Pythagorean theorem to derive an expression for the length of the line (labeled b) that runs diagonally across one of the faces of the cube in terms of the edge length (l).

c. Use the answer to part (a) and (b) along with the Pythagorean theorem to derive the expression for the edge length (l) in terms of r.

81. The unit cell in a crystal of diamond belongs to a crystal system different from any we have discussed. The volume of a unit cell of diamond is 0.0454 nm^3 and the density of diamond is 3.52 g/cm^3. Find the number of carbon atoms in a unit cell of diamond.

82. The density of an unknown metal is 12.3 g/cm^3 and its atomic radius is 0.134 nm. It has a face-centered cubic lattice. Find the atomic weight of this metal.

83. Based on the phase diagram of CO_2 shown in Figure 11.32, describe the transitions that occur when the temperature of CO_2 is increased from 190 K to 350 K at a constant pressure of (a) 1 atm, (b) 5.1 atm, (c) 10 atm, (d) 100 atm.

84. Consider a planet where the pressure of the atmosphere at sea level is 2500 mmHg. Will water behave in a way that can sustain life on this planet?

85. An unknown metal is found to have a density of 7.8748 g/cm^3 and to crystallize in a body-centered cubic lattice. The edge of the unit cell is 0.28664 nm. Calculate the atomic mass of the metal.

86. When spheres of radius r are packed in a body-centered cubic arrangement, they occupy 68.0% of the available volume. Calculate the value of a, the length of the edge of the cube in terms of r.

Challenge Problems

87. Potassium chloride crystallizes in the rock salt structure. Estimate the density of potassium chloride using the ionic radii listed in Chapter 8.

88. Butane (C_4H_{10}) has a heat of vaporization of 22.44 kJ/mol and a normal boiling point of -0.4 °C. A 250-mL sealed flask contains 0.55 g of butane at -22 °C. How much butane is present as a liquid? If the butane is warmed to 25 °C, how much is present as a liquid?

89. Liquid nitrogen can be used as a cryogenic substance to obtain low temperatures. Under atmospheric pressure, liquid nitrogen boils at 77 K, allowing low temperatures to be reached. However, if the nitrogen is placed in a sealed, insulated container connected to a vacuum pump, even lower temperatures can be reached. Why? If the vacuum pump has sufficient capacity, and is left on for an extended period of time, the liquid nitrogen will start to freeze. Explain.

90. Calculate the fraction of empty space in cubic closest packing to five significant figures.

91. A tetrahedral site in a close-packed lattice is formed by four spheres at the corners of a regular tetrahedron. This is equivalent to placing the spheres at alternate corners of a cube. In such a close-packed arrangement the spheres are in contact and if the spheres have a radius r, the diagonal of the face of the cube is $2r$. The tetrahedral hole is inside the middle of the cube. Find the length of the body diagonal of this cube and then find the radius of the tetrahedral hole.

92. Given that the heat of fusion of water is 6.02 kJ/mol, that the heat capacity of $H_2O(l)$ is 75.2 J/mol·K and that the heat capacity of $H_2O(s)$ is 37.7 J/mol·K, calculate the heat of fusion of water at -10 °C.

93. The heat of combustion of CH_4 is 890.4 kJ/mol and the heat capacity of H_2O is 75.2 J/mol·K. Find the volume of methane measured at 298 K and 1.00 atm required to convert 1.00 L of water at 298 K to water vapor at 373 K.

94. Three 1.0-L flasks, maintained at 308 K, are connected to each other with stopcocks. Initially the stopcocks are closed. One of the flasks contains 1.0 atm of N_2, the second 2.0 g of H_2O, and the third, 0.50 g of ethanol, C_2H_6O. The vapor pressure of H_2O at 308 K is 42 mmHg and that of ethanol is 102 mmHg. The stopcocks are then opened and the contents mix freely. What is the pressure?

Conceptual Problems

95. One prediction of global warming is the melting of global ice, which may result in coastal flooding. A criticism of this prediction is that the melting of icebergs does not increase ocean levels any more than the melting of ice in a glass of water increases the level of liquid in the glass. Is this a valid criticism? Does the melting of an ice cube in a cup of water raise the level of the liquid in the cup? Why or why not? In response to this criticism, scientists have asserted that they are not worried about melting icebergs, but rather the melting of ice sheets that sit on the continent of Antarctica. Would the melting of this ice increase ocean levels? Why or why not?

96. The rate of vaporization depends on the surface area of the liquid. However, the vapor pressure of a liquid does not depend on the surface area. Explain.

97. Substance A has a smaller heat of vaporization than substance B. Which of the two substances will undergo a larger change in vapor pressure for a given change in temperature?

98. The density of a substance is greater in its solid state than in its liquid state. If the triple point in the phase diagram of the substance is below 1.0 atm, then which will necessarily be at a lower temperature, the triple point or the normal melting point?

99. A substance has a heat of vaporization of ΔH_{vap} and heat of fusion of ΔH_{fus}. Express the heat of sublimation in terms of ΔH_{vap} and ΔH_{fus}.

100. Examine the heating curve for water in Section 11.7. If heat is added to the water at a constant rate, which of the three segments in which temperature is rising will have the least steep slope? Why?

101. A root cellar is an underground chamber used to store fruits, vegetables, and even meats. In extreme cold, farmers put large vats of water into the root cellar to prevent the fruits and vegetables from freezing. Explain why this works.

102. Why is heat of fusion of a substance always smaller than its heat of vaporization?

Answers to Conceptual Connections

State Changes

11.1 (a) When water boils, it simply changes state from liquid to gas. Water molecules do not decompose during boiling.

Dispersion Forces

11.2 (c) I_2 has the highest boiling point because it has the highest molar mass. Since the halogens are all similar in other ways, we expect I_2 to have the greatest dispersion forces and therefore the highest boiling point (and in fact it does).

Intermolecular Forces and Boiling Point

11.3 (a) CH_3OH. The compounds all have similar molar masses, so the dispersion forces are similar in all three. CO is polar, but since CH_3OH contains H directly bonded to O, it has hydrogen bonding, resulting in the highest boiling point.

Vapor Pressure

11.4 (b) Although the *rate of vaporization* increases with increasing surface area, the *vapor pressure* of a liquid is independent of surface area. An increase in surface increases both the rate of vaporization and the rate of condensation—the effects of surface area exactly cancel and the vapor pressure does not change.

Cooling Water with Ice

11.5 The melting of the ice. The warming of the ice from $-10\,^{\circ}C$ to $0\,^{\circ}C$ absorbs only 20.9 J/g of ice. The melting of the ice, however, absorbs about 334 J/g of ice. (This value is obtained by dividing the heat of fusion by the molar mass of water.) Therefore, the melting of the ice produces a larger temperature decrease in the water than does the warming of the ice.

Phase Diagrams

11.6 (b) The solid will sublime into a gas. Since the pressure is below the triple point the liquid state is not stable.

12 Solutions

One molecule of nonsaline substance (held in the solvent) dissolved in 100 molecules of any volatile liquid decreases the vapor pressure of this liquid by a nearly constant fraction, nearly 0.0105. —FRANCOIS-MARIE RAOULT (1830–1901)

Drinking seawater causes dehydration because seawater draws water out of body tissues.

WE LEARNED IN Chapter 1 that most of the matter we encounter is in the form of mixtures. In this chapter, we focus on homogeneous mixtures, known as solutions. Solutions are mixtures in which atoms and molecules intermingle on the molecular and atomic scale. Some common examples of solutions include ocean water, gasoline, and air. Why do solutions form? How are their properties different from the properties of the pure substances that compose them? As you read this chapter, keep in mind the great number of solutions that surround you at every moment, including those that exist within your own body.

12.1   Thirsty Solutions: Why You Should Not Drink Seawater

In a popular novel, *Life of Pi* by Yann Martel, the main character (whose name is Pi) is stranded on a lifeboat with a Bengal tiger in the middle of the Pacific Ocean for 227 days. He survives in part by distilling seawater for drinking using a solar still. However, in the first three days of his predicament, before he rigs up the still, he becomes severely dehydrated from lack of water. Drinking the seawater that surrounded him would only have made his condition worse. Why? Seawater actually draws water *out of the body* as it passes through the stomach and intestines, resulting in diarrhea and further dehydration. We can think of seawater as a *thirsty solution*—it draws water to itself. Consequently, seawater should never be consumed.

Seawater (**Figure 12.1▼**) is an example of a **solution**, a homogeneous mixture of two or more substances. A solution has at least two components. The majority component is usually called the **solvent** and the minority component is usually called the **solute**. In seawater, water is the solvent and sodium chloride is the primary solute.

The reason that seawater draws water to itself is related to nature's tendency toward spontaneous mixing, which we discuss in more detail later in this chapter and in Chapter 17. For now, we can make the observation that, unless it is highly unfavorable energetically, *substances tend to combine into uniform mixtures rather than separating into pure substances*. For example, suppose that pure water and a sodium chloride solution are enclosed in separate compartments with a removable barrier between them, as shown in **Figure 12.2(a)▼**. If the barrier is removed, the two liquids will spontaneously mix together, eventually forming a more dilute salt solution of uniform concentration, as shown in **Figure 12.2(b)▼**. The tendency

> In some cases, the concepts of solute and solvent are not useful (for example, a solution that is equal parts ethanol and water.)

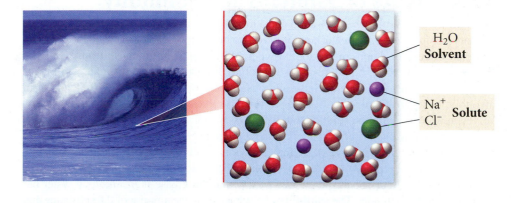

H$_2$O
Solvent

Na$^+$
Cl$^-$ **Solute**

◀ **FIGURE 12.1 A Typical Solution**
In seawater, sodium chloride is the primary solute. Water is the solvent.

▼ **FIGURE 12.2 The Tendency to Mix**   **(a)** Pure water and a sodium chloride solution are separated by a barrier. **(b)** When the barrier is removed, the two liquids spontaneously mix, producing a single solution of uniform concentration.

Spontaneous Mixing

> When the barrier is removed, spontaneous mixing occurs, producing a solution of uniform concentration.

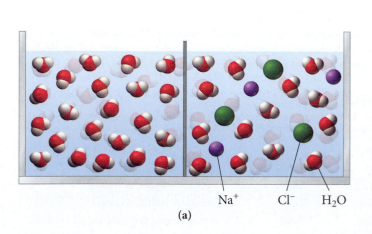

Na$^+$ Cl$^-$ H$_2$O
(a)

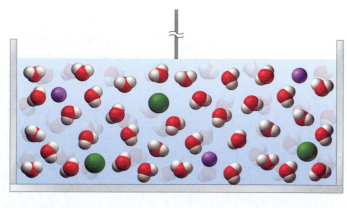

(b)

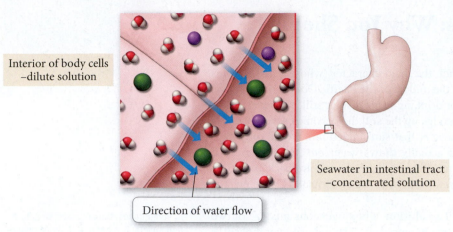

Interior of body cells –dilute solution

Direction of water flow

Seawater in intestinal tract –concentrated solution

▲ Seawater is a more concentrated solution than the fluids in body cells. As a result, when seawater flows through the digestive tract, it draws water out of the surrounding tissues.

CO_2 H_2O

▲ Club soda is a solution of carbon dioxide and water.

The general solubilities of a number of ionic compounds are described by the solubility rules in Section 4.5.

toward mixing results in a uniform concentration of the final solution.

Seawater is a *thirsty* solution because of this tendency toward mixing. As seawater moves through the intestine, it flows past cells that line the digestive tract, which consist of largely fluid interiors surrounded by membranes. Cellular fluids themselves contain dissolved ions, including sodium and chloride, but the fluids are more dilute than seawater. Nature's tendency toward mixing (which tends to produce solutions of uniform concentration), together with the selective permeability of the cell membranes (which allow water to flow in and out, but restrict the flow of dissolved solids), causes a *flow of water out of the body's cells and into the seawater*. In this way, the two solutions become more similar in concentration (as though they had mixed)—the solution in the intestine becomes somewhat more dilute than it was and the solution in the cells becomes somewhat more concentrated. The accumulation of extra fluid in the intestines causes diarrhea, and the decreased fluid in the cells causes dehydration. If Pi had drunk the seawater instead of constructing the solar still, neither he nor the Bengal tiger would have survived their ordeal.

12.2   Types of Solutions and Solubility

A solution may be composed of a solid and a liquid (such as the salt and water that are the primary components of seawater), but it may also be composed of a gas and a liquid, two different liquids, or other combinations (see Table 12.1). In **aqueous solutions** water is the solvent, and a solid, liquid, or gas is the solute. For example, solid sugar or salt readily mix with water to form aqueous solutions. Similarly, ethyl alcohol—the alcohol in alcoholic beverages—mixes with water to form a solution, and carbon dioxide gas dissolves in water to form the solution that we know as club soda.

You probably know from experience that a particular solvent, such as water, does not dissolve all possible solutes. For example, if your hands are covered with automobile grease, you cannot remove that grease with plain water. However, another solvent, such as paint thinner, easily dissolves the grease. We say that the grease is *insoluble* in water but *soluble* in the paint thinner. The **solubility** of a substance is the amount of the substance that will dissolve in a given amount of solvent. For example, the solubility of sodium chloride in water at 25 °C is 36 g NaCl per 100 g water, while the solubility of grease in water is nearly zero. The solubility of one substance in another depends both on nature's tendency toward mixing that we discussed in Section 12.1 and on the types of intermolecular forces that we discussed in Chapter 11.

Nature's Tendency toward Mixing: Entropy

So far in this book, we have seen that many chemical systems tend toward lower *potential energy*. For example, two particles with opposite charges (such as a proton and an electron or a cation and an anion) move toward each other because their potential energy is lowered

TABLE 12.1 Common Types of Solutions

Solution Phase	Solute Phase	Solvent Phase	Example
Gaseous solution	Gas	Gas	Air (mainly oxygen and nitrogen)
Liquid solution	Gas	Liquid	Club soda (CO_2 and water)
	Liquid	Liquid	Vodka (ethanol and water)
	Solid	Liquid	Seawater (salt and water)
Solid solution	Solid	Solid	Brass (copper and zinc) and other alloys

as their separation decreases according to Coulomb's law. The formation of a solution, however, *does not necessarily* lower the potential energy of its constituent particles. The clearest example of this is the formation of a homogeneous mixture—a *solution*—of two ideal gases. Suppose that neon and argon are enclosed in a container with a removable barrier between them, as shown in **Figure 12.3(a)▶**. As soon as the barrier is removed, the neon and argon mix together to form a solution, as shown in **Figure 12.3(b)▶**. *Why?*

Recall that at low pressures and moderate temperatures both neon and argon behave as ideal gases—they do not interact with each other in any way (that is, there are no significant forces between their constituent particles). When the barrier is removed, the two gases mix, but their potential energy remains unchanged. *We cannot think of the mixing of two ideal gases as lowering their potential energy.* The tendency to mix is related, rather, to a concept called *entropy*.

Entropy is a measure of energy randomization or energy dispersal in a system. Recall that a gas at any temperature above 0 K has kinetic energy due to the motion of its atoms. When neon and argon are confined to their individual compartments, their kinetic energies are also confined to those compartments. However, when the barrier between the compartments is removed, each gas—along with its kinetic energy—becomes *spread out* or *dispersed* over a larger volume. Therefore, the mixture of the two gases has greater energy dispersal, or greater *entropy* than the separated components.

The pervasive tendency for all kinds of energy to spread out, or disperse, whenever it is not restrained from doing so is the reason that two ideal gases mix. Another common example of the tendency toward energy dispersal is the transfer of thermal energy from hot to cold. If you heat one end of an iron rod, the thermal energy deposited at the end of the rod will spontaneously spread along the entire length of the rod. The tendency for energy to disperse is the reason that thermal energy flows from the hot end of the rod to the cold one, and not the other way around. Imagine a metal rod that became spontaneously hotter on one end and ice cold on the other—it simply does not happen because energy does not spontaneously concentrate itself. In Chapter 17, we will see that the dispersal of energy is actually the fundamental criterion that ultimately determines the spontaneity of all processes.

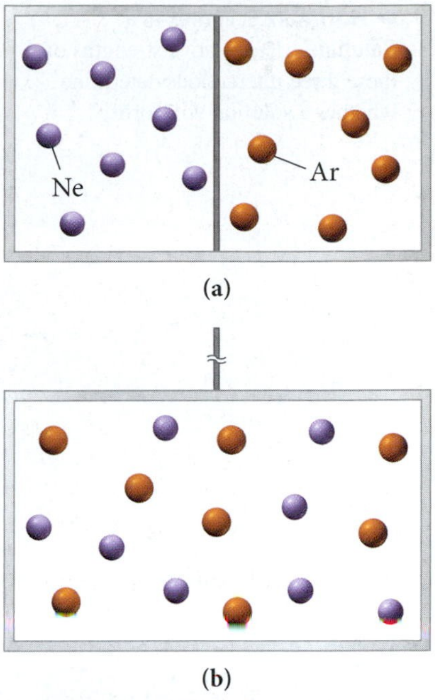

The Effect of Intermolecular Forces

We have just seen that, in the absence of intermolecular forces, two substances will spontaneously mix to form a homogeneous solution. We know from Chapter 11, however, that solids and liquids exhibit a number of different types of intermolecular forces including dispersion forces, dipole–dipole forces, hydrogen bonding, and ion–dipole forces **Figure 12.4▼**. These forces may promote the formation of a solution or prevent it, depending on the nature of the forces in the particular combination of solute and solvent.

Intermolecular forces exist between each of the following: (a) the solvent and solute particles, (b) the solvent particles themselves, and (c) the solute particles themselves, as shown in **Figure 12.5▶**.

As summarized in Table 12.2, a solution always forms if the solvent–solute interactions are comparable to, or stronger than, the solvent–solvent interactions and the

▲ **FIGURE 12.3 Spontaneous Mixing of Two Ideal Gases** (a) Neon and argon are separated by a barrier. (b) When the barrier is removed, the two gases spontaneously mix to form a uniform solution.

Intermolecular Forces

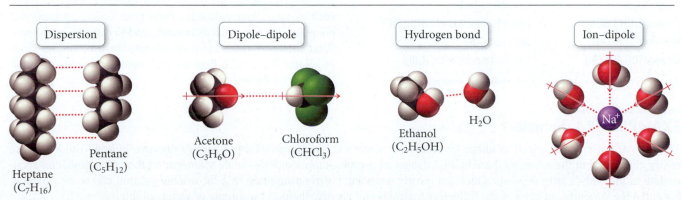

▲ **FIGURE 12.4 Intermolecular Forces Involved in Solutions**

▶ **FIGURE 12.5 Forces in a Solution**   The relative strengths of these three interactions determine whether a solution will form.

Solution Interactions

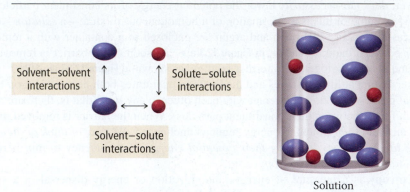

Solvent–solute interactions:	**The interactions between a solvent particle and a solute particle.**
Solvent–solvent interactions:	**The interactions between a solvent particle and another solvent particle.**
Solute–solute interactions:	**The interactions between a solute particle and another solute particle.**

TABLE 12.2 Relative Interactions and Solution Formation

Solvent–solute interactions	>	Solvent–solvent and solute–solute interactions	Solution forms
Solvent–solute interactions	=	Solvent–solvent and solute–solute interactions	Solution forms
Solvent–solute interactions	<	Solvent–solvent and solute–solute interactions	Solution may or may not form, depending on relative disparity

solute–solute interactions. For example, consider mixing the hydrocarbons pentane (C_5H_{12}) and heptane (C_7H_{16}). The intermolecular forces present within both pentane and heptane are dispersion forces. Similarly, the intermolecular forces present *between* heptane and pentane are also dispersion forces. All three interactions are of similar magnitude and the two substances are soluble in each other in all proportions—they are **miscible**. The formation of the solution is driven by the tendency toward mixing, or toward greater entropy, that we just discussed.

If the solvent–solute interactions are weaker than the solvent–solvent and solute–solute interactions—in other words, if the solvent molecules and the solute molecules each interact more strongly with molecules of their own kind than with molecules of the other kind—then a solution may still form, depending on the relative disparities between the interactions. If the disparity is small, the tendency to mix results in the formation of a solution even though the process is energetically uphill. If the disparity is large, however, a solution will not form. For example, consider mixing hexane and water. The water molecules have strong hydrogen-bonding attractions to each other but cannot hydrogen bond to hexane. The energy required to pull water molecules away from one another is too great, and too little energy is returned when the water molecules interact with hexane molecules. As a result, a solution does not form when hexane and water are mixed. Although the tendency to mix is strong, it cannot overcome the large energy disparity between the powerful solvent–solvent interactions and the weak solvent–solute interactions.

In general, we can use the rule of thumb that *like dissolves like* when predicting the formation of solutions. Polar solvents, such as water, tend to dissolve many polar or ionic solutes, and nonpolar solvents, such as hexane, tend to dissolve many nonpolar solutes. Similar kinds of solvents dissolve similar kinds of solutes. Table 12.3 lists some common polar and nonpolar laboratory solvents.

TABLE 12.3 Common Laboratory Solvents

Common Polar Solvents	Common Nonpolar Solvents
Water (H_2O)	Hexane (C_6H_{14})
Acetone (CH_3COCH_3)	Diethyl ether ($CH_3CH_2OCH_2CH_3$)*
Methanol (CH_3OH)	Toluene (C_7H_8)
Ethanol (CH_3CH_2OH)	Carbon tetrachloride (CCl_4)

*Diethyl ether can be considered to be intermediate between polar and nonpolar.

EXAMPLE 12.1 Solubility

Vitamins are often categorized as either fat soluble or water soluble. Water-soluble vitamins dissolve in body fluids and are easily eliminated in the urine, so there is little danger of overconsumption. Fat-soluble vitamins, on the other hand, can accumulate in the body's fatty deposits which are mostly nonpolar. Overconsumption of a fat-soluble vitamin can be detrimental. Examine the structures of each of the following vitamins and classify them as fat soluble or water soluble.

(a) Vitamin C

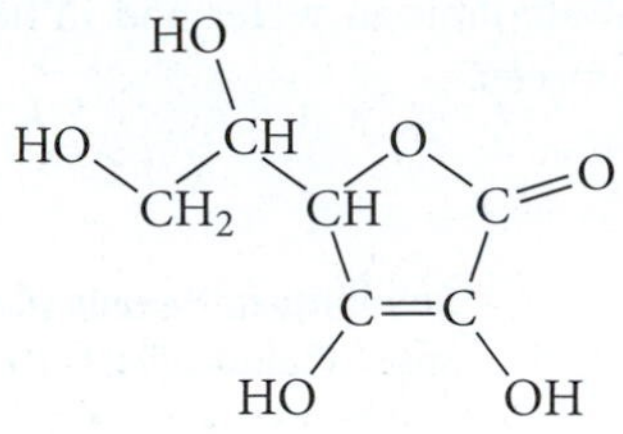

(b) Vitamin K_3

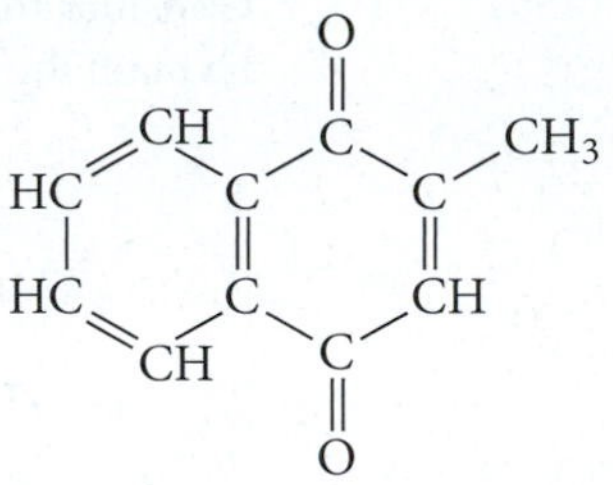

(c) Vitamin A

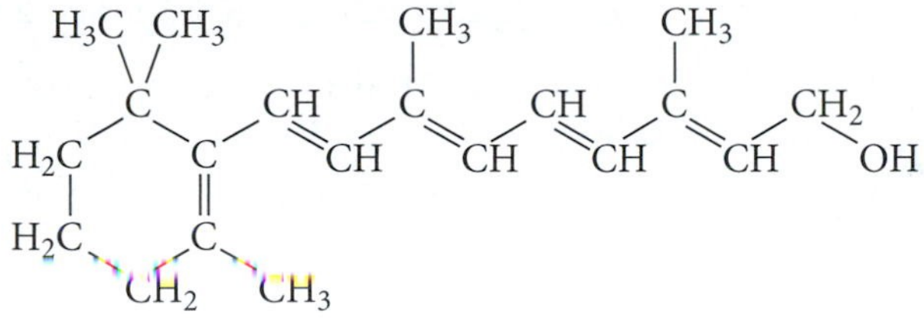

(d) Vitamin B_5

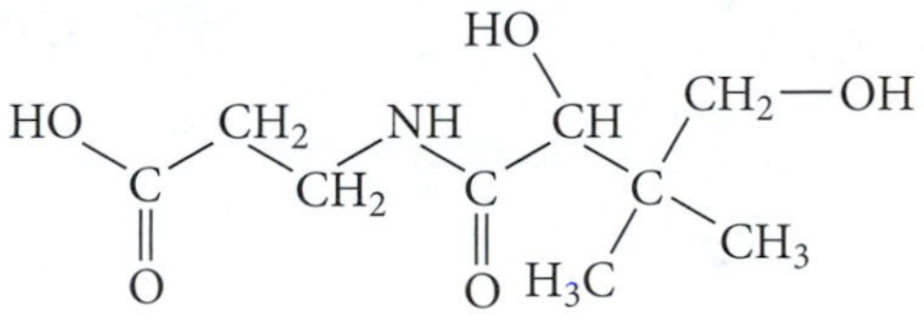

SOLUTION

(a) The four —OH bonds in vitamin C make it highly polar and give it the ability to hydrogen bond with water. Vitamin C is water soluble.	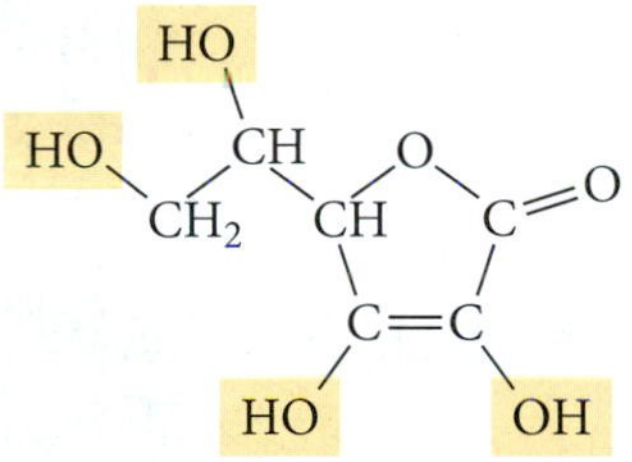
(b) The C—C bonds in vitamin K_3 are nonpolar and the C—H bonds nearly so. The C=O bonds are polar, but the bond dipoles oppose each other and therefore largely cancel, so the molecule is dominated by the nonpolar bonds. Vitamin K_3 is fat soluble.	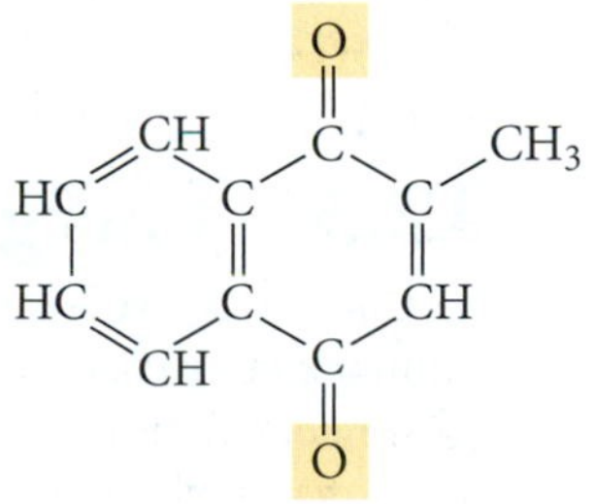
(c) The C—C bonds in vitamin A are nonpolar and the C—H bonds nearly so. The one polar —OH bond may increase the water solubility slightly, but overall vitamin A is nonpolar and therefore fat soluble.	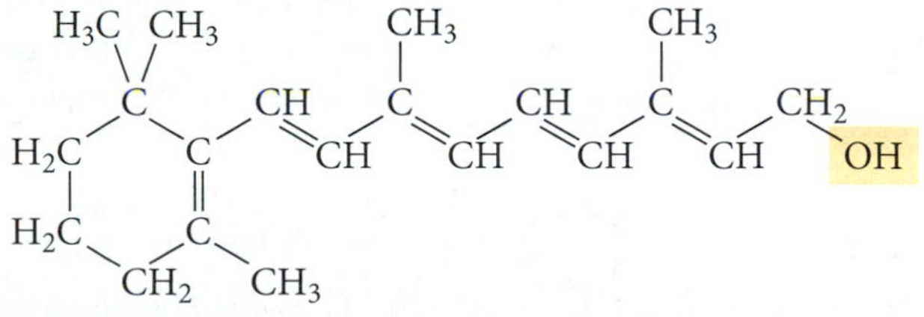
(d) The three —OH bonds and one —NH bond in vitamin B_5 make it highly polar and give it the ability to hydrogen bond with water. Vitamin B_5 is water soluble.	

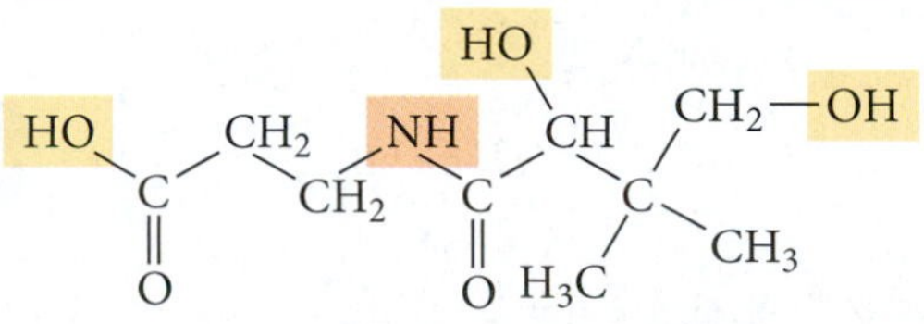

FOR PRACTICE 12.1

Determine whether each compound is soluble in hexane.

(a) water (H_2O)   **(b)** propane ($CH_3CH_2CH_3$)   **(c)** ammonia (NH_3)   **(d)** hydrogen chloride (HCl)

Conceptual Connection 12.1 Solubility

Consider this table showing the solubilities of several alcohols in water and in hexane. Explain the observed trend in terms of intermolecular forces.

Alcohol	Space-Filling Model	Solubility in H_2O (mol alcohol/100 g H_2O)	Solubility in Hexane (C_6H_{14}) (mol alcohol/100 g C_6H_{14})
Methanol (CH_3OH)		Miscible	0.12
Ethanol (CH_3CH_2OH)		Miscible	Miscible
Propanol ($CH_3CH_2CH_2OH$)		Miscible	Miscible
Butanol ($CH_3CH_2CH_2CH_2OH$)		0.11	Miscible
Pentanol ($CH_3CH_2CH_2CH_2CH_2OH$)		0.030	Miscible

12.3   Energetics of Solution Formation

In Chapter 6, we examined the energy changes associated with chemical reactions. Similar energy changes can occur upon the formation of a solution, depending on the relative interactions of the solute and solvent particles. For example, when sodium hydroxide is dissolved in water, heat is evolved—the process is *exothermic*. In contrast, when ammonium nitrate (NH_4NO_3) is dissolved in water, heat is absorbed—this process is *endothermic*. Other solutions, such as sodium chloride in water, barely absorb or evolve any heat upon formation. What causes these different behaviors?

We can understand the energy changes associated with solution formation by envisioning the process as occurring in three steps, each with an associated change in enthalpy:

1. Separating the solute into its constituent particles.

This step is always endothermic (positive ΔH) because energy is required to overcome the forces that hold the solute particles together.

2. Separating the solvent particles from each other to make room for the solute particles.

$\Delta H_{solvent} > 0$

This step is also endothermic because energy is required to overcome the intermolecular forces among the solvent particles.

3. Mixing the solute particles with the solvent particles.

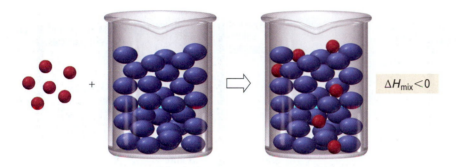

$\Delta H_{mix} < 0$

This step is exothermic, because energy is released as the solute particles interact with the solvent particles through the various types of intermolecular forces.

According to Hess's law, the overall enthalpy change upon solution formation, the **enthalpy of solution** (ΔH_{soln}) is the sum of the changes in enthalpy for each step:

$$\Delta H_{soln} = \underset{\text{endothermic (+)}}{\Delta H_{solute}} + \underset{\text{endothermic (+)}}{\Delta H_{solvent}} + \underset{\text{exothermic (–)}}{\Delta H_{mix}}$$

The first two terms are endothermic (positive ΔH) and the third term is exothermic (negative ΔH); the overall sign of ΔH_{soln} depends on the magnitudes of the individual terms, as shown in **Figure 12.6▶**.

1. *If the sum of the endothermic terms is about equal in magnitude to the exothermic term, then ΔH_{soln} is about zero.* The increasing entropy upon mixing drives the formation of a solution while the overall energy of the system remains nearly constant.

2. *If the sum of the endothermic terms is smaller in magnitude than the exothermic term, then ΔH_{soln} is negative and solution formation is exothermic.* In this case, both the tendency toward lower energy and the tendency toward greater entropy drive the formation of a solution.

3. *If the sum of the endothermic terms is greater in magnitude than the exothermic term, then ΔH_{soln} is positive and solution formation is endothermic.* In this case, as long as ΔH_{soln} is not too large, the tendency toward greater entropy will still drive the formation of a solution. If, on the other hand, ΔH_{soln} is too large, a solution will not form.

Energetics of Solution Formation

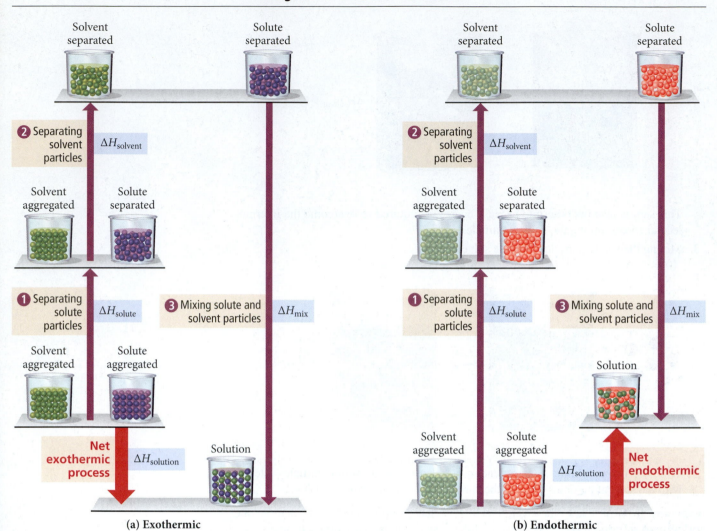

▲ **FIGURE 12.6 Energetics of the Solution Process**   (a) When ΔH_{mix} is greater in magnitude than the sum of ΔH_{solute} and $\Delta H_{solvent}$, the heat of solution is negative (exothermic). (b) When ΔH_{mix} is smaller in magnitude than the sum of ΔH_{solute} and $\Delta H_{solvent}$, the heat of solution is positive (endothermic).

Aqueous Solutions and Heats of Hydration

Many common solutions, such as the seawater in the opening example of this chapter, contain an ionic compound dissolved in water. In these aqueous solutions, $\Delta H_{solvent}$ and ΔH_{mix} can be combined into a single term called the **heat of hydration** ($\Delta H_{hydration}$) (**Figure 12.7▶** on p. 465). The heat of hydration is the enthalpy change that occurs when 1 mol of the gaseous solute ions are dissolved in water. Because the ion–dipole interactions that occur between a dissolved ion and the surrounding water molecules (**Figure 12.8▶**) are much stronger than the hydrogen bonds in water, $\Delta H_{hydration}$ is always largely negative (exothermic) for ionic compounds. Using the heat of hydration, we can write the enthalpy of solution as a sum of just two terms, one endothermic and one exothermic:

$$\Delta H_{soln} = \Delta H_{solute} + \underbrace{\Delta H_{solvent} + \Delta H_{mix}}$$

$$\Delta H_{soln} = \underset{\substack{\text{endothermic} \\ \text{(positive)}}}{\Delta H_{solute}} + \underset{\substack{\text{exothermic} \\ \text{(negative)}}}{\Delta H_{hydration}}$$

For ionic compounds, ΔH_{solute}, the energy required to separate the solute into its constituent particles is the negative of the solute's lattice energy ($\Delta H_{solute} = -\Delta H_{lattice}$), discussed in Section 9.4. For ionic aqueous solutions then, the overall enthalpy of solution

Heat of Hydration

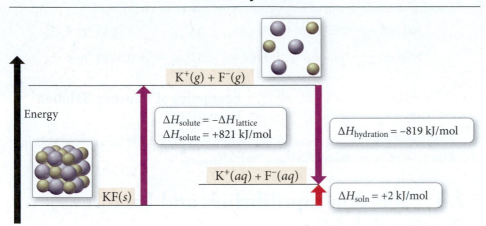

depends on the relative magnitudes of ΔH_{solute} and $\Delta H_{hydration}$, with three possible scenarios (in each case we refer to the *magnitude* or *absolute value* of ΔH):

1. $|\Delta H_{solute}| < |\Delta H_{hydration}|$. The amount of energy required to separate the solute into its constituent ions is less than the energy given off when the ions are hydrated. ΔH_{soln} is therefore negative and solution formation is exothermic. Solutes with negative enthalpies of solution include lithium bromide and potassium hydroxide. When these solutes dissolve in water, the resulting solutions feel warm to the touch.

$$LiBr(s) \xrightarrow{\text{H}_2\text{O}} Li^+(aq) + Br^-(aq) \qquad \Delta H_{soln} = -48.78 \text{ kJ/mol}$$
$$KOH(s) \xrightarrow{\text{H}_2\text{O}} K^+(aq) + OH^-(aq) \qquad \Delta H_{soln} = -57.56 \text{ kJ/mol}$$

2. $|\Delta H_{solute}| > |\Delta H_{hydration}|$. The amount of energy required to separate the solute into its constituent ions is greater than the energy given off when the ions are hydrated. ΔH_{soln} is therefore positive and the solution process is endothermic (if a solution forms at all). Solutes that form aqueous solutions with positive enthalpies of solution include ammonium nitrate and silver nitrate. When these solutes dissolve in water, the resulting solutions feel cool to the touch.

$$NH_4NO_3(s) \xrightarrow{\text{H}_2\text{O}} NH_4^+(aq) + NO_3^-(aq) \qquad \Delta H_{soln} = +25.67 \text{ kJ/mol}$$
$$AgNO_3(s) \xrightarrow{\text{H}_2\text{O}} Ag^+(aq) + NO_3^-(aq) \qquad \Delta H_{soln} = +36.91 \text{ kJ/mol}$$

3. $|\Delta H_{solute}| \approx |\Delta H_{hydration}|$. The amount of energy required to separate the solute into its constituent ions is about equal to the energy given off when the ions are hydrated. ΔH_{soln} is therefore approximately zero and the solution process is neither appreciably exothermic nor appreciably endothermic. Solutes with enthalpies of solution near zero

Ion–Dipole Interactions

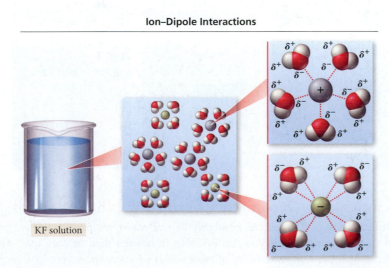

◀ **FIGURE 12.8 Ion–Dipole Interactions** Ion–dipole interactions such as those between potassium ions, fluoride ions, and water molecules cause the heat of hydration to be largely negative (exothermic).

include sodium chloride and sodium fluoride. When these solutes dissolve in water, the resulting solutions do not undergo a noticeable change in temperature.

$$NaCl(s) \xrightarrow[H_2O]{} Na^+(aq) + Cl^-(aq) \qquad \Delta H_{soln} = +3.88 \text{ kJ/mol}$$

$$NaF(s) \xrightarrow[H_2O]{} Na^+(aq) + F^-(aq) \qquad \Delta H_{soln} = +0.91 \text{ kJ/mol}$$

Conceptual Connection 12.2 Energetics of Aqueous Solution Formation

The enthalpy of solution for cesium fluoride is −36.8 kJ/mol. What can you conclude about the relative magnitudes of ΔH_{solute} and $\Delta H_{hydration}$?

12.4   Solution Equilibrium and Factors Affecting Solubility

The dissolution of a solute in a solvent is an equilibrium process similar to the equilibrium process associated with a state change (discussed in Chapter 11). Imagine, from a molecular viewpoint, the dissolving of a solid solute such as sodium chloride in a liquid solvent such as water. Initially, water molecules rapidly solvate sodium cations and chloride anions, resulting in a noticeable decrease in the amount of solid sodium chloride in the water. Over time, however, the concentration of dissolved sodium chloride in the solution increases. This dissolved sodium chloride can then begin to recrystallize as solid sodium chloride. At first, the rate of dissolution far exceeds the rate of recrystallization. But as the concentration of dissolved sodium chloride increases, the rate of recrystallization also increases. Eventually the rates of dissolution and recrystallization become equal—**dynamic equilibrium** has been reached.

$$NaCl(s) \underset{H_2O}{\rightleftharpoons} Na^+(aq) + Cl^-(aq)$$

A solution in which the dissolved solute is in dynamic equilibrium with the solid (or undissolved) solute is called a **saturated solution**. *If you add additional solute to a saturated solution, it will not dissolve.* A solution containing less than the equilibrium amount of solute is called an **unsaturated solution**. *If you add additional solute to an unsaturated solution, it will dissolve.*

Under certain circumstances, a **supersaturated solution**—one containing more than the equilibrium amount of solute—may form. Such solutions are unstable and the excess solute normally precipitates out of the solution. However, in some cases, if left undisturbed, a supersaturated solution can exist for an extended period of time. In a common classroom demonstration, a tiny piece of solid sodium acetate is added to a supersaturated solution of sodium acetate. This triggers the precipitation of the solute, which crystallizes out of solution in an often beautiful and dramatic way **Figure 12.9▼**.

▶ **FIGURE 12.9 Precipitation from a Supersaturated Solution**   When a small piece of solid sodium acetate is added to a supersaturated sodium acetate solution, the excess solid precipitates out of the solution.

The Temperature Dependence of the Solubility of Solids

The solubility of solids in water can be highly dependent on temperature. Have you ever noticed how much more sugar you can dissolve in hot tea than in cold tea? Although several exceptions exist, *the solubility of most solids in water increases with increasing*

In the case of sugar dissolving in water, the higher temperature increases both *how fast* the sugar dissolves and *how much* sugar dissolves.

temperature, as shown in **Figure 12.10▶**. For example, the solubility of potassium nitrate (KNO_3) at room temperature is about 37 g KNO_3 per 100 g of water. At 50 °C, however, the solubility rises to 88 g KNO_3 per 100 g of water.

A common way to purify a solid is a technique called **recrystallization**. In this technique, enough solid is added to water (or some other solvent) to create a saturated solution at an elevated temperature. As the solution cools, it becomes supersaturated and the excess solid precipitates out of solution. If the solution cools slowly, the solid forms crystals as it comes out of solution. The crystalline structure tends to reject impurities, resulting in a purer solid.

Factors Affecting the Solubility of Gases in Water

Solutions of gases dissolved in water are common. Club soda, for example, is a solution of carbon dioxide and water, and most liquids exposed to air contain dissolved gases from air. Fish depend on the oxygen dissolved in lake or seawater for life, and our blood contains dissolved nitrogen, oxygen, and carbon dioxide. Even tap water contains dissolved gases from air. The solubility of a gas in a liquid is affected by both temperature and pressure.

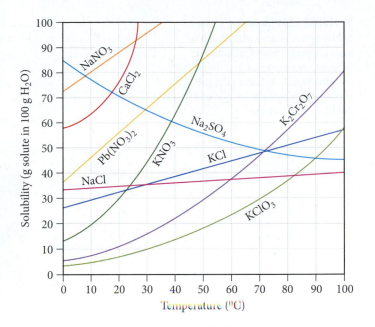

▲ **FIGURE 12.10 Solubility and Temperature** The solubility of most solids increases with increasing temperature.

The Effect of Temperature The effect of temperature on the solubility of a gas in water can be observed by heating ordinary tap water on a stove. Before the water reaches its boiling point, you will see small bubbles develop in the water. These bubbles are the dissolved air (mostly nitrogen and oxygen) coming out of solution. (Once the water boils, the bubbling becomes more vigorous—these larger bubbles are composed of water vapor.) The dissolved air comes out of solution because—unlike solids, whose solubility generally increases with increasing temperature—*the solubility of gases in liquids decreases with increasing temperature.*

The inverse relationship between gas solubility and temperature is the reason that warm soda pop bubbles more than cold soda pop when you open it and also the reason that warm beer goes flat faster than cold beer. More carbon dioxide comes out of solution at room temperature than at a lower temperature because the gas is less soluble at room temperature. The decreasing solubility of gases with increasing temperature is also the reason that fish don't bite much if the lake you are fishing in is too warm. The warm temperature results in a lower oxygen concentration. With lower oxygen levels, the fish become lethargic and tend not to strike at any lure or bait you might cast their way.

Cold soda pop   Warm soda pop

▲ Warm soda pop bubbles more than cold soda pop because carbon dioxide is less soluble in the warm solution.

Conceptual Connection 12.3 Solubility and Temperature

A solution is saturated in both nitrogen gas and potassium bromide at 75 °C. When the solution is cooled to room temperature, what is most likely to occur?

(a) Some nitrogen gas bubbles out of solution.

(b) Some potassium bromide precipitates out of solution.

(c) Some nitrogen gas bubbles out of solution *and* some potassium bromide precipitates out of solution.

(d) Nothing happens.

The Effect of Pressure The solubility of gases also depends on pressure. The higher the pressure of a gas above a liquid, the more soluble the gas is in the liquid. In a sealed can of soda pop, for example, the carbon dioxide is maintained in solution by a high pressure of carbon dioxide within the can. When the can is opened, this pressure is released and the solubility of carbon dioxide decreases, resulting in bubbling **Figure 12.11▶**.

▶ **FIGURE 12.11** **Soda Fizz** The bubbling that occurs when a can of soda is opened results from the reduced pressure of carbon dioxide over the liquid. At lower pressure, the carbon dioxide is less soluble and bubbles out of solution.

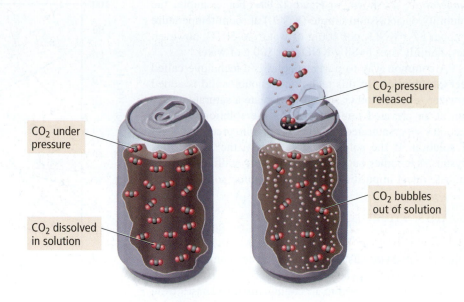

CO₂ pressure released

CO₂ under pressure

CO₂ bubbles out of solution

CO₂ dissolved in solution

The increased solubility of a gas in a liquid can be understood by considering the following cylinders containing water and carbon dioxide gas:

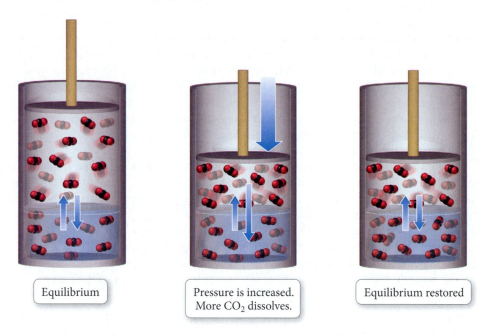

Equilibrium

Pressure is increased. More CO₂ dissolves.

Equilibrium restored

The first cylinder represents an equilibrium between gaseous and dissolved carbon dioxide—the rate of carbon dioxide molecules entering solution exactly equals the rate of molecules leaving the solution. Now imagine decreasing the volume, as shown in the second cylinder. The pressure of carbon dioxide now increases, causing the rate of molecules entering the solution to rise. The number of molecules in solution increases until equilibrium is established again, as shown in the third cylinder. However, the amount of carbon dioxide in solution is now greater.

The solubility of gases with increasing pressure can be quantified with **Henry's law**:

$$S_{gas} = k_H P_{gas}$$

where S_{gas} is the solubility of the gas (usually in M), k_H is a constant of proportionality (called the *Henry's law constant*) that depends on the specific solute and solvent and also on temperature, and P_{gas} is the partial pressure of the gas (usually in atm). The equation shows that the solubility of a gas in a liquid is directly proportional to the pressure of the gas above the liquid. Table 12.4 lists the Henry's law constants for several common gases.

TABLE 12.4 Henry's Law Constants for Several Gases in Water at 25 °C

Gas	k_H (M/atm)
O_2	1.3×10^{-3}
N_2	6.1×10^{-4}
CO_2	3.4×10^{-2}
NH_3	5.8×10^{1}
He	3.7×10^{-4}

Conceptual Connection 12.4 Henry's Law

Examine the Henry's law constants in Table 12.4. Why is the constant for ammonia bigger than the others?

EXAMPLE 12.2 Henry's Law

What pressure of carbon dioxide is required to keep the carbon dioxide concentration in a bottle of club soda at 0.12 M at 25 °C?

SORT You are given the desired solubility of carbon dioxide and asked to find the pressure required to achieve this solubility.	**GIVEN** $S_{CO_2} = 0.12$ M **FIND** P_{CO_2}
STRATEGIZE Use Henry's law to find the required pressure from the solubility. You will need the Henry's law constant for carbon dioxide.	**CONCEPTUAL PLAN**  $S_{CO_2} = k_{H,CO_2}P_{CO_2}$ **RELATIONSHIPS USED** $S_{gas} = k_H P_{gas}$ (Henry's law) $k_{H,CO_2} = 3.4 \times 10^{-2}$ M/atm (from Table 12.4)
SOLVE Solve the Henry's law equation for P_{CO_2} and substitute the other quantities to calculate it.	**SOLUTION** $S_{CO_2} = k_{H,CO_2}P_{CO_2}$ $P_{CO_2} = \dfrac{S_{CO_2}}{k_{H,CO_2}}$ $= \dfrac{0.12\,\text{M}}{3.4 \times 10^{-2}\dfrac{\text{M}}{\text{atm}}}$ $= 3.5$ atm

CHECK The answer is in the correct units and seems reasonable. A small answer (for example, less than 1 atm) would be suspect because you know that the soda is under a pressure greater than atmospheric pressure when you open it. A very large answer (for example, over 100 atm) would be suspect because an ordinary can or bottle probably could not sustain such high pressures without bursting.

FOR PRACTICE 12.2
Determine the solubility of oxygen in water at 25 °C exposed to air at 1.0 atm. Assume a partial pressure for oxygen of 0.21 atm.

12.5 Expressing Solution Concentration

As we have seen, the amount of solute in a solution is an important property of the solution. For example, the amount of sodium chloride in a solution determines whether or not the solution will cause dehydration if consumed. A **dilute solution** contains small quantities of solute relative to the amount of solvent. Drinking a dilute sodium chloride solution will not cause dehydration. A **concentrated solution** contains large quantities of solute relative to the amount of solvent. Drinking a concentrated sodium chloride solution will cause dehydration. Common ways of reporting solution concentration include molarity, molality, parts by mass, parts by volume, mole fraction, and mole percent, as summarized in Table 12.5 on the next page. We have seen two of these units before, molarity in Section 4.4 and mole fraction in Section 5.6. In the following section, we review the terms we have already covered and introduce the new ones.

TABLE 12.5 Solution Concentration Terms

Unit	Definition	Units
Molarity (M)	$\dfrac{\text{amount solute (in mol)}}{\text{volume solution (in L)}}$	$\dfrac{\text{mol}}{\text{L}}$
Molality (*m*)	$\dfrac{\text{amount solute (in mol)}}{\text{mass solvent (in kg)}}$	$\dfrac{\text{mol}}{\text{kg}}$
Mole fraction (χ)	$\dfrac{\text{amount solute (in mol)}}{\text{total amount of solute and solvent (in mol)}}$	None
Mole percent (mol %)	$\dfrac{\text{amount solute (in mol)}}{\text{total amount of solute and solvent (in mol)}} \times 100\%$	%
Parts by mass	$\dfrac{\text{mass solute}}{\text{mass solution}} \times \text{multiplication factor}$	
Percent by mass (%)	Multiplication factor = 100	%
Parts per million by mass (ppm)	Multiplication factor = 10^6	ppm
Parts per billion by mass (ppb)	Multiplication factor = 10^9	ppb
Parts by volume (%, ppm, ppb)	$\dfrac{\text{volume solute}}{\text{volume solution}} \times \text{multiplication factor*}$	

*Multiplication factors for parts by volume are identical to those for parts by mass.

Molarity

The **molarity (M)** of a solution is the amount of solute (in moles) divided by the volume of solution (in liters).

$$\text{Molarity (M)} = \frac{\text{amount solute (in mol)}}{\text{volume solution (in L)}}$$

Note that molarity is moles of solute per liter of *solution*, not liter of solvent. To make a solution of a specified molarity, you usually put the solute into a flask and then add water (or another solvent) to the desired volume of solution, as shown in **Figure 12.12▼**. Molarity is a convenient unit to use when making, diluting, and transferring solutions because it specifies the amount of solute per unit of solution transferred.

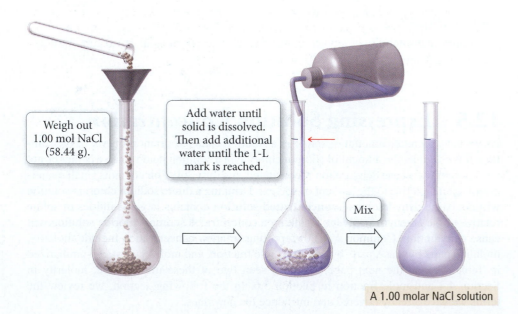

Weigh out 1.00 mol NaCl (58.44 g).

Add water until solid is dissolved. Then add additional water until the 1-L mark is reached.

Mix

A 1.00 molar NaCl solution

▶ **FIGURE 12.12 Preparing a Solution of Known Concentration** To make a 1 M NaCl solution, add 1 mol of the solid to a flask and dilute with water to make 1 L of solution.

Molality

Molarity depends on volume, and since volume varies with temperature, molarity also varies with temperature. For example, a 1 M aqueous solution at room temperature will be slightly less than 1 M at an elevated temperature because the volume of the solution is greater at the elevated temperature. A concentration unit that is independent of temperature is **molality (*m*)**, the amount of solute (in moles) divided by the mass of solvent (in kilograms).

$$\text{Molality } (m) = \frac{\text{amount solute (in mol)}}{\text{mass solvent (in kg)}}$$

Notice that molality is defined with respect to kilograms *solvent*, not kilograms solution. Molality is particularly useful when concentrations must be compared over a range of different temperatures.

> Note that molality is abbreviated with a lowercase italic *m* while molarity is abbreviated with a capital M.

Parts by Mass and Parts by Volume

It is often convenient to report a concentration as a ratio of masses. A **parts by mass** concentration is the ratio of the mass of the solute to the mass of the solution, all multiplied by a multiplication factor:

$$\frac{\text{Mass solute}}{\text{Mass solution}} \times \text{multiplication factor}$$

The particular parts by mass unit used, which determines the size of the multiplication factor, depends on the concentration of the solution. For example, for **percent by mass**, the multiplication factor is 100.

$$\text{Percent by mass} = \frac{\text{Mass solute}}{\text{Mass solution}} \times 100\%$$

Percent means *per hundred*, so a solution with a concentration of 14% by mass contains 14 g of solute per 100 g of solution.

For more dilute solutions, we might use **parts per million (ppm)**, which uses a multiplication factor of 10^6, or **parts per billion (ppb)**, which uses a multiplication factor of 10^9.

$$\text{ppm} = \frac{\text{mass solute}}{\text{mass solution}} \times 10^6$$

$$\text{ppb} = \frac{\text{mass solute}}{\text{mass solution}} \times 10^9$$

> For dilute aqueous solutions near room temperature, the units of ppm are equivalent to milligrams solute per liter of solution. This is because the density of a dilute solution near room temperature is 1.0 g/mL so that 1 L has a mass of 1000 g.

A solution with a concentration of 15 ppm by mass, for example, contains 15 g of solute per 10^6 g of solution.

Sometimes, we report concentrations as a ratio of volumes, especially for solutions in which the solute and solvent are both liquids. A **parts by volume** concentration is usually the ratio of the volume of the solute to the volume of the solution, all multiplied by a multiplication factor.

$$\frac{\text{Volume solute}}{\text{Volume solution}} \times \text{multiplication factor}$$

The multiplication factors are identical to those just described for parts by mass concentrations. For example, a 22% ethanol solution contains 22 mL of ethanol for every 100 mL of solution.

Using Parts by Mass (or Parts by Volume) in Calculations We can use the parts by mass (or parts by volume) concentration of a solution as a conversion factor between mass (or volume) of the solute and mass (or volume) of the solution. For example, for a solution containing 3.5% sodium chloride by mass, we write the following conversion factor:

$$\frac{3.5 \text{ g NaCl}}{100 \text{ g solution}} \qquad \text{converts} \qquad \boxed{\text{g solution}} \longrightarrow \boxed{\text{g NaCl}}$$

This conversion factor converts from grams solution to grams NaCl. To convert the other way, we invert the conversion factor:

$$\frac{100 \text{ g solution}}{3.5 \text{ g NaCl}} \quad \text{converts} \quad \boxed{\text{g NaCl}} \longrightarrow \boxed{\text{g solution}}$$

EXAMPLE 12.3 Using Parts by Mass in Calculations

What volume (in mL) of a soft drink that is 10.5% sucrose ($C_{12}H_{22}O_{11}$) by mass contains 78.5 g of sucrose? (The density of the solution is 1.04 g/mL.)

SORT You are given a mass of sucrose and the concentration and density of a sucrose solution, and you are asked to find the volume of solution containing that mass.	**GIVEN** 78.5 g $C_{12}H_{22}O_{11}$ 10.5% $C_{12}H_{22}O_{11}$ by mass density = 1.04 g/mL **FIND** mL
STRATEGIZE Begin with the mass of sucrose in grams. Use the mass percent concentration of the solution (written as a ratio, as shown under relationships used) to find the number of grams of solution containing this quantity of sucrose. Then use the density of the solution to convert grams to milliliters of solution.	**CONCEPTUAL PLAN** 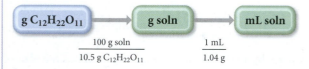 **RELATIONSHIPS USED** $\dfrac{10.5 \text{ g } C_{12}H_{22}O_{11}}{100 \text{ g soln}}$ (percent by mass written as ratio) $\dfrac{1 \text{ mL}}{1.04 \text{ g}}$ (given density of the solution)
SOLVE Begin with 78.5 g $C_{12}H_{22}O_{11}$ and multiply by the conversion factors to arrive at the volume of solution.	**SOLUTION** $78.5 \text{ g } C_{12}H_{22}O_{11} \times \dfrac{100 \text{ g soln}}{10.5 \text{ g } C_{12}H_{22}O_{11}} \times \dfrac{1 \text{ mL}}{1.04 \text{ g}} = 719 \text{ mL soln}$

CHECK The units of the answer are correct. The magnitude seems correct because the solution is approximately 10% sucrose by mass. Since the density of the solution is approximately 1 g/mL, the volume containing 78.5 g sucrose should be roughly 10 times larger, as calculated (719 ≈ 10 × 78.5).

FOR PRACTICE 12.3

How much sucrose ($C_{12}H_{22}O_{11}$), in g, is contained in 355 mL (12 ounces) of a soft drink that is 11.5% sucrose by mass? (Assume a density of 1.04 g/mL.)

FOR MORE PRACTICE 12.3

A water sample contains the pollutant chlorobenzene with a concentration of 15 ppb (by mass). What volume of this water contains 5.00×10^2 mg of chlorobenzene? (Assume a density of 1.00 g/mL.)

The mole fraction can also be defined for the solvent:

$$\chi_{solvent} = \frac{n_{solvent}}{n_{solute} + n_{solvent}}$$

Mole Fraction and Mole Percent

For some applications, especially those in which the ratio of solute to solvent can vary widely, the most useful way to express concentration is the amount of solute (in moles) divided by the total amount of solute and solvent (in moles). This ratio is the **mole fraction** (χ_{solute}):

$$\chi_{solute} = \frac{\text{amount solute (in mol)}}{\text{total amount of solute and solvent (in mol)}} = \frac{n_{solute}}{n_{solute} + n_{solvent}}$$

Also in common use is the **mole percent (mol %)**, which is the mole fraction $\times$ 100 percent.

$$\text{mol \%} = \chi_{\text{solute}} \times 100\%$$

EXAMPLE 12.4 Calculating Concentrations

A solution is prepared by dissolving 17.2 g of ethylene glycol ($C_2H_6O_2$) in 0.500 kg of water. The final volume of the solution is 515 mL. For this solution, calculate:

(a) molarity **(b)** molality **(c)** percent by mass **(d)** mole fraction **(e)** mole percent

SOLUTION

(a) To calculate molarity, first find the amount of ethylene glycol in moles from the mass and molar mass.

$$\text{mol } C_2H_6O_2 = 17.2 \text{ g } C_2H_6O_2 \times \frac{1 \text{ mol } C_2H_6O_2}{62.07 \text{ g } C_2H_6O_2} = 0.2771 \text{ mol } C_2H_6O_2$$

Then divide the amount in moles by the volume of the solution in liters.

$$\text{Molarity (M)} = \frac{\text{amount solute (in mol)}}{\text{volume solution (in L)}}$$

$$= \frac{0.2771 \text{ mol } C_2H_6O_2}{0.515 \text{ L solution}}$$

$$= 0.538 \text{ M}$$

(b) To calculate molality, use the amount of ethylene glycol in moles from part a, and divide by the mass of the water in kilograms.

$$\text{Molality (m)} = \frac{\text{amount solute (in mol)}}{\text{mass solvent (in kg)}} = \frac{0.2771 \text{ mol } C_2H_6O_2}{0.500 \text{ kg } H_2O}$$

$$= 0.554 \text{ m}$$

(c) To calculate percent by mass, divide the mass of the solute by the sum of the masses of the solute and solvent and multiply the ratio by 100%.

$$\text{Percent by mass} = \frac{\text{mass solute}}{\text{mass solution}} \times 100\%$$

$$= \frac{17.2 \text{ g}}{17.2 \text{ g} + 5.00 \times 10^2 \text{ g}} \times 100\%$$

$$= 3.33\%$$

(d) To calculate mole fraction, first determine the amount of water in moles from the mass of water and its molar mass.

$$\text{mol } H_2O = 5.00 \times 10^2 \text{ g } H_2O \times \frac{1 \text{ mol } H_2O}{18.02 \text{ g } H_2O} = 27.75 \text{ mol } H_2O$$

Then divide the amount of ethylene glycol in moles (from part a) by the total number of moles.

$$\chi_{\text{solute}} = \frac{n_{\text{solute}}}{n_{\text{solute}} + n_{\text{solvent}}}$$

$$= \frac{0.2771 \text{ mol}}{0.2771 \text{ mol} + 27.75 \text{ mol}}$$

$$= 9.89 \times 10^{-3}$$

(e) To calculate mole percent, simply multiply the mole fraction by 100%.

$$\text{mol \%} = \chi_{\text{solute}} \times 100\%$$

$$= 0.989\%$$

FOR PRACTICE 12.4

A solution is prepared by dissolving 50.4 g sucrose ($C_{12}H_{22}O_{11}$) in 0.332 kg of water. The final volume of the solution is 355 mL. For this solution, calculate:

(a) molarity **(b)** molality **(c)** percent by mass **(d)** mole fraction **(e)** mole percent

EXAMPLE 12.5 Converting between Concentration Units

What is the molarity of a 6.55% by mass glucose ($C_6H_{12}O_6$) solution? (The density of the solution is 1.03 g/mL.)

SORT You are given the concentration of a glucose solution in percent by mass and the density of the solution. Find the concentration of the solution in molarity.

GIVEN 6.55% $C_6H_{22}O_{11}$
    density = 1.03 g/mL

FIND M

STRATEGIZE Begin with the mass percent concentration of the solution written as a ratio, and separate the numerator from the denominator. Convert the numerator from g $C_6H_{12}O_6$ to mol $C_6H_{12}O_6$. Convert the denominator from g soln to mL of soln and then to L soln. Then divide the numerator (now in mol) by the denominator (now in L) to obtain molarity.

CONCEPTUAL PLAN

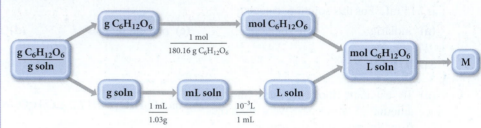

RELATIONSHIPS USED

$\dfrac{6.55\,g\,C_6H_{12}O_6}{100\,g\,soln}$ (percent by mass written as ratio)

$\dfrac{1\,mol}{180.16\,g\,C_6H_{12}O_6}$ (from molar mass of glucose)

$\dfrac{1\,mL}{1.03\,g}$ (from given density of the solution)

SOLVE Begin with the numerator (6.55 g $C_6H_{12}O_6$) and use the molar mass to convert to mol $C_6H_{12}O_6$.

Then convert the denominator (100 g solution) into mL of solution (using the density) and then to L of solution.

Finally, divide mol $C_6H_{12}O_6$ by L soln to arrive at molarity.

SOLUTION

$6.55\,g\,C_6H_{12}O_6 \times \dfrac{1\,mol\,C_6H_{12}O_6}{180.16\,g\,C_6H_{12}O_6} = 0.03636\,mol\,C_6H_{12}O_6$

$100\,g\,soln \times \dfrac{1\,mL}{1.03\,g} \times \dfrac{10^{-3}\,L}{mL} = 0.09709\,L\,soln$

$\dfrac{0.03636\,mol\,C_6H_{12}O_6}{0.09709\,L\,soln} = 0.374\,M\,C_6H_{12}O_6$

CHECK The units of the answer are correct. The magnitude seems correct. Very high molarities (especially above 25 M) should immediately appear suspect. One liter of water contains about 55 moles of water molecules, so molarities higher than 55 M are physically impossible for aqueous solutions.

FOR PRACTICE 12.5

What is the molarity of a 10.5% by mass glucose ($C_6H_{12}O_6$) solution? (The density of the solution is 1.03 g/mL.)

FOR MORE PRACTICE 12.5

What is the molality of a 10.5% by mass glucose ($C_6H_{12}O_6$) solution? (The density of the solution is 1.03 g/mL.)

12.6   Colligative Properties: Vapor Pressure Lowering, Freezing Point Depression, Boiling Point Elevation, and Osmotic Pressure

Have you ever wondered why salt is added to ice in an ice-cream maker? Or in cold climates, why salt is often scattered on icy roads? Salt actually lowers the temperature at which a salt water solution freezes. A salt and water solution will remain liquid even below 0 °C.

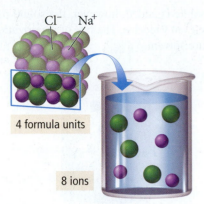

▲ FIGURE 12.13 Electrolyte Dissolution   When sodium chloride is dissolved in water, each mole of NaCl produces 2 mol of particles: 1 mol of Na^+ cations and 1 mol of Cl^- anions.

By adding salt to ice in the ice-cream maker, you form an ice/water/salt mixture that can reach a temperature of about -10 °C, causing the cream in the ice cream to freeze. On the road, the salt allows the ice to melt, even if the ambient temperature is below freezing.

▲ In winter, salt is often applied to roads to lower the melting point of ice.

The lowering of the melting point of ice by salt is an example of a **colligative property**, a property that depends on the number of particles dissolved in solution, not on the type of particle. In this section, we examine four colligative properties: vapor pressure lowering, freezing point depression, boiling point elevation, and osmotic pressure. Since these properties depend on the *number* of dissolved particles, nonelectrolytes must be treated slightly differently than electrolytes when determining colligative properties. When 1 mol of a nonelectrolyte dissolves in water, it forms 1 mol of dissolved particles. When 1 mol of an electrolyte dissolves in water, however, it normally forms more than 1 mol of dissolved particles (as shown in **Figure 12.13▲**). For example, when 1 mol of NaCl dissolves in water, it forms 1 mol of dissolved Na^+ ions and 1 mol of dissolved Cl^- ions. Therefore the resulting solution will have 2 mol of dissolved particles—the colligative properties will reflect this higher concentration of dissolved particles. In this section we examine colligative properties of nonelectrolyte solutions; we then expand the concept to include electrolyte solutions in Section 12.7.

Vapor Pressure Lowering

Recall from Section 11.5 that the vapor pressure of a liquid is the pressure of the gas above the liquid when the two are in dynamic equilibrium (that is, when the rate of vaporization equals the rate of condensation). What is the effect of a nonvolatile (i.e., not easily vaporized), nonelectrolyte solute on the vapor pressure of the liquid into which it dissolves? The basic answer to this question is that *the vapor pressure of the solution is lower than the vapor pressure of the pure solvent*. We can understand why this happens in two different ways.

The simplest explanation for the lowering of the vapor pressure of a solution relative to that of the pure solvent is related to the concept of dynamic equilibrium itself. Consider this representation of a liquid in dynamic equilibrium with its vapor. Here the rate of vaporization is equal to the rate of condensation.

Dynamic
equilibrium

When a nonvolatile solute is added, however, the solute particles interfere with the ability of the solvent particles to vaporize. The rate of vaporization is therefore diminished compared to that of the pure solvent.

Rate of vaporization reduced by solute

The change in the rate of vaporization creates an imbalance in the rates; the rate of condensation is now *greater* than the rate of vaporization. The net effect is that some of the molecules that were in the gas state condense into the liquid. As they condense, the reduced number of molecules in the gas state causes the rate of condensation to decrease. Eventually the two rates become equal again, but only after the concentration of molecules in the gas state has decreased.

Equilibrium reestablished but with fewer molecules in gas phase

The result is a lower vapor pressure for the solution compared to the pure solvent.

A more fundamental explanation of why the vapor pressure of a solution is lower than that of the pure solvent is related to the tendency toward mixing (toward greater entropy) that we discussed in Section 12.1 and 12.2. Recall from Section 12.1 that a concentrated solution is a *thirsty* solution—it has the ability to draw solvent to itself. We can see a dramatic demonstration of this tendency by placing both a beaker of a concentrated solution of a nonvolatile solute and a beaker of the pure solvent in a sealed container, as shown at the top of the next page. Over time, the volume of the pure solvent will drop and the volume of the solution will rise as molecules vaporize out of the pure solvent and condense in the solution. Notice the similarity between this observation and the dehydration caused by drinking seawater. In both cases, a concentrated solution has the ability to draw solvent to itself. The reason is nature's tendency to mix. If a pure solvent and concentrated solution are combined in a beaker, they naturally form a mixture in which the solution is less concentrated than it was initially. Similarly, if a pure solvent and concentrated solution are combined in a sealed container—even though they are in separate beakers—the two mix to form a more dilute solution.

The net transfer of solvent from the beaker containing pure solvent to the one containing the solution demonstrates that the vapor pressure of the solution is lower than that

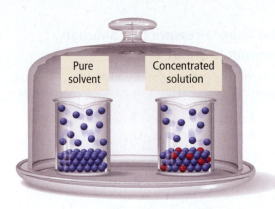

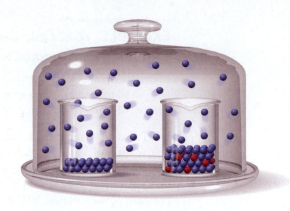

of the pure solvent. As solvent molecules vaporize, the vapor pressure in the container rises. Before dynamic equilibrium can be attained, however, the pressure in the container exceeds the vapor pressure of the solution, causing molecules to condense into the solution. Therefore, molecules constantly vaporize from the pure solvent, but the solvent's vapor pressure is never reached because molecules are constantly condensing into the solution. The result is a continuous transfer of solvent molecules from the pure solvent to the solution.

We can quantify the vapor pressure of a solution with **Raoult's law**:

$$P_{solution} = \chi_{solvent} P^{\circ}_{solvent}$$

In this equation, $P_{solution}$ is the vapor pressure of the solution, $\chi_{solvent}$ is the mole fraction of the solvent, and $P^{\circ}_{solvent}$ is the vapor pressure of the pure solvent at the same temperature. For example, suppose a water sample at 25 °C contains 0.90 mol of water and 0.10 mol of a nonvolatile solute such as sucrose. The pure water would have a vapor pressure of 23.8 torr. We calculate the vapor pressure of the solution by substituting into Raoult's law:

$$P_{solution} = \chi_{H_2O} P^{\circ}_{H_2O}$$

$$= 0.90(23.8 \text{ torr})$$

$$= 21.4 \text{ torr}$$

Notice that the vapor pressure of the solution is lowered in direct proportion to the mole composition of the solvent—since the solvent particles compose 90% of all of the particles in the solution, the vapor pressure is 90% of the vapor pressure of the pure solvent.

To arrive at an equation that shows how much the vapor pressure is lowered by a solute, we define the **vapor pressure lowering (ΔP)** as the difference in vapor pressure between the pure solvent and the solution:

$$\Delta P = P^{\circ}_{solvent} - P_{solution}$$

Then, for a two-component solution, we substitute $\chi_{solvent} = 1 - \chi_{solute}$ into Raoult's law:

$$P_{solution} = \chi_{solvent} P^{\circ}_{solvent}$$

$$P_{solution} = (1 - \chi_{solute}) P^{\circ}_{solvent}$$

$$P^{\circ}_{solvent} - P_{solution} = \chi_{solute} P^{\circ}_{solvent}$$

$$\Delta P = \chi_{solute} P^{\circ}_{solvent}$$

We can see from this last equation that the lowering of the vapor pressure is directly proportional to the mole fraction of the solute.

EXAMPLE 12.6 Calculating the Vapor Pressure of a Solution Containing a Nonionic and Nonvolatile Solute

Calculate the vapor pressure at 25 °C of a solution containing 99.5 g sucrose ($C_{12}H_{22}O_{11}$) and 300.0 mL water. The vapor pressure of pure water at 25 °C is 23.8 torr. Assume the density of water to be 1.00 g/mL.

SORT You are given the mass of sucrose and volume of water in a solution. You are also given the vapor pressure of pure water and asked to find the vapor pressure of the solution. The density of the pure water is also provided.	**GIVEN** 99.5 g $C_{12}H_{22}O_{11}$ 300.0 mL H_2O $P°_{H_2O} = 23.8$ torr at 25 °C $d_{H_2O} = 1.00$ g/mL **FIND** $P_{solution}$

STRATEGIZE Raoult's law relates the vapor pressure of a solution to the mole fraction of the solvent and the vapor pressure of the pure solvent. Begin by calculating the amount in moles of sucrose and water.

CONCEPTUAL PLAN

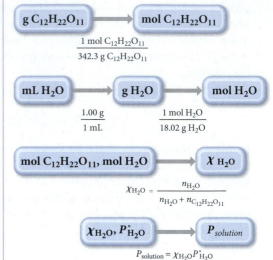

Calculate the mole fraction of the solvent from the calculated amounts of solute and solvent.

$$\chi_{H_2O} = \frac{n_{H_2O}}{n_{H_2O} + n_{C_{12}H_{22}O_{11}}}$$

Then use Raoult's law to calculate the vapor pressure of the solution.

$$P_{solution} = \chi_{H_2O} P°_{H_2O}$$

SOLVE Calculate the number of moles of each solution component.

SOLUTION

$$99.5 \text{ g } C_{12}H_{22}O_{11} \times \frac{1 \text{ mol } C_{12}H_{22}O_{11}}{342.30 \text{ g } C_{12}H_{22}O_{11}}$$

$$= 0.2907 \text{ mol } C_{12}H_{22}O_{11}$$

$$300.0 \text{ mL } H_2O \times \frac{1.00 \text{ g}}{1 \text{ mL}} \times \frac{1 \text{ mol } H_2O}{18.02 \text{ g } H_2O}$$

$$= 16.65 \text{ mol } H_2O$$

Use the number of moles of each component to calculate the mole fraction of the solvent (H_2O).

$$\chi_{H_2O} = \frac{n_{H_2O}}{n_{C_{12}H_{22}O_{11}} + n_{H_2O}}$$

$$= \frac{16.65 \text{ mol}}{0.2907 \text{ mol} + 16.65 \text{ mol}}$$

$$= 0.9828$$

Use the mole fraction of water and the vapor pressure of pure water to calculate the vapor pressure of the solution.

$$P_{solution} = \chi_{H_2O} P°_{H_2O}$$

$$= 0.9828 (23.8 \text{ torr})$$

$$= 23.4 \text{ torr}$$

CHECK The units of the answer are correct. The magnitude of the answer is about right because the calculated vapor pressure of the solution is just below that of the pure liquid, as expected for a solution with a large mole fraction of solvent.

FOR PRACTICE 12.6

Calculate the vapor pressure at 25 °C of a solution containing 55.3 g ethylene glycol ($HOCH_2CH_2OH$) and 285.2 g water. The vapor pressure of pure water at 25 °C is 23.8 torr.

FOR MORE PRACTICE 12.6

A solution containing ethylene glycol and water has a vapor pressure of 7.88 torr at 10 °C. Pure water has a vapor pressure of 9.21 torr at 10 °C. What is the mole fraction of ethylene glycol in the solution?

Vapor Pressures of Solutions Containing a Volatile (Nonelectrolyte) Solute

A solution may contain not only a volatile solvent but also a volatile solute. A solution that follows Raoult's law at all concentrations for both the solute and the solvent is called an **ideal solution** and is similar in concept to an ideal gas. Just as an ideal gas follows the ideal gas law exactly, so an ideal solution follows Raoult's law exactly. The vapor pressure of each of the solution components is given by Raoult's law throughout the entire composition range of the solution. For a two-component solution containing liquids A and B, we can write:

$$P_A = \chi_A P°_A$$

$$P_B = \chi_B P°_B$$

The total pressure above such a solution is the sum of the partial pressures of the components:

$$P_{tot} = P_A + P_B$$

Figure 12.14▼ is a plot of vapor pressure versus solution composition for an ideal two-component solution.

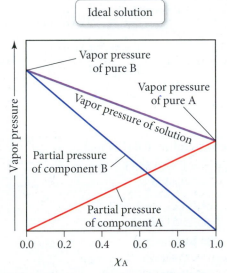

▲ FIGURE 12.14 Behavior of Ideal Solutions An ideal solution follows Raoult's law for both components.

EXAMPLE 12.7 Calculating the Vapor Pressure of a Two-Component Solution

A solution contains 3.95 g of carbon disulfide (CS_2) and 2.43 g of acetone (CH_3COCH_3). The vapor pressures at 35 °C of pure carbon disulfide and pure acetone are 515 torr and 332 torr, respectively. Assuming ideal behavior, calculate the vapor pressures of each of the components and the total vapor pressure above the solution.

SORT You are given the masses and vapor pressures of carbon disulfide and acetone and are asked to find the vapor pressures of each component in the mixture and the total pressure assuming ideal behavior.

GIVEN 3.95 g CS_2
2.43 g CH_3COCH_3
$P°_{CS_2}$ = 515 torr (at 35 °C)
$P°_{CH_3COCH_3}$ = 332 torr (at 35 °C)
FIND P_{CS_2}, $P_{CH_3COCH_3}$, P_{tot} (ideal)

STRATEGIZE This problem requires the use of Raoult's law to calculate the partial pressures of each component. In order to use Raoult's law, you must first calculate the mole fractions of the two components. Convert the masses of each component to moles and then use the definition of mole fraction to calculate the mole fraction of carbon disulfide. The mole fraction of acetone can be readily determined because the mole fractions of the two components add up to one.

CONCEPTUAL PLAN

$$3.95 \text{ g } CS_2 \longrightarrow \text{mol } CS_2$$
$$\frac{1 \text{ mol } CS_2}{76.15 \text{ g } CS_2}$$

$$2.43 \text{ g } CH_3COCH_3 \longrightarrow \text{mol } CH_3COCH_3$$
$$\frac{1 \text{ mol } CH_3COCH_3}{58.08 \text{ g } CH_3COCH_3}$$

$$\text{mol } CS_2, \text{ mol } CH_3OCH_3 \longrightarrow \chi_{CS_2}, \chi_{CH_3COCH_3}$$
$$\chi_{CS_2} = \frac{n_{CS_2}}{n_{CS_2} + n_{CH_3COCH_3}}$$

Use the mole fraction of each component along with Raoult's law to calculate the partial pressure of each component. The total pressure is simply the sum of the partial pressures.

$$P_{CS_2} = \chi_{CS_2}P°_{CS_2}$$
$$P_{CH_3COCH_3} = \chi_{CH_3COCH_3}P°_{CH_3COCH_3}$$
$$P_{tot} = P_{CS_2} + P_{CH_3COCH_3}$$

RELATIONSHIPS USED

$$\chi_A = \frac{n_A}{n_A + n_B} \text{ (mole fraction definition)}$$
$$P_A = \chi_A P°_A \text{ (Raoult's law)}$$

SOLVE Begin by converting the masses of each component to the amounts in moles.

SOLUTION

$$3.95 \text{ g } CS_2 \times \frac{1 \text{ mol } CS_2}{76.15 \text{ g } CS_2} = 0.05187 \text{ mol } CS_2$$

$$2.43 \text{ g } CH_3COCH_3 \times \frac{1 \text{ mol } CH_3COCH_3}{58.0 \text{ g } CH_3COCH_3} = 0.04184 \text{ mol } CH_3COCH_3$$

Then calculate the mole fraction of carbon disulfide.

$$\chi_{CS_2} = \frac{n_{CS_2}}{n_{CS_2} + n_{CH_3COCH_3}}$$
$$= \frac{0.05187 \text{ mol}}{0.05187 \text{ mol} + 0.04184 \text{ mol}} = 0.5535$$

Calculate the mole fraction of acetone by subtracting the mole fraction of carbon disulfide from one.

$$\chi_{CH_3COCH_3} = 1 - 0.5535$$
$$= 0.4465$$

Calculate the partial pressures of carbon disulfide and acetone by using Raoult's law and the given values of the vapor pressures of the pure substances.

$$P_{CS_2} = \chi_{CS_2}P°_{CS_2}$$
$$= 0.5535(515 \text{ torr})$$
$$= 285 \text{ torr}$$

$$P_{CH_3COCH_3} = \chi_{CH_3COCH_3}P°_{CH_3COCH_3}$$
$$= 0.4465(332 \text{ torr})$$
$$= 148 \text{ torr}$$

Calculate the total pressure by summing the partial pressures.

$$P_{tot}(\text{ideal}) = 285 \text{ torr} + 148 \text{ torr}$$
$$= 433 \text{ torr}$$

CHECK The units of the answer (torr) are correct. The magnitude seems reasonable given the partial pressures of the pure substances.

FOR PRACTICE 12.7

A solution of benzene (C_6H_6) and toluene (C_7H_8) is 25.0% benzene by mass. The vapor pressures of pure benzene and pure toluene at 25 °C are 94.2 torr and 28.4 torr, respectively. Assuming ideal behavior, calculate:

(a) The vapor pressure of each of the solution components in the mixture.

(b) The total pressure above the solution.

(c) The composition of the vapor in mass percent.

Why is the composition of the vapor different from the composition of the solution?

 Conceptual Connection 12.5 Vapor Pressure of a Two-Component Solution

A solution is equimolar in A and B (has equal number of moles of each). The vapor pressure of pure A is 100 torr and the vapor pressure of pure B is 200 torr. What is the total vapor pressure of the solution?

(a) 100 torr

(b) 150 torr

(c) 200 torr

(d) 350 torr

Freezing Point Depression and Boiling Point Elevation

The vapor pressure lowering that we just discussed occurs at all temperatures. We can see the effect of vapor pressure lowering over a range of temperatures by comparing the phase diagrams for pure water and for an aqueous solution containing a nonvolatile solute (shown at right).

Notice that the vapor pressure for the solution is shifted downward compared to that of the pure solvent. Consequently, the vapor pressure curve intersects the solid–gas curve at a lower temperature. The net effect is that the solution has a *lower melting point* and a *higher boiling point* than the pure solvent. These effects are called **freezing point depression** and **boiling point elevation**.

The freezing point of a solution containing a nonvolatile solute is lower than the freezing point of the pure solvent. For example, antifreeze, used to prevent the freezing of engine blocks in cold climates, is an aqueous solution of ethylene glycol ($C_2H_6O_2$). The more concentrated the solution, the lower the freezing point becomes.

The amount that the freezing point is lowered for solutions is given by the equation

$$\Delta T_f = m \times K_f$$

where

- ΔT_f is the change in temperature of the freezing point in degrees Celsius (relative to the freezing point of the pure solvent), usually reported as a positive number;
- m is the molality of the solution in moles solute per kilogram solvent;
- K_f is the freezing point depression constant for the solvent.

For water,

$$K_f = 1.86 \ °C/m$$

When an aqueous solution containing a dissolved solid solute freezes slowly, the solid that forms does not normally contain much of the solute. For example, when ice forms in ocean water, the ice is not salt water, but freshwater. As the ice forms, the crystal structure

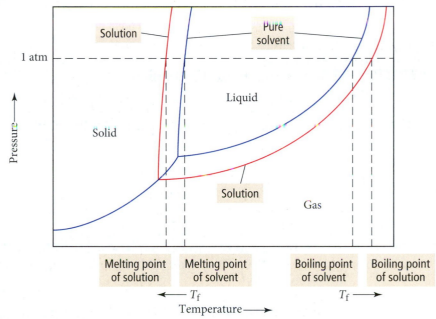

1 atm

Pressure →

Solution — Pure solvent

Liquid

Solid

Solution

Gas

Melting point of solution | Melting point of solvent | Boiling point of solvent | Boiling point of solution

← T_f T_f →

Temperature →

▲ A nonvolatile solute lowers the vapor pressure of a solution, resulting in a lower freezing point and an elevated boiling point.

▲ Antifreeze is an aqueous solution of ethylene glycol. The solution has a lower freezing point and higher boiling point than pure water.

TABLE 12.6 Freezing Point Depression and Boiling Point Elevation Constants for Several Liquid Solvents

Solvent	Normal Freezing Point (°C)	K_f (°C/m)	Normal Boiling Point (°C)	K_b (°C/m)
Benzene (C_6H_6)	5.5	5.12	80.1	2.53
Carbon tetrachloride (CCl_4)	−22.9	29.9	76.7	5.03
Chloroform ($CHCl_3$)	−63.5	4.70	61.2	3.63
Ethanol (C_2H_5OH)	−114.1	1.99	78.3	1.22
Diethyl ether ($C_4H_{10}O$)	−116.3	1.79	34.6	2.02
Water (H_2O)	0.00	1.86	100.0	0.512

of the ice tends to exclude the solute particles. You can verify this yourself by partially freezing a salt water solution in the freezer. Take out the newly formed ice, rinse it several times, and taste it. Compare the taste of the ice to the taste of the original solution. The ice will be much less salty.

Freezing point depression and boiling point elevation constants for several liquids are listed in Table 12.6. Calculating the freezing point of a solution involves substituting into the freezing point depression equation, as Example 12.8 demonstrates.

EXAMPLE 12.8 Freezing Point Depression

Calculate the freezing point of a 1.7 m aqueous ethylene glycol solution.

SORT You are given the molality of a solution and asked to find its freezing point	**GIVEN** 1.7 m solution. **FIND** freezing point (from ΔT_f)
STRATEGIZE To solve this problem, use the freezing point depression equation.	**CONCEPTUAL PLAN**  $\Delta T_f = m \times K_f$
SOLVE Substitute into the equation to calculate ΔT_f. The actual freezing point is the freezing point of pure water (0.00 °C) − ΔT_f.	**SOLUTION** $\Delta T_f = m \times K_f$ $\quad = 1.7\ m \times 1.86\ °C/m$ $\quad = 3.2\ °C$ Freezing point $= 0.00\ °C - 13.2\ °C$ $\quad\quad\quad\quad\quad\quad = -3.2\ °C$

CHECK The units of the answer are correct. The magnitude seems about right. The expected range for freezing points of an aqueous solution is anywhere from −10 °C to just below 0 °C. Any answers not within this range would be suspect.

FOR PRACTICE 12.8

Calculate the freezing point of a 2.6 m aqueous sucrose solution.

The boiling point of a solution containing a nonvolatile solute is higher than the boiling point of the pure solvent. In automobiles, antifreeze not only prevents the freezing of water within engine blocks in cold climates, it also prevents the boiling of water within engine blocks in hot climates. The amount that the boiling point is raised for solutions is given by the equation

$$\Delta T_b = m \times K_b$$

where

- ΔT_b is the change in temperature of the boiling point in degrees Celsius (relative to the boiling point of the pure solvent);
- m is the molality of the solution in moles solute per kilogram solvent;
- K_b is the boiling point elevation constant for the solvent.

For water,

$$K_b = 0.512 \,°C/m$$

The boiling point of a solution is calculated by substituting into the boiling point elevation equation, as Example 12.9 demonstrates.

 Conceptual Connection 12.6 **Freezing Point Depression**

Solution A is a 1.0 M solution with a nonionic solute and water as the solvent. Solution B is a 1.0 M solution with the same nonionic solute and ethanol as the solvent. Which solution has the greatest change in its boiling point (relative to the pure solvent)?

EXAMPLE 12.9 Boiling Point Elevation

How much ethylene glycol ($C_2H_6O_2$), in grams, must be added to 1.0 kg of water to produce a solution that boils at 105.0 °C?

SORT You are given the desired boiling point of an ethylene glycol solution containing 1.0 kg of water and asked to find the mass of ethylene glycol required to achieve the boiling point.	**GIVEN** $\Delta T_b = 5.0\,°C$, $1.0\,kg\,H_2O$ **FIND** g $C_2H_6O_2$
STRATEGIZE To solve this problem, use the boiling-point elevation equation to find the desired molality of the solution from ΔT_b. Then use the molality you just found to determine how many moles of ethylene glycol are needed per kilogram of water. Finally, calculate the molar mass of ethylene glycol and use it to convert from moles of ethylene glycol to mass of ethylene glycol.	**CONCEPTUAL PLAN**  $\Delta T_b = m \times K_b$ 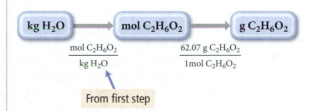 **RELATIONSHIPS USED** $C_2H_6O_2$ molar mass $= 62.07\,g/mol$ $\Delta T_b = m \times K_b$ (boiling point elevation)
SOLVE Begin by solving the boiling point elevation equation for molality and substituting the required quantities to calculate m. Then determine the mass of ethylene glycol in 1.0 kg H_2O by using the molality and the molar mass of ethylene glycol.	**SOLUTION** $\Delta T_b = m \times K_b$ $m = \dfrac{\Delta T_b}{K_b} = \dfrac{5.0\,°C}{0.512\dfrac{°C}{m}} = 9.77\,m$ $1.0\,kg\,H_2O \times \dfrac{9.77\,mol\,C_2H_6O_2}{kg\,H_2O} \times \dfrac{62.07\,g\,C_2H_6O_2}{1\,mol\,C_2H_6O_2} = 6.1 \times 10^2\,g\,C_2H_6O_2$

CHECK The units of the answer are correct. The magnitude might seem a little high initially, but the boiling point elevation constant is so small that a lot of solute is required to raise the boiling point by a small amount.

FOR PRACTICE 12.9

Calculate the boiling point of a 3.60 m aqueous sucrose solution.

Osmosis and Osmotic Pressure

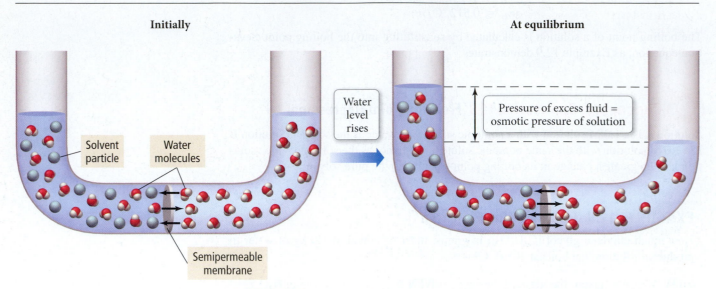

▲ **FIGURE 12.15 An Osmosis Cell**   In an osmosis cell, water flows from the pure-water side of the cell through the semipermeable membrane to the salt-water side.

Osmosis

The process discussed in the opening section of this chapter by which seawater causes dehydration is called *osmosis*. **Osmosis** can be formally defined as the flow of solvent from a solution of lower solute concentration to one of higher solute concentration. Concentrated solutions draw solvent from more dilute solutions because of nature's tendency to mix, which we discussed previously.

Figure 12.15▲ shows an osmosis cell. The left side of the cell contains a concentrated salt-water solution and the right side of the cell contains pure water. A **semipermeable membrane**—a membrane that selectively allows some substances to pass through but not others—separates the two halves of the cell. Water flows by osmosis from the pure-water side of the cell through the semipermeable membrane and into the salt-water side. Over time, the water level on the left side of the cell rises, while the water level on the right side of the cell falls. If external pressure is applied to the water in the left cell, this process can be opposed and even stopped. The pressure required to stop the osmotic flow, called the **osmotic pressure** (Π), is given by the equation:

$$\Pi = MRT$$

where M is the molarity of the solution, T is the temperature (in kelvins), and R is the ideal gas constant (0.08206 L · atm/mol · K).

EXAMPLE 12.10 Osmotic Pressure

The osmotic pressure of a solution containing 5.87 mg of an unknown protein per 10.0 mL of solution was 2.45 torr at 25 °C. Find the molar mass of the unknown protein.

SORT You are told that a solution of an unknown protein contains 5.87 mg of the protein per 10.0 mL of solution. You are also given the osmotic pressure of the solution at a particular temperature and asked to find the molar mass of the unknown protein.	**GIVEN** 5.87 mg protein 10.0 mL solution Π = 2.45 torr T = 25 °C **FIND** molar mass of protein (g/mol)

STRATEGIZE Step 1: Use the given osmotic pressure and temperature to find the molarity of the protein solution.

CONCEPTUAL PLAN

$$\Pi = MRT$$

Step 2: Use the molarity calculated in step 1 to find the number of moles of protein in 10 mL of solution.

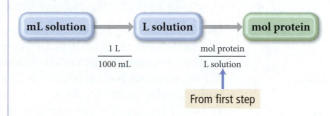

From first step

Step 3: Finally, use the number of moles of the protein calculated in step 2 and the given mass of the protein in 10.0 mL of solution to find the molar mass.

$$\text{Molar mass} = \frac{\text{mass protein}}{\text{moles protein}}$$

RELATIONSHIPS USED

$\Pi = MRT$ (osmotic pressure equation)

SOLVE Step 1: Begin by solving the osmotic pressure equation for molarity and substituting in the required quantities in the correct units to calculate M.

SOLUTION

$$\Pi = MRT$$

$$M = \frac{\Pi}{RT} = \frac{2.45 \text{ torr} \times \dfrac{1 \text{ atm}}{760 \text{ torr}}}{0.08206 \dfrac{\text{L} \cdot \text{atm}}{\text{mol} \cdot \text{K}}(298 \text{ K})}$$

$$= 1.318 \times 10^{-4} \text{ M}$$

Step 2: Begin with the given volume, convert to liters, then use the molarity to find the number of moles of protein.

$$10.0 \text{ mL} \times \frac{1 \text{ L}}{1000 \text{ mL}} \times \frac{1.318 \times 10^{-4} \text{ mol}}{\text{L}}$$

$$= 1.318 \times 10^{-6} \text{ mol}$$

Step 3: Use the given mass and the number of moles from step 2 to calculate the molar mass of the protein.

$$\text{Molar mass} = \frac{\text{mass protein}}{\text{moles protein}}$$

$$= \frac{5.87 \times 10^{-3} \text{ g}}{1.318 \times 10^{-6} \text{ mol}}$$

$$= 4.45 \times 10^{3} \text{ g/mol}$$

CHECK The units of the answer are correct. The magnitude might seem a little high initially, but proteins are large molecules and therefore have high molar masses.

FOR PRACTICE 12.10
Calculate the osmotic pressure (in atm) of a solution containing 1.50 g ethylene glycol ($C_2H_6O_2$) in 50.0 mL of solution at 25 °C.

12.7 Colligative Properties of Strong Electrolyte Solutions

At the beginning of Section 12.6, we learned that colligative properties depend on the *number* of dissolved particles, and that electrolytes must therefore be treated slightly differently than nonelectrolytes when determining colligative properties. For example, the freezing point depression of a 0.10 m sucrose solution is $\Delta T_f = 0.186$ °C. However, the

TABLE 12.7 Van't Hoff Factors at 0.05 *m* Concentration in Aqueous Solution

Solute	*i* Expected	*i* Measured
Nonelectrolyte	1	1
NaCl	2	1.9
MgSO$_4$	2	1.3
MgCl$_2$	3	2.7
K$_2$SO$_4$	3	2.6
FeCl$_3$	4	3.4

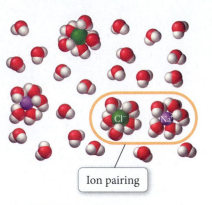

▲ FIGURE 12.16 Ion Pairing
Hydrated anions and cations may get close enough together to effectively pair, lowering the concentration of particles below what would be expected ideally.

freezing point depression of a 0.10 *m* sodium chloride solution is nearly twice this large. Why? Because 1 mol of sodium chloride dissociates into *nearly* 2 mol of ions in solution. The ratio of moles of particles in solution to moles of formula units dissolved is the **van't Hoff factor (*i*)**:

$$i = \frac{\text{moles of particles in solution}}{\text{moles of fomula units dissolved}}$$

Since 1 mol of NaCl produces 2 mol of particles in solution, we expect the van't Hoff factor to be exactly 2. In fact, this expected factor only occurs in very dilute solutions. For example, the van't Hoff factor for a 0.10 *m* NaCl solution is 1.87 and that for a 0.010 *m* NaCl in solution is 1.94. The van't Hoff factor approaches the expected value at infinite dilution (i.e., as the concentration approaches zero). Table 12.7 lists the actual and expected van't Hoff factors for a number of solutes.

The reason that the van't Hoff factors are not exactly equal to the expected values is that some ions effectively pair in solution. In other words, ideally we expect the dissociation of an ionic compound to be complete in solution. In reality, however, the dissociation is not complete—at any moment, some cations are effectively combined with the corresponding anions **Figure 12.16◄**, slightly reducing the number of particles in solution.

To calculate colligative properties of ionic solutions use the van't Hoff factor in each equation as follows:

$$\Delta T_f = im \times K_f \text{ (freezing point depression)}$$
$$\Delta T_b = im \times K_b \text{ (boiling point elevation)}$$
$$\Pi = iMRT \quad \text{(osmotic pressure)}$$

Conceptual Connection 12.7 Colligative Properties

Which solution will have the highest boiling point?

(a) 0.50 M C$_{12}$H$_{22}$O$_{11}$ **(b)** 0.50 M NaCl   **(c)** 0.50 M MgCl$_2$

EXAMPLE 12.11 Van't Hoff Factor and Freezing Point Depression

The freezing point of an aqueous 0.050 *m* CaCl$_2$ solution is −0.27 °C. What is the van't Hoff factor (*i*) for CaCl$_2$ at this concentration? How does it compare to the predicted value of *i*?

SORT You are given the molality of a solution and its freezing point. You are asked to find the value of *i*, the van't Hoff factor, and compare it to the predicted value.	**GIVEN** 0.050 *m* CaCl$_2$ solution, $\quad \Delta T_f = 0.27\,°C$ **FIND** *i*
STRATEGIZE To solve this problem, use the freezing point depression equation including the van't Hoff factor.	**CONCEPTUAL PLAN** $\Delta T_f = im \times K_f$
SOLVE Solve the freezing point depression equation for *i* and substitute in the given quantities to calculate its value.	**SOLUTION** $\Delta T_f = im \times K_f$ $i = \dfrac{\Delta T_f}{m \times K_f}$ $= \dfrac{0.27\,°C}{0.050\,m \times \dfrac{1.86\,°C}{m}}$ $= 2.9$
The expected value of *i* for CaCl$_2$ is 3 because calcium chloride forms 3 mol of ions for each mole of calcium chloride that dissolves. The experimental value is slightly less than 3, probably because of ion pairing.	

CHECK The answer has no units, as expected since *i* is a ratio. The magnitude is about right since it is close to the value you would expect upon complete dissociation of CaCl$_2$.

FOR PRACTICE 12.11
Calculate the freezing point of an aqueous 0.10 *m* FeCl$_3$ solution using a van't Hoff factor of 3.2.

Strong Electrolytes and Vapor Pressure

Just as the freezing point depression of a solution containing an electrolyte solute is greater than that of a solution containing the same concentration of a nonelectrolyte solute, so the vapor pressure lowering is greater (for the same reasons). The vapor pressure for a sodium chloride solution, for example, is lowered about twice as much as it is for a nonionic solution of the same concentration. To calculate the vapor pressure of a solution containing an ionic solute, we must account for the dissociation of the solute when we calculate the mole fraction of the solvent, as shown in Example 12.12.

EXAMPLE 12.12 Calculating the Vapor Pressure of a Solution Containing an Ionic Solute

A solution contains 0.102 mol $Ca(NO_3)_2$ and 0.927 mol H_2O. Calculate the vapor pressure of the solution at 55 °C. The vapor pressure of pure water at 55 °C is 118.1 torr. (Assume that the solute completely dissociates.)

SORT You are given the number of moles of each component of a solution and asked to find the vapor pressure of the solution. You are also given the vapor pressure of pure water at the appropriate temperature.	**GIVEN** 0.102 mol $Ca(NO_3)_2$ 0.927 mol H_2O $P°_{H_2O}$ = 118.1 torr (at 55 °C) **FIND** $P_{solution}$
STRATEGIZE To solve this problem, use Raoult's law as in Example 12.6. Calculate $\chi_{solvent}$ from the given amounts of solute and solvent.	**CONCEPTUAL PLAN** 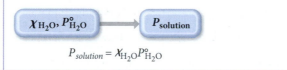 $P_{solution} = \chi_{H_2O} P°_{H_2O}$
SOLVE The key to this problem is to understand the dissociation of calcium nitrate. Write an equation showing the dissociation. Since 1 mol of calcium nitrate dissociates into 3 mol of dissolved particles, the number of moles of calcium nitrate must be multiplied by 3 when calculating the mole fraction. Use the mole fraction of water and the vapor pressure of pure water to calculate the vapor pressure of the solution.	**SOLUTION** $Ca(NO_3)_2(s) \longrightarrow Ca^{2+}(aq) + 2\,NO_3^-(aq)$ $\chi_{H_2O} = \dfrac{n_{H_2O}}{3 \times n_{Ca(NO_3)_2} + n_{H_2O}}$ $= \dfrac{0.927 \text{ mol}}{3(0.102) \text{ mol} + 0.927 \text{ mol}}$ $= 0.7518$ $P_{solution} = \chi_{H_2O} P°_{H_2O}$ $= 0.7518(118.1 \text{ torr})$ $= 88.8 \text{ torr}$

CHECK The units of the answer are correct. The magnitude also seems right because the calculated vapor pressure of the solution is significantly less than that of the pure solvent, as expected for a solution with a significant amount of solute.

FOR PRACTICE 12.12

A solution contains 0.115 mol H_2O and an unknown number of moles of sodium chloride. The vapor pressure of the solution at 30 °C is 25.7 torr. The vapor pressure of pure water at 30 °C is 31.8 torr. Calculate the number of moles of sodium chloride in the solution.

CHAPTER IN REVIEW

Key Terms

Section 12.1
solution (457)
solvent (457)
solute (457)

Section 12.2
aqueous solution (458)
solubility (458)
entropy (459)
miscible (460)

Section 12.3
enthalpy of solution (ΔH_{soln}) (463)
heat of hydration ($\Delta H_{hydration}$) (464)

Section 12.4
dynamic equilibrium (466)
saturated solution (466)

unsaturated solution (466)
supersaturated solution (466)
recrystallization (467)
Henry's law (468)

Section 12.5
dilute solution (469)
concentrated solution (469)
molarity (M) (470)
molality (m) (471)
parts by mass (471)
percent by mass (471)
parts per million (ppm) (471)
parts per billion (ppb) (471)
parts by volume (471)
mole fraction (χ_{solute}) (472)
mole percent (mol %) (473)

Section 12.6
colligative property (475)
Raoult's law (477)
vapor pressure lowering (ΔP) (477)
ideal solution (479)
freezing point depression (481)
boiling point elevation (481)
osmosis (484)
semipermeable membrane (484)
osmotic pressure (Π) (484)

Section 12.7
van't Hoff factor (i) (486)

Key Concepts

Solutions (12.1, 12.2)

▶ A solution is a homogeneous mixture of two or more substances. In a solution, the majority component is the solvent and the minority component is the solute.
▶ The tendency toward greater entropy (or greater energy dispersal) is the driving force for solution formation.
▶ Aqueous solutions contain water as a solvent and a solid, liquid, or gas as the solute.

Solubility and Energetics of Solution Formation (12.2, 12.3)

▶ The solubility of a substance is the amount of the substance that will dissolve in a given amount of solvent. The solubility of one substance in another depends on the types of intermolecular forces that exist *between* the substances as well as *within* each substance.
▶ The overall enthalpy change upon solution formation can be determined by adding the enthalpy changes for the three steps of solution formation: (1) separation of the solute particles, (2) separation of the solvent particles, and (3) mixing of the solute and solvent particles. The first two steps are both endothermic, while the last is exothermic.
▶ In aqueous solutions of an ionic compound, the change in enthalpy for steps 2 and 3 can be combined as the heat of hydration ($\Delta H_{hydration}$), which is always negative.

Solution Equilibrium (12.4)

▶ Dynamic equilibrium is the state in which the rates of dissolution and recrystallization in a solution are equal. A solution in this state is said to be saturated. Solutions containing less than or more than the equilibrium amount of solute are unsaturated or supersaturated, respectively.

▶ The solubility of most solids in water increases with increasing temperature.
▶ The solubility of gases in liquids generally decreases with increasing temperature, but increases with increasing pressure.

Concentration Units (12.5)

▶ Common units used to express solution concentration include molarity (M), molality (m), mole fraction (χ), mole percent (mol %), percent (%) by mass or volume, parts per million (ppm) by mass or volume, and parts per billion (ppb) by mass or volume. These units are summarized in Table 12.5.

Vapor Pressure Lowering, Freezing Point Depression, Boiling Point Elevation, and Osmosis (12.6, 12.7)

▶ The presence of a nonvolatile solute in a liquid results in a lower vapor pressure of the solution relative to the vapor pressure of the pure liquid. This lower vapor pressure is predicted by Raoult's law for an ideal solution.
▶ The addition of a nonvolatile solute to a liquid will result in a solution with a lower freezing point and a higher boiling point than those of the pure solvent.
▶ The flow of solvent from a solution of lower concentration to a solution of higher concentration is called osmosis.
▶ All of these phenomena are colligative properties and depend only on the number of solute particles added, not the type of solute particles.
▶ Electrolyte solutes have a greater effect on these properties than the corresponding amount of a nonelectrolyte solute as specified by the van't Hoff factor.

Key Equations and Relationships

Henry's Law: Solubility of Gases with Increasing Pressure (12.4)

$$S_{gas} = k_H P_{gas} \; (k_H \text{ is Henry's law constant})$$

Molarity (M) of a Solution (12.5)

$$M = \frac{\text{amount solute (in mol)}}{\text{volume solution (in L)}}$$

Molality (*m*) of a Solution (12.5)

$$m = \frac{\text{amount solute (in mol)}}{\text{mass solvent (in kg)}}$$

Concentration of a Solution in Parts by Mass and Parts by Volume (12.5)

$$\text{Percent by mass} = \frac{\text{mass solute} \times 100\%}{\text{mass solution}}$$

$$\text{Parts per million (ppm)} = \frac{\text{mass solute} \times 10^6}{\text{mass solution}}$$

$$\text{Parts per billion (ppb)} = \frac{\text{mass solute} \times 10^9}{\text{mass solution}}$$

$$\text{Parts by volume} = \frac{\text{volume solute} \times \text{multiplication factor}}{\text{volume solution}}$$

Concentration of a Solution in Mole Fraction (χ) and Mole Percent (12.5)

$$\chi_{solute} = \frac{n_{solute}}{n_{solute} + n_{solvent}}$$

$$\text{Mol \%} = \chi \times 100\%$$

Raoult's Law: Relationship between the Vapor Pressure of a Solution ($P_{solution}$), the Mole Fraction of the Solvent

($\chi_{solvent}$), and the Vapor Pressure of the Pure Solvent ($P^\circ_{solvent}$) (12.6)

$$P_{solution} = \chi_{solvent} P^\circ_{solvent}$$

Vapor Pressure Lowering (12.6)

$$\Delta P = \chi_{solute} P^\circ_{solvent}$$

Vapor Pressure of a Solution Containing Two Volatile Components (12.6)

$$P_A = \chi_A P^\circ_A$$
$$P_B = \chi_B P^\circ_B$$
$$P_{tot} = P_A + P_B$$

Relationship between Freezing Point Depression (ΔT_f), molality (*m*), and Freezing Point Depression Constant (K_f) (12.6)

$$\Delta T_f = i \times m \times K_f$$

Relationship between Boiling Point Elevation (ΔT_b), Molality (*m*), and Boiling Point Elevation Constant (K_b) (12.6)

$$\Delta T_b = i \times m \times K_b$$

Relationship between Osmotic Pressure (Π), Molarity (M), the Ideal Gas Constant (*R*), and Temperature (*T*, in K) (12.6)

$$\Pi = i \times MRT \; (R = 0.08206 \; L \cdot atm/mol \cdot K)$$

van't Hoff Factor (*i*): Ratio of Moles of Particles in Solution to Moles of Formula Units Dissolved (12.7)

$$i = \frac{\text{moles of particles in solution}}{\text{moles of formula units dissolved}}$$

Key Learning Objectives

Chapter Objectives	Assessment
Determining Whether a Solute Is Soluble in a Solvent (12.2)	Example 12.1 For Practice 12.1 Exercises 3–6
Using Henry's Law to Predict the Solubility of Gases with Increasing Pressure (12.4)	Example 12.2 For Practice 12.2 Exercises 21, 22
Calculating Concentrations of Solutions (12.5)	Examples 12.3, 12.4 For Practice 12.3, 12.4 For More Practice 12.3 Exercises 23–28, 35, 36
Converting between Concentration Units (12.5)	Example 12.5 For Practice 12.5 For More Practice 12.5 Exercises 37–40
Determining the Vapor Pressure of a Solution Containing a Nonionic and Nonvolatile Solute (12.6)	Example 12.6 For Practice 12.6 For More Practice 12.6 Exercises 43, 44
Determining the Vapor Pressure of a Two-Component Solution (12.6)	Example 12.7 For Practice 12.7 Exercises 45, 46
Calculating Freezing Point Depression (12.6)	Example 12.8 For Practice 12.8 Exercises 49–54, 59, 60
Calculating Boiling Point Elevation (12.6)	Example 12.9 For Practice 12.9 Exercises 49, 50, 59, 60
Determining the Osmotic Pressure (12.6)	Example 12.10 For Practice 12.10 Exercises 55–58
Determining and Using the van't Hoff Factor (12.7)	Example 12.11 For Practice 12.11 Exercises 63–68

EXERCISES

Problems by Topic

Solubility

1. Pick an appropriate solvent from Table 12.3 to dissolve each substance. State the kind of intermolecular forces that would occur between the solute and solvent in each case.
   a. motor oil (nonpolar)
   b. ethanol (polar, contains an OH group)
   c. lard (nonpolar)
   d. potassium chloride (ionic)

2. Pick an appropriate solvent from Table 12.3 to dissolve each substance.
   a. isopropyl alcohol (polar, contains an OH group)
   b. sodium chloride (ionic)
   c. vegetable oil (nonpolar)
   d. sodium nitrate (ionic)

3. Which compound would you expect to be more soluble in water, $CH_3CH_2CH_2OH$ or $HOCH_2CH_2CH_2OH$?

4. Which compound would you expect to be more soluble in water, CCl_4 or CH_2Cl_2?

5. For each molecule, would you expect greater solubility in water or in hexane? For each case, indicate the kinds of intermolecular forces that occur between the solute and the solvent in which the molecule is most soluble.

   a. glucose

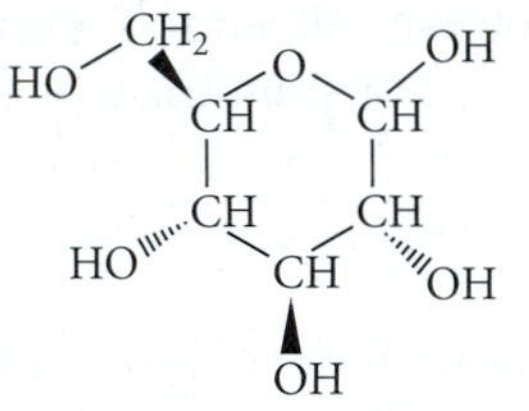

   b. naphthalene

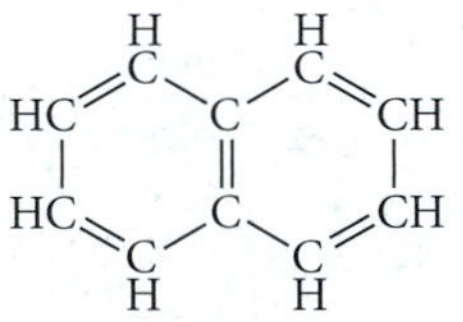

   c. dimethyl ether

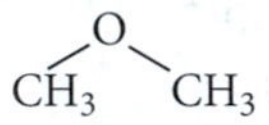

   d. alanine (an amino acid)

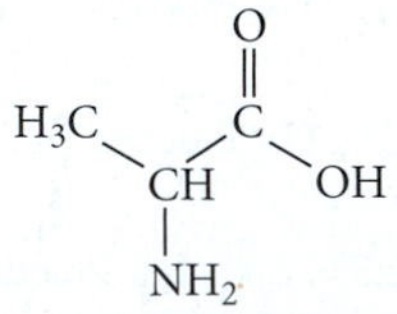

6. For each molecule, would you expect greater solubility in water or in hexane? For each case, indicate the kinds of intermolecular forces that would occur between the solute and the solvent in which the molecule is most soluble.

   a. toluene

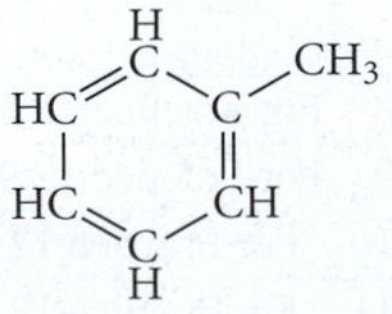

   b. sucrose (table sugar)

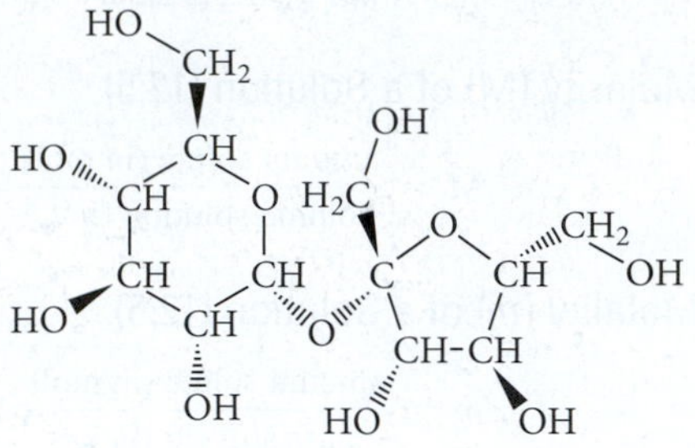

   c. isobutene

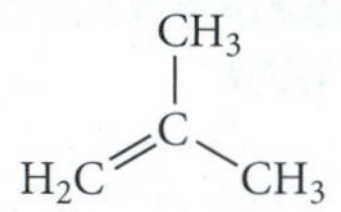

   d. ethylene glycol

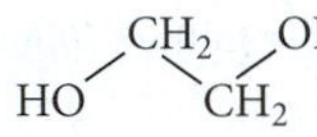

Energetics of Solution Formation

7. When ammonium chloride (NH_4Cl) is dissolved in water, the solution becomes colder.
   a. Is the dissolution of ammonium chloride endothermic or exothermic?
   b. What can you say about the relative magnitudes of the lattice energy of ammonium chloride and its heat of hydration?
   c. Sketch a qualitative energy diagram similar to Figure 12.7 for the dissolution of NH_4Cl.
   d. Why does the solution form? What drives the process?

8. When lithium iodide (LiI) is dissolved in water, the solution becomes hotter.
   a. Is the dissolution of lithium iodide endothermic or exothermic?
   b. What can you say about the relative magnitudes of the lattice energy of lithium iodide and its heat of hydration?
   c. Sketch a qualitative energy diagram similar to Figure 12.7 for the dissolution of LiI.
   d. Why does the solution form? What drives the process?

9. Silver nitrate has a lattice energy of -8.20×10^2 kJ/mol and a heat of solution of $+22.6$ kJ/mol. Calculate the heat of hydration for silver nitrate.

10. Use the data below to calculate the heats of hydration of lithium chloride and sodium chloride. Which of the two cations, lithium or sodium, has stronger ion–dipole interactions with water? How do you know?

Compound	Lattice Energy (kJ/mol)	ΔH_{soln} (kJ/mol)
LiCl	−834	−37.0
NaCl	−769	+3.88

11. Lithium iodide has a lattice energy of -7.3×10^2 (kJ/mol) and a heat of hydration of -793 kJ/mol. Find the heat of solution for lithium iodide and determine how much heat is evolved or absorbed when 15.0 g of lithium iodide completely dissolves in water.

12. Potassium nitrate has a lattice energy of -163.8 kcal/mol and a heat of hydration of -155.5 kcal/mol. How much potassium nitrate has to dissolve in water to absorb 1.00×10^2 kJ of heat?

Solution Equilibrium and Factors Affecting Solubility

13. A solution contains 25 g of NaCl per 100.0 g of water at 25 °C. Is the solution unsaturated, saturated, or supersaturated? (Use Figure 12.10.)

14. A solution contains 32 g of KNO$_3$ per 100.0 g of water at 25 °C. Is the solution unsaturated, saturated, or supersaturated? (Use Figure 12.10.)

15. A KNO$_3$ solution containing 45 g of KNO$_3$ per 100.0 g of water is cooled from 40 °C to 0 °C. What happens during cooling? (Use Figure 12.10.)

16. A KCl solution containing 42 g of KCl per 100.0 g of water is cooled from 60 °C to 0 °C. What happens during cooling? (Use Figure 12.10.)

17. Some laboratory procedures involving oxygen-sensitive reactants or products call for using preboiled (and then cooled) water. Explain why this is so.

18. A person preparing a fish tank uses boiled (and then cooled) water to fill it. When the fish is put into the tank, it dies. Explain.

19. Scuba divers breathing air at increased pressure can suffer from nitrogen narcosis—a condition resembling drunkenness—when the partial pressure of nitrogen exceeds about 4 atm. What property of gas/water solutions causes this to happen? How could the diver reverse this effect?

20. Scuba divers breathing air at increased pressure can suffer from oxygen toxicity—too much oxygen in their bloodstream—when the partial pressure of oxygen exceeds about 1.4 atm. What happens to the amount of oxygen in a diver's bloodstream when he or she breathes oxygen at elevated pressures? How can this be reversed?

21. Calculate the mass of nitrogen dissolved at room temperature in an 80.0-L home aquarium. Assume a total pressure of 1.0 atm and a mole fraction for nitrogen of 0.78.

22. Use Henry's law to determine the molar solubility of helium at a pressure of 1.0 atm and 25 °C.

Concentrations of Solutions

23. An aqueous NaCl solution is made using 145 g of NaCl diluted to a total solution volume of 1.00 L. Calculate the molarity, molality, and mass percent of the solution. (Assume a density of 1.08 g/mL for the solution.)

24. An aqueous KNO$_3$ solution is made using 72.3 g of KNO$_3$ diluted to a total solution volume of 1.50 L. Calculate the molarity, molality, and mass percent of the solution. (Assume a density of 1.05 g/mL for the solution.)

25. To what final volume should you dilute 50.0 mL of a 5.00 M KI solution so that 25.0 mL of the diluted solution contains 3.25 g of KI?

26. To what volume should you dilute 125 mL of an 8.00 M CuCl$_2$ solution so that 50.0 mL of the diluted solution contains 5.9 g CuCl$_2$?

27. Silver nitrate solutions are often used to plate silver onto other metals. What is the maximum amount of silver (in grams) that can be plated out of 4.8 L of an AgNO$_3$ solution containing 3.4% Ag by mass? Assume that the density of the solution is 1.01 g/mL.

28. A dioxin-contaminated water source contains 0.085% dioxin by mass. How much dioxin is present in 2.5 L of this water? Assume a density of 1.00 g/mL.

29. A hard water sample contains 0.0085% Ca by mass (in the form of Ca^{2+} ions). How much water (in grams) contains 1.2 g of Ca? (1.2 g of Ca is the recommended daily allowance of calcium for those between 19 and 24 years old.)

30. Lead is a toxic metal that affects the central nervous system. A Pb-contaminated water sample contains 0.0011% Pb by mass. How much of the water (in mL) contains 150 mg of Pb? (Assume a density of 1.0 g/mL.)

31. Nitric acid is usually purchased in a concentrated form that is 70.3% HNO$_3$ by mass and has a density of 1.41 g/mL. Describe exactly how you would prepare 1.15 L of 0.100 M HNO$_3$ from the concentrated solution.

32. Hydrochloric acid is usually purchased in a concentrated form that is 37.0% HCl by mass and has a density of 1.20 g/mL. Describe exactly how you would prepare 2.85 L of 0.500 M HCl from the concentrated solution.

33. Describe how to prepare each solution from the dry solute and the solvent.
   a. 1.00×10^2 mL of 0.500 M KCl
   b. 1.00×10^2 g of 0.500 m KCl
   c. 1.00×10^2 g of 5.0% KCl solution by mass

34. Describe how to prepare each solution from the dry solute and the solvent.
   a. 125 mL of 0.100 M NaNO$_3$
   b. 125 g of 0.100 m NaNO$_3$
   c. 125 g of 1.0% NaNO$_3$ solution by mass

35. A solution is prepared by dissolving 28.4 g of glucose (C$_6$H$_{12}$O$_6$) in 355 g of water. The final volume of the solution is 378 mL. Calculate the solution concentration in each unit.
   a. molarity
   b. molality
   c. percent by mass
   d. mole fraction
   e. mole percent

36. A solution is prepared by dissolving 20.2 mL of methanol (CH$_3$OH) in 100.0 mL of water at 25 °C. The final volume of the solution is 118 mL. The densities of methanol and water at this temperature are 0.782 g/mL and 1.00 g/mL, respectively. Calculate the solution concentration in each unit.
   a. molarity
   b. molality
   c. percent by mass
   d. mole fraction
   e. mole percent

37. Household hydrogen peroxide is an aqueous solution containing 3.0% hydrogen peroxide by mass. What is the molarity of this solution? (Assume a density of 1.01 g/mL.)

38. One brand of laundry bleach is an aqueous solution 4.55% sodium hypochlorite (NaOCl) by mass. What is the molarity of this solution? (Assume a density of 1.02 g/mL.)

39. An aqueous solution contains 36% HCl by mass. Calculate the molality and mole fraction of the solution.

40. An aqueous solution contains 5.0% NaCl by mass. Calculate the molality and mole fraction of the solution.

Vapor Pressure of Solutions

41. A beaker contains 100.0 mL of pure water. A second beaker contains 100.0 mL of seawater. The two beakers are left side by side on a lab bench for one week. At the end of the week, the liquid level in both beakers has decreased. However, the level has decreased more in one of the beakers than in the other. Which one and why?

42. Which solution has the highest vapor pressure?
    a. 20.0 g of glucose ($C_6H_{12}O_6$) in 100.0 mL of water
    b. 20.0 g of sucrose ($C_{12}H_{22}O_{11}$) in 100.0 mL of water
    c. 10.0 g of potassium acetate $KC_2H_3O_2$ in 100.0 mL of water

43. Calculate the vapor pressure of a solution containing 28.5 g of glycerin ($C_3H_8O_3$) in 125 mL of water at 30.0 °C. The vapor pressure of pure water at this temperature is 31.8 torr. Assume that glycerin is not volatile and dissolves molecularly (i.e., it is not ionic) and use a density of 1.00 g/mL for the water.

44. A solution contains naphthalene ($C_{10}H_8$) dissolved in hexane (C_6H_{14}) at a concentration of 10.85% naphthalene by mass. Calculate the vapor pressure at 25 °C of hexane above the solution. The vapor pressure of pure hexane at 25 °C is 151 torr.

45. A solution contains 50.0 g of heptane (C_7H_{16}) and 50.0 g of octane (C_8H_{18}) at 25 °C. The vapor pressures of pure heptane and pure octane at 25 °C are 45.8 torr and 10.9 torr, respectively. Assuming ideal behavior, calculate:
    a. the vapor pressure of each of the solution components in the mixture
    b. the total pressure above the solution
    c. the composition of the vapor in mass percent
    d. Why is the composition of the vapor different from the composition of the solution?

46. A solution contains a mixture of pentane and hexane at room temperature. The solution has a vapor pressure of 258 torr. Pure pentane and hexane have vapor pressures of 425 torr and 151 torr, respectively, at room temperature. What is the mole fraction composition of the mixture? (Assume ideal behavior.)

47. Calculate the vapor pressure at 25 °C of an aqueous solution that is 5.50% NaCl by mass.

48. An aqueous $CaCl_2$ solution has a vapor pressure of 81.6 mmHg at 50 °C. The vapor pressure of pure water at this temperature is 92.6 mmHg. What is the concentration of $CaCl_2$ in mass percent?

Freezing Point Depression, Boiling Point Elevation, and Osmosis

49. A glucose solution contains 55.8 g of glucose ($C_6H_{12}O_6$) in 455 g of water. Compute the freezing point and boiling point of the solution.

50. An ethylene glycol solution contains 21.2 g of ethylene glycol ($C_2H_6O_2$) in 85.4 mL of water. Calculate the freezing point and boiling point of the solution. (Assume a density of 1.00 g/mL for water.)

51. Calculate the freezing point and melting point of a solution containing 10.0 g of naphthalene ($C_{10}H_8$) in 100.0 mL of benzene. Benzene has a density of 0.877 g/cm³.

52. Calculate the freezing point and melting point of a solution containing 7.55 g of ethylene glycol ($C_2H_6O_2$) in 85.7 mL of ethanol. Ethanol has a density of 0.789 g/cm³.

53. An aqueous solution containing 17.5 g of an unknown molecular (nonelectrolyte) compound in 100.0 g of water was found to have a freezing point of −1.8 °C. Calculate the molar mass of the unknown compound.

54. An aqueous solution containing 35.9 g of an unknown molecular (nonelectrolyte) compound in 150.0 g of water was found to have a freezing point of −1.3 °C. Calculate the molar mass of the unknown compound.

55. Calculate the osmotic pressure of a solution containing 24.6 g of glycerin ($C_3H_8O_3$) in 250.0 mL of solution at 298 K.

56. What mass of sucrose ($C_{12}H_{22}O_{11}$) should be combined with 5.00×10^2 g of water to make a solution with an osmotic pressure of 8.55 atm at 298 K? (Assume a density of 1.0 g/mL for the solution.)

57. A solution containing 27.55 mg of an unknown protein per 25.0 mL solution has an osmotic pressure of 3.22 torr at 25 °C. What is the molar mass of the protein?

58. Calculate the osmotic pressure of a solution containing 18.75 mg of hemoglobin in 15.0 mL of solution at 25 °C. The molar mass of hemoglobin is 6.5×10^4 g/mol.

59. Calculate the freezing point and boiling point of the following aqueous solutions, assuming complete dissociation.
    a. 0.100 m K_2S
    b. 21.5 g of $CuCl_2$ in 4.50×10^2 g water
    c. 5.5% $NaNO_3$ by mass (in water)

60. Calculate the freezing point and boiling point of the following solutions, assuming complete dissociation.
    a. 10.5 g $FeCl_3$ in 1.50×10^2 g water
    b. 3.5% KCl by mass (in water)
    c. 0.150 m MgF_2

61. What mass of salt (NaCl) should you add to 1.0 L of water in an ice-cream maker to make a solution that freezes at −10.0 °C? Assume complete dissociation of the NaCl and density of 1.0 g/mL for water.

62. Determine the required concentration (in percent by mass) for an aqueous ethylene glycol ($C_2H_6O_2$) solution to have a boiling point of 104.0 °C.

63. Use the van't Hoff factors in Table 12.7 to calculate:
    a. the melting point of a 0.100 m iron(III) chloride solution
    b. the osmotic pressure of a 0.085 M potassium sulfate solution at 298 K
    c. the boiling point of a 1.22% by mass magnesium chloride solution

64. Assuming the van't Hoff factors in Table 12.7, calculate the mass of each solute required to produce each aqueous solution:
    a. a sodium chloride solution containing 1.50×10^2 g of water that has a melting point of −1.0 °C.
    b. 2.50×10^2 mL of a magnesium sulfate solution that has an osmotic pressure of 3.82 atm at 298 K.
    c. an iron(III) chloride solution containing 2.50×10^2 g of water that has a boiling point of 102 °C.

65. A 1.2 m aqueous solution of an ionic compound with the formula MX_2 has a boiling point of 101.4 °C. Calculate the van't Hoff factor for MX_2 at this concentration.

66. A 0.95 m aqueous solution of an ionic compound with the formula MX has a freezing point of −3.0 °C. Calculate the van't Hoff factor for MX at this concentration.

67. A 0.100 M ionic solution has an osmotic pressure of 8.3 atm at 25 °C. Calculate the van't Hoff factor (i) for this solution.

68. A solution contains 8.92 g of KBr in 500.0 mL of solution and has an osmotic pressure of 6.97 atm at 25 °C. Calculate the van't Hoff factor (i) for KBr at this concentration.

Cumulative Problems

69. The solubility of carbon tetrachloride (CCl_4) in water at 25 °C is 1.2 g/L. The solubility of chloroform ($CHCl_3$) at the same temperature is 10.1 g/L. Why is chloroform almost 10 times more soluble in water than is carbon tetrachloride?

70. The solubility of phenol in water at 25 °C is 8.7 g/L. The solubility of naphthol at the same temperature is only 0.074 g/L. Examine the structures of phenol and naphthol shown here and explain why phenol is so much more soluble than naphthol.

Phenol   Naphthol

71. Potassium perchlorate ($KClO_4$) has a lattice energy of −599 kJ/mol and a heat of hydration of −548 kJ/mol. Find the heat of solution for potassium perchlorate and determine the temperature change that occurs when 10.0 g of potassium perchlorate is dissolved with enough water to make 100.0 mL of solution. (Assume a heat capacity of 4.05 J/g · °C for the solution and a density of 1.05 g/mL.)

72. Sodium hydroxide (NaOH) has a lattice energy of −887 kJ/mol and a heat of hydration of −932 kJ/mol. How much solution could be heated to boiling by the heat evolved by the dissolution of 25.0 g of NaOH? (For the solution, assume a heat capacity of 4.0 J/g · °C, an initial temperature of 25.0 °C, a boiling point of 100.0 °C, and a density of 1.05 g/mL.)

73. A saturated solution was formed when 0.0537 L of argon, at a pressure of 1.0 atm and temperature of 25 °C, was dissolved in 1.0 L of water. Calculate the Henry's law constant for argon.

74. A gas has a Henry's law constant of 0.112 M/atm. How much water would be needed to completely dissolve 1.65 L of the gas at a pressure of 725 torr and a temperature of 25 °C?

75. The Safe Drinking Water Act (SDWA) sets a limit for mercury—a toxin to the central nervous system—at 0.0020 ppm by mass. Water suppliers must periodically test their water to ensure that mercury levels do not exceed this limit. Suppose water becomes contaminated with mercury at twice the legal limit (0.0040 ppm). How much of this water would a person need to consume to ingest 50.0 mg of mercury?

76. Water softeners often replace calcium ions in hard water with sodium ions. Since sodium compounds are soluble, the presence of sodium ions in water does not cause the white, scaly residues caused by calcium ions. However, calcium is more beneficial to human health than sodium because calcium is a necessary part of the human diet, while high levels of sodium intake are linked to increases in blood pressure. The U.S. Food and Drug Administration (FDA) recommends that adults ingest less than 2.4 g of sodium per day. How many liters of softened water, containing a sodium concentration of 0.050% sodium by mass, have to be consumed to exceed the FDA recommendation? (Assume a water density of 1.0 g/mL.)

77. An aqueous solution contains 12.5% NaCl by mass. What mass of water (in grams) is contained in 2.5 L of the vapor above this solution at 55 °C? The vapor pressure of pure water at 55 °C is 118 torr. (Assume complete dissociation of NaCl.)

78. The vapor above an aqueous solution contains 19.5 mg water per liter at 25 °C. Assuming ideal behavior, what is the concentration of the solute within the solution in mole percent?

79. What is the freezing point of an aqueous solution that boils at 106.5 °C?

80. What is the boiling point of an aqueous solution that has a vapor pressure of 20.5 torr at 25 °C? (Assume a nonvolatile solute.)

81. An isotonic solution (one with the same osmotic pressure as bodily fluids) contains 0.90% NaCl by mass per volume. Calculate the percent mass per volume for isotonic solutions containing each solute at 25 °C. Assume a van't Hoff factor of 1.9 for all *ionic* solutes.
  a. KCl
  b. NaBr
  c. Glucose ($C_6H_{12}O_6$)

82. Magnesium citrate, $Mg_3(C_6H_5O_7)_2$ belongs to a class of laxatives called *hyperosmotics*, which rapidly empty the bowel. When a concentrated solution of magnesium citrate is consumed, it passes through the intestines, drawing water and promoting diarrhea, usually within 6 hours. Calculate the osmotic pressure of a magnesium citrate laxative solution containing 28.5 g of magnesium citrate in 235 mL of solution at 37 °C (approximate body temperature). Assume complete dissociation of the ionic compound.

83. A solution is prepared from 4.5701 g of magnesium chloride and 43.238 g of water. The vapor pressure of water above this solution is 0.3624 atm at 348.0 K. The vapor pressure of pure water at this temperature is 0.3804 atm. Find the value of the van't Hoff factor i for magnesium chloride in this solution.

84. When HNO_2 is dissolved in water it partially dissociates according to the equation $HNO_2 \rightleftharpoons H^+ + NO_2^-$. A solution is prepared that contains 7.050 g of HNO_2 in 1.000 kg of water. Its freezing point is −0.2929 °C. Calculate the fraction of HNO_2 that has dissociated.

85. A solution of a nonvolatile solute in water has a boiling point of 375.3 K. Calculate the vapor pressure of water above this solution at 338 K. The vapor pressure of pure water at this temperature is 0.2467 atm.

86. The density of a 0.438 M solution of potassium chromate (K_2CrO_4) at 298 K is 1.063 g/mL. Calculate the vapor pressure of water above the solution. The vapor pressure of pure water at this temperature is 0.0313 atm. Assume complete dissociation.

87. The vapor pressure of carbon tetrachloride, CCl_4, is 0.354 atm and the vapor pressure of chloroform, $CHCl_3$, is 0.526 atm at 316 K. A solution is prepared from equal masses of these two compounds at this temperature. Calculate the mole fraction of the chloroform in the vapor above the solution. If the vapor above the original solution is condensed and isolated into a separate flask, what would the vapor pressure of chloroform be above this new solution?

88. Distillation is a method of purification based on successive separations and recondensations of vapor above a solution. Use the result of the previous problem to calculate the mole fraction of chloroform in the vapor above a solution obtained by three successive separations and condensations of the vapors above the original solution of carbon tetrachloride and chloroform. Show how this result explains the use of distillation as a separation method.

89. A solution of 49.0% H_2SO_4 by mass has a density of 1.39 g/cm^3 at 293 K. A 25.0-cm^3 sample of this solution is mixed with enough water to increase the volume of the solution to 99.8 cm^3. Find the molarity of sulfuric acid in this solution.

90. Find the mass of urea (CH_4N_2O) needed to prepare 50.0 g of a solution in water in which the mole fraction of urea is 0.0770.

91. A solution contains 10.05 g of unknown compound dissolved in 50.0 mL of water. (Assume a density of 1.00 g/mL for water.) The freezing point of the solution is $-3.16\,°C$. The mass percent composition of the compound is 60.97% C, 11.94% H, and the rest is O. What is the molecular formula of the compound?

92. The osmotic pressure of a solution containing 1.05 g of an unknown compound dissolved in 175.0 mL of solution at 25 °C is 1.93 atm. The combustion of 24.02 g of the unknown compound produced 28.16 g CO_2 and 8.64 g H_2O. What is the molecular formula of the compound (which contains only carbon, hydrogen, and oxygen)?

93. A 100.0 mL aqueous sodium chloride solution is 13.5% NaCl by mass and has a density of 1.12 g/mL. What would you add (solute or solvent) and what mass of it to make the boiling point of the solution 104.4 °C? (Use $i = 1.8$ for NaCl.)

94. A 50.0 mL solution is initially 1.55% $MgCl_2$ by mass and has a density of 1.05 g/mL. What is the freezing point of the solution after you add an additional 1.35 g $MgCl_2$? (Use $i = 2.5$ for $MgCl_2$.)

Challenge Problems

95. The small bubbles that form on the bottom of a water pot that is being heated (before boiling) are due to dissolved air coming out of solution. Use Henry's law and the solubilities given to calculate the total volume of nitrogen and oxygen gas that should bubble out of 1.5 L of water upon warming from 25 °C to 50 °C. Assume that the water is initially saturated with nitrogen and oxygen gas at 25 °C and a total pressure of 1.0 atm. Assume that the gas bubbles out at a temperature of 50 °C. The solubility of oxygen gas at 50 °C is 27.8 mg/L at an oxygen pressure of 1.00 atm. The solubility of nitrogen gas at 50 °C is 14.6 mg/L at a nitrogen pressure of 1.00 atm. Assume that the air above the water contains an oxygen partial pressure of 0.21 atm and a nitrogen partial pressure of 0.78 atm.

96. The vapor above a mixture of pentane and hexane at room temperature contains 35.5% pentane by mass. What is the mass percent composition of the solution? Pure pentane and hexane have vapor pressures of 425 torr and 151 torr, respectively, at room temperature.

97. A 1.10 g sample contains only glucose ($C_6H_{12}O_6$) and sucrose ($C_{12}H_{22}O_{11}$). When the sample is dissolved in water to a total solution volume of 25.0 mL, the osmotic pressure of the solution is 3.78 atm at 298 K. What is the percent composition of glucose and sucrose in the sample?

98. A solution is prepared by mixing 631 mL of methanol with 501 mL of water. The molarity of methanol in the resulting solution is 14.29 M. The density of methanol at this temperature is 0.792 g/mL. Calculate the difference in volume between this solution and the total volume of water and methanol that were mixed to prepare the solution.

99. Two alcohols, isopropyl alcohol and propyl alcohol, have the same molecular formula, C_3H_8O. A solution of the two that is two-thirds by mass isopropyl alcohol has a vapor pressure of 0.110 atm at 313 K. A solution that is one-third by mass isopropyl alcohol has a vapor pressure of 0.089 atm at 313 K. Calculate the vapor pressure of each pure alcohol at this temperature. Explain the difference given that the formula of propyl alcohol is $CH_3CH_2CH_2OH$ and that of isopropyl alcohol is $(CH_3)_2CHOH$.

100. A metal, M, of atomic weight 96 reacts with fluorine to form a salt that can be represented as MF_x. In order to determine x and therefore the formula of the salt, a boiling point elevation experiment is performed. A 9.18 g sample of the salt is dissolved in 100.0 g of water and the boiling point of the solution is 374.38 K. Find the formula of the salt. Assume complete dissociation of the salt in solution.

101. A solution of 75.0 g of benzene (C_6H_6) and 75.0 g of toluene (C_7H_8) has a total vapor pressure of 80.9 mmHg at 303 K. Another solution of 100.0 g benzene and 50.0 g toluene has a total vapor pressure of 93.9 mmHg at this temperature. Find the vapor pressure of pure benzene and pure toluene at 303 K.

102. A solution is prepared by dissolving 11.60 g of a mixture of sodium carbonate and sodium bicarbonate in 1.00 L of water. A 300.0 cm^3 sample of the solution is then treated with excess HNO_3 and boiled to remove all the dissolved gas. A total of 0.940 L of dry CO_2 is collected at 298 K and 0.972 atm. Find the molarity of the carbonate and bicarbonate in the solution.

Conceptual Problems

103. Substance A is a nonpolar liquid and has only dispersion forces among its constituent particles. Substance B is also a nonpolar liquid and has about the same magnitude of dispersion forces among its constituent particles. When substance A and B are combined, they spontaneously mix.
   a. Why do the two substances mix?
   b. Predict the sign and magnitude of ΔH_{soln}.

   c. Determine the signs and relative magnitudes of ΔH_{solute}, $\Delta H_{solvent}$, and ΔH_{mix}.

104. A power plant built on a river uses river water as a coolant. The water is warmed as it is used in heat exchangers within the plant. Should the warm water be immediately cycled back into the river? Why or why not?

105. The vapor pressure of a 1 M ionic solution is different from the vapor pressure of a 1 M nonionic solution. In both cases, the solute is nonvolatile. Which set of diagrams best represents the differences between the two solutions and their vapors?

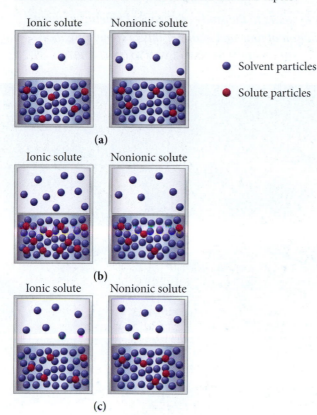

Ionic solute   Nonionic solute

(a)

Ionic solute   Nonionic solute

(b)

Ionic solute   Nonionic solute

(c)

● Solvent particles

● Solute particles

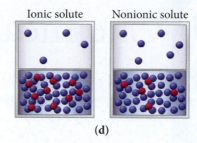

Ionic solute   Nonionic solute

(d)

106. If all of these substances cost the same amount per kilogram, which would be most cost-effective as a way to lower the freezing point of water? (Assume complete dissociation for all ionic compounds.) Explain.
   a. $HOCH_2CH_2OH$
   b. $NaCl$
   c. KCl
   d. $MgCl_2$
   e. $SrCl_2$

107. A helium balloon inflated on one day will fall to the ground by the next day. The volume of the balloon decreases somewhat overnight, but not by enough to explain why it no longer floats. (If you inflate a new balloon with helium to the same size as the balloon that fell to the ground, the newly inflated balloon floats.) Explain.

Answers to Conceptual Connections

Solubility

12.1 The first alcohol on the list is methanol, which is highly polar and forms hydrogen bonds with water. It is miscible in water and has only limited solubility in hexane, which is nonpolar. However, as the carbon chain gets longer in the series of alcohols, the OH group becomes less important relative to the growing nonpolar carbon chain. Therefore the alcohols become progressively less soluble in water and more soluble in hexane. This table demonstrates the rule of thumb, *like dissolves like*. Methanol is like water and therefore dissolves in water. It is unlike hexane and therefore has limited solubility in hexane. As you move down the list, the alcohols become increasingly like hexane and increasingly unlike water and therefore become increasingly soluble in hexane and increasingly insoluble in water.

Energetics of Aqueous Solution Formation

12.2 You can conclude that $|\Delta H_{solute}| < |\Delta H_{hydration}|$. Since ΔH_{soln} is negative, the absolute value of the negative term ($\Delta H_{hydration}$) must be greater than the absolute value of the positive term (ΔH_{solute}).

Solubility and Temperature

12.3 (b) Some potassium bromide precipitates out of solution. The solubility of most solids decreases with decreasing temperature.

However, the solubility of gases increases with decreasing temperature. Therefore, the nitrogen becomes more soluble and will not bubble out of solution.

Henry's Law

12.4 Ammonia is the only compound on the list that is polar, so we would expect its solubility in water to be greater than those of the other gases (which are all nonpolar).

Vapor Pressure of a Two Component Solution

12.5 (b) 150 torr. Since the solution is equimolar in A and B, the mole fractions of A and B are each 0.5. Therefore, the vapor pressure of A is 50 torr and the vapor pressure of B is 100 torr. The total vapor pressure is 150 torr.

Freezing Point Depression

12.6 Solution B because K_b for ethanol is greater than K_b for water.

Colligative Properties

12.7 (c) The 0.50 M $MgCl_2$ solution will have the highest boiling point because it has the highest concentration of particles. We expect 1 mol of $MgCl_2$ to form 3 mol of particles in solution (although it effectively forms slightly fewer).

13 Chemical Kinetics

*Nobody, I suppose, could devote many years to the study of chemical kinetics without being deeply conscious of the fascination of time and change: this is something that goes outside science into poetry. . . .—*SIR CYRIL N. HINSHELWOOD (1897–1967)

Pouring ice water on a lizard slows it down, making it easier to catch.

I N THE PASSAGE QUOTED here, Oxford chemistry professor Sir Cyril Hinshelwood calls attention to an aspect of chemistry often overlooked by the casual observer—the mystery of change with time. Since the opening chapter of this book, we have emphasized that the goal of chemistry is to understand the macroscopic world by examining the molecular one. In this chapter, we focus on understanding how this molecular world changes with time, a topic called chemical kinetics. The molecular world is anything but static. Thermal energy produces constant molecular motion, causing molecules to repeatedly collide with one another. In a tiny fraction of these collisions, something unique happens—the electrons on one molecule or atom are attracted to the nuclei of another. Some bonds weaken and new bonds form—a chemical reaction occurs. Chemical kinetics is the study of how these kinds of changes occur in time.

13.1   Catching Lizards

Kids (including my own) who live in my neighborhood have a unique way of catching lizards. Armed with cups of ice water, they chase the cold-blooded reptiles into a corner, and then take aim and fire—or pour, actually. They pour the cold water directly onto the lizard's body. The lizard's body temperature drops and it becomes virtually immobilized—easy prey for little hands. The kids scoop up the lizard and place it in a tub filled with sand and leaves. They then watch as the lizard warms back up and becomes active again. They usually release the lizard back into the yard within hours. I guess you could call them catch-and-release lizard hunters.

Unlike mammals, which actively regulate their body temperature through metabolic activity, lizards are *ectotherms*—their body temperature depends on their surroundings. When splashed with cold water, a lizard's body simply gets colder. The drop in body temperature slows down the lizard because its movement depends on chemical reactions that occur within its muscles, and the rates of those reactions—how fast they occur—are highly sensitive to temperature. When the temperature drops, the reactions that produce movement occur more slowly; therefore, the movement itself slows down. Cold reptiles are lethargic, unable to move very quickly. For this reason, reptiles try to maintain their body temperature within a narrow range by moving between sun and shade.

The rates of chemical reactions, and especially the ability to *control* those rates, are important not just in reptile movement but in many other phenomena as well. For example, the launching of a rocket depends on controlling the rate at which fuel burns—too quickly and the rocket can explode, too slowly and it will not leave the ground. Chemists must always consider reaction rates when synthesizing compounds. No matter how stable a compound might be, its synthesis is impossible if the rate at which it forms is too slow. As we have seen with reptiles, reaction rates are important to life. In fact, the human body's ability to switch a specific reaction on or off at a specific time is achieved largely by controlling the rate of that reaction through the use of enzymes (biological molecules that we explore more fully in Section 13.7).

The knowledge of reaction rates is not only practically important—giving us the ability to control how fast a reaction occurs—but also theoretically important. As you will see in Section 13.6, knowledge of the rate of a reaction can tell us much about how the reaction occurs on the molecular scale.

13.2   The Rate of a Chemical Reaction

The rate of a chemical reaction is a measure of how fast the reaction occurs. If a chemical reaction has a slow rate, only a relatively small fraction of molecules react to form products in a given period of time. If a chemical reaction has a fast rate, a large fraction of molecules react to form products in a given period of time.

When we measure how fast something occurs, or more specifically the *rate* at which it occurs, we usually express the measurement as a change in some quantity per unit of time. For example, we measure the speed of a car—the rate at which it travels—in *miles per hour*, and we measure how fast people lose weight in *pounds per week*. Notice that both of these rates are reported in units that represent the change in what we are measuring (distance or weight) divided by the change in time.

$$\text{Speed} = \frac{\text{change in distance}}{\text{change in time}} = \frac{\Delta x}{\Delta t} \qquad \text{Weight loss} = \frac{\text{change in weight}}{\text{change in time}} = \frac{\Delta \text{ weight}}{\Delta t}$$

Similarly, the rate of a chemical reaction is measured as a change in the amounts of reactants or products (usually in terms of concentration) divided by the change in time. Consider the following gas-phase reaction between $H_2(g)$ and $I_2(g)$:

$$H_2(g) + I_2(g) \longrightarrow 2\,HI(g)$$

$[A]$ means the concentration of A in M (mol/L).

We can define the rate of this reaction in the time interval t_1 to t_2 as follows:

$$\text{Rate} = -\frac{\Delta[H_2]}{\Delta t} = -\frac{[H_2]_{t_2} - [H_2]_{t_1}}{t_2 - t_1} \qquad [13.1]$$

In this expression, $[H_2]_{t_2}$ is the hydrogen concentration at time t_2 and $[H_2]_{t_1}$ is the hydrogen concentration at time t_1. Notice that the reaction rate is defined as *the negative* of the change in concentration of a reactant divided by the change in time. The negative sign is part of the definition when the reaction rate is defined with respect to a reactant because reactant concentrations decrease as a reaction proceeds; therefore *the change in the concentration of a reactant is negative*. The negative sign thus makes the overall *rate* positive. (By convention, reaction rates are reported as positive quantities.)

The reaction rate can also be defined with respect to the other reactant:

$$\text{Rate} = -\frac{\Delta[I_2]}{\Delta t} \qquad [13.2]$$

Since 1 mol of H_2 reacts with 1 mol of I_2, the rates are defined in the same way. The rate can also be defined with respect to the *product* of the reaction:

$$\text{Rate} = +\frac{1}{2}\frac{\Delta[HI]}{\Delta t} \qquad [13.3]$$

Notice that, because product concentrations *increase* as the reaction proceeds, the change in concentration of a product is positive. Therefore, when the rate is defined with respect to a product, we do not include a negative sign in the definition—the rate is naturally positive. Notice also the factor of $\frac{1}{2}$ in this definition. This factor is related to the stoichiometry of the reaction. In order to have a single rate for the entire reaction, the definition of the rate with respect to each reactant and product must reflect the stoichiometric coefficients of the reaction. For this particular reaction, 2 mol of HI is produced from 1 mol of H_2 and 1 mol of I_2.

Therefore the concentration of HI increases at twice the rate that the concentration of H_2 or I_2 decreases. If 100 I_2 molecules react per second, then 200 HI molecules form per second. In order for the overall rate to have the same value when defined with respect to any of the reactants or products, the change in HI concentration must be multiplied by a factor of one-half.

Consider the graph shown in **Figure 13.1▶** (on p. 499), which represents the changes in concentration for H_2 (one of the reactants) and HI (the product) versus time.

Let's examine several features of this graph individually.

Change in Reactant and Product Concentrations The reactant concentration, as expected, *decreases* with time because reactants are consumed in a reaction. The product concentration *increases* with time because products are formed in a reaction. The increase in HI concentration occurs at exactly twice the rate of the decrease in H_2 concentration because of the stoichiometry of the reaction—2 mol of HI is formed for every 1 mol of H_2 consumed.

The Average Rate of the Reaction We can calculate the average rate of the reaction for any time interval using Equation 13.1. The table shown at the top of the next page lists: H_2 concentration $([H_2])$ at various times, the change in H_2 concentration for each interval $(\Delta[H_2])$, the change in time for each interval (Δt), and the rate for each interval $(-\Delta[H_2]/\Delta t)$. The rate calculated in this way represents the average rate within the given time interval. For example, the average rate of the reaction in the time interval between 10 and 20 seconds is 0.0149 M/s, while the average rate in the time interval between 20 and 30 seconds is 0.0121 M/s. Notice that the average rate *decreases* as the reaction progresses. In other words, the reaction slows down as it proceeds. We discuss this further in the next section where we will see that, for most reactions, the rate depends on the concentrations of the reactants. As the reactants are consumed, their concentrations decrease, and the reaction slows down.

Time (s)	$[H_2]$ (M)		$\Delta[H_2]$	Δt	Rate $= -\Delta[H_2]/\Delta t$ (M/s)
00.000	1.000				
		}	-0.181	10.000	0.0181
10.000	0.819				
		}	-0.149	10.000	0.0149
20.000	0.670				
		}	-0.121	10.000	0.0121
30.000	0.549				
		}	-0.100	10.000	0.0100
40.000	0.449				
		}	-0.081	10.000	0.0081
50.000	0.368				
		}	-0.067	10.000	0.0067
60.000	0.301				
		}	-0.054	10.000	0.0054
70.000	0.247				
		}	-0.045	10.000	0.0045
80.000	0.202				
		}	-0.037	10.000	0.0037
90.000	0.165				
		}	-0.030	10.000	0.0030
100.000	0.135				

The Instantaneous Rate of the Reaction The instantaneous rate of the reaction is the rate at any one point in time, represented by the instantaneous slope of the curve at that point. We can obtain the instantaneous rate from the slope of the tangent to the curve at the point of interest. In our graph, we have drawn the tangent lines for both $[H_2]$ and $[HI]$ at 50 seconds.

We calculate the instantaneous rate at 50 seconds as follows:

Using $[H_2]$

$$\text{Instantaneous rate (at 50 s)} = -\frac{\Delta[H_2]}{\Delta t} = -\frac{-0.28\text{ M}}{40\text{ s}} = 0.0070\text{ M/s}$$

Using $[HI]$

$$\text{Instantaneous rate (at 50 s)} = +\frac{1}{2}\frac{\Delta[HI]}{\Delta t} = +\frac{1}{2}\frac{(0.56\text{ M})}{40\text{ s}} = 0.0070\text{ M/s}$$

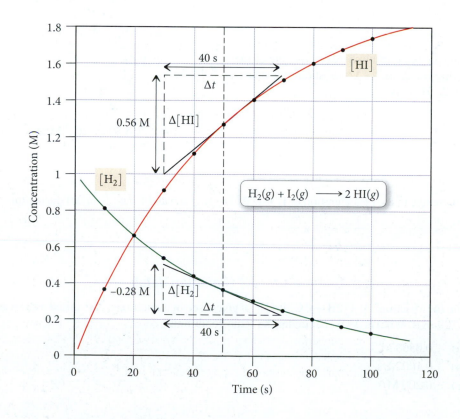

◀ **FIGURE 13.1 Reactant and Product Concentrations as a Function of Time** The concentration of I_2 has been intentionally omitted from the graph for clarity.

Notice that, as expected, the rate is the same whether we use one of the reactants or the product for the calculation. Notice also that the instantaneous rate at 50 seconds (0.0070 M/s) lies between the average rates calculated for the 10-second intervals just before and just after 50 seconds.

We can generalize our definition of reaction rates for the generic reaction:

$$aA + bB \longrightarrow cC + dD \qquad [13.4]$$

where A and B are reactants, C and D are products, and a, b, c, and d are the stoichiometric coefficients. The rate of the reaction is then defined as follows:

$$\text{Rate} = -\frac{1}{a}\frac{\Delta[A]}{\Delta t} = -\frac{1}{b}\frac{\Delta[B]}{\Delta t} = +\frac{1}{c}\frac{\Delta[C]}{\Delta t} = +\frac{1}{d}\frac{\Delta[D]}{\Delta t} \qquad [13.5]$$

Knowing the rate of change in the concentration of any one reactant or product at a point in time allows you to determine the rate of change in the concentration of any other reactant or product at that point in time (from the balanced equation). *However, predicting the rate at some future time is not possible from just the balanced equation.*

EXAMPLE 13.1 Expressing Reaction Rates

Consider the balanced chemical equation:

$$H_2O_2(aq) + 3\,I^-(aq) + 2\,H^+(aq) \longrightarrow I_3^-(aq) + 2\,H_2O(l)$$

In the first 10.0 seconds of the reaction, the concentration of I^- drops from 1.000 M to 0.868 M.

(a) Calculate the average rate of this reaction in this time interval.

(b) Predict the rate of change in the concentration of H^+ (that is, $\Delta[H^+]/\Delta t$) during this time interval.

SOLUTION

(a) Use Equation 13.5 to calculate the average rate of the reaction.	$\text{Rate} = -\dfrac{1}{3}\dfrac{\Delta[I^-]}{\Delta t}$ $\quad= -\dfrac{1}{3}\dfrac{(0.868\text{ M} - 1.000\text{ M})}{10.0\text{ s}}$ $\quad= 4.40 \times 10^{-3}\text{ M/s}$
(b) Use Equation 13.5 again to determine the relationship between the rate of the reaction and $\Delta[H^+]/\Delta t$. After solving for $\Delta[H^+]/\Delta t$, substitute the calculated rate from part (a) and calculate $\Delta[H^+]/\Delta t$.	$\text{Rate} = -\dfrac{1}{2}\dfrac{\Delta[H^+]}{\Delta t}$ $\dfrac{\Delta[H^+]}{\Delta t} = -2\,(\text{rate})$ $\quad= -2(4.40 \times 10^{-3}\text{ M/s})$ $\quad= -8.80 \times 10^{-3}\text{ M/s}$

FOR PRACTICE 13.1

For the above reaction, predict the rate of change in the concentration of H_2O_2 ($\Delta[H_2O_2]/\Delta t$) and I_3^- ($\Delta[I_3^-]/\Delta t$) during this time interval.

⬤ Conceptual Connection 13.1 Reaction Rates

For the reaction, A + 2B → C, under a given set of conditions, the initial rate was 0.100 M/s. What is $\Delta[B]/\Delta t$ under the same conditions?

(a) -0.0500 M/s

(b) -0.100 M/s

(c) -0.200 M/s

13.3   The Rate Law: The Effect of Concentration on Reaction Rate

The rate of a reaction often depends on the concentration of one or more of the reactants. Chemist Ludwig Wilhelmy noticed this effect in 1850 for the hydrolysis of sucrose. For simplicity, consider a reaction in which a single reactant, A, decomposes into products:

$$A \longrightarrow products$$

As long as the rate of the reverse reaction (the reaction in which the products return to reactants) is negligibly slow, we can write a relationship—called the **rate law**—between the rate of the reaction and the concentration of the reactant as follows:

$$Rate = k[A]^n \qquad [13.6]$$

where k is a constant of proportionality called the **rate constant** and n is a number called the **reaction order**. The value of n determines how the rate depends on the concentration of the reactant.

- If $n = 0$, the reaction is *zero order* and the rate is independent of the concentration of A.

- If $n = 1$, the reaction is *first order* and the rate is directly proportional to the concentration of A.

- If $n = 2$, the reaction is *second order* and the rate is proportional to the square of the concentration of A.

> By definition, $[A]^0 = 1$, so the rate is equal to k regardless of $[A]$.

Although other orders are possible, including noninteger (or fractional) orders, these three are the most common.

Figure 13.2▼ shows three plots illustrating how the *concentration of A changes with time* for the three common reaction orders with identical values for the rate constant (k). **Figure 13.3▼** has three plots showing the *rate of the reaction* (the slope of the lines in Figure 13.2) *as a function of the reactant concentration* for each reaction order.

Zero-Order Reaction In a zero-order reaction, the rate of the reaction is independent of the concentration of the reactant:

$$Rate = k[A]^0 = k \qquad [13.7]$$

Reactant Concentration versus Time

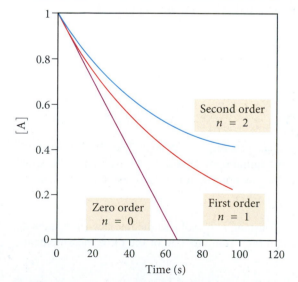

▲ **FIGURE 13.2 Reactant Concentration as a Function of Time for Different Reaction Orders**

Rate versus Reactant Concentration

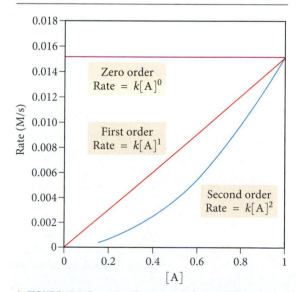

▲ **FIGURE 13.3 Reaction Rate as a Function of Reactant Concentration for Different Reaction Orders**

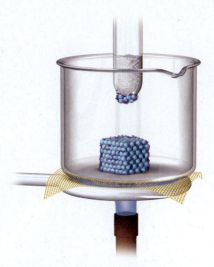

▲ FIGURE 13.4 Sublimation When a layer of particles sublimes, another identical layer is just below it. Consequently, the number of particles available to sublime at any one time does not change with the total number of particles in the sample, and the process is zero order.

Consequently, for a zero-order reaction, the concentration of the reactant decreases linearly with time, as shown in Figure 13.2. Notice the constant slope in the plot—a constant slope indicates a constant rate. The rate is constant because the reaction does not slow down as the concentration of A decreases. The graph in Figure 13.3 shows that the rate of a zero-order reaction is the same at any concentration of A. Zero-order reactions occur under conditions where the amount of reactant actually *available for reaction* is unaffected by changes in the *overall quantity of reactant*. For example, sublimation is normally zero order because only molecules at the surface can sublime, and their concentration does not change when the amount of subliming substance decreases (**Figure 13.4◄**).

First-Order Reaction In a first-order reaction, the rate of the reaction is directly proportional to the concentration of the reactant.

$$\text{Rate} = k[A]^1 \qquad [13.8]$$

Consequently, for a first-order reaction the rate slows down as the reaction proceeds because the concentration of the reactant decreases. You can see this in Figure 13.2 because the slope of the curve (the rate) becomes less steep (slower) with time. Figure 13.3 shows the rate as a function of the concentration of A. Notice the linear relationship—the rate is directly proportional to the concentration.

Second-Order Reaction In a second-order reaction, the rate of the reaction is proportional to the square of the concentration of the reactant.

$$\text{Rate} = k[A]^2 \qquad [13.9]$$

Consequently, for a second-order reaction, the rate is even more sensitive to the reactant concentration. You can see this in Figure 13.2 because the slope of the curve (the rate) flattens out more quickly than it does for a first-order reaction. Figure 13.3 shows the rate as a function of the concentration of A. Notice the quadratic relationship—the rate is proportional to the square of the concentration.

Determining the Order of a Reaction

The order of a reaction can be determined only by experiment. A common way to determine reaction order is by the *method of initial rates*. In this method, the initial rate—the rate for a short period of time at the beginning of the reaction—is measured at several different initial reactant concentrations to determine the effect of the concentration on the rate. For example, let's return to our simple reaction in which a single reactant, A, decomposes into products:

$$A \longrightarrow \text{products}$$

In an experiment, the initial rate was measured at several different initial concentrations with the following results:

[A] (M)	Initial Rate (M/s)
0.10	0.015
0.20	0.030
0.40	0.060

Notice that, in this data set, when the concentration of A doubles, the rate doubles—the initial rate is directly proportional to the initial concentration. The reaction is therefore first order in A and the rate is equal to the rate constant times the concentration of A:

$$\text{Rate} = k[A]^1$$

We can determine the value of the rate constant, k, by solving the rate law for k and substituting the concentration and the initial rate from any one of the three measurements. Using the first measurement, we get:

$$\text{Rate} = k[A]^1$$

$$k = \frac{\text{rate}}{[A]} = \frac{0.015 \text{ M/s}}{0.10 \text{ M}} = 0.15 \text{ s}^{-1}$$

Notice that the rate constant for a first-order reaction has units of s^{-1}.

The following two data sets show how measured initial rates would differ for zero-order and for second-order reactions:

Zero Order ($n=0$)		Second Order ($n=2$)	
[A] (M)	Initial Rate (M/s)	[A] (M)	Initial Rate (M/s)
0.10	0.015	0.10	0.015
0.20	0.015	0.20	0.060
0.40	0.015	0.40	0.240

For a zero-order reaction, the initial rate is independent of the reactant concentration—the rate is the same at all measured initial concentrations. For a second-order reaction, the initial rate quadruples for a doubling of the reactant concentration—the relationship between concentration and rate is quadratic. The rate constants for zero- and second-order reactions have different units than for first-order reactions. The rate constant for a zero-order reaction has units of $M \cdot s^{-1}$ and that for a second-order reaction has units of $M^{-1} \cdot s^{-1}$.

If we are unsure about how the initial rate is changing with the initial reactant concentration, or if the numbers are not as obvious as they are in these examples, we can substitute any two initial concentrations and the corresponding initial rates into a ratio of the rate laws to determine the order (n):

$$\frac{\text{rate } 2}{\text{rate } 1} = \frac{k[A]_2^n}{k[A]_1^n}$$

For example, we can substitute the last two measurements in the above data set for the second-order reaction to determine that $n = 2$:

$$\frac{0.240 \text{ M/s}}{0.060 \text{ M/s}} = \frac{k(0.40 \text{ M})^n}{k(0.20 \text{ M})^n}$$

$$4.0 = \left(\frac{0.40}{0.20}\right)^n = 2^n$$

$$\log 4.0 = \log(2^n)$$

$$= n \log 2$$

$$n = \frac{\log 4}{\log 2}$$

$$= 2$$

 Conceptual Connection 13.2 Order of Reaction

The reaction $A \longrightarrow B$ has been experimentally determined to be second order. The initial rate is 0.0100 M/s at an initial concentration of A of 0.100 M. What is the initial rate at $[A] = 0.500$ M?

(a) 0.00200 M/s
(b) 0.0100 M/s
(c) 0.0500 M/s
(d) 0.250 M/s

Reaction Order for Multiple Reactants

So far, we have considered a simple reaction with only one reactant. How is the rate law defined for reactions with more than one reactant? Consider the generic reaction:

$$a\text{A} + b\text{B} \longrightarrow c\text{C} + d\text{D}$$

As long as the reverse reaction is negligibly slow, the rate law is proportional to the concentration of A raised to the m multiplied by the concentration of B raised to the n:

$$\text{Rate} = k[A]^m[B]^n \qquad [13.10]$$

where m is the reaction order with respect to A and n is the reaction order with respect to B. The **overall order** is the sum of the exponents $(m + n)$. For example, the reaction between hydrogen and iodine has been experimentally determined to be first order with respect to hydrogen, first order with respect to iodine, and thus second order overall.

$$H_2(g) + I_2(g) \longrightarrow 2\,HI(g) \quad Rate = k[H_2]^1[I_2]^1$$

Similarly, the reaction between hydrogen and nitrogen monoxide has been experimentally determined to be first order with respect to hydrogen, second order with respect to nitrogen monoxide, and thus third order overall.

$$2\,H_2(g) + 2\,NO(g) \longrightarrow N_2(g) + 2\,H_2O(g) \quad Rate = k[H_2]^1[NO]^2$$

The rate law for any reaction must always be determined by experiment, often by the method of initial rates described previously. There is no way merely to look at a chemical equation and determine the rate law for the reaction. When the reaction has two or more reactants, the concentration of each reactant is usually varied independently of the others to determine the dependence of the rate on the concentration of that reactant. Example 13.2 uses the method of initial rates for determining the order of a reaction with multiple reactants.

EXAMPLE 13.2 Determining the Order and Rate Constant of a Reaction

Consider the reaction between nitrogen dioxide and carbon monoxide:

$$NO_2(g) + CO(g) \longrightarrow NO(g) + CO_2(g)$$

The initial rate of the reaction was measured at several different concentrations of the reactants with the following results:

$[NO_2]$ (M)	$[CO]$ (M)	Initial Rate (M/s)
0.10	0.10	0.0021
0.20	0.10	0.0082
0.20	0.20	0.0083
0.40	0.10	0.033

From the data, determine:

(a) the rate law for the reaction   **(b)** the rate constant (k) for the reaction

SOLUTION

(a) Begin by examining how the rate changes for each change in concentration. Between the first two experiments, the concentration of NO_2 doubled, the concentration of CO stayed constant, and the rate quadrupled, suggesting that the reaction is second order in NO_2.

Between the second and third experiments, the concentration of NO_2 stayed constant, the concentration of CO doubled, and the rate remained constant (the small change in the least significant figure is simply experimental error), suggesting that the reaction is zero order in CO.

Between the third and fourth experiments, the concentration of NO_2 again doubled, the concentration of CO halved, yet the rate quadrupled again, confirming that the reaction is second order in NO_2 and zero order in CO.

Write the overall rate expression.

$[NO_2]$	$[CO]$	Initial Rate (M/s)
0.10 M	0.10 M	0.0021
↓ × 2	↓ constant	↓ × 4
0.20 M	0.10 M	0.0082 M
↓ constant	↓ × 2	↓ × 1
0.20 M	0.20 M	0.0083 M
↓ × 2	↓ × ½	↓ × 4
0.40 M	0.10 M	0.033 M

$$Rate = k[NO_2]^2[CO]^0 = k[NO_2]^2$$

(b) To determine the rate constant for the reaction, solve the rate law for k and substitute the concentration and the initial rate from any one of the four measurements. In this case, we use the first measurement.

$$Rate = k[NO_2]^2$$

$$k = \frac{rate}{[NO_2]^2} = \frac{0.0021\ M/s}{(0.10\ M)^2} = 0.21\ M^{-1} \cdot s^{-1}$$

FOR PRACTICE 13.2

Consider the following reaction:

$$CHCl_3(g) + Cl_2(g) \longrightarrow CCl_4(g) + HCl(g)$$

The initial rate of the reaction was measured at several different concentrations of the reactants with the following results:

$[CHCl_3]$ (M)	$[Cl_2]$ (M)	Initial Rate (M/s)
0.010	0.010	0.0035
0.020	0.010	0.0069
0.020	0.020	0.0098
0.040	0.040	0.027

From the data, determine:

(a) the rate law for the reaction   **(b)** the rate constant (k) for the reaction

---

 Conceptual Connection 13.3 **Rate and Concentration**

This reaction was experimentally determined to be first order with respect to O_2 and second order with respect to NO.

$$O_2(g) + 2\,NO(g) \longrightarrow 2\,NO_2(g)$$

These diagrams represent reaction mixtures in which the number of each type of molecule represents its relative initial concentration. Which mixture will have the fastest initial rate?

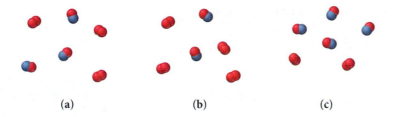

(a)   (b)   (c)

13.4 The Integrated Rate Law: The Dependence of Concentration on Time

The rate laws we have examined so far show the relationship between *the rate of a reaction and the concentration of a reactant*. However, we often want to know the relationship between *the concentration of a reactant and time*. For example, the presence of chlorofluorocarbons (CFCs) in the atmosphere threatens the ozone layer. One of the reasons that CFCs pose such a significant threat is that the reactions that consume them are so slow. Legislation has resulted in reduced CFC emissions, but even if we were to completely stop adding CFCs to the atmosphere, their concentration in the atmosphere will decrease only very slowly. Nonetheless, we would like to be able to predict how their concentration changes with time. How much will be left in 20 years? In 50 years?

The **integrated rate law** for a chemical reaction is a relationship between the concentrations of the reactants and time. For simplicity, we return to a single reactant decomposing into products:

$$A \longrightarrow products$$

The integrated rate law for this reaction depends on the order of the reaction; let's examine each of the common reaction orders individually.

First-Order Integrated Rate Law If our simple reaction is first order, the rate is equal to the rate constant multiplied by the concentration of A:

$$\text{Rate} = k[A]$$

Since $\text{Rate} = -\Delta[A]/\Delta t$, we can write

$$-\frac{\Delta[A]}{\Delta t} = k[A] \qquad [13.11]$$

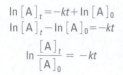

In this form, the rate law is also known as the *differential rate law*.

Although we do not show the steps here, we can use calculus (see Exercise 13.84) to integrate the differential rate law to obtain the first-order *integrated rate law*:

$$\ln[A]_t = -kt + \ln[A]_0 \qquad [13.12]$$

or

$$\ln\frac{[A]_t}{[A]_0} = -kt \qquad [13.13]$$

where $[A]_t$ is the concentration of A at any time t, k is the rate constant, and $[A]_0$ is the initial concentration of A. The two forms of the equation shown here are equivalent, as shown in the margin.

Notice that the integrated rate law shown in Equation 13.12 has the form of an equation for a straight line.

$$\ln[A]_t = -kt + \ln[A]_0$$
$$y = mx + b$$

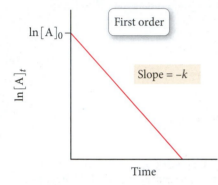

▲ **FIGURE 13.5 First-Order Integrated Rate Law**   For a first-order reaction, a plot of the natural log of the reactant concentration as a function of time yields a straight line. The slope of the line is equal to $-k$ and the y-intercept is $\ln[A]_0$.

Therefore, for a first-order reaction, a plot of the natural log of the reactant concentration as a function of time yields a straight line with a slope of $-k$ and a y-intercept of $\ln[A]_0$, as shown in **Figure 13.5◄**. (Note that the slope is negative but that the rate constant is always positive.)

EXAMPLE 13.3 The First-Order Integrated Rate Law: Using Graphical Analysis of Reaction Data

Consider the equation for the decomposition of SO_2Cl_2.

$$SO_2Cl_2(g) \longrightarrow SO_2(g) + Cl_2(g)$$

The concentration of SO_2Cl_2 is monitored at a fixed temperature as a function of time during the decomposition reaction and the following data is tabulated:

Time (s)	$[SO_2Cl_2]$ (M)	Time (s)	$[SO_2Cl_2]$ (M)
0	0.100	800	0.0793
100	0.0971	900	0.0770
200	0.0944	1000	0.0748
300	0.0917	1100	0.0727
400	0.0890	1200	0.0706
500	0.0865	1300	0.0686
600	0.0840	1400	0.0666
700	0.0816	1500	0.0647

Show that the reaction is first order and determine the rate constant for the reaction.

SOLUTION

In order to show that the reaction is first order, prepare a graph of $\ln[SO_2Cl_2]$ versus time as shown here.

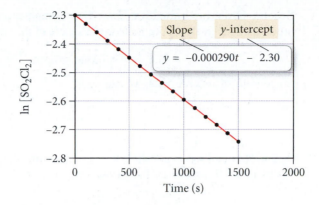

The plot is linear, confirming that the reaction is indeed first order. To obtain the rate constant, fit the data to a line. The slope of the line will be equal to $-k$. Since the slope of the best fitting line (which is most easily determined on a graphing calculator or with spreadsheet software such as Microsoft Excel) is $-2.90 \times 10^{-4} \, s^{-1}$, the rate constant is therefore $+2.90 \times 10^{-4} \, s^{-1}$.

FOR PRACTICE 13.3

Use the graph and the best fitting line in the previous example to predict the concentration of SO_2Cl_2 at 1900 s.

EXAMPLE 13.4 The First-Order Integrated Rate Law: Determining the Concentration of a Reactant at a Given Time

In Example 13.3, we learned that the decomposition of SO_2Cl_2 (under the given reaction conditions) is first order and has a rate constant of $+2.90 \times 10^{-4} \, s^{-1}$. If the reaction is carried out at the same temperature, and the initial concentration of SO_2Cl_2 is 0.0225 M, what is the SO_2Cl_2 concentration after 865 s?

SORT You are given the rate constant of a first-order reaction and the initial concentration of the reactant. You are asked to find the concentration at 865 seconds.	**GIVEN** $k = +2.90 \times 10^{-4} \, s^{-1}$ $[SO_2Cl_2]_0 = 0.0225 \, M$ **FIND** $[SO_2Cl_2]$ at $t = 865 \, s$
STRATEGIZE Use the first-order integrated rate law to determine the information you are asked to find from the given information.	**EQUATION** $\ln[A]_t = -kt + \ln[A]_0$
SOLVE Substitute the rate constant, the initial concentration, and the time into the integrated rate law. Solve the integrated rate law for the concentration of $[SOCl_2]_t$.	**SOLUTION** $\ln[SO_2Cl_2]_t = -kt + \ln[SO_2Cl_2]_0$ $\ln[SO_2Cl_2]_t = -(2.90 \times 10^{-4} s^{-1})865 \, s + \ln(0.0225)$ $\ln[SO_2Cl_2]_t = -0.251 - 3.79$ $[SO_2Cl_2]_t = e^{-4.04}$ $\quad\quad\quad\quad = 0.0175 \, M$

CHECK

The concentration is smaller than the original concentration as expected. If the concentration were larger than the initial concentration, this would indicate a mistake in the signs of one of the quantities on the right-hand side of the equation.

FOR PRACTICE 13.4

Cyclopropane rearranges to form propene in the gas phase according to the reaction:

$$\underset{\overset{|}{H_2C-CH_2}}{\overset{CH_2}{\diagup\diagdown}} \longrightarrow CH_3-CH=CH_2$$

The reaction is first order in cyclopropane and has a measured rate constant of $3.36 \times 10^{-5} \, s^{-1}$ at 720 K. If the initial cyclopropane concentration is 0.0445 M, what is the cyclopropane concentration after 235.0 minutes?

Second-Order Integrated Rate Law If our simple reaction (A $\longrightarrow$ products) is second order, the rate law is proportional to the square of the concentration of A:

$$\text{Rate} = k[A]^2$$

Since Rate $= -\Delta[A]/\Delta t$, we can write:

$$-\frac{\Delta[A]}{\Delta t} = k[A]^2 \qquad [13.14]$$

Again, although we do not show the steps here, this differential rate law can be integrated to obtain the *second-order integrated rate law*:

$$\frac{1}{[A]_t} = kt + \frac{1}{[A]_0} \qquad [13.15]$$

The second-order integrated rate law is also in the form of an equation for a straight line.

$$\frac{1}{[A]_t} = kt + \frac{1}{[A]_0}$$

$$y = mx + b$$

Notice, however, that you must now plot the inverse of the concentration of the reactant as a function of time. The plot yields a straight line with a slope of k and an intercept of $1/[A]_0$ as shown in **Figure 13.6◀**.

Zero-Order Integrated Rate Law If our simple reaction is zero order, the rate law is defined as:

$$\text{Rate} = k[A]^0$$

Since Rate $= -\Delta[A]/\Delta t$, we can write:

$$-\frac{\Delta[A]}{\Delta t} = k \qquad [13.16]$$

This differential rate law can be integrated to obtain the *zero-order integrated rate law*:

$$[A]_t = -kt + [A]_0 \qquad [13.17]$$

The zero-order integrated rate law shown in Equation 13.17 is also in the form of an equation for a straight line. A plot of the concentration of the reactant as a function of time yields a straight line with a slope of $-k$ and an intercept of $[A]_0$, as shown in **Figure 13.7◀**.

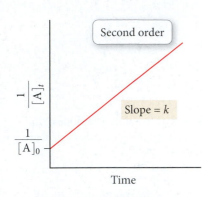

▲ **FIGURE 13.6 Second-Order Integrated Rate Law** For a second-order reaction, a plot of the inverse of the reactant concentration as a function of time yields a straight line. The slope of the line is equal to k and the y-intercept is $1/[A]_0$.

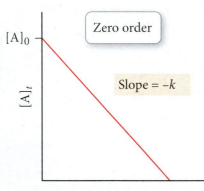

▲ **FIGURE 13.7 Zero-Order Integrated Rate Law** For a zero-order reaction, a plot of the reactant concentration as a function of time yields a straight line. The slope of the line is equal to $-k$ and the y-intercept is $[A]_0$.

EXAMPLE 13.5 The Second-Order Integrated Rate Law: Using Graphical Analysis of Reaction Data

Consider the equation for the decomposition of NO_2.

$$NO_2(g) \longrightarrow NO(g) + O(g)$$

The concentration of NO_2 is monitored at a fixed temperature as a function of time during the decomposition reaction and the data tabulated in the margin at the top right of the next page. Show by graphical analysis that the reaction is not first order and that it is second order. Determine the rate constant for the reaction.

SOLUTION

In order to show that the reaction is *not* first order, prepare a graph of $\ln[NO_2]$ versus time.

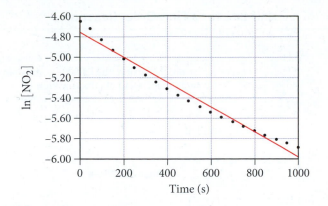

Time (s)	$[NO_2]$ (M)
0	0.01000
50	0.00887
100	0.00797
150	0.00723
200	0.00662
250	0.00611
300	0.00567
350	0.00528
400	0.00495
450	0.00466
500	0.00440
550	0.00416
600	0.00395
650	0.00376
700	0.00359
750	0.00343
800	0.00329
850	0.00316
900	0.00303
950	0.00292
1000	0.00282

The plot is *not* linear (the straight line does not fit the data points), confirming that the reaction is not first order. In order to show that the reaction is second order, prepare a graph of $1/[NO_2]$ versus time.

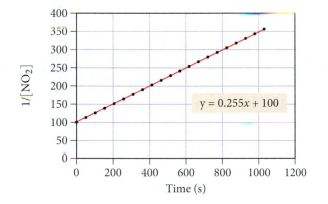

This graph is linear (the data points fit well to a straight line), confirming that the reaction is indeed second order. To obtain the rate constant, determine the slope of the best fitting line. The slope is $0.255 \text{ M}^{-1} \cdot \text{s}^{-1}$; therefore, the rate constant is $0.255 \text{ M}^{-1} \cdot \text{s}^{-1}$.

FOR PRACTICE 13.5

Use the graph and the best fitting line in Example 13.5 to predict the concentration of NO_2 at 2000 s.

The Half-Life of a Reaction

The **half-life** ($t_{1/2}$) of a reaction is the time required for the concentration of a reactant to fall to one-half of its initial value. For example, if a reaction has a half-life of 100 seconds, and if the initial concentration of the reactant is 1.0 M, then the concentration will fall to 0.50 M in 100 s. The half-life expression—which defines the dependence of half-life on the rate constant and the initial concentration—is different for different reaction orders.

First-Order Reaction Half-Life From the definition of half-life, and from the integrated rate law, we can derive an expression for the half-life. For a first-order reaction, the integrated rate law is

$$\ln \frac{[A]_t}{[A]_0} = -kt$$

At a time equal to the half-life ($t = t_{1/2}$), the concentration is exactly half of the initial concentration: ($[A]_t = \frac{1}{2}[A]_0$). Therefore, when $t = t_{1/2}$ we can write the expression:

$$\ln \frac{\frac{1}{2}[A]_0}{[A]_0} = \ln\frac{1}{2} = -kt_{1/2} \qquad [13.18]$$

▶ **FIGURE 13.8 Half-Life: Concentration versus Time for a First-Order Reaction** For this reaction, the concentration falls by one-half every 100 seconds ($t_{1/2} = 100$ s). The blue spheres represent reactant molecules (the products are omitted for clarity).

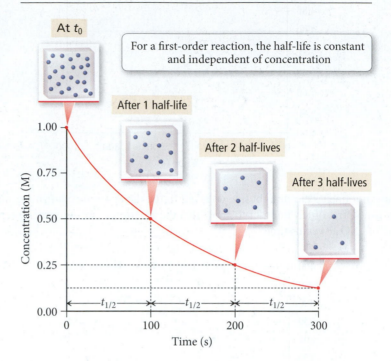

Half-Life for a First-Order Reaction

At t_0

For a first-order reaction, the half-life is constant and independent of concentration

After 1 half-life

After 2 half-lives

After 3 half-lives

Concentration (M) vs Time (s)

Solving for $t_{1/2}$, and substituting -0.693 for $\ln\frac{1}{2}$, we arrive at the following expression for the half-life of a first-order reaction:

$$t_{1/2} = \frac{0.693}{k} \qquad [13.19]$$

Notice that, for a first-order reaction, $t_{1/2}$ *is independent of the initial concentration.* For example, if $t_{1/2}$ is 100 s, and if the initial concentration is 1.0 M, the concentration falls to 0.50 M in 100 s, then to 0.25 M in another 100 s, then to 0.125 M in another 100 s, and so on (**Figure 13.8▲**). Even though the concentration is changing as the reaction proceeds, the half-life (how long it takes for the concentration to halve) is constant. A constant half-life is unique to first-order reactions, making the concept of half-life particularly useful for first-order reactions.

EXAMPLE 13.6 Half-Life

Molecular iodine dissociates at 625 K with a first-order rate constant of 0.271 s^{-1}. What is the half-life of this reaction?

SOLUTION

Since the reaction is first order, the half-life is given by Equation 13.19. Substitute the value of k into the expression and calculate $t_{1/2}$.	$t_{1/2} = \dfrac{0.693}{k}$ $= \dfrac{0.693}{0.271/\text{s}} = 2.56$ s

FOR PRACTICE 13.6

A first-order reaction has a half-life of 26.4 seconds. How long will it take for the concentration of the reactant in the reaction to fall to one-eighth of its initial value?

Second-Order Reaction Half-Life For a second-order reaction, the integrated rate law is

$$\frac{1}{[A]_t} = kt + \frac{1}{[A]_0}$$

At a time equal to the half-life ($t = t_{1/2}$), the concentration is exactly one-half of the initial concentration ($[A]_t = \frac{1}{2}[A]_0$). We can therefore write the following expression at $t = t_{1/2}$:

$$\frac{1}{\frac{1}{2}[A]_0} = kt_{1/2} + \frac{1}{[A]_0} \qquad [13.20]$$

We can then solve for $t_{1/2}$:

$$kt_{1/2} = \frac{1}{\frac{1}{2}[A]_0} - \frac{1}{[A]_0}$$

$$kt_{1/2} = \frac{2}{[A]_0} - \frac{1}{[A]_0}$$

$$t_{1/2} = \frac{1}{k[A]_0} \qquad [13.21]$$

Notice that, for a second-order reaction, the half-life depends on the initial concentration. So if the initial concentration of a reactant in a second-order reaction is 1.0 M, and the half-life is 100 s, the concentration falls to 0.50 M in 100 s. However, the time it takes for the concentration to fall to 0.25 M is now *longer than 100 s*, because the initial concentration has decreased. The half-life continues to get longer as the concentration decreases.

Zero-Order Reaction Half-Life For a zero-order reaction, the integrated rate law is

$$[A]_t = -kt + [A]_0$$

Making the substitutions ($t = t_{1/2}$; $[A]_t = \frac{1}{2}[A]_0$). We then write the following expression at $t = t_{1/2}$:

$$\frac{1}{2}[A]_0 = -kt_{1/2} + [A]_0 \qquad [13.22]$$

We then solve for $t_{1/2}$:

$$t_{1/2} = \frac{[A]_0}{2k} \qquad [13.23]$$

Notice that, for a zero-order reaction, the half-life also depends on the initial concentration but here the half-life gets shorter as the concentration decreases.

Summarizing the Integrated Rate Law (see Table 13.1):

▶ The reaction order and rate law must be determined experimentally.

▶ The differential rate law relates the *rate* of the reaction to the *concentration* of the reactant(s).

▶ The integrated rate law (which is mathematically derived from the differential rate law) relates the *concentration* of the reactant(s) to *time*.

▶ The half-life is the time it takes for the concentration of a reactant to fall to one-half of its initial value.

▶ The half-life of a first-order reaction is independent of the initial concentration.

▶ The half-lives of zero-order and second-order reactions depend on the initial concentration.

Conceptual Connection 13.4 Rate Law and Integrated Rate Law

A decomposition reaction, with a rate that is observed to slow down as the reaction proceeds, is determined to have a half-life that depends on the initial concentration of the reactant. Which statement is most likely to be true of this reaction?

(a) A plot of the natural log of the concentration of the reactant as a function of time will be linear.

(b) The half-life of the reaction increases as the initial concentration increases.

(c) A doubling of the initial concentration of the reactant results in a quadrupling of the rate.

TABLE 13.1 Rate Law Summary Table

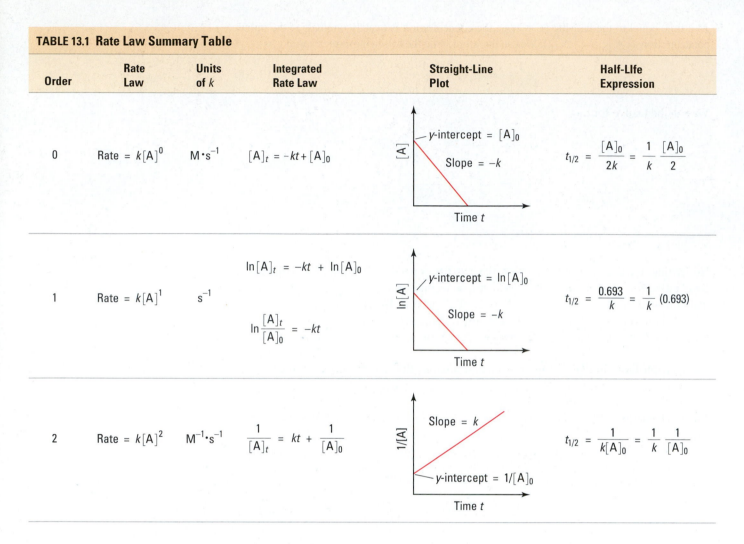

Order	Rate Law	Units of k	Integrated Rate Law	Straight-Line Plot	Half-Life Expression
0	Rate $= k[A]^0$	$M \cdot s^{-1}$	$[A]_t = -kt + [A]_0$	y-intercept $= [A]_0$ Slope $= -k$	$t_{1/2} = \dfrac{[A]_0}{2k} = \dfrac{1}{k}\dfrac{[A]_0}{2}$
1	Rate $= k[A]^1$	s^{-1}	$\ln[A]_t = -kt + \ln[A]_0$ $\ln\dfrac{[A]_t}{[A]_0} = -kt$	y-intercept $= \ln[A]_0$ Slope $= -k$	$t_{1/2} = \dfrac{0.693}{k} = \dfrac{1}{k}(0.693)$
2	Rate $= k[A]^2$	$M^{-1} \cdot s^{-1}$	$\dfrac{1}{[A]_t} = kt + \dfrac{1}{[A]_0}$	Slope $= k$ y-intercept $= 1/[A]_0$	$t_{1/2} = \dfrac{1}{k[A]_0} = \dfrac{1}{k}\dfrac{1}{[A]_0}$

13.5 The Effect of Temperature on Reaction Rate

In the opening section of this chapter, we saw that lizards become lethargic when their body temperature drops because the chemical reactions that control their muscle movement slow down at lower temperatures. The rates of chemical reactions are, in general, highly sensitive to temperature. For example, at room temperature, a 10 °C increase in temperature increases the rate of a typical biological reaction by two or three times. How do we explain this highly sensitive temperature dependence?

Recall that the rate law for a reaction is Rate $= k[A]^n$. *The temperature dependence of the reaction rate is contained in the rate constant, k* (which is actually a constant only when the temperature remains constant). An increase in temperature generally results in an increase in k, which results in a faster rate. In 1889, Swedish chemist Svante Arrhenius wrote a paper quantifying the temperature dependence of the rate constant. The modern form of the **Arrhenius equation,** which relates the rate constant (k) and the temperature in kelvins (T), is:

$$k = Ae^{\frac{-E_a}{RT}}$$

Activation energy

Frequency factor   Exponential factor

[13.24]

where R is the gas constant (8.314 J/mol $\cdot$ K), A is a constant called the *frequency factor* (or the *pre-exponential factor*), and E_a is called the *activation energy* (or *activation barrier*).

Activation Energy

$$2 H_2(g) + O_2(g) \rightleftharpoons 2 H_2O(g)$$

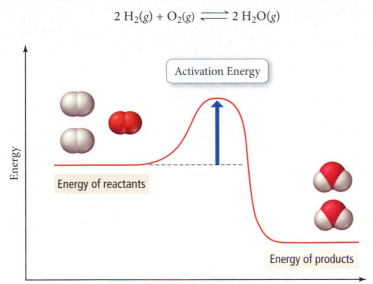

◄ **FIGURE 13.9 The Activation Energy Barrier** Even though the reaction is energetically favorable (the energy of the products is lower than that of the reactants), an input of energy is needed for the reaction to take place.

The **activation energy** (E_a) is an energy barrier or hump that must be surmounted for the reactants to be transformed into products (**Figure 13.9▲**). We will examine the frequency factor more closely in the next section; for now, we can think of the **frequency factor (A)** as the number of times that the reactants approach the activation barrier per unit time.

To understand each of these quantities better, consider the simple reaction in which CH₃NC (methyl isonitrile) rearranges to form CH₃CN (acetonitrile):

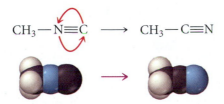

Let's examine the physical meaning of the activation energy, frequency factor, and exponential factor for this reaction.

The Activation Energy **Figure 13.10▼** shows the energy of the molecule as the reaction proceeds. The x-axis represents the progress of the reaction from left (reactant) to right (product). To get from the reactant to the product, the molecule must go through a high-energy intermediate state called the **activated complex, or transition state.** Even though

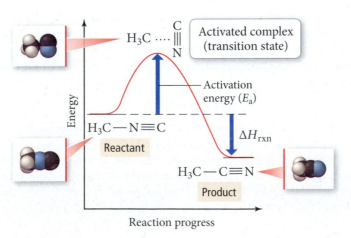

◄ **FIGURE 13.10 The Activated Complex** The reaction pathway includes a transitional state—the activated complex—that has a higher energy than either the reactant or the product.

the overall reaction is energetically downhill (exothermic), it must first go uphill to reach the activated complex because energy is required to initially weaken the H_3C—N bond and allow the NC group to begin to rotate:

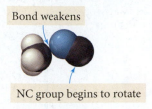

Bond weakens

NC group begins to rotate

The energy required to create the activated complex is the *activation energy. The higher the activation energy, the slower the reaction rate (at a given temperature).*

The Frequency Factor   Recall that the frequency factor represents the number of approaches to the activation barrier per unit time. Any time the NC group begins to rotate, it approaches the activation barrier. For this reaction, the frequency factor represents the rate at which the NC part of the molecule wags (or vibrates side to side). With each wag, the reactant approaches the activation barrier. However, approaching the activation barrier is not equivalent to surmounting it. Most of the approaches do not have enough total energy to make it over the activation barrier.

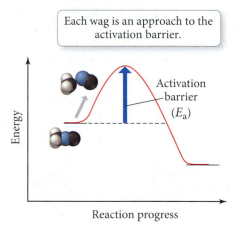

Each wag is an approach to the activation barrier.

Activation barrier (E_a)

Energy

Reaction progress

The Exponential Factor   The **exponential factor** is a number between 0 and 1 that represents the fraction of molecules that have enough energy to make it over the activation barrier on a given approach. In other words, the exponential factor is the fraction of approaches that are actually successful and result in the product. For example, if the frequency factor is 10^9/s and the exponential factor is 10^{-7} at a certain temperature, then the overall rate constant at that temperature is 10^9/s $\times$ $10^{-7} = 10^2$/s. In this case, the CN group is "wagging" at a rate of 10^9/s. With each wag, the activation barrier is approached. However, only 1 in 10^7 molecules have sufficient energy to actually make it over the activation barrier for a given wag.

The exponential factor depends on both the temperature (T) and the activation energy (E_a) of the reaction.

$$\text{Exponential factor} = e^{-E_a/RT}$$

As the temperature increases, the number of molecules having enough thermal energy to surmount the activation barrier increases. At any given temperature, a sample of molecules will have a distribution of energies, as shown in **Figure 13.11▶**. Under common circumstances, only a small fraction of the molecules have enough energy to make it over the activation barrier. Because of the shape of the energy distribution curve, however, a small change in temperature results in a large difference in the number of

Thermal Energy Distribution

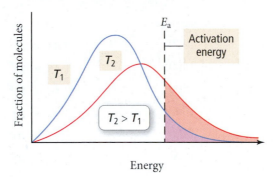

As temperature increases, the fraction of molecules with enough energy to surmount the activation energy barrier also increases.

▲ **FIGURE 13.11 Thermal Energy Distribution**   At any given temperature, the atoms or molecules in a gas sample will have a range of energies. The higher the temperature, the wider the energy distribution and the greater the average energy. The fraction of molecules with enough energy to surmount the activation energy barrier and react increases sharply as the temperature rises.

molecules having enough energy to surmount the activation barrier. This explains the sensitivity of reaction rates to temperature.

Summarizing Temperature and Reaction Rate:

▶ The frequency factor is the number of times that the reactants approach the activation barrier per unit time.

▶ The exponential factor is the fraction of approaches that are successful in surmounting the activation barrier and forming products.

▶ The exponential factor increases with increasing temperature, but decreases with an increasing value for the activation energy.

Arrhenius Plots: Experimental Measurements of the Frequency Factor and the Activation Energy

The frequency factor and activation energy are important quantities in understanding the kinetics of any reaction. To see how we can measure these factors in the laboratory, consider again Equation 13.24: $k = Ae^{-E_a/RT}$. Taking the natural log of both sides of this equation, we get:

$$\ln k = \ln\left(Ae^{-E_a/RT}\right) \qquad [13.25]$$

| Remember that $\ln(AB) = \ln A + \ln B$

$$\ln k = \ln A + \ln e^{-E_a/RT}$$

$$\ln k = \ln A - \frac{E_a}{RT}$$

| Remember that $\ln e^x = x$

$$\ln k = -\frac{E_a}{R}\left(\frac{1}{T}\right) + \ln A \qquad [13.26]$$

$$y = mx + b$$

Equation 13.26 is in the form of a straight line. *A plot of the natural log of the rate constant (ln k) versus the inverse of the temperature in kelvins (1/T) yields a straight line with a slope of $-E_a/R$ and a y-intercept of ln A. Such a plot is called an* **Arrhenius plot** and is commonly used in the analysis of kinetic data, as shown in Example 13.7.

| In an Arrhenius analysis, the pre-exponential factor (A) is assumed to be independent of temperature. Although the pre-exponential factor does depend on temperature to some degree, its temperature dependence is much less than that of the exponential factor and is often ignored.

EXAMPLE 13.7 Using an Arrhenius Plot to Determine Kinetic Parameters

The decomposition of ozone is important to many atmospheric reactions:

$$O_3(g) \longrightarrow O_2(g) + O(g)$$

A study of the kinetics of the reaction resulted in the following data:

Temperature (K)	Rate Constant ($M^{-1} \cdot s^{-1}$)	Temperature (K)	Rate Constant ($M^{-1} \cdot s^{-1}$)
600	3.37×10^3	1300	7.83×10^7
700	4.85×10^4	1400	1.45×10^8
800	3.58×10^5	1500	2.46×10^8
900	1.70×10^6	1600	3.93×10^8
1000	5.90×10^6	1700	5.93×10^8
1100	1.63×10^7	1800	8.55×10^8
1200	3.81×10^7	1900	1.19×10^9

Determine the value of the frequency factor and activation energy for the reaction.

SOLUTION

To find the frequency factor and activation energy, prepare a graph of the natural log of the rate constant ($\ln k$) versus the inverse of the temperature ($1/T$).

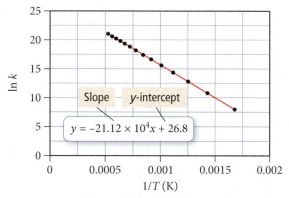

The plot is linear, as expected for Arrhenius behavior. The best fitting line has a slope of -1.12×10^4 K and a y-intercept of 26.8. Calculate the activation energy from the slope by setting the slope equal to $-E_a/R$ and solving for E_a:

$$-1.12 \times 10^4 \text{ K} = \frac{-E_a}{R}$$

$$E_a = 1.12 \times 10^4 \text{ K} \left(8.314 \frac{\text{J}}{\text{mol} \cdot \text{K}} \right)$$

$$= 9.31 \times 10^4 \text{ J/mol}$$

$$= 93.1 \text{ kJ/mol}$$

Calculate the frequency factor (A) by setting the intercept equal to $\ln A$.

$$26.8 = \ln A$$

$$A = e^{26.8}$$

$$= 4.36 \times 10^{11}$$

Since the rate constants were measured in units of $M^{-1} \cdot s^{-1}$, the frequency factor is in the same units. Consequently, we can conclude that the reaction has an activation energy of 93.1 kJ/mol and a frequency factor of 4.36×10^{11} $M^{-1} \cdot s^{-1}$.

FOR PRACTICE 13.7

For the decomposition of ozone reaction in Example 13.7, use the results of the Arrhenius analysis to predict the rate constant at 298 K.

In some cases, where either data are limited or plotting capabilities are absent, the activation energy can be calculated knowing the rate constant at just two different temperatures. The Arrhenius expression in Equation 13.25 is applied to the two different temperatures as follows:

$$\ln k_2 = -\frac{E_a}{R}\left(\frac{1}{T_2}\right) + \ln A \qquad \ln k_1 = -\frac{E_a}{R}\left(\frac{1}{T_1}\right) + \ln A$$

We can then subtract $\ln k_1$ from $\ln k_2$ as follows:

$$\ln k_2 - \ln k_1 = \left[-\frac{E_a}{R}\left(\frac{1}{T_2}\right) + \ln A \right] - \left[-\frac{E_a}{R}\left(\frac{1}{T_1}\right) + \ln A \right]$$

Rearranging, we get the two-point form of the Arrhenius equation:

$$\ln \frac{k_2}{k_1} = \frac{E_a}{R}\left(\frac{1}{T_1} - \frac{1}{T_2}\right) \qquad\qquad \text{[13.27]}$$

Example 13.8 shows how to use this equation to calculate the activation energy from experimental measurements of the rate constant at two different temperatures.

EXAMPLE 13.8   Using the Two-Point Form of the Arrhenius Equation

The reaction between nitrogen dioxide and carbon monoxide is:

$$NO_2(g) + CO(g) \longrightarrow NO(g) + CO_2(g)$$

The rate constant at 701 K is measured as $2.57 \ M^{-1} \cdot s^{-1}$ and that at 895 K is measured as $567 \ M^{-1} \cdot s^{-1}$. Find the activation energy for the reaction in kJ/mol.

SORT You are given the rate constant of a reaction at two different temperatures. You are asked to find the activation energy.	**GIVEN** $T_1 = 701 \ K, k_1 = 2.57 \ M^{-1} \cdot s^{-1}$ $T_2 = 895 \ K, k_2 = 567 \ M^{-1} \cdot s^{-1}$ **FIND** E_a
STRATEGIZE Use the two-point form of the Arrhenius equation, which relates the activation energy to the given information and R (a constant).	**EQUATION** $\ln \dfrac{k_2}{k_1} = \dfrac{E_a}{R}\left(\dfrac{1}{T_1} - \dfrac{1}{T_2}\right)$
SOLVE Substitute the two rate constants and the two temperatures into the equation.	**SOLUTION** $\ln \dfrac{567 \ M^{-1} \cdot s^{-1}}{2.57 \ M^{-1} \cdot s^{-1}} = \dfrac{E_a}{R}\left(\dfrac{1}{701 \ K} - \dfrac{1}{895 \ K}\right)$ $5.40 = \dfrac{E_a}{R}\left(\dfrac{3.09 \times 10^{-4}}{K}\right)$
Solve the equation for E_a, the activation energy, and convert to kJ/mol.	$E_a = 5.40\left(\dfrac{K}{3.09 \times 10^{-4}}\right)R$ $= 5.40\left(\dfrac{K}{3.09 \times 10^{-4}}\right)8.314 \dfrac{J}{mol \cdot K}$ $= 1.5 \times 10^5 \ J/mol$ $= 1.5 \times 10^2 \ kJ/mol$

CHECK The magnitude of the answer is reasonable. Activation energies for most reactions range from tens to hundreds of kilojoules per mole.

FOR PRACTICE 13.8

Use the results from Example 13.8 and the given rate constant of the reaction at either of the two temperatures to predict the rate constant at 525 K.

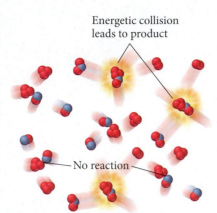

▲ **FIGURE 13.12 The Collision Model** In the collision model, two molecules react after a sufficiently energetic collision with the correct orientation to bring the reacting groups together.

Conceptual Connection 13.5 Temperature Dependence of Reaction Rate

Reaction A and reaction B have identical frequency factors. However, reaction B has a higher activation energy than reaction A. Which reaction has a faster rate at room temperature?

The Collision Model: A Closer Look at the Frequency Factor

We saw previously that the frequency factor in the Arrhenius equation represents the number of approaches to the activation barrier per unit time. We now refine that idea for a reaction involving two gas-phase reactants:

$$A(g) + B(g) \longrightarrow \text{products}$$

In the **collision model,** a chemical reaction occurs after a sufficiently energetic collision between two reactant molecules (**Figure 13.12◄**). In collision theory, therefore, each approach to the activation barrier is a collision between the reactant molecules. Consequently, the value of the frequency factor should simply be the number of collisions that occur per second. However, the frequency factors of most (though not all) gas-phase chemical reactions tend to be smaller than the number of collisions that occur per second. Why?

In the collision model, we can separate the frequency factor into two distinct parts, as shown in the following equations:

$$k = Ae^{\frac{-E_a}{RT}}$$
$$= pze^{\frac{-E_a}{RT}}$$

Orientation factor   Collision frequency

where p is the **orientation factor** and z is the **collision frequency.** The collision frequency is the number of collisions that occur per unit time, which can be calculated for a gas-phase reaction from the pressure of the gases and the temperature of the reaction mixture. Under typical conditions, a single molecule undergoes on the order of 10^9 collisions every second. To understand the orientation factor consider the following reaction:

$$NOCl(g) + NOCl(g) \longrightarrow 2\,NO(g) + Cl_2(g)$$

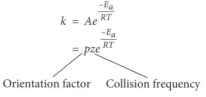

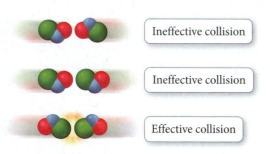

Ineffective collision

Ineffective collision

Effective collision

In order for the reaction to occur, two NOCl molecules must collide with sufficient energy. However, not all collisions with sufficient energy will lead to products because the reactant molecules must be properly oriented. Consider each of the possible orientations of the reactant molecules (shown at left) during a collision. The first two collisions, even if they occur with sufficient energy, will not result in a reaction because the reactant molecules are not oriented in a way that allows the chlorine atoms to bond. If two molecules are to react with each other, they must collide in such a way that allows the necessary bonds to break and form. For the reaction of NOCl(g), the orientation factor is $p = 0.16$. This means that only 16 out of 100 sufficiently energetic collisions are actually successful in forming the products.

Conceptual Connection 13.6 Collision Theory

Which reaction would you expect to have the smallest orientation factor?

(a) $H(g) + I(g) \longrightarrow HI(g)$

(b) $H_2(g) + I_2(g) \longrightarrow 2\,HI(g)$

(c) $HCl(g) + HCl(g) \longrightarrow H_2(g) + Cl_2(g)$

13.6 Reaction Mechanisms

Most chemical reactions do not occur in a single step, but through several steps. When we write a chemical equation to represent a chemical reaction, *we usually represent the overall reaction, not the series of individual steps by which the reaction occurs.* Consider the reaction in which hydrogen gas reacts with iodine monochloride:

$$H_2(g) + 2\,ICl(g) \longrightarrow 2\,HCl(g) + I_2(g)$$

The overall equation simply shows the substances present at the beginning of the reaction and the substances formed by the reaction—it does not show the intermediate steps that may take place. A **reaction mechanism** is a series of individual chemical steps by which an overall chemical reaction occurs. For example, the reaction between hydrogen and iodine monochloride occurs through the following proposed mechanism.

Step 1	$H_2(g) + ICl(g) \longrightarrow HI(g) + HCl(g)$
Step 2	$HI(g) + ICl(g) \longrightarrow HCl(g) + I_2(g)$

In the first step, an H_2 molecule collides with an ICl molecule and forms an HI molecule and an HCl molecule. In the second step, the HI molecule formed in the first step collides with a second ICl molecule to form another HCl molecule and an I_2 molecule. Each step in a reaction mechanism is an **elementary step.** Elementary steps cannot be broken down into simpler steps—they occur as they are written.

> An elementary step represents an actual interaction between the reactant molecules in the step. An overall reaction equation shows only the starting substances and the ending substances, not the path between them.

One of the requirements for a valid reaction mechanism is that the individual steps in the mechanism must add to the overall reaction. For example, the proposed mechanism sums to the overall reaction as shown here:

$$\begin{aligned} H_2(g) + ICl(g) &\longrightarrow \cancel{HI(g)} + HCl(g) \\ \underline{\cancel{HI(g)} + ICl(g)} &\longrightarrow \underline{HCl(g) + I_2(g)} \\ H_2(g) + 2\,ICl(g) &\longrightarrow 2\,HCl(g) + I_2(g) \end{aligned}$$

Species—such as HI—that are formed in one step of a mechanism and consumed in another are **reaction intermediates.** An intermediate is not found in the balanced equation for the overall reaction but plays a key role in the mechanism. A reaction mechanism is a complete, detailed description of the reaction at the molecular level—it specifies the individual collisions and reactions that result in the overall reaction. As such, reaction mechanisms are highly sought-after pieces of chemical knowledge.

How do we determine reaction mechanisms? Recall from the opening section of this chapter that chemical kinetics are not only practically important (allowing us to control the rate of a particular reaction), but also theoretically important because they can help us determine the mechanism of the reaction. We can piece together a reaction mechanism by measuring the kinetics of the overall reaction and working backward to write a mechanism consistent with the measured kinetics.

Rate Laws for Elementary Steps

Elementary steps are characterized by their **molecularity,** the number of reactant particles involved in the step. The molecularity of the three most common types of elementary steps are:

$A \longrightarrow$ products	**Unimolecular**
$A + A \longrightarrow$ products	**Bimolecular**
$A + B \longrightarrow$ products	**Bimolecular**

Elementary steps in which three reactant particles collide, called **termolecular** steps, are very rare because the probability of three particles simultaneously colliding is small.

Although we cannot deduce the rate law for an overall chemical reaction from the balanced chemical equation, we can deduce the rate law for an elementary step. Since we know that an elementary step occurs through the collision of the reactant particles, the rate is proportional to the product of the concentrations of those particles. For example, the rate law for the bimolecular elementary step in which A reacts with B is:

$$A + B \longrightarrow \text{products} \qquad \text{Rate} = k[A][B]$$

Similarly, the rate law for the bimolecular step in which A reacts with A is:

$$A + A \longrightarrow products \quad Rate = k[A]^2$$

Table 13.2 summarizes the rate laws for the common elementary steps, as well as those for the rare termolecular step. Notice that the molecularity of the elementary step is equal to the overall order of the step.

TABLE 13.2 Rate Laws for Elementary Step		
Elementary Step	**Molecularity**	**Rate Law**
A $\longrightarrow$ products	1	Rate $= k[A]$
A + A $\longrightarrow$ products	2	Rate $= k[A]^2$
A + B $\longrightarrow$ products	2	Rate $= k[A][B]$
A + A + A $\longrightarrow$ products	3 (rare)	Rate $= k[A]^3$
A + A + B $\longrightarrow$ products	3 (rare)	Rate $= k[A]^2[B]$
A + A + C $\longrightarrow$ products	3 (rare)	Rate $= k[A][B][C]$

Rate-Determining Steps and Overall Reaction Rate Laws

As we have noted, most chemical reactions occur through a series of elementary steps. In most cases, one of those steps—called the **rate-determining step**—is much slower than the others. The rate-determining step in a chemical reaction is analogous to the narrowest section on a freeway. If a freeway narrows from four lanes to two lanes, the rate at which cars travel along the freeway is limited by the rate at which they can travel through the narrow section (even though the rate could be much faster along the four-lane section). Similarly, the rate-determining step in a reaction mechanism limits the overall rate of the reaction (even though the other steps occur much faster) and therefore determines *the rate law for the overall reaction.*

As an example, consider the reaction between nitrogen dioxide gas and carbon monoxide gas:

$$NO_2(g) + CO(g) \longrightarrow NO(g) + CO_2(g)$$

The experimentally determined rate law for this reaction is Rate $= k[NO_2]^2$. We can see from this rate law that the reaction must not be a single-step reaction—otherwise the rate law would be Rate $= k[NO_2][CO]$. A possible mechanism for this reaction is:

$$NO_2(g) + NO_2(g) \longrightarrow NO_3(g) + NO(g) \quad Slow$$
$$NO_3(g) + CO(g) \longrightarrow NO_2(g) + CO_2(g) \quad Fast$$

Figure 13.13▶ shows the energy diagram accompanying this mechanism. The first step has a much larger activation energy than the second step. The greater activation energy results in a much smaller rate constant for the first step compared to the second step. The first step determines the overall rate of the reaction, and the predicted rate law is therefore Rate $= k[NO_2]^2$, which is consistent with the observed experimental rate law.

For a proposed reaction mechanism such as the one above to be valid—mechanisms can only be validated, not proven—two conditions must be met:

1. **The elementary steps in the mechanism must sum to the overall reaction.**
2. **The rate law predicted by the mechanism must be consistent with the experimentally observed rate law.**

We have already seen that the rate law predicted by the proposed mechanism is consistent with the experimentally observed rate law. We can check whether the elementary steps sum to the overall reaction by adding them together:

$$NO_2(g) + \cancel{NO_2(g)} \longrightarrow \cancel{NO_3(g)} + NO(g) \quad Slow$$
$$\cancel{NO_3(g)} + CO(g) \longrightarrow \cancel{NO_2(g)} + CO_2(g) \quad Fast$$
$$\overline{NO_2(g) + CO(g) \longrightarrow NO(g) + CO_2(g)} \quad Overall$$

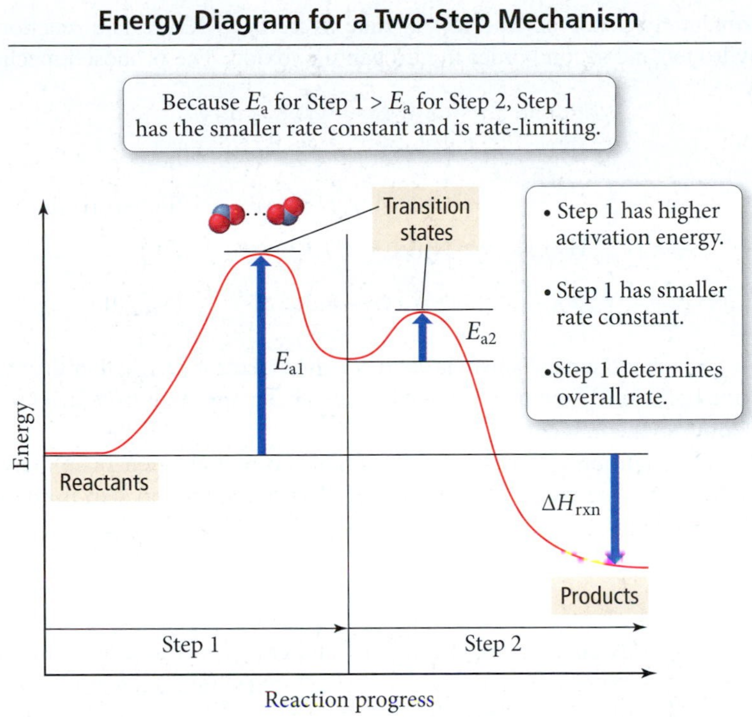

Energy Diagram for a Two-Step Mechanism

Because E_a for Step 1 > E_a for Step 2, Step 1 has the smaller rate constant and is rate-limiting.

Transition states

- Step 1 has higher activation energy.
- Step 1 has smaller rate constant.
- Step 1 determines overall rate.

E_{a1}

E_{a2}

Energy

Reactants

ΔH_{rxn}

Products

Step 1   Step 2

Reaction progress

The mechanism fulfills both of the requirements and is therefore valid. A valid mechanism is not necessarily a *proven* mechanism (because other mechanisms may also fulfill both of the requirements). We can only say that a given mechanism is consistent with kinetic observations of the reaction and therefore possible. Other types of data—such as the experimental evidence for a proposed intermediate—can further strengthen the validity of a proposed mechanism.

Mechanisms with a Fast Initial Step

When the proposed mechanism for a reaction has a slow initial step—like the one shown previously for the reaction between NO_2 and CO—the rate law predicted by the mechanism normally contains only reactants involved in the overall reaction. However, when a mechanism begins with a fast initial step, then some other subsequent step in the mechanism will be the rate-limiting step. In these cases, the rate law predicted by the rate-limiting step may contain reaction intermediates. Since reaction intermediates do not appear in the overall reaction equation, a rate law containing intermediates cannot correspond directly to the experimental rate law. Fortunately, however, we can often express the concentration of intermediates in terms of the concentrations of the reactants of the overall reaction.

In a multistep mechanism where the first step is fast, the products of the first step build up, because the rate at which they are consumed is limited by some slower step further down the line. As those products build up, they begin to react with one another to re-form the reactants. As long as the first step is fast enough in comparison to the rate-limiting step, the first-step reaction will reach equilibrium. We indicate the equilibrium as follows:

$$\text{Reactants} \underset{k_{-1}}{\overset{k_1}{\rightleftharpoons}} \text{Products}$$

The double arrows indicate that both the forward reaction and the reverse reaction occur. If equilibrium is reached, then the rate of the forward reaction equals the rate of the reverse reaction.

Consider the reaction by which hydrogen reacts with nitrogen monoxide to form water and nitrogen gas:

$$2\,H_2(g) + 2\,NO(g) \longrightarrow 2\,H_2O(g) + N_2(g)$$

The experimentally observed rate law is Rate $= k[H_2][NO]^2$. The reaction is first order in hydrogen and second order in nitrogen monoxide. The proposed mechanism is as follows:

$$2\,NO(g) \underset{k_{-1}}{\overset{k_1}{\rightleftharpoons}} N_2O_2(g) \qquad\qquad\qquad \text{Fast}$$

$$H_2(g) + N_2O_2(g) \xrightarrow{k_2} H_2O(g) + N_2O(g) \qquad \text{Slow (rate limiting)}$$

$$N_2O(g) + H_2(g) \xrightarrow{k_3} N_2(g) + H_2O(g) \qquad \text{Fast}$$

$$\overline{2\,H_2(g) + 2\,NO(g) \longrightarrow 2\,H_2O(g) + N_2(g)} \qquad \text{Overall}$$

To determine whether the mechanism is valid, we must determine whether the two conditions described previously are met. As you can see above, the steps do indeed sum to the overall reaction, so the first condition is met.

The second condition is that the rate law predicted by the mechanism must be consistent with the experimentally observed rate law. Since the second step is rate limiting, we write the following expression for the rate law:

$$\text{Rate} = k_2[H_2][N_2O_2] \qquad\qquad [13.28]$$

This rate law contains an intermediate (N_2O_2) and can therefore not be immediately reconciled with the experimentally observed rate law (which does not contain intermediates). Because of the equilibrium in the first step, however, *we can express the concentration of the intermediate in terms of the reactants of the overall equation.* Since the first step reaches equilibrium, the rate of the forward reaction in the first step equals the rate of the reverse reaction:

$$\text{Rate (forward)} = \text{Rate (backward)}$$

The rate of the forward reaction is given by

$$\text{Rate} = k_1[NO]^2$$

The rate of the reverse reaction is given by

$$\text{Rate} = k_{-1}[N_2O_2]$$

Since these two rates are equal at equilibrium, we can write the expression:

$$k_1[NO]^2 = k_{-1}[N_2O_2]$$

Rearranging, we get

$$[N_2O_2] = \frac{k_1}{k_{-1}}[NO]^2$$

We can now substitute this expression into Equation 13.28, the rate law obtained from the slow step:

$$\text{Rate} = k_2[H_2][N_2O_2]$$

$$= k_2[H_2]\frac{k_1}{k_{-1}}[NO]^2$$

$$= \frac{k_2 k_1}{k_{-1}}[H_2][NO]^2$$

If we combine the individual rate constants into one overall rate constant, we get the predicted rate law:

$$\text{Rate} = k[H_2][NO]^2 \qquad\qquad [13.29]$$

This rate law is consistent with the experimentally observed rate law so condition two is met and the proposed mechanism is valid.

EXAMPLE 13.9 Reaction Mechanisms

Ozone naturally decomposes to oxygen by the reaction:

$$2\, O_3(g) \longrightarrow 3\, O_2(g)$$

The experimentally observed rate law for this reaction is:

$$\text{Rate} = k[O_3]^2[O_2]^{-1}$$

Show that the following proposed mechanism is consistent with the experimentally observed rate law.

$$O_3(g) \underset{k_{-1}}{\overset{k_1}{\rightleftharpoons}} O_2(g) + O(g) \qquad \text{Fast}$$

$$O_3(g) + O(g) \xrightarrow{k_2} 2\, O_2(g) \qquad \text{Slow}$$

SOLUTION

To determine whether the mechanism is valid, you must first determine whether the steps sum to the overall reaction. Since the steps do indeed sum to the overall reaction, the first condition is met.	$O_3(g) \underset{k_{-1}}{\overset{k_1}{\rightleftharpoons}} O_2(g) + \cancel{O}(g)$ $O_3(g) + \cancel{O}(g) \xrightarrow{k_2} 2\, O_2(g)$ ──────────────── $2\, O_3(g) \longrightarrow 3\, O_2(g)$
The second condition is that the rate law predicted by the mechanism must be consistent with the experimentally observed rate law. Since the second step is rate limiting, write the rate law based on the second step.	$\text{Rate} = k_2[O_3][O]$
Because the rate law contains an intermediate (O), you must express the concentration of the intermediate in terms of the concentrations of the reactants of the overall reaction. To do this, set the rates of the forward reaction and the reverse reaction of the first step equal to each other. Solve the expression from the previous step for $[O]$, the concentration of the intermediate.	$\text{Rate (forward)} = \text{Rate (backward)}$ $k_1[O_3] = k_{-1}[O_2][O]$ $[O] = \dfrac{k_1[O_3]}{k_{-1}[O_2]}$
Finally, substitute $[O]$ into the rate law predicted by the slow step.	$\begin{aligned}\text{Rate} &= k_2[O_3][O] \\ &= k_2[O_3]\dfrac{k_1[O_3]}{k_{-1}[O_2]} \\ &= k_2\dfrac{k_1}{k_{-1}}\dfrac{[O_3]^2}{[O_2]} \\ &= k[O_3]^2[O_2]^{-1}\end{aligned}$

CHECK Since the two steps in the proposed mechanism sum to the overall reaction, and since the rate law obtained from the proposed mechanism is consistent with the experimentally observed rate law, the proposed mechanism is valid. The -1 reaction order with respect to $[O_2]$ indicates that the rate slows down as the concentration of oxygen increases—oxygen inhibits, or slows down, the reaction.

FOR PRACTICE 13.9

Predict the overall reaction and rate law that would result from the following two-step mechanism.

$$2\, A \longrightarrow A_2 \qquad \text{Slow}$$
$$A_2 + B \longrightarrow A_2B \qquad \text{Fast}$$

13.7 Catalysis

Throughout this chapter, we have learned ways to control the rates of chemical reactions. For example, we can speed up the rate of a reaction by increasing the concentration of the reactants or by increasing the temperature. However, these ways are not always feasible. There are limits to how concentrated we can make a reaction mixture, and increases in temperature may allow unwanted reactions—such as the decomposition of a reactant—to occur.

Alternatively, reaction rates can be increased by using a **catalyst,** a substance that increases the rate of a chemical reaction but is not consumed by the reaction. A catalyst works by providing an alternative mechanism for the reaction—one in which the rate-determining step has a lower activation energy. For example, consider the noncatalytic destruction of ozone in the upper atmosphere:

$$O_3(g) + O(g) \longrightarrow 2\,O_2(g)$$

In this reaction, an ozone molecule collides with an oxygen atom to form two oxygen molecules in a single elementary step. The reason that Earth has a protective ozone layer in the upper atmosphere is that the activation energy for this reaction is fairly high and the reaction, therefore, proceeds at a fairly slow rate. The ozone layer does not rapidly decompose into O_2. However, the addition of Cl atoms (which come from the photodissociation of man-made chlorofluorocarbons) to the upper atmosphere makes available another pathway by which O_3 can be destroyed. The first step in this pathway—called the catalytic destruction of ozone—is the reaction of Cl with O_3 to form ClO and O_2.

$$Cl + O_3 \longrightarrow ClO + O_2$$

This is followed by a second step in which ClO reacts with O, regenerating Cl.

$$ClO + O \longrightarrow Cl + O_2$$

Notice that if we add the two reactions, the overall reaction is identical to the noncatalytic reaction.

$$\cancel{Cl} + O_3 \longrightarrow \cancel{ClO} + O_2$$
$$\cancel{ClO} + O \longrightarrow \cancel{Cl} + O_2$$
$$O_3 + O \longrightarrow 2\,O_2$$

However, the activation energy for the rate-limiting step in this pathway is much smaller than for the first, uncatalyzed pathway (as shown in **Figure 13.14◀**), and therefore the reaction occurs at a much faster rate. Note that the Cl is not consumed in the overall reaction—this is characteristic of a catalyst.

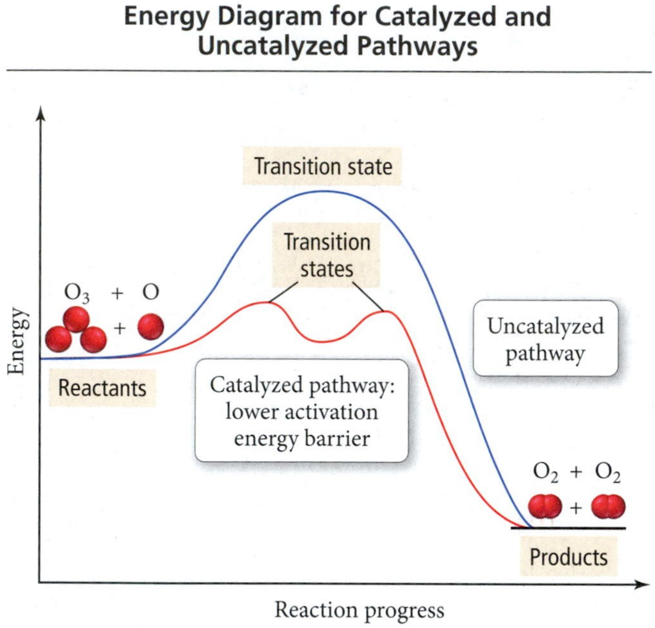

Photodissociation means *light-induced* dissociation. The energy from a photon of light can break a chemical bond and therefore dissociate—break apart—a molecule.

Energy Diagram for Catalyzed and Uncatalyzed Pathways

Transition state

Transition states

$O_3 + O$

Reactants

Uncatalyzed pathway

Catalyzed pathway: lower activation energy barrier

Energy

$O_2 + O_2$

Products

Reaction progress

▲ **FIGURE 13.14 Catalyzed and Uncatalyzed Decomposition of Ozone** In the catalytic destruction of ozone, the activation barrier for the rate-limiting step is much lower than in the uncatalyzed process.

Homogeneous and Heterogeneous Catalysis

Catalysis can be divided into two types: homogeneous and heterogeneous (**Figure 13.15▼**). In **homogeneous catalysis,** the catalyst exists in the same phase as the reactants. The catalytic destruction of ozone by Cl is an example of homogeneous catalysis—the chlorine atoms exist in the gas phase with the gas-phase reactants. In **heterogeneous catalysis,** the catalyst exists in a phase different from the reactants. The use of solid catalysts with gas-phase or solution-phase reactants is the most common type of heterogeneous catalysis.

Research has shown that heterogeneous catalysis is most likely responsible for the ozone hole over Antarctica. In 1985, when scientists discovered the dramatic drop in ozone over Antarctica, they wondered why it existed only there and not over the rest of

▶ **FIGURE 13.15 Homogeneous and Heterogeneous Catalysis** A homogeneous catalyst exists in the same phase as the reactants. A heterogeneous catalyst exists in a different phase than the reactants. Often a heterogeneous catalyst provides a solid surface on which the reaction can take place.

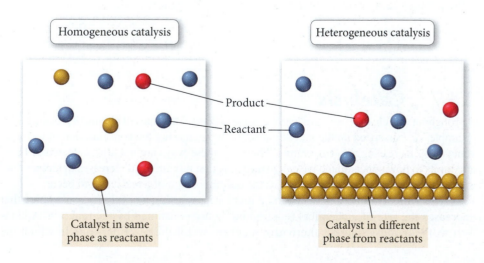

Homogeneous catalysis

Heterogeneous catalysis

Product

Reactant

Catalyst in same phase as reactants

Catalyst in different phase from reactants

the planet. After all, the chlorine from chlorofluorocarbons that catalyzes ozone destruction is evenly distributed throughout the entire atmosphere.

As it turns out, most of the chlorine that enters the atmosphere from chlorofluorocarbons gets bound up in chlorine reservoirs, substances such as $ClONO_2$ that hold chlorine and prevent it from catalyzing ozone destruction. The unique conditions over Antarctica—especially the cold isolated air mass that exists during the long dark winter—result in clouds that contain solid ice particles. These unique clouds are called polar stratospheric clouds (or PSCs), and the surfaces of the ice particles within these clouds appear to catalyze the release of chlorine from their reservoirs:

$$ClONO_2 + HCl \xrightarrow[\text{PSCs}]{} Cl_2 + HNO_3$$

When the sun rises in the Antarctic spring, the sunlight dissociates the chlorine molecules into chlorine atoms:

$$Cl_2 \xrightarrow[\text{light}]{} 2\,Cl$$

The chlorine atoms then catalyze the destruction of ozone by the mechanism discussed previously. This continues until the sun melts the stratospheric clouds, allowing chlorine atoms to be reincorporated into their reservoirs. The result is an ozone hole that forms every spring and lasts about 6–8 weeks.

▲ Polar stratospheric clouds contain ice particles that catalyze reactions by which chlorine is released from its atmospheric chemical reservoirs.

Enzymes: Biological Catalysts

Perhaps the best example of chemical catalysis is found in living organisms. Most of the thousands of reactions that must occur for an organism to survive would be too slow at normal temperatures. So living organisms rely on **enzymes,** biological catalysts that increase the rates of biochemical reactions. Enzymes are usually large protein molecules with complex three-dimensional structures. Within the enzyme structure is a specific area called the **active site.** The properties and shape of the active site are just right to bind the reactant molecule, usually called the **substrate.** The substrate fits into the active site in a manner that is analogous to a key fitting into a lock (**Figure 13.16▼**). When the substrate binds to the active site—through intermolecular forces such as hydrogen bonding and dispersion forces, or even covalent bonds—the activation energy of the reaction is greatly lowered, allowing the reaction to occur at a much faster rate. The general mechanism by which an enzyme (E) binds a substrate (S) and then reacts to form the products (P) is:

$$\begin{array}{ll} E + S \rightleftharpoons ES & \text{Fast} \\ ES \longrightarrow E + P & \text{Slow, rate limiting} \end{array}$$

Sucrase is an enzyme that catalyzes the breaking up of sucrose (table sugar) into glucose and fructose within the body. At body temperature, sucrose does not break into glucose and fructose because the activation energy is high, resulting in a slow reaction rate. However, when a sucrose molecule binds to the active site within sucrase, the bond between the glucose and fructose units weakens because glucose is forced into a

> The strategies used to speed up chemical reactions in the laboratory—high temperatures, high pressures, strongly acidic or alkaline conditions—are not available to living organisms, since they would be fatal to cells.

Enzyme–Substrate Binding

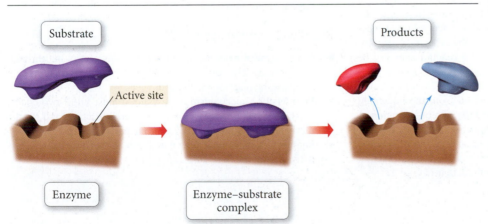

Substrate

Active site

Enzyme

Products

Enzyme–substrate complex

◀ FIGURE 13.16 Enzyme–Substrate Binding A substrate (or reactant) fits into the active site of an enzyme much as a key fits into a lock. It is held in place by intermolecular forces, forming an enzyme–substrate complex. (Sometimes temporary covalent bonding may also be involved.) After the reaction occurs, the products are released from the active site.

geometry that stresses the bond. Weakening of this bond lowers the activation energy for the reaction, increasing the reaction rate. The reaction can then proceed toward equilibrium—which favors the products—at a much lower temperature.

CHAPTER IN REVIEW

Key Terms

Section 13.3
rate law (501)
rate constant (k) (501)
reaction order (n) (501)
overall order (504)

Section 13.4
integrated rate law (505)
half-life ($t_{1/2}$) (509)

Section 13.5
Arrhenius equation (512)
activation energy (E_a) (513)
frequency factor (A) (513)
activated complex (transition
    state) (513)
exponential factor (514)
Arrhenius plot (515)
collision model (518)

orientation factor (518)
collision frequency (518)

Section 13.6
reaction mechanism (519)
elementary step (519)
reaction intermediates (519)
molecularity (519)
unimolecular (519)
bimolecular (519)

termolecular (519)
rate-determining step (520)

Section 13.7
catalyst (523)
homogeneous catalysis (524)
heterogeneous catalysis (524)
enzyme (525)
active site (525)
substrate (525)

Key Concepts

Reaction Rates, Orders, and Rate Laws (13.1–13.3)

▶ The rate of a chemical reaction is a measure of how fast a reaction occurs. The rate reflects the change in the concentration of a reactant or product per unit time, and is usually reported in units of M/s.

▶ Reaction rates generally depend on the concentration of the reactants. The rate of a first-order reaction is directly proportional to the concentration of the reactant; the rate of a second-order reaction is proportional to the square of the concentration of the reactant; and the rate of a zero-order reaction is independent of the concentration of the reactant.

▶ For a reaction with more than one reactant, the order of each reactant is, in general, independent of the order of the other reactants. The rate law shows the relationship between the rate and the concentration of each reactant.

Integrated Rate Laws and Half-Life (13.4)

▶ The rate law for a reaction gives the relationship between the rate of the reaction and the concentrations of the reactants. The integrated rate law, by contrast, gives the relationship between the concentration of a reactant and time.

▶ The integrated rate law for a zero-order reaction shows that the concentration of the reactant varies linearly with time. For a first-order reaction, the *natural log* of the concentration of the reactant varies linearly with time, and for a second-order reaction, the *inverse* of the concentration of the reactant varies linearly with time. Therefore, the order of a reaction can also be determined by plotting measurements of concentration as a function of time in these three different ways.

▶ The half-life of a reaction can be derived from the integrated rate law; it represents the time required for the concentration of a reactant to fall to one-half of its initial value. The half-life of a first-order reaction is *independent* of initial concentration of the reactant. The half-life of a zero-order or second-order reaction *depends* on the initial concentration of reactant.

The Effect of Temperature on Reaction Rate (13.5)

▶ The rate constant of a reaction generally depends on temperature and can be expressed by the Arrhenius equation, which consists of a frequency factor and an exponential factor.

▶ The frequency factor represents the number of times that the reactants approach the activation barrier per unit time.

▶ The exponential factor depends on both the temperature and the activation energy, a barrier that the reactants must overcome to become products. The exponential factor is the fraction of approaches that are successful in surmounting the activation barrier and forming products.

▶ The exponential factor increases with increasing temperature, but decreases with an increasing value of the activation energy.

▶ The frequency factor and activation energy for a reaction can be determined by measuring the rate constant at different temperatures and constructing an Arrhenius plot.

▶ For reactions in the gas phase, Arrhenius behavior can be modeled with the collision model. In this model, reactions occur as a result of sufficiently energetic collisions. The colliding molecules must be oriented in such a way that the reaction can occur. The frequency factor then contains two terms: p, which represents the fraction of collisions that have the proper orientation, and z, which represents the number of collisions per unit time.

Reaction Mechanisms (13.6)

▶ Most chemical reactions occur not in a single step, but through several steps. The series of individual steps by which a reaction occurs is the reaction mechanism.

▶ In order for a mechanism to be valid, it must fulfill two conditions: (a) the steps must sum to the overall reaction; and (b) the mechanism must predict the experimentally observed rate law.

▶ For mechanisms with a slow initial step, the predicted rate law is derived from the slow step.

▶ For mechanisms with a fast initial step, the predicted rate law is first written based on the slow step. Then equilibration of the fast steps is assumed in order to write concentrations of intermediates in terms of the reactants.

Catalysis (13.7)

▶ A catalyst is a substance that increases the rate of a chemical reaction but is not consumed by the reaction. A catalyst provides an alternative mechanism that has a lower activation energy for the rate-determining step.

▶ Catalysts can be divided into two types: homogeneous and heterogeneous. A homogeneous catalyst exists in the same phase as the reactants and forms a homogeneous mixture with them. A heterogeneous catalyst generally exists in a different phase than the reactants.

▶ Enzymes are biological catalysts capable of increasing the rate of specific biochemical reactions by many orders of magnitude.

Key Equations and Relationships

The Rate of Reaction (13.2)

For a reaction, $a\text{A} + b\text{B} \longrightarrow c\text{C} + d\text{D}$, the rate is defined as

$$\text{Rate} = -\frac{1}{a}\frac{\Delta[\text{A}]}{\Delta t} = -\frac{1}{b}\frac{\Delta[\text{B}]}{\Delta t} = +\frac{1}{c}\frac{\Delta[\text{C}]}{\Delta t} = +\frac{1}{d}\frac{\Delta[\text{D}]}{\Delta t}$$

The Rate Law (13.3)

$$\text{Rate} = k[\text{A}]^n \quad \text{(single reactant)}$$

$$\text{Rate} = k[\text{A}]^m[\text{B}]^n \quad \text{(multiple reactants)}$$

Integrated Rate Laws and Half-Life (13.4)

Reaction Order	Integrated Rate Law	Units of k	Half-Life Expression
0	$[\text{A}]_t = -kt + [\text{A}]_0$	$\text{M} \cdot \text{s}^{-1}$	$t_{1/2} = \dfrac{[\text{A}]_0}{2k}$
1	$\ln[\text{A}]_t = -kt + \ln[\text{A}]_0$	s^{-1}	$t_{1/2} = \dfrac{0.693}{k}$
2	$\dfrac{1}{[\text{A}]_t} = kt + \dfrac{1}{[\text{A}]_0}$	$\text{M}^{-1} \cdot \text{s}^{-1}$	$t_{1/2} = \dfrac{1}{k[\text{A}]_0}$

Arrhenius Equation (13.5)

$$k = Ae^{-E_a/RT}$$

$$\ln k = -\frac{E_a}{R}\left(\frac{1}{T}\right) + \ln A \qquad \text{(linearized form)}$$

$$\ln \frac{k_2}{k_1} = \frac{E_a}{R}\left(\frac{1}{T_1} - \frac{1}{T_2}\right) \qquad \text{(two-point form)}$$

$$k = pz\, e^{-E_a/RT} \qquad \text{(collision theory)}$$

Rate Laws for Elementary Steps (13.6)

Elementary Step	Molecularity	Rate Law
$\text{A} \longrightarrow \text{products}$	1	$\text{Rate} = k[\text{A}]$
$\text{A} + \text{A} \longrightarrow \text{products}$	2	$\text{Rate} = k[\text{A}]^2$
$\text{A} + \text{B} \longrightarrow \text{products}$	2	$\text{Rate} = k[\text{A}][\text{B}]$
$\text{A} + \text{A} + \text{A} \longrightarrow \text{products}$	3 (rare)	$\text{Rate} = k[\text{A}]^3$
$\text{A} + \text{A} + \text{B} \longrightarrow \text{products}$	3 (rare)	$\text{Rate} = k[\text{A}]^2[\text{B}]$
$\text{A} + \text{B} + \text{C} \longrightarrow \text{products}$	3 (rare)	$\text{Rate} = k[\text{A}][\text{B}][\text{C}]$

Key Learning Objectives

Chapter Objectives	Assessment
Expressing Reaction Rates (13.2)	Example 13.1 For Practice 13.1 Exercises 1–8
Determining the Order, Rate Law, and Rate Constant of a Reaction (13.3)	Example 13.2 For Practice 13.2 Exercises 15–18
Using Graphical Analysis of Reaction Data to Determine Reaction Order and Rate Constants (13.4)	Examples 13.3, 13.5 For Practice 13.3, 13.5 Exercises 21–26
Determining the Concentration of a Reactant at a Given Time (13.4)	Example 13.4 For Practice 13.4 Exercises 25–28
Working with the Half-Life of a Reaction (13.4)	Example 13.6 For Practice 13.6 Exercises 27–30
Using the Arrhenius Equation to Determine Kinetic Parameters (13.5)	Examples 13.7, 13.8 For Practice 13.7, 13.8 Exercises 33–40
Determining whether a Reaction Mechanism Is Valid (13.6)	Example 13.9 For Practice 13.9 Exercises 47–50

EXERCISES

Problems by Topic

Reaction Rates

1. Consider the reaction:

$$2 HBr(g) \longrightarrow H_2(g) + Br_2(g)$$

a. Express the rate of the reaction with respect to each of the reactants and products.
b. In the first 15.0 s of this reaction, the concentration of HBr dropped from 0.500 M to 0.455 M. Calculate the average rate of the reaction in this time interval.
c. If the volume of the reaction vessel in part b was 0.500 L, what amount of Br_2 (in moles) was formed during the first 15.0 s of the reaction?

2. Consider the reaction:

$$2 N_2O(g) \longrightarrow 2 N_2(g) + O_2(g)$$

a. Express the rate of the reaction with respect to each of the reactants and products.
b. In the first 10.0 s of the reaction, 0.018 mol of O_2 is produced in a reaction vessel with a volume of 0.250 L. What is the average rate of the reaction over this time interval?
c. Predict the rate of change in the concentration of N_2O over this time interval. In other words, what is $\Delta[N_2O]/\Delta t$?

3. For the reaction $2 A(g) + B(g) \longrightarrow 3 C(g)$,
a. Determine the expression for the rate of the reaction with respect to each of the reactants and products.
b. When A is decreasing at a rate of 0.100 M/s, how fast is B decreasing? How fast is C increasing?

4. For the reaction $A(g) + \frac{1}{2}B(g) \longrightarrow 2 C(g)$,
a. Determine the expression for the rate of the reaction with respect to each of the reactants and products.
b. When C is increasing at a rate of 0.025 M/s, how fast is B decreasing? How fast is A decreasing?

5. Consider the reaction:

$$C_4H_8(g) \longrightarrow 2 C_2H_4(g)$$

The following data were collected for the concentration of C_4H_8 as a function of time:

Time (s)	$[C_4H_8]$ (M)
0	1.000
10	0.913
20	0.835
30	0.763
40	0.697
50	0.637

a. What is the average rate of the reaction between 0 and 10 s? Between 40 and 50 s?
b. What is the rate of formation of C_2H_4 between 20 and 30 s?

6. Consider the reaction:

$$NO_2(g) \longrightarrow NO(g) + \frac{1}{2}O_2(g)$$

The following data were collected for the concentration of NO_2 as a function of time:

Time (s)	$[NO_2]$ (M)
0	1.000
10	0.951
20	0.904
30	0.860
40	0.818
50	0.778
60	0.740
70	0.704
80	0.670
90	0.637
100	0.606

a. What is the average rate of the reaction between 10 and 20 s? Between 50 and 60 s?
b. What is the rate of formation of O_2 between 50 and 60 s?

7. Consider the reaction:

$$H_2(g) + Br_2(g) \longrightarrow 2 HBr(g)$$

The graph below shows the concentration of Br_2 as a function of time.

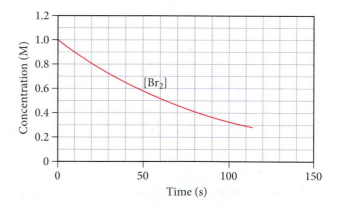

a. Use the graph to calculate:
   (i) The average rate of the reaction between 0 and 25 s.
   (ii) The instantaneous rate of the reaction at 25 s.
   (iii) The instantaneous rate of formation of HBr at 50 s.
b. Make a rough sketch of a curve representing the concentration of HBr as a function of time. Assume that the initial concentration of HBr is zero.

8. Consider the reaction:

$$2 H_2O_2(aq) \longrightarrow 2 H_2O(l) + O_2(g)$$

The graph below shows the concentration of H_2O_2 as a function of time.

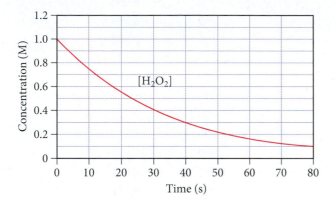

Use the graph to calculate:
a. The average rate of the reaction between 10 and 20 s.
b. The instantaneous rate of the reaction at 30 s.
c. The instantaneous rate of formation of O_2 at 50 s.
d. If the initial volume of the H_2O_2 is 1.5 L, what total amount of O_2 (in moles) is formed in the first 50 s of reaction?

The Rate Law and Reaction Orders

9. The graph that follows shows a plot of the rate of a reaction versus the concentration of the reactant A for the reaction A $\longrightarrow$ products.

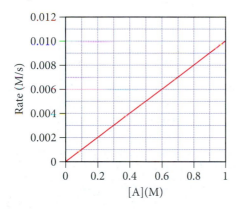

a. What is the order of the reaction with respect to A?
b. Make a rough sketch of how a plot of $[A]$ versus *time* would appear.
c. Write a rate law for the reaction including an estimate for the value of k.

10. The graph below shows a plot of the rate of a reaction versus the concentration of the reactant.

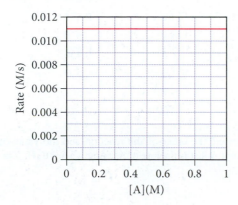

a. What is the order of the reaction with respect to A?
b. Make a rough sketch of how a plot of $[A]$ versus *time* would appear.
c. Write a rate law for the reaction including the value of k.

11. What are the units of k for each reaction order?
a. first order
b. second order
c. zero order

12. This reaction is first order in N_2O_5:

$$N_2O_5(g) \longrightarrow NO_3(g) + NO_2(g)$$

The rate constant for the reaction at a certain temperature is 0.053/s.
a. Calculate the rate of the reaction when $[N_2O_5] = 0.055$ M.
b. What would the rate of the reaction be at the same concentration as in part a if the reaction were second order? Zero order? (Assume the same *numerical* value for the rate constant with the appropriate units.)

13. A reaction in which A, B, and C react to form products is first order in A, second order in B, and zero order in C.
a. Write a rate law for the reaction.
b. What is the overall order of the reaction?
c. By what factor does the reaction rate change if $[A]$ is doubled (and the other reactant concentrations are held constant)?
d. By what factor does the reaction rate change if $[B]$ is doubled (and the other reactant concentrations are held constant)?
e. By what factor does the reaction rate change if $[C]$ is doubled (and the other reactant concentrations are held constant)?
f. By what factor does the reaction rate change if the concentrations of all three reactants are doubled?

14. A reaction in which A, B, and C react to form products is zero order in A, one-half order in B, and second order in C.
a. Write a rate law for the reaction.
b. What is the overall order of the reaction?
c. By what factor does the reaction rate change if $[A]$ is doubled (and the other reactant concentrations are held constant)?
d. By what factor does the reaction rate change if $[B]$ is doubled (and the other reactant concentrations are held constant)?
e. By what factor does the reaction rate change if $[C]$ is doubled (and the other reactant concentrations are held constant)?
f. By what factor does the reaction rate change if the concentrations of all three reactants are doubled?

15. Consider the data showing the initial rate of a reaction (A $\longrightarrow$ products) at several different concentrations of A. What is the order of the reaction? Write a rate law for the reaction including the value of the rate constant, k.

$[A]$ (M)	Initial Rate (M/s)
0.100	0.053
0.200	0.210
0.300	0.473

16. Consider the data showing the initial rate of a reaction (A $\longrightarrow$ products) at several different concentrations of A. What is the order of the reaction? Write a rate law for the reaction including the value of the rate constant, k.

[A] (M)	Initial Rate (M/s)
0.15	0.008
0.30	0.016
0.60	0.032

17. The data below were collected for the reaction:

$$2\,NO_2(g) + F_2(g) \longrightarrow 2\,NO_2F(g)$$

[NO$_2$] (M)	[F$_2$] (M)	Initial Rate (M/s)
0.100	0.100	0.026
0.200	0.100	0.051
0.200	0.200	0.103
0.400	0.400	0.411

Write an expression for the reaction rate law and calculate the value of the rate constant, k. What is the overall order of the reaction?

18. The data below were collected for the reaction:

$$CH_3Cl(g) + 3\,Cl_2(g) \longrightarrow CCl_4(g) + 3\,HCl(g)$$

[CH$_3$Cl] (M)	[Cl$_2$] (M)	Initial Rate (M/s)
0.050	0.050	0.014
0.100	0.050	0.029
0.100	0.100	0.041
0.200	0.200	0.115

Write an expression for the reaction rate law and calculate the value of the rate constant, k. What is the overall order of the reaction?

The Integrated Rate Law and Half-Life

19. Indicate the order of reaction consistent with each observation.
  a. A plot of the concentration of the reactant versus time yields a straight line.
  b. The reaction has a half-life that is independent of initial concentration.
  c. A plot of the inverse of the concentration versus time yields a straight line.

20. Indicate the order of reaction consistent with each observation.
  a. The half-life of the reaction gets shorter as the initial concentration is increased.
  b. A plot of the natural log of the concentration of the reactant versus time yields a straight line.
  c. The half-life of the reaction gets longer as the initial concentration is increased.

21. The following data show the concentration of AB versus time for the reaction:

$$AB(g) \longrightarrow A(g) + B(g)$$

Determine the order of the reaction and the value of the rate constant. Predict the concentration of AB at 25 s.

Time (s)	[AB] (M)
0	0.950
50	0.459
100	0.302
150	0.225
200	0.180
250	0.149
300	0.128
350	0.112
400	0.0994
450	0.0894
500	0.0812

22. The data below show the concentration of N_2O_5 versus time for the reaction:

$$N_2O_5(g) \longrightarrow NO_3(g) + NO_2(g)$$

Determine the order of the reaction and the value of the rate constant. Predict the concentration of N_2O_5 at 250 s.

Time (s)	[N$_2$O$_5$] (M)
0	1.000
25	0.822
50	0.677
75	0.557
100	0.458
125	0.377
150	0.310
175	0.255
200	0.210

23. The data below show the concentration of cyclobutane (C_4H_8) versus time for the reaction:

$$C_4H_8 \longrightarrow 2\,C_2H_4$$

Time (s)	[C$_4$H$_8$] (M)
0	1.000
10	0.894
20	0.799
30	0.714
40	0.638
50	0.571
60	0.510
70	0.456
80	0.408
90	0.364
100	0.326

Determine the order of the reaction and the value of the rate constant. What is the rate of reaction when $[C_4H_8] = 0.25$ M?

24. A reaction in which A $\longrightarrow$ products was monitored as a function of time and the results are shown here.

Time (s)	[A] (M)
0	1.000
25	0.914
50	0.829
75	0.744
100	0.659
125	0.573
150	0.488
175	0.403
200	0.318

Determine the order of the reaction and the value of the rate constant. What is the rate of reaction when $[A] = 0.10$ M?

25. The following reaction was monitored as a function of time:

$$A \longrightarrow B + C$$

A plot of $\ln[A]$ versus time yields a straight line with slope $-0.0045/s$.

a. What is the value of the rate constant (k) for this reaction at this temperature?
b. Write the rate law for the reaction.
c. What is the half-life?
d. If the initial concentration of A is 0.250 M, what is the concentration after 225 s?

26. The following reaction was monitored as a function of time:

$$AB \longrightarrow A + B$$

A plot of $1/[AB]$ versus time yields a straight line with slope $0.055/M \cdot s$.

a. What is the value of the rate constant (k) for this reaction at this temperature?
b. Write the rate law for the reaction.
c. What is the half-life when the initial concentration is 0.55 M?
d. If the initial concentration of AB is 0.250 M, and the reaction mixture initially contains no products, what are the concentrations of A and B after 75 s?

27. The decomposition of SO_2Cl_2 is first order in SO_2Cl_2 and has a rate constant of 1.42×10^{-4} s^{-1} at a certain temperature.

a. What is the half-life for this reaction?
b. How long will it take for the concentration of SO_2Cl_2 to decrease to 25% of its initial concentration?
c. If the initial concentration of SO_2Cl_2 is 1.00 M, how long will it take for the concentration to decrease to 0.78 M?
d. If the initial concentration of SO_2Cl_2 is 0.150 M, what is the concentration of SO_2Cl_2 after 2.00×10^2 s? After 5.00×10^2 s?

28. The decomposition of XY is second order in XY and has a rate constant of 7.02×10^{-3} M$^{-1} \cdot$ s^{-1} at a certain temperature.

a. What is the half-life for this reaction at an initial concentration of 0.100 M?
b. How long will it take for the concentration of XY to decrease to 12.5% of its initial concentration when the initial concentration is 0.100 M? When the initial concentration is 0.200 M?
c. If the initial concentration of XY is 0.150 M, how long will it take for the concentration to decrease to 0.062 M?
d. If the initial concentration of XY is 0.050 M, what is the concentration of XY after 5.0×10^1 s? After 5.50×10^2 s?

29. The half-life for the radioactive decay of U-238 is 4.5 billion years and is independent of initial concentration. How long will it take for 10% of the U-238 atoms in a sample of U-238 to decay? If a sample of U-238 initially contained 1.5×10^{18} atoms when the universe was formed 13.8 billion years ago, how many U-238 atoms will it contain today?

30. The half-life for the radioactive decay of C-14 is 5730 years and is independent of initial concentration. How long will it take for 25% of the C-14 atoms in a sample of C-14 to decay? If a sample of C-14 initially contains 1.5 mmol of C-14, how many millimoles will be left after 2255 years?

The Effect of Temperature and the Collision Model

31. This diagram shows the energy of a reaction as the reaction progresses. Label each blank box in the diagram:

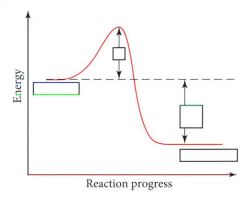

a. reactants
b. products
c. activation energy (E_a)
d. enthalpy of reaction (ΔH_{rxn})

32. A chemical reaction is endothermic and has an activation energy that is twice the value of the enthalpy of the reaction. Draw a diagram depicting the energy of the reaction as it progresses. Label the position of the reactants and products and indicate the activation energy and enthalpy of reaction.

33. The activation energy of a reaction is 56.8 kJ/mol and the frequency factor is 1.5×10^{11}/s. Calculate the rate constant of the reaction at 25 °C.

34. The rate constant of a reaction at 32 °C is measured to be 0.055/s. If the frequency factor is 1.2×10^{13}/s, what is the activation barrier?

35. The rate constant (k) for a reaction is measured as a function of temperature. A plot of $\ln k$ versus $1/T$ (in K) is linear and has a slope of -7445 K. Calculate the activation energy for the reaction.

36. The rate constant (k) for a reaction is measured as a function of temperature. A plot of $\ln k$ versus $1/T$ (in K) is linear and has a slope of -1.01×10^4 K. Calculate the activation energy for the reaction.

37. The data below were collected for the first-order reaction:

$$N_2O(g) \longrightarrow N_2(g) + O(g)$$

Use an Arrhenius plot to determine the activation barrier and frequency factor for the reaction.

Temperature (K)	Rate Constant (1/s)
800	3.24×10^{-5}
900	0.00214
1000	0.0614
1100	0.955

38. The data below show the rate constant of a reaction measured at several different temperatures. Use an Arrhenius plot to determine the activation barrier and frequency factor for the reaction.

Temperature (K)	Rate Constant (1/s)
300	0.0134
310	0.0407
320	0.114
330	0.303
340	0.757

39. The data below were collected for the second-order reaction:

$$Cl(g) + H_2(g) \longrightarrow HCl(g) + H(g)$$

Use an Arrhenius plot to determine the activation barrier and frequency factor for the reaction.

Temperature (K)	Rate Constant (L/mol · s)
90	0.00357
100	0.0773
110	0.956
120	7.781

40. The data below show the rate constant of a reaction measured at several different temperatures. Use an Arrhenius plot to determine the activation barrier and frequency factor for the reaction.

Temperature (K)	Rate Constant (1/s)
310	0.00434
320	0.0140
330	0.0421
340	0.118
350	0.316

41. A reaction has a rate constant of 0.0117/s at 400. K and 0.689/s at 450. K.
a. Determine the activation barrier for the reaction.
b. What is the value of the rate constant at 425 K?

42. A reaction has a rate constant of 0.000122/s at 27 °C and 0.228/s at 77 °C.
a. Determine the activation barrier for the reaction.
b. What is the value of the rate constant at 17 °C?

43. If a temperature increase from 10.0 °C to 20.0 °C doubles the rate constant for a reaction, what is the value of the activation barrier for the reaction?

44. If a temperature increase from 20.0 °C to 35.0 °C triples the rate constant for a reaction, what is the value of the activation barrier for the reaction?

45. Consider these two gas-phase reactions:
a. $AA(g) + BB(g) \longrightarrow 2\,AB(g)$
b. $AB(g) + CD(g) \longrightarrow AC(g) + BD(g)$

If the two reactions have identical activation barriers and are carried out under the same conditions, which one would you expect to have the faster rate?

46. Which of these two reactions would you expect to have the smaller orientation factor? Explain.
a. $O(g) + N_2(g) \longrightarrow NO(g) + N(g)$
b. $NO(g) + Cl_2(g) \longrightarrow NOCl(g) + Cl(g)$

Reaction Mechanisms

47. The reaction shown here is experimentally observed to be second order in AB and zero order in C:

$$AB + C \longrightarrow A + BC$$

Determine whether this mechanism is valid for this reaction.

$$AB + AB \xrightarrow{\;k_1\;} AB_2 + A \quad \text{Slow}$$
$$AB_2 + C \xrightarrow{\;k_2\;} AB + BC \quad \text{Fast}$$

48. The reaction shown here is experimentally observed to be second order in X and first order in Y:

$$X + Y \longrightarrow XY$$

a. Does the reaction occur in a single step in which X and Y collide?
b. Is this two-step mechanism valid?

$$2\,X \underset{k_2}{\overset{k_1}{\rightleftharpoons}} X_2 \qquad\qquad \text{Fast}$$
$$X_2 + Y \xrightarrow{\;k_3\;} XY + X \quad \text{Slow}$$

49. Consider this three-step mechanism for a reaction:

$$Cl_2(g) \underset{k_2}{\overset{k_1}{\rightleftharpoons}} 2\,Cl(g) \qquad\qquad \text{Fast}$$
$$Cl(g) + CHCl_3(g) \xrightarrow{\;k_3\;} HCl(g) + CCl_3(g) \quad \text{Slow}$$
$$Cl(g) + CCl_3(g) \xrightarrow{\;k_4\;} CCl_4(g) \qquad \text{Fast}$$

a. What is the overall reaction?
b. Identify the intermediates in the mechanism.
c. What is the predicted rate law?

50. Consider this two-step mechanism for a reaction:

$$NO_2(g) + Cl_2(g) \xrightarrow{\;k_1\;} ClNO_2(g) + Cl(g) \quad \text{Slow}$$
$$NO_2(g) + Cl(g) \xrightarrow{\;k_2\;} ClNO_2(g) \qquad \text{Fast}$$

a. What is the overall reaction?
b. Identify the intermediates in the mechanism.
c. What is the predicted rate law?

Catalysis

51. Many heterogeneous catalysts are deposited on high surface-area supports. Why is a large surface area important in heterogeneous catalysis?

52. Suppose that the reaction $A \longrightarrow$ products is exothermic and has an activation barrier of 75 kJ/mol. Sketch an energy diagram showing the energy of the reaction as a function of the progress of the reaction. Draw a second energy curve showing the effect of a catalyst.

53. Suppose that a catalyst lowers the activation barrier of a reaction from 125 kJ/mol to 55 kJ/mol. By what factor would you expect the reaction rate to increase at 25 °C? (Assume that the frequency factors for the catalyzed and uncatalyzed reactions are identical.)

54. The activation barrier for the hydrolysis of sucrose into glucose and fructose is 108 kJ/mol. If an enzyme increases the rate of the hydrolysis reaction by a factor of 1 million, how much lower does the activation barrier have to be when sucrose is in the active site of the enzyme? (Assume that the frequency factors for the catalyzed and uncatalyzed reactions are identical and a temperature of 25 °C.)

Cumulative Problems

55. The data below were collected for this reaction at 500 °C:

$$CH_3CN(g) \longrightarrow CH_3NC(g)$$

Time (h)	$[CH_3CN]$ (M)
0.0	1.000
5.0	0.794
10.0	0.631
15.0	0.501
20.0	0.398
25.0	0.316

a. Determine the order of the reaction and the value of the rate constant at this temperature.
b. What is the half-life for this reaction (at the initial concentration)?
c. How long will it take for 90% of the CH_3CN to convert to CH_3NC?

56. The data were collected for this reaction at a certain temperature:

$$X_2Y \longrightarrow 2 X + Y$$

Time (h)	$[X_2Y]$ (M)
0.0	0.100
1.0	0.0856
2.0	0.0748
3.0	0.0664
4.0	0.0598
5.0	0.0543

a. Determine the order of the reaction and the value of the rate constant at this temperature.
b. What is the half-life for this reaction (at the initial concentration)?
c. What is the concentration of X after 10.0. hours?

57. Consider the reaction:

$$A + B + C \longrightarrow D$$

which has this rate law:

$$Rate = k \frac{[A][C]^2}{[B]^{1/2}}$$

Suppose the rate of the reaction at certain initial concentrations of A, B, and C is 0.0115 M/s. What is the rate of the reaction if the concentrations of A and C are doubled and the concentration of B is tripled?

58. Consider the reaction:

$$2 O_3(g) \longrightarrow 3 O_2(g)$$

which has this rate law:

$$Rate = k \frac{[O_3]^2}{[O_2]}$$

Suppose that a 1.0-L reaction vessel initially contains 1.0 mol of O_3 and 1.0 mol of O_2. What fraction of the O_3 will have reacted when the rate falls to one-half of its initial value?

59. At 700 K acetaldehyde decomposes in the gas phase to methane and carbon monoxide. The reaction is:

$$CH_3CHO(g) \longrightarrow CH_4(g) + CO(g)$$

A sample of CH_3CHO is heated to 700 K and the pressure is measured as 0.22 atm before any reaction takes place. The kinetics of the reaction are then followed by measurements of total pressure and these data are obtained:

P_{Total}(atm)	0.22	0.24	0.27	0.31
T(s)	0	1000	3000	7000

Find the rate law, the specific rate constant, and the total pressure after 2.00×10^4 s.

60. At 400 K oxalic acid decomposes according to the reaction:

$$H_2C_2O_4(g) \longrightarrow CO_2(g) + HCOOH(g)$$

These pressure data (in mmHg) are obtained from a kinetic study:

P_{Total} at $t = 20000$ s	94.6	132	160
P $H_2C_2O_4$ at $t = 0$	65.8	92.1	111

Find the rate law of the reaction and its specific rate constant.

61. Dinitrogen pentoxide decomposes in the gas phase to form nitrogen dioxide and oxygen gas. The reaction is first order in dinitrogen pentoxide and has a half-life of 2.81 h at 25 °C. If a 1.5-L reaction vessel initially contains 745 torr of N_2O_5 at 25 °C, what partial pressure of O_2 will be present in the vessel after 215 minutes?

62. Cyclopropane (C_3H_6) reacts to form propene (C_3H_6) in the gas phase. The reaction is first order in cyclopropane and has a rate constant of 5.87×10^{-4}/s at 485 °C. If a 2.5-L reaction vessel initially contains 722 torr of cyclopropane at 485 °C, how long will it take for the partial pressure of cyclopropane to drop to below 100. torr?

63. Iodine atoms will combine to form I_2 in liquid hexane solvent with a rate constant of 1.5×10^{10} L/mol · s. The reaction is second order in I. Since the reaction occurs so quickly, the only way to study the reaction is to create iodine atoms almost

instantaneously, usually by photochemical decomposition of I_2. Suppose a flash of light creates an initial $[I]$ concentration of 0.0100 M. How long will it take for 95% of the newly created iodine atoms to recombine to form I_2?

64. The hydrolysis of sucrose ($C_{12}H_{22}O_{11}$) into glucose and fructose in acidic water has a rate constant of $1.8 \times 10^{-4}\ s^{-1}$ at 25 °C. Assuming the reaction is first order in sucrose, determine the mass of sucrose hydrolyzed when 2.55 L of a 0.150 M sucrose solution is allowed to react for 195 minutes.

65. The reaction $AB(aq) \longrightarrow A(g) + B(g)$ is second order in AB and has a rate constant of $0.0118\ M^{-1}s^{-1}$ at 25.0 °C. A reaction vessel initially contains 250.0 mL of 0.100 M AB, which is allowed to react to form the gaseous product. The product is collected over water at 25.0 °C. How much time is required to produce 200.0 mL of the products at a barometric pressure of 755.1 mmHg? (The vapor pressure of water at this temperature is 23.8 torr.)

66. The reaction $2\ H_2O_2(aq) \longrightarrow 2\ H_2O(l) + O_2(g)$ is first order in H_2O_2 and under certain conditions has a rate constant of $0.00752\ s^{-1}$ at 20.0 °C. A reaction vessel initially contains 150.0 mL of 30.0% H_2O_2 by mass solution (the density of the solution is 1.11 g/cm^3). The gaseous oxygen is collected over water at 20.0 °C as it forms. What volume of O_2 will form in 85.0 seconds at a barometric pressure of 742.5 mmHg? (The vapor pressure of water at this temperature is 17.5 torr.)

67. Consider this energy diagram showing the energy of a reaction as it progresses:

   a. How many elementary steps are involved in this reaction?
   b. Label the reactants, products, and intermediates.
   c. Which step is rate limiting?
   d. Is the overall reaction endothermic or exothermic?

68. Consider the reaction in which HCl adds across the double bond of ethane:

$$HCl + H_2C{=}CH_2 \longrightarrow H_3C{-}CH_2Cl$$

The following mechanism, with the energy diagram shown below, has been suggested for this reaction:

Step 1 $HCl + H_2C{=}CH_2 \longrightarrow H_3C{-}CH_2^{+} + Cl^{-}$

Step 2 $H_3C{-}CH_2^{+} + Cl^{-} \longrightarrow H_3C{-}CH_2Cl$

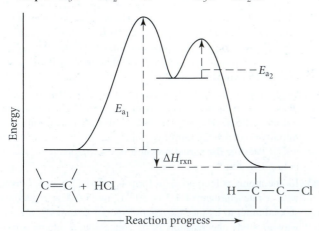

   a. Based on the energy diagram, which step is rate limiting?
   b. What is the expected order of the reaction based on the proposed mechanism?
   c. Is the overall reaction exothermic or endothermic?

69. The desorption of a single molecular layer of *n*-butane from a single crystal of aluminum oxide was found to be first order with a rate constant of 0.128/s at 150 K.
   a. What is the half-life of the desorption reaction?
   b. If the surface is initially completely covered with *n*-butane at 150 K, how long will it take for 25% of the molecules to desorb? For 50% to desorb?
   c. If the surface is initially completely covered, what fraction will remain covered after 10 s? After 20 s?

70. The evaporation of a 120-nm film of *n*-pentane from a single crystal of aluminum oxide was found to be zero order with a rate constant of 1.92×10^{13} molecules$/cm^2 \cdot$ s at 120 K.
   a. If the initial surface coverage is 8.9×10^{16} molecules$/cm^2$, how long will it take for one-half of the film to evaporate?
   b. What fraction of the film will be left after 10 s? Assume the same initial coverage as in part a.

71. The kinetics of the following reaction are studied as a function of temperature. (The reaction is first order in each reactant and second order overall.)

$$C_2H_5Br(aq) + OH^{-}(aq) \longrightarrow C_2H_5OH(l) + Br^{-}(aq)$$

Temperature (C)	$k\,(L/mol \cdot s)$
25	8.81×10^{-5}
35	0.000285
45	0.000854
55	0.00239
65	0.00633

   a. Determine the activation energy and frequency factor for the reaction.
   b. Determine the rate constant at 15 °C.
   c. If a reaction mixture is 0.155 M in C_2H_5Br, and 0.250 M in OH^{-}, what is the initial rate of the reaction at 75 °C?

72. The reaction $2\ N_2O_5 \longrightarrow 2\ N_2O_4 + O_2$ takes place at around room temperature in solvents such as CCl_4. The rate constant at 293 K is found to be $2.35 \times 10^{-4}\ s^{-1}$ and at 303 K the rate constant is found to be $9.15 \times 10^{-4}\ s^{-1}$. Calculate the frequency factor for the reaction.

73. This reaction has an activation energy of zero in the gas phase.

$$CH_3 + CH_3 \longrightarrow C_2H_6$$

   a. Would you expect the rate of this reaction to change very much with temperature?
   b. Can you think of a reason for why the activation energy is zero?
   c. What other types of reactions would you expect to have little or no activation energy?

74. Consider these two reactions:

$$O + N_2 \longrightarrow NO + N \qquad E_a = 315\ kJ/mol$$
$$Cl + H_2 \longrightarrow HCl + H \qquad E_a = 23\ kJ/mol$$

   a. Can you suggest why the activation barrier for the first reaction is so much higher than that for the second?
   b. The frequency factors for these two reactions are very close to each other in value. Assuming that they are the same, compute the ratio of the reaction rate constants for these two reactions at 25 °C.

75. Anthropologists can estimate the age of a bone or other organic matter by its carbon-14 content. The carbon-14 in a living organism is constant until the organism dies, after which carbon-14 decays with first-order kinetics and a half-life of 5730 years. Suppose a bone from an ancient human contains 19.5% of the C-14 found in living organisms. How old is the bone?

76. Geologists can estimate the age of rocks by their uranium-238 content. The uranium is incorporated in the rock as it hardens and then decays with first-order kinetics and a half-life of 4.5 billion years. A rock is found to contain 83.2% of the amount of uranium-238 that it contained when it was formed. (The amount that the rock contained when it was formed can be deduced from the presence of the decay products of U-238.) How old is the rock?

77. Consider the gas-phase reaction:

$$H_2(g) + I_2(g) \longrightarrow 2\,HI(g)$$

The reaction was experimentally determined to be first order in H_2 and first order in I_2. Consider the proposed mechanisms.
Proposed mechanism I:

$$H_2(g) + I_2(g) \longrightarrow 2\,HI(g) \quad \text{Single step}$$

Proposed mechanism II:

$$I_2(g) \underset{k_2}{\overset{k_1}{\rightleftharpoons}} 2\,I(g) \qquad\qquad \text{Fast}$$

$$H_2(g) + 2\,I(g) \xrightarrow{k_3} 2\,HI(g) \quad \text{Slow}$$

Proposed mechanism III:

$$I_2(g) \underset{k_2}{\overset{k_1}{\rightleftharpoons}} 2\,I(g) \qquad\qquad \text{Fast}$$

$$H_2(g) + I(g) \underset{k_4}{\overset{k_3}{\rightleftharpoons}} H_2I(g) \qquad \text{Fast}$$

$$H_2I(g) + I(g) \xrightarrow{k_5} 2\,HI(g) \quad \text{Slow}$$

a. Show that all three of the proposed mechanisms are valid.
b. What kind of experimental evidence might lead you to favor mechanisms II or III over mechanism I?

78. Consider the reaction:

$$2\,NH_3(aq) + OCl^-(aq) \longrightarrow N_2H_4(aq) + H_2O(l) + Cl^-(aq)$$

This three-step mechanism is proposed:

$$NH_3(aq) + OCl^-(aq) \underset{k_2}{\overset{k_1}{\rightleftharpoons}} NH_2Cl(aq) + OH^-(aq) \quad \text{Fast}$$

$$NH_2Cl(aq) + NH_3(aq) \xrightarrow{k_3} N_2H_5^+(aq) + Cl^-(aq) \quad \text{Slow}$$

$$N_2H_5^+(aq) + OH^-(aq) \xrightarrow{k_4} N_2H_4(aq) + H_2O(l) \quad \text{Fast}$$

a. Show that the mechanism sums to the overall reaction.
b. What is the rate law predicted by this mechanism?

79. A certain substance X decomposes. It is found that 50% of X remains after 100 minutes. How much X remains after 200 minutes if the reaction order with respect to X is (a) zero order, (b) first order, (c) second order?

80. The half-life for radioactive decay (a first-order process) of plutonium-239 is 24,000 years. How many years would it take for one mole of this radioactive material to decay so that just one atom remains?

81. The energy of activation for the decomposition of 2 mol of HI to H_2 and I_2 in the gas phase is 185 kJ. The heat of formation of $HI(g)$ from $H_2(g)$ and $I_2(g)$ is -5.65 kJ/mol. Find the energy of activation for the reaction of 1 mol of H_2 and 1 mol of I_2 in the gas phase.

82. Ethyl chloride vapor decomposes by the first-order reaction

$$C_2H_5Cl \longrightarrow C_2H_4 + HCl$$

The activation energy is 249 kJ/mol and the frequency factor is $1.6 \times 10^{-14}\,\text{s}^{-1}$. Find the value of the specific rate constant at 710 K. Find the fraction of the ethyl chloride that decomposes in 15 minutes at this temperature. Find the temperature at which the rate of the reaction would be twice as fast.

Challenge Problems

83. In this chapter we have seen a number of reactions in which a single reactant forms products. For example, consider the first-order reaction:

$$CH_3NC(g) \longrightarrow CH_3CN(g)$$

However, we also learned that gas-phase reactions occur through collisions.
a. One possible explanation is that two molecules of CH_3NC collide with each other and form two molecules of the product in a single elementary step. If that were the case, what reaction order would you expect?
b. Another possibility is that the reaction occurs through more than one step. For example, a possible mechanism involves one step in which the two CH_3NC molecules collide, resulting in the "activation" of one of them. In a second step, the activated molecule goes on to form the product. Write down this mechanism and determine which step must be rate determining in order for the kinetics of the reaction to be first order. Show explicitly how the mechanism predicts first-order kinetics.

84. The first-order *integrated* rate law for a reaction A $\longrightarrow$ products is derived from the rate law using calculus as follows:

$$\text{Rate} = k[A] \quad \text{(first-order rate law)}$$

$$\text{Rate} = -\frac{d[A]}{dt}$$

$$\frac{d[A]}{dt} = -k[A]$$

The above equation is a first-order, separable differential equation that can be solved by separating the variables and integrating:

$$\frac{d[A]}{[A]} = -k\,dt$$

$$\int_{[A]_0}^{[A]} \frac{d[A]}{[A]} = -\int_0^t k\,dt$$

In the above integral, $[A]_0$ is simply the initial concentration of A. We then evaluate the integral:

$$\left[\ln[A]\right]\big|_{[A]_0}^{[A]} = -k\,[t]_0^t$$

$$\ln[A] - \ln[A]_0 = -kt$$

$$\ln[A] = -kt + \ln[A]_0 \quad \text{(integrated rate law)}$$

a. Use a procedure similar to the one just shown to derive an integrated rate law for a reaction A $\longrightarrow$ products which is one-half order in the concentration of A (that is, Rate $= k[A]^{1/2}$).

b. Use the result from part a to derive an expression for the half-life of a one-half-order reaction.

85. The previous exercise shows how the first-order integrated rate law is derived from the first-order differential rate law. Begin with the second-order differential rate law and derive the second-order integrated rate law.

86. The rate constant for the first-order decomposition of $N_2O_5(g)$ to $NO_2(g)$ and $O_2(g)$ is 7.48×10^{-3} s^{-1} at a given temperature.

a. Find the length of time required for the total pressure in a system containing N_2O_5 at an initial pressure of 0.100 atm to rise to 0.145 atm.

b. To 0.200 atm.

c. Find the total pressure after 100 s of reaction.

87. Phosgene (Cl_2CO), a poison gas used in World War I, is formed by the reaction of Cl_2 and CO. The proposed mechanism for the reaction is

$$Cl_2 \rightleftharpoons 2Cl \qquad \text{(fast, equilibrium)}$$

$$Cl + CO \rightleftharpoons ClCO \qquad \text{(fast, equilibrium)}$$

$$ClCO + Cl_2 \longrightarrow Cl_2CO + Cl \quad \text{(slow)}$$

What rate law is consistent with this mechanism?

88. The rate of decomposition of $N_2O_3(g)$ to $NO_2(g)$ and $NO(g)$ is followed by measuring $[NO_2]$ at different times. The following data are obtained.

$[NO_2]$ (mol/L)	0	0.193	0.316	0.427	0.784
t (s)	0	884	1610	2460	50,000

The reaction follows a first-order rate law. Calculate the rate constant. Assume that after 50,000 s all the $N_2O_3(g)$ had decomposed.

89. At 473 K, for the elementary reaction

$$2\,NOCl(g) \underset{k_2}{\overset{k_1}{\rightleftharpoons}} 2\,NO(g) + Cl_2(g)$$

$$k_1 = 7.8 \times 10^{-2}\,\text{L/mol} \cdot \text{s} \text{ and}$$

$$k_{-1} = 4.7 \times 10^2\,\text{L}^2/\text{mol}^2 \cdot \text{s}$$

A sample of NOCl is placed in a container and heated to 473 K. When the system comes to equilibrium, $[NOCl]$ is found to be 0.12 mol/L. Find the concentrations of NO and Cl_2.

Conceptual Problems

90. Consider the reaction:

$$CHCl_3(g) + Cl_2(g) \longrightarrow CCl_4(g) + HCl(g)$$

The reaction is first order in $CHCl_3$ and one-half order in Cl_2. Which reaction mixture would you expect to have the fastest initial rate?

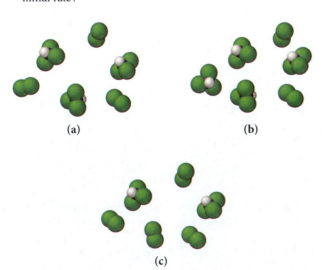

(a)   (b)

(c)

91. This graph shows the concentration of a reactant as a function of time for two different reactions. One of the reactions is first order and the other is second order. Which of the two reactions is first order? Second order? How would you change each plot to make it linear?

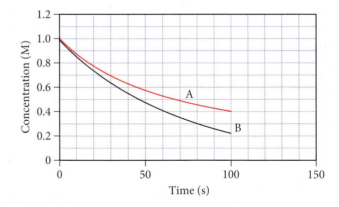

92. A particular reaction, A $\longrightarrow$ products, has a rate that slows down as the reaction proceeds. The half-life of the reaction is found to depend on the initial concentration of A. Determine whether each statement is likely to be true or false for this reaction.

a. A doubling of the concentration of A doubles the rate of the reaction.

b. A plot of $1/[A]$ versus time is linear.

c. The half-life of the reaction gets longer as the initial concentration of A increases.

d. A plot of the concentration of A versus time has a constant slope.

Answers to Conceptual Connections

Reaction Rates

13.1 (c) The rate at which B changes is twice the rate of the reaction and it is negative because B is a reactant.

Order of Reaction

13.2 (d) Since the reaction is second order, increasing the concentration of A by a factor of 5 causes the rate to increase by 5^2 or 25.

Rate and Concentration

13.3 All three mixtures have the same total number of molecules, but mixture (c) has the greatest number of NO molecules. Since the reaction is second order in NO and only first order in O_2, mixture (c) will have the fastest initial rate.

Rate Law and Integrated Rate Law

13.4 (c) The reaction is most likely second order because its rate depends on the concentration (therefore it cannot be zero order), and its half-life depends on the initial concentration (therefore it cannot be first order). For a second-order reaction, a doubling of the initial concentration results in the quadrupling of the rate.

Temperature Dependence of Reaction Rate

13.5 Reaction A because it has a lower activation energy; therefore the exponential factor will be larger at a given temperature.

Collision Theory

13.6 (c) Since the reactants in part (a) are atoms, the orientation factor should be about one. The reactants in parts (b) and (c) are both molecules, so we expect orientation factors of less than one. Since the reactants in (b) are symmetrical, we would not expect the collision to have as specific an orientation requirement as in (c), where the reactants are asymmetrical and must therefore collide in such way that a hydrogen atom is in close proximity to another hydrogen atom. Therefore, we expect (c) to have the smallest orientation factor.

14 Chemical Equilibrium

*Every system in chemical equilibrium, under the influence of a change
of any one of the factors of equilibrium, undergoes a transformation . . .
[that produces a change] . . . in the opposite direction of the factor in question.*

—Henri Le Châtelier (1850–1936)

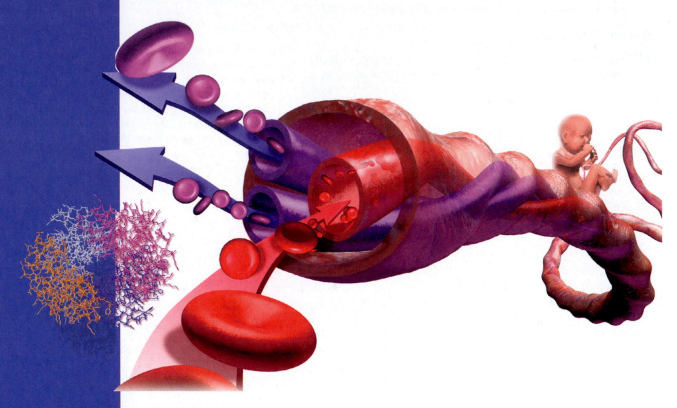

A developing fetus gets oxygen from the mother's blood because the reaction between oxygen and fetal hemoglobin has a larger equilibrium constant than the reaction between oxygen and maternal hemoglobin.

I N CHAPTER 13, we examined *how fast* a chemical reaction occurs. In this chapter we examine *how far* a chemical reaction goes. The *speed* of a chemical reaction is determined by kinetics. The *extent* of a chemical reaction is determined by thermodynamics. In this chapter, we focus on describing and quantifying how far a chemical reaction goes based on an experimentally measurable quantity called the equilibrium constant. A reaction with a large equilibrium constant proceeds nearly to completion. A reaction with a small equilibrium constant barely proceeds at all (the reactants remain as reactants, hardly forming any product). In Chapter 17, we will examine the underlying thermodynamics that determine the value of the equilibrium constant. In other words, for now we simply accept the equilibrium constant as an experimentally measurable quantity and learn how to use it to predict and quantify the extent of a reaction; in Chapter 17, we will explore the reasons underlying the magnitude of equilibrium constants.

14.1 Fetal Hemoglobin and Equilibrium

Have you ever wondered how a baby in the womb gets oxygen? Unlike you and me, a fetus does not breathe air. Yet, like you and me, a fetus needs oxygen. After we are born, we inhale air into our lungs; that air diffuses into capillaries, where it comes into contact with blood. Within red blood cells, a protein called hemoglobin (Hb) reacts with oxygen:

$$Hb + O_2 \rightleftharpoons HbO_2$$

The double arrows in this equation mean that the reaction can occur in both the forward and reverse directions and can reach chemical *equilibrium*. We have encountered this term in Chapters 11 and 12, and we define it more carefully in the next section. For now, understand that the concentrations of the reactants and products in a reaction at equilibrium are described by the *equilibrium constant, K*. A large value of K means that the reaction lies far to the right at equilibrium—a high concentration of products and a low concentration of reactants. A small value of K means that the reaction lies far to the left at equilibrium—a high concentration of reactants and a low concentration of products. In other words, the value of K is a measure of how far a reaction proceeds—the larger the value of K, the more the reaction favors the products.

The equilibrium constant for the reaction between hemoglobin and oxygen is such that hemoglobin efficiently binds oxygen at typical lung oxygen concentrations, but can also release oxygen under the appropriate conditions. Any system at equilibrium, including the hemoglobin–oxygen system, acts to maintain that equilibrium. If any of the concentrations of the reactants or products change, the reaction shifts to counteract that change. For the hemoglobin system, as blood flows through the lungs where oxygen concentrations are high, the equilibrium shifts to the right—hemoglobin binds oxygen:

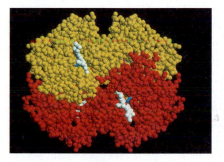

▲ Hemoglobin is the oxygen-carrying protein in red blood cells.

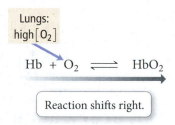

As blood flows out of the lungs and into muscles and organs where oxygen concentrations have been depleted (because muscles and organs use oxygen), the equilibrium shifts to the left—hemoglobin releases oxygen:

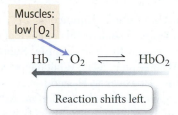

Therefore, in order to maintain equilibrium, hemoglobin binds oxygen when the surrounding oxygen concentration is high, but releases oxygen when the surrounding oxygen concentration is low. In this way, hemoglobin transports oxygen from the lungs to all parts of the body that use oxygen.

▶ **FIGURE 14.1 Oxygen Exchange between the Maternal and Fetal Circulation**   In the placenta, the blood of the fetus comes into close proximity with that of the mother without mixing. Because the reaction of fetal hemoglobin with oxygen has a larger equilibrium constant than the reaction of maternal hemoglobin with oxygen, the fetus receives oxygen from the mother's blood.

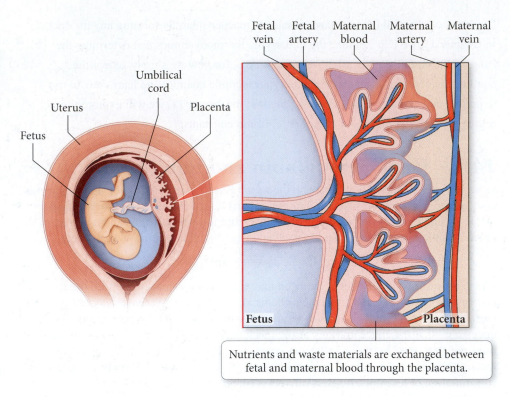

Nutrients and waste materials are exchanged between fetal and maternal blood through the placenta.

A fetus has its own circulatory system. The mother's blood never flows into the fetus's body and the fetus cannot get any air in the womb. How, then, does the fetus get oxygen? The answer lies in the properties of fetal hemoglobin (HbF), which is slightly different from adult hemoglobin. Like adult hemoglobin, fetal hemoglobin is in equilibrium with oxygen:

$$HbF + O_2 \rightleftharpoons HbFO_2$$

However, the equilibrium constant for fetal hemoglobin is larger than the equilibrium constant for adult hemoglobin, meaning that the reaction tends to go farther in the direction of the product. Consequently, fetal hemoglobin loads oxygen at a lower oxygen concentration than does adult hemoglobin. In the placenta, fetal blood flows in close proximity to maternal blood, without the two ever mixing. Because of the different equilibrium constants, the maternal hemoglobin releases oxygen which the fetal hemoglobin then binds and carries into its own circulatory system (**Figure 14.1▲**). Nature has thus evolved a chemical system through which the mother's hemoglobin can in effect *hand off* oxygen to the hemoglobin of the fetus.

14.2 The Concept of Dynamic Equilibrium

Recall from Chapter 13 that reaction rates generally increase with increasing concentration of the reactants (unless the reaction is zero order) and decrease with decreasing concentration of the reactants. With this in mind, consider the reaction between hydrogen and iodine:

$$H_2(g) + I_2(g) \rightleftharpoons 2 HI(g)$$

Nearly all chemical reactions are at least theoretically reversible. In many cases, however, the reversibility is so small that it can be ignored.

In this reaction, H_2 and I_2 react to form 2 HI molecules, but the 2 HI molecules can also react to re-form H_2 and I_2. A reaction such as this one—that can proceed in both the forward and reverse directions—is said to be **reversible.** Suppose we begin with only H_2 and I_2 in a container (**Figure 14.2(a)▶** on p. 541). What happens initially? H_2 and I_2 begin to react to form HI (**Figure 14.2(b)▶**). However, as H_2 and I_2 react their concentrations decrease, which in turn *decreases the rate of the forward reaction*. At the same time, HI begins to form. As the concentration of HI increases, the reverse reaction occurs at a faster and faster rate. Eventually the rate of the reverse reaction (which has been increasing) equals the rate of the forward reaction (which has been decreasing). At that point, **dynamic equilibrium** is reached (**Figure 14.2(c), (d)▶**).

Dynamic equilibrium for a chemical reaction is the condition in which the rate of the forward reaction equals the rate of the reverse reaction.

Dynamic Equilibrium

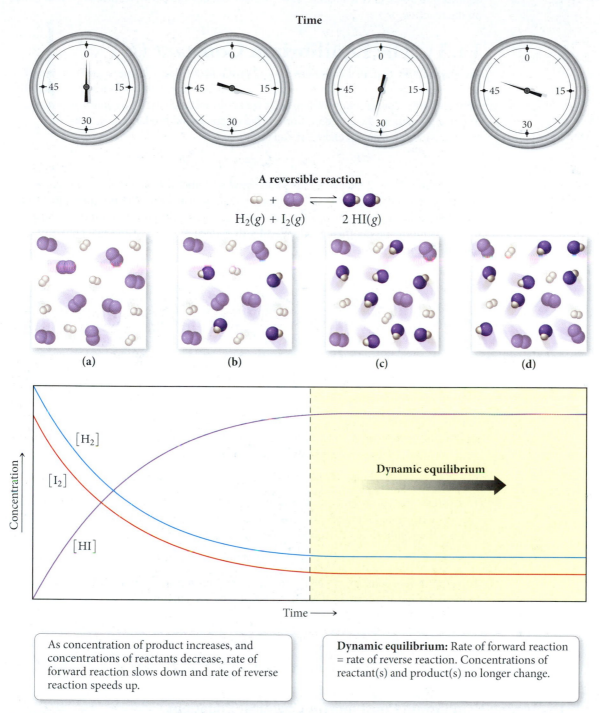

As concentration of product increases, and concentrations of reactants decrease, rate of forward reaction slows down and rate of reverse reaction speeds up.

Dynamic equilibrium: Rate of forward reaction = rate of reverse reaction. Concentrations of reactant(s) and product(s) no longer change.

▲ **FIGURE 14.2 Dynamic Equilibrium** Equilibrium is reached in a chemical reaction when the concentrations of the reactants and products no longer change. The molecular images on the top depict the progress of the reaction $H_2(g) + I_2(g) \rightleftharpoons 2\,HI(g)$. The graph on the bottom shows the concentrations of H_2, I_2, and HI as a function of time. When equilibrium is reached, both the forward and reverse reactions continue, but at equal rates, so the concentrations of the reactants and products remain constant.

Dynamic equilibrium is called "dynamic" because the forward and reverse reactions are still occurring; however, they are occurring at the same rate. When dynamic equilibrium is reached, the concentrations of H_2, I_2, and HI no longer change. They remain constant because the reactants and products are formed at the same rate that they are depleted. However, the *constancy* of the reactant and product concentrations at equilibrium *does not* imply that the concentrations are *equal* at equilibrium. Some reactions reach equilibrium only after most of the reactants have formed products. Others reach equilibrium when only a small fraction of the reactants have formed products. It depends on the reaction.

14.3   The Equilibrium Constant (K)

We just stated that the *concentrations of reactants and products* are not equal at equilibrium—rather, it is the *rates of the forward and reverse reactions* that are equal. But what about the concentrations? What can we know about them? The *equilibrium constant* is a way to quantify the concentrations of the reactants and products at equilibrium.

Consider the general chemical equation,

$$aA + bB \rightleftharpoons cC + dD$$

where A and B are reactants, C and D are products, and *a*, *b*, *c*, and *d* are the respective stoichiometric coefficients in the chemical equation. The **equilibrium constant** (K) for the reaction is the ratio—*at equilibrium*—of the concentrations of the products raised to their stoichiometric coefficients divided by the concentrations of the reactants raised to their stoichiometric coefficients.

> We distinguish between the equilibrium constant (K) and the kelvin unit of temperature (K) by italicizing the equilibrium constant.

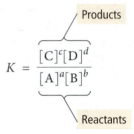

Products

$$K = \frac{[C]^c[D]^d}{[A]^a[B]^b}$$

Reactants

In this notation, $[A]$ represents the molar concentration of A. The equilibrium constant quantifies the relative concentrations of reactants and products *at equilibrium*. The relationship between the balanced chemical equation and the expression of the equilibrium constant is known as the **law of mass action**.

Expressing Equilibrium Constants for Chemical Reactions

To express an equilibrium constant for a chemical reaction, we examine the balanced chemical equation and apply the law of mass action. For example, suppose we want to express the equilibrium constant for the following reaction:

$$2\,N_2O_5(g) \rightleftharpoons 4\,NO_2(g) + O_2(g)$$

The equilibrium constant is $[NO_2]$ raised to the fourth power multiplied by $[O_2]$ raised to the first power divided by $[N_2O_5]$ raised to the second power:

$$K = \frac{[NO_2]^4[O_2]}{[N_2O_5]^2}$$

Notice that the *coefficients* in the chemical equation become the *exponents* in the expression of the equilibrium constant.

EXAMPLE 14.1   Expressing Equilibrium Constants for Chemical Equations

Express the equilibrium constant for the chemical equation.

$$CH_3OH(g) \rightleftharpoons CO(g) + 2\,H_2(g)$$

SOLUTION

The equilibrium constant is the concentrations of the products raised to their stoichiometric coefficients divided by the concentrations of the reactants raised to their stoichiometric coefficients.	$K = \dfrac{[CO][H_2]^2}{[CH_3OH]}$

FOR PRACTICE 14.1

Express the equilibrium constant for the combustion of propane in the balanced chemical equation.

$$C_3H_8(g) + 5\,O_2(g) \rightleftharpoons 3\,CO_2(g) + 4\,H_2O(g)$$

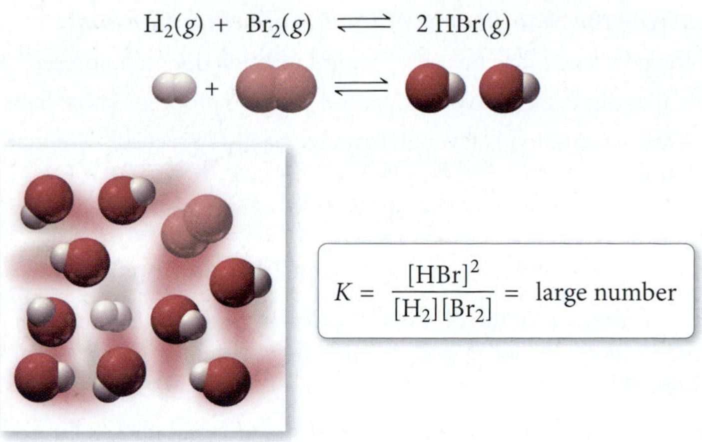

$$H_2(g) + Br_2(g) \rightleftharpoons 2\,HBr(g)$$

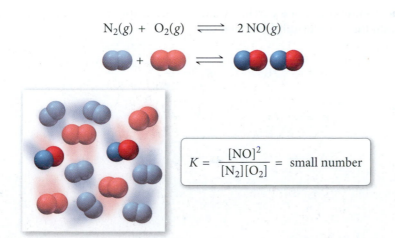

$$K = \frac{[HBr]^2}{[H_2][Br_2]} = \text{large number}$$

◀ **FIGURE 14.3 The Meaning of a Large Equilibrium Constant** If the equilibrium constant for a reaction is large, the equilibrium point of the reaction lies far to the right—the concentration of products is large and the concentration of reactants is small.

The Significance of the Equilibrium Constant

You now know how to express the equilibrium constant but what does it mean? What, for example, does a large equilibrium constant ($K \gg 1$) imply about a reaction? A large equilibrium constant indicates that the numerator of the equilibrium constant (which specifies the concentrations of products at equilibrium) is larger than the denominator (which specifies the concentrations of reactants at equilibrium). Therefore, the forward reaction is favored. For example, consider the reaction:

$$H_2(g) + Br_2(g) \rightleftharpoons 2\,HBr(g) \qquad K = 1.9 \times 10^{19}\ (\text{at } 25\,°C)$$

The equilibrium constant is large, meaning that the equilibrium point for the reaction lies far to the right—high concentrations of products, low concentrations of reactants (**Figure 14.3▲**). Remember that the equilibrium constant says nothing about *how fast* a reaction will reach equilibrium, only *how far* the reaction has proceeded once equilibrium is reached. A reaction with a large equilibrium constant may be kinetically very slow, meaning that it will take a long time to reach equilibrium.

Conversely, what does a *small* equilibrium constant ($K \ll 1$) mean? It indicates that the reverse reaction is favored; there will be more reactants than products when equilibrium is reached. For example, consider the reaction:

$$N_2(g) + O_2(g) \rightleftharpoons 2\,NO(g) \qquad K = 4.1 \times 10^{-31}\ (\text{at } 25\,°C)$$

The equilibrium constant is very small, so the equilibrium point for the reaction lies far to the left—high concentrations of reactants, low concentrations of products (**Figure 14.4▼**). This is fortunate because N_2 and O_2 are the main components of air. If this equilibrium constant were large, much of the N_2 and O_2 in air would react to form NO, a toxic gas.

$$N_2(g) + O_2(g) \rightleftharpoons 2\,NO(g)$$

$$K = \frac{[NO]^2}{[N_2][O_2]} = \text{small number}$$

◀ **FIGURE 14.4 The Meaning of a Small Equilibrium Constant** If the equilibrium constant for a reaction is small, the equilibrium point of the reaction lies far to the left—the concentration of products is small and the concentration of reactants is large.

Summarizing the Significance of the Equilibrium Constant:

▶ $K \ll 1$ Reverse reaction is favored; forward reaction does not proceed very far.

▶ $K \approx 1$ Neither direction is favored; forward reaction proceeds about halfway.

▶ $K \gg 1$ Forward reaction is favored; forward reaction proceeds essentially to completion.

Conceptual Connection 14.1 Equilibrium Constants

The equilibrium constant for the reaction $A(g) \rightleftharpoons B(g)$ is 10. A reaction mixture initially contains 11 mol of A and 0 mol of B in a fixed volume of 1 L. Which statement is true at equilibrium?

(a) The reaction mixture will contain 10 mol of A and 1 mol of B.

(b) The reaction mixture will contain 1 mol of A and 10 mol of B.

(c) The reaction mixture will contain equal amounts of A and B.

Relationships between the Equilibrium Constant and the Chemical Equation

If a chemical equation is modified in some way, then the equilibrium constant for the equation must be changed to reflect the modification. Here we list appropriate changes for common modifications.

1. **If you reverse the equation, invert the equilibrium constant:**

$$A + 2B \rightleftharpoons 3C \qquad K_{\text{forward}} = \frac{[C]^3}{[A][B]^2}$$

$$3C \rightleftharpoons A + 2B \qquad K_{\text{reverse}} = \frac{[A][B]^2}{[C]^3} = \frac{1}{K_{\text{forward}}}$$

2. **If you multiply the coefficients in the equation by a factor, raise the equilibrium constant to the same factor:**

$$A + 2B \rightleftharpoons 3C \qquad K = \frac{[C]^3}{[A][B]^2}$$

$$nA + 2nB \rightleftharpoons 3nC$$

$$K' = \frac{[C]^{3n}}{[A]^n[B]^{2n}} = \left(\frac{[C]^3}{[A][B]^2} \right)^n = K^n$$

> If n is a fractional quantity, raise K to the same fractional quantity.

3. **If you add two or more individual chemical equations to obtain an overall equation, multiply the corresponding equilibrium constants by each other to obtain the overall equilibrium constant:**

$$A \rightleftharpoons 2B \qquad K_1 = \frac{[B]^2}{[A]}$$

$$2B \rightleftharpoons 3C \qquad K_2 = \frac{[C]^3}{[B]^2}$$

$$A \rightleftharpoons 3C \qquad K_{\text{overall}} = K_1 \times K_2$$

$$= \frac{[C]^3}{[A]}$$

Conceptual Connection 14.2 Equilibrium Constant and Chemical Equation

The reaction $A(g) \rightleftharpoons 2B(g)$ has an equilibrium constant of $K = 0.010$. What is the equilibrium constant for the reaction $B(g) \rightleftharpoons \frac{1}{2}A(g)$?

(a) 1   **(b)** 10   **(c)** 100   **(d)** 0.0010

EXAMPLE 14.2 Manipulating the Equilibrium Constant to Reflect Changes in the Chemical Equation

Consider the chemical equation and equilibrium constant for the synthesis of ammonia at 25 °C:

$$N_2(g) + 3 H_2(g) \rightleftharpoons 2 NH_3(g) \qquad K = 3.7 \times 10^8$$

Calculate the equilibrium constant for the following reaction at 25 °C:

$$NH_3(g) \rightleftharpoons \tfrac{1}{2} N_2(g) + \tfrac{3}{2} H_2(g) \qquad K' = ?$$

SOLUTION

You want to manipulate the given reaction and value of K to obtain the desired reaction and value of K. You can see that the given reaction is the reverse of the desired reaction, and its coefficients are twice those of the desired reaction.

Begin by reversing the given reaction and taking the inverse of the value of K.	$N_2(g) + 3 H_2(g) \rightleftharpoons 2 NH_3(g) \qquad K = 3.7 \times 10^8$ $2 NH_3(g) \rightleftharpoons N_2(g) + 3 H_2(g) \qquad K_{rev} = \dfrac{1}{3.7 \times 10^8}$
Next, multiply the reaction by $\frac{1}{2}$ and raise the equilibrium constant to the $\frac{1}{2}$ power.	$NH_3(g) \rightleftharpoons \tfrac{1}{2} N_2(g) + \tfrac{3}{2} H_2(g)$ $K' = K_{reverse}^{1/2} = \left(\dfrac{1}{3.7 \times 10^8}\right)^{1/2}$
Calculate the value of K'.	$K' = 5.2 \times 10^{-5}$

FOR PRACTICE 14.2

Consider the chemical equation and equilibrium constant at 25 °C:

$$2 COF_2(g) \rightleftharpoons CO_2(g) + CF_4(g) \qquad K = 2.2 \times 10^6$$

Calculate the equilibrium constant for the following reaction at 25 °C:

$$2 CO_2(g) + 2 CF_4(g) \rightleftharpoons 4 COF_2(g) \qquad K' = ?$$

FOR MORE PRACTICE 14.2

Predict the equilibrium constant for the first reaction given the equilibrium constants for the second and third reactions:

$$CO_2(g) + 3 H_2(g) \rightleftharpoons CH_3OH(g) + H_2O(g) \qquad K_1 = ?$$
$$CO(g) + H_2O(g) \rightleftharpoons CO_2(g) + H_2(g) \qquad K_2 = 1.0 \times 10^5$$
$$CO(g) + 2 H_2(g) \rightleftharpoons CH_3OH(g) \qquad K_2 = 1.0 \times 10^7$$

14.4 Expressing the Equilibrium Constant in Terms of Pressure

So far, we have expressed the equilibrium constant only in terms of the *concentrations* of the reactants and products. However, for gaseous reactions, the partial pressure of a particular gas is proportional to its concentration. Therefore, we can also express the equilibrium constant in terms of the *partial pressures* of the reactants and products. For example, consider the gaseous reaction:

$$2 SO_3(g) \rightleftharpoons 2 SO_2(g) + O_2(g)$$

From this point on, we designate K_c as the equilibrium constant with respect to concentration in molarity. For the above reaction, K_c can be expressed using the law of mass action:

$$K_c = \frac{[SO_2]^2[O_2]}{[SO_3]^2}$$

We now designate K_p as the equilibrium constant with respect to partial pressures in atmospheres. *The expression for K_p takes the same form as the expression for K_c, except that we use the partial pressure of each gas in place of its concentration.* For the SO_3 reaction above, we write K_p:

$$K_p = \frac{(P_{SO_2})^2 P_{O_2}}{(P_{SO_3})^2}$$

where P_A is the partial pressure of gas A in units of atmospheres.

Since the partial pressure of a gas in atmospheres is not the same as its concentration in molarity, the value of K_p for a reaction is not necessarily equal to the value of K_c. However, as long as the gases behave ideally, we can derive a relationship between the two constants. The concentration of an ideal gas A is the number of moles of A (n_A) divided by its volume (V) in liters:

$$[A] = \frac{n_A}{V}$$

From the ideal gas law, we can relate the quantity n_A/V to the partial pressure of A:

$$P_A V = n_A RT$$

$$P_A = \frac{n_A}{V} RT$$

Since $[A] = n_A/V$, we can write

$$P_A = [A]RT \quad \text{or} \quad [A] = \frac{P_A}{RT} \qquad [14.1]$$

Now consider the general equilibrium chemical equation:

$$aA + bB \rightleftharpoons cC + dD$$

According to the law of mass action, we write K_c:

$$K_c = \frac{[C]^c[D]^d}{[A]^a[B]^b}$$

Substituting $[X] = P_X/RT$ for each concentration term, we get:

$$K_c = \frac{\left(\dfrac{P_C}{RT}\right)^c\left(\dfrac{P_D}{RT}\right)^d}{\left(\dfrac{P_A}{RT}\right)^a\left(\dfrac{P_B}{RT}\right)^b} = \frac{P_C^c P_D^d \left(\dfrac{1}{RT}\right)^{c+d}}{P_A^a P_B^b \left(\dfrac{1}{RT}\right)^{a+b}} = \frac{P_C^c P_D^d}{P_A^a P_B^b}\left(\frac{1}{RT}\right)^{c+d-(a+b)}$$

$$= K_p\left(\frac{1}{RT}\right)^{c+d-(a+b)}$$

Rearranging, we arrive at:

$$K_p = K_c(RT)^{c+d-(a+b)}$$

Finally, if we let $\Delta n = c + d - (a + b)$, which is the sum of the stoichiometric coefficients of the gaseous products minus the sum of the stoichiometric coefficients of the gaseous reactants, we get the following general result:

$$K_p = K_c(RT)^{\Delta n} \qquad [14.2]$$

Notice that if the sum of the stoichiometric coefficients for the reactants equals the sum for the products, then $\Delta n = 0$, and K_p is equal to K_c.

Units of K

Throughout this book, we express concentrations and partial pressures within the equilibrium constant in units of molarity and atmospheres, respectively. When expressing the value of the equilibrium constant, however, we have not included the units. Formally, the values of concentration or partial pressure that we substitute into the equilibrium constant expression are ratios of the concentration or pressure to a reference concentration (exactly 1 M) or a reference pressure (exactly 1 atm). For example, within the equilibrium constant expression, a pressure of 1.5 atm becomes

$$\frac{1.5 \text{ atm}}{1 \text{ atm}} = 1.5$$

Similarly, a concentration of 1.5 M becomes

$$\frac{1.5 \text{ M}}{1 \text{ M}} = 1.5$$

As long as concentration units are expressed in molarity for K_c and pressure units are expressed in atmospheres for K_p, we can skip this formality and simply enter the quantities directly into the equilibrium expression, dropping their corresponding units.

EXAMPLE 14.3 Relating K_p and K_c

Nitrogen monoxide, a pollutant in automobile exhaust, is oxidized to nitrogen dioxide in the atmosphere according to the equation:

$$2\,NO(g) + O_2(g) \rightleftharpoons 2\,NO_2(g) \qquad K_p = 2.2 \times 10^{12} \text{ at } 25\,^\circ C$$

Find K_c for this reaction.

SORT You are given K_p for the reaction and asked to find K_c.	**GIVEN** $K_p = 2.2 \times 10^{12}$ **FIND** K_c
STRATEGIZE Use Equation 14.2 to relate K_p and K_c.	**EQUATION** $K_p = K_c(RT)^{\Delta n}$
SOLVE Solve the equation for K_c. Calculate Δn. Substitute the required quantities to calculate K_c. The temperature must be in kelvins. The units are dropped when reporting K_c as described previously.	**SOLUTION** $$K_c = \frac{K_p}{(RT)^{\Delta n}}$$ $$\Delta n = 2 - 3 = -1$$ $$K_c = \frac{2.2 \times 10^{12}}{\left(0.08206 \frac{L \cdot atm}{mol \cdot K} \times 298 \text{ K}\right)^{-1}}$$ $$= 5.4 \times 10^{13}$$

CHECK The easiest way to check this answer is to substitute it back into Equation 14.2 and confirm that you get the original value for K_p.

$$K_p = K_c(RT)^{\Delta n}$$
$$= 5.4 \times 10^{13}\left(0.08206 \frac{L \cdot atm}{mol \cdot K} \times 298 \text{ K}\right)^{-1}$$
$$= 2.2 \times 10^{12}$$

FOR PRACTICE 14.3

Consider the following reaction and corresponding value of K_c:

$$H_2(g) + I_2(g) \rightleftharpoons 2\,HI(g) \qquad K_c = 6.2 \times 10^2 \text{ at } 25\,^\circ C$$

What is the value of K_p at this temperature?

Conceptual Connection 14.3 **The Relationship between K_p and K_c**

Under which circumstances will K_p and K_c be equal for the reaction,
$$aA(g) + bB(g) \rightleftharpoons cC(g) + dD(g)?$$

(a) If $a + b = c + d$.

(b) If the reaction is reversible.

(c) If the equilibrium constant is small.

14.5 Heterogeneous Equilibria: Reactions Involving Solids and Liquids

Many chemical reactions involve pure solids or pure liquids as reactants or products. Consider the reaction:

$$2\,CO(g) \rightleftharpoons CO_2(g) + C(s)$$

We might expect the expression for the equilibrium constant to be

$$K_c = \frac{[CO_2][C]}{[CO]^2} \quad \text{(incorrect)}$$

However, since carbon is a solid, its concentration is constant—its concentration does not change no matter what amount is present. The concentration of a solid does not change because a solid does not expand to fill its container. A solid's concentration, therefore, depends only on its density, which is constant as long as *some* solid is present (**Figure 14.5▼**). Consequently, pure solids—those reactants or products labeled in the chemical equation with an (s)—are not included in the equilibrium expression. The correct equilibrium expression is

$$K_c = \frac{[CO_2]}{[CO]^2} \quad \text{(correct)}$$

Similarly, the concentration of a pure liquid does not change. So, pure liquids—those reactants or products labeled in the chemical equation with an (ℓ)—are also excluded from the equilibrium expression. For example, what is the equilibrium expression for the following reaction?

$$CO_2(g) + H_2O(\ell) \rightleftharpoons H^+(aq) + HCO_3^-(aq)$$

Since $H_2O(\ell)$ is pure liquid, it is omitted from the equilibrium expression:

$$K_c = \frac{[H^+][HCO_3^-]}{[CO_2]}$$

A Heterogeneous Equilibrium

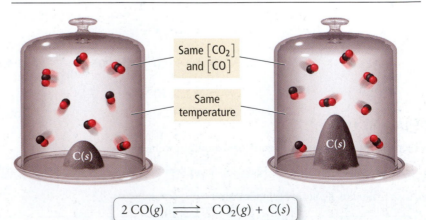

▶ **FIGURE 14.5 Heterogeneous Equilibrium** The concentration of solid carbon (the number of atoms per unit volume) is constant as long as any solid carbon is present. The same is true for pure liquids. For this reason, the concentrations of solids and pure liquids are not included in equilibrium constant expressions.

EXAMPLE 14.4 Writing Equilibrium Expressions for Reactions Involving a Solid or a Liquid

Write an expression for the equilibrium constant (K_c) for the chemical equation.

$$CaCO_3(s) \rightleftharpoons CaO(s) + CO_2(g)$$

SOLUTION

Since $CaCO_3(s)$ and $CaO(s)$ are both solids, you omit them from the equilibrium expression.	$K_c = [CO_2]$

FOR PRACTICE 14.4

Write an expression for the equilibrium constant (K_c) for the chemical equation.

$$4 HCl(g) + O_2(g) \rightleftharpoons 2 H_2O(\ell) + 2 Cl_2(g)$$

Conceptual Connection 14.4 Heterogeneous Equilibria, K_p and K_c

For which reaction is $K_p = K_c$?

(a) $2 Na_2O_2(s) + 2 CO_2(g) \rightleftharpoons 2 Na_2CO_3(s) + O_2(g)$

(b) $S(s) + CO(g) \rightleftharpoons COS(g)$

(c) $NH_4NO_3(s) \rightleftharpoons N_2O(g) + 2 H_2O(g)$

14.6 Calculating the Equilibrium Constant from Measured Equilibrium Concentrations

The most direct way to obtain an experimental value for the equilibrium constant of a reaction is to measure the concentrations of the reactants and products in a reaction mixture at equilibrium. Consider the reaction:

$$H_2(g) + I_2(g) \rightleftharpoons 2 HI(g)$$

Suppose a mixture of H_2 and I_2 is allowed to come to equilibrium at 445 °C. The measured equilibrium concentrations are $[H_2] = 0.11$ M, $[I_2] = 0.11$ M, and $[HI] = 0.78$ M. What is the value of the equilibrium constant at this temperature? The expression for K_c can be written from the balanced equation.

Since equilibrium constants depend on temperature, many equilibrium problems will state the temperature even though it has no formal part in the calculation.

$$K_c = \frac{[HI]^2}{[H_2][I_2]}$$

To calculate the value of K_c, we substitute the correct equilibrium concentrations into the expression for K_c

$$K_c = \frac{[HI]^2}{[H_2][I_2]}$$

$$= \frac{0.78^2}{(0.11)(0.11)}$$

$$= 5.0 \times 10^1$$

The concentrations within K_c should always be written in moles per liter (M); however, as noted previously in Section 14.4, we do not include units when expressing the value of the equilibrium constant, so that K_c is unitless.

For any reaction, the equilibrium *concentrations* of the reactants and products depend on the initial concentrations and, in general, will vary from one set of initial concentrations to another. However, the equilibrium *constant* is always the same at a given temperature, regardless of the initial concentrations. Table 14.1 shows several different equilibrium concentrations of H_2, I_2, and HI, each from a different set of initial concentrations. Notice that the equilibrium constant is always the same, regardless of the initial concentrations. Notice also that, whether you start with only reactants or only products, the reaction reaches equilibrium concentrations in which the equilibrium constant is the same. In other words, no matter what the initial concentrations are, the reaction always goes in a direction so that the equilibrium concentrations—when substituted into the equilibrium constant expression—give the same constant, K.

TABLE 14.1 Initial and Equilibrium Concentrations for the Reaction $H_2(g) + I_2(g) \rightleftharpoons 2\,HI(g)$ at 445 °C

Initial Concentrations			Equilibrium Concentrations			Equilibrium Constant
$[H_2]$	$[I_2]$	$[HI]$	$[H_2]$	$[I_2]$	$[HI]$	$K_c = \dfrac{[HI]^2}{[H_2][I_2]}$
0.50	0.50	0.0	0.11	0.11	0.78	$\dfrac{0.78^2}{(0.11)(0.11)} = 50$
0.0	0.0	0.50	0.055	0.055	0.39	$\dfrac{0.39^2}{(0.055)(0.055)} = 50$
0.50	0.50	0.50	0.165	0.165	1.17	$\dfrac{1.17^2}{(0.165)(0.165)} = 50$
1.0	0.50	0.0	0.53	0.033	0.934	$\dfrac{0.934^2}{(0.53)(0.033)} = 50$
0.50	1.0	0.0	0.033	0.53	0.934	$\dfrac{0.934^2}{(0.033)(0.53)} = 50$

In each of the entries in Table 14.1, we calculated the equilibrium constant from values of the equilibrium concentrations of all the reactants and products. In most cases, however, we need only know the initial concentrations of the reactant(s) and the equilibrium concentration of any *one* reactant or product. The other equilibrium concentrations can be deduced from the stoichiometry of the reaction. For example, consider the simple reaction:

$$A(g) \rightleftharpoons 2\,B(g)$$

Suppose that we have a reaction mixture in which the initial concentration of A is 1.00 M and the initial concentration of B is 0.00 M. When equilibrium is reached, the concentration of A is 0.75 M. Since $[A]$ has changed by -0.25 M, we can deduce (based on the stoichiometry) that $[B]$ must have changed by $2 \times (+0.25\text{ M})$ or $+0.50$ M. We summarize the initial conditions, the changes, and the equilibrium conditions in the following table:

	$[A]$	$[B]$
Initial	1.00	0.00
Change	−0.25	+0.50
Equilibrium	0.75	0.50

This type of table is often referred to as an ICE table (I = initial, C = change, E = equilibrium). To calculate the equilibrium constant, we use the balanced equation to write an expression for the equilibrium constant and then substitute the equilibrium concentrations from the ICE table.

$$K_c = \frac{[B]^2}{[A]} = \frac{(0.50)^2}{(0.75)} = 0.33$$

Here we show the general procedure for solving these kinds of equilibrium problems in the left column and two examples exemplifying the procedure in the center and right columns.

PROCEDURE FOR...	EXAMPLE 14.5	EXAMPLE 14.6
Finding Equilibrium Constants from Experimental Concentration Measurements To solve these types of problems, follow the procedure outlined in this column.	**Finding Equilibrium Constants from Experimental Concentration Measurements** Consider the following reaction: $CO(g) + 2\,H_2(g) \rightleftharpoons CH_3OH(g)$ A reaction mixture at 780 °C initially contains $[CO] = 0.500$ M and $[H_2] = 1.00$ M. At equilibrium, the CO concentration is found to be 0.15 M. What is the value of the equilibrium constant?	**Finding Equilibrium Constants from Experimental Concentration Measurements** Consider the following reaction: $2\,CH_4(g) \rightleftharpoons C_2H_2(g) + 3\,H_2(g)$ A reaction mixture at 1700 °C initially contains $[CH_4] = 0.115$ M. At equilibrium, the mixture contains $[C_2H_2] = 0.035$ M. What is the value of the equilibrium constant?

1. Using the balanced equation as a guide, prepare an ICE table showing the known initial concentrations and equilibrium concentrations of the reactants and products. Leave space in the middle of the table for determining the changes in concentration that occur during the reaction.

$CO(g) + 2\,H_2(g) \rightleftharpoons CH_3OH(g)$

	$[CO]$	$[H_2]$	$[CH_3OH]$
Initial	0.500	1.00	0.00
Change			
Equil	0.15		

$2\,CH_4(g) \rightleftharpoons C_2H_2(g) + 3\,H_2(g)$

	$[CH_4]$	$[C_2H_2]$	$[H_2]$
Initial	0.115	0.00	0.00
Change			
Equil		0.035	

2. Calculate the change in concentration that must have occurred for the reactant or product whose concentration is known both initially and at equilibrium.

$CO(g) + 2\,H_2(g) \rightleftharpoons CH_3OH(g)$

	$[CO]$	$[H_2]$	$[CH_3OH]$
Initial	0.500	1.00	0.00
Change	−0.35		
Equil	0.15		

$2\,CH_4(g) \rightleftharpoons C_2H_2(g) + 3\,H_2(g)$

	$[CH_4]$	$[C_2H_2]$	$[H_2]$
Initial	0.115	0.00	0.00
Change		+0.035	
Equil		0.035	

3. Use the change calculated in step 2 and the stoichiometric relationships from the balanced chemical equation to determine the changes in concentration of all other reactants and products. Since reactants are consumed during the reaction, the changes in their concentrations are negative. Since products are formed, the changes in their concentrations are positive.

$CO(g) + 2\,H_2(g) \rightleftharpoons CH_3OH(g)$

	$[CO]$	$[H_2]$	$[CH_3OH]$
Initial	0.500	1.00	0.00
Change	−0.35	−0.70	+0.35
Equil	0.15		

$2\,CH_4(g) \rightleftharpoons C_2H_2(g) + 3\,H_2(g)$

	$[CH_4]$	$[C_2H_2]$	$[H_2]$
Initial	0.115	0.00	0.00
Change	−0.070	+0.035	+0.105
Equil		0.035	

4. Sum each column for each reactant and product to determine the equilibrium concentrations.

	$[CO]$	$[H_2]$	$[CH_3OH]$
Initial	0.500	1.00	0.00
Change	−0.35	−0.70	+0.35
Equil	0.15	0.30	0.35

	$[CH_4]$	$[C_2H_2]$	$[H_2]$
Initial	0.115	0.00	0.00
Change	−0.070	+0.035	+0.105
Equil	0.045	0.035	0.105

5. Use the balanced equation to write an expression for the equilibrium constant and substitute the equilibrium concentrations to calculate K.

$$K_c = \frac{[CH_3OH]}{[CO][H_2]^2}$$
$$= \frac{0.35}{(0.15)(0.30)^2} = 26$$

$$K_c = \frac{[C_2H_2][H_2]^3}{[CH_4]^2}$$
$$= \frac{(0.035)(0.105)^3}{0.045^2} = 0.020$$

FOR PRACTICE 14.5
The reaction between CO and H_2 in Example 14.5 was carried out at a different temperature with initial concentrations of $[CO] = 0.27$ M and $[H_2] = 0.49$ M. At equilibrium, the concentration of CH_3OH was 0.11 M. Find the equilibrium constant at this temperature.

FOR PRACTICE 14.6
The reaction of CH_4 in Example 14.6 was carried out at a different temperature with an initial concentration of $[CH_4] = 0.087$ M. At equilibrium, the concentration of H_2 was 0.012 M. Find the equilibrium constant at this temperature.

14.7 The Reaction Quotient: Predicting the Direction of Change

When the reactants of a chemical reaction are mixed, they generally react to form products—the reaction proceeds to the right (toward the products). The amount of products formed when equilibrium is reached depends on the magnitude of the equilibrium constant, as we have seen. However, what if a reaction mixture that is *not at equilibrium* contains both reactants *and products*? Can we predict the direction of change for such a mixture?

To gauge the progress of a reaction with respect to equilibrium, we use a quantity called the *reaction quotient*. The definition of the reaction quotient takes the same form as the definition of the equilibrium constant, except that the reaction *need not be at equilibrium*. So, for the general reaction

$$a\text{A} + b\text{B} \rightleftharpoons c\text{C} + d\text{D}$$

we define the **reaction quotient** (Q_c) as the ratio—at any point in the reaction—of the concentrations of the products raised to their stoichiometric coefficients divided by the concentrations of the reactants raised to their stoichiometric coefficients. For gases with quantities measured in atmospheres, the reaction quotient uses the partial pressures in place of concentrations and is called Q_p.

$$Q_c = \frac{[\text{C}]^c [\text{D}]^d}{[\text{A}]^a [\text{B}]^b} \qquad Q_p = \frac{P_\text{C}^c P_\text{D}^d}{P_\text{A}^a P_\text{B}^b}$$

The difference between the reaction quotient and the equilibrium constant is that, at a given temperature, the equilibrium constant has only one value and it specifies the relative amounts of reactants and products *at equilibrium*. The reaction quotient, however, depends on the current state of the reaction and will have many different values as the reaction proceeds. For example, in a reaction mixture containing only reactants, the reaction quotient is zero ($Q_c = 0$):

$$Q_c = \frac{[0]^c [0]^d}{[\text{A}]^a [\text{B}]^b} = 0$$

In a reaction mixture containing only products, the reaction quotient is infinite ($Q_c = \infty$):

$$Q_c = \frac{[\text{C}]^c [\text{D}]^d}{[0]^a [0]^b} = \infty$$

In a reaction mixture containing both reactants and products, each at a concentration of 1 M, the reaction quotient is one ($Q_c = 1$):

$$Q_c = \frac{(1)^c (1)^d}{(1)^a (1)^b} = 1$$

The reaction quotient is useful because *the value of Q relative to K is a measure of the progress of the reaction toward equilibrium. At equilibrium, the reaction quotient is equal to the equilibrium constant.* **Figure 14.6▶** shows a plot of *Q* as a function of the concentrations of A and B for the simple reaction $\text{A}(g) \rightleftharpoons \text{B}(g)$, which has an equilibrium constant of $K = 1.45$. The following points are representative of three possible conditions.

Q	K	Predicted Direction of Reaction
0.55	1.45	To the right (toward products)
2.55	1.45	To the left (toward reactants)
1.45	1.45	No change (reaction at equilibrium)

Summarizing Direction of Change Based on Q and K:

The reaction quotient (Q) is a measure of the progress of a reaction toward equilibrium.

▶ $Q < K$ Reaction goes to the right (toward products).

▶ $Q > K$ Reaction goes to the left (toward reactants).

▶ $Q = K$ Reaction is at equilibrium—the reaction will not proceed in either direction.

Q, K, and the Direction of a Reaction

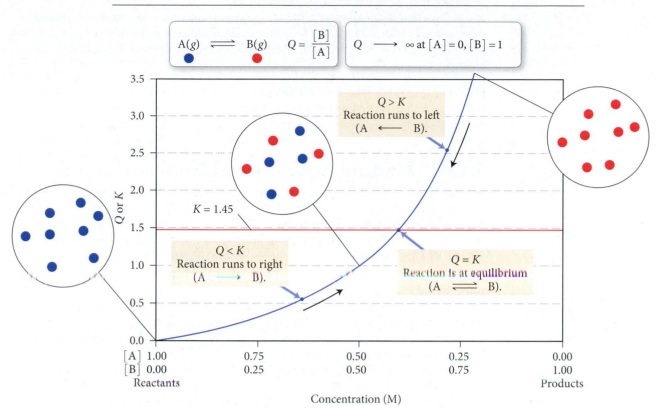

▲ **FIGURE 14.6 Q, K, and the Direction of a Reaction**   The graph shows a plot of Q as a function of the concentrations of the reactants and products in a simple reaction A ⇌ B, in which $K = 1.45$ and the sum of the reactant and product concentrations is 1 M. Therefore, the far left of the graph represents pure reactant and the far right represents pure product. The midpoint of the graph represents an equal mixture of A and B. When Q is less than K, the reaction moves in the forward direction (A ⟶ B). When Q is greater than K, the reaction moves in the reverse direction (A ⟵ B). When Q is equal to K, the reaction is at equilibrium.

EXAMPLE 14.7 Predicting the Direction of a Reaction by Comparing Q and K

Consider the reaction and its equilibrium constant:

$$I_2(g) + Cl_2(g) \rightleftharpoons 2\,ICl(g) \qquad K_p = 81.9$$

A reaction mixture contains $P_{I_2} = 0.114$ atm, $P_{Cl_2} = 0.102$ atm, and $P_{ICl} = 0.355$ atm. Is the reaction mixture at equilibrium? If not, in which direction will the reaction proceed?

SOLUTION

To determine the progress of the reaction relative to the equilibrium state, calculate Q.	$Q_p = \dfrac{P_{ICl}^2}{P_{I_2}P_{Cl_2}}$ $= \dfrac{0.355^2}{(0.114)(0.102)}$ $= 10.8$
Compare Q to K.	$Q_p = 10.8$; $K_p = 81.9$ Since $Q_p < K_p$, the reaction is not at equilibrium and will proceed to the right.

FOR PRACTICE 14.7

Consider the reaction and its equilibrium constant:

$$N_2O_4(g) \rightleftharpoons 2\,NO_2(g) \qquad K_c = 5.85 \times 10^{-3}$$

A reaction mixture contains $[NO_2] = 0.0255$ M and $[N_2O_4] = 0.0331$ M. Calculate Q_c and determine the direction in which the reaction will proceed.

Conceptual Connection 14.5   *Q* and *K*

For the reaction $N_2O_4(g) \rightleftharpoons 2 NO_2(g)$, a reaction mixture at a certain temperature initially contains both N_2O_4 and NO_2 in their standard states (see the definition of standard state in Section 6.8). If $K_p = 0.15$, which statement is true of the reaction mixture before any reaction occurs?

(a) $Q = K$; the reaction is at equilibrium.

(b) $Q < K$; the reaction will proceed to the right.

(c) $Q > K$; the reaction will proceed to the left.

14.8   Finding Equilibrium Concentrations

In Section 14.6, we demonstrated how to calculate an equilibrium constant given the equilibrium concentrations of the reactants and products. Just as commonly, we want to calculate equilibrium concentrations of reactants or products given the equilibrium constant. These kinds of calculations are important because they allow us to calculate the amount of a reactant or product at equilibrium. For example, in a synthesis reaction, we might want to know how much of the product forms when the reaction reaches equilibrium. Or for the hemoglobin–oxygen equilibrium discussed in Section 14.1, we might want to know the concentration of oxygenated hemoglobin present under certain oxygen concentrations within the lungs or muscles.

We can divide these types of problems into two categories: (1) finding equilibrium concentrations when you are given the equilibrium constant and all but one of the equilibrium concentrations of the reactants and products; and (2) finding equilibrium concentrations when you are given the equilibrium constant and only initial concentrations. The second category of problem is more difficult than the first. We examine each separately.

Finding Equilibrium Concentrations When You Are Given the Equilibrium Constant and All but One of the Equilibrium Concentrations of the Reactants and Products

We can use the equilibrium constant to calculate the equilibrium concentration of one of the reactants or products, given the equilibrium concentrations of the others. To solve this type of problem, follow our general problem-solving procedure.

EXAMPLE 14.8   Finding Equilibrium Concentrations When You Are Given the Equilibrium Constant and All but One of the Equilibrium Concentrations of the Reactants and Products

Consider the reaction.

$$2 COF_2(g) \rightleftharpoons CO_2(g) + CF_4(g) \qquad K_c = 2.00 \text{ at } 1000 \text{ °C}$$

In an equilibrium mixture, the concentration of COF_2 is 0.255 M and the concentration of CF_4 is 0.118 M. What is the equilibrium concentration of CO_2?

SORT You are given the equilibrium constant of a chemical reaction, together with the equilibrium concentrations of the reactant and one product. You are asked to find the equilibrium concentration of the other product.	**GIVEN** $[COF_2] = 0.255$ M $[CF_4] = 0.118$ M $K_c = 2.00$ **FIND** $[CO_2]$
STRATEGIZE Calculate the concentration of the product by using the given quantities and the expression for K_c.	**CONCEPTUAL PLAN** 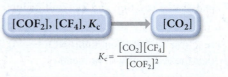 $K_c = \dfrac{[CO_2][CF_4]}{[COF_2]^2}$

SOLVE Solve the equilibrium expression for $[CO_2]$ and then substitute in the appropriate values to calculate it.	SOLUTION

$$[CO_2] = K_c \frac{[COF_2]^2}{[CF_4]}$$

$$[CO_2] = 2.00\left(\frac{0.255^2}{0.118}\right) = 1.10 \text{ M}$$

CHECK Check your answer by substituting the given values of $[COF_2]$ and $[CF_4]$ as well as the calculated value for $[CO_2]$ back into the equilibrium expression.

$$K_c = \frac{[CO_2][CF_4]}{[COF_2]^2}$$

$[CO_2]$ was found to be roughly equal to one. $[COF_2]^2 \approx 0.06$ and $[CF_4] \approx 0.12$. Therefore K_c is approximately 2, as given in the problem.

FOR PRACTICE 14.8

Molecular iodine (I_2) decomposes at high temperature to form I atoms according to the reaction:

$$I_2(g) \rightleftharpoons 2 \text{ I}(g) \qquad K_c = 0.011 \text{ at } 1200 \text{ °C}$$

In an equilibrium mixture, the concentration of I_2 is 0.10 M. What is the equilibrium concentration of I?

Finding Equilibrium Concentrations When You Are Given the Equilibrium Constant and Initial Concentrations or Pressures

More commonly, we know the equilibrium constant and only initial concentrations of reactants and need to find the *equilibrium concentrations* of the reactants or products. These kinds of problems are generally more involved than those we just examined and require a specific procedure to solve them. The procedure has some similarities to the one used in Examples 14.5 and 14.6 in that we must set up an ICE table showing the initial conditions, the changes, and the equilibrium conditions. However, unlike Examples 14.5 and 14.6, where the change in concentration of at least one reactant or product is known, here the changes in concentration are not known and are represented with the variable x. For example, consider again the simple reaction:

$$A(g) \rightleftharpoons 2 \text{ B}(g)$$

Suppose that, as before, we have a reaction mixture in which the initial concentration of A is 1.0 M and the initial concentration of B is 0.00 M. However, now we know the equilibrium constant, $K = 0.33$, and want to find the equilibrium concentrations. We set up the ICE table with the given initial concentrations and then *represent the unknown change in [A] in terms of the variable x*:

	[A]	[B]
Initial	1.0	0.00
Change	$-x$	$+2x$
Equil	$1.0-x$	$2x$

Represent changes from initial conditions with the variable x.

Notice that, due to the stoichiometry of the reaction, the change in [B] must be +2x. As before, each equilibrium concentration is the sum of the two entries above it in the ICE table. In order to find the equilibrium concentrations of A and B, we must find the value of the variable x. Since we know the value of the equilibrium constant, we can use the equilibrium expression to set up an equation in which x is the only variable.

$$K = \frac{[B]^2}{[A]} = \frac{(2x)^2}{1.0 - x} = 0.33$$

or simply,

$$\frac{4x^2}{1.0 - x} = 0.33$$

The above equation is a *quadratic* equation—it contains the variable x raised to the second power. In general, we can solve quadratic equations with the quadratic formula, which we introduce in Example 14.10. If the quadratic equation is a perfect square, however, we can solve it by simpler means, as shown in Example 14.9. For both of these examples, we give the general procedure in the left column and apply the procedure to the example problems in the center and right columns. Later in this section, we will see that quadratic equations can often be simplified by making some approximations based on our chemical knowledge.

PROCEDURE FOR...	EXAMPLE 14.9	EXAMPLE 14.10
Finding Equilibrium Concentrations from Initial Concentrations and the Equilibrium Constant To solve these types of problems, follow the procedure outlined in this column.	**Finding Equilibrium Concentrations from Initial Concentrations and the Equilibrium Constant** Consider the reaction. $$N_2(g) + O_2(g) \rightleftharpoons 2\,NO(g)$$ $$K_c = 0.10 \text{ (at 2000 °C)}$$ A reaction mixture (at 2000 °C) initially contains $[N_2] = 0.200$ M and $[O_2] = 0.200$ M. Find the equilibrium concentrations of the reactants and products at this temperature.	**Finding Equilibrium Concentrations from Initial Concentrations and the Equilibrium Constant** Consider the reaction. $$N_2O_4(g) \rightleftharpoons 2\,NO_2(g)$$ $$K_c = 0.36 \text{ (at 100 °C)}$$ A reaction mixture at 100 °C initially contains $[NO_2] = 0.100$ M. Find the equilibrium concentrations of NO_2 and N_2O_4 at this temperature.

1. Using the balanced equation as a guide, prepare a table showing the known initial concentrations of the reactants and products. Leave room in the table for the changes in concentrations and for the equilibrium concentrations.	$N_2(g) + O_2(g) \rightleftharpoons 2\,NO(g)$ 		$N_2O_4(g) \rightleftharpoons 2\,NO_2(g)$

	$[N_2]$	$[O_2]$	$[NO]$
Initial	0.200	0.200	0.00
Change			
Equil			

	$[N_2O_4]$	$[NO_2]$
Initial	0.00	0.100
Change		
Equil		

2. Use the initial concentrations to calculate the reaction quotient (Q) for the initial concentrations. Compare Q to K and predict the direction in which the reaction will proceed.	$Q_c = \dfrac{[NO]^2}{[N_2][O_2]} = \dfrac{0.00^2}{(0.200)(0.200)}$ $= 0$ $Q < K$, therefore the reaction will proceed to the right.	$Q_c = \dfrac{[NO_2]^2}{[N_2O_4]} = \dfrac{(0.100)^2}{0.00}$ $= \infty$ $Q > K$, therefore the reaction will proceed to the left.

3. Represent the change in the concentration of one of the reactants or products with the variable x. Define the changes in the concentrations of the other reactants or products in terms of x. It is usually most convenient to let x represent the change in concentration of the reactant or product with the smallest stoichiometric coefficient.

$N_2(g) + O_2(g) \rightleftharpoons 2\,NO(g)$

	$[N_2]$	$[O_2]$	$[NO]$
Initial	0.200	0.200	0.00
Change	$-x$	$-x$	$+2x$
Equil			

$N_2O_4(g) \rightleftharpoons 2\,NO_2(g)$

	$[N_2O_4]$	$[NO_2]$
Initial	0.00	0.100
Change	$+x$	$-2x$
Equil		

4. Sum each column for each reactant and product to determine the equilibrium concentrations in terms of the initial concentrations and the variable x.

$$N_2(g) + O_2(g) \rightleftharpoons 2\,NO(g)$$

	$[N_2]$	$[O_2]$	$[NO]$
Initial	0.200	0.200	0.00
Change	$-x$	$-x$	$+2x$
Equil	$0.200 - x$	$0.200 - x$	$2x$

$$N_2O_4(g) \rightleftharpoons 2\,NO_2(g)$$

	$[N_2O_4]$	$[NO_2]$
Initial	0.00	0.100
Change	$+x$	$-2x$
Equil	x	$0.100 - 2x$

5. Substitute the expressions for the equilibrium concentrations (from step 4) into the expression for the equilibrium constant. Using the given value of the equilibrium constant, solve the expression for the variable x. In some cases, such as Example 14.9 here, you can take the square root of both sides of the expression to solve for x. In other cases, such as the Example 14.10, you must solve a quadratic equation to find x.

Remember the quadratic formula:
$$ax^2 + bx + c = 0$$

$$x = \frac{-b \pm \sqrt{b^2 - 4ac}}{2a}$$

$$K_c = \frac{[NO]^2}{[N_2][O_2]}$$
$$= \frac{(2x)^2}{(0.200 - x)(0.200 - x)}$$
$$0.10 = \frac{(2x)^2}{(0.200 - x)^2}$$
$$\sqrt{0.10} = \frac{2x}{0.200 - x}$$
$$\sqrt{0.10}\,(0.200 - x) = 2x$$
$$\sqrt{0.10}\,(0.200) - \sqrt{0.10}\,x = 2x$$
$$0.063 = 2x + \sqrt{0.10}\,x$$
$$0.063 = 2.3x$$
$$x = 0.027$$

$$K_c = \frac{[NO_2]^2}{[N_2O_4]}$$
$$= \frac{(0.100 - 2x)^2}{x}$$
$$0.36 = \frac{0.0100 - 0.400x + 4x^2}{x}$$
$$0.36x = 0.0100 - 0.400x + 4x^2$$
$$4x^2 - 0.76x + 0.0100 = 0 \ (quadratic)$$
$$x = \frac{-b \pm \sqrt{b^2 - 4ac}}{2a}$$
$$= \frac{-(-0.76) \pm \sqrt{(-0.76)^2 - 4(4)(0.0100)}}{2(4)}$$
$$= \frac{0.76 \pm 0.65}{8}$$
$$x = 0.176 \quad \text{or} \quad x = 0.014$$

6. Substitute x into the expressions for the equilibrium concentrations of the reactants and products (from step 4) and calculate the concentrations. In cases where you solved a quadratic and have two values for x, choose the value for x that gives a physically realistic answer. For example, reject the value of x that results in any negative concentrations.

$$[N_2] = 0.200 - 0.027$$
$$= 0.173 \text{ M}$$
$$[O_2] = 0.200 - 0.027$$
$$= 0.173 \text{ M}$$
$$[NO] = 2(0.027)$$
$$= 0.054 \text{ M}$$

We reject the root $x = 0.176$ because it gives a negative concentration for NO_2. Using $x = 0.014$, we get the following concentrations:

$$[NO_2] = 0.100 - 2x$$
$$= 0.100 - 2(0.014) = 0.072 \text{ M}$$
$$[N_2O_4] = x$$
$$= 0.014 \text{ M}$$

7. Check your answer by substituting the calculated equilibrium values into the equilibrium expression. The calculated value of K should match the given value of K. Note that rounding errors could cause a difference in the least significant digit when comparing values of the equilibrium constant.

$$K_c = \frac{[NO]^2}{[N_2][O_2]}$$
$$= \frac{0.054^2}{(0.173)(0.173)} = 0.097$$

Since the calculated value of K_c matches the given value (to within one digit in the least significant figure), the answer is valid.

$$K_c = \frac{[NO_2]^2}{[N_2O_4]} = \frac{0.072^2}{0.014}$$
$$= 0.37$$

Since the calculated value of K_c matches the given value (to within one digit in the least significant figure), the answer is valid.

FOR PRACTICE 14.9

The reaction in Example 14.9 is carried out at a different temperature at which $K_c = 0.055$. This time, however, the reaction mixture started with only the product, $[NO] = 0.0100$ M, and no reactants. Find the equilibrium concentrations of N_2, O_2, and NO at equilibrium.

FOR PRACTICE 14.10

The reaction in Example 14.10 is carried out at the same temperature, but this time the reaction mixture initially contained only the reactant, $[N_2O_4] = 0.0250$ M, and no NO_2. Find the equilibrium concentrations of N_2O_4 and NO_2.

When the initial conditions are given in terms of partial pressures (instead of concentrations) and the equilibrium constant is given as K_p instead of K_c, use the same procedure, but substitute partial pressures for concentrations, as shown Example 14.11.

EXAMPLE 14.11 Finding Equilibrium Partial Pressures When You Are Given the Equilibrium Constant and Initial Partial Pressures

Consider the reaction.

$$I_2(g) + Cl_2(g) \rightleftharpoons 2 ICl(g) \qquad K_p = 81.9 \text{ (at } 25 \,°C)$$

A reaction mixture at 25 °C initially contains $P_{I_2} = 0.100$ atm, $P_{Cl_2} = 0.100$ atm, and $P_{ICl} = 0.100$ atm. Find the equilibrium partial pressures of I_2, Cl_2, and ICl at this temperature.

SOLUTION

Follow the procedure outlined in Examples 14.5 and 14.6 (using the pressures in place of the concentrations) to solve the problem.

1. Using the balanced equation as a guide, prepare a table showing the known initial partial pressures of the reactants and products.

$$I_2(g) + Cl_2(g) \rightleftharpoons 2 ICl(g)$$

	P_{I_2} (atm)	P_{Cl_2} (atm)	P_{ICl_2} (atm)
Initial	0.100	0.100	0.100
Change			
Equil			

2. Use the initial partial pressures to calculate the reaction quotient (Q). Compare Q to K and predict the direction in which the reaction will proceed.

$$Q_p = \frac{P_{ICl}^2}{P_{I_2}P_{Cl_2}} = \frac{0.100^2}{(0.100)(0.100)} = 1$$

$K_p = 81.9$ (given)

$Q < K$, therefore the reaction will proceed to the right.

3. Represent the change in the partial pressure of one of the reactants or products with the variable x. Define the changes in the partial pressures of the other reactants or products in terms of x.

$$I_2(g) + Cl_2(g) \rightleftharpoons 2 ICl(g)$$

	P_{I_2} (atm)	P_{Cl_2} (atm)	P_{ICl_2} (atm)
Initial	0.100	0.100	0.100
Change	$-x$	$-x$	$+2x$
Equil			

4. Sum each column for each reactant and product to determine the equilibrium partial pressures in terms of the initial partial pressures and the variable x.

$$I_2(g) + Cl_2(g) \rightleftharpoons 2 ICl(g)$$

	P_{I_2} (atm)	P_{Cl_2} (atm)	P_{ICl_2} (atm)
Initial	0.100	0.100	0.100
Change	$-x$	$-x$	$+2x$
Equil	$0.100 - x$	$0.100 - x$	$0.100 + 2x$

5. Substitute the expressions for the equilibrium partial pressures (from step 4) into the expression for the equilibrium constant. Use the given value of the equilibrium constant to solve the expression for the variable x.

$$K_p = \frac{P_{ICl}^2}{P_{I_2}P_{Cl_2}} = \frac{(0.100 + 2x)^2}{(0.100 - x)(0.100 - x)}$$

$$81.9 = \frac{(0.100 + 2x)^2}{(0.100 - x)^2} \qquad \text{(perfect square)}$$

$$\sqrt{81.9} = \frac{(0.100 + 2x)}{(0.100 - x)}$$

$$\sqrt{81.9}(0.100 - x) = 0.100 + 2x$$

$$\sqrt{81.9}(0.100) - \sqrt{81.9}x = 0.100 + 2x$$

$$\sqrt{81.9}(0.100) - 0.100 = 2x + \sqrt{81.9}x$$

$$0.805 = 11.05x$$

$$x = 0.0729$$

6. Substitute x into the expressions for the equilibrium partial pressures of the reactants and products (from step 4) and calculate the partial pressures.	$P_{I_2} = 0.100 - 0.0729 = 0.027$ atm $P_{Cl_2} = 0.100 - 0.0729 = 0.027$ atm $P_{ICl} = 0.100 + 2(0.0729) = 0.246$ atm
7. Check your answer by substituting the computed equilibrium partial pressures into the equilibrium expression. The calculated value of K should match the given value of K.	$K_p = \dfrac{P_{ICl}^2}{P_{I_2}P_{Cl_2}} = \dfrac{0.246^2}{(0.027)(0.027)}$ $= 83$ Since the computed value of K_p matches the given value (within the uncertainty indicated by the significant figures), the answer is valid.

FOR PRACTICE 14.11

The reaction between I_2 and Cl_2 (from Example 14.11) was carried out at the same temperature, but with the following initial partial pressures: $P_{I_2} = 0.150$ atm, $P_{Cl_2} = 0.150$ atm, $P_{ICl} = 0.00$ atm. Find the equilibrium partial pressures of all three substances.

Simplifying Approximations in Working Equilibrium Problems

For some equilibrium problems of the type shown in Examples 14.9, 14.10, and 14.11, there is the possibility of using an approximation that makes solving the problem much easier without significant loss of accuracy. For example, if the equilibrium constant is relatively small, the reaction will not proceed very far to the right. If the initial reactant concentration is relatively large, we can make the assumption that x is small relative to the initial concentration of reactant. To see how this approximation works, consider again the simple reaction A $\rightleftharpoons$ 2 B. Suppose that, as before, we have a reaction mixture in which the initial concentration of A is 1.0 M and the initial concentration of B is 0.0 M, and that we want to find the equilibrium concentrations. However, suppose that in this case the equilibrium constant is much smaller, say $K = 3.3 \times 10^{-5}$. The ICE table is identical to the one we set up previously:

	[A]	[B]
Initial	1.0	0.0
Change	$-x$	$+2x$
Equil	$1.0 - x$	$2x$

With the exception of the value of K, we end up with the exact quadratic equation that we had before:

$$K = \frac{[B]^2}{[A]} = \frac{(2x)^2}{1.0 - x} = 3.3 \times 10^{-5}$$

or simply,

$$\frac{4x^2}{1.0 - x} = 3.3 \times 10^{-5}$$

We can multiply this quadratic equation out and solve it using the quadratic formula. However, since we know that K is small, we know that the reaction will not proceed very far toward products; therefore, x will also be small. If x is much smaller than 1.0, then $(1.0 - x)$ (the quantity in the denominator) can be approximated by (1.0).

$$\frac{4x^2}{(1.0 - \cancel{x})} = 3.3 \times 10^{-5}$$

This approximation greatly simplifies the equation, which can then be readily solved for x as follows:

$$\frac{4x^2}{1.0} = 3.3 \times 10^{-5}$$

$$4x^2 = 3.3 \times 10^{-5}$$

$$x = \sqrt{\frac{3.3 \times 10^{-5}}{4}} = 0.0029$$

We can check the validity of this approximation by comparing the calculated value of x to the number it was subtracted from. The ratio of x to the number it is subtracted from should be less than 0.05 (or 5%) for the approximation to be valid. In this case, x was subtracted from 1.0, and the ratio of the value of x to 1.0 is calculated as follows:

$$\frac{0.0029}{1.0} \times 100\% = 0.29\%$$

The approximation is therefore valid. In Examples 14.12 and 14.13, we present two nearly identical problems—the only difference is the initial concentration of the reactant. In Example 14.12, the initial concentration of the reactant is relatively large, the equilibrium constant is small, and the x is small approximation works well. In Example 14.13, however, the initial concentration of the reactant is much smaller, and even though the equilibrium constant is the same, the x is small approximation does not work (because the initial concentration is also small). In cases like Example 14.13, we have a couple of options to solve the problem. We can either go back and solve the equation exactly (using the quadratic formula, for example), or we can use the *method of successive approximations*. In this method, which is introduced in Example 14.13, we essentially solve for x as if it were small, and then substitute the value obtained back into the equation to solve for x again. This can be repeated until the computed value of x stops changing with each iteration, an indication that we have arrived at an acceptable value for x.

Note that the x is small approximation does not imply that x is zero. If that were the case, the reactant and product concentrations would not change from their initial values. The x is small approximation simply means that when x is added or subtracted to another number, it does not change that number by very much. For example, let us compute the value of the difference $1.0 - x$ when $x = 3.0 \times 10^{-4}$:

$$1.0 - x = 1.0 - 3.0 \times 10^{-4} = 0.9997 = 1.0$$

Since the value of 1.0 is known only to two significant figures, subtracting the small x does not change the value at all. This situation is similar to weighing yourself on a bathroom scale with and without a penny in your pocket. Unless your scale is unusually precise, removing the penny from your pocket does not change the reading on the scale. This does not imply that the penny is weightless, only that its weight is small when compared to your weight. The weight of the penny can therefore be neglected in reading your weight with no detectable loss in accuracy.

PROCEDURE FOR...	**EXAMPLE 14.12**	**EXAMPLE 14.13**
Finding Equilibrium Concentrations from Initial Concentrations in Cases with a Small Equilibrium Constant	**Finding Equilibrium Concentrations from Initial Concentrations in Cases with a Small Equilibrium Constant**	**Finding Equilibrium Concentrations from Initial Concentrations in Cases with a Small Equilibrium Constant**
To solve these types of problems, follow the procedure outlined in this column.	Consider the reaction for the decomposition of hydrogen disulfide:	Consider the reaction for the decomposition of hydrogen disulfide:

$2\,H_2S(g) \rightleftharpoons 2\,H_2(g) + S_2(g)$ $K_c = 1.67 \times 10^{-7}$ at 800 °C	$2\,H_2S(g) \rightleftharpoons 2\,H_2(g) + S_2(g)$ $K_c = 1.67 \times 10^{-7}$ at 800 °C
A 0.500-L reaction vessel initially contains 0.0125 mol of H_2S at 800 °C. Find the equilibrium concentrations of H_2 and S_2.	A 0.500-L reaction vessel initially contains 1.25×10^{-4} mol of H_2S at 800 °C. Find the equilibrium concentrations of H_2 and S_2.

1. Using the balanced equation as a guide, prepare a table showing the known initial concentrations of the reactants and products. (In these examples, you must first calculate the concentration of H_2S from the given number of moles and volume.)

$$[H_2S] = \frac{0.0125\ \text{mol}}{0.500\ \text{L}} = 0.0250\ \text{M}$$

$2\,H_2S(g) \rightleftharpoons 2\,H_2(g) + S_2(g)$

	$[H_2S]$	$[H_2]$	$[S_2]$
Initial	0.0250	0.00	0.00
Change			
Equil			

$$[H_2S] = \frac{1.25 \times 10^{-4}\ \text{mol}}{0.500\ \text{L}} = 2.50 \times 10^{-4}\ \text{M}$$

$2\,H_2S(g) \rightleftharpoons 2\,H_2(g) + S_2(g)$

	$[H_2S]$	$[H_2]$	$[S_2]$
Initial	2.50×10^{-4}	0.00	0.00
Change			
Equil			

2. Use the initial concentrations to calculate the reaction quotient (Q). Compare Q to K and predict the direction that the reaction will proceed.

By inspection, $Q = 0$; the reaction will proceed to the right.	By inspection, $Q = 0$; the reaction will proceed to the right.

3. Represent the change in the concentration of one of the reactants or products with the variable x. Define the changes in the concentrations of the other reactants or products with respect to x.

$2 H_2S(g) \rightleftharpoons 2 H_2(g) + S_2(g)$

	$[H_2S]$	$[H_2]$	$[S_2]$
Initial	0.0250	0.00	0.00
Change	$-2x$	$+2x$	$+x$
Equil			

$2 H_2S(g) \rightleftharpoons 2 H_2(g) + S_2(g)$

	$[H_2S]$	$[H_2]$	$[S_2]$
Initial	2.50×10^{-4}	0.00	0.00
Change	$-2x$	$+2x$	$+x$
Equil			

4. Sum each column for each reactant and product to determine the equilibrium concentrations in terms of the initial concentrations and the variable x.

$2 H_2S(g) \rightleftharpoons 2 H_2(g) + S_2(g)$

	$[H_2S]$	$[H_2]$	$[S_2]$
Initial	0.0250	0.00	0.00
Change	$-2x$	$+2x$	$+x$
Equil	$0.0250 - 2x$	$2x$	x

$2 H_2S(g) \rightleftharpoons 2 H_2(g) + S_2(g)$

	$[H_2S]$	$[H_2]$	$[S_2]$
Initial	2.50×10^{-4}	0.00	0.00
Change	$-2x$	$+2x$	$+x$
Equil	$2.50 \times 10^{-4} - 2x$	$2x$	x

5. Substitute the expressions for the equilibrium concentrations (from step 4) into the expression for the equilibrium constant. Use the given value of the equilibrium constant to solve the expression for the variable x.

In this case, the resulting equation is cubic in x. Although cubic equations can be solved, determining the solutions is not usually easy. However, since the equilibrium constant is small, we know that the reaction does not proceed very far to the right. Therefore, x will be a small number and can be dropped from any quantities in which it is added to or subtracted from another number (as long as the number itself is not too small).

Check whether your approximation is valid by comparing the calculated value of x to the number it was added to or subtracted from. The ratio of the two numbers should be less than 0.05 (or 5%) for the approximation to be valid. If approximation is not valid, proceed to step 5a.

$$K_c = \frac{[H_2]^2[S_2]}{[H_2S]^2}$$

$$= \frac{(2x)^2(x)}{(0.0250 - 2x)^2}$$

$$1.67 \times 10^{-7} = \frac{4x^3}{(0.0250 - 2x)^2}$$

x is small.

$$1.67 \times 10^{-7} = \frac{4x^3}{(0.0250 - \cancel{2x})^2}$$

$$1.67 \times 10^{-7} = \frac{4x^3}{6.25 \times 10^{-4}}$$

$$6.25 \times 10^{-4}(1.67 \times 10^{-7}) = 4x^3$$

$$x^3 = \frac{6.25 \times 10^{-4}(1.67 \times 10^{-7})}{4}$$

$$x = 2.97 \times 10^{-4}$$

Checking the *x is small* approximation:

$$\frac{2.97 \times 10^{-4}}{0.0250} \times 100\% = 1.19\%$$

The *x is small* approximation is valid, proceed to step 6.

$$K_c = \frac{[H_2]^2[S_2]}{[H_2S]^2}$$

$$= \frac{(2x)^2 x}{(2.50 \times 10^{-4} - 2x)^2}$$

$$1.67 \times 10^{-7} = \frac{4x^3}{(2.50 \times 10^{-4} - 2x)^2}$$

x is small.

$$1.67 \times 10^{-7} = \frac{4x^3}{(2.50 \times 10^{-4} - \cancel{2x})^2}$$

$$1.67 \times 10^{-7} = \frac{4x^3}{6.25 \times 10^{-8}}$$

$$6.25 \times 10^{-8}(1.67 \times 10^{-7}) = 4x^3$$

$$x^3 = \frac{6.25 \times 10^{-8}(1.67 \times 10^{-7})}{4}$$

$$x = 1.38 \times 10^{-5}$$

Checking the *x is small* approximation:

$$\frac{1.38 \times 10^{-5}}{2.50 \times 10^{-4}} \times 100\% = 5.52\%$$

The approximation does not satisfy the $< 5\%$ rule (although it is close).

5a. If the approximation is not valid, you can either solve the equation exactly (either by hand or with your calculator), or use the method of successive approximations. In this case, we use the method of successive approximations.

$$1.67 \times 10^{-7} = \frac{4x^3}{(2.50 \times 10^{-4} - 2x)^2}$$

$$x = 1.38 \times 10^{-5}$$

$$1.67 \times 10^{-7} = \frac{4x^3}{(2.50 \times 10^{-4} - 2.76 \times 10^{-5})^2}$$

$$x = 1.27 \times 10^{-5}$$

Substitute the value obtained for x in step 5 back into the original cubic equation, but only at the exact spot where x was assumed to be negligible and then solve the equation for x again. Continue this procedure until the value of x obtained from solving the equation is the same as the one that is substituted into the equation.		If we substitute this value of x back into the cubic equation and solve it, we get $x = 1.28 \times 10^{-5}$, which is nearly identical to 1.27×10^{-5}. Therefore, we have arrived at the best approximation for x.

6. Substitute x into the expressions for the equilibrium concentrations of the reactants and products (from step 4) and calculate the concentrations.

$$[H_2S] = 0.0250 - 2(2.97 \times 10^{-4})$$
$$= 0.0244 \text{ M}$$
$$[H_2] = 2(2.97 \times 10^{-4})$$
$$= 5.94 \times 10^{-4} \text{ M}$$
$$[S_2] = 2.97 \times 10^{-4} \text{ M}$$

$$[H_2S] = 2.50 \times 10^{-4} - 2(1.28 \times 10^{-5})$$
$$= 2.24 \times 10^{-4} \text{ M}$$
$$[H_2] = 2(1.28 \times 10^{-5})$$
$$= 2.56 \times 10^{-5} \text{ M}$$
$$[S_2] = 1.28 \times 10^{-5} \text{ M}$$

7. Check your answer by substituting the calculated equilibrium concentrations into the equilibrium expression. The calculated value of K should match the given value of K. Note that the approximation method and rounding errors could cause a difference of up to about 5% when comparing values of the equilibrium constant.

$$K_c = \frac{(5.94 \times 10^{-4})^2(2.97 \times 10^{-4})}{(0.0244)^2}$$
$$= 1.76 \times 10^{-7}$$

The calculated value of K is close enough to the given value when we consider the uncertainty introduced by the approximation. Therefore the answer is valid.

$$K_c = \frac{(2.56 \times 10^{-5})^2(1.28 \times 10^{-5})}{(2.24 \times 10^{-4})^2}$$
$$= 1.67 \times 10^{-7}$$

The calculated value of K is equal to the given value. Therefore the answer is valid.

FOR PRACTICE 14.12

The reaction in Example 14.12 was carried out at the same temperature with the following initial concentrations: $[H_2S] = 0.100$ M, $[H_2] = 0.100$ M, and $[S_2] = 0.00$ M. Find the equilibrium concentration of $[S_2]$.

FOR PRACTICE 14.13

The reaction in Example 14.13 was carried out at the same temperature with the following initial concentrations: $[H_2S] = 1.00 \times 10^{-4}$ M, $[H_2] = 0.00$ M, and $[S_2] = 0.00$ M. Find the equilibrium concentration of $[S_2]$.

Conceptual Connection 14.6 **The *x is small* Approximation**

For the generic reaction, $A(g) \rightleftharpoons B(g)$, consider each value of K and initial concentration of A. For which set will the *x is small* approximation most likely apply?

(a) $K = 1.0 \times 10^{-5}$; $[A] = 0.250$ M
(b) $K = 1.0 \times 10^{-2}$; $[A] = 0.250$ M
(c) $K = 1.0 \times 10^{-5}$; $[A] = 0.00250$ M
(d) $K = 1.0 \times 10^{-2}$; $[A] = 0.00250$ M

14.9 Le Châtelier's Principle: How a System at Equilibrium Responds to Disturbances

We have seen that a chemical system not in equilibrium tends to progress toward equilibrium and that the relative concentrations of the reactants and products at equilibrium are characterized by the equilibrium constant, K. What happens when a chemical system already at equilibrium is disturbed? **Le Châtelier's principle** states that the chemical system responds to minimize the disturbance.

| Pronounced "Le-sha-te-lyay."

> **Le Châtelier's principle: When a chemical system at equilibrium is disturbed, the system shifts in a direction that minimizes the disturbance.**

In other words, a system at equilibrium tends to maintain equilibrium—it bounces back when disturbed. We can disturb a system in chemical equilibrium in several different ways, including changing the concentration of a reactant or product, changing the volume or pressure, and changing the temperature. Let's consider each of these separately.

The Effect of a Concentration Change on Equilibrium

Consider this reaction in chemical equilibrium:

$$N_2O_4(g) \rightleftharpoons 2\ NO_2(g)$$

Suppose we disturb the equilibrium by adding NO_2 to the equilibrium mixture and increasing the concentration of NO_2. What happens? According to Le Châtelier's principle, the system will shift in a direction to minimize the disturbance. The reaction goes to the left (it proceeds in the reverse direction), consuming some of the added NO_2 and bringing its concentration back down, which is shown graphically in **Figure 14.7(a)▼**.

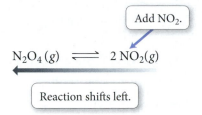

The reaction shifts to the left because the value of Q changes as follows:

- Before addition of NO_2: $Q = K$.
- Immediately after addition of NO_2: $Q > K$.
- Reaction shifts to left to reestablish equilibrium.

On the other hand, what happens if we add extra N_2O_4, increasing its concentration? In this case, the reaction shifts to the right, consuming some of the added N_2O_4 and bringing *its* concentration back down, as shown in **Figure 14.7(b)▼**.

Le Châtelier's Principle: Changing Concentration

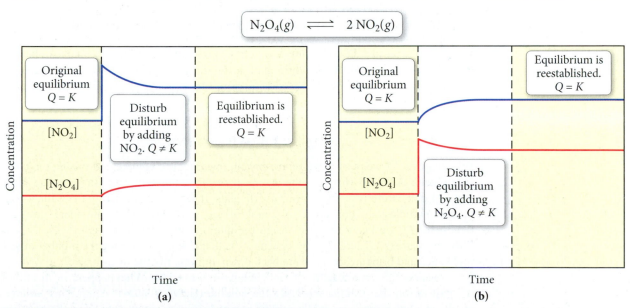

▲ **FIGURE 14.7 Le Châtelier's Principle: Changing Concentration** The graph shows the concentrations of NO_2 and N_2O_4 for the reaction $N_2O_4(g) \longrightarrow 2\ NO_2(g)$ in three distinct stages of the reaction: initially at equilibrium (left); upon disturbance of the equlibrium by addition of more NO_2 **(a)** or N_2O_4 **(b)** to the reaction mixture (center); and upon reestablishment of equilibrium (right).

The reaction shifts to the right because the value of Q changes as follows:

- Before addition of N_2O_4: $Q = K$.
- Immediately after addition of N_2O_4: $Q < K$.
- Reaction shifts to right to reestablish equilibrium.

In both of the cases, the system shifts in a direction that minimizes the disturbance. Conversely, *lowering* the concentration of a reactant (which makes $Q > K$) causes the system to shift in the direction of the reactants to minimize the disturbance. Lowering the concentration of a product (which makes $Q < K$) causes the system to shift in the direction of products.

EXAMPLE 14.14   **The Effect of a Concentration Change on Equilibrium**

Consider the following reaction at equilibrium.

$$CaCO_3(s) \rightleftharpoons CaO(s) + CO_2(g)$$

What is the effect of adding additional CO_2 to the reaction mixture? What is the effect of adding additional $CaCO_3$?

SOLUTION

Adding additional CO_2 increases the concentration of CO_2 and causes the reaction to shift to the left. Adding additional $CaCO_3$, however, does *not* increase the concentration of $CaCO_3$ because $CaCO_3$ is a solid and therefore has a constant concentration. Thus, adding additional $CaCO_3$ has no effect on the position of the equilibrium. (Note that, as we saw in Section 14.5, solids are not included in the equilibrium expression.)

FOR PRACTICE 14.14

Consider the following reaction in chemical equilibrium:
$$2\,BrNO(g) \rightleftharpoons 2\,NO(g) + Br_2(g)$$
What is the effect of adding additional Br_2 to the reaction mixture? What is the effect of adding additional BrNO?

The Effect of a Volume (or Pressure) Change on Equilibrium

In considering the effect of a change in volume, we are assuming that the change in volume is carried out at constant temperature.

How does a system in chemical equilibrium respond to a volume change? Recall from Chapter 5 that changing the volume of a gas (or a gas mixture) results in a change in pressure. Remember also that pressure and volume are inversely related: a *decrease* in volume causes an *increase* in pressure, and an *increase* in volume causes a *decrease* in pressure. So, if the volume of a reaction mixture at chemical equilibrium is changed, the pressure changes and the system shifts in a direction to minimize that change. Consider the following reaction at equilibrium in a cylinder equipped with a moveable piston:

$$N_2(g) + 3\,H_2(g) \rightleftharpoons 2\,NH_3(g)$$

What happens if we push down on the piston, lowering the volume and raising the pressure (**Figure 14.8(a)▶**)? How can the chemical system change to bring the pressure back down? Look carefully at the reaction coefficients. If the reaction shifts to the right, 4 mol of gas particles are converted to 2 mol of gas particles. From the ideal gas law ($PV = nRT$), we know that decreasing the number of moles of a gas (n) results in a lower pressure (P). Therefore, the system shifts to the right, decreasing the number of gas molecules and bringing the pressure back down, minimizing the disturbance.

Consider the same reaction mixture at equilibrium again. What happens if, this time, we pull *up* on the piston, *increasing* the volume (**Figure 14.8(b)▶**)? The higher volume results in a lower pressure and the system responds in such a way as to bring the pressure back up. It does this by shifting to the left, converting 2 mol of gas particles into 4 mol of gas particles, increasing the pressure and minimizing the disturbance.

Le Châtelier's Principle: Changing Pressure

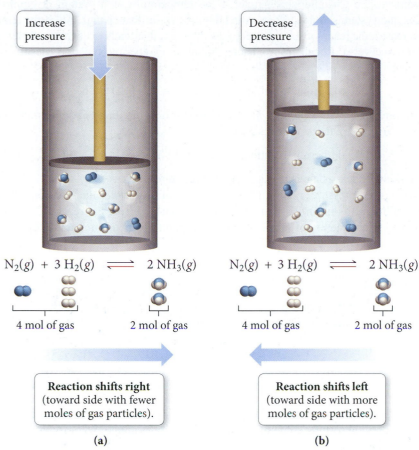

◀ **FIGURE 14.8   Le Châtelier's Principle: The Effect of a Pressure Change**   (a) Decreasing the volume increases the pressure, causing the reaction to shift to the right (fewer moles of gas, lower pressure). (b) Increasing the volume reduces the pressure, causing the reaction to shift to the left (more moles of gas, higher pressure).

$N_2(g) + 3 H_2(g) \rightleftharpoons 2 NH_3(g)$

4 mol of gas   2 mol of gas

Reaction shifts right (toward side with fewer moles of gas particles).

(a)

$N_2(g) + 3 H_2(g) \rightleftharpoons 2 NH_3(g)$

4 mol of gas   2 mol of gas

Reaction shifts left (toward side with more moles of gas particles).

(b)

Consider again the same reaction mixture at equilibrium. What happens if, this time, we keep the volume the same, but increase the pressure *by adding an inert gas* to the mixture? Although the overall pressure of the mixture increases, the partial pressures of the reactants and products do not change. Consequently, there is no effect and the reaction does not shift in either direction.

EXAMPLE 14.15   The Effect of a Volume Change on Equilibrium

Consider the following reaction at chemical equilibrium:

$$2 KClO_3(s) \rightleftharpoons 2 KCl(s) + 3 O_2(g)$$

What is the effect of decreasing the volume of the reaction mixture? Increasing the volume of the reaction mixture? Adding an inert gas at constant volume?

SOLUTION

The chemical equation has 3 mol of gas on the right and zero moles of gas on the left. Decreasing the volume of the reaction mixture increases the pressure and causes the reaction to shift to the left (toward the side with fewer moles of gas particles). Increasing the volume of the reaction mixture decreases the pressure and causes the reaction to shift to the right (toward the side with more moles of gas particles). Adding an inert gas has no effect.

FOR PRACTICE 14.15

Consider the following reaction at chemical equilibrium:

$$2 SO_2(g) + O_2(g) \rightleftharpoons 2 SO_3(g)$$

What is the effect of decreasing the volume of the reaction mixture? Increasing the volume of the reaction mixture?

The Effect of a Temperature Change on Equilibrium

In considering the effect of a change in temperature, we are assuming that the heat is added (or removed) at constant pressure.

According to Le Châtelier's principle, if the temperature of a system at equilibrium changes, the system should shift in a direction to counter that change. So, if the temperature increases, the reaction should shift in the direction that tends to decrease the temperature and vice versa. Recall from Chapter 6 that an exothermic reaction (negative ΔH) emits heat:

$$\text{Exothermic reaction:} \quad A + B \rightleftharpoons C + D + \text{heat}$$

We can think of heat as a product in an exothermic reaction. In an endothermic reaction (positive ΔH), the reaction absorbs heat.

$$\text{Endothermic reaction:} \quad A + B + \text{heat} \rightleftharpoons C + D$$

We can think of heat as a reactant in an endothermic reaction.

At constant pressure, raising the temperature of an *exothermic* reaction—think of this as adding heat—is similar to adding more product, causing the reaction to shift left. For example, the reaction of nitrogen with hydrogen to form ammonia is exothermic.

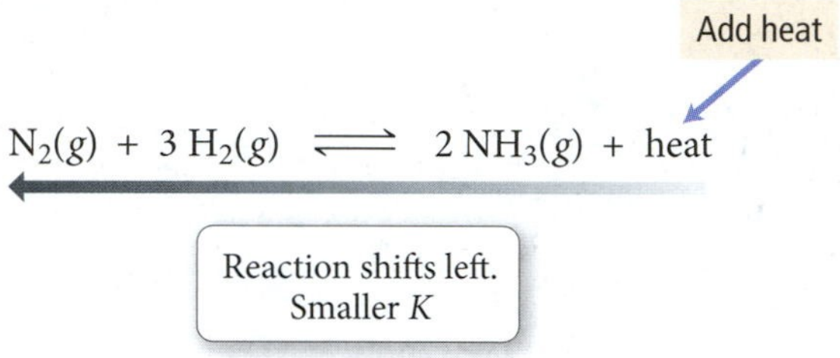

Add heat

$$N_2(g) + 3\,H_2(g) \rightleftharpoons 2\,NH_3(g) + \text{heat}$$

Reaction shifts left.
Smaller K

Adding heat favors the endothermic direction. Removing heat favors the exothermic direction.

Raising the temperature of an equilibrium mixture of these three gases causes the reaction to shift left, absorbing some of the added heat and forming fewer products and more reactants. Note that, unlike adding additional NH_3 to the reaction mixture (which does *not* change the value of the equilibrium constant), *adding heat does change the value of the equilibrium constant*. The new equilibrium mixture will have more reactants and fewer products and therefore a smaller value of K.

Conversely, lowering the temperature causes the reaction to shift right, releasing heat and producing more products because the value of K has increased.

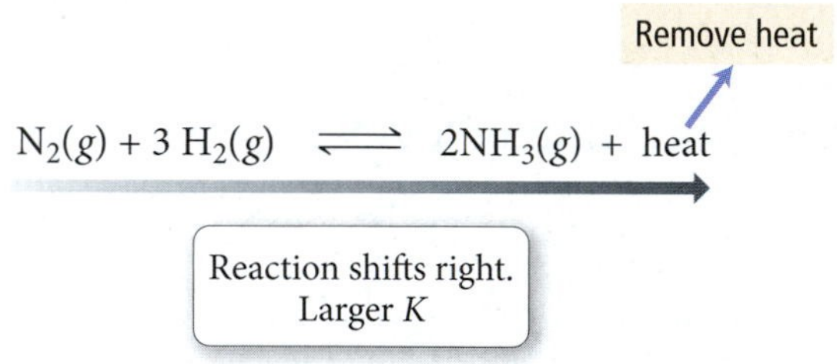

Remove heat

$$N_2(g) + 3\,H_2(g) \rightleftharpoons 2NH_3(g) + \text{heat}$$

Reaction shifts right.
Larger K

In contrast, for an *endothermic* reaction, raising the temperature (adding heat) causes the reaction to shift right, and lowering the temperature (removing heat) causes the reaction to shift left.

EXAMPLE 14.16 The Effect of a Temperature Change on Equilibrium

The following reaction is endothermic.

$$CaCO_3(s) \rightleftharpoons CaO(s) + CO_2(g)$$

What is the effect of increasing the temperature of the reaction mixture? Decreasing the temperature?

SOLUTION

Since the reaction is endothermic, we can think of heat as a reactant:

$$\text{Heat} + CaCO_3(s) \rightleftharpoons CaO(s) + CO_2(g)$$

Raising the temperature is equivalent to adding a reactant, causing the reaction to shift to the right. Lowering the temperature is equivalent to removing a reactant, causing the reaction to shift to the left.

FOR PRACTICE 14.16

The following reaction is exothermic.

$$2 SO_2(g) + O_2(g) \rightleftharpoons 2 SO_3(g)$$

What is the effect of increasing the temperature of the reaction mixture? Decreasing the temperature?

CHAPTER IN REVIEW

Key Terms

Section 14.2
reversible (540)
dynamic equilibrium (540)

Section 14.3
equilibrium constant (K) (542)
law of mass action (542)

Section 14.7
reaction quotient (Q) (552)

Section 14.9
Le Châtelier's principle (562)

Key Concepts

The Equilibrium Constant (14.1)

▶ The equilibrium constant K expresses the relative concentrations of reactants and products at equilibrium. The equilibrium constant measures how far a reaction proceeds toward products: a large K (much greater than 1) indicates a high concentration of products at equilibrium and a small K (less than 1) indicates a low concentration of products at equilibrium.

Dynamic Equilibrium (14.2)

▶ Most chemical reactions are reversible; they can proceed in either the forward or the reverse direction.

▶ When a chemical reaction is in dynamic equilibrium, the rate of the forward reaction equals the rate of the reverse reaction, so the net concentrations of reactants and products do not change. This does *not* imply that the concentrations of the reactants and the products are equal at equilibrium.

The Equilibrium Constant Expression (14.3)

▶ The equilibrium constant expression is given by the law of mass action and is equal to the concentrations of the products, raised to their stoichiometric coefficients, divided by the concentrations of the reactants, raised to their stoichiometric coefficients.

▶ When the equation for a chemical reaction is reversed, multiplied, or added to another equation, K must be modified accordingly.

The Equilibrium Constant, K (14.4)

▶ The equilibrium constant can be expressed in terms of concentrations (K_c) or in terms of partial pressures (K_p). These two constants are related by Equation 14.2. We must always express concentration in units of molarity for K_c. Partial pressures are always expressed in units of atmospheres for K_p.

States of Matter and the Equilibrium Constant (14.5)

▶ The equilibrium constant expression contains only partial pressures or concentrations of reactants and products that exist as gases or solutes dissolved in solution. Pure liquids and solids are not included in the expression for the equilibrium constant.

Calculating K (14.6)

▶ We can calculate the equilibrium constant from equilibrium concentrations or partial pressures by substituting measured values into the expression for the equilibrium constant (as obtained from the law of mass action).

▶ In most cases, we can calculate the equilibrium concentrations of the reactants and products—and therefore the value of the equilibrium constant—from the initial concentrations of the reactants and products and the equilibrium concentration of *just one* reactant or product.

The Reaction Quotient, Q (14.7)

▶ The reaction quotient Q is the ratio of the concentrations (or partial pressures) of products raised to their stoichiometric coefficients to the concentrations of reactants raised to their stoichiometric coefficients *at any point in the reaction.*

▶ Like K, we can express Q in terms of concentrations (Q_c) or partial pressures (Q_p).

▶ At equilibrium, Q is equal to K; therefore, we determine the direction in which a reaction will proceed by comparing Q to K. If $Q < K$, the reaction moves in the direction of the products; if $Q > K$, the reaction moves in the reverse direction.

Finding Equilibrium Concentrations (14.8)

▶ There are two general types of problems in which K is given and one (or more) equilibrium concentrations can be found:
  (1) K, initial concentrations, and (at least) one equilibrium concentration are given.
  (2) K and *only* initial concentrations are given.
▶ We solve the first type by rearranging the law of mass action and substituting given values.

▶ We solve the second type by using a variable x to represent the change in concentration.

Le Châtelier's Principle (14.9)

▶ When a system at equilibrium is disturbed—by a change in the amount of a reactant or product, a change in volume, or a change in temperature—the system shifts in the direction that minimizes the disturbance.

Key Equations and Relationships

Expression for the Equilibrium Constant, K_c (14.3)

$$aA + bB \rightleftharpoons cC + dD$$

$$K = \frac{[C]^c[D]^d}{[A]^a[B]^b} \quad \text{(equilibrium concentrations only)}$$

Relationship between the Equilibrium Constant and the Chemical Equation (14.3)

1. If you reverse the equation, invert the equilibrium constant.
2. If you multiply the coefficients in the equation by a factor, raise the equilibrium constant to the same factor.
3. If you add two or more individual chemical equations to obtain an overall equation, multiply the corresponding equilibrium constants by each other to obtain the overall equilibrium constant.

Expression for the Equilibrium Constant, K_p (14.4)

$$aA + bB \rightleftharpoons cC + dD$$

$$K_p = \frac{P_C^c P_D^d}{P_A^a P_B^b} \quad \text{(equilibrium partial pressures only)}$$

Relationship between the Equilibrium Constants, K_c and K_p (14.4)

$$K_p = K_c(RT)^{\Delta n}$$

The Reaction Quotient, Q_c (14.7)

$$aA + bB \rightleftharpoons cC + dD$$

$$Q_c = \frac{[C]^c[D]^d}{[A]^a[B]^b} \quad \text{(concentrations at any point in the reaction)}$$

The Reaction Quotient, Q_p (14.7)

$$aA + bB \rightleftharpoons cC + dD$$

$$Q_P = \frac{P_C^c P_D^d}{P_A^a P_B^b} \quad \text{(partial pressures at any point in the reaction)}$$

Relationship of Q to the Direction of the Reaction (14.7)

$Q < K$ Reaction goes to the right.
$Q > K$ Reaction goes to the left.
$Q = K$ Reaction is at equilibrium.

Key Learning Objectives

Chapter Objectives	Assessment
Expressing Equilibrium Constants for Chemical Equations (14.3)	Example 14.1 For Practice 14.1 Exercises 1, 2
Manipulating the Equilibrium Constant to Reflect Changes in the Chemical Equation (14.3)	Example 14.2 For Practice 14.2 For More Practice 14.2 Exercises 7–10
Relating K_p and K_c (14.4)	Example 14.3 For Practice 14.3 Exercises 11, 12
Writing Equilibrium Expressions for Reactions Involving a Solid or a Liquid (14.5)	Example 14.4 For Practice 14.4 Exercises 13, 14
Finding Equilibrium Constants from Experimental Concentration Measurements (14.6)	Examples 14.5, 14.6 For Practice 14.5, 14.6 Exercises 15, 16, 21, 22
Predicting the Direction of a Reaction by Comparing Q and K (14.7)	Example 14.7 For Practice 14.7 Exercises 25–28

EXERCISES

Problems by Topic

Equilibrium and the Equilibrium Constant Expression

1. Write an expression for the equilibrium constant for each chemical equation.
   a. $SbCl_5(g) \rightleftharpoons SbCl_3(g) + Cl_2(g)$
   b. $2\,BrNO(g) \rightleftharpoons 2\,NO(g) + Br_2(g)$
   c. $CH_4(g) + 2\,H_2S(g) \rightleftharpoons CS_2(g) + 4\,H_2(g)$
   d. $2\,CO(g) + O_2(g) \rightleftharpoons 2\,CO_2(g)$

2. Find and fix each mistake in the equilibrium constant expressions.
   a. $2\,H_2S(g) \rightleftharpoons 2\,H_2(g) + S_2(g)$

   $$K = \frac{[H_2][S_2]}{[H_2S]}$$

   b. $CO(g) + Cl_2(g) \rightleftharpoons COCl_2(g)$

   $$K = \frac{[CO][Cl_2]}{[COCl_2]}$$

3. When this reaction comes to equilibrium, are the concentrations of the reactants or products greater? Does the answer to this question depend on the initial concentrations of the reactants and products?

   $$A(g) + B(g) \rightleftharpoons 2\,C(g) \quad K_c = 1.4 \times 10^{-5}$$

4. Ethene (C_2H_4) is halogenated by the reaction:

   $$C_2H_4(g) + X_2(g) \rightleftharpoons C_2H_4X_2(g)$$

   where X_2 can be Cl_2 (green), Br_2 (brown), or I_2 (purple). Examine the three figures representing equilibrium concentrations in this reaction at the same temperature for the three different halogens. Rank the equilibrium constants for these three reactions from largest to smallest.

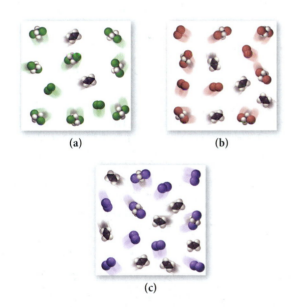

(a)   (b)

(c)

5. H_2 and I_2 are combined in a flask and allowed to react according to the reaction:

   $$H_2(g) + I_2(g) \rightleftharpoons 2\,HI(g)$$

   Examine the figures that follow (sequential in time) and answer the questions:
   a. Which figure represents the point at which equilibrium is reached?
   b. How would the series of figures change in the presence of a catalyst?
   c. Would the final figure (vi) have different amounts of reactants and products in the presence of a catalyst?

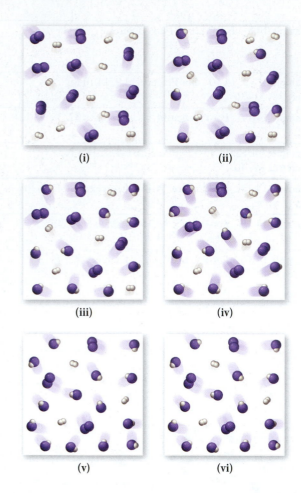

(i)　　　(ii)

(iii)　　　(iv)

(v)　　　(vi)

6. A chemist trying to synthesize a particular compound attempts two different synthesis reactions. The equilibrium constants for the two reactions are 23.3 and 2.2×10^4 at room temperature. However, upon carrying out both reactions for 15 minutes, the chemist finds that the reaction with the smaller equilibrium constant produced more of the desired product. Explain how this might be possible.

7. The reaction below has an equilibrium constant of $K_p = 2.26 \times 10^4$ at 298 K.

$$CO(g) + 2 H_2(g) \rightleftharpoons CH_3OH(g)$$

Calculate K_p for each reaction and predict whether reactants or products will be favored at equilibrium:
a. $CH_3OH(g) \rightleftharpoons CO(g) + 2 H_2(g)$
b. $\frac{1}{2}CO(g) + H_2(g) \rightleftharpoons \frac{1}{2}CH_3OH(g)$
c. $2 CH_3OH(g) \rightleftharpoons 2 CO(g) + 4 H_2(g)$

8. This reaction has an equilibrium constant
$K_p = 2.2 \times 10^6$ at 298 K.

$$2 COF_2(g) \rightleftharpoons CO_2(g) + CF_4(g)$$

Calculate K_p for each reaction and predict whether reactants or products are favored at equilibrium:
a. $COF_2(g) \rightleftharpoons \frac{1}{2}CO_2(g) + \frac{1}{2}CF_4(g)$
b. $6 COF_2(g) \rightleftharpoons 3 CO_2(g) + 3 CF_4(g)$
c. $2 CO_2(g) + 2 CF_4(g) \rightleftharpoons 4 COF_2(g)$

9. Consider these reactions and their respective equilibrium constants:
$NO(g) + \frac{1}{2}Br_2(g) \rightleftharpoons NOBr(g)$ $K_p = 5.3$
$2 NO(g) \rightleftharpoons N_2(g) + O_2(g)$ $K_p = 2.1 \times 10^{30}$

Use these reactions and their equilibrium constants to predict the equilibrium constant for the following reaction:

$$N_2(g) + O_2(g) + Br_2(g) \rightleftharpoons 2 NOBr(g).$$

10. Use the reactions that follow and their equilibrium constants to predict the equilibrium constant for the reaction $2 A(s) \rightleftharpoons 3 D(g)$.

$$A(s) \rightleftharpoons \frac{1}{2}B(g) + C(g) \quad K_1 = 0.0334$$

$$3 D(g) \rightleftharpoons B(g) + 2 C(g) \quad K_2 = 2.35$$

K_p, K_c, and Heterogeneous Equilibria

11. Calculate K_c for each reaction:
a. $I_2(g) \rightleftharpoons 2 I(g)$ $K_p = 6.26 \times 10^{-22}$ (at 298 K)
b. $CH_4(g) + H_2O(g) \rightleftharpoons CO(g) + 3 H_2(g)$
$K_p = 7.7 \times 10^{24}$ (at 298 K)
c. $I_2(g) + Cl_2(g) \rightleftharpoons 2 ICl(g)$ $K_p = 81.9$ (at 298 K)

12. Calculate K_p for each reaction:
a. $N_2O_4(g) \rightleftharpoons 2 NO_2(g)$ $K_c = 5.9 \times 10^{-3}$ (at 298 K)
b. $N_2(g) + 3 H_2(g) \rightleftharpoons 2 NH_3(g)$
$K_c = 3.7 \times 10^8$ (at 298 K)
c. $N_2(g) + O_2(g) \rightleftharpoons 2 NO(g)$
$K_c = 4.10 \times 10^{-31}$ (at 298 K)

13. Write an equilibrium expression for each chemical equation which contains one or more solid or liquid reactants or products.
a. $CO_3^{2-}(aq) + H_2O(l) \rightleftharpoons HCO_3^-(aq) + OH^-(aq)$
b. $2 KClO_3(s) \rightleftharpoons 2 KCl(s) + 3 O_2(g)$
c. $HF(aq) + H_2O(l) \rightleftharpoons H_3O^+(aq) + F^-(aq)$
d. $NH_3(aq) + H_2O(l) \rightleftharpoons NH_4^+(aq) + OH^-(aq)$

14. Find the mistake in the equilibrium constant expression and fix it.

$$PCl_5(g) \rightleftharpoons PCl_3(l) + Cl_2(g) \qquad K = \frac{[PCl_3][Cl_2]}{[PCl_5]}$$

Relating the Equilibrium Constant to Equilibrium Concentrations and Equilibrium Partial Pressures

15. Consider the reaction:

$$CO(g) + 2 H_2(g) \rightleftharpoons CH_3OH(g)$$

An equilibrium mixture of this reaction at a certain temperature was found to have $[CO] = 0.105$ M, $[H_2] = 0.114$ M, and $[CH_3OH] = 0.185$ M. What is the value of the equilibrium constant (K_c) at this temperature?

16. Consider the reaction:

$$NH_4HS(s) \rightleftharpoons NH_3(g) + H_2S(g)$$

An equilibrium mixture of this reaction at a certain temperature was found to have $[NH_3] = 0.278$ M and $[H_2S] = 0.355$ M. What is the value of the equilibrium constant (K_c) at this temperature?

17. Consider the reaction:

$$N_2(g) + 3 H_2(g) \rightleftharpoons 2 NH_3(g)$$

Complete the table. Assume that all concentrations are equilibrium concentrations in M.

T(K)	$[N_2]$	$[H_2]$	$[NH_3]$	$[K_c]$
500	0.115	0.105	0.439	_____
575	0.110	_____	0.128	9.6
775	0.120	0.140	_____	0.0584

18. Consider the reaction:

$$H_2(g) + I_2(g) \rightleftharpoons 2\,HI(g)$$

Complete the table. Assume that all concentrations are equilibrium concentrations in M.

$T(°C)$	$[H_2]$	$[I_2]$	$[HI]$	$[K_c]$
25	0.0355	0.0388	0.922	____
340	____	0.0455	0.387	9.6
445	0.0485	0.0468	____	50.2

19. Consider the reaction:

$$2\,NO(g) + Br_2(g) \rightleftharpoons 2\,NOBr(g) \quad K_p = 28.4 \text{ (at 298 K)}$$

In a reaction mixture at equilibrium, the partial pressure of NO is 118 torr and that of Br_2 is 176 torr. What is the partial pressure of NOBr in this mixture?

20. Consider the reaction:

$$SO_2Cl_2(g) \rightleftharpoons SO_2(g) + Cl_2(g)$$
$$K_p = 2.91 \times 10^3 \text{ (at 298 K)}$$

In a reaction at equilibrium, the partial pressure of SO_2 is 117 torr and that of Cl_2 is 255 torr. What is the partial pressure of SO_2Cl_2 in this mixture?

21. Consider the reaction:

$$Fe^{3+}(aq) + SCN^-(aq) \rightleftharpoons FeSCN^{2+}(aq)$$

A solution is made containing an initial $[Fe^{3+}]$ of 1.0×10^{-3} M and an initial $[SCN^-]$ of 8.0×10^{-4} M. At equilibrium, $[FeSCN^{2+}] = 1.7 \times 10^{-4}$ M. Calculate the value of the equilibrium constant (K_c).

22. Consider the reaction:

$$SO_2Cl_2(g) \rightleftharpoons SO_2(g) + Cl_2(g)$$

A reaction mixture is made containing an initial $[SO_2Cl_2]$ of 0.020 M. At equilibrium, $[Cl_2] = 1.2 \times 10^{-2}$ M. Calculate the value of the equilibrium constant (K_c).

23. Consider the reaction:

$$H_2(g) + I_2(g) \rightleftharpoons 2\,HI(g)$$

A reaction mixture in a 3.67-L flask at a certain temperature initially contains 0.763 g H_2 and 96.9 g I_2. At equilibrium, the flask contains 90.4 g HI. Calculate the equilibrium constant (K_c) for the reaction at this temperature.

24. Consider the reaction:

$$CO(g) + 2\,H_2(g) \rightleftharpoons CH_3OH(g)$$

A reaction mixture in a 5.19-L flask at a certain temperature contains 26.9 g CO and 2.34 g H_2. At equilibrium, the flask contains 8.65 g CH_3OH. Calculate the equilibrium constant (K_c) for the reaction at this temperature.

The Reaction Quotient and Reaction Direction

25. Consider the reaction:

$$NH_4HS(s) \rightleftharpoons NH_3(g) + H_2S(g)$$

At a certain temperature, $K_c = 8.5 \times 10^{-3}$. A reaction mixture at this temperature containing solid NH_4HS has $[NH_3] = 0.166$ M and $[H_2S] = 0.166$ M. Will more of the solid form or will some of the existing solid decompose as equilibrium is reached?

26. Consider the reaction:

$$2\,H_2S(g) \rightleftharpoons 2\,H_2(g) + S_2(g)$$
$$K_p = 2.4 \times 10^{-4} \text{ (at 1073 K)}$$

A reaction mixture contains 0.112 atm of H_2, 0.055 atm of S_2, and 0.445 atm of H_2S. Is the reaction mixture at equilibrium? If not, in what direction will the reaction proceed?

27. Silver sulfate dissolves in water according to the reaction:

$$Ag_2SO_4(s) \rightleftharpoons 2\,Ag^+(aq) + SO_4^{2-}(aq)$$
$$K_c = 1.1 \times 10^{-5} \text{ (at 298 K)}$$

A 1.5-L solution contains 6.55 g of dissolved silver sulfate. If additional solid silver sulfate is added to the solution, will it dissolve?

28. Nitrogen dioxide dimerizes according to the reaction:

$$2\,NO_2(g) \rightleftharpoons N_2O_4(g) \quad K_p = 6.7 \text{ (at 298 K)}$$

A 2.25-L container contains 0.055 mol of NO_2 and 0.082 mol of N_2O_4 at 298 K. Is the reaction at equilibrium? If not, in what direction will the reaction proceed?

Finding Equilibrium Concentrations from Initial Concentrations and the Equilibrium Constant

29. Consider the reaction and associated equilibrium constant:

$$aA(g) \rightleftharpoons bB(g) \quad K_c = 2.0$$

Find the equilibrium concentrations of A and B for each value of a and b. Assume that the initial concentration of A in each case is 1.0 M and that no B is present at the beginning of the reaction.

a. $a = 1; b = 1$ **b.** $a = 2; b = 2$ **c.** $a = 1; b = 2$

30. Consider the reaction and associated equilibrium constant:

$$aA(g) + bB(g) \rightleftharpoons cC(g) \quad K_c = 4.0$$

Find the equilibrium concentrations of A, B, and C for each value of a, b, and c. Assume that the initial concentrations of A and B are each 1.0 M and that no product is present at the beginning of the reaction.

a. $a = 1; b = 1; c = 2$ **b.** $a = 1; b = 1; c = 1$
c. $a = 2; b = 1; c = 1$ (set up equation for x; do not solve)

31. For the reaction, $K_c = 0.513$ at 500 K.

$$N_2O_4(g) \rightleftharpoons 2\,NO_2(g)$$

If a reaction vessel initially contains an N_2O_4 concentration of 0.0500 M at 500 K, what are the equilibrium concentrations of N_2O_4 and NO_2 at 500 K?

32. For the reaction, $K_c = 255$ at 1000 K.

$$CO(g) + Cl_2(g) \rightleftharpoons COCl_2(g)$$

If a reaction mixture initially contains a CO concentration of 0.1500 M and a Cl_2 concentration of 0.175 M at 1000 K, what are the equilibrium concentrations of CO, Cl_2, and $COCl_2$ at 1000 K?

33. Consider the reaction:

$$NiO(s) + CO(g) \rightleftharpoons Ni(s) + CO_2(g)$$
$$K_c = 4.0 \times 10^3 \text{ (at 1500 K)}$$

If a mixture of solid nickel(II) oxide and 0.10 M carbon monoxide comes to equilibrium at 1500 K, what will be the equilibrium concentration of CO_2?

34. Consider the reaction:

$$CO(g) + H_2O(g) \rightleftharpoons CO_2(g) + H_2(g)$$
$$K_c = 102 \text{ (at 500 K)}$$

If a reaction mixture initially contains 0.125 M CO and 0.125 M H_2O, what will be the equilibrium concentration of each of the reactants and products?

35. Consider the reaction:

$$HC_2H_3O_2(aq) + H_2O(l) \rightleftharpoons H_3O^+(aq) + C_2H_3O_2^-(aq)$$
$$K_c = 1.8 \times 10^{-5} \text{ at 25 °C}$$

If a solution initially contains 0.210 M $HC_2H_3O_2$, what is the equilibrium concentration of H_3O^+ at 25 °C?

36. Consider the reaction:

$$SO_2Cl_2(g) \rightleftharpoons SO_2(g) + Cl_2(g)$$
$$K_c = 2.99 \times 10^{-7} \text{ at 227 °C}$$

If a reaction mixture initially contains 0.175 M SO_2Cl_2, what is the equilibrium concentration of Cl_2 at 227 °C?

37. Consider the reaction:

$$Br_2(g) + Cl_2(g) \rightleftharpoons 2 BrCl(g)$$
$$K_p = 1.11 \times 10^{-4} \text{ at 150 K}$$

A reaction mixture initially contains a Br_2 partial pressure of 755 torr and a Cl_2 partial pressure of 735 torr at 150 K. Calculate the equilibrium partial pressure of BrCl.

38. Consider the reaction:

$$CO(g) + H_2O(g) \rightleftharpoons CO_2(g) + H_2(g)$$
$$K_p = 0.0611 \text{ at 2000 K}$$

A reaction mixture initially contains a CO partial pressure of 1344 torr and a H_2O partial pressure of 1766 torr at 2000 K. Calculate the equilibrium partial pressures of each of the products.

39. Consider the reaction:

$$A(g) \rightleftharpoons B(g) + C(g)$$

Find the equilibrium concentrations of A, B, and C for each of the different values of K_c. Assume that the initial concentration of A in each case is 1.0 M and that the reaction mixture initially contains no products. Make any appropriate simplifying assumptions.
a. $K_c = 1.0$ **b.** $K_c = 0.010$ **c.** $K_c = 1.0 \times 10^{-5}$

40. Consider the reaction:

$$A(g) \rightleftharpoons 2 B(g)$$

Find the equilibrium partial pressures of A and B for each of the different values of K_p. Assume that the initial partial pressure of B in each case is 1.0 atm and that the initial partial pressure of A is 0.0 atm. Make any appropriate simplifying assumptions.
a. $K_p = 1.0$ **b.** $K_p = 1.0 \times 10^{-4}$ **c.** $K_p = 1.0 \times 10^5$

Le Châtelier's Principle

41. Consider the reaction at equilibrium:

$$CO(g) + Cl_2(g) \rightleftharpoons COCl_2(g)$$

Predict whether the reaction will shift left, shift right, or remain unchanged upon each disturbance:
a. $COCl_2$ is added to the reaction mixture.
b. Cl_2 is added to the reaction mixture.
c. $COCl_2$ is removed from the reaction mixture.

42. Consider the reaction at equilibrium:

$$2 BrNO(g) \rightleftharpoons 2 NO(g) + Br_2(g)$$

Predict whether the reaction will shift left, shift right, or remain unchanged upon each disturbance.
a. NO is added to the reaction mixture.
b. BrNO is added to the reaction mixture.
c. Br_2 is removed from the reaction mixture.

43. Consider the reaction at equilibrium:

$$2 KClO_3(s) \rightleftharpoons 2 KCl(s) + 3 O_2(g)$$

Predict whether the reaction will shift left, shift right, or remain unchanged upon each disturbance.
a. O_2 is removed from the reaction mixture.
b. KCl is added to the reaction mixture.
c. $KClO_3$ is added to the reaction mixture.
d. O_2 is added to the reaction mixture.

44. Consider the reaction at equilibrium:

$$C(s) + H_2O(g) \rightleftharpoons CO(g) + H_2(g)$$

Predict whether the reaction will shift left, shift right, or remain unchanged upon each disturbance.
a. C is added to the reaction mixture.
b. H_2O is condensed and removed from the reaction mixture.
c. CO is added to the reaction mixture.
d. H_2 is removed from the reaction mixture.

45. Each reaction is allowed to come to equilibrium and then the volume is changed as indicated. Predict the effect (shift right, shift left, or no effect) of the indicated volume change.
a. $I_2(g) \rightleftharpoons 2 I(g)$ (volume is increased)
b. $2 H_2S(g) \rightleftharpoons 2 H_2(g) + S_2(g)$ (volume is decreased)
c. $I_2(g) + Cl_2(g) \rightleftharpoons 2 ICl(g)$ (volume is decreased)

46. Each reaction is allowed to come to equilibrium and then the volume is changed as indicated. Predict the effect (shift right, shift left, or no effect) of the indicated volume change.
a. $CO(g) + H_2O(g) \rightleftharpoons CO_2(g) + H_2(g)$ (volume is decreased)
b. $PCl_3(g) + Cl_2(g) \rightleftharpoons PCl_5(g)$ (volume is increased)
c. $CaCO_3(s) \rightleftharpoons CaO(s) + CO_2(g)$ (volume is increased)

47. This reaction is endothermic.

$$C(s) + CO_2(g) \rightleftharpoons 2 CO(g)$$

Predict the effect (shift right, shift left, or no effect) of increasing and decreasing the reaction temperature. How does the value of the equilibrium constant depend on temperature?

48. This reaction is exothermic.

$$C_6H_{12}O_6(s) + 6 O_2(g) \rightleftharpoons 6 CO_2(g) + 6 H_2O(g)$$

Predict the effect (shift right, shift left, or no effect) of increasing and decreasing the reaction temperature. How does the value of the equilibrium constant depend on temperature?

49. Coal, which is primarily carbon, can be converted to natural gas, primarily CH_4, by the exothermic reaction:

$$C(s) + 2 H_2(g) \rightleftharpoons CH_4(g)$$

Which change will favor CH_4 at equilibrium?
a. adding more C to the reaction mixture
b. adding more H_2 to the reaction mixture
c. raising the temperature of the reaction mixture
d. lowering the volume of the reaction mixture
e. adding a catalyst to the reaction mixture
f. adding neon gas to the reaction mixture

50. We can use coal to generate hydrogen gas (a potential fuel) by the endothermic reaction:

$$C(s) + H_2O(g) \rightleftharpoons CO(g) + H_2(g)$$

If this reaction mixture is at equilibrium, predict whether each change will result in the formation of additional hydrogen gas, the formation of less hydrogen gas, or have no effect on the quantity of hydrogen gas.

a. adding more C to the reaction mixture
b. adding more H_2O to the reaction mixture
c. raising the temperature of the reaction mixture
d increasing the volume of the reaction mixture
e. adding a catalyst to the reaction mixture
f. adding an inert gas to the reaction mixture

Cumulative Problems

51. Carbon monoxide replaces oxygen in oxygenated hemoglobin according to the reaction:

$$HbO_2(aq) + CO(aq) \rightleftharpoons HbCO(aq) + O_2(aq)$$

a. Use the following reactions and associated equilibrium constants at body temperature to find the equilibrium constant for the above reaction.

$$Hb(aq) + O_2(aq) \rightleftharpoons HbO_2(aq) \quad K_c = 1.8$$

$$Hb(aq) + CO(aq) \rightleftharpoons HbCO(aq) \quad K_c = 306$$

b. Suppose that an air mixture becomes polluted with carbon monoxide at a level of 0.10%. Assuming the air contains 20.0% oxygen, and that the oxygen and carbon monoxide ratios that dissolve in the blood are identical to the ratios in the air, what is the ratio of HbCO to HbO_2 in the bloodstream? Comment on the toxicity of carbon monoxide.

52. Nitrogen oxide is a pollutant in the lower atmosphere that irritates the eyes and lungs and leads to the formation of acid rain. Nitrogen oxide forms naturally in the atmosphere according to the endothermic reaction:

$$N_2(g) + O_2(g) \rightleftharpoons 2\,NO(g) \quad K_p = 4.1 \times 10^{-31} \text{ at } 298 \text{ K}$$

Use the ideal gas law to calculate the concentrations of nitrogen and oxygen present in air at a pressure of 1.0 atm and a temperature of 298 K. Assume that nitrogen composes 78% of air by volume and that oxygen composes 21% of air. Find the "natural" equilibrium concentration of NO in air in units of molecules/cm^3. How do you expect this concentration to change in an automobile engine in which combustion is occurring?

53. The reaction $CO_2(g) + C(s) \rightleftharpoons 2\,CO(g)$ has $K_p = 5.78$ at 1200 K.
a. Calculate the total pressure at equilibrium when 4.45 g of CO_2 is introduced into a 10.0-L container and heated to 1200 K in the presence of 2.00 g of graphite.
b. Repeat the calculation of part (a) in the presence of 0.50 g of graphite.

54. A mixture of water and graphite is heated to 600 K. When the system comes to equilibrium it contains 0.13 mol of H_2, 0.13 mol of CO, 0.43 mol of H_2O, and some graphite. Some O_2 is added to the system and a spark is applied so that the H_2 reacts completely with the O_2. Find the amount of CO in the flask when the system returns to equilibrium.

55. At 650 K, the reaction $MgCO_3(s) \rightleftharpoons MgO(s) + CO_2(g)$ has $K_p = 0.026$. A 10.0-L container at 650 K has 1.0 g of $MgO(s)$ and CO_2 at $P = 0.0260$ atm. The container is then compressed to a volume of 0.100 L. Find the mass of $MgCO_3$ that forms.

56. A system at equilibrium contains $I_2(g)$ at a pressure of 0.21 atm and $I(g)$ at a pressure of 0.23 atm. The system is then compressed to half its volume. Find the pressure of each gas when the system returns to equilibrium.

57. Consider the exothermic reaction:

$$C_2H_4(g) + Cl_2(g) \rightleftharpoons C_2H_4Cl_2(g)$$

If you were a chemist trying to maximize the amount of $C_2H_4Cl_2$ produced, which strategy might you try? Assume that the reaction mixture reaches equilibrium.
a. increasing the reaction volume
b. removing $C_2H_4Cl_2$ from the reaction mixture as it forms
c. lowering the reaction temperature
d. adding Cl_2

58. Consider the endothermic reaction:

$$C_2H_4(g) + I_2(g) \rightleftharpoons C_2H_4I_2(g)$$

If you were a chemist trying to maximize the amount of $C_2H_4I_2$ produced, which strategy might you try? Assume that the reaction mixture reaches equilibrium.
a. decreasing the reaction volume
b. removing I_2 from the reaction mixture
c. raising the reaction temperature
d. adding C_2H_4 to the reaction mixture

59. Consider the reaction:

$$H_2(g) + I_2(g) \rightleftharpoons 2\,HI(g)$$

A reaction mixture at equilibrium at 175 K contains $P_{H_2} = 0.958$ atm, $P_{I_2} = 0.877$ atm, and $P_{HI} = 0.020$ atm. A second reaction mixture, also at 175 K, contains $P_{H_2} = P_{I_2} = 0.621$ atm, and $P_{HI} = 0.101$ atm. Is the second reaction at equilibrium? If not, what is the partial pressure of HI when the reaction reaches equilibrium at 175 K?

60. Consider the reaction:

$$2\,H_2S(g) + SO_2(g) \rightleftharpoons 3\,S(s) + 2\,H_2O(g)$$

A reaction mixture initially containing 0.500 M H_2S and 0.500 M SO_2 contains 0.0011 M H_2O at a certain temperature. A second reaction mixture at the same temperature initially contains $[H_2S] = 0.250$ M and $[SO_2] = 0.325$ M. Calculate the equilibrium concentration of H_2O in the second mixture at this temperature.

61. Ammonia can be synthesized according to the reaction:

$$N_2(g) + 3\,H_2(g) \rightleftharpoons 2\,NH_3(g)$$

$$K_p = 5.3 \times 10^{-5} \text{ at } 725 \text{ K}$$

A 200.0-L reaction container initially contains 1.27 kg of N_2 and 0.310 kg of H_2 at 725 K. Assuming ideal gas behavior, calculate the mass of NH_3 (in g) present in the reaction mixture at equilibrium. What is the percent yield of the reaction under these conditions?

62. Hydrogen can be extracted from natural gas according to the reaction:

$$CH_4(g) + CO_2(g) \rightleftharpoons 2\,CO(g) + 2\,H_2(g)$$
$$K_p = 4.5 \times 10^2 \text{ at } 825 \text{ K}$$

An 85.0-L reaction container initially contains 22.3 kg of CH_4 and 55.4 kg of CO_2 at 825 K. Assuming ideal gas behavior, calculate the mass of H_2 (in g) present in the reaction mixture at equilibrium. What is the percent yield of the reaction under these conditions?

63. The system described by the reaction

$$CO(g) + Cl_2(g) \rightleftharpoons COCl_2(g)$$

is at equilibrium at a given temperature when $P_{CO} = 0.30$ atm, $P_{Cl_2} = 0.10$ atm, and $P_{COCl_2} = 0.60$ atm. An additional pressure of $Cl_2(g) = 0.40$ atm is added. Find the pressure of CO when the system returns to equilibrium.

64. A reaction vessel at 27 °C contains a mixture of SO_2 ($P = 3.00$ atm) and O_2 ($P = 1.00$ atm). When a catalyst is added the reaction

$$2\,SO_2(g) + O_2(g) \rightleftharpoons 2\,SO_3(g)$$

takes place. At equilibrium the total pressure is 3.75 atm. Determine the value of K_c.

65. At 70 K, CCl_4 decomposes to carbon and chlorine. The K_p for the decomposition is 0.76. Find the starting pressure of CCl_4 at this temperature that will produce a total pressure of 1.0 atm at equilibrium.

66. The equilibrium constant for the reaction $SO_2(g) + NO_2(g) \rightleftharpoons SO_3(g) + NO(g)$ is 3.0. Find the amount of NO_2 that must be added to 2.4 mol of SO_2 in order to form 1.2 mol of SO_3 at equilibrium.

67. A sample of $CaCO_3(s)$ is introduced into a sealed container of volume 0.654 L and heated to 1000 K until equilibrium is reached. The K_p for the reaction

$$CaCO_3(s) \rightleftharpoons CaO(s) + CO_2(g)$$

is 3.9×10^{-2} at this temperature. Calculate the mass of $CaO(s)$ that is present at equilibrium.

68. An equilibrium mixture contains N_2O_4, ($P = 0.28$ atm) and NO_2 ($P = 1.1$ atm) at 350 K. The volume of the container is doubled at constant temperature. Calculate the equilibrium pressures of the two gases when the system reaches a new equilibrium.

$$N_2O_4(g) \rightleftharpoons 2\,NO_2(g)$$

69. Carbon monoxide and chlorine gas react to form phosgene:

$$CO(g) + Cl_2(g) \rightleftharpoons COCl_2(g)$$
$$K_p = 3.10 \text{ at } 700 \text{ K}$$

If a reaction mixture initially contains 215 torr of CO and 245 torr of Cl_2, what is the mole fraction of $COCl_2$ when equilibrium is reached?

70. Solid carbon can react with gaseous water to form carbon monoxide gas and hydrogen gas. The equilibrium constant for the reaction at 700.0 K is $K_p = 1.60 \times 10^{-3}$. If a 1.55-L reaction vessel initially contains 145 torr of water at 700.0 K in contact with excess solid carbon, what is the percent by mass of hydrogen gas of the gaseous reaction mixture at equilibrium?

Challenge Problems

71. Consider the reaction:

$$2\,NO(g) + O_2(g) \rightleftharpoons 2\,NO_2(g)$$

a. A reaction mixture at 175 K initially contains 522 torr of NO and 421 torr of O_2. At equilibrium, the total pressure in the reaction mixture is 748 torr. Calculate K_p at this temperature.

b. A second reaction mixture at 175 K initially contains 255 torr of NO and 185 torr of O_2. What is the equilibrium partial pressure of NO_2 in this mixture?

72. Consider the reaction:

$$2\,SO_2(g) + O_2(g) \rightleftharpoons 2\,SO_3(g)$$
$$K_p = 0.355 \text{ at } 950 \text{ K}$$

A 2.75-L reaction vessel at 950 K initially contains 0.100 mol of SO_2 and 0.100 mol of O_2. Calculate the total pressure (in atmospheres) in the reaction vessel when equilibrium is reached.

73. Nitric oxide reacts with chlorine gas according to the reaction:

$$2\,NO(g) + Cl_2(g) \rightleftharpoons 2\,NOCl(g)$$
$$K_p = 0.27 \text{ at } 700 \text{ K}$$

A reaction mixture initially contains equal partial pressures of NO and Cl_2. At equilibrium, the partial pressure of NOCl is measured to be 115 torr. What were the initial partial pressures of NO and Cl_2?

74. At a given temperature a system containing $O_2(g)$ and some oxides of nitrogen can be described by the reactions:

$$2\,NO(g) + O_2(g) \rightleftharpoons 2\,NO_2(g) \quad K_p = 10^4$$
$$2\,NO_2(g) \rightleftharpoons N_2O_4(g) \quad K_p = 0.10$$

A pressure of 1 atm of $N_2O_4(g)$ is placed in a container at this temperature. Predict which, if any component (other than N_2O_4) will be present at a pressure greater than 0.2 atm at equilibrium.

75. A sample of pure NO_2 is heated to 337 °C, at which it partially dissociates according to the equation

$$2\,NO_2(g) \rightleftharpoons 2\,NO(g) + O_2(g)$$

At equilibrium the density of the gas mixture is 0.520 g/L at 0.750 atm. Calculate K_c for the reaction.

76. When $N_2O_5(g)$ is heated it dissociates into $N_2O_3(g)$ and $O_2(g)$ according to the reaction:

$$N_2O_5(g) \rightleftharpoons N_2O_3(g) + O_2(g)$$
$$K_c = 7.75 \text{ at a given temperature}$$

The $N_2O_3(g)$ dissociates to give $N_2O(g)$ and $O_2(g)$ according to the reaction:

$$N_2O_3(g) \rightleftharpoons N_2O(g) + O_2(g)$$
$$K_c = 4.00 \text{ at the same temperature}$$

When 4.00 mol of $N_2O_5(g)$ is heated in a 1.00-L reaction vessel to this temperature, the concentration of $O_2(g)$ at equilibrium is 4.50 mol/L. Find the concentrations of all the other species in the equilibrium system.

77. A sample of SO_3 is introduced into an evacuated sealed container and heated to 600 K. The following equilibrium is established:

$$2 SO_3(g) \rightleftharpoons 2 SO_2(g) + O_2(g).$$

The total pressure in the system is 3.0 atm and the mole fraction of O_2 is 0.12. Find K_p.

Conceptual Problems

78. A reaction $A(g) \rightleftharpoons B(g)$ has an equilibrium constant of 1.0×10^{-4}. For which of the initial reaction mixtures is the *x is small* approximation most likely to apply?
    a. $[A] = 0.0010$ M; $[B] = 0.00$ M
    b. $[A] = 0.00$ M; $[B] = 0.10$ M
    c. $[A] = 0.10$ M; $[B] = 0.00$ M

79. The reaction $A(g) \rightleftharpoons 2 B(g)$ has an equilibrium constant of $K_c = 1.0$ at a given temperature. If a reaction vessel contains equal initial amounts (in moles) of A and B, does the direction in which the reaction proceeds depend on the volume of the reaction vessel? Explain.

80. A particular reaction has an equilibrium constant of $K_p = 0.50$. A reaction mixture is prepared in which all the reactants and products are in their standard states. In which direction will the reaction proceed?

81. Consider the reaction:

$$aA(g) \rightleftharpoons bB(g)$$

Each of the entries in the table represents equilibrium partial pressures of A and B under different initial conditions. What are the values of *a* and *b* in the reaction?

P_A (atm)	P_B (atm)
4.0	2.0
2.0	1.4
1.0	1.0
0.50	0.71
0.25	0.50

82. Consider the simple one-step reaction:

$$A(g) \rightleftharpoons B(g)$$

Since the reaction occurs in a single step, the forward reaction has a rate of $k_{for}[A]$ and the reverse reaction has a rate of $k_{rev}[B]$. What happens to the rate of the forward reaction when we increase the concentration of A? How does this explain the reason behind Le Châtelier's principle?

Answers to Conceptual Connections

Equilibrium Constants

14.1 (b) The reaction mixture will contain 1 mol of A and 10 mol of B so that $[B]/[A] = 10$.

Relationships between the Equilibrium Constant and the Chemical Equation

14.2 (b) The reaction is reversed and divided by two. Therefore, invert the equilibrium constant and take the square root of the result. $K = (1/0.010)^{1/2} = 10$.

Relationship between K_p and K_c

14.3 (a) When $a + b = c + d$, the quantity Δn is zero so that $K_p = K_c (RT)^0$. Since $(RT)^0$ is equal to one, $K_p = K_c$.

Heterogeneous Equilibria, K_p, and K_c

14.4 (b) Since Δn for gaseous reactants and products is zero, K_p equals K_c.

Q and K

14.5 (c) Since N_2O_4 and NO_2 are both in their standard states, they each have a partial pressure of 1.0 atm. Consequently, $Q_p = 1$. Since $K_p = 0.15$, $Q_p > K_p$, and the reaction proceeds to the left.

The *x is small* Approximation

14.6 (a) The *x is small* approximation is most likely to apply to a reaction with a small equilibrium constant and an initial concentration of reactant that is not too small. The bigger the equilibrium constant and the smaller the initial concentration of reactant, the less likely that the *x is small* approximation will apply.

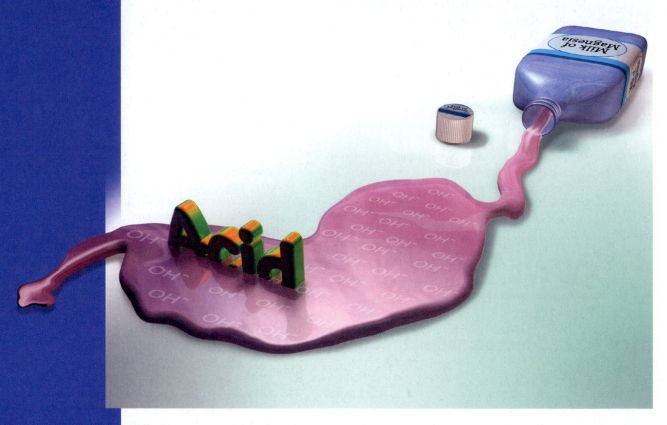

15

Acids and Bases

The differences between the various acid–base concepts are not concerned with which is right, but which is most convenient to use in a particular situation. —James E. Huheey (1935–)

Milk of magnesia contains a base that can neutralize stomach acid and relieve heartburn.

I N THIS CHAPTER, we apply the equilibrium concepts learned in the previous chapter to acid–base phenomena. Acids are common in many foods, such as limes, lemons, and vinegar, and in a number of consumer products, such as toilet cleaners and batteries. Bases are less common in foods but are found in products such as drain openers and antacids. We examine three different models for acid–base behavior, all of which define that behavior differently. In spite of their differences, the three models coexist, each being useful at explaining a particular range of acid–base phenomena. We also learn how to calculate the acidity or basicity of solutions and define a useful scale, called the pH scale, to quantify acidity and basicity. These types of calculations often involve solving the kind of equilibrium problems that we examined in Chapter 14.

15.1 Heartburn

Heartburn is a painful burning sensation in the esophagus (the tube that joins the throat to the stomach) just below the chest. The pain is caused by hydrochloric acid (HCl), which is excreted in the stomach to kill microorganisms and activate enzymes that break down food. The hydrochloric acid can sometimes back up out of the stomach and into the esophagus, a phenomenon called *acid reflux*. Recall from Section 4.8 that acids are substances that—by one definition that we elaborate on shortly—produce H^+ ions in solution. When hydrochloric acid from the stomach comes in contact with the lining of the esophagus, the H^+ ions irritate the esophageal tissues, resulting in the burning sensation. Some of the acid can work its way into the lower throat and even the mouth, producing pain in the throat and a sour taste (characteristic of acids) in the mouth. Almost everyone experiences heartburn at some time, most commonly after a large meal when the stomach is very full and the chances for reflux are greatest. Strenuous activity or lying in a horizontal position after a large meal increases the likelihood of stomach acid reflux and the resulting heartburn.

The simplest way to relieve mild heartburn is to swallow repeatedly. Saliva contains the bicarbonate ion (HCO_3^-) that acts as a base and, when swallowed, neutralizes some of the acid in the esophagus. Later in this chapter, we see how bicarbonate acts as a base. Heartburn can also be treated with antacids such as Tums, milk of magnesia, or Mylanta. These over-the-counter medications contain more base than saliva and therefore do a better job of neutralizing the esophageal acid.

For some people, heartburn becomes a chronic problem. The medical condition associated with chronic heartburn is known as gastroesophageal reflux disease (GERD). In patients with GERD, the band of muscles (called the esophageal sphincter) at the bottom of the esophagus just above the stomach does not close tightly enough, allowing the stomach contents to leak back into the esophagus on a regular basis. A wireless sensor helps to diagnose and evaluate treatment of GERD. Using a tube that goes down through the throat, a physician attaches the sensor to tissues in the patient's esophagus. The sensor reads pH—a measure of acidity we discuss in Section 15.5—and transmits the readings to a recorder worn on the patient's body. The patient goes about his or her normal business for the next few days while the recorder monitors esophageal pH. The sensor eventually falls off and is passed in the stool. The record of esophageal pH can be read by a physician to make a diagnosis or to evaluate treatment.

> The concentration of stomach acid, $[H_3O^+]$, varies from about 0.01 to 0.1 M.

In this chapter, we examine acid and base behavior. Acids and bases are not only important to our health (as we have just seen), but are also found in many household products, foods, medicines, and in nearly every chemistry laboratory. Acid–base chemistry is also central to much of biochemistry and molecular biology. The building blocks of proteins, for example, are amino acids and the molecules that carry the genetic code in DNA are bases.

15.2 The Nature of Acids and Bases

Acids have the following general properties: a sour taste; the ability to dissolve many metals; the ability to turn blue litmus paper red; and the ability to neutralize bases. Table 15.1 (on the next page) lists some common acids.

> Litmus paper contains certain dyes that change color in reponse to the presence of acids or bases.

Hydrochloric acid, nitric acid, and sulfuric acid are commonly found in most chemistry laboratories. Acetic acid is found in most people's homes; it is the active component of vinegar and is also produced in improperly stored wines. The word vinegar originates

> For a review of naming acids, see Section 3.6

$$HC_2H_3O_2$$

Acetic acid

from the French words *vin aigre*, which means sour wine. Acetic acid is an example of a **carboxylic acid**, an acid containing the following grouping of atoms:

Carboxylic acid group

Carboxylic acids are often found in substances derived from living organisms. Other carboxylic acids include citric acid, the main acid in lemons and limes, and malic acid, an acid found in apples, grapes, and wine.

TABLE 15.1 Some Common Acids	
Name	**Occurrence/Uses**
Hydrochloric acid (HCl)	Metal cleaning; food preparation; ore refining; main component of stomach acid
Sulfuric acid (H_2SO_4)	Fertilizer and explosives manufacturing; dye and glue production; automobile batteries; electroplating of copper
Nitric acid (HNO_3)	Fertilizer and explosives manufacturing; dye and glue production
Acetic acid ($HC_2H_3O_2$)	Plastic and rubber manufacturing; food preservative; active component of vinegar
Citric acid ($H_3C_6H_5O_7$)	Present in citrus fruits such as lemons and limes; used to adjust pH in foods and beverages
Carbonic acid (H_2CO_3)	Found in carbonated beverages due to the reaction of carbon dioxide with water
Hydrofluoric acid (HF)	Metal cleaning; glass frosting and etching
Phosphoric acid (H_3PO_4)	Fertilizer manufacture; biological buffering; preservative in beverages

The formula for acetic acid can also be written as CH_3COOH.

Bases have the following general properties: a bitter taste; a slippery feel; the ability to turn red litmus paper blue; and the ability to neutralize acids. Because of their bitterness, bases are less common in foods than are acids. Our aversion to the taste of bases is probably an evolutionary adaptation to warn us against **alkaloids**, organic bases found in plants that are often poisonous. (For example, the active component of hemlock—the poisonous plant that caused the death of the Greek philosopher Socrates—is the alkaloid coniine.) Nonetheless, some foods, such as coffee and chocolate (especially dark chocolate), contain small amounts of base. Many people enjoy the bitterness, but only after acquiring the taste over time.

Coffee is acidic overall, but bases present in coffee—such as caffeine—impart a bitter flavor.

Bases feel slippery because they react with oils on the skin to form soaplike substances. Some household cleaning solutions, such as ammonia, are basic and have the characteristic slippery feel of a base. Bases turn red litmus paper blue; in the laboratory, litmus paper is routinely used to test the basicity of solutions.

Some common bases are listed in Table 15.2. Sodium hydroxide and potassium hydroxide are found in most chemistry laboratories. Sodium hydroxide is the active ingredient in products such as Drāno that work to unclog drains. Sodium bicarbonate can be found in most homes as baking soda and is also an active ingredient in many antacids.

TABLE 15.2 Common Bases

Name	Occurrence/Uses
Sodium hydroxide (NaOH)	Petroleum processing; soap and plastic manufacturing
Potassium hydroxide (KOH)	Cotton processing; electroplating; soap production; batteries
Sodium bicarbonate (NaHCO$_3$)	Antacid; ingredient of baking soda; source of CO$_2$
Sodium carbonate (Na$_2$CO$_3$)	Manufacture of glass and soap; general cleanser; water softener
Ammonia (NH$_3$)	Detergent; fertilizer and explosives manufacturing; synthetic fiber production

▲ Many common household products and remedies contain bases.

15.3   Definitions of Acids and Bases

What are the main characteristics of the molecules and ions that exhibit acid and base behavior? In this chapter, we examine three different definitions: the Arrhenius definition, the Brønsted–Lowry definition, and the Lewis definition. Why are there three definitions, and which one is correct? As Huheey notes in the quotation that opens this chapter, there really is no single "correct" definition. Rather, different definitions are convenient in different situations. The Lewis definition of acids and bases is discussed in Section 15.10; here we discuss the other two.

The Arrhenius Definition

In the 1880s, Swedish chemist Svante Arrhenius proposed these molecular definitions of acids and bases:

> **Acid:** A substance that produces H$^+$ ions in aqueous solution
>
> **Base:** A substance that produces OH$^-$ ions in aqueous solution

For example, under the **Arrhenius definition**, HCl is an acid because it produces H$^+$ ions in solution (**Figure 15.1▶**):

$$HCl(aq) \longrightarrow H^+(aq) + Cl^-(aq)$$

Hydrogen chloride (HCl) is a covalent compound and does not contain ions. However, in water it *ionizes* completely to form H$^+(aq)$ ions and Cl$^-(aq)$ ions. The H$^+$ ions are highly reactive. In aqueous solution, they bond to water molecules to form H$_3$O$^+$:

$$H^+ + \; :\!\overset{H}{\underset{}{O}}\!:\!H \longrightarrow \left[H\!:\!\overset{H}{\underset{}{O}}\!:\!H \right]^+$$

The H$_3$O$^+$ ion is called the **hydronium ion**. In water, H$^+$ ions *always* associate with H$_2$O molecules to form hydronium ions and other associated species with the general formula H(H$_2$O)$_n^+$. For example, an H$^+$ ion can associate with two water molecules to form H(H$_2$O)$_2^+$, with three to form H(H$_2$O)$_3^+$, and so on. Chemists often use H$^+(aq)$ and H$_3$O$^+(aq)$ interchangeably, however, to mean the same thing—an H$^+$ ion that has been solvated by water.

NaOH is an Arrhenius base because it produces OH$^-$ ions in solution (**Figure 15.2▶**):

$$NaOH(aq) \longrightarrow Na^+(aq) + OH^-(aq)$$

NaOH is an ionic compound and therefore contains Na$^+$ and OH$^-$ ions. When NaOH is added to water, it *dissociates* or breaks apart into its component ions.

Under the Arrhenius definition, acids and bases combine to form water, neutralizing each other in the process:

$$H^+(aq) + OH^-(aq) \longrightarrow H_2O(l)$$

Arrhenius Acid

HCl

HCl$(aq) \longrightarrow$ H$^+(aq)$ + Cl$^-(aq)$

▲ **FIGURE 15.1 Arrhenius Acid** An Arrhenius acid produces H$^+$ ions in aqueous solution.

Arrhenius Base

NaOH

NaOH$(aq) \longrightarrow$ Na$^+(aq)$ + OH$^-(aq)$

▲ **FIGURE 15.2 Arrhenius Base** An Arrhenius base produces OH$^-$ ions in aqueous solution.

The Brønsted–Lowry Definition

A second, more widely applicable definition of acids and bases, called the **Brønsted–Lowry definition**, was introduced in 1923. This definition focuses on the *transfer of H^+ ions* in an acid–base reaction. Since an H^+ ion is a proton—a hydrogen atom without its electron—this definition focuses on the idea of a proton donor and a proton acceptor:

> **Acid:** proton (H^+ ion) *donor*.
> **Base:** proton (H^+ ion) *acceptor*.

Under this definition, HCl is an acid because, in solution, it donates a proton to water:

$$HCl(aq) + H_2O(l) \longrightarrow H_3O^+(aq) + Cl^-(aq)$$

> All Arrhenius acids and bases are acids and bases under the Brønsted–Lowry definition. However, some Brønsted–Lowry acids and bases cannot naturally be classified as Arrhenius acids and bases.

This definition more clearly shows what happens to the H^+ ion from an acid—it associates with a water molecule to form H_3O^+ (a hydronium ion). The Brønsted–Lowry definition also works well with bases (such as NH_3) that do not inherently contain OH^- ions. According to the Brønsted–Lowry definition, NH_3 is a base because it accepts a proton from water:

$$NH_3(aq) + H_2O(l) \rightleftharpoons NH_4^+(aq) + OH^-(aq)$$

Under the Brønsted–Lowry definition, acids (proton donors) and bases (proton acceptors) always occur together. In the reaction between HCl and H_2O, HCl is the proton donor (acid) and H_2O is the proton acceptor (base):

$$\underset{\substack{\text{acid} \\ \text{(proton donor)}}}{HCl(aq)} + \underset{\substack{\text{base} \\ \text{(proton acceptor)}}}{H_2O(l)} \longrightarrow H_3O^+(aq) + Cl^-(aq)$$

In the reaction between NH_3 and H_2O, H_2O is the proton donor (acid) and NH_3 is the proton acceptor (base).

$$\underset{\substack{\text{base} \\ \text{(proton acceptor)}}}{NH_3(aq)} + \underset{\substack{\text{acid} \\ \text{(proton donor)}}}{H_2O(l)} \rightleftharpoons NH_4^+(aq) + OH^-(aq)$$

Notice that under the Brønsted–Lowry definition, some substances—such as water in the previous two equations—can act as acids *or* bases. Substances that can act as acids or bases are termed **amphoteric**. Notice also what happens when an equation representing Brønsted–Lowry acid–base behavior is reversed:

$$\underset{\substack{\text{acid} \\ \text{(proton donor)}}}{NH_4^+(aq)} + \underset{\substack{\text{base} \\ \text{(proton acceptor)}}}{OH^-(aq)} \rightleftharpoons NH_3(aq) + H_2O(l)$$

In this reaction, NH_4^+ is the proton donor (acid) and OH^- is the proton acceptor (base). The substance that was the base (NH_3) has become the acid (NH_4^+) and vice versa. NH_4^+ and NH_3 are a **conjugate acid–base pair**, two substances related to each other by the transfer of a proton (**Figure 15.3▼**). Going back to the original forward reaction, we can identify the conjugate acid–base pairs:

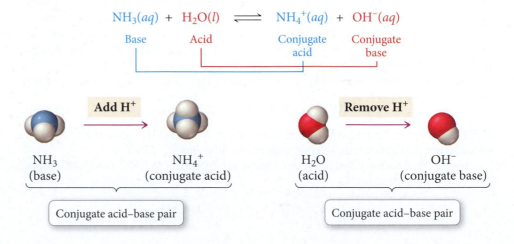

▶ **FIGURE 15.3 Conjugate Acid–Base Pairs** A conjugate acid–base pair consists of two substances related to each other by the transfer of a proton.

Summarizing Acid–Base Reactions Under the Brønsted–Lowry Definition:

▶ A base accepts a proton and becomes a conjugate acid.

▶ An acid donates a proton and becomes a conjugate base.

EXAMPLE 15.1 Identifying Brønsted–Lowry Acids and Bases and Their Conjugates

Identify the Brønsted–Lowry acid, the Brønsted–Lowry base, the conjugate acid, and the conjugate base in each reaction.

(a) $H_2SO_4(aq) + H_2O(l) \longrightarrow HSO_4^-(aq) + H_3O^+(aq)$

(b) $HCO_3^-(aq) + H_2O(l) \rightleftharpoons H_2CO_3(aq) + OH^-(aq)$

SOLUTION

(a) Since H_2SO_4 donates a proton to H_2O in this reaction, it is the acid (proton donor). After H_2SO_4 donates the proton, it becomes HSO_4^-, the conjugate base. Since H_2O accepts a proton, it is the base (proton acceptor). After H_2O accepts the proton it becomes H_3O^+, the conjugate acid.	$H_2SO_4(aq) + H_2O(l) \longrightarrow HSO_4^-(aq) + H_3O^+(aq)$ $H_2SO_4(aq)$ + $H_2O(l)$ $\longrightarrow$ $HSO_4^-(aq)$ + $H_3O^+(aq)$ Acid Base Conjugate Conjugate base acid
(b) Since H_2O donates a proton to HCO_3^- in this reaction, it is the acid (proton donor). After H_2O donates the proton, it becomes OH^-, the conjugate base. Since HCO_3^- accepts a proton, it is the base (proton acceptor). After HCO_3^- accepts the proton it becomes H_2CO_3, the conjugate acid.	$HCO_3^-(aq) + H_2O(l) \rightleftharpoons H_2CO_3(aq) + OH^-(aq)$ $HCO_3^-(aq)$ + $H_2O(l)$ $\rightleftharpoons$ $H_2CO_3(aq)$ + $OH^-(aq)$ Base Acid Conjugate Conjugate acid base

FOR PRACTICE 15.1

Identify the Brønsted–Lowry acid, the Brønsted–Lowry base, the conjugate acid, and the conjugate base in each reaction.

(a) $C_5H_5N(aq) + H_2O(l) \rightleftharpoons C_5H_5NH^+(aq) + OH^-(aq)$

(b) $HNO_3(aq) + H_2O(l) \longrightarrow H_3O^+(aq) + NO_3^-(aq)$

Conceptual Connection 15.1 Conjugate Acid–Base Pairs

Which pair is not a conjugate acid–base pair?

(a) $(CH_3)_3N$; $(CH_3)_3NH^+$

(b) H_2SO_4; H_2SO_3

(c) HNO_2; NO_2^-

15.4   Acid Strength and the Acid Ionization Constant (K_a)

The strength of an electrolyte, first discussed in Section 4.5, is determined by the extent of the dissociation of the electrolyte into its component ions in solution. A *strong electrolyte* completely dissociates into ions in solution whereas a *weak electrolyte* only partially dissociates. Strong and weak acids are defined accordingly. A **strong acid** completely ionizes in solution whereas a **weak acid** only partially ionizes. In other words, the strength of an acid depends on the equilibrium:

$$HA(aq) + H_2O(l) \rightleftharpoons H_3O^+(aq) + A^-(aq)$$

A Strong Acid

When HCl dissolves in water, it ionizes completely.

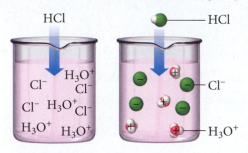

▲ **FIGURE 15.4 Ionization of a Strong Acid** When HCl dissolves in water, it completely ionizes to form H_3O^+ and Cl^-. The solution contains virtually no intact HCl.

An ionizable proton is one that ionizes in solution.

If the equilibrium lies far to the right, the acid is strong—it is virtually completely ionized. If the equilibrium lies to the left, the acid is weak—only a small percentage of the acid molecules are ionized. Of course, the range of acid strength is continuous, but for most purposes, the categories of strong and weak are useful.

Strong Acids

Hydrochloric acid (HCl) is an example of a strong acid.

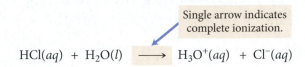

Single arrow indicates complete ionization.

$$HCl(aq) + H_2O(l) \longrightarrow H_3O^+(aq) + Cl^-(aq)$$

An HCl solution contains virtually no intact HCl; the HCl has essentially all ionized to form $H_3O^+(aq)$ and $Cl^-(aq)$ (**Figure 15.4◄**). A 1.0 M HCl solution will have an H_3O^+ concentration of 1.0 M. Abbreviating the concentration of H_3O^+ in units of molarity as $[H_3O^+]$, we say that a 1.0 M HCl solution has $[H_3O^+] = 1.0$ M.

Table 15.3 lists the six important strong acids. The first five acids in the table are **monoprotic acids**, acids containing only one ionizable proton. Sulfuric acid is a **diprotic acid**, an acid containing two ionizable protons.

TABLE 15.3 Strong Acids	
Hydrochloric acid (HCl)	Nitric acid (HNO_3)
Hydrobromic acid (HBr)	Perchloric acid ($HClO_4$)
Hydriodic acid (HI)	Sulfuric acid (H_2SO_4) (*diprotic*)

Weak Acids

In contrast to HCl, HF is an example of a weak acid, one that does not completely ionize in solution.

Equilibrium arrow indicates partial ionization.

$$HF(aq) + H_2O(l) \rightleftharpoons H_3O^+(aq) + F^-(aq)$$

The terms *strong* and *weak* acids are often confused with the terms *concentrated* and *dilute* acids. Can you articulate the difference between these terms?

An HF solution contains a lot of intact (or un-ionized) HF molecules; it also contains some $H_3O^+(aq)$ and $F^-(aq)$ (**Figure 15.5▼**). In other words, a 1.0 M HF solution has $[H_3O^+]$ that is much less than 1.0 M because only some of the HF molecules ionize to form H_3O^+.

A Weak Acid

When HF dissolves in water, only a fraction of the molecules ionize.

▶ **FIGURE 15.5 Ionization of a Weak Acid** When HF dissolves in water, only a fraction of the dissolved molecules ionize to form H_3O^+ and F^-. The solution contains many intact HF molecules.

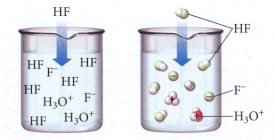

The degree to which an acid is strong or weak depends on the attraction between the anion of the acid (the conjugate base) and the hydrogen ion (relative to the attractions of these ions to water). Suppose HA is a generic formula for an acid. Then, the degree to which the following reaction proceeds in the forward direction depends on the strength of the attraction between H^+ and A^-.

$$HA(aq) + H_2O(l) \rightleftharpoons H_3O^+(aq) + A^-(aq)$$
$$\text{acid} \qquad\qquad\qquad\qquad \text{conjugate base}$$

If the attraction between H^+ and A^- is *weak*, then the reaction favors the forward direction and the acid is *strong*. If the attraction between H^+ and A^- is *strong*, then the reaction favors the reverse direction and the acid is *weak*, as shown in **Figure 15.6▶**.

For example, in HCl, the conjugate base (Cl^-) has a relatively weak attraction to H^+, meaning that the reverse reaction does not occur to any significant extent. In HF, on the other hand, the conjugate base (F^-) has a greater attraction to H^+, meaning that the reverse reaction occurs to a significant degree. *In general, the stronger the acid, the weaker the conjugate base and vice versa.* Table 15.4 lists some common weak acids.

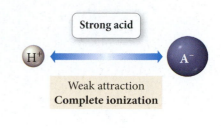
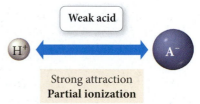

▲ FIGURE 15.6 Ionic Attraction and Acid Strength   In a strong acid, the attraction between H^+ and A^- is weak, resulting in complete ionization. In a weak acid, the attraction between H^+ and A^- is strong, resulting in only partial ionization.

TABLE 15.4 Some Weak Acids	
Hydrofluoric acid (HF)	Sulfurous acid (H_2SO_3) (*diprotic*)
Acetic acid ($HC_2H_3O_2$)	Carbonic acid (H_2CO_3) (*diprotic*)
Formic acid ($HCHO_2$)	Phosphoric acid (H_3PO_4) (*triprotic*)

The formulas for acetic acid and formic acid can also be written as CH_3COOH and HCOOH, respectively, to indicate that in these compounds the only H that ionizes is the one attached to an oxygen atom.

Notice that two of the weak acids in Table 15.4 are diprotic, meaning that they have two ionizable protons, and one is **triprotic** (three ionizable protons). Let us return to sulfuric acid for a moment. Sulfuric acid is a diprotic acid that is strong in its first ionizable proton:

$$H_2SO_4(aq) + H_2O(l) \longrightarrow H_3O^+(aq) + HSO_4^-(aq)$$

but weak in its second ionizable proton:

$$HSO_4^-(aq) + H_2O(l) \rightleftharpoons H_3O^+(aq) + SO_4^{2-}(aq)$$

Sulfurous acid and carbonic acid are weak in both of their ionizable protons, and phosphoric acid is weak in all three of its ionizable protons. We discuss polyprotic acids in more detail in Section 15.6.

The Acid Ionization Constant (K_a)

We quantify the relative strengths of weak acids with the **acid ionization constant (K_a)**, which is the equilibrium constant for the ionization reaction of a weak acid in water. As we saw in Section 14.3, for the following two equivalent reactions:

$$HA(aq) + H_2O(l) \rightleftharpoons H_3O^+(aq) + A^-(aq)$$

$$HA(aq) \rightleftharpoons H^+(aq) + A^-(aq)$$

the equilibrium constant is

$$K_a = \frac{[H_3O^+][A^-]}{[HA]} = \frac{[H^+][A^-]}{[HA]}$$

Since $[H_3O^+]$ is equivalent to $[H^+]$, both forms of the above expression are equal. Although the ionization constants for all weak acids are relatively small (otherwise the acid

Sometimes K_a is also called the acid dissociation constant.

Recall from Chapter 14 that the concentrations of pure solids or pure liquids are not included in the expression for K_c; therefore, $H_2O(l)$ is not included in the expression for K_a.

would not be a weak acid), they do vary in magnitude. The smaller the constant, the further to the left the equilibrium point for the ionization reaction lies, and the weaker the acid. Table 15.5 lists the acid ionization constants for a number of common weak acids in order of decreasing acid strength.

Conceptual Connection 15.2 Conjugate Bases

Consider these two acids and their K_a values:

$$HF \qquad K_a = 3.5 \times 10^{-4}$$
$$HClO \qquad K_a = 2.9 \times 10^{-8}$$

Which conjugate base, F^- or ClO^-, is stronger?

TABLE 15.5 Acid Ionization Constants (K_a) for Some Monoprotic Weak Acids at 25 °C

Acid	Formula	Structural Formula	Ionization Reaction	K_a
Chlorous acid	$HClO_2$	H—O—Cl=O	$HClO_2(aq) + H_2O(l) \rightleftharpoons$ $H_3O^+(aq) + ClO_2^-(aq)$	1.1×10^{-2}
Nitrous acid	HNO_2	H—O—N=O	$HNO_2(aq) + H_2O(l) \rightleftharpoons$ $H_3O^+(aq) + NO_2^-(aq)$	4.6×10^{-4}
Hydrofluoric acid	HF	H—F	$HF(aq) + H_2O(l) \rightleftharpoons$ $H_3O^+(aq) + F^-(aq)$	3.5×10^{-4}
Formic acid	$HCHO_2$		$HCHO_2(aq) + H_2O(l) \rightleftharpoons$ $H_3O^+(aq) + CHO_2^-(aq)$	1.8×10^{-4}
Benzoic acid	$HC_7H_5O_2$		$HC_7H_5O_2(aq) + H_2O(l) \rightleftharpoons$ $H_3O^+(aq) + C_7H_5O_2^-(aq)$	6.5×10^{-5}
Acetic acid	$HC_2H_3O_2$		$HC_2H_3O_2(aq) + H_2O(l) \rightleftharpoons$ $H_3O^+(aq) + C_2H_3O_2^-(aq)$	1.8×10^{-5}
Hypochlorous acid	HClO	H—O—Cl	$HClO(aq) + H_2O(l) \rightleftharpoons$ $H_3O^+(aq) + ClO^-(aq)$	2.9×10^{-8}
Hydrocyanic acid	HCN	H—C≡N	$HCN(aq) + H_2O(l) \rightleftharpoons$ $H_3O^+(aq) + CN^-(aq)$	4.9×10^{-10}
Phenol	HC_6H_5O		$HC_6H_5O(aq) + H_2O(l) \rightleftharpoons$ $H_3O^+(aq) + C_6H_5O^-(aq)$	1.3×10^{-10}

15.5 Autoionization of Water and pH

We noted earlier that water acts as a base when it reacts with HCl and as an acid when it reacts with NH_3:

Water acting as a base

$$HCl(aq) + H_2O(l) \longrightarrow H_3O^+(aq) + Cl^-(aq)$$

Acid (proton donor) **Base** (proton acceptor)

Water acting as an acid

$$NH_3(aq) + H_2O(l) \rightleftharpoons NH_4^+(aq) + OH^-(aq)$$

Base (proton acceptor) **Acid** (proton donor)

Water is *amphoteric*; it can act as either an acid or a base. Even in pure water, water acts as an acid and a base with itself, a process called **autoionization**:

Water acting as both an acid and a base

$$H_2O(l) + H_2O(l) \rightleftharpoons H_3O^+(aq) + OH^-(aq)$$

Acid (proton donor) **Base** (proton acceptor)

We can also write the autoionization reaction as:

$$H_2O(l) \rightleftharpoons H^+(aq) + OH^-(aq)$$

We quantify the autoionization of water with the equilibrium constant for the autoionization reaction.

$$K_w = [H_3O^+][OH^-] = [H^+][OH^-]$$

This equilibrium constant is the **ion product constant for water (K_w)** (sometimes called the *dissociation constant for water*). At 25 °C, $K_w = 1.0 \times 10^{-14}$. In pure water, since H_2O is the only source of these ions, the concentrations of H_3O^+ and OH^- are equal. Such a solution is said to be **neutral**. Since the concentrations are equal, they can be calculated from K_w.

$$[H_3O^+] = [OH^-] = \sqrt{K_w} = 1.0 \times 10^{-7} \text{ M} \quad \text{(in pure water at 25 °C)}$$

As you can see, in pure water, the concentrations of H_3O^+ and OH^- are *very small* (1.0×10^{-7} M) at room temperature.

An **acidic solution** contains an acid that creates additional H_3O^+ ions, causing $[H_3O^+]$ to increase. However, the *ion product constant still applies*:

$$[H_3O^+][OH^-] = K_w = 1.0 \times 10^{-14}$$

The concentration of H_3O^+ times the concentration of OH^- will always be 1.0×10^{-14} at 25 °C. If $[H_3O^+]$ increases, then $[OH^-]$ must decrease for the ion product constant to remain 1.0×10^{-14}. For example, if $[H_3O^+] = 1.0 \times 10^{-3}$ M, then we can determine $[OH^-]$ by solving the ion product constant expression for $[OH^-]$:

$$(1.0 \times 10^{-3})[OH^-] = 1.0 \times 10^{-14}$$

$$[OH^-] = \frac{1.0 \times 10^{-14}}{1.0 \times 10^{-3}} = 1.0 \times 10^{-11} \text{ M}$$

In an acidic solution $[H_3O^+] > [OH^-]$.

A **basic solution** contains a base that creates additional OH^- ions, causing $[OH^-]$ to increase and $[H_3O^+]$ to decrease but again the *ion product constant still applies*. For example, suppose $[OH^-] = 1.0 \times 10^{-2}$ M; then we can find $[H_3O^+]$ by solving the ion product constant expression for $[H_3O^+]$:

$$[H_3O^+](1.0 \times 10^{-2}) = 1.0 \times 10^{-14}$$

$$[H_3O^+] = \frac{1.0 \times 10^{-14}}{1.0 \times 10^{-2}} = 1.0 \times 10^{-12} \text{ M}$$

In a basic solution $[OH^-] > [H_3O^+]$.

Notice that changing $[H_3O^+]$ in an aqueous solution produces an inverse change in $[OH^-]$ and vice versa.

Summarizing K_w:

▶ A *neutral solution* contains $[H_3O^+] = [OH^-] = 1.0 \times 10^{-7}$ M (at 25 °C).

▶ An *acidic solution* contains $[H_3O^+] > [OH^-]$.

▶ A *basic solution* contains $[OH^-] > [H_3O^+]$.

▶ In *all aqueous solutions* both H_3O^+ and OH^- are present, with $[H_3O^+][OH^-] = K_w = 1.0 \times 10^{-14}$ (at 25 °C).

EXAMPLE 15.2 Using K_w in Calculations

Calculate $[OH^-]$ at 25 °C for each solution and determine whether the solution is acidic, basic, or neutral.

(a) $[H_3O^+] = 7.5 \times 10^{-5}$ M

(b) $[H_3O^+] = 1.5 \times 10^{-9}$ M

(c) $[H_3O^+] = 1.0 \times 10^{-7}$ M

SOLUTION

(a) To find $[OH^-]$ use the ion product constant. Substitute the given value for $[H_3O^+]$ and solve the equation for $[OH^-]$.	$[H_3O^+][OH^-] = K_w = 1.0 \times 10^{-14}$ $(7.5 \times 10^{-5})[OH^-] = 1.0 \times 10^{-14}$ $[OH^-] = \dfrac{1.0 \times 10^{-14}}{7.5 \times 10^{-5}} = 1.3 \times 10^{-10}$ M
Since $[H_3O^+] > [OH^-]$, the solution is acidic.	Acidic solution
(b) Substitute the given value for $[H_3O^+]$ and solve the acid ionization equation for $[OH^-]$.	$(1.5 \times 10^{-9})[OH^-] = 1.0 \times 10^{-14}$ $[OH^-] = \dfrac{1.0 \times 10^{-14}}{1.5 \times 10^{-9}} = 6.7 \times 10^{-6}$ M
Since $[H_3O^+] < [OH^-]$, the solution is basic.	Basic solution
(c) Substitute the given value for $[H_3O^+]$ and solve the acid ionization equation for $[OH^-]$. Since $[H_3O^+] = 1.0 \times 10^{-7}$ and $[OH^-] = 1.0 \times 10^{-7}$, the solution is neutral.	$(1.0 \times 10^{-7})[OH^-] = 1.0 \times 10^{-14}$ $[OH^-] = \dfrac{1.0 \times 10^{-14}}{1.0 \times 10^{-7}} = 1.0 \times 10^{-7}$ M Neutral solution

FOR PRACTICE 15.2

Calculate $[H_3O^+]$ at 25 °C for each solution and determine whether the solution is acidic, basic, or neutral.

(a) $[OH^-] = 1.5 \times 10^{-2}$ M

(b) $[OH^-] = 1.0 \times 10^{-7}$ M

(c) $[OH^-] = 8.2 \times 10^{-10}$ M

The pH Scale: A Way to Quantify Acidity and Basicity

The pH scale is a compact way to specify the acidity of a solution. We define **pH** as the negative of the log of the concentration of H_3O^+:

$$pH = -\log[H_3O^+]$$

A solution with $[H_3O^+] = 1.0 \times 10^{-3}$ M (acidic) has a pH of:

$$
\begin{aligned}
pH &= -\log[H_3O^+] \\
&= -\log(1.0 \times 10^{-3}) \\
&= -(-3.00) = 3.00
\end{aligned}
$$

Notice that we report the pH to two *decimal places* here. This is because only the numbers to the right of the decimal point are significant in a logarithm. Since our original value for the concentration had two significant figures, the log of that number has two decimal places.

2 significant figures | 2 decimal places

$$\log(1.0) \times 10^{-3} = 3.00$$

If the original number had three significant digits, the log would be reported to three decimal places:

3 significant figures | 3 decimal places

$$\log(1.00) \times 10^{-3} = 3.000$$

A solution with $[H_3O^+] = 1.0 \times 10^{-7}$ M (neutral) has a pH of

$$
\begin{aligned}
pH &= -\log[H_3O^+] \\
&= -\log(1.0 \times 10^{-7}) \\
&= -(-7.00) = 7.00
\end{aligned}
$$

In general, at 25 °C:

- pH < 7   The solution is *acidic*.
- pH > 7   The solution is *basic*.
- pH = 7   The solution is *neutral*.

Table 15.6 lists the pH of some common substances. Notice that, as we discussed in Section 15.2, many foods, especially fruits, are acidic and therefore have low pH values. Relatively few foods, however, are basic. The foods with the lowest pH values are limes and lemons, and they are among the sourest. Since the pH scale is a *logarithmic scale*, a change of 1 pH unit corresponds to a 10-fold change in H_3O^+ concentration (**Figure 15.7▼**). For example, a lime with a pH of 2.0 is 10 times more acidic than a plum with a pH of 3.0 and 100 times more acidic than a cherry with a pH of 4.0.

> The log of a number is the exponent to which 10 must be raised to obtain that number. Thus, $\log 10^1 = 1$; $\log 10^2 = 2$; $\log 10^{-1} = -1$; $\log 10^{-2} = -2$, etc. (see Appendix I).

> When you take the log of a quantity, the result should have the same number of decimal places as the number of significant figures in the original quantity.

TABLE 15.6 The pH of Some Common Substances

Substance	pH
Gastric juice (human stomach)	1.0–3.0
Limes	1.8–2.0
Lemons	2.2–2.4
Soft drinks	2.0–4.0
Plums	2.8–3.0
Wines	2.8–3.8
Apples	2.9–3.3
Peaches	3.4–3.6
Cherries	3.2–4.0
Beers	4.0–5.0
Rainwater (unpolluted)	5.6
Human blood	7.3–7.4
Egg whites	7.6–8.0
Milk of magnesia	10.5
Household ammonia	10.5–11.5
4% NaOH solution	14

> Concentrated acid solutions can have negative pH. For example, if $[H_3O^+] = 2.0$ M, the pH is -0.30.

The pH Scale

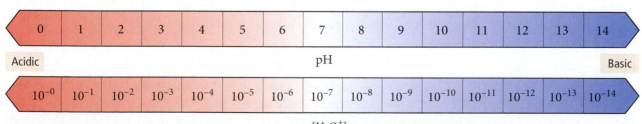

▲ FIGURE 15.7 The pH Scale An increase of 1 on the pH scale corresponds to a decrease in $[H_3O^+]$ by a factor of 10.

EXAMPLE 15.3 Calculating pH from $[H_3O^+]$ or $[OH^-]$

Calculate the pH of each solution at 25 °C and indicate whether the solution is acidic or basic.

(a) $[H_3O^+] = 1.8 \times 10^{-4}$ M   **(b)** $[OH^-] = 1.3 \times 10^{-2}$ M

SOLUTION

(a) To calculate pH, substitute the given $[H_3O^+]$ into the pH equation.	$pH = -\log[H_3O^+]$ $= -\log(1.8 \times 10^{-4})$ $= -(-3.74)$
Since pH < 7, this solution is acidic.	$= 3.74 \, (\text{acidic})$
(b) First use K_w to find $[H_3O^+]$ from $[OH^-]$.	$[H_3O^+][OH^-] = K_w = 1.0 \times 10^{-14}$ $[H_3O^+](1.3 \times 10^{-2}) = 1.0 \times 10^{-14}$ $[H_3O^+] = \dfrac{1.0 \times 10^{-14}}{1.3 \times 10^{-2}} = 7.7 \times 10^{-13}$ M
Then substitute $[H_3O^+]$ into the pH expression to find pH.	$pH = -\log[H_3O^+]$ $= -\log(7.7 \times 10^{-13})$ $= -(-12.11)$
Since pH > 7, this solution is basic.	$= 12.11 \, (\text{basic})$

FOR PRACTICE 15.3

Calculate the pH of each solution and indicate whether the solution is acidic or basic.

(a) $[H_3O^+] = 9.5 \times 10^{-9}$ M   **(b)** $[OH^-] = 7.1 \times 10^{-3}$ M

EXAMPLE 15.4 Calculating $[H_3O^+]$ from pH

Calculate the H_3O^+ concentration for a solution with a pH of 4.80.

SOLUTION

To find the $[H_3O^+]$ from pH, start with the equation that defines pH. Substitute the given value of pH and then solve for $[H_3O^+]$. Since the given pH value is reported to two decimal places, the $[H_3O^+]$ is written to two significant figures. (Remember that $10^{\log x} = x$ (see Appendix I). Some calculators use an inv log key to represent this function.)	$pH = -\log[H_3O^+]$ $4.80 = -\log[H_3O^+]$ $-4.80 = \log[H_3O^+]$ $10^{-4.80} = 10^{\log[H_3O^+]}$ $10^{-4.80} = [H_3O^+]$ $[H_3O^+] = 1.6 \times 10^{-5}$ M

FOR PRACTICE 15.4

Calculate the H_3O^+ concentration for a solution with a pH of 8.37.

pOH and Other p Scales

Notice that p is the mathematical function $-\log$; thus, $pX = -\log X$.

The pOH scale is analogous to the pH scale but is defined with respect to $[OH^-]$ instead of $[H_3O^+]$.

$$pOH = -\log[OH^-]$$

A solution having an $[OH^-]$ of 1.0×10^{-3} M (basic) has a pOH of 3.00. On the pOH scale, a pOH less than 7 is basic and a pOH greater than 7 is acidic. A pOH of 7 is neutral (**Figure 15.8▶**). We can derive a relationship between pH and pOH at 25 °C from the expression for K_w:

$$[H_3O^+][OH^-] = 1.0 \times 10^{-14}$$

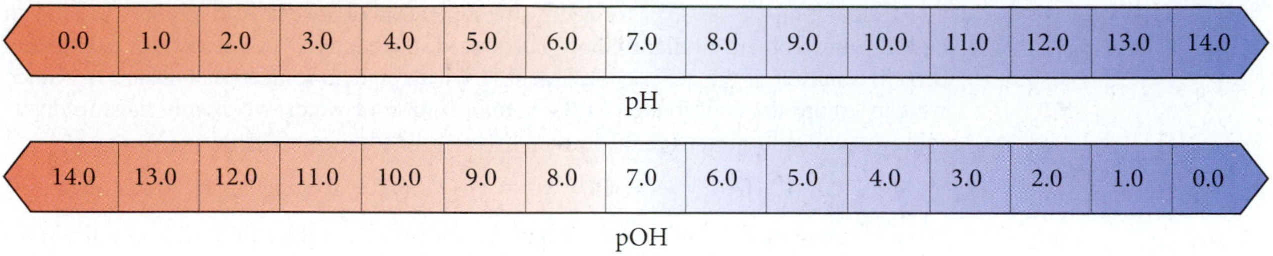

0.0	1.0	2.0	3.0	4.0	5.0	6.0	7.0	8.0	9.0	10.0	11.0	12.0	13.0	14.0

pH

14.0	13.0	12.0	11.0	10.0	9.0	8.0	7.0	6.0	5.0	4.0	3.0	2.0	1.0	0.0

pOH

▲ **FIGURE 15.8 pH and pOH**

Taking the log of both sides, we find

$$\log([H_3O^+][OH^-]) = \log(1.0 \times 10^{-14})$$
$$\log[H_3O^+] + \log[OH^-] = -14.00$$
$$-\log[H_3O^+] - \log[OH^-] = 14.00$$

$$pH + pOH = 14.00$$

The sum of pH and pOH is always equal to 14.00 at 25 °C. Therefore, a solution with a pH of 3 has a pOH of 11.

Another common p scale is pK_a, defined as follows:

$$pK_a = -\log K_a$$

The pK_a of a weak acid is just another way to quantify its strength. The smaller the pK_a, the stronger the acid. For example, chlorous acid, with a K_a of 1.1×10^{-2}, has a pK_a of 1.96 and formic acid, with a K_a of 1.8×10^{-4}, has a pK_a of 3.74.

15.6 Finding the $[H_3O^+]$ and pH of Strong and Weak Acid Solutions

In a solution containing a strong or weak acid, there are two potential sources of H_3O^+, the acid itself and the ionization of water. If we let HA be a strong or weak acid, the ionization reactions are:

$$HA(aq) + H_2O(l) \rightleftharpoons H_3O^+(aq) + A^-(aq) \quad \text{Strong or Weak Acid}$$
$$H_2O(l) + H_2O(l) \rightleftharpoons H_3O^+(aq) + OH^-(aq) \quad K_w = 1.0 \times 10^{-14}$$

Except in extremely dilute acid solutions, the ionization of water contributes a negligibly small amount of H_3O^+ compared to the ionization of a strong or weak acid. Therefore, we can focus exclusively on the amount of H_3O^+ produced by the acid.

The only exceptions would be extremely dilute ($<10^{-6}$ M) acid solutions.

Strong Acids

Because strong acids, by definition, are completely ionized in solution, and because we can (in nearly all cases) ignore the contribution of the autoionization of water, *the concentration of H_3O^+ in a strong acid solution is equal to the concentration of the strong acid.* For example, a 0.10 M HCl solution has an H_3O^+ concentration of 0.10 M and a pH of 1.00.

$$0.10 \text{ M HCl} \implies [H_3O^+] = 0.10 \text{ M} \implies pH = -\log(0.10) = 1.00$$

Weak Acids

Finding the pH of a weak acid solution is more complicated because the concentration of H_3O^+ is *not equal* to the concentration of the weak acid. For example, if you make solutions of 0.10 M HCl (a strong acid) and 0.10 M acetic acid (a weak acid) and measure the pH of each, you get the following results:

$$0.10 \text{ M HCl} \qquad pH = 1.00$$
$$0.10 \text{ M HC}_2\text{H}_3\text{O}_2 \qquad pH = 2.87$$

The pH of the acetic acid solution is higher (it is less acidic) because it is a weak acid and therefore only partially ionizes.

Calculating the $[H_3O^+]$ formed by the ionization of a weak acid requires solving an equilibrium problem similar to those introduced in Chapter 14. Consider, for example, a 0.10 M solution of the generic weak acid HA with an acid ionization constant K_a. Since we can ignore the contribution of the autoionization of water, we simply have to determine the concentration of H_3O^+ formed by the following equilibrium:

$$HA(aq) + H_2O(l) \rightleftharpoons H_3O^+(aq) + A^-(aq) \qquad K_a$$

We can summarize the initial conditions, the changes, and the equilibrium conditions in the following ICE table:

> The ICE table was first introduced in Section 14.6. The reactant $H_2O(l)$ is a pure liquid and is therefore not included in the equilibrium constant nor in the ICE table (see Section 14.5).

	[HA]	$[H_3O^+]$	$[A^-]$
Initial	0.10	≈ 0.00	0.00
Change	$-x$	$+x$	$+x$
Equilibrium	$0.10 - x$	x	x

The initial H_3O^+ concentration is listed as *approximately* zero because of the negligibly small contribution of H_3O^+ due to the autoionization of water (discussed previously). The variable x represents the amount of HA that ionizes. As discussed in Chapter 14, each *equilibrium* concentration is the sum of the two entries above it in the ICE table. In order to find the equilibrium concentration of H_3O^+, we must find the value of the variable x. We can use the equilibrium expression to set up an equation in which x is the only variable.

$$K_a = \frac{[H_3O^+][A^-]}{[HA]} = \frac{x^2}{0.10 - x}$$

As in other equilibrium problems, we arrive at a quadratic equation in x, which can be solved using the quadratic formula. In many cases, however, the *x is small* approximation discussed in Section 14.8 can be applied. Next we present the general procedure for weak acid equilibrium problems in the left column and two examples (15.5 and 15.6) of applying the procedure in the center and right columns. For both of these examples, the *x is small* approximation works well. In Example 15.7, we solve a problem in which the *x is small* approximation does not work. In such cases, we can solve the quadratic equation explicitly, or we can apply the method of successive approximations, also discussed in Section 14.8. Finally, in Example 15.8, we work a problem in which we must find the equilibrium constant of a weak acid from its pH.

PROCEDURE FOR...

Finding the pH (or $[H_3O^+]$) of a Weak Acid Solution

To solve these types of problems, follow the procedure outlined below.

1. **Write the balanced equation for the ionization of the acid and use it as a guide to prepare an ICE table showing the given concentration of the weak acid as its initial concentration.** Leave room in the table for the changes in concentrations and for the equilibrium concentrations.

(Note that the H_3O^+ concentration is listed as approximately zero because the autoionization of water produces a negligibly small amount of H_3O^+.)

EXAMPLE 15.5

Finding the $[H_3O^+]$ of a Weak Acid Solution

Find the $[H_3O^+]$ of a 0.100 M HCN solution.

$$HCN(aq) + H_2O(l) \rightleftharpoons H_3O^+(aq) + CN^-(aq)$$

	[HCN]	$[H_3O^+]$	$[CN^-]$
Initial	0.100	≈ 0.00	0.00
Change			
Equil			

EXAMPLE 15.6

Finding the pH of a Weak Acid Solution

Find the pH of a 0.200 M HNO_2 solution.

$$HNO_2(aq) + H_2O(l) \rightleftharpoons H_3O^+(aq) + NO_2^-(aq)$$

	[HNO]	$[H_3O^+]$	$[NO_2^-]$
Initial	0.200	≈ 0.00	0.00
Change			
Equil			

2. Represent the change in the concentration of H_3O^+ with the variable x. Define the changes in the concentrations of the other reactants and products in terms of x, always keeping in mind the stoichiometry of the reaction.	$HCN(aq) + H_2O(l) \rightleftharpoons$ $H_3O^+(aq) + CN^-(aq)$	$HNO_2(aq) + H_2O(l) \rightleftharpoons$ $H_3O^+(aq) + NO_2^-(aq)$

	[HCN]	$[H_3O^+]$	$[CN^-]$
Initial	0.100	≈ 0.00	0.00
Change	$-x$	$+x$	$+x$
Equil			

	[HNO]	$[H_3O^+]$	$[NO_2^-]$
Initial	0.200	≈ 0.00	0.00
Change	$-x$	$+x$	$+x$
Equil			

3. Sum each column to determine the equilibrium concentrations in terms of the initial concentrations and the variable x.	$HCN(aq) + H_2O(l) \rightleftharpoons$ $H_3O^+(aq) + CN^-(aq)$	$HNO_2(aq) + H_2O(l) \rightleftharpoons$ $H_3O^+(aq) + NO_2^-(aq)$

	[HCN]	$[H_3O^+]$	$[CN^-]$
Initial	0.100	≈ 0.00	0.00
Change	$-x$	$+x$	$+x$
Equil	$0.100 - x$	x	x

	[HNO]	$[H_3O^+]$	$[NO_2^-]$
Initial	0.200	≈ 0.00	0.00
Change	$-x$	$+x$	$+x$
Equil	$0.200 - x$	x	x

4. Substitute the expressions for the equilibrium concentrations (from step 3) into the expression for the acid ionization constant (K_a).

In many cases, you can make the approximation that x is small (as discussed in Section 14.8). **Substitute the value of the acid ionization constant (from Table 15.5) into the K_a expression and solve for x.**

$K_a = \dfrac{[H_3O^+][CN^-]}{[HCN]}$

$= \dfrac{x^2}{0.100 - \cancel{x}}$ (x is small)

$4.9 \times 10^{-10} = \dfrac{x^2}{0.100}$

$\sqrt{4.9 \times 10^{-10}} = \sqrt{\dfrac{x^2}{0.100}}$

$x = \sqrt{(0.100)(4.9 \times 10^{-10})}$

$= 7.0 \times 10^{-6}$

$K_a = \dfrac{[H_3O^+][NO_2^-]}{[HNO_2]}$

$= \dfrac{x^2}{0.200 - \cancel{x}}$ (x is small)

$4.6 \times 10^{-4} = \dfrac{x^2}{0.200}$

$\sqrt{4.6 \times 10^{-4}} = \sqrt{\dfrac{x^2}{0.200}}$

$x = \sqrt{(0.200)(4.6 \times 10^{-4})}$

$= 9.6 \times 10^{-3}$

Confirm that the x is small approximation is valid by calculating the ratio of x to the number it was subtracted from in the approximation. The ratio should be less than 0.05 (or 5%).

$\dfrac{7.0 \times 10^{-6}}{0.100} \times 100\% = 7.0 \times 10^{-3}\%$

Therefore the approximation is valid.

$\dfrac{9.6 \times 10^{-3}}{0.200} \times 100\% = 4.8\%$

Therefore the approximation is valid (but barely so).

5. Determine the H_3O^+ concentration from the calculated value of x and calculate the pH if necessary.

$[H_3O^+] = 7.0 \times 10^{-6}$ M

(pH was not asked for in this problem.)

$[H_3O^+] = 9.6 \times 10^{-3}$ M

$pH = -\log[H_3O^+]$

$= -\log(9.6 \times 10^{-3})$

$= 2.02$

6. Check your answer by substituting the calculated equilibrium values into the acid ionization expression. The calculated value of K_a should match the given value of K_a. Note that rounding errors and the x is small approximation could cause a difference in the least significant digit when comparing values of K_a.

$K_a = \dfrac{[H_3O^+][CN^-]}{[HCN]} = \dfrac{(7.0 \times 10^{-6})^2}{0.100}$

$= 4.9 \times 10^{-10}$

Since the calculated value of K_a matches the given value, the answer is valid.

$K_a = \dfrac{[H_3O^+][NO_2^-]}{[HNO_2]} = \dfrac{(9.6 \times 10^{-3})^2}{0.200}$

$= 4.6 \times 10^{-4}$

Since the calculated value of K_a matches the given value, the answer is valid.

FOR PRACTICE 15.5

Find the H_3O^+ concentration of a 0.250 M hydrofluoric acid solution.

FOR PRACTICE 15.6

Find the pH of a 0.0150 M acetic acid solution.

EXAMPLE 15.7 Finding the pH of a Weak Acid Solution in Cases Where the *x is small* Approximation Does Not Work

Find the pH of a 0.100 M $HClO_2$ solution.

SOLUTION

1. Write the balanced equation for the ionization of the acid and use it as a guide to prepare an ICE table showing the given concentration of the weak acid as its initial concentration. (Note that the H_3O^+ concentration is listed as approximately zero. Although a little H_3O^+ is present from the autoionization of water, this amount is negligibly small compared to the amount of H_3O^+ from the acid.)	$HClO_2(aq) + H_2O(l) \rightleftharpoons H_3O^+(aq) + ClO_2^-(aq)$ <table><tr><td></td><td>$[HClO_2]$</td><td>$[H_3O^+]$</td><td>$[ClO_2^-]$</td></tr><tr><td>Initial</td><td>0.100</td><td>≈0.00</td><td>0.00</td></tr><tr><td>Change</td><td></td><td></td><td></td></tr><tr><td>Equil</td><td></td><td></td><td></td></tr></table>
2. Represent the change in $[H_3O^+]$ with the variable x. Define the changes in the concentrations of the other reactants and products in terms of x.	$HClO_2(aq) + H_2O(l) \rightleftharpoons H_3O^+(aq) + ClO_2^-(aq)$ <table><tr><td></td><td>$[HClO_2]$</td><td>$[H_3O^+]$</td><td>$[ClO_2^-]$</td></tr><tr><td>Initial</td><td>0.100</td><td>≈0.00</td><td>0.00</td></tr><tr><td>Change</td><td>$-x$</td><td>$+x$</td><td>$+x$</td></tr><tr><td>Equil</td><td></td><td></td><td></td></tr></table>
3. Sum each column to determine the equilibrium concentrations in terms of the initial concentrations and the variable x.	$HClO_2(aq) + H_2O(l) \rightleftharpoons H_3O^+(aq) + ClO_2^-(aq)$ <table><tr><td></td><td>$[HClO_2]$</td><td>$[H_3O^+]$</td><td>$[ClO_2^-]$</td></tr><tr><td>Initial</td><td>0.100</td><td>≈0.00</td><td>0.00</td></tr><tr><td>Change</td><td>$-x$</td><td>$+x$</td><td>$+x$</td></tr><tr><td>Equil</td><td>$0.100 - x$</td><td>x</td><td>x</td></tr></table>
4. Substitute the expressions for the equilibrium concentrations (from step 3) into the expression for the acid ionization constant (K_a). Make the *x is small* approximation and substitute the value of the acid ionization constant (from Table 15.5) into the K_a expression. Solve for x.	$K_a = \dfrac{[H_3O^+][ClO_2^-]}{[HNO_2]}$ $= \dfrac{x^2}{0.100 - \cancel{x}}$ (*x is small*) $0.011 = \dfrac{x^2}{0.100}$ $\sqrt{0.011} = \sqrt{\dfrac{x^2}{0.100}}$ $x = \sqrt{(0.100)(0.011)}$ $= 0.033$
Check to see if the *x is small* approximation is valid by calculating the ratio of x to the number it was subtracted from in the approximation. The ratio should be less than 0.05 (or 5%).	$\dfrac{0.033}{0.100} \times 100\% = 33\%$ Therefore, the *x is small* approximation is *not valid*.
4a. If the *x is small* approximation is not valid, solve the quadratic equation explicitly or use the method of successive approximations to find x. In this case, the quadratic equation is solved.	$0.011 = \dfrac{x^2}{0.100 - x}$ $0.011(0.100 - x) = x^2$ $0.0011 - 0.011x = x^2$ $x^2 + 0.011x - 0.0011 = 0$ $x = \dfrac{-b \pm \sqrt{b^2 - 4ac}}{2a}$ $= \dfrac{-(0.011) \pm \sqrt{(0.011)^2 - 4(1)(-0.0011)}}{2(1)}$ $= \dfrac{-0.011 \pm 0.0672}{2}$ $x = -0.039 \ or \ x = 0.028$

	Since x represents the concentration of H_3O^+, and since concentrations cannot be negative, reject the negative root. $x = 0.028$
5. Determine the H_3O^+ concentration from the calculated value of x and calculate the pH (if necessary).	$[H_3O^+] = 0.028$ M $pH = -\log[H_3O^+]$ $\quad = -\log 0.028$ $\quad = 1.55$
6. Check your answer by substituting the computed equilibrium values into the acid ionization expression. The computed value of K_a should match the given value of K_a. Note that rounding errors could cause a difference in the least significant digit when comparing values of K_a.	$K_a = \dfrac{[H_3O^+][ClO_2^-]}{[HClO_2]} = \dfrac{0.028^2}{0.100 - 0.028}$ $\quad = 0.011$ Since the calculated value of K_a matches the given value, the answer is valid.

FOR PRACTICE 15.7

Find the pH of a 0.010 M HNO_2 solution.

EXAMPLE 15.8 Finding the Equilibrium Constant from pH

A 0.100 M weak acid (HA) solution has a pH of 4.25. Find K_a for the acid.

SOLUTION

Use the given pH to find the equilibrium concentration of $[H_3O^+]$. Then write the balanced equation for the ionization of the acid and use it as a guide to prepare an ICE table showing all known concentrations.	$pH = -\log[H_3O^+]$ $4.25 = -\log[H_3O^+]$ $[H_3O^+] = 5.6 \times 10^{-5}$ M

$$HA(aq) + H_2O(l) \rightleftharpoons H_3O^+(aq) + A^-(aq)$$

	[HA]	$[H_3O^+]$	[A$^-$]
Initial	0.100	≈ 0.00	0.00
Change			
Equil		5.6×10^{-5}	

Use the equilibrium concentration of H_3O^+ and the stoichiometry of the reaction to predict the changes and equilibrium concentration for all species. For most weak acids, the initial and equilibrium concentrations of the weak acid (HA) will be equal because the amount that ionizes is usually very small compared to the initial concentration.	

$$HA(aq) + H_2O(l) \rightleftharpoons H_3O^+(aq) + A^-(aq)$$

	[HA]	$[H_3O^+]$	[A$^-$]
Initial	0.100	≈ 0.00	0.00
Change	-5.6×10^{-5}	$+5.6 \times 10^{-5}$	$+5.6 \times 10^{-5}$
Equil	$(0.100 - 5.6 \times 10^{-5})$	5.6×10^{-5}	5.6×10^{-5}
	≈ 0.100		

Substitute the equilibrium concentrations into the expression for K_a and calculate its value.	$K_a = \dfrac{[H_3O^+][A^-]}{[HA]}$ $\quad = \dfrac{(5.6 \times 10^{-5})(5.6 \times 10^{-5})}{0.100}$ $\quad = 3.1 \times 10^{-8}$

FOR PRACTICE 15.8

A 0.175 M weak acid solution has a pH of 3.25. Find K_a for the acid.

Conceptual Connection 15.3 The _x is small_ Approximation

The initial concentration and K_a's of several weak acid (HA) solutions are listed below. For which of these is the _x is small_ approximation _least_ likely to work in finding the pH of the solution?

(a) initial $[\text{HA}] = 0.100$ M; $K_a = 1.0 \times 10^{-5}$
(b) initial $[\text{HA}] = 1.00$ M; $K_a = 1.0 \times 10^{-6}$
(c) initial $[\text{HA}] = 0.0100$ M; $K_a = 1.0 \times 10^{-3}$
(d) initial $[\text{HA}] = 1.0$ M; $K_a = 1.5 \times 10^{-3}$

Conceptual Connection 15.4 Strong and Weak Acids

Which solution is most acidic (that is, which one has the lowest pH)?

(a) 0.10 M HCl
(b) 0.10 M HF
(c) 0.20 M HF

Polyprotic Acids

Recall from Section 15.4 that some acids, called polyprotic acids, contain two or more ionizable protons. Sulfurous acid (H_2SO_3) is a diprotic acid containing two ionizable protons, and phosphoric acid (H_3PO_4) is a triprotic acid containing three ionizable protons. In general, a **polyprotic acid** ionizes in successive steps, each with its own K_a. For example, sulfurous acid ionizes as follows:

$$H_2SO_3(aq) \rightleftharpoons H^+(aq) + HSO_3^-(aq) \qquad K_{a_1} = 1.6 \times 10^{-2}$$
$$HSO_3^-(aq) \rightleftharpoons H^+(aq) + SO_3^{2-}(aq) \qquad K_{a_2} = 6.4 \times 10^{-8}$$

where K_{a_1} is the acid ionization constant for the first step and K_{a_2} is the acid ionization constant for the second step. Notice that K_{a_2} is smaller than K_{a_1}. This is true for all polyprotic acids and makes physical sense because the first proton must separate from a neutral molecule while the second must separate from an anion. The negative charge of the anion causes the positively charged proton to be held more tightly, making it more difficult to remove and resulting in a smaller value of K_a. Table 15.7 lists some common polyprotic acids and their acid ionization constants. Notice that in all cases, the values of K_a for each step become successively smaller. The value of K_{a_1} for sulfuric acid is listed as strong because, as we learned in Section 15.4, sulfuric acid is strong in the first step and weak in the second.

Finding the pH of a polyprotic acid solution is simpler than might be imagined because, for most polyprotic acids, K_{a_1} is much larger than K_{a_2} (or K_{a_3} for triprotic acids). Therefore, the amount of H_3O^+ contributed by the first ionization step is much larger than that contributed by the second or third ionization step (**Figure 15.9▼**). In addition, the

Dissociation of a Polyprotic Acid

$$H_2C_6H_6O_6(aq) + H_2O(l) \rightleftharpoons H_3O^+(aq) + HC_6H_6O_6^-(aq)$$

$$[H_3O^+] = 2.8 \times 10^{-3} \text{ M}$$

$$HC_6H_6O_6^-(aq) + H_2O(l) \rightleftharpoons H_3O^+(aq) + C_6H_6O_6^{2-}(aq)$$

$$[H_3O^+] = 1.6 \times 10^{-12} \text{ M}$$

0.100 M $H_2C_6H_6O_6$

$$\text{Total} [H_3O^+] = 2.8 \times 10^{-3} \text{ M} + 1.6 \times 10^{-12} \text{ M}$$

$$= 2.8 \times 10^{-3} \text{ M}$$

▶ FIGURE 15.9 Dissociation of a Polyprotic Acid A 0.100 M $H_2C_6H_6O_6$ solution contains an H_3O^+ concentration of 2.8×10^{-3} M from the first step. The amount of H_3O^+ contributed by the second step is only 1.6×10^{-12} M, which is insignificant compared to the amount produced by the first step.

TABLE 15.7 Common Polyprotic Acids and Ionization Constants

Name (Formula)	Structure	Space-filling model	K_{a_1}	K_{a_2}	K_{a_3}
Sulfuric Acid (H_2SO_4)			Strong	1.2×10^{-2}	
Oxalic Acid ($H_2C_2O_4$)			6.0×10^{-2}	6.1×10^{-5}	
Sulfurous Acid (H_2SO_3)			1.6×10^{-2}	6.4×10^{-8}	
Phosphoric Acid (H_3PO_4)			7.5×10^{-3}	6.2×10^{-8}	4.2×10^{-13}
Citric Acid ($H_3C_6H_5O_7$)			7.4×10^{-4}	1.7×10^{-5}	4.0×10^{-7}
Ascorbic Acid ($H_2C_6H_6O_6$)			8.0×10^{-5}	1.6×10^{-12}	
Carbonic Acid (H_2CO_3)			4.3×10^{-7}	5.6×10^{-11}	

production of H_3O^+ by the first step inhibits additional production of H_3O^+ by the second step (because of Le Châtelier's principle). Consequently, we can treat most polyprotic acid solutions as if the first step were the only one that contributes to the H_3O^+ concentration.

Percent Ionization of a Weak Acid

We can quantify the ionization of a weak acid based on the percentage of acid molecules that actually ionize. For instance, in Example 15.6, we found that a 0.200 M HNO_2 solution contains 9.6×10^{-3} M H_3O^+. We can define a useful quantity called the **percent ionization** of a weak acid as follows:

$$\text{Percent ionization} = \frac{\text{concentration of ionized acid}}{\text{initial concentration of acid}} \times 100\% = \frac{[H_3O^+]_{\text{equil}}}{[HA]_{\text{init}}} \times 100\%$$

Since the concentration of ionized acid is equal to the H_3O^+ concentration at equilibrium (for a monoprotic acid), we can use $[H_3O^+]_{\text{equil}}$ and $[HA]_{\text{init}}$ to calculate the percent ionization. The 0.200 M HNO_2 solution therefore has the following percent ionization:

$$\% \text{ ionization} = \frac{\left[H_3O^+\right]_{equil}}{\left[HA\right]_{init}} \times 100\%$$

$$= \frac{9.6 \times 10^{-3}\ M}{0.200\ M} \times 100\%$$

$$= 4.8\%$$

As you can see, the percent ionization is relatively small. In this case, even for a weak acid that has a relatively large K_a (HNO_2 has the second largest K_a in Table 15.5), the percent ionization indicates that less than five molecules out of one hundred ionize. For most other weak acids (with smaller K_a values), the percent ionization is even less.

In Example 15.9, we calculate the percent ionization of a more concentrated HNO_2 solution. As you read through the example, notice the calculated H_3O^+ concentration is much greater (as we would expect for a more concentrated solution), but the *percent ionization* is actually smaller.

EXAMPLE 15.9 Finding the Percent Ionization of a Weak Acid

Find the percent ionization of a 2.5 M HNO_2 solution.

SOLUTION

To find the percent ionization, you must find the equilibrium concentration of H_3O^+. Follow the procedure in Example 15.5, shown in condensed form here.	$HNO_2(aq) + H_2O(l) \rightleftharpoons H_3O^+(aq) + NO_2^-(aq)$

	$[HNO_2]$	$[H_3O^+]$	$[NO_2^-]$
Initial	2.5	≈ 0.00	0.00
Change	$-x$	$+x$	$+x$
Equil	$2.5 - x$	x	x

$$K_a = \frac{\left[H_3O^+\right]\left[NO_2^-\right]}{\left[HNO_2\right]} = \frac{x^2}{2.5 - x} \quad (x \text{ is small})$$

$$4.6 \times 10^{-4} = \frac{x^2}{2.5}$$

$$x = 0.034$$

Therefore, $\left[H_3O^+\right] = 0.034$ M.

Use the definition of percent ionization to calculate it. (Since the percent ionization is less than 5%, the *x is small* approximation is valid.)	$\% \text{ ionization} = \frac{\left[H_3O^+\right]_{equil}}{\left[HA\right]_{init}} \times 100\%$ $= \frac{0.034\ M}{2.5\ M} \times 100\%$ $= 1.4\%$

FOR PRACTICE 15.9

Find the percent ionization of a 0.250 M $HC_2H_3O_2$ solution.

Let us now compare the results of Examples 15.6 and 15.9:

$[HNO_2]$	$[H_3O^+]$	Percent Dissociation
0.200	0.0096	4.8%
2.500	0.0340	1.4%

Summarizing [H₃O⁺] and Percent Ionization:

▶ The equilibrium H_3O^+ concentration of a weak acid increases with increasing initial concentration of the acid.

▶ The percent ionization of a weak acid decreases with increasing concentration of the acid.

In other words, as the concentration of a weak acid solution increases, the concentration of the hydronium ion also increases, but the increase is not linear. The H_3O^+ concentration increases more slowly than the concentration of the acid because as the acid concentration increases, a smaller fraction of weak acid molecules ionize.

 Conceptual Connection 15.5 Percent Ionization

Which weak acid solution has the greatest percent ionization? Which solution has the lowest (most acidic) pH?

(a) 0.100 M $HC_2H_3O_2$ **(b)** 0.500 M $HC_2H_3O_2$ **(c)** 0.0100 M $HC_2H_3O_2$

15.7 Base Solutions

Strong Bases

By analogy with the definition of a strong acid, a **strong base** is one that completely dissociates in solution. NaOH, for example, is a strong base:

$$NaOH(aq) \longrightarrow Na^+(aq) + OH^-(aq)$$

An NaOH solution contains no intact NaOH—it has all dissociated to form $Na^+(aq)$ and $OH^-(aq)$ (**Figure 15.10▶**). In other words, a 1.0 M NaOH solution will have $[OH^-] = 1.0$ M and $[Na^+] = 1.0$ M. Table 15.8 lists the common strong bases.

TABLE 15.8 Strong Bases	
Lithium hydroxide (LiOH)	Strontium hydroxide $[Sr(OH)_2]$
Sodium hydroxide (NaOH)	Calcium hydroxide $[Ca(OH)_2]$
Potassium hydroxide (KOH)	Barium hydroxide $[Ba(OH)_2]$

As you can see, most strong bases are group 1A or group 2A metal hydroxides. The group 1A metal hydroxides are highly soluble in water and can form concentrated base solutions. The group 2A metal hydroxides, however, are only slightly soluble, a useful property for some applications. Notice that the general formula for the group 2A metal hydroxides is $M(OH)_2$. When they dissolve, they produce 2 mol of OH^- per mole of the base. For example, $Sr(OH)_2$ dissociates as follows:

$$Sr(OH)_2(aq) \longrightarrow Sr^{2+}(aq) + 2\,OH^-(aq)$$

Unlike diprotic acids, which ionize in two steps, bases containing two OH^- ions dissociate in one step.

Weak Bases

A **weak base** is analogous to a weak acid. Unlike strong bases that contain OH^- and dissociate in water, the most common weak bases produce OH^- by accepting a proton from water, ionizing water to form OH^- according to the general equation:

$$B(aq) + H_2O(l) \rightleftharpoons BH^+(aq) + OH^-(aq)$$

In this equation, B is a generic symbol for a weak base. Ammonia, for example, ionizes water as follows:

$$NH_3(aq) + H_2O(l) \rightleftharpoons NH_4^+(aq) + OH^-(aq)$$

The double arrow indicates that the ionization is not complete. An NH_3 solution contains mostly NH_3 with some NH_4^+ and OH^- (**Figure 15.11▶**). A 1.0 M NH_3 solution will have

A Strong Base

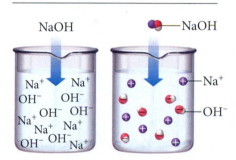

▲ **FIGURE 15.10 Ionization of a Strong Base** When NaOH dissolves in water, it dissociates completely into Na^+ and OH^-. The solution contains virtually no intact NaOH.

A Weak Base

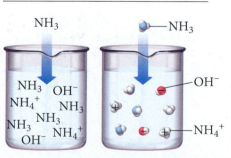

▲ **FIGURE 15.11 Ionization of a Weak Base** When NH_3 dissolves in water, it partially ionizes water to form NH_4^+ and OH^-. Most of the NH_3 molecules in solution remain as NH_3.

$[OH^-] < 1.0$ M. We quantify the extent of ionization of a weak base with the **base ionization constant, K_b.** For the general reaction in which a weak base ionizes water, we define K_b as follows:

$$B(aq) + H_2O(l) \rightleftharpoons BH^+(aq) + OH^-(aq)$$

$$K_b = \frac{[BH^+][OH^-]}{[B]}$$

By analogy with K_a, the smaller the value of K_b, the weaker the base. Table 15.9 lists some common weak bases, their ionization reactions, and values for K_b. The "p" scale can also be applied to K_b, so that $pK_b = -\log K_b$.

TABLE 15.9 Some Common Weak Bases

Weak Base	Ionization Reaction	K_b
Carbonate ion ($CO_3{}^{2-}$)*	$CO_3{}^{2-}(aq) + H_2O(l) \rightleftharpoons HCO_3{}^-(aq) + OH^-(aq)$	1.8×10^{-4}
Methylamine (CH_3NH_2)	$CH_3NH_2(aq) + H_2O(l) \rightleftharpoons CH_3NH_3{}^+(aq) + OH^-(aq)$	4.4×10^{-4}
Ethylamine ($C_2H_5NH_2$)	$C_2H_5NH_2(aq) + H_2O(l) \rightleftharpoons C_2H_5NH_3{}^+(aq) + OH^-(aq)$	5.6×10^{-4}
Ammonia (NH_3)	$NH_3(aq) + H_2O(l) \rightleftharpoons NH_4{}^+(aq) + OH^-(aq)$	1.76×10^{-5}
Pyridine (C_5H_5N)	$C_5H_5N(aq) + H_2O(l) \rightleftharpoons C_5H_5NH^+(aq) + OH^-(aq)$	1.7×10^{-9}
Bicarbonate ion ($HCO_3{}^-$)* (or hydrogen carbonate)	$HCO_3{}^-(aq) + H_2O(l) \rightleftharpoons H_2CO_3(aq) + OH^-(aq)$	1.7×10^{-9}
Aniline ($C_6H_5NH_2$)	$C_6H_5NH_2(aq) + H_2O(l) \rightleftharpoons C_6H_5NH_3{}^+(aq) + OH^-(aq)$	3.9×10^{-10}

*The carbonate and bicarbonate ions must occur with a positively charged ion such as Na^+ that serves to balance the charge but does not have any part in the ionization reaction. For example, it is the bicarbonate ion that makes sodium bicarbonate ($NaHCO_3$) basic. We look more closely at ionic bases in Section 15.8.

All but two of the weak bases listed in Table 15.9 are either ammonia or *amines*, which can be thought of as ammonia with one or more hydrocarbon groups substituted for one or more hydrogen atoms. A common molecular feature of all these bases is a nitrogen atom with a lone pair (**Figure 15.12▼**). This lone pair acts as the proton acceptor that makes the substance a base, as shown in the reactions for ammonia and methylamine:

▶ **FIGURE 15.12 Lone Pairs in Weak Bases**   Many weak bases have a nitrogen atom with a lone pair that acts as the proton acceptor.

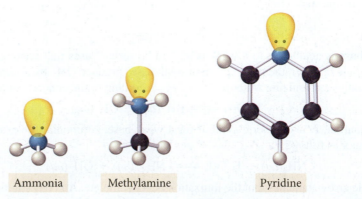

Ammonia   Methylamine   Pyridine

Finding the $[OH^-]$ and pH of Basic Solutions

Finding the $[OH^-]$ and pH of a strong base solution is relatively straightforward, as shown in Example 15.10. Just as we did when calculating the $[H_3O^+]$ in strong acid solutions, we can neglect the contribution of the autoionization of water to the $[OH^-]$ and focus solely on the strong base itself.

EXAMPLE 15.10 Finding the $[OH^-]$ and pH of a Strong Base Solution

What is the OH^- concentration and pH in each solution?

(a) 0.225 M KOH

(b) 0.0015 M $Sr(OH)_2$

SOLUTION

(a) Since KOH is a strong base, it completely dissociates into K^+ and OH^- in solution. The concentration of OH^- will therefore be the same as the given concentration of KOH.

$$KOH(aq) \longrightarrow K^+(aq) + OH^-(aq)$$
$$[OH^-] = 0.225 \text{ M}$$

Use this concentration and K_w to find $[H_3O^+]$.

$$[H_3O^+][OH^-] = K_w = 1.00 \times 10^{-14}$$
$$[H_3O^+](0.225) = 1.00 \times 10^{-14}$$
$$[H_3O^+] = 4.44 \times 10^{-14} \text{ M}$$

Then substitute $[H_3O^+]$ into the pH expression to find the pH.

$$pH = -\log[H_3O^+]$$
$$= -\log(4.44 \times 10^{-14})$$
$$= 13.353$$

(b) Since $Sr(OH)_2$ is a strong base, it completely dissociates into 1 mol of Sr^{2+} and 2 mol of OH^- in solution. The concentration of OH^- will therefore be twice the given concentration of $Sr(OH)_2$.

$$Sr^{2+}(aq) \longrightarrow Sr^{2+}(aq) + 2\,OH^-(aq)$$
$$[OH^-] = 2(0.0015) \text{ M}$$
$$= 0.0030 \text{ M}$$

Use this concentration and K_w to find $[H_3O^+]$.

$$[H_3O^+][OH^-] = K_w = 1.0 \times 10^{-14}$$
$$[H_3O^+](0.0030) = 1.0 \times 10^{-14}$$
$$[H_3O^+] = 3.3 \times 10^{-12} \text{ M}$$

Substitute $[H_3O^+]$ into the pH expression to find the pH.

$$pH = -\log[H_3O^+]$$
$$= -\log(3.3 \times 10^{-12})$$
$$= 11.48$$

FOR PRACTICE 15.10

Find the $[OH^-]$ and pH of a 0.010 M $Ba(OH)_2$ solution.

Finding the $[OH^-]$ and pH of a *weak base* solution is analogous to finding the $[H_3O^+]$ and pH of a weak acid. We can neglect the contribution of the autoionization of water to the $[OH^-]$ and focus solely on the weak base itself. We find the contribution of the weak base by preparing an ICE table showing the relevant concentrations of all species and then use the base ionization constant expression to find the $[OH^-]$. Example 15.11 illustrates how to find the $[OH^-]$ and pH of a weak base solution.

EXAMPLE 15.11 Finding the $[OH^-]$ and pH of a Weak Base Solution

Find the $[OH^-]$ and pH of a 0.100 M NH_3 solution.

SOLUTION

1. Write the balanced equation for the ionization of water by the base and use it as a guide to prepare an ICE table showing the given concentration of the weak base as its initial concentration. Leave room in the table for the changes in concentrations and for the equilibrium concentrations. (Note that we list the OH^- concentration as approximately zero. Although a little OH^- is present from the autoionization of water, this amount is negligibly small compared to the amount of OH^- formed by the base.)

$$NH_3(aq) + H_2O(l) \rightleftharpoons NH_4^+(aq) + OH^-(aq)$$

	$[NH_3]$	$[NH_4^+]$	$[OH^-]$
Initial	0.100	0.00	≈0.00
Change			
Equil			

2. Represent the change in the concentration of OH^- with the variable x. Define the changes in the concentrations of the other reactants and products in terms of x.

$$NH_3(aq) + H_2O(l) \rightleftharpoons NH_4^+(aq) + OH^-(aq)$$

	$[NH_3]$	$[NH_4^+]$	$[OH^-]$
Initial	0.100	0.00	≈0.00
Change	$-x$	$+x$	$+x$
Equil			

3. Sum each column to determine the equilibrium concentrations in terms of the initial concentrations and the variable x.

$$NH_3(aq) + H_2O(l) \rightleftharpoons NH_4^+(aq) + OH^-(aq)$$

	$[NH_3]$	$[NH_4^+]$	$[OH^-]$
Initial	0.100	0.00	≈0.00
Change	$-x$	$+x$	$+x$
Equil	$0.100 - x$	x	x

4. Substitute the expressions for the equilibrium concentrations (from step 3) into the expression for the base ionization constant.

In many cases, you can make the approximation that x *is small* (as discussed in Chapter 14).

Substitute the value of the base ionization constant (from Table 15.9) into the K_b expression and solve for x.

$$K_b = \frac{[NH_4^+][OH^-]}{[NH_3]}$$

$$= \frac{x^2}{0.100 - \cancel{x}} \quad (x \text{ is small})$$

$$1.76 \times 10^{-5} = \frac{x^2}{0.100}$$

$$\sqrt{1.76 \times 10^{-5}} = \sqrt{\frac{x^2}{0.100}}$$

$$x = \sqrt{(0.100)(1.76 \times 10^{-5})}$$

$$= 1.33 \times 10^{-3}$$

Confirm that the *x is small* approximation is valid by calculating the ratio of x to the number it was subtracted from in the approximation. The ratio should be less than 0.05 (or 5%).

$$\frac{1.33 \times 10^{-3}}{0.100} \times 100\% = 1.33\%$$

Therefore the approximation is valid.

5. Determine the OH^- concentration from the calculated value of x.	$[OH^-] = 1.33 \times 10^{-3} \text{ M}$
Use the expression for K_w to find $[H_3O^+]$.	$[H_3O^+][OH^-] = K_w = 1.00 \times 10^{-14}$ $[H_3O^+](1.33 \times 10^{-3}) = 1.00 \times 10^{-14}$ $[H_3O^+] = 7.52 \times 10^{-12} \text{ M}$
Substitute $[H_3O^+]$ into the pH equation to find pH.	$\begin{aligned} pH &= -\log[H_3O^+] \\ &= -\log(7.52 \times 10^{-12}) \\ &= 11.124 \end{aligned}$

FOR PRACTICE 15.11

Find the $[OH^-]$ and pH of a 0.33 M methylamine (CH_3NH_2) solution.

15.8 The Acid–Base Properties of Ions and Salts

We have already seen that some ions act as bases. For example, the bicarbonate ion acts as a base according to the equation:

$$HCO_3^-(aq) + H_2O(l) \rightleftharpoons H_2CO_3(aq) + OH^-(aq)$$

As we know, the bicarbonate ion does not form a stable compound by itself—charge neutrality requires that it pair with a counterion (in this case a cation) to form an ionic compound, also called a *salt*. For example, the sodium salt of bicarbonate is sodium bicarbonate. Like all soluble salts, sodium bicarbonate dissociates in solution to form a sodium cation and bicarbonate anion:

$$NaHCO_3(s) \rightleftharpoons Na^+(aq) + HCO_3^-(aq)$$

The bicarbonate ion then acts as a weak base, ionizing water as just shown to form a basic solution. The sodium ion, on the other hand, has neither acidic nor basic properties (it does not ionize water), as we will see shortly. Consequently, the pH of a sodium bicarbonate solution is above 7. In this section, we look in general at the acid–base properties of salts and the ions they contain. Some salts are pH-neutral when put into water, others are acidic, and still others are basic, depending on their constituent anions and cations. In general, anions tend to form either *basic* or neutral solutions, while cations tend to form either *acidic* or neutral solutions.

Anions as Weak Bases

We can think of any anion as the conjugate base of an acid. Consider the following anions and their corresponding acids:

This anion	is the conjugate base of	this acid
Cl^-		HCl
F^-		HF
NO_3^-		HNO_3
$C_2H_3O_2^-$		$HC_2H_3O_2$

In general, the anion A^- is the conjugate base of the acid HA. Since virtually every anion can be envisioned as the conjugate base of an acid, the ion may itself act as a base.

However, *not every anion acts as a base*—it depends on the strength of the corresponding acid. In general:

- An anion that is the conjugate base of a *weak acid* is itself a *weak base*.
- An anion that is the conjugate base of a *strong acid* is pH-*neutral* (forms solutions that are neither acidic nor basic).

For example, the Cl^- anion is the conjugate base of HCl, a strong acid. Therefore the Cl^- anion is pH-neutral (neither acidic nor basic). The F^- anion, however, is the conjugate base of HF, a weak acid. Therefore the F^- ion is itself a weak base and ionizes water according to the reaction:

$$F^-(aq) + H_2O(l) \rightleftharpoons OH^-(aq) + HF(aq)$$

We can understand why the conjugate base of a weak acid is basic by asking ourselves why an acid is weak to begin with. Hydrofluoric acid is a weak acid because this reaction lies to the left:

$$HF(aq) + H_2O(l) \rightleftharpoons H_3O^+(aq) + F^-(aq)$$

The equilibrium lies to the left because the F^- ion has a significant affinity for H^+ ions. Consequently, when F^- is put into water, its affinity for H^+ ions allows it to remove H^+ ions from water molecules, thus acting as a weak base. In general, as shown in **Figure 15.13▼**, the weaker the acid, the stronger the conjugate base (as we saw in Section 15.4). In contrast, the conjugate base of a strong acid, such as Cl^-, does not act as a base because this reaction lies far to the right:

$$HCl(aq) + H_2O(l) \longrightarrow H_3O^+(aq) + Cl^-(aq)$$

The reaction lies far to the right because the Cl^- ion has a very low affinity for H^+ ions. Consequently, when Cl^- is put into water, it does not take H^+ ions from water molecules.

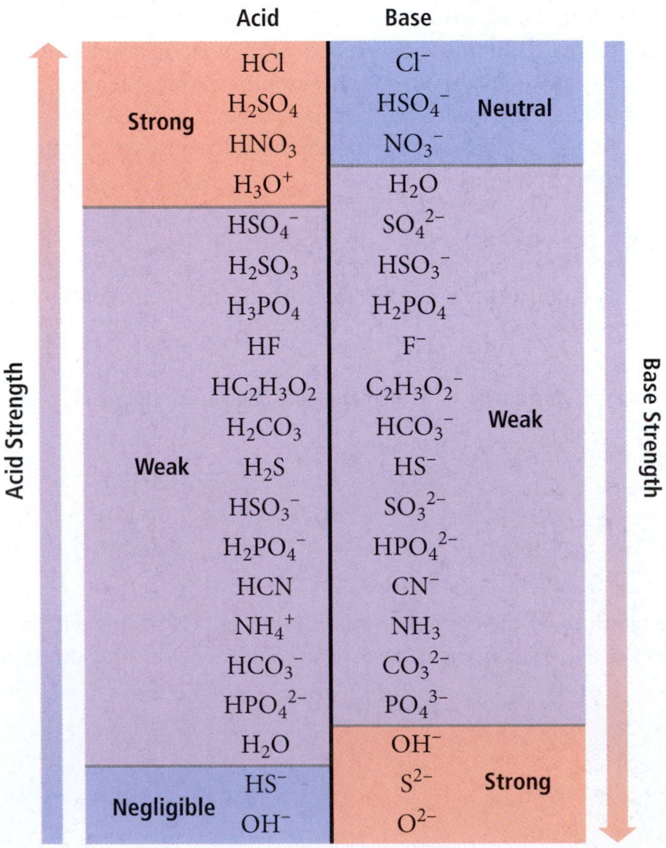

▶ **FIGURE 15.13 Strength of Conjugate Acid–Base Pairs** The stronger an acid, the weaker is its conjugate base.

EXAMPLE 15.12 Determining Whether an Anion Is Basic or pH-Neutral

Classify each anion as a weak base or pH-neutral (neither acidic nor basic):

(a) NO_3^- **(b)** NO_2^- **(c)** $C_2H_3O_2^-$

SOLUTION

(a) From Table 15.3, we can see that NO_3^- is the conjugate base of a *strong* acid (HNO_3). NO_3^- is therefore pH-neutral.

(b) From Table 15.5 (or from its absence in Table 15.3), we know that NO_2^- is the conjugate base of a weak acid (HNO_2). NO_2^- is therefore a weak base.

(c) From Table 15.5 (or from its absence in Table 15.3), we know that $C_2H_3O_2^-$ is the conjugate base of a weak acid ($HC_2H_3O_2$). $C_2H_3O_2^-$ is therefore a weak base.

FOR PRACTICE 15.12

Classify each anion as a weak base or pH-neutral:

(a) CHO_2^- **(b)** ClO_4^-

The pH of a solution containing an anion that acts as a weak base can be determined in a manner similar to that for any weak base solution. However, we need to know K_b for the anion acting as a base. We can determine the value of K_b from K_a of the corresponding acid. Recall from Section 15.4 the expression for K_a for a generic acid HA:

$$HA(aq) + H_2O(l) \rightleftharpoons H_3O^+(aq) + A^-(aq)$$

$$K_a = \frac{[H_3O^+][A^-]}{[HA]}$$

Similarly, the expression for K_b for the conjugate base (A^-) is:

$$A^-(aq) + H_2O(l) \rightleftharpoons OH^-(aq) + HA(aq)$$

$$K_b = \frac{[OH^-][HA]}{[A^-]}$$

If we multiply the expressions for K_a and K_b we get:

$$K_a \times K_b = \frac{[H_3O^+]\,\cancel{[A^-]}}{\cancel{[HA]}} \frac{[OH^-]\,\cancel{[HA]}}{\cancel{[A^-]}} = [H_3O^+][OH^-] = K_w$$

Or simply,

$$K_a \times K_b = K_w$$

The product of K_a for an acid and K_b for its conjugate base is K_w (1.0×10^{-14} at 25 °C). Consequently, we can find K_b for an anion acting as a base from the value of K_a for the conjugate acid. For example, for acetic acid ($HC_2H_3O_2$), $K_a = 1.8 \times 10^{-5}$. We can calculate the value of K_b for the conjugate base ($C_2H_3O_2^-$) as follows:

$$K_a \times K_b = K_w$$

$$K_b = \frac{K_w}{K_a} = \frac{1.0 \times 10^{-14}}{1.8 \times 10^{-5}} = 5.6 \times 10^{-10}$$

Knowing K_b, we can find the pH of a solution containing an anion acting as a base, as shown in Example 15.13.

EXAMPLE 15.13 Determining the pH of a Solution Containing an Anion Acting as a Base

Find the pH of a 0.100 M $NaCHO_2$ solution. The salt completely dissociates into $Na^+(aq)$ and $CHO_2^-(aq)$ and the Na^+ ion has no acid or base properties.

SOLUTION

<table>
<tr>
<td>

1. Since the Na^+ ion does not have any acid or base properties, you can ignore it. Write the balanced equation for the ionization of water by the basic anion and use it as a guide to prepare an ICE table showing the given concentration of the weak base as its initial concentration.

</td>
<td>

$$CHO_2^-(aq) + H_2O(l) \rightleftharpoons HCHO_2(aq) + OH^-(aq)$$

	$[CHO_2^-]$	$[HCHO_2]$	$[OH^-]$
Initial	0.100	0.00	≈ 0.00
Change			
Equil			

</td>
</tr>
<tr>
<td>

2. Represent the change in the concentration of OH^- with the variable x. Define the changes in the concentrations of the other reactants and products in terms of x.

</td>
<td>

$$CHO_2^-(aq) + H_2O(l) \rightleftharpoons HCHO_2(aq) + OH^-(aq)$$

	$[CHO_2^-]$	$[HCHO_2]$	$[OH^-]$
Initial	0.100	0.00	≈ 0.00
Change	$-x$	$+x$	$+x$
Equil			

</td>
</tr>
<tr>
<td>

3. Sum each column to determine the equilibrium concentrations in terms of the initial concentrations and the variable x.

</td>
<td>

$$CHO_2^-(aq) + H_2O(l) \rightleftharpoons HCHO_2(aq) + OH^-(aq)$$

	$[CHO_2^-]$	$[HCHO_2]$	$[OH^-]$
Initial	0.100	0.00	≈ 0.00
Change	$-x$	$+x$	$+x$
Equil	$0.100 - x$	x	x

</td>
</tr>
<tr>
<td>

4. Find K_b from K_a (for the conjugate acid which you can find in Table 15.5).

Substitute the expressions for the equilibrium concentrations (from step 3) into the expression for K_b. In many cases, you can make the approximation that x is small (as discussed in Chapter 14).

Substitute the value of K_b into the K_b expression and solve for x.

Confirm that the x is small approximation is valid by calculating the ratio of x to the number it was subtracted from in the approximation. The ratio should be less than 0.05 (or 5%).

</td>
<td>

$$K_a \times K_b = K_w$$

$$K_b = \frac{K_w}{K_a} = \frac{1.0 \times 10^{-14}}{1.8 \times 10^{-4}} = 5.6 \times 10^{-11}$$

$$K_b = \frac{[HCHO_2][OH^-]}{[CHO_2^-]}$$

$$= \frac{x^2}{0.100 - \cancel{x}}$$

$$5.6 \times 10^{-11} = \frac{x^2}{0.100}$$

$$x = 2.4 \times 10^{-6}$$

$$\frac{2.4 \times 10^{-6}}{0.100} \times 100\% = 0.0024\%$$

Therefore the approximation is valid.

</td>
</tr>
<tr>
<td>

5. Determine the OH^- concentration from the calculated value of x.
Use the expression for K_w to determine $[H_3O^+]$.

Substitute $[H_3O^+]$ into the pH equation to find pH.

</td>
<td>

$$[OH^-] = 2.4 \times 10^{-6} \text{ M}$$

$$[H_3O^+][OH^-] = K_w = 1.0 \times 10^{-14}$$

$$[H_3O^+](2.4 \times 10^{-6}) = 1.0 \times 10^{-14}$$

$$[H_3O^+] = 4.2 \times 10^{-9} \text{ M}$$

$$pH = -\log[H_3O^+]$$

$$= -\log(4.2 \times 10^{-9})$$

$$= 8.38$$

</td>
</tr>
</table>

FOR PRACTICE 15.13

Find the pH of a 0.250 M $NaC_2H_3O_2$ solution.

We can also express the relationship between K_a and K_b in terms of pK_a and pK_b. By taking the log of both sides of $K_a \times K_b = K_w$, we get:

$$\log(K_a \times K_b) = \log K_w$$

$$\log K_a + \log K_b = \log K_w$$

Since $K_w = 10^{-14}$, we can rearrange the equation to get:

$$\log K_a + \log K_b = \log 10^{-14} = -14$$

Rearranging further,

$$-\log K_a - \log K_b = 14$$

Since $-\log K = pK$, we get:

$$pK_a + pK_b = 14$$

Cations as Weak Acids

In contrast to anions, which in some cases act as weak bases, cations can in some cases act as *weak acids*. We can generally divide cations into three categories: cations that are the counterions of strong bases; cations that are the conjugate acids of *weak bases*; and cations that are small, highly charged metals. Here we examine each category individually.

Cations That Are the Counterions of Strong Bases   Strong bases such as NaOH or $Ca(OH)_2$ generally contain hydroxide ions and a counterion. In solution, a strong base completely dissociates to form $OH^-(aq)$ and the solvated counterion. Although these counterions interact with water molecules via ion–dipole forces, they do not ionize water and they do not contribute to the acidity or basicity of the solution. In general, therefore, *cations that are the counterions of strong bases are themselves pH-neutral* (they form solutions that are neither acidic nor basic). For example, Na^+, K^+, and Ca^{2+} are the counterions of the strong bases NaOH, KOH, and $Ca(OH)_2$ and are, therefore, themselves pH-neutral.

Cations That Are the Conjugate Acids of Weak Bases   A cation can be formed from any nonionic weak base by adding a proton (H^+) to its formula. The cation will be the conjugate acid of the base. For example, consider the following cations and their corresponding weak bases.

This cation	is the conjugate acid of	this weak base
NH_4^+		NH_3
$C_2H_5NH_3^+$		$C_2H_5NH_2$
$CH_3NH_3^+$		CH_3NH_2

Any of these cations, with the general formula BH^+, will act as a weak acid according to the general equation:

$$BH^+(aq) + H_2O(aq) \rightleftharpoons H_3O^+(aq) + B(aq)$$

In general, *a cation that is the conjugate acid of a weak base is a weak acid.*

We can calculate the pH of a solution containing the conjugate acid of a weak base just like that of any other weakly acidic solution. However, the value of K_a for the acid must be derived from K_b using the previously derived relationship $K_a \times K_b = K_w$.

Cations That Are Small, Highly Charged Metals *Small, highly charged metal cations such as* Al^{3+} *and* Fe^{3+} *form weakly acidic solutions.* For example, when Al^{3+} is dissolved in water, it becomes hydrated according to the equation:

$$Al^{3+}(aq) + 6\,H_2O(l) \longrightarrow Al(H_2O)_6{}^{3+}(aq)$$

The hydrated form of the ion then acts as a Brønsted–Lowry acid:

$$Al(H_2O)_6{}^{3+}(aq) \quad + \quad H_2O(aq) \quad \rightleftharpoons \quad Al(H_2O)_5(OH)^{2+}(aq) + H_3O^+(aq)$$

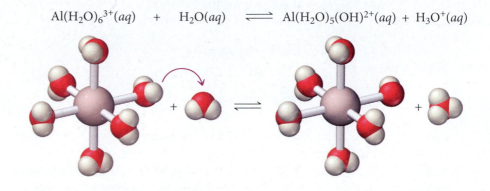

This type of behavior is not observed for alkali metals or alkaline earth metals, but is observed for the cations of most other metals.

EXAMPLE 15.14 Determining Whether a Cation Is Acidic or pH-Neutral

Classify each cation as a weak acid or pH-neutral (neither acidic nor basic):

(a) $C_5H_5NH^+$ (b) Ca^{2+} (c) Cr^{3+}

SOLUTION

(a) The $C_5H_5NH^+$ cation is the conjugate acid of a weak base. This ion is therefore a weak acid.

(b) The Ca^{2+} cation is the counterion of a strong base. This ion is therefore pH-neutral (neither acidic nor basic).

(c) The Cr^{3+} cation is a small, highly charged metal cation. It is therefore a weak acid.

FOR PRACTICE 15.14

Classify each cation as a weak acid or pH-neutral (neither acidic nor basic):

(a) Li^+ (b) $CH_3NH_3{}^+$ (c) Fe^{3+}

Classifying Salt Solutions as Acidic, Basic, or Neutral

Since salts contain both a cation and an anion, they can form acidic, basic, or neutral solutions when dissolved in water. The pH of the solution depends on the specific cation and anion involved. There are four different possibilities, and we examine examples of each individually.

1. **Salts in which neither the cation nor the anion acts as an acid or a base form pH-neutral solutions.** A salt in which the cation is the counterion of a strong base

and in which the anion is the conjugate base of strong acid will form *neutral* solutions. Salts in this category include

NaCl	Ca(NO₃)₂	KBr
sodium chloride	calcium nitrate	potassium bromide

$$NaCl \qquad Ca(NO_3)_2 \qquad KBr$$

sodium chloride   calcium nitrate   potassium bromide

Cations are pH-neutral.   Anions are conjugate bases of *strong* acids.

2. **Salts in which the cation does not act as an acid and the anion acts as a base form basic solutions.** A salt in which the cation is the counterion of a strong base and in which the anion is the conjugate base of *weak* acid will form *basic* solutions. Salts in this category include

$$NaF \qquad Ca(C_2H_3O_2)_2 \qquad KNO_2$$

sodium fluoride   calcium acetate   potassium nitrite

Cations are pH-neutral.   Anions are conjugate bases of *weak* acids.

3. **Salts in which the cation acts as an acid and the anion does not act as a base form acidic solutions.** A salt in which the cation is either the conjugate acid of a weak base or a small, highly charged metal ion and in which the anion is the conjugate base of *strong* acid will form *acidic* solutions. Salts in this category include

$$FeCl_3 \qquad Al(NO_3)_3 \qquad NH_4Br$$

iron(III) chloride   aluminum nitrate   ammonium bromide

Cations are conjugate acids
   of *weak* bases or small,   Anions are conjugate bases of *strong* acids.
highly charged metal ions.

4. **Salts in which the cation acts as an acid and the anion acts as a base form solutions in which the pH depends on the relative strengths of the acid and the base.** A salt in which the cation is either the conjugate acid of a weak base or a small, highly charged metal ion and in which the anion is the conjugate base of a *weak* acid will form a solution in which the pH depends on the relative strengths of the acid and base. Salts in this category include

$$FeF_3 \qquad Al(C_2H_3O_2)_3 \qquad NH_4NO_2$$

iron(III) fluoride   aluminum acetate   ammonium nitrite

Cations are conjugate acids
   of *weak* bases or small,   Anions are conjugate bases of *weak* acids.
highly charged metal ions.

We can determine the overall pH of a solution containing one of these salts by comparing the K_a of the acid to the K_b of the base—the ion with the higher value of K dominates and determines whether the solution will be acidic or basic, as shown in part (e) of Example 15.15. All of these possibilities are summarized in Table 15.10.

TABLE 15.10 pH of Salt Solutions

		Anion	
		Conjugate base of strong acid	**Conjugate base of weak acid**
CATION	**Conjugate acid of weak base**	*Acidic*	*Depends on relative strengths*
	Small, highly charged metal ion	*Acidic*	*Depends on relative strengths*
	Counterion of strong base	*Neutral*	*Basic*

EXAMPLE 15.15 Determining the Overall Acidity or Basicity of Salt Solutions

Determine whether the solution formed by each salt is acidic, basic, or neutral:

(a) $SrCl_2$ (b) $AlBr_3$ (c) $CH_3NH_3NO_3$ (d) $NaCHO_2$ (e) NH_4F

SOLUTION

(a) The Sr^{2+} cation is the counterion of a strong base $\left[Sr(OH)_2\right]$ and is therefore pH-neutral. The Cl^- anion is the conjugate base of a strong acid (HCl) and is therefore pH-neutral as well. The $SrCl_2$ solution will be pH-neutral (neither acidic nor basic).	$SrCl_2$ pH-neutral cation pH-neutral anion Neutral solution
(b) The Al^{3+} cation is a small, highly charged metal ion (that is not an alkali metal or an alkaline earth metal) and is therefore a weak acid. The Br^- anion is the conjugate base of a strong acid (HBr) and is therefore pH-neutral. The $AlBr_3$ solution will be acidic.	$AlBr_3$ acidic cation pH-neutral anion Acidic solution
(c) The $CH_3NH_3^+$ ion is the conjugate acid of a weak base (CH_3NH_2) and is therefore acidic. The NO_3^- anion is the conjugate base of a strong acid (HNO_3) and is therefore pH-neutral. The $CH_3NH_3NO_3$ solution will be acidic.	$CH_3NH_3NO_3$ acidic cation pH-neutral anion Acidic solution
(d) The Na^+ cation is the counterion of a strong base and is therefore pH-neutral. The CHO_2^- anion is the conjugate base of a weak acid and is therefore basic. The $NaCHO_2$ solution will be basic.	$NaCHO_2$ pH-neutral cation basic anion Basic solution
(e) The NH_4^+ ion is the conjugate acid of a weak base (NH_3) and is therefore acidic. The F^- ion is the conjugate base of a weak acid and is therefore basic. To determine the overall acidity or basicity of the solution, compare the values of K_a for the acidic cation and K_b for the basic anion. Obtain each value of K from the conjugate by using $K_a \times K_b = K_w$.	NH_4F acidic cation basic anion $K_a(NH_4^+) = \dfrac{K_w}{K_b(NH_3)} = \dfrac{1.0 \times 10^{-14}}{1.76 \times 10^{-5}}$ $= 5.68 \times 10^{-10}$ $K_b(F^-) = \dfrac{K_w}{K_a(HF)} = \dfrac{1.0 \times 10^{-14}}{3.5 \times 10^{-4}}$ $= 2.9 \times 10^{-11}$
Since K_a is greater than K_b, the solution is acidic.	$K_a > K_b$

FOR PRACTICE 15.15

Determine whether the solutions formed by each salt will be acidic, basic, or neutral:

(a) $NaHCO_3$ (b) $CH_3CH_2NH_3Cl$ (c) KNO_3 (d) $Fe(NO_3)_3$

15.9   Acid Strength and Molecular Structure

We have discussed that a Brønsted–Lowry acid is a proton (H^+) donor. However, we have not explored why some hydrogen-containing molecules act as acids while others do not, or why some acids are strong and others weak. For example, why is H_2S acidic while CH_4 is not? Or why is HF a weak acid while HCl is a strong acid? We divide our discussion about these issues into two categories: binary acids (those containing hydrogen and only one other element) and oxyacids (those containing hydrogen bonded to an oxygen atom that is bonded to another element).

Binary Acids

Consider the bond between a hydrogen atom and some other generic element (which we will call Y):

$$H—Y$$

The factors affecting the ease with which this hydrogen is donated (and therefore is acidic) are the *polarity* of the bond and the *strength* of the bond.

Bond Polarity Using the notation introduced in Chapter 9, the H—Y bond must be polarized as follows in order for the hydrogen atom to be acidic:

$$\overset{+\longrightarrow}{\delta^+ H — Y^{\delta-}}$$

This requirement makes physical sense because the hydrogen atom must be lost as a positively charged ion (H^+). Having a partial positive charge on the hydrogen atom facilitates its loss. For example, consider the following three bonds and their corresponding dipole moments:

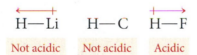

H—Li	H—C	H—F
Not acidic	Not acidic	Acidic

LiH is ionic with *the hydrogen having the negative charge*; therefore LiH is not acidic. The C—H bond is virtually nonpolar because the electronegativities of carbon and hydrogen are similar; therefore any C—H bonds in a compound are not acidic. In contrast, the H—F bond is polar with the positive charge on the hydrogen atom. As we know from this chapter, HF is a weak acid. The partial positive charge on the hydrogen atom makes it easier for the hydrogen to be lost as an H^+ ion.

Bond Strength The strength of the H—Y bond also affects the strength of the corresponding acid. As you might expect, the stronger the bond, the weaker the acid, because the more tightly the hydrogen atom is held, the less likely it is to come off. We can see the effect of bond strength by comparing the bond strengths and acidities of the following hydrogen halides.

Acid	Bond Energy (kJ/mol)	Type of Acid
H—F	565	Weak
H—Cl	431	Strong
H—Br	364	Strong

HCl and HBr, with their weaker bonds, are both strong acids while H—F, with its stronger bond, is a weak acid. This is the case despite HF's greater bond polarity.

Oxyacids

Oxyacids contain a hydrogen atom bonded to an oxygen atom which is in turn bonded to some other atom (which we will call Y):

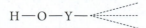

Oxyacids are sometimes called oxoacids.

Y may or may not be bonded to other atoms. The factors affecting the ease with which this hydrogen is donated (and therefore acidic) are the *electronegativity of the element Y* and the *number of oxygen atoms attached to the element Y*.

The Electronegativity of Y The more electronegative the element Y, the more it weakens and polarizes the O—H bond, resulting in greater acidity. We can see this effect by comparing the electronegativity of Y and the acid ionization constants of the following oxyacids:

Acid	Electronegativity of Y	$[K_a]$
H—O—I	2.5	2.3×10^{-11}
H—O—Br	2.8	2.0×10^{-9}
H—O—Cl	3.0	2.9×10^{-8}

Chlorine is the most electronegative of the three elements and the corresponding acid has the greatest K_a.

The Number of Oxygen Atoms Bonded to Y Oxyacids may contain additional oxygen atoms bonded to the element Y. Since these additional oxygen atoms are electronegative, they draw electron density away from the element Y, which in turn draws electron density away from the O—H bond, further weakening and polarizing it, and leading to increasing acidity. We can see this effect by comparing the following series of acid ionization constants:

Acid	Structure	K_a
HClO$_4$	$$\begin{array}{c} O \\ \parallel \\ H-O-Cl=O \\ \parallel \\ O \end{array}$$	Strong
HClO$_3$	$$\begin{array}{c} O \\ \parallel \\ H-O-Cl=O \end{array}$$	10
HClO$_2$	H—O—Cl=O	1.1×10^{-2}
HClO	H—O—Cl	2.9×10^{-8}

The greater the number of oxygen atoms bonded to Y, the stronger the acid. On this basis we would predict that H_2SO_4 is a stronger acid than H_2SO_3 and that HNO_3 is stronger than HNO_2. As we have seen in this chapter, both H_2SO_4 and HNO_3 are strong acids, while H_2SO_3 and HNO_2 are weak acids, as predicted.

Conceptual Connection 15.6 Acid Strength and Molecular Structure

Which of the protons shown in red is more acidic?

(a)
$$\begin{array}{c} O \\ \parallel \\ H-C-O-H \end{array}$$

(b)
$$\begin{array}{c} H \\ \mid \\ H-C-O-H \\ \mid \\ H \end{array}$$

15.10   Lewis Acids and Bases

We began our definitions of acids and bases with the Arrhenius model. We then saw how the Brønsted–Lowry model expanded the range of substances that could be considered acids and bases by introducing the concept of a proton donor and proton acceptor. We now introduce a third model which further broadens the range of substances that can be considered acids. This third model is the *Lewis model*, named after G. N. Lewis, the American chemist who devised the electron-dot representation of chemical bonding (Section 9.1). While the Brønsted–Lowry model focuses on the transfer of a proton, the Lewis model focuses on the transfer of an electron pair. For example, consider the simple acid–base reaction between the H^+ ion and NH_3, shown here with Lewis structures:

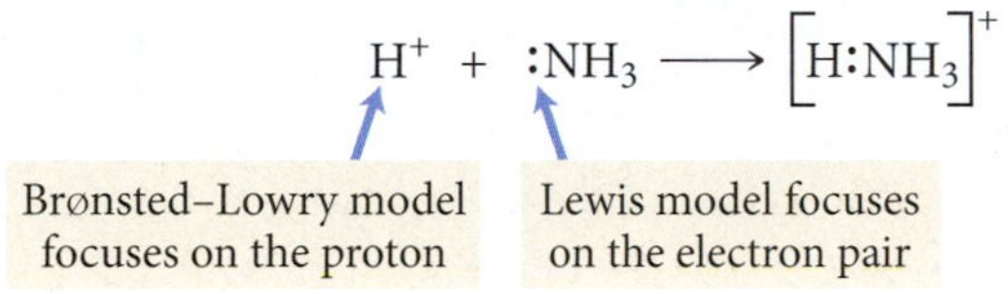

$$H^+ \ + \ :NH_3 \ \longrightarrow \ \left[H:NH_3 \right]^+$$

Brønsted–Lowry model focuses on the proton	Lewis model focuses on the electron pair

Under the Brønsted–Lowry model, the ammonia accepts a proton, thus acting as a base. Under the Lewis model, the ammonia acts as a base by *donating an electron pair*. The general definitions of acids and bases according to the Lewis model are:

Lewis acid: electron pair acceptor

Lewis base: electron pair donor

Under the Lewis definition, H^+ in the above reaction is acting as an acid because it is accepting an electron pair from NH_3. In contrast, NH_3 is acting as a Lewis base because it is donating an electron pair to H^+.

Although the Lewis model does not significantly expand the view of a base—because all proton acceptors must have an electron pair to bind the proton—it significantly broadens the view of an acid. Under the Lewis model, a substance need not even contain hydrogen to be an acid. For example, consider the gas-phase reaction between boron trifluoride and ammonia:

$$\underset{\text{Lewis acid}}{BF_3} \ + \ \underset{\text{Lewis base}}{:NH_3} \ \longrightarrow \ \underset{\text{adduct}}{F_3B:NH_3}$$

Boron trifluoride has an empty orbital that can accept the electron pair from ammonia and form the product (the product of a Lewis acid–base reaction is sometimes called an *adduct*). This reaction demonstrates an important property of Lewis acids:

> *A Lewis acid has an empty orbital (or can rearrange electrons to create an empty orbital) that can accept an electron pair.*

Consequently, the Lewis definition subsumes a whole new class of acids. Let's examine a few examples.

Molecules That Act as Lewis Acids

Since molecules with incomplete octets have empty orbitals, they can serve as Lewis acids. For example, both $AlCl_3$ and BCl_3 have incomplete octets:

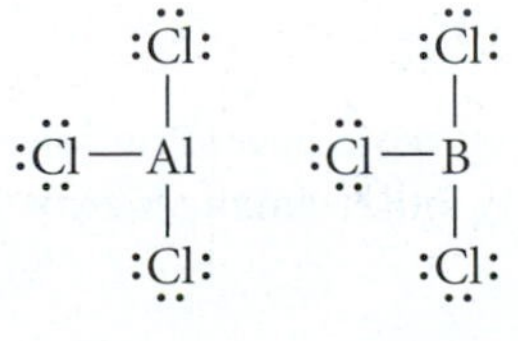

These both act as Lewis acids, as shown in these reactions:

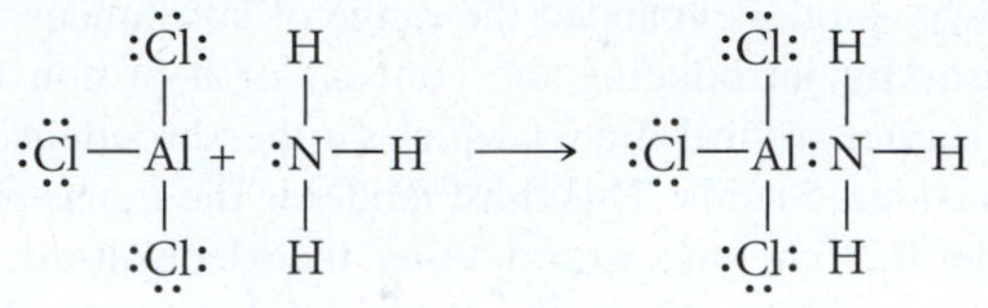

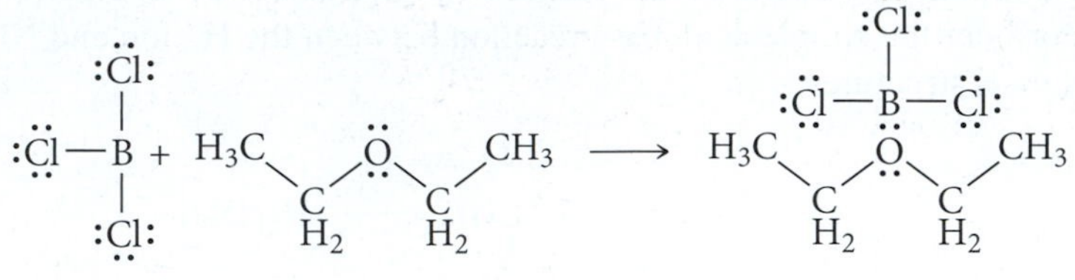

Some molecules that may not initially contain empty orbitals can rearrange their electrons to act as Lewis acids. For example, consider the reaction between carbon dioxide and water:

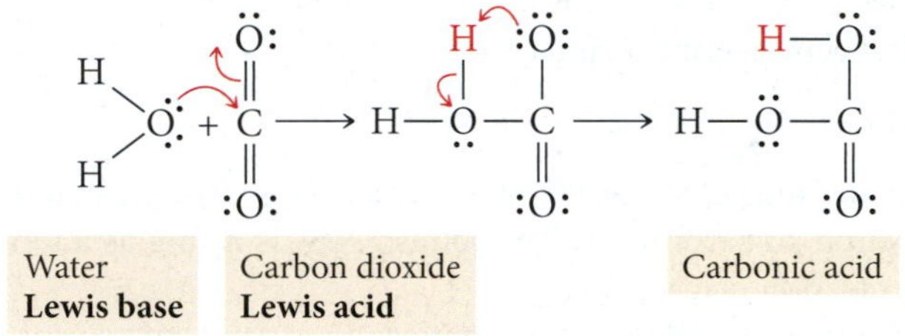

| Water | Carbon dioxide | Carbonic acid |
| **Lewis base** | **Lewis acid** | |

The electrons in the double bond on carbon move to the terminal oxygen atom, allowing carbon dioxide to act as a Lewis acid by accepting an electron pair from water. The molecule then undergoes a rearrangement in which the hydrogen atom shown in red bonds with the terminal oxygen atom instead of the internal one.

Cations That Act as Lewis Acids

Some cations, since they are positively charged and have lost some electrons, have empty orbitals that allow them to act as Lewis acids. For example, consider the hydration process of the Al^{3+} ion discussed in Section 15.8.

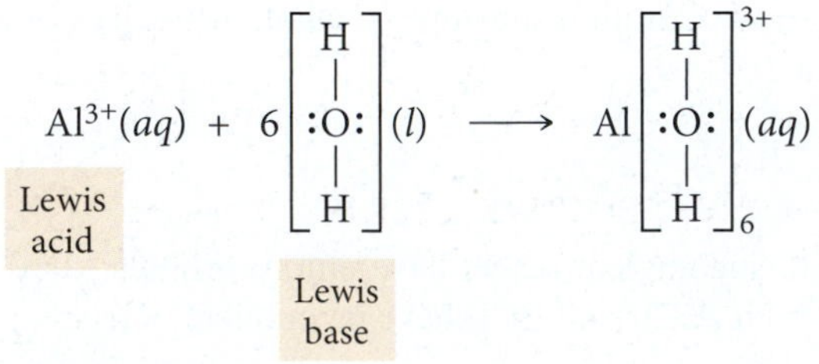

The aluminum ion acts as a Lewis acid, accepting lone pairs from six water molecules to form the hydrated ion. Many other small, highly charged metal ions also act as Lewis acids.

CHAPTER IN REVIEW

Key Terms

Section 15.2
carboxylic acid (578)
alkaloid (578)

Section 15.3
Arrhenius definitions (of acids and bases) (579)
hydronium ion (579)
Brønsted–Lowry definitions (of acids and bases) (580)

amphoteric (580)
conjugate acid–base pair (580)

Section 15.4
strong acid (581)
weak acid (581)
monoprotic acid (582)
diprotic acid (582)
triprotic acid (583)
acid ionization constant (K_a) (583)

Section 15.5
autoionization (585)
ion product constant for water (K_w) (585)
neutral (585)
acidic solution (585)
basic solution (586)
pH (587)

Section 15.6
polyprotic acid (594)
percent ionization (595)

Section 15.7
strong base (597)
weak base (597)
base ionization constant (K_b) (598)

Section 15.10
Lewis acid (611)
Lewis base (611)

Key Concepts

Heartburn (15.1)

▶ Hydrochloric acid from the stomach sometimes contacts the esophageal lining, resulting in irritation and burning, called heartburn. Heartburn is treated with antacids, bases that neutralize stomach acid.

The Nature of Acids and Bases (15.2)

▶ Acids generally taste sour, dissolve metals, turn blue litmus paper red, and neutralize bases. Common acids include hydrochloric, sulfuric, nitric, and carboxylic acids.

▶ Bases generally taste bitter, feel slippery, turn red litmus paper blue, and neutralize acids. Common bases include sodium hydroxide, sodium bicarbonate, and potassium hydroxide.

Definitions of Acids and Bases (15.3)

▶ The Arrhenius definition of acids and bases states that in an aqueous solution, an acid produces hydrogen ions and a base produces hydroxide ions.

▶ The Brønsted–Lowry definition states that an acid is a proton (hydrogen ion) donor and a base is a proton acceptor. In the Brønsted–Lowry definition two substances related by the transfer of a proton are a conjugate acid–base pair.

Acid Strength and the Acid Ionization Constant, K_a (15.4)

▶ In a solution, a strong acid completely ionizes but a weak acid only partially ionizes.

▶ Generally, the stronger the acid, the weaker the conjugate base, and vice versa.

▶ We quantify the extent of dissociation of a weak acid by the acid ionization constant, K_a, which is the equilibrium constant for the ionization of the weak acid.

Autoionization of Water and pH (15.5)

▶ In an acidic solution, the concentration of hydrogen ions is always greater than the concentration of hydroxide ions; however, $[H_3O^+]$ multiplied by $[OH^-]$ is always constant at a constant temperature.

▶ There are two types of logarithmic acid–base scales: pH and pOH. At 25 °C, the sum of the pH and pOH is always 14.

Finding the $[H_3O^+]$ and pH of Strong and Weak Acid Solutions (15.6)

▶ In a strong acid solution, the hydrogen ion concentration equals the initial concentration of the acid.

▶ In a weak acid solution, the hydrogen ion concentration—which can be determined by solving an equilibrium problem—is lower than the initial acid concentration.

▶ Polyprotic acids contain two or more ionizable protons. Generally, polyprotic acids ionize in successive steps, with the values of K_a becoming smaller for each step. In many cases, we can determine the $[H_3O^+]$ of a polyprotic acid solution by considering only the first ionization step.

▶ The percent ionization of weak acids decreases as the acid (and hydrogen ion) concentration increases.

Base Solutions (15.7)

▶ A strong base dissociates completely while a weak base does not.

▶ Most weak bases produce hydroxide ions through the ionization of water. The base ionization constant, K_b, describes the extent of ionization.

Ions as Acids and Bases (15.8)

▶ A cation will be a weak acid if it is the conjugate acid of a weak base; it will be neutral if it is the conjugate acid of a strong base.

▶ An anion is a weak base if it is the conjugate base of a weak acid; it is neutral if it is the conjugate base of a strong acid.

▶ To calculate the pH of an acidic cation or basic anion, find K_a or K_b from the equation $K_a \times K_b = K_w$.

Acid Strength and Molecular Structure (15.9)

▶ For binary acids, acid strength decreases with increasing bond energy, and increases with increasing bond polarity.

▶ For oxyacids, acid strength increases with the electronegativity of the atoms bonded to the oxygen atom and also increases with the number of oxygen atoms in the molecule.

Lewis Acids and Bases (15.10)

▶ The third model of acids and bases, the Lewis model, defines a base as an electron pair donor and an acid as an electron pair acceptor; therefore, an acid does not have to contain hydrogen.

By this definition an acid can be a compound with an empty orbital—or one that will rearrange to make an empty orbital—or a cation.

Key Equations and Relationships

Note: In all of these equations $[H^+]$ is interchangeable with $[H_3O^+]$.

Expression for the Acid Ionization Constant, K_a and Base Ionization Constant, K_b. (15.4)

$$K_a = \frac{[H_3O^+][A^-]}{[HA]} \qquad K_b = \frac{[BH^+][OH^-]}{[B]}$$

The Ion Product Constant for Water, K_w (15.5)

$$K_w = [H_3O^+][OH^-] = 1.0 \times 10^{-14} \text{ (at 25 °C)}$$

Expression for the pH Scale (15.5)

$$pH = -\log[H_3O^+]$$

Expression for the pOH Scale (15.5)

$$pOH = -\log[OH^-]$$

Relationship between pH and pOH (15.5)

$$pH + pOH = 14.00 \text{ (at 25 °C)}$$

Expression for the pK_a Scale (15.5)

$$pK_a = -\log K_a$$

Expression for Percent Ionization (15.6)

$$\text{Percent ionization} = \frac{\text{concentration of ionized acid}}{\text{initial concentration of acid}} \times 100\%$$

$$= \frac{[H_3O^+]_{equil}}{[HA]_{init}} \times 100\%$$

Relationship between K_a, K_b and K_w (15.8)

$$K_a \times K_b = K_w$$

Key Learning Objectives

Chapter Objectives	Assessment
Identifying Brønsted–Lowry Acids and Bases and Their Conjugates (15.3)	Example 15.1 For Practice 15.1 Exercises 3, 4
Using $[K_w]$ in Calculations (15.5)	Example 15.2 For Practice 15.2 Exercises 15, 16
Calculating pH from $[H_3O^+]$ or $[OH^-]$ (15.5)	Examples 15.3, 15.4 For Practice 15.3, 15.4 Exercises 17–20
Finding the pH of a Weak Acid Solution (15.6)	Examples 15.5, 15.6, 15.7 For Practice 15.5, 15.6, 15.7 Exercises 31–36
Finding the Acid Ionization Constant from pH (15.6)	Example 15.8 For Practice 15.8 Exercises 37, 38
Finding the Percent Ionization of a Weak Acid (15.6)	Example 15.9 For Practice 15.9 Exercises 39–42
Finding the $[OH^-]$ and pH of a Strong Base Solution (15.7)	Example 15.10 For Practice 15.10 Exercises 51, 52
Finding the $[OH^-]$ and pH of a Weak Base Solution (15.7)	Example 15.11 For Practice 15.11 Exercises 59, 60
Determining Whether an Anion Is Basic or Neutral (15.8)	Example 15.12 For Practice 15.12 Exercises 65, 66
Determining the pH of a Solution Containing an Anion Acting as a Base (15.8)	Example 15.13 For Practice 15.13 Exercises 67, 68
Determining Whether a Cation Is Acidic or Neutral (15.8)	Example 15.14 For Practice 15.14 Exercises 69, 70
Determining the Overall Acidity or Basicity of Salt Solutions (15.8)	Example 15.15 For Practice 15.15 Exercises 71, 72

EXERCISES

Problems by Topic

The Nature and Definitions of Acids and Bases

1. Identify each substance as an acid or a base and write a chemical equation showing how it is an acid or a base according to the Arrhenius definition.
   a. $HNO_3(aq)$ b. $NH_4^+(aq)$
   c. $KOH(aq)$ d. $HC_2H_3O_2(aq)$

2. Identify each substance as an acid or a base and write a chemical equation showing how it is an acid or a base according to the Arrhenius definition.
   a. $NaOH(aq)$ b. $H_2SO_4(aq)$
   c. $HBr(aq)$ d. $Sr(OH)_2(aq)$

3. For each reaction, identify the Brønsted–Lowry acid, the Brønsted–Lowry base, the conjugate acid, and the conjugate base.
   a. $H_2CO_3(aq) + H_2O(l) \rightleftharpoons H_3O^+(aq) + HCO_3^-(aq)$
   b. $NH_3(aq) + H_2O(l) \rightleftharpoons NH_4^+(aq) + OH^-(aq)$
   c. $HNO_3(aq) + H_2O(l) \longrightarrow H_3O^+(aq) + NO_3^-(aq)$
   d. $C_5H_5N(aq) + H_2O(l) \rightleftharpoons C_5H_5NH^+(aq) + OH^-(aq)$

4. For each reaction, identify the Brønsted–Lowry acid, the Brønsted–Lowry base, the conjugate acid, and the conjugate base.
   a. $HI(aq) + H_2O(l) \longrightarrow H_3O^+(aq) + I^-(aq)$
   b. $CH_3NH_2(aq) + H_2O(l) \rightleftharpoons CH_3NH_3^+(aq) + OH^-(aq)$
   c. $CO_3^{2-}(aq) + H_2O(l) \rightleftharpoons HCO_3^-(aq) + OH^-(aq)$
   d. $HBr(aq) + H_2O(l) \longrightarrow H_3O^+(aq) + Br^-(aq)$

5. Write the formula for the conjugate base of each acid.
   a. HCl b. H_2SO_3 c. $HCHO_2$ d. HF

6. Write the formula for the conjugate acid of each base.
   a. NH_3 b. ClO_4^- c. HSO_4^- d. CO_3^{2-}

7. $H_2PO_4^-$ is amphoteric. Write equations that demonstrate both its acidic nature and its basic nature in aqueous solution.

8. HCO_3^- is amphoteric. Write equations that demonstrate both its acidic nature and its basic nature in aqueous solution.

Acid Strength and K_a

9. Classify each acid as strong or weak. If the acid is weak, write an expression for the acid ionization constant (K_a).
   a. HNO_3 b. HCl c. HBr d. H_2SO_3

10. Classify each acid as strong or weak. If the acid is weak, write an expression for the acid ionization constant (K_a).
    a. HF b. $HCHO_2$ c. H_2SO_4 d. H_2CO_3

11. These three diagrams represent three different solutions of the binary acid HA. Water molecules have been omitted for clarity and hydronium ions (H_3O^+) are represented by hydrogen ions (H^+). Rank the acids in order of decreasing acid strength.

(a) (b) (c)

12. Rank the solutions in order of decreasing $[H_3O^+]$: 0.10 M HCl; 0.10 M HF; 0.10 M HClO; 0.10 M HC_6H_5O.

13. Pick the stronger base from each pair:
    a. F^- or Cl^- b. NO_2^- or NO_3^-
    c. F^- or ClO^-

14. Pick the stronger base from each pair:
    a. ClO_4^- or ClO_2^- b. Cl^- or H_2O
    c. CN^- or ClO^-

Autoionization of Water and pH

15. Calculate $[OH^-]$ in each aqueous solution at 25 °C, and classify each solution as acidic or basic.
    a. $[H_3O^+] = 9.7 \times 10^{-9}$ M b. $[H_3O^+] = 2.2 \times 10^{-6}$ M
    c. $[H_3O^+] = 1.2 \times 10^{-9}$ M

16. Calculate $[H_3O^+]$ in each aqueous solution at 25 °C, and classify each solution as acidic or basic.
    a. $[OH^-] = 5.1 \times 10^{-4}$ M b. $[OH^-] = 1.7 \times 10^{-12}$ M
    c. $[OH^-] = 2.8 \times 10^{-2}$ M

17. Calculate the pH and pOH of each solution.
    a. $[H_3O^+] = 1.7 \times 10^{-8}$ M b. $[H_3O^+] = 1.0 \times 10^{-7}$ M
    c. $[H_3O^+] = 2.2 \times 10^{-6}$ M

18. Calculate $[H_3O^+]$ and $[OH^-]$ for each solution.
    a. pH = 8.55 b. pH = 11.23 c. pH = 2.87

19. Complete the table. (All solutions are at 25 °C.)

$[H_3O^+]$	$[OH^-]$	pH	Acidic or Basic
——	——	3.15	——
3.7×10^{-9}	——	——	——
——	——	11.1	——
——	1.6×10^{-11}	——	——

20. Complete the table. (All solutions are at 25 °C.)

$[H_3O^+]$	$[OH^-]$	pH	Acidic or Basic
3.5×10^{-3}	——	——	——
——	3.8×10^{-7}	——	——
1.8×10^{-9}	——	——	——
——	——	7.15	——

21. Like all equilibrium constants, the value of K_w depends on temperature. At body temperature (37 °C), $K_w = 2.4 \times 10^{-14}$. What is the $[H_3O^+]$ and pH of pure water at body temperature?

22. The value of K_w increases with increasing temperature. Is the autoionization of water endothermic or exothermic?

23. Calculate the pH of each acid solution. Explain how the resulting pH values demonstrate that the pH of an acid solution should carry as many digits to the right of the decimal place as the number of significant figures in the concentration of the solution.
    a. $[H_3O^+] = 0.044$
    b. $[H_3O^+] = 0.045$
    c. $[H_3O^+] = 0.046$

24. Determine the concentration of H_3O^+, to the correct number of significant figures, in a solution with each pH. Describe how these calculations show the relationship between the number of digits to the right of the decimal place in pH and the number of significant figures in concentration.
    a. pH = 2.50   b. pH = 2.51   c. pH = 2.52

Acid Solutions

25. For each strong acid solution, determine $[H_3O^+]$, $[OH^-]$, and pH.
    a. 0.15 M HCl   b. 0.025 M HNO_3
    c. a solution that is 0.072 M in HBr and 0.015 M in HNO_3
    d. a solution that is 0.855% HNO_3 by mass (Assume a density of 1.01 g/mL for the solution.)

26. Determine the pH of each solution.
    a. 0.028 M HI   b. 0.115 M $HClO_4$
    c. a solution that is 0.055 M in $HClO_4$ and 0.028 M in HCl
    d. a solution that is 1.85% HCl by mass (Assume a density of 1.01 g/mL for the solution.)

27. What mass of HI should be present in 0.250 L of solution to obtain a solution with each pH?
    a. pH = 1.25   b. pH = 1.75   c. pH = 2.85

28. What mass of $HClO_4$ should be present in 0.500 L of solution to obtain a solution with each pH?
    a. pH = 2.50   b. pH = 1.50   c. pH = 0.50

29. What is the pH of a solution in which 224 mL of HCl(g), measured at 27.2 °C and 1.02 atm, is dissolved in 1.5 L of aqueous solution?

30. What volume of a concentrated HCl solution, which is 36.0% HCl by mass and has a density of 1.179 g/mL, should you use to make 5.00 L of an HCl solution with a pH of 1.8?

31. Determine the $[H_3O^+]$ and pH of a 0.100 M solution of benzoic acid.

32. Determine the $[H_3O^+]$ and pH of a 0.200 M solution of formic acid.

33. Determine the pH of an HNO_2 solution with each concentration. In which cases is the simplifying assumption that x is small not valid?
    a. 0.500 M   b. 0.100 M   c. 0.0100 M

34. Determine the pH of an HF solution with each concentration. In which cases is the simplifying assumption that x is small not valid?
    a. 0.250 M   b. 0.0500 M   c. 0.0250 M

35. If 15.0 mL of glacial acetic acid (pure $HC_2H_3O_2$) is diluted to 1.50 L with water, what is the pH of the resulting solution? The density of glacial acetic acid is 1.05 g/mL.

36. Calculate the pH of a formic acid solution that contains 1.35% formic acid by mass (Assume a density of 1.01 g/mL for the solution.)

37. A 0.185 M solution of a weak acid (HA) has a pH of 2.95. Calculate the acid ionization constant (K_a) for the acid.

38. A 0.115 M solution of a weak acid (HA) has a pH of 3.29. Calculate the acid ionization constant (K_a) for the acid.

39. Determine the percent ionization of a 0.125 M HCN solution.

40. Determine the percent ionization of a 0.225 M solution of benzoic acid.

41. Calculate the percent ionization of acetic acid solutions having each concentration.
    a. 1.00 M   b. 0.500 M
    c. 0.100 M   d. 0.0500 M

42. Calculate the percent ionization of formic acid solutions having each concentration.
    a. 1.00 M   b. 0.500 M
    c. 0.100 M   d. 0.0500 M

43. A 0.148 M solution of a monoprotic acid has a percent ionization of 1.55%. Determine the acid ionization constant (K_a) for the acid.

44. A 0.085 M solution of a monoprotic acid has a percent ionization of 0.59%. Determine the acid ionization constant (K_a) for the acid.

45. Find the pH and percent ionization of each HF solution:
    a. 0.250 M   b. 0.100 M   c. 0.050 M

46. Find the pH and percent ionization of a 0.100 M solution of a weak monoprotic acid having each K_a value.
    a. $K_a = 1.0 \times 10^{-5}$ b. $K_a = 1.0 \times 10^{-3}$
    c. $K_a = 1.0 \times 10^{-1}$

47. Write chemical equations and corresponding equilibrium expressions for each of the three ionization steps of phosphoric acid.

48. Write chemical equations and corresponding equilibrium expressions for each of the two ionization steps of carbonic acid.

49. Calculate the $[H_3O^+]$ and pH of each polyprotic acid solution:
    a. 0.350 M H_3PO_4 b. 0.350 M $H_2C_2O_4$

50. Calculate the $[H_3O^+]$ and pH of each polyprotic acid solution:
    a. 0.125 M H_2CO_3 b. 0.125 M $H_3C_6H_5O_7$

Base Solutions

51. For each strong base solution, determine $[OH^-]$, $[H_3O^+]$, pH, and pOH.
    a. 0.15 M NaOH
    b. 1.5×10^{-3} M $Ca(OH)_2$
    c. 4.8×10^{-4} M $Sr(OH)_2$
    d. 8.7×10^{-5} M KOH

52. For each strong base solution, determine $[OH^-]$, $[H_3O^+]$, pH, and pOH.
    a. 8.77×10^{-3} M LiOH
    b. 0.0112 M $Ba(OH)_2$
    c. 1.9×10^{-4} M KOH
    d. 5.0×10^{-4} M $Ca(OH)_2$

53. Determine the pH of a solution that is 3.85% KOH by mass. Assume that the solution has a density of 1.01 g/mL.

54. Determine the pH of a solution that is 1.55% NaOH by mass. Assume that the solution has a density of 1.01 g/mL.

55. What volume of 0.855 M KOH solution do you need to make 3.55 L of a solution with pH of 12.4?

56. What volume of a 15.0% by mass NaOH solution, which has a density of 1.116 g/mL, should you use to make 5.00 L of an NaOH solution with a pH of 10.8?

57. Write equations showing how each weak base ionizes water to form OH^-. Also write the corresponding expression for K_b.
    a. NH_3 b. HCO_3^- c. CH_3NH_2

58. Write equations showing how each weak base ionizes water to form OH^-. Also write the corresponding expression for K_b.
    a. CO_3^{2-} b. $C_6H_5NH_2$ c. $C_2H_5NH_2$

59. Determine the $[OH^-]$, pH, and pOH of a 0.15 M ammonia solution.

60. Determine the $[OH^-]$, pH, and pOH of a solution that is 0.125 M in CO_3^{2-}.

61. Caffeine ($C_8H_{10}N_4O_2$) is a weak base with a pK_b of 10.4. Calculate the pH of a solution containing a caffeine concentration of 455 mg/L.

62. Amphetamine ($C_9H_{13}N$) is a weak base with a pK_b of 4.2. Calculate the pH of a solution containing an amphetamine concentration of 225 mg/L.

63. Morphine is a weak base. A 0.150 M solution of morphine has a pH of 10.5. What is K_b for morphine?

64. A 0.135 M solution of a weak base has a pH of 11.23. Determine K_b for the base.

Acid–Base Properties of Ions and Salts

65. Which anions act as weak bases in solution? For the anions that are basic, write an equation that shows how the anion acts as a base.
   a. Br^- **b.** ClO^- **c.** CN^- **d.** Cl^-

66. Classify each anion as basic or neutral. For the anions that are basic, write an equation that shows how the anion acts as a base.
   a. $C_7H_5O_2^-$ **b.** I^- **c.** NO_3^- **d.** F^-

67. Determine the $[OH^-]$ and pH of a solution that is 0.140 M in F^-.

68. Determine the $[OH^-]$ and pH of a solution that is 0.250 M in HCO_3^-.

69. Determine whether each cation is acidic or pH-neutral. For the cations that are acidic, write an equation that shows how the cation acts as an acid.
   a. NH_4^+ **b.** Na^+ **c.** Co^{3+} **d.** $CH_3NH_3^+$

70. Determine whether each cation is acidic or pH-neutral. For the cations that are acidic, write an equation that shows how the cation acts as an acid.
   a. Sr^{2+} **b.** Mn^{3+} **c.** $C_5H_5NH^+$ **d.** Li^+

71. Determine whether each salt will form a solution that is acidic, basic, or pH-neutral.
   a. $FeCl_3$ **b.** NaF **c.** $CaBr_2$
   d. NH_4Br **e.** $C_6H_5NH_3NO_2$

72. Determine whether each salt will form a solution that is acidic, basic, or pH-neutral.
   a. $Al(NO_3)_3$ **b.** $C_2H_5NH_3NO_3$ **c.** K_2CO_3
   d. RbI **e.** NH_4ClO

73. Arrange the solutions in order of increasing acidity: $NaCl(aq)$, $NH_4Cl(aq)$, $NaHCO_3(aq)$, $NH_4ClO_2(aq)$, $NaOH(aq)$

74. Arrange the solutions in order of increasing basicity: $CH_3NH_3Br(aq)$, $KOH(aq)$, $KBr(aq)$, $KCN(aq)$, $C_5H_5NHNO_2(aq)$

75. Determine the pH of each solution:
   a. 0.10 M NH_4Cl **b.** 0.10 M $NaC_2H_3O_2$
   c. 0.10 M $NaCl$

76. Determine the pH of each solution:
   a. 0.20 M $KCHO_2$ **b.** 0.20 M CH_3NH_3I
   c. 0.20 M KI

77. Calculate the concentration of each species in a 0.15 M KF solution.

78. Calculate the concentration of each species in a 0.225 M $C_6H_5NH_3Cl$ solution.

Molecular Structure and Acid Strength

79. Based on their molecular structure, pick the stronger acid from each pair of binary acids. Explain your reasoning.
   a. HF or HCl   **b.** H_2O or HF   **c.** H_2Se or H_2S

80. Based on molecular structure, arrange the binary compounds in order of increasing acid strength. Explain your reasoning. H_2Te, HI, H_2S, NaH

81. Based on their molecular structure, pick the stronger acid from each pair of oxyacids. Explain your reasoning.
   a. H_2SO_4 or H_2SO_3 **b.** $HClO_2$ or $HClO$
   c. HClO or HBrO   **d.** CCl_3COOH or CH_3COOH

82. Based on molecular structure, arrange the oxyacids in order of increasing acid strength. Explain your reasoning. $HClO_3$, HIO_3, $HBrO_3$

83. Which is a stronger base, S^{2-} or Se^{2-}? Explain.

84. Which is a stronger base, PO_4^{3-} or AsO_4^{3-}? Explain.

Lewis Acids and Bases

85. Classify each species as either a Lewis acid or a Lewis base.
   a. Fe^{3+} **b.** BH_3 **c.** NH_3 **d.** F^-

86. Classify each species as either a Lewis acid or a Lewis base.
   a. $BeCl_2$ **b.** OH^- **c.** $B(OH)_3$ **d.** CN^-

87. Identify the Lewis acid and Lewis base from among the reactants in each equation:
   a. $Fe^{3+}(aq) + 6 H_2O(l) \rightleftharpoons Fe(H_2O)_6^{3+}(aq)$
   b. $Zn^{2+}(aq) + 4 NH_3(aq) \rightleftharpoons Zn(NH_3)_4^{2+}(aq)$
   c. $(CH_3)_3N(g) + BF_3(g) \rightleftharpoons (CH_3)_3 NBF_3(s)$

88. Identify the Lewis acid and Lewis base from among the reactants in each equation:
   a. $Ag^+(aq) + 2 NH_3(aq) \rightleftharpoons Ag(NH_3)_2^+(aq)$
   b. $AlBr_3 + NH_3 \rightleftharpoons H_3NAlBr_3$
   c. $F^-(aq) + BF_3(aq) \rightleftharpoons BF_4^-(aq)$

Cumulative Problems

89. Consider the molecular views of acid solutions. Based on the molecular view, determine whether the acid is weak or strong.

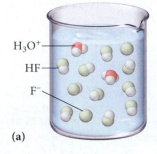

(a)

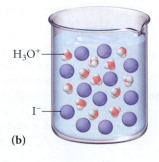

(b)

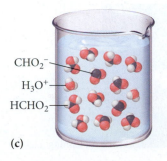

(c)

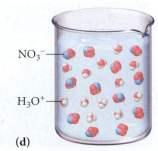

(d)

90. Consider the molecular views of base solutions. Based on the molecular view, determine whether the base is weak or strong.

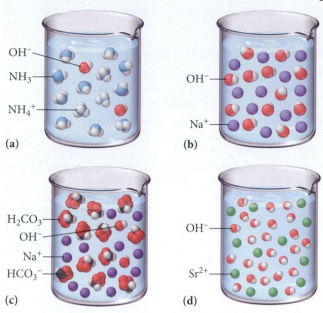

(a)

(b)

(c)

(d)

91. The binding of oxygen by hemoglobin in the blood involves the equilibrium reaction:

$$HbH^+(aq) + O_2(aq) \rightleftharpoons HbO_2(aq) + H^+(aq)$$

In this equation, Hb is hemoglobin. The pH of normal human blood is highly controlled within a range of 7.35 to 7.45. Given the above equilibrium, why is this important? What would happen to the oxygen-carrying capacity of hemoglobin if blood became too acidic (a condition known as acidosis)?

92. Carbon dioxide dissolves in water according to the equations:

$$CO_2(g) + H_2O(l) \rightleftharpoons H_2CO_3(aq)$$

$$H_2CO_3(aq) + H_2O(l) \rightleftharpoons HCO_3^-(aq) + H_3O^+(aq)$$

Carbon dioxide levels in the atmosphere have increased about 20% over the last century. Since Earth's oceans are exposed to atmospheric carbon dioxide, what effect might the increased CO_2 have on the pH of the world's oceans? What effect might this change have on the limestone structures (primarily $CaCO_3$) of coral reefs and marine shells?

93. Milk of magnesia is often taken to reduce the discomfort associated with stomach acid or heartburn. The recommended dose is 1 teaspoon, which contains 4.00×10^2 mg of $Mg(OH)_2$. What volume of an HCl solution with a pH of 1.3 can be neutralized by one dose of milk of magnesia? If the stomach contains 2.00×10^2 mL of pH 1.3 solution, will all the acid be neutralized? If not, what fraction is neutralized?

94. Lakes that have been acidified by acid rain can be neutralized by liming, the addition of limestone ($CaCO_3$). How much limestone (in kg) completely neutralizes a 4.3 billion liter lake with a pH of 5.5?

◄ Liming a lake.

95. Acid rain over the Great Lakes has a pH of about 4.5. Calculate the $[H_3O^+]$ of this rain and compare that value to the $[H_3O^+]$ of rain over the West Coast that has a pH of 5.4. How many times more concentrated is the acid in rain over the Great Lakes?

96. White wines tend to be more acidic than red wines. Find the $[H_3O^+]$ in a Sauvignon Blanc with a pH of 3.23 and a Cabernet Sauvignon with a pH of 3.64. How many times more acidic is the Sauvignon Blanc?

97. Common aspirin is acetylsalicylic acid, which has the structure shown here and a pK_a of 3.5.

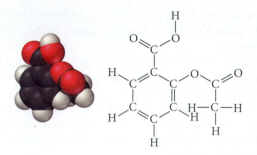

Calculate the pH of a solution in which one normal adult dose of aspirin (6.5×10^2 mg) is dissolved in 8.0 ounces of water.

98. The AIDS drug zalcitabine (also known as ddC) is a weak base with the structure shown here and a pK_b of 9.8.

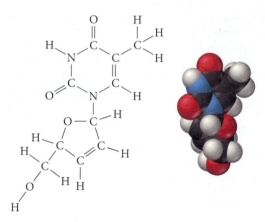

What percentage of the base is protonated in an aqueous zalcitabine solution containing 565 mg/L?

99. Determine the pH of each solution:
   a. 0.0100 M $HClO_4$ **b.** 0.115 M $HClO_2$
   c. 0.045 M $Sr(OH)_2$ **d.** 0.0852 M KCN
   e. 0.155 M NH_4Cl

100. Determine the pH of each solution:
   a. 0.0650 M HNO_3 **b.** 0.150 M HNO_2
   c. 0.0195 M KOH   **d.** 0.245 M CH_3NH_3I
   e. 0.318 M KC_6H_5O

101. Write net ionic equations for the reactions that take place when aqueous solutions of each pair of substances is mixed:
   a. sodium cyanide and nitric acid
   b. ammonium chloride and sodium hydroxide
   c. sodium cyanide and ammonium bromide
   d. potassium hydrogen sulfate and lithium acetate
   e. sodium hypochlorite and ammonia

102. Morphine has the formula $C_{17}H_{19}NO_3$. It is a base and accepts one proton per molecule. It is isolated from opium. A 0.682-g sample of opium is found to require 8.92 mL of a 0.0116 M solution of sulfuric acid for neutralization. Assuming that morphine is the only acid or base present in opium, calculate the percent morphine in the sample of opium.

103. The pH of a 1.00 M solution of urea, a weak organic base, is 7.050. Calculate the K_a of protonated urea.

104. Find the concentration of ammonia in solution made by dissolving 0.10 mol of acetic acid and 0.10 mol of ammonium chloride in enough water to make 1.0 L of solution.

105. Lactic acid is a weak acid found in milk. Its calcium salt is a source of calcium for growing animals. A saturated solution of this salt, which we can represent as Ca(Lact)$_2$ has a $[Ca^{2+}] = 0.26$ M and a pH = 8.40. Assuming the salt is completely dissociated, calculate the K_a of lactic acid.

106. A solution of 0.23 mol of the chloride salt of protonated quinine (QH$^+$), a weak organic base, in 1.0 L of solution has pH = 4.58. Find the K_b of quinine (Q).

Challenge Problems

107. A student mistakenly calculates the pH of a 1.0×10^{-7} M HI solution to be 7.0. Explain why this calculation is incorrect and calculate the correct pH.

108. When 2.55 g of an unknown weak acid (HA) with a molar mass of 85.0 g/mol is dissolved in 250.0 g of water, the freezing point of the resulting solution is $-0.257\,°C$. Calculate K_a for the unknown weak acid.

109. Calculate the pH of a solution that is 0.00115 M in HCl and 0.0100 M in HClO$_2$.

110. To what volume should you dilute 1 L of a solution of a weak acid HA to reduce the $[H^+]$ to one-half of that in the original solution? (Assume that the degree of ionization is small.)

111. HA, a weak acid, with $K_a = 1.0 \times 10^{-8}$, also forms the ion HA$_2^-$. The reaction is HA(aq) + A$^-$(aq) $\rightleftharpoons$ HA$_2^-$(aq) and its $K = 4.0$. Calculate the $[H^+]$, $[A^-]$, and $[HA_2^-]$ in a 1.0 M solution of HA.

112. We can define basicity in the gas phase as the proton affinity of the base, for example, CH$_3$NH$_2$(g) + H$^+$(g) $\rightleftharpoons$ CH$_3$NH$_3^+$(g). In the gas phase, (CH$_3$)$_3$N is more basic than CH$_3$NH$_2$ while in solution the reverse is true. Account for this observation.

113. Calculate the pH of a solution prepared from 0.200 mol of NH$_4$CN and enough water to make 1.00 L of solution.

114. We add 0.20 mol of NaF to 1.0 L of a 0.30 M solution of HClO$_2$. Calculate the $[HClO_2]$ at equilibrium.

115. A mixture of Na$_2$CO$_3$ and NaHCO$_3$ has a mass of 82.2 g. It is dissolved in 1.00 L of water and the pH is found to be 9.95. Find the mass of NaHCO$_3$ in the mixture.

116. A mixture of NaCN and NaHSO$_4$ consists of a total of 0.60 mol. When the mixture is dissolved in 1.0 L of water and comes to equilibrium the pH is found to be 9.9. Find the amount of NaCN in the mixture.

Conceptual Problems

117. Without doing any calculations, determine which solution is most acidic.
   a. 0.0100 M in HCl and 0.0100 M in KOH
   b. 0.0100 M in HF and 0.0100 M in KBr
   c. 0.0100 M in NH$_4$Cl and 0.0100 M in CH$_3$NH$_3$Br
   d. 0.100 M in NaCN and 0.100 M in CaCl$_2$

118. Without doing any calculations, determine which solution is most basic.
   a. 0.100 M in NaClO and 0.100 M in NaF
   b. 0.0100 M in KCl and 0.0100 M in KClO$_2$
   c. 0.0100 M in HNO$_3$ and 0.0100 M in NaOH
   d. 0.0100 M in NH$_4$Cl and 0.0100 M in HCN

119. Rank the acids in order of increasing acid strength.
   CH$_3$COOH   CH$_2$ClCOOH   CHCl$_2$COOH   CCl$_3$COOH

Answers to Conceptual Connections

Conjugate Acid–Base Pairs

15.1 (b) H$_2$SO$_4$ and H$_2$SO$_3$ are two different acids, not a conjugate acid–base pair.

Conjugate Bases

15.2 ClO$^-$, because the weaker the acid, the stronger the conjugate base.

The *x is small* Approximation

15.3 (c) The validity of the *x is small* approximation depends on both the value of the equilibrium constant and the initial concentration—the closer that these are to one another, the less likely the approximation will be valid.

Strong and Weak Acids

15.4 (a) A weak acid solution will usually be less than 5% dissociated. Since HCl is a strong acid, the 0.10 M solution is much more acidic than either a weak acid with the same concentration or even a weak acid that is twice as concentrated.

Percent Ionization

15.5 Solution (c) has the greatest percent ionization because percent ionization increases with decreasing weak acid concentration. Solution (b) has the lowest pH because the equilibrium H$_3$O$^+$ concentration increases with increasing weak acid concentration.

Acid Strength and Molecular Structure

15.6 (a) Since the carbon atom in (a) is bonded to another oxygen atom which draws electron density away from the O—H bond (weakening and polarizing it), and the carbon atom in (b) is bonded only to other hydrogen atoms, the proton in structure (a) is more acidic.

16 Aqueous Ionic Equilibrium

In the strictly scientific sense of the word, insolubility does not exist, and even those substances characterized by the most obstinate resistance to the solvent action of water may properly be designated as extraordinarily difficult of solution, not as insoluble. —Otto N. Witt (1853–1915)

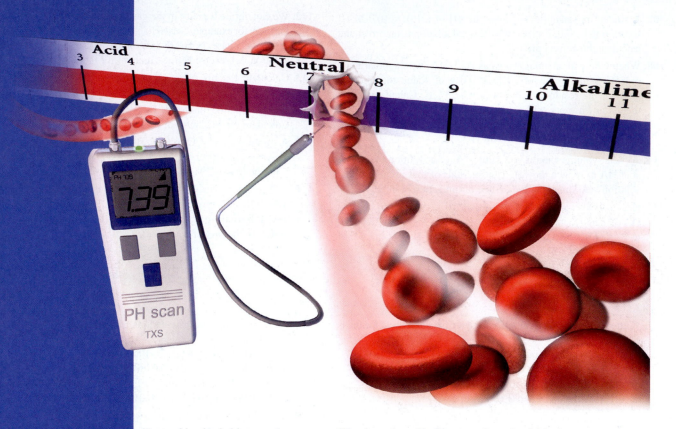

Human blood is held at nearly constant pH by the action of buffers, a main topic of this chapter.

WE HAVE ALREADY seen the importance of aqueous solutions, first in Chapters 4, 12, and 14, and most recently in Chapter 15. We now turn our attention to two additional topics involving aqueous solutions: buffers (solutions that resist pH change) and solubility equilibria (the extent to which slightly soluble ionic compounds dissolve in water). Buffers are tremendously important in biology because nearly all physiological processes must occur within a narrow pH range. Solubility equilibria are related to the solubility rules that we learned in Chapter 4. But now we find a more complicated picture: solids that we considered insoluble under the simple "solubility rules" are actually better described as being only very slightly soluble, as the chapter-opening quotation from Otto Witt suggests. Solubility equilibria are important in predicting not only solubility, but also precipitation reactions that might occur when aqueous solutions are mixed.

16.1   The Danger of Antifreeze

Every year, thousands of dogs and cats die from consuming a common household product: antifreeze that was improperly stored or that leaked out of a car radiator. Most types of antifreeze used in cars are aqueous solutions of ethylene glycol ($HOCH_2CH_2OH$), an alcohol with the structure:

Ethylene glycol has a somewhat sweet taste that often proves attractive to curious dogs and cats—and sometimes to young children, who are also vulnerable to this toxic compound. The first stage of ethylene glycol poisoning is a state resembling drunkenness. Since the compound is an alcohol, it affects the brain much as an alcoholic beverage would. Once ethylene glycol starts to be metabolized, however, a second and more deadly stage begins.

Ethylene glycol is oxidized in the liver to glycolic acid ($HOCH_2COOH$), which then enters the bloodstream. The acidity of blood is critically important, and tightly regulated, because many proteins only function properly within a narrow pH range. In human blood, for example, pH is held between 7.36 and 7.42. This nearly constant blood pH is maintained by *buffers*. We discuss buffers more carefully later, but for now know that a buffer is a chemical system that resists pH changes—it neutralizes added acid or base. An important buffer system in blood is a mixture of carbonic acid (H_2CO_3) and the bicarbonate ion (HCO_3^-). The carbonic acid neutralizes added base according to the reaction:

$$\underset{\text{added base}}{H_2CO_3(aq) + OH^-(aq)} \longrightarrow H_2O(l) + HCO_3^-(aq)$$

The bicarbonate ion neutralizes added acid according to the reaction:

$$HCO_3^-(aq) + \underset{\text{added acid}}{H^+(aq)} \longrightarrow H_2CO_3(aq)$$

With these reactions, the carbonic acid and bicarbonate ion buffering system keeps blood pH constant.

When glycolic acid first enters the bloodstream, its tendency to lower the pH of the blood is countered by the buffering action of the bicarbonate ion. However, if the original quantities of consumed antifreeze are large enough, the glycolic acid overwhelms the capacity of the buffer (the capacity of a buffer is discussed in Section 16.3), causing blood pH to drop to dangerously low levels.

Low blood pH results in *acidosis*, a condition in which the blood's ability to carry oxygen is reduced because the acid affects the equilibrium between hemoglobin (Hb) and oxygen:

$$\overset{\text{Excess } H^+}{\underset{\text{Shift left}}{HbH^+(aq) + O_2(g) \rightleftharpoons HbO_2(aq) + H^+(aq)}}$$

Excess acid causes the equilibrium to shift to the left, reducing hemoglobin's ability to carry oxygen. At this point, the cat or dog may begin hyperventilating in an effort to overcome the acidic blood's lowered oxygen-carrying capacity. If no treatment is administered, the animal will eventually go into a coma and die.

One treatment for ethylene glycol poisoning is the administration of ethyl alcohol (the alcohol found in alcoholic beverages). The two molecules are similar enough that the liver enzyme that catalyzes the metabolism of ethylene glycol also acts on ethyl alcohol, but the enzyme has higher affinity for ethyl alcohol than for ethylene glycol. Consequently, the enzyme preferentially metabolizes ethyl alcohol, allowing the unmetabolized ethylene glycol to escape through the urine. If administered early, this treatment can save the life of a dog or cat that has consumed ethylene glycol.

16.2   Buffers: Solutions That Resist pH Change

Most solutions rapidly change pH when an acid or base is added to them. As we have just discussed, however, a **buffer** resists pH change by neutralizing added acid or added base. Buffers contain significant amounts of both *a weak acid and its conjugate base* (or a weak base and its conjugate acid). For example, the buffer in blood is composed of carbonic acid (H_2CO_3) and its conjugate base, the bicarbonate ion (HCO_3^-). When additional base is added to a buffer, the weak acid reacts with the base, neutralizing it. When additional acid is added to a buffer, the conjugate base reacts with the acid, neutralizing it. In this way, a buffer can maintain a nearly constant pH.

A weak acid by itself, even though it partially ionizes to form some of its conjugate base, does not contain sufficient base to be a buffer. Similarly, a weak base by itself, even though it partially ionizes water to form some of its conjugate acid, does not contain sufficient acid to be a buffer. *A buffer contains significant amounts of both a weak acid and its conjugate base.* For example, we can make a simple buffer by dissolving acetic acid ($HC_2H_3O_2$) and sodium acetate ($NaC_2H_3O_2$) in water (**Figure 16.1▼**).

| $C_2H_3O_2^-$ is the conjugate base of $HC_2H_3O_2$.

Suppose that we add a strong base, such as NaOH, to this solution. The acetic acid neutralizes the base according to the reaction:

$$NaOH(aq) + HC_2H_3O_2(aq) \longrightarrow H_2O(l) + NaC_2H_3O_2(aq)$$

As long as the amount of NaOH we add is significantly less than the amount of $HC_2H_3O_2$ in solution, the buffer neutralizes the NaOH and the resulting pH change is small. Suppose, on the other hand, that we add a strong acid, such as HCl, to the solution. Then the conjugate base, $NaC_2H_3O_2$, neutralizes the added HCl according to the reaction:

$$HCl(aq) + NaC_2H_3O_2(aq) \longrightarrow HC_2H_3O_2(aq) + NaCl(aq)$$

Formation of a Buffer

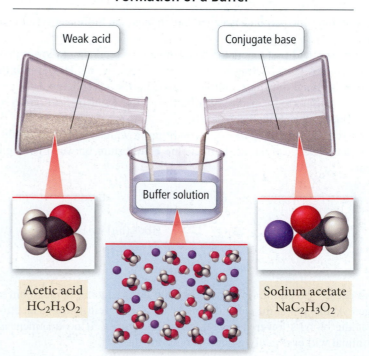

Weak acid

Conjugate base

Buffer solution

Acetic acid
$HC_2H_3O_2$

Sodium acetate
$NaC_2H_3O_2$

▶ **FIGURE 16.1   A Buffer Solution**
A buffer typically consists of a weak acid (which can neutralize added base) and its conjugate base (which can neutralize added acid).

As long as the amount of HCl added is significantly less than the amount of $NaC_2H_3O_2$ in solution, the buffer neutralizes the HCl and the resulting pH change is small.

Summarizing Characteristics of Buffers:

▶ Buffers resist pH change.

▶ Buffers contain significant amounts of both a weak acid and its conjugate base.

▶ The weak acid in a buffer neutralizes added base.

▶ The conjugate base in a buffer neutralizes added acid.

 Conceptual Connection 16.1 Buffers

Which solution is a buffer?

(a) A solution that is 0.100 M in HNO_2 and 0.100 M in HCl

(b) A solution that is 0.100 M in HNO_3 and 0.100 M in $NaNO_3$

(c) A solution that is 0.100 M in HNO_2 and 0.100 M in NaCl

(d) A solution that is 0.100 M in HNO_2 and 0.100 M in $NaNO_2$

Calculating the pH of a Buffer Solution

In Chapter 15, we learned how to calculate the pH of a solution containing either a weak acid or its conjugate base, but not both. How do we calculate the pH of a solution containing both? Consider a solution that initially contains both $HC_2H_3O_2$ and $NaC_2H_3O_2$, each at a concentration of 0.100 M. The acetic acid ionizes according to the reaction:

$$HC_2H_3O_2(aq) + H_2O(l) \rightleftharpoons H_3O^+(aq) + C_2H_3O_2^-(aq)$$

Initial concentration   0.100 M   0.100 M

However, the ionization of $HC_2H_3O_2$ in this solution is suppressed compared to its ionization in a solution that does not initially contain any $C_2H_3O_2^-$, because the presence of $C_2H_3O_2^-$ *shifts the equilibrium to the left* (as we would expect from Le Châtelier's principle). In other words, the presence of the $C_2H_3O_2^-(aq)$ ion causes the acid to ionize even less than it normally would (**Figure 16.2▼**), resulting in a less acidic solution (higher pH). This effect is known as the **common ion effect**, so named because the solution contains two substances ($HC_2H_3O_2$ and $NaC_2H_3O_2$) that share a common ion ($C_2H_3O_2^-$). To find the pH of a buffer solution containing common ions, we work an equilibrium problem in which the initial concentrations include both the acid and its conjugate base, as shown in Example 16.1.

> Le Châtelier's principle was discussed in Section 14.9.

pH = 2.9

0.100 M
$HC_2H_3O_2$

pH = 8.9

0.100 M
$NaC_2H_3O_2$

pH = 4.7

0.100 M $HC_2H_3O_2$
0.100 M $NaC_2H_3O_2$

▲ **FIGURE 16.2 The Common Ion Effect**   The pH of a 0.100 M acetic acid solution is 2.9. The pH of a 0.100 M sodium acetate solution is 8.9. The pH of a solution that is 0.100 M in acetic acid and 0.100 M in sodium acetate is 4.7.

EXAMPLE 16.1 Calculating the pH of a Buffer Solution

Calculate the pH of a buffer solution that is 0.100 M in $HC_2H_3O_2$ and 0.100 M in $NaC_2H_3O_2$.

SOLUTION

1. Write the balanced equation for the ionization of the acid and use it as a guide to prepare an ICE table showing the given concentrations of the acid and its conjugate base as the initial concentrations. Leave room in the table for the changes in concentrations and for the equilibrium concentrations.	$HC_2H_3O_2(aq) + H_2O(l) \rightleftharpoons H_3O^+(aq) + C_2H_3O_2^-(aq)$ 	

	$[HC_2H_3O_2]$	$[H_3O^+]$	$[C_2H_3O_2^-]$
Initial	0.100	≈ 0.00	0.100
Change			
Equil			

2. Represent the change in the concentration of H_3O^+ with the variable x. Express the changes in the concentrations of the other reactants and products in terms of x.	$HC_2H_3O_2(aq) + H_2O(l) \rightleftharpoons H_3O^+(aq) + C_2H_3O_2^-(aq)$

	$[HC_2H_3O_2]$	$[H_3O^+]$	$[C_2H_3O_2^-]$
Initial	0.100	≈ 0.00	0.100
Change	$-x$	$+x$	$+x$
Equil			

3. Sum each column to determine the equilibrium concentrations in terms of the initial concentrations and the variable x.	$HC_2H_3O_2(aq) + H_2O(l) \rightleftharpoons H_3O^+(aq) + C_2H_3O_2^-(aq)$

	$[HC_2H_3O_2]$	$[H_3O^+]$	$[C_2H_3O_2^-]$
Initial	0.100	≈ 0.00	0.100
Change	$-x$	$+x$	$+x$
Equil	$0.100 - x$	x	$0.100 + x$

4. Substitute the expressions for the equilibrium concentrations (from step 3) into the expression for the acid ionization constant. In most cases, you can make the approximation that x is small. (See Sections 14.8 and 15.6 to review the *x is small* approximation.) Substitute the value of the acid ionization constant (from Table 15.5) into the K_a expression and solve for x.	$K_a = \dfrac{[H_3O^+][C_2H_3O_2^-]}{[HC_2H_3O_2]}$ $= \dfrac{x(0.100 + \cancel{x})}{0.100 - \cancel{x}} \quad$ (*x is small*) $1.8 \times 10^{-5} = \dfrac{x(\cancel{0.100})}{\cancel{0.100}}$ $x = 1.8 \times 10^{-5}$
Confirm that x is small by calculating the ratio of x to the number it was subtracted from in the approximation. The ratio should be less than 0.05 (or 5%).	$\dfrac{1.8 \times 10^{-5}}{0.100} \times 100\% = 0.018\%$ Therefore the approximation is valid.
5. Determine the H_3O^+ concentration from the calculated value of x and substitute into the pH equation to find pH.	$[H_3O^+] = x = 1.8 \times 10^{-5}$ M $pH = -\log[H_3O^+]$ $\quad\;\; = -\log(1.8 \times 10^{-5})$ $\quad\;\; = 4.74$

FOR PRACTICE 16.1
Calculate the pH of a buffer solution that is 0.200 M in $HC_2H_3O_2$ and 0.100 M in $NaC_2H_3O_2$.

FOR MORE PRACTICE 16.1
Calculate the pH of the buffer that results from mixing 60.0 mL of 0.250 M $HCHO_2$ and 15.0 mL of 0.500 M $NaCHO_2$.

The Henderson–Hasselbalch Equation

Finding the pH of a buffer solution can be simplified if we derive an equation that relates the pH of the solution to the initial concentrations of the buffer components. Consider a buffer containing the generic weak acid HA and its conjugate base A^-. The acid ionization equation is as follows:

$$HA(aq) + H_2O(l) \rightleftharpoons H_3O^+(aq) + A^-(aq)$$

We can obtain an expression for the concentration of H_3O^+ from the acid ionization equation by solving it for $[H_3O^+]$.

$$K_a = \frac{[H_3O^+][A^-]}{[HA]}$$

$$[H_3O^+] = K_a \frac{[HA]}{[A^-]} \qquad [16.1]$$

If we make the *x is small* approximation that we normally make for weak acid or weak base equilibrium problems, *we can consider the equilibrium concentrations of HA and A^- to be essentially identical to the initial concentrations of HA and A^-* (see step 4 of Example 16.1). Therefore, to determine $[H_3O^+]$ for any buffer solution, we multiply K_a by the ratio of the concentrations of the acid and the conjugate base. For example, to find the $[H_3O^+]$ of the buffer in Example 16.1 (a solution that is 0.100 M in $HC_2H_3O_2$ and 0.100 M in $NaC_2H_3O_2$), we substitute the concentrations of $HC_2H_3O_2$ and $C_2H_3O_2^-$ into Equation 16.1 as follows:

Recall that the variable *x* in a weak acid equilibrium problem represents the change in the initial acid concentration. The *x is small* approximation is valid because so little of the weak acid ionizes compared to its initial concentration.

$$[H_3O^+] = K_a \frac{[HC_2H_3O_2]}{[C_2H_3O_2^-]}$$

$$= K_a \frac{0.100}{0.100}$$

$$= K_a$$

In this solution, as in any buffer solution in which the concentrations of the acid and conjugate base are equal, $[H_3O^+]$ is equal to K_a.

We can derive an equation for the pH of a buffer by taking the logarithm of both sides of Equation 16.1 as follows:

$$[H_3O^+] = K_a \frac{[HA]}{[A^-]}$$

$$\log[H_3O^+] = \log\left(K_a \frac{[HA]}{[A^-]}\right)$$

$$\log[H_3O^+] = \log K_a + \log\frac{[HA]}{[A^-]} \qquad [16.2]$$

Multiplying both sides of Equation 16.2 by -1 and rearranging, we get:

$$-\log[H_3O^+] = -\log K_a - \log\frac{[HA]}{[A^-]}$$

$$-\log[H_3O^+] = -\log K_a + \log\frac{[A^-]}{[HA]}$$

Since $pH = -\log[H_3O^+]$ and since $pK_a = -\log K_a$, we obtain the following result:

$$pH = pK_a + \log\frac{[A^-]}{[HA]}$$

Since A^- is a weak base and HA is a weak acid, we can generalize the equation:

$$pH = pK_a + \log\frac{[base]}{[acid]} \qquad [16.3]$$

Note that, as expected, the pH of a buffer increases with an increase in the amount of base relative to the amount of acid.

where the base is the conjugate base of the acid or the acid is the conjugate acid of the base. This equation, known as the **Henderson–Hasselbalch equation**, allows us to readily calculate the pH of a buffer solution from the initial concentrations of the buffer components *as long as the x is small* approximation is valid. In Example 16.2, we find the pH of a buffer in two ways: in the left column we solve a common ion effect equilibrium problem similar to the one we solved in Example 16.1; in the right column we use the Henderson–Hasselbalch equation.

EXAMPLE 16.2 Calculating the pH of a Buffer Solution as an Equilibrium Problem and with the Henderson–Hasselbalch Equation

Calculate the pH of a buffer solution that is 0.050 M in benzoic acid ($HC_7H_5O_2$) and 0.150 M in sodium benzoate ($NaC_7H_5O_2$). For benzoic acid, $K_a = 6.5 \times 10^{-5}$.

SOLUTION

Equilibrium Approach	**Henderson–Hasselbalch Approach**
Write the balanced equation for the ionization of the acid and use it as a guide to prepare an ICE table.	To find the pH of this solution, determine which component is the acid and which is the base and substitute their concentrations into the Henderson–Hasselbalch equation to calculate pH.

Equilibrium Approach:

$$HC_7H_5O_2(aq) + H_2O(l) \rightleftharpoons H_3O^+(aq) + C_7H_5O_2^-(aq)$$

	$[HC_7H_5O_2]$	$[H_3O^+]$	$[C_7H_5O_2^-]$
Initial	0.050	≈0.00	0.150
Change	−x	+x	+x
Equil	0.050−x	x	0.150 + x

Substitute the expressions for the equilibrium concentrations into the expression for the acid ionization constant. Make the *x is small* approximation and solve for *x*.

$$K_a = \frac{[H_3O^+][C_7H_5O_2^-]}{[HC_7H_5O_2]}$$

$$= \frac{x(0.150 + \cancel{x})}{0.050 - \cancel{x}} \quad (x \text{ is small})$$

$$6.5 \times 10^{-5} = \frac{x(0.150)}{0.050}$$

$$x = 2.2 \times 10^{-5}$$

Since $[H_3O^+] = x$, we compute pH as follows:

$$pH = -\log[H_3O^+]$$
$$= -\log(2.2 \times 10^{-5})$$
$$= 4.66$$

Henderson–Hasselbalch Approach:

$HC_7H_5O_2$ is the acid and $NaC_7H_5O_2$ is the base. Therefore, you can calculate the pH as follows:

$$pH = pK_a + \log\frac{[\text{base}]}{[\text{acid}]}$$

$$= -\log(6.5 \times 10^{-5}) + \log\frac{0.150}{0.050}$$

$$= 4.187 + 0.477$$

$$= 4.66$$

Confirm that the *x is small* approximation is valid by calculating the ratio of *x* to the number it was subtracted from in the approximation. The ratio should be less than 0.05 (or 5%). (See Sections 14.8 and 15.6 to review the *x is small* approximation.)

$$\frac{2.2 \times 10^{-5}}{0.050} \times 100\% = 0.044\%$$

The approximation is valid.

Confirm that the *x is small* approximation is valid by calculating the $[H_3O^+]$ from the pH. Since $[H_3O^+]$ is formed by ionization of the acid, the calculated $[H_3O^+]$ has to be less than 0.05 (or 5%) of the initial concentration of the acid.

$$pH = 4.66 = -\log[H_3O^+]$$

$$[H_3O^+] = 10^{-4.66} = 2.2 \times 10^{-5} \text{ M}$$

$$\frac{2.2 \times 10^{-5}}{0.050} \times 100\% = 0.044\%$$

The approximation is valid.

FOR PRACTICE 16.2

Calculate the pH of a buffer solution that is 0.250 M in HCN and 0.170 M in KCN. For HCN, $K_a = 4.9 \times 10^{-10}$ ($pK_a = 9.31$). Use both the equilibrium approach and the Henderson–Hasselbalch approach.

You may be wondering whether you should use the equilibrium approach or the Henderson–Hasselbalch equation when determining the pH of buffer solutions. The answer depends on the specific problem. In cases where the *x is small* approximation can be made, the Henderson–Hasselbalch equation is adequate. However, as you can see from Example 16.2, checking the *x is small* approximation is not as convenient with the Henderson–Hasselbalch equation (because the approximation is implicit). Thus, the equilibrium approach, although lengthier, gives you a better sense of the important quantities in the problem and the nature of the approximation. When first working buffer problems, use the equilibrium approach until you get a good sense for when the *x is small* approximation is adequate. Then, you can switch to the more streamlined approach in cases where the approximation applies (and only in those cases). In general, remember that the *x is small* approximation applies to problems in which both of the following are true: (a) the initial concentrations of acids (and/or bases) are not too dilute; and (b) the equilibrium constant is fairly small. Although the exact values depend on the details of the problem, for many buffer problems this means that the initial concentrations of acids and conjugate bases should be at least 10^2–10^3 times greater than the equilibrium constant (depending on the required accuracy).

 Conceptual Connection 16.2 pH of Buffer Solutions

A buffer contains the weak acid HA and its conjugate base A⁻. The weak acid has a pK_a of 4.82 and the buffer has a pH of 4.25. Which statement is true of the relative concentrations of the weak acid and conjugate base in the buffer?

(a) $[HA] > [A^-]$ **(b)** $[HA] < [A^-]$ **(c)** $[HA] = [A^-]$

Which buffer component would you add to change the pH of the buffer to 4.72?

Calculating pH Changes in a Buffer Solution

When we add an acid or a base to a buffer, the buffer tends to resist a pH change. Nonetheless, the pH does change by a small amount. To calculate the pH change we break up the problem into two parts:

1. **The stoichiometry calculation** in which we calculate how the addition changes the relative amounts of acid and conjugate base.
2. **The equilibrium calculation** in which we calculate the pH based on the new amounts of acid and conjugate base.

We demonstrate this calculation with a 1.0-L buffer solution that is 0.100 M in the generic acid HA and 0.100 M in its conjugate base A⁻. How do we calculate the pH after addition of 0.025 mol of strong acid (H⁺) to this buffer (assuming that the change in volume from adding the acid is negligible)?

The Stoichiometry Calculation

As the added acid is neutralized, it converts a stoichiometric amount of the base into its conjugate acid through the following neutralization reaction (**Figure 16.3a▶**):

$$\underset{\text{added acid}}{H^+(aq)} + \underset{\text{weak base in buffer}}{A^-(aq)} \longrightarrow HA(aq)$$

Neutralizing 0.025 mol of the strong acid (H⁺) requires 0.025 mol of the weak base (A⁻). Consequently, the amount of A⁻ *decreases* by 0.025 mol and the amount of HA *increases* by 0.025 mol because of the 1:1:1 stoichiometry of the neutralization reaction. We can track these changes in tabular form as follows:

It is best to work with amounts in moles instead of concentrations when tracking these changes, as explained later.

	$H^+(aq)$	+	$A^-(aq)$	$\longrightarrow$	$HA(aq)$
Before addition	≈0.00 mol		0.100 mol		0.100 mol
Addition	+0.025 mol		—		—
After addition	≈0.00 mol		0.075 mol		0.125 mol

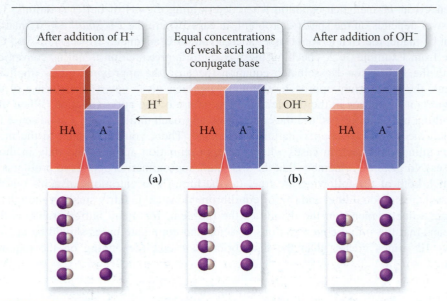

▲ **FIGURE 16.3 Buffering Action**   (a) When an acid is added to a buffer, a stoichiometric amount of the weak base is converted to the conjugate acid. (b) When a base is added to a buffer, a stoichiometric amount of the weak acid is converted to the conjugate base.

Notice that this table *is not an ICE table*. It is simply a table that tracks the stoichiometric changes that occur during the neutralization of the added acid. We write ≈ 0.00 mol for the amount of H^+ because the amount is so small compared to the amounts of A^- and HA. (Remember that weak acids ionize only to a small extent and that the presence of the common ion further suppresses the ionization.) The amount of H^+, of course, is not *exactly* zero, as we can see by completing the equilibrium part of the calculation.

The Equilibrium Calculation

We have just seen that adding a small amount of acid to a buffer is equivalent to changing the initial concentrations of the acid and conjugate base present in the buffer (in this case, since the volume is 1.0 L, [HA] increased from 0.100 M to 0.125 M and $\left[A^-\right]$ decreased from 0.100 M to 0.075 M). Once these new initial concentrations are calculated, the new pH can be calculated in the same way that we calculate the pH of any buffer: either by working a full equilibrium problem or by using the Henderson–Hasselbalch equation (see Examples 16.1 and 16.2). In this case, we work the full equilibrium problem. We begin by writing the balanced equation for the ionization of the acid and using it as a guide to prepare an ICE table. The initial concentrations for the ICE table are those calculated in the stoichiometry part of the calculation.

$$HA(aq) + H_2O(l) \rightleftharpoons H_3O^+(aq) + A^-(aq)$$

	[HA]	[H₃O⁺]	[A⁻]
Initial	0.125	≈0.00	0.075
Change	$-x$	$+x$	$+x$
Equil	$0.125 - x$	x	$0.075 + x$

From stoichiometry calculation

We then substitute the expressions for the equilibrium concentrations into the expression for the acid ionization constant. As long as K_a is sufficiently small relative to the initial concentrations, we can make the *x is small* approximation and solve for *x*, which is equal to $[H_3O^+]$.

$$K_a = \frac{[H_3O^+][A^-]}{[HA]}$$

$$= \frac{x(0.075 + \cancel{x})}{0.125 - \cancel{x}} \quad (x \text{ is small})$$

$$K_a = \frac{x(0.075)}{0.125}$$

$$x = [H_3O^+] = K_a \frac{0.125}{0.075}$$

Once $[H_3O^+]$ is calculated, pH can be calculated using $pH = -\log[H_3O^+]$.

Notice that, since the expression for *x* contains a *ratio* of concentrations $[HA]/[A^-]$, the amounts of acid and base in moles may be substituted in place of concentration because, in a single buffer solution, the volume is the same for both the acid and the base. Therefore the volumes cancel:

$$\frac{[HA]}{[A^-]} = \frac{\dfrac{n_{HA}}{\cancel{V}}}{\dfrac{n_{A^-}}{\cancel{V}}} = \frac{n_{HA}}{n_{A^-}}$$

The effect of adding a small amount of strong base to the buffer is exactly the opposite of adding acid. The added base converts a stoichiometric amount of the acid into its conjugate base through the following neutralization reaction (Figure 16.3b on p. 628):

$$OH^-(aq) + \underset{\text{weak acid in buffer}}{HA(aq)} \longrightarrow H_2O(l) + A^-(aq)$$

$\underset{\text{added base}}{}$

If we add 0.025 mol of OH^-, then the amount of A^- goes *up* by 0.025 mol and the amount of HA goes *down* by 0.025 mol as shown in the following table:

	$OH^-(aq)$	+	$HA(aq)$	$\longrightarrow$	$H_2O(l)$	+	$A^-(aq)$
Before addition	≈ 0.00 mol		0.100 mol				0.100 mol
Addition	$+0.025$ mol		—				—
After addition	≈ 0.00 mol		0.075 mol				0.125 mol

Summarizing Calculating the pH of a Buffer After Adding Small Amounts of Strong Acid or Base:

▶ Adding a small amount of strong acid to a buffer converts a stoichiometric amount of the base to the conjugate acid and decreases the pH of the buffer (adding acid decreases pH just as we would expect).

▶ Adding a small amount of strong base to a buffer converts a stoichiometric amount of the acid to the conjugate base and increases the pH of the buffer (adding base increases the pH just as we would expect).

The easiest way to remember these changes is relatively simple: adding acid creates more acid; adding base creates more base.

Example 16.3 and For Practice 16.3 involve calculating pH changes in a buffer solution after small amounts of strong acid or strong base are added.

EXAMPLE 16.3 Calculating the pH Change in a Buffer Solution After the Addition of a Small Amount of Strong Acid or Base

A 1.0-L buffer solution contains 0.100 mol $HC_2H_3O_2$ and 0.100 mol $NaC_2H_3O_2$. The value of K_a for $HC_2H_3O_2$ is 1.8×10^{-5}. Since the initial amounts of acid and conjugate base are equal, the pH of the buffer is simply equal to $pK_a = -\log(1.8 \times 10^{-5}) = 4.74$. Calculate the new pH after adding 0.010 mol of solid NaOH to the buffer. For comparison, calculate the pH after adding 0.010 mol of solid NaOH to 1.0 L of pure water. (Ignore any small changes in volume that might occur upon addition of the base.)

SOLUTION

Part I: Stoichiometry. The addition of the base converts a stoichiometric amount of acid to the conjugate base (adding base creates more base). Write an equation showing the neutralization reaction and then set up a table to track the changes.

	$OH^-(aq)$ +	$HC_2H_3O_2(aq)$	$\longrightarrow$ $H_2O(l)$ +	$C_2H_3O_2^-(aq)$
Before addition	≈ 0.00 mol	0.100 mol		0.100 mol
Addition	0.010 mol	—		—
After addition	≈ 0.00 mol	0.090 mol		0.110 mol

Part II: Equilibrium. Write the balanced equation for the ionization of the acid and use it as a guide to prepare an ICE table. Use the amounts of acid and conjugate base from part I as the initial amounts of acid and conjugate base in the ICE table.

$$HC_2H_3O_2(aq) + H_2O(l) \rightleftharpoons H_3O^+(aq) + C_2H_3O_2^-(aq)$$

	$[HC_2H_3O_2]$	$[H_3O^+]$	$[C_2H_3O_2^-]$
Initial	0.090	≈ 0.00	0.110
Change	$-x$	$+x$	$+x$
Equil	$0.090 - x$	x	$0.110 + x$

Substitute the expressions for the equilibrium concentrations of acid and conjugate base into the expression for the acid ionization constant. Make the *x is small* approximation and solve for *x*. Calculate the pH from the value of *x*, which is equal to $[H_3O^+]$.

$$K_a = \frac{[H_3O^+][C_2H_3O_2^-]}{[HC_2H_3O_2]}$$

$$= \frac{x(0.110 + \cancel{x})}{0.090 - \cancel{x}} \quad (x \text{ is small})$$

$$1.8 \times 10^{-5} = \frac{x(0.110)}{0.090}$$

$$x = [H_3O^+] = 1.47 \times 10^{-5} \text{ M}$$

$$pH = -\log[H_3O^+]$$

$$= -\log(1.47 \times 10^{-5})$$

$$= 4.83$$

Confirm that the *x is small* approximation is valid by calculating the ratio of *x* to the smallest number it was subtracted from in the approximation. The ratio should be less than 0.05 (or 5%).

$$\frac{1.47 \times 10^{-5}}{0.090} \times 100\% = 0.016\%$$

The approximation is valid.

Part II: Equilibrium Alternative (using the Henderson–Hasselbalch equation). As long at the *x is small* approximation is valid, you can simply substitute the quantities of acid and conjugate base after the addition (from part I) into the Henderson–Hasselbalch equation and calculate the new pH.

$$pH = pK_a + \log\frac{[\text{base}]}{[\text{acid}]}$$

$$= -\log(1.8 \times 10^{-5}) + \log\frac{0.110}{0.090}$$

$$= 4.74 + 0.087$$

$$= 4.83$$

The pH of 1.0 L of water after adding 0.010 mol of NaOH is calculated from the $[OH^-]$. For a strong base, $[OH^-]$ is just the number of moles of OH^- divided by the number of liters of solution.

$$[OH^-] = \frac{0.010 \text{ mol}}{1.0 \text{ L}} = 0.010 \text{ M}$$

$$pOH = -\log[OH^-] = -\log(0.010)$$

$$= 2.00$$

$$pOH + pH = 14.00$$

$$pH = 14.00 - pOH$$

$$= 14.00 - 2.00$$

$$= 12.00$$

CHECK

Notice that the buffer solution changed from pH = 4.74 to pH = 4.83 upon addition of the base (a small fraction of a single pH unit). In contrast, the pure water changed from pH = 7.00 to pH = 12.00, five whole pH units (a factor of 10^5). Notice also that even the buffer solution got slightly more basic upon addition of a base, as you would expect. To check your answer, always make sure the pH goes in the direction you expect: adding base should make the solution more basic (higher pH); adding acid should make the solution more acidic (lower pH).

FOR PRACTICE 16.3

Calculate the pH of the original buffer solution in Example 16.3 upon addition of 0.015 mol of NaOH.

FOR MORE PRACTICE 16.3

Calculate the pH in the solution upon addition of 10.0 mL of 1.00 M HCl to the original buffer in Example 16.3.

 Conceptual Connection 16.3 Adding Acid or Base to a Buffer

A buffer contains equal amounts of a weak acid and its conjugate base and has a pH of 5.25. Which would be a reasonable value to expect for the pH of the buffer after addition of a small amount of acid?

(a) 4.15   **(b)** 5.15   **(c)** 5.35   **(d)** 6.35

Buffers Containing a Base and Its Conjugate Acid

So far, we have seen examples of buffers composed of an acid and its conjugate base (where the conjugate base is an ion). However, a buffer can also be composed of a base and its conjugate acid (where the conjugate acid is an ion). For example, a solution containing significant amounts of both NH_3 and NH_4Cl will act as a buffer (**Figure 16.4▼**). The NH_3 is a weak base that neutralizes small amounts of added acid and the NH_4^+ ion is the conjugate acid that neutralizes small amounts of added base. We calculate the pH of a solution such as this in the same way we calculated the pH for a buffer containing a weak acid and its conjugate base. When using the Henderson–Hasselbalch equation, however, we must first calculate pK_a for the conjugate acid of a weak base. Recall from Section 15.8 that for a conjugate acid–base pair: $K_a \times K_b = K_w$ and $pK_a + pK_b = 14$.

Formation of a Buffer

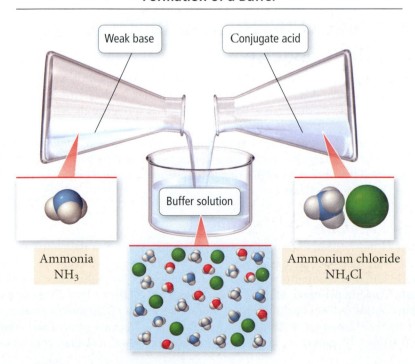

Weak base

Conjugate acid

Buffer solution

Ammonia
NH_3

Ammonium chloride
NH_4Cl

◀ **FIGURE 16.4 Buffer Containing a Base** A buffer can also consist of a weak base and its conjugate acid.

Consequently, we can find pK_a of the conjugate acid by subtracting pK_b of the weak base from 14. Example 16.4 shows how to calculate the pH of a buffer composed of a weak base and its conjugate acid.

EXAMPLE 16.4 Using the Henderson–Hasselbalch Equation to Calculate the pH of a Buffer Solution Composed of a Weak Base and Its Conjugate Acid

Use the Henderson–Hasselbalch equation to calculate the pH of a buffer solution that is 0.50 M in NH_3 and 0.20 M in NH_4Cl. For ammonia, $pK_b = 4.75$.

SOLUTION

Since K_b for NH_3 (1.8×10^{-5}) is much smaller than the initial concentrations in this problem, you can use the Henderson–Hasselbalch equation to calculate the pH of the buffer. However, you must first calculate pK_a for NH_4^+ from pK_b for NH_3.	$pK_a + pK_b = 14$ $pK_a = 14 - pK_b$ $\quad\;\; = 14 - 4.75$ $\quad\;\; = 9.25$
Now, substitute the given quantities into the Henderson–Hasselbalch equation and calculate pH.	$pH = pK_a + \log \dfrac{[\text{base}]}{[\text{acid}]}$ $\quad = 9.25 + \log \dfrac{0.50}{0.20}$ $\quad = 9.25 + 0.40$ $\quad = 9.65$

FOR PRACTICE 16.4
Calculate the pH of 1.0 L of the original buffer in Example 16.4 upon addition of 0.010 mol of solid NaOH.

FOR MORE PRACTICE 16.4
Calculate the pH of 1.0 L of the original buffer in Example 16.4 upon addition of 30.0 mL of 1.0 M HCl.

16.3 Buffer Effectiveness: Buffer Range and Buffer Capacity

An effective buffer neutralizes moderate amounts of added acid or base. Recall from the opening section of this chapter, however, that a buffer can be destroyed by the addition of too much acid or too much base. What factors influence the effectiveness of a buffer? In this section, we examine two factors: *the relative amounts of the acid and conjugate base* and *the absolute concentrations of the acid and conjugate base*. We then define the *capacity of a buffer* (how much added acid or base it can effectively neutralize) and the *range of a buffer* (the pH range over which a particular acid and its conjugate base can be effective).

Relative Amounts of Acid and Base

A buffer is most effective (most resistant to pH changes) when the concentrations of acid and conjugate base are equal. We explore this idea by considering the behavior of a generic buffer composed of HA and A^- for which $pK_a = 5.00$. Let's calculate the percent change in pH upon addition of 0.010 mol of NaOH for two different 1.0-liter solutions of this buffer system. Both solutions have exactly 0.20 mol of *total* acid and conjugate base. However, solution I has exactly equal amounts of acid and conjugate base (0.10 mol of each), while solution II has much more acid than conjugate base

(0.18 mol HA and 0.020 mol A⁻). We can calculate the initial pH values of each solution using the Henderson–Hasselbalch equation. Solution I has an initial pH of 5.00. Solution II has an initial pH of 4.05.

Solution I: 0.10 mol HA and 0.10 mol A⁻; initial pH = 5.00

	$OH^-(aq)$	+	$HA(aq)$	$\longrightarrow$	$H_2O(l)$ +	$A^-(aq)$
Before addition	≈0.00 mol		0.100 mol			0.100 mol
Addition	0.010 mol		—			—
After addition	≈0.00 mol		0.090 mol			0.110 mol

$$pH = pK_a + \log\frac{[\text{base}]}{[\text{acid}]}$$

$$= 5.00 + \log\frac{0.110}{0.090}$$

$$= 5.09$$

$$\% \text{ change} = \frac{5.09 - 5.00}{5.00} \times 100\%$$

$$= 1.8\%$$

Solution II: 0.18 mol HA and 0.020 mol A⁻; initial pH = 4.05

	$OH^-(aq)$	+	$HA(aq)$	$\longrightarrow$	$H_2O(l)$ +	$A^-(aq)$
Before addition	≈0.00 mol		0.18 mol			0.020 mol
Addition	0.010 mol		—			—
After addition	≈0.00 mol		0.17 mol			0.030 mol

$$pH = pK_a + \log\frac{[\text{base}]}{[\text{acid}]}$$

$$= 5.00 + \log\frac{0.030}{0.17}$$

$$= 4.25$$

$$\% \text{ change} = \frac{4.25 - 4.05}{4.05} \times 100\%$$

$$= 5.0\%$$

As you can see, the buffer with equal amounts of acid and conjugate base is more resistant to pH change and is, therefore, the more effective buffer. A buffer becomes less effective as the difference in the relative amounts of acid and conjugate base increases. As a guideline, we can say that an effective buffer must have a [base]/[acid] ratio in the range of 0.10 to 10. In other words, *the relative concentrations of acid and conjugate base should not differ by more than a factor of 10 in order for a buffer to be reasonably effective.*

Absolute Concentrations of the Acid and Conjugate Base

A buffer is most effective (most resistant to pH changes) when the concentrations of acid and conjugate base are highest. We explore this idea by again considering a generic buffer composed of HA and A⁻ and a pK_a of 5.00. Let's calculate the percent change in pH upon addition of 0.010 mol of NaOH for two 1.0-liter solutions of this buffer system. In this case, solution I is 10 times more concentrated in both the acid and the base than solution II. Both solutions have equal relative amounts of acid and conjugate base and therefore have the same initial pH of 5.00.

Solution I: 0.50 mol HA and 0.50 mol A⁻; initial pH = 5.00

	$OH^-(aq)$	+	$HA(aq)$	$\longrightarrow$	$H_2O(l)$ +	$A^-(aq)$
Before addition	≈0.00 mol		0.50 mol			0.50 mol
Addition	0.010 mol		—			—
After addition	≈0.00 mol		0.49 mol			0.51 mol

$$pH = pK_a + \log\frac{[\text{base}]}{[\text{acid}]}$$

$$= 5.00 + \log\frac{0.51}{0.49}$$

$$= 5.02$$

$$\% \text{ change} = \frac{5.02 - 5.00}{5.00} \times 100\%$$

$$= 0.4\%$$

Solution II: 0.050 mol HA and 0.050 mol A⁻; initial pH = 5.00

	$OH^-(aq)$	+	$HA(aq)$	$\longrightarrow$	$H_2O(l)$ +	$A^-(aq)$
Before addition	≈0.00 mol		0.050 mol			0.050 mol
Addition	0.010 mol		—			—
After addition	≈0.00 mol		0.040 mol			0.060 mol

$$pH = pK_a + \log\frac{[\text{base}]}{[\text{acid}]}$$

$$= 5.00 + \log\frac{0.060}{0.040}$$

$$= 5.18$$

$$\% \text{ change} = \frac{5.18 - 5.00}{5.00} \times 100\%$$

$$= 3.6\%$$

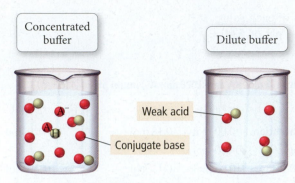

Concentrated buffer

Dilute buffer

Weak acid

Conjugate base

▲ A concentrated buffer has more weak acid and its conjugate base than does a dilute buffer. It can therefore neutralize more added acid or added base.

As you can see, the buffer with greater amounts of acid and conjugate base is most resistant to pH changes and is, therefore, the more effective buffer. The more dilute the buffer components, the less effective the buffer.

Buffer Range

Recall that the relative concentrations of acid and conjugate base should not differ by more than a factor of 10 in order for a buffer to be reasonably effective. In light of this guideline, we can calculate the pH range over which a particular acid and its conjugate base can make an effective buffer. Since the pH of a buffer is given by the Henderson–Hasselbalch equation, we can calculate the outermost points of the effective range as follows:

Lowest pH for effective buffer occurs when the base is one-tenth as concentrated as the acid.

$$pH = pK_a + \log\frac{[base]}{[acid]}$$
$$= pK_a + \log 0.10$$
$$= pK_a - 1$$

Highest pH for effective buffer occurs when the base is 10 times as concentrated as the acid.

$$pH = pK_a + \log\frac{[base]}{[acid]}$$
$$= pK_a + \log 10$$
$$= pK_a + 1$$

Therefore, the effective range for a buffering system is one pH unit on either side of pK_a. For example, we can use a buffering system with a pK_a for the weak acid of 5.0 to prepare a buffer in the range of 4.0–6.0. We can adjust the relative amounts of acid and conjugate base to achieve any pH within this range. As we noted earlier, however, the buffer is most effective at pH 5.0, because the buffer components are exactly equal at that pH.

EXAMPLE 16.5 Buffer Range

Which acid is the best choice to combine with the sodium salt of its conjugate base to make a solution buffered at pH 4.25? For the best choice, calculate the ratio of the conjugate base to the acid required to attain the desired pH.

chlorous acid ($HClO_2$)	$pK_a = 1.95$	formic acid ($HCHO_2$)	$pK_a = 3.74$
nitrous acid (HNO_2)	$pK_a = 3.34$	hypochlorous acid ($HClO$)	$pK_a = 7.54$

SOLUTION

The best choice is formic acid because its pK_a lies closest to the desired pH. You can calculate the required ratio of conjugate base (CHO_2^-) to acid ($HCHO_2$) from the Henderson–Hasselbalch equation as follows:

$$pH = pK_a + \log\frac{[base]}{[acid]}$$

$$4.25 = 3.74 + \log\frac{[base]}{[acid]}$$

$$\log\frac{[base]}{[acid]} = 4.25 - 3.74 = 0.51$$

$$\frac{[base]}{[acid]} = 10^{0.51} = 3.24$$

FOR PRACTICE 16.5

Which acid in Example 16.5 is the best choice to create a buffer with pH = 7.35? If you have 500.0 mL of a 0.10 M solution of the acid, what mass of the corresponding sodium salt of the conjugate base do you need to make the buffer (assuming no change in volume)?

Buffer Capacity

Buffer capacity is the amount of acid or base that we can add to a buffer without destroying its effectiveness. Given what we just learned about the absolute concentrations of acid and conjugate base in an effective buffer, we can conclude that the *buffer capacity increases with increasing absolute concentrations of the buffer components.* The more concentrated the weak acid and conjugate base that compose the buffer are, the higher the buffer capacity. In addition, *overall buffer capacity increases as the relative concentrations of the buffer components become closer to each other.* As the ratio of the buffer components gets closer to 1, the *overall* capacity of the buffer (the ability to neutralize added acid *and* added base) becomes greater. In some cases, however, a buffer that must neutralize primarily added acid (or primarily added base) may be overweighed in one of the buffer components. For example, the main buffering system in human blood plasma contains $[HCO_3^-] = 0.024$ M and $[H_2CO_3] = 0.0012$ M. The concentration of the conjugate base is much greater than the concentration of the acid because the substances that enter the blood are primarily acidic (such as the glycolic acid mentioned in Section 16.1) and the buffer must neutralize them.

 Conceptual Connection 16.4 **Buffer Capacity**

A 1.0-L buffer solution is 0.10 M in HF and 0.050 M in NaF. Which action will destroy the buffer?

(a) adding 0.050 mol of HCl (b) adding 0.050 mol of NaOH
(c) adding 0.050 mol of NaF (d) None of the above

16.4 Titrations and pH Curves

In an **acid–base titration**, a basic (or acidic) solution of unknown concentration reacts with an acidic (or basic) solution of known concentration. We slowly add a known solution to the unknown one while monitoring the pH with either a pH meter or an **indicator** (a substance whose color depends on the pH). As the acid and base combine, they neutralize each other. At the **equivalence point**—the point in the titration when the number of moles of base is stoichiometrically equal to the number of moles of acid—the titration is complete. When the equivalence point is reached, neither reactant is in excess and the number of moles of the reactants are related by the reaction stoichiometry (**Figure 16.5**, on the next page).

In this section, we examine acid–base titrations, concentrating on the pH changes that occur during the titration. A plot of the pH of the solution during a titration is known as a *titration curve* or *pH curve*. A pH curve for the titration of HCl with NaOH is shown in the margin. Before we add any base to the solution, the pH is low (as expected for a solution of HCl). As we add NaOH, the solution becomes less acidic as the NaOH begins to neutralize the HCl. The point of inflection in the middle of the curve is the equivalence point. Notice that the pH changes very quickly near the equivalence point (small amounts of added base cause large changes in pH). Beyond the equivalence point, the solution is basic because the HCl has been completely neutralized and excess base is being added to the solution. The exact shape of the pH curve depends on several factors, including the strength of the acid or base being titrated. In the discussions that follow, we look at several combinations individually.

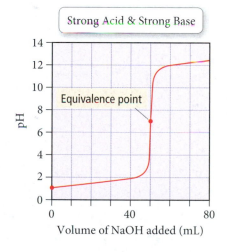

Titration Curve: Strong Acid + Strong Base This curve represents the titration of 50.0 mL of 0.100 M HCl with 0.100 M NaOH.

| The equivalence point is so named because the number of moles of acid and base are equal at this point.

The Titration of a Strong Acid with a Strong Base

Consider the titration of 25.0 mL of 0.100 M HCl with 0.100 M NaOH. We will calculate the volume of base required to reach the equivalence point, as well as the pH at several points along the way.

Volume of NaOH Required to Reach the Equivalence Point
The neutralization reaction that occurs during the titration is:

$$HCl(aq) + NaOH(aq) \longrightarrow H_2O(l) + NaCl(aq)$$

▶ **FIGURE 16.5 Acid–Base Titration**   As we add OH⁻ in a titration, it neutralizes the H⁺, forming water. At the equivalence point, the titration is complete.

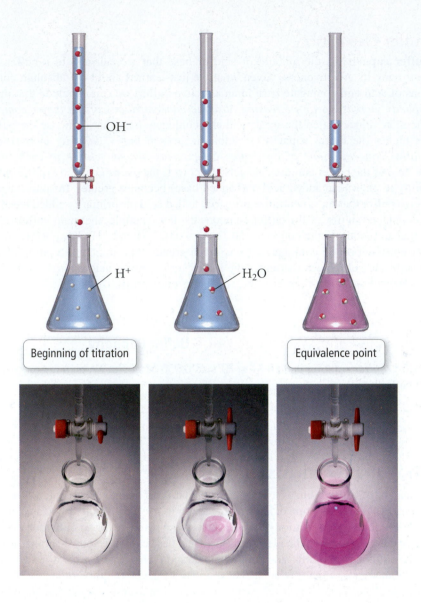

Beginning of titration

H⁺

H₂O

Equivalence point

The equivalence point is reached when the number of moles of base added equals the number of moles of acid initially in solution. We calculate the amount of acid initially in solution as follows:

$$\text{Initial mol HCl} = 0.0250 \ \cancel{L} \times \frac{0.100 \ \text{mol}}{1 \ \cancel{L}} = 0.00250 \ \text{mol HCl}$$

The amount of NaOH that we must add is therefore 0.00250 mol NaOH. We calculate the volume of NaOH required as follows:

$$\text{Volume NaOH solution} = 0.00250 \ \cancel{\text{mol}} \times \frac{1 \ \text{L}}{0.100 \ \cancel{\text{mol}}} = 0.0250 \ \text{L}$$

The equivalence point therefore is reached when 25.0 mL of NaOH has been added.

Initial pH (Before Adding Any Base)
The initial pH of the solution is simply the pH of a 0.100 M HCl solution. Since HCl is a strong acid, the concentration of H_3O^+ is also 0.100 M and the pH is 1.00.

$$pH = -\log\left[H_3O^+\right]$$
$$= -\log(0.100)$$
$$= 1.00$$

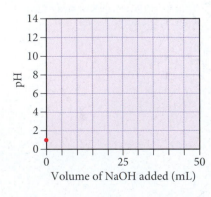

pH After Adding 5.00 mL NaOH

As NaOH is added to the solution, it neutralizes H_3O^+ according to the reaction:

$$OH^-(aq) + H_3O^+(aq) \longrightarrow 2\,H_2O(l)$$

We calculate the amount of H_3O^+ at any given point (before the equivalence point) by using the reaction stoichiometry—1 mol of NaOH neutralizes 1 mol of H_3O^+. The initial number of moles of H_3O^+ (as calculated above) is 0.00250 mol. The number of moles of NaOH added at 5.00 mL is:

$$\text{mol NaOH added} = 0.00500\,\cancel{L} \times \frac{0.100\text{ mol}}{1\,\cancel{L}} = 0.000500\text{ mol NaOH}$$

The addition of OH^- causes the amount of H^+ to decrease as shown in the following table:

	$OH^-(aq)$	$+$	$H_3O^+(aq)$	$\longrightarrow$	$2\,H_2O(l)$
Before addition	≈ 0.00 mol		0.00250 mol		
Addition	0.000500 mol		—		
After addition	≈ 0.00 mol		0.00200 mol		

We then calculate the H_3O^+ concentration by dividing the number of moles of H_3O^+ remaining by the *total volume* (initial volume plus added volume).

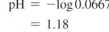

$$[H_3O^+] = \frac{0.00200\text{ mol }H_3O^+}{0.0250\text{ L} + 0.00500\text{ L}} = 0.0667\text{ M}$$

Initial volume   Added volume

The pH is therefore 1.18:

$$pH = -\log 0.0667$$
$$= 1.18$$

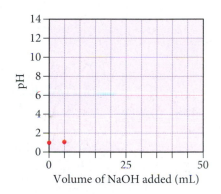

pHs After Adding 10.0, 15.0, and 20.0 mL NaOH

As more NaOH is added, it further neutralizes the H_3O^+ in the solution. We calculate the pH at each of these points in the same way as we did after 5.00 mL of NaOH was added. The results are tabulated as follows:

Volume (mL)	pH
10.0	1.37
15.0	1.60
20.0	1.95

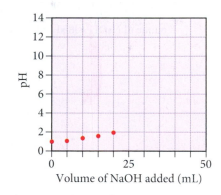

pH After Adding 25.0 mL NaOH (Equivalence Point)

The pH at the equivalence point of a strong acid–strong base titration is always 7.00 (at 25 °C). At the equivalence point, the strong base has completely neutralized the strong acid. The only source of hydronium ions then becomes the autoionization of water. The $[H_3O^+]$ at 25 °C from the autoionization of water is 1.00×10^{-7} and the pH is therefore 7.00.

pH After Adding 30.00 mL NaOH

As NaOH is added beyond the equivalence point, it becomes the excess reagent. We calculate the amount of OH^- at any given point (past the equivalence point) by subtracting the initial amount of H_3O^+ from the amount of OH^- added. The number of moles of OH^- added at 30.00 mL is

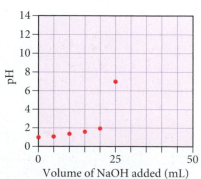

$$\text{mol }OH^-\text{ added} = 0.0300\,\cancel{L} \times \frac{0.100\text{ mol}}{1\,\cancel{L}} = 0.00300\text{ mol }OH^-$$

The number of moles of OH^- remaining after neutralization is shown in the following table:

	$OH^-(aq)$	$+$	$H_3O^+(aq)$	$\longrightarrow$	$2\,H_2O(l)$
Before addition	≈ 0.00 mol		0.00250 mol		
Addition	0.00300 mol		—		
After addition	0.00050 mol		≈ 0.00 mol		

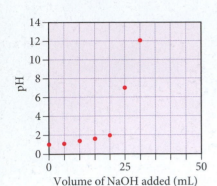

We then calculate the OH^- concentration by dividing the number of moles of OH^- remaining by the *total volume* (initial volume plus added volume).

$$[OH^-] = \frac{0.000500 \text{ mol } OH^-}{0.0250 \text{ L} + 0.0300 \text{ L}} = 0.00909 \text{ M}$$

We can then calculate the $[H_3O^+]$ and pH as follows:

$$[H_3O^+][OH^-] = 10^{-14}$$

$$[H_3O^+] = \frac{10^{-14}}{[OH^-]} = \frac{10^{-14}}{0.00909}$$

$$= 1.10 \times 10^{-12} \text{ M}$$

$$pH = -\log(1.10 \times 10^{-12})$$

$$= 11.96$$

phs After Adding 35.0, 40.0, and 50.0 mL NaOH

As more NaOH is added, it further increases the basicity of the solution. We calcaulte the pH at each of these points in the same way as we did after 30.00 mL of NaOH was added. The results are tabulated below.

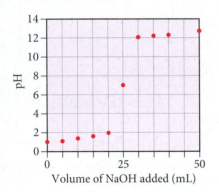

Volume (mL)	pH
35.0	12.22
40.0	12.36
50.0	12.52

The Overall pH Curve

The overall pH curve for the titration of a strong acid with a strong base has the characteristic S-shape we just calculated. The overall curve is shown here:

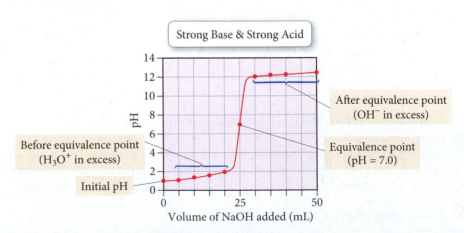

Summarizing Titration of a Strong Acid with a Strong Base:

▶ The initial pH is the pH of the strong acid solution to be titrated.

▶ Before the equivalence point, H_3O^+ is in excess. We calculate the $[H_3O^+]$ by subtracting the number of moles of added OH^- from the initial number of moles of H_3O^+ and dividing by the *total* volume.

▶ At the equivalence point, neither reactant is in excess and the pH = 7.00.

▶ Beyond the equivalence point, OH^- is in excess. We calculate the $[OH^-]$ by subtracting the initial number of moles of H_3O^+ from the number of moles of added OH^- and dividing by the *total* volume.

Figure 16.6▶ is the pH curve for the titration of a strong base with a strong acid. Calculating the points along a curve such as the one in Figure 16.7 is very similar to calculating the points along the pH curve for the titration of a strong acid with a strong base (which we just did). The main difference is that the curve starts basic and then turns acidic after the equivalence point (instead of vice versa).

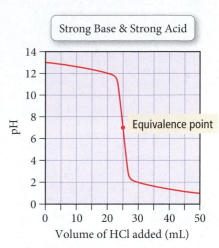

▲ **FIGURE 16.6 Titration Curve: Strong Base + Strong Acid** This curve represents the titration of 25.0 mL of 0.100 M NaOH with 0.100 M HCl.

EXAMPLE 16.6 Strong Acid–Strong Base Titration pH Curve

A 50.0-mL sample of 0.200 M sodium hydroxide is titrated with 0.200 M nitric acid. Calculate:

(a) the pH after adding 30.00 mL of HNO_3

(b) the pH at the equivalence point

Round pH values to two decimal places.

SOLUTION

(a) Begin by calculating the initial amount of NaOH (in moles) from the volume and molarity of the NaOH solution. Since NaOH is a strong base, it dissociates completely, so the amount of OH^- is equal to the amount of NaOH.

$$\text{moles NaOH} = 0.0500 \text{ L} \times \frac{0.200 \text{ mol}}{1 \text{ L}}$$
$$= 0.0100 \text{ mol}$$
$$\text{moles } OH^- = 0.0100 \text{ mol}$$

Calculate the amount of HNO_3 (in moles) added at 30.0 mL from the molarity of the HNO_3 solution.

$$\text{mol } HNO_3 \text{ added} = 0.0300 \text{ L} \times \frac{0.200 \text{ mol}}{1 \text{ L}}$$
$$= 0.00600 \text{ mol } HNO_3$$

As HNO_3 is added to the solution, it neutralizes some of the OH^-. Calculate the number of moles of OH^- remaining by setting up a table based on the neutralization reaction that shows the amount of OH^- before the addition, the amount of H_3O^+ added, and the amounts left after the addition.

	$OH^-(aq)$	+	$H_3O^+(aq)$	⟶	$2 H_2O(l)$
Before addition	0.0100 mol		≈0.00 mol		
Addition	—		0.00600 mol		
After addition	0.0040 mol		≈0.00 mol		

Calculate the OH^- concentration by dividing the amount of OH^- remaining by the *total* volume (initial volume plus added volume).

$$[OH^-] = \frac{0.0040 \text{ mol}}{0.0500 \text{ L} + 0.0300 \text{ L}}$$
$$= 0.0500 \text{ M}$$

Calculate the pOH from $[OH^-]$.

$$pOH = -\log 0.0500$$
$$= 1.30$$

Calculate the pH from the pOH using the equation pH + pOH = 14.

$$pH = 14 - pOH$$
$$= 14 - 1.30$$
$$= 12.70$$

(b) At the equivalence point, the strong base has completely neutralized the strong acid. The $[H_3O^+]$ at 25 °C from the autoionization of water is 1.00×10^{-7} and the pH is therefore 7.00.

$$pH = 7.00$$

FOR PRACTICE 16.6

Calculate the pH in the titration in Example 16.6 after you add a total of 60.0 mL of 0.200 M HNO_3.

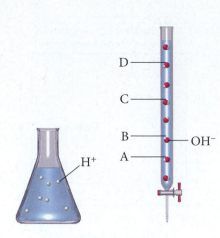

 Conceptual Connection 16.5 Titration Equivalence Point

The amount of acid in the flask on the left is to be titrated by a strong base. Which mark on the burette indicates the amount of base required to reach the equivalence point?

(a) A   **(b)** B   **(c)** C   **(d)** D

The Titration of a Weak Acid with a Strong Base

Consider the titration of 25.0 mL of 0.100 M $HCHO_2$ with 0.100 M NaOH.

$$NaOH(aq) + HCHO_2(aq) \longrightarrow H_2O(l) + NaCHO_2(aq)$$

The concentrations and the volumes here are identical to those in our previous titration, in which we calculated the pH curve for the titration of a *strong* acid with a strong base. The only difference is that $HCHO_2$ is a *weak* acid rather than a strong one. We begin by calculating the volume required to reach the equivalence point of the titration.

Volume of NaOH Required to Reach the Equivalence Point

From the stoichiometry of the equation, we can see that the equivalence point occurs when the amount (in moles) of added base equals the amount (in moles) of acid initially in solution.

$$\text{Initial mol } HCHO_2 = 0.0250 \ \cancel{L} \times \frac{0.100 \text{ mol}}{1 \cancel{L}} = 0.00250 \text{ mol } HCHO_2$$

Thus, the amount of NaOH that must be added is 0.00250 mol NaOH. The volume of NaOH required is

$$\text{Volume NaOH solution} = 0.00250 \ \cancel{\text{mol}} \times \frac{1 \text{ L}}{0.100 \ \cancel{\text{mol}}} = 0.0250 \text{ L NaOH solution}$$

The equivalence point therefore occurs at 25.0 mL of added base. Notice that the volume of NaOH required to reach the equivalence point for this weak acid is identical to that required for a strong acid. *The volume at the equivalence point in an acid–base titration does not depend on whether the acid being titrated is a strong acid or a weak acid, but only on the amount (in moles) of acid present in solution before the titration begins and on the concentration of the added base.*

Initial pH (Before Adding Any Base)

The initial pH of the solution is the pH of a 0.100 M $HCHO_2$ solution. Since $HCHO_2$ is a weak acid, we calculate the concentration of H_3O^+ and the pH by doing an equilibrium problem for the ionization of $HCHO_2$. The full procedure for solving weak acid ionization problems is shown in Examples 15.5 and 15.6. We present a highly condensed calculation here (K_a for $HCHO_2$ is 1.8×10^{-4}).

$$HCHO_2(aq) + H_2O(l) \rightleftharpoons H_3O^+(aq) + CHO_2^-(aq)$$

	$[HCHO_2]$	$[H_3O^+]$	$[CHO_2^-]$
Initial	0.100	≈ 0.00	0.00
Change	$-x$	$+x$	$+x$
Equil	$0.100 - x$	x	x

$$K_a = \frac{[H_3O^+][CHO_2^-]}{[HCHO_2]}$$

$$= \frac{x^2}{0.100 - \cancel{x}} \quad (x \text{ is small})$$

$$1.8 \times 10^{-4} = \frac{x^2}{0.100}$$

$$x = 4.2 \times 10^{-3}$$

Therefore, $[H_3O^+] = 4.2 \times 10^{-3}$ M.

$$pH = -\log(4.2 \times 10^{-3})$$

$$= 2.37$$

Notice that the pH begins at a higher value (less acidic) than it did for a strong acid of the same concentration, as we would expect because the acid is weak.

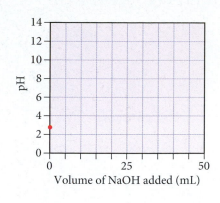

pH After Adding 5.00 mL NaOH

When titrating a *weak acid* with a strong base, the added NaOH *converts a stoichiometric amount of the acid into its conjugate base*. As we calculated previously, 5.00 mL of the 0.100 M NaOH solution contains 0.000500 mol OH$^-$. When we add the 0.000500 mol OH$^-$ to the weak acid solution, OH$^-$ reacts stoichiometrically with HCHO$_2$ causing the amount of HCHO$_2$ *to decrease* by 0.000500 mol and the amount of CHO$_2^-$ to *increase* by 0.000500 mol. This is very similar to what happens when you add strong base to a buffer, and is summarized in the following table:

	OH$^-$(aq)	+	HCHO$_2$(aq)	$\longrightarrow$	H$_2$O(l)	+	CHO$_2^-$(aq)
Before addition	$\approx$0.00 mol		0.00250 mol		—		0.00 mol
Addition	0.000500 mol		—		—		—
After addition	$\approx$0.00 mol		0.00200 mol		—		0.000500 mol

Notice that, after the addition, the solution contains significant amounts of both an acid (HCHO$_2$) and its conjugate base (CHO$_2^-$)—*the solution is now a buffer*. To calculate the pH of a buffer (when the *x is small* approximation applies, as it does here), we can use the Henderson–Hasselbalch equation and pK_a for HCHO$_2$ (which is 3.74).

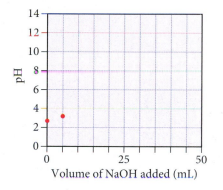

$$pH = pK_a + \log\frac{[\,base\,]}{[\,acid\,]}$$

$$= 3.74 + \log\frac{0.000500}{0.00200}$$

$$= 3.74 - 0.60$$

$$= 3.14$$

pHs After Adding 10.0, 12.5, 15.0, and 20.0 mL NaOH

As more NaOH is added, it converts more HCHO$_2$ into CHO$_2^-$. We calculate the relative amounts of HCHO$_2$ and CHO$_2^-$ at each of these volumes using the reaction stoichiometry, and then calculate the pH of the resulting buffer using the Henderson–Hasselbalch equation (as we did for the pH at 5.00 mL). The amounts of HCHO$_2$ and CHO$_2^-$ (after addition of the OH$^-$) at each volume and the corresponding pHs are tabulated as follows:

Volume (mL)	mol HCHO$_2$	mol CHO$_2^-$	pH
10.0	0.00150	0.00100	3.56
12.5	0.00125	0.00125	3.74
15.0	0.00100	0.00150	3.92
20.0	0.00050	0.00200	4.34

Notice that, as the titration proceeds, more of the HCHO$_2$ is converted to the conjugate base (CHO$_2^-$). Notice also that an added NaOH volume of 12.5 mL corresponds to exactly one-half of the equivalence point. At this volume, exactly one-half of the initial amount of HCHO$_2$ has been converted to CHO$_2^-$, resulting in *exactly equal amounts of weak acid and conjugate base*. For any buffer in which the amounts of weak acid and conjugate base are equal, the pH = pK_a as shown here:

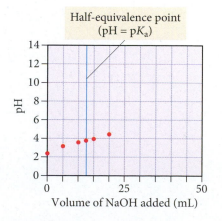

Half-equivalence point (pH = pK_a)

$$pH = pK_a + \log\frac{[\,base\,]}{[\,acid\,]}$$

$$\text{If } [\text{base}] = [\text{acid}], \text{ then } \frac{[\text{base}]}{[\text{acid}]} = 1.$$

$$\begin{aligned}
pH &= pK_a + \log 1 \\
&= pK_a + 0 \\
&= pK_a
\end{aligned}$$

pH After Adding 25.0 mL NaOH (Equivalence Point)

At the equivalence point, we have added 0.000250 mol of OH^- and have therefore converted all of the $HCHO_2$ into its conjugate base (CHO_2^-) as tabulated here.

	$OH^-(aq)$	+	$HCHO_2(aq)$	$\longrightarrow$	$H_2O(l)$	+	$CHO_2^-(aq)$
Before addition	≈ 0.00 mol		0.00250 mol		—		0.00 mol
Addition	0.00250 mol		—		—		—
After addition	≈ 0.00 mol		0.00 mol		—		0.00250 mol

The solution is no longer a buffer (it no longer contains significant amounts of both a weak acid and its conjugate base). Instead, the solution is just that of an ion (CHO_2^-) acting as a weak base. Recall that we discussed how to calculate the pH of solutions such as this in Section 15.8 (see Example 15.13) by solving an equilibrium problem involving the ionization of water by the weak base (CHO_2^-):

$$CHO_2^-(aq) + H_2O(l) \rightleftharpoons HCHO_2(aq) + OH^-(aq)$$

We calculate the initial concentration of CHO_2^- for the equilibrium problem by dividing the number of moles of CHO_2^- (0.00250 mol) by the *total* volume at the equivalence point (initial volume plus added volume).

Moles CHO_2^- at equivalence point

$$[CHO_2^-] = \frac{0.00250 \text{ mol}}{0.0250 \text{ L} + 0.0250 \text{ L}} = 0.0500 \text{ M}$$

Initial volume   Added volume at equivalence point

We then proceed to solve the equilibrium problem as shown in condensed form here:

$$CHO_2^-(aq) + H_2O(l) \rightleftharpoons HCHO_2(aq) + OH^-(aq)$$

	$[CHO_2^-]$	$[HCHO_2]$	$[OH^-]$
Initial	0.0500	0.00	≈ 0.00
Change	$-x$	$+x$	$+x$
Equil	$0.0500 - x$	x	x

Before substituting into the expression for K_b, we find the value of K_b from K_a for formic acid ($K_a = 1.8 \times 10^{-4}$) and K_w:

$$K_a \times K_b = K_w$$

$$K_b = \frac{K_w}{K_a} = \frac{1.0 \times 10^{-14}}{1.8 \times 10^{-4}} = 5.6 \times 10^{-11}$$

Now we can substitute the equilibrium concentrations from the table above into the expression for K_b as follows:

$$K_b = \frac{[HCHO_2][OH^-]}{[CHO_2^-]}$$

$$= \frac{x^2}{0.0500 - x} \quad (x \text{ is small})$$

$$5.6 \times 10^{-11} = \frac{x^2}{0.0500}$$

$$x = 1.7 \times 10^{-6}$$

Remember that x represents the concentration of the hydroxide ion. We then calculate $[H_3O^+]$ and pH:

$$[OH^-] = 1.7 \times 10^{-6} \, M$$

$$[H_3O^+][OH^-] = K_w = 1.0 \times 10^{-14}$$

$$[H_3O^+](1.7 \times 10^{-6}) = 1.0 \times 10^{-14}$$

$$[H_3O^+] = 5.9 \times 10^{-9} \, M$$

$$pH = -\log[H_3O^+]$$

$$= -\log(5.9 \times 10^{-9})$$

$$= 8.23$$

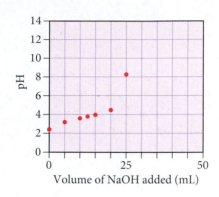

Notice that the pH at the equivalence point is *not* neutral but basic. *The titration of weak acid by a strong base will always have a basic equivalence point* because, at the equivalence point, all of the acid has been converted into its conjugate base, resulting in a weakly basic solution.

pH After Adding 30.00 mL NaOH

At this point in the titration, we have added 0.00300 mol of OH^-. NaOH has now become the excess reagent as shown in the following table:

	$OH^-(aq)$	$+$	$HCHO_2(aq)$	$\longrightarrow$	$H_2O(l)$	$+$	$CHO_2^-(aq)$
Before addition	≈0.00 mol		0.00250 mol		—		0.00 mol
Addition	0.00300 mol		—		—		—
After addition	0.00050 mol		≈0.00 mol		—		0.00250 mol

The solution becomes a mixture of a strong base (NaOH) and a weak base (CHO_2^-). The strong base completely overwhelms the weak base and we can calculate the pH by considering the strong base alone (as we did for the titration of a strong acid and a strong base). The OH^- concentration is computed by dividing the amount of OH^- remaining by the *total volume* (initial volume plus added volume).

$$[OH^-] = \frac{0.00050 \text{ mol } OH^-}{0.0250 \text{ L} + 0.0300 \text{ L}} = 0.0091 \, M$$

We can then calculate the $[H_3O^+]$ and pH:

$$[H_3O^+][OH^-] = 10^{-14}$$

$$[H_3O^+] = \frac{10^{-14}}{[OH^-]} = \frac{10^{-14}}{0.0091} = 1.1 \times 10^{-12} \, M$$

$$pH = -\log(1.1 \times 10^{-12})$$

$$= 11.96$$

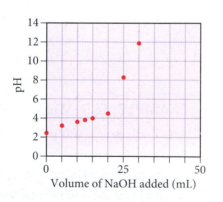

pHs After Adding 35.0, 40.0, and 50.0 mL NaOH

As more NaOH is added, the basicity of the solution increases further. We can calculate the pH at each of these volumes in the same way as we did after 30.00 mL of NaOH was added. The results are tabulated as follows:

Volume (mL)	pH
35.0	12.22
40.0	12.36
50.0	12.52

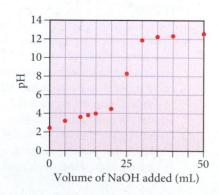

The Overall pH Curve

The overall pH curve for the titration of a weak acid with a strong base has the characteristic S-shape similar to that for the titration of a strong acid with a strong base. The main difference is that the equivalence point pH is basic (not neutral). Notice that calculating the pH in different regions throughout the titration involves working different kinds of acid–base problems, all of which we have encountered before.

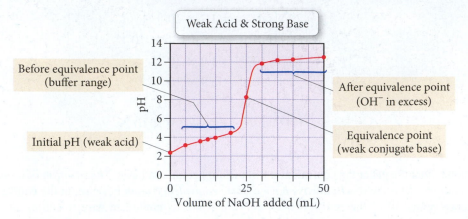

Summarizing Titration of a Weak Acid with a Strong Base:

▶ The initial pH is that of the weak acid solution to be titrated. We calculate the pH by working an equilibrium problem (similar to Examples 15.5 and 15.6) using the concentration of the weak acid as the initial concentration.

▶ Between the initial pH and the equivalence point, the solution becomes a buffer. We use the reaction stoichiometry to calculate the amounts of each buffer component and then use the Henderson–Hasselbalch equation to calculate the pH (as in Example 16.3).

▶ Halfway to the equivalence point, the buffer components are exactly equal and $pH = pK_a$.

▶ At the equivalence point, the acid has all been converted into its conjugate base. We calculate the pH by working an equilibrium problem for the ionization of water by the ion acting as a weak base (similar to Example 15.13). (We calculate the concentration of the ion acting as a weak base by dividing the number of moles of the ion by the total volume at the equivalence point.)

▶ Beyond the equivalence point, OH^- is in excess. We ignore the weak base and calculate the $[OH^-]$ by subtracting the initial number of moles of H_3O^+ from the number of moles of added OH^- and dividing by the *total* volume.

EXAMPLE 16.7 Weak Acid–Strong Base Titration pH Curve

A 40.0-mL sample of 0.100 M HNO_2 is titrated with 0.200 M KOH. Calculate:

(a) the volume required to reach the equivalence point

(b) the pH after adding 5.00 mL of KOH

(c) the pH at one-half the equivalence point

SOLUTION

(a) The equivalence point occurs when the amount (in moles) of added base equals the amount (in moles) of acid initially in the solution. Begin by calculating the amount (in moles) of acid initially in the solution. The amount (in moles) of KOH that must be added is equal to the amount of the weak acid.	$$\text{mol } HNO_2 = 0.0400 \text{ L} \times \frac{0.100 \text{ mol}}{\text{L}}$$ $$= 4.00 \times 10^{-3} \text{ mol}$$ $$\text{mol KOH required} = 4.00 \times 10^{-3} \text{ mol}$$

Calculate the volume of KOH required from the number of moles of KOH and the molarity.	volume KOH solution $= 4.00 \times 10^{-3} \, \text{mol} \times \dfrac{1 \, \text{L}}{0.200 \, \text{mol}}$

$$= 0.0200 \, \text{L KOH solution}$$
$$= 20.0 \, \text{mL KOH solution}$$

(b) Use the concentration of the KOH solution to calculate the amount (in moles) of OH^- in 5.00 mL of the solution.

$$\text{mol } OH^- = 5.00 \times 10^{-3} \, \text{L} \times \dfrac{0.200 \, \text{mol}}{1 \, \text{L}}$$
$$= 1.00 \times 10^{-3} \, \text{mol } OH^-$$

Prepare a table showing the amounts of HNO_2 and NO_2^- before and after the addition of 5.00 mL KOH. The addition of the KOH stoichiometrically reduces the concentration of HNO_2 and increases the concentration of NO_2^-.

	$OH^-(aq)$	$+$	$HNO_2(aq)$	$\longrightarrow$ $H_2O(l)$	$+$	$NO_2^-(aq)$
Before addition	≈ 0.00 mol		4.00×10^{-3} mol			0.00 mol
Addition	1.00×10^{-3} mol		—			—
After addition	≈ 0.00 mol		3.00×10^{-3} mol			1.00×10^{-3} mol

Since the solution now contains significant amounts of a weak acid and its conjugate base, use the Henderson–Hasselbalch equation and pK_a for HNO_2 (which is 3.34) to calculate the pH of the solution.

$$pH = pK_a + \log \dfrac{[\text{base}]}{[\text{acid}]}$$
$$= 3.34 + \log \dfrac{1.00 \times 10^{-3}}{3.00 \times 10^{-3}}$$
$$= 3.34 - 0.48 = 2.86$$

(c) At one-half the equivalence point, the amount of added base is exactly one-half the initial amount of acid. The base converts exactly half of the HNO_2 into NO_2^-, resulting in equal amounts of the weak acid and its conjugate base. The pH is therefore equal to pK_a.

	$OH^-(aq)$	$+$	$HNO_2(aq)$	$\longrightarrow$ $H_2O(l)$	$+$	$NO_2^-(aq)$
Before addition	≈ 0.00 mol		4.00×10^{-3} mol			0.00 mol
Addition	2.00×10^{-3} mol		—			—
After addition	≈ 0.00 mol		2.00×10^{-3} mol			2.00×10^{-3} mol

$$pH = pK_a + \log \dfrac{[\text{base}]}{[\text{acid}]}$$
$$= 3.34 + \log \dfrac{2.00 \times 10^{-3}}{2.00 \times 10^{-3}}$$
$$= 3.34 + 0 = 3.34$$

FOR PRACTICE 16.7

Determine the pH at the equivalence point for the titration between HNO_2 and KOH in Example 16.7.

The pH curve for the titration of a weak base with a strong acid is shown in **Figure 16.7▶**. Calculating the points along this curve is very similar to calculating the points along the pH curve for the titration of a weak acid with a strong base (which we just did). The main differences are that the curve starts basic and has an acidic equivalence point. We calculate the pH in the buffer region using $pH = pK_a + \log [\text{base}]/[\text{acid}]$ where the pK_a corresponds to the conjugate acid of the base being titrated.

Conceptual Connection 16.6 The Half-Equivalence Point

What is the pH at the half-equivalence point in the titration of a weak base with a strong acid? The pK_b of the weak base is 8.75.

(a) 8.75   **(b)** 7.0

(c) 5.25   **(d)** 4.37

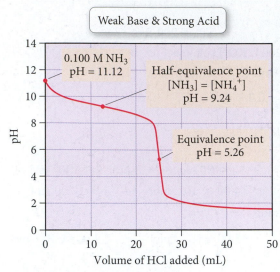

▲ **FIGURE 16.7 Titration Curve: Weak Base with Strong Acid** Curve for the titration 25.0 mL of 0.100 M NH_3 with 0.100 M HCl.

Indicators: pH-Dependent Colors

We can monitor the pH of a titration with either a pH meter or an indicator. The direct monitoring of pH with a meter yields data like the pH curves we have examined in this chapter, allowing determination of the equivalence point from the pH curve itself. With an indicator, we rely on the point where the indicator changes color—called the **endpoint**—to determine the equivalence point, as shown in **Figure 16.8▼**. With the correct indicator, the endpoint of the titration (indicated by the color change) occurs very near to the equivalence point (when the amount of acid equals the amount of base).

An indicator is itself a weak organic acid that is a different color than its conjugate base. For example, phenolphthalein is a common indicator whose acid form is colorless and conjugate base form is pink as shown in **Figure 16.9▶**. If we let HIn represent the acid form of a generic indicator and In^- the conjugate base form, we have the following equilibrium:

$$HIn(aq) + H_2O(l) \rightleftharpoons H_3O^+(aq) + In^-(aq)$$

$$\text{color 1} \qquad\qquad\qquad \text{color 2}$$

Because the color of an indicator is intense, only a small amount of indicator is required—an amount that will not affect the pH of the solution or the equivalence point of the neutralization reaction. When the $[H_3O^+]$ changes during the titration, the equilibrium shifts in response. At low pH, the $[H_3O^+]$ is high and the equilibrium lies far to the left, resulting in a solution of color 1. As the titration proceeds, however, the $[H_3O^+]$ decreases, shifting the equilibrium to the right. Since the pH change is large near the equivalence point of the titration, there is a large change in $[H_3O^+]$ near the equivalence point. Provided that the correct indicator is chosen, there will also be a correspondingly large change in color. For the titration of a strong acid with a strong base, one drop of the base near the endpoint is usually enough to change the color of the indicator from color 1 to color 2.

The color of a solution containing an indicator depends on the relative concentrations of HIn and In^-. When the pH of the solution equals the pK_a of the indicator, the solution will have an intermediate color. When the pH is 1 unit (or more) above pK_a, the indicator will be the color of In^-, and when the pH is 1 unit (or more) below pK_a, the indicator will be the color of HIn. Therefore, the indicator changes color within a range of two pH units centered at pK_a. Table 16.1 shows various indicators and their colors as a function of pH.

Using an Indicator

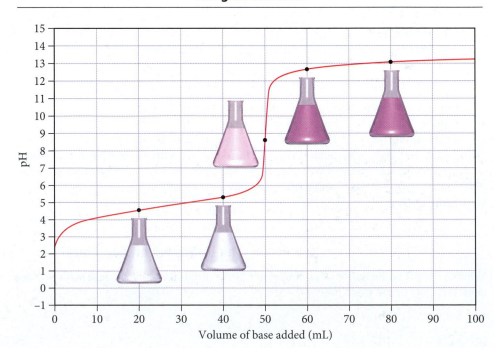

▶ **FIGURE 16.8 Monitoring the Color Change during a Titration** Titration of 50.0 mL of 0.100 M $HC_2H_3O_2$ with 0.100 M NaOH. We can detect the endpoint of a titration by a color change in an appropriate indicator (in this case, phenolphthalein).

Phenolphthalein, a Common Indicator

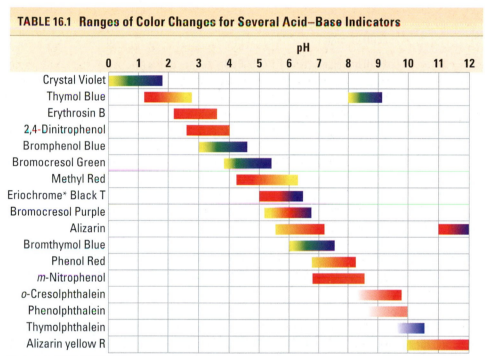

▲ **FIGURE 16.9 Phenolphthalein**   Phenolphthalein, a weakly acidic compound, is colorless. Its conjugate base is pink.

TABLE 16.1 Ranges of Color Changes for Several Acid–Base Indicators

	pH
	0 1 2 3 4 5 6 7 8 9 10 11 12
Crystal Violet	
Thymol Blue	
Erythrosin B	
2,4-Dinitrophenol	
Bromphenol Blue	
Bromocresol Green	
Methyl Red	
Eriochrome* Black T	
Bromocresol Purple	
Alizarin	
Bromthymol Blue	
Phenol Red	
m-Nitrophenol	
o-Cresolphthalein	
Phenolphthalein	
Thymolphthalein	
Alizarin yellow R	

*Trademark of CIBA GEIGY CORP.

16.5   Solubility Equilibria and the Solubility Product Constant

In Chapter 4 we learned that a compound is considered *soluble* if it dissolves in water and *insoluble* if it does not. Recall also that, through the *solubility rules* (see Table 4.1), we classified ionic compounds simply as soluble or insoluble. Now we have the tools to examine *degrees* of solubility.

We can better understand the solubility of an ionic compound by applying the concept of equilibrium to the process of dissolution. For example, we can represent the dissolution of calcium fluoride in water with the chemical equation:

$$CaF_2(s) \rightleftharpoons Ca^{2+}(aq) + 2\,F^-(aq)$$

The equilibrium expression for a chemical equation representing the dissolution of an ionic compound is called the **solubility product constant** (K_{sp}). For CaF_2, the expression of the solubility product constant is

$$K_{sp} = [Ca^{2+}][F^-]^2$$

TABLE 16.2 Selected Solubility Product Constants (K_{sp})

Compound	Formula	K_{sp}
Barium fluoride	BaF_2	2.45×10^{-5}
Barium sulfate	$BaSO_4$	1.07×10^{-10}
Calcium carbonate	$CaCO_3$	4.96×10^{-9}
Calcium fluoride	CaF_2	1.46×10^{-10}
Calcium hydroxide	$Ca(OH)_2$	4.68×10^{-6}
Calcium sulfate	$CaSO_4$	7.10×10^{-5}
Copper(II) sulfide*	CuS	1.27×10^{-36}
Iron(II) carbonate	$FeCO_3$	3.07×10^{-11}
Iron(II) hydroxide	$Fe(OH)_2$	4.87×10^{-17}
Iron(II) sulfide*	FeS	3.72×10^{-19}
Lead(II) chloride	$PbCl_2$	1.17×10^{-5}
Lead(II) bromide	$PbBr_2$	4.67×10^{-6}
Lead(II) sulfate	$PbSO_4$	1.82×10^{-8}
Lead(II) sulfide*	PbS	9.04×10^{-29}
Magnesium carbonate	$MgCO_3$	6.82×10^{-6}
Magnesium hydroxide	$Mg(OH)_2$	2.06×10^{-13}
Silver chloride	$AgCl$	1.77×10^{-10}
Silver chromate	Ag_2CrO_4	1.12×10^{-12}
Silver bromide	$AgBr$	5.35×10^{-13}
Silver iodide	AgI	8.51×10^{-17}

*Sulfide equilibrium is of the type: $MS(s) + H_2O(l) \rightleftharpoons M^{2+}(aq) + HS^-(aq) + OH^-(aq)$

Recall from Section 14.5 that solids are omitted from the equilibrium expression because the concentration of a solid is constant (it is determined by its density and does not change when we add more solid).

The value of K_{sp} is a measure of the solubility of a compound. Table 16.2 lists the values of K_{sp} for a number of ionic compounds. A more complete listing can be found in Appendix IIC.

K_{sp} and Molar Solubility

Recall from Section 12.2 that the *solubility* of a compound is the quantity of the compound that dissolves in a certain amount of liquid. The **molar solubility** is the solubility in units of moles per liter (mol/L). The molar solubility of a compound can be computed directly from K_{sp}. For example, consider silver chloride:

$$AgCl(s) \rightleftharpoons Ag^+(aq) + Cl^-(aq) \qquad K_{sp} = 1.77 \times 10^{-10}$$

First, notice that K_{sp} is *not* the molar solubility, but the *solubility product constant*. The solubility product constant has only one value at a given temperature. The solubility, however, can have different values. For example, due to the common ion effect, the solubility of AgCl in pure water is different from its solubility in an NaCl solution, even though the solubility product constant for AgCl is the same for both solutions. Second, notice that the solubility of AgCl is directly related (by the reaction stoichiometry) to the amount of Ag^+ or Cl^- present in solution when equilibrium is reached. Consequently, finding molar solubility from K_{sp} involves solving an equilibrium problem. For AgCl, we set up an ICE table for the dissolution of AgCl into its ions in pure water:

$$AgCl(s) \rightleftharpoons Ag^+(aq) + Cl^-(aq)$$

	$[Ag^+]$	$[Cl^-]$
Initial	0.00	0.00
Change	$+S$	$+S$
Equil	S	S

We let S represent the concentration of AgCl that dissolves (which is the molar solubility), and then represent the concentrations of the ions formed in terms of S. In this case, for every 1 mol of AgCl that dissolves, 1 mol of Ag^+ and 1 mol of Cl^- are produced. Therefore, the concentrations of Ag^+ or Cl^- present in solution are equal to S. Substituting the equilibrium concentrations of Ag^+ and Cl^- into the expression for the solubility product constant, we get

> Alternatively, the variable x can be used in place of S, as it was for other equilibrium calculations.

$$K_{sp} = [Ag^+][Cl^-]$$
$$= S \times S$$
$$= S^2$$

Therefore,

$$S = \sqrt{K_{sp}}$$
$$= \sqrt{1.77 \times 10^{-10}}$$
$$= 1.33 \times 10^{-5}\ M$$

So the molar solubility of AgCl is 1.33×10^{-5} mol per liter.

EXAMPLE 16.8   Calculating Molar Solubility from K_{sp}

Calculate the molar solubility of $PbCl_2$ in pure water.

SOLUTION

Begin by writing the reaction by which solid $PbCl_2$ dissolves into its constituent aqueous ions and write the corresponding expression for K_{sp}.	$PbCl_2(s) \rightleftharpoons Pb^{2+}(aq) + 2\ Cl^-(aq)$ $K_{sp} = [Pb^{2+}][Cl^-]^2$
Use the stoichiometry of the reaction to prepare an ICE table, showing the equilibrium concentrations of Pb^{2+} and Cl^- relative to S, the amount of $PbCl_2$ that dissolves.	$PbCl_2(s) \rightleftharpoons Pb^{2+}(aq) + 2\ Cl^-(aq)$ (table below)
Substitute the equilibrium expressions for $[Pb^{2+}]$ and $[Cl^-]$ from the previous step into the expression for K_{sp}.	$K_{sp} = [Pb^{2+}][Cl^-]^2$ $= S(2S)^2$ $= 4S^3$
Solve for S and substitute the numerical value of K_{sp} (from Table 16.2) to calculate S.	Therefore, $S = \sqrt[3]{\dfrac{K_{sp}}{4}}$ $S = \sqrt[3]{\dfrac{1.17 \times 10^{-5}}{4}}$ $= 1.43 \times 10^{-2}\ M$

	$[Pb^{2+}]$	$[Cl^-]$
Initial	0.00	0.00
Change	$+S$	$+2S$
Equil	S	$2S$

FOR PRACTICE 16.8
Calculate the molar solubility of $Fe(OH)_2$ in pure water.

EXAMPLE 16.9 Calculating K_{sp} from Molar Solubility

The molar solubility of Ag_2SO_4 in pure water is 1.2×10^{-5} M. Calculate K_{sp}.

SOLUTION

Begin by writing the reaction by which solid Ag_2SO_4 dissolves into its constituent aqueous ions and write the corresponding expression for K_{sp}.	$Ag_2SO_4(s) \rightleftharpoons 2\,Ag^+(aq) + SO_4^{2-}(aq)$ $K_{sp} = [Ag^+]^2[SO_4^{2-}]$
Use an ICE table to define $[Ag^+]$ and $[SO_4^{2-}]$ in terms of S, the amount of Ag_2SO_4 that dissolves.	$Ag_2SO_4(s) \rightleftharpoons 2\,Ag^+(aq) + SO_4^{2-}(aq)$

	$[Ag^+]$	$[SO_4^{2-}]$
Initial	0.00	0.00
Change	$+2S$	$+S$
Equil	$2S$	S

Substitute the expressions for $[Ag^+]$ and $[SO_4^{2-}]$ from the previous step into the expression for K_{sp}. Substitute the given value *of the molar solubility for S* and compute K_{sp}.	$\begin{aligned} K_{sp} &= [Ag^+]^2[SO_4^{2-}] \\ &= (2S)^2 S \\ &= 4S^3 \\ &= 4(1.2 \times 10^{-5})^3 \\ &= 6.9 \times 10^{-15} \end{aligned}$

FOR PRACTICE 16.9

The molar solubility of AgBr in pure water is 7.3×10^{-7} M. Calculate K_{sp}.

K_{sp} and Relative Solubility

As we have just seen, molar solubility and K_{sp} are related, and one can be calculated from the other; however, we cannot always use the K_{sp} values of two different compounds directly to compare their relative solubilities. For example, consider the following compounds, their K_{sp} values, and their molar solubilities:

Compound	K_{sp}	Solubility
$Mg(OH)_2$	2.06×10^{-13}	3.72×10^{-5} M
$FeCO_3$	3.07×10^{-11}	5.54×10^{-6} M

Magnesium hydroxide has a smaller K_{sp} than iron(II) carbonate, but a higher molar solubility. Why? The relationship between K_{sp} and molar solubility depends on the stoichiometry of the dissociation reaction. Consequently, any direct comparison of K_{sp} values for different compounds can only be made if the compounds have the same dissociation stoichiometry. Consider the following compounds with the same dissociation stoichiometry, their K_{sp} values, and their molar solubilities:

Compound	K_{sp}	Solubility
$Mg(OH)_2$	2.06×10^{-13}	3.72×10^{-5} M
CaF_2	1.46×10^{-10}	3.32×10^{-4} M

In this case, magnesium hydroxide and calcium fluoride have the same dissociation stoichiometry (1 mol of each compound produces 3 mol of dissolved ions); therefore, the K_{sp} values can be directly compared as a measure of relative solubility.

The Effect of a Common Ion on Solubility

How is the solubility of an ionic compound affected when the compound is dissolved in a solution that already contains one of its ions? For example, what is the solubility of CaF_2 in a solution that is 0.100 M in NaF? The change in solubility can be explained by the common ion effect, which we first encountered in Section 16.2. We can represent the dissociation of CaF_2 in a 0.100 M NaF solution as follows:

Common ion
0.100 M F^- *(aq)*

$CaF_2(s) \rightleftharpoons Ca^{2+}(aq) + 2\,F^-(aq)$

Equilibrium shifts left

In accordance with Le Châtelier's principle, the presence of the F^- ion in solution causes the equilibrium to shift to the left (compared to its position in pure water), which means that less CaF_2 dissolves—that is, its solubility is decreased. In general,

> **The solubility of an ionic compound is lower in a solution containing a common ion than in pure water.**

We can calculate the exact value of the solubility by working an equilibrium problem in which the concentration of the common ion is accounted for in the initial conditions, as shown in Example 16.10.

EXAMPLE 16.10 Calculating Molar Solubility in the Presence of a Common Ion

What is the molar solubility of CaF_2 in a solution containing 0.100 M NaF?

SOLUTION

Begin by writing the reaction by which solid CaF_2 dissolves into its constituent aqueous ions and write the corresponding expression for K_{sp}.	$CaF_2(s) \rightleftharpoons Ca^{2+}(aq) + 2\,F^-(aq)$ $K_{sp} = [Ca^{2+}][F^-]^2$

Use the stoichiometry of the reaction to prepare an ICE table showing the initial concentration of the common ion. Fill in the equilibrium concentrations of Ca^{2+} and F^- relative to S, the amount of CaF_2 that dissolves.

$CaF_2(s) \rightleftharpoons Ca^{2+}(aq) + 2\,F^-(aq)$

	$[Ca^{2+}]$	$[F^-]$
Initial	0.00	0.100
Change	$+S$	$+2S$
Equil	S	$0.100 + 2S$

Substitute the equilibrium expressions for $[Ca^{2+}]$ and $[F^-]$ from the previous step into the expression for K_{sp}. Since K_{sp} is small, we can make the approximation that $2S$ is much less than 0.100 and will therefore be insignificant when added to 0.100 (this is similar to the *x is small* approximation that we have made for many equilibrium problems).

$$K_{sp} = [Ca^{2+}][F^-]^2$$
$$= S(0.100 + 2S)^2 \ (S \text{ is small})$$
$$= S(0.100)^2$$

Solve for S and substitute the numerical value of K_{sp} (from Table 16.2) to calculate S.

$$K_{sp} = S(0.100)^2$$

$$S = \frac{K_{sp}}{0.0100}$$

Note that the calculated value of S is indeed small compared to 0.100, so our approximation is valid.

$$= \frac{1.46 \times 10^{-10}}{0.0100}$$
$$= 1.46 \times 10^{-8}\,\text{M}$$

For comparison, the molar solubility of CaF_2 in pure water is 3.32×10^{-4} M, which means CaF_2 is over 20,000 times more soluble in water than in the NaF solution. (Confirm this for yourself by calculating the solubility in pure water from the value of K_{sp}.)

FOR PRACTICE 16.10

Calculate the molar solubility of CaF_2 in a solution containing 0.250 M $Ca(NO_3)_2$.

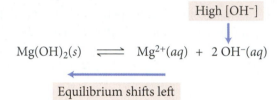 **Conceptual Connection 16.7** **Common Ion Effect**

In which solution is $BaSO_4$ most soluble?

(a) In a solution that is 0.10 M in $BaNO_3$

(b) In a solution that is 0.10 M in Na_2SO_4

(c) In a solution that is 0.10 M in $NaNO_3$

The Effect of pH on Solubility

The pH of a solution can affect the solubility of a compound in that solution. For example, consider the dissociation of $Mg(OH)_2$, the active ingredient in milk of magnesia:

$$Mg(OH)_2(s) \rightleftharpoons Mg^{2+}(aq) + 2\,OH^-(aq)$$

The solubility of this compound is highly dependent on the pH of the solution into which it dissolves. If the pH is high, then the concentration of OH^- is high. In accordance with the common ion effect, this shifts the equilibrium to the left, lowering the solubility.

High $[OH^-]$

$$Mg(OH)_2(s) \rightleftharpoons Mg^{2+}(aq) + 2\,OH^-(aq)$$

Equilibrium shifts left

If the pH is low, then the concentration of $H_3O^+(aq)$ is high. As the $Mg(OH)_2$ dissolves, these H_3O^+ ions neutralize the newly dissolved OH^- ions, driving the reaction to the right.

H_3O^+ reacts with OH^-

$$Mg(OH)_2(s) \rightleftharpoons Mg^{2+}(aq) + 2\,OH^-(aq)$$

Equilibrium shifts right

Consequently, the solubility of $Mg(OH)_2$ in an acidic solution is higher than in a pH-neutral solution. In general,

> **The solubility of an ionic compound with a strongly basic or weakly basic anion increases with increasing acidity (decreasing pH).**

Common basic anions include OH^-, S^{2-}, and CO_3^{2-}. Therefore, hydroxides, sulfides, and carbonates are more soluble in acidic water than in pure water. Since rainwater is naturally acidic due to dissolved carbon dioxide, it can dissolve rocks high in limestone ($CaCO_3$) as it flows through the ground, sometimes resulting in huge underground caverns such as those at Carlsbad Caverns National Park in New Mexico. Dripping water saturated in $CaCO_3$ within the cave creates the dramatic mineral formations known as stalagmites and stalactites.

▲ Stalactites (which hang from the ceiling) and stalagmites (which grow up from the ground) form as calcium carbonate precipitates out of water evaporating in underground caves.

EXAMPLE 16.11 The Effect of pH on Solubility

Determine if each compound is more soluble in an acidic solution than in a neutral solution.

(a) BaF_2 **(b)** AgI **(c)** $Ca(OH)_2$

SOLUTION

(a) The solubility of BaF_2 will be greater in acidic solution because the F^- ion is a weak base. (F^- is the conjugate base of the weak acid HF, and is therefore a weak base.)

(b) The solubility of AgI will not be greater in acidic solution because the I^- is *not* a base. (I^- is the conjugate base of the *strong* acid HI, and is therefore pH-neutral.)

(c) The solubility of $Ca(OH)_2$ will be greater in acidic solution because the OH^- ion is a strong base.

16.6 Precipitation

In Chapter 4, we learned that a precipitation reaction can occur upon the mixing of two solutions containing ionic compounds. The precipitation reaction occurs when one of the possible cross products—the combination of a cation from one solution and the anion from the other—is insoluble. In this chapter, however, we have seen that the terms soluble and insoluble are extremes in a continuous range of solubility—many compounds are slightly soluble and even those that we categorized as insoluble in Chapter 4 actually have some limited degree of solubility (they have very small solubility product constants).

We can better understand precipitation reactions by revisiting a concept from Chapter 14—the reaction quotient (Q). The reaction quotient for the reaction by which an ionic compound dissolves is the product of the concentrations of the ionic components raised to their stoichiometric coefficients. For example, consider the reaction by which CaF_2 dissolves:

$$CaF_2(s) \rightleftharpoons Ca^{2+}(aq) + 2\,F^-(aq)$$

The reaction quotient for this reaction is

$$Q = [Ca^{2+}][F^-]^2$$

The difference between Q and K_{sp} is that K_{sp} is the value of this product *at equilibrium only*, whereas Q is the value of the product under any conditions. We can therefore use the value of Q to compare a solution containing any concentrations of the component ions to one that is at equilibrium as follows:

- If $Q < K_{sp}$, the solution is unsaturated and more of the solid ionic compound can dissolve in the solution.
- If $Q = K_{sp}$, the solution is saturated. The solution is holding the equilibrium amount of the dissolved ions and additional solid will not dissolve in the solution.
- If $Q > K_{sp}$, the solution is supersaturated. Under most circumstances, the excess solid will precipitate out of the solution.

We can use Q to predict whether a precipitation reaction will occur upon the mixing of two solutions containing dissolved ionic compounds as shown in Example 16.12.

Na$_2$CrO$_4$

AgNO$_3$

Ag$_2$CrO$_4$

▲ In this precipitation reaction, a solution of sodium chromate is poured into a solution of silver nitrate, producing a silver chromate precipitate.

EXAMPLE 16.12 Predicting Precipitation Reactions by Comparing Q and K_{sp}

A solution containing lead(II) nitrate is mixed with one containing sodium bromide to form a solution that is 0.0150 M in $Pb(NO_3)_2$ and 0.00350 M in NaBr. Will a precipitate form in the solution?

SOLUTION

First, determine the possible cross products and their K_{sp} values (Table 16.2). Any cross products that are soluble will *not* precipitate (see Table 4.1).	Possible cross products: $NaNO_3$ soluble $PbBr_2$ $K_{sp} = 4.67 \times 10^{-6}$
Calculate Q and compare it to K_{sp}. A precipitate will only form if $Q > K_{sp}$.	$\begin{aligned} Q &= [Pb^{2+}][Br^-]^2 \\ &= (0.0150)(0.00350)^2 \\ &= 1.84 \times 10^{-7} \end{aligned}$ $Q < K_{sp}$; therefore no precipitate forms.

TABLE 16.3 Formation Constants of Selected Complex Ions in Water at 25 °C

Complex Ion	K_f
$Ag(CN)_2^-$	1×10^{21}
$Ag(NH_3)_2^+$	1.7×10^7
$Ag(S_2O_3)_2^{3-}$	3.8×10^{13}
AlF_6^{3-}	7×10^{19}
$Al(OH)_4^-$	3×10^{33}
$CdBr_4^{2-}$	5.5×10^3
CdI_4^{2-}	2×10^6
$Cd(CN)_4^{2-}$	3×10^{18}
$Co(NH_3)_6^{3+}$	2.3×10^{33}
$Co(OH)_4^{2-}$	5×10^9
$Co(SCN)_4^{2-}$	1×10^3
$Cr(OH)_4^-$	8.0×10^{29}
$Cu(CN)_4^{2-}$	1.0×10^{25}
$Cu(NH_3)_4^{2+}$	1.7×10^{13}
$Fe(CN)_6^{4-}$	1.5×10^{35}
$Fe(CN)_6^{3-}$	2×10^{43}
$Hg(CN)_4^{2-}$	1.8×10^{41}
$HgCl_4^{2-}$	1.1×10^{16}
HgI_4^{2-}	2×10^{30}
$Ni(NH_3)_6^{2+}$	2.0×10^8
$Pb(OH)_3^-$	8×10^{13}
$Sn(OH)_3^-$	3×10^{25}
$Zn(CN)_4^{2-}$	2.1×10^{19}
$Zn(NH_3)_4^{2+}$	2.8×10^9
$Zn(OH)_4^{2-}$	3×10^{15}

16.7 Complex Ion Equilibria

We have learned about several different types of equilibria so far, including acid–base equilibria and solubility equilibria. We now turn to equilibria of another type, which involve primarily transition metal ions in solution. Transition metal ions tend to be good electron acceptors (or good Lewis acids). In aqueous solutions, water molecules can act as electron donors (or Lewis bases) to hydrate transition metal ions. For example, silver ions are hydrated by water in solution to form $Ag(H_2O)_2^+(aq)$. Chemists will often write $Ag^+(aq)$ as a shorthand notation for the hydrated silver ion, but the bare ion does not really exist by itself in solution.

Species such as $Ag(H_2O)_2^+$ are known as *complex ions*. A **complex ion** contains a central metal ion bound to one or more *ligands*. A **ligand** is a neutral molecule or ion that acts as a Lewis base with the central metal ion. In $Ag(H_2O)_2^+$, water is acting as the ligand. If a stronger Lewis base is put into a solution containing $Ag(H_2O)_2^+$, the stronger Lewis base will displace the water in the complex ion. For example, ammonia reacts with $Ag(H_2O)_2^+$ according to the reaction:

$$Ag(H_2O)_2^+(aq) + 2\,NH_3(aq) \rightleftharpoons Ag(NH_3)_2^+(aq) + 2\,H_2O(l)$$

For simplicity, water is often left out of the above equation and the reaction is written as:

$$Ag^+(aq) + 2\,NH_3(aq) \rightleftharpoons Ag(NH_3)_2^+(aq) \qquad K_f = 1.7 \times 10^7$$

The equilibrium constant associated with the reaction for the formation of a complex ion, such as this one, is the **formation constant (K_f)**. The expression for K_f is determined by the law of mass action, just as for any equilibrium constant. For $Ag(NH_3)_2^+$, the expression for K_f is written as:

$$K_f = \frac{\left[Ag(NH_3)_2^+\right]}{\left[Ag^+\right]\left[NH_3\right]^2}$$

Notice that the value of K_f for $Ag(NH_3)_2^+$ is large, indicating that the formation of the complex ion is highly favored. Table 16.3 lists the formation constants for a number of common complex ions. You can see that, in general, the values of K_f are very large, indicating that the formation of complex ions is highly favored in each case. Example 16.13 demonstrates how to use K_f in calculations.

EXAMPLE 16.13 Complex Ion Equilibria

A 200.0-mL sample of a solution that is 1.5×10^{-3} M in $Cu(NO_3)_2$ is mixed with a 250.0-mL sample of a solution that is 0.20 M in NH_3. After the solution reaches equilibrium, what concentration of $Cu^{2+}(aq)$ remains?

SOLUTION

Write the balanced equation for the complex ion equilibrium that occurs and look up the value of K_f in Table 16.3. Since this is an equilibrium problem, we must create an ICE table, which requires the initial concentrations of Cu^{2+} and NH_3. Calculate those concentrations from the given values.	$Cu^{2+}(aq) + 4\,NH_3(aq) \rightleftharpoons Cu(NH_3)_4^{2+}(aq)$ $K_f = 1.7 \times 10^{13}$ $\left[Cu^{2+}\right]_{initial} = \dfrac{0.200\,\cancel{L} \times \dfrac{1.5 \times 10^{-3}\,mol}{1\,\cancel{L}}}{0.200\,L + 0.250\,L} = 6.7 \times 10^{-4}$ M $\left[NH_3\right]_{initial} = \dfrac{0.250\,\cancel{L} \times \dfrac{0.20\,mol}{1\,\cancel{L}}}{0.200\,L + 0.250\,L} = 0.11$ M

Construct an ICE table for the reaction and write down the initial concentrations of each species.	$Cu^{2+}(aq) + 4\,NH_3(aq) \rightleftharpoons Cu(NH_3)_4^{2+}(aq)$

	$[Cu^{2+}]$	$[NH_3]$	$[Cu(NH_3)_4^{2+}]$
Initial	6.7×10^{-4}	0.11	0.0
Change			
Equil			

Since the equilibrium constant is so large, and since the concentration of ammonia is much larger than the concentration of Cu^{2+}, we can assume that the reaction will be driven to the right and most of the Cu^{2+} is consumed. Unlike previous ICE tables, where we let x represent the change in concentration in going to equilibrium, here we let x represent the small amount of Cu^{2+} that remains when equilibrium is reached.	$Cu^{2+}(aq) + 4\,NH_3(aq) \rightleftharpoons Cu(NH_3)_4^{2+}(aq)$

	$[Cu^{2+}]$	$[NH_3]$	$[Cu(NH_3)_4^{2+}]$
Initial	6.7×10^{-4}	0.11	0.0
Change	$\approx(-6.7 \times 10^{-4})$	$\approx 4(-6.7 \times 10^{-4})$	$\approx(+6.7 \times 10^{-4})$
Equil	x	0.11	6.7×10^{-4}

Substitute the expressions for the equilibrium concentrations into the expression for K_f and solve for x.	$K_f = \dfrac{[Cu(NH_3)_6^{2+}]}{[Cu^{2+}][NH_3]^4}$

$$= \frac{6.7 \times 10^{-4}}{x(0.11)^4}$$

$$x = \frac{6.7 \times 10^{-4}}{K_f(0.11)^4}$$

$$= \frac{6.7 \times 10^{-4}}{1.7 \times 10^{13}(0.11)^4}$$

$$= 2.7 \times 10^{-13}$$

Confirm that x is indeed small compared to the initial concentration of the metal cation. The remaining Cu^{2+} is very small because the formation constant is very large.	Since $x = 2.7 \times 10^{-13} \ll 6.7 \times 10^{-4}$; the approximation is valid. The remaining $[Cu^{2+}] = 2.7 \times 10^{-13}$ M.

FOR PRACTICE 16.13

A 125.0-mL sample of a solution that is 0.0117 M in $NiCl_2$ is mixed with a 175.0-mL sample of a solution that is 0.250 M in NH_3. After the solution reaches equilibrium, what concentration of $Ni^{2+}(aq)$ remains?

CHAPTER IN REVIEW

Key Terms

Section 16.2
buffer (622)
common ion effect (623)
Henderson–Hasselbalch
   equation (626)

Section 16.3
buffer capacity (635)

Section 16.4
acid–base titration (635)
indicator (635)

equivalence point (635)
endpoint (646)

Section 16.5
solubility product constant
   (K_{sp}) (647)
molar solubility (648)

Section 16.7
complex ion (654)
ligand (654)
formation constant (K_f) (654)

Key Concepts

The Danger of Antifreeze (16.1)

▶ Although the pH of mammalian blood is closely regulated by buffers, the capacity of these buffers to neutralize can be overwhelmed. For example, ethylene glycol, the main component of antifreeze, is metabolized by the liver into glycolic acid. The resulting acidity can exceed the buffering capacity of blood and cause acidosis, a serious condition that results in oxygen deprivation.

Buffers: Solutions That Resist pH Change (16.2)

▶ Buffers contain significant amounts of both a weak acid and its conjugate base, enabling the buffer to neutralize added acid or added base.

▶ Adding a small amount of acid to a buffer converts a stoichiometric amount of base to the conjugate acid. Adding a small amount of base to a buffer converts a stoichiometric amount of the acid to the conjugate base.

▶ The pH of a buffer solution can be found either by solving an equilibrium problem, focusing on the common ion effect, or by using the Henderson–Hasselbalch equation.

Buffer Range and Buffer Capacity (16.3)

▶ A buffer works best when the amounts of acid and conjugate base it contains are large and approximately equal.

▶ If the relative amounts of acid and base in a buffer differ by more than a factor of 10, the ability of the buffer to neutralize added acid and added base is significantly diminished. The maximum pH range at which a buffer is effective is therefore one pH unit on either side of the acid's pK_a.

Titrations and pH Curves (16.4)

▶ A titration curve is a graph of the change in pH versus added volume of acid or base during a titration.

▶ This chapter covers two types of titration curves, representing two types of acid–base reactions: a strong acid with a strong base (or vice versa), and a weak acid with a strong base (or vice versa).

▶ The equivalence point of a titration can be made visible by an indicator, a compound that changes color at a specific pH.

Solubility Equilibria and the Solubility Product Constant (16.5)

▶ The solubility product constant (K_{sp}) is an equilibrium constant for the dissolution of an ionic compound in water.

▶ The molar solubility of an ionic compound can be determined from K_{sp} and vice versa.

▶ Although the value of K_{sp} is constant at a given temperature, the solubility of an ionic substance can depend on other factors such as the presence of common ions and the pH of the solution.

Precipitation (16.6)

▶ We can compare the magnitude of K_{sp} with the reaction quotient, Q, in order to determine the relative saturation of a solution.

Complex Ion Equilibria (16.7)

▶ A complex ion contains a central metal ion coordinated to two or more ligands.

▶ The equilibrium constant for the formation of a complex ion is called a formation constant and is usually quite large.

Key Equations and Relationships

The Henderson–Hasselbalch Equation

$$pH = pK_a + \log \frac{[\text{base}]}{[\text{acid}]}$$

Effective Buffer Range (16.3)

$$pH \text{ range} = pK_a \pm 1$$

The Relation between Q and K_{sp} (16.6)

If $Q < K_{sp}$, the solution is unsaturated. More of the solid ionic compound can dissolve in the solution.

If $Q = K_{sp}$, the solution is saturated. The solution is holding the equilibrium amount of the dissolved ions and additional solid will not dissolve in the solution.

If $Q > K_{sp}$, the solution is supersaturated. Under most circumstances, the solid will precipitate out of a supersaturated solution.

Key Learning Objectives

Chapter Objectives	Assessment
Calculating the pH of a Buffer Solution (16.2)	Example 16.1 For Practice 16.1 For More Practice 16.1 Exercises 3, 4, 7, 8
Using the Henderson–Hasselbalch Equation to Calculate the pH of a Buffer Solution (16.2)	Example 16.2 For Practice 16.2 Exercises 11–16
Calculating the pH Change in a Buffer Solution After the Addition of a Small Amount of Strong Acid or Base (16.2)	Example 16.3 For Practice 16.3 For More Practice 16.3 Exercises 21–24
Using the Henderson–Hasselbalch Equation to Calculate the pH of a Buffer Solution Composed of a Weak Base and Its Conjugate Acid (16.2)	Example 16.4 For Practice 16.4 For More Practice 16.4 Exercises 11–16
Determining Buffer Range (16.3)	Example 16.5 For Practice 16.5 Exercises 31, 32

EXERCISES

Problems by Topic

The Common Ion Effect and Buffers

1. In which of these solutions will HNO_2 ionize less than it does in pure water?
   a. 0.10 M NaCl   b. 0.10 M KNO_3
   c. 0.10 M NaOH   d. 0.10 M $NaNO_2$

2. A formic acid solution has a pH of 3.25. Which of these substances will raise the pH of the solution upon addition? Explain.
   a. HCl   b. NaBr
   c. $NaCHO_2$ d. KCl

3. Solve an equilibrium problem (using an ICE table) to calculate the pH of each solution:
   a. a solution that is 0.15 M in $HCHO_2$ and 0.10 M in $NaCHO_2$
   b. a solution that is 0.12 M in NH_3 and 0.18 M in NH_4Cl

4. Solve an equilibrium problem (using an ICE table) to calculate the pH of each solution:
   a. a solution that is 0.175 M in $HC_2H_3O_2$ and 0.110 M in $KC_2H_3O_2$
   b. a solution that is 0.195 M in CH_3NH_2 and 0.105 M in CH_3NH_3Br

5. Calculate the percent ionization of a 0.15 M benzoic acid solution in pure water and also in a solution containing 0.10 M sodium benzoate. Why is the percent ionization so different in the two solutions?

6. Calculate the percent ionization of a 0.13 M formic acid solution in pure water and also in a solution containing 0.11 M potassium formate. Explain the difference in percent ionization in the two solutions.

7. Solve an equilibrium problem (using an ICE table) to calculate the pH of each solution:
   a. 0.15 M HF
   b. 0.15 M NaF
   c. a mixture that is 0.15 M in HF and 0.15 M in NaF

8. Solve an equilibrium problem (using an ICE table) to calculate the pH of each solution:
   a. 0.18 M CH_3NH_2
   b. 0.18 M CH_3NH_3Cl
   c. a mixture that is 0.18 M in CH_3NH_2 and 0.18 M in CH_3NH_3Cl

9. A buffer contains significant amounts of acetic acid and sodium acetate. Write equations showing how this buffer neutralizes added acid and added base.

10. A buffer contains significant amounts of ammonia and ammonium chloride. Write equations showing how this buffer neutralizes added acid and added base.

11. Use the Henderson–Hasselbalch equation to calculate the pH of each of the solutions in problem 3.

12. Use the Henderson–Hasselbalch equation to calculate the pH of each of the solutions in problem 4.

13. Use the Henderson–Hasselbalch equation to calculate the pH of a solution that is
    a. 0.125 M in HClO and 0.150 M in KClO
    b. 0.175 M in $C_2H_5NH_2$ and 0.150 M in $C_2H_5NH_3Br$
    c. 10.0 g of $HC_2H_3O_2$ and 10.0 g of $NaC_2H_3O_2$ in 150.0 mL of solution

14. Use the Henderson–Hasselbalch equation to calculate the pH of a solution that is
    a. 0.155 M in propanoic acid and 0.110 M in potassium propanoate
    b. 0.15 M in C_5H_5N and 0.10 M in C_5H_5NHCl
    c. 15.0 g of HF and 25.0 g of NaF in 125 mL of solution

15. Calculate the pH of the solution that results from each mixture:
    a. 50.0 mL of 0.15 M $HCHO_2$ with 75.0 mL of 0.13 M $NaCHO_2$
    b. 125.0 mL of 0.10 M NH_3 with 250.0 mL of 0.10 M NH_4Cl

16. Calculate the pH of the solution that results from each mixture:
    a. 150.0 mL of 0.25 M HF with 225.0 mL of 0.30 M NaF
    b. 175.0 mL of 0.10 M $C_2H_5NH_2$ with 275.0 mL of 0.20 M $C_2H_5NH_3Cl$

17. Calculate the ratio of NaF to HF required to create a buffer with pH = 4.00.

18. Calculate the ratio of CH_3NH_2 to CH_3NH_3Cl concentration required to create a buffer with pH = 10.24.

19. What mass of sodium benzoate should be added to 150.0 mL of a 0.15 M benzoic acid solution in order to obtain a buffer with a pH of 4.25? (Assume no volume change.)

20. What mass of ammonium chloride should be added to 2.55 L of a 0.155 M NH_3 in order to obtain a buffer with a pH of 9.55? (Assume no volume change.)

21. A 250.0-mL buffer solution is 0.250 M in acetic acid and 0.250 M in sodium acetate.
    a. What is the initial pH of this solution?
    b. What is the pH after addition of 0.0050 mol of HCl?
    c. What is the pH after addition of 0.0050 mol of NaOH?

22. A 100.0-mL buffer solution is 0.175 M in HClO and 0.150 M in NaClO.
    a. What is the initial pH of this solution?
    b. What is the pH after addition of 150.0 mg of HBr?
    c. What is the pH after addition of 85.0 mg of NaOH?

23. For each solution, calculate the initial pH and the final pH after adding 0.010 mol of HCl.
    a. 500.0 mL of pure water
    b. 500.0 mL of a buffer solution that is 0.125 M in $HC_2H_3O_2$ and 0.115 M in $NaC_2H_3O_2$
    c. 500.0 mL of a buffer solution that is 0.155 M in $C_2H_5NH_2$ and 0.145 M in $C_2H_5NH_3Cl$

24. For each solution, calculate the initial pH and the final pH after adding 0.010 mol of NaOH.
    a. 250.0 mL of pure water
    b. 250.0 mL of a buffer solution that is 0.195 M in $HCHO_2$ and 0.275 M in $KCHO_2$
    c. 250.0 mL of a buffer solution that is 0.255 M in $CH_3CH_2NH_2$ and 0.235 M in $CH_3CH_2NH_3Cl$

25. A 350.0-mL buffer solution is 0.150 M in HF and 0.150 M in NaF. What mass of NaOH can this buffer neutralize before the pH rises above 4.00? If the same volume of the buffer was 0.350 M in HF and 0.350 M in NaF, what mass of NaOH could be handled before the pH rises above 4.00?

26. A 100.0-mL buffer solution is 0.100 M in NH_3 and 0.125 M in NH_4Br. What mass of HCl can this buffer neutralize before the pH falls below 9.00? If the same volume of the buffer were 0.250 M in NH_3 and 0.400 M in NH_4Br, what mass of HCl could be handled before the pH fell below 9.00?

27. Determine whether mixing each pair of solutions results in a buffer.
    a. 100.0 mL of 0.10 M NH_3; 100.0 mL of 0.15 M NH_4Cl
    b. 50.0 mL of 0.10 M HCl; 35.0 mL of 0.150 M NaOH
    c. 50.0 mL of 0.15 M HF; 20.0 mL of 0.15 M NaOH
    d. 175.0 mL of 0.10 M NH_3; 150.0 mL of 0.12 M NaOH
    e. 125.0 mL of 0.15 M NH_3; 150.0 mL of 0.20 M NaOH

28. Determine whether mixing each pair of solutions results in a buffer.
    a. 75.0 mL of 0.10 M HF; 55.0 mL of 0.15 M NaF
    b. 150.0 mL of 0.10 M HF; 135.0 mL of 0.175 M HCl
    c. 165.0 mL of 0.10 M HF; 135.0 mL of 0.050 M KOH
    d. 125.0 mL of 0.15 M CH_3NH_2; 120.0 mL of 0.25 M CH_3NH_3Cl
    e. 105.0 mL of 0.15 M CH_3NH_2; 95.0 mL of 0.10 M HCl

29. Blood is buffered by carbonic acid and the bicarbonate ion. Normal blood plasma is 0.024 M in HCO_3^- and 0.0012 M H_2CO_3(pK_{a_1} for H_2CO_3 at body temperature is 6.1).
    a. What is the pH of blood plasma?
    b. If the volume of blood in a normal adult is 5.0 L, what mass of HCl could be neutralized by the buffering system in blood before the pH falls below 7.0 (which would result in death)?
    c. Given the volume from part (b), what mass of NaOH could be neutralized before the pH rises above 7.8?

30. The fluids within cells are buffered by $H_2PO_4^-$ and HPO_4^{2-}.
    a. Calculate the ratio of HPO_4^{2-} to $H_2PO_4^-$ required to maintain a pH of 7.1 within a cell.
    b. Could a buffer system employing H_3PO_4 as the weak acid and $H_2PO_4^-$ as the weak base be used within cells? Explain.

31. Which buffer system is the best choice to create a buffer with pH = 7.20? For the best system, calculate the ratio of the masses of the buffer components required to make the buffer.
    $HC_2H_3O_2/KC_2H_3O_2$ $HClO_2/KClO_2$
    NH_3/NH_4Cl $HClO/KClO$

32. Which buffer system is the best choice to create a buffer with pH = 9.00? For the best system, calculate the ratio of the masses of the buffer components required to make the buffer.
    HF/KF HNO_2/KNO_2
    NH_3/NH_4Cl $HClO/KClO$

33. A 500.0-mL buffer solution is 0.100 M in HNO_2 and 0.150 M in KNO_2. Determine whether or not the addition of the given amount of each substance exceeds the capacity of the buffer to neutralize it.
    a. 250 mg NaOH
    b. 350 mg KOH
    c. 1.25 g HBr
    d. 1.35 g HI

34. A 1.0-L buffer solution is 0.125 M in HNO_2 and 0.145 M in $NaNO_2$. Determine the concentrations of HNO_2 and $NaNO_2$ after the given amount of each substance is added:
    a. 1.5 g HCl
    b. 1.5 g NaOH
    c. 1.5 g HI

Titrations, pH Curves, and Indicators

35. The graphs labeled (a) and (b) show the titration curves for two equal-volume samples of monoprotic acids, one weak and one strong. Both titrations were carried out with the same concentration of strong base.

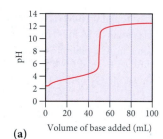
(a)

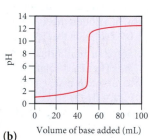
(b)

(i) What is the approximate pH at the equivalence point of each curve?
(ii) Which curve corresponds to the titration of the strong acid and which one to the titration of the weak acid?

36. Two 25.0-mL samples, one 0.100 M HCl and the other 0.100 M HF, are titrated with 0.200 M KOH. Answer each question regarding these two titrations.
    a. What is the volume of added base at the equivalence point for each titration?
    b. Is the pH at the equivalence point for each titration acidic, basic, or neutral?
    c. Which titration curve will have the lower initial pH?
    d. Make a rough sketch of each titration curve.

37. Two 20.0-mL samples, one 0.200 M KOH and the other 0.200 M CH_3NH_2, are titrated with 0.100 M HI. Answer each question regarding these two titrations.
    a. What is the volume of added acid at the equivalence point for each titration?
    b. Is the pH at the equivalence point for each titration acidic, basic, or neutral?
    c. Which titration curve will have the lower initial pH?
    d. Make a rough sketch of each titration curve.

38. The graphs labeled (a) and (b) show the titration curves for two equal-volume samples of bases, one weak and one strong. Both titrations were carried out with the same concentration of strong acid.

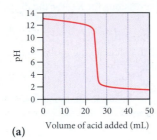
(a)

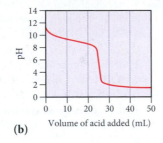
(b)

(i) What is the approximate pH at the equivalence point of each curve?

(ii) Which curve corresponds to the titration of the strong base and which one to the weak base?

39. Consider the curve for the titration of a weak monoprotic acid with a strong base and answer each question.

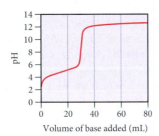

a. What is the pH and what is the volume of added base at the equivalence point?

b. At what volume of added base is the pH calculated by working an equilibrium problem based on the initial concentration and K_a of the weak acid?

c. At what volume of added base does pH = pK_a?

d. At what volume of added base is the pH calculated by working an equilibrium problem based on the concentration and K_b of the conjugate base?

e. Beyond what volume of added base is the pH calculated by focusing on the amount of excess strong based added?

40. Consider the curve for the titration of a weak base with a strong acid and answer each question.

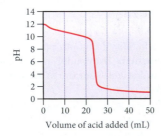

a. What is the pH and what is the volume of added acid at the equivalence point?

b. At what volume of added acid is the pH calculated by working an equilibrium problem based on the initial concentration and K_b of the weak base?

c. At what volume of added acid does pH = $14 - pK_b$?

d. At what volume of added acid is the pH calculated by working an equilibrium problem based on the concentration and K_a of the conjugate acid?

e. Beyond what volume of added acid is the pH calculated by focusing on the amount of excess strong acid added?

41. Consider the titration of a 35.0-mL sample of 0.175 M HBr with 0.200 M KOH. Determine each quantity:

a. the initial pH

b. the volume of added base required to reach the equivalence point

c. the pH at 10.0 mL of added base

d. the pH at the equivalence point

e. the pH after adding 5.0 mL of base beyond the equivalence point

42. A 20.0-mL sample of 0.125 M HNO_3 is titrated with 0.150 M NaOH. Calculate the pH for at least five different points throughout the titration curve and make a sketch of the curve. Indicate the volume at the equivalence point on your graph.

43. Consider the titration of a 25.0-mL sample of 0.115 M RbOH with 0.100 M HCl. Determine each quantity:

a. the initial pH

b. the volume of added acid required to reach the equivalence point

c. the pH at 5.0 mL of added acid

d. the pH at the equivalence point

e. the pH after adding 5.0 mL of acid beyond the equivalence point

44. A 15.0-mL sample of 0.100 M $Ba(OH)_2$ is titrated with 0.125 M HCl. Calculate the pH for at least five different points throughout the titration curve and make a sketch of the curve. Indicate the volume at the equivalence point on your graph.

45. Consider the titration of a 20.0-mL sample of 0.105 M $HC_2H_3O_2$ with 0.125 M NaOH. Determine each quantity:

a. the initial pH

b. the volume of added base required to reach the equivalence point

c. the pH at 5.0 mL of added base

d. the pH at one-half of the equivalence point

e. the pH at the equivalence point

f. the pH after adding 5.0 mL of base beyond the equivalence point

46. A 30.0-mL sample of 0.165 M propanoic acid is titrated with 0.300 M KOH. Calculate the pH at each volume of added base: 0 mL, 5 mL, 10 mL, equivalence point, one-half equivalence point, 20 mL, 25 mL. Use your calculations to make a sketch of the titration curve.

47. Consider the titration of a 25.0-mL sample of 0.175 M CH_3NH_2 with 0.150 M HBr. Determine each quantity:

a. the initial pH

b. the volume of added acid required to reach the equivalence point

c. the pH at 5.0 mL of added acid

d. the pH at one-half of the equivalence point

e. the pH at the equivalence point

f. the pH after adding 5.0 mL of acid beyond the equivalence point

48. A 25.0-mL sample of 0.125 M pyridine is titrated with 0.100 M HCl. Calculate the pH at each volume of added acid: 0 mL, 10 mL, 20 mL, equivalence point, one-half equivalence point, 40 mL, 50 mL. Use your calculations to make a sketch of the titration curve.

49. Consider the titration curves (labeled a and b) for two weak acids, both titrated with 0.100 M NaOH.

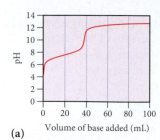

(a)

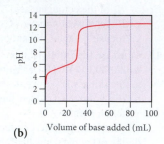

(b)

(i) Which of the two acid solutions is more concentrated?
(ii) Which of the two acids has the larger K_a?

50. Consider the titration curves (labeled a and b) for two weak bases, both titrated with 0.100 M HCl.

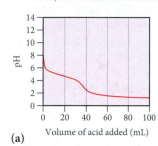

(a)

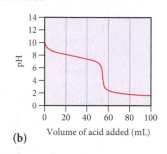

(b)

(i) Which of the two base solutions is more concentrated?
(ii) Which of the two bases has the larger K_b?

51. A 0.229-g sample of an unknown monoprotic acid was titrated with 0.112 M NaOH and the resulting titration curve is shown below. Determine the molar mass and pK_a of the acid.

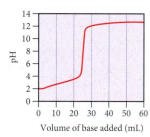

52. A 0.446-g sample of an unknown monoprotic acid was titrated with 0.105 M KOH and the resulting titration curve is shown below. Determine the molar mass and pK_a of the acid.

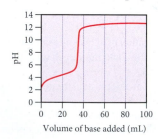

53. Using Table 16.1, pick an indicator for use in the titration of each acid with a strong base.
a. HF   **b.** HCl   **c.** HCN

54. Using Table 16.1, pick an indicator for use in the titration of each of base with a strong acid.
a. CH_3NH_2 **b.** NaOH   **c.** $C_6H_5NH_2$

Solubility Equilibria

55. Write balanced equations and expressions for K_{sp} for the dissolution of each ionic compound:
a. $BaSO_4$ **b.** $PbBr_2$ **c.** Ag_2CrO_4

56. Write balanced equations and expressions for K_{sp} for the dissolution of each ionic compound:
a. $CaCO_3$ **b.** $PbCl_2$ **c.** AgI

57. Use the K_{sp} values in Table 16.2 to calculate the molar solubility of each compound in pure water:
a. AgBr   **b.** $Mg(OH)_2$ **c.** CaF_2

58. Use the K_{sp} values in Table 16.2 to calculate the molar solubility of each compound in pure water:
a. MX ($K_{sp} = 1.27 \times 10^{-36}$)
b. Ag_2CrO_4
c. $Ca(OH)_2$

59. Use the given molar solubilities in pure water to calculate K_{sp} for each compound:
a. MX; molar solubility = 3.27×10^{-11} M
b. PbF_2; molar solubility = 5.63×10^{-3} M
c. MgF_2; molar solubility = 2.65×10^{-4} M

60. Use the given molar solubilities in pure water to calculate K_{sp} for each compound:
a. $BaCrO_4$; molar solubility = 1.08×10^{-5} M
b. Ag_2SO_3; molar solubility = 1.55×10^{-5} M
c. $Pd(SCN)_2$; molar solubility = 2.22×10^{-8} M

61. Two compounds with general formulas AX and AX_2 have $K_{sp} = 1.5 \times 10^{-5}$. Which of the compounds has the higher molar solubility?

62. Consider the compounds with the generic formulas listed here and their corresponding molar solubilities in pure water. Which compound will have the smallest value of K_{sp}?
$\qquad$ AX; molar solubility = 1.35×10^{-4} M
$\qquad$ AX_2; molar solubility = 2.25×10^{-4} M
$\qquad$ A_2X; molar solubility = 1.75×10^{-4} M

63. Use the K_{sp} value from Table 16.2 to calculate the solubility of iron(II) hydroxide in pure water in grams per 100.0 mL of solution.

64. The solubility of copper(I) chloride is 3.91 mg per 100.0 mL of solution. Calculate K_{sp} for CuCl.

65. Calculate the molar solubility of barium fluoride in:
a. pure water   **b.** 0.10 M $Ba(NO_3)_2$ **c.** 0.15 M NaF

66. Calculate the molar solubility of MX ($K_{sp} = 1.27 \times 10^{-36}$) in:
a. pure water   **b.** 0.25 M MCl_2 **c.** 0.20 M Na_2X

67. Calculate the molar solubility of calcium hydroxide in a solution buffered at each pH.
a. pH = 4   **b.** pH = 7   **c.** pH = 9

68. Calculate the solubility (in grams per 1.00×10^2 mL of solution) of magnesium hydroxide in a solution buffered at pH = 10. How does this compare to the solubility of $Mg(OH)_2$ in pure water?

69. Determine if each compound is more soluble in acidic solution than in pure water. Explain.
a. $BaCO_3$ **b.** CuS
c. AgCl   **d.** PbI_2

70. Determine if each compound is more soluble in acidic solution than in pure water. Explain.
a. Hg_2Br_2 **b.** $Mg(OH)_2$
c. $CaCO_3$ **d.** AgI

71. A solution containing sodium fluoride is mixed with one containing calcium nitrate to form a solution that is 0.015 M in NaF and 0.010 M in $Ca(NO_3)_2$. Will a precipitate form in the mixed solution? If so, identify the precipitate.

72. A solution containing potassium bromide is mixed with one containing lead acetate to form a solution that is 0.013 M in KBr and 0.0035 M in $Pb(C_2H_3O_2)_2$. Will a precipitate form in the mixed solution? If so, identify the precipitate.

73. Predict whether or not a precipitate will form upon mixing 75.0 mL of a NaOH solution with pOH = 2.58 with 125.0 mL of a 0.018 M $MgCl_2$ solution. Identify the precipitate, if any.

74. Predict whether or not a precipitate will form upon mixing 175.0 mL of a 0.0055 M KCl solution with 145.0 mL of a 0.0015 M $AgNO_3$ solution. Identify the precipitate, if any.

75. Potassium hydroxide is used to precipitate each of the cations from their respective solution. Determine the minimum concentration of KOH required for precipitation to begin in each case.

a. 0.015 M $CaCl_2$ b. 0.0025 M $Fe(NO_3)_2$
c. 0.0018 M $MgBr_2$

76. Determine the minimum concentration of the precipitating agent on the right to cause precipitation of the cation from the solution on the left.

a. 0.035 M $Ba(NO_3)_2$; NaF
b. 0.086 M CaI_2; K_2SO_4
c. 0.0018 M $AgNO_3$; RbCl

Complex Ion Equilibria

77. A solution is made that is 1.1×10^{-3} M in $Zn(NO_3)_2$ and 0.150 M in NH_3. After the solution reaches equilibrium, what concentration of $Zn^{2+}(aq)$ remains?

78. A 120.0-mL sample of a solution that is 2.8×10^{-3} M in $AgNO_3$ is mixed with a 225.0-mL sample of a solution that is 0.10 M in NaCN. After the solution reaches equilibrium, what concentration of $Ag^+(aq)$ remains?

Cumulative Problems

79. A 150.0-mL solution contains 2.05 g of sodium benzoate and 2.47 g of benzoic acid. Calculate the pH of the solution.

80. A solution is made by combining 10.0 mL of 17.5 M acetic acid with 5.54 g of sodium acetate and diluting to a total volume of 1.50 L. Calculate the pH of the solution.

81. A buffer is created by combining 150.0 mL of 0.25 M $HCHO_2$ with 75.0 mL of 0.20 M NaOH. Determine the pH of the buffer.

82. A buffer is created by combining 3.55 g of NH_3 with 4.78 g of HCl and diluting to a total volume of 750.0 mL. Determine the pH of the buffer.

83. A 1.0-L buffer solution initially contains 0.25 mol of NH_3 and 0.25 mol of NH_4Cl. In order to adjust the buffer pH to 8.75, should you add NaOH or HCl to the buffer mixture? What mass of the correct reagent should you add?

84. A 250.0-mL buffer solution initially contains 0.025 mol of $HCHO_2$ and 0.025 mol of $NaCHO_2$. In order to adjust the buffer pH to 4.10, should you add NaOH or HCl to the buffer mixture? What mass of the correct reagent should you add?

85. In analytical chemistry, bases used for titrations must often be standardized; that is, their concentration must be precisely determined. Standardization of sodium hydroxide solutions is often accomplished by titrating potassium hydrogen phthalate ($KHC_8H_4O_4$) also known as KHP, with the NaOH solution to be standardized.

a. Write an equation for the reaction between NaOH and KHP.
b. The titration of 0.5527 g of KHP required 25.87 mL of an NaOH solution to reach the equivalence point. What is the concentration of the NaOH solution?

86. A 0.5224-g sample of an unknown monoprotic acid is titrated with 0.0998 M NaOH. The equivalence point of the titration occurs at 23.82 mL. Determine the molar mass of the unknown acid.

87. A 0.25-mol sample of a weak acid with an unknown pK_a is combined with 10.0 mL of 3.00 M KOH and the resulting solution is diluted to 1.500 L. The measured pH of the solution is 3.85. What is the pK_a of the weak acid?

88. A 5.55-g sample of a weak acid with $K_a = 1.3 \times 10^{-4}$ is combined with 5.00 mL of 6.00 M NaOH and the resulting solution is diluted to 750 mL. The measured pH of the solution is 4.25. What is the molar mass of the weak acid?

89. One of the main components of hard water is $CaCO_3$. When hard water evaporates, some of the $CaCO_3$ is left behind as a white mineral deposit. If a hard water solution is saturated with calcium carbonate, what volume of the solution has to evaporate to deposit 1.00×10^2 mg of $CaCO_3$?

90. Gout—a condition that results in joint swelling and pain—is caused by the formation of sodium urate ($NaC_5H_3N_4$) crystals within tendons, cartilage, and ligaments. Sodium urate will precipitate out of blood plasma when uric acid levels become abnormally high. This could happen as a result of eating too many rich foods and consuming too much alcohol, which is why gout is sometimes referred to as the "disease of kings." If the sodium concentration in blood plasma is 0.140 M, and K_{sp} for sodium urate is 5.76×10^{-8}, what minimum concentration of urate would result in the precipitation of sodium urate?

91. Pseudogout, a condition with symptoms similar to those of gout (see previous problem), is caused by the formation of calcium diphosphate ($Ca_2P_2O_7$) crystals within tendons, cartilage, and ligaments. Calcium diphosphate will precipitate out of blood plasma when diphosphate levels become abnormally high. If the calcium concentration in blood plasma is 9.2 mg/dL, and K_{sp} for calcium diphosphate is 8.64×10^{-13}, what minimum concentration of diphosphate results in the precipitation of calcium diphosphate?

92. Calculate the solubility of silver chloride in a solution that is 0.100 M in NH_3.

93. Calculate the solubility of CuX ($K_{sp} = [Cu^{2+}][X^{2-}] = 1.27 \times 10^{-36}$) in a solution that is 0.150 M in NaCN.

94. Aniline, abbreviated ϕNH_2, where ϕ is C_6H_5, is an important organic base used in the manufacture of dyes. It has $K_b = 4.3 \times 10^{-10}$. In a certain manufacturing process it is necessary to keep the concentration of ϕNH_3^+ (its conjugate acid, called the anilinium ion) below 1.0×10^{-9} M in a solution that is 0.10 M in aniline. Find the concentration of NaOH necessary for this process.

95. The K_b of hydroxylamine, NH_2OH, is 1.10×10^{-8}. A buffer solution is prepared by mixing 100.0 mL of a 0.36 M hydroxylamine solution with 50.0 mL of a 0.26 M HCl solution. Find the pH of the resulting solution.

96. Determine the mass of sodium formate that must be dissolved in 250.0 cm^3 of a 1.4 M solution of formic acid to prepare a buffer solution with pH = 3.36.

97. What relative masses of dimethyl amine and dimethyl ammonium chloride do you need to prepare a buffer solution of pH = 10.43?

98. You are asked to prepare 2.0 L of a HCN/NaCN buffer that has a pH of 9.8 and an osmotic pressure of 1.35 atm at 298 K. What masses of HCN and NaCN should you use to prepare the buffer? (Assume complete dissociation of NaCN.)

99. What are the concentrations of benzoic acid and sodium benzoate in a solution that is buffered at a pH of 4.55 and has a freezing point of $-2.0\ °C$? (Assume complete dissociation of sodium benzoate and a density of 1.01 g/mL for the solution.)

Challenge Problems

100. Derive an equation similar to the Henderson–Hasselbalch equation for a buffer composed of a weak base and its conjugate acid. Instead of relating pH to pK_a and the relative concentrations of an acid and its conjugate base (as the Henderson–Hasselbalch equation does), the equation should relate pOH to pK_b and the relative concentrations of a base and its conjugate acid.

101. Since soap and detergent action is hindered by hard water, laundry formulations usually include water softeners—called builders—designed to remove hard water ions (especially Ca^{2+} and Mg^{2+}) from the water. A common builder used in North America is sodium carbonate. Suppose that the hard water used to do laundry contains 75 ppm $CaCO_3$ and 55 ppm $MgCO_3$ (by mass). What mass of Na_2CO_3 will remove 90.0% of these ions from 10.0 L of laundry water?

102. When excess solid $Mg(OH)_2$ is shaken with 1.00 L of 1.0 M NH_4Cl solution, the resulting saturated solution has pH = 9.00. Calculate the K_{sp} of $Mg(OH)_2$.

103. What volume of 0.100 M sodium carbonate solution is required to precipitate 99% of the Mg from 1.00 L of 0.100 M magnesium nitrate solution?

104. Determine the solubility of CuI in 0.40 M HCN solution. The K_{sp} of CuI is 1.1×10^{-12} and the K_f for the $Cu(CN)_2^-$ complex ion is 1×10^{24}.

105. Determine the pH of a solution prepared from 1.0 L of a 0.10 M solution of $Ba(OH)_2$ and excess $Zn(OH)_2(s)$. The K_{sp} of $Zn(OH)_2$ is 3×10^{-15} and the K_f of $Zn(OH)_4^{2-}$ is 2×10^{15}.

106. What amount of HCl gas must be added to 1.00 L of a buffer solution that contains [acetic acid] = 2.0 M and [acetate] = 1.0 M in order to produce a solution with pH = 4.00?

Conceptual Problems

107. Without doing any calculations, determine whether each buffer solution will have pH = pK_a, pH > pK_a, or pH < pK_a. Assume that HA is a weak monoprotic acid.
  a. 0.10 mol HA and 0.050 mol of A$^-$ in 1.0 L of solution
  b. 0.10 mol HA and 0.150 mol of A$^-$ in 1.0 L of solution
  c. 0.10 mol HA and 0.050 mol of OH$^-$ in 1.0 L of solution
  d. 0.10 mol HA and 0.075 mol of OH$^-$ in 1.0 L of solution

108. A buffer contains 0.10 mol of a weak acid and 0.20 mol of its conjugate base in 1.0 L of solution. Determine whether or not each addition exceeds the capacity of the buffer.
  a. adding 0.020 mol of NaOH
  b. adding 0.020 mol of HCl
  c. adding 0.10 mol of NaOH
  d. adding 0.010 mol of HCl

109. Consider these two solutions:
  (i) 0.10 M solution of a weak monoprotic acid
  (ii) 0.10 M solution of strong monoprotic acid

  Each solution is titrated with 0.15 M NaOH. Which quantity will be the same for both solutions?
  a. the volume required to reach the equivalence point
  b. the pH at the equivalence point
  c. the pH at one-half the equivalence point

110. Two monoprotic acid solutions (A and B) are titrated with identical NaOH solutions. The volume to reach the equivalence point for solution A is twice the volume required to reach the equivalence point for solution B, and the pH at the equivalence point of solution A is higher than the pH at the equivalence point for solution B. Which statement is true of the two acids?
  a. The acid in solution A is more concentrated than in solution B and is also a stronger acid than that in solution B.
  b. The acid in solution A is less concentrated than in solution B and is also a weaker acid than that in solution B.
  c. The acid in solution A is more concentrated than in solution B and is also a weaker acid than that in solution B.
  d. The acid in solution A is less concentrated than in solution B and is also a stronger acid than that in solution B.

111. Describe the solubility of CaF_2 in each solution compared to its solubility in pure water.
  a. in a 0.10 M NaCl solution
  b. in a 0.10 M NaF solution
  c. in a 0.10 M HCl solution

Answers to Conceptual Connections

Buffers

16.1 (d) Only this solution contains significant amounts of a weak acid and its conjugate base. (Remember that HNO_3 is a strong acid, but HNO_2 is a weak acid.)

pH of Buffer Solutions

16.2 (a) Since the pH of the buffer is less than the pK_a of the acid, the buffer must contain more acid than base ($[HA] > [A^-]$). In order to raise the pH of the buffer from 4.25 to 4.72, you must add more of the weak base (adding a base will make the buffer solution more basic).

Adding Acid or Base to a Buffer

16.3 (b) Since acid is added to the buffer, the pH will become slightly lower (slightly more acidic). Answer (a) reflects too large a change in pH for a buffer, and answers (c) and (d) shift the solution in the wrong direction.

Buffer Capacity

16.4 (a) Adding 0.050 mol of HCl will destroy the buffer because it will react with all of the NaF, leaving no conjugate base in the buffer mixture.

Titration Equivalence Point

16.5 (d) Since the flask contains 7 H^+ ions, the equivalence point is reached when 7 OH^- ions have been added.

The Half-Equivalence Point

16.6 (c) The pH at the half-equivalence point is the pK_a of the conjugate acid which is equal to $14.00 - 8.75 = 5.25$.

Common Ion Effect

16.7 (c) The sodium nitrate solution is the only one that has no common ion with barium sulfate. The other two solutions have common ions with barium sulfate; therefore the solubility of barium sulfate is lower in these solutions.

Free Energy and Thermodynamics

Die Energie der Welt ist konstant. Die Entropie der Welt strebt einem Maximum zu. (The energy of the world is constant. The entropy of the world tends towards a maximum.) —Rudolf Clausius (1822–1888)

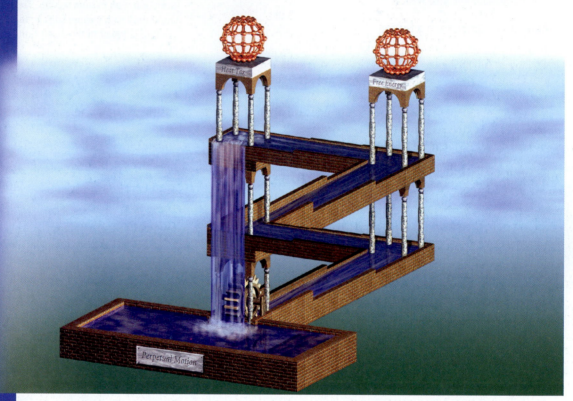

In this clever illusion, it seems that the water can perpetually flow through the canal. However, perpetual motion is forbidden by the laws of thermodynamics.

IN THIS BOOK, we have examined and learned much about chemical and physical changes. We have studied how fast chemical changes occur (kinetics) and how to predict how far they will go (through the use of equilibrium constants). We have learned that acids neutralize bases and that gases expand to fill their containers. We now turn to the following question: Why do these changes occur in the first place? What ultimately drives physical and chemical changes in matter? The answer may surprise you. The driving force behind chemical and physical change in the universe is a quantity called *entropy*, which is related to the dispersion (spreading out) of energy. Nature tends toward a state in which energy is spread out to the greatest extent possible. Although it does not seem obvious at first glance, the freezing of water below 0 °C, the dissolving of a solid into a solution, the neutralization of an acid by a base, and even the development of a person from an embryo all increase the entropy in the universe (they all result in greater energy dispersion). In our universe, entropy always increases.

17.1   Nature's Heat Tax: You Can't Win and You Can't Break Even

Energy transactions are like gambling—you walk into the casino with your pockets full of cash and (if you keep gambling long enough) you walk out empty-handed. In the long run, you lose money gambling because the casino takes a cut on each transaction. So it is with energy. Nature takes a cut—sometimes referred to as nature's heat tax—on every energy transaction so that, in the end, energy is dissipated.

Recall from Chapter 6 that, according to the first law of thermodynamics, energy is conserved in chemical processes. When we burn gasoline to run a car, for example, the amount of energy produced by the chemical reaction does not vanish, nor does any new energy appear that was not present as potential energy (within the gasoline) before the combustion. Some of the energy from the combustion reaction goes toward driving the car forward (about 20%), and the rest is dissipated into the surroundings as heat (just feel the engine). However, the total energy given off by the combustion reaction exactly equals the sum of the amount of energy driving the car forward and the amount being dissipated as heat—energy is conserved. In other words, when it comes to energy, you can't win; you cannot create more energy than what was there to begin with.

The picture becomes more interesting, however, when we consider the second law of thermodynamics. The second law—which we examine in more detail throughout this chapter—implies that not only can we not win in an energy transaction, but we cannot even break even. For example, consider a rechargeable battery. Suppose that, upon using the fully charged battery for some application, the energy from the battery does 100 kJ of work. Recharging the battery to its original state will *necessarily* (according to the second law of thermodynamics) require *more than* 100 kJ of energy. Energy is not destroyed during the cycle of discharging and recharging the battery, but some energy must be lost to the surroundings in order for the process to occur at all. The implications of the second law for energy use are significant. First of all, according to the second law, we cannot create a perpetual motion machine, that is, a machine that perpetually moves *without any energy input*. If the machine is in motion, it must pay the heat tax with each cycle of its motion—over time, it will therefore run down and stop moving.

Secondly, in most energy transactions, not only is the heat tax lost to the surroundings, but additional energy is also lost as heat because real-world processes do not achieve the theoretically possible maximum efficiency **Figure 17.1▶**. Consequently, the most efficient use of energy generally occurs with the smallest number of transactions. For example, heating your home with natural gas is generally cheaper and more efficient than heating it with electricity. Why? When you heat your home with natural gas, there is only one energy transaction—you burn the gas and the heat from the reaction warms the house. When you heat your home with electricity, however, several transactions occur. Most electricity is generated from the combustion of fossil fuels; the fuel is burned and the heat from the reaction is used to boil water. The steam generated by the boiling water then turns a turbine on a generator to create electricity. The electricity must then travel from the power plant to your home, with some of the energy lost as heat during the trip. Finally, the electricity must run the heater that creates the heat. With each transaction, energy is lost to the surroundings, resulting in a less efficient use of energy than if you had burned natural gas directly.

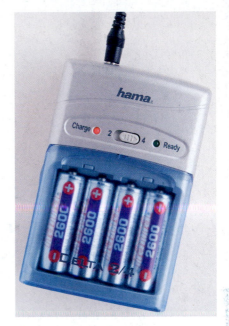

▲ A rechargeable battery always requires more energy to charge than the energy available for work during discharging because some energy is always lost to the surroundings during the charging/discharging cycle.

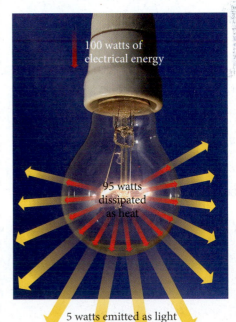

100 watts of electrical energy

95 watts dissipated as heat

5 watts emitted as light

▲ **FIGURE 17.1 Energy Loss**   In most energy transactions, some energy is lost to the surroundings, so that each transaction is only fractionally efficient.

⬤⬤ **Conceptual Connection 17.1** **Nature's Heat Tax and Diet**

Advocates of a vegetarian diet argue that the amount of cropland required for one person to maintain a meat-based diet is about 6–10 times greater than the amount required for the same person to maintain a vegetarian diet. Use the concept of nature's heat tax to explain why this might be so.

17.2 Spontaneous and Nonspontaneous Processes

A fundamental goal of thermodynamics is the prediction of *spontaneity*. For example, will rust spontaneously form when iron comes into contact with oxygen? Will water spontaneously decompose into hydrogen and oxygen? A **spontaneous process** is one that occurs *without ongoing outside intervention* (such as the performance of work by some external force). For example, when you drop a book in a gravitational field, the book spontaneously falls to the floor. When you place a ball on a slope, the ball spontaneously rolls down the slope. For simple mechanical systems, such as dropping a book or rolling a ball, the prediction of spontaneity is fairly intuitive. A mechanical system tends toward lowest potential energy, and this is usually easy to see (at least in *simple* mechanical systems). However, the prediction of spontaneity for chemical systems is not so intuitively obvious. We wish to develop a criterion for the spontaneity of chemical systems. In other words, we wish to develop a *chemical potential* that predicts the direction of a chemical system, much as mechanical potential energy predicts the direction of a mechanical system **Figure 17.2▼**.

We must not, however, confuse the *spontaneity* of a chemical reaction with the *speed* (or rate) of a chemical reaction. In thermodynamics, we study the *spontaneity* of a reaction—the direction in which and extent to which a chemical reaction proceeds. In kinetics, we study the *speed* of the reaction—how fast a reaction takes place **Figure 17.3▶**. A reaction may be thermodynamically spontaneous but kinetically slow at a given temperature. For example, the conversion of diamond to graphite is thermodynamically spontaneous. But your diamonds will not become worthless anytime soon because the process is extremely slow kinetically. Although the rate of a spontaneous process can be increased by the use of a catalyst, a nonspontaneous process cannot be made spontaneous by the use of a catalyst. Catalysts affect only the rate of a reaction, not the spontaneity.

The Concept of Chemical Potential

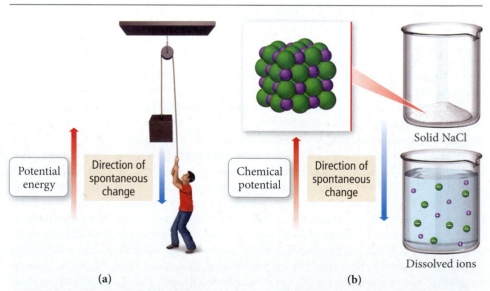

▲ **FIGURE 17.2 Mechanical Potential Energy and Chemical Potential** (a) Mechanical potential energy predicts the direction in which a mechanical system will spontaneously move. (b) We seek a chemical potential that predicts the direction in which a chemical system will spontaneously move.

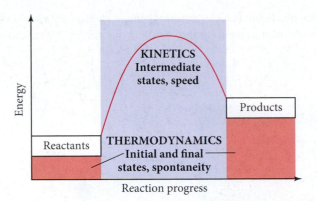

One last word about nonspontaneity—a nonspontaneous process is not *impossible*. For example, the extraction of iron metal from iron ore is a nonspontaneous process; it does not happen if the iron ore is left to itself, but that doesn't mean it is impossible. A nonspontaneous process can be made spontaneous by coupling it to another process that is spontaneous, or by supplying energy from an external source. Iron can be separated from its ore if external energy is supplied, usually by means of another reaction (that is highly spontaneous).

17.3 Entropy and the Second Law of Thermodynamics

The first candidate in our search for a chemical potential might be enthalpy, which we defined in Chapter 6. Perhaps, just as a mechanical system proceeds in the direction of lowest potential energy, so a chemical system might proceed in the direction of lowest enthalpy. If this were the case, all exothermic reactions would be spontaneous and all endothermic reactions would not. However, although *most* spontaneous reactions are exothermic, some spontaneous reactions are *endothermic*. For example, above 0 °C, ice spontaneously melts (an endothermic process). So enthalpy must not be the sole criterion for spontaneity.

See Section 6.5 for the definition of enthalpy.

We can learn more about the driving force behind chemical reactions by considering several processes (like ice melting) that involve an increase in enthalpy. These processes are energetically uphill (they are endothermic), yet they occur spontaneously. What drives them?

- the melting of ice above 0 °C
- the evaporation of liquid water to gaseous water
- the dissolution of sodium chloride in water

Each of these processes is endothermic *and* spontaneous. Do they have anything in common? Notice that, in each process, disorder or randomness increases. In the melting of ice, the arrangement of the water molecules changes from a highly ordered one (in ice) to a somewhat disorderly one (in liquid water).

The use of the word *disorder* here is only analogous to our macroscopic notions of disorder. The definition of molecular disorder, which is covered shortly, is very specific.

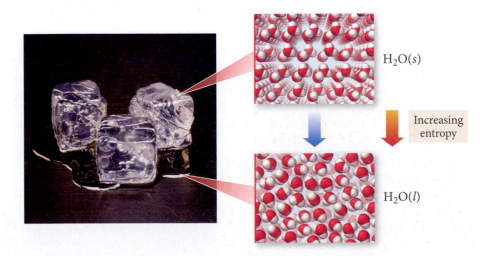

$H_2O(s)$

Increasing entropy

$H_2O(l)$

◀ When ice melts, the arrangement of water molecules changes from an orderly one to a more disorderly one.

In the evaporation of a liquid to a gas, the arrangement changes from a *somewhat* disorderly one (atoms or molecules in the liquid) to a *highly* disorderly one (atoms or molecules in the gas).

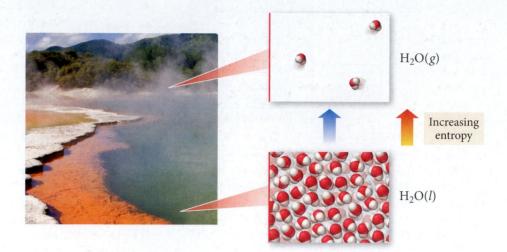

▶ When water evaporates, the arrangement of water molecules becomes still more disorderly.

In the dissolution of a salt into water, the arrangement again changes from an orderly one (in which the ions in the salt occupy regular positions in the crystal lattice) to a more disorderly one (in which the ions are randomly dispersed throughout the liquid water).

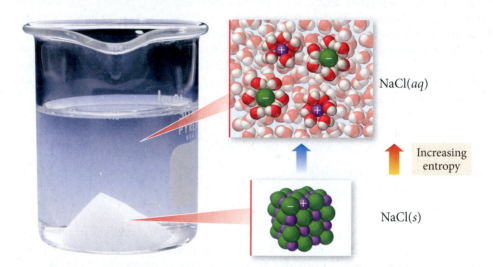

▶ When salt dissolves in water, the arrangement of the molecules and ions becomes more disorderly.

In all three of these processes, *entropy*—which is related to disorder or randomness at the molecular level—increases.

Entropy

We have hit upon the criterion for spontaneity in chemical systems: entropy. Informally, we can think of entropy as disorder or randomness. *But the concept of disorder or randomness on the macroscopic scale—such as the messiness of a drawer—is only analogous to the concept of disorder or randomness on the molecular scale.* Formally, **entropy**, abbreviated by the symbol *S*, has the following definition:

> **Entropy (*S*) is a thermodynamic function that increases with the number of *energetically equivalent* ways to arrange the components of a system to achieve a particular state.**

This definition was expressed mathematically by Ludwig Boltzmann as

$$S = k \ln W$$

where k is the Boltzmann constant (the gas constant divided by Avogadro's number, $R/N_A = 1.38 \times 10^{-23}$ J/K) and W is the number of energetically equivalent ways to arrange the components of the system. Since W is unitless (it is simply a number), the units of entropy are joules per kelvin (J/K). We will talk about the significance of the units shortly. As you can see from the equation, as W increases, entropy increases.

The key to understanding entropy is the quantity W. What does W—the number of energetically equivalent ways to arrange the components of the system—signify? Imagine a system of particles such as a fixed amount of an ideal gas. A given set of conditions (P, V, and T) define the *state* (or *macrostate*) of the system. As long as these conditions remain constant, the energy of the system also remains constant. However, exactly *where* that energy is at any given instant is anything but constant.

At any one instant, a particular gas particle may have lots of kinetic energy. However, after a very short period of time, that particle may have only a little kinetic energy (because it lost its energy through collisions with other particles). The exact internal energy distribution among the particles at any one instant is sometimes referred to as a *microstate*. You can think of a microstate as a snapshot of the system at a given instant in time. The next instant, the snapshot (the microstate) changes. However, the *macrostate*— defined by P, V, and T—remains constant. A given macrostate exists as a result of a large number of different microstates. In other words, the snapshot (or microstate) of a given macrostate is generally different from one moment to the next as the energy of the system is constantly redistributing itself among the particles of the system.

We can now conceive of W in terms of microstates. The quantity, W, is the number of *possible* microstates that can result in a given macrostate. Let's explore this concept further with a simple example. Suppose we have 2 systems (call them System A and System B) and that each is composed of two particles (one blue and one red). Both systems have a total energy of 4 Joules, but System A has only one energy level and System B has two:

▲ Boltzman's equation is engraved on his tombstone.

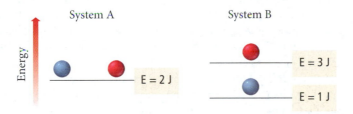

Each system has the same total energy (4 J), but System A has only one possible microstate (the red and the blue particle both occupying the 2 J energy level) while System B has a second possible microstate:

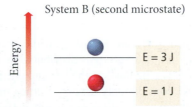

In this second microstate for System B, the blue particle has 3 J and red one has 1 J (as opposed to System B's first microstate where the energy of the particles is switched). This second microstate is not possible for System A because it has only 1 energy level. For System A, $W = 1$, but for System B, $W = 2$. In other words, System B has more microstates that result in the same 4 J macrostate. Since *W* is larger for System B than for System A, System B has greater *entropy*; it has more *energetically equivalent ways to arrange the components of the system.*

We can understand an important aspect of entropy by turning our attention to energy for a moment. The entropy of a state increases with the number of *energetically equivalent* ways to arrange the components of the system to achieve that particular state. This implies that *the state with the highest entropy also has the greatest dispersal of energy.* Returning to our previous example, the energy of System B is dispersed over two energy levels instead of being confined to just one. At the heart of entropy is the concept of energy dispersal or energy randomization. *A state in which a given amount of energy is more highly dispersed (or more highly randomized) has more entropy than a state in which the same energy is more highly concentrated.*

Although we have already alluded to the **second law of thermodynamics**, we can now formally define it:

> **For any spontaneous process, the entropy of the *universe* increases ($\Delta S_{univ} > 0$).**

The criterion for spontaneity is the entropy of the universe. Processes that increase the entropy of the universe—those that result in greater dispersal or randomization of energy—occur spontaneously. Processes that decrease the entropy of the universe do not occur spontaneously.

See the discussion of state functions in Section 6.2.

Entropy, like enthalpy, is a *state function*—its value depends only on the state of the system, not on how the system got to that state. Therefore, for any process, *the change in entropy is the entropy of the final state minus the entropy of the initial state.*

$$\Delta S = S_{final} - S_{initial}$$

Entropy determines the direction of chemical and physical change. *A chemical system proceeds in a direction that increases the entropy of the universe*—it proceeds in a direction that has the largest number of *energetically equivalent* ways to arrange its components. To better understand this tendency, let's examine the expansion of an ideal gas into a vacuum (a spontaneous process with no associated change in enthalpy). A flask containing an ideal gas is connected to another, evacuated flask by a tube equipped with a stopcock. When the stopcock is opened, the gas spontaneously expands into the evacuated flask. Since the gas is expanding into a vacuum, the pressure against which it expands is zero, and therefore the work ($w = -P_{ext}\Delta V$) is also zero.

However, even though the total energy of the gas does not change during the expansion, the entropy does change. To see this, consider the following simplified system containing only four gas atoms.

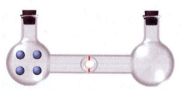

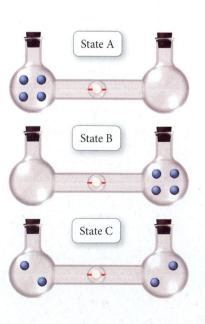

When the stopcock is opened, there are several possible energetically equivalent final states that may result, each with the four atoms distributed in a different way. For example, there could be three atoms in the flask on the left and one in the flask on the right, or vice versa. For simplicity, we consider only the possibilities shown at left.

Since the energy of any one atom is the same in either flask, and since the atoms do not interact, all three states are energetically equivalent. Now we ask the following question for each possible state: How many *internal arrangements* (microstates) give rise to the same *external arrangement* (macrostate)? To keep track of the internal arrangements we label the atoms 1–4. However, since the atoms are all the same, there is no difference between them externally. For states A and B, there is only one internal arrangement that gives the specified external arrangement—atoms 1–4 on the left side or the right side, respectively.

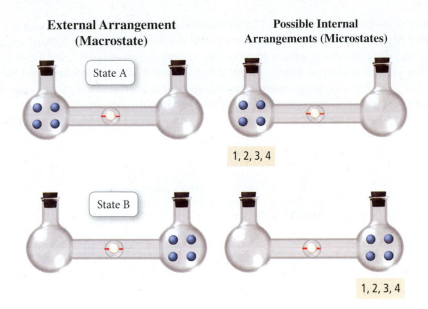

External Arrangement (Macrostate)   **Possible Internal Arrangements (Microstates)**

For state C, however, there are six possible internal arrangements that all give the same external arrangement (two atoms on each side).

External Arrangement (Macrostate)   **Possible Internal Arrangements (Microstates)**

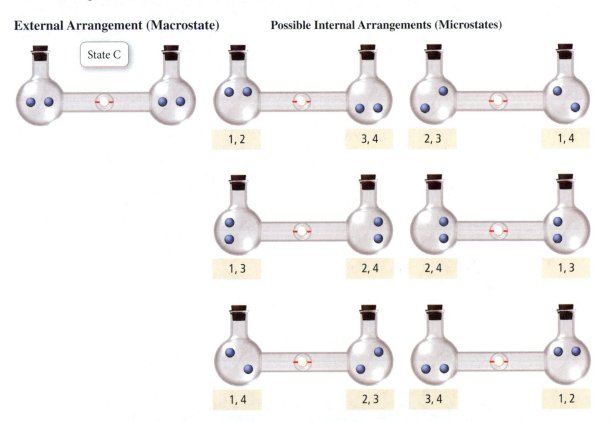

In other words, if the atoms are just randomly moving between the two flasks, the statistical probability of finding the atoms in state C is six times greater than the probability of finding the atoms in states A or B. Consequently, even for a simple system consisting of only four atoms, the atoms are most likely to be found in state C. State C has the greatest entropy—it has the greatest number of energetically equivalent ways to distribute its components.

As the number of atoms increases, the number of internal arrangements that leads to the atoms being equally distributed between the two flasks increases dramatically. For example, with 10 atoms, the number of internal arrangements leading to an equal distribution is 252 and with 20 atoms the number of internal arrangements is 184,756. Yet, the number of internal arrangements that leads to all of the atoms being on the left side does

In these drawings, the exact location of an atom within a flask is insignificant. The focus is only on whether the atom is in the left flask or the right flask.

For n particles, the number of ways to put r particles in one flask and $n - r$ particles in the other flask is $n!/[(n-r)!\, r!]$. For 10 atoms equally distributed, $n = 10$ and $r = 5$.

not increase—it is always only 1. The arrangement in which the atoms are equally distributed between the two flasks has a much larger number of possible internal arrangements and therefore much greater entropy. The system therefore tends toward that state.

The change in *entropy* in going from a state in which all of the atoms are in the left flask to the state in which the atoms are evenly distributed between both flasks is positive because the final state has a greater entropy than the initial state:

$$\Delta S = S_{final} - S_{initial}$$

Entropy of state in which atoms are distributed between both flasks

Entropy of state in which atoms are all in one flask

Since S_{final} is greater than $S_{initial}$, ΔS is positive and the process is spontaneous according to the second law. Notice that when the atoms are confined to one flask, their energy is also confined to that one flask; however, when the atoms are evenly distributed between both flasks, their energy is spread out over a greater volume. As the gas expands into the empty flask, energy is dispersed.

Heat

The second law explains many phenomena not explained by the first law. For example, in Chapter 6, we learned that heat travels from a substance at higher temperature to one at lower temperature. If we drop an ice cube into water, heat travels from the water to the ice cube—the water cools and the ice warms (and eventually melts). Why? The first law would not prohibit some heat from flowing the other way—from the ice to the water. For example, the ice could lose 10 J of heat (cooling even more) and the water could gain 10 J of heat (warming even more). The first law of thermodynamics is not violated by such a heat transfer. Imagine putting ice into water only to have the water get warmer as it absorbed thermal energy from the ice! It will never happen. Why? Because heat transfer from cold to hot violates the *second* law of thermodynamics. According to the second law, energy is dispersed, not concentrated. The transfer of heat from a substance of higher temperature to one of lower temperature results in greater energy randomization—the energy that was concentrated in the hot substance becomes dispersed between the two substances. The second law describes this pervasive tendency.

Conceptual Connection 17.2 Entropy

Consider these three changes in the distribution of six gaseous particles into interconnected boxes. Which one has a positive ΔS?

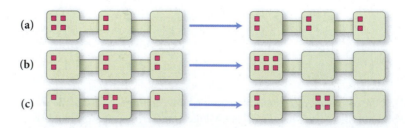

(a)

(b)

(c)

The Entropy Change Associated with a Change in State

The entropy of a sample of matter *increases* as it changes state from a solid to a liquid or from a liquid to a gas **Figure 17.4▶**. We can informally think of this increase in entropy by analogy with macroscopic disorder. The gaseous state is more disorderly than the liquid state, which is in turn more disorderly than the solid state. More formally, however, the differences in entropy are related to the number of energetically equivalent ways of

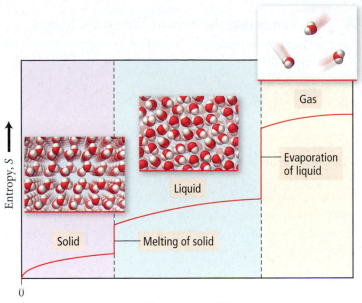

◀ **FIGURE 17.4 Entropy and Phase Change** Entropy increases in going from a solid to a liquid and in going from a liquid to a gas.

arranging the particles in each state—more in the gas than in the liquid, and more in the liquid than in the solid.

A gas has more energetically equivalent configurations because it has more ways to distribute its energy than a solid. The energy in a molecular solid consists largely of the vibrations of its molecules. If the same substance is vaporized, however, the energy can take the form of straight-line motions of the molecules (called translational energy) and rotations of the molecules (called rotational energy). In other words, when a solid vaporizes, there are new "places" to put energy **Figure 17.5▼**. The gas has more possible microstates (more energetically equivalent configurations) than the solid and therefore a greater entropy.

We can now predict the sign of ΔS for processes involving changes of state (or phase). In general, entropy increases ($\Delta S > 0$) for each transition:

- the phase transition from a solid to a liquid
- the phase transition from a solid to a gas
- the phase transition from a liquid to a gas
- an increase in the number of moles of a gas during a chemical reaction

Additional "Places" for Energy

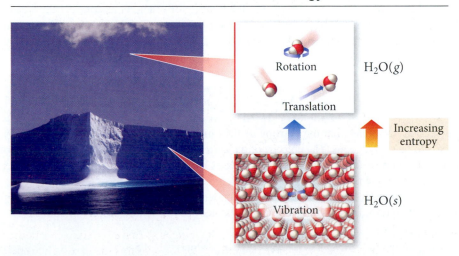

▲ **FIGURE 17.5 "Places" for Energy** In the solid phase, energy is contained largely in the vibrations between molecules. In the gas phase, energy can be contained in both the straight-line motion of molecules (translational energy) and the rotation of molecules (rotational energy).

EXAMPLE 17.1 Predicting the Sign of Entropy Change

Predict the sign of ΔS for each process:

(a) $H_2O(g) \longrightarrow H_2O(l)$

(b) Solid carbon dioxide sublimes.

(c) $2 N_2O(g) \longrightarrow 2 N_2(g) + O_2(g)$

SOLUTION

(a) Since a gas has a greater entropy than a liquid, the entropy decreases and ΔS is therefore negative.

(b) Since a solid has a lower entropy than a gas, the entropy increases and ΔS is therefore positive.

(c) Since the number of moles of gas increases, the entropy increases and ΔS is therefore positive.

FOR PRACTICE 17.1

Predict the sign of ΔS for each process:

(a) the boiling of water

(b) $I_2(g) \longrightarrow I_2(s)$

(c) $CaCO_3(s) \longrightarrow CaO(s) + CO_2(g)$

17.4 Heat Transfer and Changes in the Entropy of the Surroundings

We have now seen that the criterion for spontaneity is an increase in the entropy of the universe. However, you can probably think of several spontaneous processes in which entropy seems to decrease. For example, when water freezes at temperatures below 0 °C, the entropy of the water decreases, yet the process is spontaneous. Similarly, when water vapor in air condenses into fog on a cold night, the entropy of the water also decreases. Why are these processes spontaneous?

To answer this question, we must return to our statement of the second law: for any spontaneous process, the entropy *of the universe* increases ($\Delta S_{univ} > 0$). Even though the entropy *of the water* decreases during freezing and condensation, the entropy *of the universe* must somehow increase in order for these processes to be spontaneous. In Chapter 6, we distinguished between a thermodynamic system and its surroundings. The same distinction is helpful in our discussion of entropy. For the freezing of water, let us consider the water as the system. The surroundings are then the rest of the universe. Using these distinctions, ΔS_{sys} is the entropy change for the water itself, ΔS_{surr} is the entropy change for the surroundings, and ΔS_{univ} is the entropy change for the universe. The entropy change for the universe is just the sum of the entropy changes for the system and the surroundings:

$$\Delta S_{univ} = \Delta S_{sys} + \Delta S_{surr}$$

The second law states that the entropy of the universe must increase ($\Delta S_{univ} > 0$) for a process to be spontaneous. The entropy of the *system* can therefore decrease ($\Delta S_{sys} < 0$) as long as the entropy of the *surroundings* increases by a greater amount ($\Delta S_{surr} > -\Delta S_{sys}$), so that the overall entropy of the *universe* undergoes a net increase.

For liquid water freezing or water vapor condensing, we know that the change in entropy for the system (ΔS_{sys}) is negative because the water becomes more orderly in both cases.

$$\Delta S_{univ} = \Delta S_{sys} + \Delta S_{surr}$$

Negative   Positive

For ΔS_{univ} to be positive, therefore, ΔS_{surr} must be *positive* and greater in absolute value (or magnitude) than ΔS_{sys}. But why does the freezing or condensation of water increase the entropy of the surroundings? Both processes are *exothermic*: they give off heat to the surroundings. If we think of entropy as the dispersal or randomization of energy, then *the release of heat energy by the system disperses that energy into the surroundings, increasing the entropy of the surroundings*. The freezing of water below 0 °C and the condensation of water vapor on a cold night both increase the entropy of the universe because the heat given off to the surroundings increases the entropy of the surroundings to a sufficient degree to overcome the entropy decrease in the water.

Even though (as we saw earlier) enthalpy by itself cannot determine spontaneity, the increase in the entropy of the surroundings caused by the release of heat explains why exothermic processes are so *often* spontaneous.

The Temperature Dependence of ΔS_{surr}

We have just seen how the freezing of water increases the entropy of the surroundings by dispersing heat energy into the surroundings. However, we know that the freezing of water is not spontaneous at all temperatures. The freezing of water becomes *nonspontaneous* above 0 °C. Why? Because the magnitude of the increase in the entropy of the surroundings due to the dispersal of energy into the surroundings is *temperature dependent*.

$$\Delta S_{univ} = \Delta S_{sys} + \Delta S_{surr} \quad \text{(for water freezing)}$$

Negative   Positive, but magnitude
depends on temperature

The greater the temperature, the smaller the increase in entropy for a given amount of energy dispersed into the surroundings. Recall that the units of entropy are joules per kelvin—that is, those of energy divided by those of temperature. Entropy is a measure of energy dispersal *per unit temperature*. The higher the temperature, the smaller the increase in entropy for a given amount of energy dispersed. We can understand the temperature dependence of entropy changes due to heat flow with an analogy. Imagine that you have $1000 to give away. If you gave the $1000 to a rich man, the impact on his net worth would be negligible (because he already has so much money). If you gave the same $1000 to a poor man, however, his net worth would change substantially (because he has so little money). Similarly, if you disperse 1000 J of energy into surroundings that are hot, the entropy increase is small (because the impact of the 1000 J on surroundings that already contain a lot of energy is small). If you disperse the same 1000 J of energy into surroundings that are cold, however, the entropy increase is large (because the impact of the 1000 J on surroundings that contain little energy is great). Therefore, the impact of the heat released to the surroundings by the freezing of water depends on the temperature of the surroundings—the higher the temperature, the smaller the impact.

We can now see why the freezing of water is spontaneous at low temperature but nonspontaneous at high temperature.

$$\Delta S_{univ} = \Delta S_{sys} + \Delta S_{surr}$$

Negative   Positive and large at low temperature
Positive and small at high temperature

At low temperature, the decrease in entropy of the system (a negative quantity) is overcome by the large increase in the entropy of the surroundings (a positive quantity), resulting in a positive ΔS_{univ} and therefore a spontaneous process. At high temperature, however, the decrease in entropy of the system is not overcome by the increase in entropy of the surroundings (because the magnitude of the positive ΔS_{surr} is smaller at higher temperatures), resulting in a negative ΔS_{univ}; therefore, the freezing of water is not spontaneous at high temperature.

Quantifying Entropy Changes in the Surroundings

We have seen that when a system exchanges heat with the surroundings, it changes the entropy of the surroundings. At constant pressure, we can use q_{sys} to quantify the change in *entropy* for the surroundings (ΔS_{surr}). In general,

- A process that emits heat into the surroundings (q_{sys} negative) *increases* the entropy of the surroundings (positive ΔS_{surr}).
- A process that absorbs heat from the surroundings (q_{sys} positive) *decreases* the entropy of the surroundings (negative ΔS_{surr}).
- The magnitude of the change in entropy of the surroundings is proportional to the magnitude of q_{sys}.

We can summarize these three points with this proportionality:

$$\Delta S_{surr} \propto -q_{sys} \qquad [17.1]$$

We have also seen that, for a given amount of heat exchanged with the surroundings, the magnitude of ΔS_{surr} is inversely proportional to the temperature. In general, the higher the temperature, the lower the magnitude of ΔS_{surr} for a given amount of heat exchanged:

$$\Delta S_{surr} \propto \frac{1}{T} \qquad [17.2]$$

Combining the proportionalities in Equations 17.1 and 17.2, we get the following general expression at constant temperature:

$$\Delta S_{surr} = \frac{-q_{sys}}{T}$$

For any chemical or physical process occurring at constant temperature and pressure, the entropy change of the surroundings is equal to the energy dispersed into the surroundings ($-q_{sys}$) divided by the temperature of the surroundings in kelvins.

From this equation, we can see why exothermic processes have a tendency to be spontaneous at low temperatures—they increase the entropy of the surroundings. As temperature increases, however, a given negative q produces a smaller positive ΔS_{surr}; and exothermicity becomes less of a determining factor for spontaneity.

Under conditions of constant pressure $q_{sys} = \Delta H_{sys}$; therefore,

$$\Delta S_{surr} = \frac{-\Delta H_{sys}}{T} \qquad \text{(constant } P, T) \qquad [17.3]$$

EXAMPLE 17.2   Calculating Entropy Changes in the Surroundings

Consider the combustion of propane gas:

$$C_3H_8(g) + 5\,O_2(g) \longrightarrow 3\,CO_2(g) + 4\,H_2O(g) \quad \Delta H_{rxn} = -2044 \text{ kJ}$$

(a) Calculate the entropy change in the surroundings associated with this reaction occurring at 25 °C.

(b) Determine the sign of the entropy change for the system.

(c) Determine the sign of the entropy change for the universe. Will the reaction be spontaneous?

SOLUTION

(a) The entropy change of the surroundings is given by Equation 17.3. Substitute the value of ΔH_{rxn} and the temperature in kelvins and calculate ΔS_{surr}.	$T = 273 + 25 = 298$ K $$\Delta S_{surr} = \frac{-\Delta H_{rxn}}{T}$$ $$= \frac{-(-2044 \text{ kJ})}{298 \text{ K}}$$ $$= +6.86 \text{ kJ/K}$$ $$= +6.86 \times 10^3 \text{ J/K}$$

(b) Determine the number of moles of gas on each side of the reaction. An increase in the number of moles of gas implies a positive ΔS_{sys}.	$C_3H_8(g) + 5\,O_2(g) \longrightarrow 3\,CO_2(g) + 4\,H_2O(g)$ 6 mol gas $\qquad\qquad$ 7 mol gas ΔS_{sys} is positive.
(c) The change in entropy of the universe is the sum of the entropy changes of the system and the surroundings. If the entropy changes of the system and surroundings are both the same sign, the entropy change for the universe also has the same sign.	$\Delta S_{univ} = \Delta S_{sys} + \Delta S_{surr}$ Positive $\qquad$ Positive Therefore, ΔS_{univ} is positive and the reaction is spontaneous.

FOR PRACTICE 17.2

Consider the reaction between nitrogen and oxygen gas to form dinitrogen monoxide:

$$2\,N_2(g) + O_2(g) \longrightarrow 2\,N_2O(g) \quad \Delta H_{rxn} = +163.2 \text{ kJ}$$

(a) Calculate the entropy change in the surroundings associated with this reaction occurring at 25 °C.

(b) Determine the sign of the entropy change for the system.

(c) Determine the sign of the entropy change for the universe. Will the reaction be spontaneous?

FOR MORE PRACTICE 17.2

A reaction has $\Delta H_{rxn} = -107$ kJ and $\Delta S_{rxn} = 285$ J/K. At what temperature is the change in entropy for the reaction equal to the change in entropy for the surroundings?

 Conceptual Connection 17.3 **Entropy and Biological Systems**

Biological systems seem (at first glance) to contradict the second law of thermodynamics. By taking energy from their surroundings and synthesizing large, complex biological molecules, plants and animals tend to concentrate energy, not disperse it. How can this be so?

17.5 Gibbs Free Energy

Equation 17.3 establishes a relationship between the enthalpy change in a system and the entropy change in the surroundings. Recall that for any process the entropy change of the universe is the sum of the entropy change of the system and the entropy change of the surroundings:

$$\Delta S_{univ} = \Delta S_{sys} + \Delta S_{surr} \qquad [17.4]$$

Combining this equation with Equation 17.3 gives us the following relationship at constant temperature and pressure:

$$\Delta S_{univ} = \Delta S_{sys} - \frac{\Delta H_{sys}}{T} \qquad [17.5]$$

Notice that, by using Equation 17.5, we can calculate ΔS_{univ} while focusing only on the *system*. If we multiply Equation 17.5 by $-T$, we get:

$$-T\,\Delta S_{univ} = -T\,\Delta S_{sys} + \cancel{T}\frac{\Delta H_{sys}}{\cancel{T}}$$

$$= \Delta H_{sys} - T\,\Delta S_{sys} \qquad [17.6]$$

If we now drop the subscript *sys*—from now on ΔH and ΔS without subscripts will mean ΔH_{sys} and ΔS_{sys}—we get:

$$-T\,\Delta S_{univ} = \Delta H - T\,\Delta S \qquad [17.7]$$

The right hand side of Equation 17.7 represents the change in a thermodynamic function called the *Gibbs free energy*. The formal definition of **Gibbs free energy (G)** is

$$G = H - TS \qquad [17.8]$$

where H is enthalpy, T is the temperature in kelvins, and S is entropy. The *change* in Gibbs free energy, symbolized by ΔG, is therefore expressed as follows (at constant temperature):

$$\Delta G = \Delta H - T\,\Delta S \qquad [17.9]$$

If we combine Equations 17.7 and 17.9, we can understand the significance of ΔG:

$$\Delta G = -T\,\Delta S_{\text{univ}} \qquad (\text{constant } T, P) \qquad [17.10]$$

The change in Gibbs free energy for a process occurring at constant temperature and pressure is proportional to the negative of ΔS_{univ}. Since ΔS_{univ} is a criterion for spontaneity, ΔG is also a criterion for spontaneity (although opposite in sign). In fact, Gibbs free energy is sometimes called *chemical potential*, because it is analogous to mechanical potential energy discussed earlier. Just as mechanical systems tend toward lower potential energy, so chemical systems tend toward lower Gibbs free energy (or toward lower chemical potential) **Figure 17.6▼**.

Summarizing Gibbs Free Energy (at Constant Temperature and Pressure):

▶ ΔG is proportional to the negative of ΔS_{univ}.

▶ A decrease in Gibbs free energy ($\Delta G < 0$) corresponds to a spontaneous process.

▶ An increase in Gibbs free energy ($\Delta G > 0$) corresponds to a nonspontaneous process.

Notice that we can calculate changes in Gibbs free energy solely with reference to the system. So, to determine whether a process is spontaneous, we just have to find the change in *entropy* for the system (ΔS) and the change in *enthalpy* for the system (ΔH). We can then predict the spontaneity of the process at any temperature. In Chapter 6, we learned how to calculate changes in enthalpy (ΔH) for chemical reactions. In Section 17.6, we learn how to calculate changes in entropy (ΔS) for chemical reactions. We can use those two quantities to calculate changes in free energy (ΔG) for chemical reactions and therefore predict their spontaneity (Section 17.7). Before we move on to these matters, however, let's examine some examples that demonstrate how ΔH, ΔS, and T affect the spontaneity of chemical processes.

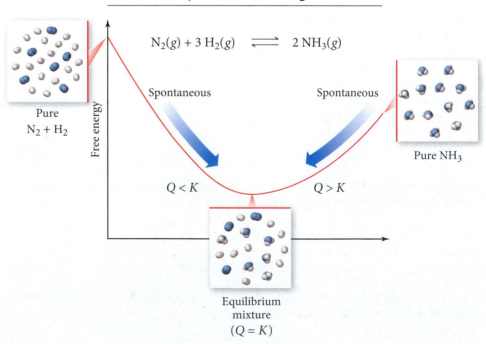

Free Energy Determines the Direction of Spontaneous Change

$$N_2(g) + 3\,H_2(g) \; \rightleftharpoons \; 2\,NH_3(g)$$

Pure N₂ + H₂

Spontaneous

Spontaneous

Pure NH₃

$Q < K$

$Q > K$

Free energy

Equilibrium mixture ($Q = K$)

▶ FIGURE 17.6 Gibbs Free Energy Gibbs free energy is also called chemical potential because it determines the direction of spontaneous change for chemical systems. The relationship between Q, K, and ΔG, (implied in this figure) is covered in detail in Sections 17.8 and 17.9.

The Effect of ΔH, ΔS, and T on Spontaneity

Case 1: ΔH Negative, ΔS Positive

If a reaction is exothermic ($\Delta H < 0$), and if the change in entropy for the reaction is positive ($\Delta S > 0$), then the change in free energy is negative at all temperatures and the reaction is therefore spontaneous at all temperatures.

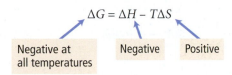

Case 2: ΔH Positive, ΔS Negative

If a reaction is endothermic ($\Delta H > 0$), and if the change in entropy for the reaction is negative ($\Delta S < 0$), then the change in free energy is positive at all temperatures and the reaction is therefore nonspontaneous at all temperatures.

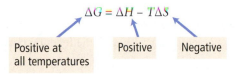

Case 3: ΔH Negative, ΔS Negative

If a reaction is exothermic ($\Delta H < 0$), and if the change in entropy for the reaction is negative ($\Delta S < 0$), then the change in free energy depends on temperature. At a low enough temperature, the heat emitted into the surroundings causes a large entropy change in the surroundings, making the process spontaneous. At high temperature, the same amount of heat is dispersed into warmer surroundings, so the positive entropy change in the surroundings is smaller, resulting in a nonspontaneous process. The reaction is spontaneous at low temperature, but nonspontaneous at high temperature.

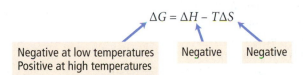

Case 4: ΔH Positive, ΔS Positive

If a reaction is endothermic ($\Delta H > 0$), and if the change in entropy for the reaction is positive ($\Delta S > 0$), then the change in free energy also depends on temperature. In this case, however, high temperature favors spontaneity because the absorption of heat from the surroundings has less effect on the entropy of the surroundings as the temperature increases. The reaction is nonspontaneous at low temperature but spontaneous at high temperature.

$$\Delta G = \Delta H - T\Delta S$$

Positive at low temperatures
Negative at high temperatures   Positive   Positive

Examples of each case are summarized in Table 17.1. Notice that when ΔH and ΔS have opposite signs, the spontaneity of the reaction does not depend on temperature. *When ΔH and ΔS have the same sign, however, the spontaneity does depend on temperature.* The temperature at which the reaction changes from being spontaneous to being nonspontaneous (or vice versa) is the temperature at which ΔG changes sign, which can be found by setting $\Delta G = 0$ and solving for T, as shown in part b of Example 17.3.

TABLE 17.1 The Effect of ΔH, ΔS, and T on Spontaneity

ΔH	ΔS	Low Temperature	High Temperature	Example	
$-$	$+$	Spontaneous ($\Delta G < 0$)	Spontaneous ($\Delta G < 0$)	$2 N_2O(g) \longrightarrow 2 N_2(g) + O_2(g)$	$\Delta H^\circ_{rxn} = -163.2$ kJ
$+$	$-$	Nonspontaneous ($\Delta G > 0$)	Nonspontaneous ($\Delta G > 0$)	$3 O_2(g) \longrightarrow 2 O_3(g)$	$\Delta H^\circ_{rxn} = +285.4$ kJ
$-$	$-$	Spontaneous ($\Delta G < 0$)	Nonspontaneous ($\Delta G > 0$)	$H_2O(l) \longrightarrow H_2O(s)$	$\Delta H^\circ = -6.01$ kJ
$+$	$+$	Nonspontaneous ($\Delta G > 0$)	Spontaneous ($\Delta G < 0$)	$H_2O(l) \longrightarrow H_2O(g)$	$\Delta H^\circ = +40.7$ kJ (at 100 °C)

EXAMPLE 17.3 Calculating Gibbs Free Energy Changes and Predicting Spontaneity from ΔH and ΔS

Consider the reaction for the decomposition of carbon tetrachloride gas:

$$CCl_4(g) \longrightarrow C(s, \text{graphite}) + 2 Cl_2(g) \quad \Delta H = +95.7 \text{ kJ}; \Delta S = +142.2 \text{ J/K}$$

(a) Calculate ΔG at 25 °C and determine whether the reaction is spontaneous.

(b) If the reaction is not spontaneous at 25 °C, determine at what temperature (if any) the reaction becomes spontaneous.

SOLUTION

(a) Use Equation 17.9 to calculate ΔG from the given values of ΔH and ΔS. The temperature must be in kelvins. Also, *be sure to express both ΔH and ΔS in the same units (usually joules).*	$T = 273 + 25 = 298$ K $\Delta G = \Delta H - T\,\Delta S$ $\quad = 95.7 \times 10^3 \text{ J} - (298 \text{ K})\,142.2 \text{ J/K}$ $\quad = 95.7 \times 10^3 \text{ J} - 42.4 \times 10^3 \text{ J}$ $\quad = +53.3 \times 10^3 \text{ J}$ Therefore the reaction is not spontaneous.
(b) Since ΔS is positive, ΔG becomes more negative with increasing temperature. To determine the temperature at which the reaction becomes spontaneous, use Equation 17.9 to find the temperature at which ΔG changes from positive to negative (that is, set $\Delta G = 0$ and solve for T). The reaction is spontaneous above this temperature.	$\Delta G = \Delta H - T\,\Delta S$ $0 = 95.7 \times 10^3 \text{ J} - (T)\,142.2 \text{ J/K}$ $T = \dfrac{95.7 \times 10^3 \text{ J}}{142.2 \text{ J/K}}$ $\quad = 673$ K

FOR PRACTICE 17.3

Consider the following reaction:

$$C_2H_4(g) + H_2(g) \longrightarrow C_2H_6(g) \quad \Delta H = -137.5 \text{ kJ}; \Delta S = -120.5 \text{ J/K}$$

Calculate ΔG at 25 °C and determine whether the reaction is spontaneous. Does ΔG become more negative or more positive as the temperature increases?

Conceptual Connection 17.4 ΔH, ΔS, and ΔG

Which statement is true for the sublimation of dry ice (solid CO_2)?

(a) ΔH is positive, ΔS is positive, and ΔG is positive at low temperature and negative at high temperature.

(b) ΔH is negative, ΔS is negative, and ΔG is negative at low temperature and positive at high temperature.

(c) ΔH is negative, ΔS is positive, and ΔG is negative at all temperatures.

(d) ΔH is positive, ΔS is negative, and ΔG is positive at all temperatures.

17.6 Entropy Changes in Chemical Reactions: Calculating ΔS°_{rxn}

In Chapter 6, we learned how to calculate standard changes in enthalpy (ΔH°_{rxn}) for chemical reactions. We now turn to calculating standard changes in *entropy* for chemical reactions. Recall from Section 6.8 that the standard enthalpy change for a reaction (ΔH°_{rxn}) is the change in enthalpy for a process in which all reactants and products are in their standard states. Recall also the definition of the standard state:

- *For a Gas*: The standard state for a gas is the pure gas at a pressure of exactly 1 atm.
- *For a Liquid or Solid*: The standard state for a liquid or solid is the pure substance in its most stable form at a pressure of 1 atm and at the temperature of interest (often taken to be 25 °C).
- *For a Substance in Solution*: The standard state for a substance in solution is a concentration of exactly 1 M.

We now define the **standard entropy change for a reaction (ΔS°_{rxn})** as the change in *entropy* for a process in which all reactants and products are in their standard states. Since entropy is a function of state, the standard change in entropy is just the standard entropy of the products minus the standard entropy of the reactants.

$$\Delta S^{\circ}_{rxn} = S^{\circ}_{products} - S^{\circ}_{reactants}$$

But how do we find the standard entropies of the reactants and products? Recall from Chapter 6 that we defined *standard molar enthalpies of formation* (ΔH°_f) to use in calculating ΔH°_{rxn}. We now want to define **standard molar entropies (S°)** to use in calculating ΔS°_{rxn}.

> The standard state has recently been changed to a pressure of 1 bar, which is very close to 1 atm (1 atm = 1.013 bar). Both standards are now in common use.

Standard Molar Entropies (S°) and the Third Law of Thermodynamics

In Chapter 6, we defined a *relative* zero for enthalpy. Recall that we assigned a value of zero to the standard enthalpy of formation for an element in its standard state. This was necessary because absolute values of enthalpy cannot be determined. In other words, for enthalpy, there is no absolute zero against which to measure all other values; therefore, we always have to rely on enthalpy changes from an arbitrarily assigned standard. For entropy, however, *there is an absolute zero*. The absolute zero of entropy is given by the **third law of thermodynamics**, which states:

The entropy of a perfect crystal at absolute zero (0 K) is zero.

A perfect crystal at a temperature of absolute zero has only one possible way ($W = 1$) to arrange its components **Figure 17.7▶**. Therefore, based on Boltzmann's definition of entropy ($S = k \ln W$), its entropy is zero ($S = k \ln 1 = 0$).

All entropy values can then be measured against the absolute zero of entropy defined by the third law. Table 17.2 lists values of standard entropies at 25 °C for some selected substances. A more complete list can be found in Appendix IIB. Notice that standard entropy values are listed in units of joules per mole per kelvin (J/mol·K). The unit of mole in the denominator is required because *entropy is an extensive property*—it depends on the amount of the substance.

At 25 °C, the standard entropy of any substance is a measure of the energy dispersed into one mole of that substance at 25 °C, which in turn depends on the number of "places" to put energy within the substance. The factors that affect the number of "places" to put energy—and therefore the standard entropy—include the state of the substance, the molar mass of the substance, the particular allotrope, the molecular complexity, and the extent of dissolution. In this section of the chapter, we examine each of these separately.

Perfect crystal at 0 K
$W = 1$ $S = 0$

▲ **FIGURE 17.7 Zero Entropy**
A perfect crystal at 0 K has only one possible way to arrange its components.

Relative Standard Entropies: Gases, Liquids, and Solids

As we saw in Section 17.3, the entropy of a gas is generally greater than the entropy of a liquid, which is in turn greater than the entropy of a solid. We can easily see these trends

> Some elements exist in two or more forms, called *allotropes*, within the same state. We discuss allotropes further later in this section.

TABLE 17.2 Standard Molar Entropy Values ($S°$) for Selected Substances at 298 K

Substance	$S°$ (J/mol · K)	Substance	$S°$ (J/mol · K)	Substance	$S°$ (J/mol · K)
Gases		**Liquids**		**Solids**	
$H_2(g)$	130.7	$H_2O(l)$	70.0	$MgO(s)$	27.0
$Ar(g)$	154.8	$CH_3OH(l)$	126.8	$Fe(s)$	27.3
$CH_4(g)$	186.3	$Br_2(l)$	152.2	$Li(s)$	29.1
$H_2O(g)$	188.8	$C_6H_6(l)$	173.4	$Cu(s)$	41.6
$N_2(g)$	191.6			$Na(s)$	51.3
$NH_3(g)$	192.8			$K(s)$	64.7
$F_2(g)$	202.8			$NaCl(s)$	72.1
$O_2(g)$	205.2			$CaCO_3(s)$	91.7
$Cl_2(g)$	223.1			$FeCl_3(s)$	142.3
$C_2H_4(g)$	219.3				

	$S°$ (J/mol · K)
$H_2O(l)$	70.0
$H_2O(g)$	188.8

in the tabulated values of standard entropies. Consider the relative standard entropies of liquid water and gaseous water at 25 °C (shown in the margin).

Gaseous water has a much greater standard entropy because, as we discussed in Section 17.3, it has more energetically equivalent ways to arrange its components, which in turn results in greater energy dispersal at 25 °C.

Relative Standard Entropies: Molar Mass
Consider the standard entropies of the noble gases at 25 °C:

	$S°$ (J/mol · K)	
$He(g)$	126.2	
$Ne(g)$	146.1	
$Ar(g)$	154.8	
$Kr(g)$	163.8	
$Xe(g)$	169.4	

The more massive the noble gas, the greater its entropy at 25 °C. A complete explanation of why entropy increases with increasing molar mass is beyond the scope of this book. Briefly, the energy states associated with the motion of heavy atoms are more closely spaced than those of lighter atoms. The more closely spaced energy states allow for greater dispersal of energy at a given temperature and therefore greater entropy. This trend holds only for elements in the same state. (The effect of a state change—from a liquid to a gas, for example—is far greater than the effect of molar mass.)

Relative Standard Entropies: Allotropes
As we mentioned previously, some elements can exist in two or more forms—called *allotropes*—that both have the same state. For example, the allotropes of carbon include diamond and graphite—both solid forms of carbon. Since the arrangement of atoms

within these forms is different, their standard molar entropies are different. The molar entropies of the allotropes of carbon are:

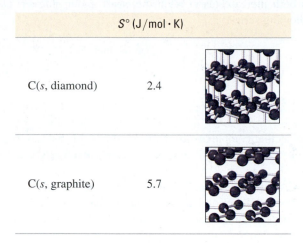

	S° (J/mol·K)	
C(s, diamond)	2.4	
C(s, graphite)	5.7	

In diamond the atoms are constrained by chemical bonds that lock them in a highly restricted three-dimensional crystal structure. In graphite the atoms bond together in sheets, but the sheets have freedom to slide past each other. The less constrained structure of graphite results in more "places" to put energy and therefore greater entropy compared to diamond.

Relative Standard Entropies: Molecular Complexity

For a given state of matter, entropy generally increases with increasing molecular complexity. Consider the standard entropies of the following two gases:

	Molar Mass (g/mol)	S° (J/mol·K)
Ar(g)	39.948	154.8
NO(g)	30.006	210.8

Ar has a greater molar mass than NO, yet it has less entropy at 25 °C. Why? Molecules generally have more "places" to put energy than do atoms. In a gaseous sample of argon, the only form that energy can take is the translational motion of the atoms. In a gaseous sample of NO, however, energy can take the form of translational motion, rotational motion, and (at high enough temperatures) vibrational motions of the molecules **Figure 17.8▼**. Consequently, for a given state, molecules will generally have a greater entropy than free atoms. Similarly, more complex molecules will generally have more entropy than simpler ones. For example, consider the standard entropies of CO and C_2H_4:

	Molar Mass (g/mol)	S° (J/mol·K)
CO(g)	28.01	197.7
C_2H_4(g)	28.05	219.3

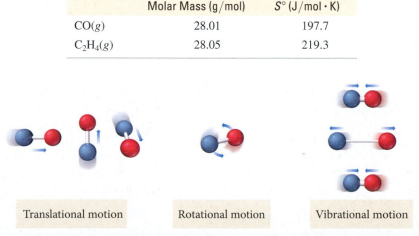

Translational motion	Rotational motion	Vibrational motion

▲ **FIGURE 17.8 "Places" for Energy in Gaseous NO**   Energy can be contained in translational motion, rotational motion, and vibrational motion.

These two substances have nearly the same molar mass, but the greater complexity of C_2H_4 compared to CO results in a greater molar entropy. When molecular complexity and molar mass both increase (as is often the case), molar entropy also increases, as shown by the following series:

	$S°$ (J/mol·K)
NO(g)	210.8
NO$_2$(g)	240.1
N$_2$O$_4$(g)	304.4

The increasing molecular complexity as you move down this list as well as the increasing molar mass results in more "places" to put energy and therefore greater entropy.

Relative Standard Entropies: Dissolution

The dissolution of a crystalline solid into solution usually results in an increase in entropy. Consider the standard entropies of solid and aqueous potassium chlorate:

	$S°$ (J/mol·K)
KClO$_3$(s)	143.1
KClO$_3$(aq)	265.7

The standard entropy for an aqueous solution is for the solution in its standard state, which is defined as having a concentration of 1 M.

When solid potassium chlorate dissolves in water, the energy that was concentrated within the crystal becomes dispersed throughout the entire solution. The greater energy dispersal results in greater entropy.

Calculating the Standard Entropy Change ($\Delta S°_{rxn}$) for a Reaction

Since entropy is a state function, and since standard entropies for many common substances are tabulated, we can calculate the standard entropy change for a chemical reaction by calculating the difference in entropy between the products and the reactants. More specifically,

> **To calculate $\Delta S°_{rxn}$, we subtract the standard entropies of the reactants multiplied by their stoichiometric coefficients from the standard entropies of the products multiplied by their stoichiometric coefficients. In the form of an equation,**

$$\Delta S°_{rxn} = \sum n_p S° \text{ (products)} - \sum n_r S° \text{ (reactants)} \qquad [17.11]$$

In Equation 17.11, n_p represents the stoichiometric coefficients of the products, n_r represents the stoichiometric coefficients of the reactants, and $S°$ represents the standard entropies. Keep in mind when using this equation that, *unlike enthalpies of formation, which are equal to zero for elements in their standard states, standard entropies are always nonzero at 25 °C.* Example 17.4 demonstrates this process.

EXAMPLE 17.4 Calculating Standard Entropy Changes ($\Delta S°_{rxn}$)

Calculate $\Delta S°_{rxn}$ for the balanced chemical equation:

$$4\,NH_3(g) + 5\,O_2(g) \longrightarrow 4\,NO(g) + 6\,H_2O(g)$$

SOLUTION

Begin by looking up (in Appendix IIB) the standard entropy for each reactant and product. Always note the correct state—(g), (l), (aq), or (s)—for each reactant and product.	

Reactant or product	$S°$ (in J/mol·K)
NH$_3$(g)	192.8
O$_2$(g)	205.2
NO(g)	210.8
H$_2$O(g)	188.8

Calculate ΔS_{rxn}° by substituting these values into Equation 17.11. Remember to include the stoichiometric coefficients in your calculation.	$\begin{aligned} \Delta S_{rxn}^{\circ} &= \sum n_p S^{\circ}\,(\text{products}) - \sum n_r S^{\circ}\,(\text{reactants}) \\ &= \left[4\left(S_{NO(g)}^{\circ}\right) + 6\left(S_{H_2O(g)}^{\circ}\right) \right] - \left[4\left(S_{NH_3(g)}^{\circ}\right) + 5\left(S_{O_2(g)}^{\circ}\right) \right] \\ &= \left[4(210.8 \text{ J/K}) + 6(188.8 \text{ J/K}) \right] - \left[4(192.8 \text{ J/K}) + 5(205.2 \text{ J/K}) \right] \\ &= 1976.0 \text{ J/K} - 1797.2 \text{ J/K} \\ &= 178.8 \text{ J/K} \end{aligned}$

CHECK

Notice that ΔS_{rxn}° is positive, as you would expect for a reaction in which the number of moles of gas increases.

FOR PRACTICE 17.4

Calculate ΔS_{rxn}° for the balanced chemical equation:

$$2\,H_2S(g) + 3\,O_2(g) \longrightarrow 2\,H_2O(g) + 2\,SO_2(g)$$

 Conceptual Connection 17.5 Standard Entropies

Arrange these gases in order of increasing standard molar entropy: SO_3, Kr, Cl_2.

17.7 Free Energy Changes in Chemical Reactions: Calculating ΔG_{rxn}°

In Section 17.6, we learned how to calculate the standard change in entropy for a chemical reaction (ΔS_{rxn}°). However, the criterion for spontaneity is the **standard change in free energy** (ΔG_{rxn}°). In this section, we examine three different ways to calculate the standard change in free energy for a reaction (ΔG_{rxn}°). In the first way, we calculate ΔH_{rxn}° and ΔS_{rxn}° from tabulated values of ΔH_f° and S°, and then use the relationship $\Delta G_{rxn}^{\circ} = \Delta H_{rxn}^{\circ} - T\,\Delta S_{rxn}^{\circ}$ to calculate ΔG_{rxn}°. In the second way, we use tabulated values of free energies of formation to calculate ΔG_{rxn}° directly. In the third way, we determine the free energy change for a stepwise reaction from the free energies of each of the steps. At the end of the section, we look at what is "free" about free energy. Remember that ΔG_{rxn}° is extremely useful because it tells us about the spontaneity of a process. The more negative that ΔG_{rxn}° is, the more spontaneous the process (the further it will go toward products to reach equilibrium).

Calculating Free Energy Changes with $\Delta G_{rxn}^{\circ} = \Delta H_{rxn}^{\circ} - T\,\Delta S_{rxn}^{\circ}$

In Chapter 6 (Section 6.8), we learned how to use tabulated values of standard enthalpies of formation to calculate ΔH_{rxn}°. In the previous section of this chapter, we learned how to use tabulated values of standard entropies to calculate ΔS_{rxn}°. We can use the values of ΔH_{rxn}° and ΔS_{rxn}° calculated in these ways to determine the standard free energy change for a reaction by using the following equation:

$$\Delta G_{rxn}^{\circ} = \Delta H_{rxn}^{\circ} - T\,\Delta S_{rxn}^{\circ} \qquad [17.12]$$

Since tabulated values of standard enthalpies of formation (ΔH_f°) and standard entropies (S°) are usually at 25 °C, the equation should (strictly speaking) be valid only when $T = 298$ K (25 °C). However, the changes in ΔH_{rxn}° and ΔS_{rxn}° over a limited temperature range are small when compared to the changes in the value of the temperature itself. Therefore, we can use Equation 17.12 to estimate changes in free energy at temperatures other than 25 °C.

EXAMPLE 17.5 Calculating the Standard Change in Free Energy for a Reaction Using $\Delta G^\circ_{rxn} = \Delta H^\circ_{rxn} - T\,\Delta S^\circ_{rxn}$

One of the possible initial steps in the formation of acid rain is the oxidation of the pollutant SO_2 to SO_3 by the reaction:

$$SO_2(g) + \tfrac{1}{2}O_2(g) \longrightarrow SO_3(g)$$

Calculate ΔG°_{rxn} at 25 °C and determine whether the reaction is spontaneous.

SOLUTION

Begin by looking up (in Appendix IIB) the standard enthalpy of formation and the standard entropy for each reactant and product.	**Reactant or product**	**ΔH°_f(kJ/mol)**	**S° (in J/mol · K)**

Reactant or product	ΔH°_f(kJ/mol)	S° (in J/mol · K)
$SO_2(g)$	−296.8	248.2
$O_2(g)$	0	205.2
$SO_3(g)$	−395.7	256.8

Calculate ΔH°_{rxn} using Equation 6.13.

$$\Delta H^\circ_{rxn} = \sum n_p\,\Delta H^\circ_f(\text{products}) - \sum n_r\,\Delta H^\circ_f(\text{reactants})$$

$$= \left[\Delta H^\circ_{f,\,SO_3(g)}\right] - \left[\Delta H^\circ_{f,\,SO_2(g)} + \tfrac{1}{2}(\Delta H^\circ_{f,\,O_2(g)})\right]$$

$$= -395.7\text{ kJ} - (-296.8\text{ kJ} + 0.0\text{ kJ})$$

$$= -98.9\text{ kJ}$$

Calculate ΔS°_{rxn} using the Equation 17.11.

$$\Delta S^\circ_{rxn} = \sum n_p S^\circ(\text{products}) - \sum n_r S^\circ(\text{reactants})$$

$$= \left[S^\circ_{SO_3(g)}\right] - \left[S^\circ_{SO_2(g)} + \tfrac{1}{2}(S^\circ_{O_2(g)})\right]$$

$$= 256.8\text{ J/K} - \left[248.2\text{ J/K} + \tfrac{1}{2}(205.2\text{ J/K})\right]$$

$$= -94.0\text{ J/K}$$

Calculate ΔG°_{rxn} using the calculated values of ΔH°_{rxn} and ΔS°_{rxn} and Equation 17.12. The temperature must be converted to kelvins.

$$T = 25 + 273 = 298\text{ K}$$

$$\Delta G^\circ_{rxn} = \Delta H^\circ_{rxn} - T\,\Delta S^\circ_{rxn}$$

$$= -98.9 \times 10^3\text{ J} - 298\text{ K}\,(-94.0\text{ J/K})$$

$$= -70.9 \times 10^3\text{ J}$$

$$= -70.9\text{ kJ}$$

The reaction is therefore spontaneous at this temperature, because ΔG_{rxn} is negative.

FOR PRACTICE 17.5
Consider the oxidation of NO to NO_2:

$$NO(g) + \tfrac{1}{2}O_2(g) \longrightarrow NO_2(g)$$

Calculate ΔG°_{rxn} at 25 °C and determine whether the reaction is spontaneous.

EXAMPLE 17.6 Estimating the Standard Change in Free Energy for a Reaction at a Temperature Other than 25 °C Using $\Delta G^\circ_{rxn} = \Delta H^\circ_{rxn} - T\,\Delta S^\circ_{rxn}$

For the reaction in Example 17.5, estimate the value of ΔG°_{rxn} at 125 °C. Does the reaction become more or less spontaneous at this elevated temperature; that is, does the value of ΔG°_{rxn} become more negative (more spontaneous) or more positive (less spontaneous)?

SOLUTION

Estimate ΔG°_{rxn} at the new temperature using the calculated values of ΔH°_{rxn} and ΔS°_{rxn} from Example 17.5. For T, use the given temperature converted to kelvins. Make sure to use the same units for ΔH°_{rxn} and ΔS°_{rxn} (usually joules).	$T = 125 + 273 = 398 \text{ K}$ $\Delta G^\circ_{rxn} = \Delta H^\circ_{rxn} - T\,\Delta S^\circ_{rxn}$ $\qquad = -98.9 \times 10^3 \text{ J} - 398 \text{ K} \, (-94.0 \text{ J/K})$ $\qquad = -61.5 \times 10^3 \text{ J}$ $\qquad = -61.5 \text{ kJ}$ Since the value of ΔG°_{rxn} at this elevated temperature is less negative (or more positive) than the value of ΔG°_{rxn} at 25 °C (which is −70.9 kJ), the reaction is less spontaneous.

FOR PRACTICE 17.6

For the reaction in For Practice 17.5, calculate the value of ΔG°_{rxn} at −55 °C. Does the reaction become more spontaneous (more negative ΔG°_{rxn}) or less spontaneous (more positive ΔG°_{rxn}) at the lower temperature?

Calculating ΔG°_{rxn} with Tabulated Values of Free Energies of Formation

Since ΔG°_{rxn} is the *change* in free energy for a chemical reaction—the difference in free energy between the products and the reactants—and since free energy is a state function, we can calculate ΔG°_{rxn} by subtracting the free energies of the reactants of the reaction from the free energies of the products of the reaction. Also, since we are interested only in *changes* in free energy (and not in absolute values of free energy itself), we are free to define the *zero* of free energy as conveniently as possible. By analogy with our definition of enthalpies of formation, we define the **free energy of formation (ΔG°_f)** as follows:

> The free energy of formation (ΔG°_f) is the change in free energy when 1 mol of a compound forms from its constituent elements in their standard states. The free energy of formation of pure elements in their standard states is zero.

We can then measure all changes in free energy relative to pure elements in their standard states. To calculate ΔG°_{rxn}, we subtract the free energies of formation of the reactants multiplied by their stoichiometric coefficients from the free energies of formation of the products multiplied by their stoichiometric coefficients. In the form of an equation:

$$\Delta G^\circ_{rxn} = \sum n_p \Delta G^\circ_f \text{ (products)} - \sum n_r \Delta G^\circ_f \text{ (reactants)} \qquad [17.13]$$

In Equation 17.13, n_p represents the stoichiometric coefficients of the products, n_r represents the stoichiometric coefficients of the reactants, and ΔG°_f represents the standard free energies of formation. Keep in mind when using this equation that elements in their standard states have $\Delta G^\circ_f = 0$. Table 17.3 shows ΔG°_f values for some selected substances. A more complete list can be found in Appendix IIB. Notice that, by definition, elements have standard free energies of formation of zero. Notice also that most compounds have negative standard free energies of formation. This means that those compounds have lower free energies than their elements in their standard states. Compounds with positive free energies of formation have higher free energies than their elements and are generally less stable.

Example 17.7 demonstrates the calculation of ΔG°_{rxn} from ΔG°_f values. *Note that this method of calculating ΔG°_{rxn} works only at the temperature for which the free energies of formation are tabulated, namely, 25 °C. Estimating ΔG°_{rxn} at other temperatures requires the use of $\Delta G^\circ_{rxn} = \Delta H^\circ_{rxn} - T\,\Delta S^\circ_{rxn}$, as demonstrated previously.*

TABLE 17.3 Standard Molar Free Energies of Formation (ΔG°_f) for Selected Substances at 298 K

Substance	ΔG°_f(kJ/mol)	Substance	ΔG°_f(kJ/mol)
$H_2(g)$	0	$CH_4(g)$	−50.5
$O_2(g)$	0	$H_2O(g)$	−228.6
$N_2(g)$	0	$H_2O(l)$	−237.1
$C(s, \text{graphite})$	0	$NH_3(g)$	−16.4
$C(s, \text{diamond})$	2.900	$NO(g)$	+87.6
$CO(g)$	−137.2	$NO_2(g)$	+51.3
$CO_2(g)$	−394.4	$NaCl(s)$	−384.1

EXAMPLE 17.7 ΔG_{rxn}° from Standard Free Energies of Formation

Ozone in the lower atmosphere is a pollutant that forms by the following reaction involving the oxidation of unburned hydrocarbons:

$$CH_4(g) + 8\,O_2(g) \longrightarrow CO_2(g) + 2\,H_2O(g) + 4\,O_3(g)$$

Use the standard free energies of formation to determine ΔG_{rxn}° for this reaction at 25 °C.

SOLUTION

Begin by looking up (in Appendix IIB) the standard free energies of formation for each reactant and product. Remember that the standard free energy of formation of a pure element in its standard state is zero.	

Reactant/product	ΔG_f° (kJ/mol)
$CH_4(g)$	-50.5
$O_2(g)$	0.0
$CO_2(g)$	-394.4
$H_2O(g)$	-228.6
$O_3(g)$	163.2

Calculate ΔG_{rxn}° by substituting into Equation 17.13.

$$\Delta G_{rxn}^{\circ} = \sum n_p \,\Delta G_f^{\circ} \text{ (products)} - \sum n_r \,\Delta G_f^{\circ} \text{ (reactants)}$$

$$= \left[\Delta G_{f,\,CO_2(g)}^{\circ} + 2(\Delta G_{f,\,H_2O(g)}^{\circ}) + 4(\Delta G_{f,\,O_3(g)}^{\circ})\right] - \left[\Delta G_{f,\,CH_4(g)}^{\circ} + 8(\Delta G_{f,\,O_2(g)}^{\circ})\right]$$

$$= \left[-394.4\text{ kJ} + 2(-228.6\text{ kJ}) + 4(163.2\text{ kJ})\right] - \left[-50.5\text{ kJ} + 8(0.0\text{ kJ})\right]$$

$$= -198.8\text{ kJ} + 50.5\text{ kJ}$$

$$= -148.3\text{ kJ}$$

FOR PRACTICE 17.7

One of the reactions that occurs within a catalytic converter in the exhaust pipe of a car is the simultaneous oxidation of carbon monoxide and reduction of NO (both of which are harmful pollutants).

$$2\,CO(g) + 2\,NO(g) \longrightarrow 2\,CO_2(g) + N_2(g)$$

Use standard free energies of formation to determine ΔG_{rxn}° for this reaction at 25 °C. Is the reaction spontaneous?

FOR MORE PRACTICE 17.7

In For Practice 17.7, you calculated ΔG_{rxn}° for the simultaneous oxidation of carbon monoxide and reduction of NO using standard free energies of formation. Calculate ΔG_{rxn}° again at 25 °C, only this time use $\Delta G_{rxn}^{\circ} = \Delta H_{rxn}^{\circ} - T\,\Delta S_{rxn}^{\circ}$. How do the two values compare? Use your results to calculate ΔG_{rxn}° at 500.0 K and explain why you could not calculate ΔG_{rxn}° at 500.0 K using tabulated standard free energies of formation.

Calculating ΔG_{rxn}° for a Stepwise Reaction from the Changes in Free Energy for Each of the Steps

Recall from Section 6.7 that, since enthalpy is a state function, ΔH_{rxn}° can be calculated for a stepwise reaction from the sum of the changes in enthalpy for each step according to Hess's law. Since free energy is also a state function, the same relationships that we covered in Chapter 6 for enthalpy also apply to free energy:

1. If a chemical equation is multiplied by some factor, then ΔG_{rxn} is also multiplied by the same factor.
2. If a chemical equation is reversed, then ΔG_{rxn} changes sign.
3. If a chemical equation can be expressed as the sum of a series of steps, then ΔG_{rxn} for the overall equation is the sum of the free energies of reactions for each step.

Example 17.8 illustrates the use of these relationships to calculate ΔG_{rxn}° for a stepwise reaction.

EXAMPLE 17.8 Calculating ΔG°_{rxn} for a Stepwise Reaction

Find ΔG°_{rxn} for the reaction:

$$3\ C(s) + 4\ H_2(g) \longrightarrow C_3H_8(g)$$

Use the following reactions with known ΔG values:

$$C_3H_8(g) + 5\ O_2(g) \longrightarrow 3\ CO_2(g) + 4\ H_2O(g) \qquad \Delta G^{\circ}_{rxn} = -2074\ kJ$$

$$C(s) + O_2(g) \longrightarrow CO_2(g) \qquad \Delta G^{\circ}_{rxn} = -394.4\ kJ$$

$$2\ H_2(g) + O_2(g) \longrightarrow 2\ H_2O(g) \qquad \Delta G^{\circ}_{rxn} = -457.1\ kJ$$

SOLUTION

To work this problem, manipulate the reactions with known ΔG°_{rxn} values in such a way as to get the reactants of interest on the left, the products of interest on the right, and other species to cancel.

Since the first reaction has C_3H_8 as a reactant, and the reaction of interest has C_3H_8 as a product, reverse the first reaction and change the sign of ΔG°_{rxn}.	$3\ CO_2(g) + 4\ H_2O(g) \longrightarrow C_3H_8(g) + 5\ O_2(g) \quad \Delta G^{\circ}_{rxn} = +2074\ kJ$
The second reaction has C as a reactant and CO_2 as a product, as required in the reaction of interest. However, the coefficient for C is 1, and in the reaction of interest, the coefficient for C is 3. Therefore, multiply this equation and its ΔG°_{rxn} by 3.	$3 \times [C(s) + O_2(g) \longrightarrow CO_2(g)] \qquad \Delta G^{\circ}_{rxn} = 3 \times (-394.4\ kJ)$ $= -1183\ kJ$
The third reaction has $H_2(g)$ as a reactant, as required. However, the coefficient for H_2 is 2, and in the reaction of interest, the coefficient for H_2 is 4. Therefore multiply this reaction and its ΔG°_{rxn} by 2.	$2 \times [2\ H_2(g) + O_2(g) \longrightarrow 2\ H_2O(g)] \qquad \Delta G^{\circ}_{rxn} = 2 \times (-457.1\ kJ)$ $= -914.2\ kJ$
Lastly, rewrite the three reactions after multiplying through by the indicated factors and show how they sum to the reaction of interest. ΔG°_{rxn} for the reaction of interest is the sum of the ΔG's for the steps.	$3\ CO_2(g) + 4\ H_2O(g) \longrightarrow C_3H_8(g) + 5\ O_2(g) \qquad \Delta G^{\circ}_{rxn} = +2074\ kJ$ $3\ C(s) + 3\ O_2(g) \longrightarrow 3\ CO_2(g) \qquad \Delta G^{\circ}_{rxn} = -1183\ kJ$ $\underline{4\ H_2(g) + 2\ O_2(g) \longrightarrow 4\ H_2O(g) \qquad \Delta G^{\circ}_{rxn} = -914.2\ kJ}$ $3\ C(s) + 4\ H_2(g) \longrightarrow C_3H_8(g) \qquad \Delta G^{\circ}_{rxn} = -23\ kJ$

FOR PRACTICE 17.8

Find ΔG°_{rxn} for the following reaction:

$$N_2O(g) + NO_2(g) \longrightarrow 3\ NO(g)$$

Use the following reactions with known ΔG values:

$$2\ NO(g) + O_2(g) \longrightarrow 2\ NO_2(g) \qquad \Delta G^{\circ}_{rxn} = -71.2\ kJ$$

$$N_2(g) + O_2(g) \longrightarrow 2\ NO(g) \qquad \Delta G^{\circ}_{rxn} = +175.2\ kJ$$

$$2\ N_2O(g) \longrightarrow 2\ N_2(g) + O_2(g) \qquad \Delta G^{\circ}_{rxn} = -207.4\ kJ$$

Why Free Energy Is "Free"

As we have discussed previously, we often want to use the energy released by a chemical reaction to do work. For example, in an automobile, we use the energy released by the combustion of gasoline to move the car forward. The change in free energy of a chemical reaction represents the maximum amount of energy available, or *free*, to do work

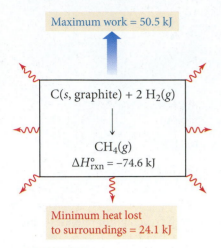

Maximum work = 50.5 kJ

$$C(s, \text{graphite}) + 2\,H_2(g)$$

$$CH_4(g)$$
$$\Delta H°_{rxn} = -74.6 \text{ kJ}$$

Minimum heat lost
to surroundings = 24.1 kJ

▲ **FIGURE 17.9 Free Energy**
Although the reaction produces 74.6 kJ
of heat, only a maximum of 50.5 kJ is
available to do work. The rest of the
energy is lost to the surroundings.

The formal definition of a reversible
reaction is: *A reversible reaction is one
that will change direction upon an
infinitesimally small change in a
variable (such as temperature or
pressure) related to the reaction.*

(if $\Delta G°_{rxn}$ is negative). For many reactions, the amount of free energy is less than the change in enthalpy for the reaction. For example, consider the reaction occurring at 25 °C:

$$C(s, \text{graphite}) + 2\,H_2(g) \longrightarrow CH_4(g)$$

$$\Delta H°_{rxn} = -74.6 \text{ kJ}$$
$$\Delta S°_{rxn} = -80.8 \text{ J/K}$$
$$\Delta G°_{rxn} = -50.5 \text{ kJ}$$

The reaction is exothermic and gives off 74.6 kJ of thermal energy. However, the maximum amount of energy available for useful work is only 50.5 kJ **Figure 17.9◄**. Why? We can see that the change in entropy of the *system* is negative. Nevertheless, the reaction is spontaneous. This is possible only if some of the emitted heat goes to increase the entropy of the surroundings by an amount sufficient to make the change in entropy of the *universe* positive. The amount of energy available to do work (the free energy) is what is left after accounting for the heat that must be lost to the surroundings.

The change in free energy for a chemical reaction represents a *theoretical limit* as to how much work can be done by the reaction. In any *real* reaction, the amount of energy available to do work is even *less* than $\Delta G°_{rxn}$ because additional energy is lost to the surroundings as heat. A reaction that achieves the theoretical limit with respect to free energy is called a **reversible reaction**. A reversible reaction occurs infinitesimally slowly, and the free energy must be drawn out in infinitesimally small increments that exactly match the amount of energy that the reaction is producing in that increment **Figure 17.10▼**.

All real reactions, however, are **irreversible reactions** and do not achieve the theoretical limit of available free energy. Let's return to our discharging battery from the opening section of this chapter as an example of this concept. A battery contains chemical reactants configured in such a way that, upon spontaneous reaction, the battery produces an electrical current. The free energy released by the reaction can be harnessed to do work. For example, an electric motor can be wired to the battery and made to turn by the flowing electrical current **Figure 17.11►**. However, owing to resistance in the wire, the flowing electrical current will also produce some heat, which is lost to the surroundings and is not available to do work. The amount of free energy lost as heat can be lowered by slowing down the rate of current flow. The slower the rate of current flow, the less free energy is lost as heat and the more is available to do work. However, only in the theoretical case of infinitesimally slow current flow will the maximum amount of work (equal to $\Delta G°_{rxn}$) be done. Any real rate of current flow will result in some loss of energy as heat. This lost energy is the heat tax that we saw in the opening section of this chapter. Recharging the battery will necessarily require more energy than was obtained as work because some of the energy is lost as heat.

Reversible Process

Weight of sand exactly matches pressure at each increment.

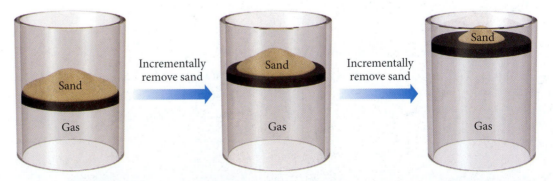

▲ **FIGURE 17.10 A Reversible Process** In a reversible process, the free energy is drawn out in infinitesimally small increments that exactly match the amount of energy that the process is producing in that increment. In this case, grains of sand are removed one at a time, resulting in a series of small expansions in which the weight of sand almost exactly matches the pressure of the expanding gas. This process is close to reversible—each sand grain would need to have an infinitesimally small mass for the process to be fully reversible.

If the change in free energy of a chemical reaction is positive, then ΔG°_{rxn} represents *the minimum amount of energy required to make the reaction occur*. Again, ΔG°_{rxn} represents a theoretical limit. Making a real nonspontaneous reaction occur always requires more energy than the theoretical limit.

17.8 Free Energy Changes for Nonstandard States: The Relationship between ΔG°_{rxn} and ΔG_{rxn}

We have learned how to calculate the *standard* free energy change for a reaction (ΔG°_{rxn}). However, the standard free energy change applies only to a very narrow set of conditions, specifically, those conditions in which the reactants and products are in their standard states. For example, consider the standard free energy change for the evaporation of liquid water to gaseous water at 25 °C:

$$H_2O(l) \rightleftharpoons H_2O(g) \qquad \Delta G^\circ_{rxn} = +8.59 \text{ kJ/mol}$$

The standard free energy change for this process is positive, so the process is nonspontaneous. But you know that if you spill water onto the floor under ordinary conditions, it eventually will evaporate spontaneously. Why? *Because ordinary conditions are not standard conditions* and ΔG°_{rxn} applies only to standard conditions. For a gas (such as the water vapor in the above reaction), standard conditions are those in which the pure gas is present at a partial pressure of 1 atmosphere. In a flask containing liquid water and water vapor under standard conditions ($P_{H_2O} = 1$ atm) at 25 °C the water would not vaporize. In fact, since ΔG°_{rxn} would be negative for the reverse reaction, the reaction would spontaneously occur in reverse—water vapor would condense.

However, in open air under ordinary circumstances, the partial pressure of water vapor is much less than 1 atm. The conditions are not standard and therefore the value of ΔG°_{rxn} does not apply. For nonstandard conditions, we need to calculate ΔG_{rxn} (as opposed to ΔG°_{rxn}) to predict spontaneity.

The Free Energy Change of a Reaction under Nonstandard Conditions

The **free energy change of a reaction under nonstandard conditions** (ΔG_{rxn}) can be calculated from ΔG°_{rxn} using the relationship

$$\Delta G_{rxn} = \Delta G^\circ_{rxn} + RT \ln Q \qquad [17.14]$$

where Q is the reaction quotient (defined in Section 14.7), T is the temperature in kelvins, and R is the gas constant in the appropriate units (8.314 J/mol·K). We demonstrate the use of this equation by applying it to the liquid–vapor water equilibrium under several different conditions, as shown in **Figure 17.12▶**. Note that by the law of mass action, for this equilibrium, $Q = P_{H_2O}$ (where the pressure is expressed in atmospheres).

$$H_2O(l) \longrightarrow H_2O(g) \qquad Q = P_{H_2O}$$

Standard Conditions
Under standard conditions, $P_{H_2O} = 1$ atm and therefore $Q = 1$. Substituting, we get the following expression:

$$\Delta G_{rxn} = \Delta G^\circ_{rxn} + RT \ln Q$$
$$= +8.59 \text{ kJ/mol} + RT \ln(1)$$
$$= +8.59 \text{ kJ/mol}$$

Under standard conditions, Q will always be equal to 1, and (since $\ln(1) = 0$) the value of ΔG_{rxn} will therefore be equal to ΔG°_{rxn}, as expected. For the liquid–vapor water equilibrium, since $\Delta G^\circ_{rxn} > 0$, the reaction is not spontaneous in the forward direction but is spontaneous in the reverse direction. As stated previously, under standard conditions water vapor condenses into liquid water.

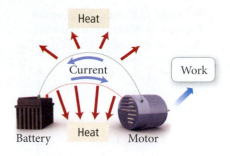

▲ **FIGURE 17.11 Energy Loss in a Battery** When current is drawn from a battery to do work, some energy is lost as heat due to resistance in the wire. Consequently, the quantity of energy required to recharge the battery will necessarily be more than the quantity of work done.

▲ Spilled water spontaneously evaporates even though ΔG° for the vaporization of water is positive. Why?

► **FIGURE 17.12 Free Energy versus Pressure for Water** The free energy change for the vaporization of water is a function of pressure.

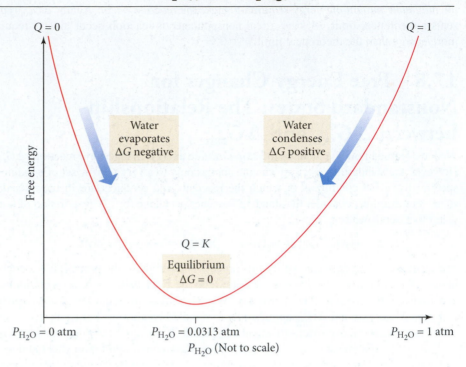

$$H_2O(l) \rightleftharpoons H_2O(g)$$

Equilibrium Conditions

At 25.00 °C, liquid water is in equilibrium with water vapor at a pressure of 0.0313 atm; therefore $Q = K_p = 0.0313$. Substituting:

$$\Delta G_{rxn} = \Delta G_{rxn}^{\circ} + RT \ln(0.0313)$$

$$= +8.59 \text{ kJ/mol} + 8.314 \frac{J}{mol \cdot K}(298.15 \text{ K}) \ln(0.0313)$$

$$= +8.59 \text{ kJ/mol} + (-8.59 \times 10^3 \text{ J/mol})$$

$$= +8.59 \text{ kJ/mol} - 8.59 \text{ kJ/mol}$$

$$= 0$$

Under equilibrium conditions, the value of $RT \ln Q$ is always equal in magnitude but opposite in sign to the value of ΔG_{rxn}°. Therefore, the value of ΔG_{rxn} is always zero. Since $\Delta G_{rxn} = 0$, the reaction is not spontaneous in either direction, as expected for a reaction at equilibrium.

Other Nonstandard Conditions

A water partial pressure of 5.00×10^{-3} atm corresponds to a relative humidity of 16% at 25 °C.

To calculate the value of ΔG_{rxn} under any other set of nonstandard conditions, calculate Q and substitute the value into Equation 17.14. For example, the partial pressure of water vapor in the air on a dry (nonhumid) day might be 5.00×10^{-3} atm, so $Q = 5.00 \times 10^{-3}$. Substituting,

$$\Delta G_{rxn} = \Delta G_{rxn}^{\circ} + RT \ln(5.00 \times 10^{-3})$$

$$= +8.59 \text{ kJ/mol} + 8.314 \frac{J}{mol \cdot K} (298 \text{ K}) \ln(5.00 \times 10^{-3})$$

$$= +8.59 \text{ kJ/mol} + (-13.1 \times 10^3 \text{ J/mol})$$

$$= +8.59 \text{ kJ/mol} - 13.1 \text{ kJ/mol}$$

$$= -4.5 \text{ kJ/mol}$$

Under these conditions, the value of $\Delta G_{rxn} < 0$, so the reaction is spontaneous in the forward direction, consistent with our experience of water evaporating when spilled on the floor.

EXAMPLE 17.9 Calculating ΔG_{rxn} under Nonstandard Conditions

Consider the following reaction at 298 K:

$$2\ NO(g) + O_2(g) \longrightarrow 2\ NO_2(g) \qquad \Delta G_{rxn}^\circ = -71.2\ kJ$$

Calculate ΔG_{rxn} under the conditions:

$$P_{NO} = 0.100\ atm;\ P_{O_2} = 0.100\ atm;\ P_{NO_2} = 2.00\ atm$$

Is the reaction more or less spontaneous under these conditions than under standard conditions?

SOLUTION

Use the law of mass action to calculate Q.	$Q = \dfrac{P_{NO_2}^2}{P_{NO}^2 P_{O_2}}$ $= \dfrac{(2.00)^2}{(0.100)^2(0.100)}$ $= 4.00 \times 10^3$
Substitute Q, T, and ΔG_{rxn}° into Equation 17.14 to calculate ΔG_{rxn}. (Since the units of R include joules, write ΔG_{rxn}° in joules.)	$\Delta G_{rxn} = \Delta G_{rxn}^\circ + RT \ln Q$ $= -71.2 \times 10^3\ J + 8.314\ \dfrac{J}{mol \cdot K}(298\ K)\ \ln(4.00 \times 10^3)$ $= -71.2 \times 10^3\ J + 20.5 \times 10^3\ J$ $= -50.7 \times 10^3\ J$ $= -50.7\ kJ$ The reaction is spontaneous under these conditions, but less spontaneous than it was under standard conditions (because ΔG_{rxn} is less negative than ΔG_{rxn}°).

CHECK

The calculated result is consistent with what you would expect based on Le Châtelier's principle; increasing the concentration of the products and decreasing the concentration of the reactants relative to standard conditions should make the reaction less spontaneous than it was under standard conditions.

FOR PRACTICE 17.9

Consider the following reaction at 298 K:

$$2\ H_2S(g) + SO_2(g) \longrightarrow 3\ S(s, rhombic) + 2\ H_2O(g) \qquad \Delta G_{rxn}^\circ = -102\ kJ$$

Compute ΔG_{rxn} under the conditions:

$$P_{H_2S} = 2.00\ atm;\ P_{SO_2} = 1.50\ atm;\ P_{H_2O} = 0.0100\ atm$$

Is the reaction more or less spontaneous under these conditions than under standard conditions?

Conceptual Connection 17.6 Free Energy Changes and Le Châtelier's Principle

According to Le Châtelier's principle and the dependence of free energy on reactant and product concentrations, which statement is true? (Assume that both the reactants and products are gaseous.)

(a) A high concentration of reactants relative to products results in a more spontaneous reaction than one in which the reactants and products are in their standard states.

(b) A high concentration of products relative to reactants results in a more spontaneous reaction than one in which the reactants and products are in their standard states.

(c) A reaction in which the reactants are in standard states, but in which no products have formed, has a ΔG_{rxn} that is more positive than ΔG_{rxn}°.

17.9 Free Energy and Equilibrium: Relating ΔG_{rxn}° to the Equilibrium Constant (K)

We have discussed throughout this chapter that ΔG_{rxn}° determines the spontaneity of a reaction when the reactants and products are in their standard states. In Chapter 14, we learned that the equilibrium constant (K) determines how far a reaction goes toward products, a measure of spontaneity. As you might expect, the standard free energy change of a reaction and the equilibrium constant are related—the equilibrium constant becomes larger as the free energy change becomes more negative. In other words, if the reactants in a particular reaction undergo a large *negative* free energy change in going to products, then the reaction has a large equilibrium constant, with products strongly favored at equilibrium. If, on the other hand, the reactants in a particular reaction undergo a large *positive* free energy change in going to products, then the reaction has a small equilibrium constant, with reactants strongly favored at equilibrium.

We can derive a relationship between ΔG_{rxn}° and K from Equation 17.14. We know that at equilibrium $Q = K$ and $\Delta G_{rxn} = 0$. Making these substitutions,

$$\Delta G_{rxn} = \Delta G_{rxn}^{\circ} + RT \ln Q$$

$$0 = \Delta G_{rxn}^{\circ} + RT \ln K$$

$$\boxed{\Delta G_{rxn}^{\circ} = -RT \ln K} \qquad [17.15]$$

| Notice that the relationship between ΔG_{rxn} and K is logarithmic—small changes in ΔG_{rxn} have a large effect on K.

We can better understand the relationship between ΔG_{rxn}° and K by considering these ranges of values for K, as summarized in **Figure 17.13▶** (on p. 695).

- When $K < 1$, $\ln K$ is negative and ΔG_{rxn}° is therefore positive. Under standard conditions (when $Q = 1$) the reaction is spontaneous in the reverse direction.
- When $K > 1$, $\ln K$ is positive and ΔG_{rxn}° is therefore negative. Under standard conditions (when $Q = 1$) the reaction is spontaneous in the forward direction.
- When $K = 1$, $\ln K$ is zero and ΔG_{rxn}° is therefore zero. The reaction happens to be at equilibrium under standard conditions.

EXAMPLE 17.10 The Equilibrium Constant and ΔG_{rxn}°

Use tabulated free energies of formation to calculate the equilibrium constant for the following reaction at 298 K:

$$N_2O_4(g) \rightleftharpoons 2 \, NO_2(g)$$

SOLUTION

Begin by looking up (in Appendix IIB) the standard free energies of formation for each reactant and product.

Reactant or product	ΔG_f° (kJ/mol)
$N_2O_4(g)$	99.8
$NO_2(g)$	51.3

Calculate ΔG_{rxn}° by substituting into Equation 17.13.

$$\Delta G_{rxn}^{\circ} = \sum n_p \Delta G_f^{\circ} \text{ (products)} - \sum n_r \Delta G_f^{\circ} \text{ (reactants)}$$

$$= 2 \left[\Delta G_f^{\circ}(NO_2) \right] - \Delta G_f^{\circ}(N_2O_4)$$

$$= 2(51.3 \text{ kJ}) - 99.8 \text{ kJ}$$

$$= 2.8 \text{ kJ}$$

Compute K from ΔG°_{rxn} by solving Equation 17.15 for K and substituting the values of ΔG°_{rxn} and temperature.	$\Delta G^{\circ}_{rxn} = -RT \ln K$
	$\ln K = \dfrac{-\Delta G^{\circ}_{rxn}}{RT}$
	$= \dfrac{-2.8 \times 10^3 \ \cancel{\text{J}}/\cancel{\text{mol}}}{8.314 \dfrac{\cancel{\text{J}}}{\cancel{\text{mol}} \cdot \cancel{\text{K}}}(298 \ \cancel{\text{K}})}$
	$= -1.13$
	$K = e^{-1.13}$
	$= 0.32$

FOR PRACTICE 17.10

Calculate ΔG°_{rxn} at 298 K for the following reaction:

$$I_2(g) + Cl_2(g) \rightleftharpoons 2 \, ICl(g) \qquad K_p = 81.9$$

Free Energy and the Equilibrium Constant

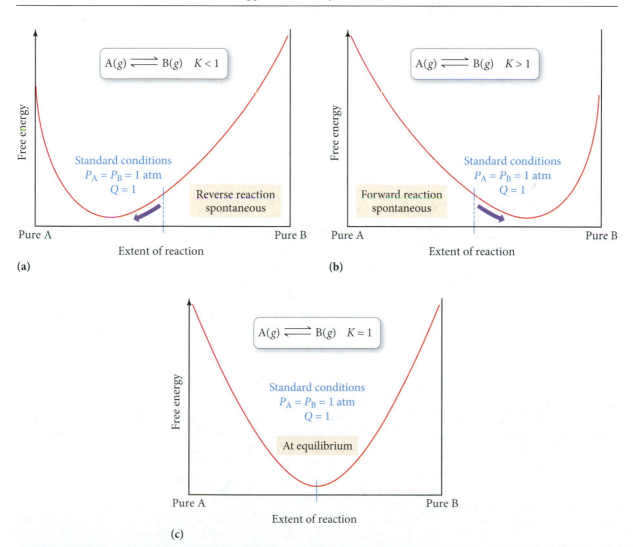

▲ **FIGURE 17.13 Free Energy and the Equilibrium Constant** (**a**) Free energy curve for a reaction with a small equilibrium constant. (**b**) Free energy curve for a reaction with a large equilibrium constant. (**c**) Free energy curve for a reaction in which $K = 1$.

Conceptual Connection 17.7 **K and ΔG_{rxn}°**

The reaction $A(g) \rightleftharpoons B(g)$ has an equilibrium constant that is less than one. What can you conclude about ΔG_{rxn}° for the reaction?

(a) $\Delta G_{rxn}^{\circ} = 0$ **(b)** $\Delta G_{rxn}^{\circ} < 0$ **(c)** $\Delta G_{rxn}^{\circ} > 0$

CHAPTER IN REVIEW

Key Terms

Section 17.2
spontaneous process (666)

Section 17.3
entropy (S) (668)
second law of
  thermodynamics (670)

Section 17.5
Gibbs free energy (G) (678)

Section 17.6
standard entropy change for a
  reaction (ΔS_{rxn}°) (681)
standard molar entropies
  (S°) (681)

third law of thermodynamics
  (681)

Section 17.7
standard free energy change
  (ΔG_{rxn}°) (685)
free energy of formation
  (ΔG_f°) (687)

reversible reaction (690)
irreversible reaction (690)

Section 17.8
free energy change of a
  reaction under nonstandard
  conditions (ΔG_{rxn}) (691)

Key Concepts

Nature's Heat Tax: You Can't Win and You Can't Break Even (17.1)

▶ The first law of thermodynamics states that energy can be neither created nor destroyed.

▶ The second law implies that for every energy transaction, some energy is lost to the surroundings; this lost energy is nature's heat tax.

Spontaneous and Nonspontaneous Processes (17.2)

▶ Both spontaneous and nonspontaneous processes can occur, but only spontaneous processes can take place without outside intervention.

▶ Thermodynamics is the study of the *spontaneity* of a reaction, *not* to be confused with kinetics, which is the study of the *rate* of a reaction.

Entropy and the Second Law of Thermodynamics (17.3)

▶ The second law of thermodynamics states that for *any* spontaneous process, the entropy of the universe increases.

▶ Entropy (S) is proportional to the number of energetically equivalent ways in which the components of a system can be arranged and is a measure of energy dispersal per unit temperature.

Heat Transfer and Changes in the Entropy of the Surroundings (17.4)

▶ For a process to be spontaneous, the total entropy of the universe (system plus surroundings) must increase.

▶ The entropy of the surroundings increases when the change in *enthalpy* of the system (ΔH_{sys}) is negative (i.e., for exothermic reactions).

▶ The change in entropy of the surroundings for a given ΔH_{sys} depends inversely on temperature—the greater the temperature, the lower the magnitude of ΔS_{surr}.

Gibbs Free Energy (17.5)

▶ Gibbs free energy, G, is a thermodynamic function that is proportional to the negative of the change in the entropy of the universe.

▶ A negative ΔG represents a spontaneous reaction and a positive ΔG represents a nonspontaneous reaction.

▶ We can calculate the value of ΔG for a reaction from the values of ΔH and ΔS for the *system* according to the equation $\Delta G = \Delta H - T \Delta S$.

Entropy Changes in Chemical Reactions: Calculating ΔS_{rxn}° (17.6)

▶ We calculate the standard change in entropy for a reaction similarly to the standard change in enthalpy for a reaction: by subtracting the sum of the standard entropies of the reactants multiplied by their stoichiometric coefficients from the sum of the standard entropies of the products multiplied by their stoichiometric coefficients.

▶ The standard entropies are *absolute*: an entropy of zero is given by the third law of thermodynamics as the entropy of a perfect crystal at absolute zero.

▶ The absolute entropy of a substance depends on factors that affect the number of energetically equivalent arrangements of the substance; these include the phase, size, and molecular complexity of the substance.

Free Energy Changes in Chemical Reactions: Calculating ΔG_{rxn}° (17.7)

▶ There are three methods of calculating ΔG_{rxn}°: (1) from ΔH° and ΔS°; (2) from free energies of formations (only at 25 °C); and (3) from the ΔG_{rxn}° values that sum to the reaction of interest.

▶ The magnitude of a negative ΔG°_{rxn} represents the theoretical amount of energy available to do work, while a positive ΔG°_{rxn} represents the minimum amount of energy required to make a nonspontaneous process occur.

Free Energy Changes for Nonstandard States: The Relationship between ΔG°_{rxn} and ΔG_{rxn} (17.8)

▶ The value of ΔG°_{rxn} applies only to standard conditions, and most real conditions are not standard.

▶ Under nonstandard conditions, ΔG_{rxn} can be calculated from the equation $\Delta G_{rxn} = \Delta G^{\circ}_{rxn} + RT \ln Q$.

Free Energy and Equilibrium: Relating ΔG°_{rxn} to the Equilibrium Constant (K) (17.9)

▶ Under standard conditions, the free energy change for a reaction is directly proportional to the negative of the natural log of the equilibrium constant, K; the more negative the free energy change (i.e., the more spontaneous the reaction), the larger the equilibrium constant.

Key Equations and Relationships

The Definition of Entropy (17.3)

$$S = k \ln W \qquad k = 1.38 \times 10^{-23} \text{ J/K}$$

Change in Entropy (17.3)

$$\Delta S = S_{final} - S_{initial}$$

Change in the Entropy of the Universe (17.4)

$$\Delta S_{univ} = \Delta S_{sys} + \Delta S_{surr}$$

Change in the Entropy of the Surroundings (17.4)

$$\Delta S_{surr} = \frac{-\Delta H_{sys}}{T} \text{ (constant } T, P)$$

Change in Gibbs Free Energy (17.5)

$$\Delta G = \Delta H - T \Delta S$$

The Relationship between Spontaneity and ΔH, ΔS, and T (17.5)

ΔH	ΔS	Low Temperature	High Temperature
−	+	Spontaneous	Spontaneous
+	−	Nonspontaneous	Nonspontaneous
−	−	Spontaneous	Nonspontaneous
+	+	Nonspontaneous	Spontaneous

Standard Change in Entropy (17.6)

$$\Delta S^{\circ}_{rxn} = \sum n_p S^{\circ} \text{ (products)} - \sum n_r S^{\circ} \text{ (reactants)}$$

Methods for Calculating the Free Energy of Formation (ΔG°_{rxn}) (17.7)

1. $\Delta G^{\circ}_{rxn} = \Delta H^{\circ}_{rxn} - T \Delta S^{\circ}_{rxn}$
2. $\Delta G^{\circ}_{rxn} = \sum n_p \Delta G^{\circ}_f \text{ (products)} - \sum n_r \Delta G^{\circ}_f \text{ (reactants)}$
3. $\Delta G^{\circ}_{rxn(overall)} = \Delta G^{\circ}_{rxn(step 1)} + \Delta G^{\circ}_{rxn(step 2)} + \Delta G^{\circ}_{rxn(step 3)} + \cdots$

The Relationship between ΔG°_{rxn} and ΔG_{rxn} (17.8)

$$\Delta G_{rxn} = \Delta G^{\circ}_{rxn} + RT \ln Q \qquad R = 8.314 \text{ J/mol} \cdot \text{K}$$

The Relationship between ΔG°_{rxn} and K (17.9)

$$\Delta G^{\circ}_{rxn} = -RT \ln K$$

Key Learning Objectives

Chapter Objectives	Assessment
Predicting the Sign of Entropy Change (17.3)	Example 17.1 For Practice 17.1 Exercises 1, 2, 5–8, 11, 12
Calculating Entropy Changes in the Surroundings (17.4)	Example 17.2 For Practice 17.2 For More Practice 17.2 Exercises 7–10
Computing Gibbs Free Energy Changes and Predicting Spontaneity from ΔH and ΔS (17.5)	Example 17.3 For Practice 17.3 Exercises 13–18
Computing Standard Entropy Changes (ΔS°_{rxn}) (17.6)	Example 17.4 For Practice 17.4 Exercises 25, 26
Calculating the Standard Change in Free Energy for a Reaction Using $\Delta G^{\circ}_{rxn} = \Delta H^{\circ}_{rxn} - T \Delta S^{\circ}_{rxn}$ (17.7)	Examples 17.5, 17.6 For Practice 17.5, 17.6 Exercises 29–32, 35, 36

Calculating ΔG°_{rxn} from Standard Free Energies of Formation (17.7)	Example 17.7 For Practice 17.7 For More Practice 17.7 Exercises 33, 34
Determining ΔG°_{rxn} for a Stepwise Reaction (17.7)	Example 17.8 For Practice 17.8 Exercises 37, 38
Calculating ΔG_{rxn} under Nonstandard Conditions (17.8)	Example 17.9 For Practice 17.9 Exercises 39–46
Relating the Equilibrium Constant and ΔG°_{rxn} (17.9)	Example 17.10 For Practice 17.10 Exercises 47, 48

EXERCISES

Problems by Topic

Entropy, the Second Law of Thermodynamics, and the Direction of Spontaneous Change

1. Which processes are spontaneous?
   a. the combustion of natural gas
   b. the extraction of iron metal from iron ore
   c. a hot drink cooling to room temperature
   d. drawing heat energy from the ocean's surface to power a ship

2. Which processes are nonspontaneous? Are the nonspontaneous processes impossible?
   a. a bike going up a hill
   b. a meteor falling to Earth
   c. obtaining hydrogen gas from liquid water
   d. a ball rolling down a hill

3. Suppose that two systems each composed of two particles represented by circles, have 20 J of total energy. Which system, A or B, has the greatest entropy? Why?

   System A

   10 J ___oo___

   System B

   12 J ___o___

   8 J ___o___

4. Suppose two systems, each composed of three particles represented by circles, have 30 J of total energy. How many energetically equivalent ways can you distribute the particles in each system? Which system has greater entropy?

   System A

   10 J ___ooo___

   System B

   12 J ___o___

   10 J ___o___

   8J ___o___

5. Without doing any calculations, predict the sign of ΔS_{sys} for each chemical reaction:
   a. $2 KClO_3(s) \longrightarrow 2 KCl(s) + 3 O_2(g)$
   b. $CH_2{=}CH_2(g) + H_2(g) \longrightarrow CH_3CH_3(g)$
   c. $Na(s) + \frac{1}{2} Cl_2(g) \longrightarrow NaCl(s)$
   d. $N_2(g) + 3 H_2(g) \longrightarrow 2 NH_3(g)$

6. Without doing any calculations, predict the sign of ΔS_{sys} for each chemical reaction:
   a. $Mg(s) + Cl_2(g) \longrightarrow MgCl_2(s)$
   b. $2 H_2S(g) + 3 O_2(g) \longrightarrow 2 H_2O(g) + 2 SO_2(g)$
   c. $2 O_3(g) \longrightarrow 3 O_2(g)$
   d. $HCl(g) + NH_3(g) \longrightarrow NH_4Cl(s)$

7. Without doing any calculations, determine the sign of ΔS_{sys} and ΔS_{surr} for each chemical reaction. In addition, predict under what temperatures (all temperatures, low temperatures, or high temperatures), if any, the reaction will be spontaneous.
   a. $C_3H_8(g) + 5 O_2(g) \longrightarrow 3 CO_2(g) + 4 H_2O(g)$
   $$\Delta H^{\circ}_{rxn} = -2044 \text{ kJ}$$
   b. $N_2(g) + O_2(g) \longrightarrow 2 NO(g) \quad \Delta H^{\circ}_{rxn} = +182.6 \text{ kJ}$
   c. $2 N_2(g) + O_2(g) \longrightarrow 2 N_2O(g) \quad \Delta H^{\circ}_{rxn} = +163.2 \text{ kJ}$
   d. $4 NH_3(g) + 5 O_2(g) \longrightarrow 4 NO(g) + 6 H_2O(g)$
   $$\Delta H^{\circ}_{rxn} = -906 \text{ kJ}$$

8. Without doing any calculations, determine the sign of ΔS_{sys} and ΔS_{surr} for each chemical reaction. In addition, predict under what temperatures (all temperatures, low temperatures, or high temperatures), if any, the reaction will be spontaneous.
   a. $2 CO(g) + O_2(g) \longrightarrow 2 CO_2(g) \quad \Delta H^{\circ}_{rxn} = -566.0 \text{ kJ}$
   b. $2 NO_2(g) \longrightarrow 2 NO(g) + O_2(g) \quad \Delta H^{\circ}_{rxn} = +113.1 \text{ kJ}$
   c. $2 H_2(g) + O_2(g) \longrightarrow 2 H_2O(g) \quad \Delta H^{\circ}_{rxn} = -483.6 \text{ kJ}$
   d. $CO_2(g) \longrightarrow C(s) + O_2(g) \quad \Delta H^{\circ}_{rxn} = +393.5 \text{ kJ}$

9. Calculate ΔS_{surr} at the indicated temperature for a reaction experiencing each change in enthalpy:
   a. $\Delta H^{\circ}_{rxn} = -287 \text{ kJ}; 298 \text{ K}$
   b. $\Delta H^{\circ}_{rxn} = -287 \text{ kJ}; 77 \text{ K}$
   c. $\Delta H^{\circ}_{rxn} = +127 \text{ kJ}; 298 \text{ K}$
   d. $\Delta H^{\circ}_{rxn} = +127 \text{ kJ}; 77 \text{ K}$

10. A reaction has $\Delta H^{\circ}_{rxn} = -127 \text{ kJ}$ and $\Delta S^{\circ}_{rxn} = 314 \text{ J/K}$. At what temperature is the change in entropy for the reaction equal to the change in entropy for the surroundings?

11. Given the values of ΔH°_{rxn}, ΔS°_{rxn}, and T, determine ΔS_{univ} and predict whether or not the reaction will be spontaneous.
    a. $\Delta H^{\circ}_{rxn} = -125 \text{ kJ}; \Delta S^{\circ}_{rxn} = +253 \text{ J/K}; T = 298 \text{ K}$
    b. $\Delta H^{\circ}_{rxn} = +125 \text{ kJ}; \Delta S^{\circ}_{rxn} = -253 \text{ J/K}; T = 298 \text{ K}$
    c. $\Delta H^{\circ}_{rxn} = -125 \text{ kJ}; \Delta S^{\circ}_{rxn} = -253 \text{ J/K}; T = 298 \text{ K}$
    d. $\Delta H^{\circ}_{rxn} = -125 \text{ kJ}; \Delta S^{\circ}_{rxn} = -253 \text{ J/K}; T = 555 \text{ K}$

12. Given the values of ΔH°_{rxn}, ΔS°_{rxn}, and T, determine ΔS_{univ} and predict whether or not the reaction will be spontaneous.
    a. $\Delta H^{\circ}_{rxn} = +85 \text{ kJ}; \Delta S_{rxn} = +147 \text{ J/K}; T = 298 \text{ K}$
    b. $\Delta H^{\circ}_{rxn} = +85 \text{ kJ}; \Delta S_{rxn} = +147 \text{ J/K}; T = 755 \text{ K}$
    c. $\Delta H^{\circ}_{rxn} = +85 \text{ kJ}; \Delta S_{rxn} = -147 \text{ J/K}; T = 298 \text{ K}$
    d. $\Delta H^{\circ}_{rxn} = -85 \text{ kJ}; \Delta S_{rxn} = +147 \text{ J/K}; T = 398 \text{ K}$

Standard Entropy Changes and Gibbs Free Energy

13. Calculate the change in Gibbs free energy for each of the sets of ΔH_{rxn}, ΔS_{rxn}, and T given in Problem 11. Predict whether or not the reaction will be spontaneous at the temperature indicated.

14. Calculate the change in Gibbs free energy for each of the sets of ΔH_{rxn}, ΔS_{rxn}, and T given in Problem 12. Predict whether or not the reaction will be spontaneous at the temperature indicated.

15. Calculate the free energy change for the reaction at 25 °C. Is the reaction spontaneous?

$$C_3H_8(g) + 5\ O_2(g) \longrightarrow 3\ CO_2(g) + 4\ H_2O(g)$$

$$\Delta H^\circ_{rxn} = -2217\ kJ;\ \Delta S^\circ_{rxn} = 101.1\ J/K$$

16. Calculate the free energy change for the reaction at 25 °C. Is the reaction spontaneous?

$$2\ Ca(s) + O_2(g) \longrightarrow 2\ CaO(s)$$

$$\Delta H^\circ_{rxn} = -1269.8\ kJ;\ \Delta S^\circ_{rxn} = -364.6\ J/K$$

17. Fill in the blanks in the table. Both ΔH and ΔS refer to the system.

ΔH	ΔS	ΔG	Low Temperature	High Temperature
−	+	−	Spontaneous	_____
−	−	Temperature dependent	_____	_____
+	+	_____	_____	Spontaneous
_____	−	_____	Nonspontaneous	Nonspontaneous

18. Predict the conditions (high temperature, low temperature, all temperatures, or no temperatures) under which each reaction is spontaneous.
    a. $H_2O(g) \longrightarrow H_2O(l)$
    b. $CO_2(s) \longrightarrow CO_2(g)$
    c. $H_2(g) \longrightarrow 2\ H(g)$
    d. $2\ NO_2(g) \longrightarrow 2\ NO(g) + O_2(g)$ (endothermic)

19. How does the molar entropy of a substance change with increasing temperature?

20. What is the molar entropy of a pure crystal at 0 K? What is the significance of the answer to this question?

21. For each pair of substances, choose the one that you expect to have the higher standard molar entropy (S°) at 25 °C. Explain your choices.
    a. $CO(g)$; $CO_2(g)$
    b. $CH_3OH(l)$; $CH_3OH(g)$
    c. $Ar(g)$; $CO_2(g)$
    d. $CH_4(g)$; $SiH_4(g)$
    e. $NO_2(g)$; $CH_3CH_2CH_3(g)$
    f. $NaBr(s)$; $NaBr(aq)$

22. For each pair of substances, choose the one that you expect to have the higher standard molar entropy (S°) at 25 °C. Explain your choices.
    a. $NaNO_3(s)$; $NaNO_3(aq)$
    b. $CH_4(g)$; $CH_3CH_3(g)$
    c. $Br_2(l)$; $Br_2(g)$
    d. $Br_2(g)$; $F_2(g)$
    e. $PCl_3(g)$; $PCl_5(g)$
    f. $CH_3CH_2CH_2CH_3(g)$; $SO_2(g)$

23. Rank each set of substances in order of increasing standard molar entropy (S°). Explain your reasoning.
    a. $NH_3(g)$; $Ne(g)$; $SO_2(g)$; $CH_3CH_2OH(g)$; $He(g)$
    b. $H_2O(s)$; $H_2O(l)$; $H_2O(g)$
    c. $CH_4(g)$; $CF_4(g)$; $CCl_4(g)$

24. Rank each set of substances in order of increasing standard molar entropy (S°). Explain your reasoning.
    a. $I_2(g)$; $F_2(g)$; $Br_2(g)$; $Cl_2(g)$
    b. $H_2O(g)$; $H_2O_2(g)$; $H_2S(g)$
    c. $C(s, graphite)$; $C(s, diamond)$; $C(s, amorphous)$

25. Use data from Appendix IIB to calculate ΔS°_{rxn} for each of the reactions. In each case, try to rationalize the sign of ΔS°_{rxn}.
    a. $C_2H_4(g) + H_2(g) \longrightarrow C_2H_6(g)$
    b. $C(s) + H_2O(g) \longrightarrow CO(g) + H_2(g)$
    c. $CO(g) + H_2O(g) \longrightarrow H_2(g) + CO_2(g)$
    d. $2\ H_2S(g) + 3\ O_2(g) \longrightarrow 2\ H_2O(l) + 2\ SO_2(g)$

26. Use data from Appendix IIB to calculate ΔS°_{rxn} for each of the reactions. In each case, try to rationalize the sign of ΔS°_{rxn}.
    a. $3\ NO_2(g) + H_2O(l) \longrightarrow 2\ HNO_3(aq) + NO(g)$
    b. $Cr_2O_3(s) + 3\ CO(g) \longrightarrow 2\ Cr(s) + 3\ CO_2(g)$
    c. $SO_2(g) + \frac{1}{2}\ O_2(g) \longrightarrow SO_3(g)$
    d. $N_2O_4(g) + 4\ H_2(g) \longrightarrow N_2(g) + 4\ H_2O(g)$

27. Find ΔS° for the formation of $CH_2Cl_2(g)$ from its gaseous elements in their standard states. Rationalize the sign of ΔS°.

28. Find ΔS° for the reaction between nitrogen gas and fluorine gas to form nitrogen trifluoride gas. Rationalize the sign of ΔS°.

29. Methanol burns in oxygen to form carbon dioxide and water. Write a balanced equation for the combustion of liquid methanol and calculate ΔH°_{rxn}, ΔS°_{rxn}, and ΔG°_{rxn} at 25 °C. Is the combustion of methanol spontaneous?

30. In photosynthesis, plants form glucose ($C_6H_{12}O_6$) and oxygen from carbon dioxide and water. Write a balanced equation for photosynthesis and calculate ΔH°_{rxn}, ΔS°_{rxn}, and ΔG°_{rxn} at 25 °C. Is photosynthesis spontaneous?

31. For each reaction, calculate ΔH°_{rxn}, ΔS°_{rxn}, and ΔG°_{rxn} at 25 °C and state whether or not the reaction is spontaneous. If the reaction is not spontaneous, would a change in temperature make it spontaneous? If so, should the temperature be raised or lowered from 25 °C?
    a. $N_2O_4(g) \longrightarrow 2\ NO_2(g)$
    b. $NH_4Cl(s) \longrightarrow HCl(g) + NH_3(g)$
    c. $3\ H_2(g) + Fe_2O_3(s) \longrightarrow 2\ Fe(s) + 3\ H_2O(g)$
    d. $N_2(g) + 3\ H_2(g) \longrightarrow 2\ NH_3(g)$

32. For each reaction, calculate ΔH°_{rxn}, ΔS°_{rxn}, and ΔG°_{rxn} at 25 °C and state whether or not the reaction is spontaneous. If the reaction is not spontaneous, would a change in temperature make it spontaneous? If so, should the temperature be raised or lowered from 25 °C?
    a. $2\ CH_4(g) \longrightarrow C_2H_6(g) + H_2(g)$
    b. $2\ NH_3(g) \longrightarrow N_2H_4(g) + H_2(g)$
    c. $N_2(g) + O_2(g) \longrightarrow 2\ NO(g)$
    d. $2\ KClO_3(s) \longrightarrow 2\ KCl(s) + 3\ O_2(g)$

33. Use standard free energies of formation to calculate ΔG° at 25 °C for each of the reactions in Problem 31. How do the values of ΔG° calculated this way compare to those calculated from ΔH° and ΔS°? Which of the two methods could be used to determine how ΔG° changes with temperature?

34. Use standard free energies of formation to calculate ΔG° at 25 °C for each of the reactions in Problem 32. How well do the values of ΔG° calculated this way compare to those calculated from ΔH° and ΔS°? Which of the two methods could be used to determine how ΔG° changes with temperature?

35. Consider the reaction:

$$2\ NO(g) + O_2(g) \longrightarrow 2\ NO_2(g)$$

Estimate ΔG° for this reaction at each temperature and predict whether or not the reaction will be spontaneous. (Assume that ΔH° and ΔS° do not change too much within the given temperature range.)
    a. 298 K
    b. 715 K
    c. 855 K

36. Consider the reaction:

$$CaCO_3(s) \longrightarrow CaO(s) + CO_2(g)$$

Estimate ΔG° for this reaction at each temperature and predict whether or not the reaction will be spontaneous. (Assume that ΔH° and ΔS° do not change too much within the given temperature range.)
a. 298 K   **b.** 1055 K   **c.** 1455 K

37. Determine ΔG° for the reaction:

$$Fe_2O_3(s) + 3\,CO(g) \longrightarrow 2\,Fe(s) + 3\,CO_2(g)$$

Use the following reactions with known ΔG_{rxn}° values:

$$2\,Fe(s) + \tfrac{3}{2}O_2(g) \longrightarrow Fe_2O_3(s) \quad \Delta G_{rxn}^\circ = -742.2 \text{ kJ}$$

$$CO(g) + \tfrac{1}{2}O_2(g) \longrightarrow CO_2(g) \quad \Delta G_{rxn}^\circ = -257.2 \text{ kJ}$$

38. Calculate ΔG_{rxn}° for the reaction:

$$CaCO_3(s) \longrightarrow CaO(s) + CO_2(g)$$

Use the following reactions and given ΔG_{rxn}° values:

$$Ca(s) + CO_2(g) + \tfrac{1}{2}O_2(g) \longrightarrow CaCO_3(s) \quad \Delta G_{rxn}^\circ = -734.4 \text{ kJ}$$

$$2\,Ca(s) + O_2(g) \longrightarrow 2\,CaO(s) \quad \Delta G_{rxn}^\circ = -1206.6 \text{ kJ}$$

Free Energy Changes, Nonstandard Conditions, and the Equilibrium Constant

39. Consider the sublimation of iodine at 25.0 °C:

$$I_2(s) \longrightarrow I_2(g)$$

a. Find ΔG_{rxn}° at 25.0 °C.
b. Find ΔG_{rxn} at 25.0 °C under these nonstandard conditions:
   (i) $P_{I_2} = 1.00$ mmHg
   (ii) $P_{I_2} = 0.100$ mmHg
c. Explain why iodine spontaneously sublimes in open air at 25.0 °C.

40. Consider the evaporation of methanol at 25.0 °C:

$$CH_3OH(l) \longrightarrow CH_3OH(g)$$

a. Find ΔG° at 25.0 °C.
b. Find ΔG at 25.0 °C under these nonstandard conditions:
   (i) $P_{CH_3OH} = 150.0$ mmHg
   (ii) $P_{CH_3OH} = 100.0$ mmHg
   (iii) $P_{CH_3OH} = 10.0$ mmHg
c. Explain why methanol spontaneously evaporates in open air at 25.0 °C.

41. Consider the reaction:

$$CH_3OH(g) \rightleftharpoons CO(g) + 2\,H_2(g)$$

Calculate ΔG for this reaction at 25 °C under these conditions:

$$P_{CH_3OH} = 0.855 \text{ atm}$$
$$P_{CO} = 0.125 \text{ atm}$$
$$P_{H_2} = 0.183 \text{ atm}$$

42. Consider the reaction:

$$CO_2(g) + CCl_4(g) \rightleftharpoons 2\,COCl_2(g)$$

Calculate ΔG for this reaction at 25 °C under these conditions:

$$P_{CO_2} = 0.112 \text{ atm}$$
$$P_{CCl_4} = 0.174 \text{ atm}$$
$$P_{COCl_2} = 0.744 \text{ atm}$$

43. Use data from Appendix IIB to calculate the equilibrium constants at 25 °C for each reaction.
a. $2\,CO(g) + O_2(g) \rightleftharpoons 2\,CO_2(g)$
b. $2\,H_2S(g) \rightleftharpoons 2\,H_2(g) + S_2(g)$

44. Use data from Appendix IIB to calculate the equilibrium constants at 25 °C for each reaction. ΔG_f° for $BrCl(g)$ is -1.0 kJ/mol.
a. $2\,NO_2(g) \rightleftharpoons N_2O_4(g)$
b. $Br_2(g) + Cl_2(g) \rightleftharpoons 2\,BrCl(g)$

45. Consider the reaction:

$$CO(g) + 2\,H_2(g) \rightleftharpoons CH_3OH(g)$$
$$K_p = 2.26 \times 10^4 \text{ at } 25\,°C$$

Calculate ΔG_{rxn} for the reaction at 25 °C under each condition:
a. standard conditions
b. at equilibrium
c. $P_{CH_3OH} = 1.0$ atm; $P_{CO} = P_{H_2} = 0.010$ atm

46. Consider the reaction:

$$I_2(g) + Cl_2(g) \rightleftharpoons 2\,ICl(g)$$
$$K_p = 81.9 \text{ at } 25\,°C$$

Calculate ΔG_{rxn} for the reaction at 25 °C under each condition:
a. standard conditions
b. at equilibrium
c. $P_{ICl} = 2.55$ atm; $P_{I_2} = 0.325$ atm; $P_{Cl_2} = 0.221$ atm

47. Estimate the value of the equilibrium constant at 525 K for each reaction in Problem 43.

48. Estimate the value of the equilibrium constant at 655 K for each reaction in Problem 44. (ΔH_f° for BrCl is 14.6 kJ/mol.)

Cumulative Problems

49. Predict the sign of ΔS_{sys} in each process:
a. water boiling
b. water freezing
c.

50. Predict the sign of ΔS_{sys} for each process:
a. dry ice subliming
b. dew forming

c.

51. Earth's atmosphere is composed primarily of nitrogen and oxygen, which coexist at 25 °C without reacting to any significant extent. However, the two gases can react to form nitrogen monoxide according to the reaction:

$$N_2(g) + O_2(g) \rightleftharpoons 2\,NO(g)$$

a. Calculate $\Delta G°$ and K_p for this reaction at 298 K. Is the reaction spontaneous?

b. Estimate $\Delta G°$ at 2000 K. Does the reaction become more spontaneous with increasing temperature?

52. Nitrogen dioxide, a pollutant in the atmosphere, can combine with water to form nitric acid. One of the possible reactions is shown. Calculate $\Delta G°$ and K_p for this reaction at 25 °C and comment on the spontaneity of the reaction.

$$3\,NO_2(g) + H_2O(l) \longrightarrow 2\,HNO_3(aq) + NO(g)$$

53. Ethene (C_2H_4) can be halogenated by the reaction:

$$C_2H_4(g) + X_2(g) \rightleftharpoons C_2H_4X_2(g)$$

where X_2 can be Cl_2, Br_2, or I_2. Use the thermodynamic data given to calculate $\Delta H°$, $\Delta S°$, $\Delta G°$, and K_p for the halogenation reaction by each of the three halogens at 25 °C. Which reaction is most spontaneous? Least spontaneous? What is the main factor driving the difference in the spontaneity of the three reactions? Does higher temperature make the reactions more spontaneous or less spontaneous?

Compound	$\Delta H_f°$(kJ/mol)	$S°$(J/mol · K)
$C_2H_4Cl_2(g)$	−129.7	308.0
$C_2H_4Br_2(g)$	−38.3	330.6
$C_2H_4I_2(g)$	+66.5	347.8

54. H_2 reacts with the halogens (X_2) according to the reaction:

$$H_2(g) + X_2(g) \rightleftharpoons 2\,HX(g)$$

where X_2 can be Cl_2, Br_2, or I_2. Use the thermodynamic data in Appendix IIB to calculate $\Delta H°$, $\Delta S°$, $\Delta G°$, and K_p for the reaction between hydrogen and each of the three halogens. Which reaction is most spontaneous? Least spontaneous? What is the main factor driving the difference in the spontaneity of the three reactions? Does higher temperature make the reactions more spontaneous or less spontaneous?

55. Consider this reaction occurring at 298 K:

$$N_2O(g) + NO_2(g) \rightleftharpoons 3\,NO(g)$$

a. Show that the reaction is not spontaneous under standard conditions by calculating $\Delta G_{rxn}°$.

b. If a reaction mixture contains only N_2O and NO_2 at partial pressures of 1.0 atm each, the reaction will be spontaneous until some NO forms in the mixture. What maximum partial pressure of NO builds up before the reaction ceases to be spontaneous?

c. Can the reaction be made more spontaneous by an increase or decrease in temperature? If so, at what temperature is the reaction spontaneous under standard conditions?

56. Consider this reaction occurring at 298 K:

$$BaCO_3(s) \rightleftharpoons BaO(s) + CO_2(g)$$

a. Show that the reaction is not spontaneous under standard conditions by calculating $\Delta G_{rxn}°$.

b. If $BaCO_3$ is placed in an evacuated flask, what is the partial pressure of CO_2 when the reaction reaches equilibrium?

c. Can the reaction be made more spontaneous by an increase or decrease in temperature? If so, at what temperature is the carbon dioxide partial pressure 1.0 atm?

57. Living organisms use energy from the metabolism of food to create an energy-rich molecule called adenosine triphosphate (ATP). The ATP acts as an energy source for a variety of reactions that the living organism must carry out to survive. ATP provides energy through its hydrolysis:

$$ATP(aq) + H_2O(l) \longrightarrow ADP(aq) + P_i(aq) \quad \Delta G_{rxn}° = -30.5 \text{ kJ}$$

where ADP represents adenosine diphosphate and P_i represents an inorganic phosphate group (such as HPO_4^{2-}).

a. Calculate the equilibrium constant, K, for this reaction at 298 K.

b. The free energy obtained from the oxidation (reaction with oxygen) of glucose ($C_6H_{12}O_6$) to form carbon dioxide and water can be used to re-form ATP by driving this reaction in reverse. Calculate the standard free energy change for the oxidation of glucose and estimate the maximum number of moles of ATP that can be formed by the oxidation of one mole of glucose.

58. The standard free energy change for the hydrolysis of ATP is given in Problem 57. In a particular cell, the concentrations of ATP, ADP, and P_i are 0.0031 M, 0.0014 M, and 0.0048 M, respectively. Calculate the free energy change for the hydrolysis of ATP under these conditions. (Assume a temperature of 298 K.)

59. These reactions are important in catalytic converters in automobiles. Calculate $\Delta G°$ for each at 298 K. Predict the effect of increasing temperature on the magnitude of $\Delta G°$.

a. $2\,CO(g) + 2\,NO(g) \longrightarrow N_2(g) + 2\,CO_2(g)$

b. $5\,H_2(g) + 2\,NO(g) \longrightarrow 2\,NH_3(g) + 2\,H_2O(g)$

c. $2\,H_2(g) + 2\,NO(g) \longrightarrow N_2(g) + 2\,H_2O(g)$

d. $2\,NH_3(g) + 2\,O_2(g) \longrightarrow N_2O(g) + 3\,H_2O(g)$

60. Calculate $\Delta G°$ at 298 K for these reactions and predict the effect on $\Delta G°$ of lowering the temperature.

a. $NH_3(g) + HBr(g) \longrightarrow NH_4Br(s)$

b. $CaCO_3(s) \longrightarrow CaO(s) + CO_2(g)$

c. $CH_4(g) + 3\,Cl_2(g) \longrightarrow CHCl_3(g) + 3\,HCl(g)$
($\Delta G_f°$ for $CHCl_3(g)$ is −70.4 kJ/mol.)

61. All the oxides of nitrogen have positive values of $\Delta G_f°$ at 298 K, but only one common oxide of nitrogen has a positive $\Delta S_f°$. Identify that oxide of nitrogen without reference to thermodynamic data and explain.

62. The trend in the $\Delta G_f°$ values of the hydrogen halides is to become less negative with increasing atomic number. The $\Delta G_f°$ of HI is slightly positive. On the other hand the trend in $\Delta S_f°$ is to become more positive with increasing atomic number. Explain these trends.

Challenge Problems

63. The hydrolysis of ATP, shown in Problem 57, is often used to drive nonspontaneous processes—such as muscle contraction and protein synthesis—in living organisms. The nonspontaneous process must be coupled to the ATP hydrolysis reaction. For example, suppose the nonspontaneous process is $A + B \longrightarrow AB$ ($\Delta G°$ positive). The coupling of a nonspontaneous reaction such as this

one to the hydrolysis of ATP is often accomplished by this mechanism.

$$A + ATP + H_2O \longrightarrow A{-}P_i + ADP$$
$$\underline{A{-}P_i + B \longrightarrow AB + P_i}$$
$$A + B + ATP + H_2O \longrightarrow AB + ADP + P_i$$

As long as ΔG°_{rxn} for the nonspontaneous reaction is less than 30.5 kJ, the reaction can be made spontaneous by coupling in this way to the hydrolysis of ATP. Suppose that ATP drives the reaction between glutamate and ammonia to form glutamine:

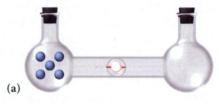

a. Calculate *K* for the reaction between glutamate and ammonia. (The standard free energy change for the reaction is +14.2 kJ/mol. Assume a temperature of 298 K.)

b. Write a set of reactions such as these showing how the glutamate and ammonia reaction can couple with the hydrolysis of ATP. What is ΔG°_{rxn} and *K* for the coupled reaction?

64. Calculate the entropy of each state and rank the states in order of increasing entropy.

(a)

(b)

(c)

65. Suppose we redefine the standard state as $P = 2$ atm. Find the new standard ΔG°_f values of each substance.
a. HCl(*g*)   **b.** N$_2$O(*g*)   **c.** H(*g*)
Explain the results in terms of the relative entropies of reactants and products of each reaction.

66. The ΔG for the freezing of H$_2$O(*l*) at $-10\ °C$ is -210 J/mol and the heat of fusion of ice at this temperature is 5610 J/mol. Find the entropy change of the universe when 1 mol of water freezes at $-10\ °C$.

67. The salt ammonium nitrate can follow three modes of decomposition: (a) to HNO$_3$(*g*) and NH$_3$(*g*), (b) to N$_2$O(*g*) and H$_2$O(*g*), and (c) to N$_2$(*g*), O$_2$(*g*), and H$_2$O(*g*). Calculate ΔG°_{rxn} for each mode of decomposition at 298 K. Explain in light of these results how it is still possible to use ammonium nitrate as a fertilizer and the precautions that should be taken when it is used.

68. A metal salt with the formula MCl$_2$ crystallizes from water to form a solid with the composition MCl$_2 \cdot 6$ H$_2$O. The vapor pressure of water above this solid at 298 K is 18.3 mmHg. What is the value of ΔG for the reaction MCl$_2 \cdot 6$ H$_2$O(*s*) $\rightleftharpoons$ MCl$_2$(*s*) $+$ 6 H$_2$O(*g*) when the pressure of water vapor is 18.3 mmHg? When the pressure of water vapor is 760 mmHg?

69. Given the following data, calculate ΔS_{vap} for each of the first four liquids. ($\Delta S_{vap} = \Delta H_{vap}/T$, where *T* is in K.)

compound	Name	BP (°C)	ΔH_{vap} (kJ/mol) at BP
C$_4$H$_{10}$O	diethyl ether	34.6	26.5
C$_3$H$_6$O	acetone	56.1	29.1
C$_6$H$_6$	benzene	79.8	30.8
CHCl$_3$	chloroform	60.8	29.4
C$_2$H$_5$OH	ethanol	77.8	38.6
H$_2$O	water	100	40.7

All four values should be close to each other. Predict whether the last two liquids in the table are expected to have ΔS_{vap} in this same range. If either does not, predict whether it should be larger or smaller and explain. Verify your prediction.

Conceptual Problems

70. Which is more efficient, a butane lighter or an electric lighter (such as can be found in most automobiles)? Explain.

71. Which statement is true?
a. A spontaneous reaction is always a fast reaction.
b. A spontaneous reaction is always a slow reaction.
c. The spontaneity of a reaction is not necessarily related to the speed of a reaction.

72. Which process is necessarily driven by an increase in the entropy of the surroundings?
a. the condensation of water
b. the sublimation of dry ice
c. the freezing of water

73. Consider these changes in the distribution of nine particles into three interconnected boxes. Which one has the most negative ΔS?

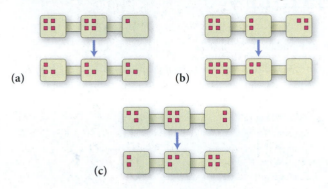

(a)　(b)　(c)

74. Which statement is true?
a. A reaction in which the entropy of the system increases can be spontaneous only if it is exothermic.
b. A reaction in which the entropy of the system increases can be spontaneous only if it is endothermic.
c. A reaction in which the entropy of the system decreases can be spontaneous only if it is exothermic.

75. One of these processes is spontaneous at 298 K. Which one?
a. $H_2O(l) \longrightarrow H_2O(g, 1 \text{ atm})$
b. $H_2O(l) \longrightarrow H_2O(g, 0.10 \text{ atm})$
c. $H_2O(l) \longrightarrow H_2O(g, 0.010 \text{ atm})$

76. The free energy change of the reaction $A(g) \longrightarrow B(g)$ is zero under certain conditions. The *standard* free energy change of the reaction is -42.5 kJ. Which statement must be true about the reaction?
a. The concentration of the product is greater than the concentration of the reactant.
b. The reaction is at equilibrium.
c. The concentration of the reactant is greater than the concentration of the product.

77. The reaction $A(g) \rightleftharpoons B(g)$ has an equiblibrium constant of 5.8 and under certain conditions has $Q = 336$. Which statement is true of the reaction under these conditions?
a. $\Delta G_{rxn}^\circ < 0; \Delta G_{rxn} < 0$
b. $\Delta G_{rxn}^\circ > 0; \Delta G_{rxn} > 0$
c. $\Delta G_{rxn}^\circ < 0; \Delta G_{rxn} > 0$
d. $\Delta G_{rxn}^\circ > 0; \Delta G_{rxn} < 0$

Answers to Conceptual Connections

Nature's Heat Tax and Diet

17.1 A person subsisting on a vegetarian diet eats fruits and vegetables and metabolizes their energy-containing molecules. A person subsisting on a meat-based diet eats the meat of an animal such as a cow and metabolizes energy-containing molecules that were part of the cow. However, the cow synthesized its energy-containing molecules from compounds that it obtained by eating and digesting plants. Since breaking down and resynthesizing biological molecules requires energy—and since the cow also needs to extract some of the energy in its food to live—a meat-based diet requires additional energy transactions in the overall process of obtaining energy for life. Therefore, due to nature's heat tax, the meat-based diet is less efficient than the vegetarian diet.

Entropy

17.2 (a) The more spread out the particles are between the three boxes, the greater the entropy. Therefore, the entropy change is positive only in scheme **(a)**.

Entropy and Biological Systems

17.3 Biological systems do not violate the second law of thermodynamics. The key to understanding this concept is that entropy changes in the system can be negative as long as the entropy change of the universe is positive. Biological systems can decrease their own entropy, but only at the expense of creating more entropy in the surroundings (which they do primarily by emitting heat generated by their metabolic processes). Thus, for any biological process, ΔS_{univ} is positive.

ΔH, ΔS, and ΔG

17.4 (a) Sublimation is endothermic (it requires energy to overcome the intermolecular forces that hold solid carbon dioxide together), so ΔH is positive. The number of moles of gas increases when the solid turns into a gas, so the entropy of the carbon dioxide increases and ΔS is positive. Since $\Delta G = \Delta H - T \Delta S$, ΔG is positive at low temperature and negative at high temperature.

Standard Entropies

17.5 $Kr < Cl_2 < SO_3$; Because krypton is a monoatomic gas; it has the least entropy. Because SO_3 is the most complex molecule; it has the most entropy. The molar masses of the three gases vary slightly, but not enough to overcome the differences in molecular complexity.

Free Energy Changes and Le Châtelier's Principle

17.6 (a) A high concentration of reactants relative to products leads to $Q < 1$, making the term $RT \ln Q$ in Equation 17.14 negative. ΔG_{rxn} is therefore more negative than ΔG_{rxn}° and the reaction is more spontaneous.

K and ΔG_{rxn}°

17.7 (c) Since the equilibrium constant is less than one, the reaction will proceed towards reactants under standard conditions (when $Q = 1$). Therefore, ΔG_{rxn}° is positive.

18 Electrochemistry

One day sir, you may tax it. —Michael Faraday (1791–1867)

[To Mr. Gladstone, the British Chancellor of the Exchequer, who asked about the practical worth of electricity.]

The MP3 player shown here is powered by a hydrogen/oxygen fuel cell, a device that generates electricity from the reaction between hydrogen and oxygen to form water.

THIS CHAPTER'S OPENING QUOTE FROM MICHAEL FARADAY illustrates an important aspect of basic research (research for the sake of understanding how nature works). The Chancellor of the Exchequer (the British cabinet minister responsible for all financial matters) was questioning the value of Michael Faraday's investigations of electricity. The Chancellor wanted to know how the apparently esoteric research would ever be useful. Faraday responded in a way that the Chancellor would understand—citing a future financial payoff. Today, of course, electricity is a fundamental form of energy that powers our entire economy. Although basic research does not always lead to useful applications, much of our technology has grown out of basic research. The history of modern science shows that it is necessary first to understand nature (the goal of basic research) before you can harness its power. In this chapter, we seek to understand oxidation–reduction reactions (first introduced in Chapter 4) and how we can harness them to generate electricity. The applications range from the batteries that power your flashlight to the fuel cells that may one day power our homes and automobiles.

18.1 Pulling the Plug on the Power Grid

The power grid distributes centrally generated electricity throughout the country to homes and businesses. When you turn on a light or electrical appliance, electricity flows from the grid, through the wires in your home, and to the light or appliance. The electrical energy is then converted into light within the light bulb or into work within the appliance. The average U.S. household consumes about 1000 kilowatt-hours (kWh) of electricity per month. The local electrical utility, of course, monitors your electricity use and bills you for it.

In the future, you may have the option of disconnecting from the power grid. Several innovative companies are developing small, fuel-cell power plants—no bigger than refrigerators—that each sit next to a home and quietly generate enough electricity to power its lights and appliances. The heat produced by the fuel cell's operation can be recaptured and used to heat water or the space within the home, eliminating the need for a hot-water heater and a furnace. Similar fuel cells could also be used to power cars. Fuel cells are highly efficient and, although many obstacles remain, they may someday supply all of our power needs while producing less pollution than fossil fuel combustion.

▲ The energy produced by this fuel cell can power an entire house.

▲ General Motors demonstrated this HydroGen3—a five-passenger fuel-cell automobile with a top speed of 100 miles per hour and a range of 250 miles on one tank of fuel—to members of the U.S. Congress.

Fuel cells are based on oxidation–reduction reactions (see Section 4.9). The most common type of fuel cell—called the hydrogen–oxygen fuel cell—is based on the reaction between hydrogen and oxygen.

$$2\,H_2(g) + O_2(g) \longrightarrow 2\,H_2O(l)$$

In this reaction, hydrogen and oxygen form covalent bonds with one another. Recall that, according to the Lewis model, a single covalent bond is a shared electron pair. However, since oxygen is more electronegative than hydrogen, the electron pair in a hydrogen–oxygen bond is not *equally* shared; oxygen gets the larger portion (see Section 9.6). In effect, oxygen has more electrons in H_2O than it does in elemental O_2—it has gained electrons in the reaction and has therefore been reduced.

In a reaction between hydrogen and oxygen, oxygen atoms gain the electrons directly from hydrogen atoms as the reaction proceeds. In a hydrogen–oxygen fuel cell, the same redox reaction occurs, but the hydrogen and oxygen are separated, forcing the electrons to move through an external wire to get from hydrogen to oxygen. These moving electrons constitute an electrical current. In effect, fuel cells employ the electron-gaining tendency of oxygen and the electron-losing tendency of hydrogen to force electrons to move through a wire, creating electricity to power a home or an electric automobile. Smaller fuel cells can replace batteries and be used to power consumer electronics such as laptop computers, cell phones, and MP3 players. The generation of electricity from spontaneous redox reactions (such as those that occur in a fuel cell) and the use of electricity to drive nonspontaneous redox reactions (such as those that occur in gold or silver plating) are both part of the field of electrochemistry, the subject of this chapter.

18.2 Balancing Oxidation–Reduction Equations

Recall from Section 4.9 that a fundamental definition of *oxidation* is the loss of electrons, and a fundamental definition of *reduction* is the gain of electrons. Recall also that we can identify oxidation–reduction reactions through changes in oxidation states: *oxidation corresponds to an increase in oxidation state and reduction corresponds to a decrease in oxidation state*. For example, consider the reaction between calcium and water:

Review Section 4.9 on assigning oxidation states.

$$Ca(s) + 2\,H_2O(l) \longrightarrow Ca(OH)_2(aq) + H_2(g)$$

Since calcium increases in oxidation state from 0 to +2, it is oxidized. Since hydrogen decreases in oxidation state from +1 to 0, it is reduced.

Balancing redox reactions can be more complicated than balancing other types of reactions because both the mass (or number of each type of atom) and the *charge* must be balanced. We can balance redox reactions occurring in aqueous solutions by using a special procedure called the *half-reaction method of balancing*. In this procedure, we separate the overall equation into two half-reactions: one for oxidation and one for reduction. We balance the half-reactions individually and then add them together. The steps differ slightly for reactions occurring in acidic and in basic solution. Examples 18.1 and 18.2 demonstrate the method used for an acidic solution, and Example 18.3 demonstrates the method used for a basic solution.

PROCEDURE FOR...	EXAMPLE 18.1	EXAMPLE 18.2
Half-Reaction Method of Balancing Aqueous Redox Equations in Acidic Solution	**Half-Reaction Method of Balancing Aqueous Redox Equations in Acidic Solution**	**Half-Reaction Method of Balancing Aqueous Redox Equations in Acidic Solution**
	Balance the redox equation.	Balance the redox equation.
GENERAL PROCEDURE	$Al(s) + Cu^{2+}(aq) \longrightarrow Al^{3+}(aq) + Cu(s)$	$Fe^{2+}(aq) + MnO_4^-(aq) \longrightarrow Fe^{3+}(aq) + Mn^{2+}(aq)$
Step 1 *Assign oxidation states to all atoms and identify the substances being oxidized and reduced.*		

Step 2	*Separate the overall reaction into two half-reactions: one for oxidation and one for reduction.*	**Oxidation:** $Al(s) \longrightarrow Al^{3+}(aq)$ **Reduction:** $Cu^{2+}(aq) \longrightarrow Cu(s)$	**Oxidation:** $Fe^{2+}(aq) \longrightarrow Fe^{3+}(aq)$ **Reduction:** $MnO_4^-(aq) \longrightarrow Mn^{2+}(aq)$
Step 3	*Balance each half-reaction with respect to mass in the following order:* • Balance all elements other than H and O. • Balance O by adding H_2O. • Balance H by adding H^+.	All elements are balanced, so proceed to next step.	All elements other than H and O are balanced, so proceed to balance H and O. $Fe^{2+}(aq) \longrightarrow Fe^{3+}(aq)$ $MnO_4^-(aq) \longrightarrow Mn^{2+}(aq) + 4\,H_2O(l)$ $8\,H^+(aq) + MnO_4^-(aq) \longrightarrow$ $\qquad Mn^{2+}(aq) + 4\,H_2O(l)$
Step 4	*Balance each half-reaction with respect to charge by adding electrons. (Make the sum of the charges on both sides of the equation equal by adding as many electrons as necessary.)*	$Al(s) \longrightarrow Al^{3+}(aq) + 3\,e^-$ $2\,e^- + Cu^{2+}(aq) \longrightarrow Cu(s)$	$Fe^{2+}(aq) \longrightarrow Fe^{3+}(aq) + 1\,e^-$ $5\,e^- + 8\,H^+(aq) + MnO_4^-(aq) \longrightarrow$ $\qquad Mn^{2+}(aq) + 4\,H_2O(l)$
Step 5	*Make the number of electrons in both half-reactions equal by multiplying one or both half-reactions by a small whole number.*	$2\big[Al(s) \longrightarrow Al^{3+}(aq) + 3\,e^-\big]$ $2\,Al(s) \longrightarrow 2\,Al^{3+}(aq) + 6\,e^-$ $3\big[2\,e^- + Cu^{2+}(aq) \longrightarrow Cu(s)\big]$ $6\,e^- + 3\,Cu^{2+}(aq) \longrightarrow 3\,Cu(s)$	$5\big[Fe^{2+}(aq) \longrightarrow Fe^{3+}(aq) + 1\,e^-\big]$ $5\,Fe^{2+}(aq) \longrightarrow 5\,Fe^{3+}(aq) + 5\,e^-$ $5\,e^- + 8\,H^+(aq) + MnO_4^-(aq) \longrightarrow$ $\qquad Mn^{2+}(aq) + 4\,H_2O(l)$
Step 6	*Add the two half-reactions together, canceling electrons and other species as necessary.*	$2\,Al(s) \longrightarrow 2\,Al^{3+}(aq) + \cancel{6\,e^-}$ $\cancel{6\,e^-} + 3\,Cu^{2+}(aq) \longrightarrow 3\,Cu(s)$ $2\,Al(s) + 3\,Cu^{2+}(aq) \longrightarrow$ $\qquad 2\,Al^{3+}(aq) + 3\,Cu(s)$	$5\,Fe^{2+}(aq) \longrightarrow 5\,Fe^{3+}(aq) + \cancel{5\,e^-}$ $\cancel{5\,e^-} + 8\,H^+(aq) + MnO_4^-(aq) \longrightarrow$ $\qquad Mn^{2+}(aq) + 4\,H_2O(l)$ $5\,Fe^{2+}(aq) + 8\,H^+(aq) + MnO_4^-(aq) \longrightarrow$ $\quad 5\,Fe^{3+}(aq) + Mn^{2+}(aq) + 4\,H_2O(l)$
Step 7	*Verify that the reaction is balanced both with respect to mass and with respect to charge.*	Reactants / Products 2 Al / 2 Al 3 Cu / 3 Cu +6 charge / +6 charge	Reactants / Products 5 Fe / 5 Fe 8 H / 8 H 1 Mn / 1 Mn 4 O / 4 O +17 charge / +17 charge
		FOR PRACTICE 18.1 Balance the redox reaction in acidic solution. $H^+(aq) + Cr(s) \longrightarrow$ $\qquad H_2(g) + Cr^{2+}(aq)$	**FOR PRACTICE 18.2** Balance the redox reaction in acidic solution. $Cu(s) + NO_3^-(aq) \longrightarrow$ $\qquad Cu^{2+}(aq) + NO_2(g)$

When a redox reaction occurs in basic solution, we balance the reaction in a similar manner, except that we must add an additional step to neutralize any H^+ with OH^-. We then combine the H^+ and OH^- to form H_2O, as shown in Example 18.3.

EXAMPLE 18.3 Balancing Redox Reactions Occurring in Basic Solution

Balance the equation occurring in basic solution:

$$I^-(aq) + MnO_4^-(aq) \longrightarrow I_2(aq) + MnO_2(s)$$

SOLUTION

To balance redox reactions occurring in basic solution, we follow the half-reaction method outlined in Examples 18.1 and 18.2, but add an extra step to neutralize the acid with OH^- as shown in step 3 below.

1. Assign oxidation states.	
2. Separate the overall reaction into two half-reactions.	**Oxidation:** $I^-(aq) \longrightarrow I_2(aq)$ **Reduction:** $MnO_4^-(aq) \longrightarrow MnO_2(s)$
3. Balance each half-reaction with respect to mass. • Balance all elements other than H and O. • Balance O by adding H_2O. • Balance H by adding H^+. • Neutralize H^+ by adding enough OH^- to neutralize each H^+. Add the same number of OH^- ions to each side of the equation.	$\begin{cases} 2\,I^-(aq) \longrightarrow I_2(aq) \\ MnO_4^-(aq) \longrightarrow MnO_2(s) \end{cases}$ $\begin{cases} 2\,I^-(aq) \longrightarrow I_2(aq) \\ MnO_4^-(aq) \longrightarrow MnO_2(s) + 2\,H_2O(l) \end{cases}$ $\begin{cases} 2\,I^-(aq) \longrightarrow I_2(aq) \\ 4\,H^+(aq) + MnO_4^-(aq) \longrightarrow MnO_2(s) + 2\,H_2O(l) \end{cases}$ $\begin{cases} 2\,I^-(aq) \longrightarrow I_2(aq) \\ 4\,H^+(aq) + 4\,OH^-(aq) + MnO_4^-(aq) \longrightarrow MnO_2(s) + 2\,H_2O(l) + 4\,OH^-(aq) \end{cases}$
4. Balance each half-reaction with respect to charge.	$2\,I^-(aq) \longrightarrow I_2(aq) + 2\,e^-$ $4\,H_2O(l) + MnO_4^-(aq) + 3\,e^- \longrightarrow MnO_2(s) + 2\,H_2O(l) + 4\,OH^-(aq)$
5. Make the number of electrons in both half-reactions equal.	$3[2\,I^-(aq) \longrightarrow I_2(aq) + 2\,e^-]$ $6\,I^-(aq) \longrightarrow 3\,I_2(aq) + 6\,e^-$ $2[4\,H_2O(l) + MnO_4^-(aq) + 3\,e^- \longrightarrow MnO_2(s) + 2\,H_2O(l) + 4\,OH^-(aq)]$ $8\,H_2O(l) + 2\,MnO_4^-(aq) + 6\,e^- \longrightarrow 2\,MnO_2(s) + 4\,H_2O(l) + 8\,OH^-(aq)$
6. Add the half-reactions together.	$6\,I^-(aq) \longrightarrow 3\,I_2(aq) + \cancel{6\,e^-}$ $\overset{4}{\cancel{8}}\,H_2O(l) + 2\,MnO_4^-(aq) + \cancel{6\,e^-} \longrightarrow 2\,MnO_2(s) + \cancel{4\,H_2O(l)} + 8\,OH^-(aq)$ $6\,I^-(aq) + 4\,H_2O(l) + 2\,MnO_4^-(aq) \longrightarrow 3\,I_2(aq) + 2\,MnO_2(s) + 8\,OH^-(aq)$
7. Verify that the reaction is balanced.	<table><tr><td>Reactants</td><td>Products</td></tr><tr><td>6 I</td><td>6 I</td></tr><tr><td>8 H</td><td>8 H</td></tr><tr><td>2 Mn</td><td>2 Mn</td></tr><tr><td>12 O</td><td>12 O</td></tr><tr><td>−8 charge</td><td>−8 charge</td></tr></table>

FOR PRACTICE 18.3

Balance the redox reaction occurring in basic solution.

$$ClO^-(aq) + Cr(OH)_4^-(aq) \longrightarrow CrO_4^{2-}(aq) + Cl^-(aq)$$

18.3 Voltaic (or Galvanic) Cells: Generating Electricity from Spontaneous Chemical Reactions

Electrical current is the flow of electrical charge. For example, electrons flowing through a wire or ions flowing through a solution both constitute electrical current. Since redox reactions involve the transfer of electrons from one substance to another, they have the potential to generate electrical current as discussed in Section 18.1.

The generation of electricity through redox reactions is typically carried out in a device called an **electrochemical cell**. A **voltaic** (or **galvanic**) **cell**, is an electrochemical cell that *produces* electrical current from a *spontaneous* chemical reaction. A second type of electrochemical cell, called an **electrolytic cell**, *consumes* electrical current to drive a *nonspontaneous* chemical reaction. We discuss voltaic cells in this section and electrolytic cells in Section 18.8.

In the voltaic cell shown in **Figure 18.1▼**, a solid strip of zinc is placed in a $Zn(NO_3)_2$ solution to form a **half-cell**. Similarly, a solid strip of copper is placed in a $Cu(NO_3)_2$ solution to form a second half-cell. The metal strips act as **electrodes**, conductive surfaces through which electrons can enter or leave the half-cells. Each metal strip reaches equilibrium with its ions in solution according to these half-reactions:

$$Zn(s) \rightleftharpoons Zn^{2+}(aq) + 2\ e^-$$
$$Cu(s) \rightleftharpoons Cu^{2+}(aq) + 2\ e^-$$

However, the positions of these equilibria are not the same for both metals. The zinc has a greater tendency to ionize than the copper, so the zinc half-reaction lies further to the right (toward production of electrons). *As a result, the zinc electrode becomes negatively charged relative to the copper electrode.*

The idea that one electrode in a voltaic cell becomes more negatively charged relative to the other electrode due to differences in ionization tendencies is central to understanding how a voltaic cell works.

A Voltaic Cell

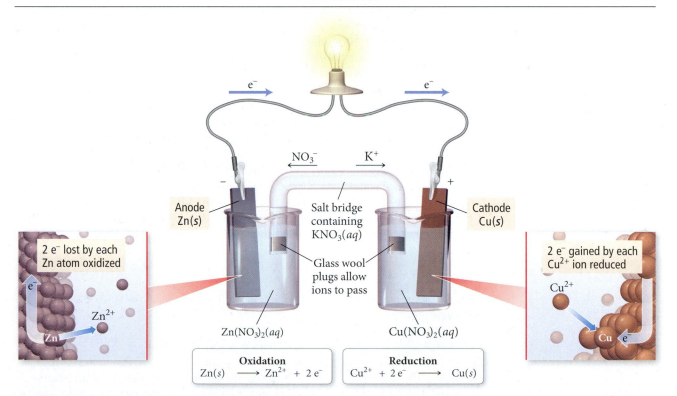

▲ **FIGURE 18.1 A Voltaic Cell** The tendency of zinc to transfer electrons to copper results in a flow of electrons through the wire, lighting the bulb. The movement of electrons from the zinc to the copper creates a positive charge buildup at the zinc half-cell and a negative charge buildup at the copper half-cell. The flow of ions within the salt bridge neutralizes this charge buildup, allowing the reaction to continue.

The ampere is often abbreviated as amp.

▲ **FIGURE 18.2 An Analogy for Electrical Current**   Just as water flows downhill in response to a difference in gravitational potential energy, electrons flow through a conductor in response to an electrical potential difference, creating an electrical current.

If the two half-cells are connected by a wire running from the zinc, through a light bulb or other electrical device, to the copper, electrons spontaneously flow from the zinc electrode (which is more negatively charged and therefore repels electrons) to the copper electrode. As the electrons flow away from the zinc electrode, the Zn/Zn^{2+} equilibrium shifts to the right (according to Le Châtelier's principle) and oxidation occurs. As electrons flow to the copper electrode, the Cu/Cu^{2+} equilibrium shifts to the left, and reduction occurs. The flowing electrons constitute an electrical current that lights the bulb.

We can understand electrical current and why it flows by analogy with water current in a stream (**Figure 18.2◄**). The *rate of electrons flowing* through a wire is analogous to the *rate of water moving* through a stream. Electrical current is measured in units of **amperes (A)**. One ampere represents the flow of one coulomb (a measure of electrical charge) per second:

$$1 \text{ A} = 1 \text{ C/s}$$

Since an electron has a charge of -1.602×10^{-19} C, 1 A corresponds to the flow of 6.242×10^{18} electrons per second.

The *driving force* for electrical current is analogous to the driving force for water current. Water current is driven by a difference in gravitational potential energy (caused by a gravitational field). Streams flow downhill, from higher to lower potential energy. Electrical current is also driven by a difference in potential energy called **potential difference** (caused by an electric field that results from the charge difference on the two electrodes). *Potential difference is a measure of the difference in potential energy (usually in joules) per unit of charge (coulombs)*. The SI unit of potential difference is the **volt (V)**, which is equal to one joule per coulomb:

$$1 \text{ V} = 1 \text{ J/C}$$

In other words, a potential difference of one volt indicates that a charge of one coulomb experiences an energy difference of one joule between the two electrodes.

A large potential difference corresponds to a large difference in charge between the two electrodes and, therefore, a strong tendency for electron flow (analogous to a steeply descending streambed). Potential difference, since it gives rise to the force that results in the motion of electrons, is also referred to as **electromotive force (emf)**.

In a voltaic cell, the potential difference between the two electrodes is called the **cell potential (E_{cell})** or **cell emf**. The cell potential depends on the relative tendencies of the reactants to undergo oxidation and reduction. Combining the oxidation of a substance with a strong tendency to undergo oxidation and the reduction of a substance with a strong tendency to undergo reduction produces a large difference in charge between the two electrodes and, therefore, a high positive cell potential.

In general, the cell potential also depends on the concentrations of the reactants and products in the cell and the temperature (which we will assume to be 25 °C unless otherwise noted). Under standard conditions (1 M concentration for reactants in solution and 1 atm pressure for gaseous reactants), the cell potential is called the **standard cell potential (E°_{cell})** or **standard emf**. The standard cell potential in the zinc and copper cell described previously is 1.10 volts:

$$Zn(s) + Cu^{2+}(aq) \longrightarrow Zn^{2+}(aq) + Cu(s) \quad E^{\circ}_{cell} = +1.10 \text{ V}$$

If the zinc is replaced with nickel (which has a lower tendency to be oxidized) the cell potential is lower:

$$Ni(s) + Cu^{2+}(aq) \longrightarrow Ni^{2+}(aq) + Cu(s) \quad E^{\circ}_{cell} = +0.62 \text{ V}$$

The cell potential is a measure of the overall tendency of the redox reaction to occur spontaneously. The greater the cell potential, the greater the tendency for the reaction to occur spontaneously. A negative cell potential indicates that the redox reaction *does not* occur spontaneously in the forward direction.

In all electrochemical cells, the electrode where oxidation occurs is the **anode** and the electrode where reduction occurs is the **cathode**. In a voltaic cell, the anode is the more negatively charged electrode and is labeled with a negative ($-$) sign. The cathode of a voltaic cell is the more positively charged electrode and is labeled with a ($+$) sign.

Note that the anode and cathode need not actually be negatively and positively charged, respectively. The anode is the electrode with the relatively *more* negative (or least positive) charge.

Electrons flow from the anode to the cathode (from negative to positive) through the wires connecting the electrodes.

As electrons flow out of the anode, positive ions (Zn^{2+} in the preceding example) form in the oxidation half-cell, resulting in a buildup of *positive charge* in the *solution*. As electrons flow into the cathode, positive ions (Cu^{2+} in the preceding example) are reduced at the reduction half-cell, resulting in a buildup of *negative charge* in the solution.

If the movement of electrons from anode to cathode were the only flow of charge, the buildup of the opposite charge in the solution would stop electron flow almost immediately. The cell needs a pathway by which counterions can flow between the half-cells without the solutions in the half-cells totally mixing. One such pathway is a **salt bridge**, an inverted, U-shaped tube that contains a strong electrolyte such as KNO_3 and connects the two half-cells (see Figure 18.1). The electrolyte is usually suspended in a gel and held within the tube by permeable stoppers. The salt bridge allows a flow of ions that neutralizes the charge buildup in the solution. *The negative ions within the salt bridge flow to neutralize the accumulation of positive charge at the anode, and the positive ions flow to neutralize the accumulation of negative charge at the cathode.* In other words, the salt bridge serves to complete the circuit, allowing electrical current to flow.

 Conceptual Connection 18.1 Voltaic Cells

In a voltaic cell, electrons flow:

(a) From the more negatively charged electrode to the more positively charged electrode

(b) From the more positively charged electrode to the more negatively charged electrode

(c) From lower potential energy to higher potential energy

Electrochemical Cell Notation

We can represent electrochemical cells with a compact notation called a *cell diagram* or *line notation*. For example, the electrochemical cell discussed previously in which Zn is oxidized to Zn^{2+} and Cu^{2+} is reduced to Cu is represented as:

$$Zn(s)\,|\,Zn^{2+}(aq)\,\|\,Cu^{2+}(aq)\,|\,Cu(s)$$

In this kind of representation,

- The oxidation half-reaction is always written on the left and the reduction on the right. A double vertical line (∥), indicating the salt bridge, separates the two half-reactions.
- Substances in different phases are separated by a single vertical line (|), which represents the boundary between the phases.
- For some redox reactions, the reactants and products of one or both of the half-reactions may be in the same phase. In these cases (explained further below), the reactants and products are separated from each other with a comma in the line diagram. Such cells use an inert electrode, such as platinum (Pt) or graphite, as the anode or cathode (or both).

Consider the redox reaction in which Fe(s) is oxidized and $MnO_4^-(aq)$ is reduced:

$$5\,Fe(s) + 2\,MnO_4^-(aq) + 16\,H^+(aq) \longrightarrow 5\,Fe^{2+}(aq) + 2\,Mn^{2+}(aq) + 8\,H_2O(l)$$

The half-reactions for this overall reaction are:

Oxidation: $Fe(s) \longrightarrow Fe^{2+}(aq) + 2\,e^-$

Reduction: $MnO_4^-(aq) + 5\,e^- + 8\,H^+(aq) \longrightarrow Mn^{2+}(aq) + 4\,H_2O(l)$

Notice that, in the reduction half-reaction, the principal species are all in the aqueous phase. However, the electron transfer needs an electrode on which to occur. In this case, an inert platinum electrode is used, and the electron transfer takes place at its surface. Using line notation, we represent the electrochemical cell corresponding to the above reaction as:

$$Fe(s)\,|\,Fe^{2+}(aq)\,\|\,MnO_4^-(aq),\,H^+(aq),\,Mn^{2+}(aq)\,|\,Pt(s)$$

The Pt(s) on the far right indicates the inert platinum electrode which acts as the cathode in this reaction, as shown in **Figure 18.3▶**.

► **FIGURE 18.3 Inert Platinum Electrode** When the participants in a half-reaction are all in the aqueous phase, a conductive surface is needed for electron transfer to take place. In such cases an inert electrode of graphite or platinum is often used. In this electrochemical cell, an iron strip acts as the anode and a platinum strip acts as the cathode. Iron is oxidized at the anode and MnO_4^- is reduced at the cathode.

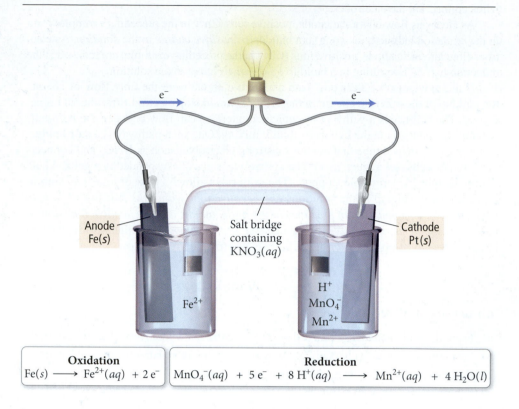

Inert Platinum Electrode

Anode Fe(s)

Salt bridge containing $KNO_3(aq)$

Cathode Pt(s)

Fe^{2+}

H^+
MnO_4^-
Mn^{2+}

Oxidation	Reduction
$Fe(s) \longrightarrow Fe^{2+}(aq) + 2 e^-$	$MnO_4^-(aq) + 5 e^- + 8 H^+(aq) \longrightarrow Mn^{2+}(aq) + 4 H_2O(l)$

18.4 Standard Electrode Potentials

As we have just seen, the standard cell potential (E°_{cell}) for an electrochemical cell depends on the specific half-reactions occurring in the half-cells, and is a measure of the potential energy difference (per unit charge) between the two electrodes. We can think of the electrode in each half-cell as having its own individual potential, called the **standard electrode potential**. The overall standard cell potential (E°_{cell}) is the difference between the two standard electrode potentials.

We can better understand this idea with an analogy. Consider two water tanks with different water levels connected by a common pipe, as shown in **Figure 18.4◄**. The water in each tank has its own level and corresponding potential energy. When the tanks are connected, water will flow from the tank with the higher level (higher potential energy) to the tank with a lower water level (lower potential energy). Similarly, each half-cell in an electrochemical cell has its own charge and corresponding electrode potential. *When the cells are connected, electrons will flow from the electrode with greater negative charge (greater potential energy) to the electrode with less negative or more positive charge (less potential energy).*

A major limitation of this analogy is that, unlike the water level in a tank, we cannot measure the electrode potential in a half-cell directly—we can only measure the overall potential that occurs when two half-cells are combined in a whole cell. However, we can arbitrarily assign a potential of zero to the electrode in a *particular* type of half-cell and then measure all other electrode potentials relative to that zero.

The half-cell electrode that is normally assigned a potential of zero is the **standard hydrogen electrode (SHE).** This half-cell consists of an inert platinum electrode immersed in 1 M HCl with hydrogen gas at 1 atm bubbling through the solution, as shown in **Figure 18.5►**. When the SHE acts as the cathode, the following half-reaction occurs:

$$2 H^+(aq) + 2 e^- \longrightarrow H_2(g) \quad E^\circ_{cathode} = 0.00 \text{ V}$$

If we connect the standard hydrogen electrode to an electrode in another half-cell of interest, we can measure the potential difference (or voltage) between the two electrodes.

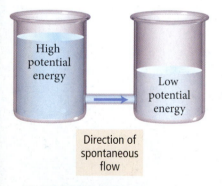

High potential energy

Low potential energy

Direction of spontaneous flow

▲ **FIGURE 18.4 An Analogy for Electrode Potential**

Since the standard hydrogen electrode is assigned a zero voltage, we can then determine the electrode potential of the other half-cell.

For example, consider the electrochemical cell shown in **Figure 18.6▼**. In this electrochemical cell, Zn is oxidized to Zn^{2+} (anode half-reaction) and H^+ is reduced to H_2 (cathode half-reaction) under standard conditions (all solutions are 1 M in concentration and all gases are 1 atm in pressure) and at 25 °C. Since electrons travel from the anode (where oxidation occurs) to the cathode (where reduction occurs) we can define E°_{cell} as *the difference in voltage between the cathode (final state) and the anode (initial state)*:

$$E^{\circ}_{cell} = E^{\circ}_{final} - E^{\circ}_{initial}$$
$$= E^{\circ}_{cathode} - E^{\circ}_{anode}$$

The measured cell potential for this cell is +0.76 V. The anode (in this case, Zn/Zn^{2+}) is at a more negative voltage (higher potential energy) than the cathode (in this case, the SHE). Therefore, electrons spontaneously flow from the anode to the cathode. We can diagram the potential energy and the voltage as follows:

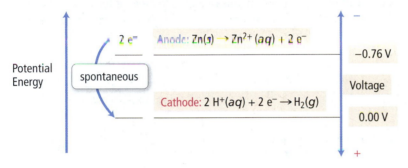

Referring back to our water tank analogy, the zinc half-cell is like the water tank with the higher water level, and electrons therefore flow from the zinc electrode to the standard hydrogen electrode.

Standard Hydrogen Electrode (SHE)

▲ FIGURE 18.5 The Standard Hydrogen Electrode The standard hydrogen electrode (SHE) is arbitrarily assigned an electrode potential of zero. All other electrode potentials are then measured relative to the SHE.

Measuring Half-Cell Potential with the SHE

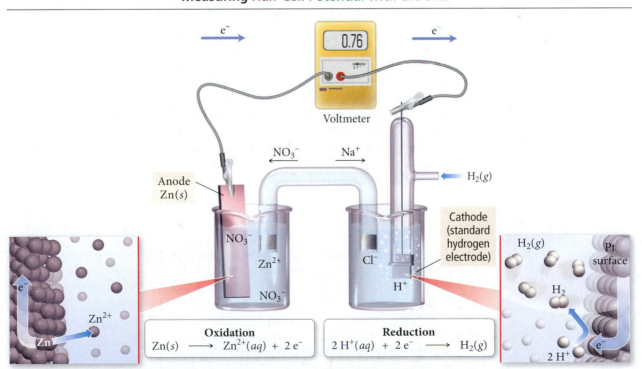

▲ FIGURE 18.6 Measuring Electrode Potential Since the electrode potential of the SHE is zero, the electrode potential for the oxidation of Zn is equal to the cell potential.

Since we assigned the SHE a potential of zero (0.00 V), we can determine the electrode potential for the Zn/Zn^{2+} half-cell (the anode):

$$E^\circ_{cell} = E^\circ_{cathode} - E^\circ_{anode}$$
$$0.76 \text{ V} = 0.00 \text{ V} - E^\circ_{anode}$$
$$E^\circ_{anode} = -0.76 \text{ V}$$

Notice that the potential for the Zn/Zn^{2+} electrode is *negative*. The negative potential indicates that an electron at the Zn/Zn^{2+} electrode has greater potential energy than it does at the SHE. *Remember the more negative the electrode potential is, the greater the potential energy of an electron at that electrode (because negative charge repels electrons).*

What would happen if we connect an electrode in which the electron has a *more positive* potential (than the standard hydrogen electrode) to the standard hydrogen electrode? That electrode would then have a more *positive* voltage (lower potential energy for an electron).

For example, suppose we connect the standard hydrogen electrode to a Cu electrode immersed in a 1 M Cu^{2+} solution. The measured cell potential for this cell is -0.34 V. The anode (defined as Cu/Cu^{2+}) is at a more positive voltage (lower potential energy) than the cathode (the SHE). Therefore, electrons will *not* spontaneously flow from the anode to the cathode. We can diagram the potential energy and the voltage of this cell as follows:

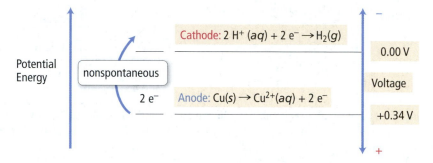

The copper half-cell is like the water tank with the lower water level, and electrons, therefore, *do not* spontaneously flow from copper electrode to the standard hydrogen electrode.

We can determine the electrode potential for the Cu/Cu^{2+} half-cell (the anode):

$$E^\circ_{cell} = E^\circ_{cathode} - E^\circ_{anode}$$
$$-0.34 \text{ V} = 0.00 \text{ V} - E^\circ_{anode}$$
$$E^\circ_{anode} = +0.34 \text{ V}$$

Notice that the potential for the Cu/Cu^{2+} electrode is *positive*. The positive potential indicates that an electron at the Cu/Cu^{2+} electrode has *lower* potential energy than it does at the SHE. The more positive the electrode potential, the lower the potential energy of an electron at that electrode (because positive charge attracts electrons).

By convention, standard electrode potentials are written for *reduction* half-reactions. We write the standard electrode potentials for the two half-reactions just discussed as:

$$Cu^{2+}(aq) + 2\,e^- \longrightarrow Cu(s) \qquad E^\circ = +0.34 \text{ V}$$
$$Zn^{2+}(aq) + 2\,e^- \longrightarrow Zn(s) \qquad E^\circ = -0.76 \text{ V}$$

A half-reaction with a positive electrode potential attracts electrons and, therefore, has a greater tendency towards reduction.

We can now clearly see that the Cu/Cu^{2+} electrode is positive relative to the SHE (and will therefore tend to draw electrons *away from* the SHE), and that the Zn/Zn^{2+} electrode is negative relative to the SHE (and will therefore tend to repel electrons *towards* the SHE). The standard electrode potentials for a number of common half-reactions are shown in Table 18.1.

Summarizing Standard Electrode Potentials:

▶ The electrode potential of the standard hydrogen electrode (SHE) is defined as exactly zero.

▶ The electrode in any half-cell with a greater tendency to undergo reduction is positively charged relative to the SHE and therefore has a positive E°.

TABLE 18.1 Standard Electrode Potentials at 25 °C

Reduction Half-Reaction		$E°$ (V)	
Stronger oxidizing agent	$F_2(g) + 2\,e^- \longrightarrow 2\,F^-(aq)$	2.87	Weaker reducing agent
	$H_2O_2(aq) + 2\,H^+(aq) + 2\,e^- \longrightarrow 2\,H_2O(l)$	1.78	
	$PbO_2(s) + 4\,H^+(aq) + SO_4^{2-}(aq) + 2\,e^- \longrightarrow PbSO_4(s) + 2\,H_2O(l)$	1.69	
	$MnO_4^-(aq) + 4\,H^+(aq) + 3\,e^- \longrightarrow MnO_2(s) + 2\,H_2O(l)$	1.68	
	$MnO_4^-(aq) + 8\,H^+(aq) + 5\,e^- \longrightarrow Mn^{2+}(aq) + 4\,H_2O(l)$	1.51	
	$Au^{3+}(aq) + 3\,e^- \longrightarrow Au(s)$	1.50	
	$PbO_2(s) + 4\,H^+(aq) + 2\,e^- \longrightarrow Pb^{2+}(aq) + 2\,H_2O(l)$	1.46	
	$Cl_2(g) + 2\,e^- \longrightarrow 2\,Cl^-(aq)$	1.36	
	$Cr_2O_7^{2-}(aq) + 14\,H^+(aq) + 6\,e^- \longrightarrow 2\,Cr^{3+}(aq) + 7\,H_2O(l)$	1.33	
	$O_2(g) + 4\,H^+(aq) + 4\,e^- \longrightarrow 2\,H_2O(l)$	1.23	
	$MnO_2(s) + 4\,H^+(aq) + 2\,e^- \longrightarrow Mn^{2+}(aq) + 2\,H_2O(l)$	1.21	
	$IO_3^-(aq) + 6\,H^+(aq) + 5\,e^- \longrightarrow \frac{1}{2}I_2(aq) + 3\,H_2O(l)$	1.20	
	$Br_2(l) + 2\,e^- \longrightarrow 2\,Br^-(aq)$	1.09	
	$VO_2^+(aq) + 2\,H^+(aq) + e^- \longrightarrow VO^{2+}(aq) + H_2O(l)$	1.00	
	$NO_3^-(aq) + 4\,H^+(aq) + 3\,e^- \longrightarrow NO(g) + 2\,H_2O(l)$	0.96	
	$ClO_2(g) + e^- \longrightarrow ClO_2^-(aq)$	0.95	
	$Ag^+(aq) + e^- \longrightarrow Ag(s)$	0.80	
	$Fe^{3+}(aq) + e^- \longrightarrow Fe^{2+}(aq)$	0.77	
	$O_2(g) + 2\,H^+(aq) + 2\,e^- \longrightarrow H_2O_2(aq)$	0.70	
	$MnO_4^-(aq) + e^- \longrightarrow MnO_4^{2-}(aq)$	0.56	
	$I_2(s) + 2\,e^- \longrightarrow 2\,I^-(aq)$	0.54	
	$Cu^+(aq) + e^- \longrightarrow Cu(s)$	0.52	
	$O_2(g) + 2\,H_2O(l) + 4\,e^- \longrightarrow 4\,OH^-(aq)$	0.40	
	$Cu^{2+}(aq) + 2\,e^- \longrightarrow Cu(s)$	0.34	
	$SO_4^{2-}(aq) + 4\,H^+(aq) + 2\,e^- \longrightarrow H_2SO_3(aq) + H_2O(l)$	0.20	
	$Cu^{2+}(aq) + e^- \longrightarrow Cu^+(aq)$	0.16	
	$Sn^{4+}(aq) + 2\,e^- \longrightarrow Sn^{2+}(aq)$	0.15	
	$2\,H^+(aq) + 2\,e^- \longrightarrow H_2(g)$	0	
	$Fe^{3+}(aq) + 3\,e^- \longrightarrow Fe(s)$	−0.036	
	$Pb^{2+}(aq) + 2\,e^- \longrightarrow Pb(s)$	−0.13	
	$Sn^{2+}(aq) + 2\,e^- \longrightarrow Sn(s)$	−0.14	
	$Ni^{2+}(aq) + 2\,e^- \longrightarrow Ni(s)$	−0.23	
	$Cd^{2+}(aq) + 2\,e^- \longrightarrow Cd(s)$	−0.40	
	$Fe^{2+}(aq) + 2\,e^- \longrightarrow Fe(s)$	−0.45	
	$Cr^{3+}(aq) + e^- \longrightarrow Cr^{2+}(aq)$	−0.50	
	$Cr^{3+}(aq) + 3\,e^- \longrightarrow Cr(s)$	−0.73	
	$Zn^{2+}(aq) + 2\,e^- \longrightarrow Zn(s)$	−0.76	
	$2\,H_2O(l) + 2\,e^- \longrightarrow H_2(g) + 2\,OH^-(aq)$	−0.83	
	$Mn^{2+}(aq) + 2\,e^- \longrightarrow Mn(s)$	−1.18	
	$Al^{3+}(aq) + 3\,e^- \longrightarrow Al(s)$	−1.66	
	$Mg^{2+}(aq) + 2\,e^- \longrightarrow Mg(s)$	−2.37	
	$Na^+(aq) + e^- \longrightarrow Na(s)$	−2.71	
	$Ca^{2+}(aq) + 2\,e^- \longrightarrow Ca(s)$	−2.76	
	$Ba^{2+}(aq) + 2\,e^- \longrightarrow Ba(s)$	−2.90	Stronger reducing agent
Weaker oxidizing agent	$K^+(aq) + e^- \longrightarrow K(s)$	−2.92	
	$Li^+(aq) + e^- \longrightarrow Li(s)$	−3.04	

▶ The electrode in any half-cell with a lesser tendency to undergo reduction (or greater tendency to undergo oxidation) is negatively charged relative to the SHE and, therefore, has a negative $E°$.

▶ The cell potential of any electrochemical cell ($E°_{cell}$) is the difference between the electrode potentials of the cathode and the anode ($E°_{cell} = E°_{cat} - E°_{an}$).

▶ $E°_{cell}$ is positive for spontaneous reactions and negative for nonspontaneous reactions.

Example 18.4 shows how to calculate the potential of an electrochemical cell from the standard electrode potentials of the half-reactions.

EXAMPLE 18.4 Calculating Standard Potentials for Electrochemical Cells from Standard Electrode Potentials of the Half-Reactions

Use tabulated standard electrode potentials to calculate the standard cell potential for the reaction occurring in an electrochemical cell at 25 °C. (The equation is balanced.)

$$Al(s) + NO_3^-(aq) + 4\,H^+(aq) \longrightarrow Al^{3+}(aq) + NO(g) + 2\,H_2O(l)$$

SOLUTION

Begin by separating the reaction into oxidation and reduction half-reactions. [In this case, you can readily see that Al(s) is oxidized. In cases where it is not so apparent, you many want to assign oxidation states to determine the correct half-reactions.]	***Oxidation:*** $\quad Al(s) \longrightarrow Al^{3+}(aq) + 3\,e^-$ ***Reduction:*** $\quad NO_3^-(aq) + 4\,H^+(aq) + 3\,e^- \longrightarrow NO(g) + 2\,H_2O(l)$
Look up the standard electrode potentials for each half-reaction. Add the half-cell reactions together to obtain the overall redox equation. Calculate the standard cell potential by subtracting the electrode potential of the anode from the electrode potential of the cathode.	***Oxidation (Anode):*** $\qquad\qquad Al(s) \longrightarrow Al^{3+}(aq) + 3\,e^- \quad E° = -1.66\text{ V}$ ***Reduction (Cathode):*** $\underline{\quad NO_3^-(aq) + 4\,H^+(aq) + 3\,e^- \longrightarrow NO(g) + 2\,H_2O(l) \quad E° = 0.96\text{ V}}$ $Al(s) + NO_3^-(aq) + 4\,H^+(aq) \longrightarrow Al^{3+}(aq) + NO(g) + 2\,H_2O(l)$ $E°_{cell} = E°_{cat} - E°_{an}$ $\qquad = 0.96\text{ V} - (-1.66\text{ V})$ $\qquad = 2.62\text{ V}$

FOR PRACTICE 18.4

Use tabulated standard electrode potentials to calculate the standard cell potential for the reaction occurring in an electrochemical cell at 25 °C. (The equation is balanced.)

$$3\,Pb^{2+}(aq) + 2\,Cr(s) \longrightarrow 3\,Pb(s) + 2\,Cr^{3+}(aq)$$

Conceptual Connection 18.2 Standard Electrode Potentials

A particular electrode has a negative electrode potential. Which statement is correct regarding the potential energy of an electron at this electrode?

(a) An electron at this electrode has a lower potential energy than it has at a standard hydrogen electrode.

(b) An electron at this electrode has a higher potential energy than it has at a standard hydrogen electrode.

(c) An electron at this electrode has the same potential energy as it has at a standard hydrogen electrode.

Predicting the Spontaneous Direction of an Oxidation–Reduction Reaction

To determine the spontaneous direction of an oxidation–reduction reaction, examine the electrode potentials of the two relevant half-reactions in Table 18.1. The half-reaction with the more *negative* electrode potential will tend to lose electrons and, therefore, undergo oxidation. (Remember that negative charge repels electrons.) The half-reaction with the more *positive* electrode potential will tend to gain electrons and, therefore, undergo reduction. (Remember that positive charge attracts electrons.)

Consider the two reduction half-reactions:

$$Ni^{2+}(aq) + 2\,e^- \longrightarrow Ni(s) \qquad E° = -0.23 \text{ V}$$

$$Mn^{2+}(aq) + 2\,e^- \longrightarrow Mn(s) \qquad E° = -1.18 \text{ V}$$

Since the manganese half-reaction has the more negative electrode potential, it repels electrons and proceeds in the reverse direction (oxidation). Since the nickel half-reaction has the more positive (or least negative) electrode potential, it attracts electrons and proceeds in the forward direction.

We can confirm this by calculating the standard electrode potential for manganese acting as the anode (oxidation) and nickel acting as the cathode (reduction).

Oxidation (*Anode*): $\qquad\qquad Mn(s) \longrightarrow Mn^{2+}(aq) + 2\,e^- \quad E° = -1.18 \text{ V}$

Reduction (*Cathode*): $Ni^{2+}(aq) + 2\,e^- \longrightarrow Ni(s) \qquad\qquad\qquad E° = -0.23 \text{ V}$

$$Ni^{2+}(aq) + Mn(s) \longrightarrow Ni(s) + Mn^{2+}(aq) \quad E°_{cell} = E°_{cat} - E°_{an}$$

$$= -0.23 \text{ V} - (-1.18 \text{ V})$$

$$= 0.95 \text{ V}$$

The overall cell potential is positive, indicating a spontaneous reaction.

Another way to predict the spontaneity of a redox reaction is to note the relative positions of the two half-reactions in Table 18.1 Since the table lists half-reactions in order of *decreasing* electrode potential, the half-reactions near the top of the table—those having large *positive* electrode potentials—attract electrons and, therefore, tend to occur in the forward direction. Half-reactions near the bottom of the table—those having large *negative* electrode potentials—repel electrons and, therefore, tend to occur in the reverse direction. In other words, as you move down Table 18.1, the half-reactions become less likely to occur in the forward direction and more likely to occur in the reverse direction. As a result, *any reduction half-reaction listed will be spontaneous when paired with the reverse of any half-reaction that appears below it in Table 18.1.*

For example, if we return to our two previous half-reactions involving manganese and nickel:

$$Ni^{2+}(aq) + 2\,e^- \longrightarrow Ni(s) \qquad E° = -0.23 \text{ V}$$

$$Mn^{2+}(aq) + 2\,e^- \longrightarrow Mn(s) \qquad E° = -1.18 \text{ V}$$

We can see that the manganese half-reaction is listed below the nickel half-reaction in Table 18.1; therefore, the nickel reaction occurs in the forward direction (reduction) and the manganese reaction occurs in the reverse direction (oxidation).

Summarizing Spontaneous Direction Prediction for Redox Reactions:

▶ The half-reaction with the more *negative* electrode potential repels electrons most strongly and will undergo oxidation. (Substances listed near the bottom of Table 18.1 tend to undergo oxidation; they are good reducing agents.)

▶ The half-reaction with the more *positive* electrode potential attracts electrons most strongly and will undergo reduction. (Substances listed at the top of Table 18.1 tend to undergo reduction; they are good oxidizing agents.)

▶ Any reduction reaction in Table 18.1 is spontaneous when paired with the *reverse* of any of the reactions listed below it.

The following mnemonic (NIO and PIR) can help you predict the spontaneous direction of redox reactions:
N.I.O.—More Negative Is Oxidation
P.I.R.—More Positive Is Reduction

EXAMPLE 18.5 Predicting Spontaneous Redox Reactions and Sketching Electrochemical Cells

Without calculating $E°_{cell}$, predict whether each redox reaction is spontaneous. If the reaction is spontaneous as written, make a sketch of the electrochemical cell in which the reaction could occur. If the reaction is not spontaneous as written, write an equation for the spontaneous direction in which the reaction would occur and make a sketch of the electrochemical cell in which the spontaneous reaction would occur. In your sketches, make sure to label the anode (which should be drawn on the left), the cathode, and the direction of electron flow.

(a) $Fe(s) + Mg^{2+}(aq) \longrightarrow Fe^{2+}(aq) + Mg(s)$

(b) $Fe(s) + Pb^{2+}(aq) \longrightarrow Fe^{2+}(aq) + Pb(s)$

SOLUTION

(a) $Fe(s) + Mg^{2+}(aq) \longrightarrow Fe^{2+}(aq) + Mg(s)$

This reaction involves the reduction of Mg^{2+}

$$Mg^{2+}(aq) + 2\,e^- \longrightarrow Mg(s) \quad E° = -2.37\ V$$

and the oxidation of Fe.

$$Fe(s) \longrightarrow Fe^{2+}(aq) + 2\,e^- \quad E° = -0.45\ V$$

The magnesium half-reaction has the more negative electrode potential, and, therefore, repels electrons more strongly and undergoes oxidation. The iron half-reaction has the more positive electrode potential and, therefore, attracts electrons more strongly and undergoes reduction. So the reaction as written is *not* spontaneous. (The reaction pairs the reduction of Mg^{2+} with the reverse of a half-reaction *above it* in Table 18.1—such pairings are not spontaneous.)

However, the reverse reaction is spontaneous.

$$Fe^{2+}(aq) + Mg(s) \longrightarrow Fe(s) + Mg^{2+}(aq)$$

The corresponding electrochemical cell is shown in **Figure 18.7▶**.

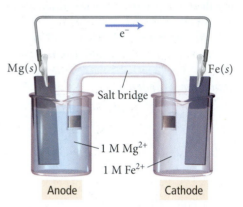

▲ **FIGURE 18.7 Mg/Fe²⁺ Electrochemical Cell**

(b) $Fe(s) + Pb^{2+}(aq) \longrightarrow Fe^{2+}(aq) + Pb(s)$

This reaction involves the reduction of Pb^{2+}

$$Pb^{2+}(aq) + 2\,e^- \longrightarrow Pb(s) \quad E° = -0.13\ V$$

and the oxidation of iron.

$$Fe(s) \longrightarrow Fe^{2+}(aq) + 2\,e^- \quad E° = -0.45\ V$$

The iron half-reaction has the more negative electrode potential; it repels electrons and undergoes oxidation. The lead half-reaction has the more positive electrode potential; it attracts electrons and undergoes reduction. Therefore, the reaction *is* spontaneous as written. (The reaction pairs the reduction of Pb^{2+} with the reverse of a half-reaction *below it* in Table 18.1—such pairings are always spontaneous.) The corresponding electrochemical cell is shown in **Figure 18.8▶**.

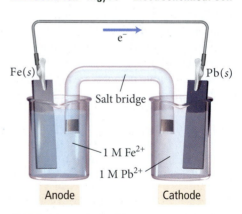

▲ **FIGURE 18.8 Fe/Pb²⁺ Electrochemical Cell**

FOR PRACTICE 18.5

Is each redox reaction spontaneous under standard conditions?

(a) $Zn(s) + Ni^{2+}(aq) \longrightarrow Zn^{2+}(aq) + Ni(s)$

(b) $Zn(s) + Ca^{2+}(aq) \longrightarrow Zn^{2+}(aq) + Ca(s)$

Conceptual Connection 18.3 Selective Oxidation

A solution contains both NaI and NaBr. Which oxidizing agent could be added to the solution to selectively oxidize $I^-(aq)$ but not $Br^-(aq)$?

(a) Cl_2 **(b)** H_2O_2 **(c)** $CuCl_2$ **(d)** HNO_3

Predicting Whether a Metal Will Dissolve in Acid

In Chapter 15, we learned that acids dissolve metals. Most acids dissolve metals by the reduction of H^+ ions to hydrogen gas and the corresponding oxidation of the metal to its ion. For example, if solid Zn is dropped into hydrochloric acid, this reaction occurs.

$$2 H^+(aq) + 2e^- \longrightarrow H_2(g)$$
$$\underline{\qquad\qquad Zn(s) \longrightarrow Zn^{2+}(aq) + 2e^-}$$
$$Zn(s) + 2 H^+(aq) \longrightarrow Zn^{2+}(aq) + H_2(g)$$

We observe the reaction as the dissolving of the zinc and the bubbling of hydrogen gas. The zinc is oxidized and the H^+ ions are reduced. Notice that this reaction involves the pairing of a reduction half-reaction (the reduction of H^+) with the reverse of a half-reaction that falls below it on Table 18.1. Therefore, this reaction is spontaneous. What would happen, however, if we paired the reduction of H^+ with the oxidation of Cu? The reaction would not be spontaneous because it involves pairing the reduction of H^+ with the reverse of a half-reaction that is listed *above it* in the table. Consequently, copper does not react with H^+ and does not dissolve in acids such as HCl. In general, *metals whose reduction half-reactions lie below the reduction of H^+ to H_2 in Table 18.1 dissolve in acids, while metals above it do not.*

An important exception to this rule is nitric acid (HNO_3), which can oxidize metals through the reduction half-reaction:

$$NO_3^-(aq) + 4 H^+(aq) + 3 e^- \longrightarrow NO(g) + 2 H_2O(l) \quad E^\circ = 0.96 \text{ V}$$

Since this half-reaction is above the reduction of H^+ in Table 18.1, HNO_3 can oxidize metals (such as copper) that can't be oxidized by HCl.

$$Zn(s) + 2 H^+(aq) \longrightarrow$$
$$Zn^{2+}(aq) + H_2(g)$$

▲ When zinc is immersed in hydrochloric acid, the zinc is oxidized, forming ions that become solvated in the solution. Hydrogen ions are reduced, forming bubbles of hydrogen gas.

 Conceptual Connection 18.4 Metals Dissolving in Acids

Which metal dissolves in HNO_3 but not in HCl?

(a) Fe   **(b)** Au   **(c)** Ag

18.5   Cell Potential, Free Energy, and the Equilibrium Constant

We have seen that a positive standard cell potential (E°_{cell}) corresponds to a spontaneous oxidation–reduction reaction. We also know (from Chapter 17) that the spontaneity of a reaction is determined by the sign of ΔG°. Therefore, E°_{cell} and ΔG° must be related. We also know from Section 17.9 that ΔG° for a reaction is related to the equilibrium constant (K) for the reaction. Since E°_{cell} and ΔG° are related, then E°_{cell} and K must also be related.

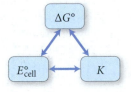

Before we look at the nature of each of these relationships in detail, let's consider the following generalizations.

For a spontaneous reaction (one that will proceed in the forward direction when all reactants and products are in their standard states):

- ΔG° is negative (< 0)
- E°_{cell} is positive (> 0)
- $K > 1$

For a nonspontaneous reaction (one that will proceed in the reverse direction when all reactants and products are in their standard states):

- $\Delta G°$ is positive (>0)
- $E°_{cell}$ is negative (<0)
- $K < 1$

The Relationship between $\Delta G°$ and $E°_{cell}$

We can derive a relationship between $\Delta G°$ and $E°_{cell}$ by briefly returning to the definition of potential difference from Section 18.3—a potential difference is a measure of the difference of potential energy per unit charge (q):

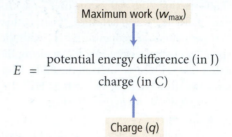

Since the potential energy difference represents the maximum amount of work that can be done by the system on the surroundings, we can write:

$$w_{max} = -qE°_{cell} \qquad [18.1]$$

The negative sign follows the convention used throughout this book that work done by the system on the surroundings is negative.

We can quantify the charge (q) that flows in an electrochemical reaction by using **Faraday's constant (F)**, which represents the charge in coulombs of 1 mol of electrons.

$$F = \frac{96,485 \text{ C}}{\text{mol e}^-}$$

The total charge is therefore $q = nF$, where n is the number of moles of electrons from the balanced chemical equation and F is Faraday's constant. Substituting $q = nF$ into Equation 18.1,

$$w_{max} = -qE°_{cell}$$
$$= -nFE°_{cell} \qquad [18.2]$$

Finally, recall from Chapter 17 that the standard change in free energy for a chemical reaction ($\Delta G°$) represents the maximum amount of work that can be done by the reaction. Therefore, $w_{max} = \Delta G°$. Making this substitution into Equation 18.2, we arrive at this important result:

$$\Delta G° = -nFE°_{cell} \qquad [18.3]$$

where $\Delta G°$ is the standard change in free energy for an electrochemical reaction, n is the number of moles of electrons transferred in the balanced equation, F is Faraday's constant, and $E°_{cell}$ is the standard cell potential. Example 18.6 illustrates how to apply this equation to calculate the standard free energy change for an electrochemical cell.

EXAMPLE 18.6 Relating $\Delta G°$ and $E°_{cell}$

Use the tabulated electrode potentials to calculate $\Delta G°$ for the reaction:

$$I_2(s) + 2\,Br^-(aq) \longrightarrow 2\,I^-(aq) + Br_2(l)$$

Is the reaction spontaneous?

SORT You are given a redox reaction and asked to find $\Delta G°$.	**GIVEN** $I_2(s) + 2\,Br^-(aq) \longrightarrow 2\,I^-(aq) + Br_2(l)$ **FIND** $\Delta G°$

STRATEGIZE Use the tabulated values of electrode potentials to calculate E°_{cell}. Then use Equation 18.3 to calculate ΔG° from E°_{cell}.	CONCEPTUAL PLAN 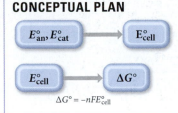 $\Delta G^\circ = -nFE^\circ_{cell}$

| SOLVE Separate the reaction into oxidation and reduction half-reactions and find the standard electrode potentials for each. Determine E°_{cell} by subtracting E_{an} from E_{cat}. | SOLUTION
 Oxidation (Anode): $\quad\quad 2\,Br^-(aq) \longrightarrow Br_2(l) + 2\,e^- \quad E^\circ = 1.09\ V$
 Reduction (Cathode): $\quad\quad I_2(s) + 2\,e^- \longrightarrow 2\,I^-(aq) \quad E^\circ = 0.54\ V$
 $I_2(s) + 2\,Br^-(aq) \longrightarrow 2\,I^-(aq) + Br_2(l)\quad E^\circ_{cell} = E^\circ_{cat} - E^\circ_{an}$
 $\quad\quad\quad\quad = -0.55\ V$ |

| Calculate ΔG° from E°_{cell}. The value of n (the number of moles of electrons) corresponds to the number of electrons that are canceled in the half-reactions. Remember that $1\ V = 1\ J/C$. | $\Delta G^\circ = -nFE^\circ_{cell}$

 $= -2\ \text{mol e}^- \left(\dfrac{96{,}485\ C}{\text{mol e}^-} \right)\left(-0.55\ \dfrac{J}{C} \right)$

 $= +1.1 \times 10^5\ J$

 Since ΔG° is positive, the reaction is not spontaneous under standard conditions. |

CHECK The answer is in the correct units (joules) and seems reasonable in magnitude ($\approx 110\ kJ$) since you have seen (in Chapter 17) that values of ΔG° are typically in the range of plus or minus tens to hundreds of kilojoules. The sign is positive, as expected for a reaction in which E°_{cell} is negative.

FOR PRACTICE 18.6

Use tabulated electrode potentials to calculate ΔG° for the reaction:

$$2\,Na(s) + 2\,H_2O(l) \longrightarrow H_2(g) + 2\,OH^-(aq) + 2\,Na^+(aq)$$

Is the reaction spontaneous?

The Relationship between E°_{cell} and K

We can derive a relationship between the standard cell potential (E°_{cell}) and the equilibrium constant for the redox reaction occurring in the cell (K) by returning to the relationship between ΔG° and K that we learned in Chapter 17. Recall from Section 17.9 that

$$\Delta G^\circ = -RT \ln K \qquad [18.4]$$

By setting Equations 18.3 and 18.4 equal to each other, we get:

$$-nFE^\circ_{cell} = -RT \ln K$$

$$E^\circ_{cell} = \frac{RT}{nF} \ln K \qquad [18.5]$$

Equation 18.5 is usually simplified for use at 25 °C by making these substitutions:

$$R = 8.314\ \frac{J}{mol \cdot K}; T = 298.15\ K; F = \left(\frac{96{,}485\ C}{mol\ e^-} \right); \text{ and } \ln K = 2.303 \log K$$

Substituting these into Equation 18.5, we arrive at this important result:

$$E^\circ_{cell} = \frac{0.0592\ V}{n} \log K \qquad [18.6]$$

where E°_{cell} is the standard cell potential, n is the number of moles of electrons transferred in the redox reaction, and K is the equilibrium constant for the balanced redox reaction at 25 °C.

EXAMPLE 18.7 Relating $E°_{cell}$ and K

Use the tabulated electrode potentials to calculate K for the oxidation of copper by H^+:

$$Cu(s) + 2\,H^+(aq) \longrightarrow Cu^{2+}(aq) + H_2(g)$$

SORT You are given a redox reaction and asked to find K.	**GIVEN** $Cu(s) + 2\,H^+(aq) \longrightarrow Cu^{2+}(aq) + H_2(g)$ **FIND** K
STRATEGIZE Use the tabulated values of electrode potentials to calculate $E°_{cell}$. Then use Equation 18.6 to calculate K from $E°_{cell}$.	**CONCEPTUAL PLAN** 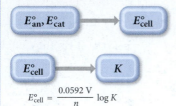 $E°_{cell} = \dfrac{0.0592\ V}{n}\log K$

SOLVE Separate the reaction into oxidation and reduction half-reactions and find the standard electrode potentials for each. Find $E°_{cell}$ by subtracting E_{an} from E_{cat}.

SOLUTION

Oxidation (Anode): $\qquad Cu(s) \longrightarrow Cu^{2+}(aq) + 2\,e^- \qquad E° = 0.34\ V$

Reduction (Cathode):

$$\underline{2\,H^+(aq) + 2\,e^- \longrightarrow H_2(g) \qquad\qquad\qquad\qquad E° = 0.00\ V}$$

$Cu(s) + 2\,H^+(aq) \longrightarrow Cu^{2+}(aq) + H_2(g) \quad E°_{cell} = E°_{cat} - E°_{an}$

$$= -0.34\ V$$

Calculate K from $E°_{cell}$. The value of n (the number of moles of electrons) corresponds to the number of electrons canceled in the half-reactions.

$$E°_{cell} = \frac{0.0592\ V}{n}\log K$$

$$\log K = E°_{cell}\,\frac{n}{0.0592\ V}$$

$$\log K = -0.34\ \cancel{V}\,\frac{2}{0.0592\ \cancel{V}}$$

$$= -11.48$$

$$K = 10^{-11.48}$$

$$= 3.3 \times 10^{-12}$$

CHECK The answer has no units, as expected for an equilibrium constant. The magnitude of the answer is small, meaning that the reaction lies far to the left at equilibrium, as expected for a reaction in which $E°_{cell}$ is negative.

FOR PRACTICE 18.7

Use the tabulated electrode potentials to calculate K for the oxidation of iron by H^+:

$$2\,Fe(s) + 6\,H^+(aq) \longrightarrow 2\,Fe^{3+}(aq) + 3\,H_2(g)$$

Notice that the fundamental quantity in the preceeding relationships is the standard change in free energy for a chemical reaction ($\Delta G°_{rxn}$). From that quantity, we can calculate both $E°_{cell}$ and K. The relationships between these three quantities is summarized in the following diagram:

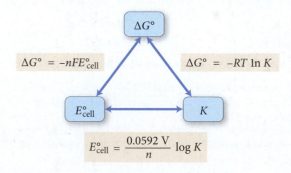

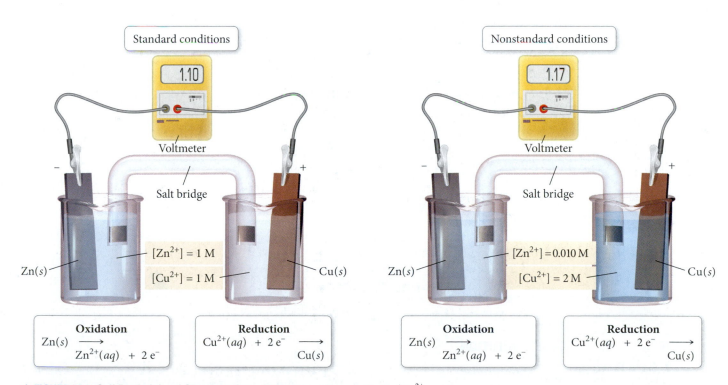

Conceptual Connection 18.5 Relating K, ΔG°_{rxn}, and E°_{cell}

A redox reaction has an equilibrium constant of $K = 1.2 \times 10^3$. Which statement is true regarding ΔG°_{rxn} and E°_{cell} for this reaction?

(a) E°_{cell} is positive and ΔG°_{rxn} is positive. **(b)** E°_{cell} is negative and ΔG°_{rxn} is negative.
(c) E°_{cell} is positive and ΔG°_{rxn} is negative. **(d)** E°_{cell} is negative and ΔG°_{rxn} is positive.

18.6 Cell Potential and Concentration

We have learned how to find E°_{cell} under standard conditions. For example, we know that when $[Cu^{2+}] = 1$ M and $[Zn^{2+}] = 1$ M, the following reaction will produce a potential of 1.10 V.

$$Zn(s) + Cu^{2+}(aq, 1\text{ M}) \longrightarrow Zn^{2+}(aq, 1\text{ M}) + Cu(s) \qquad E^\circ_{cell} = 1.10 \text{ V}$$

However, what if $[Cu^{2+}] > 1$ M and $[Zn^{2+}] < 1$ M? For example, how would the cell potential for the following conditions be different from the potential under standard conditions?

$$Zn(s) + Cu^{2+}(aq, 2\text{ M}) \longrightarrow Zn^{2+}(aq, 0.010\text{ M}) + Cu(s) \qquad E_{cell} = ?$$

Since the concentration of a reactant is greater than standard conditions, and since the concentration of product is less than standard conditions, we can use Le Châtelier's principle to predict that the reaction would have an even stronger tendency to occur in the forward direction and that E_{cell} would therefore be greater than $+1.10$ V (**Figure 18.9▼**).

We can derive an exact relationship between E_{cell} (under nonstandard conditions) and E°_{cell} by considering the relationship between the change in free energy (ΔG) and the *standard* change in free energy (ΔG°) that we learned in Section 17.8:

$$\Delta G = \Delta G^\circ + RT \ln Q \qquad\qquad [18.7]$$

▲ FIGURE 18.9 Cell Potential and Concentration   This figure compares the Zn/Cu^{2+} electrochemical cell under standard and nonstandard conditions. In this case, the nonstandard conditions consist of a higher Cu^{2+} concentration ($[Cu^{2+}] > 1$ M) at the cathode and a lower Zn^{2+} concentration at the anode ($[Zn^{2+}] < 1$ M). According to Le Châtelier's principle, the forward reaction would therefore have a greater tendency to occur under nonstandard conditions, resulting in a greater overall cell potential than the potential under standard conditions.

where R is the gas constant $(8.314 \text{ J/mol} \cdot \text{K})$, T is the temperature in kelvins, and Q is the reaction quotient corresponding to the nonstandard conditions. Since we know the relationship between ΔG and E_{cell} (Equation 18.3), we can substitute into Equation 18.7:

$$\Delta G = \Delta G° + RT \ln Q$$

$$-nFE_{cell} = -nFE°_{cell} + RT \ln Q$$

We can then divide each side by $-nF$ to arrive at:

$$E_{cell} = E°_{cell} - \frac{RT}{nF} \ln Q \qquad [18.8]$$

As we have seen, R and F are constants; at $T = 25\,°\text{C}$, $\frac{RT}{nF} \ln Q = \frac{0.0592 \text{ V}}{n} \log Q$.

Substituting into Equation 18.8, we arrive at the **Nernst equation**:

$$E_{cell} = E°_{cell} - \frac{0.0592 \text{ V}}{n} \log Q \qquad [18.9]$$

where E_{cell} is the cell potential in volts, $E°_{cell}$ is the *standard* cell potential in volts, n is the number of moles of electrons transferred in the redox reaction, and Q is the reaction quotient. Example 18.8 illustrates how to calculate the cell potential under non-standard conditions.

EXAMPLE 18.8 Calculating E_{cell} under Nonstandard Conditions

An electrochemical cell is based on these two half-reactions:

Oxidation: $Cu(s) \longrightarrow Cu^{2+}(aq, 0.010 \text{ M}) + 2\,e^-$
Reduction: $MnO_4^-(aq, 2.0\,M) + 4\,H^+(aq, 1.0\,M) + 3\,e^- \longrightarrow MnO_2(s) + 2\,H_2O(l)$

Calculate the cell potential.

SORT You are given the half-reactions for a redox reaction and the concentrations of the aqueous reactants and products. You are asked to find the cell potential.	**GIVEN** $[MnO_4^-] = 2.0 \text{ M}$; $[H^+] = 1.0 \text{ M}$; $[Cu^{2+}] = 0.010 \text{ M}$ **FIND** E_{cell}
STRATEGIZE Use the tabulated values of electrode potentials to calculate $E°_{cell}$. Then use Equation 18.9 to calculate E_{cell}.	**CONCEPTUAL PLAN** $\boxed{E°_{an}, E°_{cat}} \longrightarrow \boxed{E°_{cell}}$ $\boxed{E°_{cell},\ [MnO_4^-], [H^+], [Cu^{2+}]} \longrightarrow \boxed{E_{cell}}$ $E_{cell} = E°_{cell} - \dfrac{0.0592 \text{ V}}{n} \log Q$
SOLVE Write the oxidation and reduction half-reactions, multiplying by the appropriate coefficients to cancel the electrons. Find the standard electrode potentials for each and find $E°_{cell}$.	**SOLUTION** *Ox (Anode):* $\qquad 3[Cu(s) \longrightarrow Cu^{2+}(aq) + 2\,e^-] \qquad E° = 0.34 \text{ V}$ *Red (Cathode):* $\underline{2[MnO_4^-(aq) + 4\,H^+(aq) + 3\,e^- \longrightarrow MnO_2(s) + 2\,H_2O(l)] \quad E° = 1.68 \text{ V}}$ $3\,Cu(s) + 2\,MnO_4^-(aq) + 8\,H^+(aq) \longrightarrow 3\,Cu^{2+}(aq) + 2\,MnO_2(s) + 4\,H_2O(l)$ $E°_{cell} = E°_{cat} - E°_{an} = 1.34 \text{ V}$

Calculate E_{cell} from $E°_{cell}$. The value of n (the number of moles of electrons) corresponds to the number of electrons (in this case 6) that are canceled in the half-reactions. Determine Q based on the overall balanced equation and the given concentrations of the reactants and products. (Note that pure liquid water, solid MnO_2, and solid copper are omitted from the expression for Q.)	$\begin{aligned} E_{cell} &= E°_{cell} - \frac{0.0592\ \text{V}}{n} \log Q \\[1mm] &= E°_{cell} - \frac{0.0592\ \text{V}}{n} \log \frac{[Cu^{2+}]^3}{[MnO_4^-]^2[H^+]^8} \\[1mm] &= 1.34\ \text{V} - \frac{0.0592\ \text{V}}{6} \log \frac{(0.010)^3}{(2.0)^2(1.0)^8} \\[1mm] &= 1.34\ \text{V} - (-0.065\ \text{V}) \\[1mm] &= 1.41\ \text{V} \end{aligned}$

CHECK The answer has the correct units (V). The value of E_{cell} is larger than $E°_{cell}$, as expected based on Le Châtelier's principle because one of the aqueous reactants has a concentration greater than standard conditions and the one aqueous product has a concentration less than standard conditions. Therefore, the reaction would have a greater tendency to proceed toward products and a greater cell potential.

FOR PRACTICE 18.8
An electrochemical cell is based on these two half-reactions:

Oxidation: $Ni(s) \longrightarrow Ni^{2+}(aq, 2.0\ \text{M}) + 2\ e^-$

Reduction: $VO_2^+(aq, 0.010\ \text{M}) + 2\ H^+(aq, 1.0\ \text{M}) + e^- \longrightarrow VO^{2+}(aq, 2.0\ \text{M}) + H_2O(l)$

Calculate the cell potential.

From the previous examples, and from Equation 18.9, we can conclude:

- When a redox reaction within a voltaic cell occurs under standard conditions, $Q = 1$; therefore $E_{cell} = E°_{cell}$.

$$E_{cell} = E°_{cell} - \frac{0.0592\ \text{V}}{n} \log Q$$

$$= E°_{cell} - \frac{0.0592\ \text{V}}{n} \log 1 \qquad \overset{\displaystyle \log 1 = 0}{}$$

$$= E°_{cell}$$

- When a redox reaction within a voltaic cell occurs under conditions in which $Q < 1$, the greater concentration of reactants relative to products drives the reaction to the right, resulting in $E_{cell} > E°_{cell}$.
- When a redox reaction within an electrochemical cell occurs under conditions in which $Q > 1$, the greater concentration of products relative to reactants drives the reaction to the left, resulting in $E_{cell} < E°_{cell}$.
- When a redox reaction reaches equilibrium, $Q = K$. The redox reaction has no tendency to occur in either direction and $E_{cell} = 0$.

$$E_{cell} = E°_{cell} - \frac{0.0592\ \text{V}}{n} \log Q \qquad \overset{E°_{cell}}{\nwarrow} \qquad \text{(see Equation 18.6)}$$

$$= E°_{cell} - \frac{0.0592\ \text{V}}{n} \log K$$

$$= E°_{cell} - E°_{cell}$$

$$= 0$$

This last point explains why batteries do not last forever—as the reactants are depleted, the reaction proceeds toward equilibrium and the potential tends toward zero.

Conceptual Connection 18.6 Relating Q, K, E_{cell}, and $E°_{cell}$

In an electrochemical cell, $Q = 0.0010$ and $K = 0.10$. Which statement is true regarding E_{cell} and $E°_{cell}$?

(a) E_{cell} is positive and $E°_{cell}$ is negative.   **(b)** E_{cell} is negative and $E°_{cell}$ is positive.

(c) Both E_{cell} and $E°_{cell}$ are positive.   **(d)** Both E_{cell} and $E°_{cell}$ are negative.

Concentration Cells

Since cell potential depends not only on the half-reactions occurring in the cell, but also on the *concentrations* of the reactants and products in those half-reactions, it is possible to construct a voltaic cell in which both half-reactions are the same, but in which *a difference in concentration drives the current flow*. For example, consider the electrochemical cell shown in **Figure 18.10▼**, in which copper is oxidized at the anode and copper ions are reduced at the cathode. The second part of Figure 18.10 depicts this cell under nonstandard conditions, with $[Cu^{2+}] = 2.0$ M in one half-cell and $[Cu^{2+}] = 0.010$ M in the other:

$$Cu(s) + Cu^{2+}(aq, 2.0\ M) \longrightarrow Cu^{2+}(aq, 0.010\ M) + Cu(s)$$

The half-reactions are identical and the *standard* cell potential is, therefore, zero.

*Reduction (**Cathode**):* $Cu^{2+}(aq) + 2e^- \longrightarrow Cu(s)$ $E° = 0.34$ V

*Oxidation (**Anode**):* $Cu(s) \longrightarrow Cu^{2+}(aq) + 2e^-$ $E° = 0.34$ V

$\overline{Cu^{2+}(aq) + Cu(s) \longrightarrow Cu(s) + Cu^{2+}(aq)}$ $E°_{cell} = E°_{cat} - E°_{an}$

$= +0.00$ V

A Concentration Cell

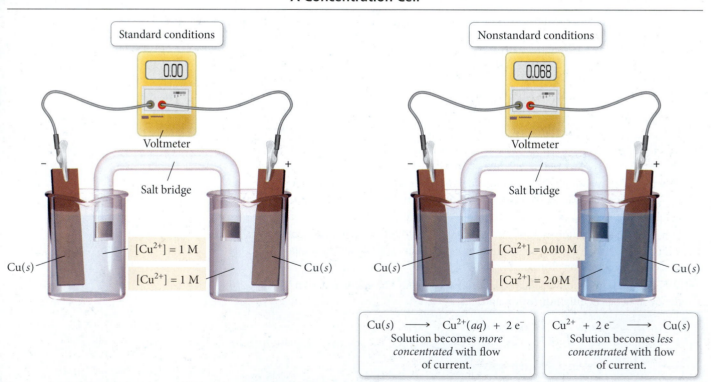

▲ **FIGURE 18.10 Cu/Cu²⁺ Concentration Cell**   If the two half-cells have the same Cu^{2+} concentration, the cell potential is zero. If one half-cell has a greater Cu^{2+} concentration than the other, a spontaneous reaction occurs. In the reaction, Cu^{2+} ions in the more concentrated cell are reduced (to solid copper), while Cu^{2+} ions in the more dilute cell are formed (from solid copper). In effect, the concentration of copper ions in the two half-cells tends toward equality.

However, because of the different concentrations in the two half-cells, the cell potential must be calculated using the Nernst equation:

$$E_{cell} = E°_{cell} - \frac{0.0592 \text{ V}}{2} \log \frac{0.010}{2.0}$$

$$= 0.000 \text{ V} + 0.068 \text{ V}$$

$$= 0.068 \text{ V}$$

The cell produces a potential of 0.068 V. *Electrons spontaneously flow from the half-cell with the lower copper ion concentration to the half-cell with the higher copper ion concentration.* You can imagine a concentration cell in the same way you think about any concentration gradient. If you mix a concentrated solution of Cu^{2+} with a dilute solution, the Cu^{2+} ions flow from the concentrated solution to the dilute one. Similarly, in a concentration cell, the transfer of electrons *from* the dilute half-cell results in the forming of Cu^{2+} ions in the dilute half-cell. The electrons flow to the concentrated cell, where they react with Cu^{2+} ions and reduce them to $Cu(s)$. Therefore, *the flow of electrons has the effect of increasing the concentration of Cu^{2+} in the dilute cell and decreasing the concentration of Cu^{2+} in the concentrated half-cell.*

18.7 Batteries: Using Chemistry to Generate Electricity

We have seen that we can combine the electron-losing tendency of one substance with the electron-gaining tendency of another to create electrical current in a voltaic cell. Batteries are voltaic cells conveniently packaged to act as portable sources of electricity. The actual oxidation and reduction reactions vary according to battery type. In this section, we examine several different types.

Dry-Cell Batteries

Common batteries, such as the ones in a flashlight, are called **dry-cell batteries** because they do not contain large amounts of liquid water. There are several familiar types of dry-cell batteries. The most inexpensive dry cells are composed of a zinc case that acts as the anode (**Figure 18.11▶**). The zinc is oxidized according to the reaction:

Oxidation (Anode): $Zn(s) \longrightarrow Zn^{2+}(aq) + 2 e^-$

The cathode is a carbon rod immersed in a moist paste of MnO_2 that also contains NH_4Cl. The MnO_2 is reduced to Mn_2O_3 according to the reaction:

Reduction (Cathode): $2 MnO_2(s) + 2 NH_4^+(aq) + 2 e^- \longrightarrow$
$Mn_2O_3(s) + 2 NH_3(g) + H_2O(l)$

These two half-reactions produce a voltage of about 1.5 V. Two or more of these batteries can be connected in series (cathode-to-anode connection) to produce higher voltages.

The more common **alkaline batteries** (Figure 18.11b) employ slightly different half-reactions in a basic medium (therefore, the name alkaline). In an alkaline battery, the reactions are:

Oxidation (Anode): $Zn(s) + 2 OH^-(aq) \longrightarrow Zn(OH)_2(s) + 2 e^-$

Reduction (Cathode): $2 MnO_2(s) + 2 H_2O(l) + 2 e^- \longrightarrow$
$2 MnO(OH)(s) + 2 OH^-(aq)$

Overall reaction: $Zn(s) + 2 MnO_2(s) + 2 H_2O(l) \longrightarrow$
$Zn(OH)_2(s) + 2 MnO(OH)(s)$

Alkaline batteries have a longer working life and a longer shelf life than their non-alkaline counterparts.

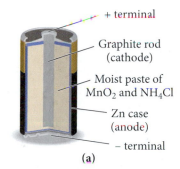

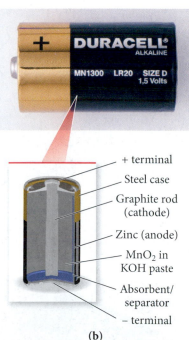

▲ **FIGURE 18.11 Dry-Cell Batteries**
(a) In a common dry-cell battery, the zinc case acts as the anode and a graphite rod immersed in a moist, slightly acidic paste of MnO_2 and NH_4Cl acts as the cathode. **(b)** The longer-lived alkaline batteries now in common use employ a graphite cathode immersed in a paste of MnO_2 and a base.

▶ **FIGURE 18.12 Lead–Acid Storage Battery** A lead–acid storage battery consists of six cells wired in series. Each cell contains a porous lead anode and a lead oxide cathode, both immersed in sulfuric acid.

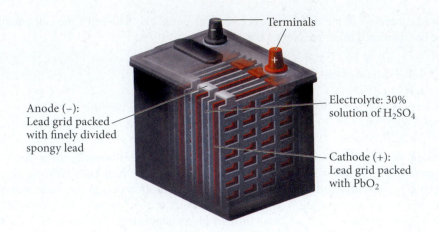

Terminals

Anode (–):
Lead grid packed
with finely divided
spongy lead

Electrolyte: 30%
solution of H_2SO_4

Cathode (+):
Lead grid packed
with PbO_2

Lead–Acid Storage Batteries

The batteries in most automobiles are **lead–acid storage batteries**. These batteries consist of six electrochemical cells wired in series (**Figure 18.12▲**). Each cell produces 2 V for a total of 12 V. Each cell contains a porous lead anode where oxidation occurs and a lead(IV) oxide cathode where reduction occurs according to these reactions:

Oxidation (Anode): $Pb(s) + HSO_4^-(aq) \longrightarrow PbSO_4(s) + H^+(aq) + 2e^-$

Reduction (Cathode): $PbO_2(s) + HSO_4^-(aq) + 23H^+(aq) + 2e^- \longrightarrow$
$$PbSO_4(s) + 2 H_2O(l)$$

Overall reaction: $Pb(s) + PbO_2(s) + 2 HSO_4^-(aq) + 2 H^+(aq) \longrightarrow$
$$2 PbSO_4(s) + 2 H_2O(l)$$

Both the anode and the cathode are immersed in sulfuric acid (H_2SO_4). As electrical current is drawn from the battery, both the anode and the cathode become coated with $PbSO_4(s)$. If the battery is run for a long time without recharging, too much $PbSO_4(s)$ develops on the surface of the electrodes and the battery goes dead. The lead–acid storage battery can be recharged if an electrical current (which must come from an external source such as an alternator in a car) causes the reaction to occur in reverse, converting the $PbSO_4(s)$ back to $Pb(s)$ and $PbSO_2(s)$.

Other Rechargeable Batteries

The ever-growing need to power electronic products such as laptops, cell phones, and digital cameras, as well as the growth in popularity of hybrid electric vehicles, has driven the development of efficient, long-lasting, rechargeable batteries. The most common types include the **nickel–cadmium (NiCad) battery**, the **nickel–metal hydride (NiMH) battery**, and the **lithium ion battery**.

The Nickel–Cadmium (NiCad) Battery

Nickel–cadmium batteries consist of an anode composed of solid cadmium and a cathode composed of $NiO(OH)(s)$. The electrolyte is usually $KOH(aq)$. During operation, the cadmium is oxidized and the $NiO(OH)$ is reduced according to the equations:

Oxidation (Anode): $Cd(s) + 2OH^-(aq) \longrightarrow Cd(OH)_2(s) + 2 e^-$

Reduction (Cathode): $2 NiO(OH)(s) + 2 H_2O(l) + 2 e^- \longrightarrow$
$$2 Ni(OH)_2(s) + 2 OH^-(aq)$$

▲ Several types of batteries, including NiCad, NiMH, and lithium ion batteries, can be recharged by chargers that use household current.

The overall reaction produces about 1.30 V. As current is drawn from the NiCad battery, solid cadmium hydroxide accumulates on the anode and solid nickel(II) hydroxide accumulates on the cathode. By running current in the opposite direction, the reactants can be regenerated from the products. A common problem in recharging NiCad and other rechargeable batteries is knowing when to stop. Once all of the products of the reaction are converted back to reactants, the charging process should ideally terminate—otherwise the electrical current will drive other, usually unwanted, reactions such as the electrolysis of water to form hydrogen and oxygen gas. These reactions will typically damage the battery and may sometimes even cause an explosion. Consequently, most commercial battery chargers have sensors that measure when the charging is complete. These sensors rely on the small changes in voltage or increases in temperature that occur once the products have all been converted back to reactants.

The Nickel–Metal Hydride (NiMH) Battery

Although NiCad batteries were the standard rechargeable battery for many years, they are being replaced by others, in part because of the toxicity of cadmium and the resulting disposal problems. One of these replacements is the nickel–metal hydride or NiMH battery. The NiMH battery uses the same cathode reaction as the NiCad battery but a different anode reaction. In the anode of a NiMH battery, hydrogen atoms held in a metal alloy are oxidized. If we let M represent the metal alloy, we can write the half-reactions as:

Oxidation (Anode): $M \cdot H(s) + OH^-(aq) \longrightarrow M(s) + H_2O(l) + e^-$

Reduction (Cathode): $NiO(OH)(s) + H_2O(l) + e^- \longrightarrow Ni(OH)_2(s) + OH^-(aq)$

In addition to being more environmentally friendly than NiCad batteries, NiMH batteries also have a greater energy density (energy content per unit battery mass), as summarized in Table 18.2. In some cases, a NiMH battery can carry twice the energy of a NiCad battery of the same mass, making NiMH batteries the most common choice for hybrid electric vehicles.

TABLE 18.2 Energy Density and Overcharge Tolerance of Several Rechargeable Batteries

Battery Type	Energy Density (W · h/kg)	Overcharge Tolerance
NiCad	45–80	Moderate
NiMH	60–120	Low
Li ion	110–160	Low
Pb storage	30–50	High

The Lithium Ion Battery

The newest and most expensive common type of rechargeable battery is the lithium ion battery. Since lithium is the least dense metal $(0.53 \, g/cm^3)$ it can be used to make batteries with high energy densities (see Table 18.2). The lithium battery works differently than the other batteries we have examined so far, and the details of its operation are beyond our current scope. Briefly, we can think of the operation of the lithium battery as being due primarily to the motion of lithium ions from the anode to the cathode. The anode is composed of graphite into which lithium ions are incorporated between layers of carbon atoms. Upon discharge, the lithium ions spontaneously migrate to the cathode, which consists of a lithium transition metal oxide such as $LiCoO_2$ or $LiMn_2O_4$. The transition metal is reduced during this process. Upon recharging, the transition metal is oxidized, forcing the lithium to migrate back into the graphite (**Figure 18.13▶**). The flow of lithium ions from the anode to the cathode causes a corresponding flow of electrons in the external circuit. Lithium ion batteries are commonly used in applications where light weight and high energy density are required. These include cell phones, laptop computers, and digital cameras.

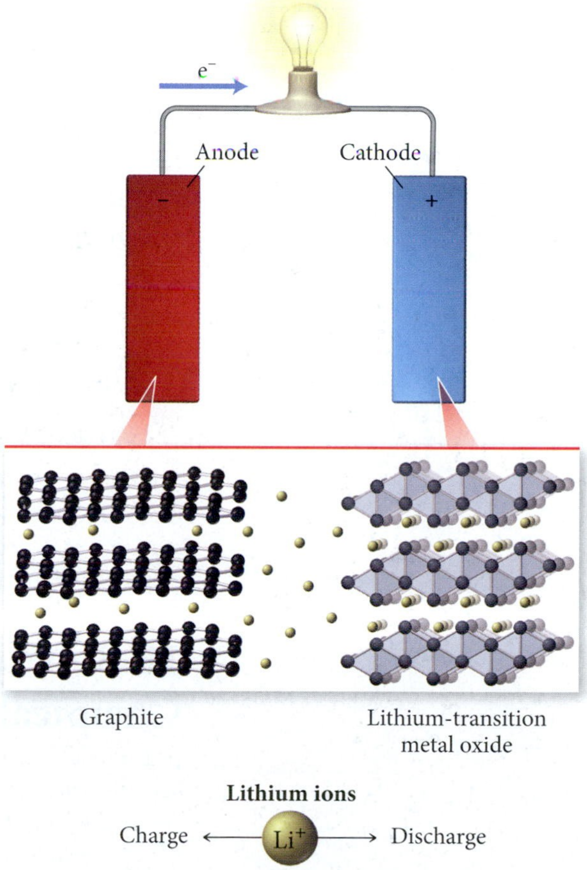

Fuel Cells

We discussed the potential for *fuel cells* in the opening section of this chapter. Fuel cells may one day replace—or at least work in combination with—centralized power grid electricity. In addition, electric vehicles powered by fuel cells may one day replace vehicles powered by internal combustion engines. Fuel cells are like batteries, but the reactants must be constantly replenished. Normal batteries lose their ability to generate voltage with use because the reactants become depleted as electrical current is drawn

▲ FIGURE 18.13 Lithium Ion Battery In the lithium ion battery, the spontaneous flow of lithium ions from the graphite anode to the lithium transition metal oxide cathode causes a corresponding flow of electrons in the external circuit.

▶ **FIGURE 18.14 Hydrogen–Oxygen Fuel Cell** In this fuel cell, hydrogen and oxygen combine to form water.

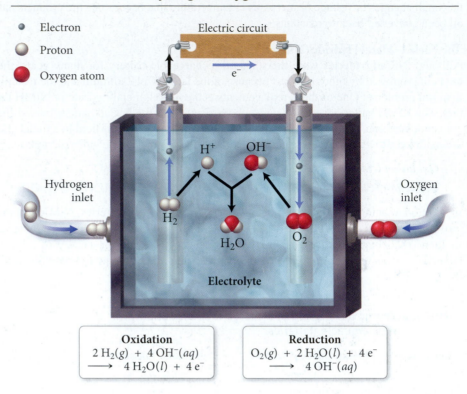

Hydrogen–Oxygen Fuel Cell

- Electron
- Proton
- Oxygen atom

Electric circuit

e^-

Hydrogen inlet

Oxygen inlet

H^+ OH^-

H_2

H_2O O_2

Electrolyte

Oxidation
$2\,H_2(g) + 4\,OH^-(aq)$
$\longrightarrow 4\,H_2O(l) + 4\,e^-$

Reduction
$O_2(g) + 2\,H_2O(l) + 4\,e^-$
$\longrightarrow 4\,OH^-(aq)$

from the battery. In a **fuel cell**, the reactants—the fuel—constantly flow through the battery, generating electrical current as they undergo a redox reaction.

The most common fuel cell is the hydrogen–oxygen fuel cell (**Figure 18.14▲**). In this cell, hydrogen gas flows past the anode (a screen coated with platinum catalyst) and undergoes oxidation:

Oxidation (Anode): $2\,H_2(g) + 4\,OH^-(aq) \longrightarrow 4\,H_2O(l) + 4\,e^-$

Oxygen gas flows past the cathode (a similar screen) and undergoes reduction:

Reduction (Cathode): $O_2(g) + 2\,H_2O(l) + 4\,e^- \longrightarrow 4\,OH^-(aq)$

The half-reactions sum to the following overall reaction:

Overall reaction: $2\,H_2(g) + O_2(g) \longrightarrow 2\,H_2O(l)$

Notice that the only product is water. In the space shuttle program, hydrogen–oxygen fuel cells provide electricity and astronauts drink the water that is produced by the reaction. In order for hydrogen-powered fuel cells to become more widely used, a more readily available source of hydrogen must be developed.

18.8 Electrolysis: Driving Nonspontaneous Chemical Reactions with Electricity

In a voltaic cell, a spontaneous redox reaction produces electrical current. In an *electrolytic cell*, electrical current drives an otherwise nonspontaneous redox reaction through a process called **electrolysis**. In a fuel cell, the reaction of hydrogen with oxygen to form water is spontaneous and can be used to produce an electrical current. By providing electrical current, we can cause the reverse reaction to occur, breaking water into hydrogen and oxygen (**Figure 18.15▶**).

$2\,H_2(g) + O_2(g) \longrightarrow 2\,H_2O(l)$	(spontaneous—produces electrical current; occurs in a voltaic cell)
$2\,H_2O\,(l) \longrightarrow 2\,H_2(g) + O_2(g)$	(nonspontaneous—consumes electrical current; occurs in an electrolytic cell)

Electrolysis of Water

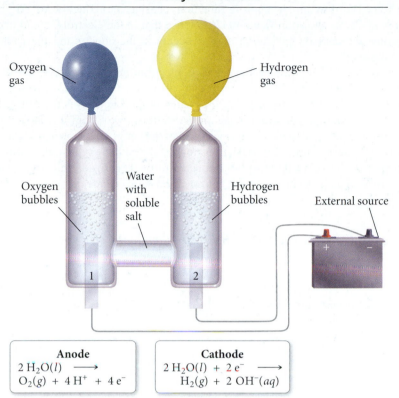

Anode	Cathode
$2\,H_2O(l) \longrightarrow$	$2\,H_2O(l) + 2\,e^- \longrightarrow$
$O_2(g) + 4\,H^+ + 4\,e^-$	$H_2(g) + 2\,OH^-(aq)$

One of the problems associated with the widespread adoption of fuel cells is the scarcity of hydrogen. Where will the hydrogen to power these fuel cells come from? One possible answer is to obtain hydrogen from water through solar-powered electrolysis. A solar-powered electrolytic cell can be used to make hydrogen from water when the sun is shining. The hydrogen can then be converted back to water to generate electricity when needed. Hydrogen made in this way could also be used to power fuel-cell vehicles.

Electrolysis also has numerous other applications. For example, most metals are found in Earth's crust as metal oxides. Converting the oxides to pure metals requires the reduction of the metal, a nonspontaneous process. Electrolysis can be used to produce these metals. Thus, sodium can be produced by the electrolysis of molten sodium chloride. Electrolysis can also be used to plate metals onto other metals. For example, silver can be plated onto a less expensive metal using the electrolytic cell shown in **Figure 18.16▶**. In this cell, a silver electrode is placed in a solution containing silver ions. An electrical current then causes the oxidation of silver at the anode (replenishing the silver ions in solution) and the reduction of silver ions at the cathode (coating the less expensive metal with solid silver).

Oxidation (Anode): $\quad Ag(s) \longrightarrow Ag^+(aq) + e^-$

Reduction (Cathode): $Ag^+(aq) + e^- \longrightarrow Ag(s)$

Since the standard cell potential of this reaction is zero, the reaction is not spontaneous under standard conditions. However, an external power source can be used to drive current flow and, therefore, cause the reaction to occur.

The voltage required to cause electrolysis depends on the specific half-reactions. For example, we have seen that the oxidation of zinc and the reduction of Cu^{2+} produces a voltage of 1.10 V under standard conditions.

Electrolytic Cell for Silver Plating

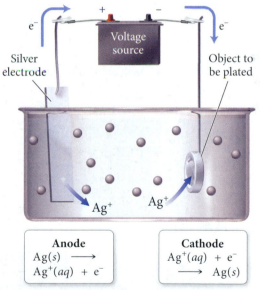

Anode	Cathode
$Ag(s) \longrightarrow$	$Ag^+(aq) + e^-$
$Ag^+(aq) + e^-$	$\longrightarrow Ag(s)$

▲ **FIGURE 18.16 Silver Plating** Silver can be plated from a solution of silver ions onto metallic objects in an electrolytic cell.

Reduction (Cathode): $\quad Cu^{2+}(aq) + 2\,e^- \longrightarrow Cu(s) \qquad E° = 0.34\ V$

Oxidation (Anode): $\qquad\qquad\quad Zn(s) \longrightarrow Zn^{2+}(aq) + 2\,e^- \quad E° = -0.76\ V$

$$Cu^{2+}(aq) + Zn(s) \longrightarrow Cu(s) + Zn^{2+}(aq)\ \ E°_{cell} = E°_{cat} - E°_{an}$$
$$= +1.10\ V$$

If a power source producing *more than 1.10 V* is inserted into the voltaic cell, electrons can be forced to flow in the opposite direction, causing the reduction of Zn^{2+} and the oxidation of Cu, as shown in **Figure 18.17▼**. Notice that in the electrolytic cell, the anode has become the cathode (oxidation always occurs at the anode) and the cathode has become the anode.

In a *voltaic cell*, the anode is the source of electrons and is, therefore, labeled with a negative charge. The cathode draws electrons and is, therefore, labeled with a positive charge. In an *electrolytic cell*, however, the source of the electrons is the external power source. The external power source must *draw electrons away* from the anode; thus, the anode must be connected to the positive terminal of the battery (as shown in Figure 18.17). Similarly, the power source drives electrons toward the cathode (where they will be used in reduction), so the cathode must be connected to the *negative* terminal of the battery. This is why the charge labels (+ and −) on an electrolytic cell are the opposite of those in a voltaic cell.

Summarizing Characteristics of Electrochemical Cell Types:

In all electrochemical cells:

▶ Oxidation occurs at the anode.

▶ Reduction occurs at the cathode.

In voltaic cells:

▶ The anode is the source of electrons and has a negative charge (anode −).

▶ The cathode draws electrons and has a positive charge (cathode +).

In electrolytic cells:

▶ Electrons are drawn away from the anode, which must, therefore, be connected to the positive terminal of the external power source (anode +).

▶ Electrons are forced to the cathode, which must, therefore, be connected to the negative terminal of the power source (cathode −).

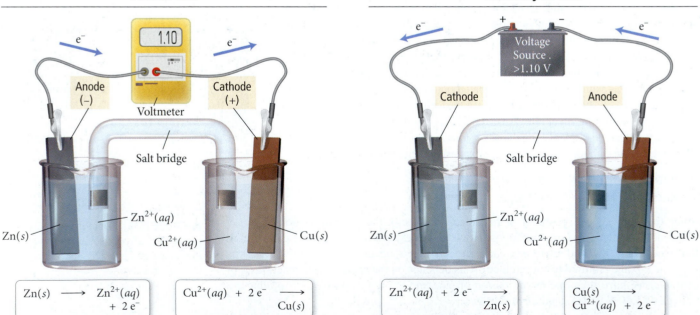

▲ **FIGURE 18.17 Voltaic versus Electrolytic Cells**   In a Zn/Cu^{2+} voltaic cell, the reaction proceeds in the spontaneous direction. In a Zn^{2+}/Cu electrolytic cell, electrical current drives the reaction in the nonspontaneous direction.

Stoichiometry of Electrolysis

In an electrolytic cell, electrical current drives a particular chemical reaction. In a sense, the electrons act as a reactant and, therefore, have a stoichiometric relationship with the other reactants and products. Unlike ordinary reactants, for which we usually measure quantity as mass, for electrons we measure quantity as charge. For example, consider an electrolytic cell used to coat copper onto metals, as shown in **Figure 18.18▼**. The half-reaction by which copper is deposited onto the metal is:

$$Cu^{2+}(aq) + 2\,e^- \longrightarrow Cu(s)$$

For every 2 mol of electrons that flow through the cell, 1 mol of solid copper is plated. We can write the stoichiometric relationship as:

$$2 \text{ mol } e^- : 1 \text{ mol } Cu(s)$$

We can determine the number of moles of electrons that have flowed in a given electrolysis cell by measuring the total charge that has flowed through the cell, which in turn depends on the *magnitude* of the current and on the *time* that the current has run. Recall from Section 18.3 that the unit of current is the ampere:

$$1\text{ A} = 1\,\frac{C}{s}$$

If we multiply the amount of current (in A) flowing through the cell by the time (in s) that the current flowed, we can determine the total charge that passed through the cell in that time:

$$\text{Current}\left(\frac{C}{s}\right) \times \text{time (s)} = \text{charge (C)}$$

The relationship between charge and the number of moles of electrons is given by Faraday's constant, which, as we saw previously, corresponds to the charge in coulombs of 1 mol of electrons.

$$F = \frac{96{,}485\text{ C}}{\text{mol } e^-}$$

These relationships can be used to solve problems involving the stoichiometry of electrolytic cells, as shown in Example 18.9.

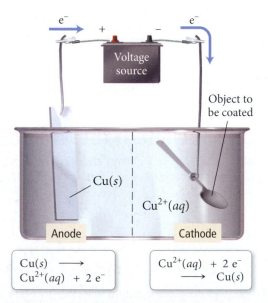

▲ **FIGURE 18.18 Electrolytic Cell for Copper Plating**   In this cell, copper ions are plated onto other metals. It takes two moles of electrons to plate one mole of copper atoms.

EXAMPLE 18.9 Stoichiometry of Electrolysis

Gold can be plated out of a solution containing Au^{3+} according to the half-reaction:

$$Au^{3+}(aq) + 3\,e^- \longrightarrow Au(s)$$

What mass of gold (in grams) is plated by the flow of 5.5 A of current for 25 minutes?

SORT You are given the half-reaction for the plating of gold, which shows the stoichiometric relationship between moles of electrons and moles of gold. You are also given the current and time. You are asked to find the mass of gold that will be deposited in that time.	**GIVEN** 3 mol e⁻ : 1 mol Au 　　　　5.5 amps 　　　　25 min **FIND** g Au

STRATEGIZE You need to find the amount of gold, which is related stoichiometrically to the number of electrons that have flowed through the cell. Begin with time in minutes and convert to seconds. Then, since current is a measure of charge per unit time, use the given current and the time to find the number of coulombs. You can then use Faraday's constant to calculate the number of moles of electrons and the stoichiometry of the reaction to find the number of moles of gold. Finally, use the molar mass of gold to convert to mass of gold.

CONCEPTUAL PLAN

$$\boxed{\text{min}} \xrightarrow{\dfrac{60\,\text{s}}{1\,\text{min}}} \boxed{\text{s}} \xrightarrow{\dfrac{5.5\,\text{C}}{1\,\text{s}}} \boxed{\text{C}}$$

$$\boxed{\text{mol}\,e^-} \xrightarrow{\dfrac{1\,\text{mol Au}}{3\,\text{mol}\,e^-}} \boxed{\text{mol Au}} \xrightarrow{\dfrac{196.97\,\text{g Au}}{1\,\text{mol Au}}} \boxed{\text{g Au}}$$

with $\dfrac{1\,\text{mol}\,e^-}{96{,}485\,\text{C}}$ between C and mol e⁻

SOLVE Follow the conceptual plan to solve the problem, canceling units to arrive at mass of gold.

SOLUTION

$$25\,\text{min} \times \frac{60\,\text{s}}{1\,\text{min}} \times \frac{5.5\,\text{C}}{1\,\text{s}} \times \frac{1\,\text{mol}\,e^-}{96{,}485\,\text{C}} \times \frac{1\,\text{mol Au}}{3\,\text{mol}\,e^-} \times \frac{196.97\,\text{g Au}}{1\,\text{mol Au}} = 5.6\,\text{g Au}$$

CHECK The answer has the correct units (g Au). The magnitude of the answer is also reasonable considering that 10 amps of current for 1 hour is the equivalent of about $\frac{1}{3}$ mol of electrons (check for yourself), which would produce $\frac{1}{9}$ mol (or about 20 g) of gold.

FOR PRACTICE 18.9

Silver can be plated out of a solution containing Ag^+ according to the half-reaction:

$$Ag^+(aq) + e^- \longrightarrow Ag(s)$$

How much time (in minutes) does it take to plate 12 g of silver using a current of 3.0 A?

18.9 Corrosion: Undesirable Redox Reactions

Corrosion is the (usually) gradual, nearly always undesired, oxidation of metals that are exposed to oxidizing agents in the environment. Notice from Table 18.1 that the reduction of oxygen in the presence of water has an electrode potential of +0.40 V:

$$O_2(g) + 2\,H_2O(l) + 4\,e^- \longrightarrow 4\,OH^-(aq) \qquad E° = +0.40\,\text{V}$$

In the presence of acid, the reduction of oxygen has an even greater electrode potential of +1.23 V.

$$O_2(g) + 4\,H^+(aq) + 4\,e^- \longrightarrow 2\,H_2O(l) \qquad E° = +1.23\,\text{V}$$

The reduction of oxygen, therefore, has a strong tendency to occur and bring about the oxidation of other substances, especially metals. Notice that the half-reactions for the reduction of most metal ions lie *below* the half-reactions for the reduction of oxygen in Table 18.1. Consequently, the oxidation (or corrosion) of those metals will be spontaneous when paired with the reduction of oxygen. Corrosion is the opposite of the process by which metals are extracted from their ores. In extraction, the free metal is reduced out from its ore. In corrosion, the metal is oxidized.

Given the ease with which metals are oxidized in the presence of oxygen, acid, and water, why are metals used so frequently as building materials in the first place? Many metals form oxides that coat the surface of the metal and prevent further corrosion. For example, bare aluminum metal, with an electrode potential of -1.66 V, is quickly oxidized in the presence of oxygen. However, the oxide that forms at the surface of aluminum is Al_2O_3. In its crystalline form, Al_2O_3 is sapphire, a highly inert and structurally solid substance. Consequently, the coating acts to protect the underlying aluminum metal, preventing further corrosion.

The oxides of iron, however, are not structurally stable, and eventually tend to flake away, exposing the underying metal to further corrosion. A significant part of the iron produced each year is used to replace rusted iron. Rusting is a redox reaction in which iron is oxidized according to the half-reaction:

$$Fe(s) \longrightarrow Fe^{2+}(aq) + 2\,e^- \qquad E° = -0.45 \text{ V}$$

This oxidation reaction tends to occur at defects on the surface of the iron—known as *anodic regions* because oxidation is occurring at these locations—as shown in **Figure 18.19▼**. The electrons produced at the anodic region then travel through the metal to areas called *cathodic regions* where they react with oxygen and H^+ ions dissolved in moisture. (The H^+ ions come from carbonic acid, which naturally forms in water when it reacts with the carbon dioxide in air.) The overall reaction has an electrode potential of $+1.68$ V and is, therefore, highly spontaneous.

$$2\,Fe(s) + O_2(g) + 4\,H^-(aq) \longrightarrow 2\,H_2O(l) + 2\,Fe^{2+}(aq) \qquad E°_{cell} = +1.68 \text{ V}$$

The Fe^{2+} ions formed in the anodic regions can migrate through moisture on the surface of the iron to cathodic regions, where they are further oxidized by reaction with more oxygen:

$$4\,Fe^{2+}(aq) + O_2(g) + (4 + 2n)\,H_2O(l) \longrightarrow 2\,Fe_2O_3 \cdot nH_2O(s) + 8\,H^+(aq)$$
$$\text{rust}$$

Rust is a hydrated form of iron(III) oxide whose exact composition depends on the conditions under which it forms.

Consider each of these important components in the formation of rust:

- *Moisture must be present for rusting to occur.* The presence of water is necessary because water is a reactant in the last reaction, and also because a charge (either electrons or ions) must be free to flow between the anodic and cathodic regions.

- *Additional electrolytes promote rusting.* The presence of an electrolyte (such as sodium chloride) on the surface of iron promotes rusting because it enhances current flow. This is why cars rust so quickly in cold climates where roads are salted, or in areas directly adjacent to beaches where salt water mist is present.

- *The presence of acids promotes rusting.* Since H^+ ions are involved in the reduction of oxygen, lower pH enhances the cathodic reaction and leads to faster rusting.

▲ Aluminum is stable because its oxide forms a protective film over the underlying metal, preventing further oxidation.

▲ A scratch in paint often allows the rusting of the underlying iron.

The Rusting of Iron

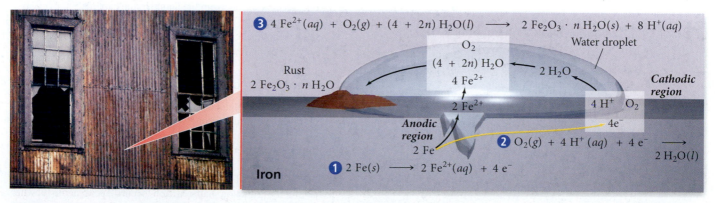

▲ **FIGURE 18.19 Corrosion of Iron: Rusting** The oxidation of iron occurs at anodic regions on the metal surface. The iron ions migrate to cathodic regions, where they react with oxygen and water to form rust.

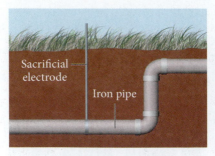

▲ If a metal more active than iron, such as magnesium or aluminum, is in electrical contact with iron, the metal rather than the iron will be oxidized. This principle underlies the use of sacrificial electrodes to prevent the corrosion of iron.

Preventing Corrosion

Preventing the rusting of iron is a major industry. The most obvious way to prevent rust is to keep iron dry. Without water, the redox reaction cannot occur. Another way of preventing rust is to coat the iron with a substance that is impervious to water. Cars, for example, are painted and sealed to prevent rust. A scratch in the paint, however, can lead to rusting of the underlying iron.

◀ In galvanized nails, a layer of zinc prevents the underlying iron from rusting. The zinc oxidizes in place of the iron, forming a protective layer of zinc oxide.

CHAPTER IN REVIEW

Key Terms

Section 18.3
electrical current (709)
electrochemical cell (709)
voltaic (galvanic) cell (709)
electrolytic cell (709)
half-cell (709)
electrode (709)
ampere (A) (710)
potential difference (710)
volt (V) (710)
electromotive force
(emf) (710)

cell potential (cell emf) (E_{cell})
(710)
standard cell potential
(standard emf)
(E°_{cell}) (710)
anode (710)
cathode (710)
salt bridge (711)

Section 18.4
standard electrode
potential (712)

standard hydrogen electrode
(SHE) (712)

Section 18.5
Faraday's constant (F) (720)

Section 18.6
Nernst equation (724)

Section 18.7
dry-cell battery (727)
alkaline battery (727)
lead–acid storage battery (728)

nickel–cadmium (NiCad)
battery (728)
nickel–metal hydride (NiMH)
battery (728)
lithium ion battery (728)
fuel cell (730)

Section 18.8
electrolysis (730)

Section 18.9
corrosion (734)

Key Concepts

Pulling the Plug on the Power Grid (18.1)

▶ In oxidation–reduction reactions electrons are transferred.
▶ If the reactants of a redox reaction are separated and connected by an external wire, electrons flow through the wire.
▶ In the most common form of fuel cell, an electrical current is created as hydrogen is oxidized and oxygen is reduced, with water the only product.

Balancing Oxidation–Reduction Equations (18.2)

▶ Oxidation is the loss of electrons and corresponds to an increase in oxidation state; reduction is the gain of electrons and corresponds to a decrease in oxidation state.
▶ We can balance redox reactions by the half-reaction method, in which the oxidation and reduction reactions are balanced separately and then added. This method differs slightly for redox reactions in acidic and in basic solutions.

Voltaic (or Galvanic) Cells: Generating Electricity from Spontaneous Chemical Reactions (18.3)

▶ A voltaic cell separates the reactants of a spontaneous redox reaction into two half-cells that are connected by a wire and a means to exchange ions, so that electricity is generated.

▶ The rate of electrons flowing through a wire is measured in amperes (A), and the cell potential is measured in volts (V).
▶ A salt bridge is commonly used to allow ions to flow between the half-cell solutions, thereby preventing the buildup of charge.
▶ In an electrochemical cell, the electrode where oxidation occurs is the anode and the electrode where reduction occurs is the cathode; electrons flow from the anode to the cathode.
▶ Cell diagram or line notation provides a technique for writing redox reactions concisely by separating the components of the reaction using lines or commas.

Standard Electrode Potentials (18.4)

▶ The electrode potentials of half-cells are measured in relation to that of a standard hydrogen electrode (SHE), which is assigned an electrode potential of zero under standard conditions (solution concentrations of 1 M, gas pressures of 1 atm, and a temperature of 25 °C).
▶ A species with a highly positive E° is an excellent oxidizing agent; its reduction half-reaction occurs spontaneously when coupled with any half-reaction that has a lower (more negative) E°, because the combined reaction has a positive standard cell potential (E°_{cell}).

Cell Potential, Free Energy, and the Equilibrium Constant (18.5)

▶ In a spontaneous reaction, E°_{cell} is positive, the change in free energy (ΔG°) is negative, and the equilibrium constant (K) is greater than one.

▶ In a nonspontaneous reaction, E°_{cell} is negative, ΔG° is positive, and K is less than one.

▶ Because E°_{cell}, ΔG°, and K all relate to spontaneity, we can derive equations relating all three quantities.

Cell Potential and Concentration (18.6)

▶ The standard cell potential (E°_{cell}) is related to the cell potential (E_{cell}) by the Nernst equation, $E_{cell} = E^{\circ}_{cell} - (0.0592\ V/n) \log Q$.

▶ As shown by this equation, E_{cell} is related to the reaction quotient (Q); since E_{cell} equals zero when Q equals K, a battery is depleted as the reaction proceeds toward equilibrium.

▶ In a concentration cell, the reactions at both electrodes are identical and electrons flow because of a difference in concentration.

Batteries: Using Chemistry to Generate Electricity (18.7)

▶ Batteries are packaged voltaic cells.

▶ Dry-cell batteries, including alkaline batteries, typically have zinc cases that act as anodes.

▶ Rechargeable batteries, such as lead–acid storage, nickel–cadmium, nickel–metal hydride, and lithium ion batteries, allow the reaction to be reversed.

▶ Fuel cells are similar to batteries except that the reactants must be continually replenished.

Electrolysis: Driving Nonspontaneous Chemical Reactions with Electricity (18.8)

▶ An electrolytic cell differs from a voltaic cell in that (1) an electrical charge is used to drive the reaction, and (2) although the anode is still the site of oxidation and the cathode the site of reduction they are represented with signs opposite those of a voltaic cell (anode +, cathode −).

▶ We can use stoichiometry to calculate the quantity of reactants consumed or products produced in an electrolytic cell.

Corrosion: Undesirable Redox Reactions (18.9)

▶ Corrosion is the undesired oxidation of metal by environmental oxidizing agents.

▶ When some metals, such as aluminum, oxidize they form a stable compound that prevents further oxidation.

▶ Iron, however, does not form a structurally stable compound when oxidized and, therefore, rust flakes off and exposes more iron to corrosion.

▶ The corrosion of iron can be prevented by keeping water out of contact with metal, minimizing the presence of electrolytes and acids, or coating the iron with a sacrificial electrode.

Key Equations and Relationships

Definition of an Ampere (18.3)

$$1\ A = 1\ C/s$$

Definition of a Volt (18.3)

$$1\ V = 1\ J/C$$

Standard Hydrogen Electrode (18.4)

$$2\ H^{+}(aq) + 2\ e^{-} \longrightarrow H_2(g) \qquad E^{\circ} = 0.00\ V$$

Equation for Cell Potential (18.4)

$$E^{\circ}_{cell} = E^{\circ}_{cathode} - E^{\circ}_{anode}$$

Relating ΔG° and E°_{cell} (18.5)

$$\Delta G^{\circ} = -nFE^{\circ}_{cell} \qquad F = \frac{96{,}485\ C}{mol\ e^{-}}$$

Relating E°_{cell} and K (18.5)

$$E^{\circ}_{cell} = \frac{0.0592\ V}{n} \log K$$

The Nernst Equation (18.6)

$$E_{cell} = E^{\circ}_{cell} - \frac{0.0592\ V}{n} \log Q$$

Key Learning Objectives

Chapter Objectives	Assessment
Half-Reaction Method of Balancing Aqueous Redox Equations in Acidic Solution (18.2)	Examples 18.1, 18.2 For Practice 18.1, 18.2 Exercises 1–4
Balancing Redox Reactions Occurring in Basic Solution (18.2)	Example 18.3 For Practice 18.3 Exercises 5, 6
Calculating Standard Potentials for Electrochemical Cells from Standard Electrode Potentials of the Half-Reactions (18.4)	Example 18.4 For Practice 18.4 Exercises 9, 10, 25, 26

Predicting Spontaneous Redox Reactions and Sketching Electrochemical Cells (18.4)	Example 18.5	For Practice 18.5	Exercises 7, 8, 11, 12, 15–18
Relating $G°$ and $E°_{cell}$ (18.5)	Example 18.6	For Practice 18.6	Exercises 29, 30
Relating $E°_{cell}$ and K (18.5)	Example 18.7	For Practice 18.7	Exercises 31–36
Calculating $E°_{cell}$ under Nonstandard Conditions (18.6)	Example 18.8	For Practice 18.8	Exercises 37–42
Stoichiometry of Electrolysis (18.8)	Example 18.9	For Practice 18.9	Exercises 57–60

EXERCISES

Problems by Topic

Balancing Redox Reactions

1. Balance each redox reaction occurring in acidic aqueous solution.
   a. $K(s) + Cr^{3+}(aq) \longrightarrow Cr(s) + K^{+}(aq)$
   b. $Al(s) + Fe^{2+}(aq) \longrightarrow Al^{3+}(aq) + Fe(s)$
   c. $BrO_3^{-}(aq) + N_2H_4(g) \longrightarrow Br^{-}(aq) + N_2(g)$

2. Balance each redox reaction occurring in acidic aqueous solution.
   a. $Zn(s) + Sn^{2+}(aq) \longrightarrow Zn^{2+}(aq) + Sn(s)$
   b. $Mg(s) + Cr^{3+}(aq) \longrightarrow Mg^{2+}(aq) + Cr(s)$
   c. $MnO_4^{-}(aq) + Al(s) \longrightarrow Mn^{2+}(aq) + Al^{3+}(aq)$

3. Balance each redox reaction occurring in acidic aqueous solution.
   a. $PbO_2(s) + I^{-}(aq) \longrightarrow Pb^{2+}(aq) + I_2(s)$
   b. $SO_3^{2-}(aq) + MnO_4^{-}(aq) \longrightarrow SO_4^{2-}(aq) + Mn^{2+}(aq)$
   c. $S_2O_3^{2-}(aq) + Cl_2(g) \longrightarrow SO_4^{2-}(aq) + Cl^{-}(aq)$

4. Balance each redox reaction occurring in acidic aqueous solution.
   a. $I^{-}(aq) + NO_2^{-}(aq) \longrightarrow I_2(s) + NO(g)$
   b. $ClO_4^{-}(aq) + Cl^{-}(aq) \longrightarrow ClO_3^{-}(aq) + Cl_2(g)$
   c. $NO_3^{-}(aq) + Sn^{2+}(aq) \longrightarrow Sn^{4+}(aq) + NO(g)$

5. Balance each redox reaction occurring in basic aqueous solution.
   a. $H_2O_2(aq) + ClO_2(aq) \longrightarrow ClO_2^{-}(aq) + O_2(g)$
   b. $Al(s) + MnO_4^{-}(aq) \longrightarrow MnO_2(s) + Al(OH)_4^{-}(aq)$
   c. $Cl_2(g) \longrightarrow Cl^{-}(aq) + ClO^{-}(aq)$

6. Balance each redox reaction occurring in basic aqueous solution.
   a. $MnO_4^{-}(aq) + Br^{-}(aq) \longrightarrow MnO_2(s) + BrO_3^{-}(aq)$
   b. $Ag(s) + CN^{-}(aq) + O_2(g) \longrightarrow Ag(CN)_2^{-}(aq)$
   c. $NO_2^{-}(aq) + Al(s) \longrightarrow NH_3(g) + AlO_2^{-}(aq)$

Voltaic Cells, Standard Cell Potentials, and Direction of Spontaneity

7. Sketch a voltaic cell for each overall redox reaction. Label the anode and cathode and indicate the half-reaction occurring at each electrode and the species present in each solution. Also indicate the direction of electron flow.
   a. $2 Ag^{+}(aq) + Pb(s) \longrightarrow 2 Ag(s) + Pb^{2+}(aq)$
   b. $2 ClO_2(g) + 2 I^{-}(aq) \longrightarrow 2 ClO_2^{-}(aq) + I_2(s)$
   c. $O_2(g) + 4 H^{+}(aq) + 2 Zn(s) \longrightarrow 2 H_2O(l) + 2 Zn^{2+}(aq)$

8. Sketch a voltaic cell for each overall redox reaction. Label the anode and cathode and indicate the half-reaction occurring at each electrode and the species present in each solution. Also indicate the direction of electron flow.
   a. $Ni^{2+}(aq) + Mg(s) \longrightarrow Ni(s) + Mg^{2+}(aq)$
   b. $2 H^{+}(aq) + Fe(s) \longrightarrow H_2(g) + Fe^{2+}(aq)$
   c. $2 NO_3^{-}(aq) + 8 H^{+}(aq) + 3 Cu(s) \longrightarrow$
      $\qquad\qquad 2 NO(g) + 4 H_2O(l) + 3 Cu^{2+}(aq)$

9. Calculate the standard cell potential for each of the electrochemical cells in Problem 7.

10. Calculate the standard cell potential for each of the electrochemical cells in Problem 8.

11. Consider the voltaic cell:

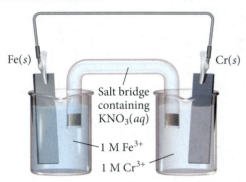

   a. Determine the direction of electron flow and label the anode and the cathode.
   b. Write a balanced equation for the overall reaction and calculate $E°_{cell}$.
   c. Label each electrode as negative or positive.
   d. Indicate the direction of anion and cation flow in the salt bridge.

12. Consider the voltaic cell:

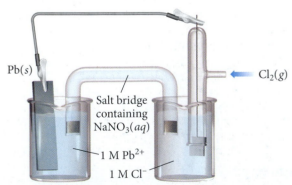

   a. Determine the direction of electron flow and label the anode and the cathode.
   b. Write a balanced equation for the overall reaction and calculate $E°_{cell}$.
   c. Label each electrode as negative or positive.
   d. Indicate the direction of anion and cation flow in the salt bridge.

13. Use line notation to represent each of the electrochemical cells in Problem 7.

14. Use line notation to represent each of the electrochemical cells in Problem 8.

15. Make a sketch of the voltaic cell represented with the line notation. Write the overall balanced equation for the reaction and calculate E_{cell}°.

$$Sn(s) \mid Sn^{2+}(aq) \mid\mid NO(g) \mid NO_3^-(aq), H^+(aq) \mid Pt(s)$$

16. Make a sketch of the voltaic cell represented with the line notation. Write the overall balanced equation for the reaction and calculate E_{cell}°.

$$Mn(s) \mid Mn^{2+}(aq) \mid\mid ClO_2^-(aq) \mid ClO_2(g) \mid Pt(s)$$

17. Determine whether or not each redox reaction occurs spontaneously in the forward direction.
    a. $Ni(s) + Zn^{2+}(aq) \longrightarrow Ni^{2+}(aq) + Zn(s)$
    b. $Ni(s) + Pb^{2+}(aq) \longrightarrow Ni^{2+}(aq) + Pb(s)$
    c. $Al(s) + 3 Ag^+(aq) \longrightarrow Al^{3+}(aq) + 3 Ag(s)$
    d. $Pb(s) + Mn^{2+}(aq) \longrightarrow Pb^{2+}(aq) + Mn(s)$

18. Determine whether or not each redox reaction occurs spontaneously in the reverse direction.
    a. $Ca^{2+}(aq) + Zn(s) \longrightarrow Ca(s) + Zn^{2+}(aq)$
    b. $2 Ag^+(aq) + Ni(s) \longrightarrow 2 Ag(s) + Ni^{2+}(aq)$
    c. $Fe(s) + Mn^{2+}(aq) \longrightarrow Fe^{2+}(aq) + Mn(s)$
    d. $2 Al(s) + 3 Pb^{2+}(aq) \longrightarrow 2 Al^{3+}(aq) + 3 Pb(s)$

19. Which metal could you use to reduce Mn^{2+} ions but not Mg^{2+} ions?

20. Which metal can be oxidized with an Sn^{2+} solution but not with an Fe^{2+} solution?

21. Decide whether or not each metal dissolves in 1 M HCl. For those metals that do dissolve, write a balanced redox reaction showing what happens when the metal dissolves.
    a. Al   b. Ag   c. Pb

22. Decide whether or not each metal dissolves in 1 M HCl. For those metals that do dissolve, write a balanced redox reaction showing what happens when the metal dissolves.
    a. Cu   b. Fe   c. Au

23. Decide whether or not each metal dissolves in 1 M HNO_3. For those metals that do dissolve, write a balanced redox reaction showing what happens when the metal dissolves.
    a. Cu   b. Au

24. Decide whether or not each metal dissolves in 1 M HIO_3. For those metals that do dissolve, write a balanced redox equation for the reaction that occurs.
    a. Au   b. Cr

25. Calculate E_{cell}° for each balanced redox reaction and determine whether the reaction is spontaneous as written.
    a. $2 Cu(s) + Mn^{2+}(aq) \longrightarrow 2 Cu^+(aq) + Mn(s)$
    b. $MnO_2(s) + 4 H^+(aq) + Zn(s) \longrightarrow$
       $\qquad Mn^{2+}(aq) + 2 H_2O(l) + Zn^{2+}(aq)$
    c. $Cl_2(g) + 2 F^-(aq) \longrightarrow F_2(g) + 2 Cl^-(aq)$

26. Calculate E_{cell}° for each balanced redox reaction and determine whether the reaction is spontaneous as written.
    a. $O_2(g) + 2 H_2O(l) + 4 Ag(s) \longrightarrow 4 OH^-(aq) + 4 Ag^+(aq)$
    b. $Br_2(l) + 2 I^-(aq) \longrightarrow 2 Br^-(aq) + I_2(s)$
    c. $PbO_2(s) + 4 H^+(aq) + Sn(s) \longrightarrow$
       $\qquad Pb^{2+}(aq) + 2 H_2O(l) + Sn^{2+}(aq)$

27. Which metal cation is the best oxidizing agent?
    a. Pb^{2+} b. Cr^{3+} c. Fe^{2+} d. Sn^{2+}

28. Which metal is the best reducing agent?
    a. Mn   b. Al   c. Ni   d. Cr

Cell Potential, Free Energy, and the Equilibrium Constant

29. Use tabulated electrode potentials to calculate ΔG_{rxn}° for each reaction at 25 °C.
    a. $Pb^{2+}(aq) + Mg(s) \longrightarrow Pb(s) + Mg^{2+}(aq)$
    b. $Br_2(l) + 2 Cl^-(aq) \longrightarrow 2 Br^-(aq) + Cl_2(g)$
    c. $MnO_2(s) + 4 H^+(aq) + Cu(s) \longrightarrow$
       $\qquad Mn^{2+}(aq) + 2 H_2O(l) + Cu^{2+}(aq)$

30. Use tabulated electrode potentials to calculate ΔG_{rxn}° for each reaction at 25 °C.
    a. $2 Fe^{3+}(aq) + 3 Sn(s) \longrightarrow 2 Fe(s) + 3 Sn^{2+}(aq)$
    b. $O_2(g) + 2 H_2O(l) + 2 Cu(s) \longrightarrow$
       $\qquad 4 OH^-(aq) + 2 Cu^{2+}(aq)$
    c. $Br_2(l) + 2 I^-(aq) \longrightarrow 2 Br^-(aq) + I_2(s)$

31. Calculate the equilibrium constant for each of the reactions in Problem 29.

32. Calculate the equilibrium constant for each of the reactions in Problem 30.

33. Calculate the equilibrium constant for the reaction between $Ni^{2+}(aq)$ and $Cd(s)$.

34. Calculate the equilibrium constant for the reaction between $Fe^{2+}(aq)$ and $Zn(s)$.

35. Calculate ΔG_{rxn}° and E_{cell}° for a redox reaction with $n = 2$ that has an equilibrium constant of $K = 25$.

36. Calculate ΔG_{rxn}° and E_{cell}° for a redox reaction with $n = 3$ that has an equilibrium constant of $K = 0.050$.

Nonstandard Conditions and the Nernst Equation

37. A voltaic cell employs the redox reaction:

$$Sn^{2+}(aq) + Mn(s) \longrightarrow Sn(s) + Mn^{2+}(aq)$$

Calculate the cell potential at 25 °C under each set of conditions:
    a. standard conditions
    b. $[Sn^{2+}] = 0.0100$ M; $[Mn^{2+}] = 2.00$ M
    c. $[Sn^{2+}] = 2.00$ M; $[Mn^{2+}] = 0.0100$ M

38. A voltaic cell employs the redox reaction:

$$2 Fe^{3+}(aq) + 3 Mg(s) \longrightarrow 2 Fe(s) + 3 Mg^{2+}(aq)$$

Calculate the cell potential at 25 °C under each set of conditions:
    a. standard conditions
    b. $[Fe^{3+}] = 1.0 \times 10^{-3}$ M; $[Mg^{2+}] = 2.50$ M
    c. $[Fe^{3+}] = 2.00$ M; $[Mg^{2+}] = 1.5 \times 10^{-3}$ M

39. An electrochemical cell is based on the two half-reactions:

Ox: $Pb(s) \longrightarrow Pb^{2+}(aq, 0.10 M) + 2 e^-$
Red: $MnO_4^-(aq, 1.50 M) + 4 H^+(aq, 2.0 M) + 3 e^- \longrightarrow$
$\qquad MnO_2(s) + 2 H_2O(l)$

Calculate the cell potential at 25 °C.

40. An electrochemical cell is based on the two half-reactions:

Ox: $Sn(s) \longrightarrow Sn^{2+}(aq, 2.00 M) + 2 e^-$
Red: $ClO_2(g, 0.100 atm) + e^- \longrightarrow ClO_2^-(aq, 2.00 M)$

Calculate the cell potential at 25 °C.

41. A voltaic cell consists of a Zn/Zn^{2+} half-cell and a Ni/Ni^{2+} half-cell at 25 °C. The initial concentrations of Ni^{2+} and Zn^{2+} are 1.50 M and 0.100 M, respectively.
    a. What is the initial cell potential?
    b. What is the cell potential when the concentration of Ni^{2+} has fallen to 0.500 M?
    c. What are the concentrations of Ni^{2+} and Zn^{2+} when the cell potential falls to 0.45 V?

42. A voltaic cell consists of a Pb/Pb^{2+} half-cell and a Cu/Cu^{2+} half-cell at 25 °C. The initial concentrations of Pb^{2+} and Cu^{2+} are 0.0500 M and 1.50 M, respectively.
    a. What is the initial cell potential?
    b. What is the cell potential when the concentration of Cu^{2+} has fallen to 0.200 M?
    c. What are the concentrations of Pb^{2+} and Cu^{2+} when the cell potential falls to 0.35 V?

43. Make a sketch of a concentration cell employing two Zn/Zn^{2+} half-cells. The concentration of Zn^{2+} in one of the half-cells is 2.0 M and the concentration in the other half-cell is 1.0 × 10^{-3} M. Label the anode and the cathode and indicate the half-reaction occurring at each electrode. Indicate the direction of electron flow.

44. Consider the concentration cell:

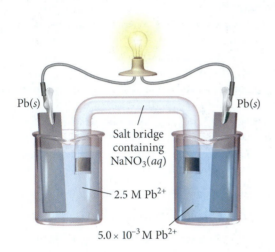

Salt bridge containing NaNO$_3$(aq)

2.5 M Pb^{2+}

5.0 × 10^{-3} M Pb^{2+}

    a. Label the anode and cathode.
    b. Indicate the direction of electron flow.
    c. Indicate what happens to the concentration of Pb^{2+} in each half-cell.

45. A concentration cell consists of two Sn/Sn^{2+} half-cells. The cell has a potential of 0.10 V at 25 °C. What is the ratio of the Sn^{2+} concentrations in the two half-cells?

46. A Cu/Cu^{2+} concentration cell has a voltage of 0.22 V at 25 °C. The concentration of Cu^{2+} in one of the half-cells is 1.5 × 10^{-3} M. What is the concentration of Cu^{2+} in the other half-cell?

Batteries, Fuel Cells, and Corrosion

47. Determine the optimum mass ratio of Zn to MnO$_2$ in an alkaline battery.

48. What mass of lead sulfate is formed in a lead–acid storage battery when 1.00 g of Pb undergoes oxidation?

49. Use the tabulated values of ΔG_f° in Appendix IIB to calculate E_{cell}° for a fuel cell that employs the reaction between methane gas (CH$_4$) and oxygen to form carbon dioxide and gaseous water.

50. Use the tabulated values of ΔG_f° in Appendix IIB to calculate E_{cell}° for the fuel cell breathalyzer, which employs the reaction below. (ΔG_f° for HC$_2$H$_3$O$_2$(g) = −374.2 kJ/mol.)

$$CH_3CH_2OH(g) + O_2(g) \longrightarrow HC_2H_3O_2(g) + H_2O(g)$$

51. Determine whether or not each metal, if coated onto iron, would prevent the corrosion of iron?
    a. Zn   b. Sn   c. Mn

52. Determine whether or not each metal, if coated onto iron, would prevent the corrosion of iron?
    a. Mg   b. Cr   c. Cu

Electrolytic Cells and Electrolysis

53. Consider the electrolytic cell:

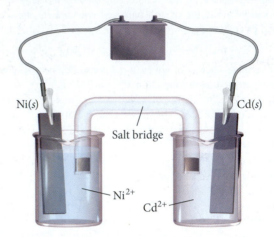

Ni(s)   *Cd(s)*

Salt bridge

Ni^{2+}   *Cd^{2+}*

    a. Label the anode and the cathode and indicate the half-reactions occurring at each.
    b. Indicate the direction of electron flow.
    c. Label the terminals on the battery as positive or negative and calculate the minimum voltage necessary to drive the reaction.

54. Draw an electrolytic cell in which Mn^{2+} is reduced to Mn and Sn is oxidized to Sn^{2+}. Label the anode and cathode, indicate the direction of electron flow, and write an equation for the half-reaction occurring at each electrode. What minimum voltage is necessary to drive the reaction?

55. Make a sketch of an electrolysis cell that might be used to electroplate copper onto other metal surfaces. Label the anode and the cathode and show the reactions that occur at each.

56. Make a sketch of an electrolysis cell that might be used to electroplate nickel onto other metal surfaces. Label the anode and the cathode and show the reactions that occur at each.

57. Copper can be electroplated at the cathode of an electrolysis cell by the half-reaction:

$$Cu^{2+}(aq) + 2\,e^- \longrightarrow Cu(s)$$

How long would it take for 225 mg of copper to be plated at a current of 7.8 A?

58. Silver can be electroplated at the cathode of an electrolysis cell by the half-reaction:

$$Ag^+(aq) + e^- \longrightarrow Ag(s)$$

What mass of silver would plate onto the cathode if a current of 5.8 A flowed through the cell for 55 min?

59. A major source of sodium metal is the electrolysis of molten sodium chloride. What magnitude of current is required to produce 1.0 kg of sodium metal in one hour?

60. What mass of aluminum metal can be produced per hour in the electrolysis of a molten aluminum salt by a current of 25 A?

Cumulative Problems

61. Consider the unbalanced redox reaction:

$$MnO_4^-(aq) + Zn(s) \longrightarrow Mn^{2+}(aq) + Zn^{2+}(aq)$$

Balance the equation and determine the volume of a 0.500 M $KMnO_4$ solution required to completely react with 2.85 g of Zn.

62. Consider the unbalanced redox reaction:

$$Cr_2O_7^{2-}(aq) + Cu(s) \longrightarrow Cr^{3+}(aq) + Cu^{2+}(aq)$$

Balance the equation and determine the volume of a 0.850 M $K_2Cr_2O_7$ solution required to completely react with 5.25 g of Cu.

63. Consider the molecular views of an Al strip and Cu^{2+} solution. Draw a similar sketch showing what happens to the atoms and ions if the Al strip is submerged in the solution for a few minutes.

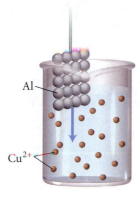

64. Consider the molecular view of an electrochemical cell involving the overall reaction:

$$Zn(s) + Ni^{2+}(aq) \longrightarrow Zn^{2+}(aq) + Ni(s)$$

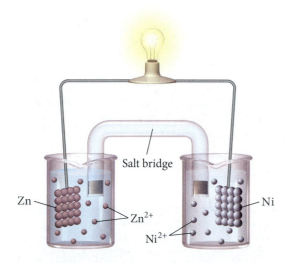

Draw a similar sketch showing how the cell might appear after it has generated a substantial amount of electrical current.

65. Determine whether or not hydroiodic acid can dissolve each metal sample. If it can, write a balanced chemical reaction showing how the metal dissolves in HI and determine the minimum volume of 3.5 M HI required to completely dissolve the sample.
a. 2.15 g Al   **b.** 4.85 g Cu   **c.** 2.42 g Ag

66. Determine whether or not nitric acid can dissolve each metal sample. If it can, write a balanced chemical reaction showing how the metal dissolves in HNO_3 and determine the minimum volume of 6.0 M HNO_3 required to completely dissolve the sample.
a. 5.90 g Au   **b.** 2.55 g Cu   **c.** 4.83 g Ni

67. The cell potential of this electrochemical cell depends on the pH of the solution in the anode half-cell:

$$Pt(s)\,|\,H_2(g, 1\ atm)\,|\,H^+(aq, ?\ M)\,||\,Cu^{2+}(aq, 1.0\ M)\,|\,Cu(s)$$

What is the pH of the solution if E_{cell} is 355 mV?

68. The cell potential of this electrochemical cell depends on the gold concentration in the cathode half-cell:

$$Pt(s)\,|\,H_2(g, 1.0\ atm)\,|\,H^+(aq, 1.0\ M)\,||\,Au^{3+}(aq, ?\ M)\,|\,Au(s)$$

What is the concentration of Au^{3+} in the solution if E_{cell} is 1.22 V?

69. A battery is constructed based on the oxidation of magnesium and the reduction of Cu^{2+}. The initial concentrations of Mg^{2+} and Cu^{2+} are 1.0×10^{-4} M and 1.5 M, respectively, in 1.0-liter half-cells.
a. What is the initial voltage of the battery?
b. What is the voltage of the battery after delivering 5.0 A for 8.0 h?
c. How long can the battery deliver 5.0 A before it dies?

70. A rechargeable battery is constructed based on a concentration cell constructed of two Ag/Ag^+ half-cells. The volume of each half-cell is 2.0 L and the concentrations of Ag^+ in the half-cells are 1.25 M and 1.0×10^{-3} M.
a. For how long can this battery deliver 2.5 A of current before it dies?
b. What mass of silver is plated onto the cathode by running at 3.5 A for 5.5 h?
c. Upon recharging, how long would it take to redissolve 1.00×10^2 g of silver at a charging current of 10.0 amps?

71. The K_{sp} of CuI is 1.1×10^{-12}. Find E_{cell} for the following cell.

$$Cu(s)\,||\,CuI(s)\,|\,I^-(aq)(1.0\ M)\,||\,Cu^+(aq)(1.0\ M)\,|\,Cu(s)$$

72. The K_{sp} of $Zn(OH)_2$ is 1.8×10^{-14}. Find E_{cell} for the following half-reaction.

$$Zn(OH)_2(s) + 2\ e^- \Longleftrightarrow Zn(s) + 2\ OH^-(aq)$$

73. Calculate ΔG°_{rxn} and K for each reaction:
a. The disproportionation of $Mn^{2+}(aq)$ to $Mn(s)$ and $MnO_2(s)$ in acid solution at 25 °C.
b. The disproportionation of $MnO_2(s)$ to $Mn^{2+}(aq)$ and $MnO_4^-(aq)$ in acid solution at 25 °C.

74. Calculate ΔG°_{rxn} and K for each reaction:
a. The reaction of $Cr^{2+}(aq)$ with $Cr_2O_7^{2-}(aq)$ in acid solution to form $Cr^{3+}(aq)$.
b. The reaction of $Cr^{3+}(aq)$ and $Cr(s)$ to form $Cr^{2+}(aq)$. [The electrode potential of $Cr^{2+}(aq)$ to $Cr(s)$ is −0.91 V.]

75. The molar mass of a metal (M) is 50.9 g/mol and it forms a chloride of unknown composition. Electrolysis of a sample of the molten chloride with a current of 6.42 A for 23.6 minutes produces 1.20 g of M at the cathode. Find the empirical formula of the chloride.

76. A metal forms the fluoride MF_3. Electrolysis of the molten fluoride by a current of 3.86 A for 16.2 minutes deposits 1.25 g of the metal. Calculate the molar mass of the metal.

77. A sample of impure tin of mass 0.535 g is dissolved in strong acid to give a solution of Sn^{2+}. The solution is then titrated with a 0.0448 M solution of NO_3^-, which is reduced to $NO(g)$. The equivalence point is reached upon the addition of 0.0344 L of the NO_3^- solution. Find the percent by mass of tin in the original sample, assuming that it contains no other reducing agents.

78. A 0.0251-L sample of a solution of Cu^+ requires 0.0322 L of 0.129 M $KMnO_4$ solution to reach the equivalence point. The products of the reaction are Cu^{2+} and Mn^{2+}. What is the concentration of the Cu^+ solution?

Challenge Problems

79. A hydrogen–oxygen fuel-cell generator is used to produce electricity for a house. Use the balanced redox reactions and the standard cell potential to predict the volume of hydrogen gas (at STP) required each month to generate the electricity needed for a typical house. Assume the home uses 1.2×10^3 kWh of electricity per month.

80. A voltaic cell designed to measure $[Cu^{2+}]$ is constructed of a standard hydrogen electrode and a copper metal electrode in the Cu^{2+} solution of interest. If you wanted to construct a calibration curve for how the cell potential varies with the concentration of copper(II), what would you plot in order to obtain a straight line? What would be the slope of the line?

81. A thin layer of gold can be applied to another material by an electrolytic process. The surface area of an object to be gold plated is 49.8 cm^2 and the density of gold is 19.3 g/cm^3. A current of 3.25 A is applied to a solution that contains gold in the +3 oxidation state. Calculate the time required to deposit an even layer of gold 1.00×10^{-3} cm thick on the object.

82. To electrodeposit all the Cu and Cd from a solution of $CuSO_4$ and $CdSO_4$ required 1.20 F of electricity (1 F = 1 mol e^-). The mixture of Cu and Cd deposited had a mass of 50.36 g. What mass of $CuSO_4$ was present in the original mixture?

83. Sodium oxalate, $Na_2C_2O_4$, in solution is oxidized to $CO_2(g)$ by MnO_4^- which is reduced to Mn^{2+}. A 50.1-mL volume of a solution of MnO_4^- is required to titrate a 0.339-g sample of sodium oxalate. This solution of MnO_4^- is then used to analyze uranium-containing samples. A 4.62-g sample of a uranium-containing material requires 32.5 mL of the solution for titration. The oxidation of the uranium can be represented by the change $UO^{2+} \longrightarrow UO_2^{2+}$. Calculate the percentage of uranium in the sample.

84. A current of 11.3 A is applied to 1.25 L of a solution of 0.552 M HBr converting some of the H^+ to $H_2(g)$, which bubbles out of solution. What is the pH of the solution after 73 minutes?

85. A 215-mL sample of a 0.500 M NaCl solution with an initial pH of 7.00 is subjected to electrolysis. After 15.0 minutes, a 10.0-mL portion (or aliquot) of the solution was removed from the cell and titrated with 0.100 M HCl solution. The endpoint in the titration was reached upon addition of 22.8 mL of HCl. Assuming constant current, what was the current (in A) running through the cell?

86. An $MnO_2(s)/Mn^{2+}(aq)$ electrode in which the pH is 10.24 is prepared. Find the $[Mn^{2+}]$ necessary to lower the potential of the half-cell to 0.00 V.

87. To what pH should you adjust a standard hydrogen electrode to get an electrode potential of -0.122 V?

Conceptual Problems

88. An electrochemical cell has a positive standard cell potential but a negative cell potential. What must be true of Q and K for the cell?
a. $K > 1; Q > K$ b. $K < 1; Q > K$
c. $K > 1; Q < K$ d. $K < 1; Q < K$

89. Which oxidizing agent will oxidize Br^- but not Cl^-?
a. $K_2Cr_2O_7$ (in acid)
b. $KMnO_4$ (in acid)
c. HNO_3

90. A redox reaction employed in an electrochemical cell has a negative ΔG°_{rxn}. Which statement is true?
a. E°_{cell} is positive; $K < 1$
b. E°_{cell} is positive; $K > 1$
c. E_{cell} is negative; $K > 1$
d. E°_{cell} is negative; $K < 1$

91. A redox reaction has an equilibrium constant of $K = 0.055$. What is true of ΔG°_{rxn} and E°_{cell} for this reaction?

Answers to Conceptual Connections

Voltaic Cells

18.1 (a) Electrons are negatively charged and, therefore, flow away from the more negatively charged electrode and towards the more positively charged electrode.

Standard Electrode Potentials

18.2 (b) A negative electrode potential indicates that an electron at that electrode has greater potential energy than it has at a standard hydrogen electrode.

Selective Oxidation

18.3 (d) The reduction of HNO_3 is found below the reduction of Br_2 and above the reduction of I_2 in Table 18.1. Since any reduction half-reaction will be spontaneous when paired with the reverse of any half-reaction below it in the table, the reduction of HNO_3 is spontaneous when paired with the oxidation of I^-, but is not spontaneous when paired with the oxidation of Br^-.

Metals Dissolving in Acids

18.4 (c) Ag falls *above* the half-reaction for the reduction of H^+ but *below* the half-reaction for the reduction of NO_3^- in Table 18.1.

Relating K, ΔG°_{rxn}, and E°_{cell}

18.5 (c) Since $K > 1$, the reaction is spontaneous under standard conditions (when $Q = 1$, the reaction proceeds toward the products). Therefore E°_{cell} is positive and ΔG°_{rxn} is negative.

Relating Q, K, E_{cell} and E°_{cell}

18.6 (a) Since $K < 1$, E°_{cell} is negative (under standard conditions, the reaction is not spontaneous). Since $Q < K$, E_{cell} is positive (the reaction is spontaneous under the nonstandard conditions of the cell).

19

Radioactivity and Nuclear Chemistry

I am among those who think that science has great beauty. A scientist in his laboratory is not only a technician; he is also a child placed before natural phenomena which impress him like a fairy tale. —Marie Curie (1867–1934)

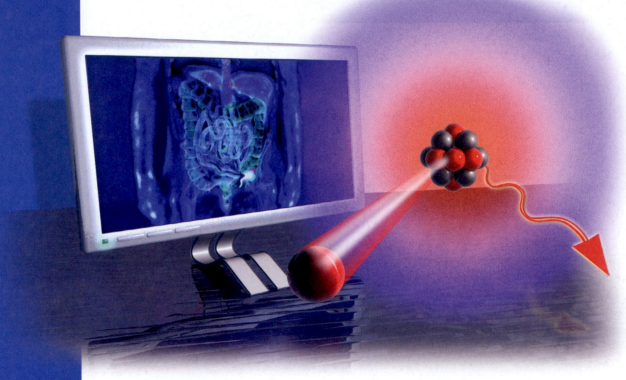

Antibodies labeled with radioactive atoms can be used to diagnose an infected appendix.

IN THIS CHAPTER, WE EXAMINE RADIOACTIVITY and nuclear chemistry, both of which are associated with changes that occur within the nuclei of atoms. Unlike ordinary chemical processes, in which elements retain their identity, nuclear processes result in one element changing into another. This process is often accompanied by the release of tremendous amounts of energy. Radioactivity has numerous applications, including the diagnosis and treatment of conditions such as cancer, thyroid disease, abnormal kidney and bladder function, and heart disease. Naturally occurring radioactivity also allows us to estimate the age of fossils, rocks, and artifacts. And radioactivity, perhaps most famously, led to the development of nuclear fission, used for electricity generation and nuclear weapons. In this chapter, we will discuss radioactivity—how it was discovered, what it is, and how we use it.

19.1   Diagnosing Appendicitis

One morning a few years ago I awoke with a dull pain on the lower right side of my abdomen that seemed to worsen by early afternoon. Since pain in this area can indicate appendicitis (inflammation of the appendix), and since I knew that appendicitis could be dangerous if left untreated, I went to the hospital emergency room. The doctor who examined me recommended a simple blood test to determine my white blood cell count. Patients with appendicitis usually have a high white blood cell count because the body is trying to fight the infection. In my case, however, the test was negative—I had a normal white blood cell count.

Although my symptoms were consistent with appendicitis, the negative blood test clouded the diagnosis. The doctor said that I could elect to have my appendix removed anyway (even though there was a chance of it being healthy) or I could submit to another test that might confirm the appendicitis. I chose the additional test, which involved *nuclear medicine*, an area of medical practice that uses *radioactivity* to diagnose and treat disease. **Radioactivity** is the emission of subatomic particles or high-energy electromagnetic radiation by the nuclei of certain atoms. Such atoms are said to be **radioactive**. Most radioactive emissions are so energetic that they can pass through many types of matter (such as skin and muscle).

To perform the test, antibodies—naturally occurring molecules that fight infection—were "labeled" with radioactive atoms and then injected into my bloodstream. Since antibodies attack infection, they migrate to areas of the body where infection is present. If my appendix was infected, the antibodies would accumulate there. After waiting about an hour, I was taken to a room and laid on a table. A photographic film was inserted in a panel above me. Radioactivity is invisible to the eye, but it exposes photographic film. If my appendix had indeed been infected, it would have by then contained a high concentration of the radioactively tagged antibodies, and the film would show an exposure spot at the location of my appendix. In this procedure, my appendix, if it were infected, would be the radiation source that would expose the film. The test, however, was negative. No radioactivity was emanating from my appendix; it was healthy. After several hours, the pain in my abdomen subsided and I went home. I never did find out what caused the pain.

19.2   Types of Radioactivity

Radioactivity was discovered in 1896 by a French scientist named Antoine-Henri Becquerel (1852–1908) and further characterized by Marie Sklodowska Curie (1867–1934), one of the first women in France to pursue doctoral work, and Ernest Rutherford (1871–1937). These scientists found that certain nuclei are unstable and spontaneously decompose, emitting small pieces of themselves to gain stability. This process of emission is called radioactivity. Natural radioactivity is divided into several different types, including *alpha (α) decay*, *beta (β) decay*, *gamma (γ) ray emission*, and *positron emission*. In addition, some unstable atomic nuclei attain greater stability by absorbing an electron, a process called *electron capture*.

In order to understand these different types of radioactivity, we must briefly review the notation for symbolizing isotopes that was first introduced in Section 2.5. Recall that any isotope can be represented with the following notation:

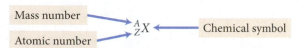

Mass number (*A*) = the sum of the number of protons and number of neutrons in the nucleus

Atomic number (*Z*) = the number of protons in the nucleus

Element 96 is named curium in honor of Marie Curie and her contributions to our understanding of radioactivity.

Since A represents the sum of the protons and neutrons, and since Z represents the number of protons, the number of neutrons in the nucleus (N) is $A - Z$.

$$N = A - Z$$

↑
Number of neutrons

For example, the symbol $^{21}_{10}\text{Ne}$ represents the neon isotope containing 10 protons and 11 neutrons. The symbol $^{20}_{10}\text{Ne}$ represents the neon isotope containing 10 protons and 10 neutrons. Remember that most elements have several different isotopes. When discussing nuclear properties, a particular isotope (or species) of an element is often referred to as a **nuclide**.

The main subatomic particles—protons, neutrons, and electrons—can all be represented with similar notation.

Proton symbol ^1_1p Neutron symbol ^1_0n Electron symbol $^{\ 0}_{-1}\text{e}$

The 1 in the lower left of the proton symbol indicates 1 proton, and the 0 in the lower left corner of the neutron symbol represents 0 protons. The -1 in the lower left corner of the electron symbol is a bit different from the other atomic numbers, but it will make sense when we see it in the context of nuclear decay a bit later in this section.

Alpha (α) Decay

Alpha (α) decay occurs when an unstable nucleus emits a particle composed of two protons and two neutrons (**Figure 19.1▼**). Since two protons and two neutrons are identical to a helium-4 nucleus, the symbol for alpha radiation is the symbol for helium-4:

Alpha (α) particle ^4_2He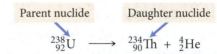

When an element emits an alpha particle, the number of protons in its nucleus changes, transforming it into a different element. We symbolize this phenomenon with a **nuclear equation**, an equation that represents nuclear processes such as radioactivity. For example, the nuclear equation for the alpha decay of uranium-238 is

Parent nuclide   Daughter nuclide

$$^{238}_{92}\text{U} \longrightarrow \ ^{234}_{90}\text{Th} + \ ^4_2\text{He}$$

In nuclear chemistry, we are primarily interested in changes within the nucleus; therefore, the 2+ charge that we would normally write for a helium nucleus is omitted for an alpha particle.

The original atom is the *parent nuclide* and the product of the decay is the *daughter nuclide*. In this case, uranium-238 becomes thorium-234. Unlike a chemical reaction, in which elements retain their identity, a nuclear reaction often results in elements changing their identity. Like a chemical equation, however, a nuclear equation must be balanced.

Alpha Decay

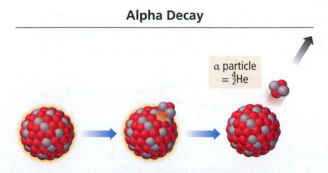

▲ **FIGURE 19.1 Alpha Decay**   In alpha decay, a nucleus emits a particle composed of two protons and two neutrons (a helium-4 nucleus).

The sum of the atomic numbers on both sides of a nuclear equation must be equal, and the sum of the mass numbers on both sides must also be equal.

$$^{238}_{92}\text{U} \longrightarrow {}^{234}_{90}\text{Th} + {}^{4}_{2}\text{He}$$

Reactants	Products
Sum of mass numbers = 238	Sum of mass numbers = 234 + 4 = 238
Sum of atomic numbers = 92	Sum of atomic numbers = 90 + 2 = 92

We can identify the symbol of the daughter nuclide in any alpha decay from the mass and atomic number of the parent nuclide. During alpha decay, the mass number decreases by 4 and the atomic number decreases by 2, as shown in Example 19.1.

EXAMPLE 19.1 Writing Nuclear Equations for Alpha Decay

Write a nuclear equation for the alpha decay of Ra-224.

SOLUTION

Begin with the symbol for Ra-224 on the left side of the equation and the symbol for an alpha particle on the right side.	$^{224}_{88}\text{Ra} \longrightarrow {}^{?}_{?}? + {}^{4}_{2}\text{He}$
Equalize the sum of the mass numbers and the sum of the atomic numbers on both sides of the equation by writing the appropriate mass number and atomic number for the unknown daughter nuclide.	$^{224}_{88}\text{Ra} \longrightarrow {}^{220}_{86}? + {}^{4}_{2}\text{He}$
Using the periodic table, deduce the identity of the unknown daughter nuclide from the atomic number and write its symbol. Since the atomic number is 86, the daughter nuclide must be radon (Rn).	$^{224}_{88}\text{Ra} \longrightarrow {}^{220}_{86}\text{Rn} + {}^{4}_{2}\text{He}$

FOR PRACTICE 19.1

Write a nuclear equation for the alpha decay of Po-216.

Alpha radiation is the 18-wheeler truck of radioactivity; the alpha particle is by far the most massive of all particles emitted by radioactive nuclei. Consequently, alpha radiation has the most potential to interact with and damage molecules, including biological ones. Highly energetic radiation interacts with other molecules and atoms by ionizing them. If radiation ionizes molecules within the cells of living organisms, those molecules may undergo damaging chemical reactions, and the cell can die or begin to reproduce abnormally. The ability of radiation to ionize molecules and atoms is called its **ionizing power**. Of all types of radioactivity, alpha radiation has the highest ionizing power.

However, in order for radiation to damage important molecules within living cells, it must penetrate into the cell. Alpha particles, because of their large size, have the lowest **penetrating power**—the ability to penetrate matter. (Imagine a semi truck trying to get through a traffic jam.) Alpha radiation does not easily penetrate into cells because it is stopped by a sheet of paper, by clothing, or even by air. Consequently, a low-level alpha emitter kept outside the body is relatively safe. If an alpha emitter is ingested, however, it becomes very dangerous because the alpha particles then have direct access to the molecules that compose organs and tissues.

Beta (β) Decay

Beta (β) decay occurs when an unstable nucleus emits an electron (**Figure 19.2▶**). How does a nucleus, which contains only protons and neutrons, emit an electron? In some unstable nuclei, a neutron changes into a proton and emits an electron in the process:

Beta decay   Neutron $\longrightarrow$ proton + emitted electron

The symbol for a beta (β) particle in a nuclear equation is

Beta (β) particle $^{0}_{-1}\text{e}$ ●

Beta Decay

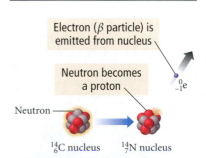

Electron (β particle) is emitted from nucleus

Neutron becomes a proton

$^{0}_{-1}\text{e}$

Neutron

$^{14}_{6}\text{C}$ nucleus $^{14}_{7}\text{N}$ nucleus

▲ **FIGURE 19.2 Beta Decay** In beta decay, a neutron emits an electron and becomes a proton.

This kind of beta radiation is also called beta minus (β^-) radiation due to its negative charge.

Thus, we can represent beta decay as

$$^1_0n \longrightarrow {}^1_1p + {}^{\ 0}_{-1}e$$

The -1 reflects the charge of the electron, which is equivalent to an atomic number of -1 in a nuclear equation. When an atom emits a beta particle, its atomic number increases by 1 because it now has an additional proton. For example, the nuclear equation for the beta decay of radium-228 is

$$^{228}_{88}Ra \longrightarrow {}^{228}_{89}Ac + {}^{\ 0}_{-1}e$$

Notice that the nuclear equation is still balanced—the sum of the mass numbers on both sides is equal and the sum of the atomic numbers on both sides is equal.

Beta radiation is the 4-door sedan of radioactivity. Beta particles are much less massive than alpha particles and consequently have a lower ionizing power. However, because of their smaller size, beta particles have a higher penetrating power and require a sheet of metal or a thick piece of wood to stop them. Consequently, a beta emitter outside of the body poses a higher risk than an alpha emitter. Inside the body, however, the beta emitter does less damage than an alpha emitter.

Gamma (γ) Ray Emission

See Section 7.2 for a review of electromagnetic radiation.

Gamma (γ) ray emission is significantly different from alpha or beta radiation. Gamma radiation is *electromagnetic* radiation. Gamma rays are high-energy (short-wavelength) photons. The symbol for a gamma ray is

Gamma (γ) ray $^0_0\gamma$ 〰〰〰〰➤

A gamma ray has no charge and no mass. When a gamma-ray photon is emitted from a radioactive atom, it does not change the mass number or the atomic number of the element. Gamma rays, however, are usually emitted in conjunction with other types of radiation. For example, the alpha emission of U-238 (discussed previously) is also accompanied by the emission of a gamma ray.

$$^{238}_{92}U \longrightarrow {}^{234}_{90}Th + {}^4_2He + {}^0_0\gamma$$

Gamma rays are the motorbikes of radioactivity. They have the lowest ionizing power, but the highest penetrating power. (Imagine a motorbike zipping through a traffic jam.) Stopping gamma rays requires several inches of lead shielding or thick slabs of concrete.

Positron Emission

Positron emission can be thought of as a type of beta emission and is sometimes referred to as beta plus emission (β^+).

Positron emission occurs when an unstable nucleus emits a positron (**Figure 19.3▼**). A **positron** is the *antiparticle* of the electron: that is, it has the same mass, but opposite charge. If a positron collides with an electron, the two particles annihilate each other, releasing energy in the form of gamma rays. In positron emission, a proton is converted into a neutron and emits a positron:

Positron emission   Proton $\longrightarrow$ neutron + emitted positron

The symbol for a positron in a nuclear equation is

Positron $^{\ 0}_{+1}e$ •

We represent positron emission as

$$^1_1p \longrightarrow {}^1_0n + {}^{\ 0}_{+1}e$$

When an atom emits a positron, its atomic number *decreases* by 1 because it now has one fewer proton. For example, the nuclear equation for the positron emission of phosphorus-30 is

$$^{30}_{15}P \longrightarrow {}^{30}_{14}Si + {}^{\ 0}_{+1}e$$

We can determine the identity and symbol of the daughter nuclide in any positron emission in a manner similar to that used for alpha and beta decay, as shown

Positron Emission

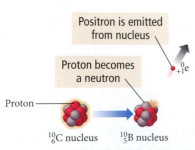

Positron is emitted from nucleus

Proton becomes a neutron

$^{\ 0}_{+1}e$

Proton

$^{10}_6C$ nucleus $^{10}_5B$ nucleus

▲ FIGURE 19.3 Positron Emission
In positron emission, a proton emits a positron and becomes a neutron.

in Example 19.2. Positrons are similar to beta particles in their ionizing and penetrating power.

Electron Capture

Unlike the forms of radioactive decay described so far, electron capture involves a particle being *absorbed by* instead of *ejected from* an unstable nucleus. **Electron capture** occurs when a nucleus assimilates an electron from an inner orbital of its electron cloud. Like positron emission, the net effect of electron capture is the conversion of a proton into a neutron:

Electron capture   Proton + electron $\longrightarrow$ neutron

We represent electron capture as

$$_1^1\text{p} + _{-1}^0\text{e} \longrightarrow _0^1\text{n}$$

When an atom undergoes electron capture, its atomic number decreases by 1 because it has one fewer proton. For example, the nuclear equation for electron capture in Ru-92 is:

$$_{44}^{92}\text{Ru} + _{-1}^0\text{e} \longrightarrow _{43}^{92}\text{Tc}$$

The different kinds of radiation are summarized in Table 19.1.

TABLE 19.1 Modes of Radioactive Decay

Decay Mode	Process	Change in: A	Z	N/Z*	Example
α	 Parent nuclide → Daughter nuclide + α particle ($_2^4\text{He}$)	−4	−2	Increase	$_{92}^{238}\text{U} \longrightarrow _{90}^{234}\text{Th} + _2^4\text{He}$
β	 Neutron; Parent nuclide → Neutron becomes a proton; Daughter nuclide + β particle ($_{-1}^0\text{e}$)	0	+1	Decrease	$_{88}^{228}\text{Ra} \longrightarrow _{89}^{228}\text{Ac} + _{-1}^0\text{e}$
γ	 Excited nuclide → Stable nuclide + Photon ($_0^0\gamma$)	0	0	None	$_{90}^{234}\text{Th} \longrightarrow _{90}^{234}\text{Th} + _0^0\gamma$
Positron emission	 Proton; Parent nuclide → Proton becomes a neutron; Daughter nuclide + Positron ($_{+1}^0\text{e}$)	0	−1	Increase	$_{15}^{30}\text{P} \longrightarrow _{14}^{30}\text{Si} + _{+1}^0\text{e}$
Electron capture	 Proton; Parent nuclide + Electron ($_{-1}^0\text{e}$) → Proton becomes a neutron; Daughter nuclide	0	−1	Increase	$_{44}^{92}\text{Ru} + _{-1}^0\text{e} \longrightarrow _{43}^{92}\text{Tc}$

* Neutron-to-proton ratio

EXAMPLE 19.2 **Writing Nuclear Equations for Beta Decay, Positron Emission, and Electron Capture**

Write a nuclear equation for each:

(a) beta decay in Bk-249   (b) positron emission in K-40

(c) electron capture in I-111

SOLUTION

(a) In beta decay, the atomic number *increases* by 1 and the mass number remains unchanged. The daughter nuclide is element number 98, californium.	$^{249}_{97}\text{Bk} \longrightarrow {}^{249}_{98}? + {}^{0}_{-1}\text{e}$ $^{249}_{97}\text{Bk} \longrightarrow {}^{249}_{98}\text{Cf} + {}^{0}_{-1}\text{e}$
(b) In positron emission, the atomic number *decreases* by 1 and the mass number remains unchanged. The daughter nuclide is element number 18, argon.	$^{40}_{19}\text{K} \longrightarrow {}^{40}_{18}? + {}^{0}_{+1}\text{e}$ $^{40}_{19}\text{K} \longrightarrow {}^{40}_{18}\text{Ar} + {}^{0}_{+1}\text{e}$
(c) In electron capture, the atomic number also *decreases* by 1 and the mass number remains unchanged. The daughter nuclide is element number 52, tellurium.	$^{111}_{53}\text{I} + {}^{0}_{-1}\text{e} \longrightarrow {}^{111}_{52}?$ $^{111}_{53}\text{I} + {}^{0}_{-1}\text{e} \longrightarrow {}^{111}_{52}\text{Te}$

FOR PRACTICE 19.2

(a) Write three nuclear equations to represent the nuclear decay sequence that begins with the alpha decay of U-235 followed by a beta decay of the daughter nuclide and then another alpha decay.

(b) Write a nuclear equation for the positron emission of Na-22.

(c) Write a nuclear equation for electron capture in Kr-76.

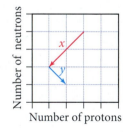

Number of neutrons / Number of protons

⬤ Conceptual Connection 19.1 Alpha and Beta Decay

Consider the graphical representation of two decays shown in the figure. The arrows labeled *x* and *y* correspond to what kinds of decay?

(a) *x* corresponds to alpha decay and *y* corresponds to positron emission.

(b) *x* corresponds to positron emission and *y* corresponds to alpha decay.

(c) *x* corresponds to alpha decay and *y* corresponds to beta decay.

(d) *x* corresponds to beta decay and *y* corresponds to alpha decay.

19.3   The Valley of Stability: Predicting the Type of Radioactivity

So far, we have described various different types of radioactivity. But we have not addressed the question of what causes a particular nuclide to be radioactive in the first place. And why do some nuclides decay via alpha decay, while others decay via beta decay or positron emission? The answers to these questions are not simple, but we can get a basic idea of the factors that influence the stability of the nucleus and the nature of its decay.

A nucleus is a collection of protons (positively charged) and neutrons (uncharged). We know that positively charged particles such as protons repel one another. So what holds the nucleus together? The binding is provided by a fundamental force of physics known as the **strong force**. All **nucleons**—protons and neutrons—are attracted to one another by the strong force. However, the strong force acts only at very short distances. So we can think of the stability of a nucleus as a balance between the *repulsive* electrostatic force among protons and the *attractive* strong force among all nucleons. The neutrons in a nucleus play an important role in stabilizing the nucleus because they attract other nucleons (through the strong force) but lack the repulsive force associated with positive charge. It might seem that adding more neutrons would *always* lead to greater stability, so that the more neutrons the better. This is not the case, however, because protons and neutrons occupy energy levels in a nucleus that are similar to those occupied by electrons in an atom. As you add more neutrons, they must occupy increasingly higher energy levels within the nucleus. At some point, the energy payback from the strong nuclear force is not enough to compensate for the high energy state that the neutron must occupy.

An important number in determining nuclear stability, therefore, is the *ratio* of neutrons to protons (N/Z). **Figure 19.4** shows a plot of the number of neutrons versus the number of protons for all known stable nuclei. The green dots along the diagonal of the graph represent stable nuclei, and this region is known as the *valley (or island) of stability*. Notice that for the lighter elements, the N/Z ratio of stable isotopes is about one (equal numbers of neutrons and protons). For example, the most abundant isotope of carbon ($Z = 6$) is carbon-12, which contains 6 protons and 6 neutrons. However, beyond about $Z = 20$, the N/Z ratio of stable nuclei begins to get larger. For example, at $Z = 40$, stable nuclei have an N/Z ratio of about 1.25 and at $Z = 80$, the N/Z ratio reaches about 1.5. Above $Z = 83$, stable nuclei do not exist—bismuth ($Z = 83$) is the heaviest element with stable (nonradioactive) isotopes.

The type of radioactivity emitted by a nuclide depends in part on the N/Z ratio:

N/Z too high: Nuclides that lie above the valley of stability have too many neutrons and will tend to convert neutrons to protons via beta decay. The process of undergoing beta decay moves the nuclide down in the plot in Figure 19.4 and closer to (or into) the valley of stability.

N/Z too low: Nuclides that lie below the valley of stability have too many protons and will tend to convert protons to neutrons via positron emission or electron capture. This moves the nuclide up in the plot in Figure 19.4 and closer to (or into) the valley of stability. (Alpha decay also raises the N/Z ratio for nuclides in which $N/Z > 1$, but the effect is smaller than for positron emission or electron capture.)

One way to decide whether N/Z for a particular nuclide is too high, too low, or generally stable is to consult Figure 19.4. Those nuclides that lie within the valley of stability are stable and those above or below the valley are unstable. Alternatively, we can compare the mass number of the nuclide in question to the atomic mass listed in the periodic table for the corresponding element. The atomic mass is an average of the masses of the stable nuclides for an element and thus represents an N/Z that is generally stable. For example, suppose we want to evaluate the N/Z ratio for Ru-112. Ruthenium has an atomic mass of 101.07 amu so we know that the nuclide with a mass number of 112 must have a N/Z that is too high; it contains too many neutrons and is likely to undergo positron emission or electron capture. Example 19.3 shows how to apply these considerations when predicting the mode of decay for a nucleus.

The Valley of Stability

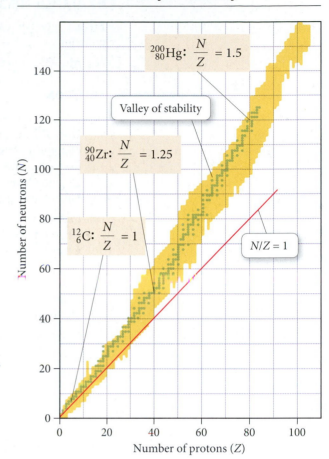

▲ **FIGURE 19.4 Stable and Unstable Nuclei** A plot of N (the number of neutrons) versus Z (the number of protons) for all known stable nuclei—represented by green dots on this graph—shows that these nuclei cluster together in a region known as the valley (or island) of stability. Nuclei with an N/Z ratio that is too high tend to undergo beta decay. Nuclei with an N/Z ratio that is too low tend to undergo positron emission or electron capture.

EXAMPLE 19.3 Predicting the Type of Radioactive Decay

Predict whether each nuclide is more likely to decay via beta decay or positron emission.

(a) Mg-28 **(b)** Mg-22 **(c)** Mo-102

SOLUTION

(a) Magnesium-28 has 16 neutrons and 12 protons, so $N/Z = 1.33$. However, for $Z = 12$, you can see in Figure 19.4 that stable nuclei should have an N/Z ratio of about 1. Alternatively, on the periodic table you can look up the atomic mass of magnesium, which is 24.31 amu; a magnesium nuclide with a mass number of 28 is too heavy to be stable because the N/Z ratio is too high. Therefore, Mg-28 undergoes *beta decay*, resulting in the conversion of a neutron to a proton.

(b) Magnesium-22 has 10 neutrons and 12 protons, so $N/Z = 0.83$ (too low). Alternatively, you can look up the atomic mass of magnesium, which is 24.31 amu; a magnesium nuclide with a mass number of 22 is too light (the N/Z ratio is too low). Therefore, Mg-22 undergoes *positron emission*, resulting in the conversion of a proton to a neutron. (Electron capture would accomplish the same thing as positron emission, but in Mg-22, positron emission is the only decay mode observed.)

(c) Molybdenum-102 has 60 neutrons and 42 protons, so $N/Z = 1.43$. However, for $Z = 42$, you can see from Figure 19.4 that stable nuclei should have an N/Z ratio of about 1.3. Alternatively, you can look up the atomic mass of molybdenum, which is 95.94 amu; a molybdenum nuclide with a mass number of 102 is too heavy to be stable (the N/Z ratio is too high). Therefore, Mo-102 undergoes *beta decay*, resulting in the conversion of a neutron to a proton.

FOR PRACTICE 19.3

Predict whether each nuclide is more likely to decay via beta decay or positron emission.

(a) Pb-192 **(b)** Pb-212 **(c)** Xe-114

Magic Numbers

In addition to the N/Z ratio, the *actual number* of protons and neutrons also affects the stability of the nucleus. Table 19.2 shows the number of nuclei with different possible combinations of even or odd nucleons. Notice that a large number of stable nuclides have an even number of protons and an even number of neutrons. Only five stable nuclides have an odd and odd combination.

The reason for this behavior is that nucleons occupy energy levels within the nucleus much as electrons occupy energy levels within an atom. Just as atoms with certain numbers of electrons are uniquely stable (in particular, the number of electrons associated with the noble gases: 2, 10, 18, 36, 54, etc.), so atoms with certain numbers of nucleons (N or $Z = 2, 8, 20, 28, 50, 82$, and $N = 126$) are uniquely stable. These numbers are often referred to as **magic numbers**. Nuclei containing a magic number of protons or neutrons are particularly stable. Since the magic numbers are even, this in part accounts for the abundance of stable nuclides with even numbers of nucleons. Moreover, nucleons also have a tendency to pair together (much as electrons pair together). This tendency and the resulting stability of paired nucleons also contribute to the abundance of stable nuclides with even numbers of nucleons.

TABLE 19.2 Number of Stable Nuclides with Even and Odd Numbers of Nucleons

Z	N	Number of Nuclides
Even	Even	157
Even	Odd	53
Odd	Even	50
Odd	Odd	5

Radioactive Decay Series

Atoms with $Z > 83$ are radioactive and decay in one or more steps involving mostly alpha and beta decay (with some gamma decay to carry away excess energy). For example, uranium (atomic number 92) is the heaviest naturally occurring element. Its most common isotope is U-238, an alpha emitter that decays to Th-234:

$$^{238}_{92}\text{U} \longrightarrow ^{234}_{90}\text{Th} + ^{4}_{2}\text{He}$$

The daughter nuclide, Th-234, is itself radioactive—it is a beta emitter that decays to Pa-234:

$$^{234}_{90}\text{Th} \longrightarrow {}^{234}_{91}\text{Pa} + {}^{0}_{-1}\text{e}$$

Protactinium-234 is also radioactive, decaying to U-234 via beta emission. Radioactive decay continues until a stable nuclide, Pb-206, is reached. The entire uranium-238 decay series is shown in **Figure 19.5▶**.

19.4 The Kinetics of Radioactive Decay and Radiometric Dating

Radioactivity is a natural component of our environment. The ground beneath you most likely contains radioactive atoms that emit radiation. The food you eat contains a residual quantity of radioactive atoms that are absorbed into your body fluids and incorporated into your tissues. Small amounts of radiation from space make it through our atmosphere and constantly bombard Earth. Humans and other living organisms have evolved in this environment and have adapted to survive in it.

One reason for the radioactivity in our environment is the instability of all atomic nuclei beyond atomic number 83 (bismuth). Every element with more than 83 protons in its nucleus is unstable and therefore radioactive. In addition, some isotopes of elements with fewer than 83 protons are also unstable and radioactive. Radioactive nuclides *persist* in our environment because new ones are constantly being formed, and because many of the existing ones decay only very slowly.

All radioactive nuclei decay via first-order kinetics, so the rate of decay in a particular sample is directly proportional to the number of nuclei present:

$$\text{Rate} = kN$$

where N is the number of radioactive nuclei and k is the rate constant. Different radioactive nuclides decay into their daughter nuclides with different rate constants. Some nuclides decay quickly while others decay slowly.

The time it takes for one-half of the parent nuclides in a radioactive sample to decay to the daughter nuclides is called the *half-life*, and is identical to the concept of half-life for chemical reactions that we covered in Chapter 13. Thus, the relationship between the half-life of a nuclide and its rate constant is given by the same expression (Equation 13.19) that we derived for a first-order reaction in Section 13.4:

$$t_{1/2} = \frac{0.693}{k} \qquad [19.1]$$

Nuclides that decay quickly have short half-lives and large rate constants—they are considered very active (many decay events per unit time). Nuclides that decay slowly have long half-lives and are less active (fewer decay events per unit time). For example, thorium-232 is an alpha emitter with a half-life of 1.4×10^{10} years, or 14 billion years. If we start with a sample of Th-232 containing one million atoms, the sample would decay to $^1/_2$ million atoms in 14 billion years and then to $^1/_4$ million in another 14 billion years and so on.

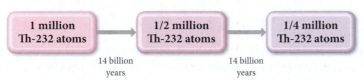

Notice that a radioactive sample does *not* decay to *zero* atoms in two half-lives—you can't add two half-lives together to get a "whole" life. The amount that remains after one half-life is always one-half of what was present at the start. The amount that remains after two half-lives is one-quarter of what was present at the start, and so on.

By contrast, radon-220 has a half-life of approximately 1 minute (**Figure 19.6▶**). If we had a 1-million-atom sample of radon-220, it would be diminished to $^1/_4$ million radon-220 atoms in just 2 minutes and to approximately 1000 atoms in 10 minutes. Table 19.3 lists several nuclides and their half-lives.

A Decay Series

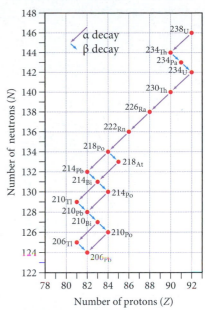

▲ **FIGURE 19.5 The Uranium-238 Radioactive Decay Series**
Uranium-238 decays via a series of steps ending in Pb-206, which is stable. Each diagonal line to the left represents an alpha decay and each diagonal line to the right represents a beta decay.

You may find it useful to review the discussion of first-order kinetics in Section 13.3.

TABLE 19.3 Selected Nuclides and Their Half-Lives

Nuclide	Half-life	Type of Decay
$^{232}_{90}\text{Th}$	1.4×10^{10} yr	Alpha
$^{238}_{92}\text{U}$	4.5×10^9 yr	Alpha
$^{14}_{6}\text{C}$	5730 yr	Beta
$^{220}_{86}\text{Rn}$	55.6 s	Alpha
$^{219}_{90}\text{Th}$	1.05×10^{-6} s	Alpha

▶ **FIGURE 19.6 The Decay of Radon-220** Radon-220 decays with a half-life of approximately 1 minute.

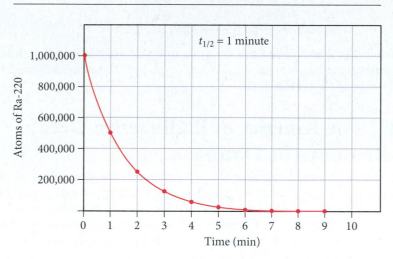

Decay of Radon-220

$t_{1/2} = 1$ minute

Conceptual Connection 19.2 Half-Life

Consider the graph representing the decay of a radioactive nuclide:

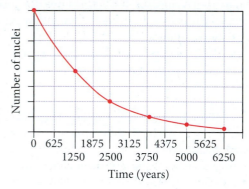

What is the half-life of the nuclide?

(a) 625 years (b) 1250 years (c) 2500 years (d) 3125 years

The Integrated Rate Law

Recall from Chapter 13 that for first-order chemical reactions, the concentration of a reactant as a function of time is given by the integrated rate law:

$$\ln \frac{[A]_t}{[A]_0} = -kt \qquad [19.2]$$

Since nuclear decay follows first-order kinetics, we can substitute the number of nuclei for concentration to arrive at the equation

$$\ln \frac{N_t}{N_0} = -kt \qquad [19.3]$$

where N_t is the number of radioactive nuclei at time t and N_0 is the initial number of radioactive nuclei. Example 19.4 demonstrates how to use this equation.

EXAMPLE 19.4 Radioactive Decay Kinetics

Plutonium-236 is an alpha emitter with a half-life of 2.86 years. If a sample initially contains 1.35 mg of Pu-236, what mass of Pu-236 is present after 5.00 years?

SORT You are given the initial mass of Pu-236 in a sample and asked to find the mass after 5.00 years.	**GIVEN** $m_{Pu-236}(\text{initial}) = 1.35$ mg; $t = 5.00$ yr; $t_{1/2} = 2.86$ yr
	FIND $m_{Pu-236}(\text{final})$

STRATEGIZE Use the integrated rate law (Equation 19.3) to solve this problem. However, you must determine the value of the rate constant (k) from the half-life expression (Equation 19.1).

Use the value of the rate constant, the initial mass of Pu-236, and the time along with integrated rate law to find the final mass of Pu-236. Since the mass of the Pu-236 (m_{Pu-236}) is directly proportional to the number of atoms (N), and since the integrated rate law contains the ratio (N_t/N_0), the initial and final masses can be substituted for the initial and final number of atoms.

CONCEPTUAL PLAN

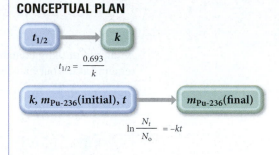

$$t_{1/2} = \frac{0.693}{k}$$

$$\ln \frac{N_t}{N_0} = -kt$$

SOLVE Follow your plan. Begin by finding the rate constant from the half-life.

SOLUTION

$$t_{1/2} = \frac{0.693}{k}$$

$$k = \frac{0.693}{t_{1/2}} = \frac{0.693}{2.86 \text{ yr}}$$

$$= 0.2423/\text{yr}$$

Solve the integrated rate law for N_t and substitute the values of the rate constant, the initial mass of Pu-236, and the time into the solved equation. Calculate the final mass of Pu-236.

$$\ln \frac{N_t}{N_0} = -kt$$

$$\frac{N_t}{N_0} = e^{-kt}$$

$$N_t = N_0 e^{-kt}$$

$$N_t = 1.35 \text{ mg} \left[e^{-(0.2423/\text{yr})(5.00\,\text{yr})} \right]$$

$$N_t = 0.402 \text{ mg}$$

CHECK The units of the answer (mg) are correct. The magnitude of the answer (0.402 mg) is about one-third of the original mass (1.35 mg) which seems reasonable given that the amount of time is between one and two half-lives. (One half-life would result in one-half of the original mass and two half-lives would result in one-fourth of the original mass.)

FOR PRACTICE 19.4

How long does it take for the 1.35-mg sample of Pu-236 in Example 19.4 to decay to 0.100 mg?

Since radioactivity is a first-order process, the rate of decay is linearly proportional to the number of nuclei in the sample. As shown below, the initial rate of decay ($rate_0$) and the rate of decay at time t ($rate_t$) can also be used in the integrated rate law.

$$\text{Rate}_t = kN_t \qquad \text{Rate}_0 = kN_0$$

$$\frac{N_t}{N_0} = \frac{\text{rate}_t/k}{\text{rate}_0/k} = \frac{\text{rate}_t}{\text{rate}_0}$$

Substituting into Equation 19.3, we get the result:

$$\ln \frac{\text{rate}_t}{\text{rate}_0} = -kt \qquad\qquad [19.4]$$

We can use Equation 19.4 to predict how the rate of decay of a radioactive sample will change with time or how much time has passed based on how the rate has changed (see the examples later in this section).

We can use the presence of radioactive isotopes in our environment, and their predictable decay with time, to estimate the age of rocks or artifacts containing those isotopes. The technique is known as **radiometric dating**, and next we examine two different types individually.

Conceptual Connection 19.3   Half Life and Amount of Radioactive Sample

A sample initially contains 1.6 moles of a radioactive isotope. How much of the isotope remains after four half-lives?

(a) 0.0 mol   **(b)** 0.10 mol   **(c)** 0.20 mol   **(d)** 0.40 mol

▲ The Dead Sea Scrolls are 2000-year-old biblical manuscripts. Their age was determined by radiocarbon dating.

Libby received the Nobel Prize in 1960 for the development of radiocarbon dating.

Radiocarbon Dating: Using Radioactivity to Measure the Age of Fossils and Artifacts

Archeologists, geologists, anthropologists, and other scientists use **radiocarbon dating**, a technique devised in 1949 by Willard Libby at the University of Chicago, to estimate the ages of fossils and artifacts. For example, in 1947, young shepherds searching for a stray goat near the Dead Sea (east of Jerusalem) entered a cave and discovered ancient scrolls stuffed into jars. These scrolls—now named the Dead Sea Scrolls—are 2000-year-old texts of the Hebrew Bible, predating other previously known manuscripts by almost a thousand years.

The Dead Sea Scrolls, like other ancient artifacts, contain a radioactive signature that reveals their age. This signature results from the presence of carbon-14 (which is radioactive) in the environment. Carbon-14 is constantly formed in the upper atmosphere by the neutron bombardment of nitrogen:

$$^{14}_{7}\text{N} + ^{1}_{0}\text{n} \longrightarrow ^{14}_{6}\text{C} + ^{1}_{1}\text{H}$$

Carbon-14 then decays back to nitrogen by beta emission with a half-life of 5730 years.

$$^{14}_{6}\text{C} \longrightarrow ^{14}_{7}\text{N} + ^{0}_{-1}\text{e} \qquad t_{1/2} = 5730 \text{ yr}$$

The continuous formation of carbon-14 in the atmosphere and its continuous decay back to nitrogen-14 produces a nearly constant equilibrium amount of atmospheric carbon-14, which is oxidized to carbon dioxide and incorporated into plants by photosynthesis. The C-14 then makes its way up the food chain and ultimately into all living organisms. As a result, living plants, animals, and humans contain the same ratio of carbon-14 to carbon-12 ($^{14}\text{C}:^{12}\text{C}$) as is found in the atmosphere. When a living organism dies, however, it stops incorporating new carbon-14 into its tissues. The $^{14}\text{C}:^{12}\text{C}$ ratio then decreases with a half-life of 5730 years. Since many artifacts, including the Dead Sea Scrolls, are made from materials that were once living—such as papyrus, wood, or other plant and animal derivatives—the $^{14}\text{C}:^{12}\text{C}$ ratio in these artifacts indicates their age. For example, suppose an ancient artifact has a $^{14}\text{C}:^{12}\text{C}$ ratio that is 25% of that found in living organisms. How old is the artifact? Since it contains one-quarter as much carbon-14 as a living organism, it must be two half-lives or 11,460 years old. The maximum age that can be estimated from carbon-14 dating is about 50,000 years—beyond that, the amount of carbon-14 becomes too low to measure accurately.

The accuracy of carbon-14 dating can be checked against objects whose ages are known from historical sources. These kinds of comparisons have revealed that ages obtained from C-14 dating can deviate from the actual ages by up to about 5%. For a 6000-year-old object, that would result in an error of about 300 years. The reason for the deviations is the variance of atmospheric C-14 levels over time.

In order to make C-14 dating more accurate, scientists have studied the carbon-14 content of western bristlecone pine trees, which can live up to 5000 years. The tree trunk contains growth rings corresponding to each year of the tree's life, and the wood laid down in each ring incorporates carbon derived from the carbon dioxide in the atmosphere at that time. The rings thus provide a record of the historical atmospheric carbon-14 content. In addition, the rings of living trees can be correlated with the rings of dead trees (if part of the lifetimes of the trees overlapped), allowing the record to be extended back about 11,000 years. Using the data from the bristlecone pine, the 5% deviations from historical dates can be corrected. In effect, the known ages of bristlecone pine trees are used to calibrate C-14 dating, resulting in more accurate results.

▲ Western bristlecone pine trees can live up to 5000 years, and their age can be precisely determined by counting the annual rings in their trunks. They can therefore be used to calibrate the timescale for radiocarbon dating.

EXAMPLE 19.5 Radiocarbon Dating

A skull believed to belong to an ancient human being is found to have a carbon-14 decay rate of 4.50 disintegrations per minute per gram of carbon (4.50 dis/min · g C). If living organisms have a decay rate 15.3 dis/min · g C, how old is the skull? (The decay rate is directly proportional to the amount of carbon-14 present.)

SORT You are given the current rate of decay for the skull and the assumed initial rate. You are asked to find the age of the skull, which is the time that passed in order for the rate to have reached its current value.	**GIVEN** $\text{rate}_t = 4.50 \text{ dis/min} \cdot \text{g C}$; $\quad\quad\quad \text{rate}_0 = 15.3 \text{ dis/min} \cdot \text{g C}$; **FIND** t
STRATEGIZE Use the expression for half-life (Equation 19.1) to find the rate constant (k) from the half-life for C-14, which is 5730 yr (Table 19.3).	**CONCEPTUAL PLAN** 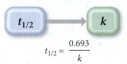 $$t_{1/2} = \frac{0.693}{k}$$
Use the value of the rate constant and the initial and current rates to find t from the integrated rate law (Equation 19.4).	$$\ln \frac{\text{rate}_t}{\text{rate}_0} = -kt$$
SOLVE Follow your plan. Begin by finding the rate constant from the half-life.	**SOLUTION** $$t_{1/2} = \frac{0.693}{k}$$ $$k = \frac{0.693}{t_{1/2}} = \frac{0.693}{4.5 \times 10^9 \text{ yr}}$$ $$= 1.209 \times 10^{-4}/\text{yr}$$
Substitute the rate constant and the initial and current rates into the integrated rate law and solve for t.	$$\ln \frac{\text{rate}_t}{\text{rate}_0} = -kt$$ $$t = -\frac{\ln \dfrac{\text{rate}_t}{\text{rate}_0}}{k} = -\frac{\ln \dfrac{4.50 \text{ dis/min} \cdot \text{g C}}{15.3 \text{ dis/min} \cdot \text{g C}}}{1.209 \times 10^{-4}/\text{yr}}$$ $$= 1.01 \times 10^4 \text{ yr}$$

CHECK The units of the answer (yr) are correct. The magnitude of the answer is about 10,000 years, which is a little less than two half-lives. This value is reasonable given that two half-lives would result in a decay rate of about 3.8 dis/min · g C.

FOR PRACTICE 19.5

An ancient scroll allegedly originates from Greek scholars in about 500 B.C. A measure of its carbon-14 decay rate gives a value that is 89% of that found in living organisms. How old is the scroll and could it be authentic?

Uranium/Lead Dating

Radiocarbon dating is limited to measuring the ages of objects that were once living and that are relatively young (<50,000 years). Other radiometric dating techniques can measure the ages of prehistoric objects that were never alive. The most dependable technique relies on the ratio of uranium-238 to lead-206 within igneous rocks (rocks of volcanic origin). This technique measures the time that has passed since the rock solidified (at which point the "radiometric clock" was reset).

Since U-238 decays into Pb-206 with a half-life of 4.5×10^9 years, the relative amounts of U-238 and Pb-206 in a uranium-containing rock reveal its age. For example, if a rock originally contained U-238 and currently contains equal amounts of U-238 and Pb-206, it would be 4.5 billion years old, assuming that the rock did not contain any Pb-206 when it was formed. The latter assumption can be tested because the lead that results from the decay of uranium has a different isotopic composition than the lead that would have been deposited in rocks at the time of their formation. Example 19.6 shows how the relative amounts of Pb-206 and U-238 in a rock are used to estimate its age.

EXAMPLE 19.6 Using Uranium/Lead Dating to Estimate the Age of a Rock

A meteor contains 0.556 g of Pb-206 to every 1.00 g of U-238. Assuming that the meteor did not contain any Pb-206 at the time of its formation, determine the age of the meteor. Uranium-238 decays to lead-206 with a half-life of 4.5 billion years.

SORT You are given the current masses of Pb-206 and U-238 in a rock and asked to find its age. You are also given the half-life of U-238.	**GIVEN** $m_{U\text{-}238} = 1.00$ g; $m_{Pb\text{-}206} = 0.556$ g; $t_{1/2} = 4.5 \times 10^9$ yr **FIND** t

STRATEGIZE Use the integrated rate law (Equation 19.3) to solve this problem. However, you must first determine the value of the rate constant (k) from the half-life expression (Equation 19.1).

Before substituting into the integrated rate law, you will also need the ratio of the current amount of U-238 to the original amount (N_t/N_0). The current mass of uranium is simply 1.00 g. The initial mass includes the current mass (1.00 g) plus the mass that has decayed into lead-206, which can be determined from the current mass of Pb-206.

Use the value of the rate constant and the initial and current amounts of U-238 along with integrated rate law to find t.

CONCEPTUAL PLAN

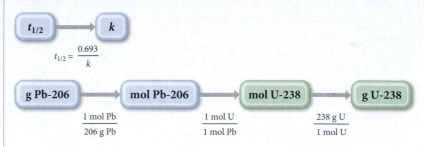

$$\ln \frac{N_t}{N_0} = -kt$$

SOLVE Follow your plan. Begin by finding the rate constant from the half-life.

$$t_{1/2} = \frac{0.693}{k}$$

$$k = \frac{0.693}{t_{1/2}} = \frac{0.693}{4.5 \times 10^9 \text{ yr}} = 1.54 \times 10^{-10}/\text{yr}$$

Determine the mass in grams of U-238 that would have been required to form the given mass of Pb-206.

$$0.556 \text{ g Pb-206} \times \frac{1 \text{ mol Pb-206}}{206 \text{ g Pb-206}} \times \frac{1 \text{ mol U-238}}{1 \text{ mol Pb-206}} \times \frac{238 \text{ g U-238}}{1 \text{ mol U-238}}$$
$$= 0.6424 \text{ g U-238}$$

Substitute the rate constant and the initial and current masses of U-238 into the integrated rate law and solve for t. (The initial mass of U-238 is the sum of the current mass and the mass that would have been required to form the given mass of Pb-206.)

$$\ln \frac{N_t}{N_0} = -kt$$

$$t = -\frac{\ln \frac{N_t}{N_0}}{k} = -\frac{\ln \frac{1.00 \text{ g}}{1.00 \text{ g} + 0.6424 \text{ g}}}{1.54 \times 10^{-10}/\text{yr}}$$
$$= 3.2 \times 10^9 \text{ yr}$$

CHECK The units of the answer are correct. The magnitude of the answer is about 3.2 billion years, which is less than one half-life. This value is reasonable given that less than half of the uranium has decayed into lead.

FOR PRACTICE 19.6

A rock contains Pb-206 to U-238 mass ratio of 0.145 : 1.00. Assuming that the rock did not contain any Pb-206 at the time of its formation, determine the age of the rock.

The Age of Earth

The uranium/lead radiometric dating technique as well as other radiometric dating techniques (such as the decay of potassium-40 to argon-40) have been widely used to measure the ages of rocks on Earth and have produced highly consistent results. Rocks with ages greater than 3.5 billion years have been found on every continent. The oldest rocks have an age of approximately 4.0 billion years, establishing a lower bound for Earth's age (Earth must be at least as old as its oldest rocks). The ages of about 70 meteorites that have struck Earth have also been extensively studied and have been found to be about 4.5 billion years old. Since the meteorites were formed at the same time as our solar system (which includes Earth), the best estimate for Earth's age is therefore about 4.5 billion years. That age is consistent with the estimated age of our universe—about 13.7 billion years.

> The age of the universe is estimated from its expansion rate, which can be measured by examining changes in the wavelength of light from distant galaxies.

19.5   The Discovery of Fission: The Atomic Bomb and Nuclear Power

In the mid-1930s Enrico Fermi (1901–1954), an Italian physicist, tried to synthesize a new element by bombarding uranium—the heaviest known element at that time—with neutrons. Fermi hypothesized that if a neutron were to be incorporated into the nucleus of a uranium atom, the nucleus might undergo beta decay, converting a neutron into a proton. If that happened, a new element, with atomic number 93, would be synthesized for the first time. The nuclear equation for the process is

> The element with atomic number 100 is named fermium in honor of Enrico Fermi.

$$\underset{\text{Neutron}}{{}^{238}_{92}\text{U} + {}^{1}_{0}\text{n}} \longrightarrow {}^{239}_{92}\text{U} \longrightarrow \underset{\text{Newly synthesized element}}{{}^{239}_{93}\text{X} + {}^{0}_{-1}\text{e}}$$

Fermi performed the experiment and detected the emission of beta particles. However, his results were inconclusive. Had he synthesized a new element? Fermi never chemically examined the products to determine their composition and therefore could not say with certainty that he had.

Subsequently, three researchers in Germany—Lise Meitner (1878–1968), Fritz Strassmann (1902–1980), and Otto Hahn (1879–1968)—repeated Fermi's experiments, and then performed careful chemical analysis of the products. What they found in the products—several elements *lighter* than uranium—would change the world forever. On January 6, 1939, Meitner, Strassmann, and Hahn reported that the neutron bombardment of uranium resulted in **nuclear fission**—the splitting of the uranium atom. The nucleus of the neutron-bombarded uranium atom had been split into barium, krypton, and other smaller products. They also noted that the process emitted enormous amounts of energy. A nuclear equation for a fission reaction, showing how uranium breaks apart into the daughter nuclides, is shown here.

$${}^{235}_{92}\text{U} + {}^{1}_{0}\text{n} \longrightarrow {}^{140}_{56}\text{Ba} + {}^{93}_{36}\text{Kr} + 3\,{}^{1}_{0}\text{n} + \text{energy}$$

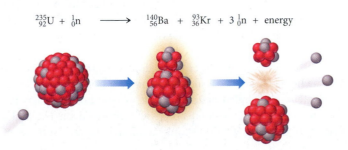

Notice that the initial uranium atom is the U-235 isotope, which constitutes less than 1% of all naturally occurring uranium. U-238, the most abundant uranium isotope, does not undergo fission. Notice also that the process produces three neutrons, which have the potential to initiate fission in three other U-235 atoms.

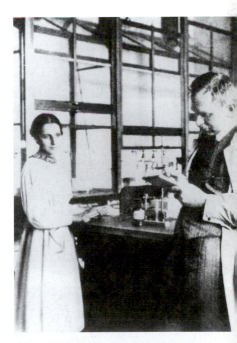

▲ Lise Meitner in Otto Hahn's Berlin laboratory. Together with Hahn and Fritz Strassmann, Meitner determined that U-235 could undergo nuclear fission. The element with atomic number 109 is named meitnerium in honor of Lise Meitner.

▶ **FIGURE 19.7 A Self-Amplifying Chain Reaction**   The fission of one U-235 nucleus emits neutrons which can then initiate fission in other U-235 nuclei, resulting in a chain reaction that can release enormous amounts of energy.

Fission Chain Reaction

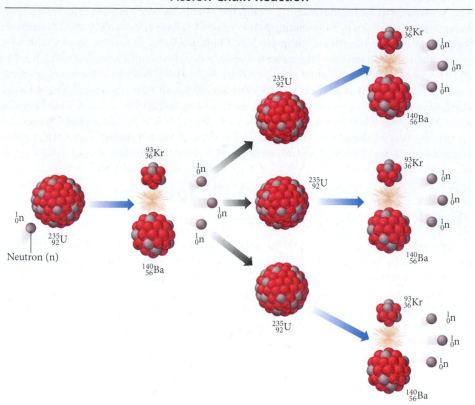

▲ On July 16, 1945, in the New Mexico desert, the world's first atomic bomb was detonated. It had the power of 18,000 tons of dynamite.

Scientists quickly realized that a sample rich in U-235 could undergo a **chain reaction** in which neutrons produced by the fission of one uranium nucleus would induce fission in other uranium nuclei (**Figure 19.7 ▲**) . The result would be a self-amplifying reaction capable of producing an enormous amount of energy—an atomic bomb. However, to make a bomb, a **critical mass** of U-235—enough U-235 to produce a self-sustaining reaction—would be necessary. Fearing that Nazi Germany would develop such a bomb, several U.S. scientists persuaded Albert Einstein, the most famous scientist of the time, to write a letter to President Franklin Roosevelt warning of this possibility. Einstein wrote, "… and it is conceivable—though much less certain—that extremely powerful bombs of a new type may thus be constructed. A single bomb of this type, carried by boat and exploded in a port, might very well destroy the whole port together with some of the surrounding territory."

Roosevelt was convinced by Einstein's letter, and in 1941 he assembled the resources to begin the costliest scientific project ever attempted. The top-secret endeavor was called the *Manhattan Project* and its main goal was to build an atomic bomb before the Germans did. The project was led by physicist J. R. Oppenheimer (1904–1967) at a high-security research facility in Los Alamos, New Mexico. Four years later, on July 16, 1945, the world's first nuclear weapon was successfully detonated at a test site in New Mexico. The first atomic bomb exploded with a force equivalent to 18,000 tons of dynamite. Ironically, the Germans—who had *not* made a successful nuclear bomb—had already been defeated by this time. Instead, the United States used the atomic bomb on Japan. One bomb was dropped on Hiroshima and a second bomb was dropped on Nagasaki. Together, the bombs killed approximately 200,000 people and led to Japan's surrender.

Nuclear Power: Using Fission to Generate Electricity

Nuclear reactions, such as fission, generate enormous amounts of energy. In a nuclear bomb, the energy is released all at once. However, the energy can also be released more slowly and used for peaceful purposes such as electricity generation. In the United States, about 20% of electricity is generated by nuclear fission. In some other countries, as much as 70% of electricity is generated by nuclear fission. To get an idea of the amount of energy released during fission, imagine a hypothetical nuclear-powered car. Suppose the fuel for such a car was a uranium cylinder about the size of a pencil. The energy content of the uranium cylinder would be equivalent to about 1000 twenty-gallon tanks of gasoline. If you refuel your gasoline-powered car once a week, your nuclear-powered car could go 1000 weeks—almost 20 years—before refueling.

Similarly, a nuclear-powered electricity generation plant can produce a lot of electricity from a small amount of fuel. Such plants exploit the heat created by fission. The heat is used to boil water and make steam, which then turns the turbine on a generator to produce electricity (**Figure 19.8 ▼**). The fission reaction itself occurs in the nuclear core of the power plant. The core consists of uranium fuel rods—enriched to about 3.5% U-235—interspersed between retractable neutron-absorbing control rods. When the control rods are fully retracted from the fuel rod assembly, the chain reaction can occur. When the control rods are fully inserted into the fuel assembly, however, they absorb the neutrons that would otherwise induce fission, shutting down the chain reaction. By retracting or inserting the control rods, the operator can increase or decrease the rate of fission. In this way, the fission reaction is controlled to produce the right amount of heat needed for electricity generation. In case of a power failure, the control rods automatically drop into the fuel rod assembly, shutting down the fission reaction.

A typical nuclear power plant generates enough electricity for a city of about 1 million people and uses about 50 kg of fuel per day. In contrast, a coal-burning power plant uses about 2,000,000 kg of fuel to generate the same amount of electricity. Furthermore, a nuclear power plant generates no air pollution and no greenhouse gases. A coal-burning power plant emits pollutants such as carbon monoxide, nitrogen oxides, and sulfur oxides. Coal-burning power plants also emit carbon dioxide, a greenhouse gas.

Nuclear power generation, however, is not without problems. Foremost among them is the danger of nuclear accidents. In spite of safety precautions, the fission reaction occurring in nuclear power plants can overheat. The most famous examples of this occurred in Chernobyl, in the former Soviet Union, on April 26, 1986, and at the Fukushima Daiichi Nuclear Power Plant in Japan in March of 2011.

In the Chernobyl incident, operators of the plant were performing an experiment designed to reduce maintenance costs. In order to perform the experiment, many of the safety features of the reactor core were disabled. The experiment failed with disastrous results. The nuclear core, composed partly of graphite, overheated and began to burn. The accident caused 31 immediate deaths and produced a fire that scattered radioactive debris into the atmosphere, making much of the surrounding land (within about a 32-kilometer

Reactor cores in the United States are not made of graphite and could not burn in the way that the Chernobyl core did.

Nuclear Reactor

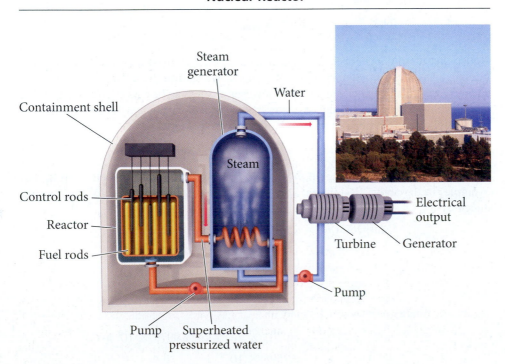

◀ **FIGURE 19.8 A Nuclear Reactor**
The fission of U-235 in the core of a nuclear power plant generates heat that creates steam, which is used to turn a turbine on an electrical generator. The control rods are raised or lowered to control the fission reaction.

▲ In 1986, the reactor core at Chernobyl (in what is now Ukraine) overheated, exploded, and destroyed part of the containment structure. The release of radioactive nuclides into the environment forced the government to relocate over 335,000 people. It is estimated that there may eventually be several thousand additional cancer deaths among the exposed populations.

radius) uninhabitable. It is important to understand, however, that a nuclear power plant *cannot* become a nuclear bomb. The uranium fuel used in electricity generation is not sufficiently enriched in U-235 to produce a nuclear detonation.

In the Japanese accident, a 9.0 magnitude earthquake triggered a tsunami that flooded the coastal plant and caused the cooling system pumps to fail. Several of the nuclear cores within the plant dramatically overheated and at least one of the cores experienced a partial meltdown (in which the fuel gets so hot that it melts). The accident was intensified by the loss of water in the fuel storage ponds (pools of water used to keep spent and future fuel cool), which caused the fuel stored in the ponds to also overheat. The release of radiation into the environment, while significant, was lower than that of Chernobyl. As of press time, no radioactivity-related deaths have been reported at the Fukushima plant. The cleanup of the site, however, will continue for many years.

A second problem associated with nuclear power is waste disposal. Although the amount of fuel used in electricity generation is small compared to other fuels, the products of the reaction are radioactive and have long half-lives. What do we do with this waste? Currently, in the United States, nuclear waste is stored on site at the nuclear power plants. A single permanent disposal site was being developed in Yucca Mountain, Nevada; however, in the Spring of 2010, the Obama administration halted the development of this project. The administration formed the Blue Ribbon Commission on America's Nuclear Future to explore alternatives.

19.6 Converting Mass to Energy: Mass Defect and Nuclear Binding Energy

Nuclear fission produces large amounts of energy. But where does the energy come from? We can answer this question by carefully examining the masses of the reactants and products in the fission equation from Section 19.5.

$$^{235}_{92}U + ^{1}_{0}n \longrightarrow ^{140}_{56}Ba + ^{93}_{36}Kr + 3\,^{1}_{0}n$$

	Mass Reactants		Mass Products
$^{235}_{92}U$	235.04392 amu	$^{140}_{56}Ba$	139.910581 amu
$^{1}_{0}n$	1.00866 amu	$^{93}_{36}Kr$	92.931130 amu
		$3\,^{1}_{0}n$	3(1.00866) amu
Total	**236.05258 amu**		**235.86769 amu**

In a chemical reaction, there are also mass changes associated with the emission or absorption of energy. Because the energy involved in chemical reactions is so much smaller than that of nuclear reactions, however, these mass changes are completely negligible.

Notice that the products of the nuclear reaction have less mass than the reactants. The missing mass is converted to energy. In Chapter 1, we learned that matter is conserved in chemical reactions. In nuclear reactions, however, matter can be converted to energy. The relationship between the amount of matter that is lost and the amount of energy formed is given by Einstein's famous equation relating the two quantities,

$$E = mc^2$$

where E is the energy produced, m is the mass lost, and c is the speed of light. For example, in the above fission reaction, we can calculate the quantity of energy produced as

$$\text{Mass lost } (m) = 236.05258 \text{ amu} - 235.86769 \text{ amu}$$

$$= 0.18489 \text{ amu} \times \frac{1.66054 \times 10^{-27} \text{ kg}}{1 \text{ amu}}$$

$$= 3.0702 \times 10^{-28} \text{ kg}$$

$$\text{Energy produced } (E) = mc^2$$

$$= 3.0702 \times 10^{-28} \text{ kg } (2.9979 \times 10^8 \text{ m/s})^2$$

$$= 2.7593 \times 10^{-11} \text{ J}$$

The result is the energy produced when one nucleus of U-235 undergoes fission. To compare to a chemical reaction, we can calculate the energy produced *per mole* of U-235:

$$2.7593 \times 10^{-11} \frac{J}{\text{U-235 atom}} \times \frac{6.0221 \times 10^{23} \text{ U-235 atom}}{1 \text{ mol U-235}}$$

$$= 1.6617 \times 10^{13} \text{ J/mol U-235}$$

The energy produced by the fission of 1 mol of U-235 is therefore about 17 billion kJ. In contrast, a highly exothermic chemical reaction might produce 1000 kJ per mole of reactant. Fission produces over a million times more energy per mole than chemical processes.

Mass Defect

We can view the formation of a stable nucleus from its component particles as a nuclear reaction in which mass is converted to energy. For example, consider the formation of helium-4 from its components:

$$2 \, {}_{1}^{1}\text{H} + 2 \, {}_{0}^{1}\text{n} \longrightarrow {}_{2}^{4}\text{He}$$

	Mass Reactants		Mass Products
$2 \, {}_{1}^{1}\text{H}$	2(1.00783) amu	${}_{2}^{4}\text{He}$	4.00260 amu
$2 \, {}_{0}^{1}\text{n}$	2(1.00866) amu		
Total	**4.03298 amu**		**4.00260 amu**

A helium-4 atom has less mass than the sum of the masses of its separate components. This difference in mass, known as the **mass defect**, exists in all stable nuclei. The energy corresponding to the mass defect—obtained by substituting the mass defect into the equation $E = mc^2$—is the **nuclear binding energy**, the amount of energy that would be required to break apart the nucleus into its component nucleons.

Although chemists typically report energies in joules, nuclear physicists often use the electron volt (eV) or megaelectron volt (MeV): 1 MeV = 1.602×10^{-13} J. Unlike energy in joules, which is usually reported per mole, energy in electron volts is usually reported per nucleus. A particularly useful conversion for calculating and reporting nuclear binding energies is the relationship between amu (mass units) and MeV (energy units):

$$1 \text{ amu} = 931.5 \text{ MeV}$$

In other words, a mass defect of 1 amu, when substituted into the equation $E = mc^2$, gives an energy of 931.5 MeV. Using this conversion factor, we can readily calculate the binding energy of the helium nucleus:

$$\text{Mass defect} = 4.03298 \text{ amu} - 4.00260 \text{ amu}$$

$$= 0.03038 \text{ amu}$$

$$\text{Nuclear binding energy} = 0.03038 \text{ amu} \times \frac{931.5 \text{ MeV}}{1 \text{ amu}}$$

$$= 28.30 \text{ MeV}$$

The binding energy of the helium nucleus is 28.30 MeV. In order to compare the binding energy of one nucleus to that of another, we calculate the *binding energy per nucleon*, which is the nuclear binding energy of a nuclide divided by the number of nucleons in the nuclide. For helium-4, we calculate the binding energy per nucleon as follows:

$$\text{Binding energy per nucleon} = \frac{28.30 \text{ MeV}}{4 \text{ nucleons}}$$

$$= 7.075 \text{ MeV per nucleon}$$

The electrons are contained on the left side in the two ${}_{1}^{1}\text{H}$, and on the right side in ${}_{2}^{4}\text{He}$. If you write the equation using only two protons on the left (${}_{1}^{1}\text{p}$), you must also add two electrons to the left.

An electron volt is defined as the kinetic energy of an electron that has been accelerated through a potential difference of 1 V.

We can calculate the binding energy per nucleon for other nuclides in the same way. For example, the nuclear binding energy of carbon-12 is 7.680 MeV per nucleon. Since the binding energy per nucleon of carbon-12 is greater than that of helium-4, we can conclude the carbon-12 nuclide is more *stable* (it has lower potential energy).

EXAMPLE 19.7 Calculating Mass Defect and Nuclear Binding Energy

Calculate the mass defect and nuclear binding energy per nucleon (in MeV) for C-16, a radioactive isotope of carbon with a mass of 16.014701 amu.

SOLUTION

Calculate the mass defect as the difference between the mass of one C-16 atom and the sum of the masses of 6 hydrogen atoms and 10 neutrons.	Mass defect = $6(\text{mass } {}^{1}_{1}\text{H}) + 10(\text{mass } {}^{1}_{0}\text{n}) - \text{mass } {}^{16}_{6}\text{C}$ $= 6(1.00783) \text{ amu} + 10(1.00866) \text{ amu} - 16.014701 \text{ amu}$ $= 0.118879 \text{ amu}$
Calculate the nuclear binding energy by converting the mass defect (in amu) into MeV. (Use 1 amu = 931.5 MeV.)	$0.118879 \text{ amu} \times \dfrac{931.5 \text{ MeV}}{\text{amu}} = 110.74 \text{ MeV}$
Calculate the nuclear binding energy per nucleon by dividing by the number of nucleons in the nucleus.	Nuclear binding energy per nucleon $= \dfrac{110.74 \text{ MeV}}{16 \text{ nucleons}}$ $= 6.921 \text{ MeV/nucleon}$

FOR PRACTICE 19.7

Calculate the mass defect and nuclear binding energy per nucleon (in MeV) for U-238, which has a mass of 238.050784 amu.

Figure 19.9▼ shows the binding energy per nucleon plotted as a function of mass number (*A*). Notice that the binding energy per nucleon is relatively low for small mass numbers and increases until about *A* = 60, where it reaches a maximum. This means that nuclides with mass numbers of about 60 are among the most stable. Beyond *A* = 60, the binding energy per nucleon decreases again. Figure 19.9 clearly shows why nuclear fission is a highly exothermic process. When a heavy nucleus, such as U-235, breaks up into smaller nuclei, such as Ba-140 and Kr-93, the binding energy per nucleon increases.

The Curve of Binding Energy

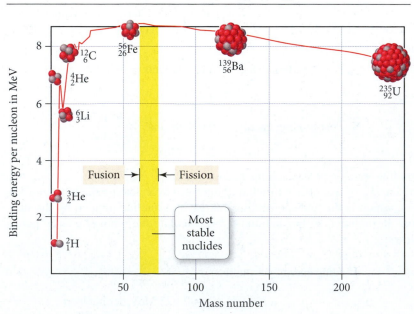

▶ FIGURE 19.9 **Nuclear Binding Energy per Nucleon** The nuclear binding energy per nucleon (a measure of the stability of a nucleus) reaches a maximum at Fe-56. Energy can be obtained either by breaking a heavy nucleus up into lighter ones (fission) or by combining lighter nuclei into heavier ones (fusion).

This is analogous to a chemical reaction in which weak bonds break and strong bonds form. In both cases, the process is exothermic. Notice also, however, that Figure 19.9 reveals that the combining of two lighter nuclei (below $A = 60$) to form a heavier nucleus should also be exothermic. This process is called *nuclear fusion, which we discuss in more detail in the next section of the chapter.*

19.7   Nuclear Fusion: The Power of the Sun

Nuclear fission is the *splitting* of a heavy nucleus to form two or more lighter ones. **Nuclear fusion**, by contrast, is the *combination* of two light nuclei to form a heavier one. Both fusion and fission emit large amounts of energy because, as we have just seen in Section 19.6, they both form daughter nuclides with greater binding energies per nucleon than the parent nuclides. Nuclear fusion is the energy source of stars, including our sun. In stars, hydrogen atoms fuse together to form helium atoms, emitting energy in the process.

Nuclear fusion is also the basis of modern nuclear weapons called hydrogen bombs, which have up to 1000 times the explosive force of the first atomic bombs. Hydrogen bombs employ the following fusion reaction:

$$^2_1H + ^3_1H \longrightarrow ^4_2He + ^1_0n$$

In this reaction, deuterium (the isotope of hydrogen with one neutron) and tritium (the isotope of hydrogen with two neutrons) combine to form helium-4 and a neutron (**Figure 19.10▶**). Because fusion reactions require two positively charged nuclei (which repel each other) to fuse together, extremely high temperatures are required. In a hydrogen bomb, a small fission bomb is detonated first, creating temperatures high enough for fusion to proceed.

Nuclear fusion has been intensely investigated as a way to produce electricity. Because of the higher energy potential—fusion provides about 10 times more energy per gram of fuel than does fission—and because the products of the reaction are less problematic than those of fission, fusion holds promise as a future energy source. However, in spite of intense efforts, the generation of electricity by fusion remains elusive. One of the main problems is the high temperature required for fusion to occur—no material can withstand those temperatures. Using powerful magnetic fields or laser beams, scientists have succeeded in compressing and heating nuclei to the point where fusion has been initiated, and even sustained for brief periods of time (**Figure 19.11▶**). To date, however, the amount of energy generated by fusion reactions has been less than the amount required to get it to occur. After years of spending billions of dollars on fusion research, the U.S. Congress has reduced funding for these projects. Whether fusion will ever be a viable energy source remains uncertain.

19.8   The Effects of Radiation on Life

As we discussed in Section 19.2, the energy associated with radioactivity can ionize molecules. When radiation ionizes important molecules in living cells, problems can develop. The ingestion of radioactive materials—especially alpha and beta emitters—is particularly dangerous because the radioactivity once inside the body can do substantial damage. The effects of radiation can be divided into three different types: acute radiation damage, increased cancer risk, and genetic effects.

Acute Radiation Damage

Acute radiation damage results from exposure to large amounts of radiation in a short period of time. The main sources of this kind of exposure are nuclear bombs and exposed nuclear reactor cores. These high levels of radiation kill large numbers of cells. Cells that divide rapidly, such as those in the immune system and the intestinal

Deuterium-Tritium Fusion Reaction

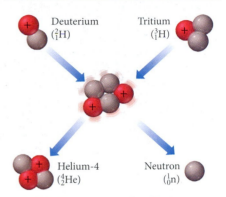

▲ **FIGURE 19.10 A Nuclear Fusion Reaction**   In this particular reaction, two heavy isotopes of hydrogen, deuterium (hydrogen-2) and tritium (hydrogen-3), fuse to form helium-4 and a neutron.

Tokamak Fusion Reactor

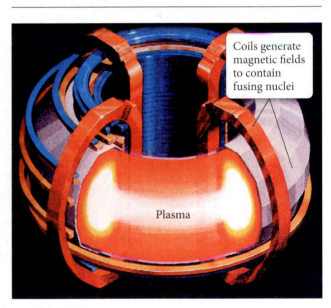

Coils generate magnetic fields to contain fusing nuclei

Plasma

▲ **FIGURE 19.11 Tokamak Fusion Reactor**   A tokamak uses powerful magnetic fields to confine nuclear fuel at the enormous temperatures needed for fusion.

lining, are most susceptible. Consequently, people exposed to high levels of radiation have weakened immune systems and a lowered ability to absorb nutrients from food. In milder cases, recovery is possible with time. In more extreme cases, death results, often from unchecked infection.

Increased Cancer Risk

Lower doses of radiation over extended periods of time can increase cancer risk. Radiation increases cancer risk because it can damage DNA, the molecules in cells that carry instructions for cell growth and replication. When the DNA within a cell is damaged, the cell normally dies. Occasionally, however, changes in DNA cause cells to grow abnormally and to become cancerous. These cancerous cells grow into tumors that can spread and, in some cases, cause death. Cancer risk increases with increasing radiation exposure. However, cancer is so prevalent and can be attributed to so many causes that it is difficult to determine an exact threshold for increased cancer risk from radiation exposure.

Genetic Defects

Another possible effect of radiation exposure is genetic defects in offspring. If radiation damages the DNA of reproductive cells—such as eggs or sperm—then the offspring that develops from those cells may have genetic abnormalities. Genetic defects of this type have been observed in laboratory animals exposed to high levels of radiation. However, such genetic defects—with a clear causal connection to radiation exposure—have yet to be verified in humans, even in studies of Hiroshima survivors.

Measuring Radiation Exposure

We can measure radiation exposure in a number of different ways. One method is to simply measure the number of decay events to which a person is exposed. The unit used for this type of exposure measurement is called the *curie* (Ci), defined as 3.7×10^{10} decay events per second. A person exposed to a curie of radiation from an alpha emitter is bombarded by 3.7×10^{10} alpha particles per second. Different kinds of radiation produce different effects. As we have learned, alpha radiation has a much greater ionizing power than beta radiation. Consequently, a certain number of alpha decays occurring within a person's body (due to the ingestion of an alpha emitter) would do more damage than the same number of beta decays. If the alpha emitter and beta emitter are external to the body, however, the radiation from the alpha emitter would largely be stopped by clothing or the skin (due to the low penetrating power of alpha radiation), while the radiation from the beta emitter could penetrate the skin and therefore cause more damage.

A better way to assess radiation exposure is to measure the amount of energy actually absorbed by body tissue. The units used for this type of exposure measurement are called the *gray* (Gy), which corresponds to 1 J of energy absorbed per kilogram of body tissue, and the *rad* (for *rad*iation *a*bsorbed *d*ose), which corresponds to 0.01 Gy.

$$1 \text{ gray (Gy)} = 1 \text{ J/kg body tissue}$$
$$1 \text{ rad} = 0.01 \text{ J/kg body tissue}$$

Although these units measure the actual energy absorbed by bodily tissues, they still do not account for the amount of damage to biological molecules caused by that energy absorption, which differs from one type of radiation to another and from one type of biological tissue to another. For example, when a gamma ray passes through biological tissue, the energy absorbed is spread out over the long distance that the radiation travels through the body, resulting in a low ionization density within the tissue. When an alpha particle passes through biological tissue, however, the energy is absorbed over a much shorter distance, resulting in a much higher ionization density. The higher ionization density results in greater damage, even though the amount of energy absorbed might be the same.

Consequently, a correction factor, called the **biological effectiveness factor**, or **RBE** (for *R*elative *B*iological *E*ffectiveness), is usually multiplied by the dose in rads to obtain the dose in a unit called the **rem** for *r*oentgen *e*quivalent *m*an.

$$\text{Dose in rads} \times \text{biological effectiveness factor} = \text{dose in rems}$$

The SI unit that corresponds to the rem is the sievert (Sv). However, the rem is still commonly used in the United States. The conversion factor is 1 rem = 0.01 Sv.

TABLE 19.4 Exposure by Source for Persons Living in the United States

Source	Dose
Natural Radiation	
A 5-hour jet airplane ride	2.5 mrem/trip (0.5 mrem/hr at 39,000 feet) (whole body dose)
Cosmic radiation from outer space	27 mrem/yr (whole body dose)
Terrestrial radiation	28 mrem/yr (whole body dose)
Natural radionuclides in the body	35 mrem/yr (whole body dose)
Radon gas	200 mrem/yr (lung dose)
Diagnostic Medical Procedures	
Chest X-ray	8 mrem (whole body dose)
Dental X-rays (panoramic)	30 mrem (skin dose)
Dental X-rays (two bitewings)	80 mrem (skin dose)
Mammogram	138 mrem per image
Barium enema (X-ray portion only)	406 mrem (bone marrow dose)
Upper gastrointestinal tract	244 mrem (X-ray portion only) (bone marrow dose)
Thallium heart scan	500 mrem (whole body dose)
Consumer Products	
Building materials	3.5 mrem/year (whole body dose)
Luminous watches (H-3 and Pm-147)	0.04–0.1 mrem/year (whole body dose)
Tobacco products (to smokers of 30 cigarettes per day)	16,000 mrem/year (bronchial epithelial dose)

Source: **Department of Health and Human Services,** National Institutes of Health.

The biological effectiveness factor for alpha radiation, for example, is much higher than that for gamma radiation.

On average, each of us is exposed to approximately 360 mrem of radiation per year from sources shown in Table 19.4. The majority of this exposure comes from natural sources, especially radon, one of the products in the uranium decay series. As you can see from Table 19.4, however, some medical procedures also involve exposure levels similar to those received from natural sources.

It takes much more than the average radiation dose to produce significant health effects in humans. The first measurable effect, a decreased white blood cell count, occurs at instantaneous exposures of approximately 20 rem (Table 19.5). Exposures of 100 rem produce a definite increase in cancer risk, and exposures of over 500 rem often result in death.

TABLE 19.5 Effects of Radiation Exposure

Approximate Dose (rem)	Probable Outcome
20–100	Decreased white blood cell count; possible increase in cancer risk
100–400	Radiation sickness including vomiting and diarrhea; skin lesions; increase in cancer risk
500	Death (often within 2 months)
1000	Death (often within 2 weeks)
2000	Death (within hours)

Conceptual Connection 19.4 Radiation Exposure

A person ingests equal amounts of two nuclides, both of which are beta emitters (of roughly equal energy). Nuclide A has a half-life of 8.5 hours and Nuclide B has a half-life of 15.0 hours. Both nuclides are eliminated from the body within 24 hours of ingestion. Which of the two nuclides produces the greater radiation exposure?

19.9   Radioactivity in Medicine

Radioactivity is often perceived as dangerous; however, it is also immensely useful to physicians in the diagnosis and treatment of disease and has numerous other valuable applications. The use of radioactivity in medicine can be broadly divided into *diagnostic techniques* (used to diagnose disease) and *therapeutic techniques* (used to treat disease).

Diagnosis in Medicine

The use of radioactivity in diagnosis usually involves a **radiotracer**, a radioactive nuclide that has been attached to a compound or introduced into a mixture in order to track the movement of the compound or mixture within the body. Tracers are useful in the diagnosis of disease primarily because of two factors: (1) the sensitivity with which radioactivity can be detected, and (2) the identical chemical behavior of a radioactive nucleus and its nonradioactive counterpart. For example, the thyroid gland naturally concentrates iodine. When a patient is given small amounts of iodine-131 (a radioactive isotope of iodine), the radioactive iodine accumulates in the thyroid, just as nonradioactive iodine does. However, the radioactive iodine emits radiation, which can be detected with great sensitivity and used to measure the rate of iodine uptake by the thyroid, and to image the gland.

Different elements are taken up preferentially by different organs or tissues, so various radiotracers can be used to monitor metabolic activity and image a variety of organs and structures, including the kidneys, heart, brain, gallbladder, bones, and arteries, as shown in Table 19.6. Radiotracers can also be used to locate infections or cancers within the body. To locate an infection, antibodies are "labeled" or "tagged" with a radioactive nuclide, such as technetium-99m (where "m" means metastable), and administered to the patient. The antibodies then aggregate at the infected site, as described in Section 19.1. Cancerous tumors can be detected because they naturally concentrate phosphorus. When a patient is given phosphorus-32 (a radioactive isotope of phosphorus) or a phosphate compound incorporating another radioactive isotope such as Tc-99m, the tumors concentrate the radioactive substance and can be detected (**Figure 19.12◀**).

A specialized imaging technique known as **positron emission tomography (PET)** employs positron-emitting nuclides, such as fluorine-18. The fluorine-18 is attached to a metabolically active substance such as glucose and administered to the patient. As the glucose travels through the bloodstream and to the heart and brain, it carries the radioactive fluorine, which decays with a half-life of just under 2 hours. When a fluorine-18 nuclide decays, it emits a positron which immediately combines with any nearby electrons. Since a positron and an electron are antiparticles, they annihilate one other, producing two gamma rays that travel in exactly opposing directions. The gamma rays are detected by an array of detectors that can locate the point of origin with great accuracy. The result is a set of highly detailed images that show both the rate of glucose metabolism and structural features of the imaged organ (**Figure 19.13◀**).

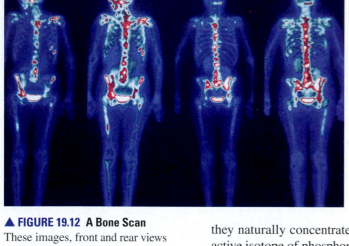

▲ FIGURE 19.12 A Bone Scan
These images, front and rear views of the human body, were created by the gamma ray emissions of Tc-99m. Such scans are often used to locate cancer that has metastasized (or spread) to the bones from a primary tumor elsewhere.

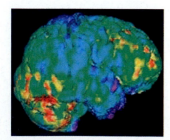

▲ FIGURE 19.13 A PET Scan   The colored areas indicate regions of high metabolic activity in the brain of a schizophrenic patient experiencing hallucinations.

TABLE 19.6 Common Radiotracers

Nuclide	Type of Emission	Half-Life	Part of Body Studied
Technetium-99m	Gamma (primarily)	6.01 hours	Various organs, bones
Iodine-131	Beta	8.0 days	Thyroid
Iron-59	Beta	44.5 days	Blood, spleen
Thallium-201	Electron capture	3.05 days	Heart
Fluorine-18	Positron emission	1.83 hours	PET studies of heart, brain
Phosphorus-32	Beta	14.3 days	Tumors in various organs

Radiotherapy in Medicine

Because radiation kills cells, and because it is particularly effective at killing rapidly dividing cells, it is often used as a therapy for cancer (cancer cells reproduce much faster than normal cells). During radiotherapy, gamma rays are focused on internal tumors to kill them. The gamma ray beam is usually moved in a circular path around the tumor (**Figure 19.14▶**), maximizing the exposure of the tumor while minimizing the exposure of the surrounding healthy tissue. Nonetheless, cancer patients receiving such treatment usually develop the symptoms of radiation sickness, which include vomiting, skin burns, and hair loss.

Some people wonder why radiation—which is known to cause cancer—can also be used to treat cancer. The answer lies in risk analysis. A cancer patient is normally exposed to radiation doses of about 100 rem. Such a dose increases cancer risk by about 1%. However, if the patient has a significant chance of dying from the cancer that he already has, such a risk becomes acceptable, especially when there is a proven chance of curing the cancer.

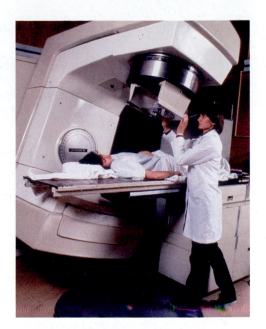

▶ **FIGURE 19.14 Radiotherapy for Cancer** This treatment involves exposing a malignant tumor to gamma rays generated by nuclides such as cobalt-60.

CHAPTER IN REVIEW

Key Terms

Section 19.1
radioactivity (745)
radioactive (745)

Section 19.2
nuclide (746)
alpha (α) decay (746)
alpha (α) particle (746)
nuclear equation (746)
ionizing power (747)
penetrating power (747)
beta (β) decay (747)

beta (β) particle (747)
gamma (γ) ray
    emission (748)
gamma (γ) ray (748)
positron emission (748)
positron (748)
electron capture (749)

Section 19.3
strong force (751)
nucleon (751)
magic numbers (752)

Section 19.4
radiometric dating (755)
radiocarbon dating (756)

Section 19.5
nuclear fission (759)
chain reaction (760)
critical mass (760)

Section 19.6
mass defect (763)
nuclear binding energy (763)

Section 19.7
nuclear fusion (765)

Section 19.8
biological effectiveness factor
    (RBE) (766)
rem (766)

Section 19.9
radiotracer (768)
positron emission tomography
    (PET) (768)

Key Concepts

Diagnosing Appendicitis (19.1)

▶ Radioactivity is the emission of subatomic particles or energetic electromagnetic radiation by the nuclei of certain atoms.

▶ Because some of these emissions can pass through matter, radioactivity is useful in medicine and many other areas of study.

Types of Radioactivity (19.2)

▶ The major types of natural radioactivity are alpha (α) decay, beta (β) decay, gamma (γ) ray emission, and positron emission.

▶ Alpha radiation consists of helium nuclei. Beta particles are electrons. Gamma rays are electromagnetic radiation of very high energy. Positrons are the antiparticles of electrons.

▶ A nucleus may also absorb one of its orbital electrons in a process called electron capture.

▶ We can represent each radioactive process with a nuclear equation in which the parent nuclide changes into the daughter nuclide. In a nuclear equation, although the specific types of atoms may not balance, the atomic numbers and mass numbers must.

▶ Each type of radioactivity has a different ionizing and penetrating power. These values are inversely related; a particle with a higher ionizing power has a lower penetrating power. Alpha particles are the most massive and they therefore have the highest ionizing power, followed by beta particles and positrons, which are equivalent in their ionizing power. Gamma rays have the lowest ionizing power.

The Valley of Stability: Predicting the Type of Radioactivity (19.3)

▶ The stability of a nucleus, and therefore the probability that it will undergo radioactive decay, depends largely on two factors. The first is the ratio of neutrons to protons (N/Z), because neutrons provide a strong force which overcomes the electromagnetic repulsions between the positive protons. This ratio is one for smaller elements, but becomes greater than one for larger elements. The second factor in nuclei stability is a concept known as magic numbers; certain numbers of nucleons are more stable than others.

The Kinetics of Radioactive Decay and Radiometric Dating (19.4)

▶ All radioactive elements decay according to first-order kinetics (Chapter 13); the half-life equation and the integrated rate law for radioactive decay are derived from the first-order rate laws.

▶ The kinetics of radioactive decay can be used to date objects and artifacts. The age of materials that were once part of living organisms is measured by carbon-14 dating. The age of ancient rocks and even Earth itself is determined by uranium/lead dating.

The Discovery of Fission: The Atomic Bomb and Nuclear Power (19.5)

▶ Fission is the splitting of an atom, such as uranium-235, into two atoms of lesser atomic weight.

▶ Because the fission of one uranium-235 atom releases enormous amounts of energy and produces neutrons that can split other uranium-235 atoms, the energy from these collective reactions can be harnessed in an atomic bomb or nuclear reactor.

▶ Nuclear power produces no air pollution and requires little mass to release lots of energy; however, there is always a danger of accidents, and it is difficult to dispose of nuclear waste.

Converting Mass to Energy: Mass Defect and Nuclear Binding Energy (19.6)

▶ In a nuclear fission reaction, a substantial amount of mass is converted into energy.

▶ The difference in mass between a nuclide and the individual protons and neutrons that compose it is the mass defect, and the corresponding energy, calculated from Einstein's equation $E = mc^2$, is the nuclear binding energy.

▶ The stability of a nucleus is determined by the binding energy per nucleon, which increases up to mass number 60 and then decreases.

Nuclear Fusion: The Power of the Sun (19.7)

▶ Stars produce their energy by a process that is the opposite of fission: nuclear fusion, the combination of two light nuclei to form a heavier one.

▶ Modern nuclear weapons employ fusion. Although fusion has been examined as a possible method to produce electricity, experiments with hydrogen fusion have thus far been more costly than productive.

The Effects of Radiation on Life (19.8)

▶ The effects of radiation can be grouped into three categories. Acute radiation damage is caused by a substantial exposure to radiation for a short period of time. Lower radiation exposures may result in increased cancer risk because of damage to DNA. Genetic defects are caused by damage to the DNA of reproductive cells.

▶ The most effective unit of measurement for the amount of radiation absorbed is the rem, which takes into account the different penetrating and ionizing powers of the various types of radiation.

Radioactivity in Medicine (19.9)

▶ Radioactivity is central to the diagnosis of medical problems by means of radiotracers and positron emission tomography (PET). Both of these techniques can provide data about the appearance and metabolic activity of an organ, or help locate a tumor.

▶ Radiation is also used to treat cancer because it can kill cells. Although this treatment has unpleasant side effects and increases the risk of a new cancer developing, the risk is usually acceptable compared to the risk of death from an established cancer.

Key Equations and Relationships

The First-Order Rate Law (19.4)

$$\text{Rate} = kN$$

The Half-Life Equation (19.4)

$$t_{1/2} = \frac{0.693}{k} \quad k = \text{rate constant}$$

The Integrated Rate Law (19.4)

$$\ln \frac{N_t}{N_0} = -kt \quad N_t = \text{number of radioactive nuclei at time } t$$
$$N_0 = \text{initial number of radioactive nuclei}$$

Einstein's Energy–Mass Equation (19.6)

$$E = mc^2$$

Key Learning Objectives

Chapter Objectives	Assessment		
Writing Nuclear Equations for Alpha Decay (19.2)	Example 19.1	For Practice 19.1	Exercises 1–6
Writing Nuclear Equations for Beta Decay, Positron Emission, and Electron Capture (19.2)	Example 19.2 Exercises 1–6	For Practice 19.2	For More Practice 19.2
Predicting the Type of Radioactive Decay (19.3)	Example 19.3	For Practice 19.3	Exercises 11, 12
Using Radioactive Decay Kinetics (19.4)	Example 19.4	For Practice 19.4	Exercises 15–22
Using Radiocarbon Dating (19.4)	Example 19.5	For Practice 19.5	Exercises 23, 24
Using Uranium/Lead Dating to Estimate the Age of a Rock (19.4)	Example 19.6	For Practice 19.6	Exercises 25, 26
Determining the Mass Defect and Nuclear Binding Energy (19.6)	Example 19.7	For Practice 19.7	Exercises 33–40

EXERCISES

Problems by Topic

Radioactive Decay and Nuclide Stability

1. Write a nuclear equation for the indicated decay of each nuclide.
   a. U-234 (alpha)   **b.** Th-230 (alpha)
   c. Pb-214 (beta)   **d.** N-13 (positron emission)
   e. Cr-51 (electron capture)

2. Write a nuclear equation for the indicated decay of each nuclide.
   a. Po-210 (alpha)   **b.** Ac-227 (beta)
   c. Tl-207 (beta)   **d.** O-15 (positron emission)
   e. Pd-103 (electron capture)

3. Write a partial decay series for Th-232 undergoing the sequential decays: $\alpha, \beta, \beta, \alpha$.

4. Write a partial decay series for Rn-220 undergoing the sequential decays: $\alpha, \alpha, \beta, \alpha$.

5. Fill in the missing particles in each nuclear equation.

   a. _____ $\longrightarrow$ $^{217}_{85}At + ^{4}_{2}He$

   b. $^{241}_{94}Pu \longrightarrow ^{241}_{95}Am +$ _____

   c. $^{19}_{11}Na \longrightarrow ^{19}_{10}Ne +$ _____

   d. $^{75}_{34}Se +$ _____ $\longrightarrow ^{75}_{33}As$

6. Fill in the missing particles in each nuclear equation.

   a. $^{241}_{95}Am \longrightarrow ^{237}_{93}Np +$ _____

   b. _____ $\longrightarrow ^{233}_{92}U + ^{0}_{-1}e$

   c. $^{237}_{93}Np \longrightarrow$ _____ $+ ^{4}_{2}He$

   d. $^{75}_{35}Br \longrightarrow$ _____ $+ ^{0}_{+1}e$

7. Determine whether or not each nuclide is likely to be stable. State your reasons.
   a. Mg-26   **b.** Ne-25   **c.** Co-51   **d.** Te-124

8. Determine whether or not each nuclide is likely to be stable. State your reasons.
   a. Ti-48   **b.** Cr-63   **c.** Sn-102   **d.** Y-88

9. The first six elements of the first transition series have the following number of stable isotopes:

Element	Number of Stable Isotopes
Sc	1
Ti	5
V	1
Cr	3
Mn	1
Fe	4

   Explain why Sc, V, and Mn each has only one stable isotope while the other elements have several.

10. Neon and magnesium each has three stable isotopes while sodium and aluminum each has only one. Explain why this might be so.

11. Predict a likely mode of decay for each unstable nuclide.
    a. Mo-109   **b.** Ru-90   **c.** P-27   **d.** Rn-196

12. Predict a likely mode of decay for each unstable nuclide.
    a. Sb-132   **b.** Te-139   **c.** Fr-202   **d.** Ba-123

13. Which member of each pair of nuclides would you expect to have the longest half-life?
    a. Cs-113 or Cs-125   **b.** Fe-62 or Fe-70

14. Which member of each pair of nuclides would you expect to have the longest half-life?
    a. Cs-149 or Cs-139   **b.** Fe-45 or Fe-52

The Kinetics of Radioactive Decay and Radiometric Dating

15. One of the nuclides in spent nuclear fuel is U-235, an alpha emitter with a half-life of 703 million years. How long does it take for the U-235 to reach one-eighth of its initial amount?

16. A patient is given 0.050 mg of technetium-99m, a radioactive isotope with a half-life of about 6.0 hours. How long until the radioactive isotope decays to 6.3×10^{-3} mg? (Assume no excretion of the nuclide from the body.)

17. A radioactive sample contains 1.55 g of an isotope with a half-life of 3.8 days. What mass of the isotope remains after 5.5 days? (Assume no excretion of the nuclide from the body.)

18. At 8:00 a.m., a patient receives a 58-mg dose of I-131 to obtain an image of her thyroid. If the nuclide has a half-life of 8 days, what mass of the nuclide remains in the patient at 5:00 p.m. the next day? (Assume no excretion of the nuclide from the body.)

19. A sample of F-18 has an initial decay rate of $1.5 \times 10^5/s$. How long will it take for the decay rate to fall to $1.0 \times 10^2/s$? (F-18 has a half-life of 1.83 hours.)

20. A sample of Tl-201 has an initial decay rate of $5.88 \times 10^4/s$. How long will it take for the decay rate to fall to $55/s$? (Tl-201 has a half-life of 3.042 days.)

21. A wooden boat discovered just south of the Great Pyramid in Egypt has a carbon-14/carbon-12 ratio that is 72.5% of that found in living organisms. How old is the boat?

22. A layer of peat beneath the glacial sediments of the last ice age has a carbon-14/carbon-12 ratio that is 22.8% of that found in living organisms. How long ago was this ice age?

23. An ancient skull has a carbon-14 decay rate of 0.85 disintegrations per minute per gram of carbon ($0.85 \text{ dis/min} \cdot \text{g C}$). How old is the skull? (Assume that living organisms have a carbon-14 decay rate of $15.3 \text{ dis/min} \cdot \text{g C}$ and that carbon-14 has a half-life of 5730 yr.)

24. A mammoth skeleton has a carbon-14 decay rate of 0.48 disintegrations per minute per gram of carbon ($0.48 \text{ dis/min} \cdot \text{g C}$). When did the mammoth live? (Assume that living organisms have a carbon-14 decay rate of $15.3 \text{ dis/min} \cdot \text{g C}$ and that carbon-14 has a half-life of 5730 yr.)

25. A rock from Australia contains 0.438 g of Pb-206 to every 1.00 g of U-238. Assuming that the rock did not contain any Pb-206 at the time of its formation, how old is the rock?

26. A meteor has a Pb-206:U-238 mass ratio of 0.855:1.00. What is the age of the meteor? (Assume that the meteor did not contain any Pb-206 at the time of its formation.)

Fission and Fusion

27. Write a nuclear reaction for the neutron-induced fission of U-235 to form Xe-144 and Sr-90. How many neutrons are produced in the reaction?

28. Write a nuclear reaction for the neutron-induced fission of U-235 to produce Te-137 and Zr-97. How many neutrons are produced in the reaction?

29. Write a nuclear equation for the fusion of two H-2 atoms to form He-3 and one neutron.

30. Write a nuclear equation for the fusion of H-3 with H-1 to form He-4.

31. A breeder nuclear reactor is a reactor in which U-238 (which does not undergo fission) is converted into Pu-239 (which does undergo fission). The process involves bombardment of U-238 by neutrons to form U-239 which then undergoes two sequential beta decays. Write nuclear equations to represent this process.

32. Write a series of nuclear equations to represent the bombardment of Al-27 with a neutron to form a product that then undergoes an alpha decay followed by a beta decay.

Energetics of Nuclear Reactions, Mass Defect, and Nuclear Binding Energy

33. If 1.0 g of matter were converted to energy, how much energy would be formed?

34. A typical home uses approximately 1.0×10^3 kWh of energy per month. If the energy came from a nuclear reaction, what mass would have to be converted to energy per year to meet the energy needs of the home?

35. Calculate the mass defect and nuclear binding energy per nucleon of each of the nuclides.
    a. O-16 (atomic mass = 15.994915 amu)
    b. Ni-58 (atomic mass = 57.935346 amu)
    c. Xe-129 (atomic mass = 128.904780 amu)

36. Calculate the mass defect and nuclear binding energy per nucleon of each of the nuclides.
    a. Li-7 (atomic mass = 7.016003 amu)
    b. Ti-48 (atomic mass = 47.947947 amu)
    c. Ag-107 (atomic mass = 106.905092 amu)

37. Calculate the quantity of energy produced per gram of U-235 (atomic mass = 235.043922 amu) for the neutron-induced fission of U-235 to form Xe-144 (atomic mass = 143.9385 amu) and Sr-90 (atomic mass = 89.907738 amu) (discussed in Problem 27).

38. Calculate the quantity of energy produced per mole of U-235 (atomic mass = 235.043922 amu) for the neutron-induced fission of U-235 to produce Te-137 (atomic mass = 136.9253 amu) and Zr-97 (atomic mass = 96.910950 amu) (discussed in Problem 28).

39. Calculate the quantity of energy produced per gram of reactant for the fusion of two H-2 (atomic mass = 2.014102 amu) atoms to form He-3 (atomic mass = 3.016029 amu) and one neutron (discussed in Problem 29).

40. Calculate the quantity of energy produced per gram of reactant for the fusion of H-3 (atomic mass = 3.016049 amu) with H-1 (atomic mass = 1.007825 amu) to form He-4 (atomic mass = 4.002603 amu) (discussed in Problem 30).

Effects and Applications of Radioactivity

41. A 75-kg human is exposed to 32.8 rad of radiation. How much energy is absorbed by the person's body? Compare this energy to the amount of energy absorbed by a person's body if he or she jumped from a chair to the floor (assume that all of the energy from the fall is absorbed by the person).

42. If a 55-gram laboratory mouse is exposed to 20.5 rad of radiation, how much energy was absorbed by the mouse's body?

43. PET studies require fluorine-18, which decays with a half-life of 1.83 hours from the time it is produced. Assuming that the F-18 can be transported at 60.0 miles/hour, how close must the hospital be to the production site of F-18 if 65% of the F-18 produced is to make it to the hospital?

44. A patient ingests 155 mg of I-131, a beta emitter with a half-life of 8.0 days. Assuming that none of the I-131 is eliminated from the person's body in the first 4.0 hours of treatment, what is the exposure (in Ci) during those first four hours?

Cumulative Problems

45. Write a nuclear equation for the most likely mode of decay for each unstable nuclide.
    a. Ru-114   b. Ra-216   c. Zn-58   d. Ne-31

46. Write a nuclear equation for the most likely mode of decay for each unstable nuclide.
    a. Kr-74   b. Th-221   c. Ar-44   d. Nb-85

47. Bismuth-210 is a beta emitter with a half-life of 5.0 days. If a sample contains 1.2 g of Bi-210 (atomic mass = 209.984105 amu), how many beta emissions occur in 13.5 days? If a person's body intercepts 5.5% of those emissions, to what dose of radiation (in Ci) is the person exposed?

48. Polonium-218 is an alpha emitter with a half-life of 3.0 minutes. If a sample contains 55 mg of Po-218 (atomic mass = 218.008965 amu), how many alpha emissions occur in 25.0 minutes? If the polonium is ingested by a person, to what dose of radiation (in Ci) is the person exposed?

49. Radium-226 (atomic mass = 226.025402 amu) decays to radon-224 (a radioactive gas) with a half-life of 1.6×10^3 years. What volume of radon gas (at 25.0 °C and 1.0 atm) is produced by 25.0 g of radium in 5.0 days? (Report your answer to two significant digits.)

50. In one of the neutron-induced fission reactions of U-235 (atomic mass = 235.043922 amu), the products are Ba-140 and Kr-93 (a radioactive gas). What volume of Kr-93 (at 25.0 °C and 1.0 atm) is produced when 1.00 g of U-235 undergoes this fission reaction?

51. When a positron and an electron annihilate one another, the resulting mass is completely converted to energy. Calculate the energy associated with this process in kJ/mol.

52. A typical nuclear reactor produces about 1.0 MW of power per day. What is the minimum rate of mass loss required to produce this much energy?

53. Determine the binding energy in an atom of ^{3}He, that has a mass of 3.016030 amu.

54. The overall hydrogen burning reaction in stars can be represented as the conversion of four protons to one α particle. Use the data for the mass of H-1 and He-4 to calculate the energy released by this process.

55. The half-life of ^{238}U is 4.5×10^9 yr. A sample of rock of mass 1.6 g produces 29 dis/s. Assuming all the radioactivity is due to ^{238}U, find the percent by mass of ^{238}U in the rock.

56. The half-life of ^{232}Th is 1.4×10^{10} yr. Determine the number of disintegrations per hour emitted by 1.0 mol of ^{232}Th in 1 minute.

57. A 1.50-L gas sample at 745 mm Hg and 25.0° C contains 3.55% radon-220 by volume. Radon-220 is an alpha-emitter with a half-life of 55.6 s. How many alpha particles are emitted by the gas sample in 5.00 minutes?

58. A 228-mL sample of an aqueous solution contains 2.35% $MgCl_2$ by mass. Exactly one-half of the magnesium ions are Mg-28, a beta emitter with a half-life of 21 hours. What is the decay rate of Mg-28 in the solution after 4.00 days? (Assume a density of 1.02 g/mL for the solution.)

59. When a positron and an electron collide and annihilate each other, two photons of equal energy are produced. What is the wavelength of these photons? (See problem 51.)

60. The half-life of ^{235}U, an alpha emitter, is 7.1×108 yr. Calculate the number of alpha particles emitted by 1.0 mg of this nuclide in 1.0 minute.

61. Given that the energy released in the fusion of two deuterons to a ^{3}He and a neutron is 3.3 MeV and in the fusion to tritium and a proton it is 4.0 MeV, calculate the energy change for the process ^{3}He + ^{1}n $\longrightarrow$ ^{3}H + ^{1}p. Suggest an explanation for why this process occurs at much lower temperatures than either of the first two.

62. The nuclide ^{18}F decays by both electron capture and β^+ decay. Find the difference in the energy released by these two processes. The atomic masses are ^{18}F = 18.000950 and ^{18}O = 17.9991598.

Challenge Problems

63. The space shuttle carries about 72,500 kg of solid aluminum fuel, which is oxidized with ammonium perchlorate according to the reaction:

$$10\,Al(s) + 6\,NH_4ClO_4(s) \longrightarrow$$
$$4\,Al_2O_3(s) + 2\,AlCl_3(s) + 12\,H_2O(g) + 3\,N_2(g)$$

The space shuttle also carries about 608,000 kg of oxygen (which reacts with hydrogen to form gaseous water).
  a. Assuming that aluminum and oxygen are the limiting reactants, determine the total energy produced by these fuels. (ΔH_f° for solid ammonium perchlorate is -295 kJ/mol.)
  b. Suppose that a future space shuttle is powered by matter–antimatter annihilation. The matter could be normal hydrogen (containing a proton and an electron) and the antimatter could be antihydrogen (containing an antiproton and a positron). What mass of antimatter would be required to produce

the energy equivalent of the aluminum and oxygen fuel currently carried on the space shuttle?

64. An 85.0-gram laboratory animal ingests 10.0 mg of a substance that contained 2.55% by mass Pu-239, an alpha emitter with a half-life of 24,110 years.
  a. What is the animal's initial radiation exposure in Curies?
  b. If all of the energy from the emitted alpha particles is absorbed by the animal's tissues, and if the energy of each emission is 7.77×10^{-12} J what is the dose in rads to the animal in the first 4.0 hours following the ingestion of the radioactive material? Assuming a biological effectiveness factor of 20, what is the 4.0-hour dose in rems?

65. In addition to the natural radioactive decay series that begins with U-238 and ends with Pb-206, there are natural radioactive decay series that begin with U-235 and Th-232. Both of these series end with nuclides of Pb. Predict the likely end product of each series and the number of α decay steps that occur.

Conceptual Problems

66. Examine the diagram representing the beta decay of fluorine-21 and identify the missing nucleus.

$$^{21}_{9}F \longrightarrow \quad ? \quad 1 \quad ^{0}_{-1}e$$

67. Approximately how many half-lives must pass for the amount of radioactivity in a substance to decrease to below 1% of its initial level?

68. A person is exposed for three days to identical amounts of two different nuclides that emit positrons of roughly equal energy. The half-life of nuclide A is 18.5 days and the half-life of nuclide B is 255 days. Which of the two nuclides poses the greater health risk to the person?

69. Identical amounts of two different nuclides, an alpha emitter and a gamma emitter, with roughly equal half-lives are spilled in a building adjacent to your bedroom. Which of the two nuclides poses the greater health threat to you while you sleep in your bed? If you accidentally wander into the building and ingest equal amounts of the two nuclides, which one poses the greater health threat?

Answers to Conceptual Connections

Alpha and Beta Decay

19.1 (c) The arrow labeled x shows a decrease of 2 neutrons and 2 protons, indicative of alpha decay. The arrow labeled y shows a decrease of 1 neutron and an increase of 1 proton, indicative of beta decay.

Half-Life

19.2 (b) The half-life is the time it takes for the number of nuclei to decay to one-half of their original number.

Half Life and Amount of Radioactive Sample

19.3 (b) 0.10 mol. The sample loses one-half of the number of moles per half-life; so for four half-lives, the amount falls to 0.10 mol.

Radiation Exposure

19.4 Nuclide A. Because nuclide A has a shorter half-life, more of the nuclides will decay, and therefore produce radiation, before they exit the body.

20

Organic Chemistry

Organic chemistry just now is enough to drive one mad. It gives one the impression of a primeval, tropical forest full of the most remarkable things. . . . —Friedrich Wöhler (1800–1882)

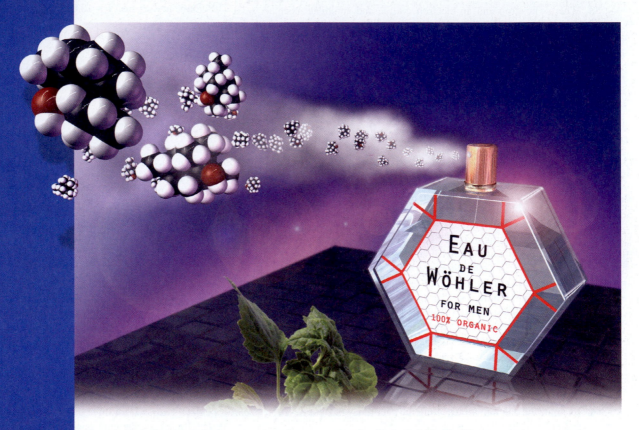

About half of all men's colognes contain at least some patchouli alcohol ($C_{15}H_{26}O$), an organic compound (pictured here) derived from the patchouli plant. Patchouli alcohol has a pungent, musty, earthy fragrance.

ORGANIC CHEMISTRY IS THE STUDY of carbon-containing compounds. Carbon is unique in the sheer number of compounds that it can form. Millions of organic compounds are known, and new ones are discovered every day. Carbon is also unique in the diversity of compounds that it can form. In most cases, a fixed number of carbon atoms can combine with a fixed number of atoms of another element to form a large number of different compounds. For example, 10 carbon atoms and 22 hydrogen atoms form 75 distinctly different compounds. It is not surprising that life is based on the chemistry of carbon because life needs diversity to exist, and organic chemistry is nothing if it is not diverse. With carbon as the backbone, nature can take the same combination of atoms and bond them together in slightly different ways to produce a huge variety of substances. In this chapter, we will peer into Friedrich Wöhler's "primeval tropical forest" (see the chapter-opening quotation) and discover the most remarkable things.

20.1 Fragrances and Odors

Have you ever ridden an elevator with someone wearing too much perfume? Or accidentally gotten too close to a skunk? Or caught a whiff of rotting fish? What causes these fragrances and odors? Smells are caused by odorants, molecules that bind to olfactory receptors in our noses when we inhale. The interaction sends a nerve signal to the brain that we experience as a smell. Some smells, such as that of perfume, are pleasant (when not in excess). Other smells, such as that of the skunk or rotting fish, are unpleasant. Our sense of smell helps us identify food, people, and other organisms, and alerts us to dangers such as polluted air or spoiled food. Smell (or olfaction) is one way we probe the environment around us.

Odorants, if they are to reach our noses, must be volatile. However, many volatile substances have no scent at all. Nitrogen, oxygen, water, and carbon dioxide molecules, for example, are constantly passing through our noses, yet they produce no smell because they do not bind to olfactory receptors. Most common smells are caused by **organic molecules**, molecules containing carbon combined with several other elements, such as hydrogen, nitrogen, oxygen, and sulfur. Organic molecules are responsible for the smells of flowers, vanilla, cinnamon, almond, jasmine, body odor, and rotting fish. When you wander into a rose garden, you experience the sweet smell caused in part by geraniol, an organic compound emitted by roses. Men's colognes often contain patchouli alcohol, an earthy-smelling organic compound that can be extracted from the patchouli plant. If you have been in the vicinity of skunk spray (or have been unfortunate enough to be sprayed yourself), you are familiar with 2-butene-1-thiol and 3-methyl-1-butanethiol, two particularly odoriferous compounds present in the secretion that skunks use to defend themselves.

The study of compounds containing carbon combined with one or more of the elements mentioned previously (that is, hydrogen, nitrogen, oxygen, and sulfur), including the properties and reactions of these compounds, is known as **organic chemistry**. Besides composing much of what we smell, organic compounds are prevalent in foods, drugs, petroleum products, and pesticides. Organic chemistry is also the basis for living organisms. Life has evolved based on carbon-containing compounds, making organic chemistry of utmost importance to any person interested in understanding living organisms.

$CH_3CH\!=\!CHCH_2SH$
2-Butene-1-thiol

CH_3
|
$CH_3CHCH_2CH_2SH$
3-Methyl-1-butanethiol

▲ The smell of skunk spray is due primarily to the molecules shown here.

20.2 Carbon: A Unique Element

Why did life evolve based on the chemistry of carbon? Why is life not based on some other element? The answer may not be simple, but we know that life—in order to exist—must have complexity. We also know that carbon chemistry is complex. The number of compounds containing carbon is greater than the number of compounds made up of all the rest of the elements combined. The reasons for this include carbon's ability to form four covalent bonds, its ability to form double and triple bonds, and its tendency to *catenate* (that is, to form chains).

Carbon's Tendency to Form Four Covalent Bonds

Carbon—with its four valence electrons—can form four covalent bonds. Consider the Lewis structure and space-filling models of two simple carbon compounds, methane and ethane.

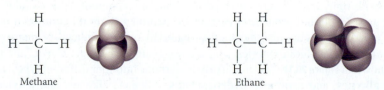

Methane

Ethane

The geometry of a carbon atom forming four single bonds is tetrahedral, as shown in the figure of methane. Carbon's ability to form four bonds, and to form those bonds with a number of different elements, results in the potential to form many different compounds. As you learn to draw structures for organic compounds, always remember to draw carbon with four bonds.

Carbon's Ability to Form Double and Triple Bonds

Carbon atoms can also form double bonds (trigonal planar geometry) and triple bonds (linear geometry), adding even more diversity to the number of compounds that the element can form.

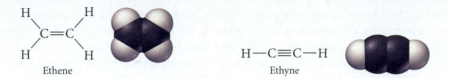

Ethene   Ethyne

In contrast, silicon (the element in the periodic table with properties closest to that of carbon) does not readily form double or triple bonds because the greater size of silicon atoms results in a Si—Si bond that is too long for much overlap between nonhybridized p orbitals.

Carbon's Tendency to Catenate

Carbon, more than any other element, can bond to itself to form chain, branched, and ring structures.

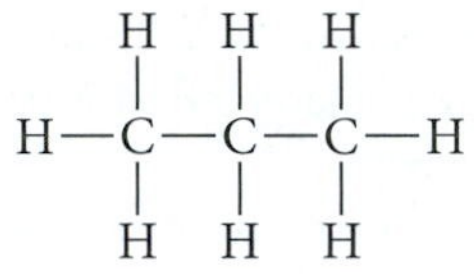

Propane 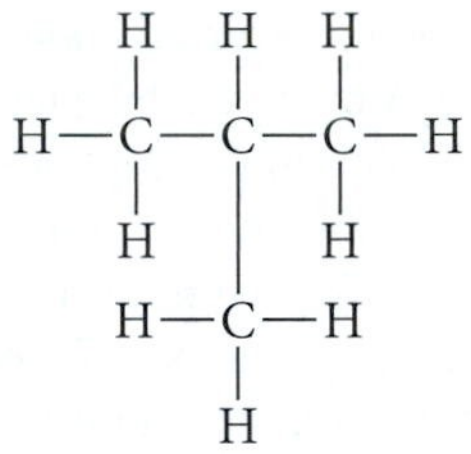   Isobutane 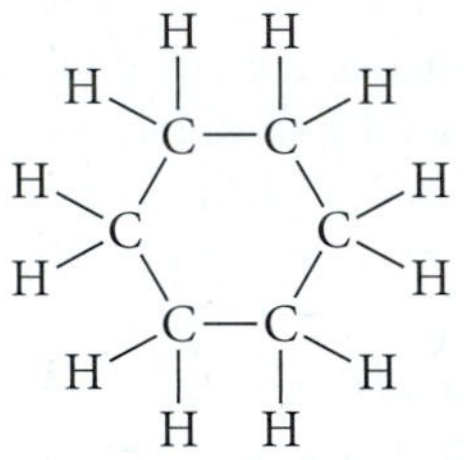   Cyclohexane

Although other elements can form chains, none surpasses carbon at this ability. Silicon, for example, can form chains with itself. However, silicon's affinity for oxygen (the Si—O bond is 142 kJ/mol stronger than the Si—Si bond) coupled with the prevalence of oxygen in our atmosphere means that silicon–silicon chains are readily oxidized to form silicates (the silicon–oxygen compounds that compose a significant proportion of minerals). By contrast, the C—C bond (347 kJ/mol) and the C—O bond (359 kJ/mol) are nearly the same strength, allowing carbon chains to exist relatively peacefully in an oxygen-rich environment. In other words, silicon's affinity for oxygen robs it of the rich diversity that catenation provides to carbon.

20.3   Hydrocarbons: Compounds Containing Only Carbon and Hydrogen

Hydrocarbons—compounds that contain only carbon and hydrogen—are the simplest organic compounds. However, because of the uniqueness of carbon just discussed, many different kinds of hydrocarbons exist. Hydrocarbons are commonly used as fuels. Candle wax, oil, gasoline, liquid propane (LP) gas, and natural gas are all composed of hydrocarbons. Hydrocarbons are also the starting materials in the synthesis of many different consumer products including fabrics, soaps, dyes, cosmetics, drugs, plastic, and rubber.

As shown in **Figure 20.1▶**, we classify hydrocarbons into four different types: **alkanes**, **alkenes**, **alkynes**, and **aromatic hydrocarbons**. Alkanes, alkenes, and alkynes—also called **aliphatic hydrocarbons**—are differentiated based on the kinds of bonds between

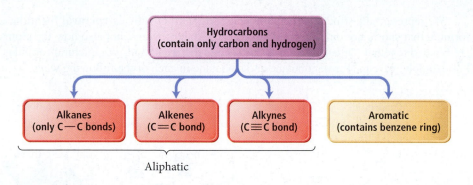

◀ **FIGURE 20.1   Four Types of Hydrocarbons**

carbon atoms. (Aromatic hydrocarbons will be discussed in more detail in Section 20.7.) As shown in Table 20.1, alkanes have only single bonds between carbon atoms, alkenes have at least one double bond, and alkynes have at least one triple bond.

TABLE 20.1   Alkanes, Alkenes, Alkynes

Type of Hydrocarbon	Type of bonds	Generic Formula*	Example
Alkane	All single	C_nH_{2n+2}	Ethane
Alkenes	One (or more) double	C_nH_{2n}	Ethene
Alkynes	One (or more) triple	C_nH_{2n-2}	$H-C\equiv C-H$ Ethyne

* *n* is the number of carbon atoms. These formulas apply only to noncyclic structures containing no more than one multiple bond.

Drawing Hydrocarbon Structures

Throughout this book, we have relied primarily on molecular formulas as the simplest way to represent compounds. In organic chemistry, however, molecular formulas are insufficient because, as we have already discussed, the same atoms can bond together in different ways to form different compounds. For example, consider an alkane with 4 carbon atoms and 10 hydrogen atoms. Two different structures, named butane and isobutane, are possible.

Butane   Isobutane

Butane and isobutane are **structural isomers**, molecules with the same molecular formula but different structures. Because of their different structures, they have different properties—indeed they are different compounds. Isomerism is ubiquitous in organic chemistry. Butane has 2 structural isomers. Pentane (C_5H_{12}) has 3, hexane (C_6H_{14}) has 5, and decane ($C_{10}H_{22}$) has 75!

We represent the structure of a particular hydrocarbon with a **structural formula**, a formula that shows not only the numbers of each kind of atoms, but also how the atoms are bonded together. Organic chemists use several different kinds of structural formulas. For example, we can represent butane and isobutane in each of the following ways:

Structural formula	Condensed structural formula	Carbon skeleton formula	Ball-and-stick model	Space-filling model

Butane

$$\begin{array}{c}H\;\;\;H\;\;\;H\;\;\;H\\ |\;\;\;\;|\;\;\;\;|\;\;\;\;|\\ H-C-C-C-C-H\\ |\;\;\;\;|\;\;\;\;|\;\;\;\;|\\ H\;\;\;H\;\;\;H\;\;\;H\end{array}$$

$CH_3-CH_2-CH_2-CH_3$

Isobutane

$$\begin{array}{c}H\\ |\\ H\;\;H-C-H\;\;H\\ |\;\;\;\;|\;\;\;\;|\\ H-C-\;\;\;\;C-\;\;\;\;C-H\\ |\;\;\;\;|\;\;\;\;|\\ H\;\;\;\;H\;\;\;\;H\end{array}$$

$$\begin{array}{c}CH_3\\ |\\ CH_3-CH-CH_3\end{array}$$

The structural formula shows all of the carbon and hydrogen atoms in the molecule and how they are bonded together. The condensed structural formula groups the hydrogen atoms together with the carbon atom to which they are bonded. Condensed structural formulas may show some of the bonds (as above) or none at all. For example, the condensed structural formula for butane can also be written as $CH_3CH_2CH_2CH_3$. The carbon skeleton formula (also called a line formula) shows the carbon–carbon bonds only as lines. Each end or bend of a line represents a carbon atom bonded to as many hydrogen atoms as necessary to form a total of four bonds.

Note that structural formulas are generally not three-dimensional representations of the molecule—as space-filling or ball-and-stick models are—but rather two-dimensional representations that show how atoms are bonded together. As such, the most important feature of a structural formula is the *connectivity* of the atoms, not the exact way the formula is drawn. For example, consider the two condensed structural formulas for butane and the corresponding space-filling models below them:

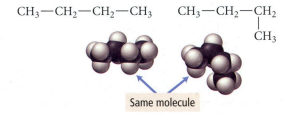

$CH_3-CH_2-CH_2-CH_3$ $\begin{array}{c}CH_3-CH_2-CH_2\\ |\\ CH_3\end{array}$

Same molecule

Since rotation about the single bond is relatively unhindered at room temperature, the two structural formulas are identical, even though they are drawn differently.

We represent double and triple bonds in structural formulas by double or triple lines. For example, the structural formulas for C_3H_6 (propene) and C_3H_4 (propyne) are drawn as:

Structural formula	Condensed structural formula	Carbon skeleton formula	Ball-and-stick model	Space-filling model

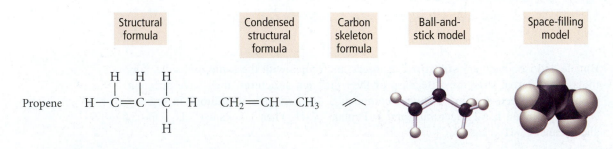

Propene

$$\begin{array}{c}H\;\;\;H\;\;\;H\\ |\;\;\;\;|\;\;\;\;|\\ H-C=C-C-H\\ |\\ H\end{array}$$

$CH_2=CH-CH_3$

Propyne $H-C\equiv C-\overset{\overset{\displaystyle H}{|}}{\underset{\underset{\displaystyle H}{|}}{C}}-H$ $CH\equiv C-CH_3$ $\equiv -$

The kind of structural formula we use depends on how much information we want to portray. Example 20.1 shows how to write structural formulas for a compound.

EXAMPLE 20.1 Writing Structural Formulas for Hydrocarbons

Write the structural formulas and carbon skeleton formulas for the five isomers of C_6H_{14} (hexane).

SOLUTION

To start, draw the carbon backbone of the straight-chain isomer.	$C-C-C-C-C-C$
Next, determine the carbon backbone structure of the other isomers by arranging the carbon atoms in four other unique ways.	
Fill in all the hydrogen atoms so that each carbon has four bonds.	
Write the carbon skeleton formulas by using lines to represent each carbon–carbon bond. Remember that each end or bend represents a carbon atom.	

FOR PRACTICE 20.1

Write the structural formulas and carbon skeleton formulas for the three isomers of C_5H_{12} (pentane).

Conceptual Connection 20.1 Organic Structures

Which of these structures is an *isomer* of $CH_3-CH(-CH_3)-CH_2-CH(-CH_3)-CH_3$ (and not just the same structure)?

$$CH_3-\underset{\underset{\displaystyle CH_3}{|}}{CH}-CH_2-\underset{\underset{\displaystyle CH_3}{|}}{CH}-CH_3$$

(a) $CH_3-\underset{\underset{\displaystyle CH_3}{|}}{CH}-CH_2-\underset{\underset{\displaystyle CH_3}{|}}{CH}-CH_3$

(b) $CH_3-\underset{\underset{\displaystyle CH_3}{|}}{CH}-CH_2-\underset{\overset{\overset{\displaystyle CH_3}{|}}{\underset{\displaystyle CH_3}{|}}}{CH}$

(c) $\underset{\underset{\displaystyle CH_3}{|}}{\overset{\overset{\displaystyle CH_3}{|}}{CH}}-CH_2-\underset{\underset{\displaystyle CH_3}{|}}{CH}$

(d) $CH_3-\underset{\overset{\displaystyle CH_3}{|}}{CH}-CH_2-CH_2-\underset{\underset{\displaystyle CH_3}{|}}{CH_2}$

Stereoisomerism and Optical Isomerism

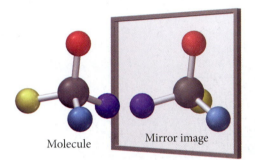

Molecule   Mirror image

Stereoisomers are molecules in which the atoms have the same connectivity, but have a different spatial arrangement. Stereoisomers can be of two types: geometric (cis–trans) isomers, and optical isomers. We discuss geometric isomers in Section 20.5. **Optical isomers** are two molecules that are nonsuperimposable mirror images of one another. For example, consider the molecule shown at left with its mirror image.

The molecule cannot be superimposed onto its mirror image. If you swing the mirror image around to try to superimpose the two, you find that there is no way to get all four substituent atoms to align together.

Optical isomers are similar to your right and left hands (**Figure 20.2◄**). The two are mirror images of one another, but you cannot superimpose one on the other. For this reason, a right-handed glove does fit on your left hand and vice versa.

Any carbon atom with four different substituents in a tetrahedral arrangement will exhibit optical isomerism. (A substituent is an atom or group of atoms that has been substituted for a hydrogen atom in an organic compound.) For example, consider 3-methylhexane:

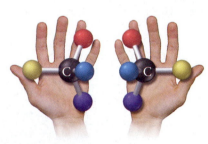

▲ FIGURE 20.2 Mirror Images
Your left and right hands are nonsuperimposable mirror images, similar to optical isomers.

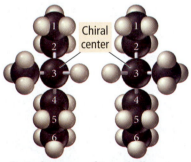

Chiral center

Optical isomers of 3-methylhexane

The two molecules shown here are nonsuperimposable mirror images and are, therefore, optical isomers of one other; optical isomers are also called **enantiomers**. Any molecule, such as 3-methylhexane, that exhibits optical isomerism is said to be **chiral**, from the Greek word *cheir*, which means "hand." Many of the physical and chemical properties of entantiomers are indistinguishable from one another. For example, the optical isomers of 3-methylhexane have identical freezing points, melting points, and densities. However, the properties of enantiomers differ from one another in two important ways: (1) in the direction in which they rotate polarized light and (2) in their chemical behavior in a chiral environment.

Rotation of Polarized Light

Plane-polarized light is light whose electric field waves oscillate in only one plane. When a beam of plane-polarized light is directed through a sample containing only one of two optical isomers, the plane of polarization of the light is rotated. One of the two optical isomers rotates the polarization of the light clockwise and is called the **dextrorotatory** isomer (or the *d* isomer). The other isomer rotates the polarization of the light counterclockwise and is called the **levorotatory** isomer (or the *l* isomer). An equimolar mixture of both optical isomers does not rotate the polarization of light at all and is called a **racemic mixture**.

> *Dextrorotatory* means turning clockwise or to the right. *Levorotatory* means turning counterclockwise or to the left.

Chemical Behavior in a Chiral Environment

Optical isomers also exhibit different chemical behavior when they are in a chiral environment (a chiral environment is one that is not superimposable on its mirror image). For example, enzymes are large biological molecules that catalyze reactions in living organisms and provide chiral environments. Consider the following simplified picture of two enantiomers in a chiral environment.

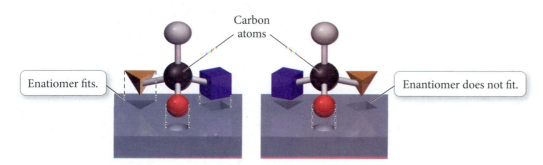

One of the enantiomers fits the template, but the other does not, no matter how it is rotated. An enzyme can catalyze the reaction of one enantiomer because it fits the "template" but not the other. Most biological molecules are chiral and for these biological molecules usually only one or the other enantiomer is active in biological systems. For example, glucose, the primary fuel of cells is chiral. Only one of the enantiomers of glucose has that familar sweet taste and only that enantiomer can fuel our cellular functioning; the other enantiomer is not even metabolized by our bodies.

Conceptual Connection 20.2 Optical Isomers

Which of these organic molecules can exhibit optical isomerism?

(a) H—C—Cl with H above and Br below   (b) Br—C—C—H with H,H above and Cl,H below   (c) Br—C—C—Cl with H,H above and H,H below   (d) Br—C—C—Cl with H,Cl above and H,H below

20.4 Alkanes: Saturated Hydrocarbons

As we have seen, alkanes are hydrocarbons containing only single bonds. Alkanes are often called **saturated hydrocarbons** because they are saturated (loaded to capacity) with hydrogen. The simplest hydrocarbons are methane (CH_4), the main component of natural gas, ethane (C_2H_6), a minority component in natural gas, and propane (C_3H_8), the main component of liquified petroleum (LP) gas.

Alkanes containing four or more carbon atoms may have either straight or branched carbon chains (as we saw in Section 20.3). The straight-chain isomers are often called normal alkanes, or *n*-alkanes. As the number of carbon atoms increases in the *n*-alkanes, so does their boiling point (as shown in Table 20.2). Methane, ethane, propane, and *n*-butane are all gases at room temperature, but the next *n*-alkane in the series, pentane, is a liquid at room temperature. Pentane is a component of gasoline. Table 20.3 summarizes the *n*-alkanes

TABLE 20.2 *n*-Alkane Boiling Points	
n-Alkane	Boiling point (°C)
Methane	−161.5
Ethane	−88.6
Propane	−42.1
n-Butane	−0.5
n-Pentane	36.0
n-Hexane	68.7
n-Heptane	98.5
n-Octane	125.6

TABLE 20.3 n-Alkanes

n	Name	Molecular Formula C_nH_{2n+2}	Structural Formula	Condensed Structural Formula
1	Methane	CH_4		CH_4
2	Ethane	C_2H_6		CH_3CH_3
3	Propane	C_3H_8		$CH_3CH_2CH_3$
4	n-Butane	C_4H_{10}		$CH_3CH_2CH_2CH_3$
5	n-Pentane	C_5H_{12}		$CH_3CH_2CH_2CH_2CH_3$
6	n-Hexane	C_6H_{14}		$CH_3CH_2CH_2CH_2CH_2CH_3$
7	n-Heptane	C_7H_{16}		$CH_3CH_2CH_2CH_2CH_2CH_2CH_3$
8	n-Octane	C_8H_{18}		$CH_3CH_2CH_2CH_2CH_2CH_2CH_2CH_3$
9	n-Nonane	C_9H_{20}		$CH_3CH_2CH_2CH_2CH_2CH_2CH_2CH_2CH_3$
10	n-Decane	$C_{10}H_{22}$		$CH_3CH_2CH_2CH_2CH_2CH_2CH_2CH_2CH_2CH_3$

TABLE 20.4 Uses of Hydrocarbons

Number of Carbon Atoms	State	Major Uses
1–4	Gas	Heating fuel, cooking fuel
5–7	Low-boiling liquids	Solvents, gasoline
6–18	Liquids	Gasoline
12–24	Liquids	Jet fuel, portable-stove fuel
18–50	High-boiling liquids	Diesel fuel, lubricants, heating oil
50+	Solids	Petroleum jelly, paraffin wax

through decane, which contains 10 carbon atoms. Like pentane, hexane through decane are all components of gasoline. Table 20.4 summarizes the many uses of hydrocarbons.

Naming Alkanes

Many organic compounds have common names that we can learn only through familiarity. Because of the sheer number of organic compounds, however, we need a systematic method of nomenclature. In this book, we adopt the nomenclature system, which is recommended by the IUPAC (International Union of Pure and Applied Chemistry) and used throughout the world. In this system, the longest continuous chain of carbon atoms—called the base chain—determines the base name of the compound. The root of the base name depends on the number of carbon atoms in the base chain, as shown in Table 20.5. Base names for alkanes always have the ending *-ane*. Groups of carbon atoms branching off the base chain are called alkyl groups and are named as substituents. Remember that a *substituent* is an atom or group of atoms that has been substituted for a hydrogen atom in an organic compound. Common alkyl groups are shown in Table 20.6.

The procedure shown in Examples 20.2 and 20.3 allows us to systematically name many alkanes.

TABLE 20.5 Prefixes for Base Names of Alkane Chains

Number of Carbon Atoms	Prefix
1	meth-
2	eth-
3	prop-
4	but-
5	pent-
6	hex-
7	hept-
8	oct-
9	non-
10	dec-

TABLE 20.6 Common Alkyl Groups

Condensed Structural Formula	Name	Condensed Structural Formula	Name
$-CH_3$	Methyl	$-CHCH_3$ $\quad\vert$ $\quad CH_3$	Isopropyl
$-CH_2CH_3$	Ethyl	$-CH_2CHCH_3$ $\qquad\vert$ $\qquad CH_3$	Isobutyl
$-CH_2CH_2CH_3$	Propyl	$-CHCH_2CH_3$ $\quad\vert$ $\quad CH_3$	*sec*-Butyl
$-CH_2CH_2CH_2CH_3$	Butyl	$\qquad CH_3$ $\qquad\vert$ $-CCH_3$ $\qquad\vert$ $\qquad CH_3$	*tert*-Butyl

PROCEDURE FOR... Naming Alkanes	EXAMPLE 20.2 Naming Alkanes	EXAMPLE 20.3 Naming Alkanes
	Name the alkane. $$CH_3-CH_2-CH-CH_2-CH_3$$ with CH_2 and CH_3 branches below the third carbon.	Name the alkane. $$CH_3-CH-CH_2-CH-CH_2-CH_2-CH-CH_3$$ with CH_3, CH_2–CH_3, and CH_3 branches below.
1. Count the number of carbon atoms in the longest continuous carbon chain to determine the base name of the compound. Find the prefix corresponding to this number of atoms in Table 20.5 and add the ending -*ane* to form the base name.	**SOLUTION** This compound has five carbon atoms in its longest continuous chain. $$CH_3-CH_2-CH-CH_2-CH_3$$ with CH_2 and CH_3 below. The correct prefix from Table 20.5 is *pent-*. The base name is pentane.	**SOLUTION** This compound has eight carbon atoms in its longest continuous chain. $$CH_3-CH-CH_2-CH-CH_2-CH_2-CH-CH_3$$ with CH_3, CH_2–CH_3, and CH_3 below. The correct prefix from Table 20.5 is *oct-*. The base name is octane.
2. Consider every branch from the base chain to be a substituent. Name each substituent according to Table 20.6.	This compound has one substituent named *ethyl*. $$CH_3-CH_2-CH-CH_2-CH_3$$ ethyl → CH_2 / CH_3	This compound has one substituent named *ethyl* and two named *methyl*. ethyl → CH_2–CH_3; methyl → CH_3 groups
3. Beginning with the end closest to the branching, number the base chain and assign a number to each substituent. (If two substituents occur at equal distances from each end, go to the next substituent to determine from which end to start numbering.)	Number the base chain as follows: $$\overset{1}{CH_3}-\overset{2}{CH_2}-\overset{3}{CH}-\overset{4}{CH_2}-\overset{5}{CH_3}$$ with CH_2 / CH_3 below carbon 3. Assign the ethyl substituent the number 3.	Number the base chain as follows: $$\overset{1}{CH_3}-\overset{2}{CH}-\overset{3}{CH_2}-\overset{4}{CH}-\overset{5}{CH_2}-\overset{6}{CH_2}-\overset{7}{CH}-\overset{8}{CH_3}$$ Assign the ethyl substituent the number 4 and assign the two methyl substituents the numbers 2 and 7.
4. Write the name of the compound in the format: (substituent number)-(substituent name)(base name) If there are two or more substituents, give each one a number and list them alphabetically with hyphens between words and numbers.	The name of the compound is 3-ethylpentane	The basic form of the name of the compound is 4-ethyl-2,7-methyloctane List ethyl before methyl because substituents are listed in alphabetical order.

5. If a compound has two or more identical substituents, designate the number of identical substituents with the prefix *di-* (2), *tri-* (3), or *tetra-* (4) before the substituent's name. Separate the numbers indicating the positions of the substituents relative to each other with a comma. The prefixes are not taken into account when alphabetizing.	Does not apply to this compound.	This compound has two methyl substituents; therefore, the name of the compound is 4-ethyl-2,7-dimethyloctane

FOR PRACTICE 20.2 Name the alkane. 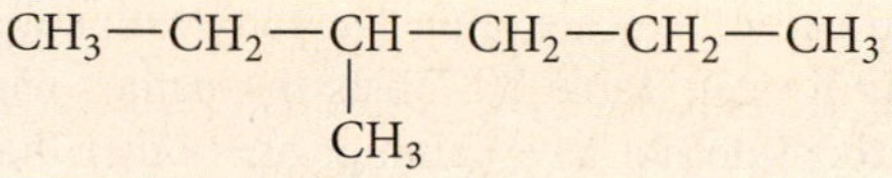 $CH_3-CH_2-CH-CH_2-CH_2-CH_3$ $\qquad\qquad\quad\ \ \|$ $\qquad\qquad\quad CH_3$	**FOR PRACTICE 20.3** Name the alkane. 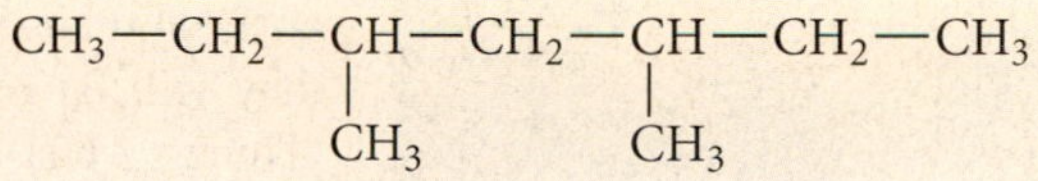 $CH_3-CH_2-CH-CH_2-CH-CH_2-CH_3$ $\qquad\qquad\quad\ \ \|\qquad\qquad\ \|$ $\qquad\qquad\quad CH_3\qquad\quad CH_3$

EXAMPLE 20.4 Naming Alkanes

Name the alkane:

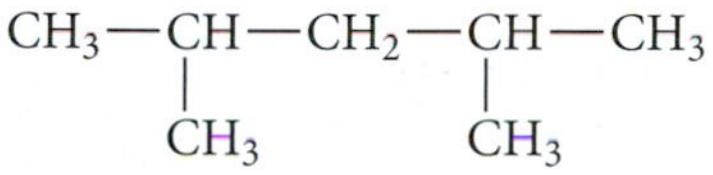

$CH_3-CH-CH_2-CH-CH_3$
$\qquad\ \ \|\qquad\qquad\ \|$
$\qquad CH_3\qquad\quad CH_3$

SOLUTION

1. The longest continuous carbon chain has five atoms. Therefore the base name is pentane.	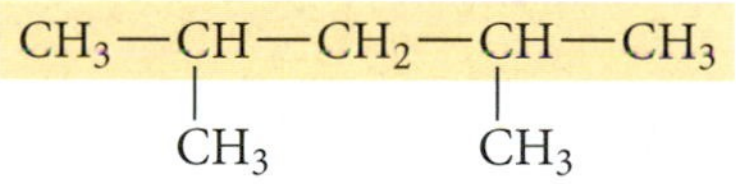 $CH_3-CH-CH_2-CH-CH_3$ $\qquad\ \ \|\qquad\qquad\ \|$ $\qquad CH_3\qquad\quad CH_3$
2. This compound has two substituents, both of which are named methyl.	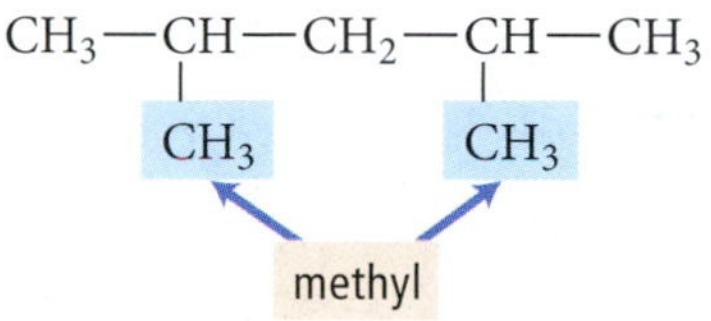 $CH_3-CH-CH_2-CH-CH_3$ $\qquad\ \ \|\qquad\qquad\ \|$ $\qquad CH_3\qquad\quad CH_3$ methyl
3. Since the two substituents are equidistant from the ends, it does not matter from which end you start numbering.	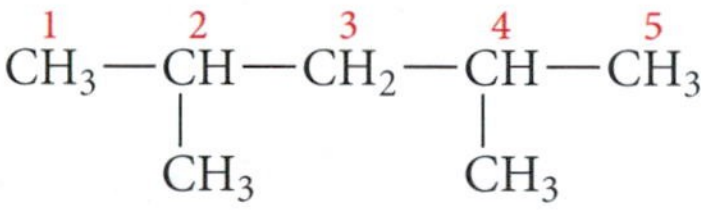 $\overset{1}{C}H_3-\overset{2}{C}H-\overset{3}{C}H_2-\overset{4}{C}H-\overset{5}{C}H_3$ $\qquad\ \ \|\qquad\qquad\ \|$ $\qquad CH_3\qquad\quad CH_3$
4, 5. Use the general form for the name: (substituent number)-(substituent name)(base name) Since this compound contains two identical substituents, step 5 from the naming procedure applies and you use the prefix *di-*. You also indicate the position of each substituent with a number separated by a comma.	2,4-dimethylpentane

FOR PRACTICE 20.4

Name the alkane:

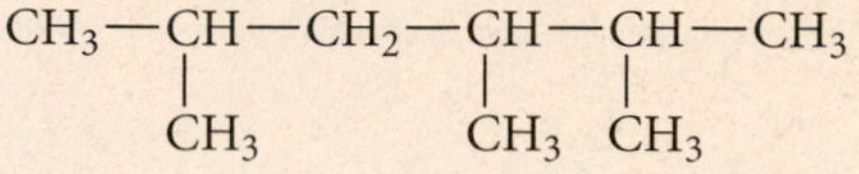

$CH_3-CH-CH_2-CH-CH-CH_3$
$\qquad\ \ \|\qquad\qquad\ \|\quad\ \|$
$\qquad CH_3\qquad\quad CH_3\ CH_3$

20.5 Alkenes and Alkynes

Alkenes are hydrocarbons containing at least one double bond between carbon atoms; alkynes contain at least one triple bond. Because of the double or triple bond, alkenes and alkynes have fewer hydrogen atoms than the corresponding alkane and are therefore called **unsaturated hydrocarbons**—they are not loaded to capacity with hydrogen. As we learned earlier, noncyclic alkenes have the formula C_nH_{2n} and noncyclic alkynes have the formula C_nH_{2n-2}. The simplest alkene is ethene (C_2H_4), also called ethylene.

The formulas shown here for alkenes and alkynes assume only one multiple bond.

Ethene or ethylene C_2H_4

Formula   Structural formula   Space-filling model

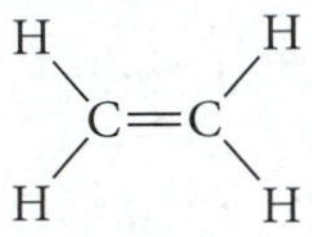

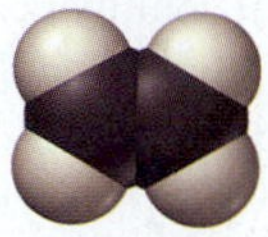

The geometry about each carbon atom in ethene is trigonal planar, making ethene a flat, rigid molecule (see the the valence bond model of ethane in Example 10.8). Ethene is a ripening agent in fruit. For example, when a banana within a cluster of bananas begins to ripen, it emits ethene. The ethene then causes other bananas in the cluster to ripen. Banana farmers usually pick bananas green for ease of shipping. When the bananas arrive at their destination, they are often "gassed" with ethene to initiate ripening so that they will be ready to sell. Table 20.7 lists the names and structures of several other alkenes. Most of them do not have familiar uses other than their presence as minority components of fuels.

TABLE 20.7 Alkenes

n	Name	Molecular Formula C_nH_{2n}	Structural Formula	Condensed Structural Formula
2	Ethene	C_2H_4	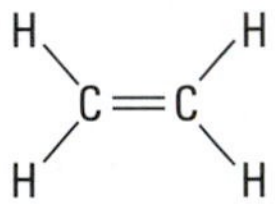	$CH_2{=}CH_2$
3	Propene	C_3H_6	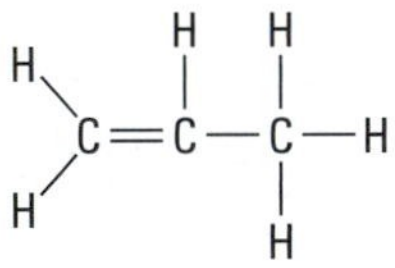	$CH_2{=}CHCH_3$
4	1-Butene*	C_4H_8	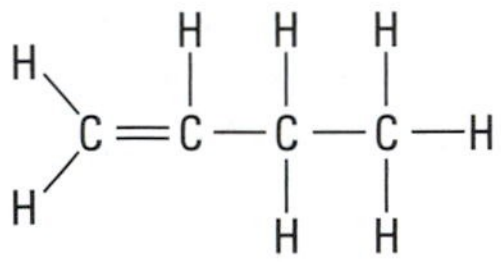	$CH_2{=}CHCH_2CH_3$
5	1-Pentene*	C_5H_{10}	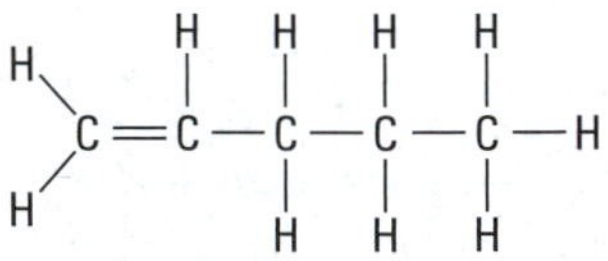	$CH_2{=}CHCH_2CH_2CH_3$
6	1-Hexene*	C_6H_{12}	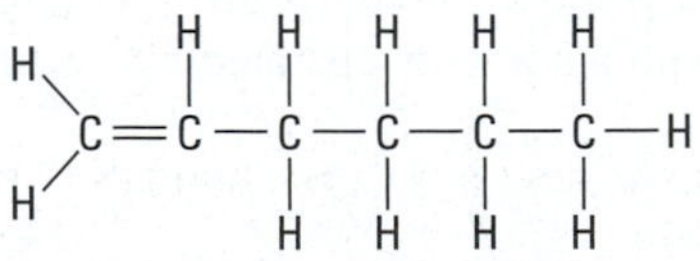	$CH_2{=}CHCH_2CH_2CH_2CH_3$

* These alkenes have one or more isomers depending on the position of the double bond. The isomers shown here have the double bond in the 1 position, meaning the first carbon–carbon bond of the chain.

TABLE 20.8   Alkynes

n	Name	Molecular Formula C_nH_{2n-2}	Structural Formula	Condensed Structural Formula
2	Ethyne	C_2H_2	H—C≡C—H	CH≡CH
3	Propyne	C_3H_4	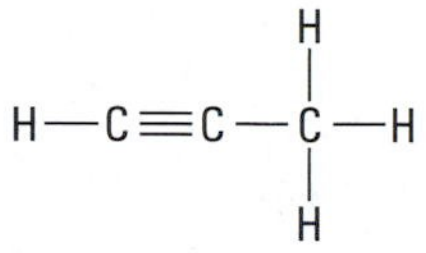 H—C≡C—C—H (with H above and below C)	CH≡CCH$_3$
4	1-Butyne*	C_4H_6	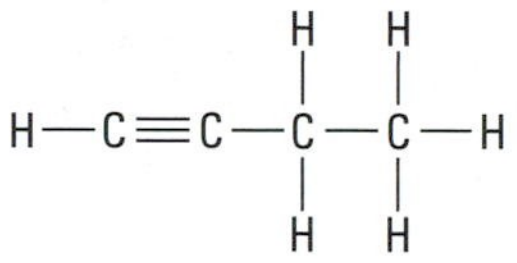 H—C≡C—C—C—H (with H's above and below)	CH≡CCH$_2$CH$_3$
5	1-Pentyne*	C_5H_8	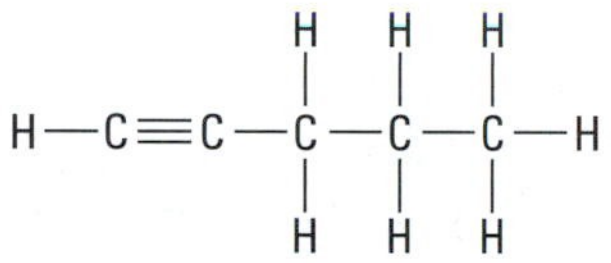 H—C≡C—C—C—C—H (with H's above and below)	CH≡CCH$_2$CH$_2$CH$_3$
6	1-Hexyne*	C_6H_{10}	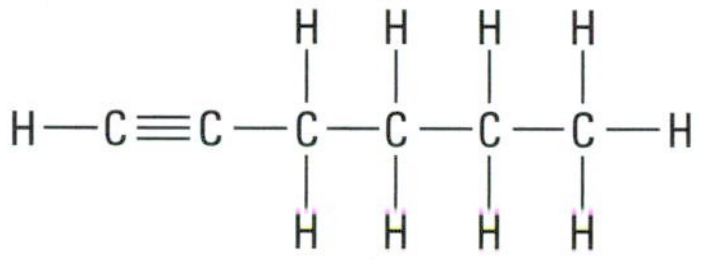 H—C≡C—C—C—C—C—H (with H's above and below)	CH≡CCH$_2$CH$_2$CH$_2$CH$_3$

* These alkynes have one or more isomers depending on the position of the triple bond. The isomers shown here have the triple bond in the 1 position, meaning the first carbon–carbon bond of the chain.

The simplest alkyne is ethyne, C_2H_2, also called acetylene.

Ethyne or acetylene C_2H_2 H—C≡C—H

Formula   Structural formula   Space-filling model

The geometry about each carbon atom in ethyne is linear, making ethyne a linear molecule. Ethyne (or acetylene) is commonly used as fuel for welding torches. The names and structures of several other alkynes are shown in Table 20.8. Like alkenes, the alkynes do not have familiar uses, other than being present as minority components of gasoline.

Naming Alkenes and Alkynes

We name alkenes and alkynes in the same way as alkanes with the following exceptions.

- The base chain is the longest continuous carbon chain *that contains the double or triple bond.*
- We give the base name the ending *-ene* for alkenes and *-yne* for alkynes.
- We number the base chain to give the double or triple bond the lowest possible number.
- We insert a number indicating the position of the double or triple bond (lowest possible number) just before the base name.

▲ Welding torches often burn ethyne in pure oxygen to produce the very hot flame needed for melting metals.

For example, the alkenes shown here are named 2-methyl-2-pentene and 1-butyne:

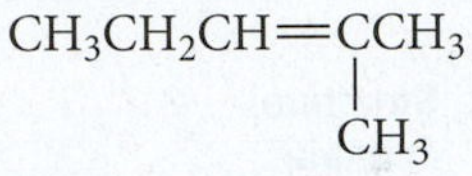

$$CH_3CH_2CH{=}CCH_3$$
$$\overset{|}{\underset{CH_3}{}}$$

$$CH{\equiv}CCH_2CH_3$$

2-Methyl-2-pentene   1-Butyne

EXAMPLE 20.5 Naming Alkenes and Alkynes

Name each compound:

(a) 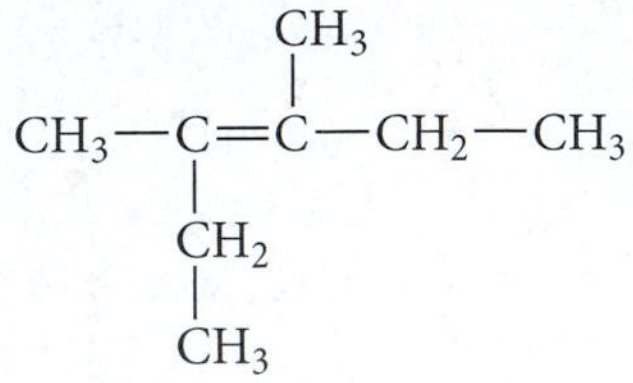
$$CH_3{-}\overset{\overset{\displaystyle CH_3}{|}}{C}{=}\underset{\underset{\displaystyle CH_2}{\underset{\underset{\displaystyle CH_3}{|}}{|}}}{C}{-}CH_2{-}CH_3$$

(b) 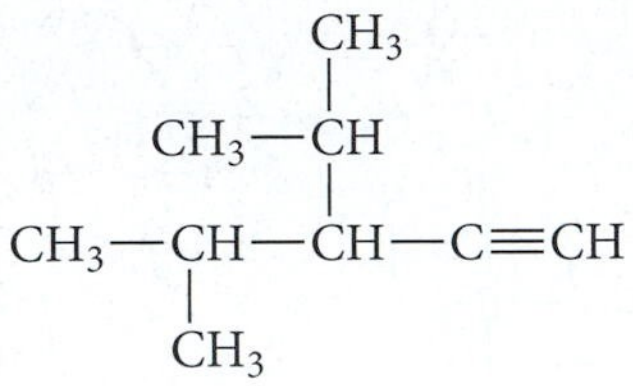
$$CH_3{-}\underset{\underset{\displaystyle CH_3}{|}}{CH}{-}\overset{\overset{\displaystyle CH_3{-}\overset{\displaystyle CH_3}{|}}{CH}}{CH}{-}C{\equiv}CH$$

SOLUTION

(a) 1. The longest continuous carbon chain containing the double bond has six carbon atoms. The base name is therefore *hexene*.	$$CH_3{-}\overset{\overset{\displaystyle CH_3}{|}}{C}{=}\underset{\underset{\displaystyle CH_2}{\underset{\underset{\displaystyle CH_3}{|}}{|}}}{C}{-}CH_2{-}CH_3$$
2. The two substituents are both methyl.	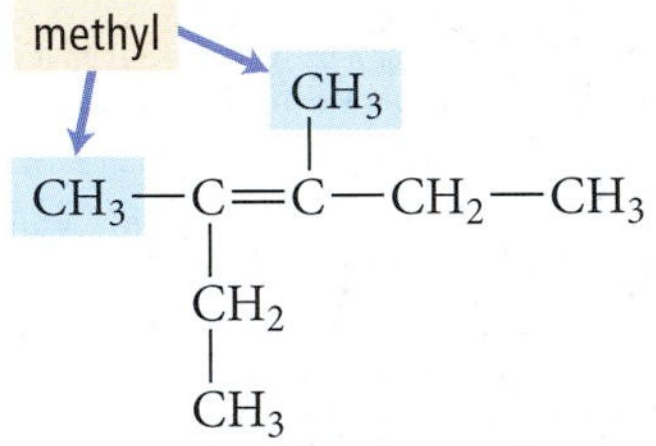 methyl $$CH_3{-}\overset{\overset{\displaystyle CH_3}{|}}{C}{=}\underset{\underset{\displaystyle CH_2}{\underset{\underset{\displaystyle CH_3}{|}}{|}}}{C}{-}CH_2{-}CH_3$$
3. One of the exceptions listed previously states that, in naming alkenes, you should number the chain so that the *double bond* has the lowest number. In this case, the double bond is equidistant from the ends. Assign the double bond the number 3. The two methyl groups are, therefore, at positions 3 and 4.	$$CH_3{-}\overset{\overset{\displaystyle CH_3}{|}}{\underset{3}{C}}{=}\underset{\underset{\displaystyle 2\,CH_2}{\underset{\underset{\displaystyle CH_3}{|}}{\underset{\displaystyle 1}{|}}}}{\underset{4}{C}}{-}\underset{5}{CH_2}{-}\underset{6}{CH_3}$$
4, 5. Use the general form for the name: (substituent number)-(substituent name)(base name) Since this compound contains two identical substituents, step 5 of the naming procedure applies, so use the prefix *di-*. In addition, indicate the position of each substituent with a number separated by a comma. Since this compound is an alkene, specify the position of the double bond, isolated by hyphens, just before the base name.	3,4-dimethyl-3-hexene
(b) 1. The longest continuous carbon chain containing the triple bond is five carbons long; therefore the base name is *pentyne*.	$$CH_3{-}\underset{\underset{\displaystyle CH_3}{|}}{CH}{-}\overset{\overset{\displaystyle CH_3{-}\overset{\displaystyle CH_3}{|}}{CH}}{CH}{-}C{\equiv}CH$$

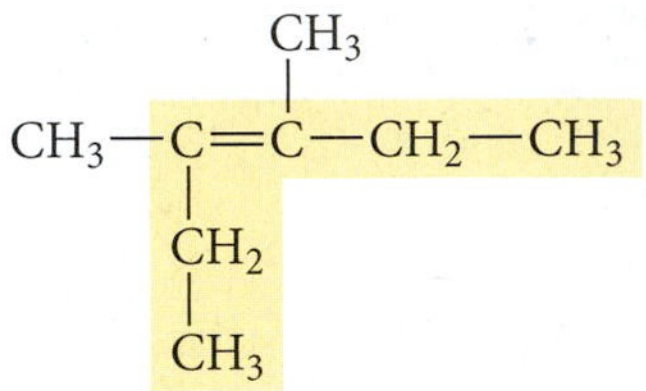

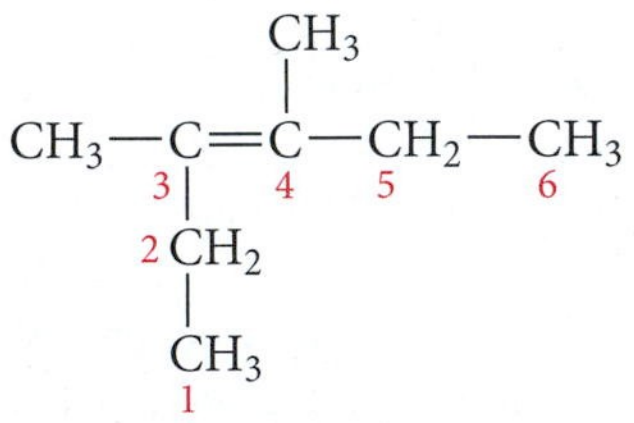

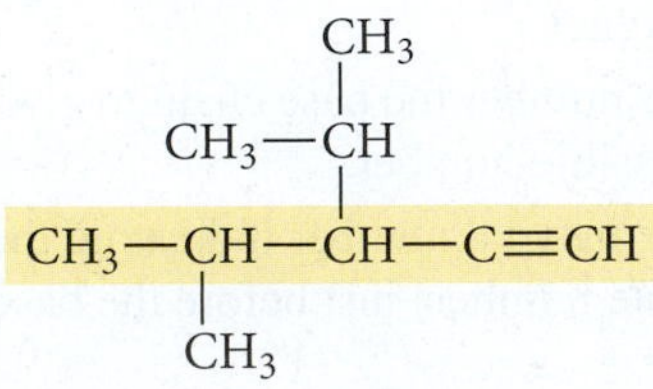

2. There are two substituents; one is a methyl group and the other an isopropyl group.

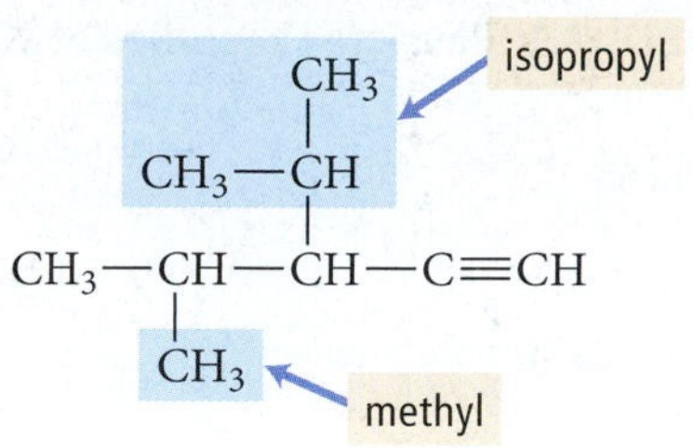

3. Number the base chain, giving the triple bond the lowest number (1). Give the isopropyl and methyl groups the numbers 3 and 4, respectively.

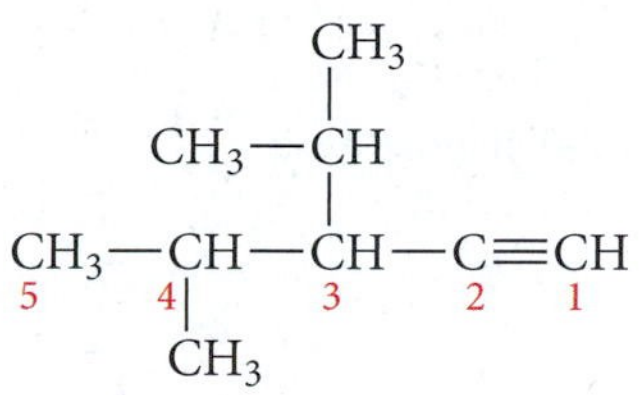

4. Use the general form for the name:

(substituent number)-(substituent name)(base name)

Since there are two substituents, list both of them in alphabetical order. Because this compound is an alkyne, specify the position of the triple bond with a number isolated by hyphens just before the base name.

3-isopropyl-4-methyl-1-pentyne

FOR PRACTICE 20.5

Name each compound:

(a) 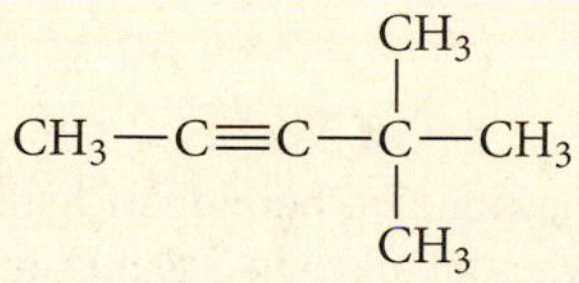 $CH_3-C\equiv C-\underset{\underset{CH_3}{|}}{\overset{\overset{CH_3}{|}}{C}}-CH_3$

(b) 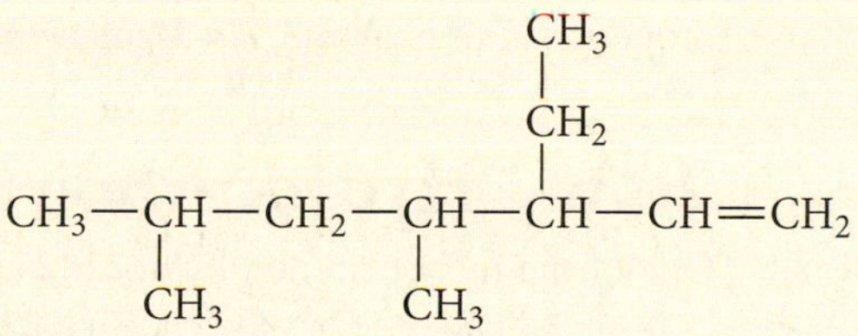 $CH_3-\underset{\underset{CH_3}{|}}{CH}-CH_2-\underset{\underset{CH_3}{|}}{CH}-\underset{\underset{|}{CH_2}}{\overset{\overset{CH_3}{|}}{CH}}-CH=CH_2$

Geometric (Cis–Trans) Isomerism in Alkenes

A major difference between a single bond and a double bond is the degree to which rotation occurs about the bond. As discussed in Section 10.7, rotation about a double bond is highly restricted due to the overlap between unhybridized p orbitals on the adjacent carbon atoms. Consider, for example, the difference between 1,2-dichloroethane and 1,2-dichloroethene:

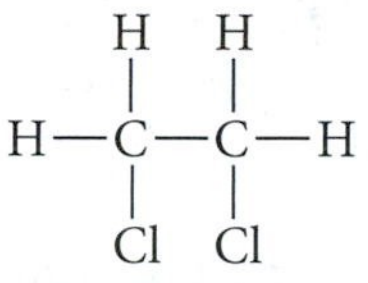

1, 2-Dichloroethane 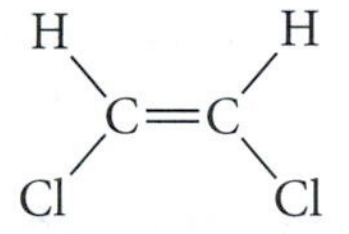   1, 2-Dichloroethene

The hybridization of the carbon atoms in 1,2-dichloroethane is sp^3, resulting in relatively free rotation about the sigma single bond. Consequently, there is no difference between these two structures at room temperature because they quickly interconvert:

In contrast, rotation about the double bond (sigma + pi) in 1,2-dichloroethene is restricted, so that, at room temperature, 1,2-dichloroethene can exist in the two isomeric forms shown in Table 20.9. These two forms of 1,2-dichloroethene are different compounds with

TABLE 20.9 Physical Properties of *cis*- and *trans*-1,2-Dichloroethene

Name	Structure	Space-filling Model	Density (g/mL)	Melting Point (°C)	Boiling Point (°C)
cis-1,2-Dichloroethene			1.284	−80.5	60.1
trans-1,2-Dichloroethene			1.257	−49.4	47.5

different properties. This kind of isomerism is a type of stereoisomerism (see Section 20.3) called **geometric (or cis–trans) isomerism**. We distinguish between the two isomers with the designations *cis* (meaning "same side") and *trans* (meaning "opposite sides"). Cis–trans isomerism is common in alkenes. Consider *cis*- and *trans*-2-butene.

cis-2-Butene   *trans*-2-Butene

Like the two isomers of 1,2-dichloroethene, these two isomers have different physical properties. For example, *cis*-2-butene boils at 3.7 °C, and *trans*-2-butene boils at 0.9 °C.

20.6 Hydrocarbon Reactions

One of the most common hydrocarbon reactions is combustion, the burning of hydrocarbons in the presence of oxygen. Alkanes, alkenes, and alkynes all undergo combustion. In a combustion reaction, the hydrocarbon reacts with oxygen to form carbon dioxide and water.

$$CH_3CH_2CH_3(g) + 5\,O_2(g) \longrightarrow 3\,CO_2(g) + 4\,H_2O(g) \quad \text{Alkane combustion}$$
$$CH_2{=}CHCH_2CH_3(g) + 6\,O_2(g) \longrightarrow 4\,CO_2(g) + 4\,H_2O(g) \quad \text{Alkene combustion}$$
$$CH{\equiv}CCH_3(g) + 4\,O_2(g) \longrightarrow 3\,CO_2(g) + 2\,H_2O(g) \quad \text{Alkyne combustion}$$

Hydrocarbon combustion reactions are highly exothermic and are commonly used to warm homes and buildings, to generate electricity, and to power the engines of cars, ships, and airplanes.

Reactions of Alkanes

In addition to combustion, alkanes also undergo substitution reactions, in which one or more hydrogen atoms on an alkane are replaced by one or more other types of atoms. The most common substitution reaction is halogen substitution. For example, methane reacts with chlorine gas in the presence of heat or light to form chloromethane:

$$CH_4(g) + Cl_2(g) \xrightarrow{\text{heat or light}} CH_3Cl(g) + HCl(g)$$

Methane   Chlorine   Chloromethane

Ethane reacts with chlorine gas to form chloroethane:

$$CH_3CH_3(g) + Cl_2(g) \xrightarrow{\text{heat or light}} CH_3CH_2Cl(g) + HCl(g)$$

Ethane   Chlorine   Chloroethane

Multiple halogenation reactions can occur because halogens can replace more than one of the hydrogen atoms on an alkane. The general form for halogen substitution reactions is:

$$R{-}H + X_2 \xrightarrow{\text{heat or light}} R{-}X + HX$$

Alkane   Halogen   Haloalkane   Hydrogen
                                      halide

The R in this chemical equation is generic for a hydrocarbon group.

Reactions of Alkenes and Alkynes

Alkenes and alkynes undergo addition reactions in which molecules add across (on either side of) the multiple bond. For example, ethene reacts with chlorine gas to form dichloroethane.

$$\underset{H}{\overset{H}{>}}C=C\underset{H}{\overset{H}{<}} \ + \ Cl-Cl \ \longrightarrow \ H-\underset{|}{\overset{|}{C}}-\underset{|}{\overset{|}{C}}-H$$

Catalysts are often labeled over the reaction arrow.

Notice that the addition of chlorine converts the carbon–carbon double bond into a single bond because each carbon atom now has a new bond to a chlorine atom. Alkenes and alkynes can also add hydrogen in hydrogenation reactions. For example, in the presence of an appropriate catalyst, propene reacts with hydrogen gas to form propane.

$$H-\underset{\underset{H}{|}}{\overset{\overset{H}{|}}{C}}-C=C\underset{H}{\overset{H}{<}} \ + \ H-H \ \xrightarrow{\text{catalyst}} \ H-\underset{\underset{H}{|}}{\overset{\overset{H}{|}}{C}}-\underset{\underset{H}{|}}{\overset{\overset{H}{|}}{C}}-\underset{\underset{H}{|}}{\overset{\overset{H}{|}}{C}}-H$$

Hydrogenation reactions convert unsaturated hydrocarbons into saturated hydrocarbons. Hydrogenation reactions are also used to convert unsaturated vegetable oils into saturated fats. Most vegetable oils are unsaturated because their carbon chains contain double bonds. The double bonds put bends into the carbon chains that result in less efficient packing of molecules; thus vegetable oils are liquids at room temperature while saturated fats are solids at room temperature. When hydrogen is added to the double bonds of vegetable oil, the unsaturated fat is converted into a saturated fat, turning the liquid oil into a solid at room temperature. The words "hydrogenated vegetable oil" on a label indicate a food product that contains saturated fats made via hydrogenation reactions.

▲ Hydrogenated vegetable oil is a saturated fat that is made by hydrogenating unsaturated fats.

EXAMPLE 20.6 Alkene Addition Reactions

Determine the products of the reaction:

$$CH_3CH_2CH=CH_2 + Br_2 \longrightarrow$$

SOLUTION

The reaction of 1-butene with bromine is an example of a symmetric addition. The bromine adds across the double bond so that each carbon forms a single bond to a bromine atom.

$$H-\underset{\underset{H}{|}}{\overset{\overset{H}{|}}{C}}-\underset{\underset{H}{|}}{\overset{\overset{H}{|}}{C}}-C=C-H \ + \ Br-Br \ \longrightarrow$$

$$H-\underset{\underset{H}{|}}{\overset{\overset{H}{|}}{C}}-\underset{\underset{H}{|}}{\overset{\overset{H}{|}}{C}}-\underset{\underset{Br}{|}}{\overset{\overset{H}{|}}{C}}-\underset{\underset{H}{|}}{\overset{\overset{Br}{|}}{C}}-H$$

FOR PRACTICE 20.6

Determine the products of the reaction:

$$CH_3-\underset{\underset{H}{|}}{\overset{\overset{CH_3}{|}}{C}}-\underset{\underset{H}{|}}{C}=CH_2 + H_2 \xrightarrow{\text{catalyst}}$$

20.7   Aromatic Hydrocarbons

As you might imagine, determining the structure of organic compounds has not always been easy. In the mid-1800s chemists were trying to determine the structure of a particularly stable organic compound named benzene (C_6H_6). In 1865, Friedrich August Kekulé (1829–1896) had a dream in which he envisioned chains of carbon atoms as snakes. The snakes danced before him, and one of them twisted around and bit its tail. Based on that vision, Kekulé proposed the following structure for benzene.

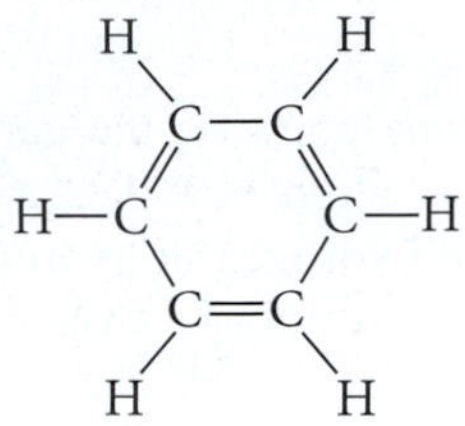

This structure features alternating single and double bonds. When we examine the bond lengths in benzene, however, we find that all the bonds are the same length. The structure of benzene is better represented by the following resonance structures.

Resonance structures were defined in Section 9.8. Recall that the actual structure of a molecule represented by resonance structures is intermediate between the two resonance structures and is called a *resonance hybrid*.

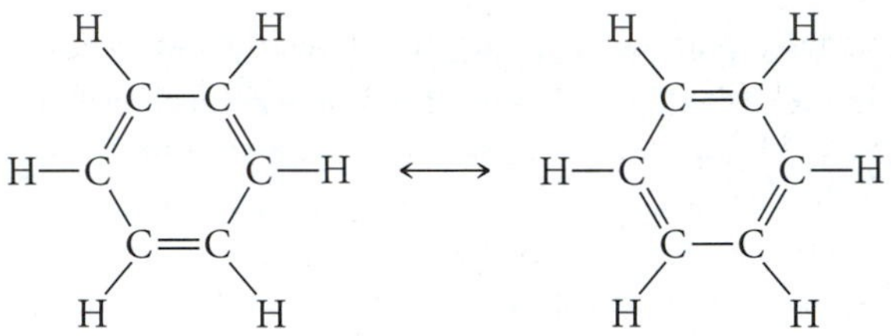

The true structure of benzene is a hybrid of the two resonance structures. Benzene is often represented with the following carbon skeletal formula (or line formula).

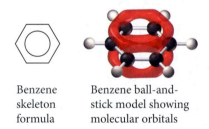

Benzene   Benzene ball-and-
skeleton   stick model showing
formula   molecular orbitals

The ring in the carbon skeletal formula represents the delocalized π electrons that occupy the molecular orbitals superimposed on the ball-and-stick structure. When drawing benzene rings, either by themselves or as parts of other compounds, organic chemists often use either the carbon skeletal structure shown here or just one of the resonance structures with alternating single and double bonds. Both representations indicate the same thing—a benzene ring.

The benzene ring structure occurs in many organic compounds. An atom or group of atoms can be substituted for one or more of the six hydrogen atoms on the ring to form substituted benzenes. Chlorobenzene and phenol are two examples of substituted benzenes.

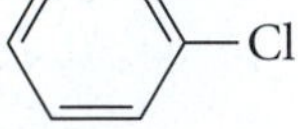

Chlorobenzene 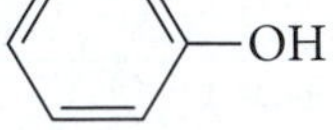   Phenol

Since many compounds containing benzene rings have pleasant aromas, benzene rings are also called aromatic rings, and compounds containing them are referred to as aromatic compounds. The pleasant smells of cinnamon, vanilla, and jasmine all result from aromatic compounds.

Naming Aromatic Hydrocarbons

Monosubstituted benzenes—benzenes in which only one of the hydrogen atoms has been substituted—are often named as derivatives of benzene.

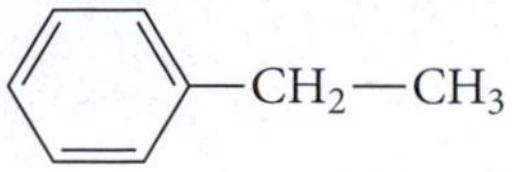

Ethylbenzene 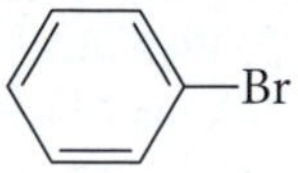   Bromobenzene

These names take the following general form:

(name of substituent)benzene

However, many monosubstituted benzenes have common names that can only be learned through familiarity. Four examples are shown here:

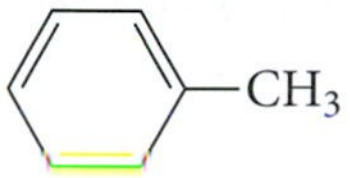

Toluene 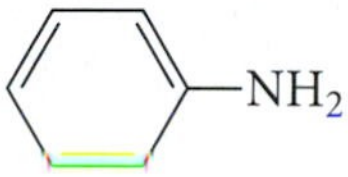   Aniline 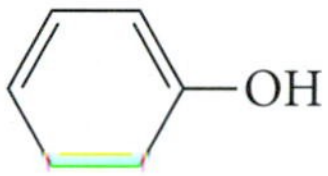   Phenol 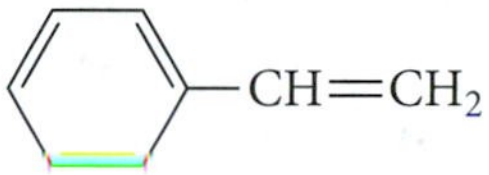   Styrene

Disubstituted benzenes—benzenes in which two hydrogen atoms have been substituted—are numbered and the substituents are listed alphabetically. The order of numbering on the ring is also determined by the alphabetical order of the substituents.

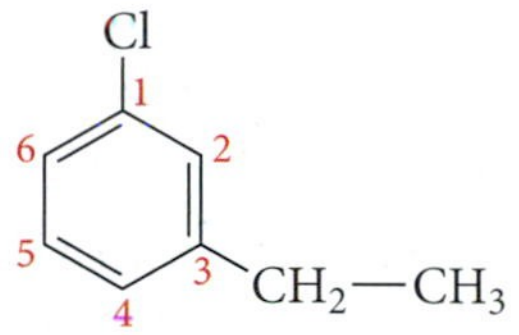

1-Chloro-3-ethylbenzene    1-Bromo-2-chlorobenzene

When the two substituents are identical, we use the prefix *di-*.

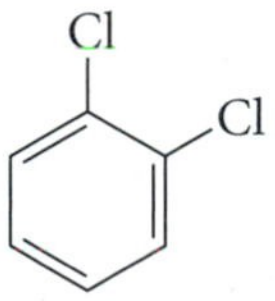

1,2-Dichlorobenzene   1,3-Dichlorobenzene   1,4-Dichlorobenzene
ortho-Dichlorobenzene   *meta*-Dichlorobenzene   *para*-Dichlorobenzene

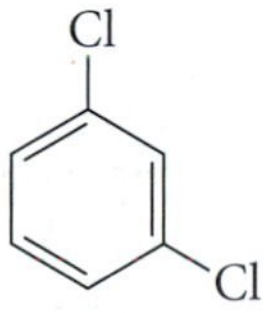

Also in common use, in place of numbering, are the prefixes ortho (1,2 disubstituted), meta (1,3 disubstituted), and para (1,4 disubstituted).

Compounds containing fused aromatic rings are called polycyclic aromatic hydrocarbons. Naphthalene, the substance that composes mothballs, and pyrene, a carcinogen found in cigarette smoke are common examples and are shown in **Figure 20.3▶**.

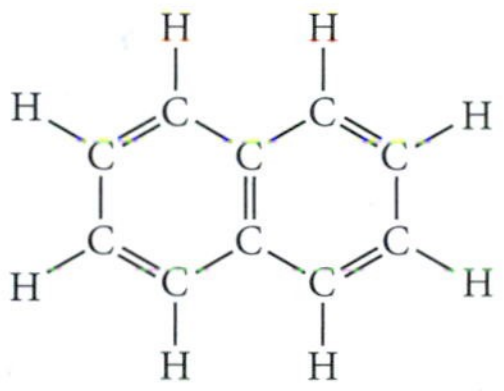

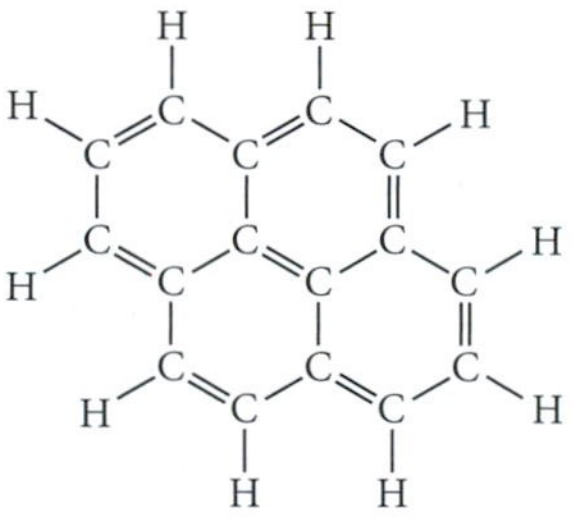

▲ **FIGURE 20.3 Polycyclic Aromatic Compounds** The structures of some common polycyclic aromatic compounds contain fused rings.

20.8 Functional Groups

Most other families of organic compounds can be thought of as hydrocarbons with a **functional group**—a characteristic atom or group of atoms—inserted into the hydrocarbon. A group of organic compounds with the same functional group forms a **family**. For example, the members of the family of alcohols have an —OH functional group and the general formula R—OH, where R represents a hydrocarbon group. (That is, we refer to

TABLE 20.10   **Some Common Functional Groups**

Family	General Formula*	Condensed General Formula	Example	Name
Alcohols	R—OH	ROH	CH_3CH_2OH	Ethanol (ethyl alcohol)
Ethers	R—O—R	ROR	CH_3OCH_3	Dimethyl ether
Aldehydes	$R-\overset{\displaystyle O}{\overset{\displaystyle \|}{C}}-H$	RCHO	$CH_3-\overset{\displaystyle O}{\overset{\displaystyle \|}{C}}-H$	Ethanal (acetaldehyde)
Ketones	$R-\overset{\displaystyle O}{\overset{\displaystyle \|}{C}}-R$	RCOR	$CH_3-\overset{\displaystyle O}{\overset{\displaystyle \|}{C}}-CH_3$	Propanone (acetone)
Carboxylic acids	$R-\overset{\displaystyle O}{\overset{\displaystyle \|}{C}}-OH$	RCOOH	$CH_3-\overset{\displaystyle O}{\overset{\displaystyle \|}{C}}-OH$	Ethanoic acid (acetic acid)
Esters	$R-\overset{\displaystyle O}{\overset{\displaystyle \|}{C}}-OR$	RCOOR	$CH_3-\overset{\displaystyle O}{\overset{\displaystyle \|}{C}}-OCH_3$	Methyl acetate
Amines	$R-\overset{\displaystyle R}{\overset{\displaystyle \|}{N}}-R$	R_3N	$CH_3CH_2-\overset{\displaystyle H}{\overset{\displaystyle \|}{N}}-H$	Ethylamine

*In ethers, ketones, esters, and amines, the two R groups may be the same or different.

the hydrocarbon group as an "R group.") Some specific examples of alcohols include methanol and isopropyl alcohol (also known as rubbing alcohol).

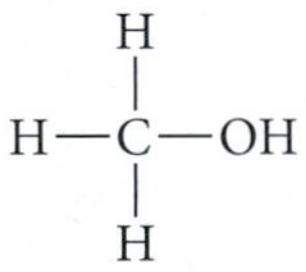

$$\underset{\text{Methanol}}{\overset{\displaystyle H}{\underset{\displaystyle H}{H-C-OH}}} \qquad \underset{\text{Isopropyl alcohol}}{\overset{\displaystyle H \quad CH_3}{\underset{\displaystyle H \quad H}{H-C-C-OH}}}$$

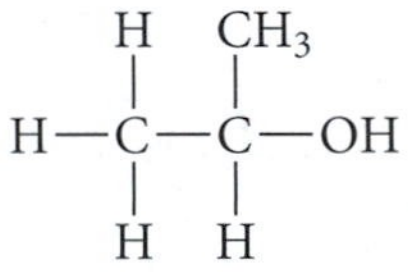

The insertion of a functional group into a hydrocarbon alters the properties of the compound significantly. For example, methanol—which can be thought of as methane with an —OH group substituted for one of the hydrogen atoms—is a polar, hydrogen-bonded liquid at room temperature. Methane, in contrast, is a nonpolar gas. Although each member of a family is unique and different, their common functional group also gives them some similarities in both their physical and chemical properties. Table 20.10 lists some common functional groups, their general formulas, and an example of each.

Alcohols

Alcohols are organic compounds containing the —OH functional group, or **hydroxyl group**, and they have the general formula R—OH. In addition to methanol and isopropyl alcohol, other common alcohols include ethanol and 1-butanol shown here.

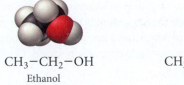

CH_3-CH_2-OH
Ethanol

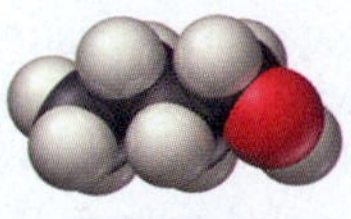

$CH_3-CH_2-CH_2-CH_2-OH$
1-Butanol

Naming Alcohols

We name alcohols similarly to the way we name alkanes with the following differences:

- The base chain is the longest continuous carbon chain that contains the —OH functional group.
- We give the base name the ending *-ol*.
- We number the base chain to give the —OH group the lowest possible number.
- We insert a number indicating the position of the —OH group just before the base name. For example,

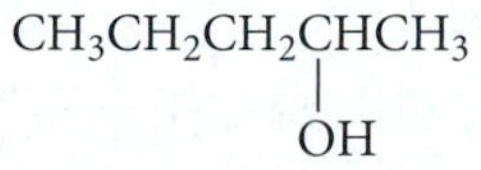

$$CH_3CH_2CH_2CHCH_3$$
$$|$$
$$OH$$
2-Pentanol

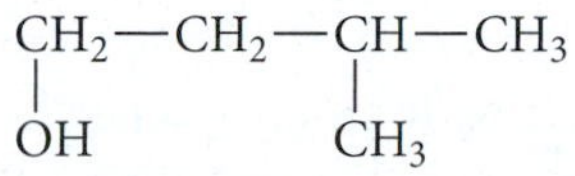

$$CH_2\!-\!CH_2\!-\!CH\!-\!CH_3$$
$$| \qquad\qquad |$$
$$OH \qquad\quad CH_3$$
3-Methyl-1-butanol

About Alcohols

Among the most familiar alcohols is ethanol, the alcohol in alcoholic beverages. Ethanol is commonly formed by the yeast fermentation of sugars, such as glucose, from fruits and grains.

$$C_6H_{12}O_6 \xrightarrow{\text{yeast}} 2\ CH_3CH_2OH + 2\ CO_2$$
Glucose $\qquad\qquad$ Ethanol

Alcoholic beverages contain ethanol, water, and a few other components that provide flavor and color. Ethanol is also used as a gasoline additive because it fosters complete combustion, reducing the levels of certain pollutants such as carbon monoxide and ozone precursors. Isopropyl alcohol (or 2-propanol) can be purchased at any drug store as rubbing alcohol. It is used as a disinfectant for wounds and to sterilize medical instruments. Isopropyl alcohol should never be consumed internally, as it is highly toxic. A few ounces of isopropyl alcohol can cause death. A third common alcohol is methanol, also called wood alcohol. Methanol is used as a laboratory solvent and as a fuel additive. Like isopropyl alcohol, methanol is toxic and should never be consumed.

Alcohol Reactions

Alcohols undergo a number of reactions including substitution, elimination (or dehydration), and oxidation, as shown in these examples.

Substitution $\qquad\qquad\qquad$ $ROH + HBr \longrightarrow R\!-\!Br + H_2O$

Elimination (or Dehydration)
$$CH_2\!-\!CH_2$$
$$| \qquad\quad | \qquad \xrightarrow{H_2SO_4} \quad CH_2\!=\!CH_2 + H_2O$$
$$H \qquad\; OH$$

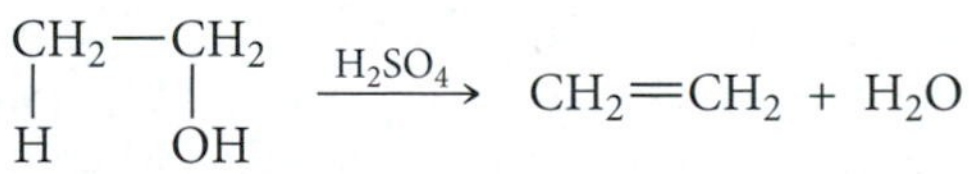

Oxidation $\qquad\qquad$ $CH_3CH_2OH \xrightarrow[H_2SO_4]{Na_2Cr_2O_7} CH_3COOH$

Aldehydes and Ketones

Aldehydes and **ketones** have the general formulas:

$$O$$
$$\|$$
$$R\!-\!C\!-\!H$$
Aldehyde

$$O$$
$$\|$$
$$R\!-\!C\!-\!R$$
Ketone

The condensed structural formula for aldehydes is RCHO and that for ketones is RCOR.

Both aldehydes and ketones contain a **carbonyl group**:

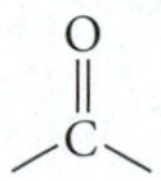

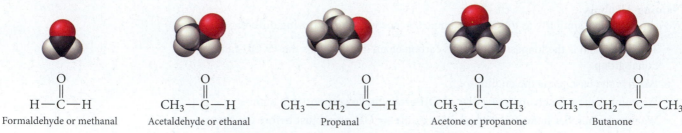

$$H-\overset{\displaystyle O}{\overset{\|}{C}}-H$$
Formaldehyde or methanal

$$CH_3-\overset{\displaystyle O}{\overset{\|}{C}}-H$$
Acetaldehyde or ethanal

$$CH_3-CH_2-\overset{\displaystyle O}{\overset{\|}{C}}-H$$
Propanal

$$CH_3-\overset{\displaystyle O}{\overset{\|}{C}}-CH_3$$
Acetone or propanone

$$CH_3-CH_2-\overset{\displaystyle O}{\overset{\|}{C}}-CH_3$$
Butanone

▲ **FIGURE 20.4 Common Aldehydes and Ketones**

Ketones have an R group attached to both sides of the carbonyl, while aldehydes have one R group and a hydrogen atom. (An exception is formaldehyde, which is an aldehyde with two H atoms attached to the carbonyl group.) Common aldehydes and ketones are shown in **Figure 20.4▲**.

Naming Aldehydes and Ketones

Many aldehydes and ketones have common names that can be learned only by becoming familiar with them. We systematically name simple aldehydes according to the number of carbon atoms in the longest continuous carbon chain that contains the carbonyl group. We form the base name from the name of the corresponding alkane by dropping the -*e* and adding the ending -*al*.

$$CH_3-CH_2-CH_2-\overset{\displaystyle O}{\overset{\|}{C}}-H$$
Butanal

$$CH_3-CH_2-CH_2-CH_2-\overset{\displaystyle O}{\overset{\|}{C}}-H$$
Pentanal

We systematically name simple ketones according to the longest continuous carbon chain containing the carbonyl group. We form the base name from the name of the corresponding alkane by dropping the letter -*e* and adding the ending -*one*. For ketones, we number the chain to give the carbonyl group the lowest possible number.

$$CH_3-CH_2-CH_2-\overset{\displaystyle O}{\overset{\|}{C}}-CH_3$$
2-Pentanone

$$CH_3-CH_2-CH_2-\overset{\displaystyle O}{\overset{\|}{C}}-CH_2-CH_3$$
3-Hexanone

About Aldehydes and Ketones

The most familiar aldehyde is probably formaldehyde. Formaldehyde is a gas with a pungent odor. It is often mixed with water to make formalin, a preservative and disinfectant. Formaldehyde is also found in wood smoke, which is one reason that smoking foods preserves them—the formaldehyde kills bacteria. Aromatic aldehydes, those that also contain a benzene ring, have pleasant aromas. Aldehydes are responsible for the smell of vanilla, cinnamon, and almonds. The most familiar ketone is acetone, the main component of nail polish remover. Many ketones also have pleasant aromas. For example, ketones are largely responsible for the smell of spearmint, cloves, and raspberries.

Aldehyde and Ketone Reactions

The carbonyl group in aldehydes and ketones is unsaturated, much like the double bond in an alkene. Therefore, the most common reactions of aldehydes and ketones are **addition reactions**. For example, HCN adds across the carbonyl double bond in formaldehyde:

$$H-\overset{\displaystyle O}{\overset{\|}{C}}-H \;+\; H-C\equiv N \;\xrightarrow{NaCN}\; N\equiv C-\overset{\displaystyle\overset{O-H}{|}}{\underset{|}{C}}-H$$

Carboxylic Acids and Esters

Carboxylic acids and **esters** have the following general formulas:

$$
\begin{matrix}
O & & O \\
\parallel & & \parallel \\
R-C-OH & & R-C-OR \\
\text{Carboxylic acid} & & \text{Ester}
\end{matrix}
$$

The structures of some common carboxylic acids and esters are shown in **Figure 20.5▼**.

> The condensed structural formula for carboxylic acids is RCOOH and that for esters is RCOOR.

Naming Carboxylic Acids and Esters

We name carboxylic acids systematically according to the number of carbon atoms in the longest chain containing the —COOH functional group. We form the base name by dropping the -*e* from the name of the corresponding alkane and adding the ending -*oic acid*.

$$
\begin{matrix}
O & & O \\
\parallel & & \parallel \\
CH_3-CH_2-C-OH & & CH_3-CH_2-CH_2-CH_2-C-OH \\
\text{Propanoic acid} & & \text{Pentanoic acid}
\end{matrix}
$$

We systematically name esters as if they were derived from a carboxylic acid by replacing the H on the OH with an alkyl group. The R group from the parent acid forms the base name of the compound. We change the -*ic* on the name of the corresponding carboxylic acid to -*ate*, and drop *acid*. The R group that replaced the H on the carboxylic acid is named as an alkyl group with the ending -*yl*, as shown in these examples.

$$
\begin{matrix}
O & & O \\
\parallel & & \parallel \\
CH_3-CH_2-C-OCH_3 & & CH_3-CH_2-CH_2-CH_2-C-OCH_2CH_3 \\
\text{Methyl propanoate} & & \text{Ethyl pentanoate}
\end{matrix}
$$

About Carboxylic Acids and Esters

Like all acids, carboxylic acids taste sour. The most familiar carboxylic acid is ethanoic acid, which is better known by its common name, acetic acid. Acetic acid is the active ingredient in vinegar. It can be formed by the oxidation of ethanol, which is why wines left open to air become sour. Some yeasts and bacteria also form acetic acid when they metabolize sugars in bread dough. These are added to bread dough to make sourdough bread. Carboxylic acids are also responsible for the sour taste of limes, lemons, and oranges. Esters are best known for their sweet smells. For example, methyl butanoate is largely responsible for the smell and taste of apples, and ethyl butanoate is largely responsible for the smell and taste of pineapples.

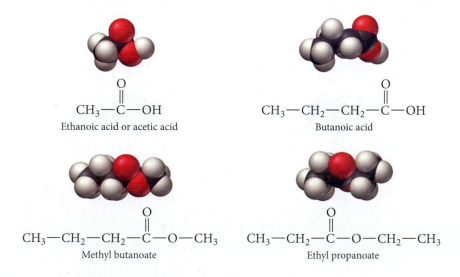

$$
\begin{matrix}
O & & O \\
\parallel & & \parallel \\
CH_3-C-OH & & CH_3-CH_2-CH_2-C-OH \\
\text{Ethanoic acid or acetic acid} & & \text{Butanoic acid}
\end{matrix}
$$

$$
\begin{matrix}
O & & O \\
\parallel & & \parallel \\
CH_3-CH_2-CH_2-C-O-CH_3 & & CH_3-CH_2-C-O-CH_2-CH_3 \\
\text{Methyl butanoate} & & \text{Ethyl propanoate}
\end{matrix}
$$

◀ **FIGURE 20.5 Common Carboxylic Acids and Esters**

Carboxylic Acid and Ester Reactions

Carboxylic acids act as weak acids in solution according to the equation:

$$RCOOH(aq) + H_2O(l) \rightleftharpoons H_3O^+(aq) + RCOO^-(aq)$$

Like all acids, carboxylic acids react with strong bases via neutralization reactions. A carboxylic acid reacts with an alcohol to form an ester via a **condensation reaction**, a reaction in which two (or more) organic compounds are joined, often with the loss of water (or some other small molecule).

Ethers

Ethers are organic compounds with the general formula ROR. The R groups may be the same or different. Some common ethers are shown in **Figure 20.6◄**.

CH₃—O—CH₃
Dimethyl ether

CH₃—O—CH₂—CH₃
Ethyl methyl ether

CH₃—CH₂—O—CH₂—CH₃
Diethyl ether

▲ **FIGURE 20.6 Ethers**

Naming Ethers

Common names for ethers have the format:

(R group 1) (R group 2) ether

If the two R groups are different, we use each of their names in alphabetical order. If the two R groups are the same, we use the prefix *di-*. Some examples include

H₃C—CH₂—CH₂—O—CH₂—CH₂—CH₃
Dipropyl ether

H₃C—CH₂—O—CH₂—CH₂—CH₃
Ethyl propyl ether

About Ethers

The most common ether is diethyl ether. Diethyl ether is a common laboratory solvent because it can dissolve many organic compounds and it has a low boiling point (34.6 °C). The low boiling point allows for easy removal of the solvent when necessary. Diethyl ether was also used as a general anesthetic for many years. When inhaled, diethyl ether depresses the central nervous system, causing unconsciousness and insensitivity to pain. Its use as an anesthetic, however, has decreased in recent years because other compounds have the same anesthetic effect with fewer side effects (such as nausea).

Amines

An amine is an organic compound that contains nitrogen. The simplest nitrogen-containing compound is ammonia (NH_3). **Amines** are derivatives of ammonia with one or more of the hydrogen atoms replaced by alkyl groups. Like ammonia, amines are weak bases. We systematically name them according to the hydrocarbon groups attached to the nitrogen and give them the ending *-amine*.

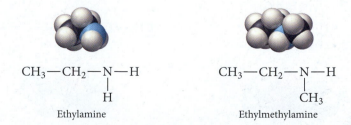

CH₃—CH₂—N—H
            |
            H
Ethylamine

CH₃—CH₂—N—H
            |
            CH₃
Ethylmethylamine

Amines are known for their awful odors. When a living organism dies, bacteria that feast on its proteins emit amines. Amines are responsible for the smell of rotten fish and of decaying animal flesh.

Amine Reactions

Just as carboxylic acids act as weak acids, amines act as weak bases:

$$RNH_2(aq) + H_2O(l) \rightleftharpoons RNH_3^+(aq) + OH^-(aq)$$

An important amine reaction is the condensation reaction between a carboxylic acid and an amine.

$$CH_3COOH(aq) + HNHR(aq) \longrightarrow CH_3CONHR(aq) + HOH(l)$$

This reaction is responsible for the formation of proteins from amino acids.

20.9 Polymers

Polymers are long, chainlike molecules composed of repeating units called monomers. Synthetic polymers compose many frequently encountered plastic products such as PVC tubing, styrofoam coffee cups, nylon rope, and plexiglass windows. Polymer materials are common in our everyday lives; they are found in everything from computers to toys to packaging materials. The simplest synthetic polymer is probably polyethylene. The polyethylene monomer is ethene (also called ethylene).

$$H_2C{=}CH_2$$ \[Monomer\]

Ethene or ethylene

Ethene monomers can react with each other, breaking the double bond between carbons and adding together to make a long polymer chain:

$$\cdots CH_2{-}CH_2{-}CH_2{-}CH_2{-}CH_2{-}CH_2{-}CH_2{-}CH_2{-}CH_2\cdots$$

\[Polymer\]

Polyethylene

Polyethylene is the plastic in bottles, juice containers, and garbage bags. It is an example of an **addition polymer**, a polymer in which the monomers simply link together without the elimination of any atoms.

An entire class of polymers can be thought of as substituted polyethylenes. For example, polyvinyl chloride (PVC)—the plastic used to make certain kinds of pipes and plumbing fixtures—is composed of monomers in which a chlorine atom has been substituted for one of the hydrogen atoms in ethene (**Figure 20.7▶**). These monomers link together to form PVC.

$$HC{=}CH_2$$
$$|$$
$$Cl$$

\[Monomer\]

Chloroethene

$$\cdots CH{-}CH_2{-}CH{-}CH_2{-}CH{-}CH_2{-}CH{-}CH_2{-}CH\cdots$$
$$|\phantom{CH_2{-}}|\phantom{CH_2{-}}|\phantom{CH_2{-}}|\phantom{CH_2{-}}|$$
$$Cl\phantom{CH_2{-}}Cl\phantom{CH_2{-}}Cl\phantom{CH_2{-}}Cl\phantom{CH_2{-}}Cl$$

\[Polymer\]

Polyvinyl chloride (PVC)

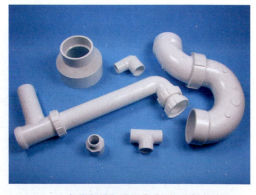

▲ **FIGURE 20.7 Polyvinyl Chloride** Polyvinyl chloride is used for many plastic plumbing supplies, such as pipes and connectors.

Table 20.11 shows several other substituted polyethylene polymers.

TABLE 20.11   Polymers of Commercial Importance

Polymer	Structure	Uses
Addition Polymers		
Polyethylene	$-(CH_2-CH_2)_n$	Films, packaging, bottles
Polypropylene	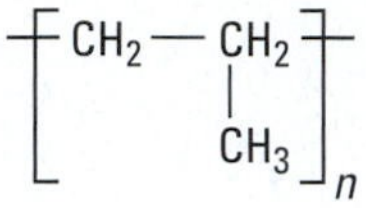 $\left[CH_2-CH_2 \atop CH_3 \right]_n$	Kitchenware, fibers, appliances
Polystyrene	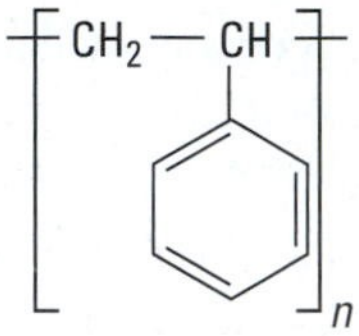 $\left[CH_2-CH \right]_n$ (with phenyl)	Packaging, disposable food containers, insulation
Polyvinyl chloride	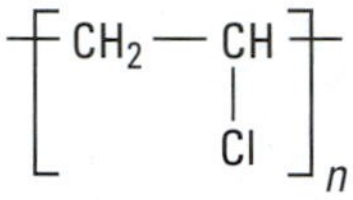 $\left[CH_2-CH \atop Cl \right]_n$	Pipe fittings, clear film for meat packaging
Condensation Polymers		
Polyurethane	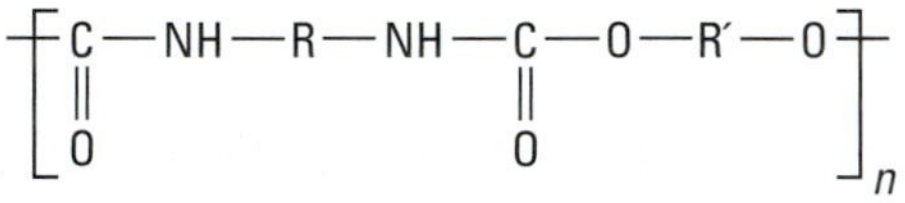 $\left[\underset{O}{C}-NH-R-NH-\underset{O}{C}-O-R'-O \right]_n$ R, R' = $-CH_2-CH_2-$ (for example)	"Foam" furniture stuffing, spray-on insulation, automotive parts, footwear, water-protective coatings
Polyethylene terephthalate (a polyester)	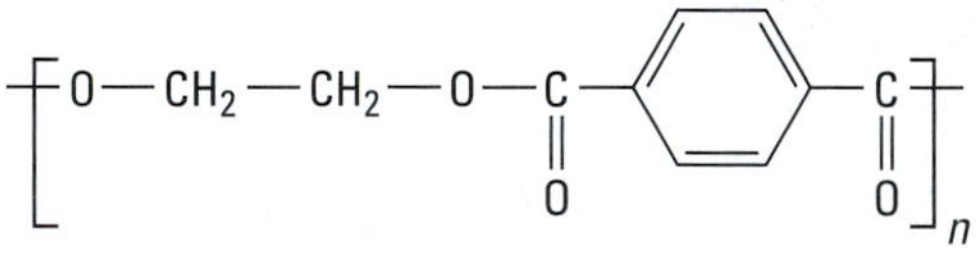 $\left[O-CH_2-CH_2-O-\underset{O}{C}-\text{(ring)}-\underset{O}{C} \right]_n$	Tire cord, magnetic tape, apparel, soda bottles
Nylon 6,6	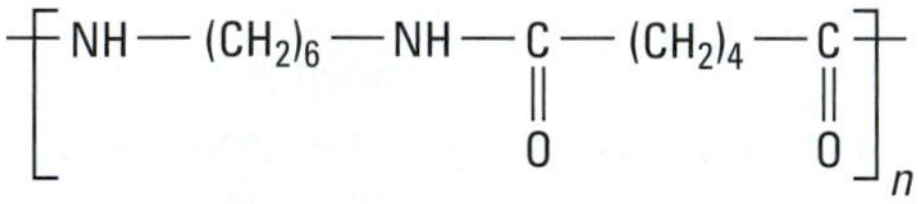 $\left[NH-(CH_2)_6-NH-\underset{O}{C}-(CH_2)_4-\underset{O}{C} \right]_n$	Home furnishings, apparel, carpet fibers, fish line, polymer blends

Some polymers—called copolymers—consist of two different kinds of monomers. For example, the monomers that compose nylon 6,6 are hexamethylenediamine and adipic acid. These two monomers add together via a condensation reaction:

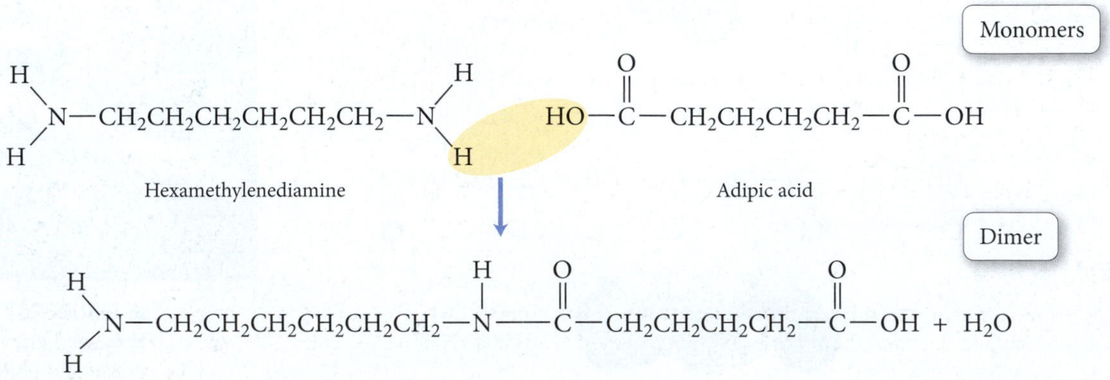

Monomers

$$\underset{H}{\overset{H}{N}}-CH_2CH_2CH_2CH_2CH_2CH_2-\underset{H}{\overset{H}{N}} \qquad HO-\underset{O}{\overset{\parallel}{C}}-CH_2CH_2CH_2CH_2-\underset{O}{\overset{\parallel}{C}}-OH$$

Hexamethylenediamine   Adipic acid

Dimer

$$\underset{H}{\overset{H}{N}}-CH_2CH_2CH_2CH_2CH_2CH_2-\underset{}{\overset{H}{N}}-\underset{O}{\overset{\parallel}{C}}-CH_2CH_2CH_2CH_2-\underset{O}{\overset{\parallel}{C}}-OH + H_2O$$

The product that forms between the reaction of two monomers is called a **dimer**. The polymer (nylon 6,6) forms as the dimer continues to add more monomers. Polymers that eliminate an atom or a small group of atoms during polymerization are called **condensation polymers**. Nylon 6,6 and other similar nylons can be drawn into fibers and used to make consumer products such as panty hose, carpet fibers, and fishing line. Table 20.11 shows other condensation polymers.

CHAPTER IN REVIEW

Key Terms

Section 20.1
organic molecule (775)
organic chemistry (775)

Section 20.3
hydrocarbon (776)
alkane (776)
alkene (776)
alkyne (776)
aromatic hydrocarbon (776)
aliphatic hydrocarbon (776)
structural isomers (777)
structural formula (778)

stereoisomers (780)
optical isomers (780)
enantiomers (780)
chiral (780)
dextrorotatory (781)
levorotatory (781)
racemic mixture (781)

Section 20.4
saturated hydrocarbon (781)

Section 20.5
unsaturated hydrocarbon (786)

geometric (cis–trans)
   isomerism (790)

Section 20.7
disubstituted benzene (793)

Section 20.8
functional group (793)
family (793)
alcohol (794)
hydroxyl group (794)
aldehyde (795)
ketone (795)

carbonyl group (795)
addition reaction (796)
carboxylic acid (797)
esters (797)
condensation reaction (798)
ether (798)
amine (798)

Section 20.9
addition polymer (799)
dimer (801)
condensation polymer (801)

Key Concepts

Fragrances and Odors (20.1)

▶ Organic chemistry is the study of organic compounds, those that contain carbon (and other elements including hydrogen, oxygen, and nitrogen). These compounds produce many common odors.

Carbon (20.2)

▶ Carbon forms more compounds than all the other elements combined.
▶ Carbon's four valence electrons (combined with its size) allow it to form four bonds (in the form of single, double, or triple bonds).
▶ Carbon also has the capacity to catenate, to form long chains, because of the strength of the carbon–carbon bond.

Hydrocarbons (20.3)

▶ Organic compounds containing only carbon and hydrogen are called hydrocarbons, most commonly known as the key components of our world's fuels.
▶ Hydrocarbons can be divided into four different types: alkanes, alkenes, alkynes, and aromatic hydrocarbons.
   Stereoisomers are molecules with the same atoms bonded in the same order, but arranged differently in space. Optical isomerism, a type of stereoisomerism, occurs when two molecules are nonsuperimposable mirror images of one another.

Alkanes (20.4)

▶ Alkanes are saturated hydrocarbons—they contain only single bonds and can therefore be represented by the generic formula C_nH_{2n+2}. Alkane names always end in -ane.

Alkenes and Alkynes (20.5)

▶ Alkenes and alkynes are unsaturated hydrocarbons—they contain double bonds (alkenes) or triple bonds (alkynes) and are represented by the generic formula C_nH_{2n} and C_nH_{2n-2}, respectively.
▶ Alkene names always end in -ene and alkynes end in -yne.
   Because rotation about a double bond is severely restricted, geometric (or cis-trans) isomerism occurs in alkenes.

Hydrocarbon Reactions (20.6)

▶ The most common hydrocarbon reaction is probably combustion in which hydrocarbons react with oxygen to form carbon dioxide and water; this reaction is exothermic and provides most of our society's energy.
▶ Alkanes can also undergo substitution reactions, in which heat or light causes another atom, commonly a halogen such as bromine, to be substituted for a hydrogen atom.
▶ Unsaturated hydrocarbons undergo addition reactions.

Aromatic Hydrocarbons (20.7)

▶ Aromatic hydrocarbons contain six-membered benzene rings represented with alternating single and double bonds that become equivalent through resonance. These compounds are called aromatic because they often produce pleasant fragrances.

Functional Groups (20.8)

▶ Characteristic groups of atoms, such as hydroxyl (—OH), that occur within organic compounds are called functional groups.
▶ Molecules that contain the same functional group have similar chemical and physical properties, and are referred to as families.

▶ Common functional groups include alcohols, aldehydes, ketones, carboxylic acids, esters, ethers, and amines.

Polymers (20.9)

▶ Polymers are long, chainlike molecules made up of repeating units called monomers.

▶ Polymers can be natural or synthetic; an example of a common synthetic polymer is polyethylene, the plastic in soda bottles and garbage bags. Polyethylene is an addition polymer, a polymer formed without the elimination of any atoms.

▶ Condensation polymers, such as nylon, are formed by the elimination of small groups of atoms.

Key Equations and Relationships

Halogen Substitution Reactions in Alkanes (20.6)

$$R\text{—}H \ + \ X_2 \ \xrightarrow{\text{heat or light}} \ R\text{—}X \ + \ HX$$

alkane halogen   haloalkane hydrogen halide

Common Functional Groups (20.8)

Family	General Formula	Condensed General Formula	Example	Name
Alcohols	R—OH	ROH	CH_3CH_2OH	Ethanol (ethyl alcohol)
Ethers	R—O—R	ROR	CH_3OCH_3	Dimethyl ether
Aldehydes	R—C(=O)—H	RCHO	H_3C—C(=O)—H	Ethanal (acetaldehyde)
Ketones	R—C(=O)—R	RCOR	H_3C—C(=O)—CH_3	Propanone (acetone)
Carboxylic acids	R—C(=O)—OH	RCOOH	H_3C—C(=O)—OH	Acetic acid
Esters	R—C(=O)—OR	RCOOR	H_3C—C(=O)—OCH_3	Methyl acetate
Amines	R—N(R)—R	R_3N	H_3CH_2C—N(H)—H	Ethylamine

Alcohol Reactions (20.8)

Substitution $ROH + HBr \longrightarrow R\text{—}Br + H_2O$

Elimination $R\text{—}CH_2\text{—}\underset{\underset{OH}{|}}{CH_2} \xrightarrow{H_2SO_4} R\text{—}CH{=}CH_2 + H_2O$

Oxidation $R\text{—}CH_2\text{—}\underset{\underset{OH}{|}}{CH_2} \xrightarrow[H_2SO_4]{Na_2Cr_2O_7} R\text{—}CH_2\text{—}\underset{\underset{O}{||}}{C}\text{—}OH$

     Alcohol   Carboxylic Acid

Carboxylic Acid Condensation Reactions (20.8)

$$R-\overset{\overset{\displaystyle O}{\|}}{C}-OH + HO-R' \xrightarrow{H_2SO_4} R-\overset{\overset{\displaystyle O}{\|}}{C}-O-R' + H_2O$$

Acid   Alcohol   Ester   Water

Amine–Carboxylic Acid Condensation Reactions (20.8)

$$CH_3COOH(aq) + HNHR(aq) \longrightarrow CH_3CONHR(aq) + H_2O(l)$$

Key Learning Objectives

Chapter Objectives	Assessment
Writing Structural Formulas for Hydrocarbons (20.3)	Example 20.1 For Practice 20.1 Exercises 3, 4
Naming Alkanes (20.4)	Examples 20.2, 20.3, 20.4 For Practice 20.2, 20.3, 20.4 Exercises 9, 10
Naming Alkenes and Alkynes (20.5)	Example 20.5 For Practice 20.5 Exercises 19–22
Writing Reactions: Addition Reactions (20.6)	Example 20.6 For Practice 20.6 Exercises 25–28

EXERCISES

Problems by Topic

Hydrocarbons

1. Based on the molecular formula, determine whether each compound is an alkane, alkene, or alkyne. (Assume that the hydrocarbons are noncyclical and there is no more than one multiple bond.)
   a. C_5H_{12} b. C_3H_6 c. C_7H_{12} d. $C_{11}H_{22}$

2. Based on the molecular formula, determine whether each compound is an alkane, alkene, or alkyne. (Assume that the hydrocarbons are noncyclical and there is no more than one multiple bond.)
   a. C_8H_{16} b. C_4H_6 c. C_7H_{16} d. C_2H_2

3. Write structural formulas for each of the nine structural isomers of heptane.

4. Write structural formulas for any 6 of the 18 structural isomers of octane.

5. Determine whether each molecule exhibits optical isomerism:
   a. CCl_4

   b. $CH_3-CH_2-\underset{\underset{\displaystyle CH_3}{|}}{CH}-CH_2-CH_2-CH_2-CH_3$

   c. $CH_3-\underset{\underset{\displaystyle NH_2}{|}}{\overset{\overset{\displaystyle H}{|}}{C}}-Cl$

   d. $CH_3CHClCH_3$

6. Determine whether each molecule exhibits optical isomerism:
   a. $CH_3CH_2CHClCH_3$ b. $CH_3CCl_2CH_3$

   c.

   d. $CH_3-\underset{\underset{\underset{\underset{\displaystyle CH_2}{\|}}{\displaystyle CH}}{|}}{\overset{\overset{\overset{\overset{\displaystyle CH_3}{|}}{\displaystyle CH_2O}}{\|}}{C}}-\overset{\overset{\displaystyle O}{\|}}{C}-OH$

7. Determine whether each pair of molecules are identical or enantiomers.

   a.

   b.

   c.

8. Determine whether each pair of molecules are identical or enantiomers.

a.

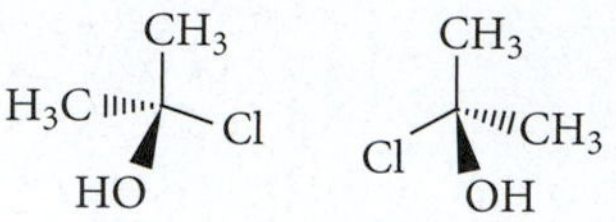

b.

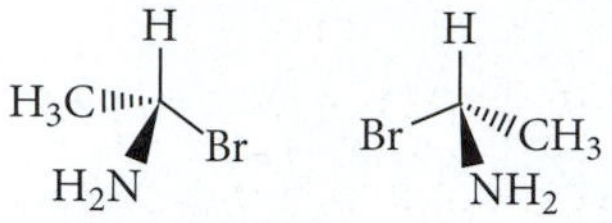

c.

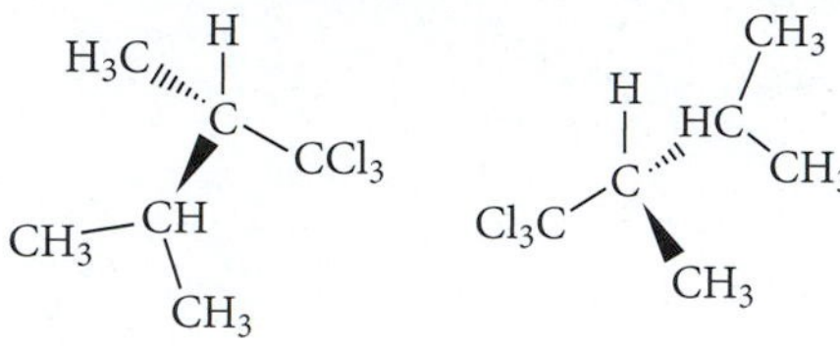

Alkanes

9. Name each alkane.

a. $CH_3-CH_2-CH_2-CH_2-CH_3$

b. 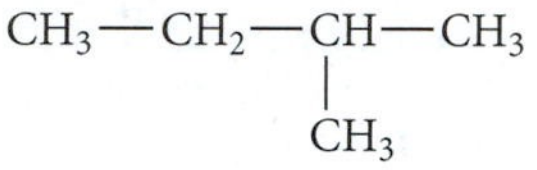 $CH_3-CH_2-CH-CH_3$ with CH_3 branch

c. 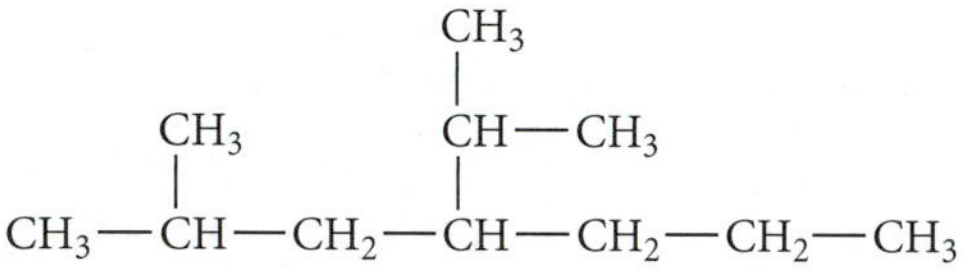 $CH_3-CH-CH_2-CH-CH_2-CH_2-CH_3$ with CH_3 branch on first and $CH-CH_3$ (CH_3) branch on second

d. 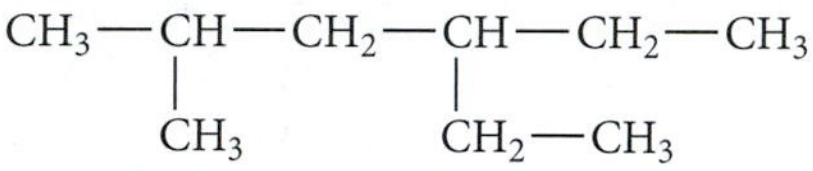 $CH_3-CH-CH_2-CH-CH_2-CH_3$ with CH_3 branch and CH_2-CH_3 branch

10. Name each alkane.

a. 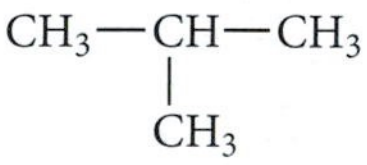 $CH_3-CH-CH_3$ with CH_3 branch

b. 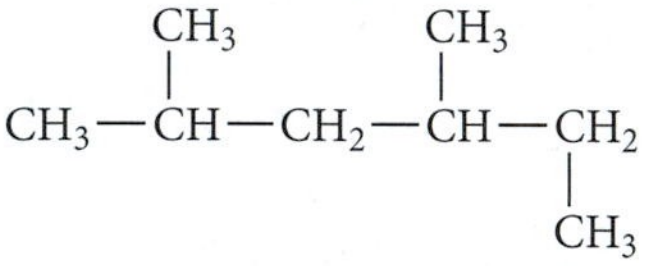 $CH_3-CH-CH_2-CH-CH_2$ with CH_3 branches and CH_3

c. 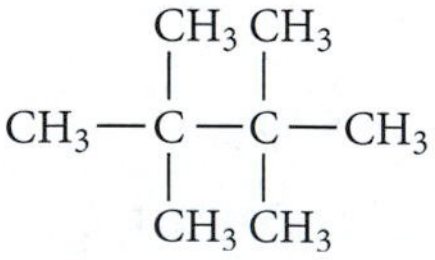 $CH_3-C-C-CH_3$ with CH_3 CH_3 above and CH_3 CH_3 below

d. 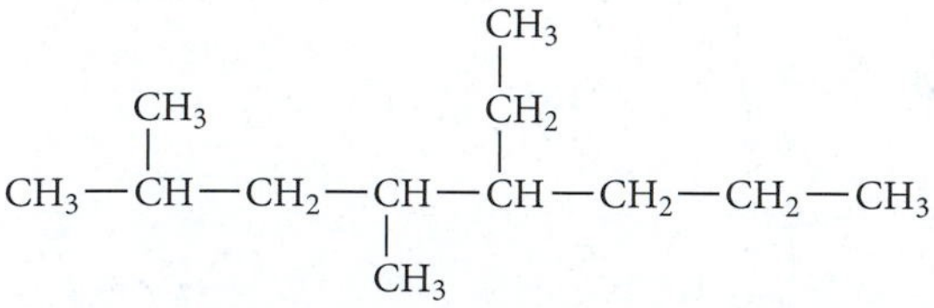 $CH_3-CH-CH_2-CH-CH-CH_2-CH_2-CH_3$ with CH_3 branch, CH_2/CH_3 branch, CH_3

11. Draw a structure for each alkane.

a. 3-ethylhexane

b. 3-ethyl-3-methylpentane

c. 2,3-dimethylbutane

d. 4,7-diethyl-2,2-dimethylnonane

12. Draw a structure for each alkane.

a. 2,2-dimethylpentane

b. 3-isopropylheptane

c. 4-ethyl-2,2-dimethylhexane

d. 4,4-diethyloctane

13. Complete and balance each hydrocarbon combustion reactions.

a. $CH_3CH_2CH_3 + O_2 \longrightarrow$

b. $CH_3CH_2CH=CH_2 + O_2 \longrightarrow$

c. $CH\equiv CH + O_2 \longrightarrow$

14. Complete and balance each hydrocarbon combustion reaction.

a. $CH_3CH_2CH_2CH_3 + O_2 \longrightarrow$

b. $CH_2=CHCH_3 + O_2 \longrightarrow$

c. $CH\equiv CCH_2CH_3 + O_2 \longrightarrow$

15. List all the possible products for each alkane substitution reaction. (Assume monosubstitution.)

a. $CH_3CH_3 + Br_2 \longrightarrow$

b. $CH_3CH_2CH_3 + Cl_2 \longrightarrow$

c. $CH_2Cl_2 + Br_2 \longrightarrow$

d. 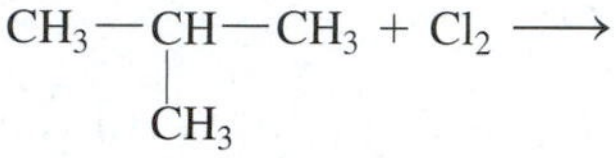 $CH_3-CH-CH_3 + Cl_2 \longrightarrow$ with CH_3 branch

16. List all the possible products for each alkane substitution reaction. (Assume monosubstitution.)

a. $CH_4 + Cl_2 \longrightarrow$

b. $CH_3CH_2Br + Br_2 \longrightarrow$

c. $CH_3CH_2CH_2CH_3 + Cl_2 \longrightarrow$

d. $CH_3CHBr_2 + Br_2 \longrightarrow$

Alkenes and Alkynes

17. Write structural formulas for all of the possible isomers of *n*-hexene that can be formed by moving the position of the double bond.

18. Write structural formulas for all of the possible isomers of *n*-pentyne that can be formed by moving the position of the triple bond.

19. Name each alkene.

a. $CH_2=CH-CH_2-CH_3$

b. 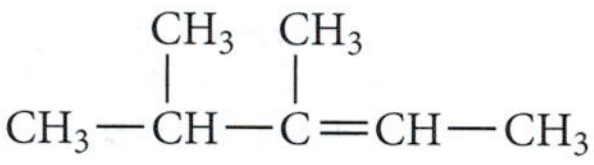 $CH_3-CH-C=CH-CH_3$ with CH_3 CH_3 branches

c. $CH_2=HC-CH-CH_2-CH_2-CH_3$ with CH_3-CH (CH_3) branch

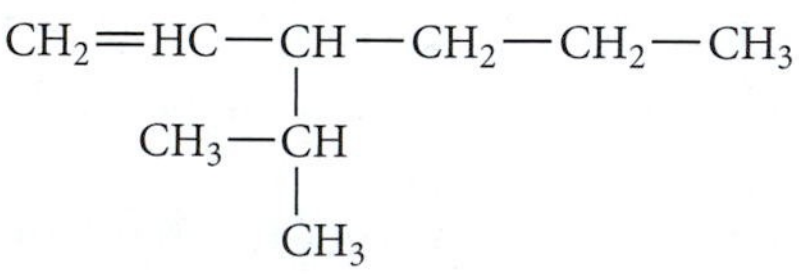

d. 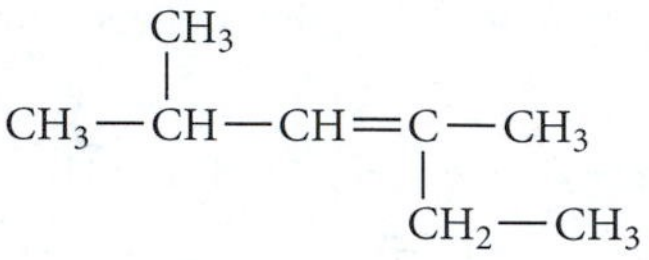 $CH_3-CH-CH=C-CH_3$ with CH_3 branch and CH_2-CH_3 branch

20. Name each alkene.

a. $CH_3-CH_2-CH=CH-CH_2-CH_3$

b. 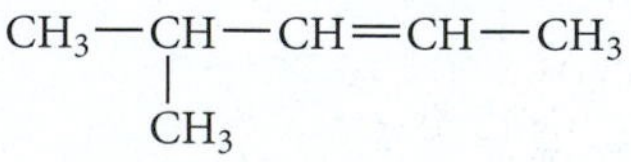 $CH_3-CH-CH=CH-CH_3$ with CH_3 branch

c. 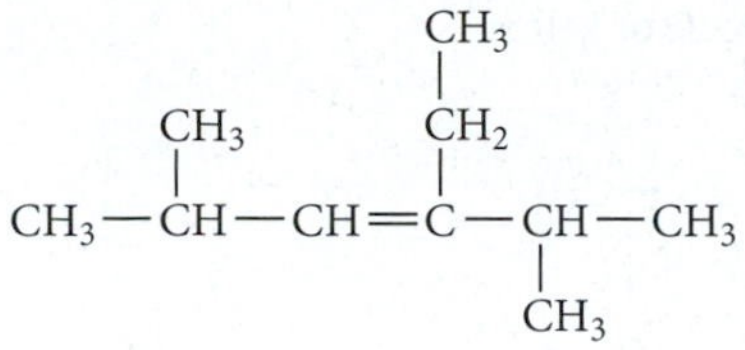 $CH_3-CH-CH=C-CH-CH_3$ with CH_3 on the left CH, CH_2 (and CH_3) above the middle C, and CH_3 below the right CH

d. 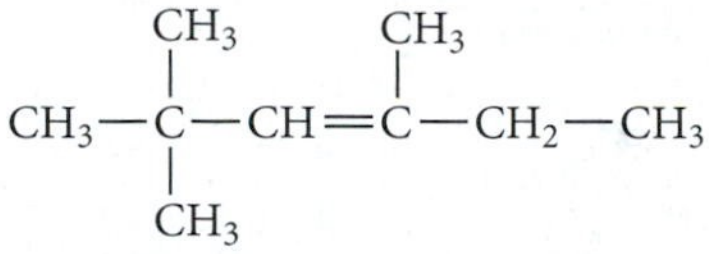 $CH_3-C-CH=C-CH_2-CH_3$ with two CH_3 groups above (on C and on the =C) and a CH_3 below the C

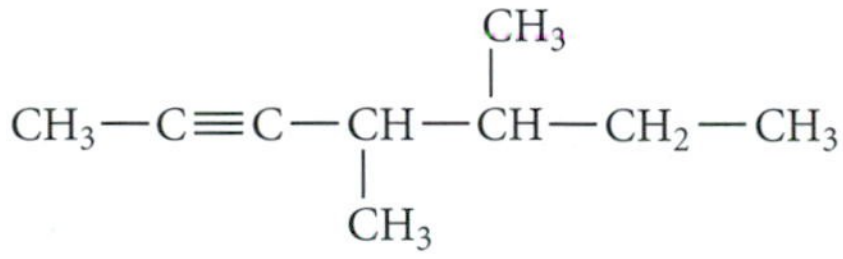

21. Name each alkyne.

a. $CH_3-C\equiv C-CH_3$

b. 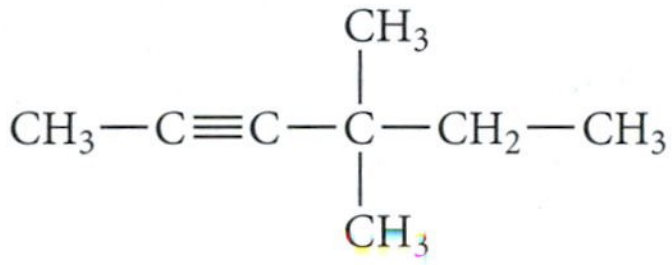 $CH_3-C\equiv C-\overset{\overset{\displaystyle CH_3}{|}}{\underset{\underset{\displaystyle CH_3}{|}}{C}}-CH_2-CH_3$

c. $CH\equiv C-CH-CH_2-CH_2-CH_3$ with $CH-CH_3$ below the first CH, and CH_3 below that

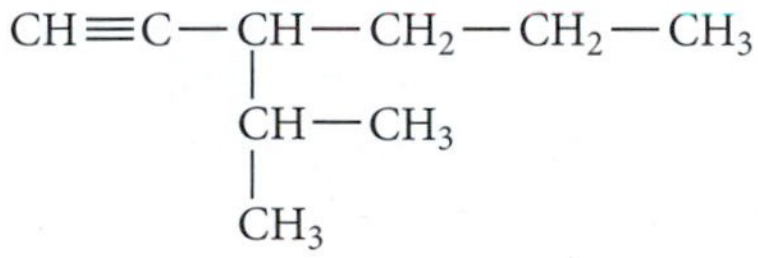

d. 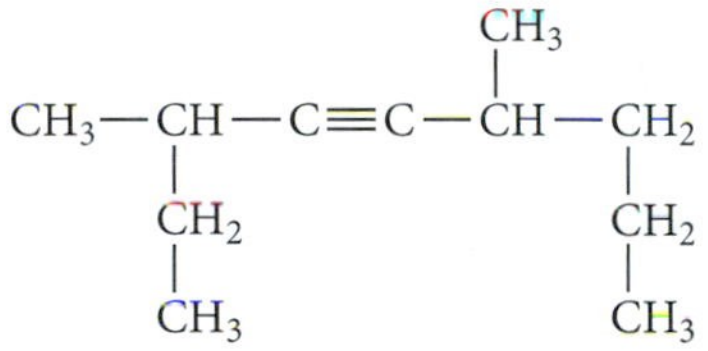 $CH_3-CH-C\equiv C-CH-CH_2$ with CH_2 and CH_3 below the first CH, CH_3 above the right CH, and CH_2, CH_3 below the CH_2

22. Name each alkyne.

a. 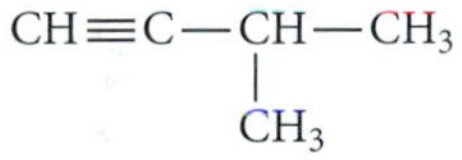 $CH\equiv C-CH-CH_3$ with CH_3 below the CH

b. $CH_3-C\equiv C-CH-CH-CH_2-CH_3$ with CH_3 above the middle CH and CH_3 below

c. $CH\equiv C-\overset{\overset{\displaystyle CH_3}{|}}{\underset{\underset{\underset{\displaystyle CH_3}{|}}{CH_2}}{C}}-CH_2-CH_3$

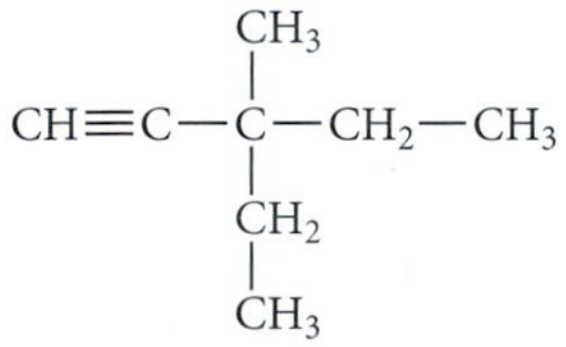

d. 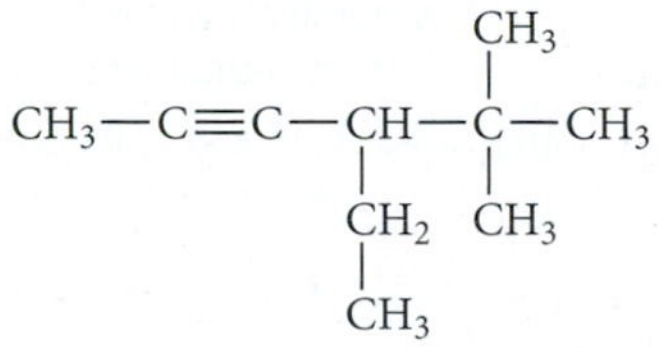 $CH_3-C\equiv C-CH-\overset{\overset{\displaystyle CH_3}{|}}{\underset{\underset{\underset{\displaystyle CH_3}{|}}{CH_2\ \ CH_3}}{C}}-CH_3$

23. Provide the correct structure for each compound.
a. 4-octyne
b. 3-nonene
c. 3,3-dimethyl-1-pentyne
d. 5-ethyl-3,6-dimethyl-2-heptene

24. Provide the correct structure for each compound.
a. 2-hexene
b. 1-heptyne
c. 4,4-dimethyl-2-hexene
d. 3-ethyl-4-methyl-2-pentene

25. Determine the products of each alkene addition reaction.

a. $CH_3-CH=CH-CH_3 + Cl_2 \longrightarrow$

b. $CH_3-CH_2-CH=CH-CH_3 + Br_2 \longrightarrow$

26. Determine the products of each alkene addition reaction.

a. 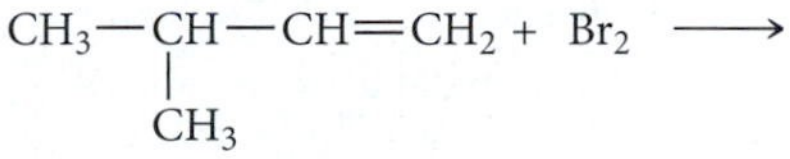 $CH_3-CH-CH=CH_2 + Br_2 \longrightarrow$ with CH_3 below the first CH

b. $CH_2=CH-CH_3 + Cl_2 \longrightarrow$

27. Complete each hydrogenation reaction.

a. $CH_2=CH-CH_3 + H_2 \xrightarrow{\text{catalyst}}$

b. 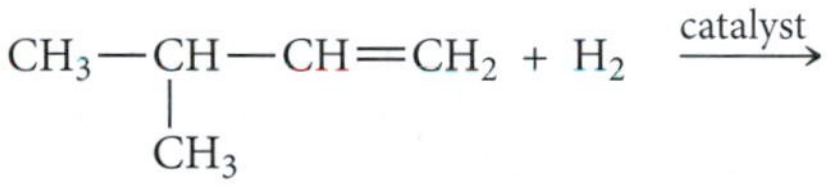 $CH_3-CH-CH=CH_2 + H_2 \xrightarrow{\text{catalyst}}$ with CH_3 below the first CH

c. 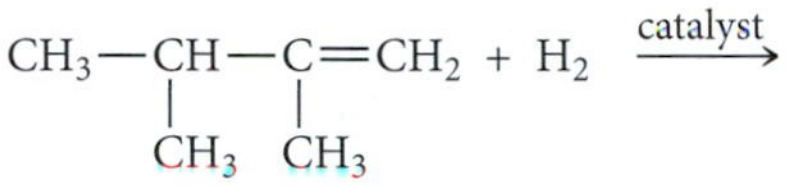 $CH_3-CH-C=CH_2 + H_2 \xrightarrow{\text{catalyst}}$ with CH_3 and CH_3 below

28. Complete each hydrogenation reaction.

a. $CH_3-CH_2-CH=CH_2 + H_2 \xrightarrow{\text{catalyst}}$

b. 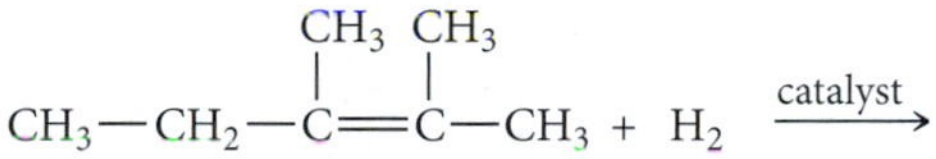 $CH_3-CH_2-C=C-CH_3 + H_2 \xrightarrow{\text{catalyst}}$ with CH_3 and CH_3 above

c. $CH_3-CH_2-C=CH_2 + H_2 \xrightarrow{\text{catalyst}}$ with CH_3 below

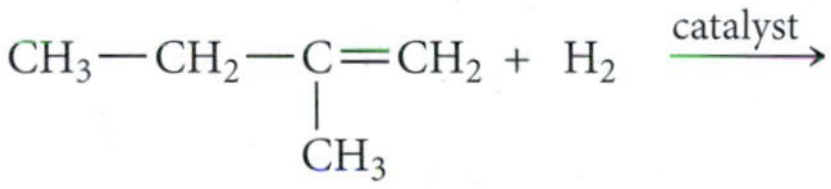

Aromatic Hydrocarbons

29. Name each monosubstituted benzene.

a. 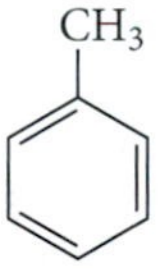 benzene ring with CH_3
b. 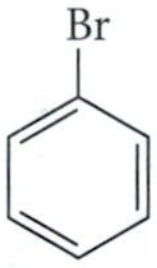 benzene ring with Br
c. 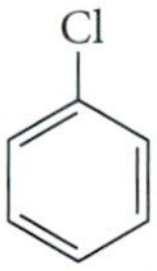 benzene ring with Cl

30. Name each monosubstituted benzene.

a. 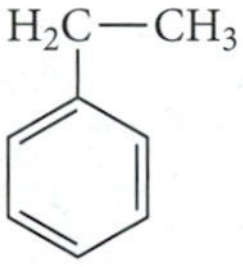 benzene ring with H_2C-CH_3
b. 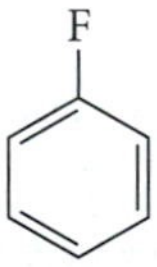 benzene ring with F
c. 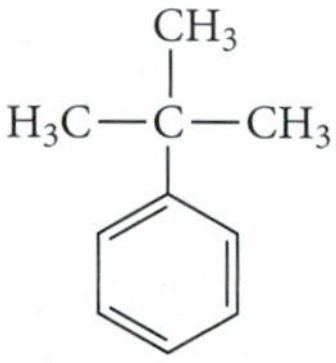 benzene ring with $H_3C-\overset{\overset{\displaystyle CH_3}{|}}{C}-CH_3$

31. Name each disubstituted benzene.

a.  benzene ring with Br top and Br bottom
b. 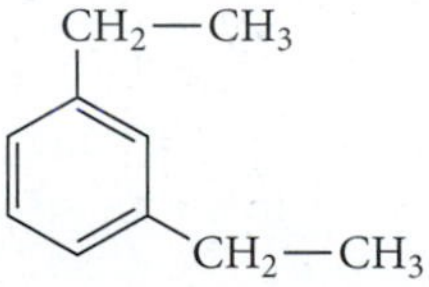 benzene ring with CH_2-CH_3 and CH_2-CH_3
c. 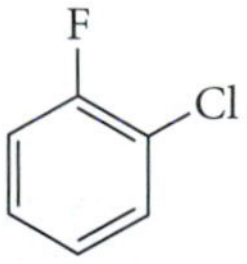 benzene ring with F and Cl

32. Name each disubstituted benzene.

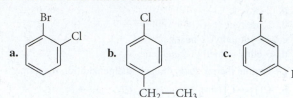

a.   b.   c.

33. Draw a structure for each compound.
   a. isopropylbenzene
   b. *meta*-dibromobenzene
   c. 1-chloro-4-methylbenzene

34. Draw a structure for each compound.
   a. ethylbenzene
   b. 1-iodo-2-methylbenzene
   c. *para*-diethylbenzene

Functional Groups

35. Name each alcohol.

   a. $CH_3-CH_2-CH_2-OH$

   b. $CH_3-\underset{\underset{OH}{|}}{CH}-CH_2-\underset{\overset{|}{CH_2-CH_3}}{CH}-CH_3$

   c. $CH_3-\underset{\overset{|}{CH_3}}{CH}-CH_2-\underset{\underset{OH}{|}}{CH}-CH_2-\underset{\overset{|}{CH_3}}{CH}-CH_3$

   d. $H_3C-CH_2-\underset{\underset{H_3C}{|}}{\overset{\overset{HO}{|}}{C}}-CH_2-CH_3$

36. Draw a structure for each alcohol.
   a. 2-butanol   **b.** 2-methyl-1-propanol
   c. 3-ethyl-1-hexanol   **d.** 2-methyl-3-pentanol

37. What are the products of each alcohol reaction?

   a. $CH_3-CH_2-CH_2-OH + HBr \longrightarrow$

   b. $CH_3-\underset{\overset{|}{CH_3}}{CH}-CH_2-OH \xrightarrow{H_2SO_4}$

   c. $CH_3-\underset{\underset{CH_3}{|}}{\overset{\overset{CH_3}{|}}{C}}-CH_2-CH_2-OH \xrightarrow[H_2SO_4]{Na_2Cr_2O_7}$

38. What are the products of each alcohol reaction?

   a. $CH_3-\underset{\underset{CH_3}{|}}{\overset{\overset{CH_3}{|}}{C}}-OH \xrightarrow{H_2SO_4}$

   b. $CH_3-\underset{\overset{|}{CH_3}}{CH}-CH_2-CH_2-OH \xrightarrow[H_2SO_4]{Na_2Cr_2O_7}$

   c. $CH_3-CH_2-OH + HCl \longrightarrow$

39. Name each aldehyde or ketone.

   a. $CH_3-\overset{\overset{O}{||}}{C}-CH_2-CH_3$

   b. $CH_3-CH_2-CH_2-CH_2-\overset{\overset{O}{||}}{CH}$

   c. $CH_3-\underset{\underset{CH_3}{|}}{\overset{\overset{CH_3}{|}}{C}}-CH_2-\underset{\overset{|}{CH_3}}{CH}-CH_2-\overset{\overset{O}{||}}{C}-H$

   d. $CH_3-\underset{\underset{CH_2-CH_3}{|}}{CH}-CH_2-\overset{\overset{O}{||}}{C}-CH_3$

40. Draw the structure of each aldehyde or ketone.
   a. hexanal   **b.** 2-pentanone
   c. 2-methylbutanal   **d.** 4-heptanone

41. Determine the product of the addition reaction:

   $CH_3-CH_2-CH_2-\overset{\overset{O}{||}}{CH} + H-C\equiv N \xrightarrow{NaCN}$

42. Determine the product of the addition reaction:

   $CH_3-\overset{\overset{O}{||}}{C}-CH_2-CH_3 + HCN \xrightarrow{NaCN}$

43. Name each carboxylic acid or ester.

   a. $CH_3-CH_2-CH_2-\overset{\overset{O}{||}}{C}-O-CH_3$

   b. $CH_3-CH_2-\overset{\overset{O}{||}}{C}-OH$

   c. $CH_3-\underset{\overset{|}{CH_3}}{CH}-CH_2-CH_2-CH_2-\overset{\overset{O}{||}}{C}-OH$

   d. $CH_3-CH_2-CH_2-CH_2-\overset{\overset{O}{||}}{C}-O-CH_2-CH_3$

44. Draw the structure of each carboxylic acid or ester.
   a. pentanoic acid   **b.** methyl hexanoate
   c. 3-ethylheptanoic acid   **d.** butyl ethanoate

45. Determine the products of the carboxylic acid reaction:

   $CH_3-CH_2-CH_2-CH_2-\overset{\overset{O}{||}}{C}-OH +$

   $CH_3-CH_2-OH \xrightarrow{H_2SO_4}$

46. Determine the products of the carboxylic acid reaction:

   $CH_3-CH_2-CH_2-\overset{\overset{O}{||}}{C}-OH +$

   $CH_3-CH_2-CH_2-OH \xrightarrow{H_2SO_4}$

47. Name each ether:
   a. $CH_3—CH_2—CH_2—O—CH_2—CH_3$
   b. $CH_3—CH_2—CH_2—CH_2—CH_2—O—CH_2—CH_3$
   c. $CH_3—CH_2—CH_2—O—CH_2—CH_2—CH_3$
   d. $CH_3—CH_2—O—CH_2—CH_2—CH_2—CH_3$

48. Draw a structure for each ether:
   a. ethyl propyl ether
   b. dibutyl ether
   c. methyl hexyl ether
   d. dipentyl ether

49. Name each amine.

a. 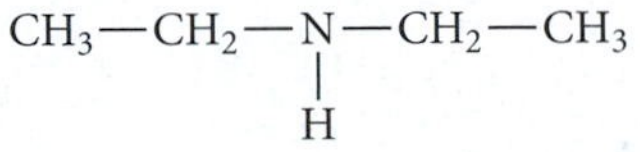

b. 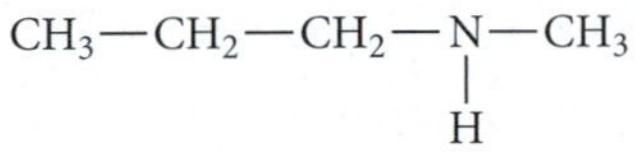

c. 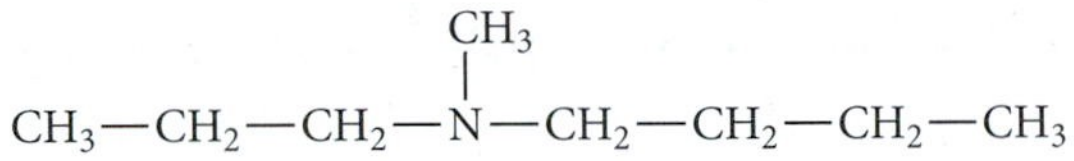

50. Draw the structure of each amine.
   a. isopropylamine
   b. triethylamine
   c. butylethylamine

51. Determine the products of the amine reaction:
   $CH_3CH_2NH_2 + CH_3CH_2COOH \longrightarrow$

52. Determine the products of the amine reaction:

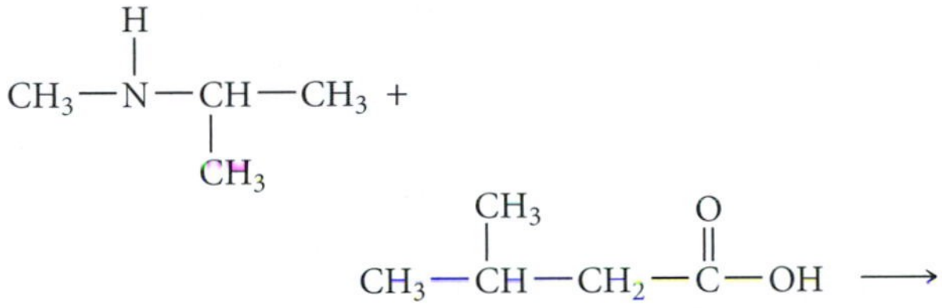

Polymers

53. Teflon is an addition polymer formed from the monomer shown here. Draw the structure of the polymer.

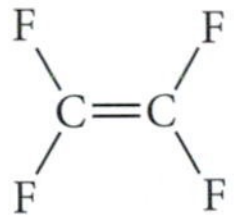

54. Saran, the polymer used to make saran wrap, is an addition polymer formed from the two monomers, vinylidene chloride and vinyl chloride. Draw the structure of the polymer. (Hint: The monomers alternate.)

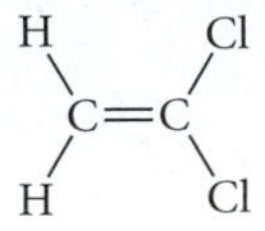

Vinylidene chloride

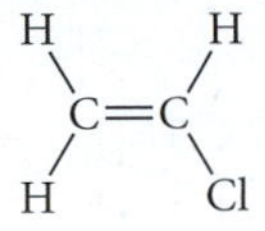

Vinyl chloride

55. One kind of polyester is a condensation copolymer formed between terephthalic acid and ethylene glycol. Draw the structure of the dimer and circle the ester functional group. [Hint: Water (circled) is eliminated when the bond between the monomers forms.]

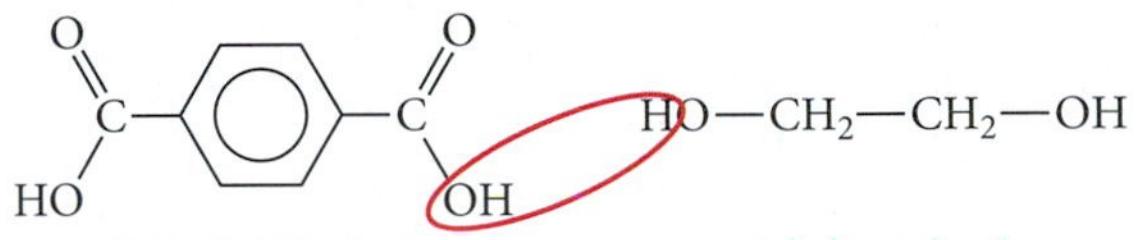

Terephthalic acid   Ethylene glycol

56. Nomex, a condensation copolymer used by firefighters because of its flame-resistant properties, is formed between isophthalic acid and *m*-aminoaniline. Draw the structure of the dimer. (Hint: Water is eliminated when the bond between the monomers forms.)

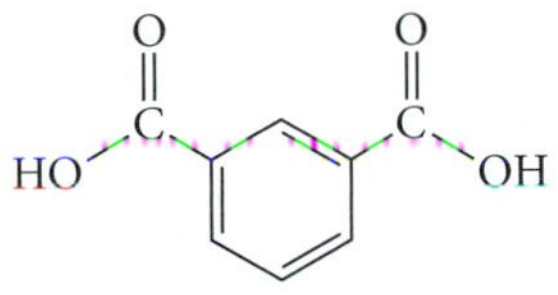

Isophthalic acid

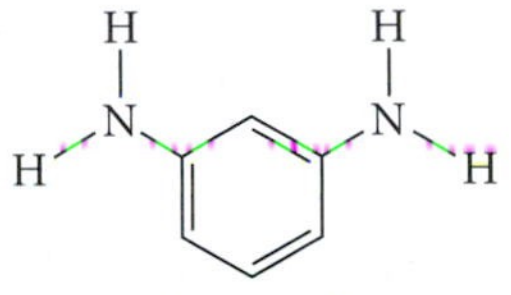

m-Aminoaniline

Cumulative Problems

57. Identify each organic compound as an alkane, alkene, alkyne, aromatic hydrocarbon, alcohol, ether, aldehyde, ketone, carboxylic acid, ester, or amine, and provide a name for the compound:

a. 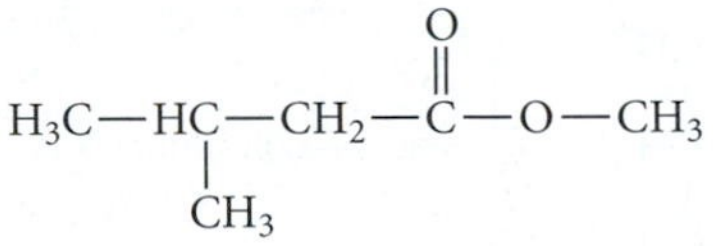

b. 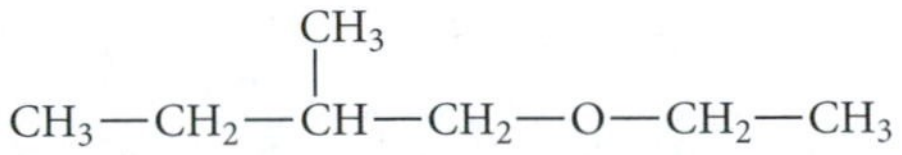

c. 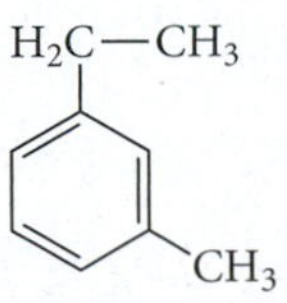

d. 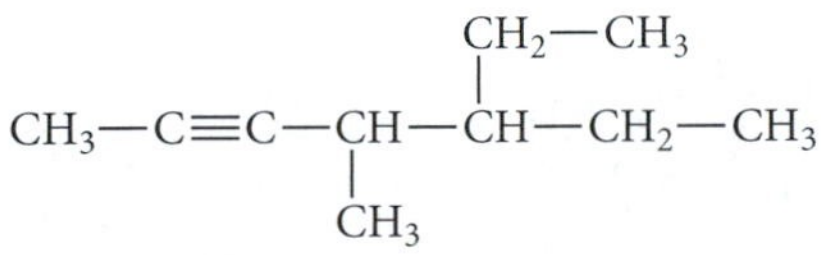

e. 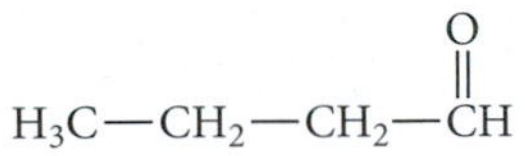

f. 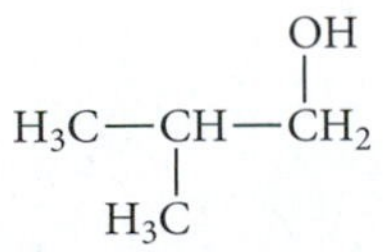

58. Identify each organic compound as an alkane, alkene, alkyne, aromatic hydrocarbon, alcohol, ether, aldehyde, ketone, carboxylic acid, ester, or amine, and provide a name for the compound:

a. 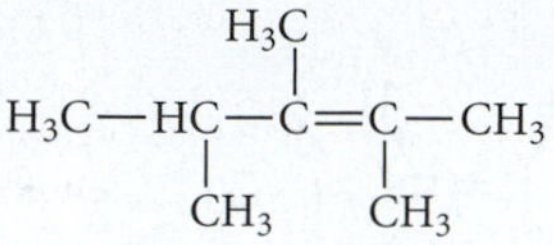
$$H_3C—HC—C=C—CH_3$$
with H_3C above the second carbon, and CH_3, CH_3 below

b.
$$CH_3—C—CH_2—CH—CH_2—CH_3$$
with CH_3, CH_3 above and CH_3 below the first C

c.
$$CH_3—CH_2—CH—CH_2—C—OH$$
with CH_3 above, and O double bonded to the C next to OH

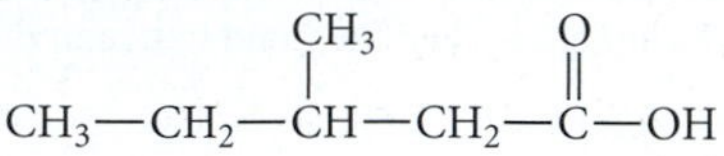

d. 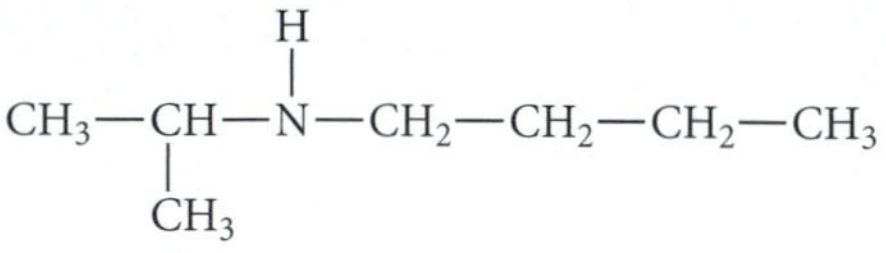
$$CH_3—CH—N—CH_2—CH_2—CH_2—CH_3$$
with H above N and CH_3 below first CH

e. 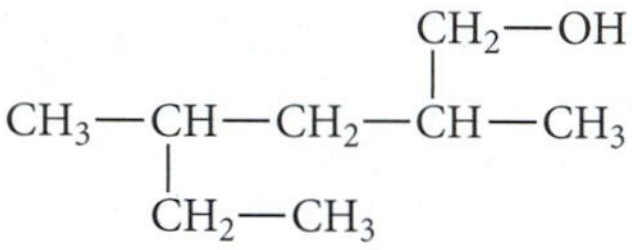
$$CH_3—CH—CH_2—CH—CH_3$$
with $CH_2—OH$ above and $CH_2—CH_3$ below

f. 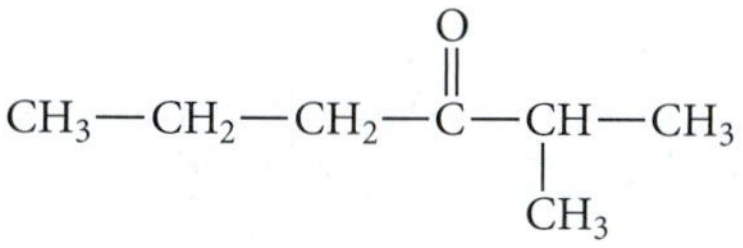
$$CH_3—CH_2—CH_2—C—CH—CH_3$$
with O double bonded to C, and CH_3 below CH

59. Name each compound.

a. 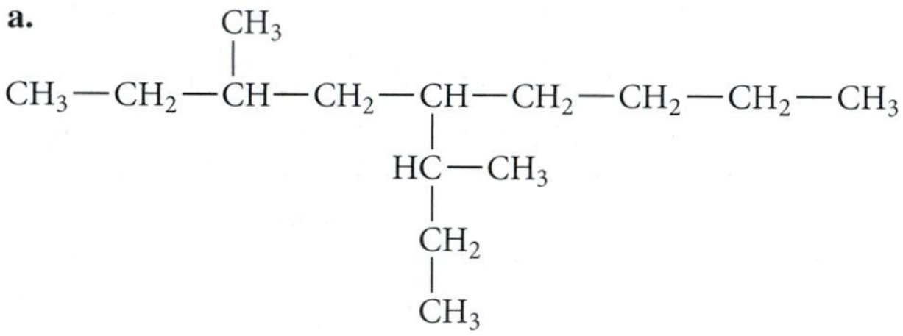
$$CH_3—CH_2—CH—CH_2—CH—CH_2—CH_2—CH_2—CH_3$$
with CH_3 above third carbon; below fifth carbon: $HC—CH_3$, then CH_2, then CH_3

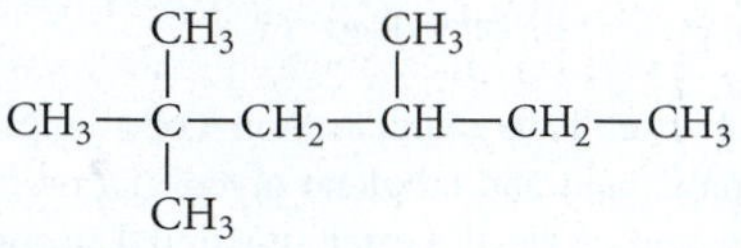

b. 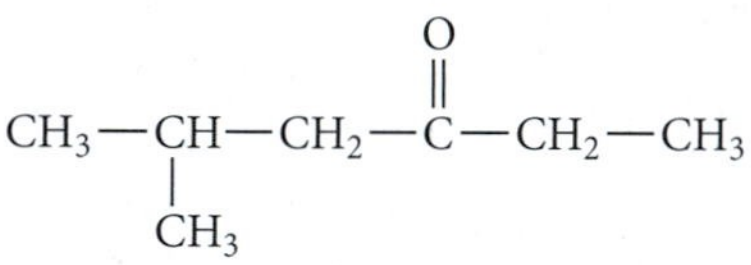
$$CH_3—CH—CH_2—C—CH_2—CH_3$$
with O double bonded to C and CH_3 below first CH

c.
$$CH_3—CH—CH—CH_3$$
with OH above and CH_3 below

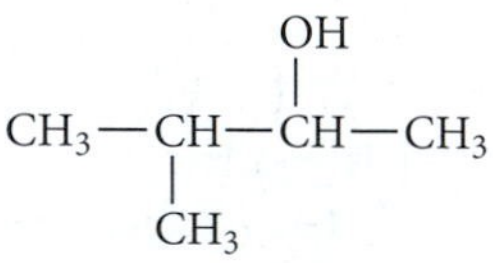

d. 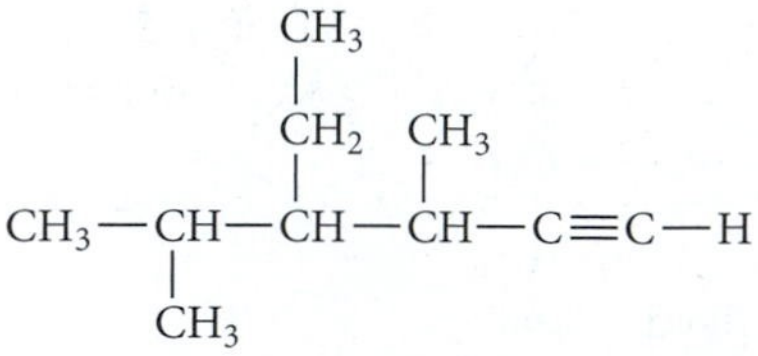
$$CH_3—CH—CH—CH—C≡C—H$$
with CH_3 below first CH; above third CH: CH_2 then CH_3, with CH_3 also shown

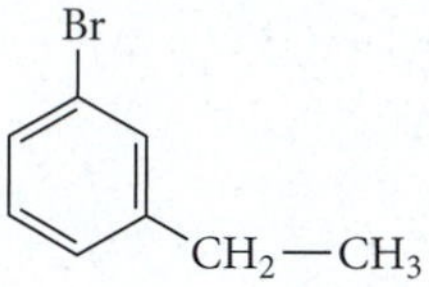

60. Name each compound.

a. 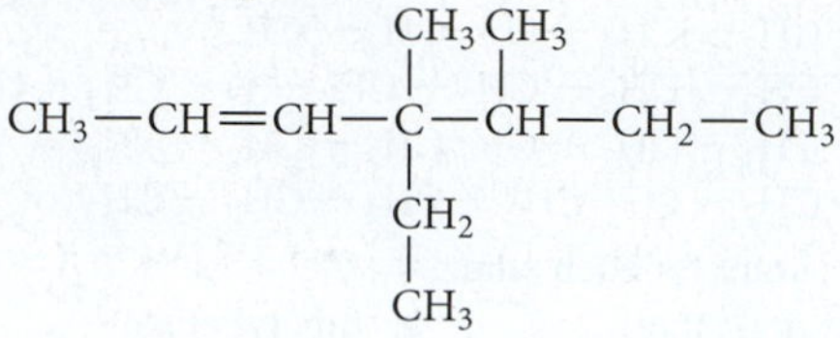
$$CH_3—CH=CH—C—CH—CH_2—CH_3$$
with CH_3, CH_3 above; below the C: CH_2 then CH_3

b.
benzene ring with Br and $CH_2—CH_3$

c.
$$CH_3—CH_2—CH—CH_2—C—O—CH—CH_3$$
with CH_3 above third C, O double bonded to middle C, and CH_3 above the final CH

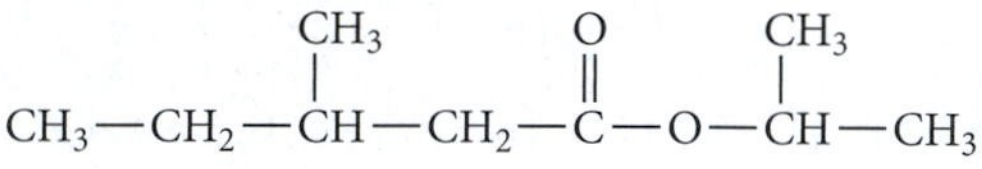

d. 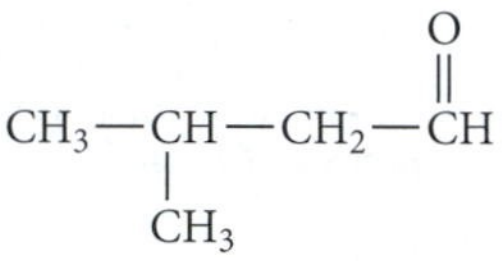
$$CH_3—CH—CH_2—CH$$
with O double bonded to last CH and CH_3 below first CH

61. Determine whether each pair of structures are isomers or the same molecule drawn in two different ways:

a. 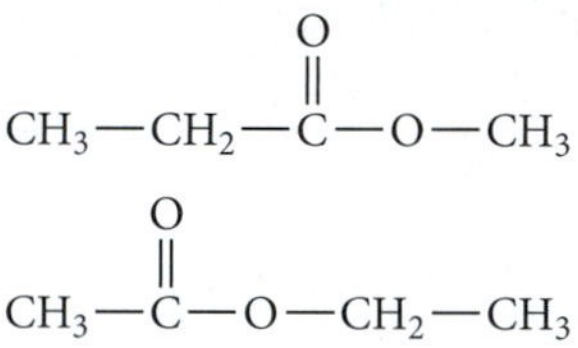
$$CH_3—CH_2—C—O—CH_3$$
with O double bonded to C
$$CH_3—C—O—CH_2—CH_3$$
with O double bonded to C

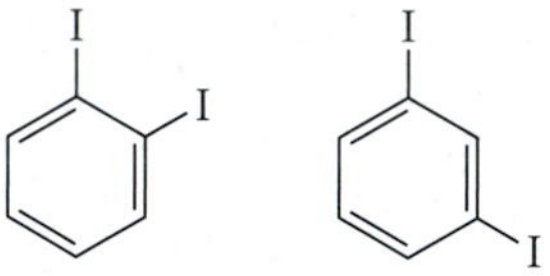

b.
benzene ring with two I (ortho) and benzene ring with two I (meta)

c.
$$CH_3—HC—CH—CH_3$$
with CH_3 and CH_3 below

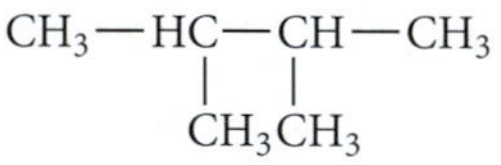

d. 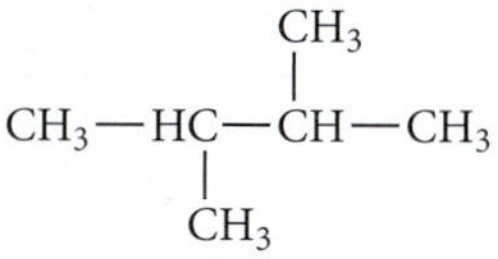
$$CH_3—HC—CH—CH_3$$
with CH_3 above and CH_3 below

62. Determine whether each pair of structures are isomers or the same molecule drawn two different ways:

a. 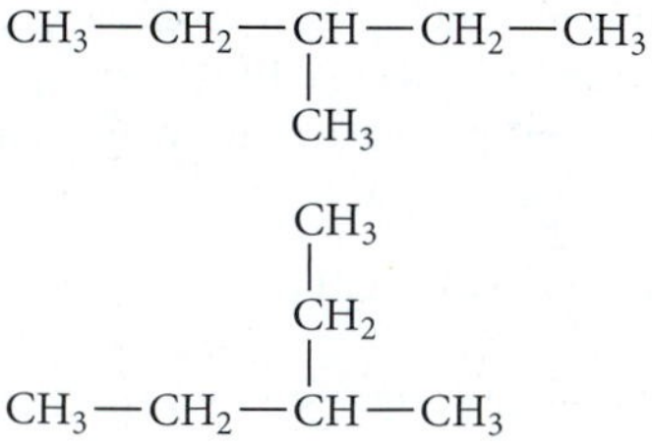
$$CH_3—CH_2—CH—CH_2—CH_3$$
with CH_3 below

$$CH_3—CH_2—CH—CH_3$$
with CH_3 then CH_2 above

$$CH_3$$
$$|$$
b. 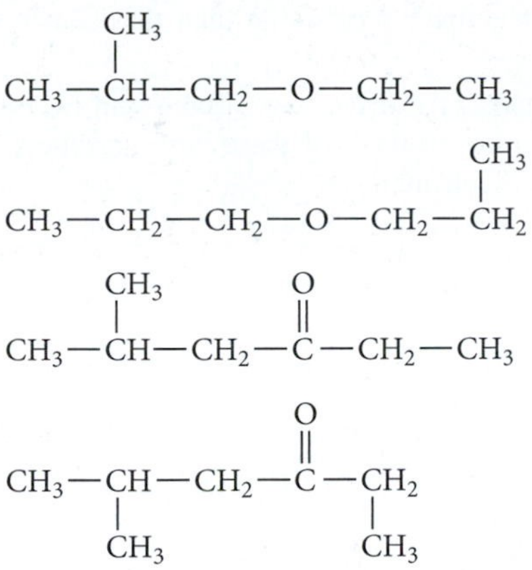 $CH_3-CH-CH_2-O-CH_2-CH_3$

$$CH_3$$
$$|$$
$CH_3-CH_2-CH_2-O-CH_2-CH_2$

$$CH_3 \quad\quad O$$
$$| \quad\quad\quad ||$$
$CH_3-CH-CH_2-C-CH_2-CH_3$

$$O$$
$$||$$
$CH_3-CH-CH_2-C-CH_2$
$$|\quad\quad\quad\quad\quad\quad\quad |$$
$$CH_3 \quad\quad\quad\quad\quad CH_3$$

63. What is the minimum amount of hydrogen gas, in grams, required to completely hydrogenate 15.5 kg of 2-butene?

64. How many kilograms of CO_2 are produced by the complete combustion of 3.8 kg of *n*-octane?

65. Draw a structure corresponding to each name and indicate which structures can exist as stereoisomers.

a. 3-methyl-1-pentene **b.** 3,5-dimethyl-2-hexene
c. 3-propyl-2-hexene

66. Two of these compounds correspond to structures that display stereoisomerism. Identify the two compounds and draw their structures.
a. 3-methyl-3-pentanol **b.** 2-methyl-2-pentanol
c. 3-methyl-2-pentanol **d.** 2-methyl-3-pentanol
e. 2,4-dimethyl-3-pentanol

67. There are 11 structures (ignoring stereoisomerism) with the formula C_4H_8O that have no carbon branches. Draw the structures and identify the functional groups in each.

68. There are seven structures with the formula C_3H_7NO in which the O is part of a carbonyl group. Draw the structures and identify the functional groups in each.

69. Explain why carboxylic acids are much stronger acids than alcohols.

70. The hydrogen at C-1 of 1-butyne is much more acidic than the one at C-1 in 1-butene. Explain.

Challenge Problems

71. Determine the one or two steps it would take to get from the starting material to the product using the reactions found in this chapter.

a. 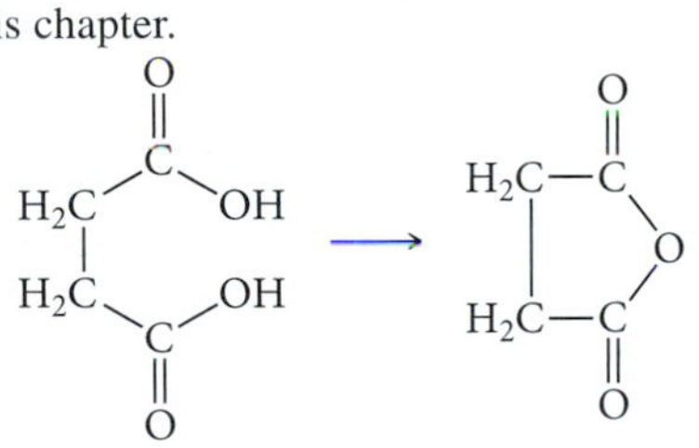
$$O$$
$$||$$
$$H_2C \quad C$$
$$\quad\quad\quad OH$$
$$H_2C$$
$$\quad\quad OH$$
$$C$$
$$||$$
$$O$$
$\longrightarrow$
$$O$$
$$||$$
$$H_2C-C$$
$$\quad\quad\quad\quad O$$
$$H_2C-C$$
$$||$$
$$O$$

b. 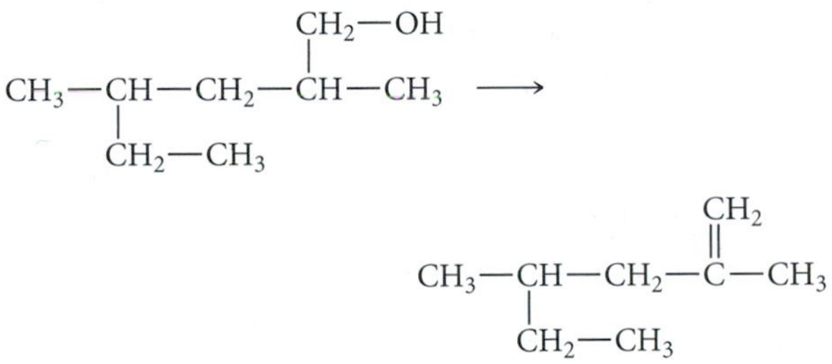
$$CH_2-OH$$
$$|$$
$CH_3-CH-CH_2-CH-CH_3 \longrightarrow$
$$|$$
$$CH_2-CH_3$$

$$CH_2$$
$$||$$
$CH_3-CH-CH_2-C-CH_3$
$$|$$
$$CH_2-CH_3$$

72. Given the synthesis of ethyl 3-chloro-3-methylbutanoate shown here, fill in the missing intermediates or reactants.

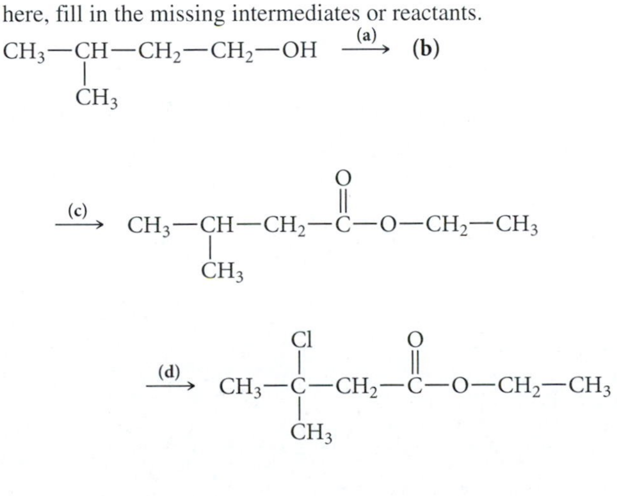

$CH_3-CH-CH_2-CH_2-OH \xrightarrow{\text{(a)}}$ **(b)**
$$|$$
$$CH_3$$

$$O$$
$$||$$
$\xrightarrow{\text{(c)}} CH_3-CH-CH_2-C-O-CH_2-CH_3$
$$|$$
$$CH_3$$

$$Cl \quad\quad O$$
$$| \quad\quad\quad ||$$
$\xrightarrow{\text{(d)}} CH_3-C-CH_2-C-O-CH_2-CH_3$
$$|$$
$$CH_3$$

73. For the chlorination of propane two isomers are possible:

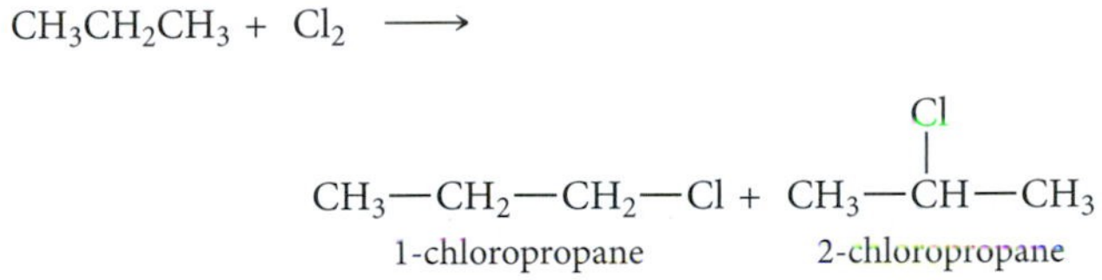

$CH_3CH_2CH_3 + Cl_2 \longrightarrow$

$$Cl$$
$$|$$
$CH_3-CH_2-CH_2-Cl + CH_3-CH-CH_3$
1-chloropropane   2-chloropropane

Propane has six hydrogen atoms on terminal carbon atoms—called primary (1°) hydrogen atoms—and two hydrogen atoms on the interior carbon atom—called secondary (2°) hydrogen atoms.
a. If the two different types of hydrogen atoms were equally reactive, what ratio of 1-chloropropane to 2-chloropropane would we expect as monochlorination products?
b. The result of a reaction yields 55% 2-chloropropane and 45% 1-chloropropane. What can you conclude about the relative reactivity of the two different kinds of hydrogen atoms? Determine a ratio of the reactivity of one type of hydrogen atom to the other.

74. There are two isomers of C_4H_{10}. Suppose that each isomer is treated with Cl_2 and the products that have the composition $C_4H_8Cl_2$ are isolated. Find the number of different products that form from each of the original C_4H_{10} compounds. Do not consider optical isomerism.

75. Identify the compounds formed in the previous problem that are chiral.

76. Nitromethane has the formula CH_3NO_2, with the N bonded to the C and without O—O bonds. Draw its two most important contributing structures.
a. What is the hybridization of the C and how many hybrid orbitals are in the molecule?
b. What is the shortest bond?
c. Between which two atoms is the strongest bond found?
d. Predict whether the HCH bond angles are greater or less than 109.5° and justify your prediction.

77. Free radical fluorination of methane is uncontrollably violent and free radical iodination of methane is a very poor reaction. Explain these observations in light of bond energies.

78. There are two compounds with the formula C_3H_6, one of which does not have a multiple bond. Draw its structure and explain why it is much less stable than the isomer with the double bond.

79. Consider molecules that have two carbons and two chlorines. Draw the structures of three of these with no dipole moment and two with a dipole moment.

Conceptual Problems

80. Draw the structure and name a compound with the formula C_8H_{18} that forms only one product with the formula $C_8H_{17}Br$ when it is treated with Br_2.

81. Determine whether each structure is chiral.

a.
$$\overset{\displaystyle Cl}{\underset{\displaystyle CH_3}{\overset{|}{\underset{|}{HC}}}{-}CH_3}$$

b.
$$CH_3{-}\underset{\underset{\displaystyle Cl}{|}}{CH}{-}\underset{\underset{\displaystyle CH_3}{|}}{CH}{-}CH_3$$

c. $CH_3{-}CH_2{-}OH$

d.
$$\underset{\underset{\displaystyle CH_3}{|}}{\overset{\overset{\displaystyle Cl}{|}}{HC}}{-}\underset{\underset{\displaystyle Br}{|}}{CH_2}$$

82. Pick the more oxidized structure from each pair.

a. $CH_3{-}\overset{\overset{\displaystyle O}{\|}}{CH}$ or $CH_3{-}CH_2{-}OH$

b. $CH_3{-}CH_2{-}OH$ or $CH_3{-}CH_3$

c. $CH_3{-}CH_2{-}\overset{\overset{\displaystyle O}{\|}}{CH}$ or $CH_3{-}CH_2{-}\overset{\overset{\displaystyle O}{\|}}{C}{-}OH$

Answers to Conceptual Connections

Organic Structures

20.1 (d) The others are just the same structure drawn in slightly different ways.

Optical Isomers

20.2 (b) This structure is the only one that contains a carbon atom (the one on the left) with four different groups attached (a Br atom, a Cl atom, an H atom, and a CH_3 group).

Common Mathematical Operations in Chemistry

A. Scientific Notation

A number written in scientific notation consists of a **decimal part**, a number that is usually between 1 and 10, and an **exponential part**, 10 raised to an **exponent**, n.

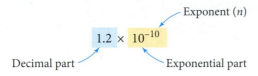

Exponent (n)

1.2×10^{-10}

Decimal part   Exponential part

Each of the following numbers is written in both scientific and decimal notation.

$$1.0 \times 10^5 = 100{,}000 \qquad 1.0 \times 10^{-5} = 0.000001$$
$$6.7 \times 10^3 = 6700 \qquad 6.7 \times 10^{-3} = 0.0067$$

A positive exponent means 1 multiplied by 10 n times.

$$10^0 = 1$$
$$10^1 = 1 \times 10$$
$$10^2 = 1 \times 10 \times 10 = 100$$
$$10^3 = 1 \times 10 \times 10 \times 10 = 1000$$

A negative exponent $(-n)$ means 1 divided by 10 n times.

$$10^{-1} = \frac{1}{10} = 0.1$$
$$10^{-2} = \frac{1}{10 \times 10} = 0.01$$
$$10^{-3} = \frac{1}{10 \times 10 \times 10} = 0.001$$

To convert a number to scientific notation, we move the decimal point to obtain a number between 1 and 10 and then multiply by 10 raised to the appropriate power. For example, to write 5983 in scientific notation, we move the decimal point to the left three places to get 5.983 (a number between 1 and 10) and then multiply by 1000 to make up for moving the decimal point.

$$5983 = 5.983 \times 1000$$

Since 1000 is 10^3, we write

$$5983 = 5.983 \times 10^3$$

We can do this in one step by counting how many places we move the decimal point to obtain a number between 1 and 10 and then writing the decimal part multiplied by 10 raised to the number of places we moved the decimal point.

$$5983 = 5.983 \times 10^3$$

If the decimal point is moved to the left, as in the previous example, the exponent is positive. If the decimal is moved to the right, the exponent is negative.

$$0.00034 = 3.4 \times 10^{-4}$$

To express a number in scientific notation:

1. **Move the decimal point to obtain a number between 1 and 10.**
2. **Write the result from step 1 multiplied by 10 raised to the number of places you moved the decimal point.**

- *The exponent is positive if you moved the decimal point to the left.*
- *The exponent is negative if you moved the decimal point to the right.*

Consider the following additional examples:

$$290,809,000 \qquad = 2.90809 \times 10^{8}$$

$$0.000000000070 \text{ m} = 7.0 \times 10^{-11} \text{ m}$$

Multiplication and Division

To multiply numbers expressed in scientific notation, multiply the decimal parts and add the exponents.

$$(A \times 10^{m})(B \times 10^{n}) = (A \times B) \times 10^{m+n}$$

To divide numbers expressed in scientific notation, divide the decimal parts and subtract the exponent in the denominator from the exponent in the numerator.

$$\frac{(A \times 10^{m})}{(B \times 10^{n})} = \left(\frac{A}{B}\right) \times 10^{m-n}$$

Consider the following example involving multiplication:

$$(3.5 \times 10^{4})(1.8 \times 10^{6}) = (3.5 \times 1.8) \times 10^{4+6}$$
$$= 6.3 \times 10^{10}$$

Consider the following example involving division:

$$\frac{(5.6 \times 10^{7})}{(1.4 \times 10^{3})} = \left(\frac{5.6}{1.4}\right) \times 10^{7-3}$$
$$= 4.0 \times 10^{4}$$

Addition and Subtraction

To add or subtract numbers expressed in scientific notation, rewrite all the numbers so that they have the same exponent, then add or subtract the decimal parts of the numbers. The exponents remained unchanged.

$$
\begin{array}{r}
A \times 10^{n} \\
\pm B \times 10^{n} \\
\hline
(A \pm B) \times 10^{n}
\end{array}
$$

Notice that the numbers *must have* the same exponent. Consider the following example involving addition:

$$
\begin{array}{r}
4.82 \times 10^{7} \\
+3.4 \;\; \times 10^{6}
\end{array}
$$

First, express both numbers with the same exponent. In this case, we rewrite the lower number and perform the addition as follows:

$$
\begin{array}{r}
4.82 \times 10^{7} \\
+0.34 \times 10^{7} \\
\hline
5.16 \times 10^{7}
\end{array}
$$

Consider the following example involving subtraction.

$$
\begin{array}{r}
7.33 \times 10^5 \\
-1.9 \times 10^4
\end{array}
$$

First, express both numbers with the same exponent. In this case, we rewrite the lower number and perform the subtraction as follows:

$$
\begin{array}{r}
7.33 \times 10^5 \\
-0.19 \times 10^5 \\
\hline
7.14 \times 10^5
\end{array}
$$

Powers and Roots

To raise a number written in scientific notation to a power, raise the decimal part to the power and multiply the exponent by the power:

$$
\begin{aligned}
(4.0 \times 10^6)^2 &= 4.0^2 \times 10^{6 \times 2} \\
&= 16 \times 10^{12} \\
&= 16 \times 10^{13}
\end{aligned}
$$

To take the nth root of a number written in scientific notation, take the nth root of the decimal part and divide the exponent by the root:

$$
\begin{aligned}
(4.0 \times 10^6)^{1/3} &= 4.0^{1/3} \times 10^{6/3} \\
&= 1.6 \times 10^2
\end{aligned}
$$

B. Logarithms

Common (or Base 10) Logarithms

The common or base 10 logarithm (abbreviated log) of a number is the exponent to which 10 must be raised to obtain that number. For example, the log of 100 is 2 because 10 must be raised to the second power to get 100. Similarly, the log of 1000 is 3 because 10 must be raised to the third power to get 1000. The logs of several multiples of 10 are shown below.

$$\log 10 = 1$$
$$\log 100 = 2$$
$$\log 1000 = 3$$
$$\log 10,000 = 4$$

Because $10^0 = 1$ by definition, $\log 1 = 0$.

The log of a number smaller than one is negative because 10 must be raised to a negative exponent to get a number smaller than one. For example, the log of 0.01 is -2 because 10 must be raised to -2 to get 0.01. Similarly, the log of 0.001 is -3 because 10 must be raised to -3 to get 0.001. The logs of several fractional numbers are shown below.

$$\log 0.1 = -1$$
$$\log 0.01 = -2$$
$$\log 0.001 = -3$$
$$\log 0.0001 = -4$$

The logs of numbers that are not multiples of 10 can be computed on your calculator. See your calculator manual for specific instructions.

Inverse Logarithms

The inverse logarithm, or invlog, function is exactly the opposite of the log function. For example, the log of 100 is 2 and the inverse log of 2 is 100. The log function and the invlog function undo one another.

$$\log 100 = 2$$
$$\text{invlog } 2 = 100$$
$$\text{invlog}(\log 100) = 100$$

The inverse log of a number is simply 10 rasied to that number.

$$\text{invlog } x = 10^x$$
$$\text{invlog } 3 = 10^3 = 1000$$

The inverse logs of numbers can be computed on your calculator. See your calculator manual for specific instructions.

Natural (or Base e) Logarithms

The natural (or base e) logarithm (abbreviated ln) of a number is the exponent to which e (which has the value of 2.71828...) must be raised to obtain that number. For example, the ln of 100 is 4.605 because e must be raised to 4.605 to get 100. Similarly, the ln of 10.0 is 2.303 because e must be raised to 2.303 to get 10.0.

The inverse natural logarithm or invln function is exactly the opposite of the ln function. For example, the ln of 100 is 4.605 and the inverse ln of 4.605 is 100. The inverse ln of a number is simply e rasied to that number.

$$\text{invln } x = e^x$$
$$\text{invln } 3 = e^3 = 20.1$$

The invln of a number can be computed on your calculator. See your calculator manual for specific instructions.

Mathematical Operations Using Logarithms

Because logarithms are exponents, mathematical operations involving logarithms are similar to those involving exponents as follows:

$$\log(a \times b) = \log a + \log b \qquad \ln(a \times b) = \ln a + \ln b$$

$$\log \frac{a}{b} = \log a - \log b \qquad \ln \frac{a}{b} = \ln a - \ln b$$

$$\log a^n = n \log a \qquad \ln a^n = n \ln a$$

C. Quadratic Equations

A quadratic equation contains at least one term in which the variable x is raised to the second power (and no terms in which x is raised to a higher power). A quadratic equation has the following general form:

$$ax^2 + bx + c = 0$$

A quadratic equation can be solved for x using the quadratic formula:

$$x = \frac{-b \pm \sqrt{b^2 - 4ac}}{2a}$$

Quadratic equations are often encountered when solving equilibrium problems. Below we show how to use the quadratic formula to solve a quadratic equation for x.

$$3x^2 - 5x + 1 = 0 \quad (\textit{quadratic equation})$$

$$x = \frac{-b \pm \sqrt{b^2 - 4ac}}{2a}$$

$$= \frac{-(-5) \pm \sqrt{(-5)^2 - 4(3)(1)}}{2(3)}$$

$$= \frac{5 \pm 3.6}{6}$$

$$x = 1.43 \quad \text{or} \quad x = 0.233$$

As you can see, the solution to a quadratic equation usually has two values. In any real chemical system, one of the values can be eliminated because it has no physical significance. (For example, it may correspond to a negative concentration, which does not exist.)

D. Graphs

Graphs are often used to visually show the relationship between two variables. For example, in Chapter 5 we show the following relationship between the volume of a gas and its pressure:

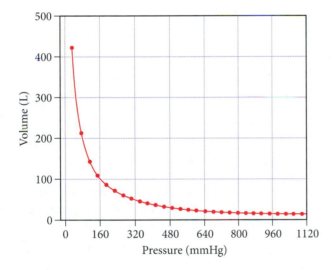

Volume versus Pressure A plot of the volume of a gas sample—as measured in a J-tube—versus pressure. The plot shows that volume and pressure are inversely related.

The horizontal axis is the x-axis and is normally used to show the independent variable. The vertical axis is the y-axis and is normally used to show how the other variable (called the dependent variable) varies with a change in the independent variable. In this case, the graph shows that as the pressure of a gas sample increases, its volume decreases.

Many relationships in chemistry are *linear*, which means that if you change one variable by a factor of n the other variable will also change by a factor of n. For example, the volume of a gas is linearly related to the number of moles of gas. When two quantities are linearly related, a graph of one versus the other produces a straight line. For example, the graph below shows how the volume of an ideal gas sample depends on the number of moles of gas in the sample:

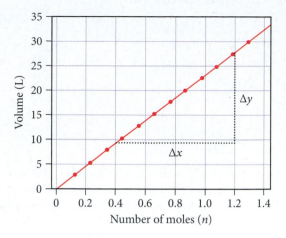

Volume versus Number of Moles   The volume of a gas sample increases linearly with the number of moles of gas in the sample.

A linear relationship between any two variables x and y can be expressed by the following equation:

$$y = mx + b$$

where m is the slope of the line and b is the y-intercept. The slope is the change in y divided by the change in x.

$$m = \frac{\Delta y}{\Delta x}$$

For the graph above, we can estimate the slope by simply estimating the changes in y and x for a given interval. For example, between $x = 0.4$ mol and 1.2 mol, $\Delta x = 0.80$ mol and we can estimate that $\Delta y = 18$ L. Therefore the slope is

$$m = \frac{\Delta y}{\Delta x} = \frac{18\ \text{L}}{0.80\ \text{mol}} = 23\ \text{mol/L}$$

In several places in this book, logarithmic relationships between variables can be plotted in order to obtain a linear relationship. For example, the variables $[A]_t$ and t in the following equation are not linearly related, but the natural logarithm of $[A]_t$ and t. are linearly related.

$$\ln[A]_t = -kt + \ln[A]_0$$
$$y = mx + b$$

A plot of $\ln[A]_t$ versus t will therefore produce a straight line with slope $= -k$ and y-intercept $= \ln[A]_0$.

Useful Data

A. Atomic Colors

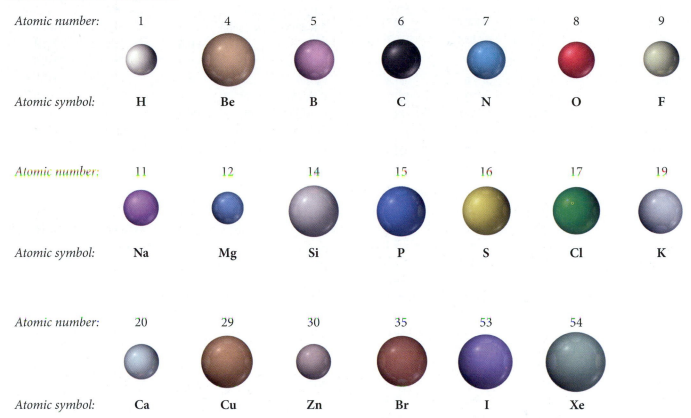

Atomic number:	1	4	5	6	7	8	9
Atomic symbol:	H	Be	B	C	N	O	F

Atomic number:	11	12	14	15	16	17	19
Atomic symbol:	Na	Mg	Si	P	S	Cl	K

Atomic number:	20	29	30	35	53	54
Atomic symbol:	Ca	Cu	Zn	Br	I	Xe

B. Standard Thermodynamic Quantities for Selected Substances at 25 °C

Substance	ΔH_f°(kJ/mol)	ΔG_f°(kJ/mol)	S°(J/mol · K)
Aluminum			
Al(s)	0	0	28.32
Al(g)	330.0	289.4	164.6
Al^{3+}(aq)	−538.4	−483	−325
AlCl$_3$(s)	−704.2	−628.8	109.3
Al$_2$O$_3$(s)	−1675.7	−1582.3	50.9
Barium			
Ba(s)	0	0	62.5
Ba(g)	180.0	146.0	170.2
Ba^{2+}(aq)	−537.6	−560.8	9.6
BaCO$_3$(s)	−1213.0	−1134.4	112.1
BaCl$_2$(s)	−855.0	−806.7	123.7
BaO(s)	−548.0	−520.3	72.1
Ba(OH)$_2$(s)	−944.7		
BaSO$_4$(s)	−1473.2	−1362.2	132.2

Substance	ΔH_f°(kJ/mol)	ΔG_f°(kJ/mol)	S°(J/mol · K)
Beryllium			
Be(s)	0	0	9.5
BeO(s)	−609.4	−580.1	13.8
Be(OH)$_2$(s)	−902.5	−815.0	45.5
Bismuth			
Bi(s)	0	0	56.7
BiCl$_3$(s)	−379.1	−315.0	177.0
Bi$_2$O$_3$(s)	−573.9	−493.7	151.5
Bi$_2$S$_3$(s)	−143.1	−140.6	200.4
Boron			
B(s)	0	0	5.9
B(g)	565.0	521.0	153.4
BCl$_3$(g)	−403.8	−388.7	290.1
BF$_3$(g)	−1136.0	−1119.4	254.4
B$_2$H$_6$(g)	36.4	87.6	232.1

(*continued on the next page*)

Substance	ΔH_f°(kJ/mol)	ΔG_f°(kJ/mol)	S°(J/mol · K)
$B_2O_3(s)$	−1273.5	−1194.3	54.0
$H_3BO_3(s)$	−1094.3	−968.9	90.0
Bromine			
$Br(g)$	111.9	82.4	175.0
$Br_2(l)$	0	0	152.2
$Br_2(g)$	30.9	3.1	245.5
$Br^-(aq)$	−121.4	−102.8	80.71
$HBr(g)$	−36.3	−53.4	198.7
Cadmium			
$Cd(s)$	0	0	51.8
$Cd(g)$	111.8	77.3	167.7
$Cd^{2+}(aq)$	−75.9	−77.6	−73.2
$CdCl_2(s)$	−391.5	−343.9	115.3
$CdO(s)$	−258.4	−228.7	54.8
$CdS(s)$	−161.9	−156.5	64.9
$CdSO_4(s)$	−933.3	−822.7	123.0
Calcium			
$Ca(s)$	0	0	41.6
$Ca(g)$	177.8	144.0	154.9
$Ca^{2+}(aq)$	−542.8	−553.6	−53.1
$CaC_2(s)$	−59.8	−64.9	70.0
$CaCO_3(s)$	−1207.6	−1129.1	91.7
$CaCl_2(s)$	−795.4	−748.8	108.4
$CaF_2(s)$	−1228.0	−1175.6	68.5
$CaH_2(s)$	−181.5	−142.5	41.4
$Ca(NO_3)_2(s)$	−938.2	−742.8	193.2
$CaO(s)$	−634.9	−603.3	38.1
$Ca(OH)_2(s)$	−985.2	−897.5	83.4
$CaSO_4(s)$	−1434.5	−1322.0	106.5
$Ca_3(PO_4)_2(s)$	−4120.8	−3884.7	236.0
Carbon			
$C(s, graphite)$	0	0	5.7
$C(s, diamond)$	1.88	2.9	2.4
$C(g)$	716.7	671.3	158.1
$CH_4(g)$	−74.6	−50.5	186.3
$CH_3Cl(g)$	−81.9	−60.2	234.6
$CH_2Cl_2(g)$	−95.4		270.2
$CH_2Cl_2(l)$	−124.2	−63.2	177.8
$CHCl_3(l)$	−134.1	−73.7	201.7
$CCl_4(g)$	−95.7	−62.3	309.7
$CCl_4(l)$	−128.2	−66.4	216.4
$CH_2O(g)$	−108.6	−102.5	218.8
$CH_2O_2(l, formic acid)$	−425.0	−361.4	129.0
$CH_3NH_2(g, methylamine)$	−22.5	32.7	242.9
$CH_3OH(l)$	−238.6	−166.6	126.8
$CH_3OH(g)$	−201.0	−162.3	239.9
$C_2H_2(g)$	227.4	209.9	200.9
$C_2H_4(g)$	52.4	68.4	219.3
$C_2H_6(g)$	−84.68	−32.0	229.2
$C_2H_5OH(l)$	−277.6	−174.8	160.7
$C_2H_5OH(g)$	−234.8	−167.9	281.6

Substance	ΔH_f°(kJ/mol)	ΔG_f°(kJ/mol)	S°(J/mol · K)
$C_2H_3Cl(g, vinyl\ chloride)$	37.2	53.6	264.0
$C_2H_4Cl_2$ (l, dichloroethane)	−166.8	−79.6	208.5
$C_2H_4O(g, acetaldehyde)$	−166.2	−133.0	263.8
$C_2H_4O_2(l, acetic\ acid)$	−484.3	−389.9	159.8
$C_3H_8(g)$	−103.85	−23.4	270.3
$C_3H_6O(l, acetone)$	−248.4	−155.6	199.8
$C_3H_7OH(l, isopropanol)$	−318.1		181.1
$C_4H_{10}(l)$	−147.3	−15.0	231.0
$C_4H_{10}(g)$	−125.7	−15.71	310.0
$C_6H_6(l)$	49.1	124.5	173.4
$C_6H_5NH_2(l, aniline)$	31.6	149.2	191.9
$C_6H_5OH(s, phenol)$	−165.1	−50.4	144.0
$C_6H_{12}O_6(s, glucose)$	−1273.3	−910.4	212.1
$C_8H_{18}(l)$	−250.1		
$C_{10}H_8(s, naphthalene)$	78.5	201.6	167.4
$C_{12}H_{22}O_{11}(s, sucrose)$	−2226.1	−1544.3	360.24
$CO(g)$	−110.5	−137.2	197.7
$CO_2(g)$	−393.5	−394.4	213.8
$CO_2(aq)$	−413.8	−386.0	117.6
$CO_3^{2-}(aq)$	−677.1	−527.8	−56.9
$HCO_3^-(aq)$	−692.0	−586.8	91.2
$H_2CO_3(aq)$	−699.7	−623.2	187.4
$CN^-(aq)$	151	166	118
$HCN(l)$	108.9	125.0	112.8
$HCN(g)$	135.1	124.7	201.8
$CS_2(l)$	89.0	64.6	151.3
$CS_2(g)$	116.7	67.1	237.8
$COCl_2(g)$	−219.1	−204.9	283.5
$C_{60}(s)$	2327.0	2302.0	426.0
Cesium			
$Cs(s)$	0	0	85.2
$Cs(g)$	76.5	49.6	175.6
$Cs^+(aq)$	−258.0	−292.0	132.1
$CsBr(s)$	−400	−387	117
$CsCl(s)$	−438	−414	101.2
$CsF(s)$	−553.5	−525.5	92.8
$CsI(s)$	−342	−337	127
Chlorine			
$Cl(g)$	121.3	105.3	165.2
$Cl_2(g)$	0	0	223.1
$Cl^-(aq)$	−167.1	−131.2	56.6
$HCl(g)$	−92.3	−95.3	186.9
$HCl(aq)$	−167.2	−131.2	56.5
$ClO_2(g)$	102.5	120.5	256.8
$Cl_2O(g)$	80.3	97.9	266.2
Chromium			
$Cr(s)$	0	0	23.8
$Cr(g)$	396.6	351.8	174.5
$Cr^{3+}(aq)$	−1971		
$CrO_4^{2-}(aq)$	−872.2	−717.1	44
$Cr_2O_3(s)$	−1139.7	−1058.1	81.2
$Cr_2O_7^{2-}(aq)$	−1476	−1279	238

Substance	ΔH_f°(kJ/mol)	ΔG_f°(kJ/mol)	S°(J/mol · K)
Cobalt			
Co(s)	0	0	30.0
Co(g)	424.7	380.3	179.5
CoO(s)	−237.9	−214.2	53.0
Co(OH)$_2$(s)	−539.7	−454.3	79.0
Copper			
Cu(s)	0	0	33.2
Cu(g)	337.4	297.7	166.4
Cu$^+$(aq)	51.9	50.2	−26
Cu^{2+}(aq)	64.9	65.5	−98
CuCl(s)	−137.2	−119.9	86.2
CuCl$_2$(s)	−220.1	−175.7	108.1
CuO(s)	−157.3	−129.7	42.6
CuS(s)	−53.1	−53.6	66.5
CuSO$_4$(s)	−771.4	−662.2	109.2
Cu$_2$O(s)	−168.6	−146.0	93.1
Cu$_2$S(s)	−79.5	−86.2	120.9
Fluorine			
F(g)	79.38	62.3	158.75
F$_2$(g)	0	0	202.79
F$^-$(aq)	−335.35	−278.8	−13.8
HF(g)	−273.3	−275.4	173.8
Gold			
Au(s)	0	0	47.4
Au(g)	366.1	326.3	180.5
Helium			
He(g)	0	0	126.2
Hydrogen			
H(g)	218.0	203.3	114.7
H$^+$(aq)	0	0	0
H$^+$(g)	1536.3	1517.1	108.9
H$_2$(g)	0	0	130.7
Iodine			
I(g)	106.76	70.2	180.79
I$_2$(s)	0	0	116.14
I$_2$(g)	62.42	19.3	260.69
I$^-$(aq)	−56.78	−51.57	106.45
HI(g)	26.5	1.7	206.6
Iron			
Fe(s)	0	0	27.3
Fe(g)	416.3	370.7	180.5
Fe^{2+}(aq)	−87.9	−84.94	113.4
Fe^{3+}(aq)	−47.69	−10.54	293.3
FeCO$_3$(s)	−740.6	−666.7	92.9
FeCl$_2$(s)	−341.8	−302.3	118.0
FeCl$_3$(s)	−399.5	−334.0	142.3
FeO(s)	−272.0	−255.2	60.75
Fe(OH)$_3$(s)	−823.0	−696.5	106.7
FeS$_2$(s)	−178.2	−166.9	52.9

Substance	ΔH_f°(kJ/mol)	ΔG_f°(kJ/mol)	S°(J/mol · K)
Fe$_2$O$_3$(s)	−824.2	−742.2	87.4
Fe$_3$O$_4$(s)	−1118.4	−1015.4	146.4
Lead			
Pb(s)	0	0	64.8
Pb(g)	195.2	162.2	175.4
Pb^{2+}(aq)	0.92	−24.4	18.5
PbBr$_2$(s)	−278.7	−261.9	161.5
PbCO$_3$(s)	−699.1	−625.5	131.0
PbCl$_2$(s)	−359.4	−314.1	136.0
PbI$_2$(s)	−175.5	−173.6	174.9
Pb(NO$_3$)$_2$(s)	−451.9		
PbO(s)	−217.3	−187.9	68.7
PbO$_2$(s)	−277.4	−217.3	68.6
PbS(s)	−100.4	−98.7	91.2
PbSO$_4$(s)	−920.0	−813.0	148.5
Lithium			
Li(s)	0	0	29.1
Li(g)	159.3	126.6	138.8
Li$^+$(aq)	−278.47	−293.3	12.24
LiBr(s)	−351.2	−342.0	74.3
LiCl(s)	−408.6	−384.4	59.3
LiF(s)	−616.0	−587.7	35.7
LiI(s)	−270.4	−270.3	86.8
LiNO$_3$(s)	−483.1	−381.1	90.0
LiOH(s)	−487.5	−441.5	42.8
Li$_2$O(s)	−597.9	−561.2	37.6
Magnesium			
Mg(s)	0	0	32.7
Mg(g)	147.1	112.5	148.6
Mg^{2+}(aq)	−467.0	−455.4	−137
MgCl$_2$(s)	−641.3	−591.8	89.6
MgCO$_3$(s)	−1095.8	−1012.1	65.7
MgF$_2$(s)	−1124.2	−1071.1	57.2
MgO(s)	−601.6	−569.3	27.0
Mg(OH)$_2$(s)	−924.5	−833.5	63.2
MgSO$_4$(s)	−1284.9	−1170.6	91.6
Mg$_3$N$_2$(s)	−461	−401	88
Manganese			
Mn(s)	0	0	32.0
Mn(g)	280.7	238.5	173.7
Mn^{2+}(aq)	−219.4	−225.6	−78.8
MnO(s)	−385.2	−362.9	59.7
MnO$_2$(s)	−520.0	−465.1	53.1
MnO$_4^-$(aq)	−529.9	−436.2	190.6
Mercury			
Hg(l)	0	0	75.9
Hg(g)	61.4	31.8	175.0
Hg^{2+}(aq)	170.21	164.4	−36.19
Hg$_2^{2+}$(aq)	166.87	153.5	65.74
HgCl$_2$(s)	−224.3	−178.6	146.0

(continued on the next page)

Substance	ΔH_f°(kJ/mol)	ΔG_f°(kJ/mol)	S°(J/mol·K)
HgO(s)	−90.8	−58.5	70.3
HgS(s)	−58.2	−50.6	82.4
Hg$_2$Cl$_2$(s)	−265.4	−210.7	191.6
Nickel			
Ni(s)	0	0	29.9
Ni(g)	429.7	384.5	182.2
NiCl$_2$(s)	−305.3	−259.0	97.7
NiO(s)	−239.7	−211.7	37.99
NiS(s)	−82.0	−79.5	53.0
Nitrogen			
N(g)	472.7	455.5	153.3
N$_2$(g)	0	0	191.6
NF$_3$(g)	−132.1	−90.6	260.8
NH$_3$(g)	−45.9	−16.4	192.8
NH$_3$(aq)	−80.29	−26.50	111.3
NH$_4^+$(aq)	−133.26	−79.31	111.17
NH$_4$Br(s)	−270.8	−175.2	113.0
NH$_4$Cl(s)	−314.4	−202.9	94.6
NH$_4$CN(s)	0.4		
NH$_4$F(s)	−464.0	−348.7	72.0
NH$_4$HCO$_3$(s)	−849.4	−665.9	120.9
NH$_4$I(s)	−201.4	−112.5	117.0
NH$_4$NO$_3$(s)	−365.6	−183.9	151.1
NH$_4$NO$_3$(aq)	−339.9	−190.6	259.8
HNO$_3$(g)	−133.9	−73.5	266.9
HNO$_3$(aq)	−207	−110.9	146
NO(g)	91.3	87.6	210.8
NO$_2$(g)	33.2	51.3	240.1
NO$_3^-$(aq)	−206.85	−110.2	146.70
NOBr(g)	82.2	82.4	273.7
NOCl(g)	51.7	66.1	261.7
N$_2$H$_4$(l)	50.6	149.3	121.2
N$_2$H$_4$(g)	95.4	159.4	238.5
N$_2$O(g)	81.6	103.7	220.0
N$_2$O$_4$(l)	−19.5	97.5	209.2
N$_2$O$_4$(g)	9.16	99.8	304.4
N$_2$O$_5$(s)	−43.1	113.9	178.2
N$_2$O$_5$(g)	13.3	117.1	355.7
Oxygen			
O(g)	249.2	231.7	161.1
O$_2$(g)	0	0	205.2
O$_3$(g)	142.7	163.2	238.9
OH$^-$(aq)	−230.02	−157.3	−10.90
H$_2$O(l)	−285.8	−237.1	70.0
H$_2$O(g)	−241.8	−228.6	188.8
H$_2$O$_2$(l)	−187.8	−120.4	109.6
H$_2$O$_2$(g)	−136.3	−105.6	232.7
Phosphorus			
P(s, white)	0	0	41.1
P(s, red)	−17.6	−12.1	22.8
P(g)	316.5	280.1	163.2

Substance	ΔH_f°(kJ/mol)	ΔG_f°(kJ/mol)	S°(J/mol·K)
P$_2$(g)	144.0	103.5	218.1
P$_4$(g)	58.9	24.4	280.0
PCl$_3$(l)	−319.7	−272.3	217.1
PCl$_3$(g)	−287.0	−267.8	311.8
PCl$_5$(s)	−443.5		
PCl$_5$(g)	−374.9	−305.0	364.6
PF$_5$(g)	−1594.4	−1520.7	300.8
PH$_3$(g)	5.4	13.5	210.2
POCl$_3$(l)	−597.1	−520.8	222.5
POCl$_3$(g)	−558.5	−512.9	325.5
PO$_4^{3-}$(aq)	−1277.4	−1018.7	−220.5
HPO$_4^{2-}$(aq)	−1292.1	−1089.2	−33.5
H$_2$PO$_4^-$(aq)	−1296.3	−1130.2	90.4
H$_3$PO$_4$(s)	−1284.4	−1124.3	110.5
H$_3$PO$_4$(aq)	−1288.3	−1142.6	158.2
P$_4$O$_6$(s)	−1640.1		
P$_4$O$_{10}$(s)	−2984	−2698	228.9
Platinum			
Pt(s)	0	0	41.6
Pt(g)	565.3	520.5	192.4
Potassium			
K(s)	0	0	64.7
K(g)	89.0	60.5	160.3
K$^+$(aq)	−252.14	−283.3	101.2
KBr(s)	−393.8	−380.7	95.9
KCN(s)	−113.0	−101.9	128.5
KCl(s)	−436.5	−408.5	82.6
KClO$_3$(s)	−397.7	−296.3	143.1
KClO$_4$(s)	−432.8	−303.1	151.0
KF(s)	−567.3	−537.8	66.6
KI(s)	−327.9	−324.9	106.3
KNO$_3$(s)	−494.6	−394.9	133.1
KOH(s)	−424.6	−379.4	81.2
KOH(aq)	−482.4	−440.5	91.6
KO$_2$(s)	−284.9	−239.4	116.7
K$_2$CO$_3$(s)	−1151.0	−1063.5	155.5
K$_2$O(s)	−361.5	−322.1	94.14
K$_2$O$_2$(s)	−494.1	−425.1	102.1
K$_2$SO$_4$(s)	−1437.8	−1321.4	175.6
Rubidium			
Rb(s)	0	0	76.8
Rb(g)	80.9	53.1	170.1
Rb$^+$(aq)	−251.12	−283.1	121.75
RbBr(s)	−394.6	−381.8	110.0
RbCl(s)	−435.4	−407.8	95.9
RbClO$_3$(s)	−392.4	−292.0	152
RbF(s)	−557.7		
RbI(s)	−333.8	−328.9	118.4
Scandium			
Sc(s)	0	0	34.6
Sc(g)	377.8	336.0	174.8

Substance	ΔH_f°(kJ/mol)	ΔG_f°(kJ/mol)	S°(J/mol · K)
Selenium			
Se(s, gray)	0	0	42.4
Se(g)	227.1	187.0	176.7
$H_2Se(g)$	29.7	15.9	219.0
Silicon			
Si(s)	0	0	18.8
Si(g)	450.0	405.5	168.0
$SiCl_4(l)$	−687.0	−619.8	239.7
$SiF_4(g)$	−1615.0	−1572.8	282.8
$SiH_4(g)$	34.3	56.9	204.6
$SiO_2(s, quartz)$	−910.7	−856.3	41.5
$Si_2H_6(g)$	80.3	127.3	272.7
Silver			
Ag(s)	0	0	42.6
Ag(g)	284.9	246.0	173.0
$Ag^+(aq)$	105.79	77.11	73.45
AgBr(s)	−100.4	−96.9	107.1
AgCl(s)	−127.0	−109.8	96.3
AgF(s)	−204.6	−185	84
AgI(s)	−61.8	−66.2	115.5
$AgNO_3(s)$	−124.4	−33.4	140.9
$Ag_2O(s)$	−31.1	−11.2	121.3
$Ag_2S(s)$	−32.6	−40.7	144.0
$Ag_2SO_4(s)$	−715.9	−618.4	200.4
Sodium			
Na(s)	0	0	51.3
Na(g)	107.5	77.0	153.7
$Na^+(aq)$	−240.34	−261.9	58.45
NaBr(s)	−361.1	−349.0	86.8
NaCl(s)	−411.2	−384.1	72.1
NaCl(aq)	−407.2	−393.1	115.5
$NaClO_3(s)$	−365.8	−262.3	123.4
NaF(s)	−576.6	−546.3	51.1
$NaHCO_3(s)$	−950.8	−851.0	101.7
$NaHSO_4(s)$	−1125.5	−992.8	113.0
NaI(s)	−287.8	−286.1	98.5
$NaNO_3(s)$	−467.9	−367.0	116.5
$NaNO_3(aq)$	−447.5	−373.2	205.4
NaOH(s)	−425.8	−379.7	64.4
NaOH(aq)	−470.1	−419.2	48.2
$NaO_2(s)$	−260.2	−218.4	115.9
$Na_2CO_3(s)$	−1130.7	−1044.4	135.0
$Na_2O(s)$	−414.2	−375.5	75.1
$Na_2O_2(s)$	−510.9	−447.7	95.0
$Na_2SO_4(s)$	−1387.1	−1270.2	149.6
$Na_3PO_4(s)$	−1917	−1789	173.8
Strontium			
Sr(s)	0	0	55.0
Sr(g)	164.4	130.9	164.6
$Sr^{2+}(aq)$	−545.51	−557.3	−39
$SrCl_2(s)$	−828.9	−781.1	114.9

Substance	ΔH_f°(kJ/mol)	ΔG_f°(kJ/mol)	S°(J/mol · K)
$SrCO_3(s)$	−1220.1	−1140.1	97.1
SrO(s)	−592.0	−561.9	54.4
$SrSO_4(s)$	−1453.1	−1340.9	117.0
Sulfur			
S(s, rhombic)	0	0	32.1
S(s, monoclinic)	0.3	0.096	32.6
S(g)	277.2	236.7	167.8
$S_2(g)$	128.6	79.7	228.2
$S_8(g)$	102.3	49.7	430.9
$S^{2-}(aq)$	41.8	83.7	22
$SF_6(g)$	−1220.5	−1116.5	291.5
$HS^-(aq)$	−17.7	12.4	62.0
$H_2S(g)$	−20.6	−33.4	205.8
$H_2S(aq)$	−39.4	−27.7	122
$SOCl_2(l)$	−245.6		
$SO_2(g)$	−296.8	−300.1	248.2
$SO_3(g)$	−395.7	−371.1	256.8
$SO_4^{2-}(aq)$	−909.3	−744.6	18.5
$HSO_4^-(aq)$	−886.5	−754.4	129.5
$H_2SO_4(l)$	−814.0	−690.0	156.9
$H_2SO_4(aq)$	−909.3	−744.6	18.5
$S_2O_3^{2-}(aq)$	−648.5	−522.5	67
Tin			
Sn(s, white)	0	0	51.2
Sn(s, gray)	−2.1	0.1	44.1
Sn(g)	301.2	266.2	168.5
$SnCl_4(l)$	−511.3	−440.1	258.6
$SnCl_4(g)$	−471.5	−432.2	365.8
SnO(s)	−280.7	−251.9	57.2
$SnO_2(s)$	−577.6	−515.8	49.0
Titanium			
Ti(s)	0	0	30.7
Ti(g)	473.0	428.4	180.3
$TiCl_4(l)$	−804.2	−737.2	252.3
$TiCl_4(g)$	−763.2	−726.3	353.2
$TiO_2(s)$	−944.0	−888.8	50.6
Tungsten			
W(s)	0	0	32.6
W(g)	849.4	807.1	174.0
$WO_3(s)$	−842.9	−764.0	75.9
Uranium			
U(s)	0	0	50.2
U(g)	533.0	488.4	199.8
$UF_6(s)$	−2197.0	−2068.5	227.6
$UF_6(g)$	−2147.4	−2063.7	377.9
$UO_2(s)$	−1085.0	−1031.8	77.0
Vanadium			
V(s)	0	0	28.9
V(g)	514.2	754.4	182.3

(*continued on the next page*)

Substance	ΔH_f°(kJ/mol)	ΔG_f°(kJ/mol)	S°(J/mol · K)
Zinc			
Zn(s)	0	0	41.6
Zn(g)	130.4	94.8	161.0
Zn²⁺(aq)	−153.39	−147.1	−109.8

Substance	ΔH_f°(kJ/mol)	ΔG_f°(kJ/mol)	S°(J/mol · K)
$ZnCl_2(s)$	−415.1	−369.4	111.5
$ZnO(s)$	−350.5	−320.5	43.7
$ZnS(s$, zinc blende)	−206.0	−201.3	57.7
$ZnSO_4(s)$	−982.8	−871.5	110.5

C. Aqueous Equilibrium Constants

1. Dissociation Constants for Acids at 25 °C

Name	Formula	K_{a_1}	K_{a_2}	K_{a_3}
Acetic	$HC_2H_3O_2$	1.8×10^{-5}		
Acetylsalicylic	$HC_9H_7O_4$	3.3×10^{-4}		
Adipic	$H_2C_6H_8O_4$	3.9×10^{-5}	3.9×10^{-6}	
Arsenic	H_3AsO_4	5.5×10^{-3}	1.7×10^{-7}	5.1×10^{-12}
Arsenous	H_3AsO_3	5.1×10^{-10}		
Ascorbic	$H_2C_6H_6O_6$	8.0×10^{-5}	1.6×10^{-12}	
Benzoic	$HC_7H_5O_2$	6.5×10^{-5}		
Boric	H_3BO_3	5.4×10^{-10}		
Butanoic	$HC_4H_7O_2$	1.5×10^{-5}		
Carbonic	H_2CO_3	4.3×10^{-7}	5.6×10^{-11}	
Chloroacetic	$HC_2H_2O_2Cl$	1.4×10^{-3}		
Chlorous	$HClO_2$	1.1×10^{-2}		
Citric	$H_3C_6H_5O_7$	7.4×10^{-4}	1.7×10^{-5}	4.0×10^{-7}
Cyanic	$HCNO$	2×10^{-4}		
Formic	$HCHO_2$	1.8×10^{-4}		
Hydrazoic	HN_3	2.5×10^{-5}		
Hydrocyanic	HCN	4.9×10^{-10}		
Hydrofluoric	HF	3.5×10^{-4}		
Hydrogen chromate ion	$HCrO_4^-$	3.0×10^{-7}		
Hydrogen peroxide	H_2O_2	2.4×10^{-12}		
Hydrogen selenate ion	$HSeO_4^-$	2.2×10^{-2}		
Hydrosulfuric	H_2S	8.9×10^{-8}	1×10^{-19}	
Hydrotelluric	H_2Te	2.3×10^{-3}	1.6×10^{-11}	

Name	Formula	K_{a_1}	K_{a_2}	K_{a_3}
Hypobromous	$HBrO$	2.8×10^{-9}		
Hypochlorous	$HClO$	2.9×10^{-8}		
Hypoiodous	HIO	2.3×10^{-11}		
Iodic	HIO_3	1.7×10^{-1}		
Lactic	$HC_3H_5O_3$	1.4×10^{-4}		
Maleic	$H_2C_4H_2O_4$	1.2×10^{-2}	5.9×10^{-7}	
Malonic	$H_2C_3H_2O_4$	1.5×10^{-3}	2.0×10^{-6}	
Nitrous	HNO_2	4.6×10^{-4}		
Oxalic	$H_2C_2O_4$	5.9×10^{-2}	6.4×10^{-5}	
Paraperiodic	H_5IO_6	2.8×10^{-2}	5.3×10^{-9}	
Phenol	HC_6H_5O	1.3×10^{-10}		
Phosphoric	H_3PO_4	7.5×10^{-3}	6.2×10^{-8}	4.2×10^{-13}
Phosphorous	H_3PO_3	5×10^{-2}	2.0×10^{-7}	
Propanoic	$HC_3H_5O_2$	1.3×10^{-5}		
Pyruvic	$HC_3H_3O_3$	4.1×10^{-3}		
Pyrophosphoric	$H_4P_2O_7$	1.2×10^{-1}	7.9×10^{-3}	2.0×10^{-7}
Selenous	H_2SeO_3	2.4×10^{-3}	4.8×10^{-9}	
Succinic	$H_2C_4H_4O_4$	6.2×10^{-5}	2.3×10^{-6}	
Sulfuric	H_2SO_4	Strong acid	1.2×10^{-2}	
Sulfurous	H_2SO_3	1.7×10^{-2}	6.4×10^{-8}	
Tartaric	$H_2C_4H_4O_6$	1.0×10^{-3}	4.6×10^{-5}	
Trichloroacetic	$HC_2Cl_3O_2$	2.2×10^{-1}		
Trifluoroacetic acid	$HC_2F_3O_2$	3.0×10^{-1}		

2. Dissociation Constants for Hydrated Metal Ions at 25 °C

Cation	Hydrated Ion	K_a
Al³⁺	$Al(H_2O)_6^{3+}$	1.4×10^{-5}
Be²⁺	$Be(H_2O)_6^{2+}$	3×10^{-7}
Co²⁺	$Co(H_2O)_6^{2+}$	1.3×10^{-9}
Cr³⁺	$Cr(H_2O)_6^{3+}$	1.6×10^{-4}
Cu²⁺	$Cu(H_2O)_6^{2+}$	3×10^{-8}
Fe²⁺	$Fe(H_2O)_6^{2+}$	3.2×10^{-10}

Cation	Hydrated Ion	K_a
Fe³⁺	$Fe(H_2O)_6^{3+}$	6.3×10^{-3}
Ni²⁺	$Ni(H_2O)_6^{2+}$	2.5×10^{-11}
Pb²⁺	$Pb(H_2O)_6^{2+}$	3×10^{-8}
Sn²⁺	$Sn(H_2O)_6^{2+}$	4×10^{-4}
Zn²⁺	$Zn(H_2O)_6^{2+}$	2.5×10^{-10}

3. Dissociation Constants for Bases at 25 °C

Name	Formula	K_b		Name	Formula	K_b
Ammonia	NH_3	1.76×10^{-5}		Ketamine	$C_{13}H_{16}ClNO$	3×10^{-7}
Aniline	$C_6H_5NH_2$	3.9×10^{-10}		Methylamine	CH_3NH_2	4.4×10^{-4}
Bicarbonate ion	HCO_3^-	1.7×10^{-9}		Morphine	$C_{17}H_{19}NO_3$	1.6×10^{-6}
Carbonate ion	CO_3^{2-}	1.8×10^{-4}		Nicotine	$C_{10}H_{14}N_2$	1.0×10^{-6}
Codeine	$C_{18}H_{21}NO_3$	1.6×10^{-6}		Piperidine	$C_5H_{10}NH$	1.33×10^{-3}
Diethylamine	$(C_2H_5)_2NH$	6.9×10^{-4}		Propylamine	$C_3H_7NH_2$	3.5×10^{-4}
Dimethylamine	$(CH_3)_2NH$	5.4×10^{-4}		Pyridine	C_5H_5N	1.7×10^{-9}
Ethylamine	$C_2H_5NH_2$	5.6×10^{-4}		Strychnine	$C_{21}H_{22}N_2O_2$	1.8×10^{-6}
Ethylenediamine	$C_2H_8N_2$	8.3×10^{-5}		Triethylamine	$(C_2H_5)_3N$	5.6×10^{-4}
Hydrazine	H_2NNH_2	1.3×10^{-6}		Trimethylamine	$(CH_3)_3N$	6.4×10^{-5}
Hydroxylamine	$HONH_2$	1.1×10^{-8}				

4. Solubility Product Constants for Compounds at 25 °C

Compound	Formula	K_{sp}		Compound	Formula	K_{sp}
Aluminum hydroxide	$Al(OH)_3$	1.3×10^{-33}		Iron(III) hydroxide	$Fe(OH)_3$	2.79×10^{-39}
Aluminum phosphate	$AlPO_4$	9.84×10^{-21}		Lanthanum fluoride	LaF_3	2×10^{-19}
Barium carbonate	$BaCO_3$	2.58×10^{-9}		Lanthanum iodate	$La(IO_3)_3$	7.50×10^{-12}
Barium chromate	$BaCrO_4$	1.17×10^{-10}		Lead(II) bromide	$PbBr_2$	4.67×10^{-6}
Barium fluoride	BaF_2	2.45×10^{-5}		Lead(II) carbonate	$PbCO_3$	7.40×10^{-14}
Barium hydroxide	$Ba(OH)_2$	5.0×10^{-3}		Lead(II) chloride	$PbCl_2$	1.17×10^{-5}
Barium oxalate	BaC_2O_4	1.6×10^{-6}		Lead(II) chromate	$PbCrO_4$	2.8×10^{-13}
Barium phosphate	$Ba_3(PO_4)_2$	6×10^{-39}		Lead(II) fluoride	PbF_2	3.3×10^{-8}
Barium sulfate	$BaSO_4$	1.07×10^{-10}		Lead(II) hydroxide	$Pb(OH)_2$	1.43×10^{-20}
Cadmium carbonate	$CdCO_3$	1.0×10^{-12}		Lead(II) iodide	PbI_2	9.8×10^{-9}
Cadmium hydroxide	$Cd(OH)_2$	7.2×10^{-15}		Lead(II) phosphate	$Pb_3(PO_4)_2$	1×10^{-54}
Cadmium sulfide	CdS	8×10^{-28}		Lead(II) sulfate	$PbSO_4$	1.82×10^{-8}
Calcium carbonate	$CaCO_3$	4.96×10^{-9}		Lead(II) sulfide	PbS	9.04×10^{-29}
Calcium chromate	$CaCrO_4$	7.1×10^{-4}		Magnesium carbonate	$MgCO_3$	6.82×10^{-6}
Calcium fluoride	CaF_2	1.46×10^{-10}		Magnesium fluoride	MgF_2	5.16×10^{-11}
Calcium hydroxide	$Ca(OH)_2$	4.68×10^{-6}		Magnesium hydroxide	$Mg(OH)_2$	2.06×10^{-13}
Calcium hydrogen phosphate	$CaHPO_4$	1×10^{-7}		Magnesium oxalate	MgC_2O_4	4.83×10^{-6}
Calcium oxalate	CaC_2O_4	2.32×10^{-9}		Manganese(II) carbonate	$MnCO_3$	2.24×10^{-11}
Calcium phosphate	$Ca_3(PO_4)_2$	2.07×10^{-33}		Manganese(II) hydroxide	$Mn(OH)_2$	1.6×10^{-13}
Calcium sulfate	$CaSO_4$	7.10×10^{-5}		Manganese(II) sulfide	MnS	2.3×10^{-13}
Chromium(III) hydroxide	$Cr(OH)_3$	6.3×10^{-31}		Mercury(I) bromide	Hg_2Br_2	6.40×10^{-23}
Cobalt(II) carbonate	$CoCO_3$	1.0×10^{-10}		Mercury(I) carbonate	Hg_2CO_3	3.6×10^{-17}
Cobalt(II) hydroxide	$Co(OH)_2$	5.92×10^{-15}		Mercury(I) chloride	Hg_2Cl_2	1.43×10^{-18}
Cobalt(II) sulfide	CoS	5×10^{-22}		Mercury(I) chromate	Hg_2CrO_4	2×10^{-9}
Copper(I) bromide	$CuBr$	6.27×10^{-9}		Mercury(I) cyanide	$Hg_2(CN)_2$	5×10^{-40}
Copper(I) chloride	$CuCl$	1.72×10^{-7}		Mercury(I) iodide	Hg_2I_2	5.2×10^{-29}
Copper(I) cyanide	$CuCN$	3.47×10^{-20}		Mercury(II) hydroxide	$Hg(OH)_2$	3.1×10^{-26}
Copper(II) carbonate	$CuCO_3$	2.4×10^{-10}		Mercury(II) sulfide	HgS	1.6×10^{-54}
Copper(II) hydroxide	$Cu(OH)_2$	2.2×10^{-20}		Nickel(II) carbonate	$NiCO_3$	1.42×10^{-7}
Copper(II) phosphate	$Cu_3(PO_4)_2$	1.40×10^{-37}		Nickel(II) hydroxide	$Ni(OH)_2$	5.48×10^{-16}
Copper(II) sulfide	CuS	1.27×10^{-36}		Nickel(II) sulfide	NiS	3×10^{-20}
Iron(II) carbonate	$FeCO_3$	3.07×10^{-11}		Silver bromate	$AgBrO_3$	5.38×10^{-5}
Iron(II) hydroxide	$Fe(OH)_2$	4.87×10^{-17}		Silver bromide	$AgBr$	5.35×10^{-13}
Iron(II) sulfide	FeS	3.72×10^{-19}				

(*continued on the next page*)

Compound	Formula	K_{sp}
Silver carbonate	Ag_2CO_3	8.46×10^{-12}
Silver chloride	$AgCl$	1.77×10^{-10}
Silver chromate	Ag_2CrO_4	1.12×10^{-12}
Silver cyanide	$AgCN$	5.97×10^{-17}
Silver iodide	AgI	8.51×10^{-17}
Silver phosphate	Ag_3PO_4	8.89×10^{-17}
Silver sulfate	Ag_2SO_4	1.20×10^{-5}
Silver sulfide	Ag_2S	6×10^{-51}
Strontium carbonate	$SrCO_3$	5.60×10^{-10}

Compound	Formula	K_{sp}
Strontium chromate	$SrCrO_4$	3.6×10^{-5}
Strontium phosphate	$Sr_3(PO_4)_2$	1×10^{-31}
Strontium sulfate	$SrSO_4$	3.44×10^{-7}
Tin(II) hydroxide	$Sn(OH)_2$	5.45×10^{-27}
Tin(II) sulfide	SnS	1×10^{-26}
Zinc carbonate	$ZnCO_3$	1.46×10^{-10}
Zinc hydroxide	$Zn(OH)_2$	3×10^{-17}
Zinc oxalate	ZnC_2O_4	2.7×10^{-8}
Zinc sulfide	ZnS	2×10^{-25}

5. Complex Ion Formation Constants in Water at 25 °C

Complex Ion	K_f
$[Ag(CN)_2]^-$	1×10^{21}
$[Ag(EDTA)]^{3-}$	2.1×10^7
$[Ag(en)_2]^+$	5.0×10^7
$[Ag(NH_3)_2]^+$	1.7×10^7
$[Ag(SCN)_4]^{3-}$	1.2×10^{10}
$[Ag(S_2O_3)_2]^-$	2.8×10^{13}
$[Al(EDTA)]^-$	1.3×10^{16}
$[AlF_6]^{3-}$	7×10^{19}
$[Al(OH)_4]^-$	3×10^{33}
$[Al(ox)_3]^{3-}$	2×10^{16}
$[CdBr_4]^{2-}$	5.5×10^3
$[Cd(CN)_4]^{2-}$	3×10^{18}
$[CdCl_4]^{2-}$	6.3×10^2
$[Cd(en)_3]^{2+}$	1.2×10^{12}
$[CdI_4]^{2-}$	2×10^6
$[Co(EDTA)]^{2-}$	2.0×10^{16}
$[Co(EDTA)]^-$	1×10^{36}
$[Co(en)_3]^{2+}$	8.7×10^{13}
$[Co(en)_3]^{3+}$	4.9×10^{48}
$[Co(NH_3)_6]^{2+}$	1.3×10^5
$[Co(NH_3)_6]^{3+}$	2.3×10^{33}
$[Co(OH)_4]^{2-}$	5×10^9
$[Co(ox)_3]^{4-}$	5×10^9
$[Co(ox)_3]^{3-}$	1×10^{20}
$[Co(SCN)_4]^{2-}$	1×10^3
$[Cr(EDTA)]^-$	1×10^{23}
$[Cr(OH)_4]^-$	8.0×10^{29}
$[CuCl_3]^{2-}$	5×10^5
$[Cu(CN)_4]^{2-}$	1.0×10^{29}
$[Cu(EDTA)]^{2-}$	5×10^{18}
$[Cu(en)_2]^{2+}$	1×10^{20}
$[Cu(NH_3)_4]^{2+}$	1.7×10^{13}
$[Cu(ox)_2]^{2-}$	3×10^8
$[Fe(CN)_6]^{4-}$	1.5×10^{35}

Complex Ion	K_f
$[Fe(CN)_6]^{3-}$	2×10^{43}
$[Fe(EDTA)]^{2-}$	2.1×10^{14}
$[Fe(EDTA)]^-$	1.7×10^{24}
$[Fe(en)_3]^{2+}$	5.0×10^9
$[Fe(ox)_3]^{4-}$	1.7×10^5
$[Fe(ox)_3]^{3-}$	2×10^{20}
$[Fe(SCN)]^{2+}$	8.9×10^2
$[Hg(CN)_4]^{2-}$	1.8×10^{41}
$[HgCl_4]^{2-}$	1.1×10^{16}
$[Hg(EDTA)]^{2-}$	6.3×10^{21}
$[Hg(en)_2]^{2+}$	2×10^{23}
$[HgI_4]^{2-}$	2×10^{30}
$[Hg(ox)_2]^{2-}$	9.5×10^6
$[Ni(CN)_4]^{2-}$	2×10^{31}
$[Ni(EDTA)]^{2-}$	3.6×10^{18}
$[Ni(en)_3]^{2+}$	2.1×10^{18}
$[Ni(NH_3)_6]^{2+}$	2.0×10^8
$[Ni(ox)_3]^{4-}$	3×10^8
$[PbCl_3]^-$	2.4×10^1
$[Pb(EDTA)]^{2-}$	2×10^{18}
$[PbI_4]^{2-}$	3.0×10^4
$[Pb(OH)_3]^-$	8×10^{13}
$[Pb(ox)_2]^{2-}$	3.5×10^6
$[Pb(S_2O_3)_3]^{4-}$	2.2×10^6
$[PtCl_4]^{2-}$	1×10^{16}
$[Pt(NH_3)_6]^{2+}$	2×10^{35}
$[Sn(OH)_3]^-$	3×10^{25}
$[Zn(CN)_4]^{2-}$	2.1×10^{19}
$[Zn(EDTA)]^{2-}$	3×10^{16}
$[Zn(en)_3]^{2+}$	1.3×10^{14}
$[Zn(NH_3)_4]^{2+}$	2.8×10^9
$[Zn(OH)_4]^{2-}$	2×10^{15}
$[Zn(ox)_3]^{4-}$	1.4×10^8

D. Standard Reduction Half-Cell Potentials at 25 °C

Half-Reaction	$E°$ (V)
$F_2(g) + 2\,e^- \longrightarrow 2\,F^-(aq)$	2.87
$O_3(g) + 2\,H^+(aq) + 2\,e^- \longrightarrow O_2(g) + H_2O(l)$	2.08
$Ag^{2+}(aq) + e^- \longrightarrow Ag^+(aq)$	1.98
$Co^{3+}(aq) + e^- \longrightarrow Co^{2+}(aq)$	1.82
$H_2O_2(aq) + 2\,H^+(aq) + 2\,e^- \longrightarrow 2\,H_2O(l)$	1.78
$PbO_2(s) + 4\,H^+(aq) + SO_4^{2-}(aq) + 2\,e^- \longrightarrow$ $PbSO_4(s) + 2\,H_2O(l)$	1.69
$MnO_4^-(aq) + 4\,H^+(aq) + 3\,e^- \longrightarrow MnO_2(s) + 2\,H_2O(l)$	1.68
$2\,HClO(aq) + 2\,H^+(aq) + 2\,e^- \longrightarrow Cl_2(g) + 2\,H_2O(l)$	1.61
$MnO_4^-(aq) + 8\,H^+(aq) + 5\,e^- \longrightarrow Mn^{2+}(aq) + 4\,H_2O(l)$	1.51
$Au^{3+}(aq) + 3\,e^- \longrightarrow Au(s)$	1.50
$2\,BrO_3^-(aq) + 12\,H^+(aq) + 10\,e^- \longrightarrow Br_2(l) + 6\,H_2O(l)$	1.48
$PbO_2(s) + 4\,H^+(aq) + 2\,e^- \longrightarrow Pb^{2+}(aq) + 2\,H_2O(l)$	1.46
$Cl_2(g) + 2\,e^- \longrightarrow 2\,Cl^-(aq)$	1.36
$Cr_2O_7^{2-}(aq) + 14\,H^+(aq) + 6\,e^- \longrightarrow$ $2\,Cr^{3+}(aq) + 7\,H_2O(l)$	1.33
$O_2(g) + 4\,H^+(aq) + 4\,e^- \longrightarrow 2\,H_2O(l)$	1.23
$MnO_2(s) + 4\,H^+(aq) + 2\,e^- \longrightarrow Mn^{2+}(aq) + 2\,H_2O(l)$	1.21
$IO_3^-(aq) + 6\,H^+(aq) + 5\,e^- \longrightarrow \frac{1}{2}I_2(aq) + 3\,H_2O(l)$	1.20
$Br_2(l) + 2\,e^- \longrightarrow 2\,Br^-(aq)$	1.09
$AuCl_4^-(aq) + 3\,e^- \longrightarrow Au(s) + 4\,Cl^-(aq)$	1.00
$VO_2^+(aq) + 2\,H^+(aq) + e^- \longrightarrow VO^{2+}(aq) + H_2O(l)$	0.99
$HNO_2(aq) + H^+(aq) + e^- \longrightarrow NO(g) + 2\,H_2O(l)$	0.98
$NO_3^-(aq) + 4\,H^+(aq) + 3\,e^- \longrightarrow NO(g) + 2\,H_2O(l)$	0.96
$ClO_2(g) + e^- \longrightarrow ClO_2^-(aq)$	0.95
$2\,Hg^{2+}(aq) + 2\,e^- \longrightarrow 2\,Hg_2^{2+}(aq)$	0.92
$Ag^+(aq) + e^- \longrightarrow Ag(s)$	0.80
$Hg_2^{2+}(aq) + 2\,e^- \longrightarrow 2\,Hg(l)$	0.80
$Fe^{3+}(aq) + e^- \longrightarrow Fe^{2+}(aq)$	0.77
$PtCl_4^{2-}(aq) + 2\,e^- \longrightarrow Pt(s) + 4\,Cl^-(aq)$	0.76
$O_2(g) + 2\,H^+(aq) + 2\,e^- \longrightarrow H_2O_2(aq)$	0.70
$MnO_4^-(aq) + e^- \longrightarrow MnO_4^{2-}(aq)$	0.56
$I_2(s) + 2\,e^- \longrightarrow 2\,I^-(aq)$	0.54
$Cu^+(aq) + e^- \longrightarrow Cu(s)$	0.52
$O_2(g) + 2\,H_2O(l) + 4\,e^- \longrightarrow 4\,OH^-(aq)$	0.40

Half-Reaction	$E°$ (V)
$Cu^{2+}(aq) + 2\,e^- \longrightarrow Cu(s)$	0.34
$BiO^+(aq) + 2\,H^+(aq) + 3\,e^- \longrightarrow Bi(s) + H_2O(l)$	0.32
$Hg_2Cl_2(s) + 2\,e^- \longrightarrow 2\,Hg(l) + 2\,Cl^-(aq)$	0.27
$AgCl(s) + e^- \longrightarrow Ag(s) + Cl^-(aq)$	0.22
$SO_4^{2-}(aq) + 4\,H^+(aq) + 2\,e^- \longrightarrow H_2SO_3(aq) + H_2O(l)$	0.20
$Cu^{2+}(aq) + e^- \longrightarrow Cu^+(aq)$	0.16
$Sn^{4+}(aq) + 2\,e^- \longrightarrow Sn^{2+}(aq)$	0.15
$S(s) + 2\,H^+(aq) + 2\,e^- \longrightarrow H_2S(g)$	0.14
$AgBr(s) + e^- \longrightarrow Ag(s) + Br^-(aq)$	0.071
$2\,H^+(aq) + 2\,e^- \longrightarrow H_2(g)$	0.00
$Fe^{3+}(aq) + 3\,e^- \longrightarrow Fe(s)$	−0.036
$Pb^{2+}(aq) + 2\,e^- \longrightarrow Pb(s)$	−0.13
$Sn^{2+}(aq) + 2\,e^- \longrightarrow Sn(s)$	−0.14
$AgI(s) + e^- \longrightarrow Ag(s) + I^-(aq)$	−0.15
$N_2(g) + 5\,H^+(aq) + 4\,e^- \longrightarrow N_2H_5^+(aq)$	−0.23
$Ni^{2+}(aq) + 2\,e^- \longrightarrow Ni(s)$	−0.23
$Co^{2+}(aq) + 2\,e^- \longrightarrow Co(s)$	−0.28
$PbSO_4(s) + 2\,e^- \longrightarrow Pb(s) + SO_4^{2-}(aq)$	−0.36
$Cd^{2+}(aq) + 2\,e^- \longrightarrow Cd(s)$	−0.40
$Fe^{2+}(aq) + 2\,e^- \longrightarrow Fe(s)$	−0.45
$2\,CO_2(g) + 2\,H^+(aq) + 2\,e^- \longrightarrow H_2C_2O_4(aq)$	−0.49
$Cr^{3+}(aq) + e^- \longrightarrow Cr^{2+}(aq)$	−0.50
$Cr^{3+}(aq) + 3\,e^- \longrightarrow Cr(s)$	−0.73
$Zn^{2+}(aq) + 2\,e^- \longrightarrow Zn(s)$	−0.76
$2\,H_2O(l) + 2\,e^- \longrightarrow H_2(g) + 2\,OH^-(aq)$	−0.83
$Mn^{2+}(aq) + 2\,e^- \longrightarrow Mn(s)$	−1.18
$Al^{3+}(aq) + 3\,e^- \longrightarrow Al(s)$	−1.66
$H_2(g) + 2\,e^- \longrightarrow 2\,H^-(aq)$	−2.23
$Mg^{2+}(aq) + 2\,e^- \longrightarrow Mg(s)$	−2.37
$La^{3+}(aq) + 3\,e^- \longrightarrow La(s)$	−2.38
$Na^+(aq) + e^- \longrightarrow Na(s)$	−2.71
$Ca^{2+}(aq) + 2\,e^- \longrightarrow Ca(s)$	−2.76
$Ba^{2+}(aq) + 2\,e^- \longrightarrow Ba(s)$	−2.90
$K^+(aq) + e^- \longrightarrow K(s)$	−2.92
$Li^+(aq) + e^- \longrightarrow Li(s)$	−3.04

E. Vapor Pressure of Water at Various Temperatures

T (°C)	P (torr)	T (°C)	P (torr)	T (°C)	P (torr)	T (°C)	P (torr)
0	4.58	21	18.65	35	42.2	92	567.0
5	6.54	22	19.83	40	55.3	94	610.9
10	9.21	23	21.07	45	71.9	96	657.6
12	10.52	24	22.38	50	92.5	98	707.3
14	11.99	25	23.76	55	118.0	100	760.0
16	13.63	26	25.21	60	149.4	102	815.9
17	14.53	27	26.74	65	187.5	104	875.1
18	15.48	28	28.35	70	233.7	106	937.9
19	16.48	29	30.04	80	355.1	108	1004.4
20	17.54	30	31.82	90	525.8	110	1074.6

Answers to Selected Exercises

Chapter 1

1. **a.** theory   **b.** observation   **c.** law   **d.** observation
3. Several answers possible.
5. **a.** mixture, homogeneous
   b. pure substance, compound
   c. pure substance, element
   d. mixture, heterogeneous

7.
Substance	Pure or Mixture	Type
Aluminum	Pure	Element
Apple juice	Mixture	Homogeneous
Hydrogen peroxide	Pure	Compound
Chicken soup	Mixture	Heterogeneous

9. **a.** pure substance, compound   **b.** mixture, heterogeneous
   c. mixture, homogeneous   **d.** pure substance, element
11. physical, chemical, physical, physical, physical
13. **a.** chemical   **b.** physical   **c.** physical   **d.** chemical
15. **a.** chemical   **b.** physical   **c.** chemical   **d.** chemical
17. **a.** physical   **b.** chemical   **c.** physical
19. **a.** $0\,°C$ **b.** $-321\,°F$ **c.** $-78.3\,°C$ **d.** $310.2\,K$
21. $-62.2\,°C$, $210.9\,K$
23. **a.** 38 ns   **b.** 5.7 pg   **c.** 59 ML   **d.** .93 Gm

25.
1245 kg	1.245×10^6 g	1.245×10^9 mg
515 km	5.15×10^6 dm	5.15×10^7 cm
122.355 s	1.22355×10^5 ms	0.122355 ks
3.345 kJ	3.345×10^3 J	3.345×10^6 mJ

27. 10,000 1-cm squares
29. no
31. $1.26\,\text{g/cm}^3$
33. **a.** 463 g   **b.** 3.7 L
35. **a.** 73.8 mL   **b.** $88.3\,°C$ **c.** 643 mL
37. **a.** 1,050,501   **b.** 0.0020
    c. 0.0000000000000002   **d.** 0.001090
39. **a.** 3
    b. ambiguous, without more information assume 3 significant figures
    c. 3
    d. 5
    e. ambiguous, without more information assume 1 significant figure
41. **a.** not exact   **b.** exact   **c.** not exact   **d.** exact
43. **a.** 156.9   **b.** 156.8   **c.** 156.8   **d.** 156.9
45. **a.** 1.84   **b.** 0.033   **c.** 0.500   **d.** 34
47. **a.** 41.4   **b.** 133.5   **c.** 73.0   **d.** 0.42
49. **a.** 391.3   **b.** 1.1×10^4 **c.** 5.96   **d.** 5.93×10^4
51. **a.** $2.78 \times 10^4\,\text{cm}^3$ **b.** 1.898×10^{-3} kg
    c. 1.98×10^7 cm
53. **a.** 89.8 in   **b.** 5.62 lb   **c.** 2.55 qt   **d.** 6.18 in
55. 5.0×10^1

57. 33 mi/gal
59. **a.** $1.95 \times 10^{-4}\,\text{km}^2$ **b.** $1.95 \times 10^4\,\text{dm}^2$
    c. $1.95 \times 10^6\,\text{cm}^2$
61. $0.680\,\text{mi}^2$
63. 0.95 mL
65. 3.1557×10^7 s/solar year
67. **a.** extensive   **b.** intensive   **c.** intensive
    d. intensive   **e.** extensive
69. $-34°$
71. **a.** 2.2×10^{-6} **b.** 0.0159   **c.** 6.9×10^4
73. **a.** mass of can of gold $= 1.9 \times 10^4$ g
    mass of can of sand $= 3.0 \times 10^3$ g
    b. Yes, the thief sets off the trap because the can of sand is lighter than the gold cylinder.
75. $21\,\text{in}^3$
77. $7.6\,\text{g/cm}^3$
79. 3.11×10^5 lb
81. 3.3×10^2 km
83. 6.8×10^{-15}
85. $0.669\,\Omega$
87. 0.492
89. $7.3 \times 10^{11}\,\text{g/cm}^3$
91. **a.** $1.6 \times 10^4\,\text{nm}^3$ **b.** 1.3×10^{-18} g oxygen
    c. 1.7×10^2 g oxygen   **d.** 1.3×10^{20} nanocontainers
    e. 2.0 L, not feasible when total blood volume in an adult is 5 L (this is almost a 40% increase in blood volume)
93. 49%
95. (c)
97. 343
99. **a.** The darker colored block
    b. the lighter colored block
    c. cannot be determined

Chapter 2

1. 13.5 g
3. These results are not consistent with the law of definite proportions because sample 1 is composed of 11.5 parts Cl to 1 part C and sample 2 is composed of 9.05 parts Cl to 1 part C. The law of definite proportions states that a given compound always contains exactly the same proportion of elements by mass.
5. 23.8 g
7. For the law of multiple proportions to hold, the ratio of the masses of O combining with 1 g of Os in the compound should be a small whole number. $0.3369/0.168 = 2.00$
9. Sample 1: $1.00\,\text{g}\,O_2/1.00\,\text{g S}$;
   sample 2: $1.50\,\text{g}\,O_2/1.00\,\text{g S}$
   Sample 2/sample 1 $= 1.50/1.00 = 1.50$
   3 O atoms/2 O atoms $= 1.5$

11. **a.** not consistent
   b. consistent: Dalton's atomic theory states that the atoms of a given element are identical.
   c. consistent: Dalton's atomic theory states that atoms combine in simple whole-number ratios to form compounds.
   d. not consistent
13. **a.** consistent: Rutherford's nuclear model states that the atom is largely empty space.
   b. consistent: Rutherford's nuclear model states that most of the atom's mass is concentrated in a tiny region called the nucleus.
   c. not consistent
   d. not consistent
15. -2.3×10^{-19} C
17. 9.4×10^{13} excess electrons, 8.5×10^{-17} kg
19. **a, b, c**
21. **a.** $^{23}_{11}$Na **b.** $^{16}_{8}$O **c.** $^{27}_{13}$Al **d.** $^{127}_{53}$I
23. **a.** 7^1_1p and 7^0_1n **b.** 11^1_1p and 12^0_1n
   c. 86^1_1p and 136^0_1n **d.** 82^1_1p and 126^0_1n
25. 6^1_1p and 8^0_1n, $^{14}_{6}$C
27. **a.** 28^1_1p and 26 e$^-$ **b.** 16^1_1p and 18 e$^-$
   c. 35^1_1p and 36 e$^-$ **d.** 24^1_1p and 21 e$^-$
29. **a.** $2-$ **b.** $1+$ **c.** $3+$ **d.** $1+$
31.

Symbol	Ion Formed	Number of Electrons in Ion	Number of Protons in Ion
Ca	Ca^{2+}	18	20
Be	Be^{2+}	2	4
Se	Se^{2-}	36	34
In	In^{3+}	46	49

33. **a.** sodium, metal **b.** magnesium, metal
   c. bromine, nonmetal **d.** nitrogen, nonmetal
   e. arsenic, metalloid
35. **a** and **b** are main group elements
37. **a.** alkali metal **b.** halogen
   c. alkaline earth metal **d.** alkaline earth metal
   e. noble gas
39. Cl and F because they are in the same group or family. Elements in the same group or family have similar chemical properties.
41. 85.47 amu, mass spectrum will have a peak at 86.91 amu and another peak about 2.5 times larger at 84.91 amu
43. 121.8 amu, Sb
45. 1.6×10^{24} atoms
47. **a.** 0.295 mol Ar **b.** 0.0543 mol Zn
   c. 0.144 mol Ta **d.** 0.0304 mol Li
49. 1.42×10^{22} atoms
51. **a.** 1.01×10^{23} atoms **b.** 6.78×10^{21} atoms
   c. 5.39×10^{21} atoms **d.** 5.6×10^{20} atoms
53. 4.2×10^{21} atoms
55. 3.239×10^{-22} g
57. 1.50 g
59. C_2O_3
61. 4.82245×10^7 C/kg

63. ^{237}Pa, ^{238}U, ^{239}Np, ^{240}Pu, ^{235}Ac, ^{234}Ra, etc.
65.

Symbol	Z	A	#p	#e$^-$	#n	Charge
O^{2-}	8	16	8	10	8	$2-$
Ca^{2+}	20	40	20	18	20	$2+$
Mg^{2+}	12	25	12	10	13	$2+$
N^{3-}	7	14	7	10	7	$3-$

67. $V_n = 8.2 \times 10^{-8}$ pm^3, $V_a = 1.4 \times 10^6$ pm^3, 5.9×10^{-12}%
69. 6.022×10^{21} dollars total, 8.9×10^{11} dollars per person, billionaires
71. 15.985 amu
73. 4.76×10^{24} atoms
75. Li $- 6 = 7.494$%, Li $- 7 = 92.506$%
77. 75.0% Au, 25.0% Pb
79. 1.7×10^{22}
81. 1×10^{78} atoms/universe
83. 0.0903
85. N_2O_4, 20; 7
87. **c.** The law of multiple proportions states that when two elements form different compounds, the masses of element B that combine with 1 gram of element A can be expressed in whole number ratios.
89. Li $- 6$: [image] Li $- 7$: [image]
   Li $- 6$: 925 atoms Li $- 7$: 75 atoms
91. If the amu and mole were not based on the same isotope, the numerical values obtained for an atom of material and a mole of material would not be the same.
93. The different isotopes of the same element have the same number of protons and electrons, so the attractive forces between the nucleus and the electrons is constant, and there is no difference in the radii of the isotopes. (Since the nucleus is so small, the extra neutron does not make the atom larger.) Ions, on the other hand, have a different number of electrons than the parent atom from which they are derived. Cations have fewer electrons than the parent atom. The attractive forces are greater because there is a larger positive charge in the nucleus than the negative charge in the electron cloud. So, cations are smaller than the parent atom from which they are derived. Anions have more electrons than the parent. The electron cloud has a greater negative charge than the nucleus, so the anions have a larger radii than the parent.

Chapter 3

1. **a.** 3 Ca, 2 P, 8 O **b.** 1 Sr, 2 Cl
   c. 1 K, 1 N, 3 O **d.** 1 Mg, 2 N, 4 O
3. **a.** NH_3 **b.** C_2H_6 **c.** SO_3
5. **a.** atomic **b.** molecular **c.** atomic **d.** molecular
7. **a.** molecular **b.** ionic **c.** ionic **d.** molecular
9. **a.** molecular element **b.** molecular compound
   c. atomic element
11. **a.** MgS **b.** BaO **c.** $SrBr_2$ **d.** $BeCl_2$
13. **a.** $Ba(OH)_2$ **b.** $BaCrO_4$ **c.** $Ba_3(PO_4)_2$ **d.** $Ba(CN)_2$
15. **a.** magnesium nitride **b.** potassium fluoride
   c. sodium oxide **d.** lithium sulfide
17. **a.** tin(IV) chloride **b.** lead(II) iodide
   c. iron(III) oxide **d.** copper(II) iodide
19. **a.** tin(II) oxide **b.** chromium(III) sulfide
   c. rubidium iodide **d.** barium bromide

21. a. copper(I) nitrite   **b.** magnesium acetate
    c. barium nitrate   **d.** lead(II) acetate
    e. potassium chlorate   **f.** lead(II) sulfate

23. a. $NaHSO_3$ **b.** $LiMnO_4$ **c.** $AgNO_3$ **d.** K_2SO_4
    e. $RbHSO_4$ **f.** $KHCO_3$

25. a. cobalt(II) sulfate heptahydrate   **b.** $IrBr_3 \cdot 4H_2O$
    c. magnesium bromate hexahydrate   **d.** $K_2CO_3 \cdot 2H_2O$

27. a. carbon monoxide   **b.** nitrogen triiodide
    c. silicon tetrachloride   **d.** tetranitrogen tetraselenide
    e. diiodine pentaoxide

29. a. PCl_3 **b.** ClO **c.** S_2F_4 **d.** PF_5 **e.** P_2S_5

31. a. hydroiodic acid   **b.** nitric acid
    c. carbonic acid   **d.** acetic acid

33. a. HF **b.** HBr **c.** H_2SO_3

35. a. 46.01 amu   **b.** 58.12 amu
    c. 180.16 amu   **d.** 238.03 amu

37. a. 0.471 mol CCl_4 **b.** 0.0362 mol $C_{12}H_{22}O_{11}$
    c. 968 mol C_2H_2 **d.** 0.279 mol N_2O

39. a. 1.2×10^{23} molecules   **b.** 4.61×10^{23} molecules
    c. 3.44×10^{23} molecules   **d.** 1.53×10^{23} molecules

41. 2.992×10^{-23} g

43. 3.04×10^{-7} mol, 0.104 mg

45. a. 74.87% C   **b.** 79.89% C
    c. 92.26% C   **d.** 37.23% C

47. NH_3 : 82.27% N $CO(NH_2)_2$: 46.65% N
    NH_4NO_3 : 35.00% N $(NH_4)_2SO_4$: 21.20% N
    NH_3 has the highest N content.

49. 27.1 g F

51. 196 μg KI

53. a. 2:1   **b.** 4:1   **c.** 6:2:1

55. a. 0.885 mol H   **b.** 5.2 mol H
    c. 29 mol H   **d.** 33.7 mol H

57. a. 3.3 g Na   **b.** 3.6 g Na
    c. 1.4 g Na   **d.** 1.7 g Na

59. a. Ag_2O **b.** $Co_3As_2O_8$ **c.** $SeBr_4$

61. a. C_5H_7N **b.** $C_4H_5N_2O$

63. NCl_3

65. a. $C_{12}H_{14}N_2$ **b.** $C_6H_3Cl_3$ **c.** $C_{10}H_{20}N_2S_4$

67. CH_2

69. C_2H_4O

71. $2\,SO_2(g) + O_2(g) + 2\,H_2O(l) \longrightarrow 2\,H_2SO_4(aq)$

73. $2\,Na(s) + 2\,H_2O(l) \longrightarrow H_2(g) + 2\,NaOH(aq)$

75. $C_{12}H_{22}O_{11}(aq) + H_2O(l) \longrightarrow 4\,C_2H_5OH(aq) + 4\,CO_2(g)$

77. a. $PbS(s) + 2\,HBr(aq) \longrightarrow PbBr_2(s) + H_2S(g)$
    b. $CO(g) + 3\,H_2(g) \longrightarrow CH_4(g) + H_2O(l)$
    c. $4\,HCl(aq) + MnO_2(s) \longrightarrow$
               $MnCl_2(aq) + 2\,H_2O(l) + Cl_2(g)$
    d. $C_5H_{12}(l) + 8\,O_2(g) \longrightarrow 5\,CO_2(g) + 6\,H_2O(l)$

79. $Na_2CO_3(aq) + CuCl_2(aq) \longrightarrow CuCO_3(s) + 2\,NaCl(aq)$

81. a. $2\,CO_2(g) + CaSiO_3(s) + H_2O(l) \longrightarrow$
               $SiO_2(s) + Ca(HCO_3)_2(aq)$
    b. $2\,Co(NO_3)_3(aq) + 3\,(NH_4)_2S(aq) \longrightarrow$
               $Co_2S_3(s) + 6\,NH_4NO_3(aq)$
    c. $Cu_2O(s) + C(s) \longrightarrow 2\,Cu(s) + CO(g)$
    d. $H_2(g) + Cl_2(g) \longrightarrow 2\,HCl(g)$

83. a. inorganic   **b.** organic
    c. organic   **d.** inorganic

85. 1.50×10^{24} molecules C_2H_5OH

87. a. K_2CrO_4, 40.27% K, 26.78% Cr, 32.95% O
    b. $Pb_3(PO_4)_2$, 76.60% Pb, 7.632% P, 15.77% O
    c. H_2SO_3, 2.46% H, 39.07% S, 58.47% O
    d. $CoBr_2$, 26.94% Co, 73.06% Br

89. 230 g Cl/yr

91. M = Fe

93. estradiol = $C_{18}H_{24}O_2$

95. $C_{18}H_{20}O_2$

97. 7 H_2O

99. C_6H_9BrO

101. 1.87×10^{21} atoms

103. 92.93 amu

105. Ca

107. 8.66 kg

109. XZ_2, X_2Z_5

111. 19.8 g S_2Cl_2

113. 0.078 metric ton H_2SO_4

115. $C_6H_7SNO_2$

117. XY_3

119. The sphere in the molecular models represents the electron cloud of the atom. On this scale, the nucleus would be too small to see.

121. The statement is incorrect because a chemical formula is based on the ratio of atoms combined not the ratio of grams combined. The statement should read: "The chemical formula for ammonia (NH_3) indicates that ammonia contains three hydrogen atoms to each nitrogen atom."

123. % mass O > % mass S > % mass H

Chapter 4

1. $2\,C_6H_{14}(g) + 19\,O_2(g) \longrightarrow$
               $12\,CO_2(g) + 14\,H_2O(g)$, 47 mol O_2

3. a. 2.6 mol   **b.** 12 mol   **c.** 0.194 mol   **d.** 28.7 mol

5.

mol SiO_2	mol C	mol SiC	mol CO
3	9	3	6
2	6	2	4
5	15	5	10
2.8	8.4	2.8	5.6
0.517	1.55	0.517	1.03

7. a. 14 g HBr, 0.17 g H_2

9. a. 3.8 g   **b.** 4.5 g   **c.** 4.1 g   **d.** 4.7 g

11. a. 4.42 g HCl   **b.** 8.25 g HNO_3 **c.** 4.24 g H_2SO_4

13. a. Na   **b.** Na   **c.** Br_2 **d.** Na

15. 3 molecules Cl_2

17. a. 2 mol   **b.** 7 mol   **c.** 9.40 mol

19. a. 2.5 g   **b.** 31.1 g   **c.** 1.16 g

21. limiting reactant: Pb^{2+}, theoretical yield: 34.5 g $PbCl_2$, percent yield: 85.3%

23. limiting reactant: NH_3, theoretical yield: 240.5 kg CH_4N_2O, percent yield: 70.01%

25. a. 1.5 M LiCl   **b.** 0.116 M $C_6H_{12}O_6$
    c. 5.05×10^{-3} M NaCl

27. a. 1.3 mol   **b.** 1.5 mol   **c.** 0.211 mol

29. 55 g

31. 0.27 M

33. 6.0 L

35. 37.1 mL

37. 2.1 L

39. Barium Nitrate, 2.81 g, 87.2%

41. a. yes   **b.** no   **c.** yes   **d.** no

43. a. soluble Ag^+, NO_3^- **b.** soluble Pb^{2+}, $C_2H_3O_2^-$
   c. soluble K^+, NO_3^- **d.** soluble NH_4^+, S^{2-}

45. a. NO REACTION   **b.** NO REACTION
   c. $CrBr_2(aq) + Na_2CO_3(aq) \longrightarrow CrCO_3(s) + 2\,NaBr(aq)$
   d. $3\,NaOH(aq) + FeCl_3(aq) \longrightarrow$
$$Fe(OH)_3(s) + 3\,NaCl(aq)$$

47. a. $K_2CO_3(aq) + Pb(NO_3)_2(aq) \longrightarrow$
$$PbCO_3(s) + 2\,KNO_3(aq)$$
   b. $Li_2SO_4(aq) + Pb(C_2H_3O_2)_2(aq) \longrightarrow$
$$PbSO_4(s) + 2\,LiC_2H_3O_3(aq)$$
   c. $Cu(NO_3)_2(aq) + MgS(aq) \longrightarrow$
$$CuS(s) + Mg(NO_3)_2(aq)$$
   d. NO REACTION

49. a. Complete:
$$H^+(aq) + Cl^-(aq) + Li^+(aq) + OH^-(aq) \longrightarrow$$
$$H_2O(l) + Li^+(aq) + Cl^-(aq)$$
    Net: $H^+(aq) + OH^-(aq) \longrightarrow H_2O(l)$
   b. Complete:
$$Mg^{2+}(aq) + S^{2-}(aq) + Cu^{2+}(aq) + 2\,Cl^-(aq) \longrightarrow$$
$$CuS(s) + Mg^{2+}(aq) + 2\,Cl^-(aq)$$
    Net: $Cu^{2+}(aq) + S^{2-}(aq) \longrightarrow CuS(s)$
   c. Complete:
$$Na^+(aq) + OH^-(aq) + H^+(aq) + NO_3^-(aq) \longrightarrow$$
$$H_2O(l) + Na^+(aq) + NO_3^-(aq)$$
    Net: $H^+(aq) + OH^-(aq) \longrightarrow H_2O(l)$
   d. Complete:
$$6\,Na^+(aq) + 2\,PO_4^{3-}(aq) + 3\,Ni^{2+}(aq) + 6\,Cl^-(aq) \longrightarrow$$
$$Ni_3(PO_4)_2(s) + 6\,Na^+(aq) + 6\,Cl^-(aq)$$
    Net: $3\,Ni^{2+}(aq) + 2\,PO_4^{3-}(aq) \longrightarrow Ni_3(PO_4)_2(s)$

51. Complete:
$$Hg_2^{2+}(aq) + 2\,NO_3^-(aq) + 2\,Na^+(aq) + 2\,Cl^-(aq) \longrightarrow$$
$$Hg_2Cl_2(s) + 2\,Na^+(aq) + 2\,NO_3^-(aq)$$
   Net: $Hg_2^{2+}(aq) + 2\,Cl^-(aq) \longrightarrow Hg_2Cl_2(s)$

53. Molecular: $HBr(aq) + KOH(aq) \longrightarrow H_2O(l) + KBr(aq)$
   Net ionic: $H^+(aq) + OH^-(aq) \longrightarrow H_2O(l)$

55. a. $H_2SO_4(aq) + Ca(OH)_2(aq) \longrightarrow 2\,H_2O(l) + CaSO_4(s)$
   b. $HClO_4(aq) + KOH(aq) \longrightarrow H_2O(l) + KClO_4(aq)$
   c. $H_2SO_4(aq) + 2\,NaOH(aq) \longrightarrow 2\,H_2O(l) + Na_2SO_4(aq)$

57. a. $2\,HBr(aq) + NiS(s) \longrightarrow H_2S(g) + NiBr_2(aq)$
   b. $NH_4I(aq) + NaOH(aq) \longrightarrow$
$$H_2O(l) + NH_3(g) + NaI(aq)$$
   c. $2\,HBr(aq) + Na_2S(aq) \longrightarrow H_2S(g) + 2\,NaBr(aq)$
   d. $2\,HClO_4(aq) + Li_2CO_3(aq) \longrightarrow$
$$H_2O(l) + CO_2(g) + 2\,LiClO_4(aq)$$

59. a. Ag: 0   **b.** Ag: +1
   c. Ca: +2, F: −1   **d.** H: +1, S: −2
   e. C: +4, O: −2   **f.** Cr: +6, O: −2

61. a. +2   **b.** +6   **c.** +3

63. a. redox reaction, oxidizing agent: O_2, reducing agent: Li
   b. redox reaction, oxidizing agent: Fe^{2+}, reducing agent: Mg
   c. not a redox reaction
   d. not a redox reaction

65. a. $S(s) + O_2(g) \longrightarrow SO_2(g)$
   b. $2\,C_3H_6(g) + 9\,O_2(g) \longrightarrow 6\,CO_2(g) + 6\,H_2O(g)$
   c. $2\,Ca(s) + O_2(g) \longrightarrow 2\,CaO(g)$
   d. $C_5H_{12}S(l) + 9\,O_2(g) \longrightarrow$
$$5\,CO_2(g) + SO_2(g) + 6\,H_2O(g)$$

67. 3.32 M

69. 1.1 g

71. 3.1 kg

73. limiting reactant: $C_7H_6O_3$, theoretical yield: 1.63 g $C_9H_8O_4$, percent yield: 74.8%

75. b

77. a. $2\,HCl(aq) + Hg_2(NO_3)_2(aq) \longrightarrow$
$$Hg_2Cl_2(s) + 2\,HNO_3(aq)$$
   b. $KHSO_3(aq) + HNO_3(aq) \longrightarrow$
$$H_2O(l) + SO_2(g) + KNO_3(aq)$$
   c. $2\,NH_4Cl(aq) + Pb(NO_3)_2(aq) \longrightarrow$
$$PbCl_2(s) + 2\,NH_4NO_3(aq)$$
   d. $2\,NH_4Cl(aq) + Ca(OH)_2(aq) \longrightarrow$
$$2\,NH_3(g) + 2\,H_2O(g) + CaCl_2(aq)$$

79. 22 g

81. 6.9 g

83. $NaNO_3$ is more economical.

85. Br is the oxidizing agent, Au is the reducing agent, 38.8 g $KAuF_4$

87. Ca^{2+} and Cu^{2+} present in the original solution
   Net ionic for first precipitate:
$$Ca^{2+}(aq) + SO_4^{2-}(aq) \longrightarrow CaSO_4(s)$$
   Net ionic for second precipitate:
$$Cu^{2+}(aq) + CO_3^{2-}(aq) \longrightarrow CuCO_3(s)$$

89. 30.8 kg

91. 3.4×10^4 kg

93. 2.0 mg

95. 96.9 g

97. d. The mass ratio must be at least $(4 \times 39.09)/32 = 4.88$ or K will be the limiting reactant. The mass ratio of 1.5/0.38 is 3.94, which is less than 4.88.

99. a. Add 0.5 mol solute
   b. Add 1 L solvent
   c. Add 0.33 L solvent

101. $A_2X(aq) + BY_2(aq) \longrightarrow AY(s) + BX(aq)$

Chapter 5

1. a. 0.832 atm   **b.** 632 mmHg
   c. 12.2 psi   **d.** 8.43×10^4 Pa

3. a. 809.0 mmHg   **b.** 1.064 atm
   c. 809.0 torr   **d.** 107.9 kPa

5. 5.70×10^2 mmHg

7. 44.6 mL

9. 4.33 L

11. 3.5 L

13. 2.1 mol

15. Yes, the final gauge pressure is 43.5 psi which exceeds the maximum rating.

17. 16.2 L

19. b

21. 4.76 atm

23. 16.7 L

25. 9.43 g/L

27. 44.0 g/mol

29. 4.00 g/mol

31. $P_{tot} = 0.879$ atm; $mass_{N_2} = 1.47$ g; $mass_{O_2} = 0.496$ g; $mass_{He} = 0.101$ g

33. 1.84 atm

35. $\chi_{N_2} = 0.627$, $\chi_{O_2} = 0.373$, $P_{N_2} = 0.687$ atm, $P_{O_2} = 0.409$ atm

37. $P_{H_2} = 0.921$ atm, $mass_{H_2} = 0.0539$ g

39. 7.47×10^{-2} g

41. 38 L

43. $V_{H_2} = 48.2$ L, $V_{CO} = 24.1$ L

45. 22.8 g NaN_3

47. 60.5%

49. Flourine, 4.84 g

51. **a.** yes   **b.** no

  c. No. Even though the argon atoms are more massive than the helium atoms, both have the same kinetic energy at a given temperature. The argon atoms, therefore, move more slowly, and so exert the same pressure as the helium atoms.

  d. He

53. F_2: $u_{rms} = 442$ m/s,$KE_{avg} = 3.72 \times 10^3$ J; Cl_2: $u_{rms} = 324$ m/s,$KE_{avg} = 3.72 \times 10^3$ J; Br_2: $u_{rms} = 216$ m/s,$KE_{avg} = 3.72 \times 10^3$ J; rankings: u_{rms}: $Br_2 < Cl_2 < F_2$, KE_{avg}: $Br_2 = Cl_2 = F_2$, rate of effusion: $Br_2 < Cl_2 < F_2$

55. rate $^{238}UF_6$/rate $^{235}UF_6 = 0.99574$

57. krypton

58. A has the higher molar mass, B has the higher rate of effusion.

61. That the volume of gas particles is small compared to the space between them breaks down under conditions of high pressure. At high pressure the particles themselves occupy a significant portion of the total gas volume.

63. 0.05826 L (ideal); 0.0708 L (V.D.W.); Difference because of high pressure, at which Ne no longer acts ideally.

65. 97.7%

67. 27.8 g/mol

69. C_4H_{10}

71. 4.70 L

73. $2 HCl(aq) + K_2S(s) \longrightarrow$
$H_2S(g) + 2 KCl(aq), 0.191$ g $K_2S(s)$

75. 11.7 L

77. $mass_{air} = 8.56$ g, $mass_{He} = 1.20$ g, mass difference $= 7.36$ g

79. 4.76 L/s

81. total force $= 6.15 \times 10^3$ pounds; no, the can cannot withstand this force

83. 5.8×10^3 balloons

85. 4.0 cm

87. 77.7%

89. 7.3×10^{-3} mol

91. 311 K

93. 5.0 g

95. C_3H_8

97. 74.0 mmHg

99. $P_{CH_4} = 7.30 \times 10^{-2}$ atm, $P_{O_2} = 4.20 \times 10^{-1}$ atm, $P_{NO} = 2.79 \times 10^{-3}$ atm, $P_{CO_2} = 5.03 \times 10^{-3}$ atm, $P_{H_2O} = 5.03 \times 10^{-3}$ atm, $P_{NO_2} = 2.51 \times 10^{-2}$ atm, $P_{OH} = 1.01 \times 10^{-2}$ atm, $P_{tot} = 0.533$ atm

101. 0.42

103. 1.1 atm

105. 0.46 g/L

107. Because helium is less dense than air, the balloon moves in a direction opposite the direction the air inside the car is moving due to the acceleration and deceleration of the car.

109. Since each gas will occupy 22.414 L/mole at STP and we have 2 moles of gas, we will have a volume of 44.828 L.

111. Br_2 would deviate the most from ideal behavior since it is the largest of the three.

113. c

Chapter 6

1. **a.** 8.48×10^3 cal   **b.** 4.289×10^6 J
   c. 8.48×10^4 cal   **d.** 4.50×10^8 J

3. **a.** 9560×10^6 J   **b.** 9560×10^3 kJ   **c.** 2.66 kWh

5. d

7. **a.** heat, +   **b.** work, −   **c.** heat, +

9. -7.30×10^2 kJ

11. 311 kJ

13. The drinks that went into cooler B had more thermal energy than the refrigerated drinks that went into cooler A. The temperature difference between the drinks in cooler B and the ice was greater than the difference between the drinks and the ice in cooler A. More thermal energy was exchanged between the drinks and the ice in cooler B, which resulted in more melting.

15. 5.4×10^5 J

17. **a.** 7.6×10^2 °C   **b.** 4.3×10^2 °C
   c. 1.3×10^2 °C   **d.** 49 °C

19. -2.80×10^2 J

21. 489 J

23. $\Delta E = -3463$ J, $\Delta H = -3452$ kJ

25. **a.** exothermic, −   **b.** endothermic, +   **c.** exothermic, −

27. -4.30×10^3 kJ

29. 6.46×10^4 kJ

31. 9.5×10^2 g CO_2

33. 77.1 g Ag

35. 28.4 °C

37. 1.10 J/g·°C

39. Measurement B corresponds to conditions of constant pressure. Measurement A corresponds to conditions of constant volume. When a fuel is burned under constant pressure some of the energy released does work on the atmosphere by expanding against it. Less energy is manifest as heat due to this work. When a fuel is burned under constant volume, all of the energy released by the combustion reaction is evolved as heat.

41. -6.3×10^3 kJ/mol

43. -1.6×10^5 J

45. **a.** $-\Delta H_1$ **b.** $2 \Delta H_1$ **c.** $-\frac{1}{2}\Delta H_1$

47. -23.9 kJ

49. 87.8 kJ

51. **a.** $N_2(g) + 3 H_2(g) \longrightarrow 2 NH_3(g)$, $\Delta H_f^\circ = -45.9$ kJ/mol
   b. $C(s, \text{graphite}) + O_2(g) \longrightarrow$
   $CO_2(g), \Delta H_f^\circ = -393.5$ kJ/mol
   c. $2 Fe(s) + 3/2 O_2(g) \longrightarrow$
   $Fe_2O_3(s), \Delta H_f^\circ = -824.2$ kJ/mol
   d. $C(s, \text{graphite}) + 2 H_2(g) \longrightarrow$
   $CH_4(g), \Delta H_f^\circ = -74.6$ kJ/mol

53. -382.1 kJ/mol

55. a. -137.1 kJ   **b.** -41.2 kJ
   c. -137 kJ   **d.** 290.7 kJ

57. $6\,CO_2(g) + 6\,H_2O(l) \longrightarrow$
$$C_6H_{12}O_6(s) + 9\,O_2(g),\ \Delta H^\circ_{rxn} = 2803 \text{ kJ}$$

59. -113.0 kJ/mol

61. $\Delta E = -1.7$ J, $q = -0.5$ J, $w = -1.2$ J

63. 78 g

65. $\Delta H = 6.0$ kJ/mol, 1.1×10^2 g

67. $26.3\,^\circ C$

69. palmitic acid: 9.9378 Cal/g, sucrose: 3.938 Cal/g, fat contains more Cal/g than sugar

71. $\Delta H = \Delta E + nR\,\Delta T$

73. 5.7 Cal/g

75. $\Delta E = 0$, $\Delta H = 0$, $q = -w = 3.0 \times 10^3$ J

77. 94.1 kJ

79. 7.3×10^3 g H_2SO_4

81. 7.2×10^2 g

83. $78.2\,^\circ C$

85. $C_v = \dfrac{3}{2}R$, $C_p = \dfrac{5}{2}R$

87. $q = 1030$ kJ, $w = -78.1$ kJ, $\Delta E = 9\,50$ kJ, $\Delta H = 1030$ kJ

89. d

91. a. At constant pressure, heat can be added and work can be done on the system. $\Delta E = q + w$, therefore, $q = \Delta E - w$.

Chapter 7

1. 499 s

3. (i) d, c, b, a   (ii) a, b, c, d

5. a. 4.74×10^{14} Hz   **b.** 5.96×10^{14} Hz
   c. 5.8×10^{18} Hz

7. a. 3.14×10^{-19} J   **b.** 3.95×10^{-19} J
   c. 3.8×10^{-15} J

9. 1.31×10^{16} photons

11. a. 79.8 kJ/mol   **b.** 239 kJ/mol   **c.** 798 kJ/mol

13.

15. 6.33 nm

17. 1.1×10^{-34} m. The wavelength of a baseball is negligible with respect to its size.

19. 2s

21. a. $l = 0$ **b.** $l = 0, 1$
   c. $l = 0, 1, 2$ **d.** $l = 0, 1, 2, 3$

23. l can equal 0, 1, and 2. For $l = 0$, m_l can equal only 0; For $l = 1$, m_l can equal -1, 0, and 1. For $l = 2$, m_l can equal $-2, -1, 0, 1$, and 2; 9 orbitals

25. $+1/2, -1/2$

27. c

29. See Figures 7.19 and 7.20. The 2s and 3p orbitals would, on average, be farther from the nucleus and have more nodes than the 1s and 2p orbitals.

31. $n = 1$

33. $2p \longrightarrow 1s$

35. a. 122 nm, UV   **b.** 103 nm, UV
   c. 486 nm, visible   **d.** 434 nm, visible

37. $n = 2$

39. 344 nm

41. 6.4×10^{17} photons/s

43. 0.0547 nm

45. 91.2 nm

47. a. 4   **b.** 9   **c.** 16

49. $n = 4 \longrightarrow n = 3$, $n = 5 \longrightarrow n = 3$, $n = 6 \longrightarrow n = 3$, respectively

51. 4.84×10^{14} s^{-1}

53. 11 m

55. 6.79×10^{-3} J

57. 632 nm

59. a. $E_1 = 2.51 \times 10^{-18}$ J, $E_2 = 1.00 \times 10^{-17}$ J, $E_3 = 2.26 \times 10^{-17}$ J
   b. 26.5 nm, UV; 15.8 nm, UV

61.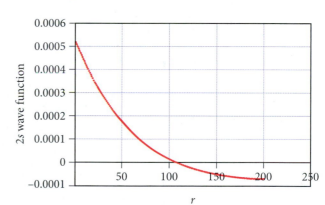

2s:

The plot for the 2s wave function extends below the x-axis. The x-intercept represents the radial node of the orbital.

63. 7.39×10^5 m/s

65. 7.5×10^{-7}

67. In the Bohr model, electrons exist in specific orbits encircling the atom. In the quantum mechanical model, electrons exist in orbitals that are really probability density maps of where the electron is likely to be found. The Bohr model is inconsistent with Heisenberg's uncertainty principle.

69. a

Chapter 8

1. a. $1s^2\,2s^2\,2p^6\,3s^2\,3p^3$ **b.** $1s^2\,2s^2\,2p^2$
   c. $1s^2\,2s^2\,2p^6\,3s^1$ **d.** $1s^2\,2s^2\,2p^6\,3s^2\,3p^6$

3. a.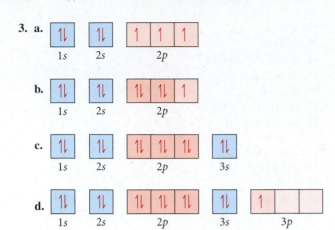

b.

c.

d.

5. a. $[\text{Ne}]\,3s^2\,3p^3$ **b.** $[\text{Ar}]\,4s^2\,3d^{10}\,4p^2$
 c. $[\text{Kr}]\,5s^2\,4d^2$ **d.** $[\text{Kr}]\,5s^2\,4d^{10}\,5p^5$

7. a. 1   **b.** 10   **c.** 5   **d.** 2

9. a. V, As   **b.** Se   **c.** V   **d.** Kr

11. a. 2   **b.** 1   **c.** 10   **d.** 6

13. reactive metal: **a**, reactive nonmetal: **c**

15. C

17. The valence electrons of nitrogen will experience a greater effective nuclear charge. The valence electrons of both atoms are screened by two core electrons but N has a greater number of protons and therefore a greater net nuclear charge.

19. a. 1+   **b.** 2+   **c.** 6+   **d.** 4+

21. a. In   **b.** Si   **c.** Pb   **d.** C

23. F, S, Si, Ge, Ca, Rb

25. a. $[\text{Ne}]$ **b.** $[\text{Kr}]$ **c.** $[\text{Kr}]$
 d. $[\text{Ar}]\,3d^6$ **e.** $[\text{Ar}]\,3d^9$

27. a. $[\text{Ar}]$ Diamagnetic

b. $[\text{Ar}]$  Paramagnetic

c. $[\text{Ar}]$ Paramagnetic

d. $[\text{Ar}]$ Paramagnetic

29. a. Li   **b.** I^- **c.** Cr   **d.** O^{2-}

31. O^{2-}, F^-, Ne, Na^+, Mg^{2+}

33. a. Br   **b.** Na
 c. cannot tell based on periodic trends   **d.** P

35. In, Si, N, F

37. a. second and third   **b.** fifth and sixth
 c. sixth and seventh   **d.** first and second

39. a. Na   **b.** S   **c.** C   **d.** F

41. a. Sr   **b.** Bi
 c. cannot tell based on periodic trends   **d.** As

43. S, Se, Sb, In, Ba, Fr

45. Br: $1s^2\,2s^2\,2p^6\,3s^2\,3p^6\,4s^2\,3d^{10}\,4p^5$
 Kr: $1s^2\,2s^2\,2p^6\,3s^2\,3p^6\,4s^2\,3d^{10}\,4p^6$
 Krypton's outer electron shell is filled, giving it chemical stability. Bromine is missing an electron from its outer shell and subsequently has a high electron affinity. Bromine tends to be easily reduced by gaining an electron, giving the bromide ion

stability due to the filled p subshell which corresponds to krypton's chemically stable electron configuration.

47. V: $[\text{Ar}]\,4s^2\,3d^3$ V^{3+}: $[\text{Ar}]\,3d^2$
 Both V and V^{3+} contain unpaired electrons in their $3d$ orbitals.

49. A substitute for K^+ would need to exhibit a 1+ electric charge and have similar mass and atomic radius. Na^+ and Rb^+ would not be good substitutes because their radii are significantly smaller and larger, respectively. Based on mass, Ca^+ and Ar^+ are the closest to K^+. Because the first ionization energy of Ca^+ is closest to that of K^+, Ca^+ is the best choice for a substitute. The difficulty lies in Ca's low second ionization energy, making it easily oxidized.

51. Si, Ge

53. a. N: $[\text{He}]\,2s^2\,2p^3$, Mg: $[\text{Ne}]\,3s^2$, O: $[\text{He}]\,2s^2\,2p^4$,
 F: $[\text{He}]\,2s^2\,2p^5$, Al: $[\text{Ne}]\,3s^2\,3p^1$
 b. Mg, Al, N, O, F   **c.** Al, Mg, O, N, F
 d. Aluminum's first ionization energy is lower than Mg because its $3p$ electron is shielded by the $3s$ orbital. Oxygen's first ionization energy is lower than that of N because its fourth $2p$ electron experiences electron–electron repulsion by the other electron in its orbital.

55. For main-group elements, atomic radii decrease across a period because the addition of a proton in the nucleus and an electron in the outermost energy level increases Z_{eff}. This does not happen in the transition metals because the electrons are added to the $n_{\text{highest}-1}$ orbital and the Z_{eff} stays roughly the same.

57. Noble gases are exceptionally unreactive due to the stability of their completely filled outer quantum levels and their high ionization energies. The ionization energies of Kr, Xe, and Rn are low enough to form some compounds.

59. 6A: $ns^2\,np^4$, 7A: $ns^2\,np^5$, group 7A elements require only one electron to achieve a noble gas configuration. Since group 6A elements require two electrons, their affinity for one electron is less negative, because one electron will merely give them an np^5 configuration.

61. 85

63. 1.390×10^{-6} J; 86.06 nm

65. 120, 170

67. a. $d_{\text{Ar}} \approx 2$ g/L, $d_{\text{Xe}} \approx 6.5$ g/L
 b. $d_{118} \approx 13$ g/L
 c. mass $= 3.35 \times 10^{-23}$ g/Ne atom, density of Ne atom $= 2.3 \times 10^4$ g/L. The separation of Ne atoms relative to their size is immense.
 d. Kr: 2.69×10^{22} atoms/L, Ne: 2.69×10^{22} atoms/L. It seems Ar will also have 2.69×10^{22} atoms/L. $d_{\text{Ar}} = 1.78$ g/L. This corresponds to accepted values.

69. Density increases to the right, because, though electrons are added successively across the period, they are added to the $3d$ subshell which is not a part of the outermost principal energy level. As a result, the atomic radius does not increase significantly across the period while mass does.

71. Longest λ: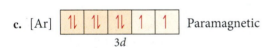

Next longest λ:

Next longest λ:

73. 168, noble gas

75. A relatively high effective nuclear charge is found in gallium with its completed $3d$ subshell and in thallium with its completed $4f$ subshell, accounting for the relatively high first ionization energies of these elements.

77. The second electron affinity requires the addition of an electron to something that is already negatively charged. The monoanions of both of these elements have relatively high electron density in a relatively small volume. As we shall see in Chapter 9 the dianions of these elements do exist in many compounds because they are stabilized by chemical bonding.

79. **a.** any group 6A element   **b.** any group 5A element
 c. any group 1A element

81. **a.** True   **b.** True   **c.** False   **d.** True

83. The $4s$ electrons in calcium have relatively low ionization energies because they are valence electrons. The energy cost of losing a third electron is extraordinarily higher. Similarly the electron affinity of fluorine to gain one electron is highly exothermic because the added electron completes fluorine's valence shell. Therefore, we expect calcium and fluoride to combine in a 1:2 ratio.

Chapter 9

1. $1s^2 2s^2 2p^3$ $\cdot\ddot{N}\!:$

3. **a.** $\cdot\dot{Al}\cdot$ **b.** Na^+ **c.** $\cdot\ddot{Cl}\!:$ **d.** $\left[:\ddot{Cl}\!:\right]^-$

5. **a.** $Na^+\left[:\ddot{F}\!:\right]^-$ **b.** $Ca^{2+}\left[:\ddot{O}\!:\right]^{2-}$

 c. $Sr^{2+}\,2\left[:\ddot{Br}\!:\right]^-$ **d.** $2\,K^+\left[:\ddot{O}\!:\right]^{2-}$

7. **a.** SrSe   **b.** $BaCl_2$ **c.** Na_2S **d.** Al_2O_3

9. As the size of the alkaline earth metal ions increases, so does the distance between the metal cations and oxygen anions. Therefore, the magnitude of the lattice energy decreases accordingly because the potential energy decreases as the distance increases.

11. One factor of lattice energy is the product of the charges of the two ions. The product of the ion charges for CsF is -1 while that for BaO is -4. Because this product is four times greater, the lattice energy is also four times greater.

13. **a.** H:H, filled duets, 0 formal charge on both atoms

 b. $:\ddot{Cl}\!:\!\ddot{Cl}\!:$, filled octets, 0 formal charge on both atoms

 c. $\ddot{O}\!=\!\ddot{O}$, filled octets, 0 formal charge on both atoms

 d. $:N\!\equiv\!N\!:$, filled octets, 0 formal charge on both atoms

15. **a.** $H\!-\!\overset{\displaystyle H}{\underset{\displaystyle |}{\overset{|}{P}}}\!-\!H$ **b.** $:\ddot{S}\!-\!\ddot{Cl}\!:$ with $:\ddot{Cl}\!:$ below

 c. $H\!-\!\ddot{I}\!:$ **d.** $H\!-\!\overset{\displaystyle H}{\underset{\displaystyle H}{\overset{|}{\underset{|}{C}}}}\!-\!H$

17. **a.** pure covalent   **b.** polar covalent
 c. pure covalent   **d.** ionic bond

19. $:C\!\!\overset{\longrightarrow}{=}\!\!O\!:$, 25%

21. **a.** $:\ddot{I}\!-\!\overset{\displaystyle :\ddot{I}:}{\underset{\displaystyle :\ddot{I}:}{\overset{|}{\underset{|}{C}}}}\!-\!\ddot{I}\!:$ **b.** $:N\!\equiv\!N\!-\!\ddot{O}\!:$ **c.** $H\!-\!\overset{\displaystyle H}{\underset{\displaystyle H}{\overset{|}{\underset{|}{Si}}}}\!-\!H$

 d. $:\ddot{Cl}\!-\!\overset{\displaystyle :O:}{\overset{\|}{C}}\!-\!\ddot{Cl}\!:$ **e.** $H\!-\!\overset{\displaystyle \ddot{O}\!-\!H}{\underset{\displaystyle H}{\overset{|}{\underset{|}{C}}}}\!-\!H$ **f.** $\left[:\ddot{O}\!-\!H\right]^-$

 g. $\left[:\ddot{Br}\!-\!\ddot{O}\!:\right]^-$

23. **a.** $\underset{-1}{:\ddot{O}}\!-\!\underset{+1}{\ddot{Se}}\!=\!\underset{0}{\ddot{O}}\!: \longleftrightarrow \underset{0}{:\ddot{O}}\!=\!\underset{+1}{\ddot{Se}}\!-\!\underset{-1}{\ddot{O}}\!:$

 b. $\left[\,^{-1}\!:\!\ddot{O}\!-\!\overset{\displaystyle :O:^0}{\overset{\|}{C}}\!-\!\ddot{O}\!:^{-1}\,\right]^{2-} \longleftrightarrow \left[\,^{-1}\!:\!\ddot{O}\!-\!\overset{\displaystyle :\ddot{O}:^{-1}}{\overset{\|}{C}}\!-\!\ddot{O}\!:^0\,\right]^{2-} \longleftrightarrow$

 $\left[\,^0\!:\!\ddot{O}\!-\!\overset{\displaystyle :\ddot{O}:^{-1}}{\overset{\|}{C}}\!-\!\ddot{O}\!:^{-1}\,\right]^{2-}$

 c. $\left[:\ddot{Cl}\underset{0}{}\!-\!\ddot{O}\!:_{-1}\right]^-$

 d. $\left[\underset{0}{:\ddot{O}}\!=\!\underset{0}{\ddot{N}}\!-\!\underset{-1}{\ddot{O}}\!:\right]^- \longleftrightarrow \left[\underset{-1}{:\ddot{O}}\!-\!\underset{0}{\ddot{N}}\!=\!\underset{0}{\ddot{O}}\!:\right]^-$

25. $H\!-\!\underset{0}{\overset{\displaystyle}{C}}\!=\!\underset{0}{\ddot{S}}\qquad H\!-\!\underset{+2}{\ddot{S}}\!=\!\underset{-2}{\ddot{C}}\;$ H_2CS is the better structure

27. $:O\!\equiv\!C\!-\!\ddot{O}\!:$ does not provide a significant contribution to the resonance hybrid as it has a $+1$ formal charge on a very electronegative atom (oxygen).

29. **a.** $\overset{\displaystyle :\ddot{Cl}:}{\underset{\displaystyle :\ddot{Cl}\quad\ddot{Cl}:}{B}}$ **b.** $\ddot{O}\!=\!\dot{N}\!-\!\ddot{O}\!: \longleftrightarrow :\ddot{O}\!-\!\dot{N}\!=\!\ddot{O}\!:$

 c. $\overset{\displaystyle H}{\underset{\displaystyle H\quad H}{\overset{|}{B}}}$

31. **a.** $\left[\,^{-1}\!:\!\ddot{O}\!-\!\overset{\displaystyle :O:^0}{\underset{\displaystyle :\ddot{O}:^{-1}}{\overset{\|}{\underset{|}{P}}}}\!-\!\ddot{O}\!:^{-1}\,\right]^{3-} \longleftrightarrow \left[\,^{-1}\!:\!\ddot{O}\!-\!\overset{\displaystyle :\ddot{O}:^{-1}}{\underset{\displaystyle :\ddot{O}:^{-1}}{\overset{|}{\underset{|}{P}}}}\!=\!\ddot{O}^0\,\right]^{3-} \longleftrightarrow$

 $\left[\,^{-1}\!:\!\ddot{O}\!-\!\overset{\displaystyle :\ddot{O}:^{-1}}{\underset{\displaystyle :\ddot{O}:^0}{\overset{|}{\underset{\|}{P}}}}\!-\!\ddot{O}\!:^{-1}\,\right]^{3-} \longleftrightarrow \left[\,^0\ddot{O}\!=\!\overset{\displaystyle :\ddot{O}:^{-1}}{\underset{\displaystyle :O:^{-1}}{\overset{|}{\underset{|}{P}}}}\!-\!\ddot{O}\!:^{-1}\,\right]^{3-}$

b. $[:C\equiv N:]^-$ (charges −1, 0)

c. $[:\ddot{O}-\overset{\overset{\displaystyle :O:}{\|}}{S}-\ddot{O}:]^{2-} \longleftrightarrow [:\ddot{O}-\overset{\overset{\displaystyle :\ddot{O}:^{-1}}{\|}}{S}=\ddot{O}]^{2-} \longleftrightarrow [:\ddot{O}-\overset{\overset{\displaystyle :\ddot{O}:^{-1}}{|}}{\underset{}{S}}-\ddot{O}:]^{2-}$

d. $[:\ddot{O}-\overset{+1}{\ddot{C}l}-\ddot{O}:]^-$ (charges −1, +1, −1)

33. a. PF₅ structure with five F around P

b. $[:\ddot{I}-\ddot{I}-\ddot{I}:]^-$

c. $:\ddot{F}-\overset{\overset{\displaystyle :\ddot{F}:}{|}}{\underset{\underset{\displaystyle :\ddot{F}:}{|}}{S}}-\ddot{F}:$

d. $:\ddot{F}-\overset{\overset{\displaystyle :\ddot{F}:}{|}}{Ge}-\ddot{F}:$ (with :F: below)

35. H₃CCH₃, H₂CCH₂, HCCH

37. −128 kJ

39. a. $:\ddot{I}-\overset{\overset{\displaystyle :\ddot{I}:}{|}}{B}-\ddot{I}:$ **b.** $2\ K^+[:\ddot{S}:]^{2-}$

c. $H-\overset{\overset{\displaystyle :O:}{\|}}{C}-\ddot{F}:$ **d.** $:\ddot{B}r-\overset{\overset{\displaystyle :\ddot{B}r:}{|}}{P}-\ddot{B}r:$

41. a. $Ba^{2+}[:\ddot{O}-\overset{\overset{\displaystyle :O:}{\|}}{C}-\ddot{O}:]^{2-} \longleftrightarrow$

$Ba^{2+}[:\ddot{O}-\overset{\overset{\displaystyle :\ddot{O}:}{|}}{C}=\ddot{O}]^{2-} \longleftrightarrow Ba^{2+}[\ddot{O}=\overset{\overset{\displaystyle :\ddot{O}:}{|}}{C}-\ddot{O}:]^{2-}$

b. $Ca^{2+}\ 2[:\ddot{O}-H]^-$

c. $K^+[:\ddot{O}-\overset{\overset{\displaystyle :O:}{\|}}{N}=\ddot{O}]^- \longleftrightarrow K^+[\ddot{O}=\overset{\overset{\displaystyle :\ddot{O}:}{|}}{N}=\ddot{O}]^- \longleftrightarrow$

$K^+[\ddot{O}=\overset{\overset{\displaystyle :O:}{\|}}{N}-\ddot{O}:]^-$

d. $Li^+[:\ddot{I}-\ddot{O}:]^-$

43. a. cyclobutane ring:
H—C(H)—C(H)—H ring with H—C—C—H

b.
H H C=C H H ⟷ H H C=C H H (cyclobutadiene resonance)

c. cyclohexane ring (chair/structure with CH₂ groups)

d. benzene resonance structures

45. CH_2O_2, $H-\overset{\overset{\displaystyle :O:}{\|}}{C}-\ddot{O}-H$

47. The reaction is exothermic due to the energy released when the Al₂O₃ lattice forms.

49. $H-\ddot{O}-\overset{\overset{\displaystyle :O:^{0}}{|}}{N}-\ddot{O}: \longleftrightarrow H-\ddot{O}-\overset{\overset{\displaystyle :\ddot{O}:^{-1}}{|}}{N}=\ddot{O} \longleftrightarrow$ (charges 0, 0, +1, −1 and 0, 0, +1, 0)

Most important (points to first two structures)

$H-\ddot{O}=\overset{\overset{\displaystyle :\ddot{O}:^{-1}}{|}}{N}-\ddot{O}:$ (charges 0, +1, +1, −1)

51. $[\ddot{C}=N=\ddot{O}]^- \longleftrightarrow [:C\equiv \ddot{N}-\ddot{O}:]^-$ (charges −2, +1, 0 and −1, +1, −1)

The fulminate ion is less stable because nitrogen is more electronegative than carbon and should, therefore, be terminal to accommodate the negative formal charge.

53.

a. $H-\overset{\overset{\displaystyle H}{|}}{\underset{\underset{\displaystyle H}{|}}{C}}-\overset{\overset{\displaystyle H}{|}}{\underset{\underset{\displaystyle H}{|}}{C}}-\overset{\overset{\displaystyle H}{|}}{\underset{\underset{\displaystyle H}{|}}{C}}-H$

d. $H-\overset{\overset{\displaystyle H}{|}}{\underset{\underset{\displaystyle H}{|}}{C}}-\overset{\overset{\displaystyle \cdot\ddot{O}\cdot}{\|}}{C}-\ddot{O}-H$

b. $H-\overset{\overset{\displaystyle H}{|}}{\underset{\underset{\displaystyle H}{|}}{C}}-\ddot{O}-\overset{\overset{\displaystyle H}{|}}{\underset{\underset{\displaystyle H}{|}}{C}}-H$

e. $H-\overset{\overset{\displaystyle H}{|}}{\underset{\underset{\displaystyle H}{|}}{C}}-\overset{\overset{\displaystyle \ddot{O}:}{\nearrow}}{C}\diagdown H$

c. $H-\overset{\overset{\displaystyle H}{|}}{\underset{\underset{\displaystyle H}{|}}{C}}-\overset{\overset{\displaystyle \cdot\ddot{O}\cdot}{\|}}{C}-\overset{\overset{\displaystyle H}{|}}{\underset{\underset{\displaystyle H}{|}}{C}}-H$

55.

Nonpolar — :S: H — Polar

H—C—N—H

Nonpolar   Polar

57. a. $\left[\ddot{\text{O}}{=}\ddot{\text{O}}\cdot\right]^{-}$ **b.** $\left[\cdot\ddot{\text{O}}\colon\right]^{-}$

c. $\colon\ddot{\text{O}}{-}\text{H}$ **d.** $\text{H}{-}\overset{\text{H}}{\underset{\text{H}}{\text{C}}}{-}\ddot{\text{O}}{-}\ddot{\text{O}}\cdot$

59. $\Delta H_{rxn(H_2)} = -243 \text{ kJ/mol} = -121 \text{ kJ/g}$
$\Delta H_{rxn(CH_4)} = -802 \text{ kJ/mol} = -50.0 \text{ kJ/g}$
CH_4 yields more energy per mole while H_2 yields more energy per gram.

61. a. $\ddot{\text{O}}{=}\overset{\colon\text{O}\colon}{\text{Cl}}{-}\ddot{\text{O}}{-}\overset{\colon\text{O}\colon}{\text{Cl}}{=}\ddot{\text{O}}$ **b.** $\text{H}{-}\overset{\colon\text{O}\colon}{\underset{\underset{\text{H}}{\colon\text{O}\colon}}{\text{P}}}{-}\ddot{\text{O}}{-}\text{H}$

c. $\text{H}{-}\ddot{\text{O}}{-}\overset{\colon\text{O}\colon}{\underset{\underset{\text{H}}{\colon\text{O}\colon}}{\text{As}}}{-}\ddot{\text{O}}{-}\text{H}$

63. Na^+F^-, Na^+O^{2-}, $Mg^{2+}F^-$, $Mg^{2+}O^{2-}$, $Al^{3+}O^{2-}$

65. $\text{H}{-}\text{C}{\equiv}\text{C}{-}\text{H}$

67. $\ddot{\text{O}}{=}\ddot{\text{S}}{=}\ddot{\text{O}} + \colon\ddot{\text{O}}{-}\text{H} \longrightarrow \text{H}{-}\ddot{\text{O}}{-}\overset{\colon\text{O}\colon}{\text{S}}{-}\ddot{\text{O}}\colon$

$\text{H}{-}\ddot{\text{O}}{-}\overset{\colon\text{O}\colon}{\text{S}}{-}\ddot{\text{O}}\colon + \ddot{\text{O}}{=}\ddot{\text{O}} \longrightarrow$

$\colon\ddot{\text{O}}{-}\overset{\colon\text{O}\colon}{\text{S}}{-}\ddot{\text{O}}\colon + \text{H}{-}\ddot{\text{O}}{-}\ddot{\text{O}}\cdot$

$\colon\ddot{\text{O}}{-}\overset{\colon\text{O}\colon}{\text{S}}{-}\ddot{\text{O}}\colon + \text{H}{-}\ddot{\text{O}}{-}\text{H} \longrightarrow \colon\overset{\text{H}}{\underset{\underset{\text{H}-\ddot{\text{O}}\colon}{\colon\text{O}}}{\text{O}}}{-}\overset{\colon\ddot{\text{O}}}{\text{S}}{=}\ddot{\text{O}}\colon$

$\Delta H_{rxn} = -172 \text{ kJ}$

69. $r_{HCl} = 113 \text{ pm}$
$r_{HF} = 84 \text{ pm}$
These values are close to the accepted values.

71. $\colon\text{P}{-}\overset{\ddot{\text{P}}}{\underset{\ddot{\text{P}}}{|}}{-}\text{P}\colon$

73. Both H: +1
S (bonded to H): −1
S (bonded only to S): 0

75. 536.2 kJ/mol
77. The compounds are energy rich because a great deal of energy is released when these compounds undergo a reaction that breaks weak bonds and forms strong ones.
79. The theory is successful because it allows us to predict and account for many chemical observations. The theory is limited because electrons cannot be treated as localized "dots."

Chapter 10

1. 4
3. a. 4 e⁻ groups, 4 bonding groups, 0 lone pairs
   b. 5 e⁻ groups, 3 bonding groups, 2 lone pairs
   c. 6 e⁻ groups, 5 bonding groups, 1 lone pair
5. a. e⁻ geometry: tetrahedral molecular geometry: trigonal pyramidal idealized bond angle: 109.5°, deviation
   b. e⁻ geometry: tetrahedral molecular geometry: bent idealized bond angle: 109.5°, deviation
   c. e⁻ geometry: tetrahedral molecular geometry: tetrahedral idealized bond angle: 109.5°, deviation (due to large size of Cl compared to H)
   d. e⁻ geometry: linear molecular geometry: linear idealized bond angle: 180°
7. H_2O has a smaller bond angle due to lone pair–lone pair repulsions, the strongest electron group repulsion.

9. a. seesaw,   **b.** T-shape, Cl—F

   c. linear, F—I—F   **d.** square planar, Br—I—Br

11. a. linear, H—C≡C—H
   b. trigonal planar,
   c. tetrahedral, H—C—C—H

13. a. The lone pair will cause lone pair–bonding pair repulsions, pushing the three bonding pairs out of the same plane. The correct molecular geometry is trigonal pyramidal.
   b. The lone pair should take an equatorial position to minimize 90° bonding pair interactions. The correct molecular geometry is seesaw.
   c. The lone pairs should take positions on opposite sides of the central atom to reduce lone pair–lone pair interactions. The correct molecular geometry is square planar.

15. a. C: tetrahedral

   O: bent

b. C's: tetrahedral

O: bent

c. O's: bent

17. The vectors of the polar bonds in both CO_2 and CCl_4 oppose each other with equal magnitude and sum to 0.

19. PF_3, polar SBr_2, nonpolar

$CHCl_3$, polar CS_2, nonpolar

21. a. polar   **b.** polar   **c.** nonpolar

23. a. 0   **b.** 3   **c.** 1

25. P:
3s   3p

H$_1$:
1s

H$_2$:
1s

H$_3$:
1s

Expected bond angle = 90°

Valence bond theory is compatible with experimentally determined bond angle of 93.3° without hybrid orbitals.

27.
2s   2p   Hybridization sp^3

29. sp^2

31. a. sp^3

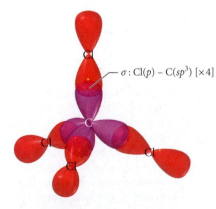

σ : Cl(p) – C(sp^3) [×4]

b. sp^3

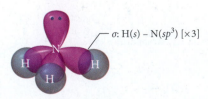

σ: H(s) – N(sp^3) [×3]

c. sp^3

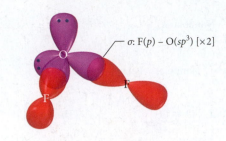

σ: F(p) – O(sp^3) [×2]

d. sp

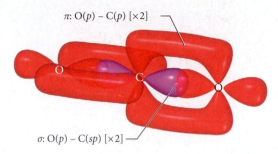

π: O(p) – C(p) [×2]

σ: O(p) – C(sp) [×2]

33. a. sp^2

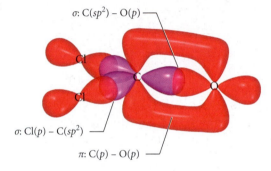

σ: C(sp^2) – O(p)

σ: Cl(p) – C(sp^2)

π: C(p) – O(p)

b. sp^3d^2

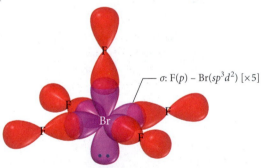

σ: F(p) – Br(sp^3d^2) [×5]

c. sp^3d

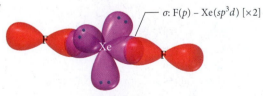

σ: F(p) – Xe(sp^3d) [×2]

d. sp^3d

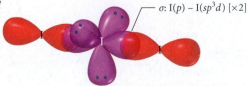

σ: I(p) – I(sp^3d) [×2]

35. a. N's: sp^2

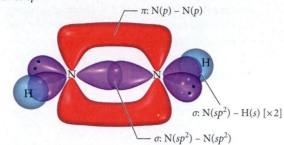
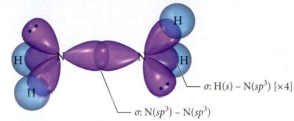

π: N(p) – N(p)

σ: N(sp^2) – H(s) [×2]

σ: N(sp^2) – N(sp^2)

b. N's: sp^3

σ: H(s) – N(sp^3) [×4]

σ: N(sp^3) – N(sp^3)

c. C: sp^3 N: sp^3

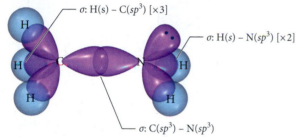

σ: H(s) – C(sp^3) [×3]

σ: H(s) – N(sp^3) [×2]

σ: C(sp^3) – N(sp^3)

37.

sp^3

sp^3 sp^2 sp^3

39.

Constructive interference

41. bond order Be$_2^+$ = 1/2   Be$_2^+$

$\underline{1}\ \sigma_{2s}^*$

$\underline{1\!\!1}\ \sigma_{2s}$

bond order Be$_2^-$ = 1/2   Be$_2^-$

$\underline{1}\ \sigma_{2p}$

$\underline{1\!\!1}\ \sigma_{2s}^*$

$\underline{1\!\!1}\ \sigma_{2s}$

Both will exist in gas phase.

43.   Bonding

Antibonding

45. a.

$\underline{}\ \sigma_{2p}^*$

$\underline{}\ \ \underline{}\ \pi_{2p}^*$

$\underline{}\ \sigma_{2p}$

$\underline{}\ \ \underline{}\ \pi_{2p}$

$\underline{1\!\!1}\ \sigma_{2s}^*$

$\underline{1\!\!1}\ \sigma_{2s}$

bond order = 0 diamagnetic

b.

$\underline{}\ \sigma_{2p}^*$

$\underline{}\ \ \underline{}\ \pi_{2p}^*$

$\underline{}\ \sigma_{2p}$

$\underline{1}\ \ \underline{1}\ \pi_{2p}$

$\underline{1\!\!1}\ \sigma_{2s}^*$

$\underline{1\!\!1}\ \sigma_{2s}$

bond order = 1 paramagnetic

c.

$\underline{}\ \sigma_{2p}^*$

$\underline{}\ \ \underline{}\ \pi_{2p}^*$

$\underline{}\ \sigma_{2p}$

$\underline{1\!\!1}\ \ \underline{1\!\!1}\ \pi_{2p}$

$\underline{1\!\!1}\ \sigma_{2s}^*$

$\underline{1\!\!1}\ \sigma_{2s}$

bond order = 2 diamagnetic

d.

$\underline{}\ \sigma_{2p}^*$

$\underline{}\ \ \underline{}\ \pi_{2p}^*$

$\underline{1}\ \sigma_{2p}$

$\underline{1\!\!1}\ \ \underline{1\!\!1}\ \pi_{2p}$

$\underline{1\!\!1}\ \sigma_{2s}^*$

$\underline{1\!\!1}\ \sigma_{2s}$

bond order = 2.5 paramagnetic

47. a. not stable   **b.** not stable
c. stable   **d.** not stable

49. C$_2^-$ has the highest bond order, the highest bond energy, and the shortest bond length.

51. a. trigonal planar polar

C: sp^2 :Ö:
               ‖
          :F̈—C—F̈:

b. bent polar

S's sp^3 :C̈l—S̈—S̈—C̈l:

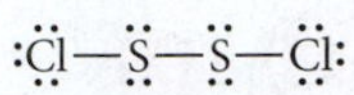

c. seesaw polar

$$\begin{array}{c} \ddot{F}: \\ | \\ :\ddot{F}—\overset{..}{S}—\ddot{F}: \\ | \\ :\ddot{F}: \end{array}$$

S: sp^3d

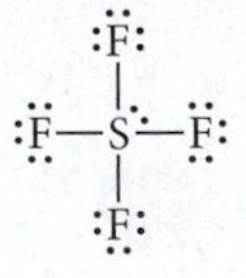

53. a.

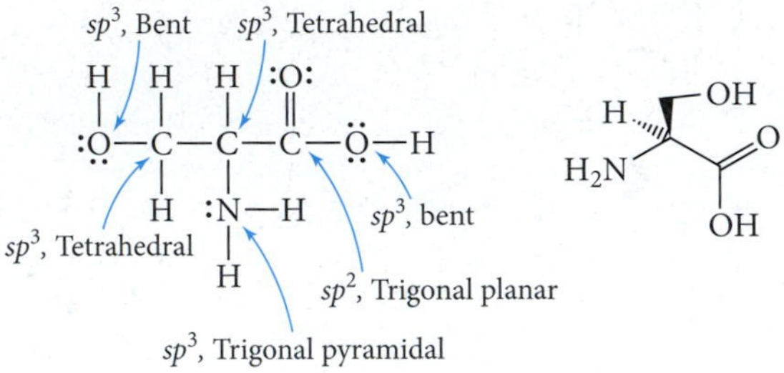

sp^3, Bent sp^3, Tetrahedral
sp^3, Tetrahedral
sp^3, bent
sp^2, Trigonal planar
sp^3, Trigonal pyramidal

b.

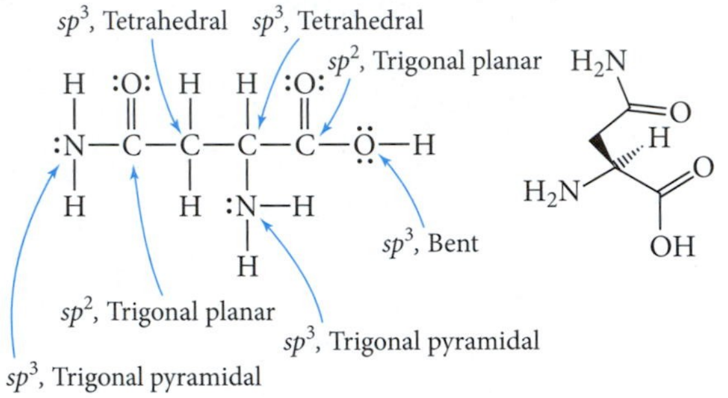

sp^3, Tetrahedral sp^3, Tetrahedral
sp^2, Trigonal planar
sp^3, Bent
sp^2, Trigonal planar
sp^3, Trigonal pyramidal
sp^3, Trigonal pyramidal

c.

sp^3, Bent sp^3, Tetrahedral
sp^3, Bent
sp^3, Tetrahedral
sp^2, Trigonal planar
sp^3, Trigonal pyramidal

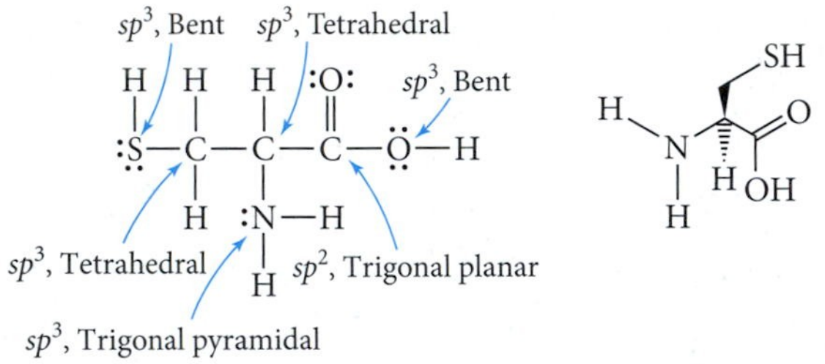

55. σ bonds: 25
π bonds: 4
lone pairs: on O's and N (without methyl group): sp^2 orbitals
    on N's (with methyl group): sp^3 orbitals

57. a. water soluble   **b.** fat soluble
c. water soluble   **d.** fat soluble

59. BrF, unhybridized, linear

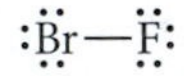

:B̈r—F̈:

BrF$_2^-$ has two bonds and three lone pairs on the central atom. The hybridization is sp^3d. The electron geometry is trigonal bipyramidal with the three lone pairs equatorial. The molecular geometry is linear.

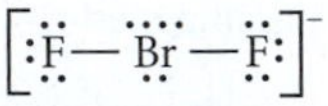

$$\left[:\ddot{F}—\overset{..}{Br}—\ddot{F}: \right]^-$$

BrF$_3$ has three bonds and two lone pairs on the central atom. The hybridization is sp^3d. The electron geometry is trigonal bipyramidal with the two lone pairs equatorial. The molecular geometry is T-shaped.

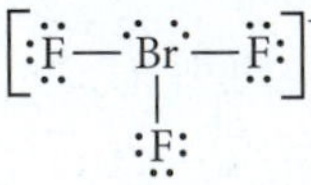

$$\left[:\ddot{F}—\overset{..}{Br}—\ddot{F}: \right]^-$$
$$\begin{array}{c} | \\ :\ddot{F}: \end{array}$$

BrF$_4^-$ has four bonds and two lone pairs on the central atom. The hybridization is sp^3d^2. The electron geometry is octahedral with the two lone pairs on the same axis. The molecular geometry is square planar.

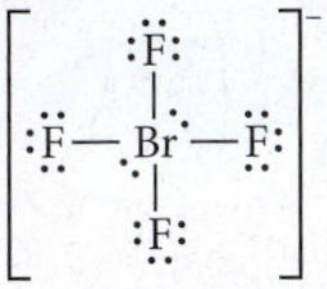

$$\left[\begin{array}{c} :\ddot{F}: \\ | \\ :\ddot{F}—\overset{..}{Br}—\ddot{F}: \\ | \\ :\ddot{F}: \end{array} \right]^-$$

BrF$_5$ has five bonds and one lone pair on the central atom. The hybridization is sp^3d^2. The electron geometry is octahedral. The molecular geometry is square pyramidal.

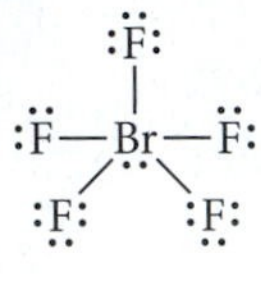

$$\begin{array}{c} :\ddot{F}: \\ | \\ :\ddot{F}—\overset{..}{Br}—\ddot{F}: \\ :\ddot{F}: \quad :\ddot{F}: \end{array}$$

61.

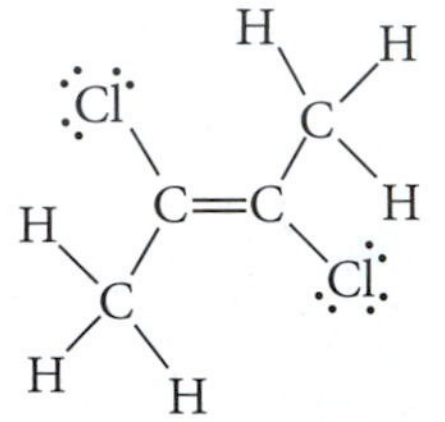

63. According to valence bond theory, CH$_4$, NH$_3$, and H$_2$O are all sp^3 hybridized. This hybridization results in a tetrahedral electron group configuration with a 109.5° bond angle. NH$_3$ and H$_2$O deviate from this idealized bond angle because their lone electron pairs exist in their own sp^3 orbitals. The presence of lone pairs lowers the tendency for the central atom's orbitals to hybridize. As a result, as lone pairs are added, the bond angle moves further from the 109.5° hybrid angle and closer to the 90° unhybridized angle.

65. In NO$_2^+$, the central N has two electron groups, so the hybridization is sp and the ONO angle is 180°. In NO$_2^-$ the central N has three electron groups, two bonds and one lone pair. The ideal hybridization is sp^2 but the ONO bond angle should close down a bit because of the lone pair. A bond angle around 115° is a good guess. In NO$_2$ there are three electron groups, but one group is a lone pair. Again the ideal hybridization would be sp^2, but since one unpaired electron must be much smaller than a lone pair or even a bonding pair, we predict that the ONO bond angle will spread and be greater than 120°. As a guess the angle is probably significantly greater than 120°.

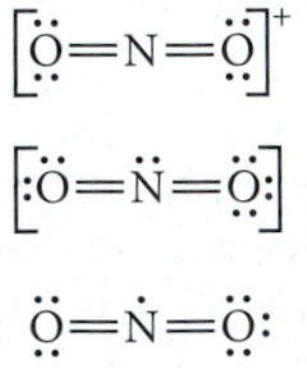

$$\left[\ddot{O}=N=\ddot{O} \right]^+$$
$$\left[:\ddot{O}=\ddot{N}=\ddot{O}: \right]$$
$$\ddot{O}=\dot{N}=\ddot{O}:$$

67. From left to right: C$_1$ tetrahedral, C$_2$ trigonal planer, N trigonal pyramidal

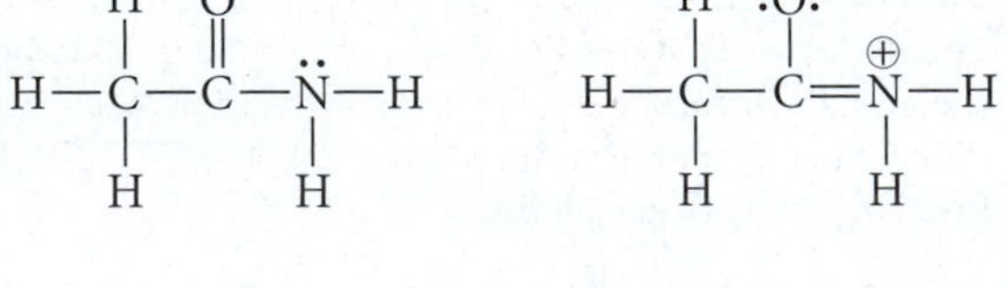

69. a. This is the best.
  b. This statement is similar to **a** but leaves out non-bonding lone-pair electron groups.
  c. Molecular geometries are not determined by overlapping orbitals, but rather by the number and type of electron groups around each central atom.

71. Lewis theory defines a single bond, double bond, and triple bond as a sharing of two electrons, four electrons, and six electrons respectively between two atoms. Valence bond theory defines a single bond as a sigma overlap of two orbitals, a double bond as a single sigma bond combined with a pi bond, and a triple bond as a double bond with an additional pi bond. Molecular orbital theory defines a single bond, double bond, and triple bond as a bond order of 1, 2, or 3 respectively between two atoms.

Chapter 11

1. a. dispersion forces
  b. dispersion forces, dipole–dipole forces, hydrogen bonding
  c. dispersion forces, dipole–dipole forces
  d. dispersion forces

3. a. dispersion forces, dipole–dipole forces
  b. dispersion forces, dipole–dipole forces, hydrogen bonding
  c. dispersion forces
  d. dispersion forces

5. a, b, c, d. Boiling point increases with increasing intermolecular forces. The molecules increase in their intermolecular forces as follows: **a,** dispersion forces; **b,** stronger dispersion forces (broader electron cloud); **c,** dispersion forces and dipole–dipole interactions; **d,** dispersion forces, dipole–dipole interactions, and hydrogen bonding.

7. a. CH_3OH, hydrogen bonding
  b. CH_3CH_2OH, hydrogen bonding
  c. CH_3CH_3, greater mass, broader electron cloud causes greater dispersion forces

9. a. Br_2, smaller mass results in weaker dispersion forces
  b. H_2S, lacks hydrogen bonding
  c. PH_3, lacks hydrogen bonding

11. a. not homogeneous
  b. homogeneous, dispersion, dipole–dipole, hydrogen bonding, ion–dipole
  c. homogeneous, dispersion
  d. homogeneous, dispersion, dipole–dipole, hydrogen bonding

13. Water. Surface tension increases with increasing intermolecular forces, and water can hydrogen bond while acetone cannot.

15. compound A

17. When the tube is clean, water experiences adhesive forces with glass that are stronger than its cohesive forces, causing it to climb the surface of a glass tube. Water does not experience strong intermolecular forces with oil, so if the tube is coated in oil, the water's cohesive forces will be greater and it will not be attracted to the surface of the tube.

19. The water in the 12-cm dish will evaporate more quickly. The vapor pressure does not change but the surface area does. The water in the dish evaporates more quickly because the greater surface area allows for more molecules to obtain enough energy at the surface and break free.

21. Water is more volatile than vegetable oil. When the water evaporates, the endothermic process results in cooling.

23. 0.471 L

25. 86 °C

27. $\Delta H_{vap} = 24.7$ kJ/mol, bp $= 239$ K

29. 41 torr

31. 15.9 kJ

33. 2.7 °C

35. 30.5 kJ

37. a. solid   **b.** liquid   **c.** gas
  d. supercritical fluid   **e.** solid/liquid
  f. liquid/gas   **g.** solid/liquid/gas

39. N_2 has a stable liquid phase at 1 atm.

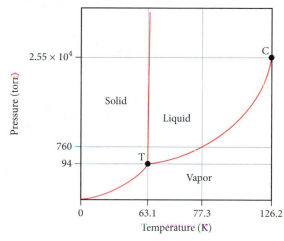

41. a. 0.027 mmHg   **b.** rhombic

43. Water has strong intermolecular forces. It is polar and experiences hydrogen bonding.

45. Water's exceptionally high specific heat capacity has a moderating effect on Earth's climate. Also, its high ΔH_{vap} causes water evaporation and condensation to have a strong effect on temperature.

47. a. 1   **b.** 2   **c.** 4

49. $l = 393$ pm, $d = 21.3$ g/cm³

51. 134.5 pm

53. 6.0×10^{23} atoms/mol

55. a. atomic   **b.** molecular   **c.** ionic   **d.** atomic

57. LiCl(*s*). The other three solids are held together by intermolecular forces while LiCl is held together by stronger coulombic interactions between the cations and anions of the crystal lattice.

59. a. $TiO_2(s)$, ionic solid
  b. $SiCl_4(s)$, larger, stronger dispersion forces
  c. Xe(*s*), larger, stronger dispersion forces
  d. CaO, ions have greater charge, and, therefore, stronger coulombic forces

61. TiO_2

63. Cs: 1(1) = 1
  Cl: 8(1/8) = 1
  1:1
  CsCl
  Ba: 8(1/8) + 6(1/2) = 4
  Cl: 8(1) = 8
  4:8 = 1:2
  $BaCl_2$

65. a

67. The general trend is that melting point increases with increasing mass. This is due to the fact that the electrons of the larger molecules are held more loosely and a stronger dipole moment can be induced more easily. HF is the exception to the rule. It has a relatively high melting point due to hydrogen bonding.

69. yes, 1.22 g

71. gas $\longrightarrow$ liquid $\longrightarrow$ solid

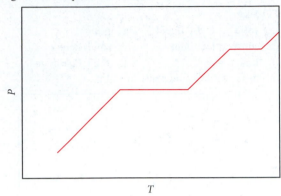

73. 26 °C

75. 3.4×10^3 g H_2O

77. CsCl has a higher melting point than AgI because of its higher coordination number. In CsCl, one anion bonds to eight cations (and vice versa) while in AgI, one anion bonds only to four cations.

79. **a.** $4r$

b. $c^2 = a^2 + b^2$
$c = 4r, a = l, b = l$
$(4r)^2 = l^2 + l^2$
$16r^2 = 2l^2$
$8r^2 = l^2$
$l = \sqrt{8r^2}$
$l = 2\sqrt{2}r$

81. 8 atoms/unit

83. **a.** $CO_2(s) \longrightarrow CO_2(g)$ at 195 K

b. $CO_2(s) \longrightarrow$ triple point at 216 K $\longrightarrow CO_2(g)$ just above 216 K

c. $CO_2(s) \longrightarrow CO_2(l)$ at somewhat above 216 K $\longrightarrow CO_2(g)$ at around 250 K

d. $CO_2(s) \longrightarrow CO_2(g) \longrightarrow$ supercritical fluid

85. 55.843 g/mol

87. 2.00 g/cm^3

89. Decreasing the pressure will decrease the temperature of liquid nitrogen. Because the nitrogen is boiling, its temperature must be constant at a given pressure. As the pressure decreases, the boiling point decreases, and therefore so does the temperature. If the pressure drops below the pressure of the triple point, the phase change will shift from vaporization to sublimation and the liquid nitrogen will become solid.

91. body diagonal $= \sqrt{6}r$, radius $= (\sqrt{3} - \sqrt{2})/\sqrt{2} = 0.2247r$

93. 70.7 L

95. This is a valid criticism because icebergs displace the same volume of water regardless of whether or not they melt. The melting of ice sheets on Antarctica would increase ocean levels because that ice is not currently displacing any water.

97. A

99. $\Delta H_{sub} = \Delta H_{vap} + \Delta H_{fus}$

101. The vats of water will prevent the cellar from reaching freezing temperatures. Because the heat capacity of water is so high, the water will lose a lot of heat before the temperature is low enough to freeze everything.

Chapter 12

1. **a.** hexane, toluene, or CCl_4; dispersion forces

b. water, methanol; dispersion, dipole–dipole, hydrogen bonding

c. hexane, toluene, or CCl_4; dispersion forces

d. water, acetone, methanol, ethanol; dispersion, ion–dipole

3. $HOCH_2CH_2CH_2OH$

5. **a.** water; dispersion, dipole–dipole, hydrogen bonding

b. hexane; dispersion

c. water; dispersion, dipole–dipole

d. water; dispersion, dipole–dipole, hydrogen bonding

7. **a.** endothermic

b. The lattice energy is greater in magnitude than the heat of hydration.

c.

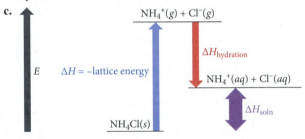

d. The solution forms because chemical systems tend toward greater entropy.

9. -797 kJ/mol

11. $\Delta H_{soln} = -6 \times 10^1$ kJ/mol, -7 kJ of energy evolved

13. unsaturated

15. About 31 g will precipitate.

17. Boiling water releases any O_2 dissolved in it. The solubility of gases decreases with increasing temperature.

19. As pressure increases, nitrogen will more easily dissolve in blood. To reverse this process, divers should ascend to lower pressures.

21. 1.1 g

23. 2.48 M, 2.65 m, 13.4% by mass

25. 319 mL

27. 1.6×10^2 g Ag

29. 1.4×10^4 g

31. Add water to 7.31 mL of concentrated solution until a total volume of 1.15 L is acquired.

33. **a.** Add water to 3.73 g KCl to a volume of 100 mL.

b. Add 3.59 g KCl to 96.41 g H_2O.

c. Add 5.0 g KCl to 95 g H_2O.

35. **a.** 0.417 M   **b.** 0.444 m **c.** 7.41% by mass

d. 0.00794   **e.** 0.794% by mole

37. 0.89 M H_2O_2

39. 15. m, 0.22

41. The level has decreased more in the beaker filled with pure water. The dissolved salt in the seawater decreases the vapor pressure and subsequently lowers the rate of vaporization.

43. 30.4 torr

45. a. $P_{hep} = 24.4$ torr, $P_{oct} = 5.09$ torr   **b.** 29.5 torr
   c. 80.8% heptane by mass, 19.2% octane by mass
   d. The vapor is richer in the more volatile component.

47. 23.0 torr

49. freezing point (fp) $= -1.27\,°C$, bp $= 100.349\,°C$

51. fp $= 1.0\,°C$, bp $= 82.4\,°C$

53. 1.8×10^2 g/mol

55. 26.1 atm

57. 6.36×10^3 g/mol

59. a. fp $= -0.558\,°C$, bp $= 100.154\,°C$
   b. fp $= -1.98\,°C$, bp $= 100.546\,°C$
   c. fp $= -2.5\,°C$, bp $= 100.70\,°C$

61. 157.1 g

63. a. $-0.632\,°C$ **b.** 5.4 atm   **c.** $100.18\,°C$

65. 2.3

67. 3.4

69. Chloroform is polar and has stronger solute–solvent interactions than nonpolar carbon tetrachloride.

71. $\Delta H_{soln} = 51$ kJ/mol, $-8.7\,°C$

73. 2.2×10^{-3} M/atm

75. 1.3×10^4 L

77. 0.24 g

79. $-24\,°C$

81. a. 1.1% by mass/V
   b. 1.58% by mass/V
   c. 5.3% by mass/V

83. 2.484

85. 0.227 atm

87. χ_{CHCl_3}(original) $= 0.657$, P_{CHCl_3}(condensed) $= 0.346$ atm

89. 1.74 M

91. $C_6H_{14}O_2$

93. 12g

95. 6.4×10^{-3} L

97. 22.4% glucose by mass, 77.6% sucrose by mass

99. $P_{iso} = 0.131$ atm, $P_{pro} = 0.068$ atm. The major intermolecular attractions are between the OH groups. The OH group at the end of the chain in propyl alcohol is more accessible than the one in the middle of the chain in isopropyl alcohol. In addition, the molecular shape of propyl alcohol is a straight chain of carbon atoms, while that of isopropyl alcohol is a branched chain and is more like a ball. The contact area between two ball-like objects is smaller than that of two chain-like objects. The smaller contact area in isopropyl alcohol means the molecules don't attract each other as strongly as do those of propyl alcohol. As a result of both of these factors, the vapor pressure of isopropyl alcohol is higher.

101. 37.2 mmHg, 118 mmHg

103. a. The two substances mix because their intermolecular forces between themselves are roughly equal to the forces between each other.
   b. 0
   c. ΔH_{solute} and $\Delta H_{solvent}$ are positive, ΔH_{mix} is negative and equals the sum of the first two.

105. d

107. The balloon not only loses He, it also takes in N_2 and O_2 from the air (due to the tendency towards mixing), increasing the density of the balloon.

Chapter 13

1. a. Rate $= -\dfrac{1}{2}\dfrac{\Delta[HBr]}{\Delta t} = \dfrac{\Delta[H_2]}{\Delta t} = \dfrac{\Delta[Br_2]}{\Delta t}$
   b. 1.5×10^{-3} M/s   **c.** 0.011 mol Br_2

3. a. Rate $= -\dfrac{1}{2}\dfrac{\Delta[A]}{\Delta t} = -\dfrac{\Delta[B]}{\Delta t} = \dfrac{1}{3}\dfrac{\Delta[C]}{\Delta t}$
   b. $\dfrac{\Delta[B]}{\Delta t} = -0.0500$ M/s, $\dfrac{\Delta[C]}{\Delta t} = 0.150$ M/s

5. a. $0 \longrightarrow 10$ s: Rate $= 8.7 \times 10^{-3}$ M/s
   $40 \longrightarrow 50$ s: Rate $= 6.0 \times 10^{-3}$ M/s
   b. 1.4×10^{-2} M/s

7. a. (i) 1.0×10^{-2} M/s (ii) 8.5×10^{-3} M/s (iii) 0.013 M/s
   b.

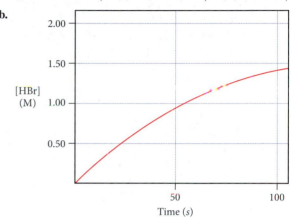

9. a. first order
   b.

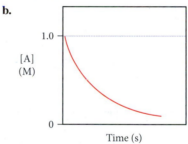

   c. Rate $= k[A]^1$, $k = 0.010$ s^{-1}

11. a. s^{-1} **b.** M^{-1}s^{-1}
   c. M·s^{-1}

13. a. Rate $= k[A][B]^2$ **b.** third order
   c. 2   **d.** 4
   e. 1   **f.** 8

15. second order, Rate $= 5.25$ M^{-1}s$^{-1}[A]^2$

17. Rate $= k[NO_2][F_2]$, $k = 2.57$ M^{-1}s^{-1}, second order

19. a. zero order **b.** first order **c.** second order

21. second order, $k = 2.25 \times 10^{-2}$ M^{-1}s^{-1}, $[AB]$ at 25 s $= 0.619$ M

23. first order, $k = 1.12 \times 10^{-2}$ s^{-1}, Rate $= 2.8 \times 10^{-3}$ M/s

25. a. 4.5×10^{-3} s^{-1} **b.** Rate $= 4.5 \times 10^{-3}$ s$^{-1}[A]$
   c. 1.5×10^2 s   **d.** $[A] = 0.0908$ M

27. a. 4.88×10^3 s   **b.** 9.8×10^3 s
   c. 1.8×10^3 s   **d.** 0.146 M at 200 s, 0.140 M at 500 s

29. 6.8×10^8 yrs; 1.8×10^{17} atoms

31.

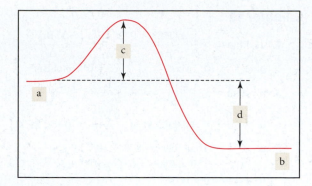

33. $17 \, \text{s}^{-1}$

35. $61.90 \, \text{kJ/mol}$

37. $E_a = 251 \, \text{kJ/mol}, A = 7.93 \times 10^{11} \, \text{s}^{-1}$

39. $E_a = 23.0 \, \text{kJ/mol}, A = 8.05 \times 10^{10} \, \text{s}^{-1}$

41. a. $122 \, \text{kJ/mol}$ **b.** $0.101 \, \text{s}^{-1}$

43. $47.85 \, \text{kJ/mol}$

45. a

47. The mechanism is valid.

49. a. $Cl_2(g) + CHCl_3(g) \longrightarrow HCl(g) + CCl_4(g)$

  b. $Cl(g), CCl_3(g)$ **c.** Rate $= k[Cl_2]^{1/2}[CHCl_3]$

51. Heterogeneous catalysts require a large surface area because catalysis can only happen at the surface. A greater surface area means greater opportunity for the substrate to react, which results in a faster reaction.

53. 10^{12}

55. a. first order, $k = 0.0462 \, \text{hr}^{-1}$ **b.** $15 \, \text{hr}$

  c. $5.0 \times 10^1 \, \text{hr}$

57. $0.0531 \, \text{M/s}$

59. $0.074 \, \text{atm}$

61. $219 \, \text{torr}$

63. $1 \times 10^{-7} \, \text{s}$

65. $160 \, \text{s}$

67. a. 2

  b.

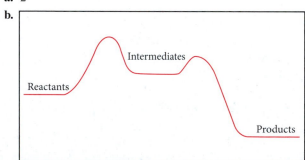

  c. first step   **d.** exothermic

69. a. $5.41 \, \text{s}$

  b. $2.2 \, \text{s}$ for 25%, $5.4 \, \text{s}$ for 50%

  c. 0.28 at $10 \, \text{s}$, 0.077 at $20 \, \text{s}$

71. a. $E_a = 89.5 \, \text{kJ/mol}, A = 4.22 \times 10^{11} \, \text{s}^{-1}$

  b. $2.5 \times 10^{-5} \, \text{M}^{-1} \, \text{s}^{-1}$

  c. $6.0 \times 10^{-4} \, \text{M/s}$

73. a. No

  b. No bond is broken and the two radicals attract each other.

  c. Formation of diatomic gases from atomic gases

75. $1.35 \times 10^4 \, \text{years}$

77. a. All are valid. For each, all steps sum to overall reaction and the predicted rate law is consistent with experimental data.

  b. Buildup of $I(g)$ and/or $H_2I(g)$.

79. a. 0% **b.** 25% **c.** 33%

81. $174 \, \text{kJ}$

83. a. second order

$$CH_3NC + CH_3NC^* \underset{k_2}{\overset{k_1}{\rightleftharpoons}} CH_3NC^* + CH_3NC \quad \text{(fast)}$$

$$CH_3NC^* \overset{k_3}{\rightarrow} CH_3NC \qquad \text{(slow)}$$

Rate $= k_3[CH_3NC^*]$

$k_1[CH_3NC]^2 = k_2[CH_3NC^*][CH_3NC]$

$$[CH_3NC^*] = \frac{k_1}{k_2}[CH_3NC]$$

$$\text{Rate} = k_3 \times \frac{k_1}{k_2}[CH_3NC]$$

Rate $= k[CH_3NC]$

85. $d[A]/dt = -k[A]^2 \longrightarrow d[A]/[A]^2 = -k \, dt \longrightarrow$ integrate from $[A]_0$ to $[A]$ and 0 to $t \longrightarrow -1/[A] - (-1/[A]_0) = -kt - (-k \times 0) \longrightarrow 1/[A] = kt + 1/[A]_0$

87. Rate $= k[CO][Cl_2]^{3/2}$

89. Cl_2: $0.0084 \, \text{M}$

NO: $0.017 \, \text{M}$

91. B is first order and A is second order.

B will be linear if you plot $\ln[B]$ vs. time, A will be linear if you plot $1/[A]$ vs. time.

Chapter 14

1. a. $K = \dfrac{[SbCl_3][Cl_2]}{[SbCl_5]}$ **b.** $K = \dfrac{[NO]^2[Br_2]}{[BrNO]^2}$

  c. $K = \dfrac{[CS_2][H_2]^4}{[CH_4][H_2S]^2}$ **d.** $K = \dfrac{[CO_2]^2}{[CO]^2[O_2]}$

3. The concentration of the reactants will be greater. No, this is not dependent on initial concentrations; it is dependent on the value of K_c.

5. a. figure v

  b. The change in the decrease of reactants and increase of products would be faster.

  c. No, catalysts affect kinetics, not equilibrium.

7. a. 4.42×10^{-5}, reactants favored

  b. 1.50×10^2, products favored

  c. 1.96×10^{-9}, reactants favored

9. 1.3×10^{-29}

11. a. 2.56×10^{-23}

  b. 1.3×10^{22}

  c. 81.9

13. a. $K_c = \dfrac{[HCO_3^-][OH^-]}{[CO_3^{2-}]}$ **b.** $K_c = [O_2]^3$

  c. $K_c = \dfrac{[H_3O^+][F^-]}{[HF]}$ **d.** $K_c = \dfrac{[NH_4^+][OH^-]}{[NH_3]}$

15. 136

17.

$T \, (K)$	$[N_2]$	$[H_2]$	$[NH_3]$	K_c
500	0.115	0.105	0.439	1.45×10^3
575	0.110	0.249	0.128	9.6
775	0.120	0.140	4.39×10^{-3}	0.0584

19. 303 torr

21. 3.3×10^2

23. 764

25. More solid will form.

27. Additional solid will not dissolve.

29. **a.** $[A] = 0.33$ M, $[B] = 0.67$ M

   b. $[A] = 0.41$ M, $[B] = 0.59$ M

   c. $[A] = 0.50$ M, $[B] = 1.0$ M

31. $[N_2O_4] = 0.0115$ M, $[NO_2] = 0.0769$ M

33. 0.10 M

35. 1.9×10^{-3} M

37. 7.84 torr

39. **a.** $[A] = 0.38$ M, $[B] = 0.62$ M, $[C] = 0.62$ M

   b. $[A] = 0.90$ M, $[B] = 0.095$ M, $[C] = 0.095$ M

   c. $[A] = 1.0$ M, $[B] = 3.2 \times 10^{-3}$ M, $[C] = 3.2 \times 10^{-3}$ M

41. **a.** shift left   **b.** shift right

   c. shift right

43. **a.** shift right   **b.** no effect

   c. no effect   **d.** shift left

45. **a.** shift right   **b.** shift left

   c. no effect

47. Increase temperature $\longrightarrow$ shift right, decrease temperature $\longrightarrow$ shift left. Increasing the temperature will increase the equilibrium constant.

49. **b, d**

51. **a.** 1.7×10^2

   b. $\dfrac{[Hb-CO]}{[Hb-O_2]} = 0.85$ or $17/20$

   CO is highly toxic, as it blocks O_2 uptake by hemoglobin. CO at a level of 0.1% will replace nearly half of the O_2 in blood.

53. **a.** 1.67 atm   **b.** 1.41 atm

55. 0.407 g $MgCO_3$

57. **b, c, d**

59. 0.0144 atm

61. 3.1×10^2 g, 20% yield

63. 0.12 atm

65. 0.72 atm

67. 0.017 g

69. 0.226

71. **a.** 29.3   **b.** 86.3 torr

73. $P_{NO} = P_{Cl_2} = 429$ torr

75. 2.52×10^{-3}

77. Yes, because the volume affects Q.

79. a = 1, b = 2

Chapter 15

1. **a.** acid, $HNO_3(aq) \longrightarrow H^+(aq) + NO_3^-(aq)$

   b. acid, $NH_4^+(aq) \rightleftharpoons H^+(aq) + NH_3(aq)$

   c. base, $KOH(aq) \longrightarrow K^+(aq) + OH^-(aq)$

   d. acid, $HC_2H_3O_2(aq) \rightleftharpoons H^+(aq) + C_2H_3O_2^-(aq)$

3. **a.** $H_2CO_3(aq) + H_2O(l) \rightleftharpoons H_3O^+(aq) + HCO_3^-(aq)$
   acid   base   conj. acid   conj. base

   b. $NH_3(aq) + H_2O(l) \rightleftharpoons NH_4^+(aq) + OH^-(aq)$
   base   acid   conj. acid   conj. base

   c. $HNO_3(aq) + H_2O(l) \longrightarrow H_3O^+(aq) + NO_3^-(aq)$
   acid   base   conj. acid   conj. base

   d. $C_5H_5N(aq) + H_2O(l) \rightleftharpoons C_5H_5NH^+(aq) + OH^-(aq)$
   base   acid   conj. acid   conj. base

5. **a.** Cl^- **b.** HSO_3^- **c.** CHO_2^- **d.** F^-

7. $H_2PO_4^-(aq) + H_2O(l) \rightleftharpoons HPO_4^{2-}(aq) + H_3O^+(aq)$
   $H_2PO_4^-(aq) + H_2O(l) \rightleftharpoons H_3PO_4(aq) + OH^-(aq)$

9. **a.** strong   **b.** strong

   c. strong   **d.** weak, $K_a = \dfrac{[H_3O^+][HSO_3^-]}{[H_2SO_3]}$

11. **a, b, c**

13. **a.** F^- **b.** NO_2^- **c.** ClO^-

15. **a.** 1.0×10^{-6}, basic   **b.** 4.5×10^{-9}, acidic

   c. 8.3×10^{-6}, basic

17. **a.** pH = 7.77, pOH = 6.23

   b. pH = 7.00, pOH = 7.00

   c. pH = 5.66, pOH = 8.34

19.

$[H_3O^+]$	$[OH^-]$	pH	Acidic or Basic
7.1×10^{-4}	1.4×10^{-11}	3.15	Acidic
3.7×10^{-9}	2.7×10^{-6}	8.43	Basic
7.9×10^{-12}	1.3×10^{-3}	11.1	Basic
6.3×10^{-4}	1.6×10^{-11}	3.20	Acidic

21. $[H_3O^+] = 1.5 \times 10^{-7}$ M, pH = 6.81

23. pH = 1.36
   pH = 1.35
   pH = 1.34
   A difference of 1 in the second significant digit in a concentration value produces a difference of 0.01 in pH. Therefore, the second significant digit in value of the concentration corresponds to the hundredths place in a pH value.

25. **a.** $[H_3O^+] = 0.15$ M, $[OH^-] = 6.7 \times 10^{-14}$ M, pH = 0.82

   b. $[H_3O^+] = 0.025$ M, $[OH^-] = 4.0 \times 10^{-13}$ M, pH = 1.60

   c. $[H_3O^+] = 0.087$ M, $[OH^-] = 1.1 \times 10^{-13}$ M, pH = 1.06

   d. $[H_3O^+] = 0.137$ M, $[OH^-] = 7.30 \times 10^{-14}$ M, pH = 0.863

27. **a.** 1.8 g   **b.** 0.57 g   **c.** 0.045 g

29. pH = 2.21

31. $[H_3O^+] = 2.5 \times 10^{-3}$ M, pH = 2.59

33. **a.** 1.82 (approximation valid)

   b. 2.18 (approximation breaks down)

   c. 2.72 (approximation breaks down)

35. 2.75

37. 6.8×10^{-6}

39. 0.0063%

41. **a.** 0.42%   **b.** 0.60%   **c.** 1.3%   **d.** 1.9%

43. 3.61×10^{-5}

45. **a.** pH = 2.03, percent ionization = 3.7%

   b. pH = 2.24, percent ionization = 5.7%

   c. pH = 2.40, percent ionization = 8.0%

47. $H_3PO_4(aq) + H_2O(l) \rightleftharpoons H_2PO_4^-(aq) + H_3O^+(aq)$,

$$K_{a_1} = \frac{[H_3O^+][H_2PO_4^-]}{[H_3PO_4]}$$

$H_2PO_4^-(aq) + H_2O(l) \rightleftharpoons HPO_4^{2-}(aq) + H_3O^+(aq)$,

$$K_{a_2} = \frac{[H_3O^+][HPO_4^{2-}]}{[H_2PO_4^-]}$$

$HPO_4^{2-}(aq) + H_2O(l) \rightleftharpoons PO_4^{3-}(aq) + H_3O^+(aq)$,

$$K_{a_3} = \frac{[H_3O^+][PO_4^{3-}]}{[HPO_4^{2-}]}$$

49. a. $[H_3O^+] = 0.048$ M, pH $= 1.32$
  b. $[H_3O^+] = 0.12$ M, pH $= 0.92$
51. a. $[OH^-] = 0.15$ M, $[H_3O^+] = 6.7 \times 10^{-14}$ M,
      pH $= 13.17$, pOH $= 0.83$
  b. $[OH^+] = 0.003$ M, $[H_3O^+] = 3.3 \times 10^{-12}$ M,
      pH $= 11.48$, pOH $= 2.52$
  c. $[OH^-] = 9.6 \times 10^{-4}$ M, $[H_3O^+] = 1.0 \times 10^{-11}$ M,
      pH $= 10.98$, pOH $= 3.02$
  d. $[OH^-] = 8.7 \times 10^{-5}$ M, $[H_3O^+] = 1.1 \times 10^{-10}$ M,
      pH $= 9.93$, pOH $= 4.07$
53. 13.842
55. 0.104 L
57. a. $NH_3(aq) + H_2O(l) \rightleftharpoons NH_4^+(aq) + OH^-(aq)$,
$$K_b = \frac{[NH_4^+][OH^-]}{[NH_3]}$$
  b. $HCO_3^-(aq) + H_2O(l) \rightleftharpoons$
$$H_2CO_3(aq) + OH^-(aq), K_b = \frac{[H_2CO_3][OH^-]}{[HCO_3^-]}$$
  c. $CH_3NH_2(aq) + H_2O(l) \rightleftharpoons$
$$CH_3NH_3^+(aq) + OH^-(aq), K_b = \frac{[CH_3NH_3^+][OH^-]}{[CH_3NH_2]}$$
59. $[OH^-] = 1.6 \times 10^{-3}$ M, pOH $= 2.79$, pH $= 11.21$
61. 7.48
63. 6.7×10^{-7}
65. a. neutral
  b. basic, $ClO^-(aq) + H_2O(l) \rightleftharpoons HClO(aq) + OH^-(aq)$
  c. basic, $CN^-(aq) + H_2O(l) \rightleftharpoons HCN(aq) + OH^-(aq)$
  d. neutral
67. $[OH^-] = 2.0 \times 10^{-6}$ M, pH $= 8.30$
69. a. acidic, $NH_4^+(aq) + H_2O(l) \rightleftharpoons NH_3(aq) + H_3O^+(aq)$
  b. neutral
  c. acidic, $Co(H_2O)_6^{3+}(aq) + H_2O(l) \rightleftharpoons$
$$Co(H_2O)_5(OH)^{2+}(aq) + H_3O^+(aq)$$
  d. acidic, $CH_2NH_3^+(aq) + H_2O(l) \rightleftharpoons$
$$CH_2NH_2(aq) + H_3O^+(aq)$$
71. a. acidic   **b.** basic   **c.** neutral
  d. acidic   **e.** acidic
73. NaOH, $NaHCO_3$, NaCl, NH_4ClO_2, NH_4Cl
75. a. 5.13   **b.** 8.87   **c.** 7.0
77. $[K^+] = 0.15$ M, $[F^-] = 0.15$ M, $[HF] = 2.1 \times 10^{-6}$ M,
      $[OH^-] = 2.1 \times 10^{-6}$ M; $[H_3O^+] = 4.8 \times 10^{-9}$ M
79. a. HCl, weaker bond   **b.** HF, bond polarity
  c. H_2Se, weaker bond
81. a. H_2SO_4, more oxygen atoms bonded to S
  b. $HClO_2$, more oxygen atoms bonded to Cl
  c. HClO, Cl has higher electronegativity
  d. CCl_3COOH, Cl has higher electronegativity
83. S^{2-}, its conjugate acid (H_2S), is a weaker acid than H_2S
85. a. Lewis acid   **b.** Lewis acid
  c. Lewis base   **d.** Lewis base
87. a. acid: Fe^{3+}, base: H_2O **b.** acid: Zn^{2+}, base: NH_3
  c. acid: BF_3, base: $(CH_3)_3N$
89. a. weak   **b.** strong   **c.** weak   **d.** strong
91. If blood became acidic, the H^+ concentration would increase.
      According to Le Châtelier's principle, equilibrium would be
      shifted to the left and the concentration of oxygenated Hb
      would decrease.

93. All acid will be neutralized.
95. $[H_3O^+]$(Great Lakes) $= 3 \times 10^{-5}$ M,
      $[H_3O^+]$(West Coast) $= 4 \times 10^{-6}$ M. The rain over the
      Great Lakes is about 8 times more concentrated.
97. 2.7
99. a. 2.000   **b.** 1.52   **c.** 12.95
  d. 11.12   **e.** 5.03
101. a. $CN^-(aq) + H^+(aq) \rightleftharpoons HCN(aq)$
  b. $NH_4^+(aq) + OH^-(aq) \rightleftharpoons NH_3(aq) + H_2O(l)$
  c. $CN^-(aq) + NH_4^+(aq) \rightleftharpoons HCN(aq) + NH_3(aq)$
  d. $HSO_4^-(aq) + C_2H_3O_2^-(aq) \rightleftharpoons$
$$SO_4^{2-}(aq) + HC_2H_3O_2(aq)$$
  e. no reaction between the major species
103. 0.794
105. 8.2×10^{-4}
107. 6.79
109. 2.14
111. $[A^-] = 4.5 \times 10^{-5}$ M
      $[H^+] = 2.2 \times 10^{-4}$ M
      $[HA_2^-] = 1.8 \times 10^{-4}$ M
113. pH $= 11.30$
115. 51 g
117. b
119. $CH_3COOH < CH_2ClCOOH < CHCl_2COOH < CCl_3COOH$

Chapter 16

1. d
3. a. 3.57   **b.** 9.08
5. pure water: 2.1%, in $NaC_7H_5O_2$: 0.065%. The percent ion-
    ization in the sodium benzoate solution is much smaller
    because the presence of the benzoate ion shifts the equilib-
    rium to the left.
7. a. 2.14   **b.** 8.32   **c.** 3.46
9. $HCl + NaC_2H_3O_2 \longrightarrow HC_2H_3O_2 + NaCl$
    $NaOH + HC_2H_3O_2 \longrightarrow NaC_2H_3O_2 + H_2O$
11. a. 3.57   **b.** 9.07
13. a. 7.62   **b.** 10.82   **c.** 4.61
15. a. 3.86   **b.** 8.95
17. 3.5
19. 3.7 g
21. a. 4.74   **b.** 4.68   **c.** 4.81
23. a. initial 7.00 after 1.70   **b.** initial 4.71 after 4.56
  c. initial 10.78 after 10.66
25. 1.2 g; 2.7 g
27. a. yes   **b.** no   **c.** yes
  d. no   **e.** no
29. a. 7.4   **b.** 0.3 g   **c.** 0.14 g
31. KClO/HClO $= 0.79$
33. a. does not exceed capacity
  b. does not exceed capacity
  c. does not exceed capacity
  d. does not exceed capacity
35. i. (a) pH $= 8$, (b) pH $= 7$ **ii.** (a) weak acid, (b) strong acid
37. a. 40.0 mL HI for both
  b. KOH: neutral, CH_3NH_2: acidic

c. CH_3NH_2

d. Titration of KOH with HI:

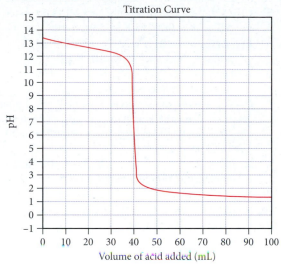

Titration Curve

Titration of CH_3NH_2 with HI:

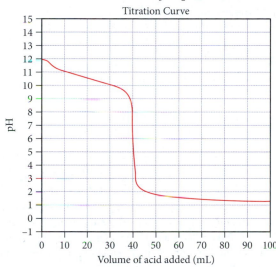

Titration Curve

39. a. pH = 9, added base = 30 mL   **b.** 0 mL

  c. 15 mL   **d.** 30 mL   **e.** 30 mL

41. a. 0.757   **b.** 30.6 mL   **c.** 1.038

  d. 7   **e.** 12.15

43. a. 13.06   **b.** 28.8 mL   **c.** 12.90

  d. 7   **e.** 2.07

45. a. 2.86   **b.** 16.8 mL   **c.** 4.37

  d. 4.74   **e.** 8.75   **f.** 12.17

47. a. 11.94   **b.** 29.2 mL   **c.** 11.33

  d. 10.64   **e.** 5.87   **f.** 1.90

49. i. (a)   **ii.** (b)

51. $pK_a = 3$, 82 g/mol

53. a. phenol red, *m*-nitrophenol

  b. alizarin, bromothymol blue, phenol red

  c. alizarin yellow R

55. a. $BaSO_4(s) \rightleftharpoons Ba^{2+}(aq) + SO_4^{2-}(aq)$,

$$K_{sp} = [Ba^{2+}][SO_4^{2-}]$$

  b. $PbBr_2(s) \rightleftharpoons Pb^{2+}(aq) + 2\ Br^-(aq)$,

$$K_{sp} = [Pb^{2+}][Br^-]^2$$

  c. $Ag_2CrO_4(s) \rightleftharpoons 2\ Ag^+(aq) + CrO_4^{2-}(aq)$,

$$K_{sp} = [Ag^+]^2[CrO_4^{2-}]$$

57. a. 7.31×10^{-7} M   **b.** 3.72×10^{-5} M

  c. 3.32×10^{-4} M

59. a. 1.07×10^{-21} **b.** 7.14×10^{-7}

  c. 7.44×10^{-11}

61. AX_2

63. 2.07×10^{-5} g/100 mL

65. a. 0.0183 M   **b.** 0.00755 M   **c.** 0.00109 M

67. a. 5×10^{14} M   **b.** 5×10^8 M

  c. 5×10^4 M

69. a. more soluble, CO_3^{2-} is basic

  b. more soluble, S^{2-} is basic

  c. not, neutral   **d.** not, neutral

71. precipitate will form, CaF_2

73. precipitate will form, $Mg(OH)_2$

75. a. 0.018 M   **b.** 1.4×10^{-7} M

  c. 1.1×10^{-5} M

77. 8.7×10^{-10} M

79. 4.03

81. 3.57

83. HCl, 4.7 g

85. a. $NaOH(aq) + KHC_8H_4O_4(aq) \longrightarrow Na^+(aq) +$
$$K^+(aq) + C_8H_4O_4^{2-}(aq) + H_2O(l)$$

  b. 0.1046 M

87. 4.73

89. 14.2 L

91. 1.6×10^{-7} M

93. 8.0×10^{-8} M

95. 6.3

97. 3.6 g $(CH_3)_2NH_2Cl$/g $(CH_3)_2NH$ or 0.28 g$(CH_3)_2NH$/g $(CH_3)_2NH_2Cl$

99. 0.18 M $HC_7H_5O_2$
  0.41 M $C_7H_5O_2Na$

101. 51.6 g

103. 1.37 L

105. 13.0

107. a. pH < pK_a **b.** pH > pK_a
  c. pH = pK_a **d.** pH > pK_a

109. a

111. a. no difference   **b.** less soluble
  c. more soluble

Chapter 17

1. a, c

3. System B has the greatest entropy. There is only one energetically equivalent arrangement for System A. However, the particles of System B may exchange positions for a second energetically equivalent arrangement.

5. a. $\Delta S > 0$ **b.** $\Delta S < 0$ **c.** $\Delta S < 0$ **d.** $\Delta S < 0$

7. a. $\Delta S_{sys} > 0$, $\Delta S_{surr} > 0$, spontaneous at all temperatures

  b. $\Delta S_{sys} < 0$, $\Delta S_{surr} < 0$, nonspontaneous at all temperatures

  c. $\Delta S_{sys} < 0$, $\Delta S_{surr} < 0$, nonspontaneous at all temperatures

  d. $\Delta S_{sys} > 0$, $\Delta S_{surr} > 0$, spontaneous at all temperatures

9. a. 963 J/K   **b.** 3.73×10^3 J/K

  c. -426 J/K   **d.** -1.65×10^3 J/K

11. a. 672 J/K, spontaneous

  b. -672 J/K, nonspontaneous

  c. 166 J/K, spontaneous

  d. -28 J/K, nonspontaneous

13. a. -2.00×10^2 kJ, spontaneous
   b. -2.00×10^2 kJ, nonspontaneous
   c. -5.0×10^1 kJ, spontaneous
   d. 15 kJ, nonspontaneous

15. -2.247×10^6 J, spontaneous

17.

ΔH	ΔS	ΔG	Low Temperature	High Temperature
$-$	$+$	$-$	Spontaneous	Spontaneous
$-$	$-$	Temperature dependent	Spontaneous	Nonspontaneous
$+$	$+$	Temperature dependent	Nonspontaneous	Spontaneous
$+$	$-$	$+$	Nonspontaneous	Nonspontaneous

19. It increases.

21. a. $CO_2(g)$, greater molar mass and complexity
   b. $CH_3OH(g)$, gas phase
   c. $CO_2(g)$, greater molar mass and complexity
   d. $SiH_4(g)$, greater molar mass
   e. $CH_3CH_2CH_3(g)$, greater molar mass and complexity
   f. $NaBr(aq)$, aqueous

23. a. He, Ne, SO_2, NH_3, CH_3CH_2OH. From He to Ne there is an increase in molar mass; beyond that, the molecules increase in complexity.
   b. $H_2O(s)$, $H_2O(l)$, $H_2O(g)$; increase in entropy in going from solid to liquid to gas phase.
   c. CH_4, CF_4, CCl_4; increasing entropy with increasing molar mass.

25. a. -120.8 J/K, decrease in moles of gas
   b. 133.9 J/K, increase in moles of gas
   c. -42.0 J/K, small change because moles of gas stay constant
   d. -390.8 J/K, decrease in moles of gas

27. -89.3 J/K, decrease in moles of gas

29. $\Delta H^\circ_{rxn} = -638.5$ kJ, $\Delta S^\circ_{rxn} = 259.4$ J/K, $\Delta G^\circ_{rxn} = -7.158 \times 10^3$ kJ; yes

31. a. $\Delta H^\circ_{rxn} = 57.2$ kJ, $\Delta S^\circ_{rxn} = 175.8$ J/K, $\Delta G^\circ_{rxn} = 4.8 \times 10^3$ J/mol; nonspontaneous, becomes spontaneous at high temperatures
   b. $\Delta H^\circ_{rxn} = 176.2$ kJ, $\Delta S^\circ_{rxn} = 285.1$ J/K, $\Delta G^\circ_{rxn} = 91.2$ kJ; nonspontaneous, becomes spontaneous at high temperatures
   c. $\Delta H^\circ_{rxn} = 98.8$ kJ, $\Delta S^\circ_{rxn} = 141.5$ J/K, $\Delta G^\circ_{rxn} = 56.6$ kJ; nonspontaneous, becomes spontaneous at high temperatures
   d. $\Delta H^\circ_{rxn} = -91.8$ kJ, $\Delta S^\circ_{rxn} = -198.1$ J/K, $\Delta G^\circ_{rxn} = -32.8$ kJ; spontaneous

33. a. 2.8 kJ   **b.** 91.2 kJ   **c.** 56.4 kJ   **d.** -32.8 kJ
   Values are comparable. The method using ΔH° and ΔS° can be used to determine how ΔG° changes with temperature.

35. a. -72.5 kJ, nonspontaneous
   b. -11.4 kJ, spontaneous
   c. 9.1 kJ, nonspontaneous

37. -29.4 kJ

39. a. 19.3 kJ   **b.** (i) 2.9 kJ
                 (ii) -2.9 kJ
   c. The partial pressure of iodine is very low.

41. 11.9 kJ

43. a. 1.48×10^{90} **b.** 2.09×10^{-26}

45. a. -24.8 kJ   **b.** 0   **c.** 9.4 kJ

47. a. 4.32×10^{26} **b.** 1.51×10^{-13}

49. a. $+$ **b.** $-$ **c.** $-$

51. a. $\Delta G^\circ = 175.2$ kJ, $K = 1.95 \times 10^{-31}$, nonspontaneous
   b. 133 kJ, yes

53. Cl_2: $\Delta H^\circ_{rxn} = -182.1$ kJ, $\Delta S^\circ_{rxn} = -134.4$ J/K,
     $\Delta G^\circ_{rxn} = -142.0$ kJ $K = 7.94 \times 10^{24}$
   Br_2: $\Delta H^\circ_{rxn} = -121.6$ kJ, $\Delta S^\circ_{rxn} = -134.2$ J/K,
     $\Delta G^\circ_{rxn} = -81.6$ kJ $K = 2.01 \times 10^{14}$
   I_2: $\Delta H^\circ_{rxn} = -48.3$ kJ, $\Delta S^\circ_{rxn} = -132.2$ J/K,
     $\Delta G^\circ_{rxn} = -8.9$ kJ $K = 36.3$
   Cl_2 is the most spontaneous, I_2 is the least. Spontaneity is determined by the standard enthalpy of formation of the dihalogenated ethane. Higher temperatures make the reactions less spontaneous.

55. a. 107.8 kJ   **b.** 5.0×10^{-7} atm
   c. spontaneous at higher temperatures, $T = 923.4$ K

57. a. 2.22×10^5 **b.** 94.4 mol

59. a. $\Delta G^\circ = -689.6$ kJ, ΔG° becomes less negative
   b. $\Delta G^\circ = -665.2$ kJ, ΔG° becomes less negative
   c. $\Delta G^\circ = -632.4$ kJ, ΔG° becomes less negative
   d. $\Delta G^\circ = -549.3$ kJ, ΔG° becomes less negative

61. With one exception, the formation of any oxide of nitrogen at 298 K requires more moles of gas as reactants than are formed as products. For example, 1 mol of N_2O requires 0.5 mol of O_2 and 1 mol of N_2, 1 mol of N_2O_3 requires 1 mol of N_2 and 1.5 mol of O_2, and so on. The exception is NO, where 1 mol of NO requires 0.5 mol of O_2 and 0.5 mol of

$$N_2\text{:} \quad \frac{1}{2} N_2(g) + \frac{1}{2} O_2(g) \longrightarrow NO(g)$$

This reaction has a positive ΔS because what is essentially mixing of the N and O has taken place in the product.

63. a. 3.24×10^{-3}

   b. $NH_3 + ATP + H_2O \longrightarrow NH_3{-}P_i + ADP$
     $\underline{NH_3{-}P_i + C_5H_8O_4N^- \longrightarrow C_5H_9O_3N_2 + P_i + H_2O}$
     $NH_3 + C_5H_8O_4N^- + ATP \longrightarrow C_5H_9O_3N_2 + ADP + P_i$
              $\Delta G^\circ = -16.3$ kJ, $K = 7.20 \times 10^2$

65. a. -95.3 kJ/mol. Since the number of moles of reactants and products are the same, the decrease in volume affects the entropy of both equally, so there is no change in ΔG.
   b. 102.8 J/mol. The entropy of the reactants (1.5 mol) is decreased more than the entropy of the product (1 mol). Since the product is relatively more favored at lower volume, ΔG is less positive.
   c. 204.2 kJ/mol. The entropy of the product (1 mol) is decreased more than the entropy of the reactant (0.5 mol). Since the product is relatively less favored, ΔG is more positive.

67. $NH_4NO_3(s) \longrightarrow HNO_3(g) + NH_3(g) \quad \Delta G^\circ_{rxn} = +94.0$ kJ
   $NH_4NO_3(s) \longrightarrow N_2O(g) + 2\,H_2O(g)$
                    $\Delta G^\circ_{rxn} = -169.6$ kJ
   $NH_4NO_3(s) \longrightarrow N_2(g) + \frac{1}{2} O_2(g) + 2\,H_2O(g)$
                    $\Delta G^\circ_{rxn} = -273.3$ kJ

The second and third reactions are spontaneous and so we would expect decomposition products of N_2O, N_2, O_2, and H_2O in the gas phase. It is still possible for ammonium nitrate to remain as a solid because the thermodynamics of the reaction say nothing of the kinetics of the reaction (reaction can be extremely slow). Since all of the products are gases, the decomposition of ammonium nitrate will result in a large increase in volume (explosion). Some of the products aid in combustion, which could facilitate the combustion of materials near the ammonium nitrate. Also N_2O is known as laughing gas, which has anesthetic and toxic effects on humans. The solid should not be kept in tightly sealed containers.

69. $C_4H_{10}O$: 0.0861 kJ/K

C_3H_6O: 0.883 kJ/K

C_6H_6O: 0.0873 kJ/K

$CHC1_3$: 0.0880 kJ/K

C_2H_5OH: 0.100 kJ/K

H_2O: 0.109 kJ/K

71. c

73. b

75. c

77. c

Chapter 18

1. **a.** $3 K(s) + Cr^{3+}(aq) \longrightarrow Cr(s) + 3 K^+(aq)$

b. $2 Al(s) + 3 Fe^{2+}(aq) \longrightarrow 2 Al^{3+}(aq) + 3 Fe(s)$

c. $2 BrO_3^-(aq) + 3 N_2H_4(g) \longrightarrow$
$\qquad 2 Br^-(aq) + 3 N_2(g) + 6 H_2O(l)$

3. **a.** $PbO_2(s) + 2 I^-(aq) + 4 H^+(aq) \longrightarrow$
$\qquad Pb^{2+}(aq) + I_2(s) + 2 H_2O(l)$

b. $5 SO_3^{2-}(aq) + 2 MnO_4^-(aq) + 6 H^+(aq) \longrightarrow$
$\qquad 5 SO_4^{2-}(aq) + 2 Mn^{2+}(aq) + 3 H_2O(l)$

c. $S_2O_3^{2-}(aq) + 4 Cl_2(g) + 5 H_2O(l) \longrightarrow$
$\qquad 2 SO_4^{2-}(aq) + 8 Cl^-(aq) + 10 H^+(aq)$

5. **a.** $H_2O_2(aq) + 2 ClO_2(aq) + 2 OH^-(aq) \longrightarrow$
$\qquad O_2(g) + 2 ClO_2^-(aq) + 2 H_2O(l)$

b. $Al(s) + MnO_4^-(aq) + 2 H_2O(l) \longrightarrow$
$\qquad Al(OH)_4^-(aq) + MnO_2(s)$

c. $Cl_2(g) + 2 OH^-(aq) \longrightarrow Cl^-(aq) + ClO^-(aq) + H_2O(l)$

7. **a.**

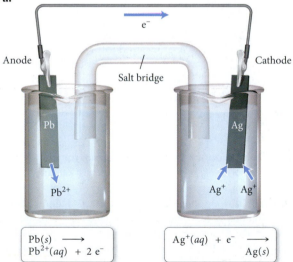

$Pb(s) \longrightarrow$
$Pb^{2+}(aq) + 2 e^-$

$Ag^+(aq) + e^- \longrightarrow$
$\qquad Ag(s)$

b.

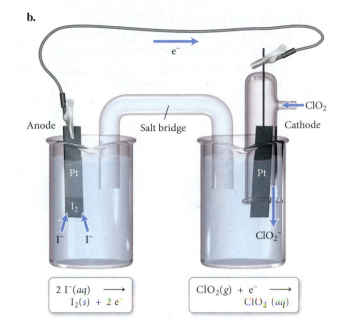

$2 I^-(aq) \longrightarrow$
$\quad I_2(s) + 2 e^-$

$ClO_2(g) + e^- \longrightarrow$
$\qquad ClO_2^-(aq)$

c.

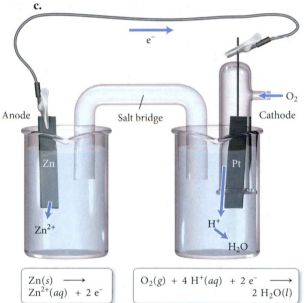

$Zn(s) \longrightarrow$
$Zn^{2+}(aq) + 2 e^-$

$O_2(g) + 4 H^+(aq) + 2 e^- \longrightarrow$
$\qquad 2 H_2O(l)$

9. **a.** 0.93 V   **b.** 0.41 V   **c.** 1.99 V

11. **a, c, d**

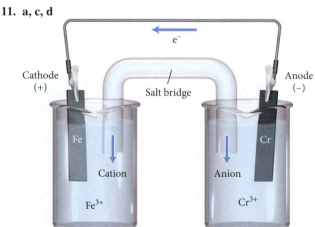

b. $Cr(s) + Fe^{3+}(aq) \longrightarrow Cr^{3+}(aq) + Fe(s)$, $E^\circ_{cell} = 0.69$ V

13. **a.** $Pb(s)|Pb^{2+}(aq)||Ag^+(aq)|Ag(s)$
   b. $Pt(s), I_2(s)|I^-(aq)||ClO_2^-(aq)|ClO_2(g)|Pt(s)$
   c. $Zn(s)|Zn^{2+}(aq)||H_2O(l)|H^+(aq)|O_2(g)|Pt(s)$

15.

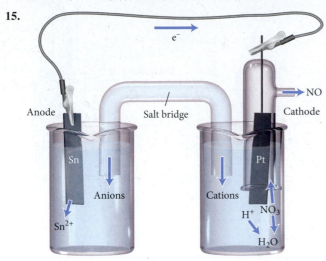

$3\,Sn(s) + 2\,NO_3^-(aq) + 8\,H^+(aq) \longrightarrow$
$\qquad 3\,Sn^{2+}(aq) + 2\,NO(g) + 4\,H_2O(l),\ E^\circ_{cell} = 1.10\ V$

17. **b, c**

19. aluminum

21. **a.** yes, $2\,Al(s) + 6\,H^+(aq) \longrightarrow 2\,Al^{3+}(aq) + 3\,H_2(g)$
   b. no   **c.** yes, $Pb(s) + 2\,H^+(aq) \longrightarrow Pb^{2+}(aq) + H_2(g)$

23. **a.** yes, $3\,Cu(s) + 2\,NO_3^-(aq) + 8\,H^+(aq) \longrightarrow$
   $\qquad 3\,Cu^{2+}(aq) + 2\,NO(g) + 4\,H_2O(l)$
   b. no

25. **a.** -1.70 V, nonspontaneous   **b.** 1.97 V, spontaneous
   c. -1.51, nonspontaneous

27. **a**

29. **a.** -432 kJ   **b.** 52 kJ   **c.** -1.7×10^2 kJ

31. **a.** 5.31×10^{75} **b.** 7.7×10^{-10}
   c. 6.3×10^{29}

33. 5.54×10^5

35. $\Delta G^\circ = -7.97$ kJ, $E^\circ_{cell} = 0.041$ V

37. **a.** 1.04 V   **b.** 0.97 V   **c.** 1.11 V

39. 1.87 V

41. **a.** 0.56 V   **b.** 0.52 V
   c. $[Ni^{2+}] = 0.003$ M, $[Zn^{2+}] = 1.60$ M

43.

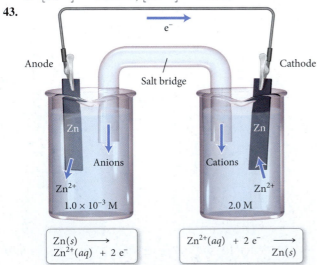

45. $\dfrac{[Sn^{2+}](ox)}{[Sn^{2+}](red)} = 4.2 \times 10^{-4}$

47. 0.3762

49. 1.038 V

51. **a, c**

53.

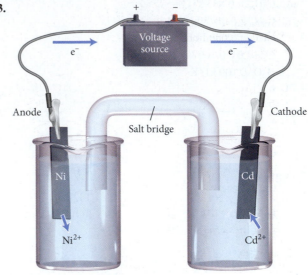

minimum voltage $= 0.17$ V

55.

57. 88 s

59. 1.2×10^3 A

61. $2\,MnO_4^-(aq) + 5\,Zn(s) + 16\,H^+(aq) \longrightarrow$
   $\qquad 2\,Mn^{2+}(aq) + 5\,Zn^{2+}(aq) + 8\,H_2O(l)$
   34.9 mL

63. The drawing should show that several Al atoms dissolve into solution as Al^{3+} ions and that several Cu^{2+} ions are deposited on the Al surface as solid Cu.

65. **a.** 68.3 mL   **b.** cannot be dissolved
   c. cannot be dissolved

67. 0.25

69. **a.** 2.83 V   **b.** 2.71 V   **c.** 16 hr

71. 0.71 V

73. **a.** $\Delta G^\circ = 461$ kJ, $K = 1.4 \times 10^{-81}$
   b. $\Delta G^\circ = 2.7 \times 10^2$ kJ, $K = 2.0 \times 10^{-48}$

75. MCl_4

77. 51.3%

79. 4.1×10^5 L

81. 435 s

83. 8.39% U

85. 5.26 A

87. pH = 2.06

89. a

91. Since $K < 1$, the reaction must be nonspontaneous under standard conditions. Therefore, E°_{cell} is negative and ΔG°_{rxn} is positive.

Chapter 19

1. a. $^{234}_{92}U \longrightarrow {}^4_2He + {}^{230}_{90}Th$ **b.** $^{230}_{90}Th \longrightarrow {}^4_2He + {}^{226}_{88}Ra$

c. $^{214}_{82}Pb \longrightarrow {}^0_{-1}e + {}^{214}_{83}Bi$ **d.** $^{13}_7N \longrightarrow {}^0_{+1}e + {}^{13}_6C$

e. $^{51}_{24}Cr + {}^0_{-1}e \longrightarrow {}^{51}_{23}V$

3. $^{232}_{90}Th \longrightarrow {}^4_2He + {}^{228}_{88}Ra$

$^{232}_{88}Ra \longrightarrow {}^0_{-1}e + {}^{228}_{89}Ac$

$^{228}_{89}Ac \longrightarrow {}^0_{-1}e + {}^{228}_{90}Th$

$^{228}_{90}Th \longrightarrow {}^4_2He + {}^{224}_{88}Ra$

5. a. $^{221}_{87}Fr$ **b.** $^0_{-1}e$ **c.** $^0_{+1}e$ **d.** $^0_{-1}e$

7. a. stable, N/Z ratio is close to 1, acceptable for low Z atoms

b. not stable, N/Z ratio much too high for low Z atom

c. not stable, N/Z ratio is less than 1, much too low

d. stable, N/Z ratio is acceptable for this Z

9. Sc, V, and Mn, each have odd numbers of protons. Atoms with an odd number of protons typically have less stable isotopes than those with an even number of protons.

11. a. beta decay   **b.** positron emission

c. positron emission   **d.** positron emission

13. a. Cs-125   **b.** Fe-62

15. 2.11×10^9 yr

17. 0.57 g

19. 19 hr

21. 2.66×10^3 yr

23. 2.4×10^4 yr

25. 2.7×10^9 yr

27. $^{235}_{92}U + {}^1_0n \longrightarrow {}^{144}_{54}Xe + {}^{90}_{38}Sr + 2{}^1_0n$

29. $^2_1H + {}^2_1H \longrightarrow {}^3_2He + {}^1_0n$

31. $^{238}_{92}U + {}^1_0n \longrightarrow {}^{239}_{92}U$

$^{239}_{92}U \longrightarrow {}^{239}_{93}Np + {}^0_{-1}e$

$^{239}_{93}Np \longrightarrow {}^{239}_{94}Pu + {}^0_{-1}e$

33. 9.0×10^{13} J

35. a. mass defect = 0.13701 amu

binding energy = 7.976 MeV/nucleon

b. mass defect = 0.54369 amu

binding energy = 8.732 MeV/nucleon

c. mass defect = 1.16754 amu

binding energy = 8.437 MeV/nucleon

37. 7.228×10^{10} J/g U-235

39. 7.84×10^{10} J/g H-2

41. radiation: 25.5 J, fall: 370 J

43. 68 mi

45. a. $^{114}_{44}Ru \longrightarrow {}^0_{-1}e + {}^{114}_{45}Rh$

b. $^{216}_{88}Ra \longrightarrow {}^0_{+1}e + {}^{216}_{87}Fr$

c. $^{58}_{30}Zn \longrightarrow {}^0_{+1}e + {}^{58}_{29}Cu$

d. $^{31}_{10}Ne \longrightarrow {}^0_{-1}e + {}^{31}_{11}Na$

47. 2.9×10^{21} beta emissions, 3700 Ci

49. 1.6×10^{-5} L

51. 4.93×10^7 kJ/mol

53. 7.72 MeV

55. 0.15%

57. 1.25×10^{21} particles

59. 2.43 pm

61. The energy change is much less and there is no coulombic barrier for the collision with the neutron, so the process can occur at lower temperatures.

63. a. 1.164×10^{10} kJ   **b.** 0.1295 g

65. U-235 forms Pb-207 in seven α-decays and four β-decays and Th-232 forms Pb-208 in six α-decays and four β-decays.

67. 7

69. The gamma emitter is the greater health threat while you sleep. The alpha emitter is the greater health threat if ingested.

Chapter 20

1. a. alkane   **b.** alkene

c. alkyne   **d.** alkene

3. $CH_3-CH_2-CH_2-CH_2-CH_2-CH_2-CH_3$

$CH_3-CH-CH_2-CH_2-CH_2-CH_3$
$\quad\quad\;\; | $
$\quad\quad CH_3$

$CH_3-CH_2-CH-CH_2-CH_2-CH_3$
$\quad\quad\quad\quad\; | $
$\quad\quad\quad\; CH_3$

$\quad\quad\quad CH_3$
$\quad\quad\quad\; | $
$CH_3-CH-CH-CH_2-CH_3$
$\quad\quad\quad\quad\; | $
$\quad\quad\quad\; CH_3$

$\quad\quad\quad\quad CH_3$
$\quad\quad\quad\quad\; | $
$CH_3-CH_2-C-CH_2-CH_3$
$\quad\quad\quad\quad\; | $
$\quad\quad\quad\; CH_3$

$\quad\quad\quad CH_3$
$\quad\quad\quad\; | $
$H_3C-C-CH_2-CH_2-CH_3$
$\quad\quad\; | $
$\quad\; CH_3$

$H_3C-CH-CH_2-CH-CH_3$
$\quad\quad\; | \quad\quad\quad | $
$\quad\; CH_3 \quad\quad CH_3$

$\quad\quad\quad CH_3$
$\quad\quad\quad\; | $
$H_3C-C-CH-CH_3$
$\quad\quad\; | \quad | $
$\quad CH_3 \; CH_3$

$H_3C-CH_2-CH-CH_2-CH_3$
$\quad\quad\quad\quad | $
$\quad\quad\quad CH_2$
$\quad\quad\quad\quad | $
$\quad\quad\quad CH_3$

5. a. no   **b.** yes   **c.** yes   **d.** no

7. a. enantiomers   **b.** same   **c.** enantiomers

9. a. pentane   **b.** 2-methylbutane
   c. 4-isopropyl-2-methylheptane
   d. 4-ethyl-2-methylhexane

11. a. $CH_3-CH_2-CH-CH_2-CH_2-CH_3$
   with CH_2-CH_3

b. $CH_3-CH_2-C-CH_2-CH_3$
   with CH_3 above and CH_2-CH_3 below

c. $CH_3-CH-CH-CH_3$
   with CH_3 CH_3

d. $CH_3-C-CH_2-CH-CH_2-CH_2-CH-CH_2-CH_3$
   with CH_3 (top left), CH_3 (bottom left), CH_2-CH_3 (middle bottom), CH_2-CH_3 (top right)

13. a. $CH_3CH_2CH_3 + 5 O_2 \longrightarrow 3 CO_2 + 4 H_2O$
   b. $CH_3CH_2CH=CH_2 + 6 O_2 \longrightarrow 4 CO_2 + 4 H_2O$
   c. $2 CH\equiv CH + 5 O_2 \longrightarrow 4 CO_2 + 2 H_2O$

15. a. CH_3CH_2Br
   b. $CH_3CH_2CH_2Cl,\ CH_3CHClCH_3$
   c. $CHCl_2Br$

d. $CH_3-\overset{H}{\underset{CH_3}{C}}-CH_2-Cl$

   $CH_3-\overset{Cl}{\underset{CH_3}{C}}-CH_3$

17. $CH_2=CH-CH_2-CH_2-CH_2-CH_3$
   $CH_3-CH=CH-CH_2-CH_2-CH_3$
   $CH_3-CH_2-CH=CH-CH_2-CH_3$

19. a. 1-butene   **b.** 3,4-dimethyl-2-pentene
   c. 3-isopropyl-1-hexene   **d.** 2,4-dimethyl-3-hexene

21. a. 2-butyne   **b.** 4,4-dimethyl-2-hexyne
   c. 3-isopropyl-1-hexyne   **d.** 3,6-dimethyl-4-nonyne

23. a. $CH_3-CH_2-CH-C\equiv C-CH_2-CH_2-CH_3$
   b. CH_3-CH_2-CH
      $\parallel$
      $CH-CH_2-CH_2-CH_2-CH_2-CH_3$

c. $CH\equiv C-\overset{CH_3}{\underset{CH_3}{C}}-CH_2-CH_2-CH_3$

d. $CH_3-CH=\overset{CH_3}{C}-CH_2-CH-\overset{CH_3}{CH}-CH_3$
   with CH_2-CH_3 below

25. a. $CH_3-\underset{Cl}{CH}-\underset{Cl}{CH}-CH_3$

b. $CH_3-CH_2-\underset{Br}{CH}-\underset{Br}{CH}-CH_3$

27. a. $CH_2=CH-CH_3 + H_2 \longrightarrow CH_3-CH_2-CH_3$
   b.
   $CH_3-\underset{CH_3}{CH}-CH=CH_2 + H_2 \longrightarrow$

   $CH_3-\underset{CH_3}{CH}-CH_2-CH_3$

   c.
   $CH_3-\underset{CH_3}{CH}-\underset{CH_3}{C}=CH_2 + H_2 \longrightarrow$

   $CH_3-\underset{CH_3}{CH}-\underset{CH_3}{CH}-CH_3$

29. a. methylbenzene or toluene
   b. bromobenzene   **c.** chlorobenzene
31. a. 1,4-dibromobenzene or p-dibromobenzene
   b. 1,3-diethylbenzene or m-diethylbenzene
   c. 1-chloro-2-fluorobenzene or o-chlorofluorobenzene
33. a. $CH_3-CH-CH_3$ (with benzene ring below)   **b.** (benzene ring with Br and Br)   **c.** (benzene ring with Cl and CH_3)

35. a. 1-propanol   **b.** 4-methyl-2-hexanol
   c. 2,6-dimethyl-4-heptanol   **d.** 3-methyl-3-pentanol
37. a. $CH_3CH_2CH_2Br + H_2O$
   b. $CH_3-\underset{CH_3}{C}=CH_2 + H_2O$

   c. $CH_3-\underset{CH_3}{\overset{CH_3}{C}}-CH_2-\overset{O}{\overset{\parallel}{C}}-OH$

39. a. butanone   **b.** pentanal
   c. 3,5,5-trimethylhexanal   **d.** 4-methyl-2-hexanone

41. $CH_3-CH_2-CH_2-\overset{OH}{\underset{H}{C}}-C\equiv N$

43. a. methylbutanoate   **b.** propanoioc acid
   c. 5-methylhexanoic acid   **d.** ethylpentanoate

45. $CH_3-CH_2-CH_2-CH_2-\overset{O}{\overset{\parallel}{C}}-O-CH_2-CH_3 + H_2O$

47. a. ethyl propyl ether   **b.** ethyl pentyl ether
   c. dipropyl ether   **d.** butyl ethyl ether
49. a. diethylamine   **b.** methylpropylamine
   c. butylmethylpropylamine
51. condensation, $CH_3CH_2CONHCH_2CH_3(aq) + H_2O$

53. ... (fluorinated polymer chain structure) ...

55. $HO-\overset{O}{\overset{\parallel}{C}}-$(benzene ring)$-\overset{O}{\overset{\parallel}{C}}-O-CH_2-CH_2-OH$

57. a. ester, methyl 3-methylbutanoate
   b. ether, ethyl 2-methylbutyl ether
   c. aromatic, 1-ethyl-3-methylbenzene or *m*-ethylmethylbenzene
   d. alkyne, 5-ethyl-4-methyl-2-heptyne
   e. aldehyde, butanal
   f. alcohol, 2-methyl-1-propanol

59. a. 5-isobutyl-3-methylnonane
   b. 5-methyl-3-hexanone
   c. 3-methyl-2-butanol
   d. 4-ethyl-3,5-dimethyl-1-hexyne

61. a. isomers   **b.** isomers   **c.** same

63. 558 g

65. a. 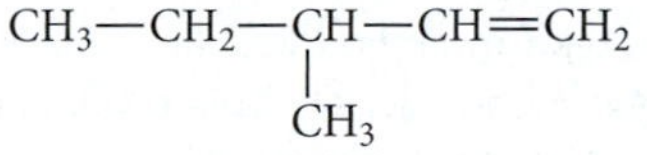 $CH_3—CH_2—CH—CH=CH_2$
   $|$
   CH_3
   Can exist as a stereoisomer

   b. 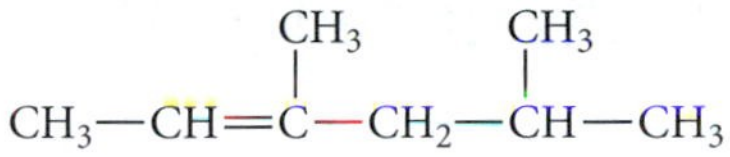 $CH_3—CH=C—CH_2—CH—CH_3$
   with CH_3 groups
   Can exist as a stereoisomer

   c. 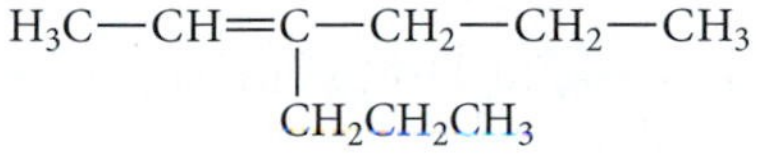 $H_3C—CH=C—CH_2—CH_2—CH_3$
   $|$
   $CH_2CH_2CH_3$
   Can exist as a stereoisomer

67. 1. 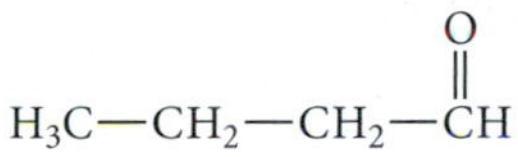 $H_3C—CH_2—CH_2—\overset{O}{\overset{\|}{C}}H$
   Aldehyde

   2. 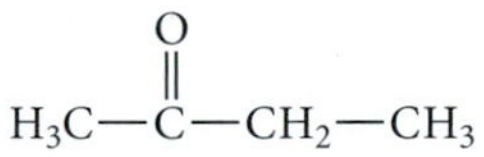 $H_3C—\overset{O}{\overset{\|}{C}}—CH_2—CH_3$
   Ketone

   3. $H_3C—CH=CH—O—CH_3$
   Alkene, ether

   4. $H_2C=CH—O—CH_2—CH_3$
   Alkene, ether

   5. $H_2C=CH—CH_2—O—CH_3$
   Alkene, ether

   6. $H_3C—CH=CH—CH_2—OH$
   Alkene, alcohol

   7. 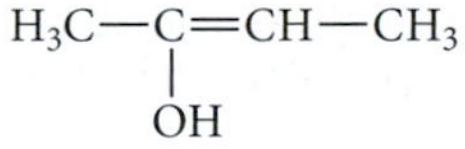 $H_3C—C=CH—CH_3$
   $|$
   OH
   Alkene, alcohol

   8. 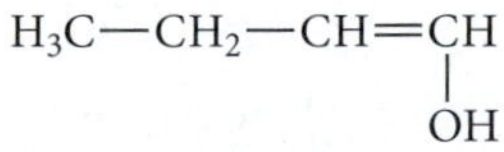 $H_3C—CH_2—CH=CH$
   $|$
   OH
   Alkene, alcohol

   9. 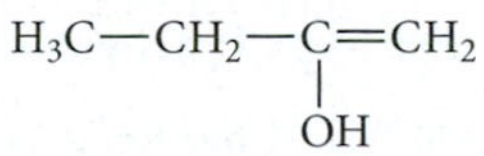 $H_3C—CH_2—C=CH_2$
   $|$
   OH
   Alkene, alcohol

   10. 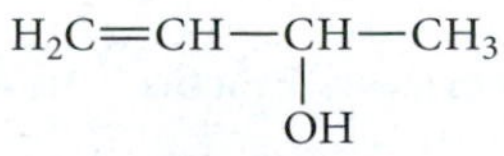 $H_2C=CH—CH—CH_3$
   $|$
   OH
   Alkene, alcohol

   11. $H_2C=CH—CH_2—CH_2—OH$
   Alkene, alcohol

69. In the acid form of the carboxylic acid, electron withdrawal by the C=O bond enhances acidity. The conjugate base, the carboxylate anion, is stabilized by resonance so the two O atoms are equivalent and bear the negative charge equally. This resonance stability is not possible in an alchol, where the carbonyl group is absent.

71. a. 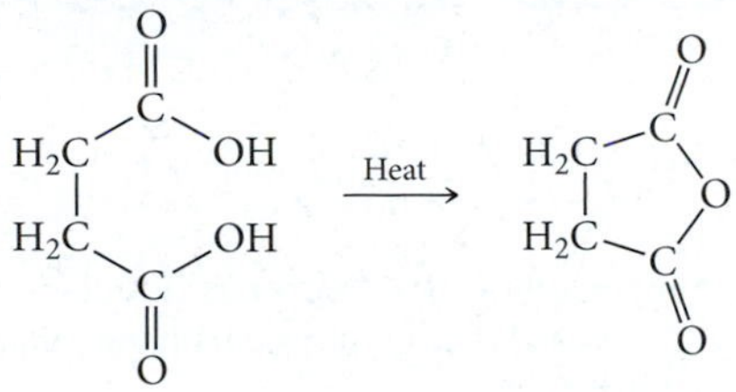 [structure: dicarboxylic acid with Heat arrow forming cyclic anhydride]

   b. 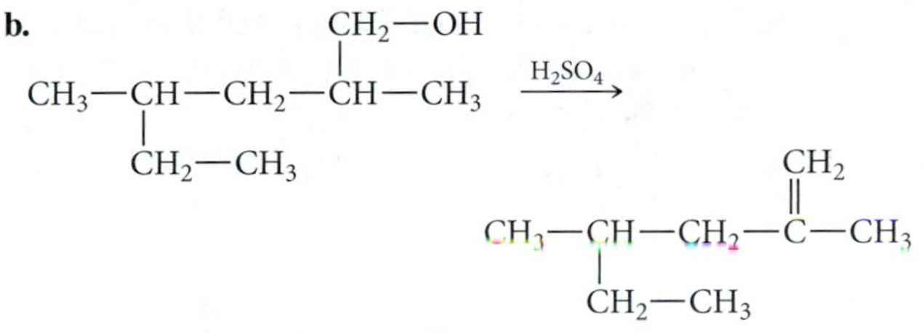
   $CH_3—CH—CH_2—CH—CH_3$ with $CH_2—OH$ and $CH_2—CH_3$ groups $\xrightarrow{H_2SO_4}$

   $CH_3—CH—CH_2—\overset{CH_2}{\overset{\|}{C}}—CH_3$ with $CH_2—CH_3$ group

73. a. 3:1
   b. 2° hydrogen atoms are more reactive. The reactivity of 2° hydrogens to 1° hydrogens is 11:3.

75. 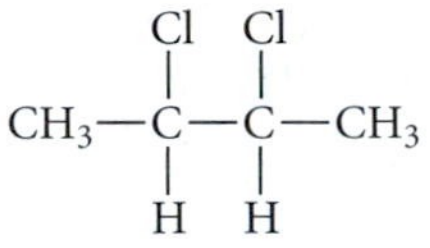 $CH_3—\overset{Cl}{\underset{H}{C}}—\overset{Cl}{\underset{H}{C}}—CH_3$
   Chiral

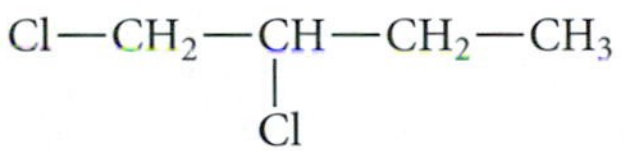

   $Cl—CH_2—\overset{}{\underset{Cl}{CH}}—CH_2—CH_3$
   Chiral

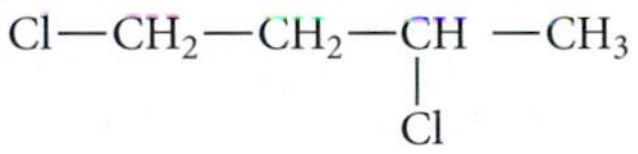

   $Cl—CH_2—CH_2—\overset{}{\underset{Cl}{CH}}—CH_3$
   Chiral

77. The first propagation step for F is a very rapid and exothermic because of the strength of the H—F bond that forms. For I the first propagation step is endothermic and slow because the H—I bond that forms is relatively weak.

79. No dipole moment:

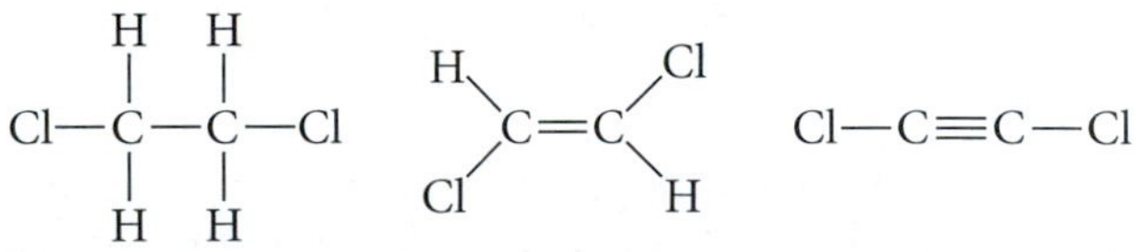

[structures: $Cl—C(H)(H)—C(H)(H)—Cl$; $H(Cl)C=C(Cl)H$; $Cl—C≡C—Cl$]

With dipole moment:

[structures: $H—C(H)(Cl)—C(H)(H)—Cl$; $H(H)C=C(Cl)Cl$; $Cl(H)C=C(Cl)H$]

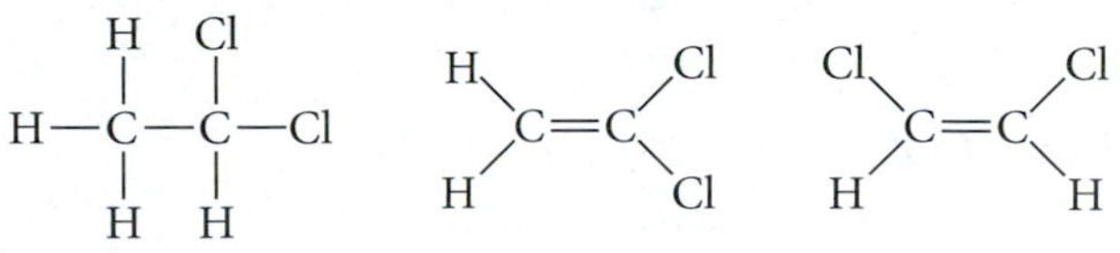

81. b, d

Answers to In-Chapter Practice Problems

Chapter 1

1.1. a. The composition of the copper is not changing; thus, being hammered flat is a physical change that signifies a physical property.

   b. The dissolution and color change of the nickel indicate that it is undergoing a chemical change and exhibiting a chemical property.

   c. Vaporization is a physical change indicative of a physical property.

   d. When a match ignites, a chemical change begins as the match reacts with oxygen to form carbon dioxide and water. Flammability is a chemical property.

1.2. a. $29.8\,°C$ **b.** $302.9\,K$

1.3. $21.4\,g/cm^3$ This matches the density of platinum.

1.3. For More Practice $4.50\,g/cm^3$ The metal is titanium.

1.4. The thermometer shown has markings every $1\,°F$; thus, the first digit of uncertainty is 0.1. The answer is $103.1\,°F$.

1.5. a. Each figure in this number is significant by rule 1: three significant figures.

   b. This is a defined quantity that has an unlimited number of significant figures.

   c. Both 1's are significant (rule 1) and the interior zero is significant as well (rule 2): three significant figures.

   d. Only the two 9's are significant, the leading zeroes are not (rule 3): two significant figures.

   e. There are five significant figures because the 1, 4, and 5 are nonzero (rule 1) and the trailing zeroes are after a decimal point so they are significant as well (rule 4).

   f. The number of significant figures is ambiguous because the trailing zeroes occur before an implied decimal point (rule 4). Assume two significant figures.

1.6. a. 0.381   **b.** 121.0   **c.** 1.174   **d.** 8

1.7. 3.15 yd   **1.8.** 2.446 gal   **1.9.** $1.61 \times 10^6\,cm^3$

1.9. For More Practice $3.23 \times 10^3\,kg$

1.10. 1.03 kg   **1.10. For More Practice** $2.9 \times 10^{-2}\,cm^3$

1.11. 0.855 cm   **1.12.** $2.70\,g/cm^3$

Chapter 2

2.1. For the first sample:

$$\frac{\text{mass of oxygen}}{\text{mass of carbon}} = \frac{17.2\,g\,O}{12.9\,g\,C} = 1.33 \text{ or } 1.33\!:\!1$$

For the second sample:

$$\frac{\text{mass of oxygen}}{\text{mass of carbon}} = \frac{10.5\,g\,O}{7.88\,g\,C} = 1.33 \text{ or } 1.33\!:\!1$$

The ratios of oxygen to carbon are the same in the two samples of carbon monoxide, so these results are consistent with the law of definite proportions.

2.2.

$$\frac{\text{mass of hydrogen to 1 g of oxygen in hydrogen peroxide}}{\text{mass of hydrogen to 1 g of oxygen in water}}$$

$$= \frac{0.250}{0.125} = 2.00$$

The ratio of the mass of hydrogen from one compound to the mass of hydrogen in the other is equal to 2. This is a simple whole number and, therefore, consistent with the law of multiple proportions.

2.3. a. $Z = 6, A = 13, {}^{13}_{6}C$ **b.** 19 protons, 20 neutrons

2.4. a. N^{3-} **b.** Rb^+ **2.5.** 24.31 amu

2.5. For More Practice 70.92 amu

2.6. $4.65 \times 10^{-2}\,mol\,Ag$ **2.7.** 0.563 mol Cu

2.7. For More Practice 22.6 g Ti

2.8. $1.3 \times 10^{22}\,C$ atoms **2.8. For More Practice** 6.87 g W

2.9. $l = 1.72\,cm$

2.9. For More Practice $2.90 \times 10^{24}\,Cu$ atoms

Chapter 3

3.1. a. C_5H_{12} **b.** $HgCl$ **c.** CH_2O

3.2. a. molecular element   **b.** molecular compound

   c. atomic element   **d.** ionic compound

   e. ionic compound

3.3. K_2S **3.4.** AlN **3.5.** silver nitride

3.5. For More Practice Rb_2S

3.6. iron(II) sulfide   **3.6. For More Practice** RuO_2

3.7. tin(II) chlorate   **3.7. For More Practice** $Co_3(PO_4)_2$

3.8. dinitrogen pentoxide   **3.8. For More Practice** PBr_3

3.9. hydrofluoric acid

3.10. nitrous acid   **3.10. For More Practice** $HClO_4$

3.11. 164.10 amu   **3.12.** $5.839 \times 10^{20}\,C_{13}H_{18}O_2$ molecules

3.12. For More Practice 1.06 g H_2O

3.13. 53.29% **3.13. For More Practice** 74.19% Na

3.14. 4.0 g O **3.14. For More Practice** 3.60 g C

3.15. CH_2O **3.16.** $C_{13}H_{18}O_2$

3.17. C_6H_6 **3.17. For More Practice** $C_2H_8N_2$

3.18. C_2H_5 **3.19.** C_2H_4O

3.20. $SiO_2(s) + 3\,C(s) \longrightarrow SiC(s) + 2\,CO(g)$

3.21. $2\,C_2H_6(g) + 7\,O_2(g) \longrightarrow 4\,CO_2(g) + 6\,H_2O(g)$

3.22. $PbCl_2(aq) + 2\,KCl(aq) \longrightarrow PbCl_2(s) + 2\,KNO_3(aq)$

Chapter 4

4.1. 4.08 g HCl **4.2.** 22 kg HNO_3

4.3. H_2 is the limiting reagent, since it produces the least amount of NH_3. Therefore, 29.4 kg NH_3 is the theoretical yield.

4.4. CO is the limiting reagent, since it only produces 114 g Fe. Therefore, 114 g Fe is the theoretical yield: percentage yield = 63.4% yield

4.5. 0.214 M $NaNO_3$ **4.5. For More Practice** 44.6 g KBr

4.6. 402 g $C_{12}H_{22}O_{11}$

4.6. For More Practice 221 mL of KCl solution

4.7. 667 mL **4.7. For More Practice** 0.105 L

4.8. 51.4 mL HNO_3 solution

4.8. For More Practice 0.170 g CO_2

4.9. a. Insoluble. **b.** Insoluble.
c. Soluble. **d.** Soluble.

4.10. $NH_4Cl(aq) + Fe(NO_3)_3(aq) \longrightarrow$ NO REACTION

4.11. $2\,NaOH(aq) + CuBr_2(aq) \longrightarrow$
$Cu(OH)_2(s) + 2\,NaBr(aq)$

4.12. $2\,H^+(aq) + 2\,I^-(aq) + Ba^{2+}(aq) + 2\,OH^-(aq) \longrightarrow$
$2\,H_2O(l) + Ba^{2+}(aq) + 2\,I^-(aq)$
$H^+(aq) + OH^-(aq) \longrightarrow H_2O(l)$

4.12. For More Practice

$2\,Ag^+(aq) + 2\,NO_3^-(aq) + Mg^{2+}(aq) + 2\,Cl^-(aq) \longrightarrow$
$2\,AgCl(s) + Mg^{2+}(aq) + 2\,NO_3^-(aq)$
$Ag^+(aq) + Cl^-(aq) \longrightarrow AgCl(s)$

4.13. $H_2SO_4(aq) + 2\,LiOH(aq) \longrightarrow 2\,H_2O(l) + Li_2SO_4(aq)$
$H^+(aq) + OH^-(aq) \longrightarrow H_2O(aq)$

4.14. $2\,HBr(aq) + K_2SO_3(aq) \longrightarrow$
$H_2O(l) + SO_2(g) + 2\,KBr(aq)$

4.14. For More Practice $2\,H^+(aq) + S^{2-}(aq) \longrightarrow H_2S(g)$

4.15. a. Cr = 0. **b.** Cr^{3+} = +3.
c. Cl^- = −1, C = +4. **d.** Br = −1, Sr = +2.
e. O = −2, S = +6. **f.** O = −2, N = +5.

4.16. Sn is oxidized and N is reduced.

4.16. For More Practice
b. Reaction b is the only redox reaction. Al is oxidized and O is reduced.

4.17. a. This is a redox reaction in which Li is the reducing agent (it is oxidized) and Cl_2 is the oxidizing reagent (it is reduced).
b. This is a redox reaction in which Al is the reducing agent and Sn^{2+} is the oxidizing agent.
c. This is not a redox reaction because no oxidation states change.
d. This is a redox reaction in which C is the reducing agent and O_2 is the oxidizing agent.

4.18. $2\,C_2H_5SH(l) + 9\,O_2(g) \longrightarrow$
$4\,CO_2(g) + 2\,SO_2(g) + 6\,H_2O(g)$

Chapter 5

5.1. 15.0 psi **5.1. For More Practice** 80.6 kPa

5.2. 2.1 atm at a depth of approximately 11 m.

5.3. 123 mL **5.4.** 11.3 L

5.5. 1.63 atm, 23.9 psi **5.6.** 16.1 L

5.6. For More Practice 976 mmHg

5.7. $d = 4.91$ g/L **5.7. For More Practice** 44.0 g/mol

5.8. 70.7 g/mol **5.9.** 0.0610 mol H_2

5.10. 4.2 atm **5.11.** 12.0 mg H_2

5.12. 82.3 g Ag_2O **5.12. For More Practice** 7.10 g Ag_2O

5.13. 6.53 L O_2 **5.14.** $u_{rms} = 238$ m/s

5.15. $\dfrac{\text{rate}_{H_2}}{\text{rate}_{Kr}} = 6.44$

Chapter 6

6.1. $\Delta E = 71$ J **6.2.** $C_s = 0.38\,\dfrac{J}{g \cdot {}^\circ C}$

The specific heat capacity of gold is 0.128 J/g·°C; therefore, the rock cannot be pure gold.

6.2. For More Practice $T_f = 42.1\,^\circ C$

6.3. 37.8 g Cu

6.4. −122 J

6.4. For More Practice $\Delta E = -998$ J

6.5. $\Delta E_{rxn} = -3.91 \times 10^3 \dfrac{kJ}{\text{mol } C_6H_{14}}$

6.5. For More Practice $C_{cal} = 4.55\,\dfrac{kJ}{^\circ C}$

6.6. a. endothermic, positive ΔH.
b. endothermic, positive ΔH.
c. exothermic, positive ΔH.

6.7. -2.06×10^3 kJ

6.7. For More Practice 33 g C_4H_{10} 99 g CO_2

6.8. $\Delta H_{rxn} = -68$ kJ

6.9. $N_2O(g) + NO_2(g) \longrightarrow 3\,NO(g)$, $\Delta H_{rxn} = +157.6$ kJ

6.9. For More Practice

$3\,H_2(g) + O_3(g) \longrightarrow 3\,H_2O(g)$, $\Delta H = -868.1$ kJ

6.10. a. $Na(s) + \dfrac{1}{2}Cl_2(g) \longrightarrow NaCl(s)$,

$$\Delta H_f^\circ = -411.2 \text{ kJ/mol}$$

b. $Pb(s) + N_2(g) + 3\,O_2(g) \longrightarrow$
$$Pb(NO_3)_2(s), \Delta H_f^\circ = -451.9 \text{ kJ/mol}$$

6.11. $\Delta H_{rxn}^\circ = -851.5$ kJ

6.12. $\Delta H_{rxn}^\circ = -1648.4$ kJ   111 kJ emitted (-111 kJ)

Chapter 7

7.1. $5.83 \times 10^{14}\,s^{-1}$

7.2. 2.64×10^{20} photons

7.2. For More Practice   435 nm

7.3. a. blue < green < red.
  b. red < green < blue.
  c. red < green < blue.

7.4. 6.1×10^6 m/s

7.5. For the $5d$ orbitals:

$n = 5 \quad l = 2 \quad m_l = -2, -1, 0, 1, 2$

The 5 integer values for m_l signify that there are five $5d$ orbitals.

7.6. a. l cannot equal 3 if $n = 3$. $l = 2$
  b. m_l cannot equal -2 if $l = -1$. Possible values for $m_l = -1, 0,$ or 1
  c. l cannot be 1 if $n = 1$. $l = 0$.

7.7. 397 nm   **7.7. For More Practice** $n = 1$

Chapter 8

8.1. a. $Cl\ 1s^2 2s^2 2p^6 3s^2 3p^5$ or $[Ne]\ 3s^2 3p^5$
  b. $Si\ 1s^2 2s^2 2p^6 3s^2 3p^2$ or $[Ne]\ 3s^2 3p^2$
  c. $Sr\ 1s^2 2s^2 2p^6 3s^2 3p^6 4s^2 3d^{10} 4p^6 5s^2$ or $[Kr]\ 5s^2$
  d. $O\ 1s^2 2s^2 2p^4$ or $[He]\ 2s^2 2p^4$

8.2. There are no unpaired electrons.

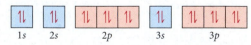
$1s \quad 2s \quad\quad 2p \quad\quad 3s \quad\quad 3p$

8.3. $1s^2 2s^2 2p^6 3s^2 3p^3$ or $[Ne]\ 3s^2 3p^3$. The five electrons in the $3s^2 3p^3$ orbitals are the valence electrons, while the 10 electrons in the $1s^2 2s^2 2p^6$ orbitals belong to the core.

8.4. $Bi\ [Xe]\ 6s^2 4f^{14} 5d^{10} 6p^3$

8.4. For More Practice $I\ [Kr]\ 5s^2 4d^{10} 5p^5$

8.5. a. Sn   **b.** cannot predict   **c.** W   **d.** Se

8.5. For More Practice   Rb > Ca > Si > S > F.

8.6. a. $[Ar]\ 4s^0 3d^7$. Co^{2+} is paramagnetic.

Co^{2+} [Ar]   ▢ 1↓ 1↓ 1 1 1
$\quad\quad\quad\quad 4s \quad\quad\quad 3d$

b. $[He]\ 2s^2 2p^6$. N^{3-} is diamagnetic.

N^{3-} [He]   1↓ 1↓ 1↓ 1↓
$\quad\quad\quad\quad 2s \quad\quad 2p$

c. $[Ne]\ 3s^2 3p^6$. Ca^{2+} is diamagnetic.

Ca^{2+} [Ne]   1↓ 1↓ 1↓ 1↓
$\quad\quad\quad\quad 3s \quad\quad 3p$

8.7. a. K   **b.** F^- **c.** Cl^-

8.7. For More Practice $Cl^- > Ar > Ca^{2+}$.

8.8. a. I   **b.** Ca   **c.** cannot predict   **d.** F

8.8. For More Practice   F > S > Si > Ca > Rb.

8.9. a. Sn
  b. cannot predict based on simple trends (Po is larger.)
  c. Bi   **d.** B

8.9. For More Practice   Cl < Si < Na < Rb.

Chapter 9

9.1. Mg_3N_2

9.2. KI < LiBr < CaO.

9.2. For More Practice $MgCl_2$

9.3. a. pure covalent   **b.** ionic   **c.** polar covalent

9.4. $:C{\equiv}O:$

9.5.
$\quad\quad :O:$
$\quad\quad \|$
$H{-}C{-}H$

9.6. $\left[:\overset{..}{\underset{..}{Cl}}:\overset{..}{\underset{..}{O}}:\right]^-$

9.7. $\left[:\overset{..}{O}{=}\overset{..}{N}{-}\overset{..}{\underset{..}{O}}:\right]^- \longleftrightarrow \left[:\overset{..}{\underset{..}{O}}{-}\overset{..}{N}{=}\overset{..}{O}:\right]^-$

9.8.

Structure	A			B			C		
	$:\overset{..}{N}{=}N{=}\overset{..}{O}:$			$:N{\equiv}N{-}\overset{..}{\underset{..}{O}}:$			$:\overset{..}{\underset{..}{N}}{-}N{\equiv}O:$		
number of valence e^-	5	5	6	5	5	6	5	5	6
number of lone pair e^-	-4	-0	-4	-2	-0	-6	-6	-0	-2
$^1/_2$ (number of bonding e^2)	-2	-4	-2	-3	-4	-1	-1	-4	-3
Formal charge	-1	$+1$	0	0	$+1$	-1	-2	$+1$	$+1$

Structure B contributes the most to the correct overall structure of N_2O.

9.8. For More Practice   The nitrogen is $+1$, the singly bonded oxygen atoms are -1, and the double-bonded oxygen atom has no formal charge.

9.9.

$$:\overset{\displaystyle :\ddot{F}:}{\underset{\displaystyle :\ddot{F}:}{:\ddot{F}-\overset{|}{\underset{|}{\ddot{X}e}}-\ddot{F}:}}$$

9.9. For More Practice

$$H-\ddot{O}-\overset{\displaystyle H}{\underset{\displaystyle \overset{\|}{:\ddot{O}:^{0}}}{\overset{|}{\underset{|}{P^{0}}}-\ddot{O}-H}}$$

9.10. $CH_3OH(g) + \dfrac{3}{2}O_2(g) \longrightarrow CO_2(g) + 2\,H_2O(g).$

$\Delta H_{rxn} = -641$ kJ

9.10. For More Practice $\Delta H_{rxn} = -8.0 \times 10^1$ kJ

Chapter 10

10.1. tetrahedral

$$:\overset{\displaystyle :\ddot{C}l:}{\underset{\displaystyle :\ddot{C}l:}{:\ddot{C}l-\overset{|}{\underset{|}{C}}-\ddot{C}l:}}$$

10.2. bent   **10.3.** linear

10.4.

Atom	Number of Electron Groups	Number of Lone Pairs	Molecular Geometry
Carbon (left)	4	0	Tetrahedral
Carbon (right)	3	0	Trigonal planar
Oxygen	4	2	Bent

10.5. The molecule is nonpolar.

10.6. The xenon atom has six electron groups and therefore has an octahedral electron geometry. An octahedral electron geometry corresponds to sp^3d^2 hybridization (refer to Table 10.3).

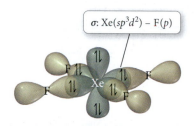

σ: Xe(sp^3d^2) – F(p)

10.7. Since there are only two electron groups around the central atom (C), the electron geometry is linear. According to Table 10.3, the corresponding hybridization on the carbon atom is sp.

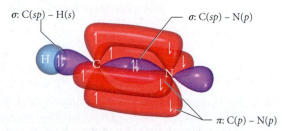

σ: C(sp) – H(s)   σ: C(sp) – N(p)

π: C(p) – N(p)

10.8. Since there are only two electron groups about the central atom (C) the electron geometry is linear. The hybridization on C is sp (refer to Table 10.3).

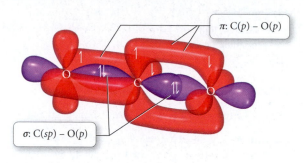

π: C(p) – O(p)

σ: C(sp) – O(p)

10.8. For More Practice There are five electron groups about the central atom (I); therefore, the electron geometry is trigonal bipyramidal and the corresponding hybridization of I is sp^3d (refer to Table 10.3).

10.9. H_2^+ bond order $= +\dfrac{1}{2}$

Since the bond order is positive, the H_2^+ ion should be stable; however, the bond order of H_2^+ is lower than the bond order of H_2 (bond order $= 1$). Therefore, the bond in H_2^+ is weaker than in H_2.

10.10. The bond order of N_2^+ is 2.5, which is lower than that of the N_2 molecule (bond order $= 3$), therefore, the bond is weaker. The MO diagram shows that the N_2^+ ion has one unpaired electron and is therefore paramagnetic.

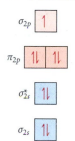

10.10. For More Practice The bond order of Ne_2 is 0, which indicates that dineon does not exist.

Chapter 11

11.1. b, c

11.2. HF has a higher boiling point than HCl because, unlike HCl, HF is able to form hydrogen bonds. The hydrogen bond is the strongest of the intermolecular forces and requires more energy to break.

11.3. 5.83×10^3 kJ

11.3. For More Practice 49 °C

11.4. 33.8 kJ/mol   **11.5.** 7.04×10^3 torr   **11.6.** $7.18\,\dfrac{g}{cm^3}$

Chapter 12

12.1. **a.** not soluble   **b.** soluble

c. not soluble   **d.** not soluble

12.2. 2.7×10^{-4} M   **12.3.** 42.5 g $C_{12}H_{22}O_{11}$

12.3. For More Practice 3.3×10^4 L

12.4. **a.** $M = 0.415$ M   **b.** $m = 0.443$ m
c. % by mass $= 13.2$%   **d.** $\chi C_{12}H_{22}O_{11} = 0.00793$
e. mole percent $= 0.793$%

12.5. 0.600 M $C_6H_{12}O_6$

12.5. For More Practice 0.651 m $C_6H_{12}O_6$

12.6. 22.5 torr   **12.6. For More Practice** 0.144

12.7. **a.** $P_{benzene} = 26.6$ torr $P_{toluene} = 20.4$ torr
b. 47.0 torr   **c.** 52.5% benzene; 47.5% toluene

The vapor will be richer in the more volatile component, which in this case is benzene.

12.8. $T_f = -4.8\,°C$ **12.9.** 101.84 °C

12.10. 11.8 atm   **12.11.** $-0.60\,°C$ **12.12.** 0.014 mols

Chapter 13

13.1. $\dfrac{\Delta[H_2O_2]}{\Delta t} = -4.40 \times 10^{-3}$ M/s

$\dfrac{\Delta[I_3^-]}{\Delta t} = 4.40 \times 10^{-3}$ M/s

13.2. **a.** Rate $= k[CHCl_3][Cl_2]^{1/2}$. (Fractional-order reactions are not common but are occasionally observed.)
b. 3.5 $M^{-1/2} \cdot s^{-1}$

13.3. 5.78×10^{-2} M   **13.4.** 0.0277 M

13.5. 1.64×10^{-3} M   **13.6.** 79.2 s

13.7. $2.07 \times 10^{-5} \dfrac{L}{mol \cdot s}$ **13.8.** $6.13 \times 10^{-4} \dfrac{L}{mol \cdot s}$

13.9. $2\,A + B \longrightarrow A_2B$ Rate $= k[A]^2$

Chapter 14

14.1. $K = \dfrac{[CO_2]^3[H_2O]^4}{[C_3H_8][O_2]^5}$ **14.2.** 2.1×10^{-13}

14.2. For More Practice 1.4×10^2

14.3. 6.2×10^2 **14.4.** $K_c = \dfrac{[Cl_2]^2}{[HCl]^4[O_2]}$

14.5. 9.4   **14.6.** 1.1×10^{-6}

14.7. $Q_c = 0.0196$
Reaction proceeds to the left.

14.8. 0.033 M

14.9. $[N_2] = 4.45 \times 10^{-3}$ M $[O_2] = 4.45 \times 10^{-3}$ M
$[NO] = 1.1 \times 10^{-3}$ M

14.10. $[N_2O_4] = 0.005$ M $[NO_2] = 0.041$ M

14.11. $P_{I_2} = 0.027$ atm $P_{Cl_2} = 0.027$ atm
$P_{ICl} = 0.246$ atm

14.12. 1.67×10^{-7} M   **14.13.** 6.78×10^{-6} M

14.14. Adding Br_2 increases the concentration of Br_2, causing a shift to the left (away from the Br_2). Adding BrNO increases the concentration of BrNO, causing a shift to the right.

14.15. Decreasing the volume causes the reaction to shift right. Increasing the volume causes the reaction to shift left.

14.16. If we increase the temperature, the reaction shifts to the left. If we decrease the temperature, the reaction shifts to the right.

Chapter 15

15.1. **a.** H_2O donates a proton to C_5H_5N, making it the acid. The conjugate base is therefore OH^-. Since C_5H_5N accepts the proton, it is the base and becomes the conjugate acid $C_5H_5NH^+$.

b. Since HNO_3 donates a proton to H_2O, it is the acid, making NO_3^- the conjugate base. Since H_2O is the proton acceptor, it is the base and becomes the conjugate acid, H_3O^+.

15.2. **a.** $[H_3O^+] = 6.7 \times 10^{-13}$ M
Since $[H_3O^+] < [OH^-]$, the solution is basic.

b. $[H_3O^+] = 1.0 \times 10^{-7}$ M
Neutral solution.

c. $[H_3O^+] = 1.2 \times 10^{-5}$ M
Since $[H_3O^+] > [OH^-]$, the solution is acidic.

15.3. **a.** 8.02 (basic)   **b.** 11.85 (basic)

15.4. 4.3×10^{-9} M   **15.5.** 9.4×10^{-3} M

15.6. 3.28   **15.7.** 2.72   **15.8.** 1.8×10^{-6}

15.9. 0.85%   **15.10.** $[OH^-] = 0.020$ M   pH $= 12.30$

15.11. $[OH^-] = 1.2 \times 10^{-2}$ M   pH $= 12.08$

15.12. **a.** weak base   **b.** pH-neutral

15.13. 9.07

15.14. **a.** pH-neutral   **b.** weak acid   **c.** weak acid

15.15. **a.** basic   **b.** acidic
c. pH-neutral   **d.** acidic

Chapter 16

16.1. 4.44   **16.1. For More Practice** 3.44

16.2. 9.14   **16.3.** 4.87

16.3. For More Practice 4.65

16.4. 9.68   **16.4. For More Practice** 9.56

16.5. hypochlorous acid (HClO); 2.4 g NaClO

16.6. 1.74   **16.7.** 7.99   **16.8.** 2.30×10^{-6} M

16.9. 5.3×10^{-13} **16.10.** 1.21×10^{-5} M

16.11. $FeCO_3$ will be more soluble in an acidic solution than $PbBr_2$ because the CO_3^{2-} ion is a basic anion, whereas Br^- is the conjugate base of a strong acid (HBr) and is, therefore, pH-neutral.

16.12. $Q > K_{sp}$; therefore, a precipitate forms.

16.13. 9.6×10^{-6} M.

Chapter 17

17.1. **a.** positive   **b.** negative
    c. positive

17.2. **a.** -548 J/K   **b.** ΔS_{sys} is negative.
    c. ΔS_{univ} is negative, and the reaction is not spontaneous.

17.2. **For More Practice** 375 K

17.3. $\Delta G = -101.6 \times 10^3$ J

Therefore, the reaction is spontaneous. Since both ΔH and ΔS are negative, as the temperature increases ΔG will become more positive.

17.4. -153.2 J/K

17.5. $\Delta G^{\circ}_{rxn} = -36.3$ kJ

Since ΔG°_{rxn} is negative, the reaction is spontaneous at this temperature.

17.6. $\Delta G^{\circ}_{rxn} = -42.1$ kJ

Since the value of ΔG°_{rxn} at the lowered temperature is more negative (or less positive) (which is -36.3 kJ), the reaction is more spontaneous.

17.7. $\Delta G^{\circ}_{rxn} = -689.6$ kJ

Since ΔG°_{rxn} is negative, the reaction is spontaneous at this temperature.

17.7. **For More Practice** $\Delta G^{\circ}_{rxn} = -689.7$ kJ (at 25° C)

The value calculated for ΔG°_{rxn} from the tabulated values (-689.6 kJ) is the same, to within 1 in the least significant digit, as the value calculated using the equation for ΔG°_{rxn}. $\Delta G^{\circ}_{rxn} = -649.7$ kJ (at 500.0 K)

You could not calculate ΔG°_{rxn} at 500.0 K using tabulated ΔG°_{f} values because the tabulated values of free energy are calculated at a standard temperature of 298 K, much lower than 500 K.

17.8. $+107.1$ kJ

17.9. $\Delta G_{rxn} = -129$ kJ

The reaction is more spontaneous under these conditions than under standard conditions because ΔG_{rxn} is more negative than ΔG°_{rxn}.

17.10. -10.9 kJ

Chapter 18

18.1. $2 \, Cr(s) + 4 \, H^{+}(aq) \longrightarrow 2 \, Cr^{2+}(aq) + 2 \, H_2(g)$

18.2. $Cu(s) + 4 \, H^{+}(aq) + 2 \, NO_3^{-}(aq) \longrightarrow$
$$Cu^{2+}(aq) + 2 \, NO_2(g) + 2 \, H_2O(l)$$

18.3. $3 \, ClO^{-}(aq) + 2 \, Cr(OH)_4^{-}(aq) + 2 \, OH^{-}(aq) \longrightarrow$
$$3 \, Cl^{-}(aq) + 2 \, CrO_4^{2-}(aq) + 5 \, H_2O(l)$$

18.4. $+0.60$ V

18.5. **a.** The reaction *will* be spontaneous as written.
    b. The reaction *will not* be spontaneous as written.

18.6. $\Delta G^{\circ} = -3.63 \times 10^5$ J

Since ΔG° is negative, the reaction is spontaneous.

18.7. 4.5×10^3 **18.8.** 1.09 V   **18.9.** 6.0×10^1 min

Chapter 19

19.1. $^{216}_{84}Po \longrightarrow \, ^{212}_{82}Pb + \, ^{4}_{2}He$

19.2. **a.** $^{235}_{92}U \longrightarrow \, ^{231}_{90}Th + \, ^{4}_{2}He$

$^{231}_{90}Th \longrightarrow \, ^{231}_{91}Pa + \, ^{0}_{-1}e$

$^{231}_{91}Pa \longrightarrow \, ^{227}_{89}Ac + \, ^{4}_{2}He$

    b. $^{22}_{11}Na \longrightarrow \, ^{22}_{10}Ne + \, ^{0}_{+1}e$

    c. $^{76}_{36}Kr + \, ^{0}_{-1}e \longrightarrow \, ^{76}_{35}Br$

19.3. **a.** positron emission   **b.** beta decay
    c. positron emission

19.4. 10.7 yr

19.5. $t = 964$ yr
No, the C-14 content suggests that the scroll is from about A.D. 1000, not 500 B.C.

19.6. 1.0×10^9 yr

19.7. Mass defect $= 1.934$ amu
Nuclear binding energy $= 7.569$ MeV/nucleon

Chapter 20

20.1.

or

20.2. 3-methylhexane

20.3. 3,5-dimethylheptane

20.4. 2,3,5-trimethylhexane

20.5. **a.** 4,4-dimethyl-2-pentyne

    b. 3-ethyl-4,6-dimethyl-1-heptene

20.6. 2-methylbutane

Glossary

accuracy   A term that refers to how close a measured value is to the actual value. (1.7)

acid   A molecular compound that is able to donate an H^+ ion (proton) when dissolved in water, thereby increasing the concentration of H^+. (3.6)

acid ionization constant (K_a)   The equilibrium constant for the ionization reaction of a weak acid; used to compare the relative strengths of weak acids. (15.4)

acid–base reaction (neutralization reaction)   A reaction in which an acid reacts with a base and the two neutralize each other, producing water. (4.8)

acid–base titration   A laboratory procedure in which a basic (or acidic) solution of unknown concentration is reacted with an acidic (or basic) solution of known concentration, in order to determine the concentration of the unknown. (16.4)

acidic solution   A solution containing an acid that creates additional H_3O^+ ions, causing $[H_3O^+]$ to increase. (15.5)

activated complex (transition state)   A high-energy intermediate state between reactant and product. (13.5)

activation energy   An energy barrier in a chemical reaction that must be overcome for the reactants to be converted into products. (13.5)

active site   The specific area of an enzyme at which catalysis occurs. (13.7)

actual yield   The amount of product actually produced by a chemical reaction. (4.3)

addition polymer   A polymer in which the monomers simply link together without the elimination of any atoms. (20.9)

addition reaction   A type of organic reaction in which two substituents are added across a double bond. (20.8)

alcohol   A member of the family of organic compounds that contain a hydroxyl functional group (—OH). (20.8)

aldehyde   A member of the family of organic compounds that contain a carbonyl functional group (C=O) bonded to two R groups, one of which is a hydrogen atom. (20.8)

aliphatic hydrocarbons   Organic compounds in which carbon atoms are joined in straight or branched chains. (20.3)

alkali metals   Highly reactive metals in group 1A of the periodic table. (2.6)

alkaline battery   A dry-cell battery that employs the oxidation of zinc and the reduction of manganese(IV) oxide in a basic medium. (18.7)

alkaline earth metals   Fairly reactive metals in group 2A of the periodic table. (2.6)

alkaloid   Organic bases found in plants; they are often poisonous. (15.2)

alkane   A hydrocarbon containing only single bonds. (20.3)

alkene   A hydrocarbon containing one or more carbon–carbon double bonds. (20.3)

alkyne   A hydrocarbon containing one or more carbon–carbon triple bonds. (20.3)

alpha (α) decay   The form of radioactive decay that occurs when an unstable nucleus emits a particle composed of two protons and two neutrons. (19.2)

alpha (α) particle   A low-energy particle released during alpha decay; equivalent to a He-4 nucleus. (19.2)

amine   An organic compound containing nitrogen and derived from ammonia with one or more of the hydrogen atoms replaced with alkyl groups. (20.8)

amorphous solid   A solid in which atoms or molecules do not have any long-range order. (1.3, 11.2)

ampere (A)   The SI unit for electrical current; $1\ A = 1\ C/s$. (18.3)

amphoteric   Able to act as either an acid or a base. (15.3)

amplitude   The vertical height of a crest (or depth of a trough) of a wave; a measure of wave intensity. (7.2)

angular momentum quantum number (l)   An integer that determines the shape of an orbital. (7.5)

anion   A negatively charged ion. (2.5)

anode   The electrode in an electrochemical cell where oxidation occurs; electrons flow away from the anode. (18.3)

antibonding orbital   A molecular orbital that is higher in energy than any of the atomic orbitals from which it was formed. (10.8)

aqueous solution   A solution in which water acts as the solvent. (4.4, 12.2)

aromatic hydrocarbon   Unusually stable cyclic conjugated hydrocarbons such as those that contain a benzene ring. (20.1, 20.7)

Arrhenius definitions (of acids and bases)   The definitions of an acid as a substance that produces H^+ ions in aqueous solution and a base as a substance that produces OH^- ions in aqueous solution. (4.8, 15.3)

Arrhenius equation   An equation which relates the rate constant of a reaction to the temperature, the activation energy, and the frequency factor; $k = Ae^{-E_a/RT}$. (13.5)

Arrhenius plot   A plot of the natural log of the rate constant ($\ln k$) versus the inverse of the temperature in kelvins ($1/T$) that yields a straight line with a slope of $-E_a/R$ and a y-intercept of $\ln A$. (13.5)

atmosphere (atm)   A unit of pressure based on the average pressure of air at sea level; $1\ atm = 101,325\ Pa$. (5.2)

atom   A submicroscopic particle that constitutes the fundamental building block of ordinary matter; the smallest identifiable unit of an element. (1.1)

atomic element   Those elements that exist in nature with single atoms as their basic units. (3.4)

atomic mass (atomic weight)   The average mass in amu of the atoms of a particular element based on the relative abundance of the various isotopes; it is numerically equivalent to the mass in grams of one mole of the element. (2.7)

atomic mass unit (amu)   A unit used to express the masses of atoms and subatomic particles, defined as $1/12$ the mass of a carbon atom containing six protons and six neutrons. (2.5)

atomic number (Z)   The number of protons in an atom; the atomic number defines the element. (2.5)

atomic radius   Refers to a set of bonding radii determined from measurements on a large number of elements and compounds. (8.6)

atomic solids   Solids whose composite units are atoms; they include nonbonding atomic solids, metallic atomic solids, and network covalent solids. (11.11)

atomic theory   The theory that each element is composed of tiny indestructible particles called atoms, that all atoms of a given element have the same mass and other properties, and that atoms combine in simple, whole-number ratios to form compounds. (2.2)

aufbau principle The principle that indicates the pattern of orbital filling in an atom. (8.3)

autoionization The process by which water acts as an acid and a base with itself. (15.5)

Avogadro's law The law that states that the volume of a gas is directly proportional to its amount in moles ($V \propto n$). (5.3)

Avogadro's number The number of ^{12}C atoms in exactly 12 g of ^{12}C; equal to 6.0221421×10^{23}. (2.8)

balanced chemical equation see *chemical equation* (3.10)

ball-and-stick model A representation of the arrangement of atoms in a molecule that shows how the atoms are bonded to each other and the overall shape of the molecule. (3.3)

band gap An energy gap that exists between the valence band and conduction band of semiconductors and insulators. (11.12)

band theory A model for bonding in atomic solids that comes from molecular orbital theory in which atomic orbitals combine and become delocalized over the entire crystal. (11.12)

barometer An instrument used to measure atmospheric pressure. (5.2)

base ionization constant (K_b) The equilibrium constant for the ionization reaction of a weak base; used to compare the relative strengths of weak bases. (15.7)

basic solution A solution containing a base that creates additional OH^- ions, causing the $[OH^-]$ to increase. (15.5)

beta (β) decay The form of radioactive decay that occurs when an unstable nucleus emits an electron. (19.2)

beta (β) particle A medium-energy particle released during beta decay; equivalent to an electron. (19.2)

bimolecular An elementary step in a reaction that involves two particles, either the same species or different, that collide and go on to form products. (13.6)

binary acid An acid composed of hydrogen and a nonmetal. (3.6)

binary compound A compound that contains only two different elements. (3.5)

biological effectiveness factor (RBE) A correction factor multiplied by the dose of radiation exposure in rad to obtain the dose rem. (19.8)

body-centered cubic A unit cell that consists of a cube with one atom at each corner and one atom at the center of the cube. (11.10)

boiling point The temperature at which the vapor pressure of a liquid equals the external pressure. (11.5)

boiling point elevation The effect of a solute that causes a solution to have a higher boiling point than the pure solvent. (12.6)

bomb calorimeter A piece of equipment designed to measure ΔE_{rxn} for combustion reactions at constant volume. (6.5)

bond energy The energy required to break 1 mol of the bond in the gas phase. (9.10)

bond length The average length of a bond between two particular atoms in a variety of compounds. (9.10)

bond order For a molecule, the number of electrons in bonding orbitals minus the number of electrons in nonbonding orbitals divided by two; a positive bond order implies that the molecule is stable. (10.8)

bonding orbital A molecular orbital that is lower in energy than any of the atomic orbitals from which it was formed. (10.8)

bonding pair A pair of electrons shared between two atoms. (9.5)

Boyle's law The law that states that volume of a gas is inversely proportional to its pressure $\left(V \propto \dfrac{1}{P}\right)$. (5.3)

Brønsted–Lowry definitions (of acids and bases) The definitions of an acid as a proton (H^+ ion) donor and a base as a proton acceptor. (15.3)

buffer A solution containing significant amounts of both a weak acid and its conjugate base (or a weak base and its conjugate acid) that resists pH change by neutralizing added acid or added base. (16.2)

buffer capacity The amount of acid or base that can be added to a buffer without destroying its effectiveness. (16.3)

calorie (cal) A unit of energy defined as the amount of energy required to raise one gram of water 1 °C; equal to 4.184 J. (6.2)

Calorie (Cal) Shorthand notation for the kilocalorie (kcal), or 1000 calories; also called the nutritional calorie, the unit of energy used on nutritional labels. (6.2)

calorimetry The experimental procedure used to measure the heat evolved in a chemical reaction. (6.5)

capillary action The ability of a liquid to flow against gravity up a narrow tube due to adhesive and cohesive forces. (11.4)

carbonyl group A functional group consisting of a carbon atom double-bonded to an oxygen atom ($C=O$). (20.8)

carboxylic acid An organic acid containing the functional group $-COOH$. (15.2, 20.8)

catalyst A substance that is not consumed in a chemical reaction, but increases the rate of the reaction by providing an alternate mechanism in which the rate-determining step has a smaller activation energy. (13.7)

cathode The electrode in an electrochemical cell where reduction occurs; electrons flow toward the cathode. (18.3)

cathode rays A stream of electrons produced when a high electrical voltage is applied between two electrodes within a partially evacuated tube. (2.3)

cathode ray tube A partially evacuated tube equipped with two electrodes and designed to create cathode rays. (2.3)

cation A positively charged ion. (2.5)

cell potential (cell emf) (E_{cell}) The potential difference between the cathode and the anode in an electrochemical cell. (18.3)

Celsius (°C) scale The temperature scale most often used by scientists (and by most countries other than the United States), on which pure water freezes at 0 °C and boils at 100 °C (at atmospheric pressure). (1.6)

chain reaction A series of reactions in which previous reactions cause future ones; in a fission bomb, neutrons produced by the fission of one uranium nucleus induce fission in other uranium nuclei. (19.5)

Charles's law The law that states that the volume of a gas is directly proportional to its temperature ($V \propto T$). (5.3)

chemical bond The sharing or transfer of electrons to attain stable electron configurations for the bonding atoms. (9.3)

chemical change A change that alters the molecular composition of a substance; see also *chemical reaction*. (1.4)

chemical energy The energy associated with the relative positions of electrons and nuclei in atoms and molecules. (6.2)

chemical equation A symbolic representation of a chemical reaction; a balanced equation contains equal numbers of the atoms of each element on both sides of the equation. (3.10)

chemical formula A symbolic representation of a compound which indicates the elements present in the compound and the relative number of atoms of each. (3.3)

chemical property A property that a substance displays only by changing its composition via a chemical change. (1.4)

chemical reaction A process by which one or more substances are converted to one or more different substances; see also *chemical change*. (3.10)

chemical symbol A one- or two-letter abbreviation for an element that is listed directly below its atomic number on the periodic table. (2.5)

chemistry The science that seeks to understand the behavior of matter by studying the behavior of atoms and molecules. (1.1)

chiral molecule A molecule that is not superimposable on its mirror image, and thus exhibits optical isomerism. (20.3)

cis–trans isomerism Another term for geometric isomerism; *cis*-isomers have the same functional group on the same side of a bond and *trans*-isomers have the same functional group on opposite sides of a bond. (20.5)

Clausius–Clapeyron equation An equation that displays the exponential relationship between vapor pressure and temperature;

$$\ln{(P_{vap})} = \frac{-\Delta H_{vap}}{R}\left(\frac{1}{T}\right) + \ln{\beta}. \text{ (11.5)}$$

coffee-cup calorimeter A piece of equipment designed to measure ΔH_{rxn} for reactions at constant pressure. (6.7)

colligative property A property that depends on the amount of a solute but not on the type. (12.6)

collision frequency In the collision model, the number of collisions that occur per unit of time represented by the symbol z. (13.5)

collision model A model of chemical reactions in which a reaction occurs after a sufficiently energetic collision between two reactant molecules. (13.5)

combustion analysis A method of obtaining empirical formulas for unknown compounds, especially those containing carbon and hydrogen, by burning a sample of the compound in pure oxygen and analyzing the products of the combustion reaction. (3.9)

combustion reaction A type of chemical reaction in which a substance combines with oxygen to form one or more oxygen-containing compounds; the reaction often causes the evolution of heat and light in the form of a flame. (3.10)

common ion effect The tendency for a common ion to decrease the solubility of an ionic compound or to decrease the ionization of a weak acid or weak base. (16.2)

common name A traditional name of a compound that gives little or no information about its chemical structure; for example, the common name of $NaHCO_3$ is "baking soda." (3.5)

complementary properties Those properties that exclude one another, i.e., the more you know about one, the less you know about the other. For example, the wave nature and particle nature of the electron are complementary. (7.4)

complete ionic equation An equation which lists individually all of the ions present as either reactants or products in a chemical reaction. (4.7)

complex ion An ion that contains a central metal ion bound to one or more ligands. (16.7)

composition The kinds and amounts of substances that compose matter. (1.3)

compound A substance composed of two or more elements in fixed, definite proportions. (1.3)

concentrated solution A solution that contains a large amount of solute relative to the amount of solvent. (4.4, 12.5)

condensation The phase transition from gas to liquid. (11.5)

condensation polymer A polymer formed by elimination of an atom or small group of atoms (usually water) between pairs of monomers during polymerization. (20.9)

condensation reaction A reaction in which two or more organic compounds are joined, often with the loss of water or some other small molecule. (20.8)

conjugate acid–base pair Two substances related to each other by the transfer of a proton. (15.3)

constructive interference The interaction of waves from two sources that align with overlapping crests, resulting in a wave of greater amplitude. (7.2)

conversion factor A factor used to convert between two different units; a conversion factor can be constructed from any two quantities known to be equivalent. (1.8)

coordination number The number of atoms with which each atom in a crystal lattice is in direct contact. (11.10)

core electrons Those electrons in a complete principal energy level and those in complete *d* and *f* sublevels. (8.4)

corrosion The gradual, nearly always undesired oxidation of metals that occurs when they are exposed to oxidizing agents in the environment. (18.9)

coulomb's law A law that states that the potential energy of two charged particles is directly proportional to the product of their charges and inversely proportional to the distance between them. (8.3)

covalent bond A chemical bond in which two atoms share electrons that interact with the nuclei of both atoms, lowering the potential energy of each through electrostatic interactions. (3.2, 9.2)

covalent radius (bonding atomic radius) Defined in nonmetals as one-half the distance between two atoms bonded together, and in metals as one-half the distance between two adjacent atoms in a crystal of the metal. (8.6)

critical mass The necessary amount of a radioactive isotope required to produce a self-sustaining fission reaction. (19.5)

critical point The temperature and pressure above which a supercritical fluid exists. (11.8)

critical pressure The pressure required to bring about a transition to a liquid at the critical temperature. (11.5)

critical temperature The temperature above which a liquid cannot exist, regardless of pressure. (11.5)

crystalline lattice The regular arrangement of atoms in a crystalline solid. (11.10)

crystalline solid (crystal) A solid in which atoms, molecules, or ions are arranged in patterns with long-range, repeating order. (1.3, 11.2)

cubic closest packing A closest-packed arrangement in which the third layer of atoms is offset from the first; the same structure as the face-centered cubic. (11.10)

Dalton's law of partial pressures The law stating that the sum of the partial pressures of the components in a gas mixture must equal the total pressure. (5.6)

de Broglie relation The observation that the wavelength of a particle is inversely proportional to its momentum $\lambda = \frac{h}{mv}$. (7.4)

degenerate A term describing two or more electron states with the same energy. (8.3)

density (*d*) The ratio of an object's mass to its volume. (1.6)

deposition   The phase transition from gas to solid. (11.6)

derived unit   A unit that is a combination of other base units. For example, the SI unit for speed is meters per second (m/s), a derived unit. (1.6)

destructive interference   The interaction of waves from two sources aligned so that the crest of one overlaps the trough of the other, resulting in cancellation. (7.2)

deterministic   A characteristic of the classical laws of motion, which imply that present circumstances determine future events. (7.4)

dextrorotatory   Capable of rotating the plane of polarization of light clockwise. (20.3)

diamagnetic   The state of an atom or ion that contains only paired electrons and is, therefore, slightly repelled by an external magnetic field. (8.7, 10.8)

diffraction   The phenomena by which a wave emerging from an aperture spreads out to form a new wave front. (7.2)

diffusion   The process by which a gas spreads through a space occupied by another gas. (5.9)

dilute solution   A solution that contains a very small amount of solute relative to the amount of solvent. (4.4, 12.5)

dimensional analysis   The use of units as a guide to solving problems. (1.8)

dimer   The product that forms from the reaction of two monomers. (20.9)

dipole moment   A measure of the separation of positive and negative charge in a molecule. (9.6)

dipole–dipole force   An intermolecular force exhibited by polar molecules that results from the uneven charge distribution. (11.3)

diprotic acid   An acid that contains two ionizable protons. (4.8, 15.4)

dispersion force (London force)   An intermolecular force exhibited by all atoms and molecules that results from fluctuations in the electron distribution. (11.3)

disubstituted benzene   A benzene in which two hydrogen atoms have been replaced by other atoms. (20.7)

double bond   The bond that forms when two electrons are shared between two atoms. (9.5)

dry-cell battery   A battery that does not contain a large amount of liquid water, often using the oxidation of zinc and the reduction of MnO_2 to provide the electrical current. (18.7)

duet   A Lewis structure with two dots, signifying a filled outer electron shell for the elements H and He. (9.3)

dynamic equilibrium   The point at which the rate of the reverse reaction or process equals the rate of the forward reaction or process. (11.5, 12.4, 14.2)

effective nuclear charge (Z_{eff})   The actual nuclear charge experienced by an electron, defined as the charge of the nucleus plus the charge of the shielding electrons. (8.3)

effusion   The process by which a gas escapes from a container into a vacuum through a small hole. (5.9)

electrical charge   A fundamental property of certain particles that causes them to experience a force in the presence of electric fields. (2.3)

electrical current   The flow of electric charge. (18.3)

electrochemical cell   A device in which a chemical reaction either produces or is carried out by an electrical current. (18.3)

electrode   A conductive surface that allows electrons to enter or leave half-cells. (18.3)

electrolysis   The process by which electrical current is used to drive an otherwise nonspontaneous redox reaction. (18.8)

electrolyte   A substance that dissolves in water to form solutions that conduct electricity. (4.5)

electrolytic cell   An electrochemical cell which uses electrical current to drive a nonspontaneous chemical reaction. (18.3)

electromagnetic radiation   A form of energy embodied in oscillating electric and magnetic fields. (7.2)

electromagnetic spectrum   The range of the wavelengths of all possible electromagnetic radiation. (7.2)

electromotive force (emf)   The force that results in the motion of electrons due to a difference in potential. (18.3)

electron   A negatively charged, low mass particle found outside the nucleus of all atoms that occupies most of the atom's volume but contributes almost none of its mass. (2.3)

electron affinity (EA)   The energy change associated with the gaining of an electron by an atom in its gaseous state. (8.8)

electron capture   The form of radioactive decay that occurs when a nucleus assimilates an electron from an inner orbital. (19.2)

electron configuration   A notation that shows the particular orbitals that are occupied by electrons in an atom. (8.3)

electron geometry   The geometrical arrangement of electron groups in a molecule. (10.3)

electron groups   A general term for lone pairs, single bonds, multiple bonds, or lone electrons in a molecule. (10.2)

electron spin   A fundamental property of electrons; spin can have a value of $\pm 1/2$. (8.3)

electronegativity   The ability of an atom to attract electrons to itself in a covalent bond. (9.6)

element   A substance that cannot be chemically broken down into simpler substances. (1.3)

elementary step   An individual step in a reaction mechanism. (13.6)

emission spectrum   The range of wavelengths emitted by a particular element; used to identify the element. (7.3)

empirical formula   A chemical formula that shows the simplest whole number ratio of atoms in the compound. (3.3)

empirical formula molar mass   The sum of the masses of all the atoms in an empirical formula. (3.9)

enantiomers (optical isomers)   Two molecules that are nonsuperimposable mirror images of one another. (20.3)

endothermic reaction   A chemical reaction that absorbs heat from its surroundings; for an endothermic reaction, $\Delta H > 0$. (6.6)

endpoint   The point of pH change where an indicator changes color. (16.4)

energy   The capacity to do work. (1.5, 6.2)

English system   The system of units used in the United States and various other countries in which the inch is the unit of length, the pound is the unit of force, and the ounce is the unit of mass. (1.6)

enthalpy (H)   The sum of the internal energy of a system and the product of its pressure and volume. (6.6)

enthalpy of solution (ΔH_{soln})   The enthalpy change associated with a substance dissolving into a solvent. (12.3)

enthalpy (heat) of reaction (ΔH_{rxn})   The enthalpy change associated with a chemical reaction.

entropy   A thermodynamic function that is proportional to the number of energetically equivalent ways to arrange the components of a system to achieve a particular state; a measure of the energy randomization or energy dispersal in a system. (12.2, 17.3)

enzyme A biochemical catalyst made of protein that increases the rates of biochemical reactions. (13.7)

equilibrium constant (K) The ratio, at equilibrium, of the concentrations of the products of a reaction raised to their stoichiometric coefficients to the concentrations of the reactants raised to their stoichiometric coefficients. (14.3)

equivalence point The point in a titration at which the added reactant completely reacts with the reactant present in the solution; for acid–base titrations, the point at which the amount of acid is stoichiometrically equal to the amount of base in solution. (16.4)

ester A family of organic compounds with the general structure R—COO—R. (20.8)

ether A member of the family of organic compounds of the form R—O—R′. (20.8)

exact numbers Numbers that have no uncertainty and thus do not limit the number of significant figures in any calculation. (1.7)

exothermic reaction A chemical reaction that releases heat to its surroundings; for an exothermic reaction, $\Delta H < 0$. (6.6)

experiment A highly controlled procedure designed to generate observations that may support a hypothesis or prove it wrong. (1.2)

exponential factor A number between 0 and 1 that represents the fraction of molecules that have enough energy to make it over the activation barrier on a given approach. (13.5)

extensive property A property that depends on the amount of a given substance, such as mass. (1.6)

face-centered cubic A crystal structure whose unit cell consists of a cube with one atom at each corner and one atom in the center of every face. (11.10)

Fahrenheit (°F) scale The temperature scale that is most familiar in the United States, on which pure water freezes at 32 °F and boils at 212 °F. (1.6)

family A group of organic compounds with the same functional group. (20.8)

family (group) Columns within the main group elements in the periodic table that contain elements that exhibit similar chemical properties. (2.6)

Faraday's constant (F) The charge in coulombs of 1 mol of electrons: $F = \dfrac{96,485\ \text{C}}{\text{mol e}^-}$. (18.5)

first law of thermodynamics The law stating that the total energy of the universe is constant. (6.3)

formal charge The charge that an atom in a Lewis structure would have if all the bonding electrons were shared equally between the bonded atoms. (9.8)

formation constant (K_f) The equilibrium constant associated with reactions for the formation of complex ions. (16.7)

formula mass The average mass of a molecule of a compound in amu. (3.7)

formula unit The smallest, electrically neutral collection of ions in an ionic compound. (3.4)

free energy of formation (ΔG_f°) The change in free energy when 1 mol of a compound forms from its constituent elements in their standard states. (17.7)

free radical A molecule or ion with an odd number of electrons in its Lewis structure. (9.9)

freezing The phase transition from liquid to solid. (11.6)

freezing point depression The effect of a solute that causes a solution to have a lower melting point than the pure solvent. (12.6)

frequency (ν) For waves, the number of cycles (or complete wavelengths) that pass through a stationary point in one second. (7.2)

frequency factor The number of times that reactants approach the activation energy per unit time. (13.5)

fuel cell A voltaic cell in which the reactants continuously flow through the cell. (18.7)

functional group A characteristic atom or group of atoms that imparts certain chemical properties to an organic compound. (20.8)

gamma (γ) rays The form of electromagnetic radiation with the shortest wavelength and highest energy. (7.2, 19.2)

gamma (γ) ray emission The form of radioactive decay that occurs when an unstable nucleus emits extremely high frequency electromagnetic radiation. (19.2)

gas A state of matter in which atoms or molecules have a great deal of space between them and are free to move relative to one another; lacking a definite shape or volume, a gas conforms to those of its container. (1.3)

gas-evolution reaction A reaction in which two aqueous solutions are mixed and a gas forms, resulting in bubbling. (4.8)

geometric isomerism A form of stereoisomerism involving the orientation of functional groups in a molecule that contains bonds incapable of rotating. (20.5)

Gibbs free energy (G) A thermodynamic state function related to enthalpy and entropy by the equation $G = H - TS$; chemical systems tend towards lower Gibbs free energy, also called the *chemical potential*. (17.5)

ground state The lowest energy state of atom, molecule, or ion. (8.3)

graham's law of effusion A law stating that the ratio of the rates of effusion of two gases is proportional to the square root of the inverse of the ratio of their molar masses. (5.9)

half-cell One half of an electrochemical cell where either oxidation or reduction occurs. (18.3)

half-life ($t_{1/2}$) The time required for the concentration of a reactant or the amount of a radioactive isotope to fall to one-half of its initial value. (13.4)

halogens Highly reactive nonmetals in group 7A of the periodic table. (2.6)

heat (q) The flow of energy caused by a temperature difference. (6.2)

heat capacity (C) The quantity of heat required to change a system's temperature by 1 °C. (6.4)

heat of fusion (ΔH_{fus}) The amount of heat required to melt 1 mole of a solid. (11.6)

heat of hydration ($\Delta H_{hydration}$) The enthalpy change that occurs when 1 mole of gaseous solute ions are dissolved in water. (12.3)

heat of reaction (ΔH_{rxn}) The enthalpy change for a chemical reaction. (6.5)

heat of vaporization (ΔH_{vap}) The amount of heat required to vaporize one mole of a liquid to a gas. (11.5)

Heisenberg's uncertainty principle The principle stating that due to the wave-particle duality, it is fundamentally impossible to precisely determine both the position and velocity of a particle at a given moment in time. (7.4)

Henderson–Hasselbalch equation An equation used to easily calculate the pH of a buffer solution from the initial concentrations of the buffer components, assuming that the "*x is small*" approximation is valid: $\text{pH} = \text{p}K_a + \log\dfrac{[\text{base}]}{[\text{acid}]}$. (16.2)

Henry's law An equation that expresses the relationship between solubility of a gas and pressure: $S_{gas} = k_H P_{gas}$. (12.4)

Hess's law The law stating that if a chemical equation can be expressed as the sum of a series of steps, then ΔH_{rxn} for the overall equation is the sum of the heats of reactions for each step. (6.8)

heterogeneous catalysis Catalysis in which the catalyst and the reactants exist in different phases. (13.7)

heterogeneous mixture A mixture in which the composition varies from one region to another. (1.3)

hexagonal closest packing A closest packed arrangement in which the atoms of the third layer align exactly over those in the first layer. (11.10)

homogeneous catalysis Catalysis in which the catalyst exists in the same phase as the reactants. (13.7)

homogeneous mixture A mixture with the same composition throughout. (1.3)

Hund's rule The principle stating that when electrons fill degenerate orbitals they first fill them singly with parallel spins. (8.3)

hybrid orbitals Orbitals formed from the combination of standard atomic orbitals that correspond more closely to the actual distribution of electrons in a chemically bonded atom. (10.7)

hybridization A mathematical procedure in which standard atomic orbitals are combined to form new, hybrid orbitals. (10.7)

hydrate An ionic compound that contains a specific number of water molecules associated with each formula unit. (3.5)

hydrocarbon An organic compound that contains only carbon and hydrogen. (3.11, 20.3)

hydrogen bond A strong dipole–dipole attractive force between a hydrogen bonded to O, N, or F and one of these electronegative atoms on a neighboring molecule. (11.3)

hydronium ion H_3O^+, the ion formed from the association of a water molecule with an H^+ ion donated by an acid. (4.8, 15.3)

hydroxyl group Refers to the -OH functional group. (20.8)

hypothesis A tentative interpretation or explanation of an observation. A good hypothesis is *falsifiable*. (1.2)

ideal gas A hypothetical gas that follows the ideal gas law. (5.4)

ideal gas constant The proportionality constant of the ideal gas law, R, equal to 8.314 J/mol · K or 0.08206 L · atm/mol · K. (5.4)

ideal gas law The law that combines the relationships of Boyle's, Charles's, and Avogadro's laws into one comprehensive equation of state with the proportionality constant R in the form $PV = nRT$. (5.4)

ideal solution A solution that follows Raoult's law at all concentrations for both solute and solvent. (12.6)

indeterminacy The principle that present circumstances do not necessarily determine future events in the quantum-mechanical realm. (7.4)

indicator A dye whose color depends on the pH of the solution it is dissolved in; often used to detect the endpoint of a titration. (16.4)

infrared (IR) radiation Electromagnetic radiation emitted from warm objects, with wavelengths slightly larger than those of visible light. (7.2)

insoluble Incapable of dissolving in water or being extremely difficult of solution. (4.5)

integrated rate law A relationship between the concentrations of the reactants in a chemical reaction and time. (13.4)

intensive property A property such as density that is independent of the amount of a given substance. (1.6)

interference The superposition of two or more waves overlapping in space, resulting in either an increase in amplitude (constructive interference) or a decrease in amplitude (destructive interfence). (7.2)

internal energy (E) The sum of the kinetic and potential energies of all of the particles that compose a system. (6.3)

International System of Units (SI) The standard unit system used by scientists, based on the metric system. (1.6)

ion An atom or molecule with a net charge caused by the loss or gain of electrons. (2.5)

ion product constant for water (K_w) The equilibrium constant for the autoionization of water. (15.5)

ion–dipole force An intermolecular force between an ion and the oppositely charged end of a polar molecule. (11.3)

ionic bond A chemical bond formed between two oppositely charged ions, generally a metallic cation and a nonmetallic anion, that are attracted to one another by electrostatic forces. (3.2, 9.2)

ionic compound A compound composed of cations and anions bound together by electrostatic attraction. (3.4)

ionic solids Solids whose composite units are ions; they generally have high melting points. (11.11)

ionization energy (IE) The energy required to remove an electron from an atom or ion in its gaseous state. (8.7)

ionizing power The ability of radiation to ionize other molecules and atoms. (19.2)

irreversible reaction A reaction that does not achieve the theoretical limit of available free energy. (17.7)

isotopes Atoms of the same element with the same number of protons but different numbers of neutrons and consequently different masses. (2.5)

joule (J) The SI unit for energy: equal to 1 kg · m^2/s^2. (6.2)

kelvin (K) The SI standard unit of temperature. (1.6)

Kelvin scale The temperature scale that assigns 0 K (-273 °C or -459 °F) to the coldest temperature possible, absolute zero, the temperature at which molecular motion virtually stops. 1 K = 1 °C. (1.6)

ketone A member of the family of organic compounds that contain a carbonyl functional group (C = O) bonded to two R groups, neither of which is a hydrogen atom. (20.8)

kilogram (kg) The SI standard unit of mass defined as the mass of a block of metal kept at the International Bureau of Weights and Measures at Sèvres, France. (1.6)

kilowatt-hour (kWh) An energy unit used primarily to express large amounts of energy produced by the flow of electricity; equal to 3.60×10^6 J. (6.2)

kinetic energy The energy associated with motion of an object. (1.5, 6.2)

kinetic molecular theory A model of an ideal gas as a collection of point particles in constant motion undergoing completely elastic collisions. (5.8)

lattice energy The energy associated with forming a crystalline lattice from gaseous ions. (9.4)

law see *scientific law*

law of conservation of energy A law stating that energy can neither be created nor destroyed, only converted from one form to another. (1.5, 6.2)

law of conservation of mass A law stating that matter is neither created nor destroyed in a chemical reaction. (2.2)

law of definite proportions   A law stating that all samples of a given compound have the same proportions of their constituent elements. (2.2)

law of mass action   The relationship between the balanced chemical equation and the expression of the equilibrium constant. (14.3)

law of multiple proportions   A law stating that when two elements (A and B) form two different compounds, the masses of element B that combine with one gram of element A can be expressed as a ratio of small whole numbers. (2.2)

Le Châtelier's principle   The principle stating that when a chemical system at equilibrium is disturbed, the system shifts in a direction that minimizes the disturbance. (14.9)

lead–acid storage battery   A battery that uses the oxidation of lead and the reduction of lead(IV) oxide in sulfuric acid to provide electrical current. (18.7)

levorotatory   Capable of rotating the polarization of light counterclockwise. (20.3)

Lewis acid   An atom, ion, or molecule that is an electron pair acceptor. (15.10)

Lewis base   An atom, ion, or molecule that is an electron pair donor. (15.10)

Lewis electron-dot structures (Lewis structures)   A drawing that represents chemical bonds between atoms as shared or transferred electrons; the valence electrons of atoms are represented as dots. (9.1)

Lewis model (theory)   A simple model of chemical bonding using diagrams that represent bonds between atoms as lines or pairs of dots. In this theory, atoms bond together to obtain stable octets (8 valence electrons). (9.1)

ligand   A neutral molecule or an ion that acts as a Lewis base with the central metal ion in a complex ion. (16.7)

limiting reactant   The reactant that has the smallest stoichiometric amount in a reactant mixture and consequently limits the amount of product in a chemical reaction. (4.3)

linear geometry   The molecular geometry of three atoms with a 180° bond angle due to the repulsion of two electron groups. (10.2)

liquid   A state of matter in which atoms or molecules pack about as closely as they do in solid matter but are free to move relative to each other, giving a fixed volume but not a fixed shape. (1.3)

liter (L)   A unit of volume equal to 1000 cm^3 or 1.057 qt. (1.6)

lithium-ion battery   A rechargeable battery that produces electrical current from the motion of lithium ions from the anode to the cathode. (18.7)

lone pair   A pair of electrons associated with only one atom. (9.5)

magic numbers   Certain numbers of nucleons (N or Z = 2, 8, 20, 28, 50, 82, and N = 126) that confer unique stability. (19.3)

magnetic quantum number (m_l)   An integer that specifies the orientation of an orbital. (7.5)

main-group elements   Those elements found in the s or p blocks of the periodic table, whose properties tend to be predictable based on their position in the table. (2.6)

mass   A measure of the quantity of matter making up an object. (1.6)

mass defect   The difference in mass between the nucleus of an atom and the sum of the separated particles that make up that nucleus. (19.6)

mass number (A)   The sum of the number of protons and neutrons in an atom. (2.5)

mass percent composition (mass percent)   An element's percentage of the total mass of a compound containing the element. (3.8)

matter   Anything that occupies space and has mass. (1.3)

mean free path   The average distance that a molecule in a gas travels between collisions. (5.9)

melting (fusion)   The phase transition from solid to liquid. (11.6)

melting point   The temperature at which the molecules of a solid have enough thermal energy to overcome intermolecular forces and become a liquid. (11.6)

metal   A large class of elements that are generally good conductors of heat and electricity, malleable, ductile, lustrous, and tend to lose electrons during chemical changes. (2.6)

metallic atomic solids   Atomic solids held together by metallic bonds; they have variable melting points. (11.11)

metallic bonding   The type of bonding that occurs in metal crystals, in which metal atoms donate their electrons to an electron sea, delocalized over the entire crystal lattice. (9.2)

metalloids   A category of elements found on the boundary between the metals and nonmetals of the periodic table, with properties intermediate between those of both groups; also called *semimetals*. (2.6)

meter (m)   The SI standard unit of length; equivalent to 39.37 inches. (1.6)

metric system   The system of measurements used in most countries in which the meter is the unit of length, the kilogram is the unit of mass, and the second is the unit of time. (1.6)

microwaves   Electromagnetic radiation with wavelengths slightly longer than those of infrared radiation; used for radar and in microwave ovens. (7.2)

milliliter (mL)   A unit of volume equal to 10^{-3} L or 1 cm^3. (1.6)

millimeter of mercury (mmHg)   A common unit of pressure referring to the air pressure required to push a column of mercury to a height of 1 mm in a barometer; 760 mmHg = 1 atm. (5.2)

miscibility   The ability to mix without separating into two phases. (11.3)

miscible   The ability of two or more substances to be soluble in each other in all proportions. (12.2)

mixture   A substance composed of two or more different types of atoms or molecules that can be combined in variable proportions. (1.3)

molality (m)   A means of expressing solution concentration as the number of moles of solute per kilogram of solvent. (12.5)

molar heat capacity   The amount of heat required to raise the temperature of one mole of a substance by 1 °C. (6.4)

molar mass   The mass in grams of one mole of atoms of an element; numerically equivalent to the atomic mass of the element in amu. (2.8)

molar solubility   The solubility of a compound in units of moles per liter. (16.5)

molar volume   The volume occupied by one mole of a gas; the molar volume of an ideal gas at STP is 22.4 L. (5.5)

molarity (M)   A means of expressing solution concentration as the number of moles of solute per liter of solution. (4.4, 12.5)

mole (mol)   A unit defined as the amount of material containing 6.0221421×10^{23} (Avogadro's number) particles. (2.8)

mole fraction (χ_a)   The number of moles of a component in a mixture divided by the total number of moles in the mixture. (5.6)

mole fraction (χ_{solute})   A means of expressing solution concentration as the number of moles of solute per moles of solution. (12.5)

mole percent   A means of expressing solution concentration as the mole fraction multiplied by 100%. (12.5)

molecular compound Compounds composed of two or more covalently bonded nonmetals. (3.4)

molecular element Those elements that exist in nature with diatomic or polyatomic molecules as their basic unit. (3.4)

molecular equation An equation showing the complete neutral formula for each compound in a reaction. (4.7)

molecular formula A chemical formula that shows the actual number of atoms of each element in a molecule of a compound. (3.3)

molecular geometry The geometrical arrangement of atoms in a molecule. (10.3)

molecular orbital theory An advanced model of chemical bonding in which electrons reside in molecular orbitals delocalized over the entire molecule. In the simplest version, the molecular orbitals are simply linear combinations of atomic orbitals. (10.8)

molecular solids Solids whose composite units are molecules; they generally have low melting points. (11.11)

molecularity The number of reactant particles involved in an elementary step. (13.6)

molecule Two or more atoms joined chemically in a specific geometrical arrangement. (1.1)

monoprotic acid An acid that contains only one ionizable proton. (15.4)

natural abundance The relative percentage of a particular isotope in a naturally occurring sample with respect to other isotopes of the same element. (2.5)

Nernst equation The equation relating the cell potential of an electrochemical cell to the standard cell potential and the reaction quotient; $E_{cell} = E°_{cell} - \dfrac{0.0592 \text{ V}}{n} \log Q$. (18.6)

net ionic equation An equation that shows only the species that actually change during the reaction. (4.7)

network covalent atomic solids Atomic solids held together by covalent bonds; they have high melting points. (11.11)

neutral The state of a solution where the concentrations of H_3O^+ and OH^- are equal. (15.5)

neutron An electrically neutral subatomic particle found in the nucleus of an atom, with a mass almost equal to that of a proton. (2.4)

nickel–cadmium (NiCad) battery A battery that consists of an anode composed of solid cadmium and a cathode composed of $NiO(OH)(s)$ in a KOH solution. (18.7)

nickel–metal hydride (NiMH) battery A battery that uses the same cathode reaction as the NiCad battery but a different anode reaction, the oxidation of hydrogens in a metal alloy. (18.7)

noble gases The group 8A elements, which are largely unreactive (inert) due to their stable filled p orbitals. (2.6)

node A point where the wave function (ψ), and therefore the probability density (ψ^2) and radial distribution function, all go through zero. (7.6)

nonbonding atomic solids Atomic solids held together by dispersion forces; they have low melting points. (11.11)

nonelectrolyte A compound that does not dissociate into ions when dissolved in water. (4.5)

nonmetal A class of elements that tend to be poor conductors of heat and electricity and usually gain electrons during chemical reactions. (2.6)

nonvolatile Not easily vaporized. (11.5)

normal boiling point The temperature at which the vapor pressure of a liquid equals 1 atm. (11.5)

nuclear binding energy The amount of energy that would be required to break apart the nucleus into its component nucleons. (19.6)

nuclear equation An equation that represents nuclear processes such as radioactivity. (19.2)

nuclear fission The splitting of the nucleus of an atom, resulting in a tremendous release of energy. (19.7)

nuclear fusion The combination of two light nuclei to form a heavier one. (19.7)

nuclear theory The theory that most of the atom's mass and all of its positive charge is contained in a small, dense nucleus. (2.4)

nucleons The particles that compose the nucleus which are protons and neutrons. (19.3)

nucleus The very small, dense core of the atom that contains most of the atom's mass and all of its positive charge; it is composed of protons and neutrons. (2.4)

nuclide A particular isotope of an atom. (19.2)

octahedral arrangement The molecular geometry of seven atoms with 90° bond angles. (10.2)

octet A Lewis structure with eight dots, signifying a filled outer electron shell for s and p block elements. (9.3)

octet rule The tendency for most bonded atoms to possess or share eight electrons in their outer shell to obtain stable electron configurations and lower their potential energy. (9.3)

optical isomers Two molecules that are nonsuperimposable mirror images of one another. (20.3)

orbital A probability distribution map, based on the quantum mechanical model of the atom, used to describe the likely position of an electron in an atom; also an allowed energy state for an electron. (7.5)

orbital diagram A diagram which gives information similar to an electron configuration, but symbolizes an electron as an arrow in a box representing an orbital, with the arrow's direction denoting the electron's spin. (8.3)

organic chemistry The study of carbon-based compounds. (20.1)

organic compound Compounds composed of carbon, hydrogen, and several other possible elements including oxygen, sulfur, and nitrogen. (3.11)

organic molecule A molecule containing carbon combined with several other elements including hydrogen, nitrogen, oxygen, or sulfur. (20.1)

orientation factor In the collision model, the percentage of sufficiently energetic collisions that succeed in forming the product based upon the orientation of the molecules; represented by the symbol p. (13.5)

osmosis The flow of solvent from a solution of lower solute concentration to one of higher solute concentration. (12.6)

osmotic pressure The pressure required to stop osmotic flow. (12.6)

overall order The sum of the orders of all reactants in a chemical reaction. (13.3)

oxidation The loss of one or more electrons; also the gaining of oxygen or the loss of hydrogen. (4.9)

oxidation state (oxidation number) A positive or negative whole number that represents the "charge" an atom in a compound would have if all shared electrons were assigned to the atom with a greater attraction for those electrons. (4.9)

oxidation–reduction (redox) reaction Reactions in which electrons are transferred from one reactant to another and the oxidation states of certain atoms are changed. (4.9)

oxidizing agent   A substance that causes the oxidation of another substance; an oxidizing agent gains electrons and is reduced. (4.9)

oxyacid   An acid composed of hydrogen and an oxyanion. (3.6)

oxyanion   A polyatomic anion containing a nonmetal covalently bonded to one or more oxygen atoms. (3.5)

packing efficiency   The percentage of volume of a unit cell occupied by the atoms, assumed to be spherical. (11.10)

paramagnetic   The state of an atom or ion that contains unpaired electrons and is, therefore, attracted by an external magnetic field. (8.7, 10.8)

partial pressure (P_n)   The pressure due to any individual component in a gas mixture. (5.6)

parts by mass   A unit for expressing solution concentration as the mass of the solute divided by the mass of the solution multiplied by a multiplication factor. (12.5)

parts by volume   A unit for expressing solution concentration as the volume of the solute divided by the volume of the solution multiplied by a multiplication factor. (12.5)

parts per billion (ppb)   A unit for expressing solution concentration in parts by mass where the multiplication factor is 10^9. (12.5)

parts per million (ppm)   A unit for expressing solution concentration in parts by mass where the multiplication factor is 10^6. (12.5)

pascal (Pa)   The SI unit of pressure, defined as $1 \text{ N}/\text{m}^2$. (5.2)

Pauli exclusion principle   The principle that no two electrons in an atom can have the same four quantum numbers. (8.3)

penetrating power   The ability of radiation to penetrate matter. (19.2)

penetration   The phenomenon of some higher-level atomic orbitals having significant amounts of probability within the space occupied by orbitals of lower energy level. For example, the $2s$ orbital penetrates into the $1s$ orbital. (8.3)

percent by mass   A unit for expressing solution concentration in parts by mass with a multiplication factor of 100%. (12.5)

percent ionic character   The ratio of a bond's actual dipole moment to the dipole moment it would have if the electron were transferred completely from one atom to the other, multiplied by 100%. (9.6)

percent ionization   The concentration of ionized acid in a solution divided by the initial concentration of acid multiplied by 100%. (15.6)

percent yield   The percentage of the theoretical yield of a chemical reaction that is actually produced; the ratio of the actual yield to the theoretical yield multiplied by 100%. (4.3)

periodic law   A law based on the observation that when the elements are arranged in order of increasing mass, certain sets of properties recur periodically. (2.6)

periodic property   A property of an element that is predictable based on an element's position in the periodic table. (8.1)

permanent dipole   A permanent separation of charge; a molecule with a permanent dipole always has a slightly negative charge at one end and a slightly positive charge at the other. (11.3)

pH   The negative log of the concentration of H_3O^+ in a solution; the pH scale is a compact way to specify the acidity of a solution. (15.5)

phase   With regard to waves and orbitals, the phase is the sign of the amplitude of the wave, which can be positive or negative. (7.6)

phase diagram   A map of the phase of a substance as a function of pressure and temperature. (11.8)

photoelectric effect   The observation that many metals emit electrons when light falls upon them. (7.2)

photon (quantum)   The smallest possible packet of electromagnetic radiation with an energy equal to $h\nu$. (7.2)

physical change   A change that alters only the state or appearance of a substance but not its chemical composition. (1.4)

physical property   A property that a substance displays without changing its chemical composition. (1.4)

pi (π) bond   The bond that forms between two p orbitals that overlap side to side. (10.7)

polar covalent bond   A covalent bond between two atoms with significantly different electronegativities, resulting in an uneven distribution of electron density. (9.6)

polyatomic ion   An ion composed of two or more atoms. (3.4)

polyprotic acid   An acid that contains more than one ionizable proton and releases them sequentially. (4.8, 15.6)

positron   The particle released in positron emission; equal in mass to an electron but opposite in charge. (19.2)

positron emission   The form of radioactive decay that occurs when an unstable nucleus emits a positron. (19.2)

positron emission tomography (PET)   A specialized imaging technique that employs positron-emitting nuclides, such as fluorine-18, as a radiotracer. (19.9)

potential difference   A measure of the difference in potential energy usually in joules per unit of charge (coulombs). (18.3)

potential energy   The energy associated with the position or composition of an object. (1.5, 6.2)

precipitate   A solid, insoluble ionic compound that forms in, and separates from, a solution. (4.6)

precipitation reaction   A reaction in which a solid, insoluble product forms upon mixing two solutions. (4.6)

precision   A term that refers to how close a series of measurements are to one another or how reproducible they are. (1.7)

prefix multipliers   Multipliers that change the value of the unit by powers of 10. (1.6)

pressure   A measure of force exerted per unit area; in chemistry, most commonly the force exerted by gas molecules as they strike the surfaces around them. (5.1)

pressure–volume work   The work that occurs when a volume change takes place against an external pressure. (6.4)

principal level (shell)   The group of orbitals with the same value of n. (7.5)

principal quantum number (n)   An integer that specifies the overall size and energy of an orbital. The higher the quantum number n, the greater the average distance between the electron and the nucleus and the higher its energy. (7.5)

probability density   The probability (per unit volume) of finding the electron at a point in space as expressed by a three-dimensional plot of the wave function squared (ψ^2). (7.6)

products   The substances produced in a chemical reaction; they appear on the right-hand side of a chemical equation. (3.10)

proton   A positively charged subatomic particle found in the nucleus of an atom. (2.4)

pure substance   A substance composed of only one type of atom or molecule. (1.3)

quantum number   One of four interrelated numbers that determine the shape and energy of orbitals, as specified by a solution of the Schrödinger equation. (7.5)

quantum-mechanical model   A model that explains the behavior of absolutely small particles such as electrons and photons. (7.1)

racemic mixture   An equimolar mixture of two optical isomers that does not rotate the plane of polarization of light at all. (20.3)

radial distribution function   A plot representing the total probability of finding an electron within a thin spherical shell at a distance r from the nucleus. (7.6)

radio waves   The form of electromagnetic radiation with the longest wavelengths and smallest energy. (7.2)

radioactive   The state of those unstable atoms that emit subatomic particles or high-energy electromagnetic radiation. (19.1)

radioactivity   The emission of subatomic particles or high-energy electromagnetic radiation by the unstable nuclei of certain atoms. (2.4, 19.1)

radiocarbon dating   A form of radiometric dating based on the C-14 isotope. (19.4)

radiometric dating   A technique used to estimate the age of rocks, fossils, or artifacts that depends on the presence of radioactive isotopes and their predictable decay with time. (19.4)

radiotracer   A radioactive nuclide that has been attached to a compound or introduced into a mixture in order to track the movement of the compound or mixture within the body. (19.9)

random error   Error that has equal probability of being too high or too low. (1.7)

Raoult's law   An equation used to determine the vapor pressure of a solution; $P_{soln} = X_{solv} P°_{solv}$. (12.6)

rate constant (k)   A constant of proportionality in the rate law. (13.3)

rate law   A relationship between the rate of a reaction and the concentration of the reactants. (13.3)

rate-determining step   The step in a reaction mechanism that occurs much more slowly than any of the other steps. (13.6)

reactants   The starting substances of a chemical reaction; they appear on the left-hand side of a chemical equation. (3.10)

reaction intermediates   Species that are formed in one step of a reaction mechanism and consumed in another. (13.6)

reaction mechanism   A series of individual chemical steps by which an overall chemical reaction occurs. (13.6)

reaction order (n)   A value in the rate law that determines how the rate depends on the concentration of the reactants. (13.3)

reaction quotient (Q_c)   The ratio, at any point in the reaction, of the concentrations of the products of a reaction raised to their stoichiometric coefficients to the concentrations of the reactants raised to their stoichiometric coefficients. (14.7)

recrystallization   A technique used to purify solids in which the solid is put into hot solvent until the solution is saturated; when the solution cools, the purified solute comes out of solution. (12.4)

reducing agent   A substance that causes the reduction of another substance; a reducing agent loses electrons and is oxidized. (4.9)

reduction   The gaining of one or more electrons; also the gaining of hydrogen or the loss of oxygen. (4.9)

rem   A unit of the dose of radiation exposure that stands for roentgen equivalent man, where a roentgen is defined as the amount of radiation that produces 2.58×10^{-4} C of charge per kg of air. (19.8)

resonance hybrid   The actual structure of a molecule that is intermediate between two or more resonance structures. (9.8)

resonance structures   Two or more valid Lewis structures that are shown with double-headed arrows between them to indicate that the actual structure of the molecule is intermediate between them. (9.8)

reversible   As applied to a reaction, the ability to proceed in either the forward or the reverse direction. (14.2)

reversible reaction   A reaction that achieves the theoretical limit with respect to free energy and will change direction upon an infinitesimally small change in a variable (such as temperature or pressure) related to the reaction. (17.7)

salt   An ionic compound formed in a neutralization reaction by the replacement of an H^+ ion from the acid with a cation from the base. (4.8)

salt bridge   An inverted, U-shaped tube containing a strong electrolyte such as KNO_3 that connects the two half-cells, allowing a flow of ions that neutralizes the charge buildup. (18.3)

saturated hydrocarbon   A hydrocarbon containing no double bonds in the carbon chain. (20.4)

saturated solution   A solution in which the dissolved solute is in dynamic equilibrium with any undissolved solute; any added solute will not dissolve. (12.4)

scientific law   A brief statement or equation that summarizes past observations and predicts future ones. (1.2)

scientific method   An approach to acquiring knowledge about the natural world that begins with observations and leads to the formation of testable hypotheses. (1.2)

second (s)   The SI standard unit of time, defined as the duration of 9,192,631,770 periods of the radiation emitted from a certain transition in a cesium-133 atom. (1.6)

second law of thermodynamics   A law stating that for any spontaneous process, the entropy of the universe increases ($\Delta S_{univ} > 0$). (17.3)

seesaw geometry   The molecular geometry of a molecule with trigonal bipyramidal electron geometry and one lone pair in an axial position. (10.3)

semiconductor   A material with intermediate electrical conductivity that can be changed and controlled. (2.6)

semipermeable membrane   A membrane that selectively allows some substances to pass through but not others. (12.6)

shielding   The effect on an electron of repulsion by electrons in lower-energy orbitals that screen it from the full effects of nuclear charge. (8.3)

sigma (σ) bond   The resulting bond that forms between a combination of any two s, p, or hybridized orbitals that overlap end to end. (10.7)

significant figures (significant digits)   In any reported measurement, the non-place-holding digits that indicate the precision of the measured quantity. (1.7)

simple cubic   A unit cell that consists of a cube with one atom at each corner. (11.10)

solid   A state of matter in which atoms or molecules are packed close to one another in fixed locations with definite volume. (1.3)

solubility   The amount of a substance that will dissolve in a given amount of solvent. (12.2)

solubility product constant (K_{sp})   The equilibrium expression for a chemical equation representing the dissolution of a slightly to moderately soluble ionic compound. (16.5)

soluble   Able to dissolve to a significant extent, usually in water. (4.5)

solute   The minority component of a solution. (4.4, 12.1)

solution   A homogenous mixture of two substances. (4.4, 12.1)

solvent   The majority component of a solution. (4.4, 12.1)

space-filling molecular model   A representation of a molecule that shows how the atoms fill the space between them. (3.3)

specific heat capacity (C_s)   The amount of heat required to raise the temperature of 1 g of a substance by 1 °C. (6.4)

spectator ion   Ions in a complete ionic equation that do not participate in the reaction and therefore remain in solution. (4.7)

spin quantum number, m_s   The fourth quantum number, which denotes the electron's spin as either $\frac{1}{2}(\uparrow)$ or $-\frac{1}{2}(\downarrow)$. (8.3)

spontaneous process   A process that occurs without ongoing outside intervention. (17.2)

square planar geometry   The molecular geometry of a molecule with octahedral electron geometry and two lone pairs. (10.3)

square pyramidal geometry   The molecular geometry of a molecule with octahedral electron geometry and one lone pair. (10.3)

standard cell potential (standard emf) (E°_{cell})   The cell potential for a system in standard states (solute concentration of 1 M and gaseous reactant partial pressure of 1 atm). (18.3)

standard change in free energy (ΔG°_{rxn})   The change in free energy for a process when all reactants and products are in their standard states. (17.7)

standard electrode potential   The standard cell potential of an individual half-cell. (18.4)

standard enthalpy change (ΔH°)   The change in enthalpy for a process when all reactants and products are in their standard states. (6.9)

standard enthalpy of formation (ΔH°_f)   The change in enthalpy when 1 mol of a compound forms from its constituent elements in their standard states. (6.9)

standard entropy change (ΔS_{rxn})   The change in entropy for a process when all reactants and products are in their standard states. (17.6)

standard entropy change for a reaction (ΔS°_{rxn})   The change in entropy for a process in which all reactants and products are in their standard states. (17.6)

Standard Hydrogen Electrode (SHE)   The half-cell consisting of an inert platinum electrode immersed in 1 M HCl with hydrogen gas at 1 atm bubbling through the solution; used as the standard of a cell potential of zero. (18.4)

standard molar entropy (S°)   A measure of the energy dispersed into one mole of a substance at a particular temperature. (17.6)

standard state   For a gas, the standard state is the pure gas at a pressure of exactly 1 atm; for a liquid or solid, the standard state is the pure substance in its most stable form at a pressure of 1 atm and the temperature of interest (often taken to be 25 °C); for a substance in solution, the standard state is a concentration of exactly 1 M. (6.9)

standard temperature and pressure (STP)   The conditions of $T = 0$ °C (273 K) and $P = 1$ atm; used primarily in reference to a gas. (5.5)

state   A classification of the form of matter as a solid, liquid, or gas. (1.3)

state function   A function whose value depends only on the state of the system, not on how the system got to that state. (6.3)

stereoisomers   Molecules in which the atoms are bonded in the same order, but have a different spatial arrangement. (20.3)

steroid   A lipid composed of four fused hydrocarbon rings. (21.2)

stock solution   A highly concentrated form of a solution used in laboratories to make less concentrated solutions via dilution. (4.4)

stoichiometry   The numerical relationships between amounts of reactants and products in a balanced chemical equation. (4.2)

strong acid   An acid that completely ionizes in solution. (4.5, 15.4)

strong base   A base that completely dissociates in solution. (15.7)

strong electrolyte   A substance that completely dissociates into ions when dissolved in water. (4.5)

strong force   Of the four fundamental forces of physics, the one that is the strongest but acts over the shortest distance; the strong force is responsible for holding the protons and neutrons together in the nucleus of an atom. (19.3)

structural formula   A molecular formula that shows how the atoms in a molecule are connected or bonded to each other. (3.3, 20.3)

structural isomers   Molecules with the same molecular formula but different structures. (20.3)

sublevel (subshell)   Those orbitals in the same principal level with the same value of n and l. (7.5)

sublimation   The phase transition from solid to gas. (11.6)

substance   A specific instance of matter. (1.3)

substrate   The reactant molecule of a biochemical reaction that binds to an enzyme at the active site. (13.7)

supersaturated solution   An unstable solution in which more than the equilibrium amount of solute is dissolved. (12.4)

surface tension   The energy required to increase the surface area of a liquid by a unit amount; responsible for the tendency of liquids to minimize their surface area, giving rise to a membrane-like surface. (11.4)

surroundings   In thermodynamics, everything in the universe which exists outside the system under investigation. (6.2)

system   In thermodynamics, the portion of the universe which is singled out for investigation. (6.2)

systematic error   Error that tends towards being consistently either too high or too low. (1.7)

systematic name   An official name for a compound, based on well-established rules, that can be determined by examining its chemical structure. (3.5)

temperature   A measure of the average kinetic energy of the atoms or molecules that compose a sample of matter. (1.6)

termolecular   An elementary step of a reaction in which three particles collide and go on to form products. (13.6)

tetrahedral geometry   The molecular geometry of five atoms with 109.5° bond angles. (10.2)

theoretical yield   The greatest possible amount of product that can be made in a chemical reaction based on the amount of limiting reactant. (4.3)

theory   A proposed explanation for observations and laws based on well-established and tested hypotheses, that presents a model of the way nature works and predicts behavior beyond the observations and laws on which it was based. (1.2)

thermal energy   A type of kinetic energy associated with the temperature of an object, arising from the motion of individual atoms or molecules in the object; see also *heat*. (1.5, 6.2)

thermal equilibrium   The point at which there is no additional net transfer of heat between a system and its surroundings. (6.4)

thermochemistry   The study of the relationship between chemistry and energy. (6.1)

thermodynamics   The general study of energy and its interconversions. (6.3)

third law of thermodynamics   The law stating that the entropy of a perfect crystal at absolute zero (0 K) is zero. (17.6)

torr   A unit of pressure equal to 1 mm Hg. (5.2)

transition elements (transition metals)   Those elements found in the d block of the periodic table whose properties tend to be less predictable based simply on their position in the table. (2.6)

trigonal bipyramidal geometry   The molecular geometry of six atoms with 120° bond angles between the three equatorial electron groups and 90° bond angles between the two axial electron groups and the trigonal plane. (10.2)

trigonal planar geometry   The molecular geometry of four atoms with 120° bond angles in a plane. (10.2)

trigonal pyramidal geometry   The molecular geometry of a molecule with tetrahedral electron geometry and one lone pair. (10.3)

triple bond   The bond that forms when three electron pairs are shared between two atoms. (9.5)

triple point   The unique set of conditions at which all three phases of a substance are equally stable and in equilibrium. (11.8)

triprotic acid   An acid that contains three ionizable protons. (15.4)

T-shaped geometry   The molecular geometry of a molecule with trigonal bipyramidal electron geometry and two lone pairs in axial positions. (10.3)

ultraviolet (UV) radiation   Electromagnetic radiation with slightly smaller wavelengths than visible light. (7.2)

unimolecular   Describes a reaction that involves only one particle that goes on to form products. (13.6)

unit cell   The smallest divisible unit of a crystal that, when repeated in three dimensions, reproduces the entire crystal lattice. (11.10)

units   Standard quantities used to specify measurements. (1.6)

unsaturated hydrocarbon   A hydrocarbon that includes one or more double or triple bonds. (20.5)

unsaturated solution   A solution containing less than the equilibrium amount of solute; any added solute will dissolve until equilibrium is reached. (12.4)

valence bond theory   An advanced model of chemical bonding in which electrons reside in quantum-mechanical orbitals localized on individual atoms that are a hybridized blend of standard atomic orbitals; chemical bonds result from an overlap of these orbitals. (10.6)

valence electrons   Those electrons that are important in chemical bonding. For main-group elements, the valence electrons are those in the outermost principal energy level. (8.4)

valence shell electron pair repulsion (VSEPR) theory   A theory that allows prediction of the shapes of molecules based on the idea that electrons—either as lone pairs or as bonding pairs—repel one another. (10.2)

van der Waals equation   The extrapolation of the ideal gas law that considers the effects of intermolecular forces and particle volume in a nonideal gas: $P + a\left(\dfrac{n}{V}\right)^2 \times (V - nb) = nRT$. (5.10)

van der Waals radius (nonbonding atomic radius)   Defined as one-half the distance between the centers of adjacent, nonbonding atoms in a crystal. (8.6)

van't Hoff factor (i)   The ratio of moles of particles in a solution to moles of formula units dissolved. (12.7)

vapor pressure   The partial pressure of a vapor in dynamic equilibrium with its liquid. (5.6, 11.5)

vapor pressure lowering (ΔP)   The difference in vapor pressure between a pure solvent and its solution. (12.6)

vaporization   The phase transition from liquid to gas. (11.5)

viscosity   A measure of the resistance of a liquid to flow. (11.4)

visible light   Those frequencies of electromagnetic radiation that can be detected by the human eye. (7.2)

volt   The SI derived unit for potential difference and equal to 1 J/C. (18.3)

volatile   Tending to vaporize easily. (11.5)

voltaic (galvanic) cell   An electrochemical cell which produces electrical current from a spontaneous chemical reaction. (18.3)

volume (V)   A measure of space. Any unit of length, when cubed (raised to the third power), becomes a unit of volume. (1.6)

wave function (ψ)   A mathematical function that describes the wavelike nature of the electron. (7.5)

wavelength (λ)   The distance between adjacent crests of a wave. (7.2)

weak acid   An acid that does not completely ionize in water. (4.5, 15.4)

weak base   A base that only partially ionizes in water. (15.7)

weak electrolyte   A substance that does not completely ionize in water and only weakly conducts electricity in solution. (4.5)

work (w)   The result of a force acting through a distance. (1.5, 6.2)

X-rays   Electromagnetic radiation with wavelengths slightly longer than those of gamma rays; used to image bones and internal organs. (7.2)

Photo Credits

Chapter 1 Page 4: Tomas Abad/Alamy, Page 7 center: Fotolia, Page 8 top, from left to right: Karl R. Martin/Shutterstock, Fotolia, iStockphoto, Page 9 left: Michael Dalton/Fundamental Photographs, NYC, Page 9 center: Siede Preis/Getty Images, 10 left, from top to bottom: Charles D. Winters/Photo Researchers, Inc., Renn Sminkey/Pearson Education, Lon C. Diehl/Photo Edit, Page 13 bottom: NASA Earth Observing System, Page 14 top left: Icon Sports Media 465/Icon Sports Media/Newscom, Page 14 top right: Siede Preis/Getty Images, Page 14 bottom left: Richard Megna/Fundamental Photographs, NYC, Page 15 bottom, from left to right: Marty Honig/Getty Images, iStockphoto, Shutterstock, Page 20 top & bottom left, top right: Richard Megna/Fundamental Photographs, NYC, Page 20 bottom right: maxwellartandphoto.com/Pearson Education, Page 34 top & center: Richard Megna/Fundamental Photographs, NYC, Page 34 bottom: iStockphoto, Page 35 (Ex. 35, a, b, c, Ex. 36, b): maxwellartandphoto.com/Pearson Education, Page 35 (Ex. 36, a, c): Warren Rosenberg/Fundamental Photographs, Page 37 NASA. **Chapter 2** Page 42 left: Veeco Instruments, Inc., Page 42 right: IBM Research, Almaden Research Center, Page 43 bottom, left to right: PH College/Pearson Education, Charles D. Winter/Photo Researchers, Inc., Shutterstock, Page 46 top right: Richard Megna/Fundamental Photographs, NYC, Page 50 top left: Jeremy Woodhouse/Getty Images, Page 50 bottom left: Steve Cole/Getty Images, Page 50 bottom right: Modern Curriculum Press/Pearson Learning/Pearson Education, Page 51: Library of Congress, Page 54: Shutterstock, Page 55 from left to right: Courtesy of Wikipedia, Charles D. Winters/Photo Researchers, Inc., Shutterstock, Clive Streeter/Dorling Kindersly Media Library, Tom Bochsler/PH College/Pearson Education, Courtesy of Wikipedia, Harry Taylor/Dorling Kindersley Media Library, Mark A. Schneider/Photo Researchers, Inc., iStockphoto, Charles D. Winters/Photo Researchers, Inc., Perennou Nuridsany/Photo Researchers, Inc., Page 59 top & bottom: maxwellartandphoto.com/Pearson Education, Page 68: IBM Communications Media Relations. **Chapter 3** Page 74 center: iStockphoto, Page 75 bottom, left to right: PH College/Pearson Education, Charles Falco/Photo Researchers, Inc., Charles D. Winter/Photo Researchers, Inc., Page 77 bottom: Charles D. Winter/Photo Researchers, Inc., Page 79 right: Charles Falco/Photo Researchers, Inc., Page 81 bottom left & right: Renn Sminkey/Pearson Education, Page 80: maxwellartandphoto.com/Pearson Education, Page 81 top left: Sydney Moulds/Photo Researchers, Inc., Page 81 top right: Basement Stock/Alamy, Page 87: Richard Megna/Fundamental Photographs, NYC, Page 89: Shutterstock, Page 94: Goddard Space Flight Center/NASA. **Chapter 4** Page 140 bottom left: Richard Megna/Fundamental Photographs, NYC, Page 142 top left: Martyn F. Chillmaid/Photo Researchers, Inc., Page 142 bottom left: Dorling Kindersly Media Library, Page 147 top right: Chip Clarke, Page 147 bottom right: Eric Schrader/Pearson Education, Page 148 top left: Eric Schrader/Pearson Education, Page 148 bottom left: Richard Megna/Fundamental Photographs, NYC, Page 150 top left: Richard Megna/Fundamental Photographs, NYC, Page 151 left: Tom Bochsler/PH College/Pearson Education, Page 151 right: Charles D. Winters/Photo Researchers, Inc., Page 152 left & right: Richard Megna/Fundamental Photographs, NYC. **Chapter 5** Page 176 top left: Shutterstock, Page 176 bottom left: Dorling Kindersley Media Library, Page 180: maxwellartandphoto.com/Pearson Education, Page 209 left & right: Tom Bochsler/PH College/Pearson Education. **Chapter 6** Page 222: Larry Brownstein/Getty Images, Page 231 left: Tom Bochsler/PH College/Pearson Education, Page 231 right: Stan Fellerman/Corbis, Page 241: Richard Megna/Fundamental Photographs, NYC, Page 249: Charles D. Winter/Photo Researchers, Inc. **Chapter 7** Page 255 top right: Thinkstock Images/Getty Images, Page 255 bottom right: Shutterstock, Page 257 right, from top to bottom: iStockphoto, Bob Richards/Sierra Pacific Innovations Corporation, Shutterstock, Page 262 top: Shutterstock, Page 262 bottom: Tom Bochsler/PH College/Pearson Education, Page 263 left: Fundamental Photographs, Page 267 bottom: AIP Emilio Segrè Visual Archives/Segre Collection, Page 269 top: John Sommers/Icon SMI/Newscom, Page 269 bottom: Bill Greenblatt/UPI/Newscom. **Chapter 8** Page 288 top left: University of Pennsylvania, Van Pelt Library, Page 288 bottom left: Charles D. Winters/Photo Researchers, Inc., Page 288 bottom right: Richard Megna/Fundamental Photographs, NYC, Page 316 counterclockwise from left: Courtesy of Wikipedia, Richard Megna/Fundamental Photographs, NYC, Charles D. Winters/Photo Researchers, Inc., Courtesy of Wikipedia, Charles D. Winters/Photo Researchers, Inc., Steve Gorton/Dorling Kindersley Media Library, Charles D. Winters/Photo Researchers, Inc., Shutterstock, Harry Taylor/Dorling Kindersly Media Library, Charles D. Winters/Photo Researchers, Inc., Dorling Kindersley Media Library. **Chapter 9** Page 325: Lawrence Berkeley National Laboratory/Photo Researchers, Page 326 top left: Shutterstock, Page 326 top right: PhotoDisc/Getty Images, Page 326 bottom: Dorling Kindersley Media Library, Page 332 left & right: Richard Megna/Fundamental Photographs, NYC. **Chapter 10** Page 363 top left & right: Renn Sminkey/Pearson Education, Page 364 top & bottom left: Renn Sminkey/Pearson Education, Page 376 center & right: maxwellartandphoto.com/Pearson Education, Page 400 top left: Richard Megna/Fundamental Photographs, NYC. **Chapter 11** Page 408: Eric Isselée/iStockphoto, Page 409: Andrew Syred/Photo Researchers, Inc., Page 419 bottom right: Douglas Allen/iStockphoto, Page 420 left: NASA/Johnson Space Center, Page 421 top: Iris Sample Processing, Page 421 bottom: Sinclair Stammers/Photo Researchers, Inc., Page 426 bottom left: Michael Dalton/Fundamental Photographs, NYC, Page 430 top left to right: Dr. P.A. Hamley, University of Nottingham, Page 431 top: Elizabeth Coelfen/Alamy, Page 431 bottom: Photolibrary, Page 433 bottom, from left to right: Pearson Education, Tim Ridley/Dorling Kindersley Media Library, Shutterstock, Page 435: NASA/JPL, Page 436: Renn Sminkey/Pearson Education, Page 437 top left: Shutterstock, Page 437 top right: Ted Kinsman/Photo Researchers, Inc., Page 442 bottom, from left to right: Photos.com, Andrew Syred/Photo Researchers, Inc., Dr. Mark J. Winter, Shutterstock, Page 444 bottom left: Charles D. Winters/Photo Researchers, Inc., Page 444 bottom right: iStockphoto, Page 445 left: Harry Taylor/Dorling Kindersley Media Library, Page 445 right: Courtesy of Wikipedia, Page 450 (Ex. 14, a): Photolibrary, Page 450 (Ex. 14, b): Harry Taylor/Dorling Kindersley Media Library, Page 450 (Ex. 17): Professor Nivaldo Jose Tro. **Chapter 12** Page 457 bottom right: Shutterstock, Page 458 bottom left: Photolibrary, Page 466 left to right: Richard Megna/Fundamental Photographs, NYC, Page 467: maxwellartandphoto.com/

Index

A

Absolute scale (Kelvin scale), 14–16
Absolute zero, 175
  of entropy, 681
Accuracy, 23–24
Acetaldehyde (ethanal), 796
Acetate, 86
Acetic acid, 91, 140, 148, 577, 583, 622, 623, 797
  acid ionization constant for, 584
Acetone, 422, 459, 460, 796
  boiling point of, 423
  heat of fusion of, 432
  heat of vaporization of, 423
Acetonitrile, 513
Acetylene, 386, 787
  common uses of, 107
  representations of, 78, 107
Acetylsalicylic acid, 405, 618
Acid(s), 140, 576–619. *See also* **Acid–base chemistry;**
    Acid–base titration
  binary, 90
  dilution of, 134
  diprotic, 147, 582, 594
  ionization of, 140
  metals dissolved in, 89–90, 719
  monoprotic, 582, 584, 595
  naming, 89–91
  polyprotic, 147
  and rusting, 35
  strong, 140
  triprotic, 583, 594
  weak, 140
Acid–base chemistry, 147–49, 576–619, 635–47. *See also*
    Buffers; pH
  acid–base properties of ions, 601–6
  acid ionization constant (K_a), 583–84
  acid strength, 581–84, 609–10
  addition to buffer, 627–31
  anions as weak bases, 601–5
  Arrhenius model of, 147, 579
  autoionization of water, 585–89
  base solutions, 597–601
  Brønsted–Lowry model of, 580–81, 609, 611
  cations as weak acids, 605
  definitions of, 579
  equations, 149
  heartburn and, 576–77
  hydroxide ion concentration and pH of, 599–601
  Lewis model of, 611–12
  molecular structure, 609–10
    binary acids, 609
    oxyacids, 609–10

  nature of, 577–79
  net ionic, 149
  pOH and other p scales, 588–89
  polyprotic acids, 594–95
    acid ionization constants for, 595
    dissociation of, 594
    ionization of, 594–95
    pH of, 594–95
  salt solutions as acidic, basic, or neutral, 606–8
  strong acid, 582
  strong bases, 597
  weak acids, 582–83
  weak bases, 597–98
Acid–base titration, 635–47
  endpoint of, 646
  indicators, 646–47
  of strong acid with strong base, 635–40
    equivalence point, 635–36
    overall pH curve, 638
  titration curve or pH curve, 635
  of weak acid with strong base, 640–45
    equivalence point, 640
    overall pH curve, 644
Acid dissociation constant. *See* **Acid ionization**
    constant(s) (K_a)
Acidic solution, 585–86
Acid ionization constant(s) (K_a), 584, 597
  for polyprotic acids, 595
  for weak acids, 590
Acidity of blood, 621
Acidosis, 621
Acid rain, 114, 124
Acid reflux, 577
Actinides (inner transition elements), 297, 298
  electron configurations of, 300
Activated complex (transition state), 513–14
Activation energy (activation barrier), 513, 515–17
  Arrhenius plots of, 515–18
  catalysis and, 523–24
  enzymes and, 525
Active site, 325, 361, 525
Actual yield, 126
Acute radiation damage, 765–66
Addition polymer, 799, 800
Addition reactions
  of aldehydes and ketones, 796
  of alkenes and alkynes, 791
Adduct, 611
Adhesive forces, 421
Adipic acid, 800
Aerosol cans, 179–80
Air, composition of dry, 186

Conversion Factors and Relationships

Length

SI unit: meter (m)

$1 \text{ m} = 1.0936 \text{ yd}$
$1 \text{ cm} = 0.39370 \text{ in}$
$1 \text{ in} = 2.54 \text{ cm (exactly)}$
$1 \text{ km} = 0.62137 \text{ mi}$
$1 \text{ mi} = 5280 \text{ ft}$
$= 1.6093 \text{ km}$
$1 \text{ Å} = 10^{-10} \text{ m}$

Temperature

SI unit: kelvin (K)

$0 \text{ K} = -273.15 \,°\text{C}$
$= -459.67 \,°\text{F}$
$\text{K} = °\text{C} + 273.15$
$°\text{C} = \dfrac{(°\text{F} - 32)}{1.8}$
$°\text{F} = 1.8 \, (°\text{C}) + 32$

Energy (derived)

SI unit: joule (J)

$1 \text{ J} = 1 \text{ kg} \cdot \text{m}^2/\text{s}^2$
$= 0.23901 \text{ cal}$
$= 1 \text{ C} \cdot \text{V}$
$= 9.4781 \times 10^{-4} \text{ Btu}$
$1 \text{ cal} = 4.184 \text{ J}$
$1 \text{ eV} = 1.6022 \times 10^{-19} \text{ J}$

Pressure (derived)

SI unit: pascal (Pa)

$1 \text{ Pa} = 1 \text{ N/m}^2$
$= 1 \text{ kg}/(\text{m} \cdot \text{s}^2)$
$1 \text{ atm} = 101,325 \text{ Pa}$
$= 760 \text{ torr}$
$= 14.70 \text{ lb/in}^2$
$1 \text{ bar} = 10^5 \text{ Pa}$
$1 \text{ torr} = 1 \text{ mmHg}$

Volume (derived)

SI unit: cubic meter (m^3)

$1 \text{ L} = 10^{-3} \text{ m}^3$
$= 1 \text{ dm}^3$
$= 10^3 \text{ cm}^3$
$= 1.0567 \text{ qt}$
$1 \text{ gal} = 4 \text{ qt}$
$= 3.7854 \text{ L}$
$1 \text{ cm}^3 = 1 \text{ mL}$
$1 \text{ in}^3 = 16.39 \text{ cm}^3$
$1 \text{ qt} = 32 \text{ fluid oz}$

Mass

SI unit: kilogram (kg)

$1 \text{ kg} = 2.2046 \text{ lb}$
$1 \text{ lb} = 453.59 \text{ g}$
$= 16 \text{ oz}$
$1 \text{ amu} = 1.66053873 \times 10^{-27} \text{ kg}$
$1 \text{ ton} = 2000 \text{ lb}$
$= 907.185 \text{ kg}$
$1 \text{ metric ton} = 1000 \text{ kg}$
$= 2204.6 \text{ lb}$

Geometric Relationships

$\pi = 3.14159\ldots$
$\text{Circumference of a circle} = 2\pi r$
$\text{Area of a circle} = \pi r^2$
$\text{Surface area of a sphere} = 4\pi r^2$
$\text{Volume of a sphere} = \dfrac{4}{3}\pi r^3$
$\text{Volume of a cylinder} = \pi r^2 h$

Fundamental Constants

Atomic mass unit	1 amu	$= 1.66053873 \times 10^{-27} \text{ kg}$
	1 g	$= 6.02214199 \times 10^{23} \text{ amu}$
Avogadro's number	N_A	$= 6.0221421 \times 10^{23}/\text{mol}$
Bohr radius	a_0	$= 5.29177211 \times 10^{-11} \text{ m}$
Boltzmann's constant	k	$= 1.38065052 \times 10^{-23} \text{ J/K}$
Electron charge	e	$= 1.60217653 \times 10^{-19} \text{ C}$
Faraday's constant	F	$= 9.64853383 \times 10^4 \text{ C/mol}$
Gas constant	R	$= 0.08205821 \text{ (L} \cdot \text{atm}/(\text{mol} \cdot \text{K})$
		$= 8.31447215 \text{ J}/(\text{mol} \cdot \text{K})$
Mass of an electron	m_e	$= 5.48579909 \times 10^{-4} \text{ amu}$
		$= 9.10938262 \times 10^{-31} \text{ kg}$
Mass of a neutron	m_n	$= 1.00866492 \text{ amu}$
		$= 1.67492728 \times 10^{-27} \text{ kg}$
Mass of a proton	m_p	$= 1.00727647 \text{ amu}$
		$= 1.67262171 \times 10^{-27} \text{ kg}$
Planck's constant	h	$= 6.62606931 \times 10^{-34} \text{ J} \cdot \text{s}$
Speed of light in vacuum	c	$= 2.99792458 \times 10^8 \text{ m/s (exactly)}$

SI Unit Prefixes

a	f	p	n	μ	m	c	d	k	M	G	T	P	E
atto	femto	pico	nano	micro	milli	centi	deci	kilo	mega	giga	tera	peta	exa
10^{-18}	10^{-15}	10^{-12}	10^{-9}	10^{-6}	10^{-3}	10^{-2}	10^{-1}	10^3	10^6	10^9	10^{12}	10^{15}	10^{18}